GCSE INTERMEDIATE MATHS

Stanley Thornes (Publishers) Ltd

First published in 1999 by:
Stanley Thornes (Publishers) Ltd
Ellenborough House
Wellington Street
CHELTENHAM
GL50 1YW

A catalogue record for this book is available from the British Library.

ISBN 0 7487 3674 3

99 00 01 02 03 / 10 9 8 7 6 5 4 3 2 1

All photographs, including cover, from STP Archive
Illustrations by Oxford Designers & Illustrators
Typeset by Tech Set Limited, Gateshead, Tyne and Wear
Printed and bound in Spain by Mateu Cromo

Acknowledgements

The authors and publishers would like to acknowledge, with thanks, the assistance of Alison Gee, June Haighton and Joan Wilson in developing a number of questions and answers in this book.

The authors and publishers are grateful to the following examination boards/Unitary Authorities for permission to reproduce questions from their past examination papers:

- [L] London Examinations (The Edexcel Foundation)
- [MEG] Midland Examining Group (OCR)
- [NEAB] Northern Examinations and Assessment Board (AQA)
- [NI] Northern Ireland Council for the Curriculum, Examinations and Assessment (NICCEA)
- [SEG] The Associated Examining Board (AQA)
- [WJEC] Welsh Joint Education Committee

Questions are followed by the initials of the board, as shown above in square brackets, and 'p' denotes a part question. Any answers and worked solutions are the responsibility of the authors and have not been provided by the examining boards.

Every effort has been made to contact copyright holders. The authors and publishers apologise to anyone whose rights have been overlooked, and will be happy to rectify any errors or omissions.

CONTENTS

INTRODUCTION

All you need for complete success in **GCSE Intermediate Maths** is provided in this book. It has been designed to be both a course book and a revision guide. It contains a number of features that will help you to build your confidence and achieve thorough and effective learning, revision and examination preparation.

ORGANISATION OF THE BOOK

The book is split into three stages and this **staged approach** helps you to progress in your learning by building on your knowledge and understanding from previous topics.

Each stage is broadly targeted at grade bands in the Intermediate Tier: you will be able to target grades E and D by working through Stage 1, grades D and C by continuing through Stage 2 and grades C and B on completion of Stage 3.

LAYOUT OF THE BOOK

The topics in each chapter are shown in **coloured guides** at the top of each page. These show you how you are progressing through the chapter and they will also help you to locate quickly other topics in the book.

A **green 'T'** tool box icon and a **green** tint identify the 'tools' you will need to do the job, that is the key skills, facts and formulas to solve problems. Each chapter begins with a 'tool box' to highlight the key objectives of the chapter.

A **blue 'open book'** icon is used to provide you with a clear summary of key words. This icon also refers you back to previous work that is useful in your understanding of the topic.

Green 'Do' icons provide reminders, hints and helpful explanations.

Red 'Do not' icons warn you against common pitfalls and mistakes.

KEY FEATURES OF THE BOOK

- **Fact Sheets**, shown in a **purple** tint, are given in Stage 1 to provide you with concise reminders about some of your Key Stage 3 work. These will be useful to refer to as you progress through the book.
- **Sample Questions** are presented in a **yellow** tint throughout each chapter, with **yellow** bullet points that take you through the main steps to obtaining the answer.
- **Exercises**, shown in a **blue** tint, are provided at the end of each chapter topic to help you to consolidate the ideas just learnt. Answers to these exercises are at the back of the book; use these wisely and only check the answer **after** you have done a question! Remember that, in an examination, you gain marks for the working, not just for the answer.
- At the end of each chapter, there is a **Worked Exam Question**, with a commentary suggesting how marks are allocated, followed by **Exam Questions** for you to do, all in a **pink** tint. These will give you plenty of practice and experience at tackling a variety of chapter topics and will help to build your confidence about taking your actual GCSE examinations.
- A **Summary** is given at the end of each stage, providing you with a revision list of the key skills and topics, all shown in a **yellow** tint.
- **Mixed Exam Questions** are also included at the end of each stage to give you more practice from all the examination boards. These are shown in a **pink** tint.
- Throughout the text, **calculator sequences** are given to help you to use your calculator efficiently. There are several types of calculator so you need to refer to the instructions for yours, which might be one of these:

  Type 1: Put in a function before the number, e.g. [√] [9] [=] gives 3
  Type 2: Put in a function after the number, e.g. [9] [√] gives 3

We hope that you enjoy using this book and wish you every success in your mathematics studies and examinations.

Janet Crawshaw and Paul Langley

1 PERIMETERS

In this chapter you will learn how to
- **work out the perimeters of simple shapes**
- **solve problems using $C = \pi d$ and $C = 2\pi r$ to find lengths**
- **calculate perimeters involving composite rectangular shapes and circles**

In everyday language the word **perimeter** is used to mean 'boundary' as in the following:
- The security cameras can see all the way around the perimeter of the car park.
- The gardener was asked to plant marigolds around the perimeter of each flower bed.
- A fence is to be built around the perimeter of the sports stadium.

In mathematics, if you are asked to find the **perimeter of a shape**, you would find the **total distance around the boundary** of the shape.

Perimeters of simple shapes

Sample Question 1 Find the perimeter of the field labelled A.

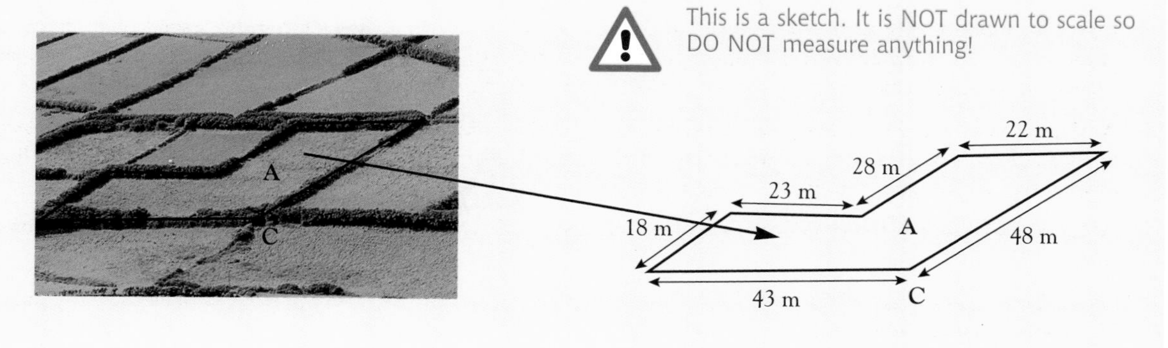

Answer
- Work out the distance you would walk if you started at corner C and walked anti-clockwise around the field.

Perimeter = 48 + 22 + 28 + 23 + 18 + 43

Perimeter = 182 **m**

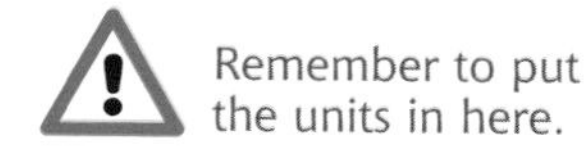

Before you start adding, check that all the lengths are given in the same units. If they are different, e.g. cm and m, change them all to the same unit.

PERIMETERS OF SIMPLE SHAPES

Sample Question 2

Chantal is tiling her kitchen. She tiles this surface.

Chantal wants to put a plastic strip around the perimeter of this surface. How many metres of plastic strip will she need?

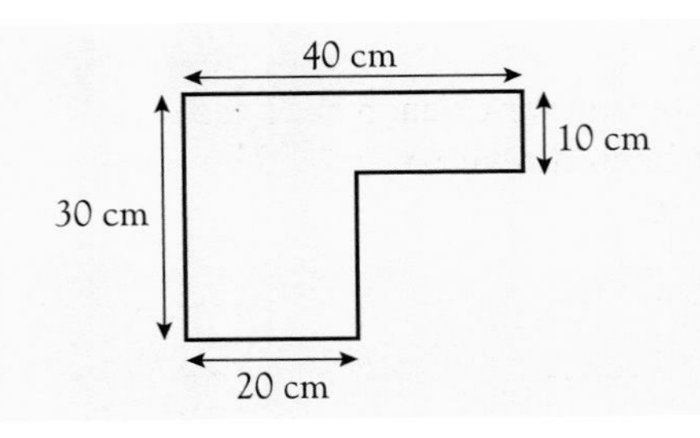

[SEG, p]

Answer

DO NOT just add up the numbers on the diagram. Some lengths are missing.

- Find the **missing lengths**
- Work out the distance around the surface.
- Change the units to metres.

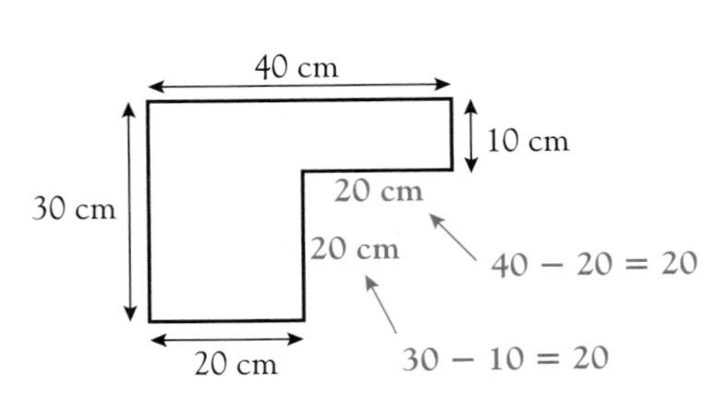

Write P for perimeter.

$P = 40 + 10 + 20 + 20 + 20 + 30$
$= 140$ cm
$\underline{P = 1.4 \text{ m}}$

To change cm to m, divide by 100

Sample Question 3

An anniversary cake is made in the shape of a **regular** hexagon. Each side is 15 cm long. A fancy trim is to be put around the cake.

What length of trim will be needed to fit exactly around the perimeter of the cake?

Answer

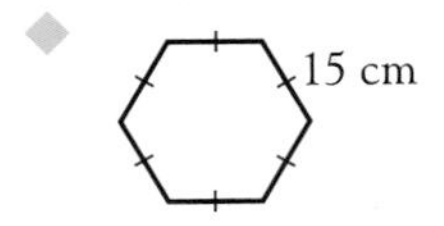

Regular means that all the sides are the same length. To show this put the same mark on each side of the shape.

$P = 15 + 15 + 15 + 15 + 15 + 15$
$= 90$
Length of trim needed = 90 cm

This is the same as 6 lots of 15 so you could find 6×15

TASK

1 Draw accurately as many rectangles as you can which have a perimeter of 24 cm.

2 Draw accurately three **different** shapes that are not rectangles or squares but have the same perimeter as each other.

PERIMETERS OF SIMPLE SHAPES

Exercise 1.1

1 Find the perimeter of each of these shapes. They have not been drawn to scale. The lengths given are in centimetres.

a

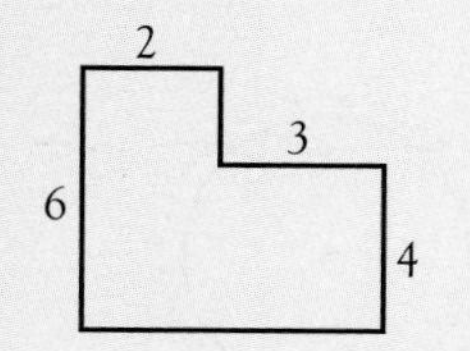

b

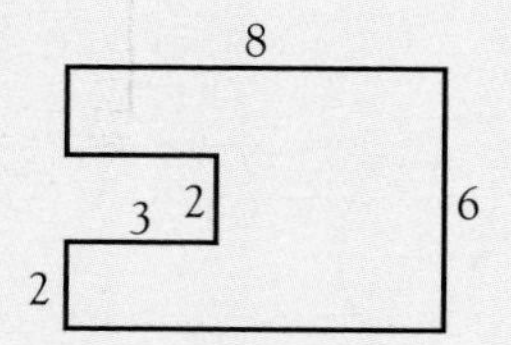

c

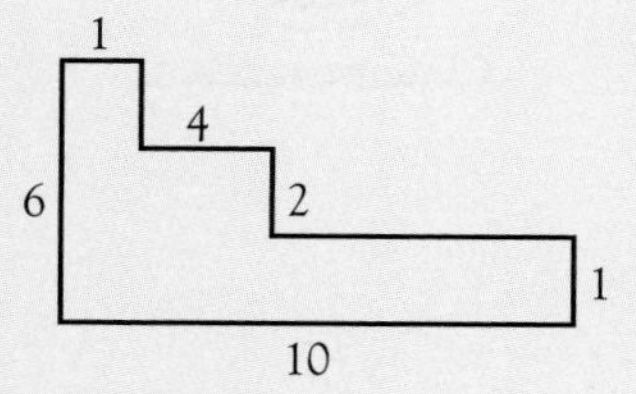

d

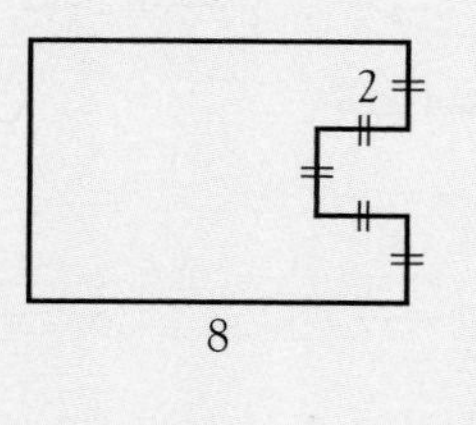

2 Find the perimeter of this quadrilateral by measuring. Give your answer in millimetres.

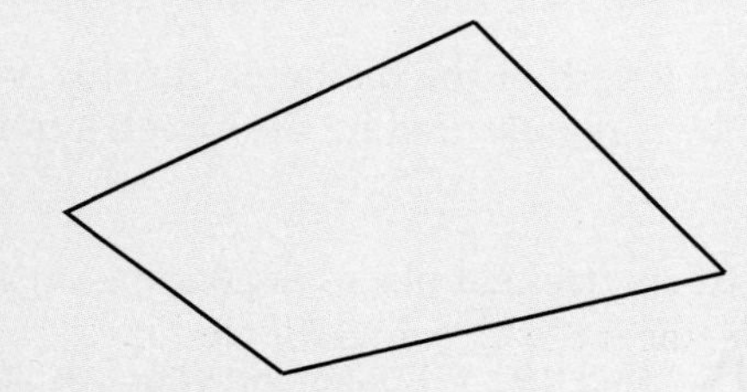

3 Calculate the perimeter of each of these shapes.

a

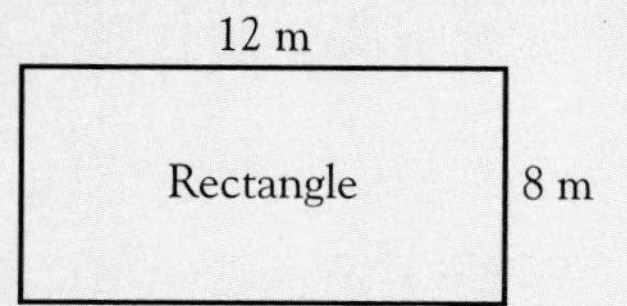

b

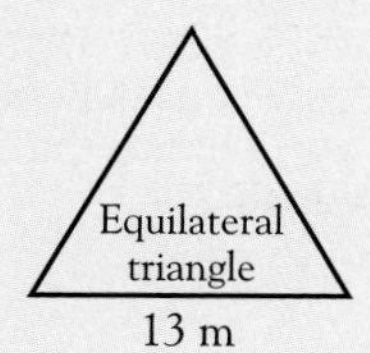

c

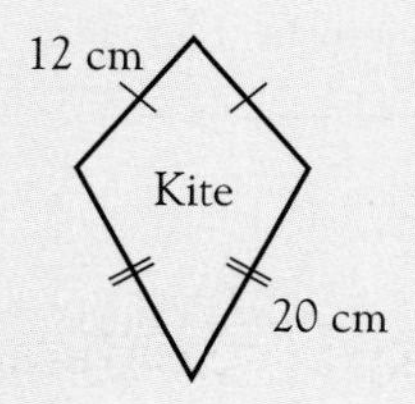

d

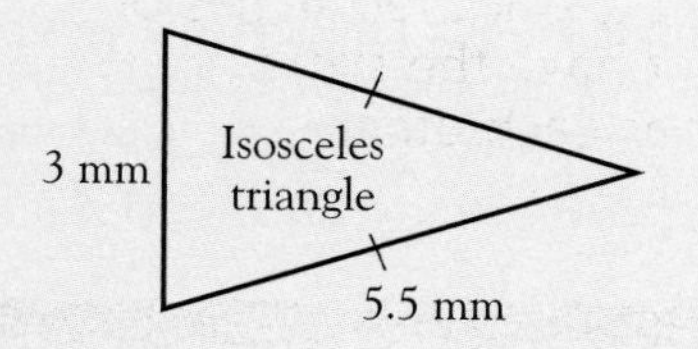

4 One of the sides of a regular pentagon is 4.6 cm long. What is the perimeter of the pentagon?

5 Lace costs 50p a metre. Vicky buys just enough to edge a rectangular tablecloth which measures 150 cm by 100 cm.

a What length of lace does she buy?

b How much does she pay for the lace?

6 On Friday night the security guard walked around the perimeter of the factory five times. How far did the security guard walk?

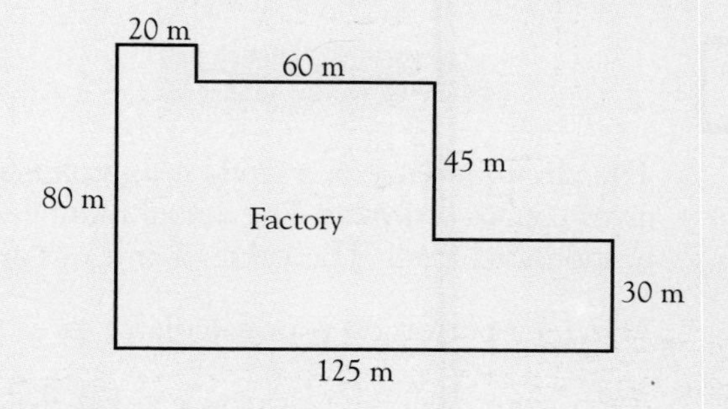

7 A square has a perimeter of 40 cm. What is the length of a side of the square?

8 A rectangle has a length of 6 cm and a perimeter of 20 cm. What is the width of the rectangle?

9 In the following diagrams, work out the length marked by a letter.

a

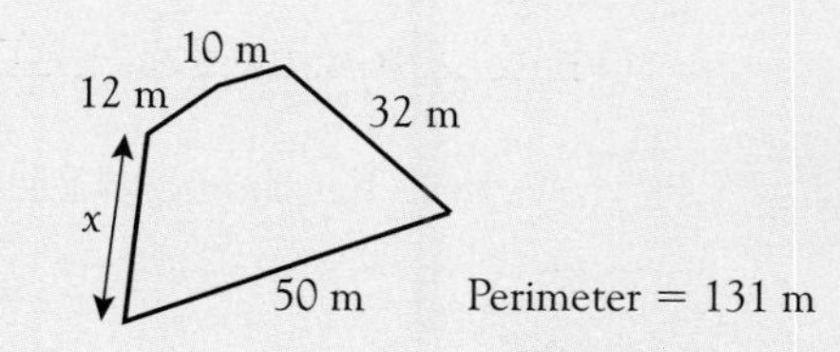

b

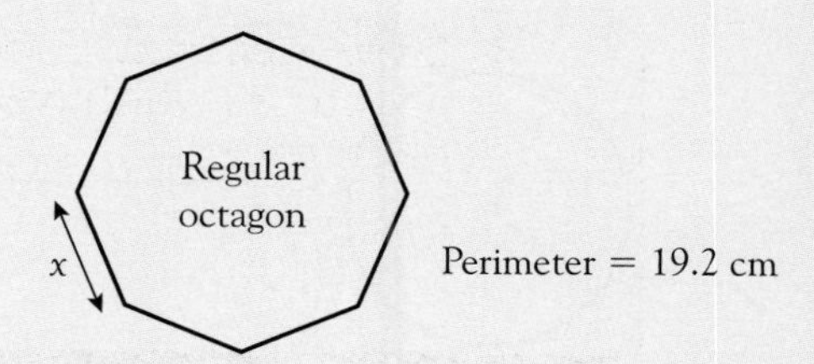

CIRCUMFERENCE OF A CIRCLE

Circumference of a circle

The **radius** of a circle is the distance from the centre of the circle to the boundary.

The **diameter** of a circle is the distance across the circle through the centre of the circle.

The **perimeter** of a circle is called the **circumference**.

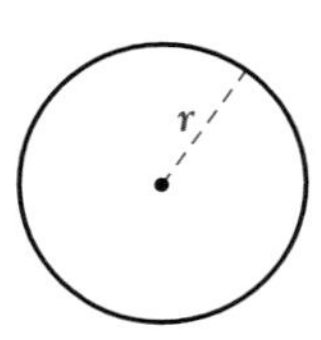

Radius, r

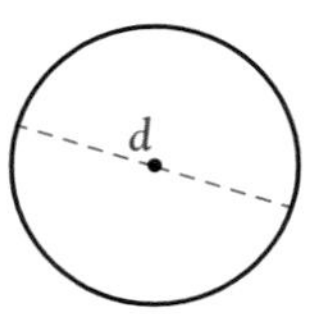

Diameter, d

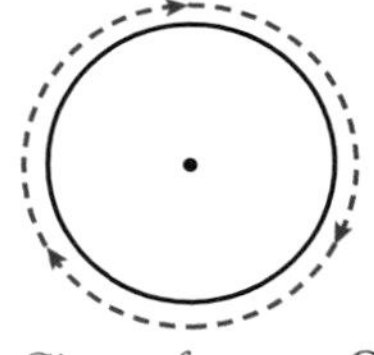

Circumference, C

The **diameter** of a circle is **twice** the radius.

This is written

$d = 2r$

Remember, $2r$ means $2 \times r$

The circumference of a circle is approximately three times the length of the diameter but this only gives you an estimate. The actual number that you multiply by is represented by the Greek letter 'π', pronounced 'pie'. The value of π is not an exact number.

The [π] button on your calculator gives the value to at least six decimal places depending on the size of your calculator display, e.g. 3.1415927

Find the [π] button on **your** calculator.

The formula for calculating the circumference of a circle is

$C = \pi d$

Remember, πd means $\pi \times d$

The value of π is often shortened to 3.14 or 3.142. You will usually be told what to use. If not, you can use the [π] button on your calculator.

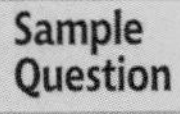

Sample Question 4 A circle has diameter 25 cm. Calculate the circumference of the circle. Use $\pi = 3.14$.

Answer

- Use the formula $C = \pi d$.

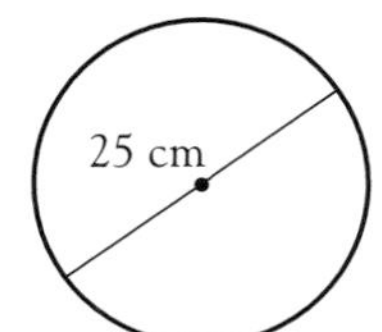

$C = 3.14 \times 25$

$\underline{C = 78.5 \text{ cm}}$

If you use the [π] button on your calculator the answer will be 78.539...

You should round this off to a sensible degree of accuracy, for example, one decimal place. The answer would then be

$C = 78.539... = 78.5$ cm (1 d.p.)

CIRCUMFERENCE OF A CIRCLE

Sometimes you are told the radius of the circle instead of the diameter. If this is the case, either

- work out the diameter using $d = 2 \times r$
- then use the formula $C = \pi d$ to work out the circumference

or

- replace d with $2 \times r$ in the formula $C = \pi d$ giving

$$C = \pi \times 2 \times r$$
$$= 2 \times \pi \times r$$

Remember that the order in which you multiply does not matter.

- then use the new formula

$$C = 2\pi r$$

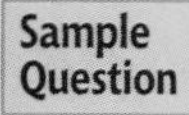

5 A bicycle wheel has a radius of 30 cm.

a How far in metres has the bicycle travelled when the wheel has turned 150 times?

b How many complete turns does the wheel make if the bicycle travels 50 m?

Answer

a The distance travelled in **one** turn (or revolution) is the same as the circumference of the wheel.

- Calculate the circumference of the wheel using the formula $C = 2\pi r$ because you are told the **radius** of the wheel.

$$C = 2 \times \pi \times 30$$
$$= 2 \times 3.14 \times 30$$
$$C = 188.4 \text{ cm}$$

- Calculate the total distance travelled by multiplying the circumference by 150.

$$\text{Distance travelled} = 150 \times 188.4$$
$$= 28\,260 \text{ cm}$$

- Change from **cm to m.**

$$\text{Distance travelled} = 28\,260 \div 100$$
$$\underline{\text{Distance travelled} = 282.6 \text{ m}}$$

To change cm to m, divide by 100

b

- Write the distance travelled in **cm.**

$$50 \text{ m} = 50 \times 100 = 5000 \text{ cm}$$

To change m to cm, multiply by 100

- Work out how many turns of the wheel there will be by dividing the distance travelled by the circumference of the wheel.

1st turn	2nd turn	
188.4 cm	188.4 cm	etc.

$$5000 \div 188.4 = 26.539\ldots$$

DO NOT round up!
The wheel has not completed the 27th turn.

<u>The wheel makes 26 complete turns when the bicycle travels 50 m</u>

CIRCUMFERENCE OF A CIRCLE

Parts of a circle

Part of the circumference of a circle is called an **arc**
The following are special arcs:

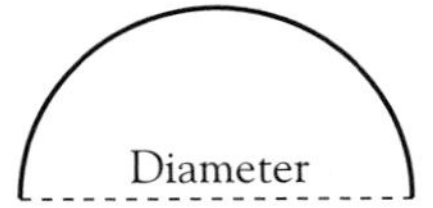

Half-circle
or **semi-circle**

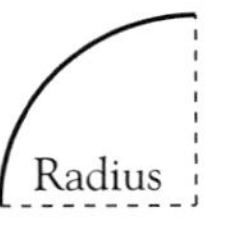

Quarter-circle
or **quadrant**

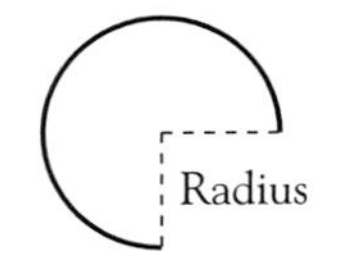

Three-quarter circle

6 A tile is in the shape of a quarter-circle, with radius 6 cm. Find:

a the length of the arc,

b the perimeter of the tile.

(Use $\pi = 3.14$)

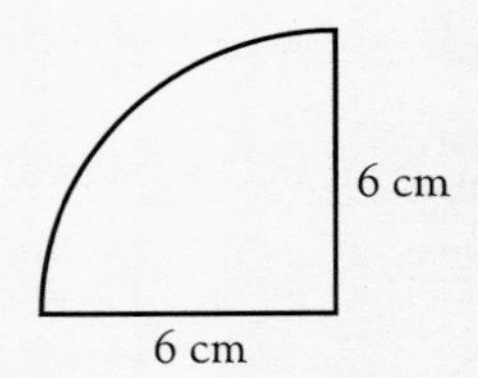

Answer

a

- Work out the circumference of the whole circle using $C = 2\pi r$.

$$C = 2 \times 3.14 \times 6$$
$$= 37.68 \text{ cm}$$

- Work out $\frac{1}{4}$ of this.

To find $\frac{1}{4}$, divide by 4

$$37.68 \div 4 = 9.42 \text{ cm}$$

<u>Length of arc = 9.42 cm</u>

b

- Add together the length of the arc and the straight edges.

$$9.42 + 6 + 6 = 21.42 \text{ cm}$$

<u>Perimeter of tile = 21.42 cm</u>

Remember to include the straight edges.

Exercise 1.2

Use $\pi = 3.14$ and give your answer to 1 decimal place where appropriate.

1 Work out the circumference of each of these circles.

a

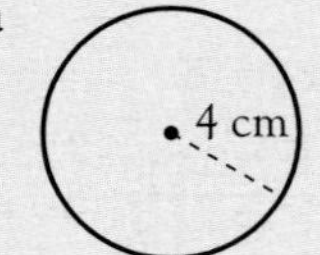

b

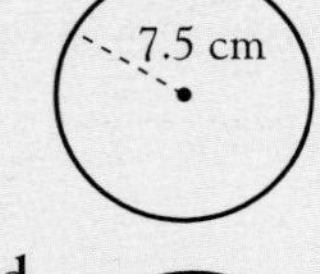

c

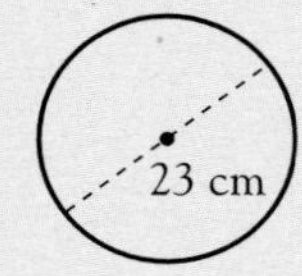

d

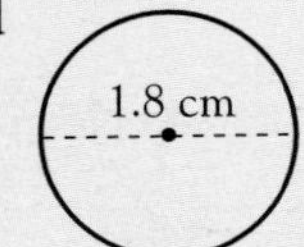

2 Work out the length of each of these arcs.

a

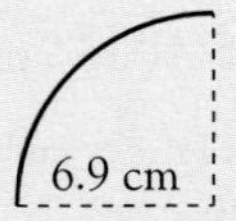

b

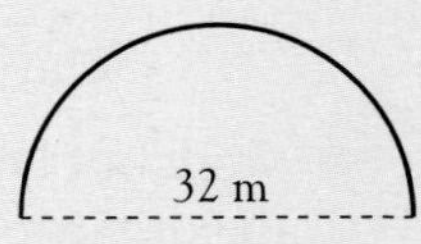

c

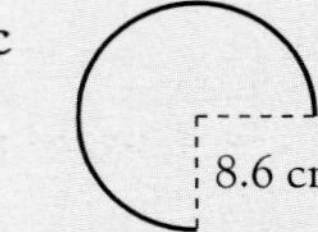

d 45 mm

CIRCUMFERENCE OF A CIRCLE

3 A circular pond has a diameter of 5 m. Around the circumference of the pond is a circular path 70 cm wide.

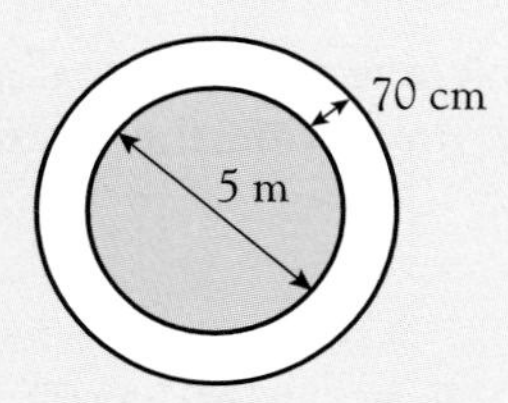

a Work out the circumference of the pond.

b What length of fencing would be needed to go around the outside of the path?

4 A semi-circular tablecloth is to be edged with a strip of lace. The diameter of the tablecloth is 2.8 m.

What length of lace is needed to go around the tablecloth?

5

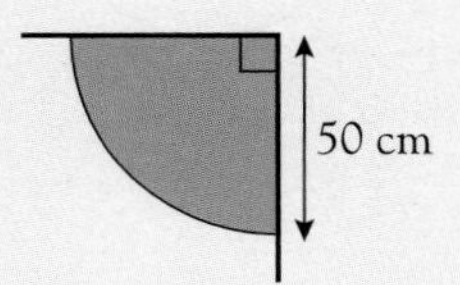

The diagram shows two fence panels meeting at a corner. They meet at right angles.

A quadrant (quarter of a circle) is cut to make a flower bed between the fences.

What length of lawn edging is needed to go around the curved part of the flower bed?

6 The diagram shows the frame of a mirror. It is in the shape of a rectangle with a semi-circular top.

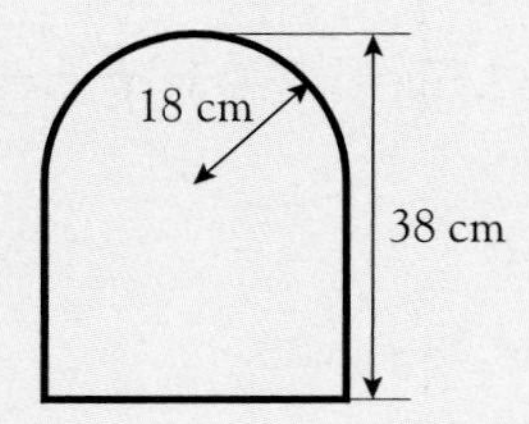

The radius of the semi-circle is 18 cm and the total height of the mirror is 38 cm.

What length of plastic trim would be needed to go around the frame of the mirror?

7 The arch above a door is in the shape of a semi-circle. The diameter of the arch is 1.4 metres.

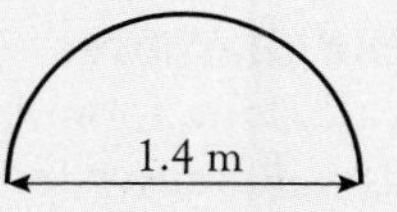

The arch is made of curved stone pieces each 31.4 cm long. How many of these stone pieces are needed to make the arch?

8 A template for part of a sewing pattern is made by cutting a quarter-circle with radius 2 cm from each of three corners of a square of side 12 cm.

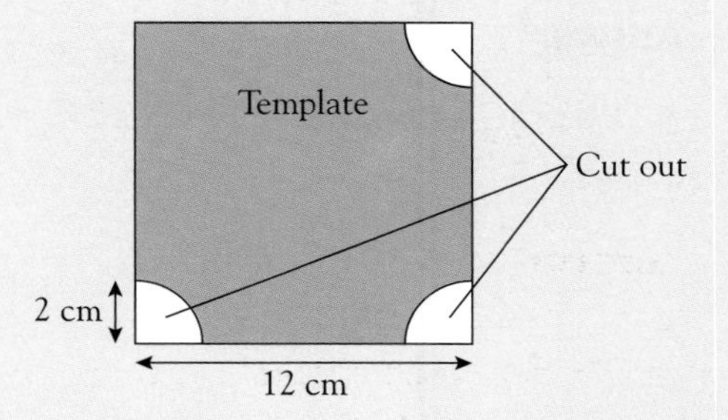

Work out the perimeter of the template.

9 A piece of jewellery is made from wire that is bent into two circles as shown in the diagram.

The radius of the larger circle is 3 cm and the diameter of the smaller circle is 1 cm.

What length of wire is needed to make 100 pieces of jewellery using this design?

(Give your answer correct to the nearest metre.)

10

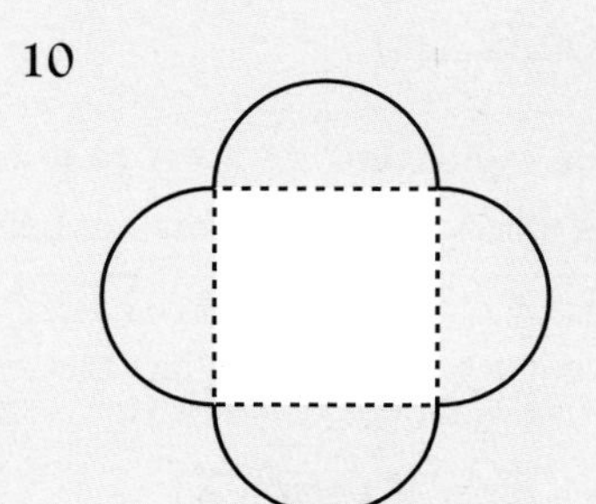

Semi-circular tiles are fitted around a square of side 6 cm. Find the perimeter of the final shape.

CIRCUMFERENCE OF A CIRCLE

Finding the diameter and radius from the circumference

You can work out the diameter or radius of a circle if you know what the circumference is.

To find the diameter

You know that the formula for working out the circumference of a circle is $C = \pi d$, that is, you **multiply** the **diameter** by π. So to work back to find the diameter you **divide** the **circumference** by π. This gives the formula

$$d = \frac{C}{\pi}$$

Remember, $\frac{C}{\pi}$ is $C \div \pi$

Sample Question 7 Find the diameter of a circle with circumference 20 cm. (Use $\pi = 3.14$)

Answer

- Use the formula $d = \frac{C}{\pi}$ where $C = 20$ and $\pi = 3.14$.

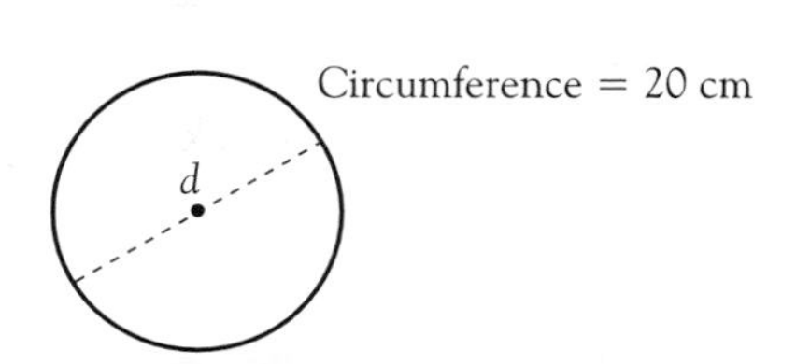

$d = 20 \div 3.14$
$= 6.369\ldots$
$\underline{d = 6.4 \text{ cm (1 d.p.)}}$

To find the radius

Using the radius of the circle, the formula for the circumference is $C = 2\pi r$.

If you rearrange this formula you get

$$r = \frac{C}{2\pi}$$

Remember, $\frac{C}{2\pi}$ means $C \div (2 \times \pi)$

Sample Question 8 Work out the radius of a circle whose circumference is 35 cm. (Use $\pi = 3.14$)

Answer

- Use the formula $r = \frac{C}{2\pi}$ where $C = 35$ and $\pi = 3.14$.

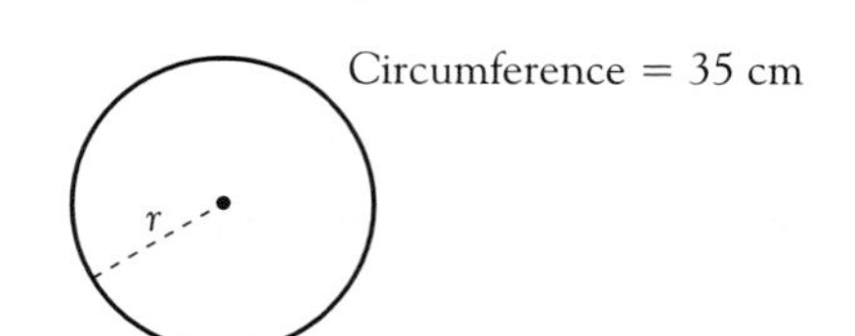

$r = 35 \div (2 \times 3.14)$
$= 5.573\ldots$
$\underline{r = 5.6 \text{ cm (1 d.p.)}}$

When calculating you could work out 2×3.14 first. This gives 6.28, so $r = 35 \div 6.28$.

It is, however, more efficient to use the bracket buttons on your calculator, e.g.

[3] [5] [÷] [(] [2] [×] [3] [·] [1] [4] [)] [=]

An even more efficient way is to use the fact that $\div 2\pi$ is the same as $\div 2$ and $\div \pi$ in any order.
On your calculator this would be keyed in as:

[3] [5] [÷] [2] [÷] [3] [·] [1] [4] [=]

CIRCUMFERENCE OF A CIRCLE

Exercise 1.3

Use $\pi = 3.14$

1 The diagram shows a circular speed-skating track.

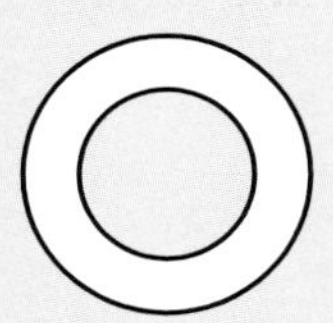

The circumference of the inside of the track is 150 m and the circumference of the outside of the track is 175 m.

Calculate, giving answers correct to 1 decimal place:

a the radius of the circle forming the inside of the track,

b the radius of the circle forming the outside of the track,

c the width of the track.

2

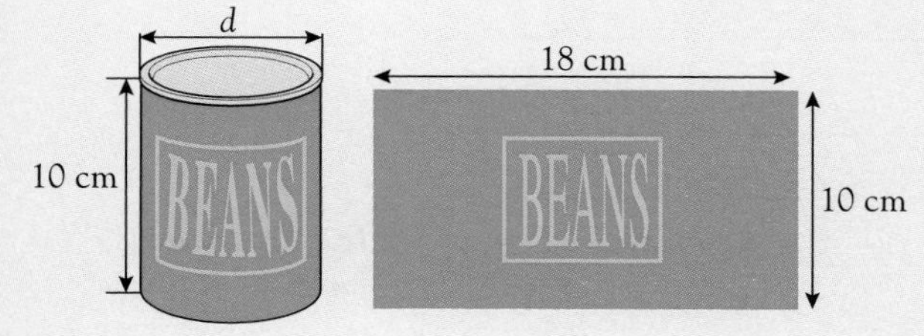

A tin of beans is in the shape of a cylinder. When the label is cut off the tin it is in the shape of a rectangle measuring 18 cm by 10 cm.

Calculate the diameter, d, of the tin.

3

Atul has to paint a design similar to the Olympic Rings on the school playing field. The rings are to be white so that they can be clearly seen from the air.

He has enough white-lining liquid to paint a total distance of 300 m.

If he has to paint five complete identical circles, calculate the largest radius he could use for the circles.

4 Lorinda is trying to work out the diameters of the boys' necks in her class. To do this she measures the circumference of each boy's neck. She then uses her formula to work out the diameter.

a Write down the formula she uses.

b Work out the diameter of the necks of which she measures the circumference as:
i 17 inches
ii $15\frac{1}{2}$ inches
iii 32 cm

5 Tony is taking part in a target golf competition. The target is in the middle of a circular pond. Tony needs to know how far the target is away so that he can select the correct golf club to use.

He estimates that the circumference of the pond is 300 yards.

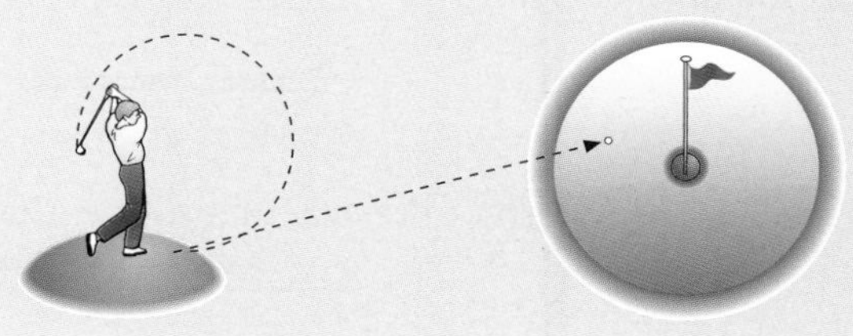

Using Tony's estimate for the circumference, calculate the distance from the edge of the pond to the target.
(Give your answer to the nearest 5 yards.)

6 Box

Mirror

A circular mirror is packed tightly into a square box, as shown. The mirror has a circumference of 50.24 cm. Calculate the length of the side of the box.

7 The area marked out for throwing a javelin is in the shape of the sector that is $\frac{1}{8}$ of a circle.

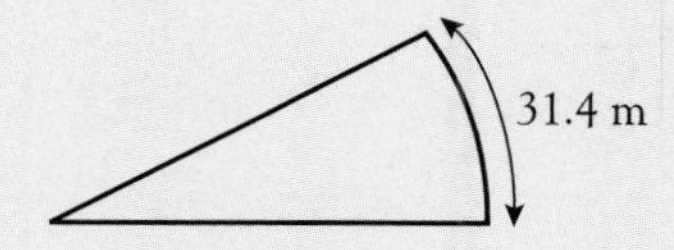

The diagram shows this sector at Amy's school. The curved part of the sector measures 31.4 m.

a Calculate the radius of the sector of the circle.

b If Amy walks around the sector, how far will she walk?

8 Calculate the diameter of a circle that has a circumference of 26.69 m.

EXAM QUESTIONS

Worked Exam Question

[SEG]

COMMENTS

John uses a piece of string to measure the perimeter of shapes. It fits exactly round a rectangle 10 cm by 8 cm.

a He fits it exactly round a square. How long is one side of the square?

◆ Find the length of the piece of string by working out the perimeter of the rectangle.

Draw a sketch. It is often helpful.

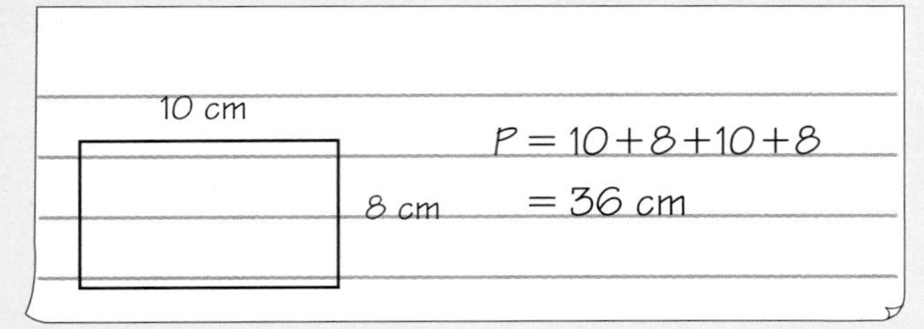

M1 for evidence of adding the lengths of the four sides of the rectangle

◆ Work out the length of the side of a square that has a perimeter of 36 cm.

Draw another sketch.

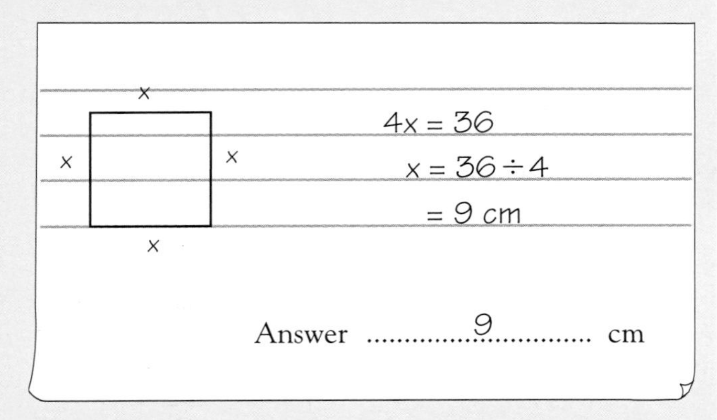

M1 for showing 36 ÷ 4

A1 for correct answer

3 marks

b He uses another piece of string of length 40 cm to form a circle. Calculate the radius of the circle.

◆ Rearrange $C = 2\pi r$ into $r = \dfrac{C}{2\pi}$.

◆ Use $r = \dfrac{C}{2\pi}$ with $C = 40$ and $\pi = 3.14$.

M1 for ÷2
M1 for $\div\pi$
(M1 for dividing only by π)

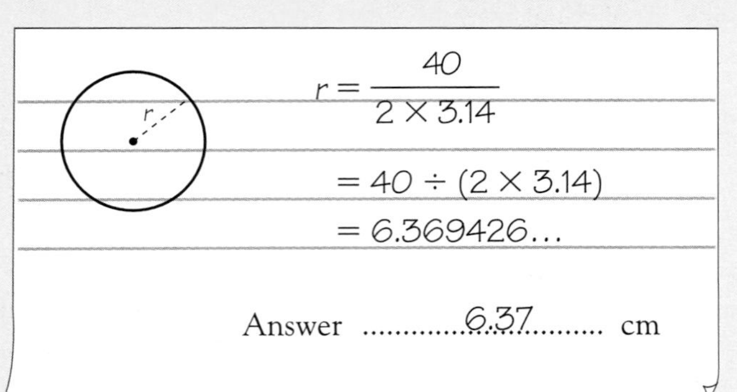

A1 for correct answer (6.4, 6.37 or other correct rounding is acceptable)

Approximate sensibly.

3 marks

Exam Questions

1

6 m
2 m
3.5 m
4 m
Not to scale

The diagram shows the floor plan of the lounge which Peter Paste is going to redecorate.
The walls are to be papered to a height of 2.4 m.

a Calculate the perimeter of the room.

The table below is used to calculate the number of rolls of wallpaper needed to paper different rooms.

Height (m)	**Perimeter** (m) 13	14	15	16	17	18	19	20
2.0	5	5	6	6	6	7	7	8
2.2	5	6	6	6	7	7	8	8
2.4	6	6	7	7	8	8	9	9
2.6	6	7	7	8	8	9	9	10
2.8	7	7	8	8	9	10	10	11

b Use the table to find how many rolls of wallpaper Peter needs to paper the lounge.

c Peter bought wallpaper which cost £7.54 a roll. How much did he spend? [MEG]

2 Janet measured the diameter and circumference of four objects:

2p coin

Can of beans

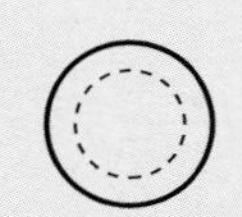
Dinner plate

Bicycle wheel

Her results are shown below:

Object	**Diameter**	**Circumference**
2p coin	2.5 cm	8 cm
Tin of beans	8 cm	25 cm
Dinner plate	23 cm	64 cm
Bicycle wheel	65 cm	205 cm

Janet has measured one of the circumferences incorrectly.
Which one?
Give a reason for your answer. [NEAB]

3

Chris wants to stick braid around the circles at the top and the bottom of this lampshade.
The top diameter is 40 cm.
The bottom diameter is 48 cm.

a What length is needed around the top?

b What length is needed around the bottom?

Braid is sold in lengths measured to the nearest 10 cm.

c What length does Chris need to buy?

d The braid costs £1.20 a metre.
How much does Chris pay? [MEG]

4

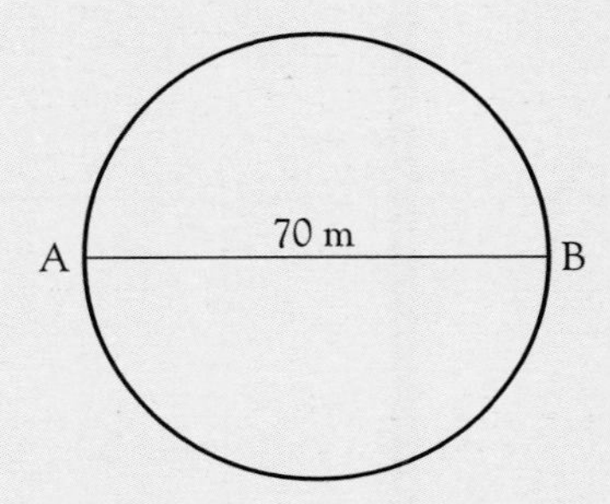

The diagram represents a circular training track.
The diameter of the track, AB, is 70 metres.
Alisa and Bryony have a race.
Alisa runs along the diameter from A to B and back again.
Bryony starts at A and runs all the way round the track to A again.

Work out how much further Bryony runs than Alisa. [L]

5 Andre is rolling a hoop along the ground.

The hoop has a diameter of 90 cm.

What is the minimum number of complete turns the hoop has to make to cover a distance of 5 m? [SEG]

EXAM QUESTIONS

6

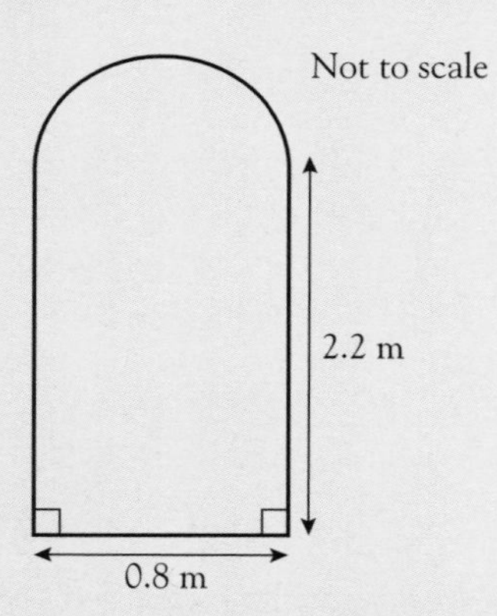

The diagram shows a wooden frame for a doorway. The frame consists of three sides of a rectangle and a semicircle.

a **i** Calculate the length of the wood that is in the shape of a semi-circle of diameter 0.8 m.

ii Calculate the perimeter of the frame.

[MEG, p]

7

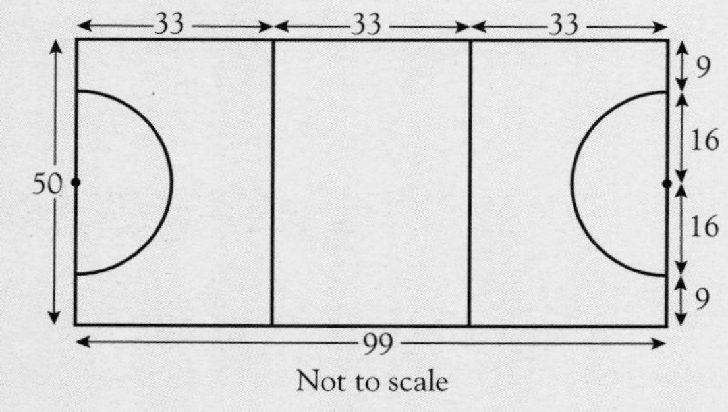

The diagram shows the dimensions in feet of a netball court. The court is a rectangle.
The markings consist of straight lines and two semi-circles of radius 16 feet.
The two straight lines inside the court are parallel to the end lines of the court.

a Calculate the total length of the straight line markings.

b Calculate the total length of the two semi-circular markings. (Take $\pi = 3.14$)

[MEG]

8 Take π as 3.142.

a Calculate the circumference of a circle of diameter of 30 cm.

b

The circumference of a trundle wheel is 100 centimetres. Work out the diameter of the trundle wheel, to the nearest centimetre.

[MEG]

9 A garden roller has a radius of 25 cm.

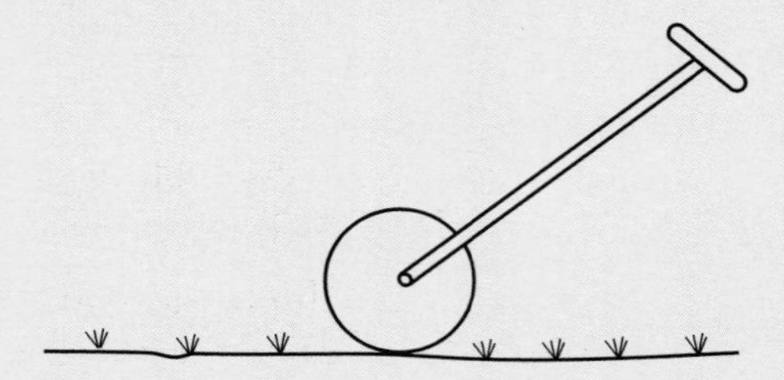

a What is the circumference of the roller? (You may take $\pi = 3.14$)

b How many revolutions are required to roll the edge of a lawn 78.5 m long?

[NI]

10 The penny-farthing is a bicycle with a large front wheel and a small rear wheel.

The radius of the large front wheel is 75 cm.

a The front wheel makes one full turn. Calculate the distance that the bicycle moves.

b The small rear wheel has a radius of 25 cm. How many full turns will the small wheel make for each full turn of the front wheel?

[SEG]

FACT SHEET 1: FRACTIONS, DECIMALS AND PERCENTAGES

Read through these important facts, using your calculator to help you to understand.

Proper fractions

Four fifths, written $\frac{4}{5}$, is a **proper fraction**. The **numerator** (top number) is smaller than the **denominator** (bottom number).

A proper fraction is part of a whole, so it is smaller than 1.

Key in	Display
4 a b/c 5	4⌟5 $\frac{4}{5}$
	or 4⌐5

Equivalent fractions

Equivalent fractions are different ways of writing the same fraction.

$\frac{2}{3}$ shaded

$\frac{8}{12}$ shaded

The same fraction of the strip has been shaded, showing that $\frac{2}{3} = \frac{8}{12}$.

You can multiply, or divide, both numerator and denominator by the same number to obtain an equivalent fraction:

$$\frac{3}{4} = \frac{3 \times 5}{4 \times 5} = \frac{15}{20} \qquad \frac{3}{4} = \frac{3 \times 7}{4 \times 7} = \frac{21}{28}$$

Key in	Display
8 a b/c 1 2	8⌟12 $\frac{8}{12}$

Try

8 a b/c 1 2 =	2⌟3 $\frac{2}{3}$

'Cancelling down'

$$\frac{12}{32} = \frac{12 \div 2}{32 \div 2} = \frac{6}{16} = \frac{6 \div 2}{16 \div 2} = \frac{3}{8}$$

Try

1 2 a b/c 3 2 =	3⌟8 $\frac{3}{8}$

The calculator gives the simplest form straight away when you press =

Simplest form or lowest terms

The fraction $\frac{3}{8}$ is in its **simplest form** (or **lowest terms**) as there are no numbers that divide exactly into 3 and 8.

Improper fractions

In an **improper** (or **top-heavy**) **fraction**, the numerator (top) is bigger than the denominator (bottom). The fraction is bigger than 1, for example:

$$\frac{11}{8} = 1\tfrac{3}{8}$$

This form is called a mixed number.

Key in

1 1 a b/c 8 =	1⌟3⌟8 $1\frac{3}{8}$

Writing as a mixed number

To do the above without a calculator:

$$\frac{11}{8} = 11 \div 8 = 1 \text{ remainder } 3 = 1\tfrac{3}{8}$$

Since you are dividing by 8, the remainder 3 means three-eighths.

Writing a mixed number as a top-heavy fraction

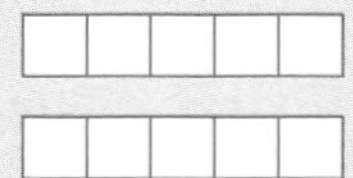

10 (fifths) 3 (fifths)

↓ ↓

$$2\tfrac{3}{5} = \frac{2 \times 5 + 3}{5} = \frac{13}{5}$$

Key in:

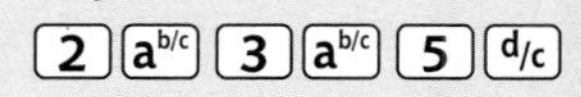

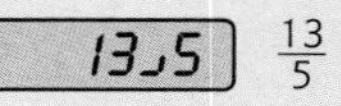

$\frac{13}{5}$

FACT SHEET 1: FRACTIONS, DECIMALS AND PERCENTAGES

Changing a fraction to a decimal

$\frac{3}{4} = 3 \div 4 = 0.75$

Key in: 3 ÷ 4 = → 0.75

Try 3 $a^{b/c}$ 4 = $a^{b/c}$

What happens if you now press $a^{b/c}$ again?

$\frac{12}{5} = 12 \div 5 = 2.4$

Key in: 1 2 ÷ 5 = → 2.4

Try 1 2 $a^{b/c}$ 5 = $a^{b/c}$

Now press $a^{b/c}$ again and again …

What happened at this stage?

Changing a decimal to a fraction

$0.7 = \frac{7}{10}$ You cannot get your calculator to tell you this. You must know it yourself!

$1.5 = 1\frac{5}{10} = 1\frac{1}{2}$ You can write this in simplest form by cancelling it yourself or using the calculator.

$5.93 = 5\frac{93}{100}$ $0.002 = \frac{2}{1000} = \frac{1}{500}$ $0.35 = \frac{35}{100} = \frac{7}{20}$

There are some decimals which are difficult to change into a fraction. They may have more figures than the display shows on the calculator or they may recur. Some never end or recur and it is impossible to write them as fractions.

Learn these common recurring ones:

$0.\dot{1} = \frac{1}{9}$, $0.\dot{2} = \frac{2}{9}$, $0.\dot{3} = \frac{3}{9} = \frac{1}{3}$, $0.\dot{4} = \frac{4}{9}$, $0.\dot{5} = \frac{5}{9}$, $0.\dot{6} = \frac{6}{9} = \frac{2}{3}$, $0.\dot{7} = \frac{7}{9}$, $0.\dot{8} = \frac{8}{9}$

(What about $0.\dot{9}$?)

Writing percentages as fractions or decimals

(i) 37% means 37 out of 100 or 37 hundredths

so $37\% = \frac{37}{100} = 0.37$

(ii) $45\% = \frac{45}{100} = \frac{9}{20}$ (cancelled down as a fraction in its simplest form)

or $45\% = \frac{45}{100} = 0.45$ (as a decimal)

Writing fractions or decimals as percentages

To change a fraction or a decimal to a percentage, **multiply by 100**

$\frac{11}{20} = \frac{11}{20} \times 100\% = 55\%$

$0.82 = 0.82 \times 100\% = 82\%$

Check these on your calculator and LEARN THEM.

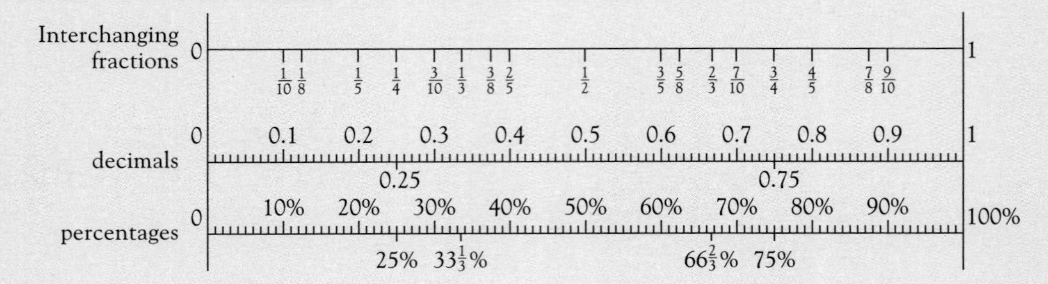

FACT SHEET 1: FRACTIONS, DECIMALS AND PERCENTAGES

Writing a number as a proportion (fraction, decimal, percentage) of another number

(i) If 25 out of 60 pupils in Yr 11 go to a concert,

- fraction is $\frac{25}{60} = \frac{5}{12}$ (cancel down)
- proportion is $\frac{25}{60} = 25 \div 60 = 0.4166\ldots \approx 0.42$

≈ means approximately equal to.

- percentage is $\frac{25}{60} \times 100 = 41\frac{2}{3}\% = 41.666\ldots\% \approx 42\%$

(ii) Write 6.2 cm as a percentage of 25 cm.

$\frac{6.2}{25} \times 100 = 24.8\%$

You cannot use your fraction button as the top is not a whole number.

Calculating a proportion (fraction, decimal or percentage) of something

(i) Find two thirds of 270.

Method 1:

$\frac{2}{3} \times 270 = \underline{180}$

Key in:
2 a^b/c 3 × 2 7 0 =
or 2 ÷ 3 × 2 7 0 =

Method 2:

This way is often easier when calculating mentally.

- Find $\frac{1}{3}$ of 270 by dividing by 3.
  $270 \div 3 = 90$
- Then multiply by 2 to find $\frac{2}{3}$.
  $90 \times 2 = \underline{180}$

Key in:
2 7 0 ÷ 3 × 2 =
↑ The calculator display shows 90 at this stage.

(ii) Find 82% of 350.

Method 1 (using fractions):

- Write 82% as a fraction, then multiply by 350.

  $82\% \text{ of } 350 = \frac{82}{100} \times 350 = \underline{287}$

Key in:
8 2 a^b/c 1 0 0 × 3 5 0 =
or 8 2 ÷ 1 0 0 × 3 5 0 =

Method 2 (using decimals):

- Write 82% as a decimal, then multiply by 350.

  $82\% \text{ of } 350 = 0.82 \times 350 = \underline{287}$

Key in:
0 . 8 2 × 3 5 0 =

This is a very useful method for later work.

Method 3:

- Find 1% of 350 by dividing by 100, then find 82% by multiplying by 82.

  $350 \div 100 = 3.5$

  $3.5 \times 82 = \underline{287}$

Key in:
3 5 0 ÷ 1 0 0 × 8 2 =
↑ 3.5 shows on display at this stage.

2 NUMBER I

In this chapter you will
- **practise solving mixed problems using fractions, decimals and percentages, using a calculator**
- **learn quick ways of calculating without a calculator**
- **make estimates by approximating numbers to one significant figure**

Solving problems using fractions, decimals and percentages

You may use a calculator for these examples.
If you have forgotten a method, look it up in Fact Sheet 1 (pages 13–15). References are given in the examples.

Sample Question 1

Pupils on a school trip were given a choice between cola and lemonade.

a There were 260 girls on the trip. 65% of them chose cola. How many girls chose cola?

b There were 210 boys on the trip. Two-fifths of them chose cola. How many boys chose cola?

c There were 470 pupils altogether on the trip. What percentage of them chose cola?

[MEG]

Answer

a Find 65% of 260 using one of the following methods. Decide which method you prefer.

- Use fractions:

$$\frac{65}{100} \times 260$$

[6] [5] [a b/c] [1] [0] [0] [×] [2] [6] [0] [=]

- Use decimals:

[0] [.] [6] [5] [×] [2] [6] [0] [=]

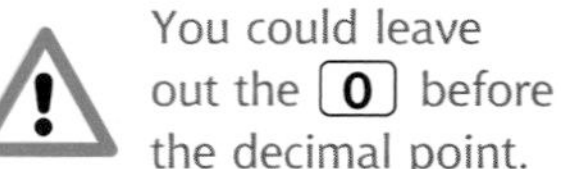

$$0.65 \times 260$$

Remember that $\frac{65}{100} = 0.65$

- Divide by 100 to find 1% then multiply by 65 to find 65%.

[2] [6] [0] [÷] [1] [0] [0] [×] [6] [5] [=]

65% of 260 = 169

<u>169 girls chose cola.</u>

CALCULATING

b Find two-fifths of 210. Choose your method.

- Multiply by $\frac{2}{5}$.   [2] [a b/c] [5] [×] [2] [1] [0] [=]

or [2] [÷] [5] [×] [2] [1] [0] [=]

FACT SHEET 1, Calculating a proportion (fraction) of something

or

- Divide by 5 to find $\frac{1}{5}$ then multiply by 2 to get $\frac{2}{5}$.   [2] [1] [0] [÷] [5] [×] [2] [=]

  $\frac{2}{5}$ of 210 = 84

  84 boys chose cola.

c

- Add your two answers to find the total number choosing cola.
- Write this as a fraction of the total number of pupils.
- Change this fraction to a percentage by multiplying by 100.

Number choosing cola $= 169 + 84 = 253$

Fraction choosing cola $= \dfrac{253}{470}$

FACT SHEET 1, Writing a number as a proportion (percentage) of another number

Percentage choosing cola $= \dfrac{253}{470} \times 100\%$

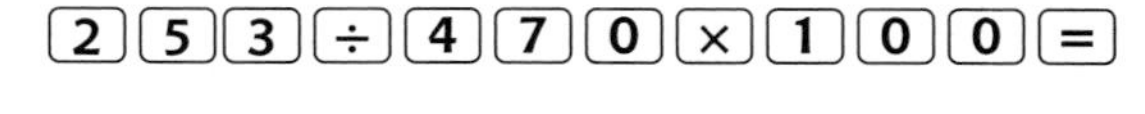

$= 53.829\ldots$
$= 53.8\%$ **(1 d.p.)**

Approximate sensibly.

53.8% of the pupils chose cola.

If you use the fraction key [2] [5] [3] [a b/c] [4] [7] [0] [×] [1] [0] [0] [=] you get $53\frac{39}{47}$.
Though this answer is exact, you would usually write it as a decimal (press [a b/c] again) and approximate it, because it is difficult to visualise $\frac{39}{47}$.

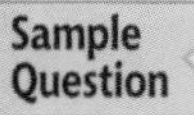

Sample Question 2

Paul and Sue got the two highest marks in a test.
Who came first, and by how many marks?

Paul $\frac{112}{140}$ Sue 75%

Answer

- Write $\frac{112}{140}$ as a percentage and compare it with 75%.

  $\dfrac{112}{140} \times 100\% = 80\%$

FACT SHEET 1, Writing fractions or decimals as percentages

  Paul came first.

- Find out how many marks Sue got by finding 75% of 140.

  Check this.

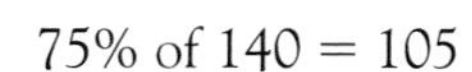

  75% of 140 = 105

FACT SHEET 1, Calculating a proportion (percentage) of something

  Sue got 105 marks, Paul got 112 marks

  Paul came first by 7 marks.

CALCULATING

Sample Question 3

Write these numbers in order of size, with the smallest first.

$\frac{2}{3}$, 0.66, 68%, $\frac{3}{5}$

Answer

- Change all the numbers to the same type, for example percentages.

$\frac{2}{3} = \frac{2}{3} \times 100\% = \mathbf{66.666\ldots\%}$

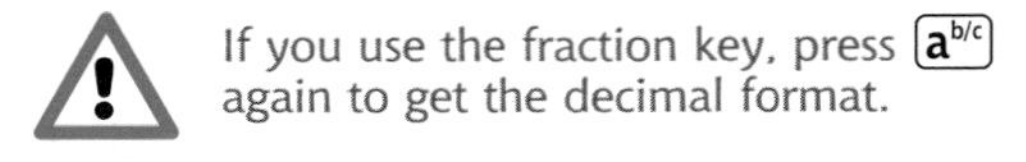
If you use the fraction key, press $a^{b/c}$ again to get the decimal format.

$0.66 = 0.66 \times 100\% = 66\%$

$\frac{3}{5} = \frac{3}{5} \times 100\% = 60\%$

So 60% is the smallest ($\frac{3}{5}$), then 66% (0.66), then 66.666 …% ($\frac{2}{3}$), then 68%.

Order is $\frac{3}{5}$, 0.66, $\frac{2}{3}$, 68%

Make sure that you write the numbers in the form given in the question.

You could have changed all the numbers to decimals.
Changing them all to fractions would make them difficult, though not impossible, to compare.

Sample Question 4

Karl did a survey to find out how pupils travelled to school. He found that 30% came by car, 45% came by bus and the remaining five people walked.

a What fraction walked?
b How many came by bus?

Answer

a

- Find the percentage that walked, then change this to a fraction.

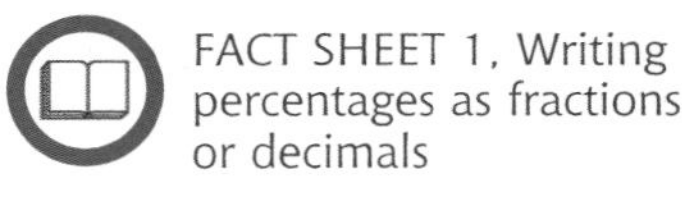
FACT SHEET 1, Writing percentages as fractions or decimals

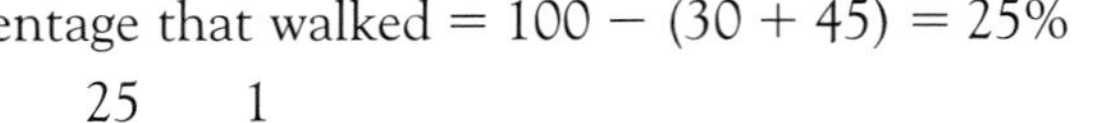
Percentage that walked = $100 - (30 + 45) = 25\%$

$$25\% = \frac{25}{100} = \frac{1}{4}$$

$\frac{1}{4}$ of the pupils walked.

b

- Find out how many people were in the survey, then find 45% of this number.

$\frac{1}{4}$ of the total is 5, so number in survey $= 4 \times 5 = 20$

45% of $20 = 9$

9 people came by bus.

Sample Question 5

12 identical wine glasses are filled to the same level using all the wine from a $4\frac{1}{2}$ litre container. What fraction of a litre of wine is in each glass?

CALCULATING

Answer

- To find out how much wine is in each glass, divide $4\frac{1}{2}$ by 12.

$4\frac{1}{2} \div 12 = \frac{3}{8}$ [4] [a b/c] [1] [a b/c] [2] [÷] [1] [2] [=]

There is $\frac{3}{8}$ litre in each glass.

If you decide to write $4\frac{1}{2}$ as a decimal first, this gives $4.5 \div 12 = 0.375$, [4] [.] [5] [÷] [1] [2] [=] so there is 0.375 litres in each glass. You now have to change this to a fraction.

Remember that $0.375 = \frac{375}{1000}$. Now cancel it down.

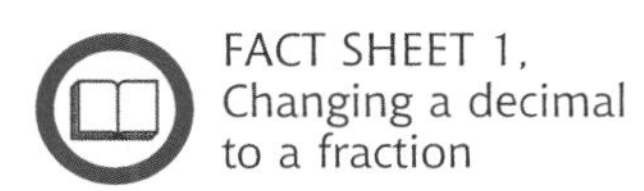

FACT SHEET 1, Changing a decimal to a fraction

On your calculator: [3] [7] [5] [a b/c] [1] [0] [0] [0] [=]

Some calculators may not be able to display all the digits here – check yours!

Without a calculator, find a number that divides exactly into the top and the bottom:

$$\frac{375}{1000} = \frac{375 \div 5}{1000 \div 5} = \frac{75}{200}$$

Do not leave it in this form. It can be simplified further.

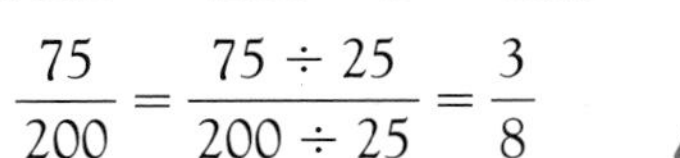

$$\frac{75}{200} = \frac{75 \div 25}{200 \div 25} = \frac{3}{8}$$

You could have divided by 5, then by 5 again, but try to spot larger numbers that divide into both. In fact, 125 goes into 375 and 1000!

$0.375 = \frac{3}{8}$

TASK

SPECIAL OFFERS!

Here are three special offers that you might see in a supermarket. Compare the percentage reductions if you take up the offer. Which is the best offer?

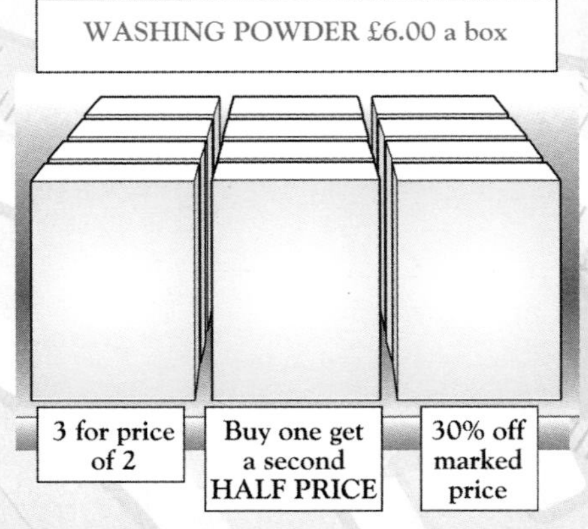

List other special offers that you see and work out the percentage reduction on the items if you take up the offer.

CALCULATING

Exercise 2.1

1 In a mixed box of apples and oranges containing 120 fruits, 60% were apples. How many apples were there?

2 Write these numbers in order, with the smallest first: $\frac{2}{5}$, 45%, 0.44, $\frac{3}{10}$

3 Jane had a film in her camera that would take 24 snaps. She had used 8 so far. Tony had a film in his camera that would take 36 snaps. He had used 9 so far.

a Who had the larger fraction of snaps left?

b Express the fractions of snaps left as percentages.

4 At the school sports day, ice cream was sold.

a It was found that a 2 litre tub of ice cream made 20 wafers. What fraction of a litre did each wafer contain?

b The ice creams were selling so well at 20p each that the teachers decided to put up the price by 5p. What percentage increase did this represent?

5 Sam and Jo sat at a rectangular table. Sam measured the length of the table top and found it was 0.9 m long. Jo measured the width and said it was $\frac{3}{4}$ m wide. What was the perimeter of the table?

6 In a sale, the prices of items were advertised as 'one-third off'.

a Mrs Rogers saw a skirt originally priced at £27. What was the sale price?

As a further incentive, if two items were bought, the shop reduced the total bill by 20%. Mrs Rogers decided to buy the skirt and a shirt for her husband, originally priced at £15.

b What was the total she had to pay?

7 Due to a heavy snowfall, only 55% of pupils out of a total of 800 managed to get to school.

a How many pupils arrived?

The teachers lived much further away and only 30% of the 50 staff arrived. The headteacher said that the school would close if she did not have at least one teacher for every 30 pupils.

b Was everyone sent home?

8 Mrs Lock hired a canal boat for a holiday with friends. It cost £750 for the week. She paid a deposit of 20%. How much was this?

9 Emma inherited £2500 from her great aunt. She spent 48% of it on a computer, gave $\frac{3}{20}$ of it to charity and saved the rest.

a How much did she spend on the computer?

b How much did she give to charity?

c How much did she save?

d What percentage of the total did she save?

10 Copy out and fill in this table, writing fractions in their simplest form.

Fraction	$\frac{3}{5}$			$\frac{1}{3}$	$\frac{9}{20}$		
Decimal			0.2				0.51
Percentage		85%				12%	

11 Write these numbers in order of size, with the smallest first: $\frac{2}{5}$, 39%, $\frac{3}{7}$, 0.405, $\frac{1}{3}$, 30%

12 Mr Brown went to a garden centre sale where plants were advertised as 'half the marked price'. He bought a clematis marked at £5.

a How much was the sale price?

Because he had a loyalty card he received a further reduction of 10% of the sale price.

b How much did he pay for the clematis?

c What percentage of the marked price did he pay?

13 Of the 200 members of the Golf Club, only 65% voted in the election of a new treasurer.

a How many voted?

To be elected, a candidate must get at least two-thirds of the votes cast. Anthony obtained 90 votes.

b Was he successful in being elected? (Show all your working.)

14 A regular hexagon has a perimeter of $10\frac{1}{2}$ cm. What is the length of each side?

15 A large tin of cat food contained $1\frac{1}{2}$ kg. How many **grams** were left after using one-third of it?

16 A washing machine is advertised for sale at £350 + VAT at $17\frac{1}{2}$%.

a How much is the VAT on the washing machine?

b What is the cost of the washing machine, including VAT?

NO CALCULATOR

Quick ways of calculating without a calculator

Your most useful tool is a knowledge of the multiplication tables.

Try to practise them!

Finding a fraction of something

It is easy to find $\frac{1}{2}$ (by dividing by 2), $\frac{1}{3}$ (by dividing by 3), $\frac{1}{4}$ (by dividing by 4), $\frac{1}{5}$ (by dividing by 5), etc. From these you can find more complicated fractions, such as $\frac{2}{3}$, $\frac{3}{4}$, $\frac{4}{5}$, etc.

◆ To find $\frac{2}{3}$ of 27

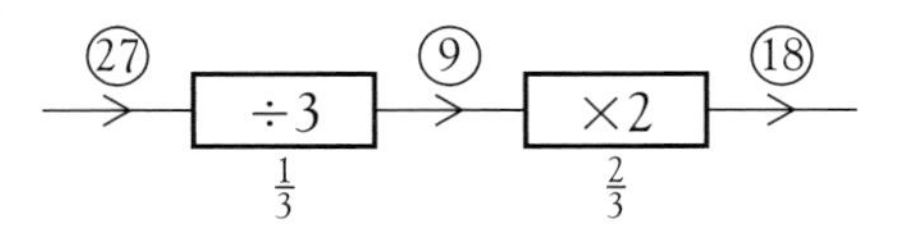

$\frac{2}{3}$ of 27 = 18

◆ To find $\frac{3}{4}$ of 28

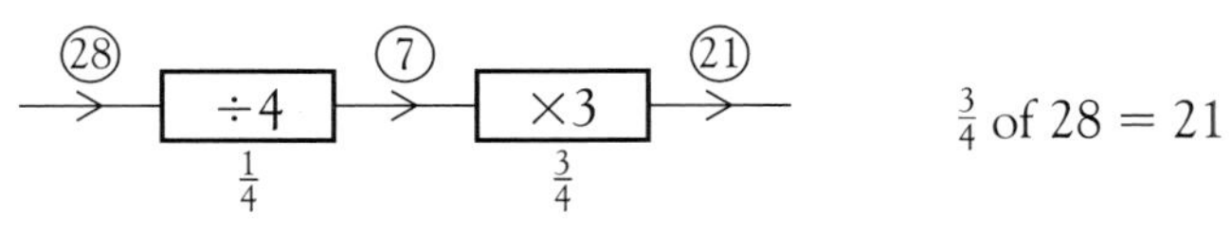

$\frac{3}{4}$ of 28 = 21

◆ To find $\frac{4}{5}$ of 20

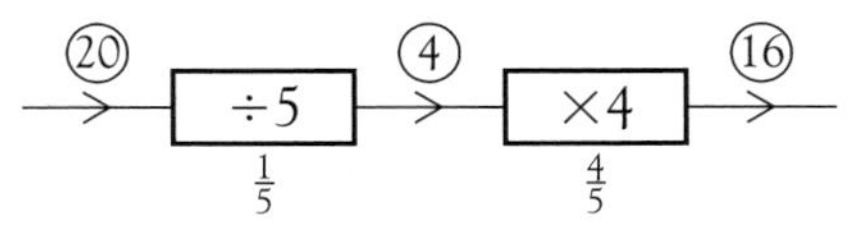

$\frac{4}{5}$ of 20 = 16

Exercise 2.2

Try these without a calculator.

1 $\frac{1}{5}$ of 40
2 $\frac{1}{3}$ of 24
3 $\frac{2}{5}$ of 55
4 $\frac{1}{2}$ of 82
5 $\frac{2}{3}$ of 12
6 $\frac{3}{5}$ of 30
7 $\frac{1}{7}$ of 35
8 $\frac{2}{9}$ of 18
9 $\frac{3}{4}$ of 200
10 $\frac{3}{8}$ of 32
11 $\frac{4}{5}$ of 100
12 $\frac{4}{9}$ of 27
13 $\frac{1}{4}$ of 480
14 $\frac{2}{3}$ of 45
15 $\frac{3}{7}$ of 49
16 $\frac{2}{11}$ of 88

◆ To find $\frac{1}{10}$ of something, **divide by 10.**

$320 \div 10 = 32$
$45 \div 10 = 4.5$
$23.6 \div 10 = 2.36$
$5400 \div 10 = 540$
$5408 \div 10 = 540.8$
$0.4 \div 10 = 0.04$

You do not need a calculator to divide by 10.

Some people remember to move the decimal point one place to the left. Others remember to move all the figures one place to the right.

Remember, in a whole number the decimal point is at the right-hand end, but is not usually written.

Exercise 2.3

DO NOT use a calculator to work out the following.

1 $\frac{1}{10}$ of 27
2 $\frac{1}{10}$ of 3400
3 $6.31 \div 10$
4 $0.324 \div 10$
5 $\frac{1}{10}$ of 540
6 $0.04 \div 10$
7 $275 \div 10$
8 $3 \div 10$
9 $120.3 \div 10$
10 $\frac{1}{10}$ of 6.4
11 $\frac{1}{10}$ of 200 + $\frac{1}{10}$ of 60
12 $0.000351 \div 10$
13 $\frac{3}{10}$ of 21
14 $\frac{7}{10}$ of 90
15 $\frac{9}{10}$ of 2000
16 $\frac{3}{10}$ of 5

NO CALCULATOR

Finding a percentage of something

FACT SHEET 1, Calculating a proportion (percentage) of something

$50\% = \frac{50}{100} = \frac{1}{2}$ → ÷2 (50%) →

$25\% = \frac{25}{100} = \frac{1}{4}$ → ÷4 (25%) →

$75\% = \frac{75}{100} = \frac{3}{4}$ → ÷4 (25%) → ×3 (75%) →

$33\frac{1}{3}\% = \frac{33\frac{1}{3}}{100} = \frac{1}{3}$ → ÷3 ($33\frac{1}{3}\%$) →

$66\frac{2}{3}\% = \frac{66\frac{2}{3}}{100} = \frac{2}{3}$ → ÷3 ($33\frac{1}{3}\%$) → ×2 ($66\frac{2}{3}\%$) →

$10\% = \frac{10}{100} = \frac{1}{10}$ → ÷10 (10%) →

Once you know 10% it is easy to find 20% (double it), 30% (multiply it by 3), 40% (multiply it by 4), 5% (halve it), etc.

◆ To find 30% of 400

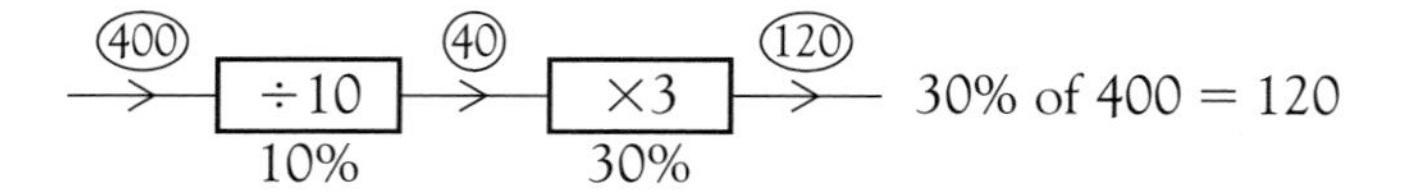

◆ To find 5% of 24

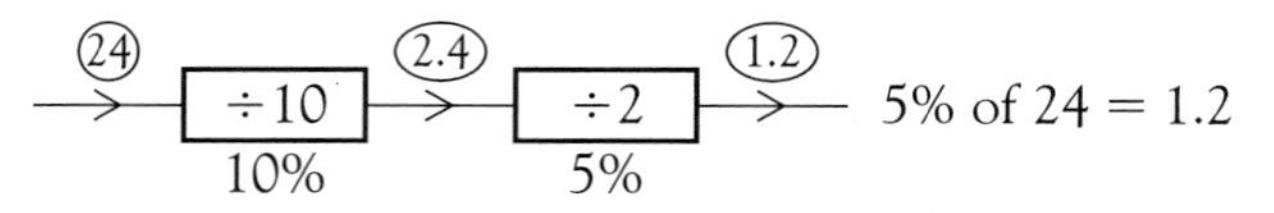

Once you get the idea, you can find more complicated percentages such as 15% or $17\frac{1}{2}\%$.

◆ To find 15% of 600

(600) → ÷10 (10%) → (60)

10%	=	60
5%	=	30
15%		90

15% of 600 = 90

◆ To find $17\frac{1}{2}\%$ of 3200

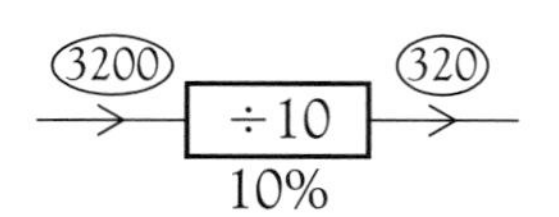

10%	=	320
5%	=	160
$2\frac{1}{2}\%$	=	80
$17\frac{1}{2}\%$		560

$17\frac{1}{2}\%$ of 3200 = 560

$17\frac{1}{2} = 10 + 5 + 2\frac{1}{2}$
This is useful when you need to find VAT when the rate is set at $17\frac{1}{2}\%$.

Exercise 2.4

DO NOT use a calculator to work out the following.

1 50% of 32
2 25% of 24
3 20% of 35
4 75% of 200
5 5% of 300
6 80% of 30
7 70% of 40
8 15% of 160
9 $17\frac{1}{2}\%$ of 40
10 35% of 800
11 $33\frac{1}{3}\%$ of 24
12 $66\frac{2}{3}\%$ of 300
13 30% of 30
14 45% of 220
15 65% of 160
16 $12\frac{1}{2}\%$ of 480

NO CALCULATOR

More calculations with fractions without using a calculator

Sample Question 

Put these fractions in order of size, smallest first.

$\frac{2}{3}, \frac{1}{2}, \frac{5}{12}, \frac{11}{24}, \frac{3}{4}$

Answer

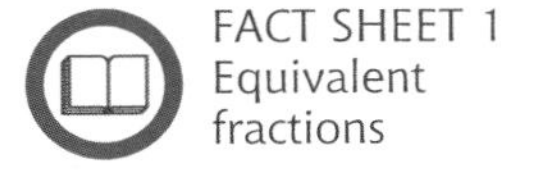

- Write all the fractions in an equivalent form, with the same denominator.

$$\frac{2}{3} = \frac{2 \times 8}{3 \times 8} = \frac{16}{24}, \quad \frac{1}{2} = \frac{12}{24}$$

$$\frac{5}{12} = \frac{5 \times 2}{12 \times 2} = \frac{10}{24}, \quad \frac{3}{4} = \frac{3 \times 6}{4 \times 6} = \frac{18}{24}$$

Look for a number that can be divided by all the denominators. 24 is the smallest number in this case.

- Put all the numbers in order:

$$\frac{10}{24}, \frac{11}{24}, \frac{12}{24}, \frac{16}{24}, \frac{18}{24}$$

- Write them in the format given in the question. The order is then:

$$\frac{5}{12}, \frac{11}{24}, \frac{1}{2}, \frac{2}{3}, \frac{3}{4}$$

Using a calculator, you could change them to decimals or percentages to compare them.

To add or subtract fractions, change them to equivalent fractions with the same denominator.

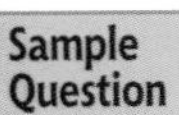 

The three classes in Year 9 raised money for charity. 9P raised $\frac{1}{2}$ of the total, 9S raised $\frac{2}{5}$ and 9J raised the rest. What fraction did 9J raise?

Answer

- Find the fraction raised by 9P and 9S by working out $\frac{1}{2} + \frac{2}{5}$.
- Change $\frac{1}{2}$ and $\frac{2}{5}$ into equivalent fractions with the same denominator. The **smallest number** that can be divided by both 2 and 5 is 10.

$$\frac{1}{2} = \frac{5}{10}, \quad \frac{2}{5} = \frac{4}{10}$$

You do not have to find the smallest number, but it is normally easier if you do!

so $$\frac{1}{2} + \frac{2}{5} = \frac{5}{10} + \frac{4}{10} = \mathbf{\frac{9}{10}}$$

5 tenths + 4 tenths = 9 tenths

- Subtract $\frac{9}{10}$ from 1 to find the amount raised by 9J.

$1 - \frac{9}{10} = \frac{1}{10}$

9J raised $\frac{1}{10}$ of the total.

NO CALCULATOR

Exercise 2.5

Do not use a calculator to work out the following.

1 a $\frac{1}{2} + \frac{1}{4}$

b $\frac{3}{4} + \frac{1}{6}$

c $\frac{2}{7} + \frac{3}{5}$

d $\frac{1}{8} + \frac{2}{3}$

e $\frac{2}{3} + \frac{1}{4}$

f $\frac{7}{12} - \frac{1}{3}$

g $\frac{2}{3} - \frac{1}{6}$

h $\frac{1}{2} - \frac{2}{7}$

i $\frac{5}{8} - \frac{2}{5}$

j $\frac{3}{5} - \frac{1}{2}$

2 Put these fractions in order of size, with the smallest first.

a $\frac{7}{20}$, $\frac{1}{4}$, $\frac{3}{10}$, $\frac{2}{5}$

b $\frac{4}{9}$, $\frac{2}{3}$, $\frac{3}{4}$, $\frac{1}{2}$, $\frac{11}{18}$

3 Add $\frac{1}{4}$ and $\frac{1}{3}$, then subtract the result from $\frac{5}{6}$.

Exercise 2.6

This is a mixed exercise involving fractions, decimals and percentages. Do not use a calculator. You may write down your working.

1 Cotswold Creameries delivered milk to $\frac{2}{5}$ of the homes in a village and Chipping Dairies delivered to $\frac{1}{3}$ of the homes. What fraction of homes in the village were covered by these deliveries?

2 A jug holds 750 ml when it is full. How many ml does it hold when it is two-thirds full?

3 20% of oranges in a bag of 20 were bad. How many oranges were bad?

4 What is the decimal equivalent of:

a $\frac{4}{10}$ b $\frac{73}{100}$ c $\frac{3}{5}$

d $\frac{7}{20}$ e $\frac{3}{25}$ f $\frac{13}{50}$

5 A train company is increasing all its fares by 10%.

a What will be the increase for a journey at present costing £2.80?

b How much will the new fare be if the present fare is £25.00?

6 A bus that can carry 60 passengers was surveyed on each of six journeys and found to be the following parts full:

$\frac{3}{4}$, 50%, $\frac{1}{3}$, 0.9, 60%, 0.4

How many people was it carrying on each occasion?

7 In a test, Abi got $\frac{3}{5}$ of the marks. What was her percentage mark?

8 Find $66\frac{2}{3}$% of £15.

9 How much is 28% of £1.00?

10 50 g of nails cost £1.00. Find the cost of $\frac{1}{4}$ kg of nails.

11 $\frac{1}{4}$ of my salary is spent on my mortgage and $\frac{2}{3}$ on food and travelling. What fraction is left?

12 Paula earned £150. She spent 40% of it and saved the rest.

a How much did she spend?

b What fraction did she save?

13 Three-quarters of the 80 houses in Bell Road were sent a questionnaire in a survey. One-third of the households completed the questionnaire and returned it. How many questionnaires were returned?

14 A pack of paper was priced at £4.00, excluding VAT at $17\frac{1}{2}$%.

a How much was the VAT?

b What was the total cost of the pack of paper including the VAT?

Estimating

It is often useful to estimate the answer to a calculation, especially when you only need a rough idea of the answer or when you want to check that your calculator answer is reasonable.

Rounding to one significant figure (1 s.f.)

Although you can use other approximations, you will often need to round numbers to 1 significant figure. The following table may be useful.

For numbers between ...	Rounding to 1 s.f. is the same as rounding ...	For example ...
0.1 and 1	to the nearest tenth	0.814 = 0.8 (1 s.f.)
1 and 10	to the nearest whole number	6.93 = 7 (1 s.f.)
10 and 100	to the nearest ten	27.1 = 30 (1 s.f.)
100 and 1000	to the nearest hundred	241 = 200 (1 s.f.)
1000 and 10 000	to the nearest thousand	2899 = 3000 (1 s.f.)

Always check that your approximation is the same size as the original number, for example 3120 to 1 s.f. should read '3 thousand'.

Using 1 s.f. approximations and estimating without a calculator

Sample Question 8

Estimate the cost of 28 books at £5.90 each.

Answer

- Round to 1 s.f., then use the rounded figures.

  $28 = 30$ (1 s.f.), $5.90 = 6$ (1 s.f.)

  $28 \times 5.90 \approx 30 \times 6 = 180$

  Cost ≈ £180

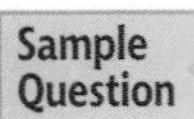

9 Estimate the number of drawing pins in 82 packets, each containing 185 pins.

Answer

- Round to 1 s.f., then use the rounded figures.

  $82 = 80$ (1 s.f.), $185 = 200$ (1 s.f.)

  $82 \times 185 \approx 80 \times 200$

  $= 16\,000$

  The number of drawing pins ≈ 16 000

Quick way:

80 × 200 = 16 000 (8 × 2)

80: 1 zero; 200: 2 zeros; 16 000: (1 + 2 = 3 zeros)

ESTIMATING

10 Estimate $2896 \div 52$.

Answer

- Round to 1 s.f.

  $2896 = 3000$ (1 s.f.), $\quad 52 = 50$ (1 s.f.)

- Then divide, using 1 s.f. values.

  $$2896 \div 52 \approx 3000 \div 50 = \frac{300\not{0}}{5\not{0}} = 60$$

<u>$2896 \div 52 \approx 60$</u>

Exercise 2.7

1 Work out approximate answers to the following, by approximating the numbers to 1 significant figure. Do not use a calculator.

a 49×9.32

b 63×27

c 2.05×68.3

d 394×809

e $693 \div 71$

f $89.41 \div 9.27$

g $28962 \div 176$

h $393.5 \div 41.7$

Now do the calculations using your calculator. Were your approximations close to the actual answers?

2 Find the approximate cost of 32 books at £11.50 a book. Is your answer an overestimate or an underestimate?

3 A season ticket for Rovers Football Club costs £621. If this allows you to go to 22 games, approximately how much are you paying for each game?

4 Each day Joshua paid £2.85 for a return bus ticket to work. If he worked 18 days in March, approximately how much did he pay on his fares that month?

5 An 800 g loaf of bread was cut into 28 slices. What was the approximate weight of each slice?

6 For which of these examples is

a 100 a good approximation,

b 10 a good approximation?

i 10×0.09

ii 10×9.9

iii 1.9×10

iv 10×0.99

v 0.9×10

7 A bicycle wheel has radius 11.3 cm. Approximately how many turns of the wheel are needed to cover a distance of 3 km? (Use $\pi \approx 3$)

8 A rectangle has length 72.4 cm and width 21.2 cm. Estimate its perimeter.

9 Estimate the perimeter of a regular octagon with sides of length 28.9 cm.

Worked Exam Question
[SEG]

COMMENTS

When 15 oranges are bought individually the total cost is £1.20.
When 15 oranges are bought in a pack the cost is £1.14.

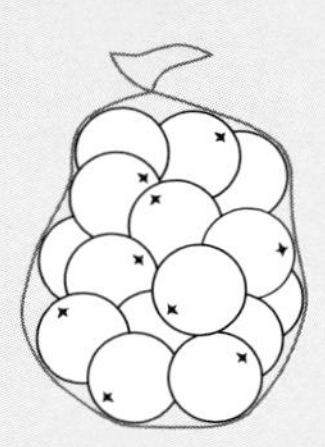

a What is the percentage saving by buying the pack?

◆ Find the amount saved. It is easier to work in pence.

◆ Write this as a percentage of the individual cost.

If you work in pounds you will need to find $\frac{0.06}{1.20} \times 100\%$

Amount saved = 6p

$\%$ saving $= \frac{6}{120} \times 100\%$

$= 5\%$

Answer ....5%..........

M1 for finding saving

M1 for expressing your answer as a % of individual cost

A1 for correct answer

3 marks

b A special offer pack of these oranges has 20% extra free.
How many extra oranges are in the special offer pack?

◆ Find 20% of 15.

Write down what you are trying to find.

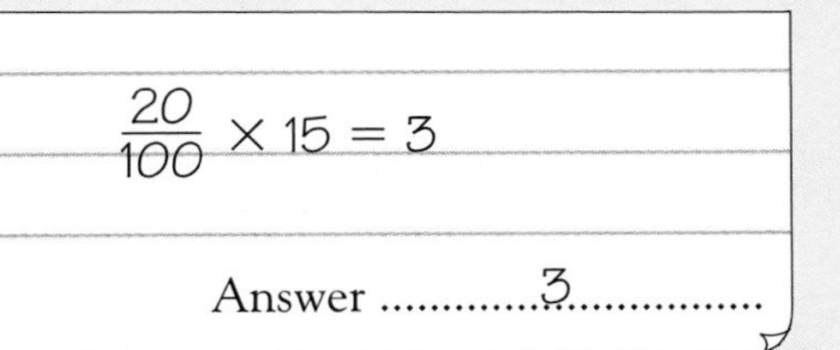

$\frac{20}{100} \times 15 = 3$

Answer ....3..........

M1 for attempting to find 20% of 15

A1 for correct answer

2 marks

c What fraction of the oranges in the special offer pack are free?

◆ Find how many there are in the special offer pack.

◆ Write 3 as a fraction of this total number.

Remember to simplify the fraction.

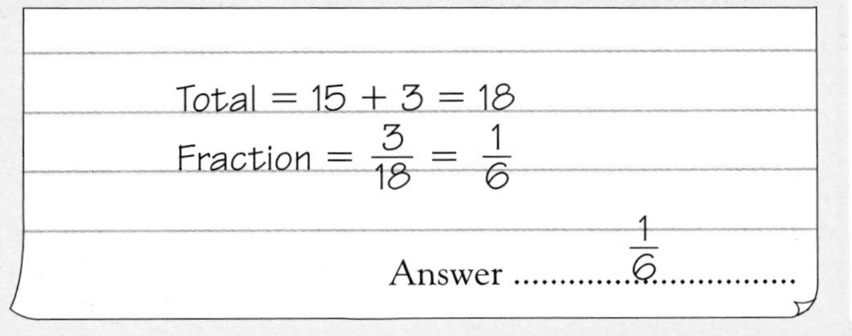

Total = 15 + 3 = 18

Fraction $= \frac{3}{18} = \frac{1}{6}$

Answer ....$\frac{1}{6}$..........

M1 for writing your **b** as a fraction of your **b** + 15

A1 for correct answer

2 marks

Exam Questions A

You must not use a calculator for these questions. You should show all your working.

1 Beacon School is organising a trip to a theme park. 307 pupils have each paid £19 for the trip.

Gina wants to know roughly how much the pupils have paid altogether. Write down the calculation Gina could do in her head. [MEG, p]

2 Freebird sells free-range eggs to supermarkets. One week they have 996 eggs. They pack the eggs in boxes. There are 12 eggs in each box.

How many boxes do they need? [MEG]

3 The *Roberts Library* is planning to buy 104 books which cost £18 each.

a **Roughly** how much will the books cost altogether? Write a calculation which you could do in your head.

b Work out the **exact** cost of the books.

c The library has £400 to spend on computer programs. Each program costs £55.
How many programs can the library buy? [MEG]

4 Flour costs 48p per kilogram. Brett bought 205 kg and shared it equally among 14 people. He calculated that each person should pay £0.72.

Without using your calculator, use a rough estimate to check whether this answer is about the right size. You must show all your working. [SEG]

5 You go into a shop and buy three pens at 49 pence each and five pads of paper at £1.99 each. The shop assistant asks you for £12.42.

Write down a calculation you can do in your head to show that the shop assistant has made a mistake. [MEG]

6

WANTED
Temporary Clerk
WAGE
£135 per week

Leon got this job.

a How much will he earn in 15 weeks?

b Deductions from his wages were £25 each week and he gave his mother £30 each week. He saved $\frac{2}{5}$ of the remainder. How much did he save each week? [MEG]

Exam Questions B

You may use a calculator for these questions.

1 Mr Donald won the first prize of £48 000 in the lottery.

a He saved $\frac{2}{3}$ of the money.
How much did he save?

b He spent £9600 on a new car.
What percentage of his winnings did he spend on the car? [MEG]

2 $\frac{3}{8}$ of the Highlands of Scotland is covered in forest.

a i Change $\frac{3}{8}$ to a decimal.

ii Write down your answer to part **i** correct to two decimal places.

b Work out the percentage of the Highlands of Scotland NOT covered in forest.

Here is a list of fractions, decimals and percentages:

67%, $\frac{1}{2}$, 0.6, 25%, 0.3, $\frac{3}{8}$

c Rewrite the list in order of size, starting with the smallest first. [L]

3 Mary earns £4.20 per hour.

a Calculate how much she earns in a 38 hour week.

b For each hour over 38 hours she works, Mary is paid overtime.
The overtime rate is 'time and a quarter'.
How much is she paid for each hour of overtime?

c Mary is given a rise of 5%.
Calculate how much she earns in a 38 hour week after the rise. [MEG]

4 In a General Election a candidate loses her/his deposit if s/he does not get at least 5% of the total votes cast.

The votes were counted in the General Election in Bradworth.

6540 people voted for Mary Ashworth,
5235 people voted for John Barnard,
425 people voted for Bill Crowther.

Calculate the percentage of the vote that Bill Crowther got.
State whether he lost his deposit. [NEAB]

EXAM QUESTIONS

5 The Standard monthly payment for an insurance scheme for Tom is £7.20.
This is reduced for the Discount monthly payment to £6.12.
Work out the percentage reduction. [L]

6 a When a pendant is carved from a block of wood, 45% of the wood is cut away. The original block of wood weighs 220 grams. What weight of wood is cut away?

b When another pendant is chipped from a stone, $\frac{4}{9}$ of the stone is chipped away. The original stone weighs 180 grams. What weight of stone is chipped away? [MEG]

7 a Change $\frac{4}{5}$ to a decimal.

b Write these numbers in order of size, smallest first.

0.805, 0.85, $\frac{4}{5}$, 0.096 [SEG]

8 The number of new cars registered in August 1994 was 460 000.

a One quarter of the cars were red. Find the number of red cars.

b 36% of the cars were blue. Find the number of blue cars.

c There were 2400 black cars. What fraction of the cars were black? [MEG, p]

9 A geography test is marked out of 80 marks.

a Georgina gets 60% of the marks.
How many marks does she get?

b Alfie gets 36 marks.
What percentage does he get? [SEG]

10

W.E. Sellem
Estate Agents
Our fee is 2% of the selling price of your house

What fee is charged on a house sold for £60 250? [NEAB]

11 In a nine carat gold ring $\frac{9}{24}$ of the weight is pure gold. What percentage of the weight of the ring is pure gold? [SEG]

12 The number of people living in cities is increasing. In 1975 about 2 people in every ten in the world lived in a city.

a What percentage of the world lived in cities in 1975?

b By the year 2000, the population in the world will be about 9000 million.
About 60% of these people will live in cities.
About how many millions will live in cities? [MEG]

13 A compact disc is normally sold for £18. In a sale the price is reduced by 20%. How much does Harry save when he buys this disc? [SEG]

14 Tom earned £40 working in a supermarket.
He spent £6.40 of this on a tee-shirt.
What is £6.40 as a percentage of £40? [MEG]

15 A club is holding a sponsored swim. It plans to give $\frac{1}{5}$ of the money to a local charity and the rest to the club's minibus fund.

The sponsored swim raises £350.

a How much is given to charity?

b How much is given to the minibus fund? [MEG]

16 a Complete the following.

$$0.14 = \frac{}{100} = \frac{}{50}$$

b Write 85% as a fraction in its simplest form.

c Write 97% as a decimal. [MEG]

17 a Change $\frac{11}{17}$ into a decimal.
Give your answer correct to 3 decimal places.

b Place the following numbers in order of size, starting with the smallest.

$\frac{11}{17}$, 65%, $\frac{3}{5}$, 0.63 [MEG]

3 PROBABILITY

In this chapter you will learn about
- **the language of probability**
- **calculating probabilities**
- **mutually exclusive events**
- **methods of estimating probabilities including relative frequency**

Language of probability

Probability involves the study of the laws of chance and is a measure of the likelihood of something happening. The thing you want to happen is called an **event**.

If you are carrying out an experiment using equipment such as a spinner, dice, or cards, each spin, roll or pick is called a **trial**.

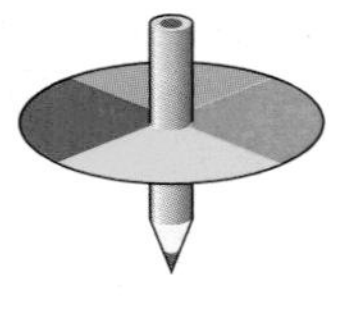

The result of a trial is called an **outcome**.

Sometimes each outcome has exactly the same chance of happening. You would say that the outcomes are **equally likely**. If you pick a card, without looking, from an ordinary pack of playing cards you are picking the card at **random**. Each card is as likely to be picked as any other.

A coin is **fair** if it is equally likely to land on a head or a tail. If the coin is not fair, that is, it is more likely to land on one side than the other, the coin is **biased**. (It is possible for a coin to land and stay on its edge! This happens so infrequently that you tend to ignore this outcome.)

Some outcomes must happen and you would say that they are **certain**. For example, when you roll a fair die with the numbers 1, 2, 3, 4, 5 and 6 on the faces the outcome must be a number between 1 and 6. However, it is **impossible** for the outcome to be 9.

Comparing probabilities

Rather than use words to describe the chance of an event happening, you can give probability as a number, usually written as a fraction or decimal, between 0 and 1.

- If it is **impossible** for an event to happen, the probability is **0**.
- If an event is **certain** to happen, the probability is **1**.
- All other probabilities are greater than 0 but less than 1.

Sometimes you will see probability written as a percentage; in this case the probability will be between 0 and 100%.

An impossible event will have a probability of 0% and an event that is certain to happen will have a probability of 100%. All other probabilities will be greater than 0% but less than 100%.

PROBABILITY

To compare probabilities you must **compare the relative sizes of the fractions, decimals or percentages.** If you have forgotten how to do this, refer to Fact Sheet 1 (pages 13–15).

You can place probabilities on a **probability line** like the one below.

The more likely an event is to happen, the higher the probability, so the larger the fraction, decimal or percentage will be.

The less likely an event is to happen, the lower the probability, so the smaller the fraction, decimal or percentage will be.

Exercise 3.1

1 Draw a probability line. Consider the likelihood of these events happening and indicate on the line where you might put each event. Label the event with its letter.

A You will throw a number bigger than 2 on an ordinary die.
B It will rain tomorrow in the Sahara Desert.
C I am 8 years old and I have my own credit card.
D There will be snow in February next year.
E You will get over 80% in your next Science test.
F The colour of the first car you see when you leave school today will be a shade of blue or white.
G The goalkeeper will score the first goal in next week's Year 10 football match.
H A fair coin will land on a 'tail'.
I The next bus to arrive at the bus station will be either early or on time.

2 For homework the pupils in a class were asked to work out the probability of each of five events happening. They were then asked to rank them in order of likelihood, starting with the least likely.

Jody worked out the probabilities correctly but did not rank them in order as asked. Here are her probabilities:

Event	1	2	3	4	5
Probability	$\frac{3}{5}$	0.25	78%	$\frac{1}{3}$	30%

By comparing the relative sizes of these probabilities, write down the events as Jody should have written them down.

3 Some students were asked what they intended to do at the end of Year 11. Different people worked out the results of the survey and they gave the results in different ways. Here are the results:

Do nothing	0.15
Go into Year 12	60%
Go to college	$\frac{1}{25}$
Get a job	12%
Don't know	$\frac{9}{100}$

Rearrange the results in order of popularity of option, starting with the least popular choice.

4 You can use these words to describe the likelihood of an event happening:

impossible, very unlikely, unlikely, evens, likely, very likely, certain

A large bag contains the following packets of flavoured crisps:

28 Salt & Vinegar, 5 Prawn Cocktail, 21 Beef & Onion, 3 Spring Onion, 83 Ready Salted

Using the words from the list above, decide which word best suits the likelihood of picking, at random, from the bag:

a a packet of Ready Salted **or** Salt & Vinegar flavour,
b a packet of Tomato Sauce flavour,
c a packet of Spring Onion **or** Prawn Cocktail flavour,
d a packet which is not Spring Onion flavour.

PROBABILITY

Listing outcomes

This wheel consists of six identical triangles. It can stop on any one of the six numbers shown, so there are six outcomes for the score – 10, 15, 20, 25, 30 and 35.

If you want the event 'the score is an odd number' to happen, then you would be interested in the numbers 15, 25 and 35 coming up. These numbers are said to be **favourable outcomes**.

If the event is 'the score is greater than 30', the only favourable outcome is a score of 35.

It might be that you have more than one piece of equipment in your experiment, as in the following Sample Question.

Sample Question 1

a The diagram shows a possible outcome from tossing a fair coin and rolling a fair die at the same time. List all the possible outcomes.

b How many of these outcomes include a head and an even number?

Possible outcome

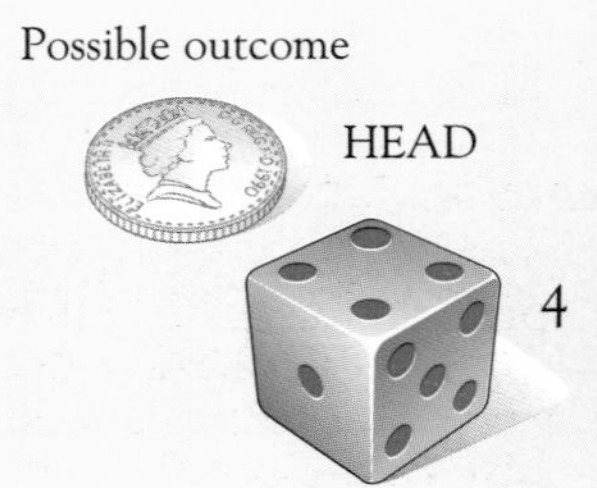

Answer

a

- To show all the outcomes you could make a table with two columns, one for the outcome of tossing the coin, the other for the outcome of rolling the die. Then you could list the outcomes for the die when a **head** (H) is thrown, and then those when a **tail** (T) is thrown.

COIN	DIE	
H	1	
H	2	*
H	3	
H	4	*
H	5	
H	6	*
T	1	
T	2	
T	3	
T	4	
T	5	
T	6	

Be systematic!

b

- Mark with an asterisk the outcomes that include a head and an even number.

There are three outcomes including a head and an even number.

An alternative method is to use a **possibility space**.

		DIE OUTCOMES					
		1	2	3	4	5	6
COIN OUTCOMES	H	H, 1	(H, 2)	H, 3	(H, 4)	H, 5	(H, 6)
	T	T, 1	T, 2	T, 3	T, 4	T, 5	T, 6

This is a good method because from the possibility space you can see that no other outcomes are possible.

PROBABILITY

Exercise 3.2

1 A bag contains three colouring pencils. One is yellow, another is green and the third is purple.

Laura chooses a pencil from the bag, Jamie chooses one of the remaining pencils and Amy is left with the remaining pencil.

List all the possible outcomes when Laura, Jamie and Amy select the three pencils. The first one is done for you.

Laura	**Jamie**	**Amy**
Purple	Yellow	Green

2 Brendan has bought a new CD single for his CD player.

The single has 4 tracks, 1, 2, 3 and 4. Brendan can programme his CD player to play two tracks at random. It will not play the same track twice.

List all the possibilities for the two tracks. [MEG]

3 In a box there are 15 tins of dog food. There are:

5 tins of Original flavour
7 tins of Chicken flavour
1 tin of Beef flavour
2 tins of Lamb flavour

Rubin selects two tins at random.
List all the possible **different** outcomes of his selections.

Note: a selection of Original and Beef is the same as Beef and Original. He can pick the same flavours in two selections.

The first two selections have been made for you.

Original	Original
Original	Chicken

4 In my left-hand pocket I have a 50p coin, a 20p coin and a 10p coin.

In my right-hand pocket I have a 2p coin, a 1p coin and a 5p coin.

I choose one coin, at random, from each pocket. List all the possible total values for the two coins selected.

5 I roll two fair six-sided dice. To get my score I must multiply the value that one die shows by the value the other die shows.

The score on these dice is $6 \times 4 = 24$.

Using a possibility space, list all the possible scores.

6 The diagram shows a flag.
The design on the flag consists of a rectangle and a triangle.
The design and the outer part of the flag are to be different colours.
The design is to be red or blue or white.
The outer part is to be red or blue or green.
List the possible pairs of colours of the two parts of the flag.

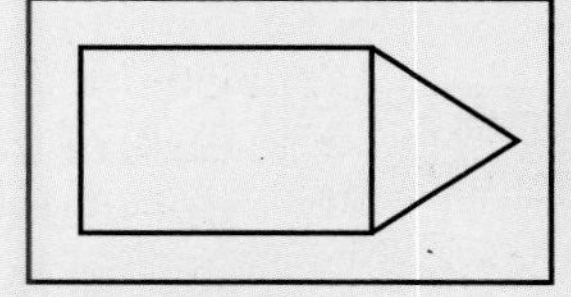

Design	Outer part

[MEG, p]

7 A bag contains 2 red marbles, 1 blue marble and 1 yellow marble.
A second bag contains 1 red marble, 2 blue marbles and 1 yellow marble.
A marble is drawn from each bag.
Copy and complete the table showing all the possible pairs of colours.

Marble from second bag (columns); Marble from first bag (rows)

	R	B	B	Y
R	RR	RB	RB	RY
R	RR			
B	BR			
Y	YR			

[MEG, p]

8 A bag contains just three balls.
Each ball is a different colour: a white, a red and a black.
Anil chooses a ball.
Carl chooses one of the remaining two balls.
Mark has the ball that remains.
Copy and complete the list of all the possible outcomes.

Anil	Carl	Mark
white	red	black

[MEG, p]

CALCULATING

Calculating probabilities

If, in an experiment, each outcome is as likely to occur as any other outcome then you have what are called **equally likely outcomes.** It is possible to calculate the probability of an event happening using the following formula:

$$\text{Probability of an event happening} = \frac{\text{Number of favourable outcomes}}{\text{Total number of outcomes}}$$

Sample Question 2

Each of five cards has a letter of the alphabet on one side of the card as shown. The cards are shuffled and laid face down on the table. You choose a card at random.

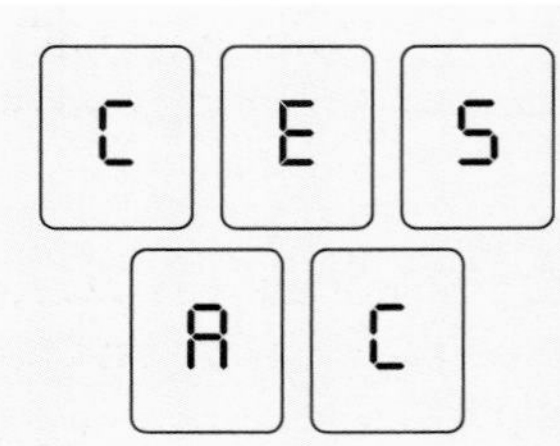

a What is the probability that you choose the letter 'A'?

b What is the probability that you choose the letter 'C'?

Answer

a

- There are five cards altogether so there are five equally likely outcomes. There is one card with the letter 'A' on it so there is one favourable outcome.

$$\text{Probability of selecting 'A'} = \frac{1}{5}$$

You would use shorthand and write $P(A) = \frac{1}{5}$

b

- There are two cards with the letter 'C' so the number of favourable outcomes is two.

$$P(C) = \frac{2}{5}$$

Sample Question 3

In a bag there are three green balls, four red balls and five blue balls. All the balls are the same size and weight. If you choose a ball, **at random**, from the bag, what is the probability that it will be:

a green, b white, c red, d not blue?

(Give each answer as a fraction in its simplest form.)

Remember that choosing 'at random' means that you are equally likely to choose any ball.

Answer

There are 12 balls altogether in the bag so there are 12 possible equally likely outcomes.

a Number of favourable outcomes (green balls) = 3

$$P(\text{green ball}) = \frac{3}{12} = \frac{1}{4}$$

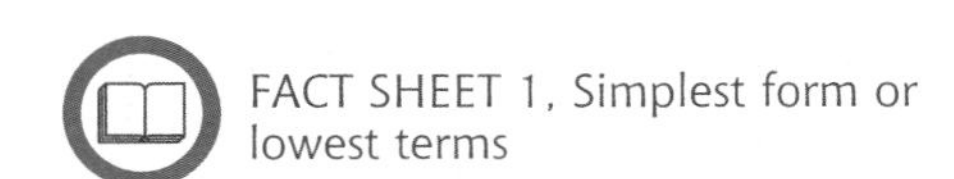

b Number of favourable outcomes (white balls) = 0
It is impossible to select a white ball.

P(white ball) = 0

Using the formula,
$P(\text{white ball}) = \frac{0}{12}$
DO NOT leave it like this. You must write 0.

c Number of favourable outcomes (red balls) = 4

$$P(\text{red ball}) = \frac{4}{12} = \frac{1}{3}$$

d If the ball is not blue then it must be either green or red.
Number of favourable outcomes (green or red balls) = 3 + 4 = 7.

$$P(\text{not blue ball}) = \frac{7}{12}$$

Sample Question 4

For a two-course meal you can choose one main course and one pudding from this menu.

a List all the different two-course meals that you can choose.

b What is the probability that, if you choose a meal at random from the menu, it will contain either cod or ice-cream but not both?

MAIN COURSE	PUDDING
Beef Stew	*Treacle Pudding*
Fillet of Cod	*Luxury Ice-cream*
Nut Cutlet	

Answer

a

- Draw a possibility space using the letters B(Beef Stew), C (Fillet of Cod), N (Nut Cutlet), T (Treacle Pudding) and I (Ice-cream).

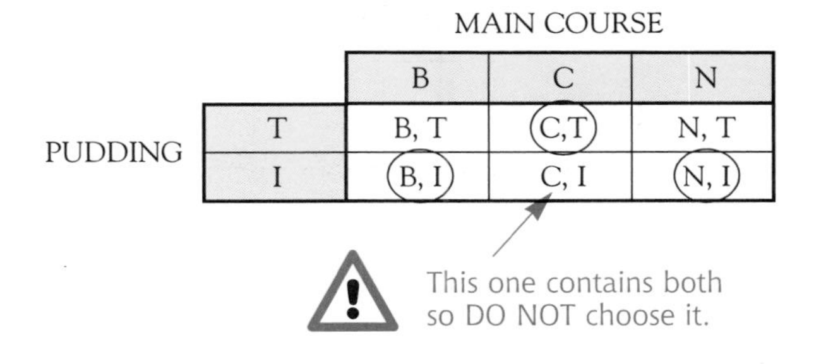

MAIN COURSE

PUDDING	B	C	N
T	B, T	(C,T)	N, T
I	(B, I)	C, I	(N, I)

b

- Decide which meals would contain cod or ice-cream **but not both.**
Number of favourable outcomes is 3.
Total number of outcomes (different meals) is 6.

$$P(\text{C or I but not both}) = \frac{3}{6} = \frac{1}{2}$$

Remember to put the fraction in its simplest form.

Mutually exclusive events

Mutually exclusive events are events that cannot happen at the same time.

If you roll a die the outcome will be a score of 1, 2, 3, 4, 5 or 6 but only **one** of these can occur. For example, the score cannot be a 2 and a 3 at the same time. The outcomes 1, 2, 3, 4, 5 or 6 are therefore mutually exclusive.

CALCULATING

Sample Question 5

The picture shows a board used for a game at a school fair. It has the numbers 1, 2 and 3 on it and is divided into identical coloured sectors. The pointer shows a score of 2 on a purple sector. The pointer is spun.

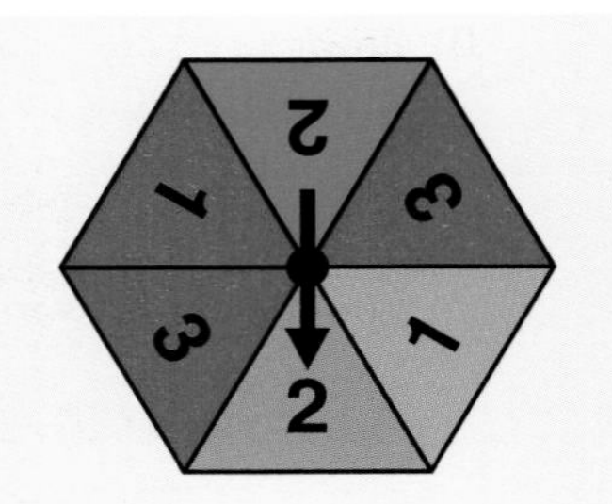

a What is the probability that the pointer will stop on the number 3?

b What is the probability that the pointer will stop on the number 2 or a red sector?

Answer

a

- As the sectors are identical, there are equally likely outcomes.
Number of favourable outcomes is 2 as there are two sectors labelled 3.
Total number of outcomes is 6.

$$P(3) = \frac{2}{6} = \frac{1}{3}$$

Give the fraction in its simplest form.

b

- There are three red sectors and two sectors numbered with a 2.
Number of favourable outcomes is 5.
Total number of outcomes is 6.

There is no sector that is red **and** numbered 2 so the events are **mutually exclusive**.

$$P(\text{red or } 2) = \frac{5}{6}$$

Notice that $P(\text{red}) = \frac{3}{6}$, $P(2) = \frac{2}{6}$ and $P(\text{red or } 2) = \frac{5}{6}$.

$$\frac{3}{6} + \frac{2}{6} = \frac{5}{6}$$

For any two **mutually exclusive** events A and B,
P(A or B) = P(A) + P(B)

Note, if the outcome in the above example is to be either **green** or a score of **2**, these two outcomes **can** happen at the same time. So the events 'a green sector and a score of 2' are not mutually exclusive. The rule for mutually exclusive events **cannot** be used.

Sample Question 6

A bag contains three red balls, five white balls, two green balls and four blue balls.
A ball is chosen at random from the bag.
What is the probability that the chosen ball is

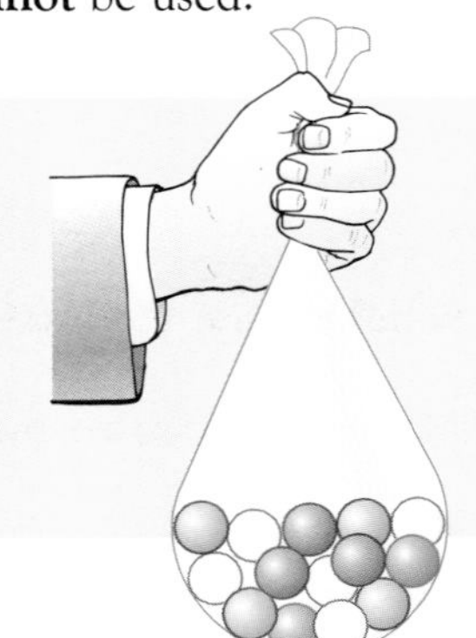

a red,

b white, green or blue?

Answer

a

- Total number of balls in the bag = 14
  Number of favourable outcomes (red balls) = 3

$$\underline{\text{P(red ball)} = \frac{3}{14}}$$

b

- The events 'white', 'green' and 'blue' are **mutually exclusive events**, so use the rule:

$$\text{P(white, green or blue)} = \text{P(white)} + \text{P(green)} + \text{P(blue)}$$

$$= \frac{5}{14} + \frac{2}{14} + \frac{4}{14}$$

$$\underline{\text{P(white, green or blue)} = \frac{11}{14}}$$

'To happen or not to happen, that is the question'

In Sample Question 6, part **b**, selecting a white, green or blue ball is the same as selecting a ball **that is not red**.

$$\text{P(not a red ball)} = \frac{11}{14}$$

From part **a** of the question you know that P(red ball) $= \frac{3}{14}$.

You can see that P(red ball) + P(not red ball) $= \frac{3}{14} + \frac{11}{14} = \frac{14}{14} = 1$

Alternatively, P(not red ball) = 1 − P(red ball)

In general,

P (event does not happen) = 1 − P (event happens)

$$\text{P}(\bar{\text{E}}) = 1 - \text{P(E)}$$

$\bar{\text{E}}$ is pronounced 'E bar' and is shorthand for 'event E does not happen'. Somethines it is written E'.

CALCULATING

Sample Question 7

Joanne plays golf every Saturday. The probability that she will lose a golf ball during the round is $\frac{5}{8}$.

What is the probability that next Saturday Joanne will complete her round of golf without losing a golf ball?

Answer

◆ Use P(event does not happen) = 1 − P(event happens)

$$\begin{aligned} P(\text{not lose ball}) &= 1 - P(\text{lose ball}) \\ &= 1 - \frac{5}{8} \\ &= \underline{\frac{3}{8}} \end{aligned}$$

Sample Question 8

In a game, Louise has to choose a card, at random, from Tony's hand, without looking at the cards. The probability that she will pick the ace of diamonds is 0.2.

What is the probability that she will not pick the ace of diamonds?

Answer

◆ P(not ace of diamonds) = 1 − P(ace of diamonds)

$$\begin{aligned} &= 1 - 0.2 \\ &= \underline{0.8} \end{aligned}$$

Exercise 3.3

1 A bag contains 20 fruit-flavoured sweets. There are four strawberry, five lemon, two blackcurrant, three orange and six raspberry. Roger picks a sweet from the bag without looking.

  a What is the probability that Roger picks either a lemon or blackcurrant sweet?

  Roger does not like orange or lemon sweets. He likes all the other flavours.

  b What is the probability that he picks a sweet that he likes?

2 Siôn and Siân cannot decide where to go on holiday this year. They have narrowed it down to a choice between Majorca (£149), Crete (£225), Portugal (£208), Norway (£175) or Turkey (£110).

  They put the name of each country on a postcard and Siôn selects one without looking. What is the probability that:

  a Siôn selects the cheapest holiday,

  b Siôn selects a holiday that costs more than £200?

3 At the end of the show, a local amateur theatre company picks, at random, the number of a programme and gives the owner of the programme two free tickets to a future production.

  You and your three friends have programmes numbered 121, 122, 123 and 124. That night 240 programmes are sold.

  What is the probability that either you or one of your friends will win the tickets?

CALCULATING

4 The students in class 11D worked out that the probability of their Maths teacher being late for the lesson is 0.6 and the probability that she is early is 0.25.

What is the probability that 11D's Maths teacher will arrive on time for the lesson?

5 Sasha is asked to write down a 2-digit number. The digits can be repeated but she cannot use 0.

What is the probability that the number she writes down:

a is greater than 75,

b is less than 40,

c has a repeated digit, e.g. 11, 55,

d does not start with an odd number?

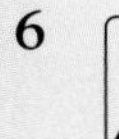

6

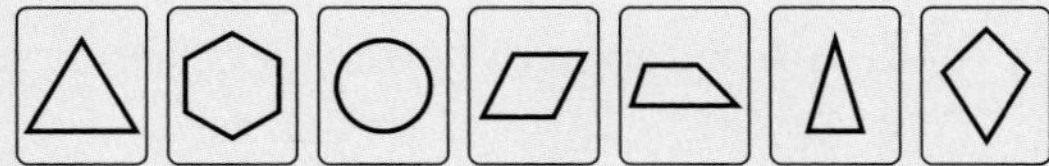

These cards are used in a Maths game. The cards are shuffled and placed face down on the table. The top card is turned over.

What is the probability that this card shows a shape that:

a has 4 sides,

b does not have any straight edges,

c has only 3 sides,

d is either a trapezium, a parallelogram or a kite,

e does not have any parallel sides?

7 In a multiple-choice science test Geoff does not know the answer to questions 3 and 8. He has the choice of five answers (A, B, C, D or E) for each question. For question 8 he knows for certain that answers B and E are incorrect. He must guess from the rest of the answers. For question 3 he must guess from all five of the answers. He is equally likely to choose any of the answers.

a Show on a possibility space all the possible pairs of answers he could pick for these two questions.

The correct answer to question 3 is E and for question 8 it is A. What is the probability that he gets:

b both questions correct,

c question 3 correct or question 8 correct but not both correct,

d at least one question correct?

8 An eight-faced die has the numbers 1 to 8 on its faces. It is rolled together with a normal six-faced die. Both dice are fair. The score on the eight-faced die is multiplied by the score on the six-faced die to give the number of 'points' for that roll. A score of 2 and 5 gives 10 'points'.

a Draw a possibility space of all the possible 'points' that can be scored.

b What is the probability that when the dice are rolled together the 'points' scored will be:

i greater than 20,

ii less than 10,

iii a prime number,

iv an odd number?

9 A badly made die does not have equally likely outcomes. The probability of the die landing on a particular number is shown in the table.

Score	1	2	3	4	5	6
Prob.	0.15	0.2	0.05	0.1	0.4	0.1

a Which number is the die most likely to land on?

b What is the probability that the die will land on:

i an even number,

ii a number other than 5,

iii the number 2 or 6,

iv a square number?

10 Class 10H has 28 pupils; 12 are boys and the rest are girls. Following a survey of the class the teacher finds the following results.

	Left-handed	Blonde hair
Boys	5	4
Girls	3	7

No student is left-handed **and** has blonde hair.

The headteacher chooses, at random, a student from the class. What is the probability that the chosen student:

a is a girl,

b does not have blonde hair,

c is either left-handed or has blonde hair?

The headteacher chooses a girl.

d What is the probability that the girl is left-handed?

Using relative frequency to estimate probabilities

A consumer group carries out a survey on the reliability of different makes of batteries used to power a personal CD player. To do this they played a CD continuously with a new set of batteries until the CD player stopped working. They tested four leading makes of battery. Here are the results of the survey:

Make of battery	Number of sets of batteries tested	Number of sets giving less than 2 hours playing time
Longalife	1240	31
Everlast	850	34
Superbat	1160	40
Cellendure	2100	42

From the results of the survey you can see that the *Cellendure* make of battery had the highest number of battery sets that lasted less than two hours and the *Longalife* had the lowest number. To compare the batteries in this way, however, would be very unfair as different numbers of each make of battery were tested.

To compare the batteries fairly you need to work out the **relative frequency** of battery sets lasting less than two hours for each make.

$$\text{Relative frequency of an event} = \frac{\text{Total number of times event happened}}{\text{Total number of trials}}$$

The relative frequency can be used as an estimate of probability.

9 Estimate, using relative frequency, the probability that a set of *Everlast* batteries, chosen at random, will work for less than two hours of continuous play.
(Use the survey results above.)

Answer

- Work out the proportion of *Everlast* batteries that lasted less than two hours in the trial; write this as a decimal. Use this as an estimate for the required probability.

$$\text{Proportion lasting less than 2 hours} = \frac{34}{850} = \mathbf{0.04}$$

Divide 34 by 850

P(*Everlast* lasts less than 2 hours) = 0.04

Exercise 3.4

1 For each of the remaining makes of battery in the survey above work out, as a decimal, the proportion of batteries lasting less than two hours.

2 Which type of batteries, based on the results of the survey, do you think it would be best to choose if you play your CD player for at least two hours a day?

3 If a packet of *Longalife* batteries is selected at random, what is the probability that they will last for two or more hours if your CD player is used continuously?

Expectation using relative frequency

You can use relative frequency to estimate how often you might expect an event to happen. This is sometimes called the **expectation** of an event.

Expectation of an event = Relative frequency of the event × Number of trials

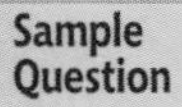

Sample Question 10 In an experiment, Joe threw a drawing pin into the air and recorded which way up it landed, either point up or point down.

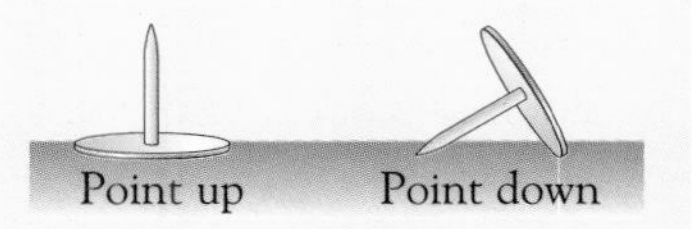

He threw the pin a total of 300 times and it landed point up 210 times.

a What is the relative frequency of the pin landing point up?

b If he were to continue the experiment by throwing the pin a total of 1500 times, how many times would he **expect** the pin to land point up?

Answer

a

- Work out the proportion of times the pin landed point up. This is the relative frequency.

$$\text{Proportion point up} = \frac{210}{300} = 0.7$$

Relative frequency = 0.7

b

- Multiply the relative frequency by the total number of throws.

$$0.7 \times 1500 = 1050$$

Expected number of point ups = 1050

The greater the number of trials, the more reliable the relative frequency becomes as an estimate for probability.

Exercise 3.5

1 Belinda did an experiment to test whether her new four-faced die was fair or whether it was more likely to land on one or more of the four numbers.

Here are the results of her first 100 rolls of the die:

Score on die	1	2	3	4
Number of times landed on	28	19	16	37

a Based on Belinda's results, do you think that the die is fair or biased? Give a reason for your answer.

b Belinda continued the test for another 400 rolls. How many more times would she expect the die to land on the number 2?

RELATIVE FREQUENCY

2 A survey of the main colour of cars entering the local supermarket car park in one hour gave these results:

Colour	Red	Blue	Green	Black	White	Other
Number of cars	245	172	179	44	110	205

If a total of 4600 cars entered the car park that afternoon, how many of them do you expect to be black or white? (Give your answer to the nearest whole car!)

3 From each class in the upper school 10 students were selected at random. They were asked how they usually travelled to school in the morning and gave these answers:

Bus	65
Car	33
Walk	72
Bicycle	26
Other	4

There are a total of 620 students in the upper school. Based on the answers of the selected students, how many students in the upper school would you expect to come to school by bus or by car or by walking?

4 It is estimated that $\frac{5}{8}$ of the tourists visiting Hardup Hall come from the USA. Last month, there were 1648 visitors to the Hall. How many of these would you expect to have come from the USA?

5 In the school's mock general election there are three candidates, one from each of the Labour, Conservative and No Homework parties. In an opinion poll of 100 students, 12 said they would vote Conservative, 42 Labour, 4 didn't know and the rest said they would vote for the No Homework party.

a Based on the opinion poll result, what is the probability that a student selected at random would vote for the No Homework party?

b In the election a total of 600 students voted. How many students would you have expected to vote for the Conservative candidate?

6 It is estimated that the probability of it raining in Waterton on any day in June is $\frac{4}{7}$. You are going to spend 21 days there visiting a relative. How many days can you expect it not to rain?

7 When I threw a drawing pin into the air 50 times I noticed that on 38 occasions it landed point upwards. If I throw the pin a total of 750 times, how many times should I expect it to land point upwards?

8 A light bulb manufacturer tests samples of bulbs to find out how many are faulty. 500 bulbs are selected at random and it is found that 30 are faulty. The sample was taken from a total of 50 000 bulbs.

Based on the result of the sample test, how many of the bulbs would you expect to have been faulty?

9 A traffic survey is carried out at a busy crossroads. In one hour it is estimated that the probability of a car:

turning right is 0.24
turning left is 0.63
going straight ahead is 0.13

If 550 cars used the junction during the hour, how many of them turned right?

10 Based on last season's results, it is estimated that the probabilities of the best four golfers in the club winning their matches this season are:

Tom	58%
Anthony	96%
James	67%
Mark	42%

This season there are 40 games.
How many games should

a Anthony,

b Mark,

expect to win?

Methods of estimating probabilities

You now have two methods for finding, or estimating, the probability of an event happening. They are:

1 using equally likely outcomes,

2 using relative frequency.

Sometimes, you are unable to use either of these methods.

A third method is to use records of previous occurrences, that is, to look at what has happened before.

You might be asked, for example, to estimate the probability that it will snow next 25th of December. You certainly cannot use equally likely outcomes and it would be very difficult (and time consuming!) to carry out a survey.

The best method, be it not very reliable as many gamblers will know to their cost, is to look back over previous years and see how many times it has actually snowed on the 25th of December.

Sample Question 11 The three ways of estimating probability are:

Method 1 Use equally likely outcomes
Method 2 Carry out an experiment and use relative frequency
Method 3 Use historical evidence

Which method would you use to find an estimate for the probability of the following events happening?

a The next student to buy a drink from the drinks machine will buy a can of diet cola.
b If there is no rain on May 1st it will not rain for the next seven days.
c The school's Year 10 netball team will win their next tournament.
d The first ball picked in the National Lottery will begin with the number 3.

Answer

a Method 3 – Find out the proportion of each drink sold since the machine was installed.
or
Method 2 – Carry out a survey over a set time to see which drinks students are buying.

b Method 3 – Look back over a period of time and see how often and for how long it has not rained when May 1st has been dry.

c Method 3 – Check the team's recent form to see how well they have been playing and how many tournaments they have won this year.

d Method 1 – Use equally likely outcomes, by finding the number of balls that begin with the number 3 (favourable numbers).

Exercise 3.6

For this exercise, the ways of estimating probability are:

Method 1 = using equally likely outcomes
Method 2 = carrying out an experiment
Method 3 = using historical evidence

1 For each of the situations below, state which method of estimating probability is most appropriate.
  a A card, selected from a normal pack of 52 cards, will be a picture card.
  b The next time you are late for school it will be because the bus is late.
  c When it is raining there are more calls to the Automobile Association.
  d During this cricket season I will score over 1000 runs.
  e The next person to buy a cake from the canteen will buy a doughnut.
  f When a biased six-faced die is rolled the score will be prime or even.
  g There will be no rain on July 1st.
  h The next person leaving the supermarket will have bought a loaf of bread.
  i The next Prime Minister of the UK will be a woman.
  j If I buy one ticket in the club raffle, I will win the 1st prize.

EXAM QUESTIONS

Worked Exam Question
[SEG]

COMMENTS

a Joyce buys a spinner with seven edges.
When it is spun it has an equally likely chance of landing on any one of the edges.

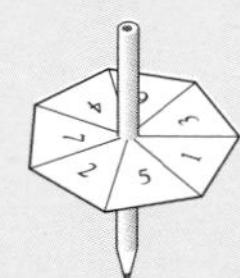

1 favourable outcome,
7 possible outcomes.

i What is the probability that it lands on the edge marked 5?

Answer ...$\frac{1}{7}$...

B1 for correct answer

Mandeep makes a seven-sided spinner. He spins it 30 times.
The table shows the results.

SCORE	1	2	3	4	5	6	7
FREQUENCY	3	2	4	5	11	3	2

ii Is Mandeep's spinner fair? Give a reason for your answer.

Answer ...*There are far more 5s than other numbers so Mandeep's spinner is not fair*...

B1 for 'not fair'

B1 for mention of too many 5s

3 marks

b A game is played with two fair spinners.
Both are spun and the numbers are added to get the score.

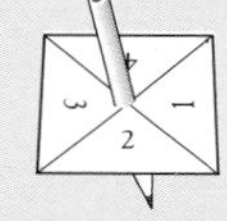
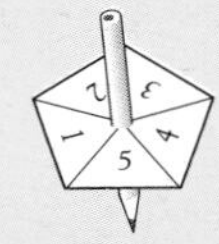

e.g. $2 + 5 = 7$

The score is shown on the table of results below.

i Complete the table to show all the possible scores.

Add the scores on each spinner.

4 favourable outcomes
20 possible outcomes

	1	2	3	4	5
1	2	3	4	5	(6)
2	3	4	5	(6)	7
3	4	5	(6)	7	8
4	5	(6)	7	8	9

B1 for all correct

ii What is the probability of scoring a 6?

$P(6) = \frac{4}{20} = \frac{1}{5}$

Take care to cancel correctly.

Answer ...$\frac{1}{5}$...

B1 for 4
B1 for 20
($\frac{4}{20}$, $\frac{1}{5}$, 0.2, 20% allowed)

The probability of scoring 2 is 0.05.
The probability of scoring 8 is 0.1.
To start the game you need to score either 2 or 8.

iii What is the probability that you will start the game first time?
Show how you worked it out.

A score of 2 or 8 cannot happen together so the events are mutually exclusive.

2 and 8 are mutually exclusive events

$P(2 \text{ or } 8) = P(2) + P(8)$

$= 0.05 + 0.1 = 0.15$

Answer ...0.15...

M1 for adding the probabilities

A1 for correct answer

5 marks

Exam Questions

1 a Giving reasons for your answers, write down an estimate for the probability that

  i it will snow in the Sahara Desert tomorrow,

  ii when I pour 300 cm^3 of water into a 200 cm^3 container, the water will overflow.

b A game uses two dice. In the game, only combined scores of 10 or more are recorded.

When two dice are rolled, two outcomes which will give a score of 10 are (6, 4) or (4, 6).

Write down all of the other possible combined scores which may be recorded in the game.

c The events A and B are described below.

A When I toss a coin it comes down heads.

B I will win the first prize in a raffle in which I only bought one of the 5000 raffle tickets sold.

Write A and B at approximate positions of their probabilities on a probability scale like the one given below.

[WJEC]

2 A firm found that the probability that a parcel will be delivered within 4 days of being posted is $\frac{5}{8}$.

What is the probability that it will be delivered more than 4 days after being posted? [MEG]

3 In a game, discs numbered 1 to 100 are picked at random from a bag and not replaced. The numbers picked are crossed off on cards.

John and Rosie are playing with one card each. No number appears on both cards.

2	14	33	50	71	92
5	19	37	62	73	95
6	26	41	65	79	96
8	29	46	67	85	99

JOHN'S CARD

1	11	30	52	74	91
4	16	34	63	77	94
7	25	47	66	80	97
9	28	49	69	86	98

ROSIE'S CARD

What is the probability that the first number picked is

a one of the numbers on John's card?

b one of the even numbers on Rosie's card?

c one of the numbers on John's card or on Rosie's card? [NI]

4 Asma has a fair coin and a fair dice with its four faces numbered 1, 2, 3 and 4.

a She throws the dice once.
What is the probability that it lands on a 3?

b She throws the coin and the dice together.
Complete the table to show all the possible outcomes.

Coin	Dice
Head	1
Head	2

c i What is the probability that she throws a head and a 4?

  ii What is the probability that she throws a tail and an even number? [MEG]

5 Ruth made a spinner with three colours, green, blue and red.

She tested it by spinning it 500 times.

Her results were 227 landed on green
176 landed on blue
97 landed on red

a Estimate the probability of the spinner landing on blue.

b In a game, the spinner is used 100 times.
How many times would you expect the spinner to land on blue? [NEAB]

6 A fair spinner is labelled as shown.

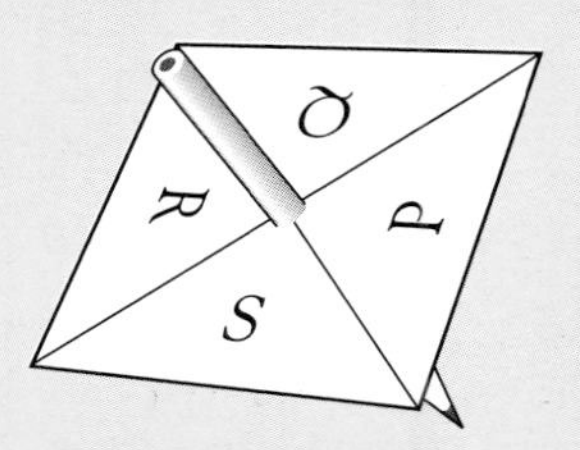

The results of the first ten spins are

$P \quad Q \quad R \quad P \quad Q \quad S \quad R \quad R \quad S \quad Q$

a Write down the relative frequency of the letter R for these results.

b As the number of results increases, what do you expect to happen to the relative frequency of the letter R? [SEG]

EXAM QUESTIONS

7 John collects model racing cars.

He has 3 Ferraris, 2 Benettons, 2 Tyrells and 1 McLaren.

They are all in a box and he takes one out without looking.

a What is the probability that it is

i a McLaren,

ii a Ferrari?

b John takes two of his cars to playgroup. Complete the list to show all his possible choices.
'Ferrari & Benetton' is the same as 'Benetton & Ferrari'.
Do not write the same pair more than once.

Choices		
Ferrari	and	Ferrari
Ferrari	and	Benetton
	and	

John picks the two cars without looking.

c Explain why the probability that he picks a Ferrari and a Benetton is not

$$\frac{1}{\text{total number of choices in the table}}$$

d What is the probability that John takes two McLaren cars with him?
Explain your answer. [MEG]

8 Jenny is conducting a survey by asking passers-by which TV programmes they watched the night before. After asking 100 people, these were her results.

Time shown	Programme	Number who watched
7.00– 7.30	*Wipe Out*	15
7.00– 7.30	*Catchphrase*	22
7.30– 8.00	*Tomorrow's World*	31
7.30– 8.00	*Coronation Street*	46
9.00– 9.30	*Nine O'clock News*	28
10.00–10.30	*News at Ten*	42

a What is the probability that the next person she asks

i watched *Tomorrow's World*,

ii watched either *Tomorrow's World* or *Coronation Street*?

b Jenny wants to work out the probability that the next person she asks watched either the *Nine O'clock News* or *News at Ten*.

Explain why this cannot be done from this information. [MEG]

9 Zaheda conducted a probability experiment using a packet of 20 sweets.
She counted the number of sweets of each colour.

Her results are shown in the table.

Red	Green	Orange
12	3	5

Zaheda is going to take one sweet at random from the packet.

Write down the probability

a that Zaheda will take a green sweet from the packet,

b that the sweet Zaheda takes will **not** be red. [L]

10 In a game at a school fete contestants throw two fair dice and add the two numbers showing. Contestants who score 10 or more win a prize.

a Copy and complete the following table to show all the possible outcomes for the game.

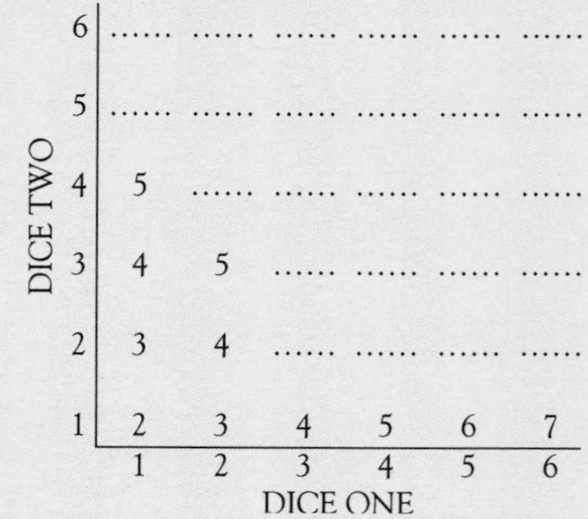

b Jeremy has one turn at the game.
What is the probability that he wins a prize?

c At the fete, 270 people each play the game once.
Approximately how many are likely to win a prize?

d The cost of one turn at the game is 10p. The prize for scoring 10 or more is 50p. How much profit is the game likely to make if 270 people each have one turn? [WJEC]

FACT SHEET 2: THE LANGUAGE AND SHORTHAND OF ALGEBRA

The language of algebra

Here is an alphabetical list of words or phrases that are associated with or used in algebra. You may already know some of them but others will be new to you. Alongside them is their meaning. Most will be used in this course, some very often, others not so often.

Use this list to help you when you come across a word that you do not know or when you have forgotten the meaning. The more you refer to the list, the more you will become familiar with the contents and hence, the more you will remember.

Word or Phrase	Meaning	How or where it might occur
Common (or constant) difference	When the difference between the numbers in a sequence is always the same.	In the sequence 3, 7, 11, 15, 19, 23, … the difference between consecutive numbers is always 4, i.e. the **common (or constant) difference** is 4.
Common factor	Numbers, letters or expressions that are common to all, or part, of an expression.	In the expression $3ab + 3bc$ the **factors** 3 and b are **common** to both parts of the expression.
Evaluate	To find the value of an expression or formula by substituting numbers for the variables. This always gives a numerical answer.	**Evaluate** $a + 2b - c$ when $a = 2$, $b = 4$ and $c = 3$.
Expression, algebraic	A calculation using a combination of letters and numbers, combined by at least one of the operations $+, -, \times, \div$.	$2a + 3c$ $3a^2 - 4b$ $5ab^2$ $\frac{a + 2b}{c}$
Flow diagram	An alternative way of writing an expression, sequence, formula or equation by giving a series of instructions to be carried out in a specified order.	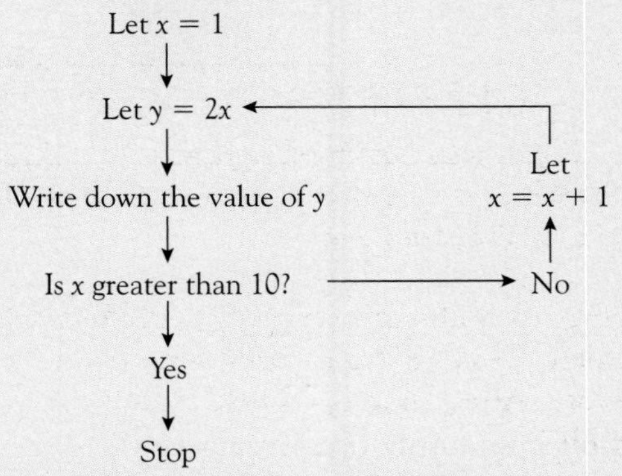 Let $x = 1$ ↓ Let $y = 2x$ ← (from Let $x = x + 1$) ↓ Write down the value of y ↓ Is x greater than 10? → No → Let $x = x + 1$ ↓ Yes ↓ Stop
Formula	A statement that describes the relationship between two or more variables. A formula must contain an = sign.	The **formula** for changing °C to °F is $F = \frac{9}{5}C + 32$ where F is the temperature in °F and C is the temperature in °C.
Integer	A positive or negative whole number or zero.	-1, 4, 6, … etc. There is no fractional or decimal part.
Sequence	A list of numbers each formed by using a set rule.	The terms of the **sequence** 1, 2, 4, 8, 16, 32, … are formed by 'doubling the value of the previous term'.
Simplest form	An expression that cannot be written in a simpler form.	$2a + b - a + 3b$ can be made simpler by adding and subtracting **like terms** to obtain $a + 4b$. $4a + 2b - c$ cannot be made simpler so is in its **simplest form**.

FACT SHEET 2: THE LANGUAGE AND SHORTHAND OF ALGEBRA

Word or Phrase	Meaning	How or where it might occur
Simplify	Make simpler by combining terms where possible.	$3a + 4a - b + 6b$ can be simplified to $7a + 5b$. $2a^2b \times 3ab^2c$ can be simplified to $6a^3b^3c$.
Solve	To calculate the value of an unknown variable in an equation.	**Solve** the equation $3x + 2 = 8$ means find the numerical value of x.
Subject of a formula	The letter that appears on its own on one side of a formula. This letter will not appear on the other side of the formula.	In the formula $s = ut + \frac{1}{2}at^2$, s is the **subject of the formula.**
Substitute	Replace a letter in a formula or expression with the numerical value of that letter.	If $x = 2$ and $y = 3$, find the value of z in the formula $z = 4x - 2y + 5$ by substituting for x and y.
Term of a sequence	A number in a sequence.	In the sequence 2, 5, 8, 11, 14, ... the individual numbers are **terms of the sequence.** 8 is the third term.
Term of an expression	Part of an expression consisting of a number, letter(s), power of a letter or any combination of these.	In the expression $3 + 2a - 3b^2$ each part separated by the operations + and − are the **terms of the expression**, e.g. $2a$ is a **term**.
Term, constant	A number on its own in an expression, equation or formula.	In $a + 2$ and $y = 3x + 1$ the numbers on their own, i.e. 2 and 1, are the **constant terms.**
Term, *n*th	The rule that will generate any term in a particular sequence.	The ***n*th term** of the sequence 4, 7, 10, 13, 16, ... is generated by $3n + 1$, where n is the position of the term in the sequence.
Terms, like	Terms of an expression that are of the same type and can be added to or subtracted from each other.	In the expression $2a + 3b - a + 4b^2$ $2a$ and a are **like terms** but neither is of the same type as the other terms in the expression.
Terms, unlike	Terms of an expression that are not of the same type so cannot be added to or subtracted from each other.	In the expression $2a + 3b - a^2 + 4b^2$ all the terms are **unlike** each other as none of the terms is of the same type as any other.
Variable	A letter that can stand for different numbers.	In the equation $y = x - 6$ x and y are **variables.** When x takes different values, the value of y changes.

Symbols commonly used in algebra

Symbol	Meaning
$+$	Add
$-$	Subtract
$\times$	Multiply
$\div$	Divide
$=$	Equals
$>$	Greater than (or bigger than)
$<$	Less than (or smaller than)
$\geqslant$	Greater than or equal to
$\leqslant$	Less than or equal to
$\equiv$	Identical to, e.g. $\frac{a}{4} \equiv \frac{1}{4}a$ for any value of a.

These symbols are the ones most commonly used in algebra.
Some you should be very familiar with, others will be new to you.

FACT SHEET 2: THE LANGUAGE AND SHORTHAND OF ALGEBRA

Directed numbers

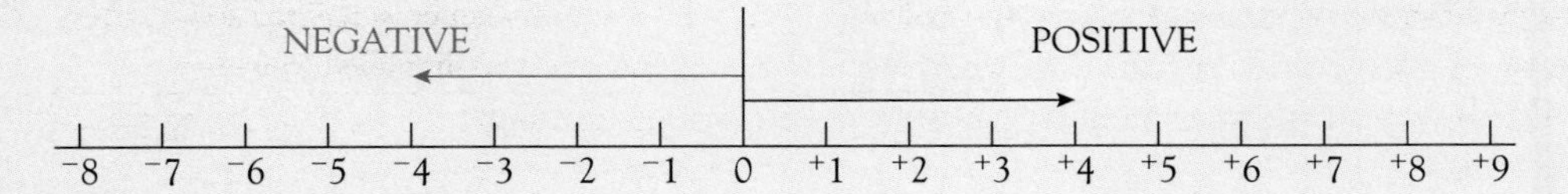

Numbers that are greater than 0 are positive, those that are less than 0 are negative. Positive and negative numbers are often referred to as **directed numbers**. You will have seen positive and negative numbers before, for example, when looking at temperature: the temperature might be $^{-}18$°C in Moscow and 23°C in Miami. Positive numbers often have the $^{+}$ sign missing but we accept that a number without a sign in front is positive. Another way of identifying directed numbers is by using brackets, e.g. (-18), $(+23)$.

You should already have met the rules for adding, subtracting, multiplying and dividing directed numbers. Here is a reminder of these rules:

Operation	Examples
Adding a positive number ($+^{+}$) is the same as **addition**	$^{+}6 + {}^{+}5 = 6 + 5 = 11$ $(-3) + (+5) = -3 + 5 = 2$
Adding a negative number ($+^{-}$) is the same as **subtraction**	$^{+}6 + {}^{-}5 = 6 - 5 = 1$ $(-3) + (-5) = -3 - 5 = -8$
Subtracting a positive number ($-^{+}$) is the same as **subtraction**	$^{-}2 - {}^{+}8 = {}^{-}2 - 8 = {}^{-}10$ $(+8) - (+3) = 8 - 3 = 5$
Subtracting a negative number ($-^{-}$) is the same as **addition**	$^{+}5 - {}^{-}4 = 5 + 4 = 9$ $(-7) - (-5) = -7 + 5 = -2$
Multiplying a $^{+}$ number by a $^{+}$ number gives a **+ result**	$^{+}5 \times {}^{+}6 = {}^{+}30$ $(+3) \times (+4) = (+12)$
Multiplying a $^{-}$ number by a $^{-}$ number gives a **+ result**	$^{-}3 \times {}^{-}6 = {}^{+}18$ $(-4) \times (-2) = (+8)$
Multiplying a $^{+}$ number by a $^{-}$ number gives a **– result**	$^{+}2 \times {}^{-}3 = {}^{-}6$ $(+10) \times (-5) = (-50)$
Multiplying a $^{-}$ number by a $^{+}$ number gives a **– result**	$^{-}4 \times {}^{+}7 = {}^{-}28$ $(-8) \times (+7) = (-56)$
Dividing a $^{+}$ number by a $^{+}$ number gives a **+ result**	$^{+}6 \div {}^{+}2 = {}^{+}3$ $(+8) \div (+4) = (+2)$
Dividing a $^{-}$ number by a $^{-}$ number gives a **+ result**	$^{-}10 \div {}^{-}2 = {}^{+}5$ $(-20) \div (-4) = (+5)$
Dividing a $^{+}$ number by a $^{-}$ number gives a **– result**	$^{+}15 \div {}^{-}3 = {}^{-}5$ $(+100) \div (-20) = (-5)$
Dividing a $^{-}$ number by a $^{+}$ number gives a **– result**	$^{-}40 \div {}^{+}5 = {}^{-}8$ $(-63) \div (+9) = (-7)$

A quick way to decide whether the result of **multiplying or dividing two numbers** is positive or negative is to remember that if the numbers have **the same sign** the answer will be **positive,** otherwise the answer will be negative.

FACT SHEET 2: THE LANGUAGE AND SHORTHAND OF ALGEBRA

Directed numbers and your calculator

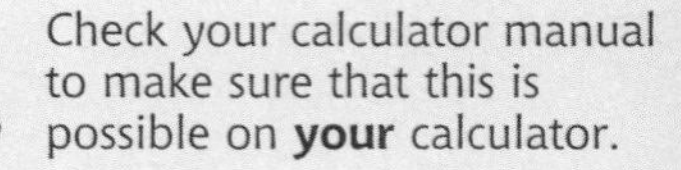
Check your calculator manual to make sure that this is possible on **your** calculator.

Your calculator has a special button that might let you change a number from **positive to negative** and **negative to positive**. This button **will** certainly allow you to enter negative numbers into your calculator. The button will look like this: [+/-]

You will need to check the type of calculator that you have to find out whether to press this button before or after you enter the number. If you have an MDF (Multiplication and Division First) or VPAM calculator you must press it **after** you enter the number. If your calculator is a DAL model then you will have to press it **before** you enter the number.

For example, to do $5 + {}^{-}4$ you would key in the numbers in this order:

MDF/VPAM [5] [+] [4] [+/-] [=]

DAL [5] [+] [+/-] [4] [=]

The answer should be 1; try it on your calculator and remember the order you must use.

The shorthand of algebra

Algebra tends to be a little frightening because the way in which it is written can appear strange. This need not be the case if you understand the shorthand; here are some of the most common types. Remember that the letters a and b are only used as examples; any letters, capital or lower case, can be used.

Shorthand	What does it mean?
a	'one lot of a', $1 \times a$, $a \times 1$, $1a$
$-a$	$-1 \times a$, $a \times -1$, $-1a$
$3a$	'3 lots of a', $3 \times a$, $a \times 3$, $a + a + a$
$\frac{a}{3}$	$\frac{1}{3}$ of a, $a \div 3$, $\frac{1}{3} \times a$
ab	$a \times b$, $b \times a$, $1a \times 1b$, $1b \times 1a$
$2ab$	'2 lots of ab', $2 \times a \times b$, $2 \times b \times a$, $ab + ab$
a^2	$a \times a$
a^3	$a \times a \times a$
$3a^2$	'3 lots of a^2', $3 \times a^2$, $3 \times a \times a$, $a^2 + a^2 + a^2$
$(3a)^2$	$3a \times 3a$, $3 \times a \times 3 \times a$, $3 \times 3 \times a \times a$
$3ab^2$	'3 lots of ab^2', $3 \times a \times b^2$, $3 \times a \times b \times b$, $ab^2 + ab^2 + ab^2$

a^2 and a^3 must not be confused with $2 \times a$ and $3 \times a$; they are different.

You might find it useful to make your own 'dictionary' of words, phrases, symbols and shorthand, adding new ones as you meet them. Students who do this on small cards or using a small notebook, taking the 'dictionary' to lessons or to homework/revision sessions, often say how useful it is. When they come across a word that they do not remember they can look it up quickly.

4 ALGEBRA I

Look out for the tools you need

In this chapter you will learn how to
- **write and simplify an algebraic expression**
- **create and interpret a formula**
- **substitute into a formula**
- **form and solve simple linear equations**

Algebraic expressions

FACT SHEET 2 pages 47–50

An **algebraic expression** is a way of writing down a calculation using letters to represent **variable** values and numbers to represent **constant** values. It does not contain an = sign as part of the expression.

Sample Question 1

If I want to hire a hedge trimmer from the local garden centre I must pay £6 for each day, or part of a day, that I wish to keep it.

a How much will it cost me to hire a hedge trimmer for four days?

b If I want to hire it for d days, write an expression to represent the total cost of the hire.

Answer

a

- Multiply the cost of hire per day by the number of days.

$6 \times 4 = 24$

Cost of hire is £24

b

- Multiply the cost of hire per day by d days.

$6 \times d = 6d$

Cost of hire, in £, is $6d$

Sample Question 2

A special relay running race is to be held to raise funds for a local charity.

In each team four athletes will each run 100 m, n athletes will each run 200 m and t athletes will each run 400 m.

a How many athletes are there in each team?

b How far altogether, in metres, will the team run?

c How far altogether, in kilometres, will the team run?

EXPRESSIONS

Answer

a <u>Number of athletes in a team is $4 + n + t$</u>

b

- Multiply the number of athletes by the distance each will run.

$$\left.\begin{aligned} 4 \times 100 &= 400 \\ n \times 200 &= 200n \\ t \times 400 &= 400t \end{aligned}\right\}$$ Add together these distances and write an expression for the total distance run in metres.

<u>Total distance run by the team, in metres, is $400 + 200n + 400t$</u>

c

- To find the distance in km, divide the answer to **a** by 1000.

1000 m = 1 km

<u>Total distance run by the team, in kilometres, is $\dfrac{400 + 200n + 400t}{1000}$</u>

Exercise 4.1

Write an expression for each of the following:

1 a Bread rolls cost 22 pence each.
  i How much, in pence, will eight rolls cost?
  ii How much, in pence, will n rolls cost?

  b Bread rolls cost b pence each.
  i How much, in pence, will n rolls cost?
  ii How much, in pounds, will n rolls cost?

2 A newsagent buys 15 copies of a daily newspaper at a cost of 12 pence each.

  a How much, in pence, does the newsagent pay for the 15 newspapers?

  The newsagent also buys c copies of a different newspaper costing 20 pence each.

  b How much, in pence, does he pay for these newspapers?

  c How much altogether, in pounds, does the newsagent pay for the 15 copies of the first and c copies of the second newspaper?

3 When full a small box contains 10 computer discs. A large box contains 25 discs when it is full. How many discs do I have altogether if I have n full large boxes and m full small boxes of discs?

4 The recommended time for cooking pork is 50 minutes per kilogram plus an extra 35 minutes, whatever the weight of the pork.

  a How many minutes should it take to cook a joint of pork weighing:
  i 2 kg
  ii w kg?

  b Write your last answer in hours.

5 A first class stamp costs f pence and a second class stamp costs s pence. How much, in pence, will 15 first class and 35 second class stamps cost?

6 A bookshop started the week with 20 copies of a new book. On Monday the shop sold s copies of the book. How many copies did the shop have left at the end of Monday?

7 A computer service engineer charges a call-out fee of £30 plus a fee of £12 per hour, or part of an hour, to repair the computer. How much, in £, does he charge for a repair that takes h hours to complete?

EXPRESSIONS

8 A coal delivery firm charges £5 for each bag of coal it delivers to its customers. It actually costs the firm £45 a day to run the lorry that is used to deliver the coal and to pay the driver, no matter how many bags of coal are delivered.

a If 80 bags of coal are delivered in a day, how much money is charged for delivering the coal?

b If N bags of coal are delivered in a day, how much money, in £, is made from the delivery?

c Any money left over after making the deduction for the lorry costs and driver's wages is profit. How much profit, in £, is there if N bags of coal are delivered in a day?

9 A school buys n boxes of duplicating paper. Each box contains five packets of paper.

a How many packets of paper does the school buy altogether?

b If the 12 subject departments in the school share the packets of paper equally, how many packets of paper does each department receive?

c There are 500 sheets of paper in a packet. How many sheets of paper does each department receive?

10 A sweet shop buys 30 boxes of chocolate eggs for Easter. There are E eggs in each box.

a How many eggs does the shop buy?

When the boxes are opened it is found that B eggs are broken and it is too late to send them back to get replacements. The unbroken eggs are shared equally between 50 baskets.

b How many chocolate eggs are there in each basket?

Simplifying an expression

Sometimes it is possible to **simplify** an expression by adding/subtracting **like terms**, by multiplying/dividing terms, or by a combination of these operations.

FACT SHEET 2
pages 47–50

Only like terms in an expression can be added or subtracted.

Remember that terms such as $2a$ and $3a$ are like terms but $2a$ and $3a^2$ are **unlike terms**.

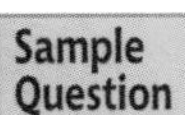

3 Simplify the following expressions.

a $2a + 3b + 3a + 4b$ b $2a^2 - 3a + a^2 - a$

Answer

- Re-arrange the expressions so that like terms are next to each other, then add or subtract the like terms.

a $\underbrace{2a + 3a}_{5a} + \underbrace{4b + 3b}_{7b}$

$\underline{2a + 3b + 3a + 4b = 5a + 7b}$

These steps can be done mentally as you become more confident.

b $\underbrace{2a^2 + a^2}_{3a^2} \underbrace{- 3a - a}_{- 4a}$

$\underline{2a^2 - 3a + a^2 - a = 3a^2 - 4a}$

The − signs belong to the $3a$ and to the a so must stay immediately in front of them when rearranging.

EXPRESSIONS

Multiplying terms. When multiplying terms of an expression, the terms do not need to be of the same type. It is a good idea to multiply numbers first, then letters.

Remember that

- $a \times b$ is the same as ab, which is also the same as ba or $b \times a$
- $a \times a$ is a^2; $a \times a \times a$ is a^3; $a \times a \times a \times a$ is a^4; etc.

Sample Question 4 Simplify these expressions.

a $2a \times 3b$ b $ab \times 4 \times b \times 3ba$ c $a^3 \times 2a \times 5$

Answer

- Re-arrange the expressions putting numbers first, then letters.

a $2a \times 3b = \underbrace{2 \times 3}\times\underbrace{a \times b}$

$= 6 \times ab$

When you become more confident you should be able to do this part in your head.

$\underline{2a \times 3b = 6ab}$

b $ab \times 4 \times b \times 3ba = \underbrace{4 \times 3} \times \underbrace{a \times a} \times \underbrace{b \times b \times b}$

$= 12 \times a^2 \times b^3$

$\underline{ab \times 4 \times b \times 3ba = 12a^2b^3}$

c $a^3 \times 2a \times 5 = \underbrace{2 \times 5} \times \underbrace{a^3 \times a}$

$= 10 \times a \times a \times a \times a$

$= 10 \times a^4$

$\underline{a^3 \times 2a \times 5 = 10a^4}$

Exercise 4.2

Simplify each of the following expressions where possible.
If it is not possible to simplify an expression, write NOT POSSIBLE.

1 $3a + 4a$

2 $7b - 2b$

3 $10c + 2c - 3c$

4 $d + d^2$

5 $2e + 5e - 6e$

6 $3a + 2b + 3b + 4a$

7 $10a + 3c - 2a - c$

8 $5f + 3g - 2f - g + 3f + 2$

9 $3ab + 2ab + 5ba$

10 $2d - e + 3d + 4e$

11 $3x^2 + 5x - x^2 + x$

12 $2a^2 + a^3$

13 $4c^2 + 3 + 2c + 7 - c^2$

14 $4cd + 2dc - 6cd$

15 $2 \times a \times 5$

16 $4b \times 6$

17 $2a \times 4c$

18 $a \times a^2$

19 $c \times 2e \times c \times 4e$

20 $3 \times a \times 4 \times a$

21 $3a \times -5b$

22 $3b \times 10c \times 2d$

23 $-4e^2 \times -d$

24 $7h \times 4k^4 \times 2hk$

EXPRESSIONS

FACT SHEET 2

Substitution into an expression

When you are told the values of the variables in an expression you can **evaluate** the expression by substituting the values into it.

Sample Question 5 If $a = 3$, $b = {}^-4$ and $c = 5$, evaluate each of the following expressions.

a $a + c$ **b** abc **c** $3a + 2b$ **d** $a^2 - c$ **e** $ac + b^2$

Answer

a $a + c = 3 + 5$
$\underline{a + c = 8}$

b $abc = a \times b \times c$
$= 3 \times -4 \times 5$
$\underline{abc = -60}$

c $3a + 2b = (3 \times a) + (2 \times b)$
$= (3 \times 3) + (2 \times {}^-4)$
$= 9 + (-8)$
$\underline{3a + 2b = 1}$

d $a^2 - c = (a \times a) - c$
$= (3 \times 3) - 5$
$= 9 - 5$
$\underline{a^2 - c = 4}$

If you are using a calculator to do this, use the x^2 button.

e $ac + b^2 = (a \times c) + (b \times b)$
$= (3 \times 5) + (-4 \times -4)$
$= 15 + 16$
$\underline{ac + b^2 = 31}$

Exercise 4.3

Do not use a calculator.

1 If $p = 4$, $q = 2$, $r = 10$, $s = 1$ and $t = 0$ find the value of each of the following expressions.

a $p + q + s$
b $3r - 2q$
c $rs + pt$
d $q^2 - 3s$
e $4p + 2q^2$
f $2r - p^2 + 4q$
g p^2qt
h $3p - 2s + 2q^2$
i $pqr - r^2s$

2 In the following questions $a = 3$, $b = -2$, $c = 6$, $d = 7$ and $e = -1$.
For each question first simplify the expression and then substitute the appropriate values into the **simplified expression** and evaluate it.

a $2a + b + 3a + b$
b $4c + 2d - 2c + d$
c $2ab + e - ab + 4e$
d $5a + c - a + 2b - a + c$
e $2 \times 4a \times d$
f $a \times 3b \times a \times e$
g $5c \times c \times 2a$
h $3ac \times 2ab$
i $5a^2 \times 7e^2$

FORMULAE

Writing and evaluating a formula

A **formula** is a statement using letters and often numbers as well, which describes a relationship between two or more variables.

Unlike an expression, a formula **must** have an = sign in it.

Sample Question 6

A holiday cottage in Berkshire costs a basic £200 a week to rent. There is also an extra charge of £25 per week for each person staying at the cottage.

a Write a formula to calculate the total cost C (in £) for a group of P people to rent the cottage for a week.

b Use your formula to calculate the total cost of six people renting the cottage for two weeks.

Answer

a

- Multiply the charge per person by the number of people.

$$\text{Extra charge for } P \text{ people} = 25 \times P$$
$$= 25P$$

- Add this to the basic rent.

$$\underline{C = 25P + 200}$$

DO NOT put the units in a formula.

b

- Substitute the value $P = 6$ into the formula to calculate the cost for one week.

$$C = (25 \times 6) + 200$$
$$= 150 + 200$$
$$= 350$$

- Multiply this answer by 2 to calculate the rent for two weeks.

$$\text{Total cost} = 2 \times 350$$
$$\underline{\text{Total cost} = £700}$$

Put the units in here.

Sample Question 7

The formula used for calculating the average speed, s mph, for a journey is

$$s = \frac{d}{t}$$

where d is the distance travelled in miles and t is the time taken in hours.

Calculate the average speed of a car that travels a distance of 216 miles in 4 hours 30 minutes.

Answer

- Substitute the known values into the formula.

$$s = \frac{216}{4.5}$$

Remember, 30 minutes is 0.5 hours.

$$\underline{s = 48 \text{ mph}}$$

FORMULAE

Exercise 4.4

1 The number of boys in a school is B. The number of girls in the school is G. The total number of students in the school is T.

Complete the formulae beginning:

a $B =$

b $G =$

c $T =$

2 A container weighing J grams is filled with F grams of flour. The total weight of the container and the flour is T grams.

Complete the formulae beginning:

a $T =$

b $F =$

c $J =$

3 A packet of biscuits costs B pence. I buy P packets of biscuits. The total cost of the biscuits is T pence.

Complete the formulae beginning:

a $P =$

b $B =$

c $T =$

4 I go into a cake shop and spend a total of C pence. I buy A cakes costing X pence each and B rolls costing Y pence each.

Write a formula starting $C = \ldots$

5 Write a formula to calculate the total cost, £T, of B Maths books each costing £P and B Physics books each costing twice as much as each Maths book.

In the next five questions of this exercise, four formulae are given for each situation. **Only two are correct**. For each situation choose and write down the two correct formulae.

6 The number of red rose bushes in a garden is R. The number of white rose bushes is W. The total number of rose bushes in the garden is T.

$T = R + W$ $\quad$ $W = R - T$

$R = T - W$ $\quad$ $R = W + T$

7 A furniture shop sells two armchairs for £A each and a sofa-bed for £B. Altogether the chairs and the sofa-bed cost £S.

$B = 2A - S$ $\quad$ $S = B + A + 2$

$S = B + 2A$ $\quad$ $B = S - 2A$

8 A box of matches contains M matches. There are b boxes of matches. The total number of matches is T.

$M = T - b$ $\quad$ $T = bM$

$b = T \div M$ $\quad$ $T = b + M$

9 A train consists of C carriages. Each carriage can carry P people. The total number of people that the train can carry is T.

$T = C + P$ $\quad$ $P = C \div T$

$P = T \div C$ $\quad$ $T = CP$

10 There are f 50 pence coins in a bag. When they are changed into 10 pence coins there are t 10 pence coins.

$t = f \div 5$ $\quad$ $f = t \div 5$

$t = 5f$ $\quad$ $f = 5t$

FORMULAE

Exercise 4.5

1 To calculate the total cost, C pence, of printing a set of worksheets, the school office uses the formula

$$C = 25 + 3n$$

where n is the number of copies required.

Use the formula to calculate the cost of printing:

a 30 copies of a worksheet,

b 100 copies of a worksheet.

2 A delivery firm charges its clients according to how many miles are travelled by road and how many by rail. It charges £3 a mile by road and £4 a mile by rail. To calculate the total charge the firm uses the formula

$$C = 4r + 3v$$

where C is the total charge, in £, r is the number of rail miles, v is the number of road miles.

Calculate the total charge made to a client if it takes 50 rail miles and 175 road miles to make the delivery.

3 The voltage, V volts, in an electrical circuit is calculated using the formula $V = IR$, where R is the resistance in ohms and I is the current in amps.

Calculate the voltage in a circuit when the current is 5 amps and the resistance is 400 ohms.

4 The approximate relationship between the temperature in degrees Fahrenheit (F) and degrees Celsius (C) is given by the formula

$$F = 2C + 30$$

Using the formula, work out the temperature in °F if the temperature is given as $-3°$ C.

5

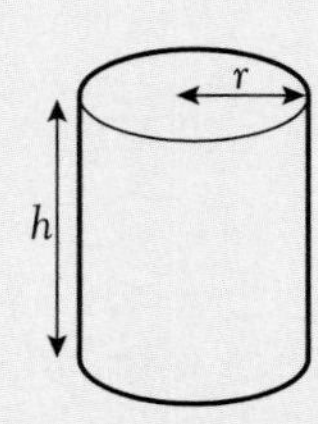

The volume, V cm^3, of a cylinder is calculated using the formula

$$V = \pi r^2 h$$

where r is the radius of the base, in cm, h is the height, in cm, and $\pi = 3.14$.

Calculate the volume of a cylinder with height 25 cm and the base radius 5 cm.

6 The density, D g/cm^3, of a material is calculated using the formula

$$D = \frac{m}{V}$$

where m is the mass of the material in g, and V is the volume of the material in cm^3.

Calculate the density of a material that has a mass of 200 g and a volume of 45 cm^3.

7 The acceleration, a m/s^2, of a particle is calculated using the formula

$$a = \frac{v - u}{t}$$

where v is the final velocity in m/s, u is the initial velocity in m/s, and t is the time taken in s.

Calculate the acceleration of a particle which has an initial velocity of 16 m/s and a final velocity of 24 m/s, in a time of 2 s.

8 The power, P watts, of a machine is calculated using the formula

$$P = \frac{Fd}{t}$$

where F is the force exerted in newtons, d is the distance moved in metres in the direction of the force, and t is the time taken in seconds.

Calculate the power of a machine if a force of 10 newtons is exerted over a distance of 50 metres for 8 seconds.

9 The distance travelled by an object in a given time can be found using the formula

$$s = ut + \tfrac{1}{2}at^2$$

where s is the distance travelled in m, u is the initial velocity in m/s, a is the acceleration of the object in m/s^2, and t is the time taken in s.

Calculate the distance travelled by an object that has an initial velocity of 40 m/s and accelerated at a rate of -3 m/s^2 for 8 seconds.

A negative acceleration means that the object is slowing down!

10 Using the formula

$$R = \frac{rs}{r + s}$$

calculate the value of R when $r = 15$ and $s = -25$.

Number puzzles and equations

'I think of a number. I multiply the number by 3 and add 1 to the result. The answer is 13. What number did I think of?'

One way to solve this number puzzle is to work backwards from the answer.

- To get the answer 13, the last operation was *'add 1'*. Do the reverse of **add 1** which is **subtract 1**.

  $13 - 1 = 12$

- The operation before this was *'multiply by 3'*. The reverse of × is ÷, so divide by 3.

  $12 \div 3 = 4$

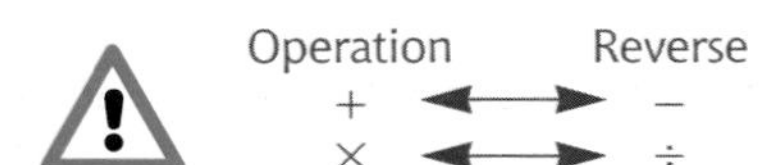

So the number originally thought of was 4

You can write this number puzzle in algebraic form.

- Let the number thought of be called x.
- Multiplying x by 3 gives $3x$.
- Adding 1 gives $3x + 1$.
- Putting in the answer 13 gives $3x + 1 = 13$. This is an **equation**.

You can use a flow diagram to solve an equation.

Sample Question 8

a Write the following number puzzle as a flow diagram and use it to solve the puzzle.

'I think of a number. I multiply the number by 3 and take 5 away from the result. The answer is 7. What number am I thinking of?'

b Write this puzzle as an equation.

Answer

a

- Write the puzzle as a series of instructions in the form of a flow diagram, letting the number being thought of be called x.

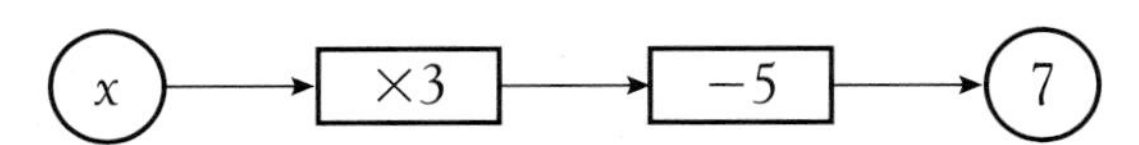

- Write in a second line to the diagram, reversing the operations.

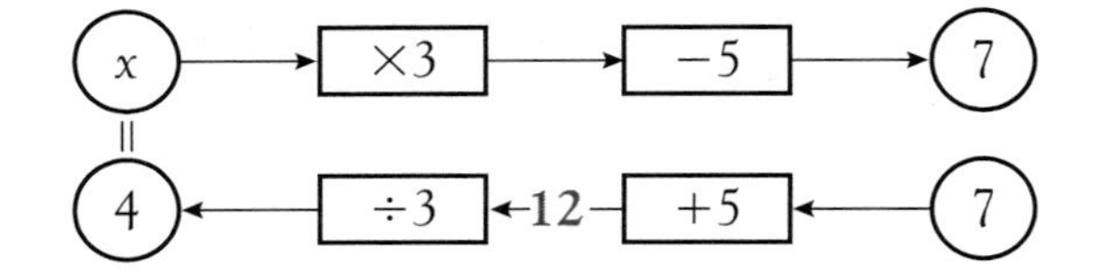

This is the answer to 7 + 5

The number thought of is 4

b

- Writing the instructions as an equation gives

  $3x - 5 = 7$

EQUATIONS

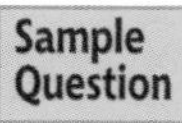

9 Write this number puzzle as an equation and solve it to find the number:

'I think of a number, multiply it by 4 and add 7 to the result. The answer is 27. What number am I thinking of?'

Answer

- Write the puzzle as an equation, using x for the number.

$$\underbrace{4 \times \text{number}}_{4x} \quad \underbrace{\text{add } 7}_{+7} \quad \underbrace{\text{answer } 27}_{= 27}$$

The equation is $4x + 7 = 27$.

- Solve the equation by reversing the order of the operations in the instructions.

$27 - 7 = 20$

$20 \div 4 = 5$

The number thought of is 5

TASK

Work in groups of three or four. Each member of the group should write five number puzzles. Members of the group must try to solve one another's puzzles and write each puzzle as an equation. (Make sure you know the answer to your own puzzles!)

Balancing

You can also think of an equation as a **balance**. Each equation has a left-hand side (LHS) and a right-hand side (RHS), separated by an = sign. The balance is unique to each equation.

Look at the equation $3x + 5 = 14$

There are two sides to the equation, balanced by the = sign. You can show this in a picture.

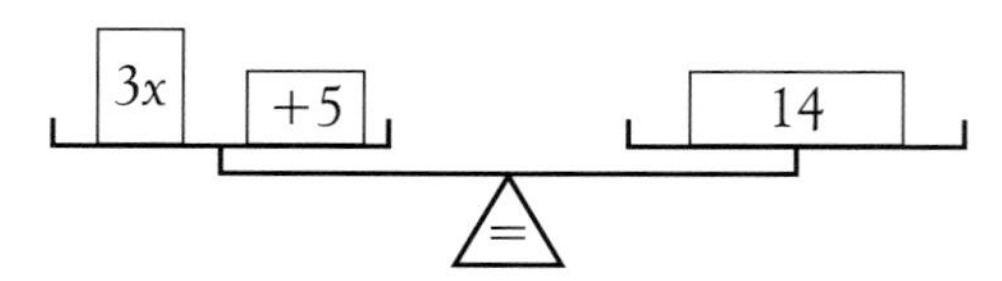

You can add to (+), remove from (−), expand (×) or shrink (÷) items on either side of the balance, as long as you follow one simple rule:

If you do something to one side of an equation, you must do exactly the same to the other side!

If you do not do this you are changing the original balance.

When solving an equation you are trying to find the value of x, so you would like the x to be on its own on one side of the balance. In this equation you have to remove the $+5$ from the left-hand side. To do this you must subtract 5 from **both** sides of the balance.

$+5 - 5 = 0$

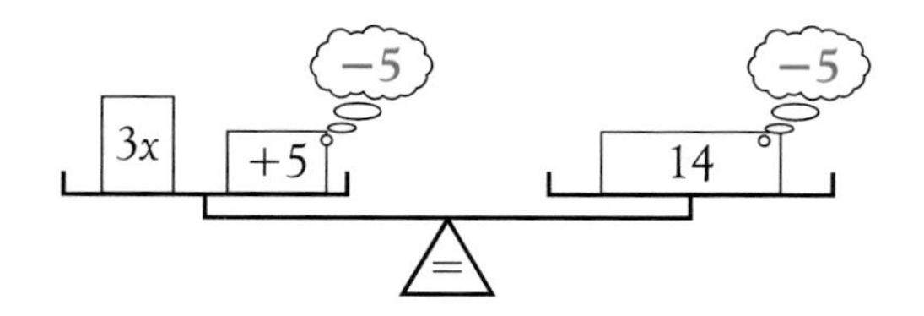

This gives a new balance:

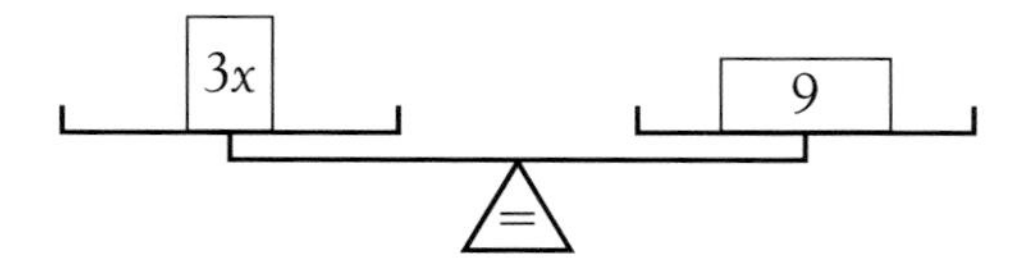

Now you need to 'shrink' the $3x$ to $1x$ by dividing it by 3, **not forgetting to divide the right-hand side by 3 as well.**

$3x \div 3 = 1x = x$

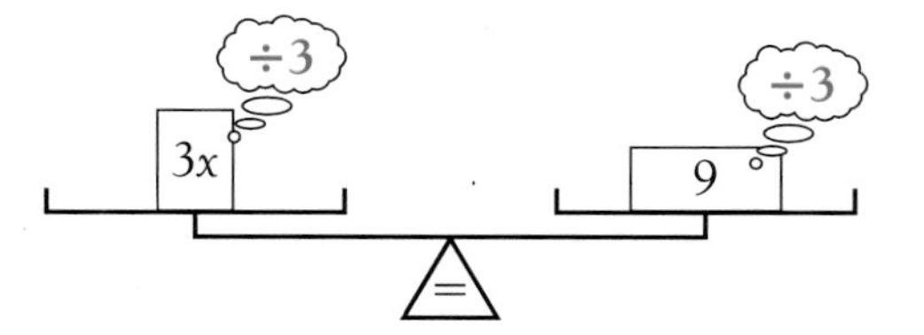

This leaves you with the required result:

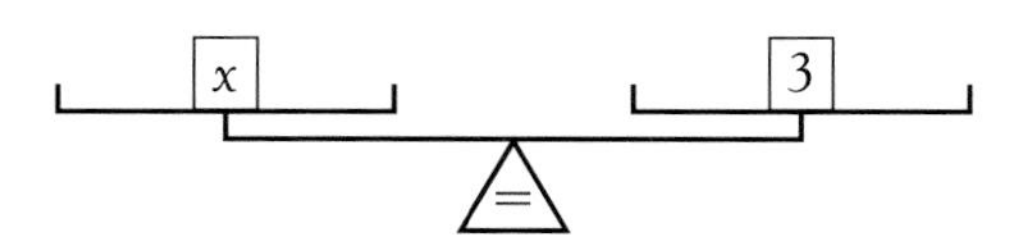

So, if $3x + 5 = 14$, then $\underline{x = 3}$

Sample Question 10 Using 'balance diagrams' solve the equation $4x - 3 = 13$.

Answer

- Isolate the $4x$ on the left-hand side by adding 3 to both sides.

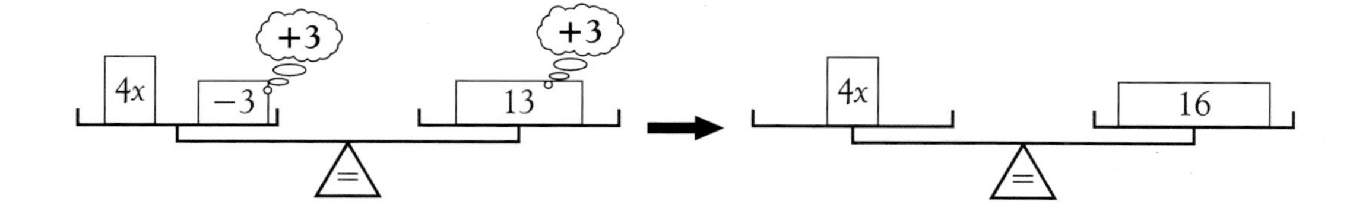

- 'Shrink' the $4x$ to $1x$ by dividing both sides by 4.

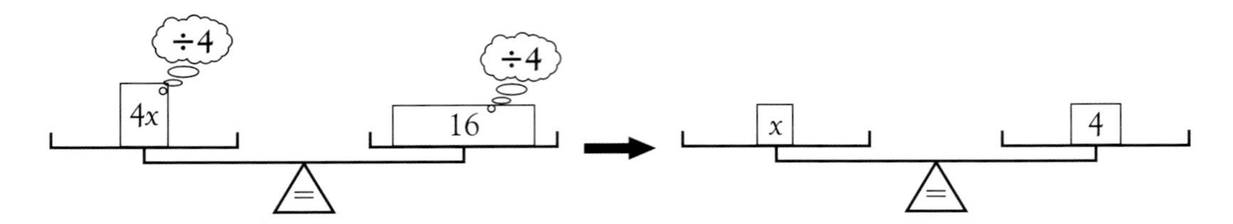

If $4x - 3 = 13$, then $\underline{x = 4}$

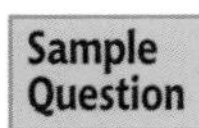

11 Solve the equation $\frac{1}{2}x + 1 = 8$.

Answer

- Subtract 1 from both sides of the equation.

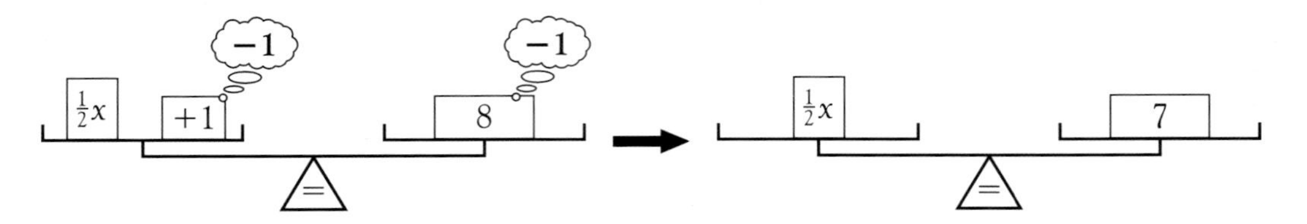

- 'Enlarge' the $\frac{1}{2}x$ to $1x$ by multiplying both sides by 2.

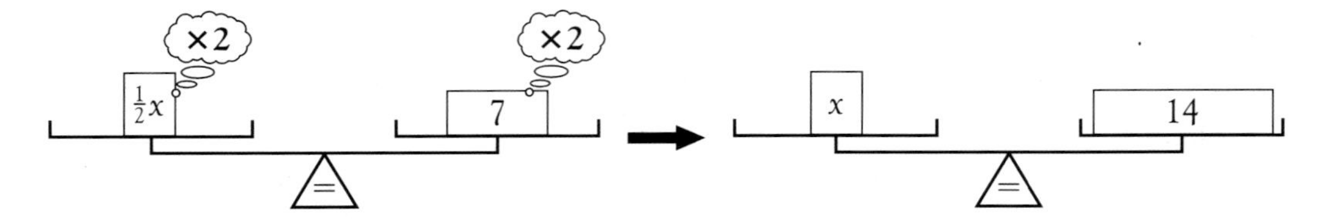

- If $\frac{1}{2}x + 1 = 8$, then $\underline{x = 14}$

It is very time consuming to draw the 'balance' every time, so instead you can write what is on the balance and what is happening like this.

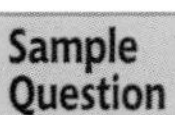

Solve the equation $4x + 3 = 11$.

Answer

- Subtract 3 from **both** sides of the equation.
- Divide **both** sides of the equation by 4.

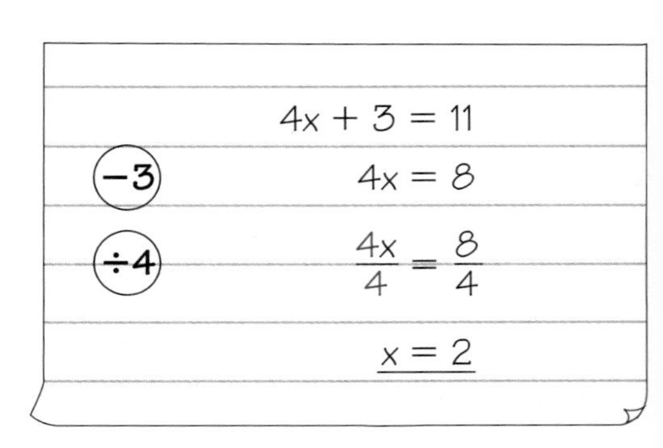

Remember, $\frac{4}{4} = 1$. Any number or letter divided by itself is always 1

EQUATIONS

Sample Question 13 Solve the equation $\frac{x}{4} + 7 = -15$.

Answer

- Subtract 7 from **both** sides of the equation.
- Multiply **both** sides of the equation by 4.

$\frac{x}{4} \times 4$ is the same as $\frac{4x}{4}$ or $\frac{4}{4} \times x$ which is $1x$

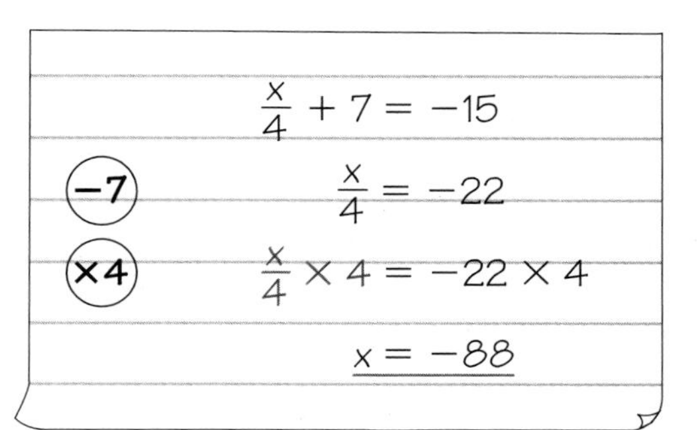

$$\frac{x}{4} + 7 = -15$$

$$(-7) \quad \frac{x}{4} = -22$$

$$(\times 4) \quad \frac{x}{4} \times 4 = -22 \times 4$$

$$x = -88$$

Exercise 4.6

1 For each number puzzle write the puzzle as an equation, and then solve the equation to find the number.
Use whichever method you are most confident with.

a I think of a number, add 8 to it and the answer is 14. What number am I thinking of?

b I think of a number, multiply it by 3 and the answer is 21. What number am I thinking of?

c I think of a number, divide it by 2 and the answer is 15. What number am I thinking of?

d I think of a number, multiply it by 6 and add 5 to the result. The answer is 23.
What number am I thinking of?

e I think of a number, divide it by 2 and subtract 3 from the result. The answer is 5.
What number am I thinking of?

2 Solve the following equations.

a $6x = 36$

b $5x = 75$

c $3x - 1 = 8$

d $9x + 2 = 29$

e $2x - 3 = 47$

f $\frac{x}{5} = 15$

g $\frac{x}{10} - 7 = 2$

3 Simplify and solve the following equations.

a $3x + 2 + 5x - 1 = 15$

b $7 - x - 2 + 4x = 20$

c $x + 1 + x + 2 + x + 3 = 180$

d $10x - 2x + 4 + x = 85$

In the following, solve the problems by forming an equation in x and solving the equation.

4 In my pocket I have x 10p coins, $2x$ 20p coins and a 50p coin.

a Write, in terms of x, the total value of the 10p coins in my pocket.

b Write, in terms of x, the total value of the 20p coins in my pocket.

Altogether I have £3 in my pocket.

c Write an equation in x and solve it to find how many 10p coins I have in my pocket.

5 The length of a rectangular playground is x metres. The width of the playground is 30 metres shorter than the length.

a Write, in terms of x, the width of the playground.

b Write an expression, in its simplest form, for the perimeter of the playground.

The perimeter of the playground is 400 metres.

c Write an equation, in terms of x, and solve it to find the length and width of the playground.

6 Rajesh is x years old. Beth is twice as old as Rajesh. Their combined age is 63 years. Write an equation, in terms of x, and solve it to find Beth's age.

EXAM QUESTIONS

Worked Exam Question
[MEG]

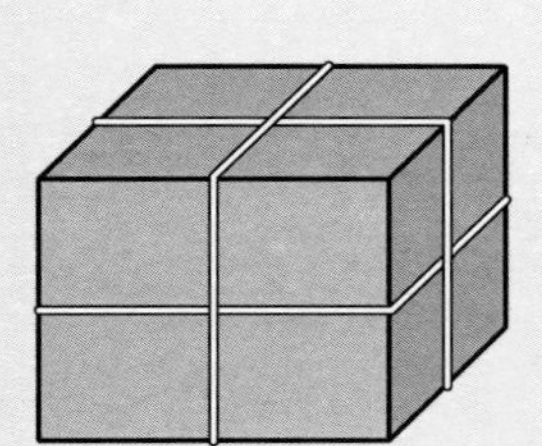

PARCEL A

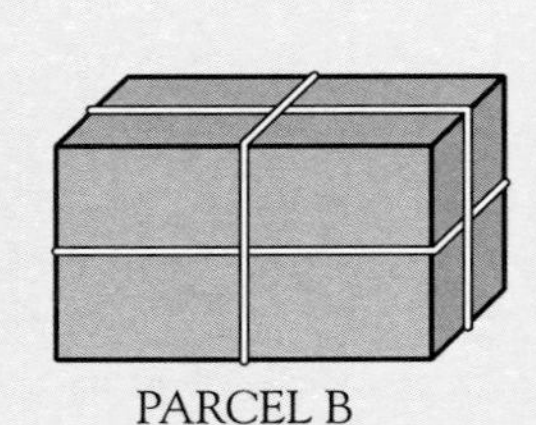

PARCEL B

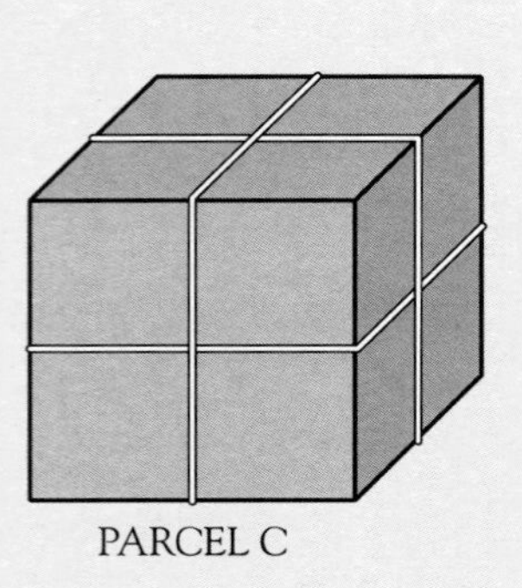

PARCEL C

COMMENTS

Steven weighed three parcels before posting them.

Parcel A weighed x grams.
Parcel B was 50 grams lighter than parcel A.
Parcel C was three times as heavy as parcel A.

a Write down, in terms of x,

i the weight of parcel B,

$x - 50$

Answer ....$x - 50$........ g

M1

ii the weight of parcel C.

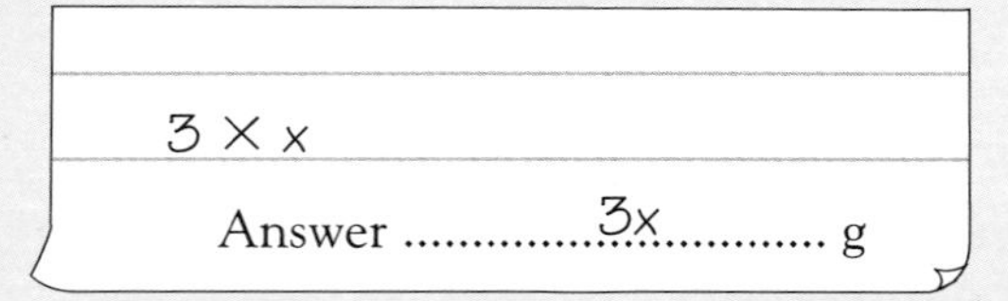

M1 (3 × x would be allowed)

2 marks

b The total weight of the three parcels was 840 grams.
Write down an equation in terms of x.

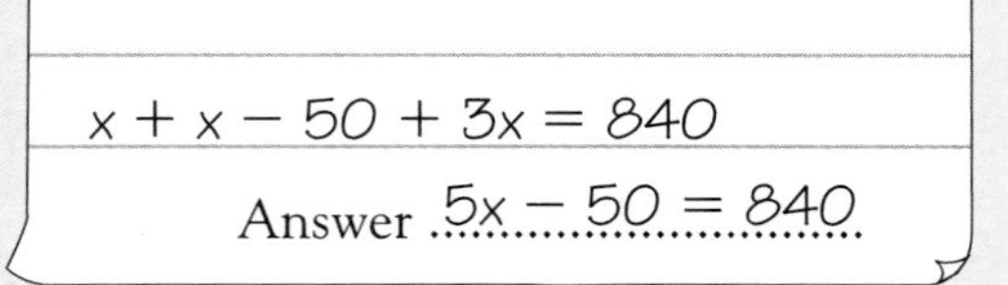

Remember to simplify your equation.

M1 for correct equation

1 mark

c Solve your equation to find the weight of parcel A.

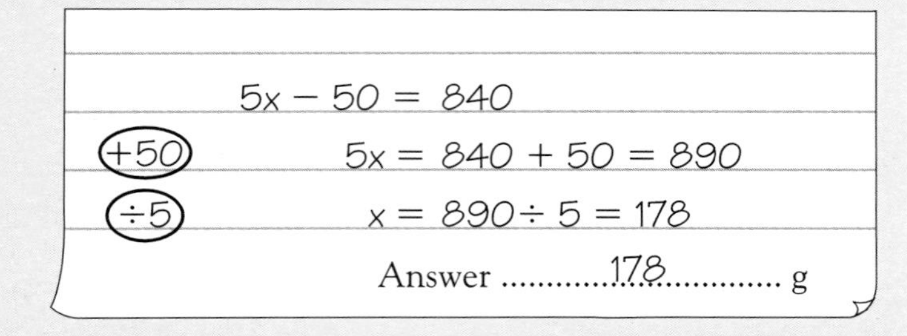

M1 for showing +50, ÷5

A1 for answer

2 marks

Exam Questions

1 The cost, C pence, of advertising in a local newspaper is worked out using the formula $C = 20n + 30$, where n is the number of words in the advertisement.

a Annalise puts in an advertisement of 15 words. Work out the cost.

b The cost of Debbie's advertisement is 250 pence.

i Use the formula to write down an equation in n.

ii Solve the equation to find the number of words in Debbie's advertisement. [MEG]

2 A minibus owner runs a daily service to London. The profit (£T) depends on the number of passengers (n). The formula connecting T and n is

$$T = 12n - 80$$

a Find the profit when the number of passengers is 13.

b i Use the formula to find the value of T when $n = 5$.

ii What does your answer indicate? [MEG]

3 Jake buys a television. He pays £125 deposit and 12 monthly instalments of £29.62.

a Work out the total amount that Jake pays.

b Anna buys a car. She pays £d deposit and 24 monthly instalments of £m each. Write down an expression for the total amount that Anna pays. [L]

4 In this triangle all the measurements are in centimetres.

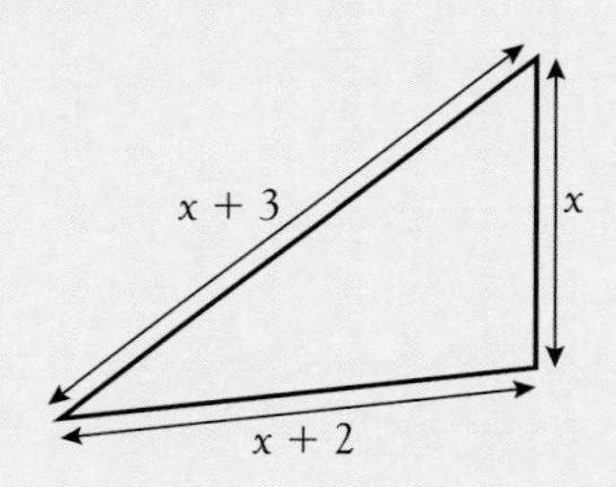

a Write a formula for the perimeter, P, of this triangle.

b The perimeter of the triangle is 38 cm.

i Write an equation in x.

ii Solve this equation to find the lengths of the sides of the triangle. [MEG]

5 Solve the equations.

a $x + 7 = 18$

b $5y = 45$

c $3a + 2 = 14$ [MEG]

6 In a certain rectangle the length of each of the longer sides are 3 cm more than the length of each of the shorter sides. Let x cm denote the length of each of the shorter sides.

a Write down, in terms of x, the length of each of the longer sides.

b Write down, in terms of x, the perimeter of the rectangle. Simplify your answer as far as possible.

c The perimeter of the rectangle is 32 cm. Write down an equation in x. Solve this equation to find the value of x. [WJEC]

7 a Solve the equation $3x + 2 = 17$.

b Use the formula $v = at$ to work out the value of v when $a = 3$ and $t = 8$.

c Given that $4y - 3 = 15$, work out the value of $2y - 3$. [L]

8 The cost, S pounds, of a chest of drawers with d drawers may be calculated using the formula

$$S = 29 + 15d$$

a Calculate the cost of a chest of drawers with 3 drawers.

Another chest of drawers costs £119.

b Calculate the number of drawers this chest has. [L]

EXAM QUESTIONS

9

Only 60p plus
50p for every window cleaned.

a Kevin has 8 windows cleaned. How much does Wendy charge?

b Betty has W windows cleaned. Which of the following expressions represents Wendy's charge, in pence, for cleaning Betty's windows?

$60 + 50W$ $\quad$ $110W$ $\quad$ $60W + 50$

c Wendy charges Ali C pence to clean his windows. Use your expression in **b** to form an equation for this charge.

d Bert pays Wendy £5.10 to clean his windows. How many windows does Wendy clean?

[NEAB]

10 A building supplier hires out cement mixers. He calculates the hire charge using this formula:

Five pounds per day plus a fixed charge of seven pounds.

a How much would it cost to hire the cement mixer for 3 days?

b A builder paid £52 altogether for hiring a cement mixer. For how many days did he hire it?

The building supplier also hires out compressors. He calculates the hire charge for compressors using this formula:

£7.50 a day plus a fixed charge of £12.

Using C for the total hire charge and d for the number of days hired, write a formula for C in terms of d.

c Use your formula to calculate the cost of hiring a compressor for 4 days. [SEG]

11 A coach has x passengers upstairs and y passengers downstairs.

a Write down an expression, in terms of x and y, for the total number of passengers on the coach.

Tickets for the journey on the coach cost £5 each.

b Write down an expression, in terms of x and y, for the total amount of money paid by the passengers on the coach. [L]

12 Hassan is twice as old as Ali. Their ages add up to 39 years. How old is

a Ali,

b Hassan? [NEAB]

13 Mrs Brown is making a collection of china jugs. Each jug costs £15.

a Write down an expression for the cost, in £, of n jugs.

Mrs Brown displays the jugs in a cabinet costing £68.

b i Write down an expression, in terms of n, for the total value of the cabinet and n jugs.

ii The total value of the cabinet and n jugs is £188. Use your answer to **b i** to form an equation.

iii Solve your equation to find the number of jugs in Mrs Brown's collection. [MEG]

14 a If 2 metres is cut off a rope 8 metres long, how much rope is left?

b If x metres is cut off a rope 10 metres long, how much is left?

c If z metres is cut off a rope y metres long, how much is left? [NEAB]

15 a Write down and simplify an algebraic expression for the perimeter of this triangle.

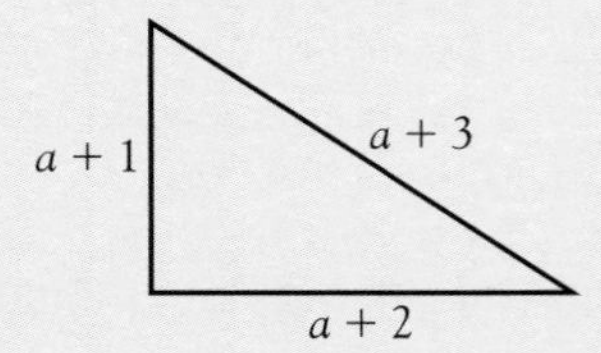

b Write down and simplify an algebraic expression for the perimeter of this rectangle.

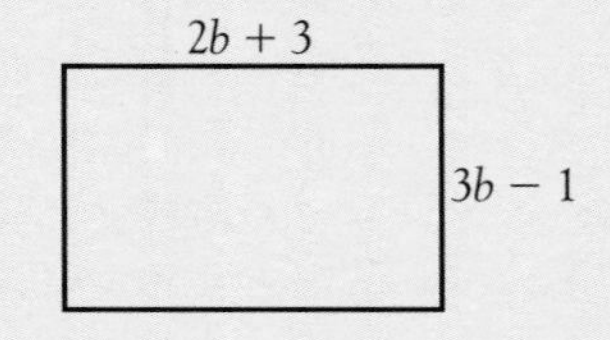

[L]

FACT SHEET 3: THE LANGUAGE OF GEOMETRY

Angles

An **angle** measures the amount of turning between two lines and is usually measured in **degrees** (°).

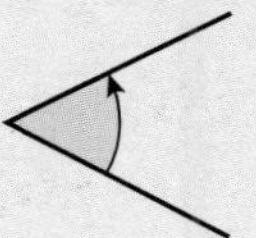

Fractions of a complete turn

A complete turn is called a **revolution** and is divided into 360°.

A quarter turn is 90° and is called a **right angle.**

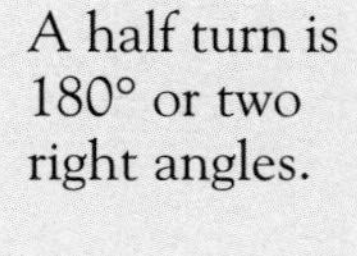

A half turn is 180° or two right angles.

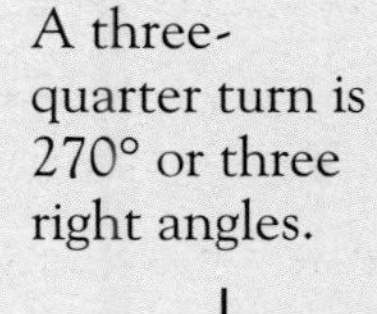

A three-quarter turn is 270° or three right angles.

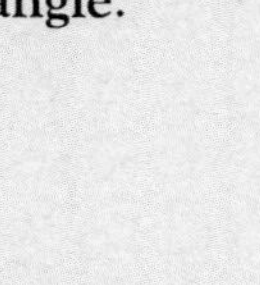

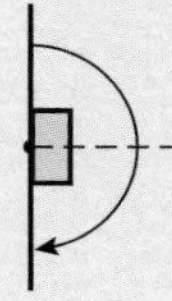

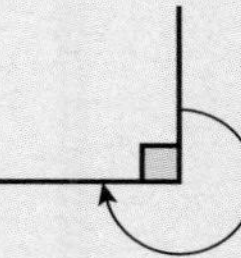

Make a square to show 90°

Types of angles

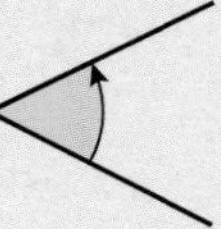

An **acute** angle is between 0 and 90°.

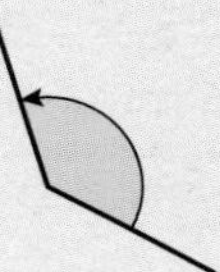

An **obtuse** angle is between 90 and 180°.

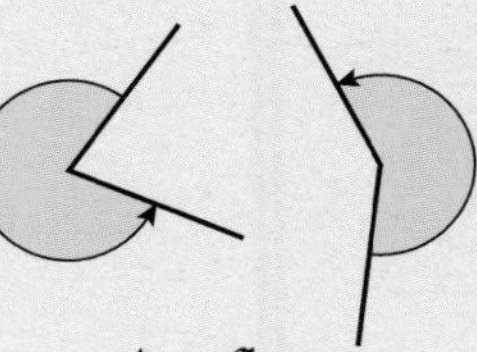

A **reflex** angle is between 180 and 360°.

Labelling angles

①

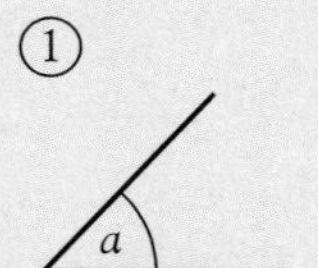

Use a small letter and write it inside the angle, angle a.

②

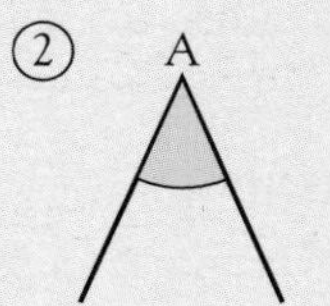

Label the point with a capital letter outside the angle. Write $\hat{A}$ for the angle.

③ Label the ends of the lines with capital letters.

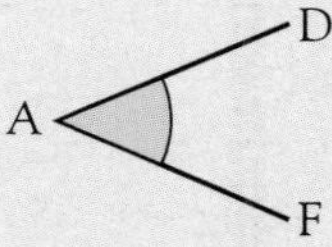

The shaded angle is written as angle DAF or $D\hat{A}F$.

Notice that A is in the middle of the three letters.

FACT SHEET 3: THE LANGUAGE OF GEOMETRY

Perpendicular lines

Two lines that meet or cross making an angle of 90° are called **perpendicular** lines.

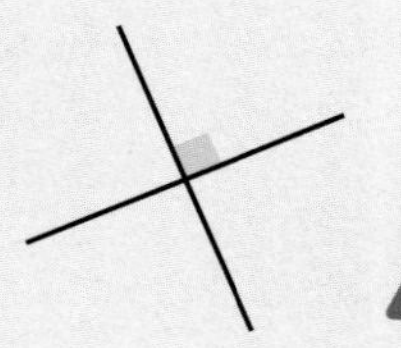

Put a 90° 'corner' on the diagram to show the right angle.

Parallel lines

Parallel lines are always the same distance apart. You show that they are parallel by putting equal numbers of arrows on the lines, sometimes one on each, sometimes two or more if there is more than one set of parallel lines in the diagram.

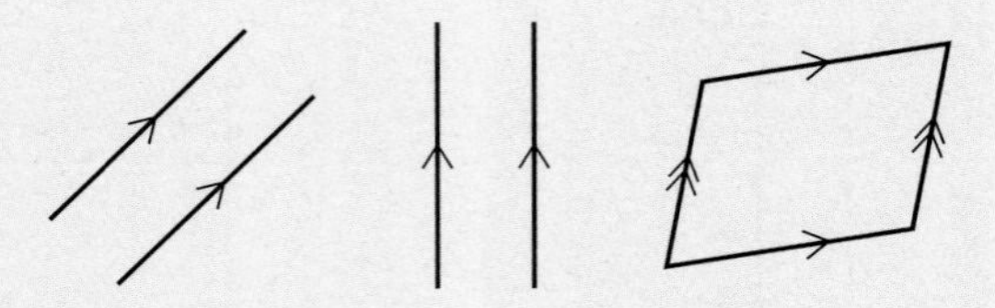

Polygons

A **polygon** is a flat shape with three or more straight sides, for example:

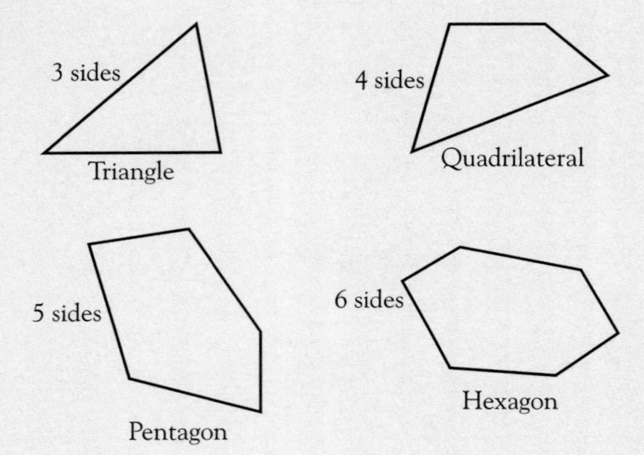

8 sides: octagon
9 sides: nonagon
10 sides: decagon
12 sides: dodecagon

Regular polygons

If all the sides of a polygon are the same length, it is **regular**. You mark the sides to show that they are equal.

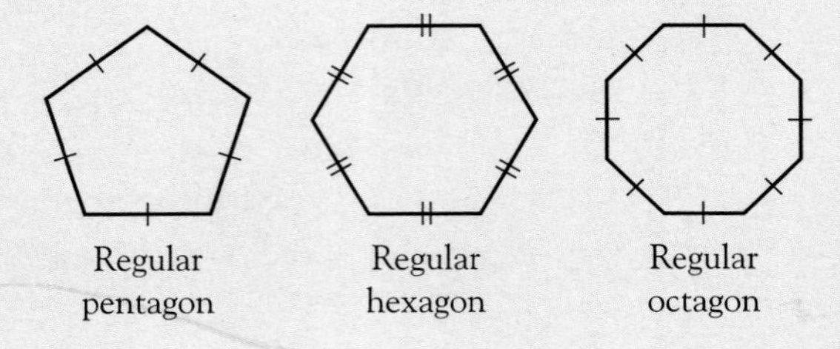

FACT SHEET 3: THE LANGUAGE OF GEOMETRY

Types of angle in a polygon

Interior angles

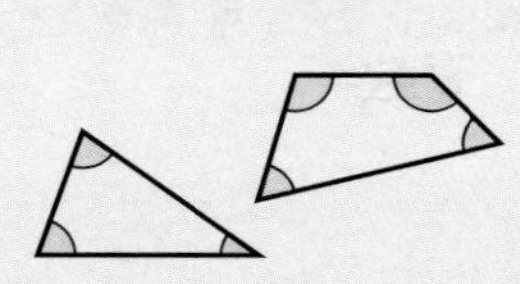

Exterior angles

Special triangles

Equilateral

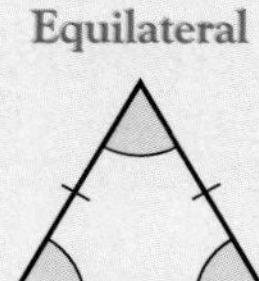

3 sides equal, 3 angles equal, (each 60°).

Isosceles

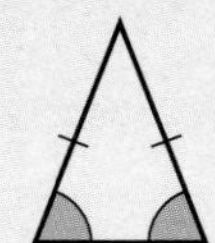

2 sides equal, 2 angles equal (called **base** angles).

Right-angled

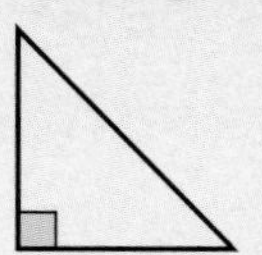

One of the angles is 90°.

Scalene

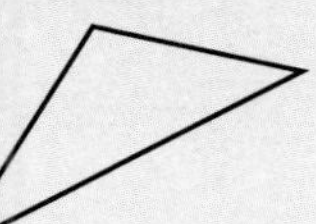

No sides or angles are the same.

An equilateral triangle is a **regular** triangle.

Special quadrilaterals

Parallelogram

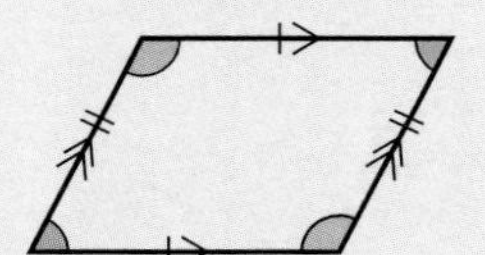

Each pair of opposite sides is parallel and equal in length.
Opposite angles are equal.

Rhombus

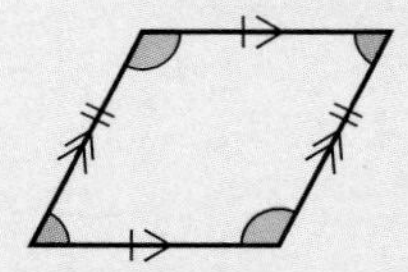

A parallelogram that has all the sides the same length.

Square

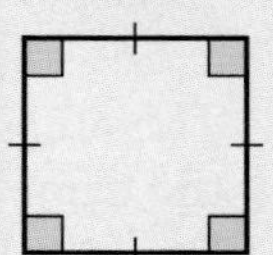

A rhombus that has each interior angle equal to 90°.

A square is a **regular** quadrilateral.

Rectangle

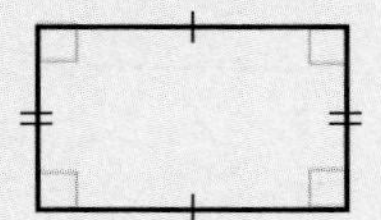

A parallelogram that has each interior angle equal to 90°.

Kite

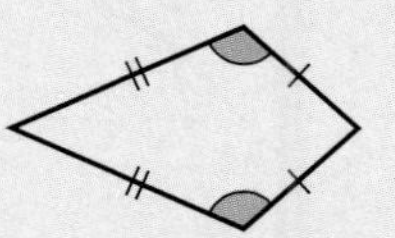

A parallelogram that has two pairs of adjacent sides that are equal in length.

Trapezium

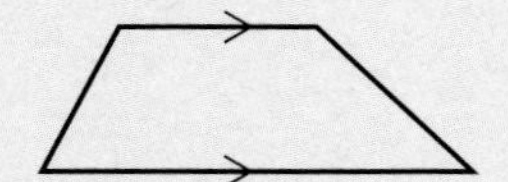

A parallelogram that has only one pair of parallel sides. These are not equal in length.

Isosceles trapezium

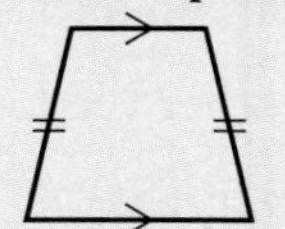

A trapezium in which the non-parallel sides are equal in length.

5 GEOMETRY I

In this chapter you will
- **learn how to use angle facts on a straight line, at a point and on parallel lines**
- **learn how to use angle facts in triangles and quadrilaterals**
- **practise drawing accurately using a ruler, compasses, protractor and set square**

Angle facts on a straight line and at a point

$a + b = 180°$

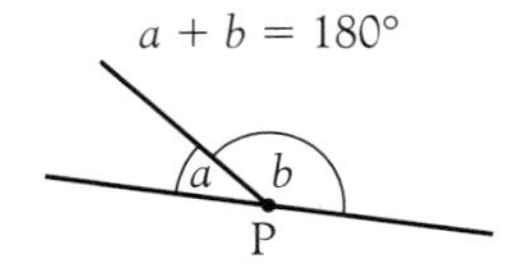

Angles on a straight line add up to 180°.

The angles must be at **the same point** on the line.

$a + b + c + d = 360°$

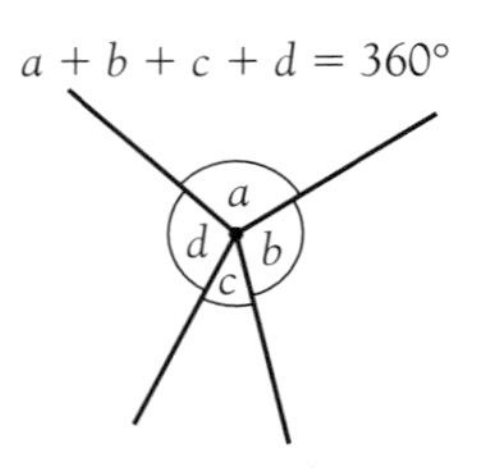

Angles at a point add up to 360°.

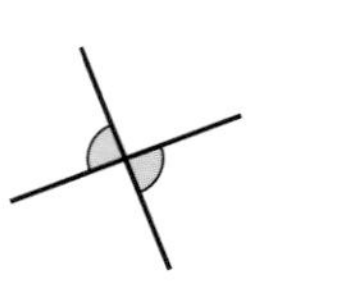

Opposite angles at a point are equal.

These are sometimes known as 'vertically opposite angles' or 'scissor' angles.

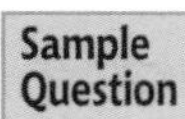

1 Work out the missing angle in each of these sketches.

a

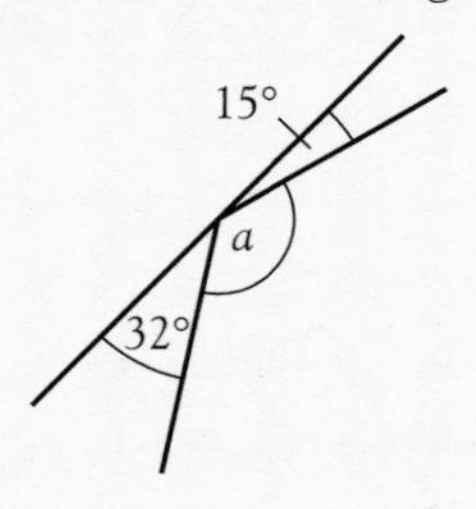

b

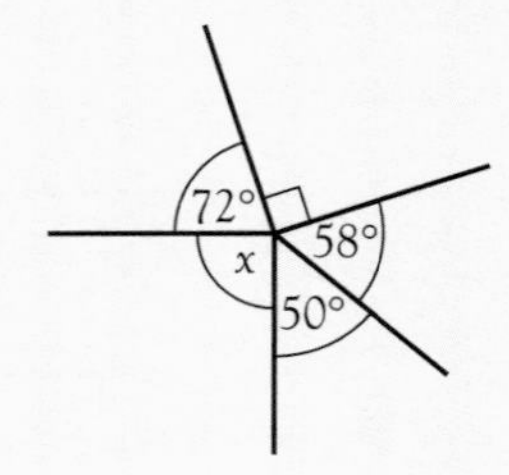

c

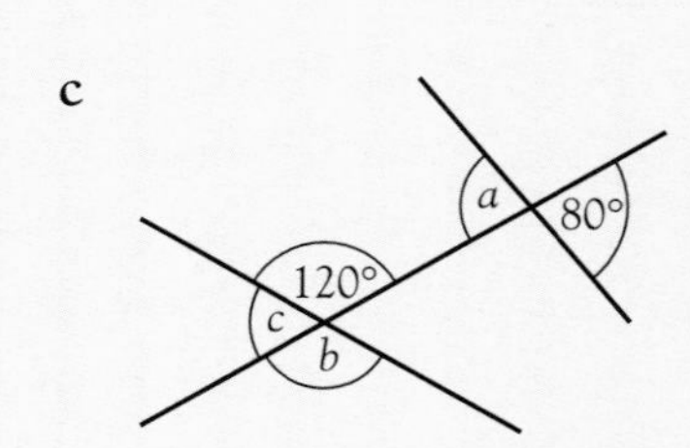

Answer

a

- Use angles on a straight line add up to 180°.
- Add together the known angles 32° and 15°.

  $32 + 15 = 47$

- Take this answer away from 180° to find a.

  $a = 180 - 47$

  $\underline{a = 133°}$

You can write your working as an equation and then solve it to find a.

$a + 32 + 15 = 180$

Simplify: $a + 47 = 180$

-47 $\quad \underline{a = 133°}$

ANGLES AND PARALLELS

b

- Use angles at a point add up to 360°.
- Add together the known angles.

  $72 + 90 + 58 + 50 = 270$
- Take this answer away from 360°.

  $x = 360 - 270$

  $\underline{x = 90°}$

$x + 72 + 90 + 58 + 50 = 360$

Simplify: $x + 270 = 360$

-270 $\underline{x = 90°}$

c

- To find a, use opposite angles at a point are equal.

  $\underline{a = 80°}$
- To find b, use opposite angles at a point are equal.

  $\underline{b = 120°}$
- To find c, use angles on a straight line add up to 180°.

  $c + 120 = 180$

  $\underline{c = 60°}$

Exercise 5.1

Find the missing angles in each of the following.
Draw a sketch each time and say which angle fact you are using.

1

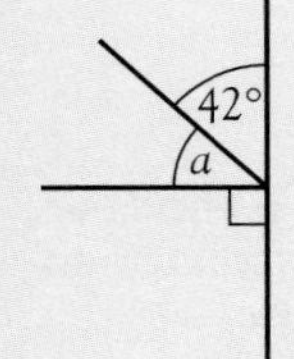

4

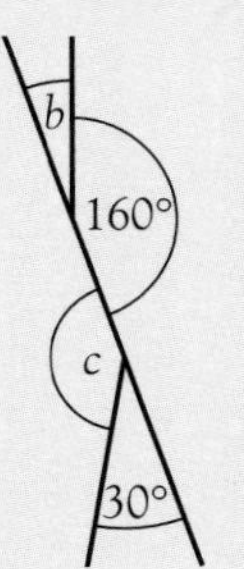

7

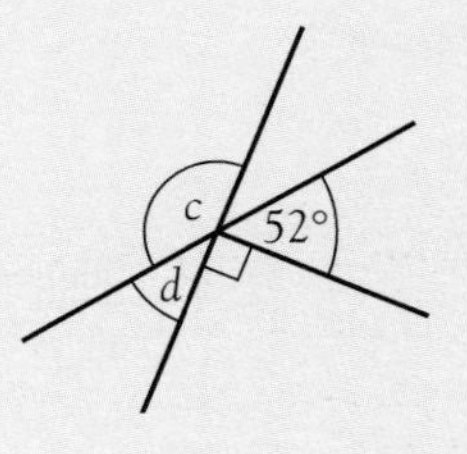

10

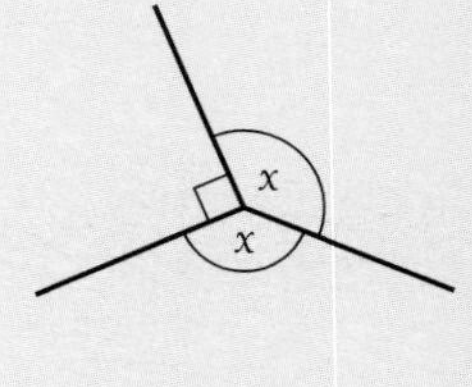

2

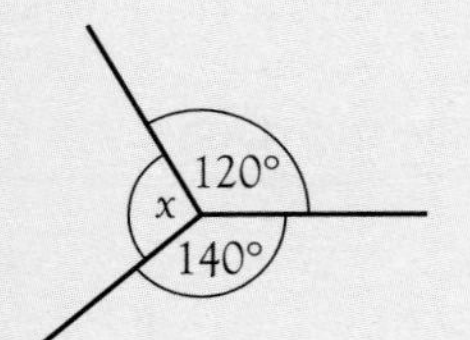

5

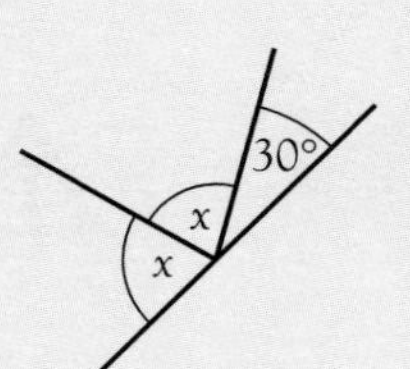

8

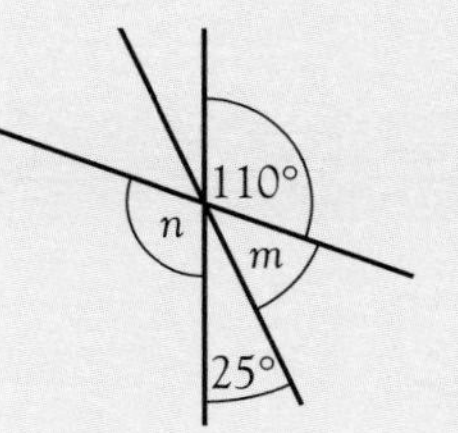

11

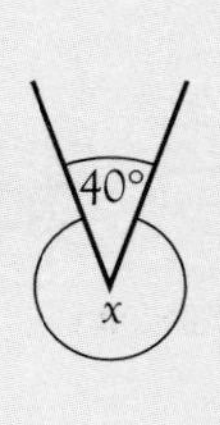

3

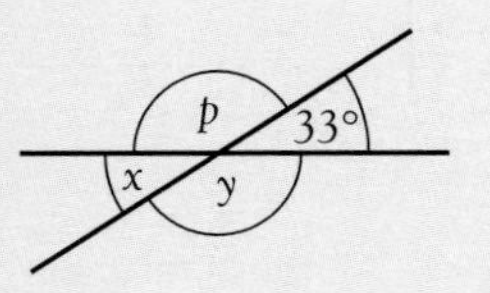

6

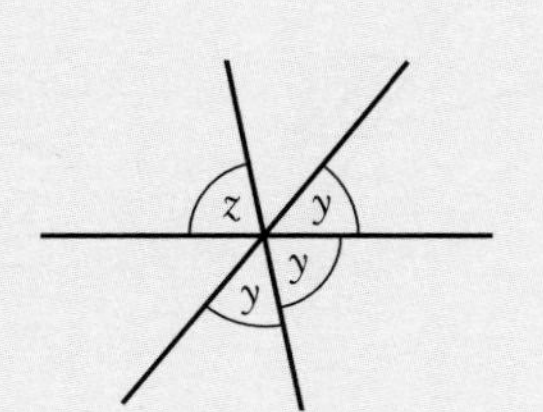

9

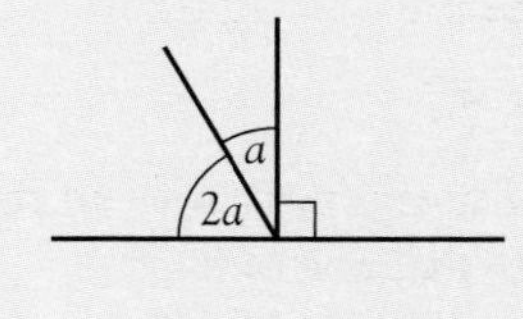

12

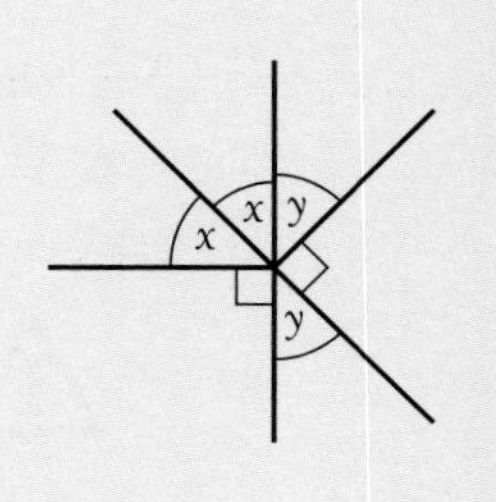

ANGLES AND PARALLELS

Angle facts on parallel lines

Corresponding angles on parallel lines are equal.

These are sometimes called 'F' angles.

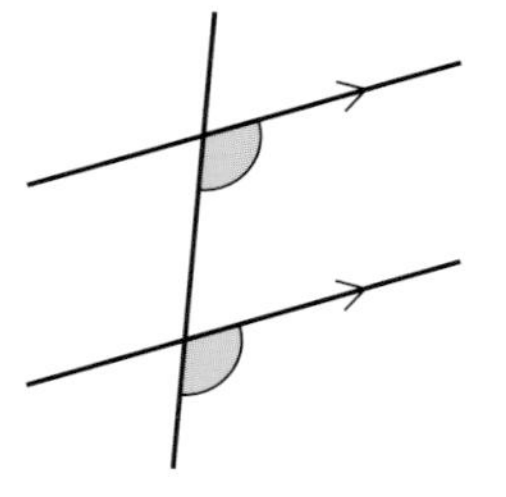

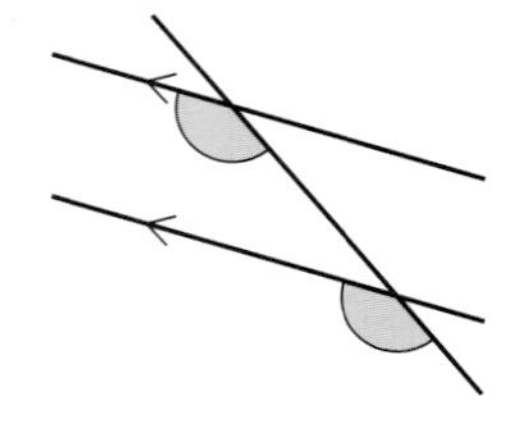

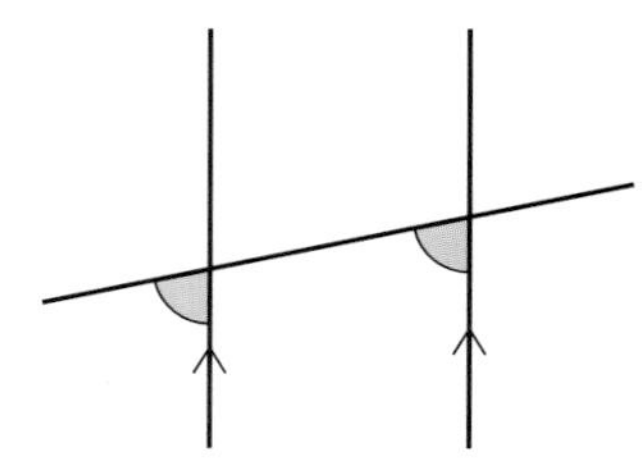

Alternate angles on parallel lines are equal.

These are sometimes called 'Z' angles.

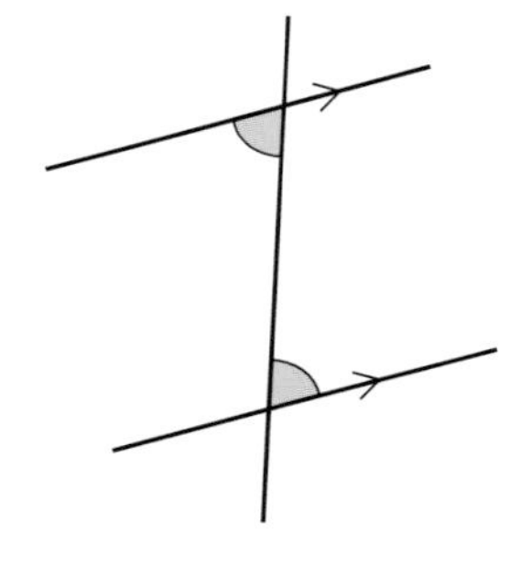

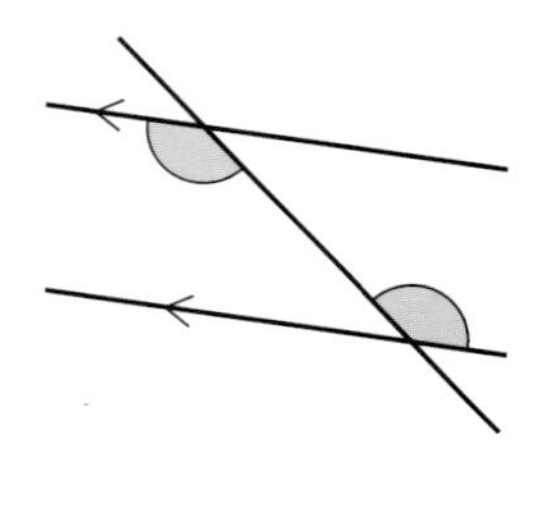

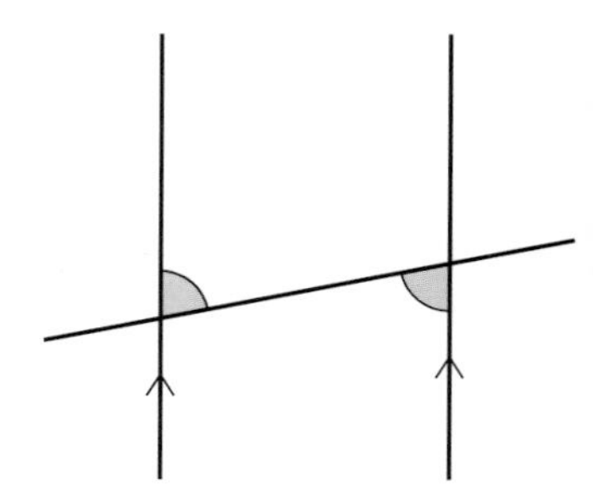

This is particularly useful for work on bearings.

Interior angles on parallel lines add up to 180°.

These are sometimes called '$\sqsubset$' angles.

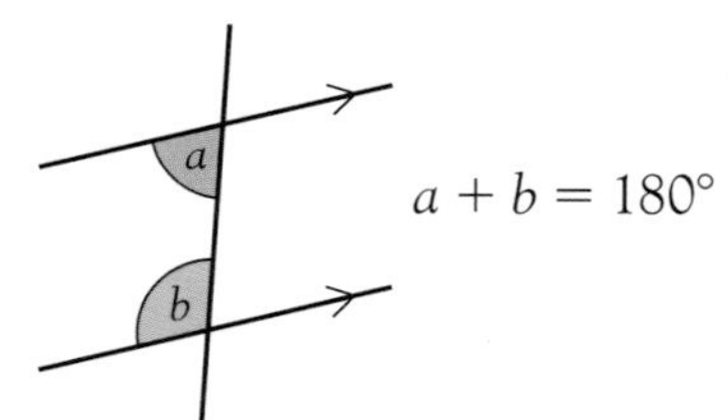

$a + b = 180°$

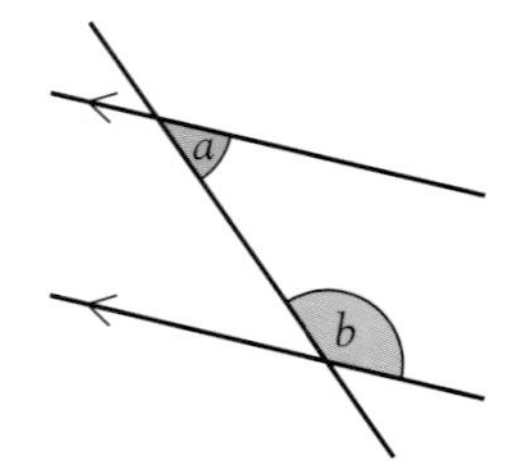

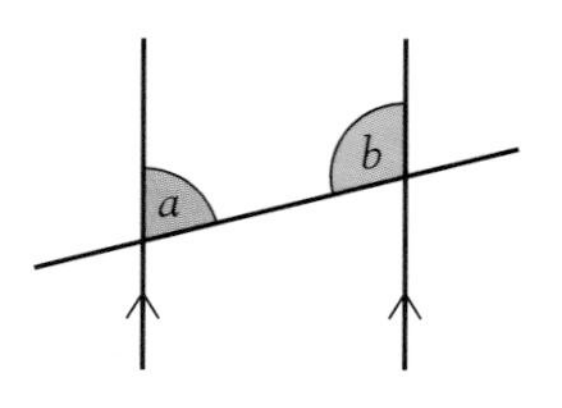

Notice that one of the angles is acute and the other is obtuse.

All these facts are very useful. You should LEARN them.

ANGLES AND PARALLELS

Sample Question 2

Find the angles marked with letters.
Give reasons for your answers.

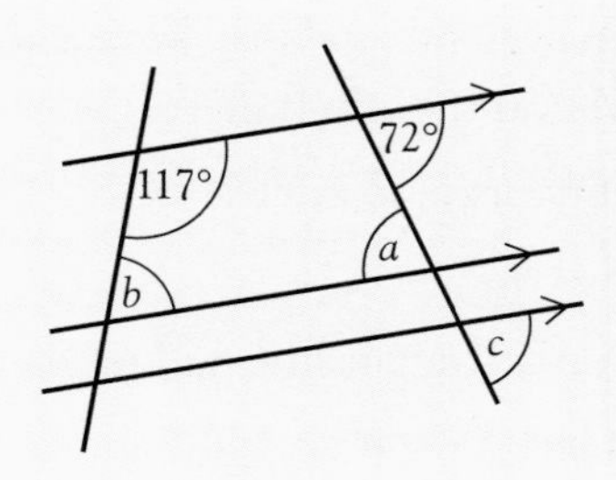

Answer

- There are parallel lines so look for Z angles, F angles or ⊏ angles.
  Using **Z** angles

  $a = 72°$ (alternate angles on parallel lines are equal)

Z angles:

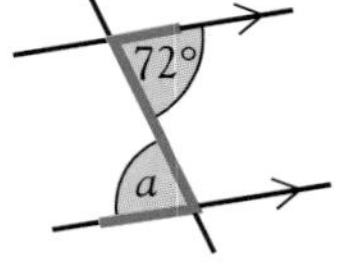

Using **F** angles

$c = 72°$ (corresponding angles on parallel lines are equal)

F angles:

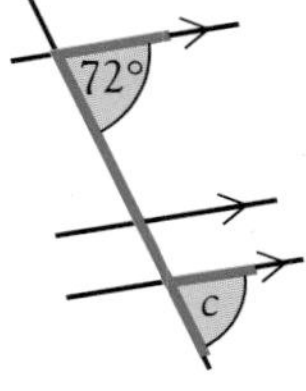

Look for ⊏ angles

$b = 180 - 117$
$b = 63°$ (interior angles on parallel lines add up to 180°)

⊏ angles:

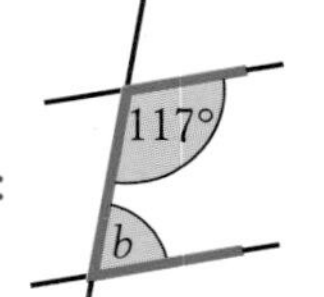

Remember that these are NOT equal; they add up to 180°.

Sample Question 3

$A\hat{B}G = 40°$. Find:

a $C\hat{B}D$,

b $F\hat{G}H$.

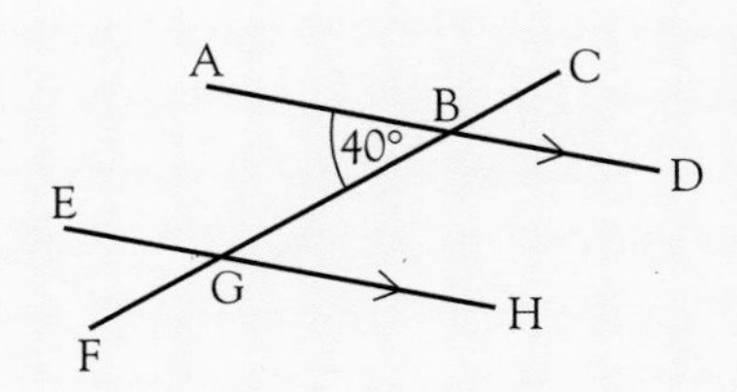

Answer

a

- Identify $C\hat{B}D$ on the diagram.
- Use opposite angles at a point are equal.

  $C\hat{B}D = 40°$

Remember that $C\hat{B}D$ is the angle turned through in going from C to D via B.

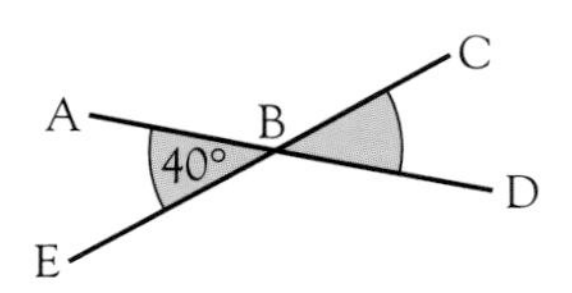

$C\hat{B}D$ may be written as angle CBD.

b

- Identify $F\hat{G}H$ on the diagram
- Find $B\hat{G}H$ first using Z angles.

  $B\hat{G}H = 40°$
- Use angles on a straight line add up to 180°.

  $F\hat{G}H = 180 - 40 = \underline{140°}$

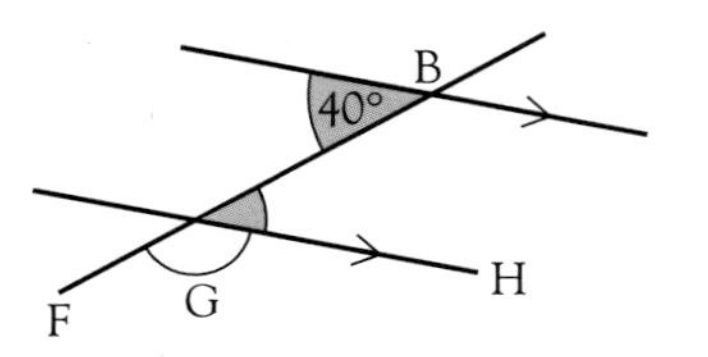

Exercise 5.2

1 Find the angles marked with letters and give reasons for your answers. Remember that these are sketches. They have not been drawn accurately, so do not measure anything.

a

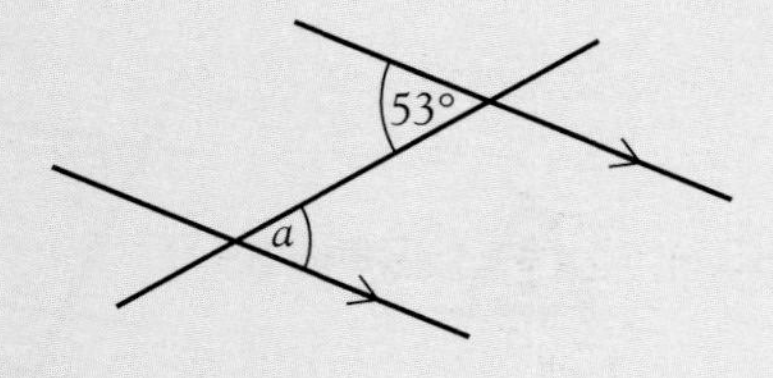

b

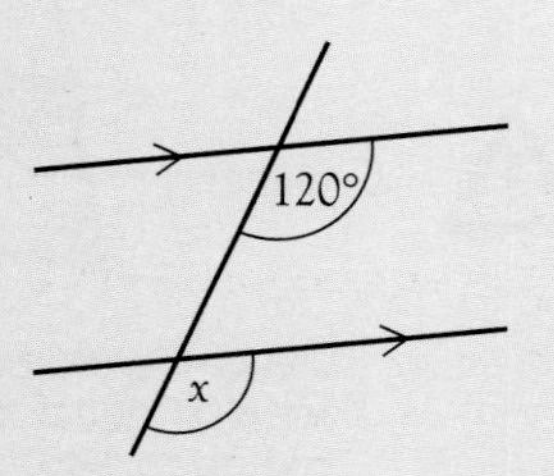

c

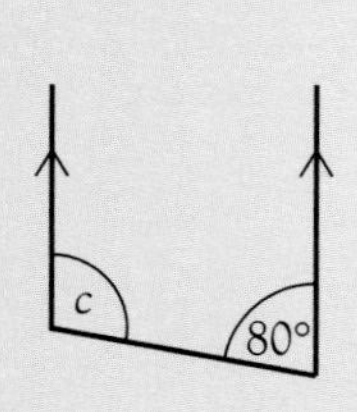

d

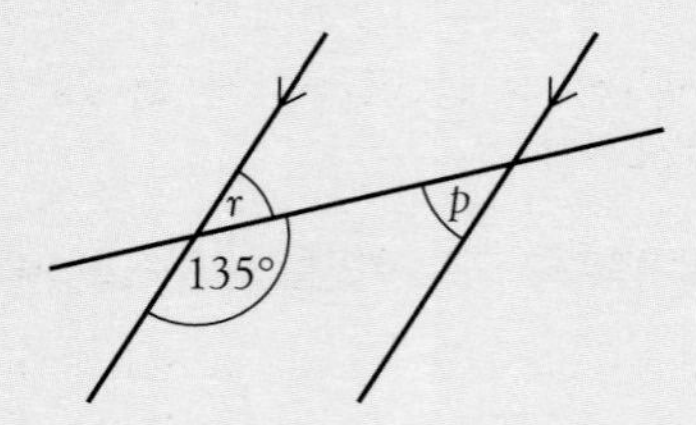

e

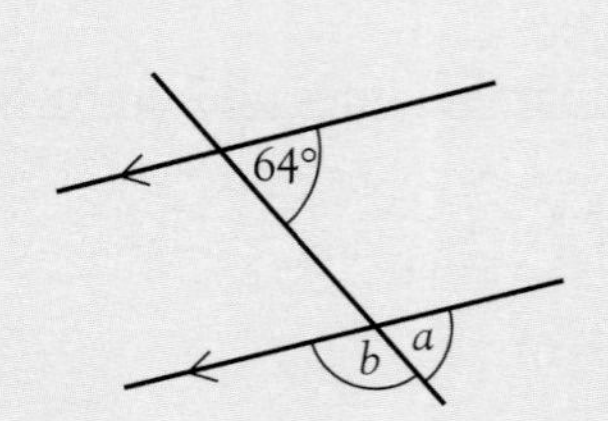

f

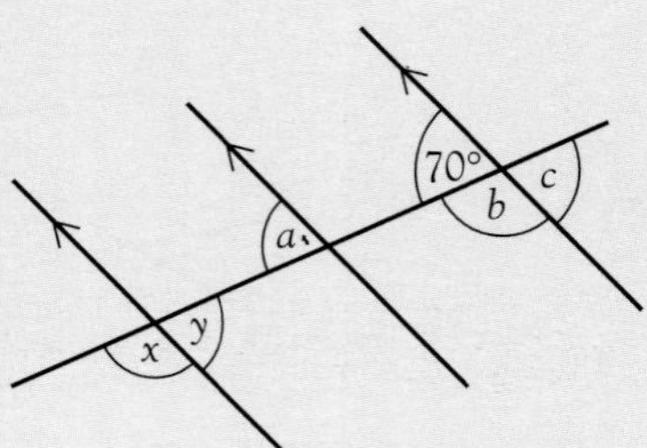

g

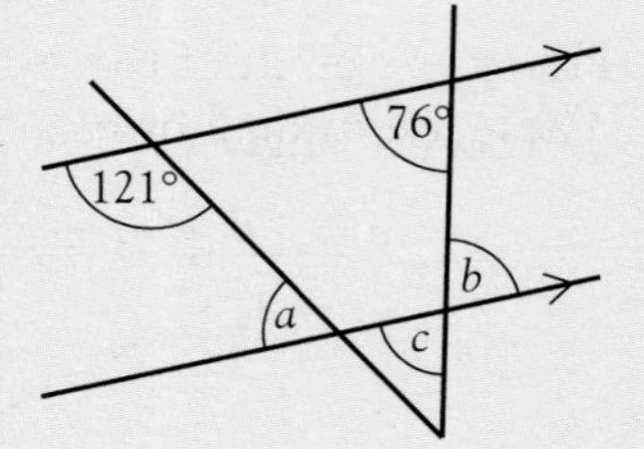

h

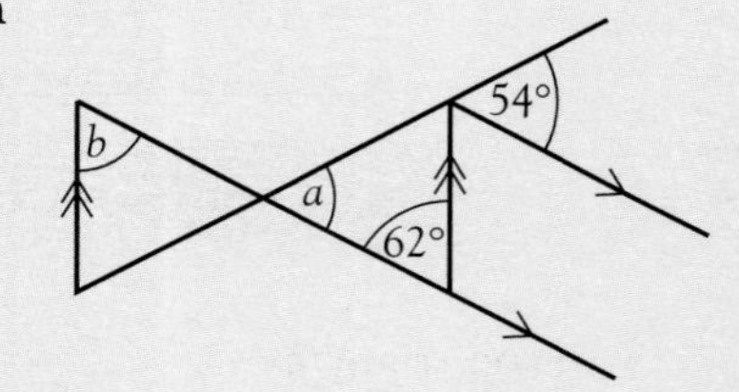

i

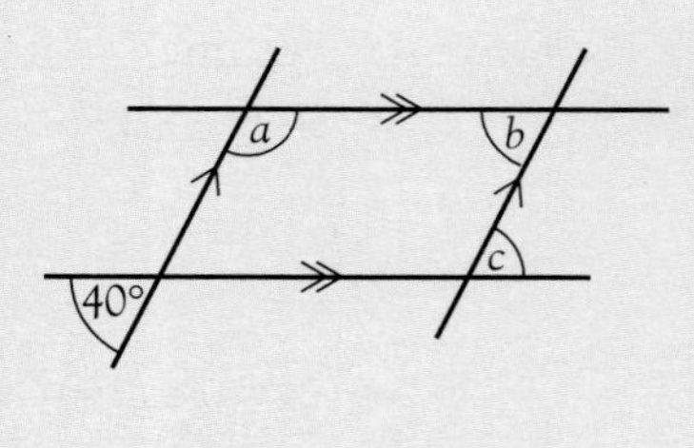

2 Find the following angles:

a $B\hat{C}F$

b $D\hat{C}F$

c $A\hat{B}F$

d $E\hat{F}C$

Give reasons for your answers.

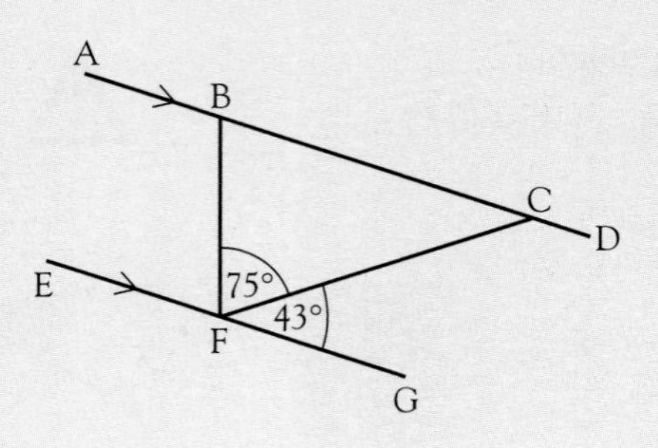

Triangles

A triangle is a polygon with three sides.

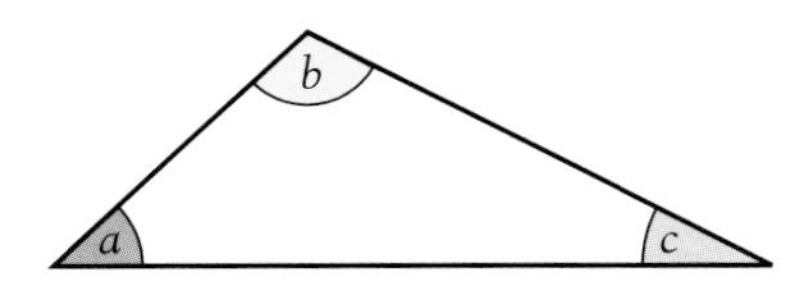

The angles marked a, b and c are called the **interior angles** of the triangle.

If you tear off the three corners and fit the angles together, they fit on a straight line. Angles on a straight line add up to 180°, so

$$a + b + c = 180°$$

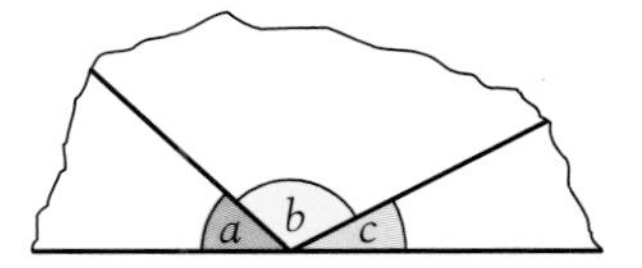

The interior angles of a triangle add up to 180°.

Here is a sketch of the triangle, with the **exterior angles** d, e and f shown as well. These are formed by extending the sides.

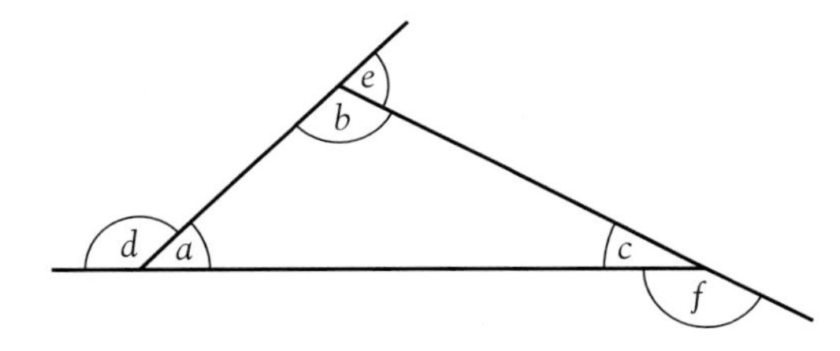

Using angles on a straight line add up to 180°

$$a + d = 180°$$
$$b + e = 180°$$
$$c + f = 180°$$

Interior angle + exterior angle = 180°

This fact is true for **any** polygon.

In the triangle

$$a + \mathbf{b} + \mathbf{c} = 180°$$

On the straight line

$$a + \mathbf{d} = 180°$$

So $\mathbf{d} = \mathbf{b} + \mathbf{c}$

It is also true that $e = a + c$ and $f = a + b$

An exterior angle of a triangle is equal to the sum of the interior opposite angles.

Remember that you obtain the **sum** by adding.

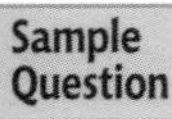

Sample Question 4 Find the angles marked with letters.

a

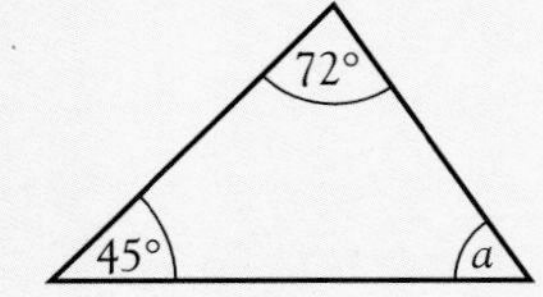

b

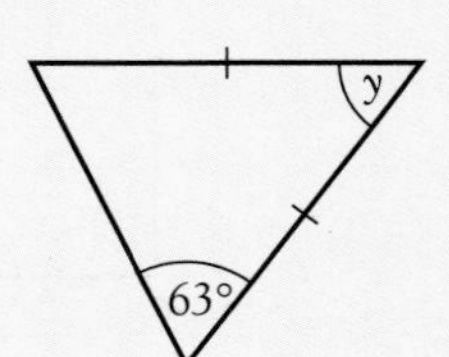

c

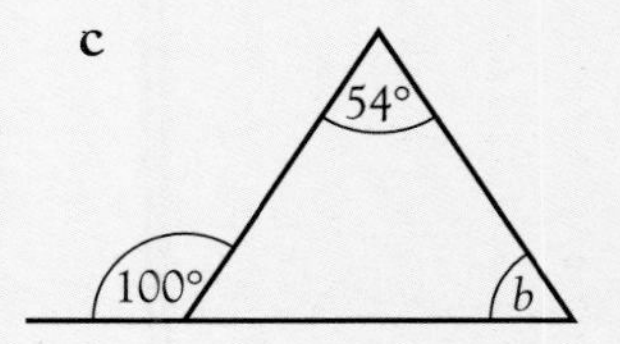

TRIANGLES

Answer

a

- Use interior angles of a triangle add up to 180°.
- Add together the two known angles.

  $45 + 72 = 117$

- Take this answer away from 180.

  $a = 180 - 117$
  $\underline{a = 63°}$

Forming an equation:
$a + 45 + 72 = 180$
Simplify: $a + 117 = 180$
(−117) $\underline{a = 63}$

b

- The triangle is **isosceles** so use the fact that the base angles are equal to find the unmarked angle.
- Add together the base angles.

  $63 + 63 = 126$

- Take this answer away from 180°.

  $y = 180 - 126$
  $\underline{y = 54°}$

The two 'base angles' are on the line that is a different length from the other two.
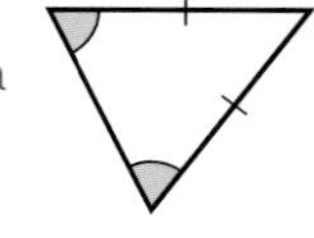

FACT SHEET 3, pages 67–69
Special triangles

$y + 63 + 63 = 180$
Simplify: $y + 126 = 180$
(−126) $\underline{y = 54}$

c

- Use exterior angle of a triangle is equal to the sum of the interior opposite angles.

  $b + 54 = 100$
  $b = 100 - 54$
  $\underline{b = 46°}$

Sample Question 5 Find:

a $B\hat{C}A$

b $C\hat{D}A$

c $B\hat{A}D$

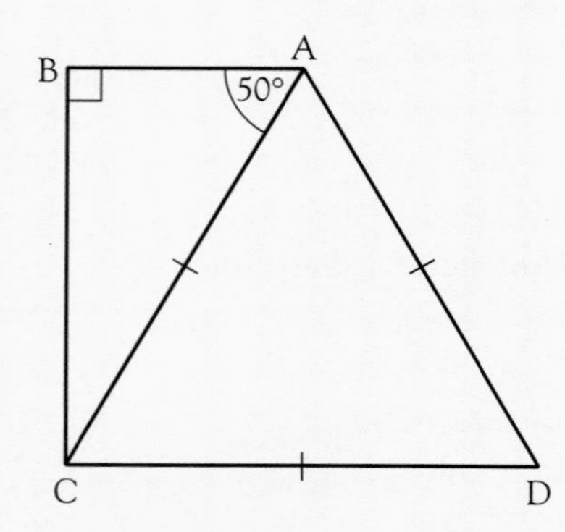

TRIANGLES

Answer

a

- Look at $\triangle ABC$ and identify $B\hat{C}A$.
- $A\hat{B}C$ is 90°, so the other two angles add up to 90°.

$B\hat{C}A = 90 - 50$

$\underline{B\hat{C}A = 40°}$

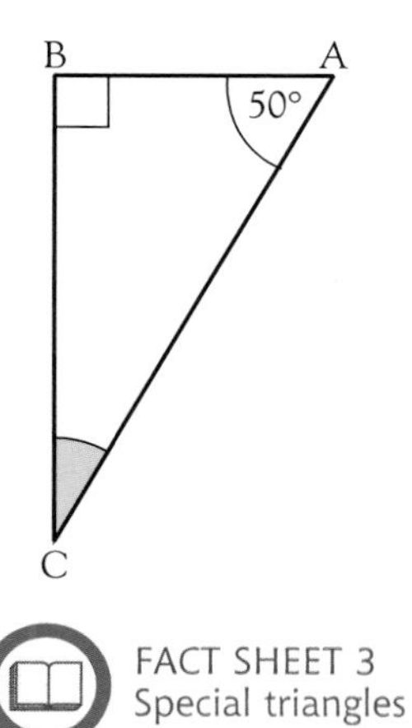

FACT SHEET 3
Special triangles

b

- Use the fact that $\triangle ACD$ is **equilateral**, so each interior angle is the same.

$C\hat{D}A = 180 \div 3$

$\underline{C\hat{D}A = 60°}$

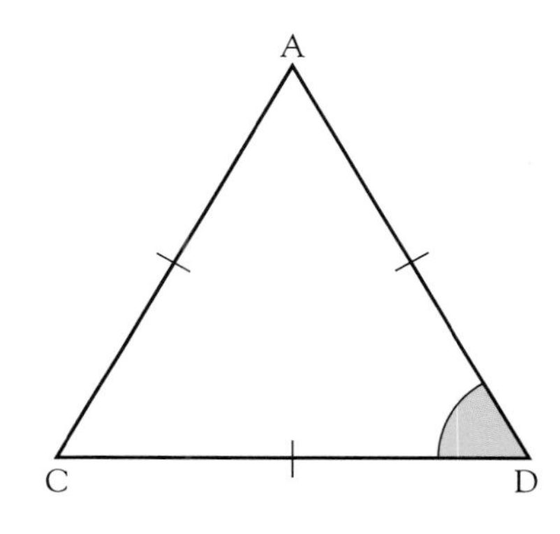

c

- Use $B\hat{A}D = B\hat{A}C + C\hat{A}D$

$B\hat{A}D = 50 + 60$

$\underline{B\hat{A}D = 110°}$

$C\hat{A}D$ is in the same equilateral triangle as $C\hat{D}A$ so it is 60°

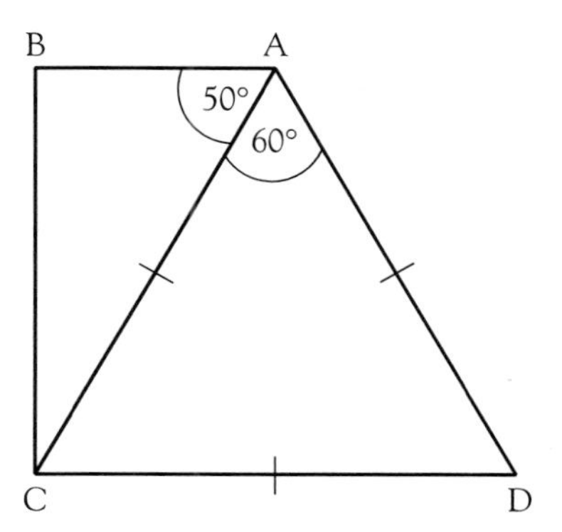

QUADRILATERALS

Quadrilaterals

A quadrilateral is a four-sided polygon.

If you tear off the corners and re-arrange them, they fit together at a point:

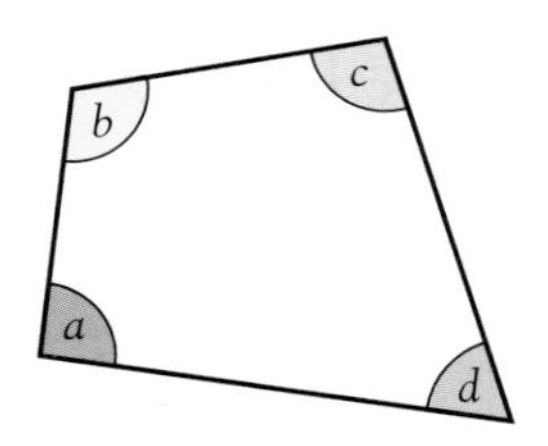

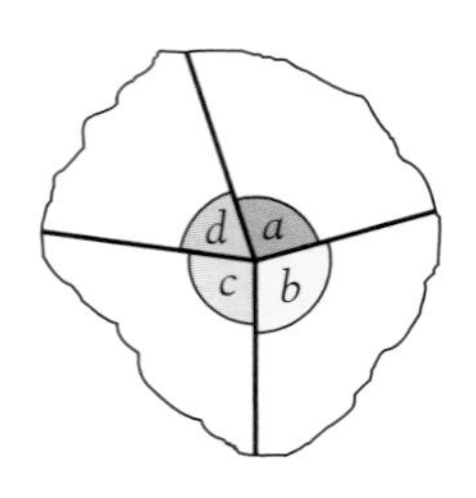

Angles at a point add up to 360°, so

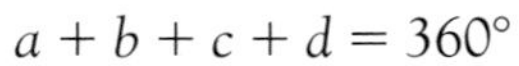

$$a + b + c + d = 360°$$

The interior angles of a quadrilateral add up to 360°.

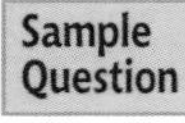

Sample Question 6

Find x and hence find the four interior angles of this quadrilateral.

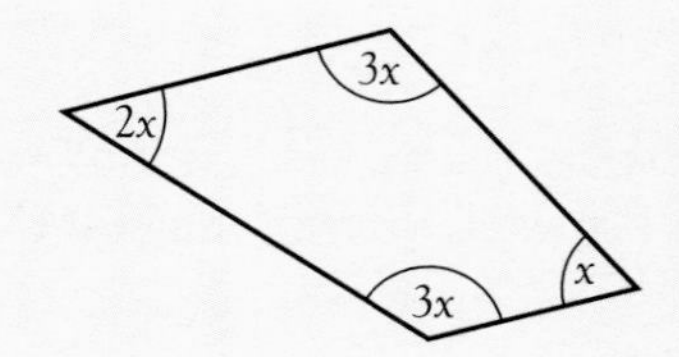

Answer

- Use interior angles of a quadrilateral add up to 360° to form an equation in x.
- Simplify it, then solve it to find x.
- Work out the four angles.

$$2x + 3x + x + 3x = 360$$

Simplify: $9x = 360$

(÷9) $x = 40$

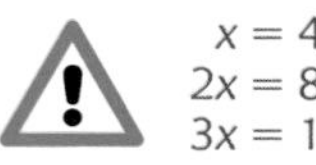

$x = 40°$
$2x = 80°$
$3x = 120°$

The angles are 40°, 80°, 120°, 120°

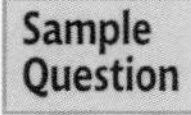

Sample Question 7

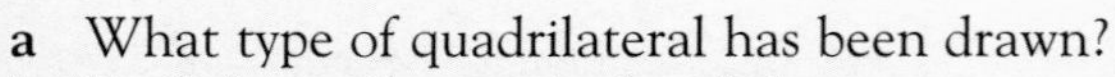

a What type of quadrilateral has been drawn?

b Find the angles marked with letters.

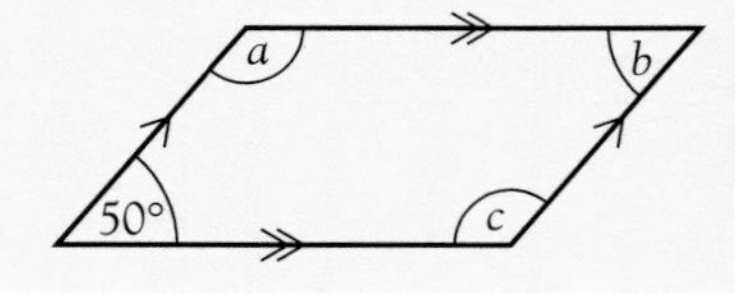

Answer

a

- The quadrilateral is a **parallelogram**.

FACT SHEET 3
Special quadrilaterals

b

- Use ⊏ angles.

$a = 180 - 50 = 130°$
$b = 180 - 130 = 50°$
$c = 180 - 50 = 130°$

$a = 130°, b = 50°, c = 130°$

Notice that opposite angles in a parallelogram are equal.

MISSING ANGLES

Exercise 5.3

When writing your answers, always give your reasons. It is also helpful to draw your own diagram, especially when the question is complicated.

1 Find the angles marked with letters.

a 20°, x, 27°

b

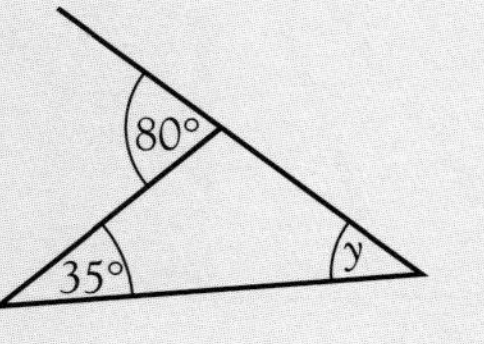

c a, 130°, 110°

d

e

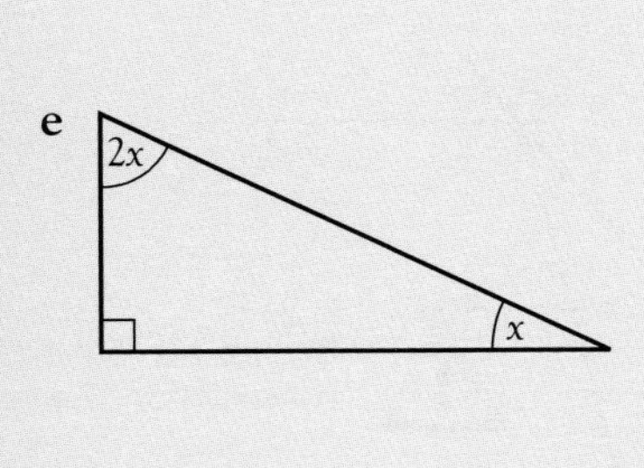

f

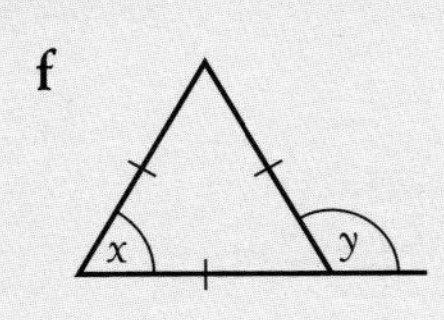

g

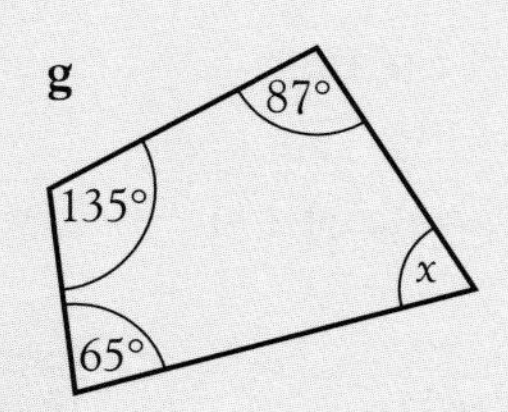

h

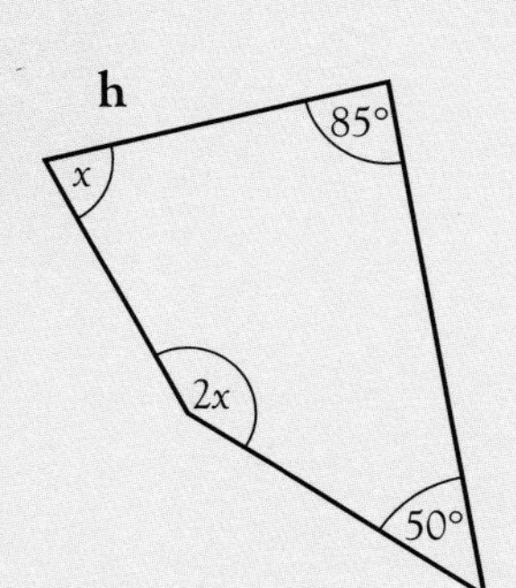

i

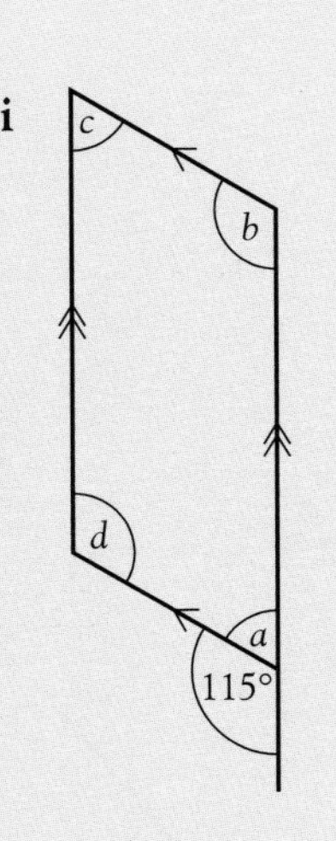

2 $\triangle ABC$ is isosceles with $A\hat{B}C = A\hat{C}B$.

a If $B\hat{A}C = 70°$, find $A\hat{B}C$.

b If $A\hat{B}C = 70°$, find $B\hat{A}C$.

3 PQRS is a rectangle. Calculate the value of a, b, c.

4

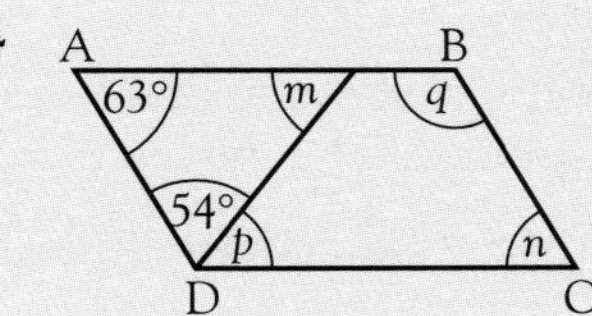

ABCD is a parallelogram. Calculate the value of m, n, p and q.

5 In a triangle, the angles are x, $x + 30°$ and $x + 24°$. Find the size of each angle.

6 Find the angles marked with letters.

a

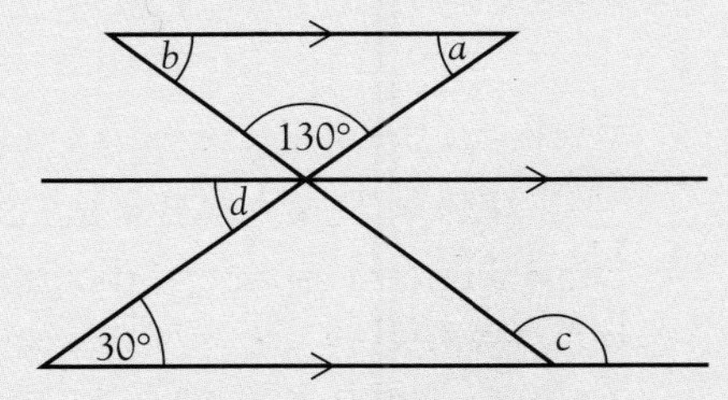

b

c

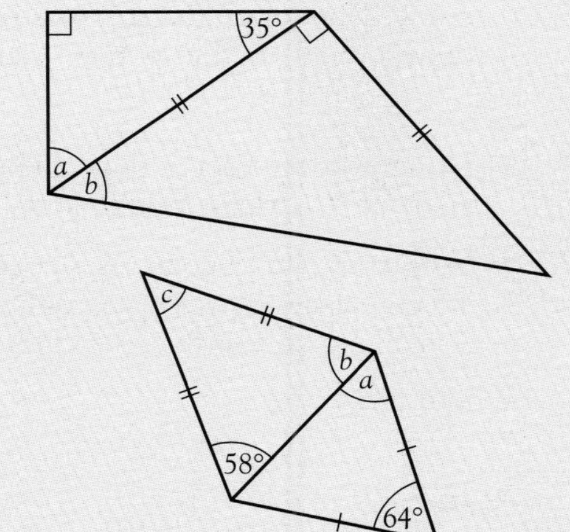

7 ABCD is a rectangle. Find the value of x, y and z.

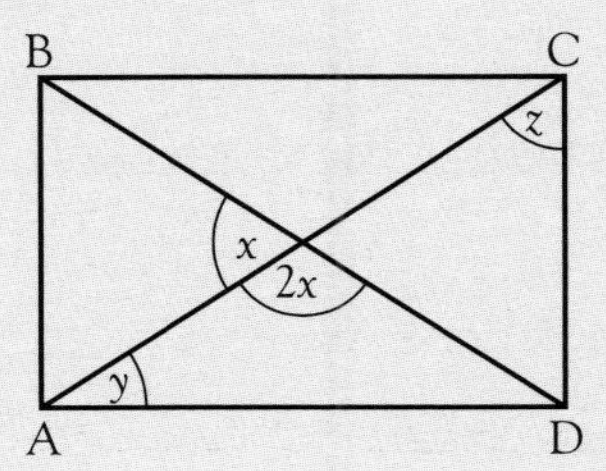

Drawing accurately

To draw accurately you need a sharp HB pencil, a ruler, a pair of compasses, an angle measurer or protractor and a set square.

Sample Question 8

AB and AC are two scaffolding poles attached to a vertical wall at A and fixed to the horizontal ground at B and C.

The longer pole AB is 4.5 m and is fixed at an angle of 50° to the ground. The distance BC is 2 m.

a Draw triangle ABC accurately, using a scale of 1 cm to 1 m.

b How long is the shorter pole?

c Measure $B\hat{A}C$, the angle between the poles.

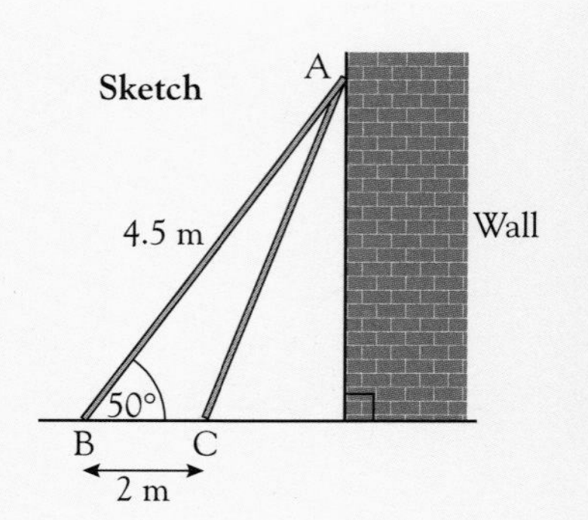

Answer

a

- Work out the scale, using 1 cm to 1 m.

  BC = 2 cm, AB = 4.5 cm

- Draw a line to represent the ground, labelling B at the end.
- To find C, with your pair of compasses opened out at 2 cm, place the compass point at B and draw a small arc on the line, 2 cm from B.

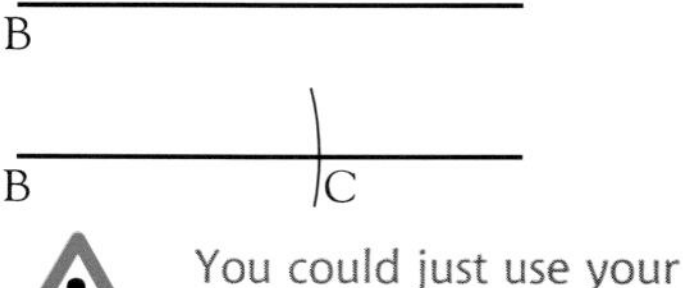

You could just use your ruler to find C but this method is more accurate.

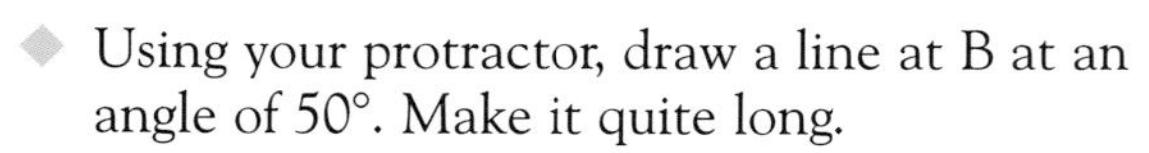

- Using your protractor, draw a line at B at an angle of 50°. Make it quite long.
- With your pair of compasses opened out at 4.5 cm, place the compass point at B and draw a small arc on this line, 4.5 cm from B. Label it A.
- Join AC.

A

B C

Do not rub out the construction marks.

b

- Measure length AC in cm.

  AC = 3.5 cm

- Convert it to metres, using your scale.

  Length of the shorter pole = 3.5 m

Remember to convert back to the original units.

c

- Measure $B\hat{A}C$.

  $B\hat{A}C = 26°$

You may need to make line AB or line AC longer in order to measure the angle.

DRAWING ACCURATELY

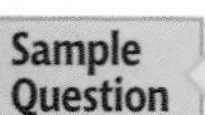

9 Draw an accurate diagram of a triangle ABC in which AB = 5.5 cm, BC = 3.5 cm and AC = 3 cm.

Measure the shortest distance from C to AB.

Answer

- Draw a rough sketch first to give an idea of the drawing.

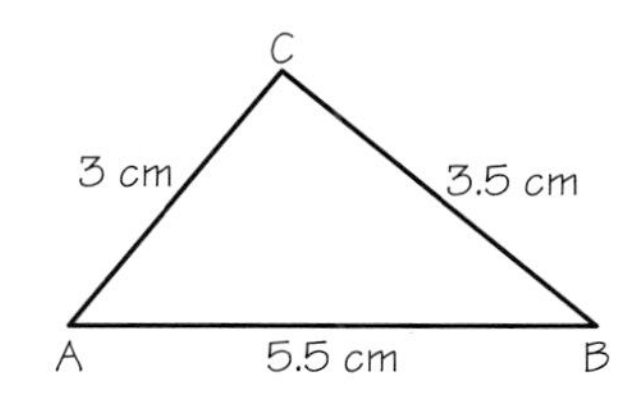

- Draw one of the sides, say AB, accurately with ruler and compasses. Leave room above it for the triangle.

- Open out your compasses to 3 cm. Placing the point at A, make an arc above the line AB.

You do not know where C is yet; draw a long arc.

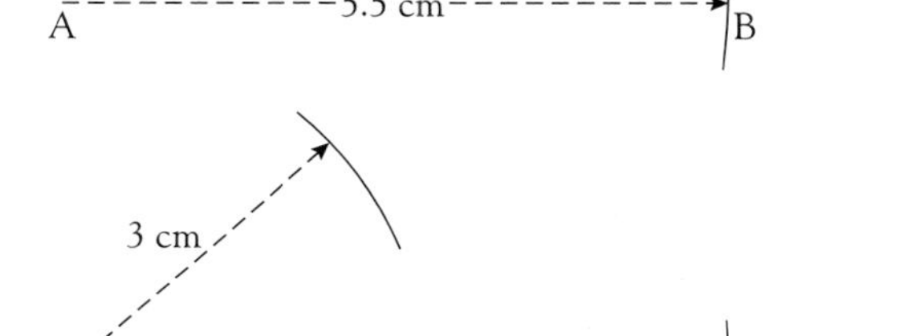

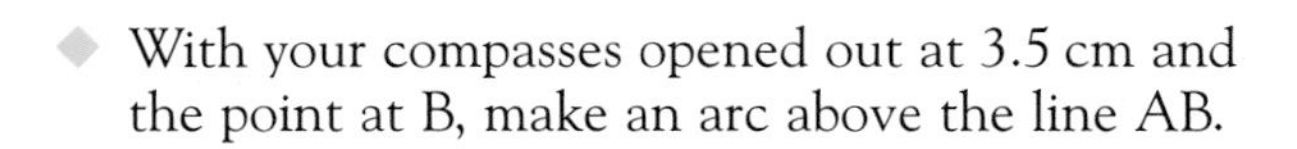

- With your compasses opened out at 3.5 cm and the point at B, make an arc above the line AB.

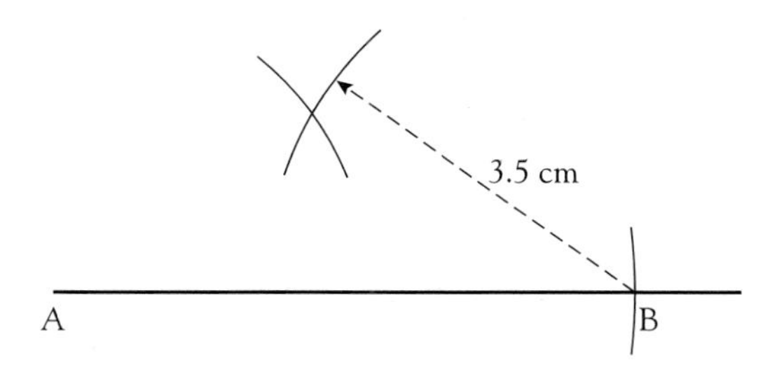

- Where the arcs cross is the point C.

  Draw AC and BC.

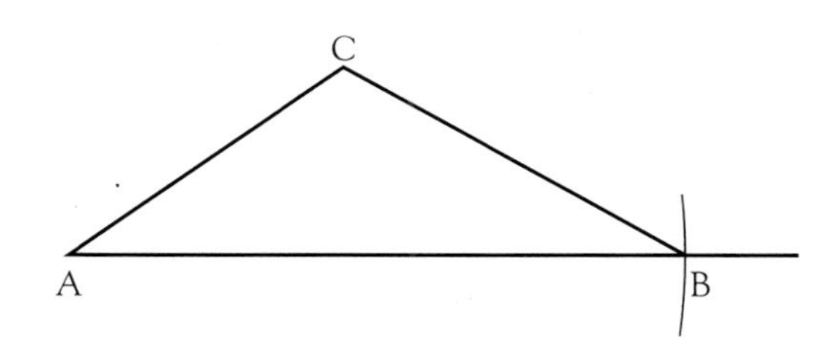

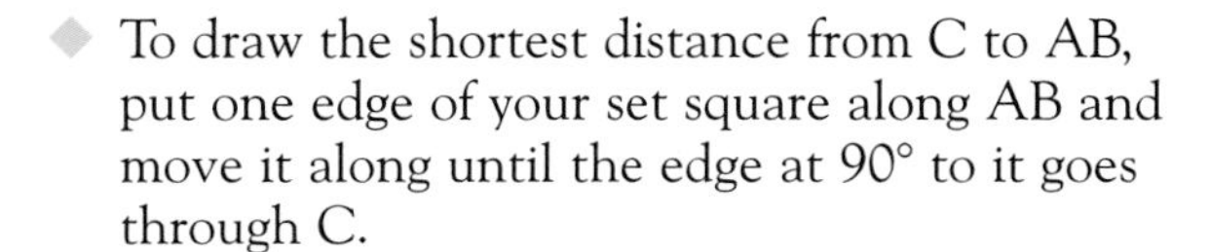

- To draw the shortest distance from C to AB, put one edge of your set square along AB and move it along until the edge at 90° to it goes through C.

  Measure the length CX.

  CX = 1.7 cm

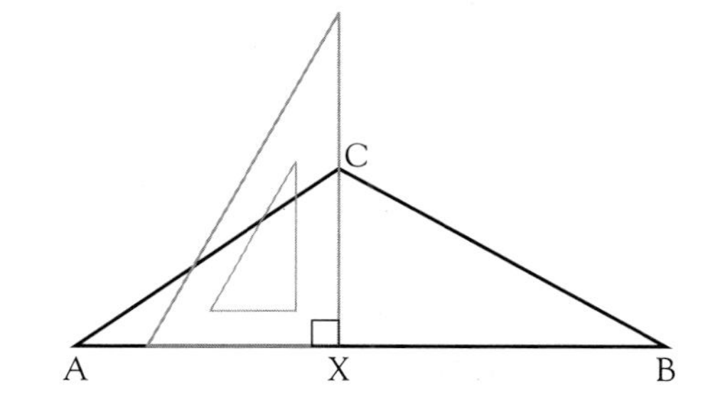

Exercise 5.4

1 Make accurate drawings of these triangles and measure the lengths and angles indicated. You may need to calculate an angle first.

a

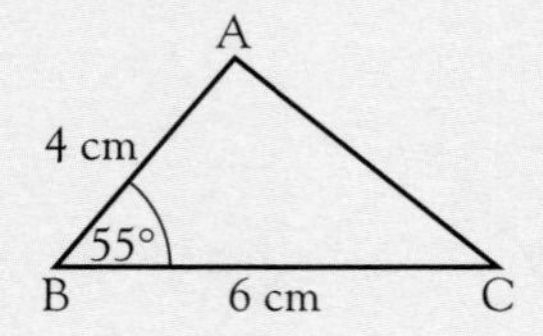

Measure AC.

b

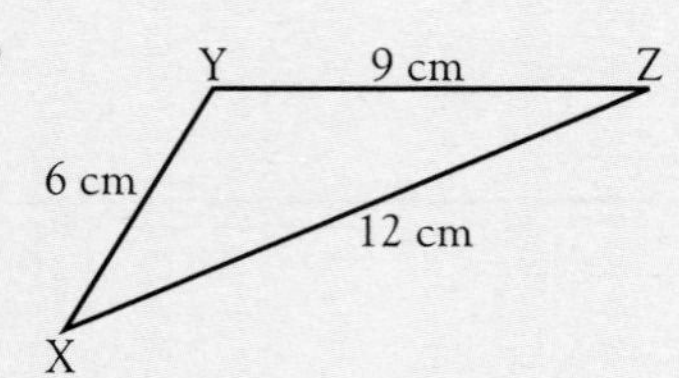

Measure $X\hat{Y}Z$.

c

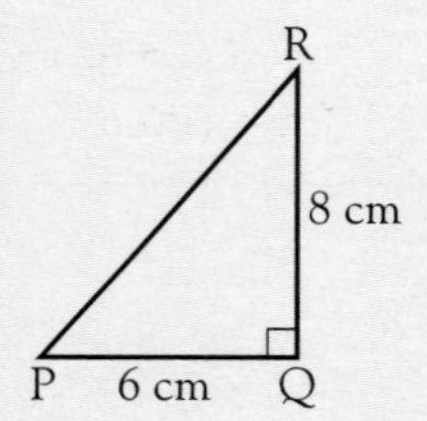

Measure PR and $R\hat{P}Q$.

d

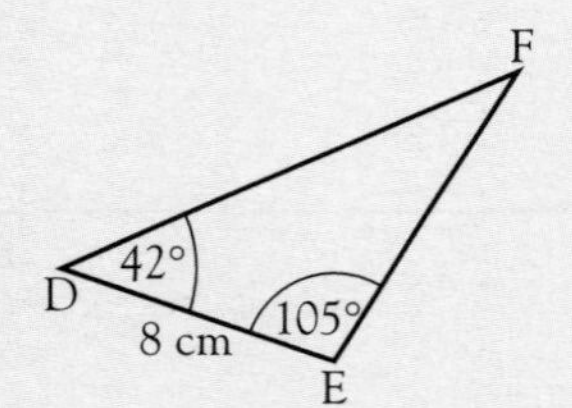

Measure DF.

e

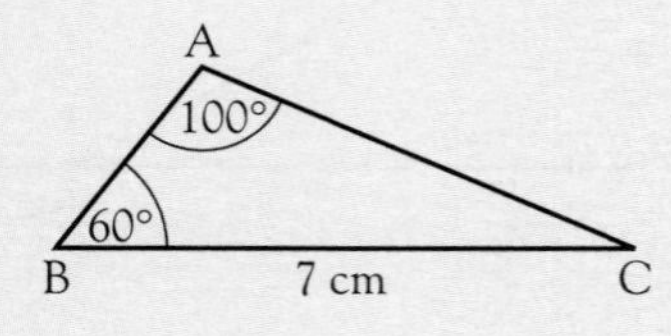

Measure AB and AC.

f

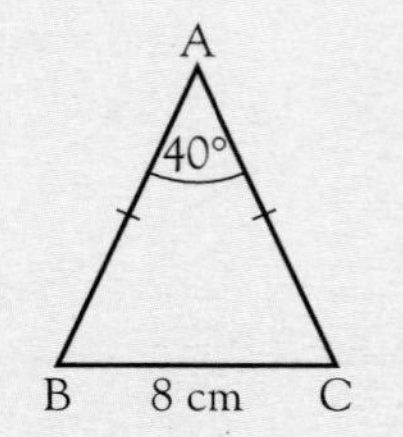

Measure AB.

2 This is a sketch of a plot of land.

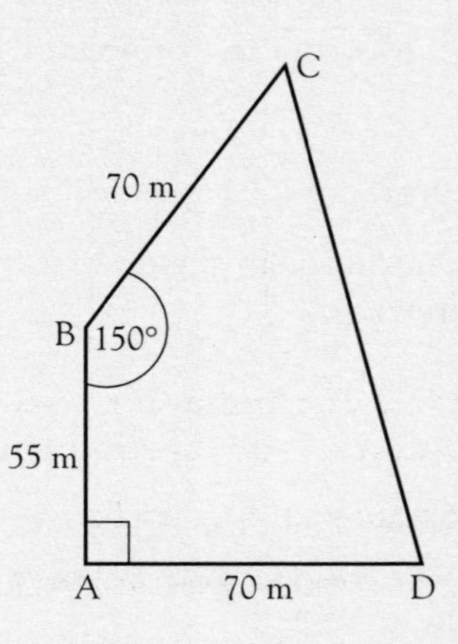

a Draw the plot accurately, using a scale of 1 cm to 10 m.

b Use your scale diagram to find the actual length of CD.

3 The diagram shows a sketch of part of a map.

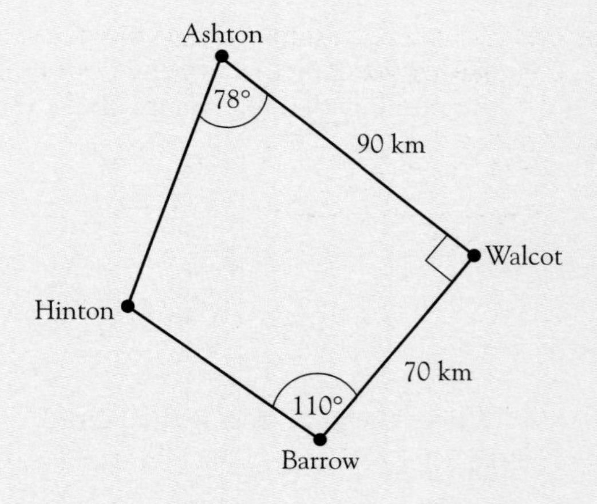

a Make an accurate scale drawing, using a scale of 1 cm to 10 km.

b Use your scale drawing to find the distance between Hinton and Ashton.

c What is the direct distance between Hinton and Walcot?

4 Using a scale of 1 cm to represent 2 m, draw a diagram to show the positions of three girls where each girl is 14 m from the other two.

Worked Exam Question
[MEG]

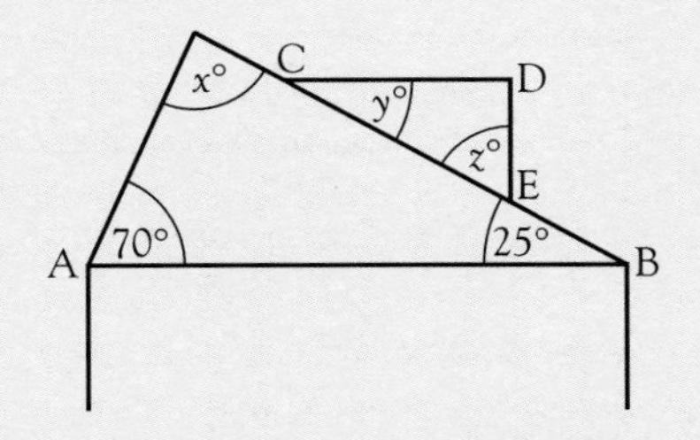

Not to scale

The diagram represents the side view of the roof of a house.
The lines AB and CD are horizontal and the line DE is vertical.

Calculate the values of *x*, *y* and *z*.

◆ Find *x* using interior angles in a triangle add up to 180°.

Look at this triangle.

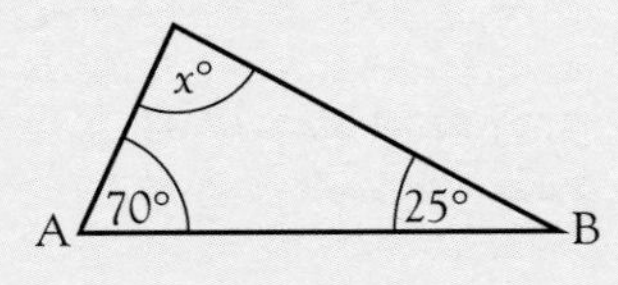

$70 + 25 = 95$

$180 - 95 = 85$

Answer ....x = 85........

◆ AB and CD are horizontal, so they are parallel.
Use Z angles to find *y*.

Look at this part of the diagram.

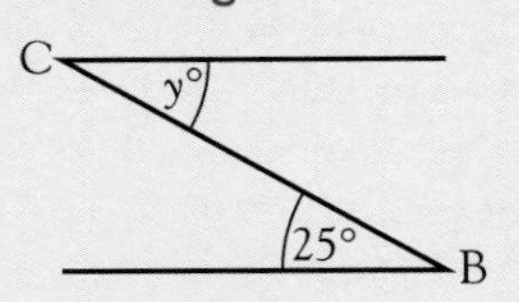

Z angles

Answer ....y = 25........

◆ CD is horizontal and DE is vertical, so $\hat{CDE} = 90°$.

◆ To find *z* use interior angles in a triangle add up to 180°;
since $\hat{D} = 90°$, *z* and 25° add up to 90°.

Look at △CDE and write in the value of $\hat{C}$ found above.

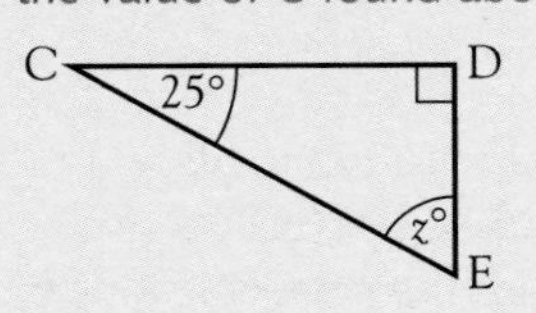

$z = 90 - 25$

$= 65$

Answer ....z = 65........

COMMENTS

M1 for adding 70 + 25

M1 for subtracting from 180

A1 for correct answer

M1 for 90 − *y* or equivalent

4 marks

Exam Questions

1 Find the sizes of the angles marked by letters in these diagrams (which are not drawn to scale).

a

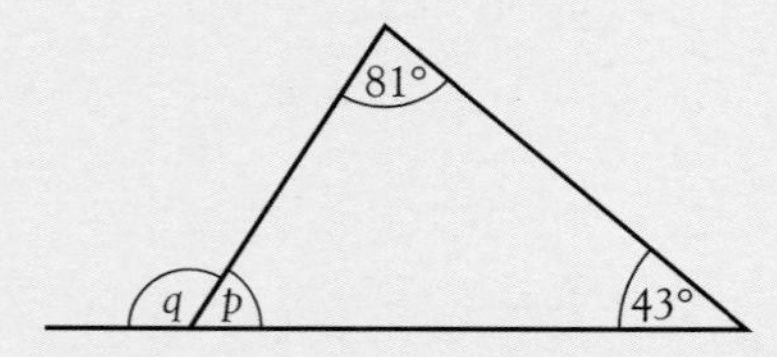

b

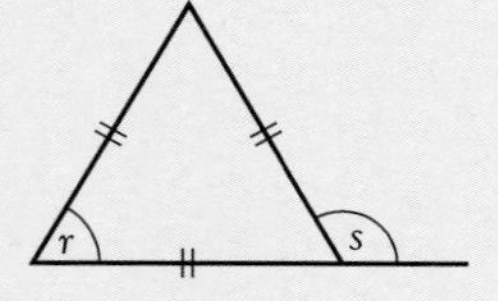

c

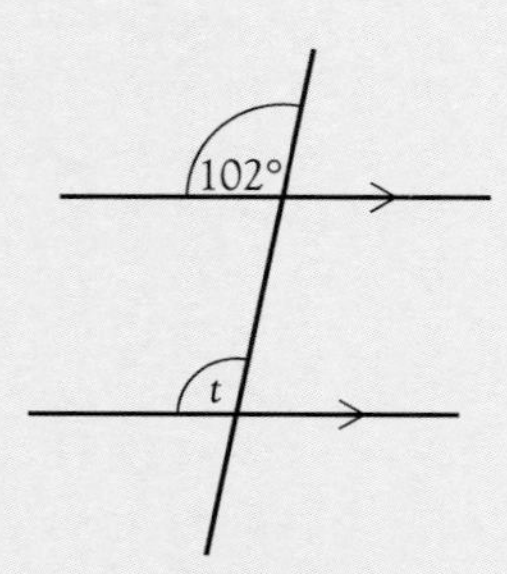

[NEAB]

2

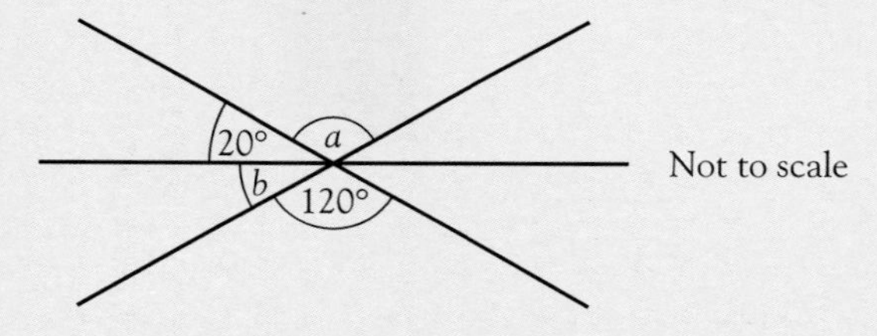

This diagram shows three straight lines which cross at the same point.

Work out the sizes of the angles marked with letters and give a reason in each case. [MEG]

3

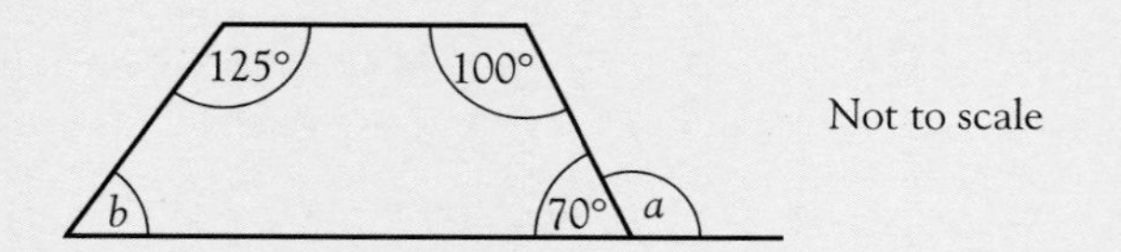

Work out the size of the angles marked with letters. Give reasons for your answers. [MEG]

4

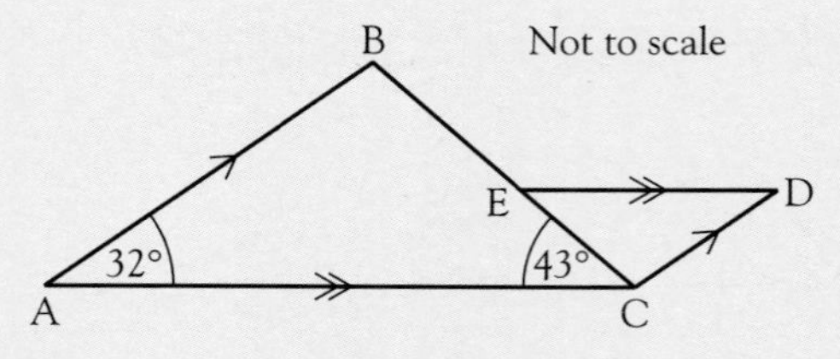

Name an obtuse angle in the diagram.

Write down the size of angle CED.

Calculate angle ACD. [MEG]

5

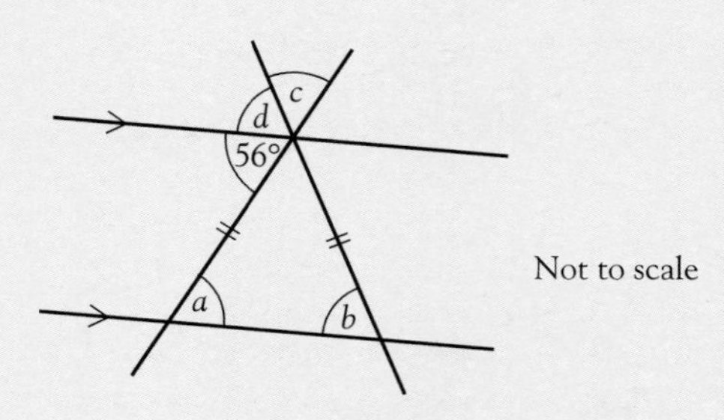

The diagram shows a pair of parallel lines and an isosceles triangle. Write down the size of each angle marked. You must give a reason for each answer. [NEAB]

6

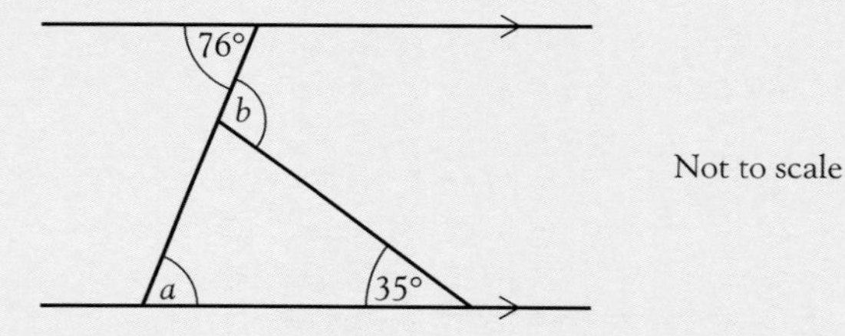

Work out the sizes of the angles marked a and b. [MEG]

7

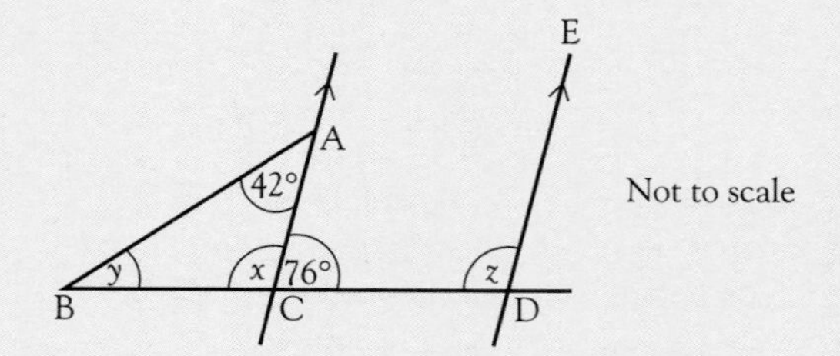

In the diagram, CA is parallel to DE.
Angle BAC = 42° and angle ACD = 76°.
Calculate x, y and z, giving a reason for each answer. [MEG]

EXAM QUESTIONS

8

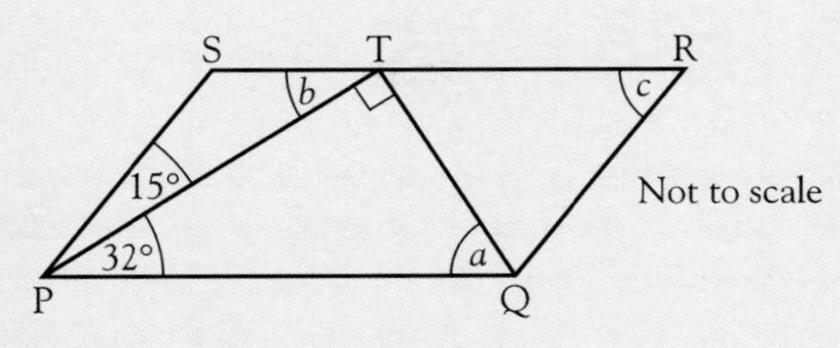

PQRS is a parallelogram.

T is a point on SR such that angle PTQ = 90°.

Calculate the size of the angles marked *a*, *b* and *c*. [MEG]

9 The three angles of a triangle are *x*, 3*x* and 5*x*. Calculate the value of *x*. [NEAB]

10 a

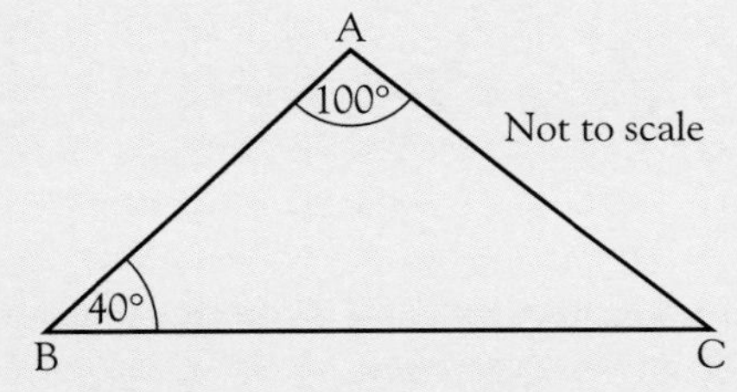

i Work out the size of angle C. Give a reason for your answer.

ii What type of triangle is ABC?

b

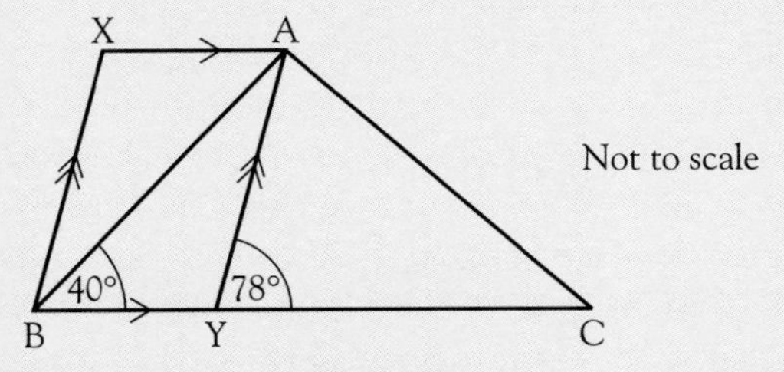

XAYB is a parallelogram.

i Find angle AYB. Give a reason for your answer.

ii Find angle XAB. Give a reason for your answer. [MEG]

11 In the diagram the lines RS and TU are parallel.

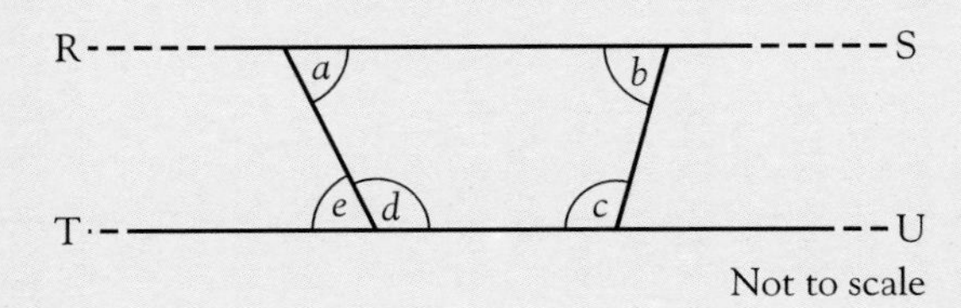

a What is the sum of the angles marked *a*, *b*, *c* and *d*?

b Angle $a = 67°$ and $c = 115°$.

i What is the size of angle *e*?

ii Work out the size of angle *b*. [SEG]

12 This is a sketch of a triangle.

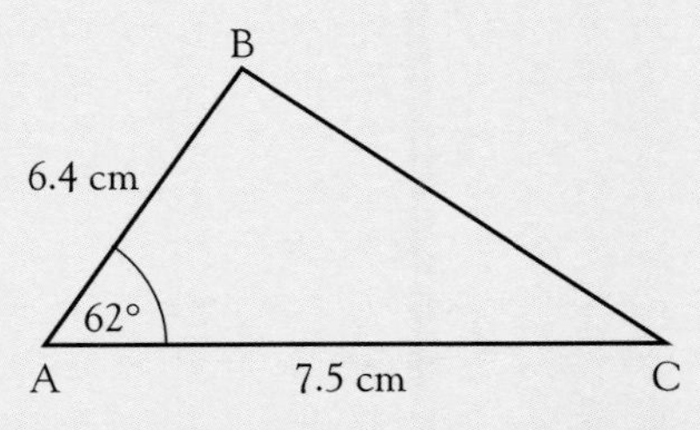

a Make a full size accurate drawing of the triangle.

b Measure the length of BC. [MEG]

13

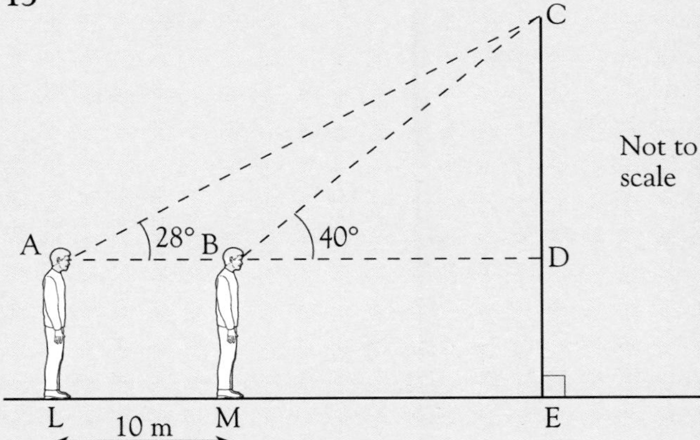

A vertical flag pole CDE stands on horizontal ground. When Pat stands at L, the angle of elevation of C from A is 28°. When Pat stands at M, 10 metres closer to the pole, the angle of elevation of C from B is 40°.

a Using a scale of 1 cm to represent 2 m, draw an accurate scale diagram showing A, B, C and D.

b Use your diagram to obtain the length, in metres, of CD.

c Pat is 16 years old and of average height. Estimate the height, in metres, of the flagpole.

[MEG]

FACT SHEET 4: THE LANGUAGE OF GRAPHS

Graphs

A **graph** is a diagram that represents the relationship between two quantities (or variables). These quantities may represent real-life situations.

A graph is usually drawn on squared paper as this helps you to identify points on the graph accurately. Sketch graphs, however, need not be drawn on squared paper as they are only an indication of the relationship between the quantities.

Axes

Each quantity is represented by an **axis**, one is **horizontal**, the other **vertical**.

This diagram is sometimes called a **grid**.

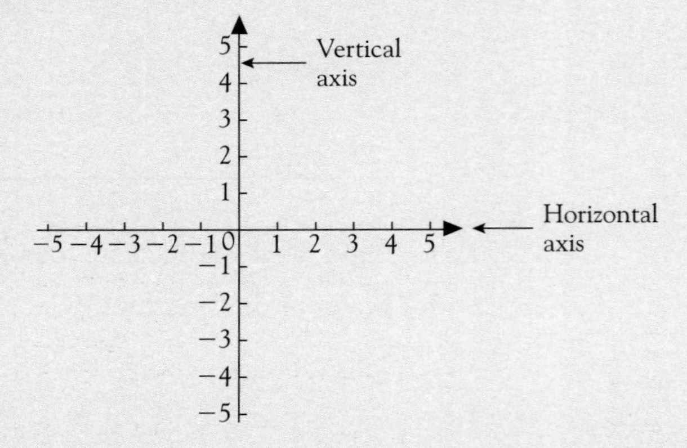

You often see axes represented by the letters x and y.

Normally, the **horizontal axis** is referred to as the ***x*-axis** and the **vertical axis** as the ***y*-axis**.

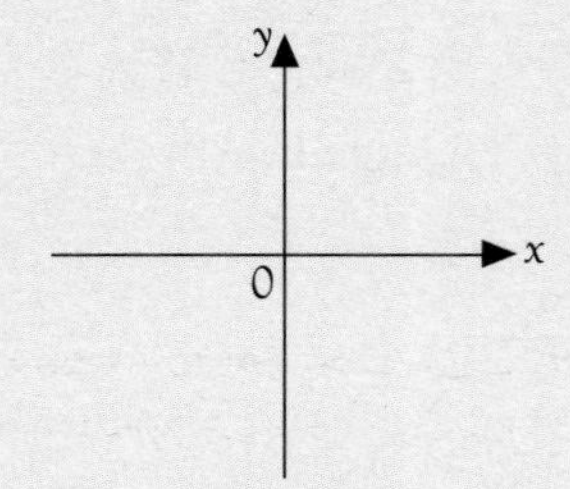

These are not the only letters that can be used. It depends on the situation that is being represented. Here are some examples.

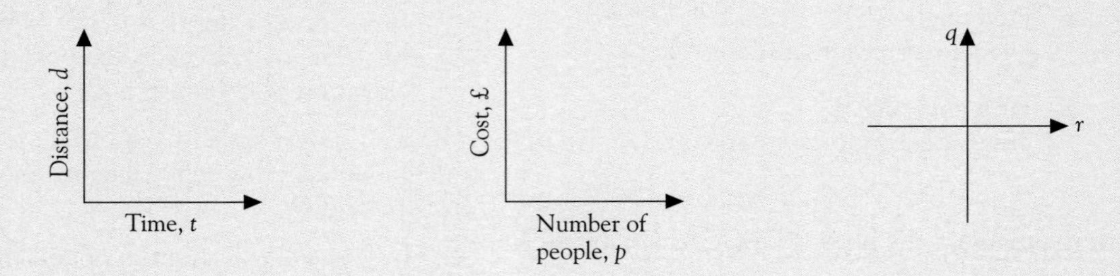

Sometimes the grid will show all four quadrants and have both positive and negative scales shown. At other times only the positive parts of the axes are needed.

FACT SHEET 4: THE LANGUAGE OF GRAPHS

Co-ordinates Points on a graph are represented using **co-ordinates**. Co-ordinates have a horizontal part (x) and a vertical part (y) and are written in a bracket, (x, y).

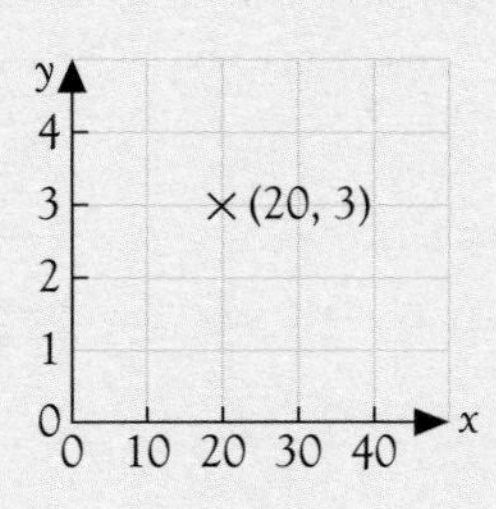

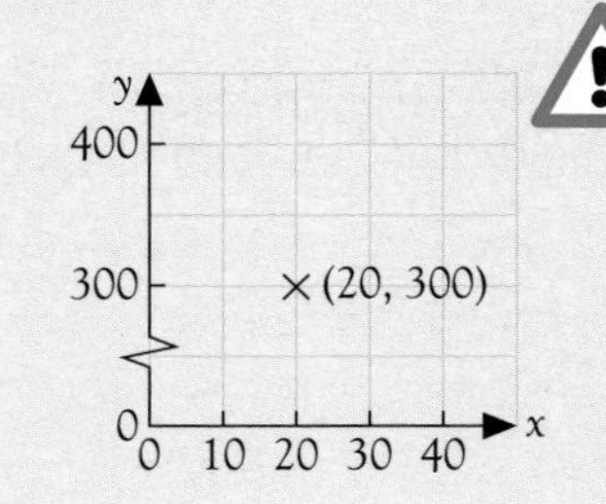

Make sure you get the co-ordinates in the right order! Remember that the letters x and y are in alphabetical order or that 'you go in the door (→) then up the stairs (↑)'.

is the symbol for a 'broken axis'. It indicates that some numbers have been left out.

Origin The point (0, 0) is called the **origin**.

Quadrants (or regions) of a graph

A grid is made up of 4 **quadrants**.

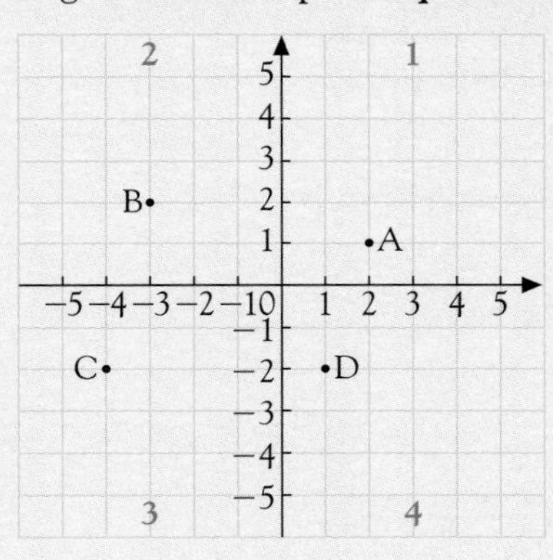

In quadrant **1**, x and y are both **positive**. Point A has co-ordinates (2, 1).

In quadrant **2**, x is **negative**, y is **positive**. Point B has co-ordinates (−3, 2).

In quadrant **3**, x and y are both **negative**. Point C has co-ordinates (−4, −2).

In quadrant **4**, x is **positive**, y is **negative**. Point D has co-ordinates (1, −2).

Scale Each axis must have a **scale**. The scale tells you how much a unit of measurement represents. For example, 1 cm might represent 1 second of time, 10 miles, 100 g and so on.

The scale on an axis **must be consistent**. If 1 cm represents 100 km then every cm represents 100 km on that axis.

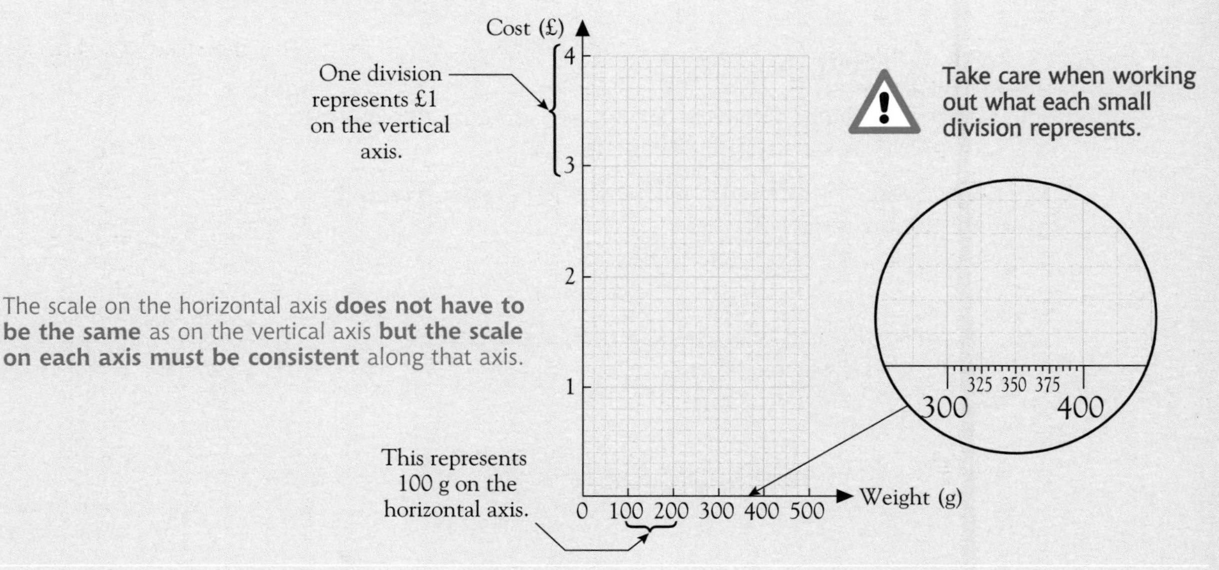

FACT SHEET 4: THE LANGUAGE OF GRAPHS

Families of graphs

There are 'families' of graphs that you need to recognise. These graphs can be drawn and recognised from their equations. You will learn more about these later on in the course.

The x family

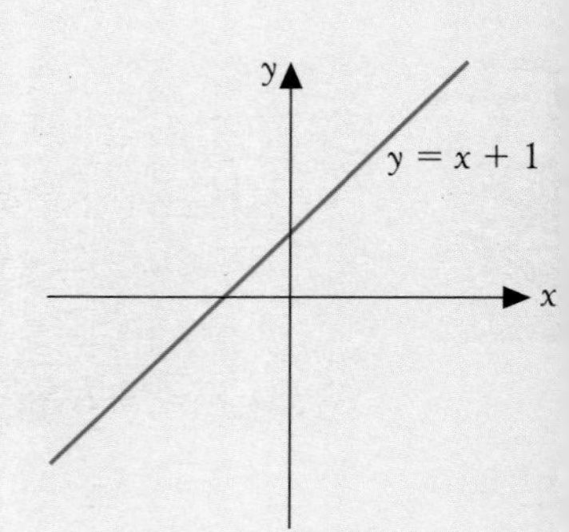

Graphs in this family are **straight lines.**
Graphs that are straight lines are said to be **linear.**

Their equations will have x (or x^1) as the highest power of x. Examples of lines in this family are

$$y = x,\quad y = 2x + 1,\quad y = -3x - 2,\quad y = \tfrac{1}{2}x$$

Their general form is

$$y = ax + b$$

where a and b are **constants.**

The x^2 family

This is a family of **curves.**

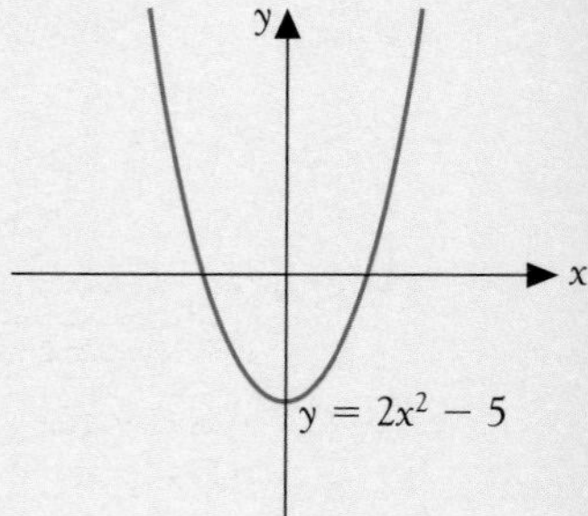

Their equations will have powers of x up to x^2 and may include lower powers of x. Examples in this family are

$$y = 2x^2,\quad y = x^2 + x - 3,\quad y = -\tfrac{1}{2}x^2 - 3$$

These are called **quadratic** curves. Their general form is

$$y = ax^2 + bx + c$$

where a, b and c are constants (a cannot be zero).

The x^3 family

This is also a family of curves.

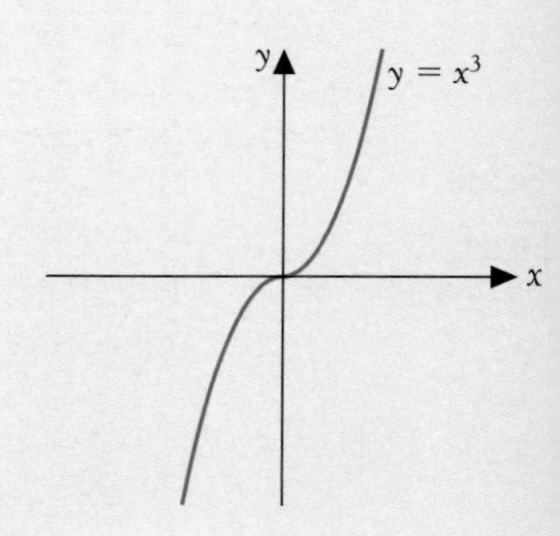

Their equations will have powers of x up to x^3 and may include lower powers of x. Examples in this family are

$$y = x^3 + 2x^2 - x + 3,\quad y = -2x^3 + x - 4$$

These are called **cubic** curves. Their general form is

$$y = ax^3 + bx^2 + cx + d$$

where a, b, c and d are constants (a cannot be zero).

The $\dfrac{1}{x}$ family

This is a strange family of curves, as each curve is split into two parts as shown in the diagram.

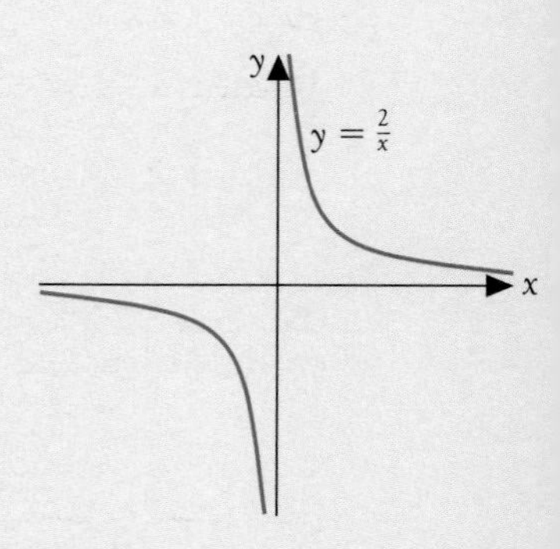

Its equation is a fraction with x at the bottom. Examples in this family are

$$y = \frac{4}{x},\quad y = \frac{-5}{x}$$

A curve of this type is called a **reciprocal curve.**
The general form is

$$y = \frac{a}{x}$$

where a is a constant which cannot be zero.

Note that neither x nor y can be zero.

LINEAR GRAPHS I

Look out for the tools you need

In this chapter you will learn how to
- **draw a linear graph given its equation**
- **identify special linear graphs from their equations**
- **use linear graphs in everyday situations**

Drawing a graph of a linear relationship

Remember that a graph is a diagram that represents the relationship between two quantities or variables. This relationship may be given as a 'rule' in the form of an equation.

A **linear** relationship is one that produces a **straight-line** graph.

FACT SHEET 4 pages 86–88

If the variables are x and y, the equation of a linear relationship will have powers of x and y no bigger than x^1 or y^1.

x^1 is the same as x. y^1 is the same as y.

Examples of linear relationships are

$$y = x, \quad y = 8 - 2x, \quad x + y = 4, \quad x = 4, \quad y = 2$$

To draw the graph of a linear relationship from its equation, create a table of values and plot these values on a grid.

Creating a table of values

Choose values for one of the quantities, or variables, and apply the 'rule' or equation to find corresponding values for the other quantity or variable.

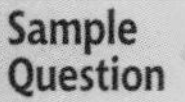

Sample Question 1 For the equation $y = 4x + 1$, complete the following table of values.

x	-3	-2	-1	0	1	2	3
y							

As in this example, you are often told what values of x to use.

Answer

- In turn, substitute each value of x into the equation to calculate the corresponding value of y.

$$\begin{aligned}
\text{For } x &= -3 & y &= (4 \times -3) + 1 = -12 + 1 = -11 \\
x &= -2 & y &= (4 \times -2) + 1 = -8 + 1 = -7 \\
x &= -1 & y &= (4 \times -1) + 1 = -4 + 1 = -3 \\
x &= 0 & y &= (4 \times 0) + 1 = 0 + 1 = 1 \\
x &= 1 & y &= (4 \times 1) + 1 = 4 + 1 = 5 \\
x &= 2 & y &= (4 \times 2) + 1 = 8 + 1 = 9 \\
x &= 3 & y &= (4 \times 3) + 1 = 12 + 1 = 13
\end{aligned}$$

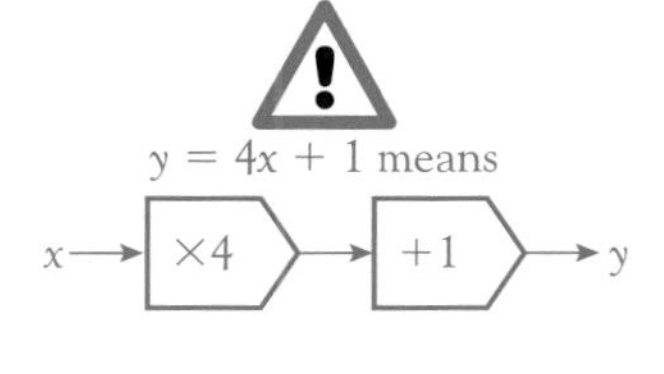

x	-3	-2	-1	0	1	2	3
y	-11	-7	-3	1	5	9	13

This table also represents a set of points (x, y), such as $(-2, -7)$, $(1, 5)$.

DRAWING A LINEAR GRAPH

Plotting the relationship on a grid

From Sample Question 1, you now have a set of points to plot on a grid.

These points are $(-3, -11)$, $(-2, -7)$, $(-1, -3)$, $(0, 1)$ $(1, 5)$, $(2, 9)$, $(3, 13)$.

The next step is to draw an **appropriate grid**.

In an exam the grid is often done for you, with the appropriate axes and scale already marked.

Looking at the points that you are going to plot,

- the smallest value of x is -3, the largest is 3,
- the smallest value of y is -11, the largest is 13.

Draw the grid, plot points and draw lines in pencil.
Do the labelling in pen.

Set your scales accordingly. On your axes, mark the scales:

- on the x-axis, just below the line,
- on the y-axis, just to the left of the line.

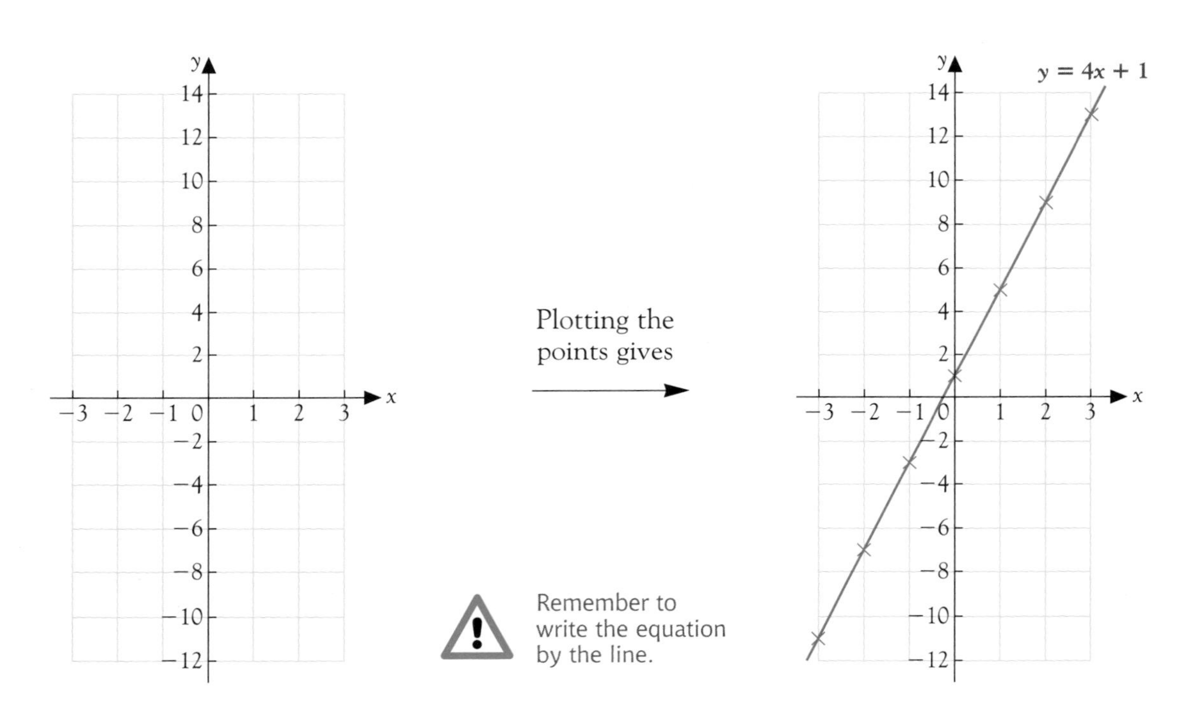

Here are some very important points to remember when drawing a graph.

- Label the axes with the appropriate letters (or description).
- Plot points accurately using a sharpened pencil.
- Get the order right for the co-ordinates: x first, then y.
- Draw a straight line through the points with a sharp pencil and using a ruler.
- **Label** your graph with **the equation of the line**, especially if you have more than one line on the same grid.
- If your line takes any unexpected changes in direction, CHECK YOUR CALCULATIONS FOR THE SUSPECT POINTS.

Draw the graph using a SHARP pencil and an UNDAMAGED ruler!

Write the equation close to the line.

DRAWING A LINEAR GRAPH

Sample Question 2

a Copy and complete this table of values for the equation $y = 2x - 3$, for values of x from -2 to 4.

x	-2	-1	0	1	2	3	4
y		-5			1		

b Draw a grid using a suitable scale on each axis. Label the x and y axes.

c On your grid draw the graph of $y = 2x - 3$.

Answer

- Substitute values for x into the equation $y = 2x - 3$.
  Notice that $x = -1$ and $x = 2$ have already been done for you.

For $x = -2$ $\quad y = (2 \times -2) - 3 = -7$
For $x = 0$ $\quad y = (2 \times 0) - 3 = -3$
For $x = 1$ $\quad y = (2 \times 1) - 3 = -1$
For $x = 3$ $\quad y = (2 \times 3) - 3 = 3$
For $x = 4$ $\quad y = (2 \times 4) - 3 = 5$

It is not necessary to write the calculations down every time. You can work them out on your calculator, or preferably in your head, and write the answers straight into your table.

x	-2	-1	0	1	2	3	4
y	-7	-5	-3	-1	1	3	5

- Draw the grid as requested and plot the points carefully. Remember to label your line with its equation.

DO NOT use a pen to draw ANY points or lines.

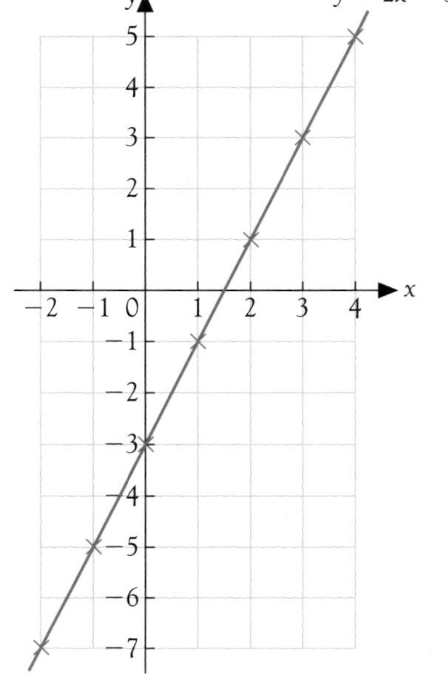

Remember that the points must be plotted accurately. You will lose marks in an exam if they are not.

Sample Question 3

a Is the point (4, 3) a point on the line $y = x - 1$?

b Is the point (1, 0) a point on the line $y = 2x - 3$?

DRAWING A LINEAR GRAPH

Answer

a

- To check whether the point (4, 3) is a point on the line, substitute the value $x = 4$ into the equation $y = x - 1$. If it does **lie on the line**, the resulting y value should be 3.

  When $x = 4$ $\quad y = x - 1 = 4 - 1 = 3$

  As the value of y **does** equal 3 when $x = 4$,
  point (4, 3) is a point on the line $y = x - 1$.

If a point is on a line you can say that the point 'lies on the line'.

b

- Substitute $x = 1$ into $y = 2x - 3$.

  When $x = 1$ $\quad y = 2x - 3 = 2 \times 1 - 3 = 2 - 3 = -1$

  As the value of y **does not** equal 0 when $x = 1$,
  point (1, 0) is not a point on the line $y = 2x - 3$.

You can draw the graph of a **straight line** by finding the co-ordinates of only **three points**.

You could draw a line using only two points but the third point is needed as a check.

- Find the co-ordinates of any three points using the equation of the line.
- Plot the points on a suitable set of axes.
- Draw the line that passes through the points.

Sample Question 4 Draw the graph of the line with equation $y = 3x - 1$.

Answer

- Find the co-ordinates of three points on the line. For example, use the values $x = -1$, 0 and 1.

You can use any values of x that you like but these three make the calculations relatively easy.

When $x = -1$ $\quad y = 3 \times -1 - 1 = -3 - 1 = -4$
When $x = 0$ $\quad y = 3 \times 0 - 1 = 0 - 1 = -1$
When $x = 1$ $\quad y = 3 \times 1 - 1 = 3 - 1 = 2$

You can draw a table for your three values if you need to.

x	−1	0	1
y	−4	−1	2

Your three points are (−1, −4), (0, −1), (1, 2).

DRAWING A LINEAR GRAPH

- Draw a set of axes on a grid that will enable you to plot your three points and to extend the line beyond the points in both directions.
- Plot your three points on the grid.
- Draw a straight line through the three points, extending it in both directions to show more of the line.

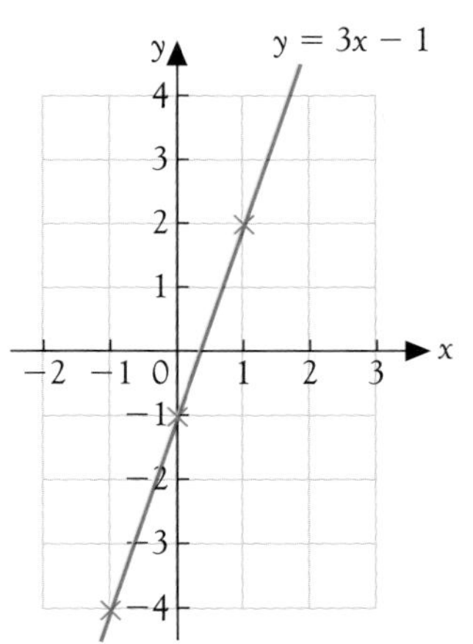

Remember **all** the labels, including the line.

Exercise 6.1

1 For each of these equations, complete a table of values for x from -3 to 3. **Do not draw any graphs.**

a $y = 5x - 2$

b $y = \frac{1}{2}x + 3$

c $y = -2x + 1$

d $y = 5 - x$

e $y = 3x + 4$

2 For each of the following, say whether the point lies on the given line. **Do not draw any graphs.**

a $(6, 3)$; $y = x - 1$

b $(0, 4)$; $y = 3x + 4$

c $(5, 2)$; $y = 7 - x$

d $(1, 1)$; $y = -2x + 3$

e $(0, 0)$; $y = 6x$

3 a Copy and complete the table of values for the equation $y = x + 4$.

x	0	1	2	3	4
y			6		

b Draw a grid using 1 cm to represent 1 unit on both the x and y axes. Using your table of values, draw accurately the graph of $y = x + 4$.

4 a Copy and complete the table of values for the equation $y = -x + 1$.

x	−3	−2	−1	0	1	2	3
y			2				−2

b Draw axes labelling the x axis from -3 to 3 and the y axis from -3 to 5.

c On your axes plot the points from the table of values and draw the graph of the line $y = -x + 1$.

5 a For the equation $y = \frac{1}{2}x - 2$, do a table of values for x and y with values of x from -2 to 6.

b Using the values from your table, draw the graph of $y = \frac{1}{2}x - 2$.

6 Draw the graph of each of the following equations **by choosing only three points.**

Use a different set of axes for each graph.

a $y = 3x - 5$

b $y = 2x + 3$

c $y = 4 - 2x$

d $y = 6x + 1$

e $y = \frac{1}{4}x$

f $y = -x - 2$

g $y = -7x + 3$

h $y = 4x + 6$

SPECIAL LINES

Special linear relationships

The graph shows a straight line that is parallel to the x axis. Three points, A, B and C, have been marked on the line.

The co-ordinates of point A are $(-3, 4)$.
The co-ordinates of point B are $(0, 4)$.
The co-ordinates of point C are $(2, 4)$.

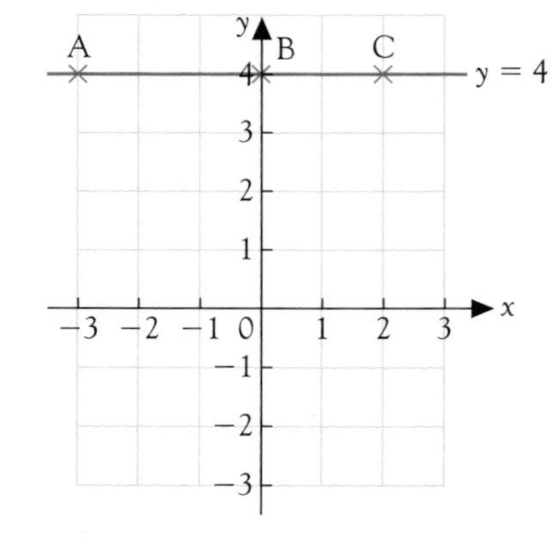

Notice that for all three points on the line, the y co-ordinate is always 4. In fact, **all points on this line have a y co-ordinate of 4.** The x co-ordinate is the one that changes.

For this reason the equation of this line is given as $\mathbf{y = 4}$.

The line on this graph is parallel to the y axis.

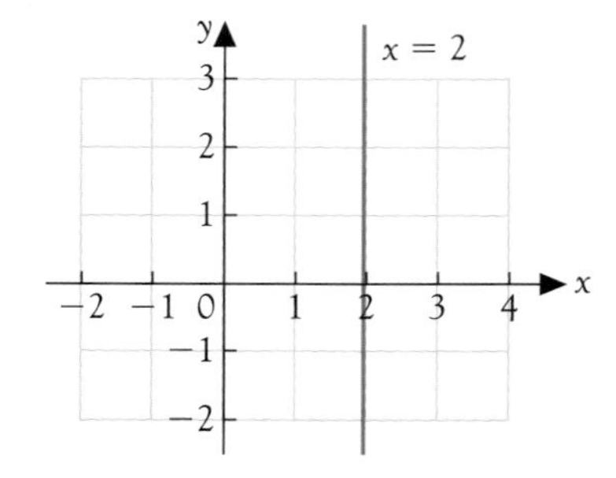

This time it is the y co-ordinate that changes for different points on the line. The x co-ordinate of **any** point on the line is **always 2.**

For this reason, the equation of this line is given as $\mathbf{x = 2}$.

The **x axis** is a line that has the equation $\mathbf{y = 0}$.
This is because, for all points on the line, only the x co-ordinate varies.
The y co-ordinate is always 0.

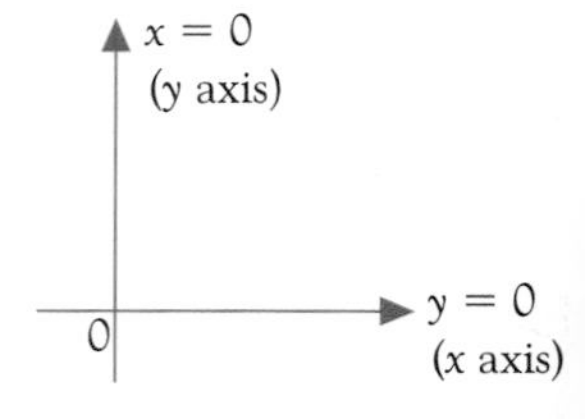

The **y axis** is a line that has the equation $\mathbf{x = 0}$.
This is because, for all points on the line, only the y co-ordinate varies.
The x co-ordinate is always 0.

In general:

- Lines that are parallel to the **x axis** are given the equation $\mathbf{y = a}$, where a is a constant and is the value of the y co-ordinate where the line crosses the y axis.
- Lines that are parallel to the **y axis** are given the equation $\mathbf{x = a}$, where a is a constant and is the value of the x co-ordinate where the line crosses the x axis.
- $y = 0$ is the equation of the x axis.
- $x = 0$ is the equation of the y axis.

SPECIAL LINES

Look at the points marked on this graph.
The co-ordinates of these points are:

Point A: (2, 2)
Point B: (−3, −3)
Point C: (0, 0)
Point D: (4, 4)

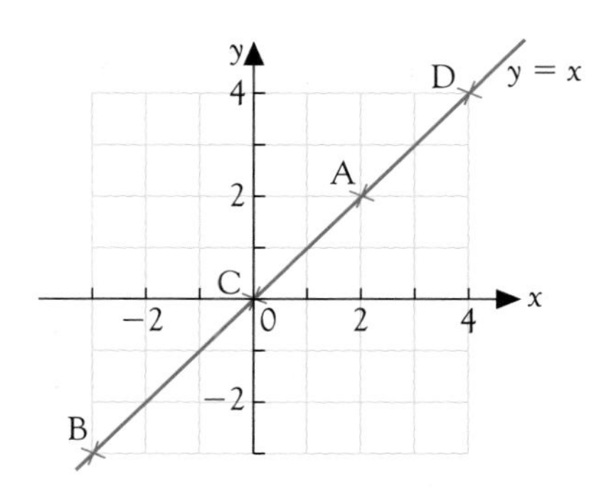

For all these points the x co-ordinate is the same as the y co-ordinate. The points all lie on the same straight line. The equation of this line is $\mathbf{y = x}$.

The co-ordinates of the points marked on this graph are:

Point E: (2, −2)
Point F: (−3, 3)
Point G: (4, −4)
Point H: (−1, 1)

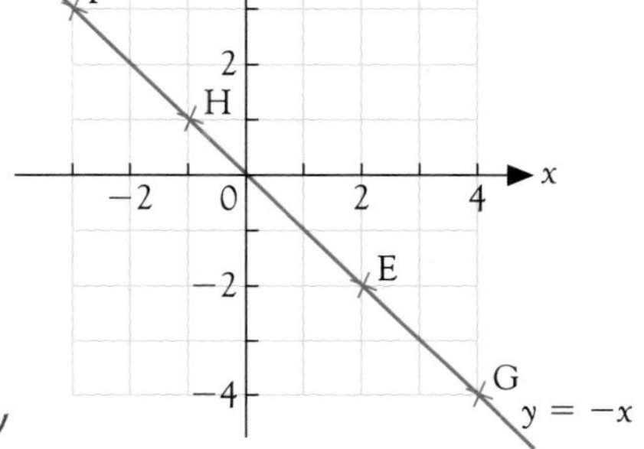

For all these points the y-co-ordinate is the negative of the x co-ordinate. The points all lie on the same straight line. The equation of this line is $\mathbf{y = -x}$.

$y = -x$ is the same as $x = -y$ or $y + x = 0$

- $y = x$ is the equation of the line with points whose x and y co-ordinates have the same value.
- $y = -x$ is the equation of the line with points whose y co-ordinate is the negative of its x co-ordinate.

Exercise 6.2

1 Identify, from the graphs, the equation of each of the following lines:

a

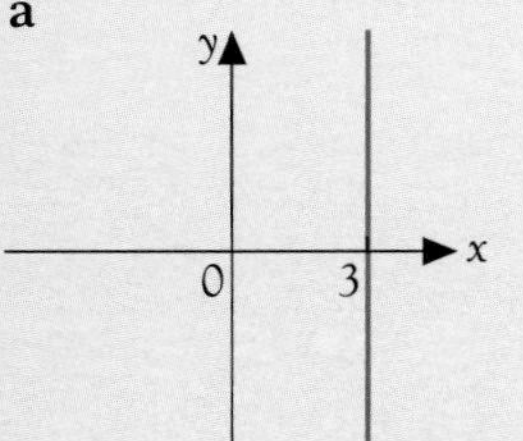

b

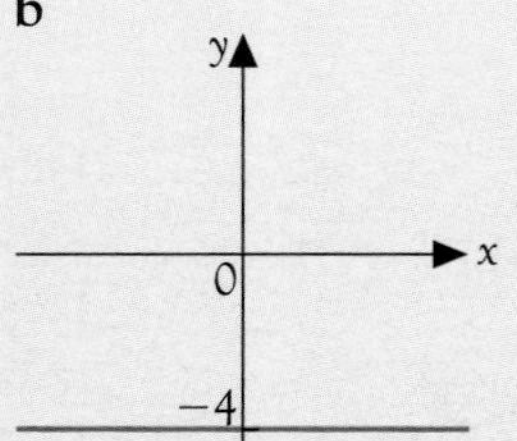

c

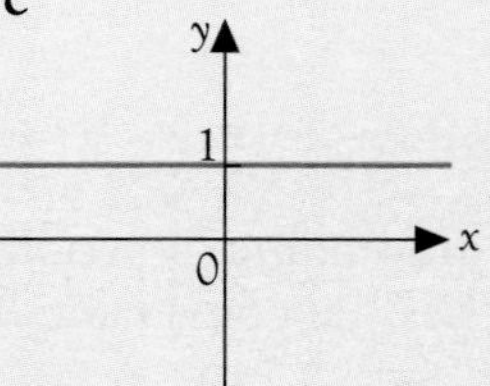

d

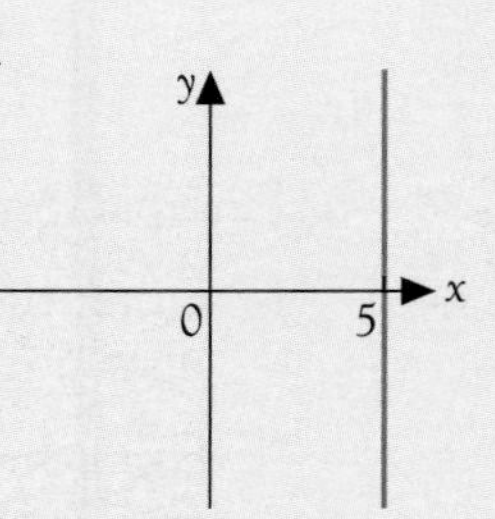

2 Draw and label each of the following lines on the same grid. Do not make a table of values or plot any points. You must show any information necessary to identify the line.

a $y = -3$ **b** $x = 6$ **c** $y = 2\frac{1}{2}$ **d** $x = 2$ **e** $y = 5\frac{1}{2}$ **f** $x = -4$

Graphs in practical situations

Use can be made of straight-line graphs in a variety of real-life situations. The axes represent practical quantities, e.g. time, distance, £, French francs, rather than unknown variables like x and y. The line represents the relationship between the two quantities concerned. A common application is to use a graph as an aid to converting one quantity to another, e.g. £ to French francs.

Sometimes the graph is drawn for you and you are asked to use it to make a conversion. At other times you might be asked to draw the graph, given some data, and then to use it.

Sample Question 5

A delivery firm uses this graph to calculate how much to charge a customer for making a delivery of goods.

a How much does the firm charge for making a delivery of 80 miles?

b A charge of £30 was made for a delivery. How many miles away was the recipient?

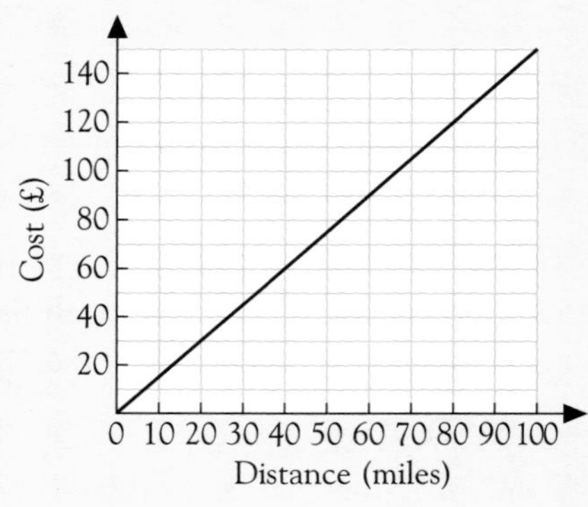

Answer

a

- Find 80 on the 'Distance' axis.

  i From this point, draw a line vertically up until it reaches the conversion line.

  ii From there, draw a line horizontally to the left until it reaches the 'Cost' axis.

  iii Read off the cost.

  <u>The firm charges £120 for the delivery.</u>

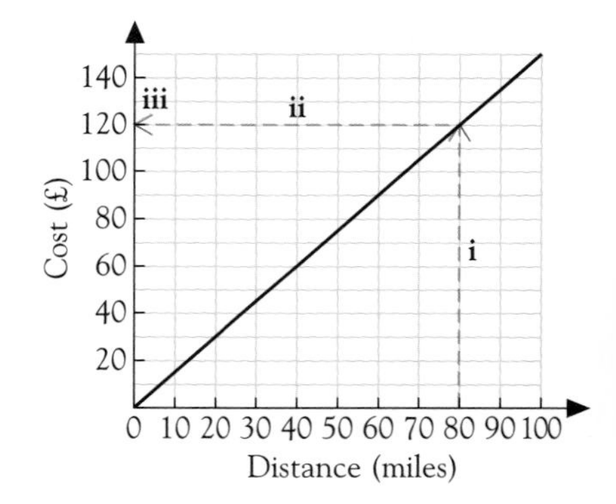

b

- Find £30 on the 'Cost' axis.

  i From this point, draw a line horizontally to the right until it reaches the conversion line.

  ii From there, draw a line vertically downwards until it reaches the 'Distance' axis.

  iii Read off the distance.

  <u>The distance was 20 miles.</u>

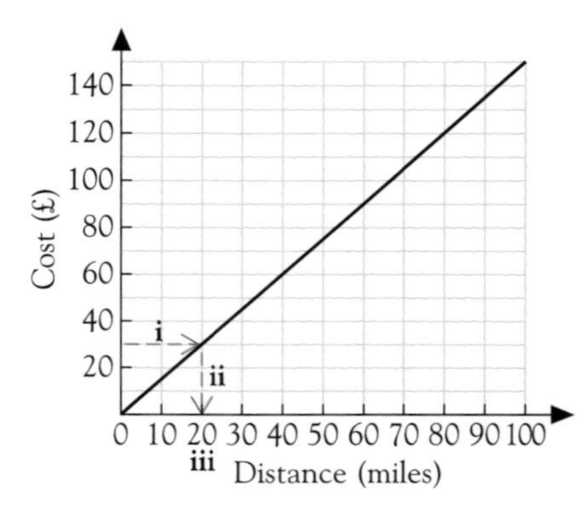

PRACTICAL SITUATIONS

Sample Question 6

On a sponsored walk Brenda walked at a constant rate. The table shows how long it took her to reach three of the checkpoints.

Distance walked (km)	2	8	12
Time taken (minutes)	25	100	150

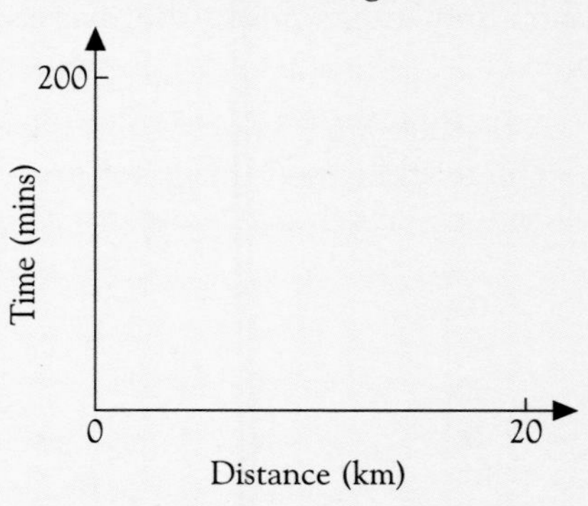

a On graph paper, copy this set of axes, choosing a suitable scale. Represent the table as a straight-line graph.

b Use your graph to find:

i the distance Brenda walked in 50 minutes

ii how long, in minutes, it took Brenda to walk 10 km.

c If it took Brenda 200 minutes to complete the sponsored walk, use your graph to find how far the walk was altogether.

When you represent this table as a graph, you are plotting the **time taken** against **distance walked**.

Answer

a

- Draw and label the axes and plot the points. Join them up to form a line.

Make sure that you extend your line beyond the points in both directions, as you will need more of the line than the table gives you.

b i

- Draw a horizontal line passing through 50 minutes on the 'Time' axis, as far as the line that you drew in **a**.
- From there, draw a vertical line downwards until it reaches the 'Distance' axis and read off the distance.

In 50 minutes Brenda walked 4 km.

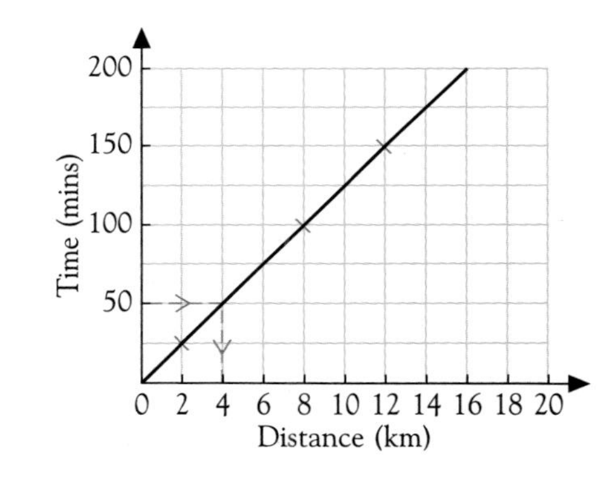

ii

- Repeat the process as described in **b i**, starting with a vertical line passing through 10 on the 'Distance' axis.

It took Brenda 125 minutes to walk 10 km.

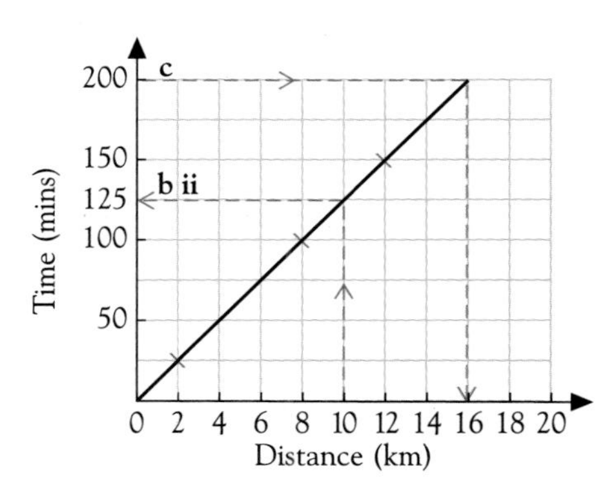

c

- In part **a** you should have extended your line.
Repeat the steps in **b ii**, starting with a horizontal line passing through 200 on the 'Time' axis.

The sponsored walk was 16 km long.

PRACTICAL SITUATIONS

Exercise 6.3

When drawing graphs in this exercise use cm graph paper.

1 A mathematics examination is marked out of 138. The teacher used this graph to convert the students' marks to a percentage.

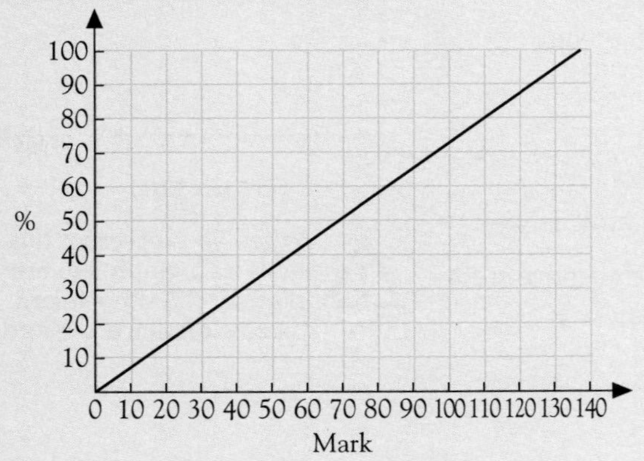

Copy the diagram onto graph paper using a scale of 1 cm to represent 10 on both axes. Notice that the line goes through (0, 0) and (138, 100). Use the graph to find:

a the mark out of 138 for a student who scored 75%,

b the percentage given to a student who scored 26 out of 138.

2 Packets of paper can be bought from a mail order company. The cost C (in £) depends on p, the number of packets bought, where $C = 5p + 2.5$.

Copy and complete this table.

p	1	5	7	10
C			37.5	

Plot the points obtained from your table on a grid with axes as shown.

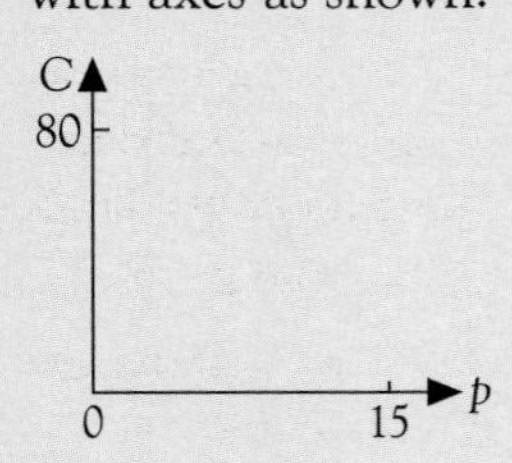

Use your graph to find the cost of the following amounts of paper:

a 3 packets

b 6 packets

c 12 packets

3 A car hire company charges a fixed fee of £40 plus 25p for every mile travelled. Copy and complete the table, where m is the number of miles travelled and C is the total cost in pounds.

m	0	100	200	300
C		65	90	

Draw a graph of C against m, using scales of 2 cm for 50 miles on the horizontal axis and 1 cm to £10 on the vertical axis.

Use your graph to find:

a the total cost of travelling 136 miles,

b how many miles were travelled if the total cost was £94.

4 The table shows how three amounts of pounds (£) convert to New Zealand dollars ($NZ).

£	10	30	60
$NZ	27.5	82.5	165

Using a scale of 1 cm to represent £10 on the horizontal axis and 1 cm to represent $NZ10 on the vertical axis, plot the three points given in the table and draw a line through the points.

Use your graph to convert:

a £25 to $NZ,

b $NZ110 to £,

c £50 to $NZ,

d $NZ55 to £.

5 A small family car, when driven at a steady speed of 56 mph, costs p pence to run when it travels m miles. Some values of m and p are given in the table.

Number of miles, m	8	13	21	25
Cost, p pence	56	91	147	175

Draw the graph of p against m with p on the vertical axis and m on the horizontal axis. Use your graph to find:

a the number of miles the car travels for a cost of 120 pence,

b the cost of running the car for 15 miles.

6 A train reaches a steady speed of 80 km/h. At this speed the train covers 80 km in 1 hour and 240 km in 3 hours.

Using this information, draw a graph to enable you to work out the distance travelled, at this steady speed, over differing periods of time.

Use a scale of 2 cm to represent 1 hour on the horizontal axis and 2 cm to represent 100 km on the vertical scale.

a Use your graph to find how far the train has travelled in:

i $2\frac{1}{2}$ hours

ii 1 hour 48 minutes

b How long does it take the train to travel:

i 168 km

ii 64 km?

EXAM QUESTIONS

Worked Exam Question
[MEG]

'Print-a-Word' charges £3 to design a business card and 5p for every card printed.

The cost, c pence, of x cards can be written as $c = 5x + 300$.

a i Complete the table of values for this equation.

x	0	50	100
c	300	550	800

Substitute $x = 0$ and $x = 100$ into the equation to find the missing values of c. Do this in your head or use your calculator.

ii Use these values to draw the graph of
$c = 5x + 300$
on the axes below.

Plot the points that you worked out in part **a**: (0, 300), (50, 550) and (100, 800). You must do this carefully.

Draw a **straight** line **through** the points, making sure to extend it beyond the point (100, 800). Label the line.

DO NOT go lower than the point (0, 300). x cannot be negative.

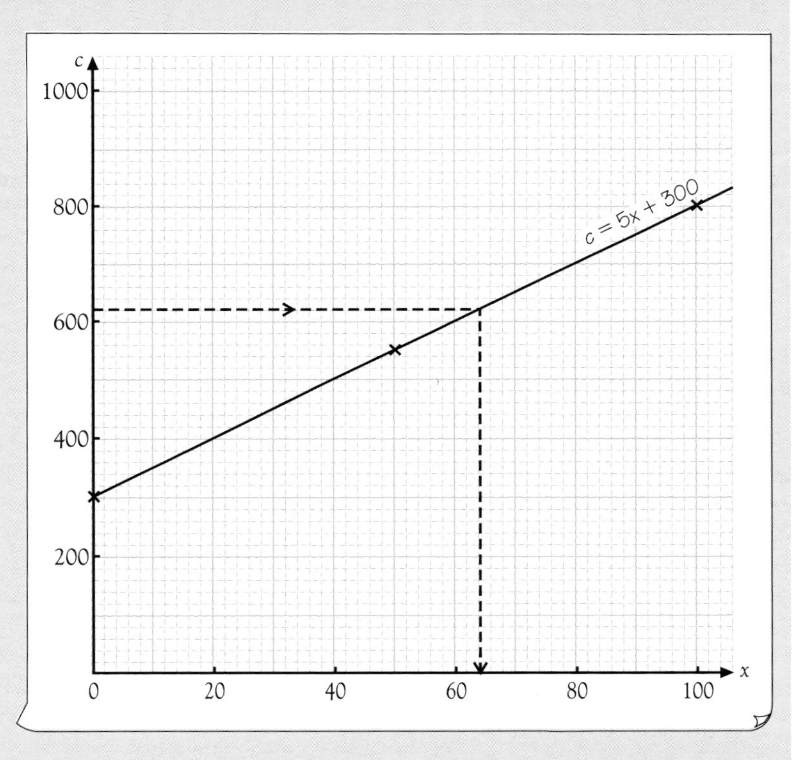

b Damian was charged £6.20 for some cards. How many cards did he buy?

DO NOT use the equation to calculate the answer.

1. Change £6.20 into pence because the 'c' axis scale is in pence.
2. Draw a horizontal line through 620 on the 'c' axis, as far as your line.
3. From there draw a vertical line down to the 'x' axis.
4. Read off the value of x.

£6.20 = 620 p

Answer ....64 cards........

COMMENTS

A1 for each answer

2 marks

A2 for plotting 3 points correctly (**A1** for 2 points correct)

A1 for joining points with a straight line, extending the line beyond the point (100, 800)

3 marks

M1 for showing appropriate vertical and horizontal lines drawn on graph

A1 for correct answer from graph

(**A0** for calculating answer)

2 marks

EXAM QUESTIONS

Exam Questions

1 Jenny weighs three objects in both kilograms and pounds.
The table shows her results.

Object	A	B	C
Weight in kilograms	2	5	10
Weight in pounds	4.4	11	22

a Use Jenny's results to draw a conversion graph between kilograms and pounds, using axes as shown.

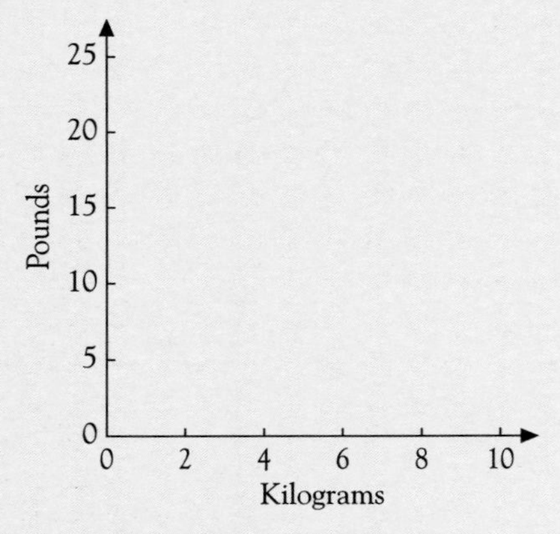

b A fourth object weighs 15 pounds.
Use your graph to find its weight in kilograms.

[SEG]

2 a On a grid with axes as shown, draw the graph of $y = \frac{1}{2}x - 1$.

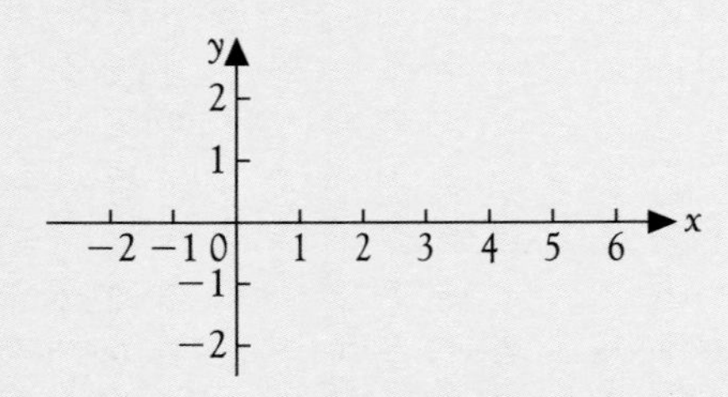

b Use your graph to solve the equation $\frac{1}{2}x - 1 = 1.7$.

[MEG]

3 a Mark suitable scales on a pair of axes and draw the graph of

$$y = 4 - \tfrac{1}{2}x$$

b The equation in part **a** gives the depth of water in a tank (y metres) after a time (x minutes).

i When is the depth 1.5 m?

ii What happens when $x = 8$?

[MEG]

4 a For the equation

$$y = \frac{x}{5} + 4$$

choose three values of x in the range 0 to 60 and work out the values of y.

b Draw the graph of $y = \frac{x}{5} + 4$ for values of x from 0 to 60.

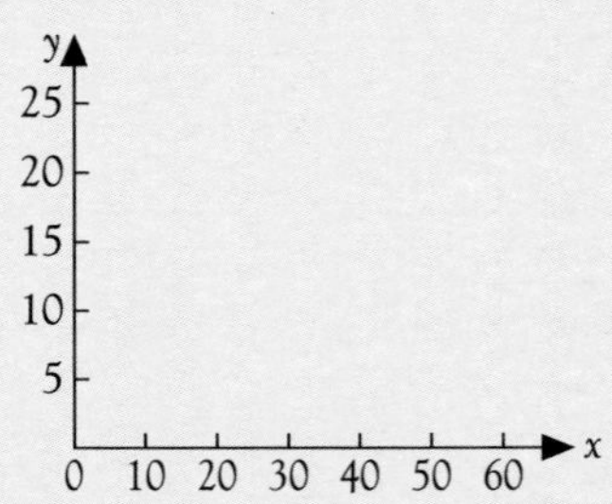

This graph shows the relationship between the perimeter, x inches, of a picture frame and the cost, £y.

A frame costs £14.

c Use the graph to find its perimeter.

[MEG]

5 The diagram shows a conversion graph between pounds (£) and German Deutschmarks (DM).

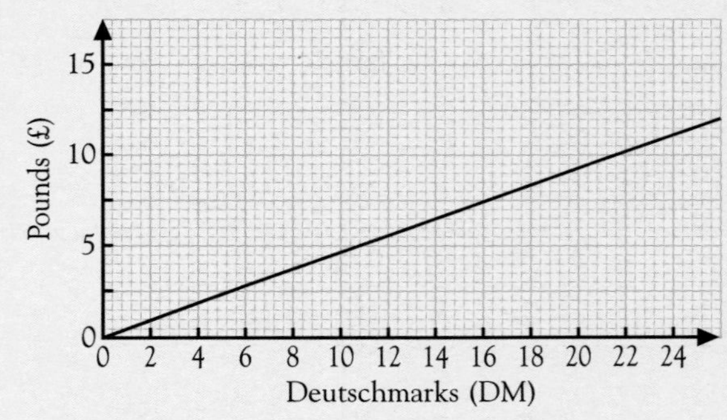

Use the graph to write down how many:

a Deutschmarks can be exchanged for £10,

b pounds can be exchanged for 14 Deutschmarks.

[L]

6 a Complete this table of values for $y = 3x - 1$.

x	−2	−1	0	1	2	3
y			−1			8

b Draw the graph of $y = 3x - 1$ on a grid, with the x axis extending from -5 to 5 and the y axis from -8 to 8.

c Use your graph to find:

i the value of x when $y = 3.5$,

ii the value of y when $x = -1.5$.

[L]

EXAM QUESTIONS

7 a Use the data to draw the conversion graph from £ (sterling) to IR£ (punts) on graph paper, using the axes as shown. Use a scale of 2 cm to 10 on both axes.

EXCHANGE

£ (sterling)	10.00	20.00	50.00
IR£ (punts)	10.80	21.60	54.00

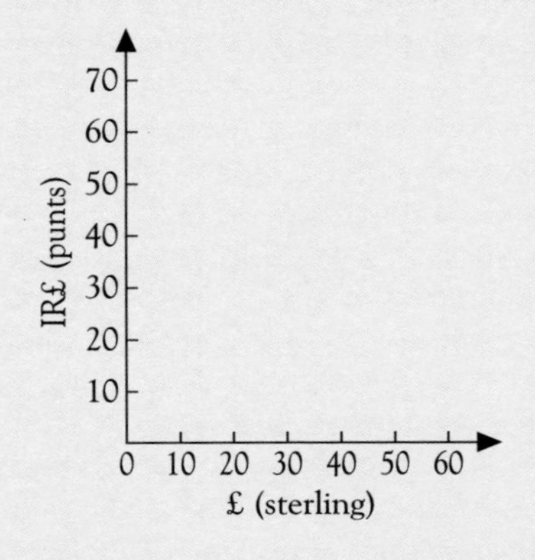

b Marking your readings clearly, use the graph to find:

i the value in punts of £38 (sterling),

ii the value of £ (sterling) of 27 punts. [NI]

8 a Complete this table of values for $y = x - 3$.

x	−3	−2	−1	0	1	2	3	4	5
$y = x - 3$	−6	−5	−4	−3	−2				

b Plot the points and draw the graph for $y = x - 3$ for these values of x and y. [L]

9 The road distances between London and three other towns are shown on the diagram.

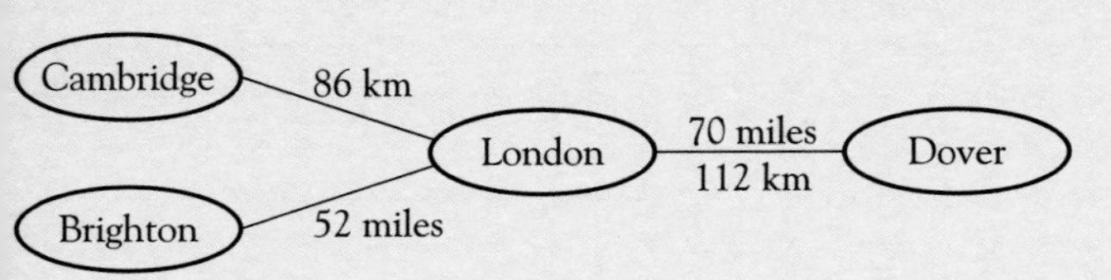

a Using the data from the London–Dover route draw a graph for converting miles into kilometres, using the axes as shown. Use a scale of 2 cm to 20 on both axes.

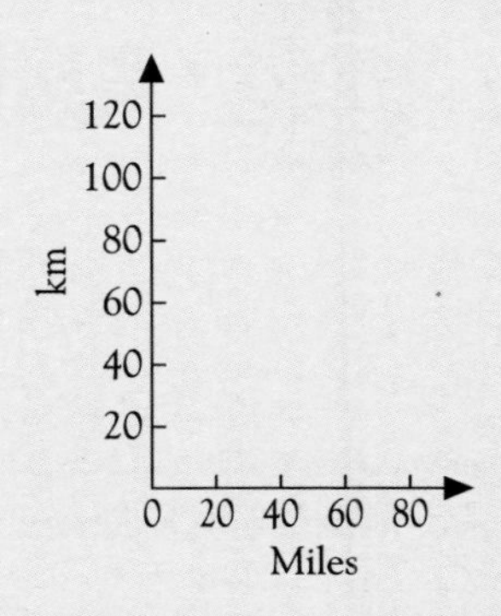

b How many kilometres is it from London to Brighton?

c How many miles is it from London to Cambridge? [SEG]

10 The table below shows the repayments required on loans of different amounts, for 1 year.

Amount of loan (£)	500	750	1500	2250	3000
Monthly repayment (£)	60	85	160	235	310

a Plot these pairs of values on a grid with axes as shown. Join them with a straight line. Use a scale of 2 cm to 500 on the horizontal axis and 2 cm to 100 on the vertical axis.

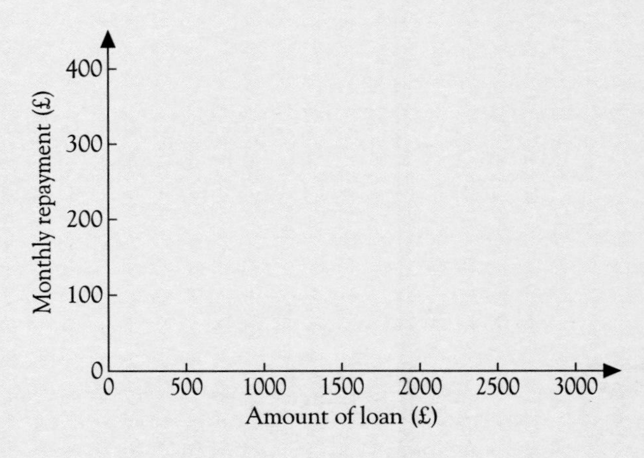

b I can afford to repay £180 a month. Use the graph to find out the largest amount I could borrow.

c Use your graph to find the monthly repayment on a loan of £1000.

d Phil borrows £1500.
Altogether his 12 monthly repayments amount to more than £1500. How much more? [MEG]

7 AREA AND VOLUME I

Look out for the tools you need

In this chapter you will learn how to
- **work out the areas of rectangles and triangles**
- **work out the area of a circle using $A = \pi r^2$**
- **work out the volume and surface area of a cuboid**

Area of simple shapes

The amount of space inside the boundary of a flat shape is called the **area** of the shape. Area is measured in **square units**.

1 cm | 1 cm

This square has an area of 1 square cm, which is written as 1 cm^2.

cm^2 is read as 'centimetres squared'.

Area = 6 cm^2

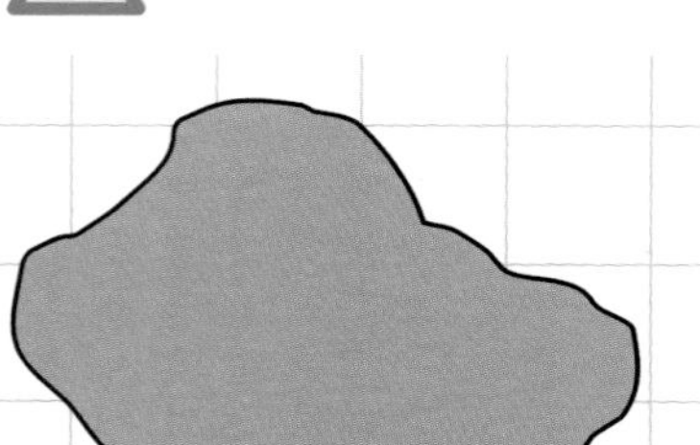

Area $\approx$ 9.5 cm^2

You have to estimate the area of this shape.

Remember that $\approx$ means 'approximately equal to'.

The unit of area that you use depends on what you are measuring. For example, you could measure the area of a forest in square km (km^2), the area of a football pitch in square m (m^2), the area of a leaf in square cm (cm^2) and the area of a bacteria colony on a petri dish in square mm (mm^2).

Area of a rectangle

Area of a rectangle = length × width

$= l \times w$

w

l

A square is a special rectangle in which $l = w$.

Area of a square $= l \times l$

$= l^2$

l

l

AREA OF SIMPLE SHAPES

Sample Question 1

A gardener marked out a plot of ground which she wanted to turn into a lawn.

She bought a 1 kg pack of grass seed. Did she have enough if she followed the instruction to use 50 g of seed for every square metre?

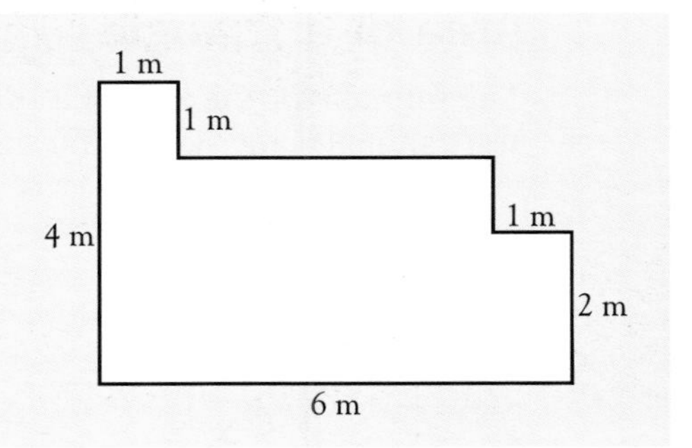

Answer

- Split the plot into rectangles and work out any **missing** lengths.
- Work out the area of each rectangle and add them up to find the total area of the plot.

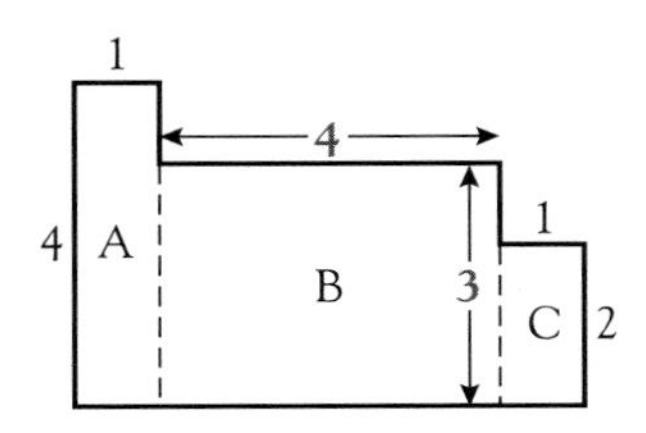

Area of A $= 4 \times 1 = 4\ \text{m}^2$
Area of B $= 4 \times 3 = 12\ \text{m}^2$
Area of C $= 2 \times 1 = 2\ \text{m}^2$

Total area $= 4 + 12 + 2 = 18\ \text{m}^2$

- Find the amount of grass seed needed to cover $18\ \text{m}^2$.

  $50 \times 18 = 900$ g

- Make your conclusion.

  Yes, she did have enough grass seed if she bought 1 kg

1 kg = 1000 g

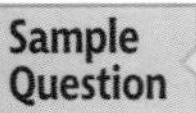

2 A square blue tile with area $256\ \text{cm}^2$ is fitted next to a rectangular green tile with area $176\ \text{cm}^2$, as shown in the diagram.

Work out the dimensions of each tile.

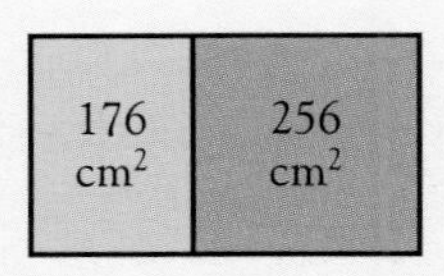

Answer

- The blue tile is a square. If the length of a side is l then

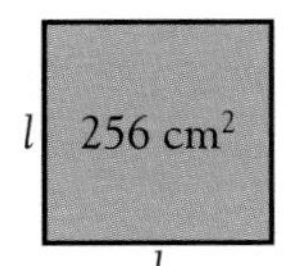

$l^2 = 256$

$l = \sqrt{256}$

$l = 16$

Make sure you know how to find the square root $\sqrt{\ }$ of a number on your calculator.

Dimensions of blue tile = 16 cm by 16 cm

- The green tile has length 16 cm. If the width is w then

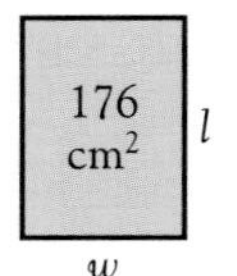

$16w = 176$

$w = 176 \div 16$

$w = 11$

Dimensions of green tile = 16 cm by 11 cm

AREA OF SIMPLE SHAPES

Area of a triangle

The formula for the area of a triangle uses the words 'base' and 'height'.
You can make any side of the triangle the base:

The height is easier to see if you turn your book so that the base is horizontal.

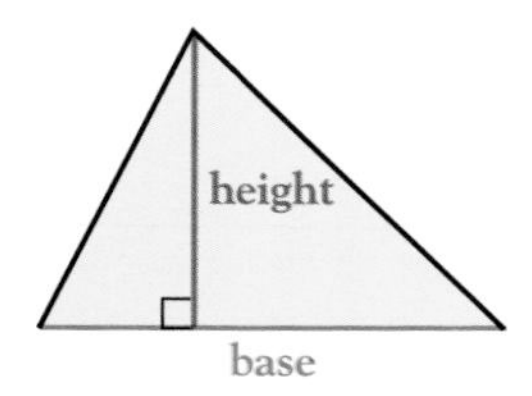

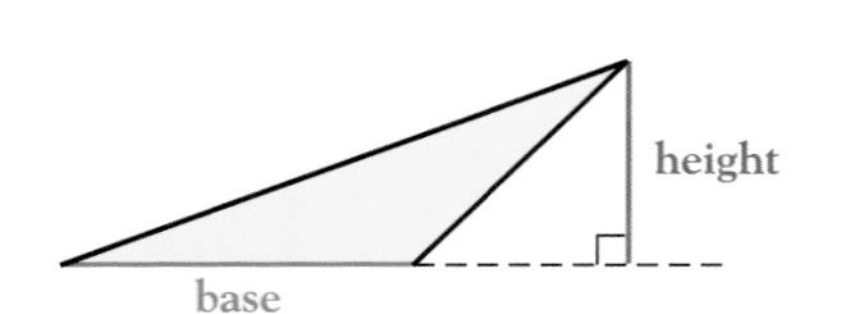

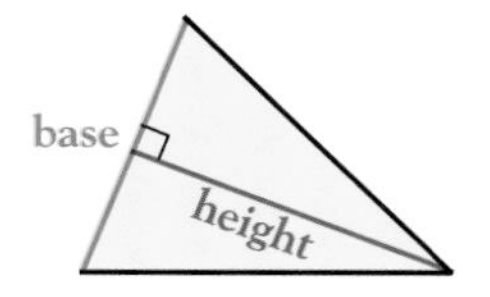

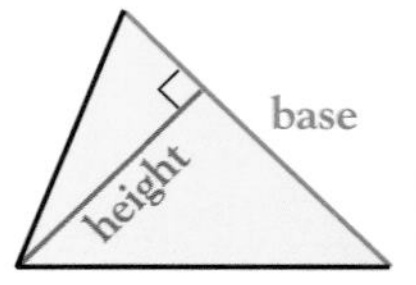

Notice that the height is at right angles (90°) to the base. It is called the **perpendicular height**.

To work out the formula for the area of a triangle, picture a box fitting around it:

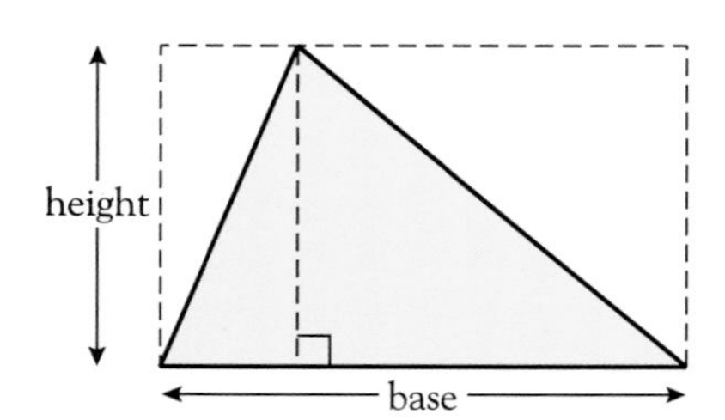

You can see that the area of the coloured triangle is **half** the area of the rectangle.

Area of rectangle = base × height, so to find the area of the triangle you need to divide by 2.

Here are some of the ways that people remember the formula:

$A = \frac{b \times h}{2}$

$A = \frac{1}{2}bh$

$A = \frac{1}{2}$ of base × height

$$\text{Area of triangle} = \frac{\text{base} \times \text{height}}{2}$$

Sample Question 3

Find the area of this sign.

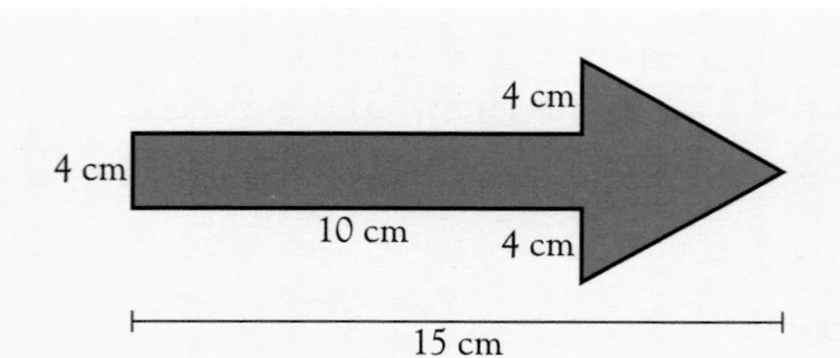

Answer

- Split the shape into a rectangle and a triangle.
- Work out each area separately, then add the areas.

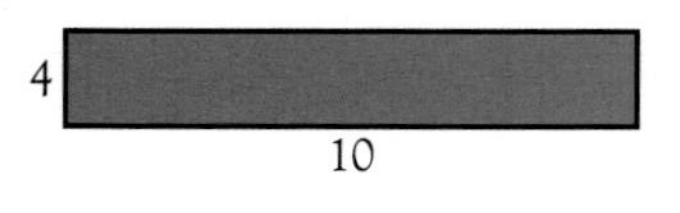

Area $= l \times w$
$= 4 \times 10$
$= 40 \text{ cm}^2$

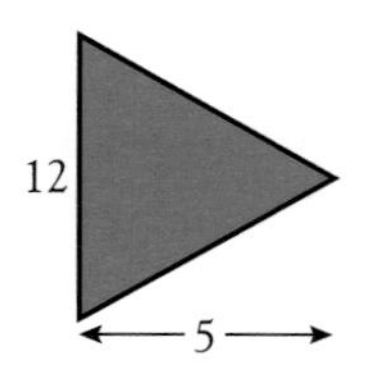

Base $= 4 + 4 + 4 = 12$ cm
Height $= 15 - 10 = 5$ cm

Area $= \frac{b \times h}{2}$
$= \frac{12 \times 5}{2}$
$= 30 \text{ cm}^2$

Total area $= 40 + 30 = \underline{70 \text{ cm}^2}$

AREA OF SIMPLE SHAPES

A band is cut from a piece of square material of side 12 cm, as shown. Find the area of the band.

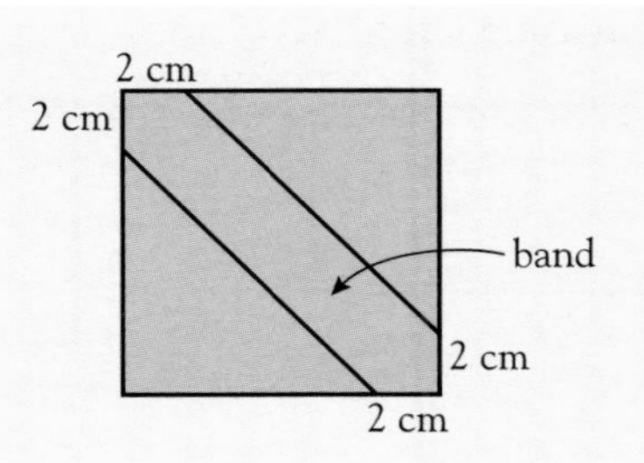

Answer

- Work out the area of the square.
- Work out the area of each triangle left when the band has been cut.
- Subtract the area of the two triangles from the area of the square.

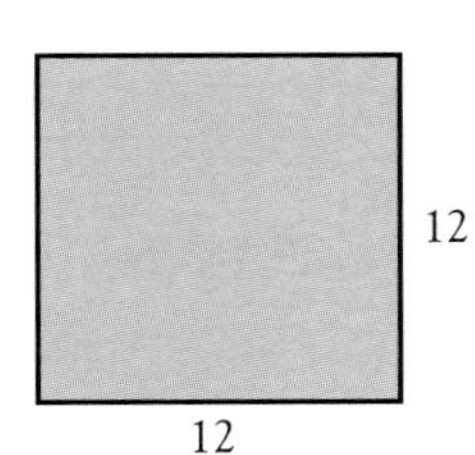

$$\begin{aligned}\text{Area} &= l^2\\ &= 12 \times 12\\ &= 144 \text{ cm}^2\end{aligned}$$

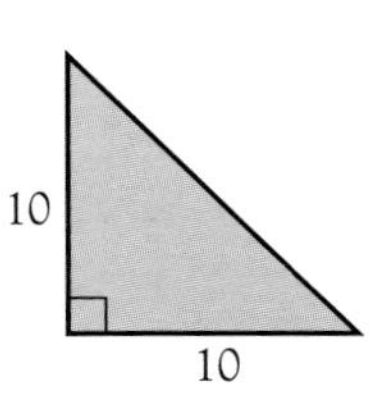

$b = 10, h = 10$

$$\begin{aligned}\text{Area} &= \frac{b \times h}{2}\\ &= \frac{10 \times 10}{2}\\ &= 50 \text{ cm}^2\end{aligned}$$

Area of two triangles $= 50 + 50 = 100$ cm^2

Area of band $= 144 - 100$

Area of band $= 44$ cm^2

Exercise 7.1

1 Tom and Harry decided to put their initials in kitchen foil on the wall in their bedroom. Their mother gave them each a rectangular sheet of foil 40 cm by 30 cm. They designed the letters with the following dimensions.

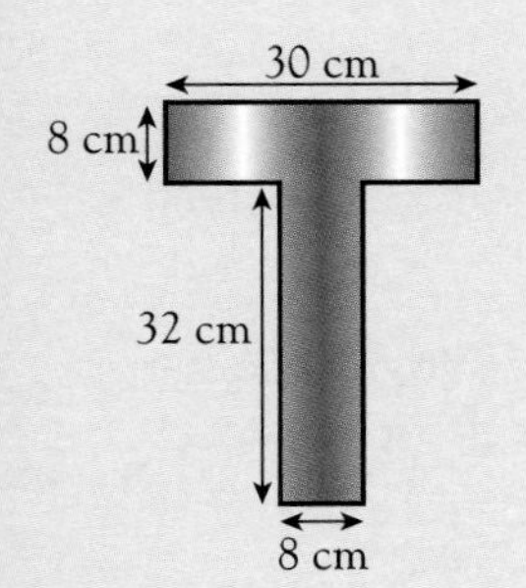

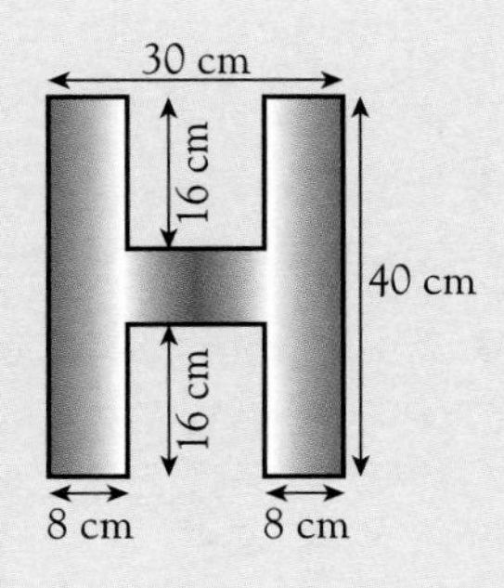

a Find the area, in cm^2, of each letter.

b Find the area of foil wasted by each of the boys.

2 A rectangular mirror 70 cm by 35 cm is surrounded by a wooden border 5 cm wide.

a Find the area of the mirror.

b Find the area of the border.

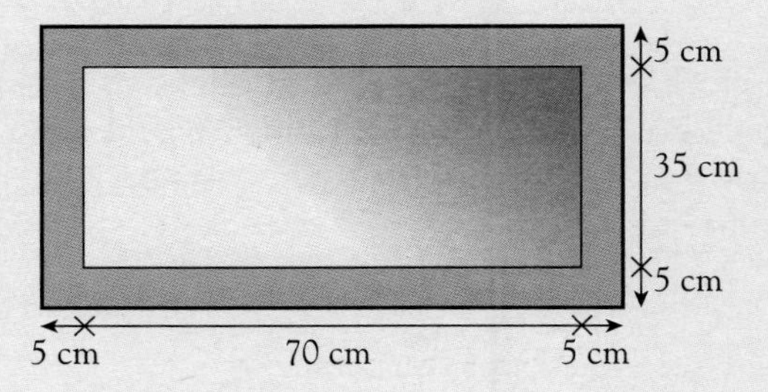

3 This rectangle has a perimeter of 34 cm. Find the area of the rectangle.

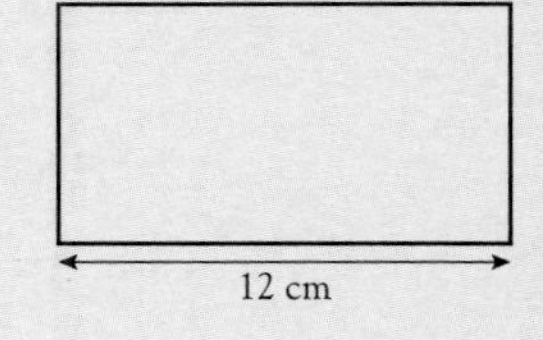

AREA OF SIMPLE SHAPES

4 Find the missing lengths.

a

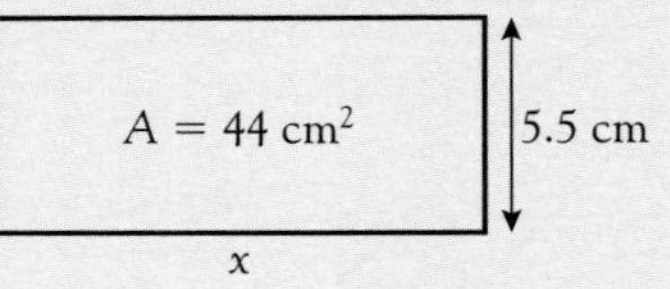

b

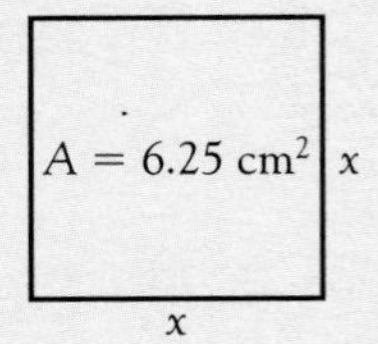

5 A square has an area of 81 cm^2.

a What is the length of a side of the square?

b What is the perimeter of the square?

6 Find the area of each triangle.

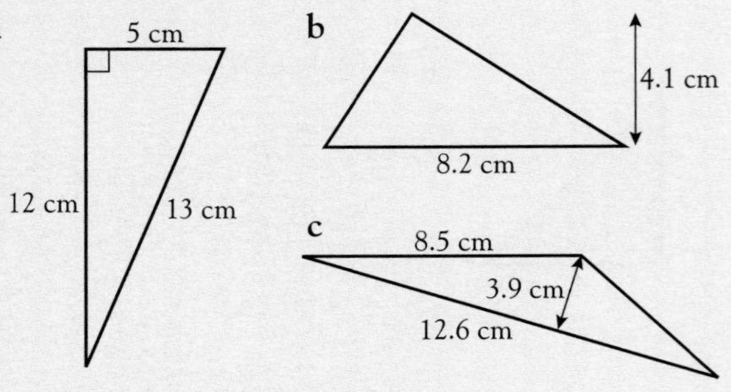

7 The diagram shows a triangle drawn inside a square of side 6 cm.

Find the area of the triangle shown shaded in the diagram.

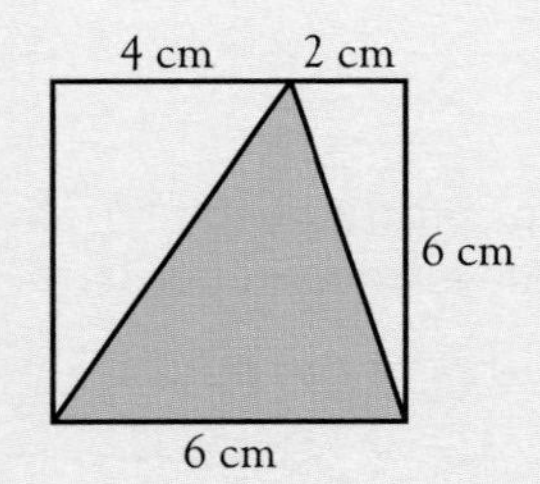

8 A piece of perspex is cut in the shape of a right-angled triangle with sides 1.5 ft, 2 ft and 2.5 ft as shown in the diagram. Perspex costs £2.70 per square foot.

a Find the area of the triangle, in square feet.

b Find the cost of the perspex.

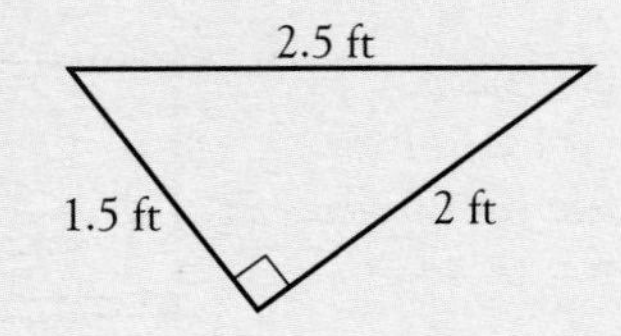

9 Work out the area of this kite.

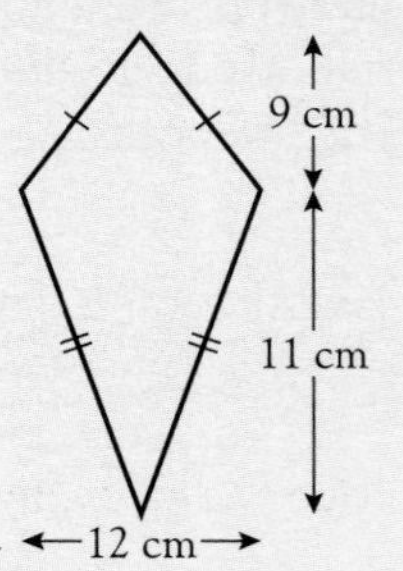

10 The diagram shows the end wall of a house.
Find its area.

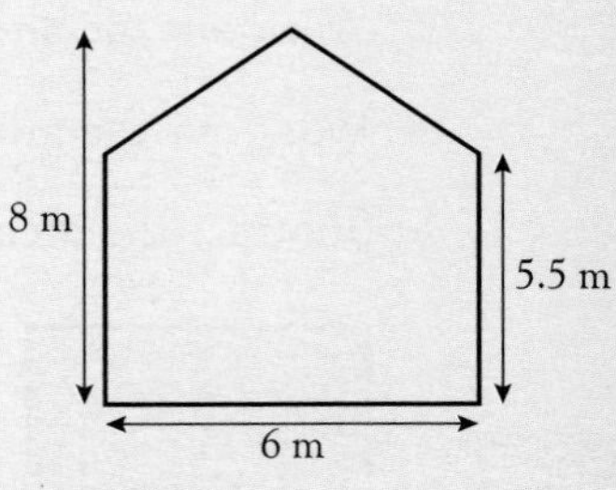

11 This triangle has an area of 48 cm^2.

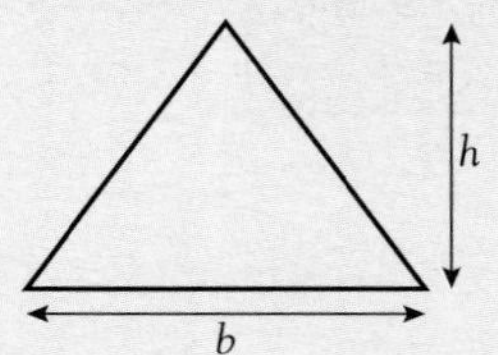

Find:

a its height h, if its base length b is 8 cm,

b its base length b, if its height h is 10 cm.

12 On squared paper, draw a grid, using a scale of 1 cm to 1 unit, with both x and y axes going from 0 to 8.

Plot each triangle on the grid, labelling it carefully. Find the area of each triangle.

Triangle A (1, 1), (3, 1), (2, 4)
Triangle B (0.5, 6), (0.5, 7.5), (3.5, 6)
Triangle C (4, 3), (4, 7), (6, 6)
Triangle D (4, 2), (7, 1), (5, 2)

13 Find the area of a rectangular stamp which is 3 cm long and 20 mm wide.

14 Susie wants to concrete a drive outside her house. Find the area of the drive.

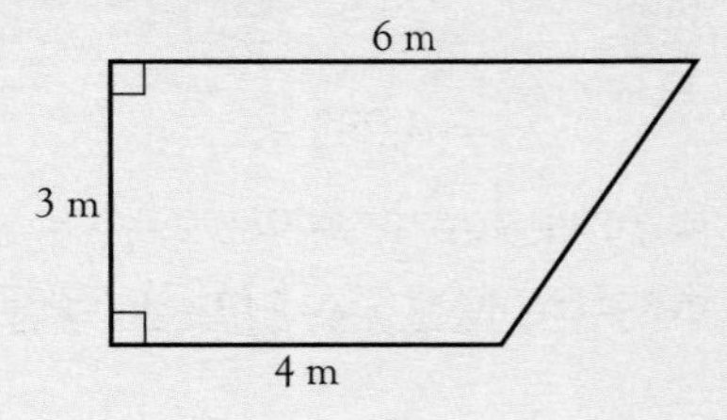

AREA OF A CIRCLE

Area of a circle

To find the area of a circle you need to know the radius, r.

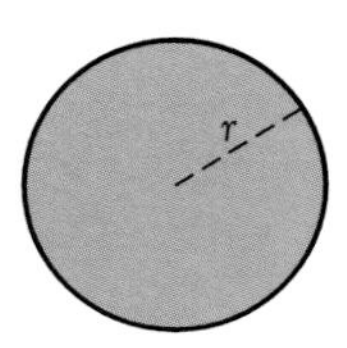

Area of a circle = πr^2

πr^2 means $\pi \times r^2$ which is $\pi \times r \times r$

For a circle with radius 4 cm

$$\text{Area} = \pi r^2 = 3.14 \times 4^2 = 50.24$$

Area = 50.24 **cm²**

Remember the units are cm²

On your calculator, key in

[3] [.] [1] [4] [×] [4] [x²] [=]

or

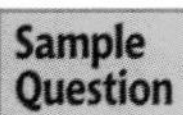

A circular pond has a path running around it. The diameter of the pond is 10 m and the width of the path is 0.5 m. Find:

a the area of the pond,

b the area of the path.

(Take $\pi = 3.14$)

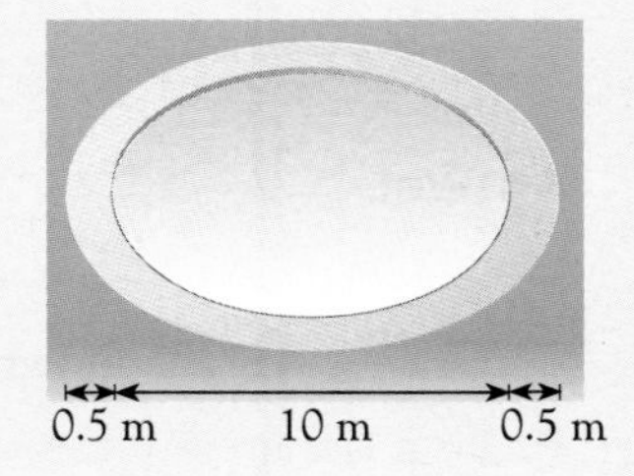

Answer

a

- Work out the radius of the pond.

  $d = 10$ m, so $\mathbf{r = 5}$ **m**

- Use $A = \pi r^2$ to find the area.

  $$A = \pi r^2 = 3.14 \times 5^2 = 78.5$$

  Area of pond = 78.5 m²

Always find the radius if the diameter is given.

b

- Work out the area of the pond and the path together.
- Subtract the area of the pond from this answer.

  For the path and pond, $r = 5.5$ m.

  $$A = \pi r^2 = 3.14 \times 5.5^2 = 94.985 \text{ m}^2$$

  Area of path = 94.985 − 78.5 = 16.485

  Area of path = 16.5 m² (1 d.p.)

A sketch is helpful here.

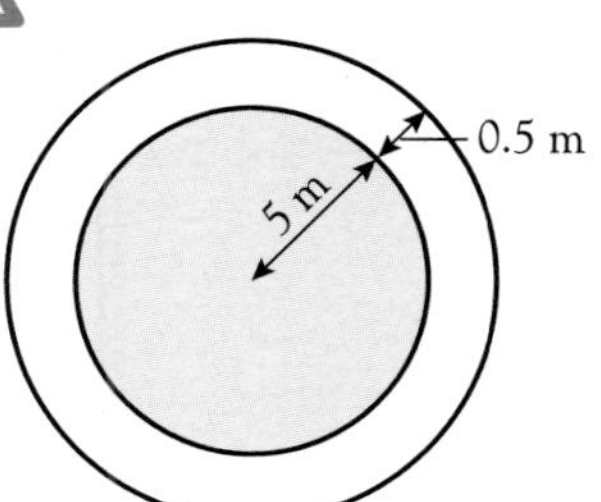

Approximate your final answer sensibly.

AREA OF A CIRCLE

Sample Question 6

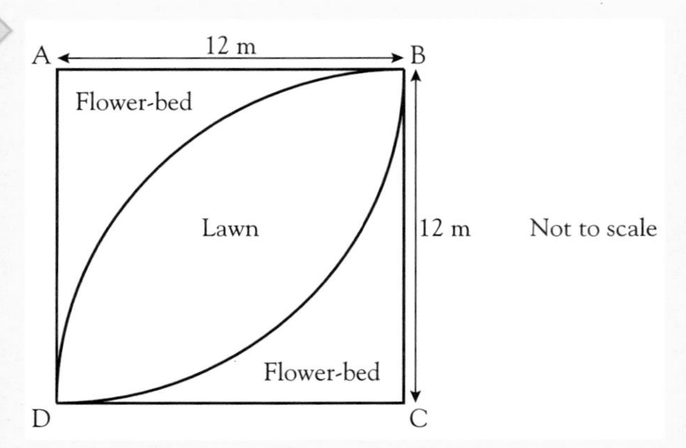

The diagram shows a square garden, ABCD, with sides of length 12 metres. It is divided into a lawn and two flower-beds by circular arcs of radius 12 m, with centres at A and C.
Taking π to be 3.14, calculate, giving your answers correct to the nearest square metre, the area, in square metres, of:

a one flower-bed,

b the lawn. [MEG]

Answer

a

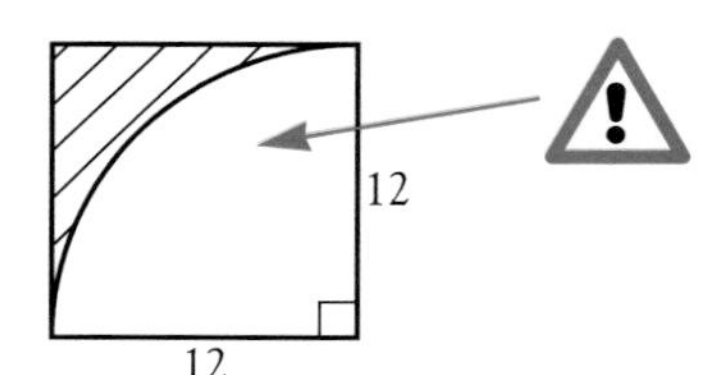

- Draw a sketch, showing one flower-bed.
- Find the area of the square.
- Find the area of a circle with radius 12 cm, then find $\frac{1}{4}$ of it.
- Subtract this from the area of the square.

Area of square $= 12 \times 12 = 144\ \text{m}^2$

Area of complete circle $= \pi r^2$
$= 3.14 \times 12^2$
$= 452.16$

Area of quarter circle $= 452.16 \div 4$
$= 113.04\ \text{m}^2$

Area of flower-bed $= 144 - 113.04$
$= 30.96\ \text{m}^2$

Round your answer to the nearest square metre, as requested.

<u>Area of flower-bed $= 31\ \text{m}^2$</u> (nearest square m)

b

- Find the area of the two flower-beds.
- Subtract this from the area of the square.

Area of two flower-beds $= 2 \times 30.96$
$= 61.2\ \text{m}^2$

Area of lawn $= 144 - 61.2$
$= 82.8\ \text{m}^2$

<u>Area of lawn $= 83\ \text{m}^2$</u> (nearest square m)

AREA OF A CIRCLE

Exercise 7.2

(Take $\pi = 3.14$)

1 The radius of a circular garden pond is 3 m. Calculate the area of the pond.

2 A circular table top has a diameter of 70 cm. Find the area of the table top.

3 Four circles are cut from a square sheet of paper of side 24 cm.

a What is the radius of each circle?

b Find the area of one circle.

c Find the wasted area.

d Find the percentage of area wasted.

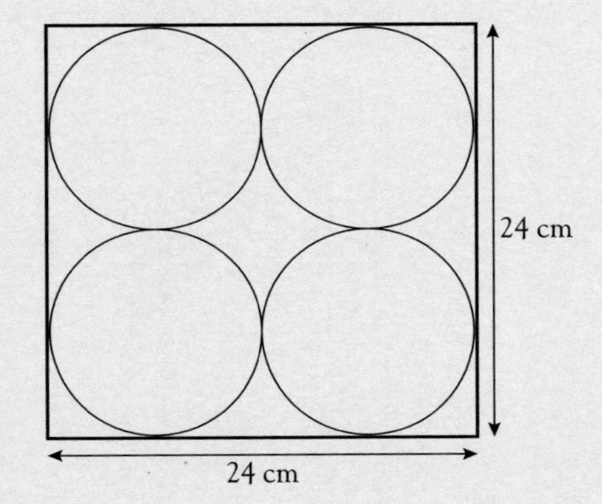

4 A wooden pendant is in the shape of a circle of diameter 40 mm, with two circles, of radius 6 mm and 4 mm, cut from it.

Find the area of the pendant, in mm^2.

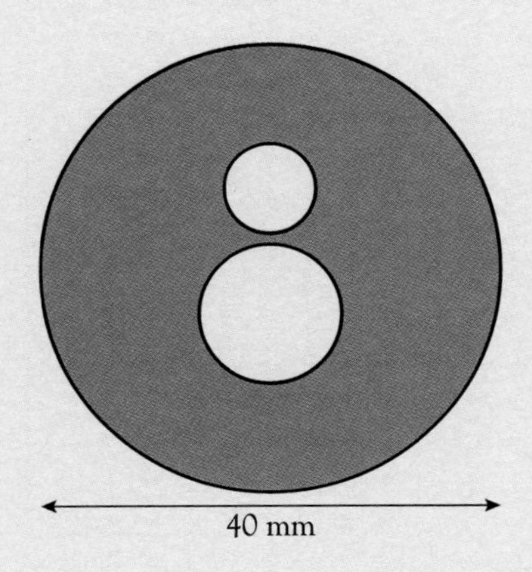

5 Which of these two diagrams shows the larger area shaded, and by how much?

a

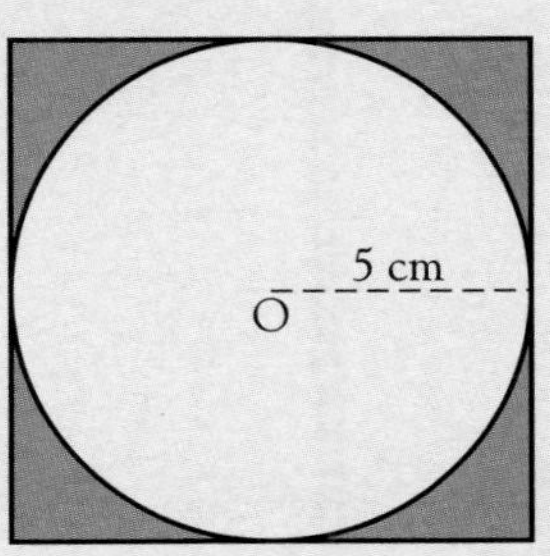

b

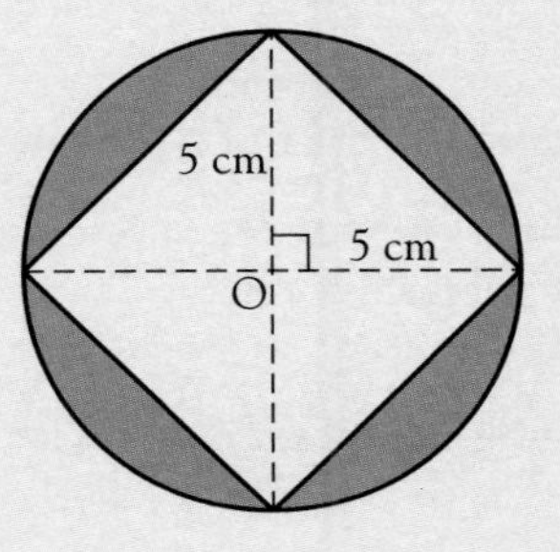

6 This semi-circular rug has diameter 1.3 metres.

a Find its area, in centimetres squared.

b Find the perimeter of the rug, in metres.

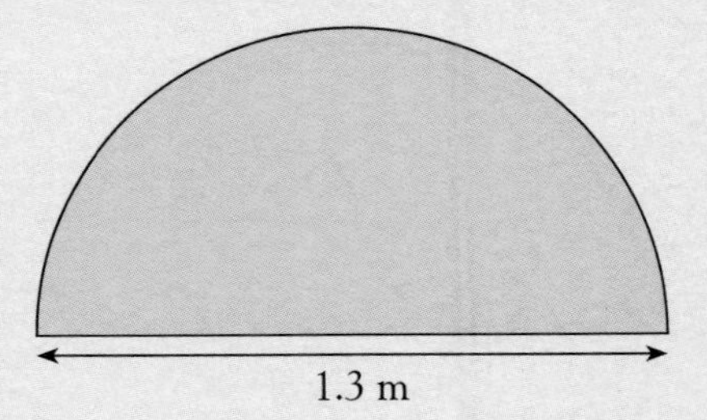

7 ABC is a triangle in which AB = 5 cm, AC = 4 cm and BC = 3 cm. Three semi-circles are drawn with AB, AC and BC as diameters.

Find the total area of the triangle and the three semi-circles.

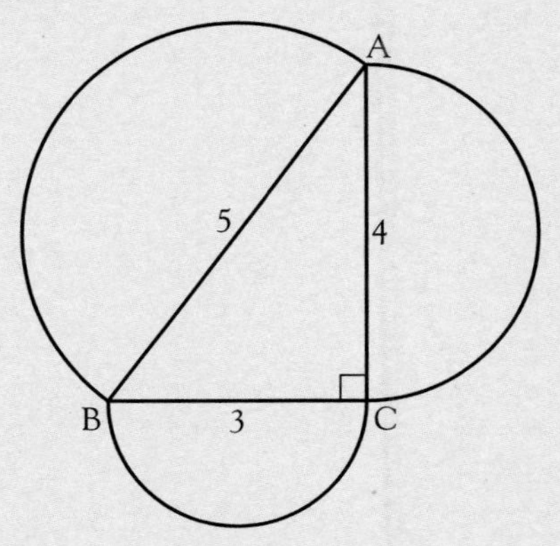

AREA OF A CIRCLE

8 A wooden coffee table has a tile inlay at each corner in the shape of a quarter circle of radius 10 cm.

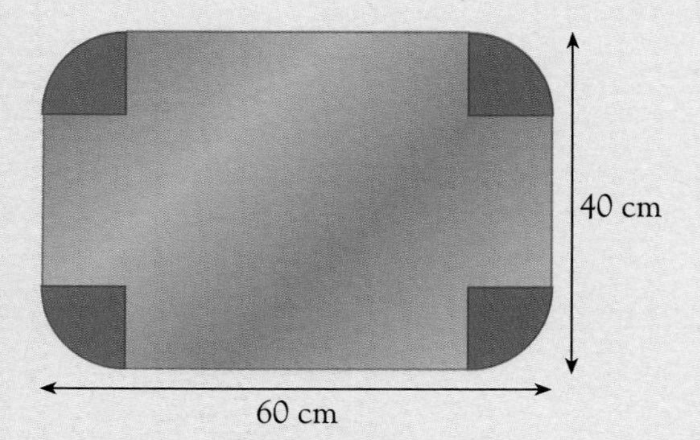

a What is the area of the four tiles?

b Find the area of the surface of the table top not covered by tiles.

9 The diagram shows a circle with centre O. The diameter is 25 cm. The shaded parts are both semi-circles.

a Find the area shaded.

b What fraction of the complete diagram is shaded?

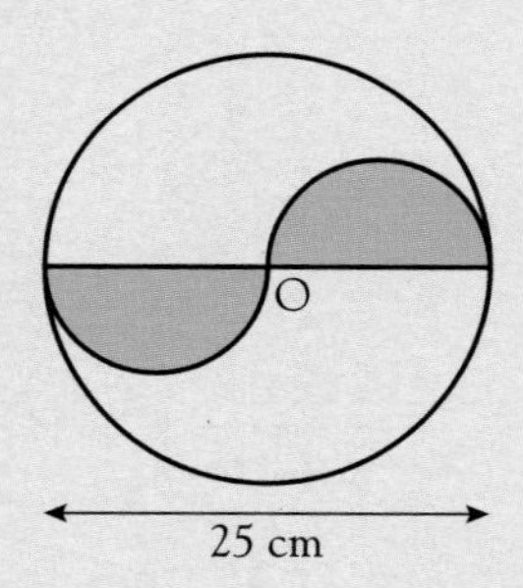

10 An isosceles triangle of height 12 cm and a semi-circle of radius 4 cm are joined together to make the shape in the diagram. Calculate the area of the complete shape.

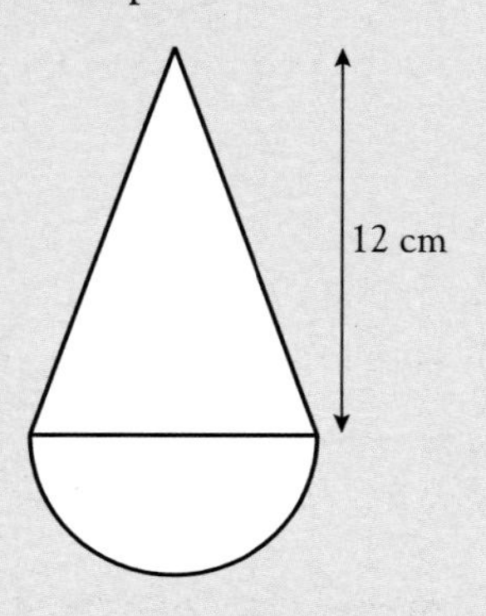

11 Three circles are drawn, each with the same centre. The radius of the smallest circle is 24 cm. The radius of the second circle is 12 cm more than the first circle and 12 cm less than the third circle. Find the shaded area.

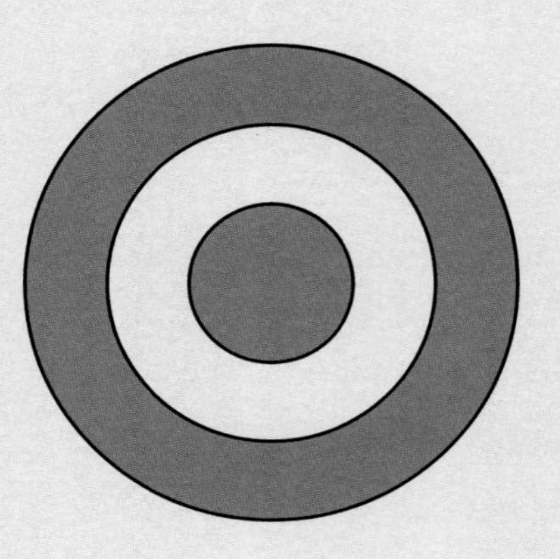

12 Find the shaded areas in the following diagrams.

a

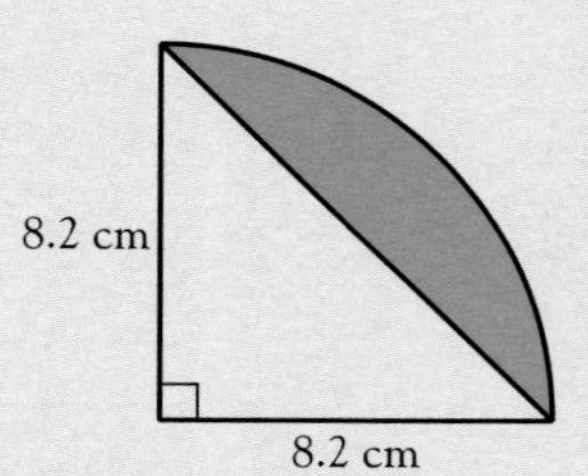

b

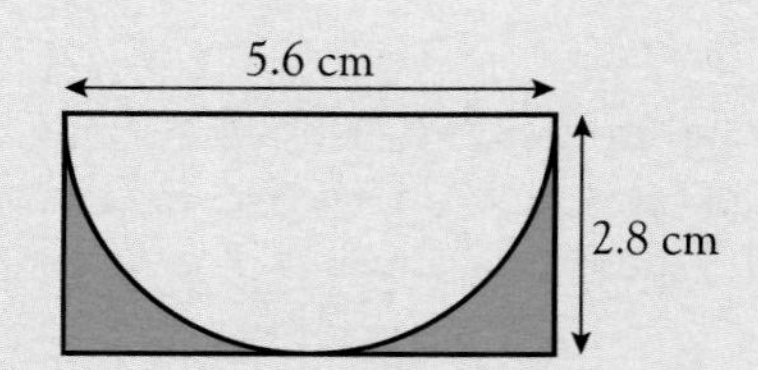

c

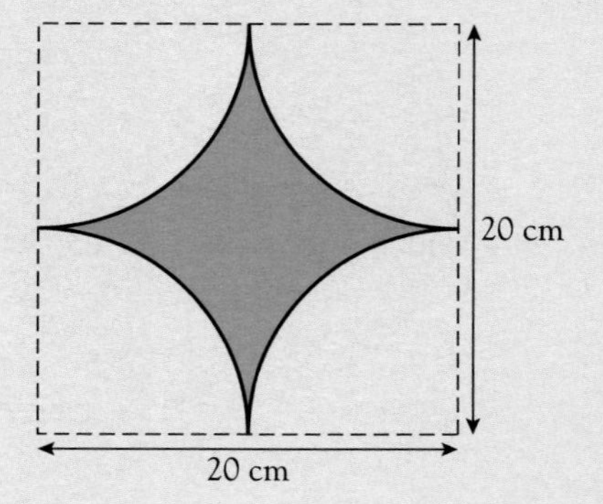

AREA OF A CIRCLE

TASK

A Garden Centre held a competition for pupils in the local school to design signs to be hung where various products could be found.

Rules of competition

1. The design must be made using only rectangles, triangles, trapezia and circles.
2. The design must be painted in green on a piece of white card measuring 30 cm by 35 cm.
3. Each entry is to be accompanied by the instructions necessary to draw the design.

The entries were as follows:

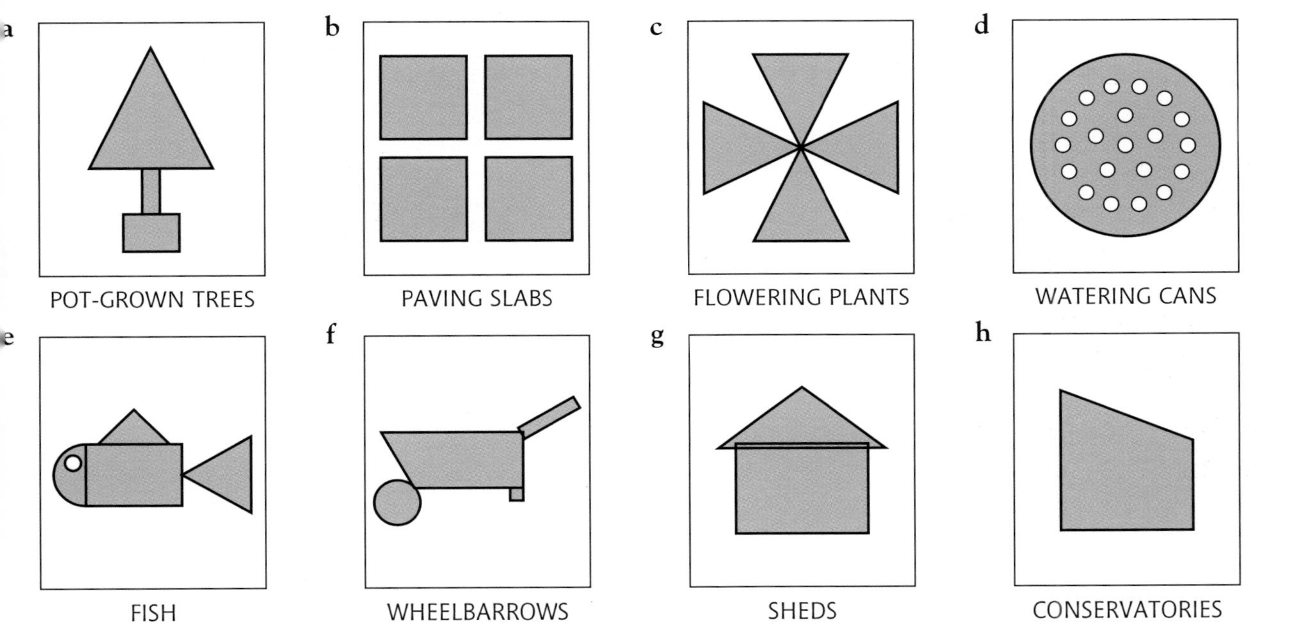

Descriptions (all units in cm)

a Isosceles triangle $b = 16$, $h = 17$; rectangles 2 by 6 and 7 by 5

b Rectangles 11 by 11

c Isosceles triangle $b = 12$, $h = 13$

d Large circle $d = 25$; small circles $d = 2$

e Semi-circle $r = 4$; eye (left white) circle $d = 2$; fin triangle $b = 9$, $h = 5$; tail triangle $b = 10$, $h = 9$; body rectangle 13 by 8

f Circle $d = 6$; trapezium top edge 19, bottom edge 14, height 7; rectangles 2 by 2 and 2 by 9

g Rectangle 17 by 12; triangle $b = 22$, $h = 9$

h Back height 20; front height 12; width 18

1. Find the area painted green in each sign.
2. What percentage of the area is painted green in each sign?
3. Design some signs of your own, using only rectangles, triangles, trapezia and circles, to advertise other products, for example for a sports shop, or a toy shop, or a vegetable shop.

   Make sure that you include instructions.

   Exchange your designs with others to work out the areas and percentages, and to mark one another's answers.

Volume of a cuboid

The volume of a solid shape is the amount of space it occupies. Volume is measured in cubic units.

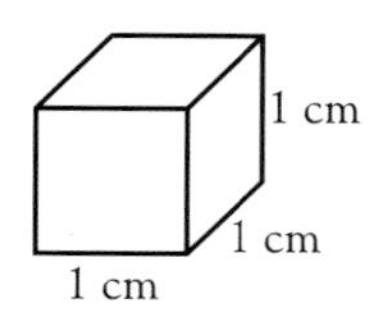

This cube has a volume of 1 cubic cm, which is written as 1 cm^3.

cm^3 is read as 'centimetres cubed'.

In a cuboid 2 cm by 5 cm by 3 cm, there are 10 cubes in each of the three layers.

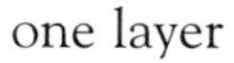

one layer

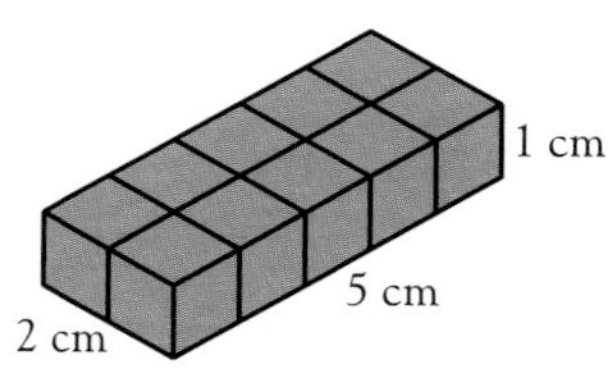

Volume = 10 cm^3

three layers

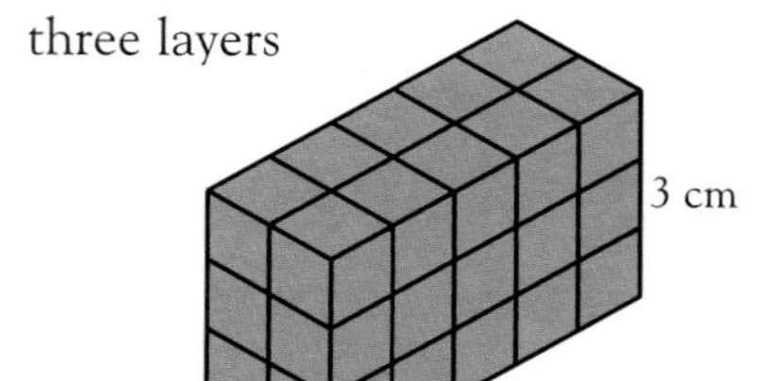

Volume = $5 \times 2 \times 3 = 30$ cm^3

Volume of a cuboid = length × width × height
$= l \times w \times h$

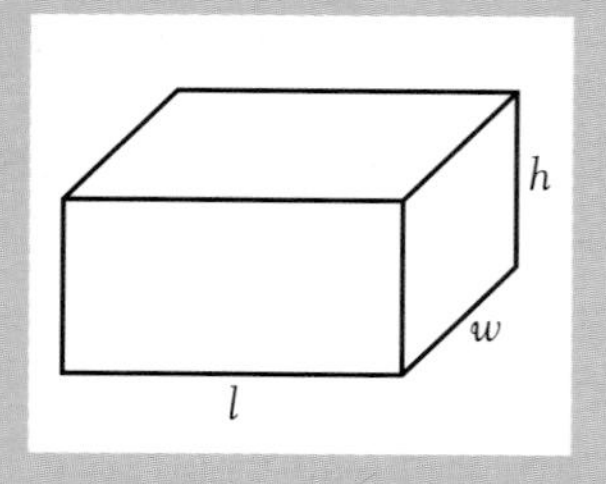

A **cube** is a special cuboid in which the length, width and height are all equal to each other.

Volume of a cube $= l \times l \times l$
$= l^3$

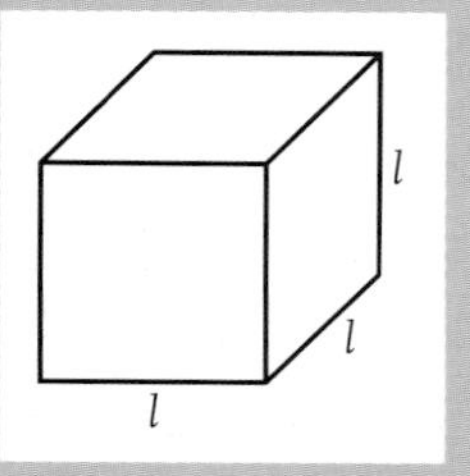

VOLUME OF A CUBOID

Sample Question 7 These two cuboids have the same volume. Find the height of cuboid **b**.

a

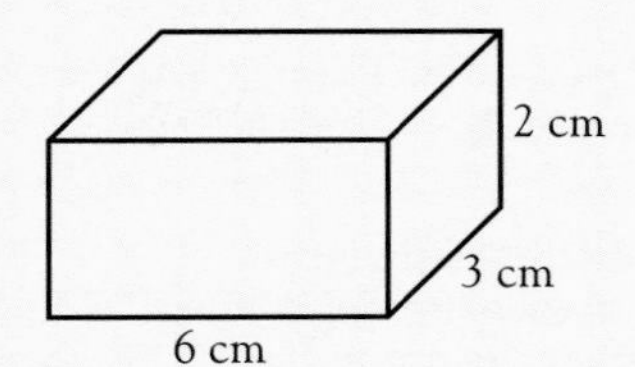

b

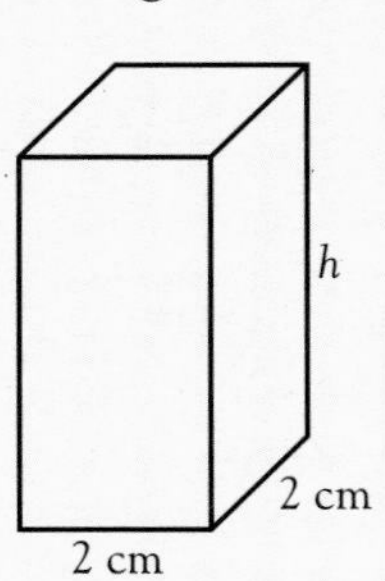

Answer

- Find the volume of cuboid **a**.

$$\begin{aligned} V &= l \times w \times h \\ &= 6 \times 3 \times 2 \\ &= 36 \text{ cm}^3 \end{aligned}$$

- Put the volume of cuboid **b** equal to this answer and work backwards to find h.

$$\begin{aligned} 2 \times 2 \times h &= 36 \\ 4h &= 36 \\ h &= 36 \div 4 \\ &= 9 \end{aligned}$$

You are solving an equation here.

Height of the second cuboid = 9 cm

Sample Question 8 A water storage tank is 2.5 m long, 2 m wide and 150 cm deep.

a What is its volume in cm^3?

b What is the capacity, in litres, of a full tank?

Answer

a

- The volume is needed in cm^3, so change all the lengths to cm.

length = 250 cm, width = 200 cm, depth = 150 cm

To change m to cm, multiply by 100

- Work out the volume.

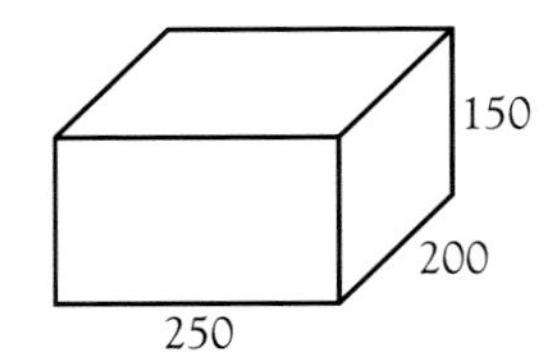

$$\begin{aligned} V &= l \times w \times h \\ &= 250 \times 200 \times 150 \end{aligned}$$

$V = 7\,500\,000 \text{ cm}^3$

b

- Change this volume into litres using the relationship **1 litre = 1000 cm^3**

$7\,500\,000 \div 1000 = 7500$

Capacity of a full tank = 7500 litres

To change cm^3 to litres, divide by 1000

Surface area of a cuboid

Sample Question 9

Work out the surface area of this cuboid.

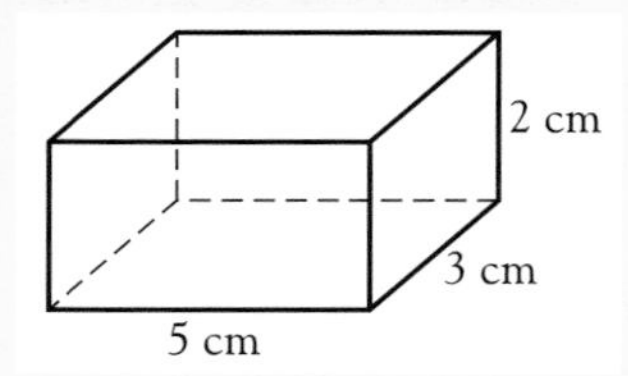

A cuboid has six faces.
The surface area of a cuboid is found by adding together the areas of each of these six faces.

Answer

- Work out the areas of the front, side and top.

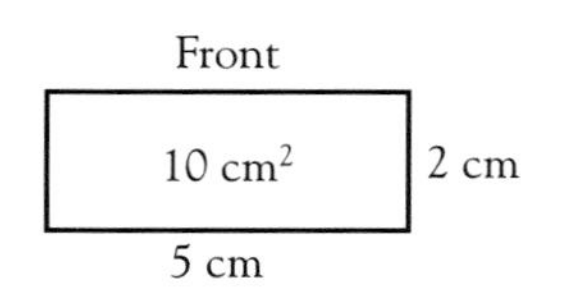

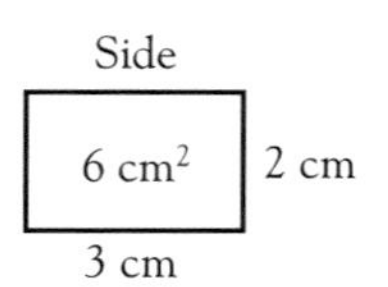

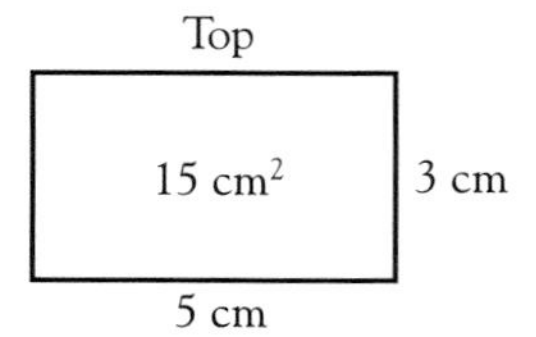

- Add together **two lots** of each area (front and back, two 'sides', top and bottom).

Surface area $= 2 \times 10 + 2 \times 6 + 2 \times 15$

$= 20 + 12 + 30$

Surface area $= 62 \text{ cm}^2$

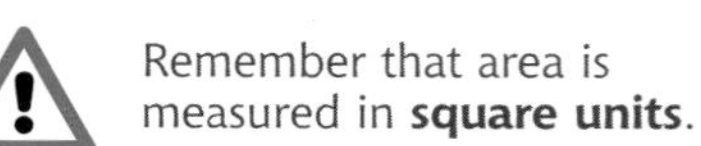

Sample Question 10

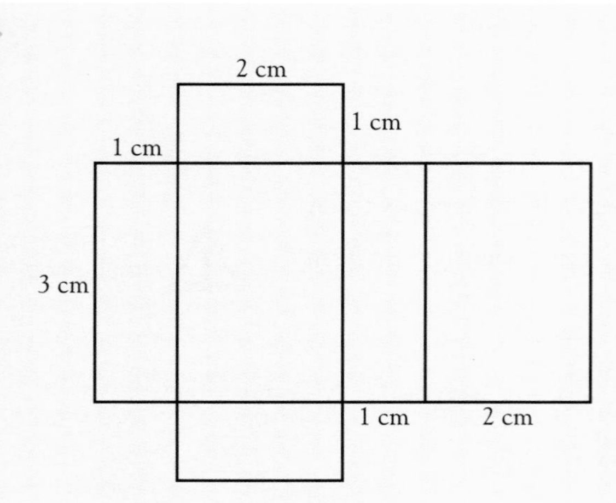

This is the net of a cuboid.

a Work out the surface area of the cuboid from the net.

b Draw the cuboid, full size, on isometric paper.

c Work out the volume of the cuboid.

Answer

a

- Work out the area of the net.

  You could look at the six separate rectangles, or put some together to make the working quicker, for example:

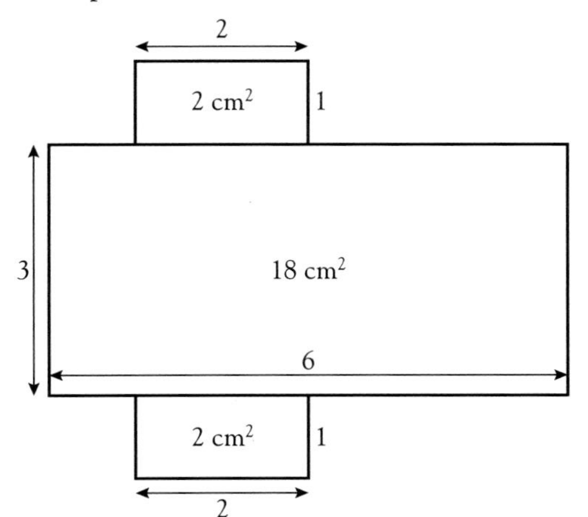

Surface area = 18 + 2 + 2

Surface area = 22 cm²

b

- Start the drawing by showing width 2 cm, length 3 cm and height 1 cm.

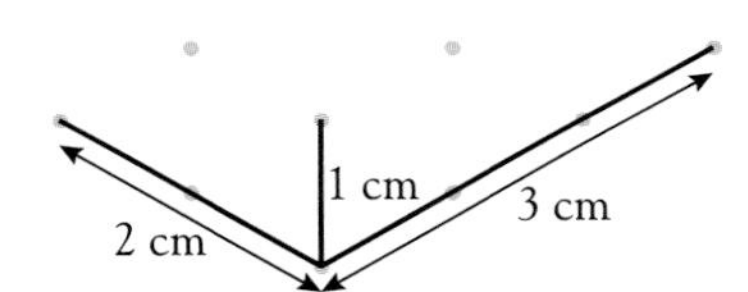

- Complete the diagram.

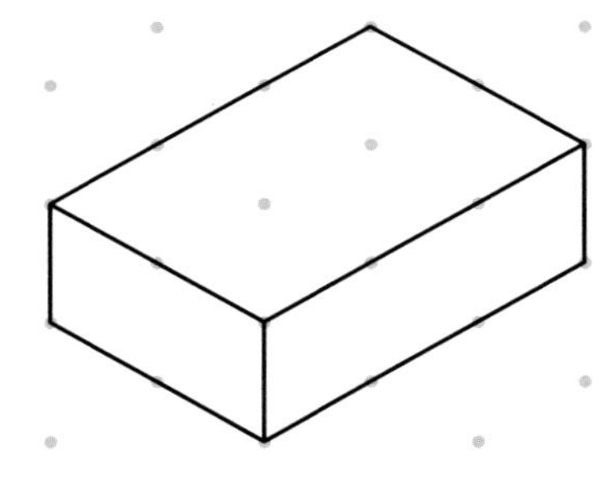

c Volume of cuboid = length × width × height

$= 3 \times 2 \times 1$

Volume of cuboid = 6 cm^3

Remember that volume is in **cubic units**.

TASK

Design and make as many cuboids as you can with a volume of 24 cm^3.

Which of your cuboids has the least surface area?

VOLUME OF A CUBOID

Exercise 7.3

1 These boxes are both cuboids. They have the same volume.

a Find the volume of Box A.

b Find the height of Box B.

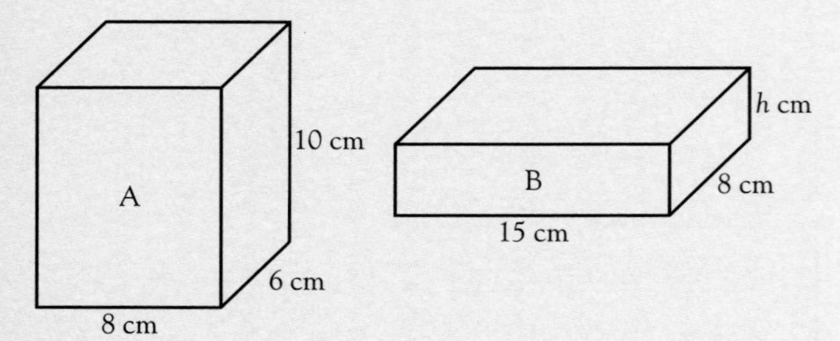

2 A carton of fruit juice is 18 cm tall and its base measures 6 cm by 12 cm. The carton is filled so that there is a 1 cm gap at the top. Find the volume of fruit juice in the carton.

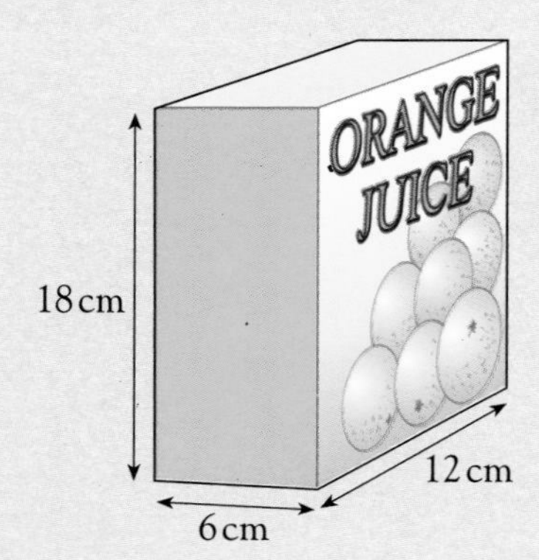

3 A gardener dug out a pond 2 m long, 1 m wide and 0.5 m deep. The pond was then filled with water.

a How many cubic metres of water were needed to fill the pond completely?

A concrete block was placed in the pond ready to position a pump for a fountain. The block had a square base of side 25 cm and height 40 cm.

b How many litres of water did it displace?

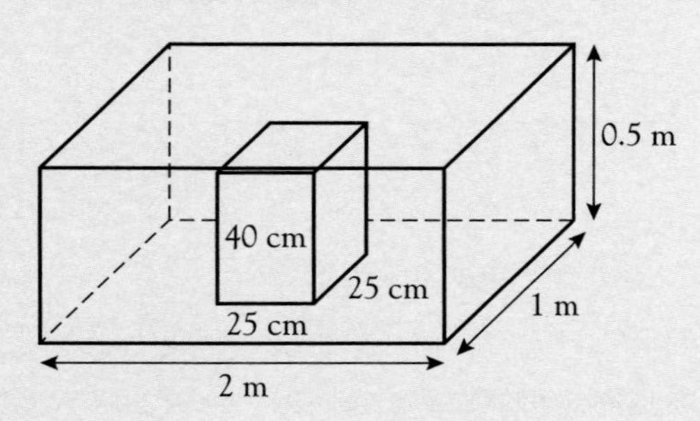

4 A header tank in a loft measures 80 cm by 50 cm by 50 cm. It can never be more than four-fifths full because of the overflow pipe. What is the maximum volume, in litres, that it can hold?

5 The concrete base of a double garage is 5 m by 5 m and laid to a depth of 20 cm. Calculate:

a the area of the garage floor,

b the volume of the concrete used.

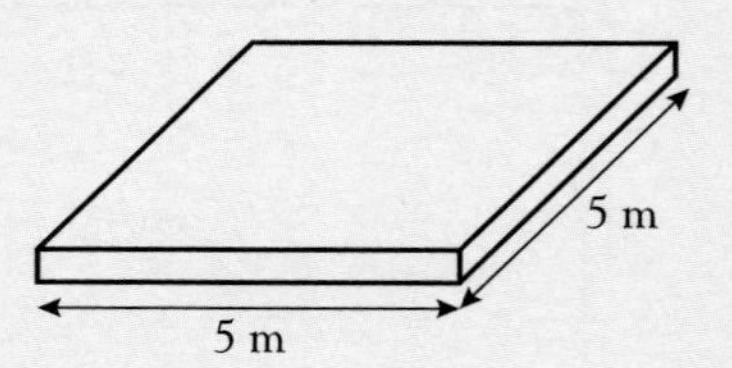

6 A children's sand pit is 3 m by 4 m and 0.5 m deep. The sand is 0.3 m deep.

a Calculate the volume of:

i the sand pit,

ii the sand.

b Calculate the percentage volume of the pit occupied by the sand.

7 Which of these nets fold up into a cube?

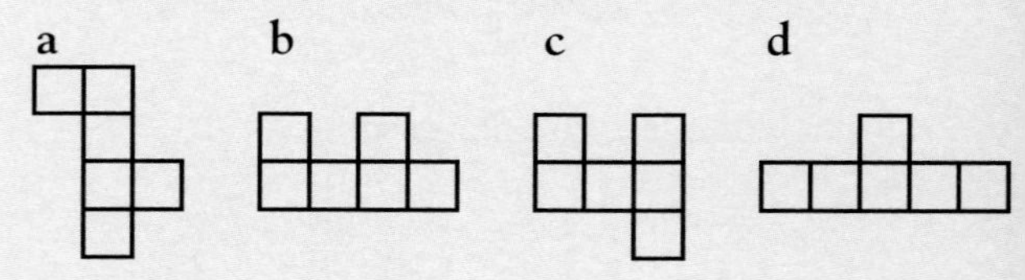

8 This is the net that Simon used to make a cardboard mould. He then used the mould to make a miniature concrete beam.

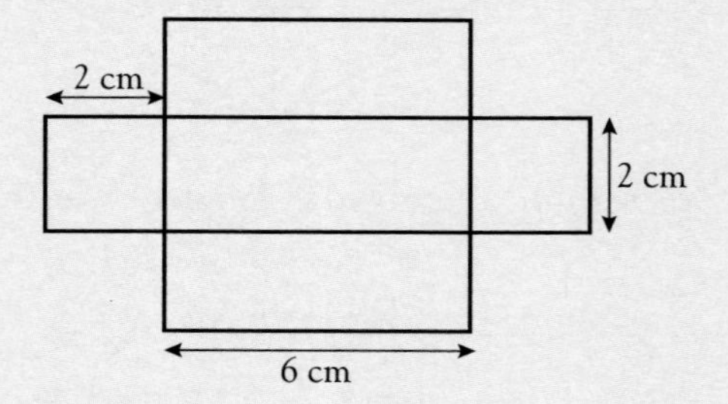

a Calculate the area of the net.

b Draw a sketch of the mould.

c Calculate the volume of the mould.

d Give the dimensions of a different cardboard mould that would hold the same volume.

e Sketch the net of this mould and work out its surface area.

VOLUME OF A CUBOID

9 An old cottage has a doorway which is only 1.75 m high. The renovators want to increase the height to 2 m. The width of the doorway is 75 cm and the walls are 30 cm thick. What volume of wall must be knocked out?

10 A square piece of cardboard, of side 12 cm, is used to make an open box.

Squares of side 2 cm are cut from each corner and the card folded to form the box shown in the diagram. Calculate the volume of the box.

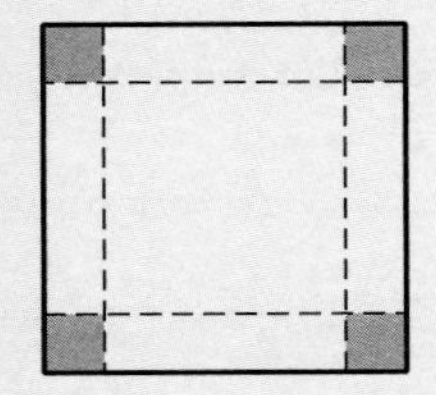

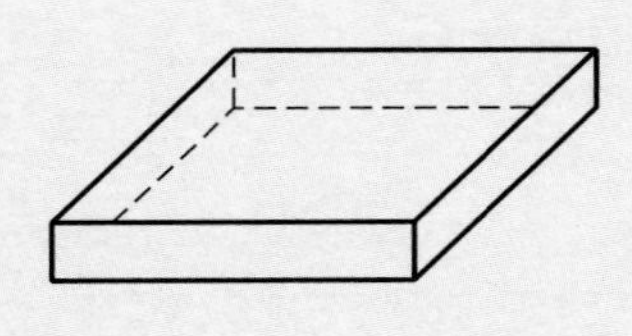

11 Both of these boxes are cuboids. What is the difference between the volumes of the two boxes?

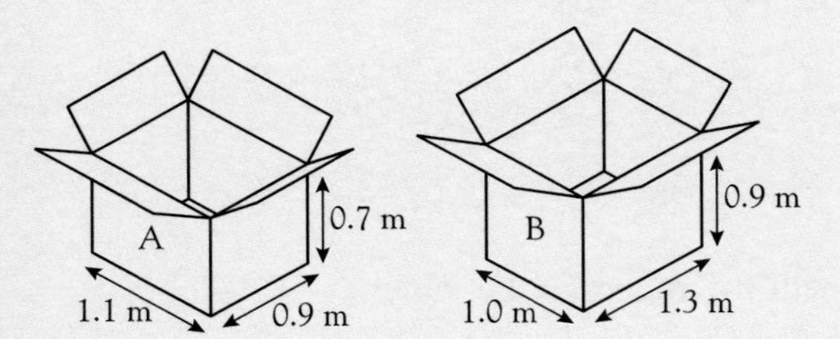

12 The diagram shows the net of a small open box, with no top face.

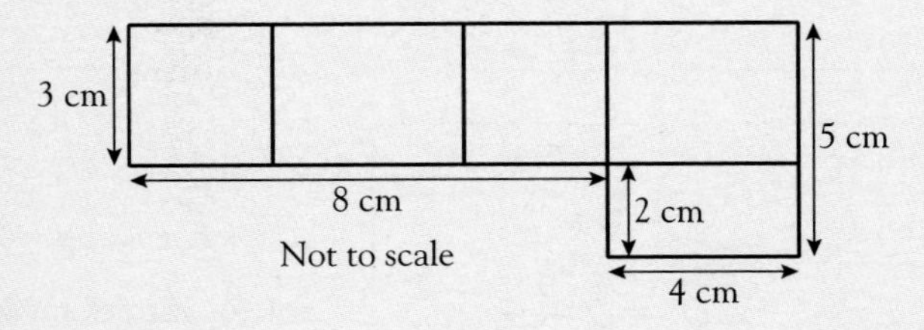

a Find the perimeter of the net.

b Calculate the area of the net.

c Sketch the net and add one more rectangle in a suitable position to change the diagram to the net of a closed box.

d Write down the length, width and height of the box (in any order).

e Calculate the volume of the box.

f Draw an isometric view of the closed box on isometric dotty paper. [MEG]

13 A present is wrapped in paper and ribbon.

a Calculate the volume of the parcel.

b An extra 20 cm of ribbon is needed to make a bow. Ribbon is sold in 1.5 m rolls. Will one roll be enough to wrap the parcel?

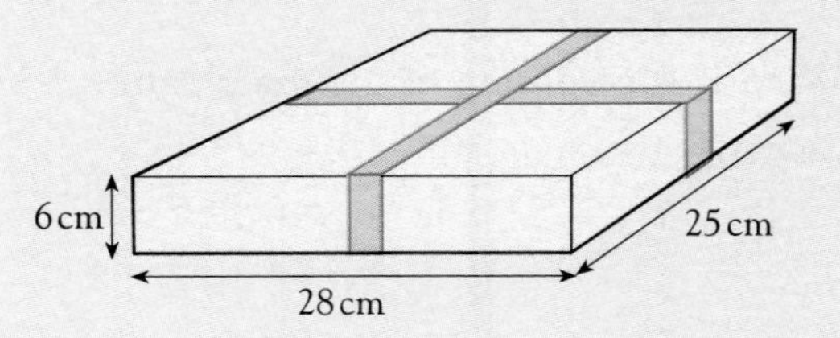

14 This is the net for a rectangular box.

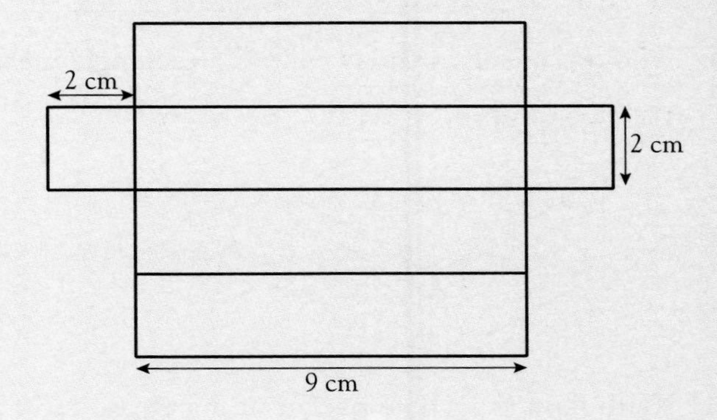

a Work out the volume enclosed by the box.

b What is the total area of cardboard needed to make the box?

15 The diagram shows the winners' rostrum at a swimming gala. It is made from six cuboids, each 50 cm by 50 cm by 20 cm.

The boxes have only their visible faces painted in the appropriate colours of gold, silver and bronze. What is the area of each colour painted?

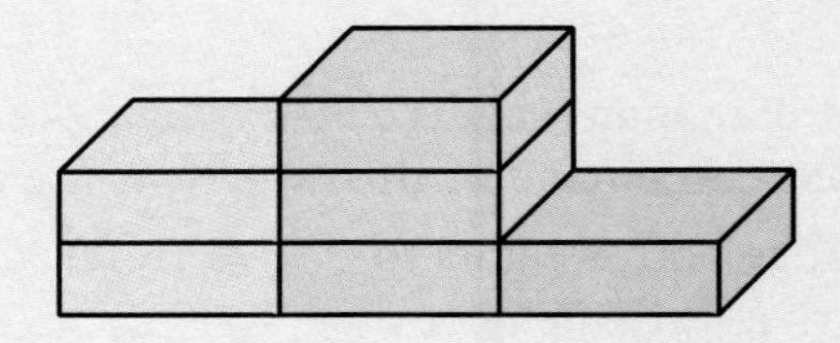

16 A cuboid has a square base of side x cm and a height of 12 cm. The volume of the cuboid is 972 cm^2. Find x.

Worked Exam Question
[SEG]

COMMENTS

Keith buys a packet of marzipan for a birthday cake. The packet measures 3.5 cm by 6 cm by 11 cm.

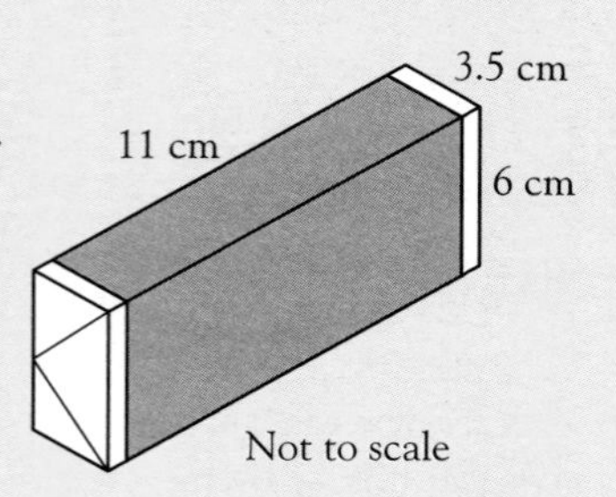

a What is the volume of the marzipan?

◆ Use volume of cuboid $= l \times w \times h$

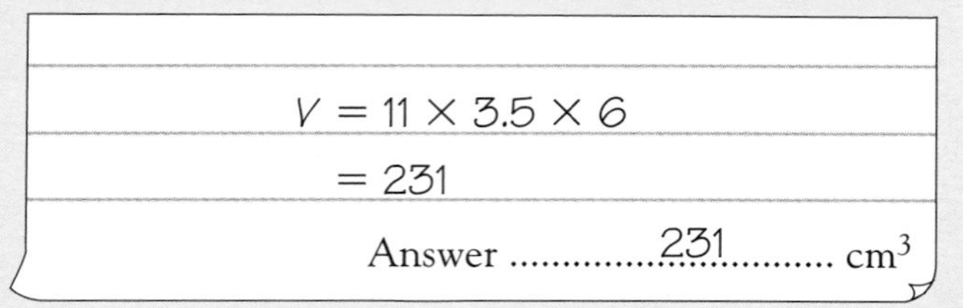

$V = 11 \times 3.5 \times 6$

$= 231$

Answer ......231...... cm³

M1 for multiplying $l \times w \times h$

A1 for correct answer

2 marks

b The top of the birthday cake is a rectangle measuring 25 cm by 20 cm.

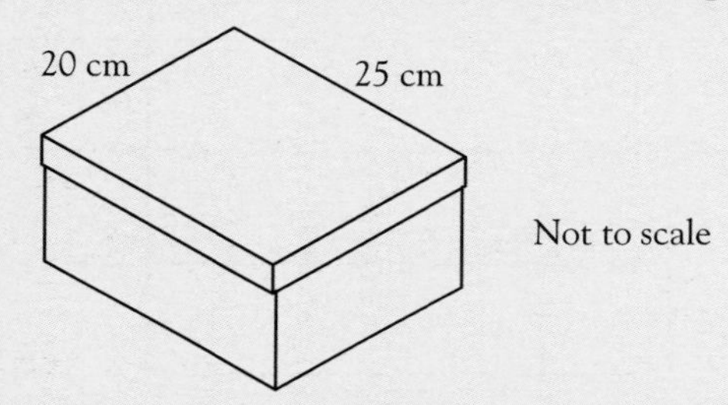

Not to scale

Keith uses the whole packet of marzipan. How thick will the marzipan be when it is rolled out to cover the top of the cake?

◆ The marzipan is in the shape of a cuboid, $l = 25$ cm, $w = 20$ cm, h is not known. The volume of this cuboid is 231 cm³. Work backwards from the volume formula to find h.

Do not be surprised that this is smaller than 1 cm

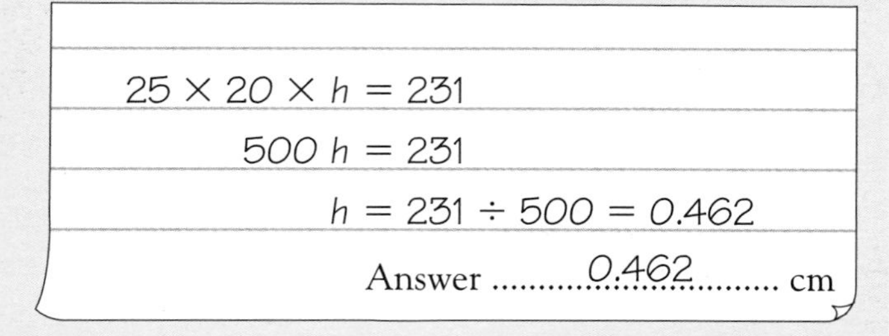

$25 \times 20 \times h = 231$

$500\,h = 231$

$h = 231 \div 500 = 0.462$

Answer ......0.462...... cm

M1 for $25 \times 20 \times h$

M1 for dividing

A1 for correct answer

3 marks

c The marzipan weighs 250 g. The cake and marzipan together weigh 1.6 kg. What percentage of this weight is the marzipan?

◆ Change both weights to g (or to kg). Remember that 1.6 kg = 1600 g.

◆ Express 250 as a percentage of 1600.

Approximate sensibly.

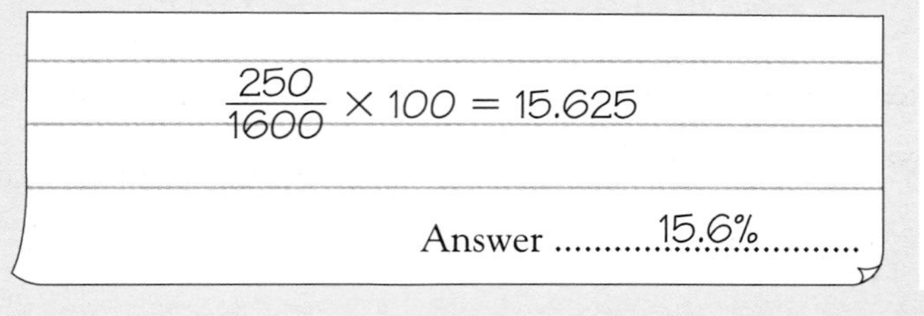

$\frac{250}{1600} \times 100 = 15.625$

Answer ......15.6%......

B1 for expressing both weights in the same unit

M1 for correct method

A1 for correct answer

3 marks

Exam Questions

In these questions think carefully about whether you are finding a length, an area or a volume, and use the correct units.

1

A, B, C, D, E
50 m
70 m
40 m
40 m

Not to scale

In this question you must write down the units with your answers.

Work out the area of

a the square EBCD,

b the triangle ABE. [L]

2

90 m
B A
C 70 m 70 m F
D E
90 m

The diagram shows a running track. BA and DE are parallel and straight. They are each of length 90 m. BCD and EFA are semi-circular. They each have a diameter of length 70 m.

a Calculate the perimeter of the track.

b Calculate the total area inside the track. [NEAB]

3

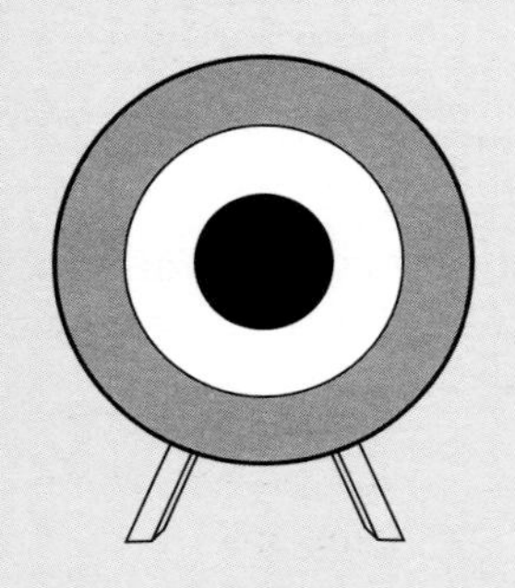

A toy archery target is made of three circles. Each circle has the same centre. The radius of the smallest circle is 8 cm.

a Find the circumference of the smallest circle.

The other two circles have radius 16 cm and radius 24 cm.

b Find the area of the outer shaded ring. [MEG]

4 A number of small cubes of side 1 cm are placed together to form a large cube of side 3 cm, as shown.

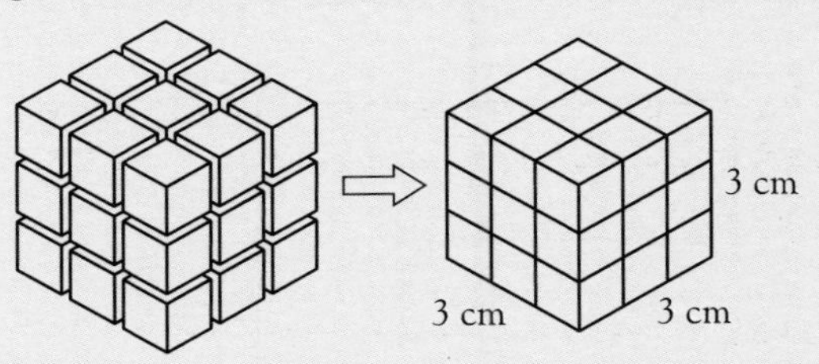

a What is the volume of the large cube?

b Draw accurately a net for the large cube.

c What is the area of the net? [SEG]

5 Boxes of drawing pins measure 4 cm by 5 cm by 2 cm. They are packed into a large box 20 cm by 15 cm by 8 cm.

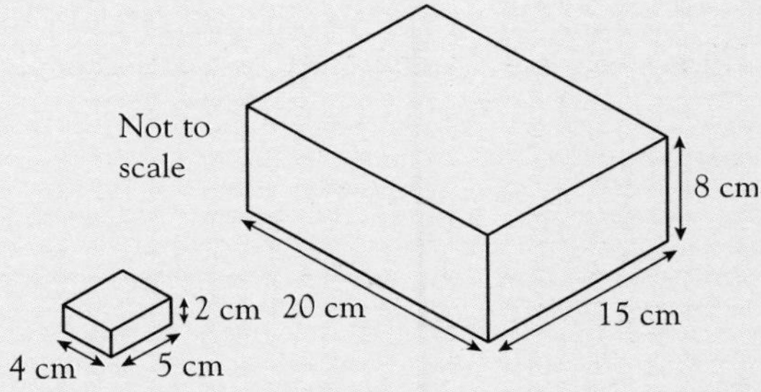

a How many boxes of drawing pins are needed so that they fill the large box exactly?

b Calculate the volume of the large box. [MEG]

6 A letter 'V' is cut from a rectangle measuring 9 cm by 8 cm, as shown.

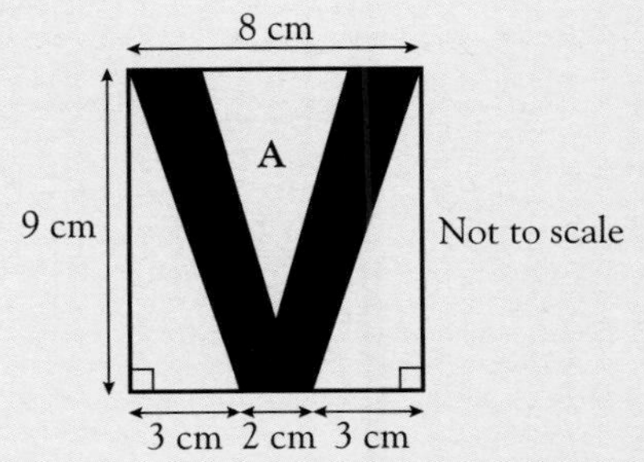

The area of the triangle marked **A** is 12 cm^2.
What is the area of the 'V' shape? [SEG]

7 The volume of a cuboid is 2565 cm^3.
The base measures 10 cm by 19 cm.
Work out *h*, the height of the cuboid.

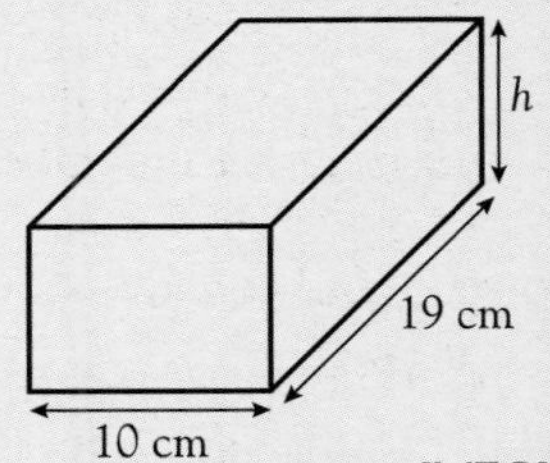

Not to scale

[MEG]

8 The diagram shows the plan of a concrete drive that Mr Fraser intends to lay in his garden.

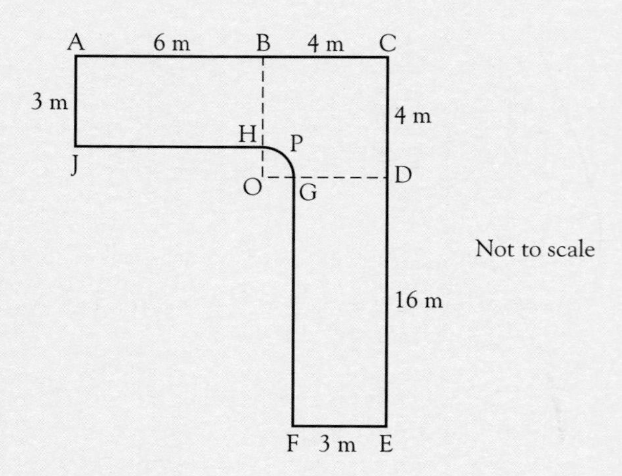

ABC and CDE are perpendicular straight lines and both DEFG and ABHJ are rectangles.
The arc HPG is part of a circle, of radius 1 m, with centre at O.

AJ = EF = 3 m
AB = 6 m, BC = CD = 4 m and DE = 16 m
OB = OD = 4 m

Taking π to be 3.14, find the total area of the drive, correct to the nearest square metre. [MEG]

9 This is a sketch plan of the walls of part of a French château:

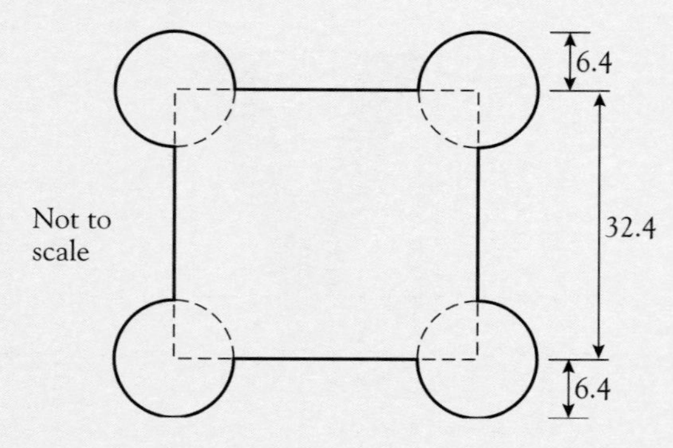

The plan has order 4 rotational symmetry. The tower at each corner is circular, with centre at the corner of the square. The lengths marked are in metres.

a Find the area of one of the complete circles.

b Find the total area shown on the plan within the walls. Include the towers. [MEG]

10 The diagram below shows the net for a rectangular box.

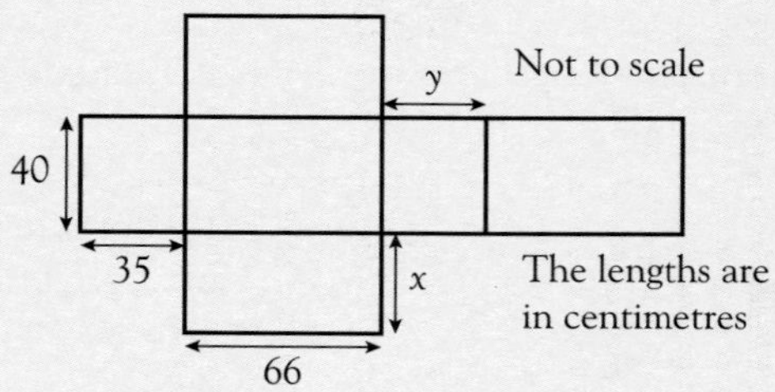

a Write down
  i the value of x,
  ii the value of y.

b Write down the number of edges of the box.

c Calculate the volume, in cubic centimetres, of the box. [MEG]

11

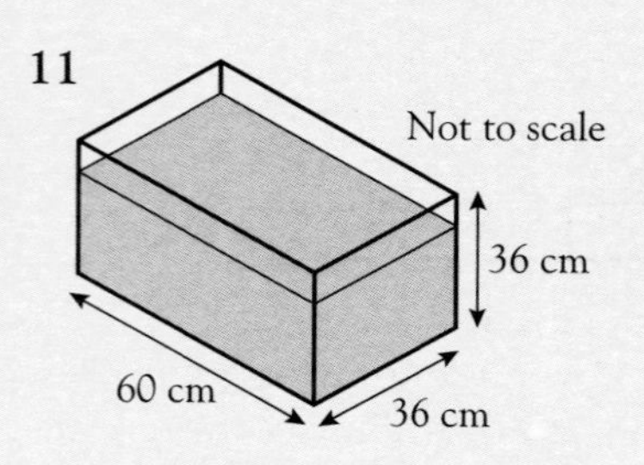

A rectangular tank has length = 60 cm, breadth = 36 cm and height = 36 cm.
It does not have a lid.

a Draw a sketch of a net of the tank. Show the dimensions of the tank in your sketch.

b Water is poured into the tank until it is three-quarters full.
  i Calculate the depth of water in the tank.
  ii Calculate, in cm^3, the volume of water in the tank.
  iii Express your answer to part **b ii** in litres. [MEG]

12

The stand for the medal winners at an athletics meeting is made up of cubical blocks, as shown in the diagram. Each block is of side 1 foot.

a What is the volume of
  i the section marked 3,
  ii the section marked 2,
  iii the section marked 1?

b Calculate the total volume of the stand.

c Work out the area of the front of the whole stand. [MEG]

TRANSFORMATIONS I

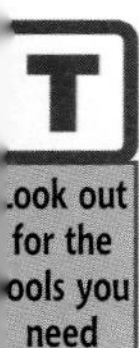

In this chapter you will learn about these transformations:
- **translation**
- **reflection**
- **rotation**
- **enlargement**

A **transformation** happens to a shape when its position, shape, size, or any combination of these, is changed.

Translations, **reflections** and **rotations** are types of transformation that change the position of a shape, but do not change the size. An **enlargement** changes the size of a shape.

There are some words associated with transformations that you should know:

Object – the original shape before you do anything to it.
Image – the shape after a transformation has been carried out.
Map or **mapped** – the word often used instead of 'move(d)' or 'transform(ed)'.
Displacement – the amount a shape has been moved.
Congruent – the object and image are congruent if they are exactly the same size and shape.

You need to be able to identify the transformations and also be able to describe them so that they can be carried out.

Translation

If every point in a shape is moved exactly the same distance in the same direction, a **translation** has been carried out. The image is identical to the object and so the shapes are **congruent**.

The shape has not been turned or 'flipped' over.

To describe a translation you need to give its direction and say by how much it has moved.

This is most easily done using a grid and indicating how far left/right and up/down the points of the shape have moved.

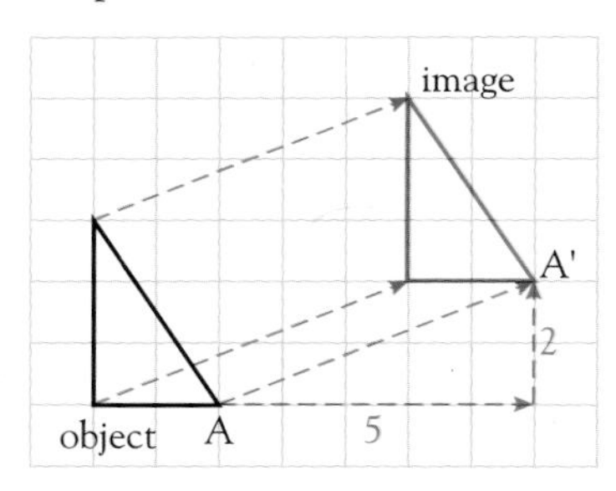

The point A on the triangle has moved 5 squares to the right and 2 squares up to A′.

Every other point has moved the same amount, as shown by the blue arrows.

A move to the **right** is given as a **positive** number, a move to the **left** as a **negative** number.

A move **up** is **positive**, **down** is **negative**.

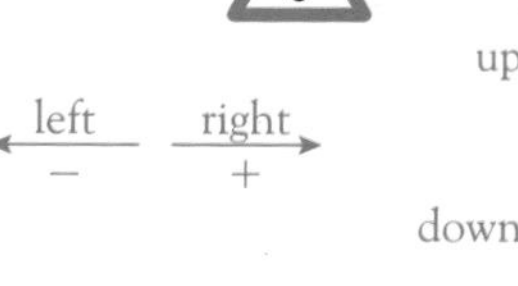

TRANSLATION

Instead of writing a sentence to describe a translation, it is often given in the form of a **column vector**.

A column vector is written like a pair of co-ordinates, the x value first then the y value, but instead of being written horizontally it is written as a **column**, with the left/right displacement on top and the up/down displacement underneath.

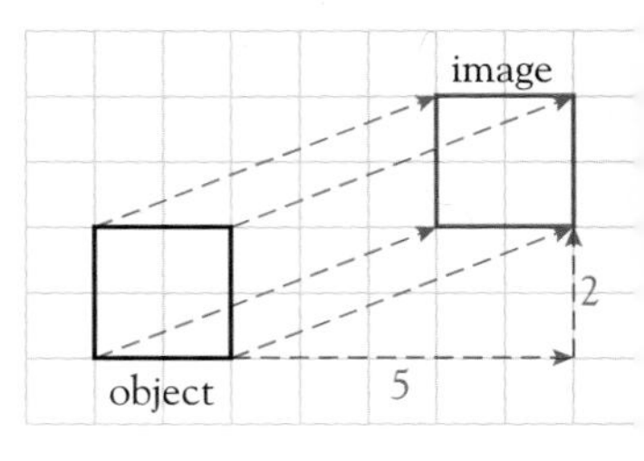

$\begin{pmatrix} 5 \\ 2 \end{pmatrix}$ is a translation of **5** to the **right** and **2 up**.

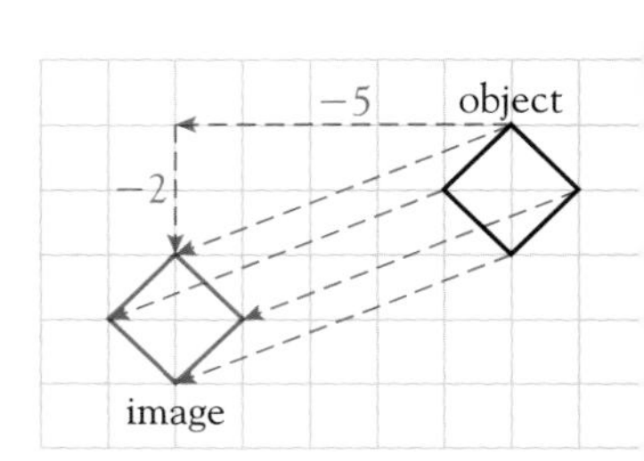

$\begin{pmatrix} -5 \\ -2 \end{pmatrix}$ is a translation of **5** to the **left** and **2 down**.

- A translation is a transformation in which all points in the shape being translated move exactly the same distance in the same direction.
- A translation is written as a column vector $\begin{pmatrix} a \\ b \end{pmatrix}$, where a is the movement in the x direction and b is the movement in the y direction. (a and b can be positive, negative or zero.)
- In a translation the image is identical to the object. The shapes are congruent. Also the object is neither turned nor flipped over.
- All lines in the image are equal in length and parallel to the corresponding lines in the object.

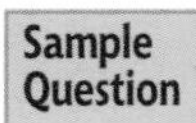

1 a The shape ABCD has been translated to position PQRS as shown on the grid. Write down the column vector that describes the translation.

b PQRS is then translated to WXYZ. Write down the column vector for this translation.

c Write down the column vector that would translate ABCD directly to WXYZ.

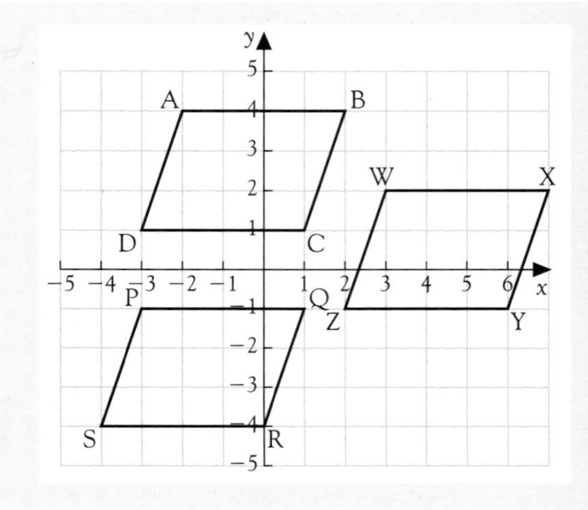

TRANSLATION

Answer

a

- Point A has co-ordinates $(-2, 4)$. It has been translated to point P whose co-ordinates are $(-3, -1)$. The x co-ordinate has changed from -2 to -3. This is a move to the **left** of 1 unit. The y co-ordinate has changed from 4 to -1. This is a move **down** of 5 units.
- Points B, C and D have all moved the same distance as point A, in the same direction.

  The translation ABCD to PQRS is a displacement with column vector $\begin{pmatrix}-1\\-5\end{pmatrix}$

b

- Point P has co-ordinates $(-3, -1)$ and has moved to point W with co-ordinates $(3, 2)$. This is a translation of 6 to the **right** and 3 **up**.

  The translation PQRS to WXYZ is a displacement with column vector $\begin{pmatrix}6\\3\end{pmatrix}$

c

- Point A $(-2, 4)$ would move to point W $(3, 2)$. This is a move of 5 to the **right** and 2 **down**.

  The translation ABCD to WXYZ is a displacement with column vector $\begin{pmatrix}5\\-2\end{pmatrix}$

Sample Question 2

For each of the following transformations of the triangle A, state whether the new position is a translation or not and, if it is, give the column vector that describes the translation.

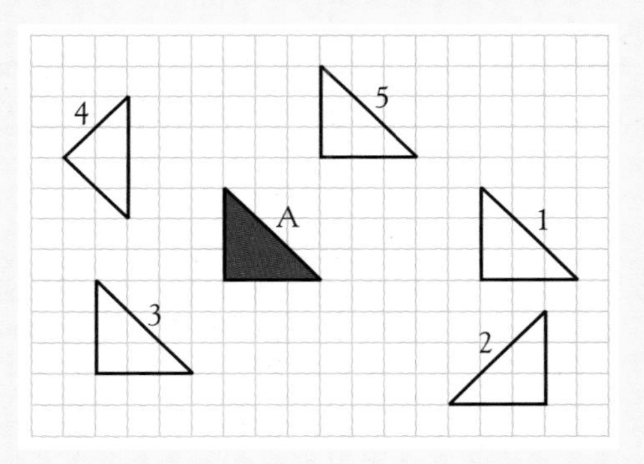

Answer

- Position 1: Each point on A is displaced 8 in the x direction and 0 in the y direction. The triangle has not been flipped or turned.

  This is a translation with column vector $\begin{pmatrix}8\\0\end{pmatrix}$

Remember to put the x displacement on top and the y displacement underneath.

- Position 2: The triangle has been flipped. This is not a translation.
- Position 3: Each point on A has been displaced -4 in the x direction and -3 in the y direction. The triangle has not been flipped or turned.

  This is a translation with column vector $\begin{pmatrix}-4\\-3\end{pmatrix}$

- Position 4: The triangle has been turned, so this is not a translation.
- Position 5: Each point on A has been displaced 3 in the x direction and 4 in the y direction. The triangle has not been flipped or turned.

  This is a translation with column vector $\begin{pmatrix}3\\4\end{pmatrix}$

TRANSLATION

Exercise 8.1

1 Make a copy of the grid on squared paper and draw the pentagon ABCDE.

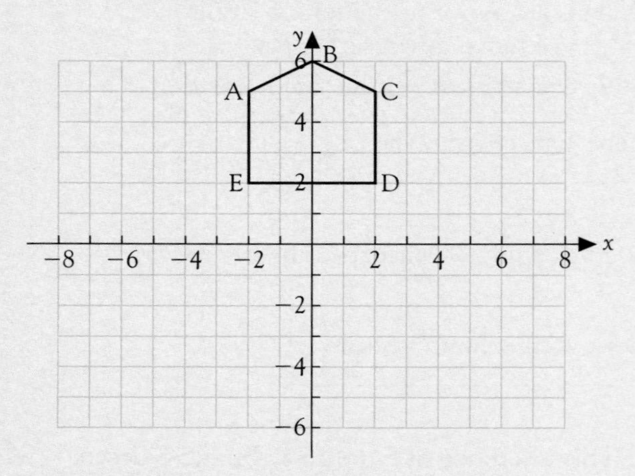

a Draw the image of ABCDE under the translation $\begin{pmatrix} 3 \\ -6 \end{pmatrix}$. Label the image PQRST and write down the co-ordinates of point R.

b ABCDE is translated so that E moves to $(-8, -4)$. Draw the image and write down the column vector of the translation.

2 Triangle ABC has been mapped onto triangle PQR by a translation.

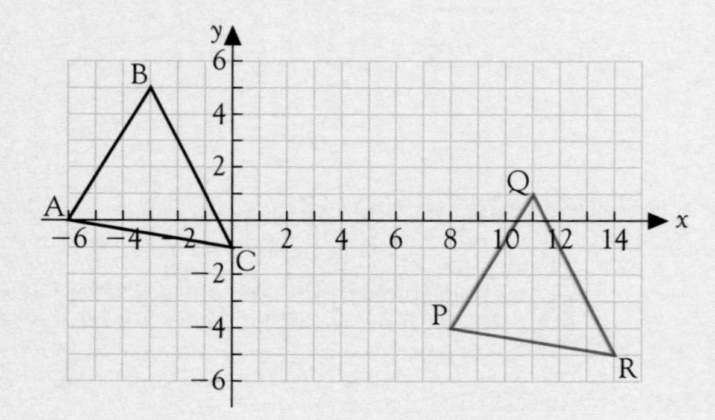

a Write down the column vector that maps ABC onto PQR.

b Copy the diagram and draw the image of PQR under the translation $\begin{pmatrix} -8 \\ 3 \end{pmatrix}$. Label it XYZ.

c Write down the column vector used to translate XYZ to ABC.

3 Use the grid below to help you decide which of the following transformations are translations and which are not.

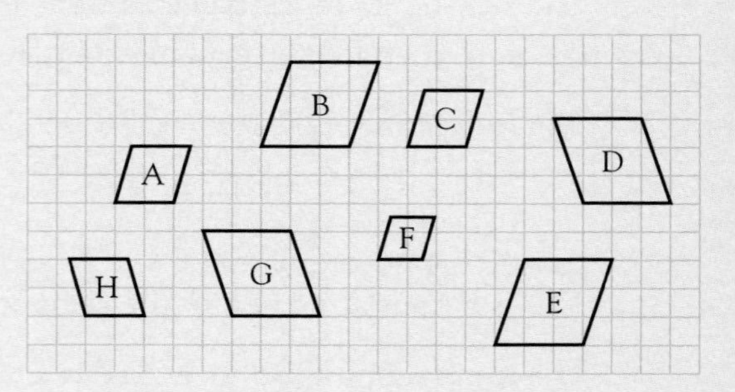

Copy down each transformation and write 'Yes' if it is a translation, giving its column vector. Write 'No' if the mapping is not a translation.

a A → C b B → H c F → A d D → G
e B → E f H → D g G → D h H → A

4 Using a grid to help you, write down the co-ordinates of the image of the following object points, after being translated by the column vector given.

a $(2, 1)$ $\begin{pmatrix} 2 \\ 0 \end{pmatrix}$

b $(4, 7)$ $\begin{pmatrix} -1 \\ 2 \end{pmatrix}$

c $(-1, 3)$ $\begin{pmatrix} 3 \\ 3 \end{pmatrix}$

d $(-1, -1)$ $\begin{pmatrix} -3 \\ -8 \end{pmatrix}$

e $(10, -5)$ $\begin{pmatrix} 0 \\ 6 \end{pmatrix}$

f $(-3, -7)$ $\begin{pmatrix} -2 \\ 9 \end{pmatrix}$

5 Using a grid to help you, write down the column vector that translates:

a $(3, -1)$ to $(4, -5)$

b $(0, 2)$ to $(-3, 6)$

c $(1, 1)$ to $(0, 7)$

d $(-3, -4)$ to $(5, -6)$

e $(7, -2)$ to $(-7, 2)$

REFLECTION

Reflection

A reflection occurs when a shape is folded or 'flipped' over a mirror line. The image is identical to the object in shape and size, so the object and image are congruent.

Here are two examples of reflections. Notice that each point on the image is the same perpendicular distance behind the mirror line as the corresponding point is in front of it.

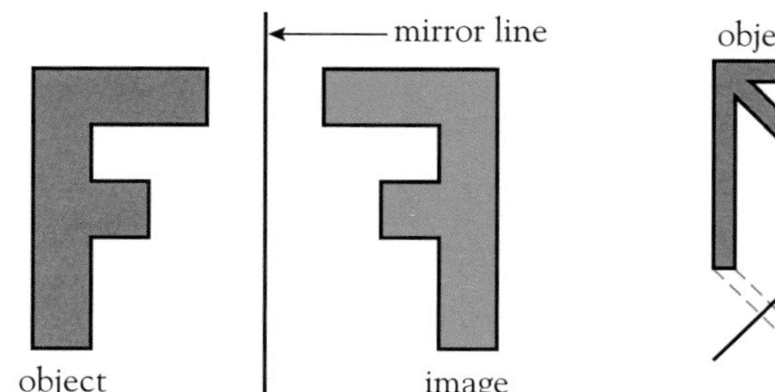

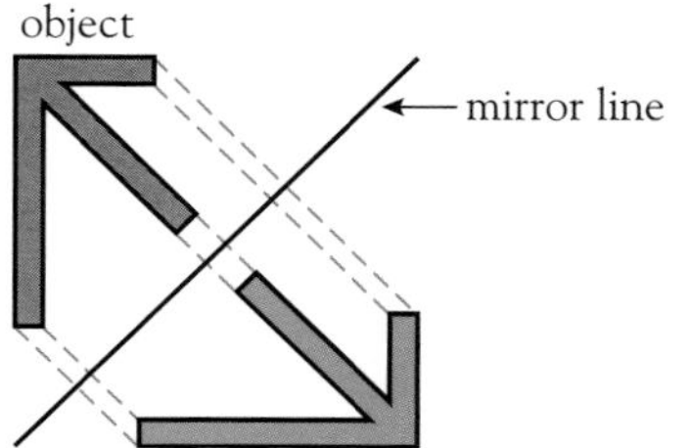

The mirror line is a line of symmetry.

To describe a reflection you must give a description of the mirror line.
If you have a mirror line on a grid, you can give the equation of the mirror line.

Look back at Chapter 6 to remind yourself about straight lines.

- In a reflection, all points on the image are the same **perpendicular** distance behind the mirror line as the corresponding points on the object are in front of the mirror line.
- To describe a reflection you must describe the mirror line, usually by giving its **equation.**
- In a reflection, only the position of the object is changed. There is no effect on the shape or size of the object. The object and the image are congruent.

FACT SHEET 3
pages 67–69

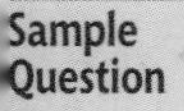

Sample Question 3

On the grid, draw the image of the triangle ABC after a reflection in the line $y = 2$.

Label the image A′B′C′.

Answer

- Draw the line $y = 2$.
- Point A is 1 unit from the line $y = 2$, so A′ (the image of A) will be 1 unit from $y = 2$ but on the other side of the mirror line.
- Similarly, point C′ is 1 unit from the mirror line on the opposite side of it.
- Point B is 4 units from the mirror line, so B′ is 4 units from the line on the other side.
- Plot A′, B′ and C′, and join the points to form the image triangle A′B′C′.

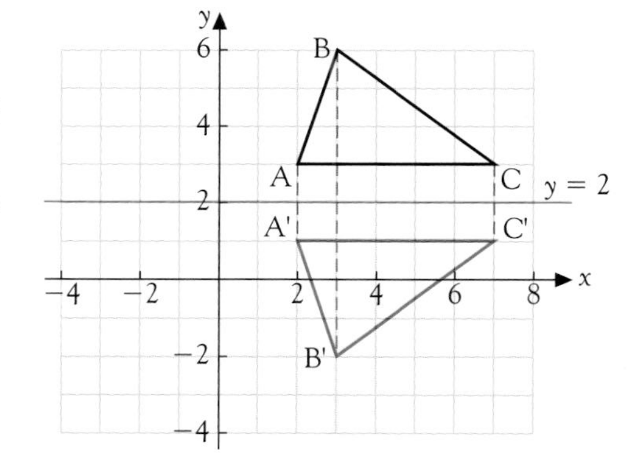

REFLECTION

Sample Question 4

On a grid, plot the points with co-ordinates P(−1, 2), Q(−3, 4), R(−2, −2) and S(1, −3). Join the points, in order, with straight lines to form the shape PQRS.

Draw and label the line $x = 1$.

Reflect the shape PQRS in the line $x = 1$. Label the image with the letters P′Q′R′S′.

Answer

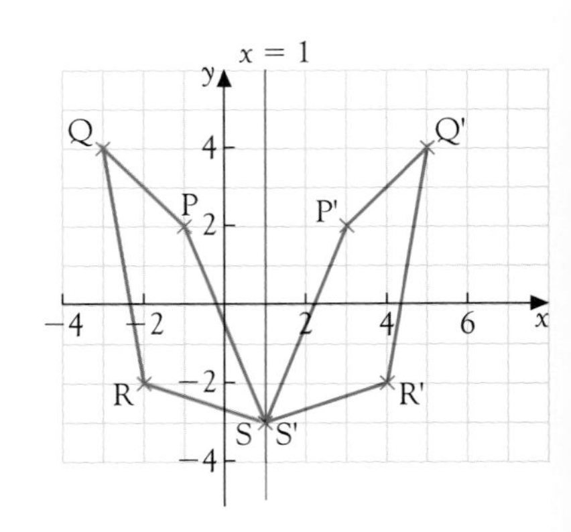

- Draw a suitable grid to enable you to plot the points.
- Plot the points carefully, labelling each one with its letter. Join the points in order PQRS and back to P.
- Draw the line $x = 1$ (drawn in red here).
- In turn, draw the image of each point. Label each point as you draw it.
- Join the image points with straight lines, in order.

Sample Question 5

The triangle ABC has been reflected in a mirror line. The image of ABC is triangle A′B′C′.

Identify the mirror line and write down its equation.

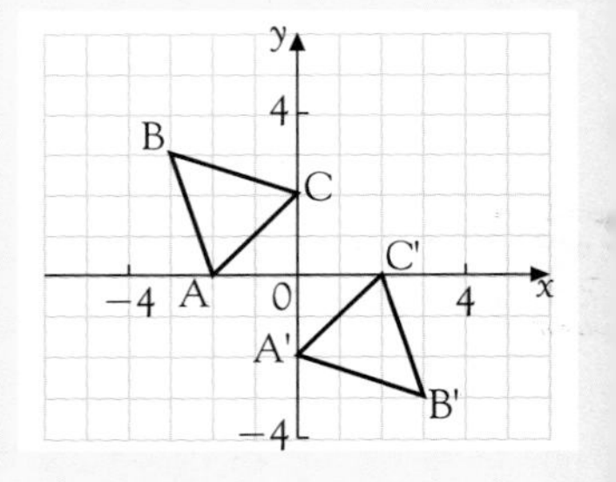

Answer

A **perpendicular bisector** cuts a line in half at right angles.

- Using straight lines, join A to A′, B to B′ and C to C′.
- Mark half-way between each pair of points. Join these half-way marks with a straight line. This is the mirror line. It is the **perpendicular bisector** of the lines joining the corresponding points.
- The co-ordinates of the half-way marks are (−1, −1), (0, 0) and (1, 1). Notice that for all three points the x co-ordinate is the same as the y co-ordinate. This means that the equation of this line is $y = x$.

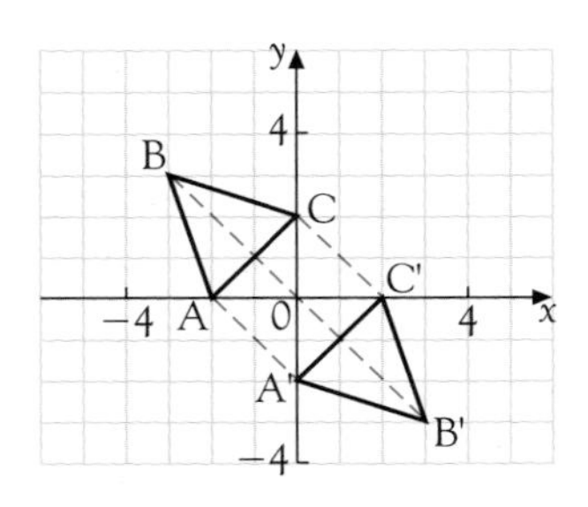

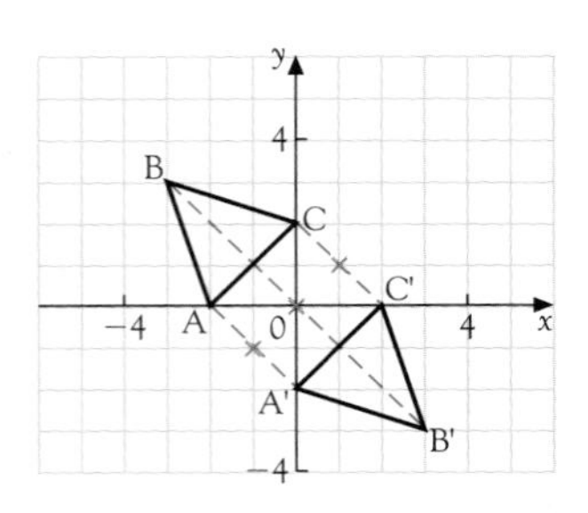

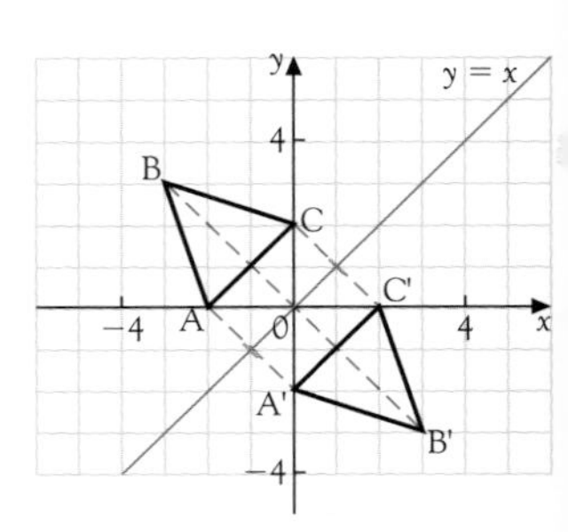

T

- The mirror line is the perpendicular bisector of the line joining corresponding object and image points.

REFLECTION

Exercise 8.2

Copy each shape onto squared paper and reflect it in the mirror line, shown dotted.

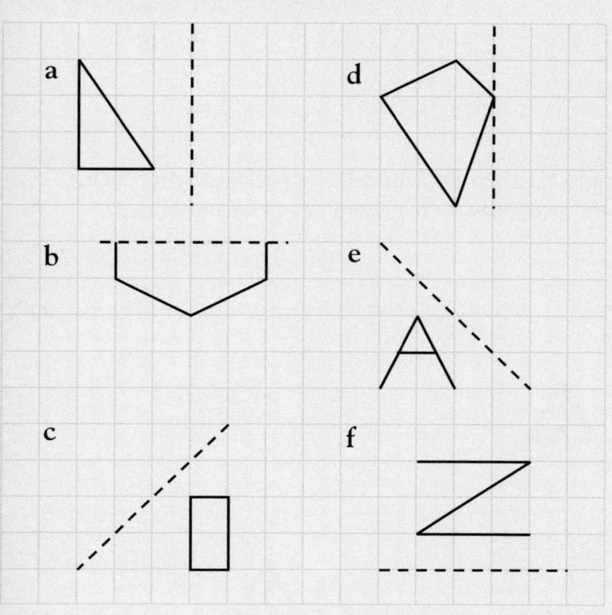

Copy each shape onto squared paper and draw dotted lines to show all the lines of symmetry.

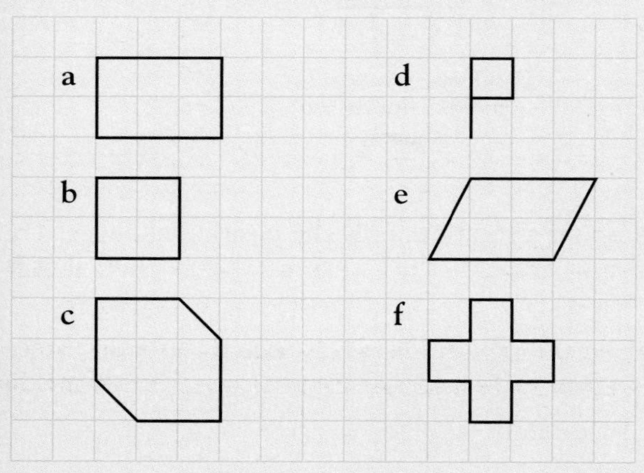

Copy the grid below onto squared paper. Draw the shape labelled T.

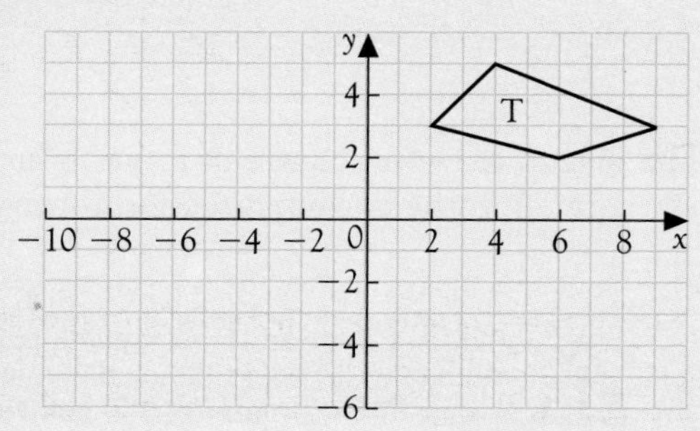

a Reflect the shape T in
  i the y axis and label the image T_1
  ii the x axis and label the image T_2.

b Reflect T_1 in the line $y = 2$. Label the image T_3.

c Reflect T_2 in the line $x = 1$. Label the image T_4.

4 Copy the grid below onto squared paper. Draw the shape labelled S.

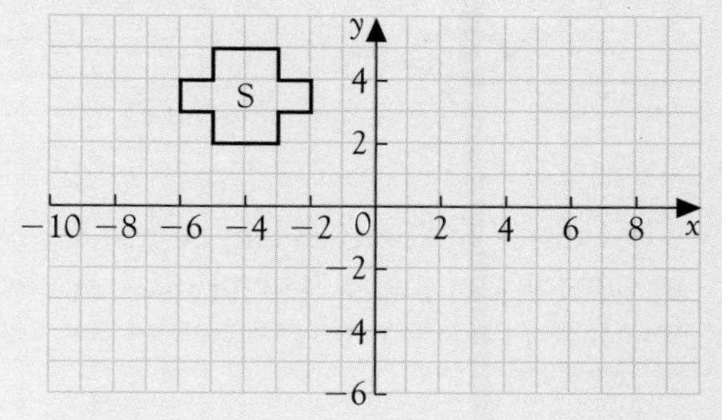

a Reflect the shape S in the line $y = 1$. Label the image P.

b Reflect the shape P in the line $y = -x$. Label the image Q.

c Reflect the shape Q in the y axis. Label the image R.

d Reflect the shape R in the line $x = 3$. Label the image T.

e Reflect the shape T in the x axis. Label the image V.

5

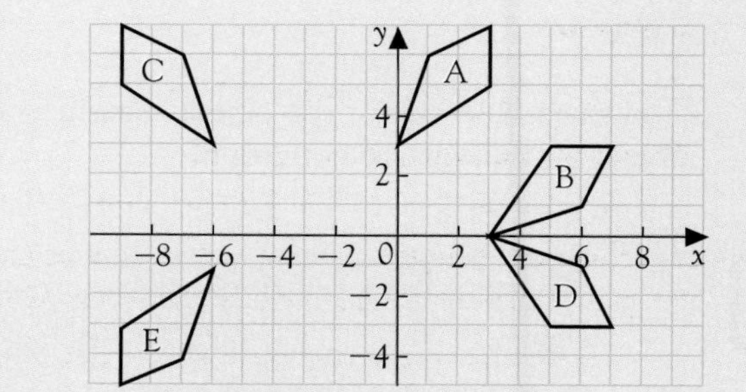

Write down the equation of the mirror line for the following reflections:

a Shape A mapped onto shape B.

b Shape A mapped onto shape C.

c Shape B mapped onto shape D.

d Shape C mapped onto shape E.

6 Using a grid to help you if necessary, write down the co-ordinates of the image of the point (1, 4) after a reflection in each of the following lines:

a $y = x$ **b** $x = 2$ **c** $y = -x$

d $y = 3$ **e** $y = 4$ **f** $x = 0$

ROTATION

Rotation

A rotation is a transformation in which an object turns through an angle about a fixed point called the **centre of rotation**.

This is an example of a rotation.

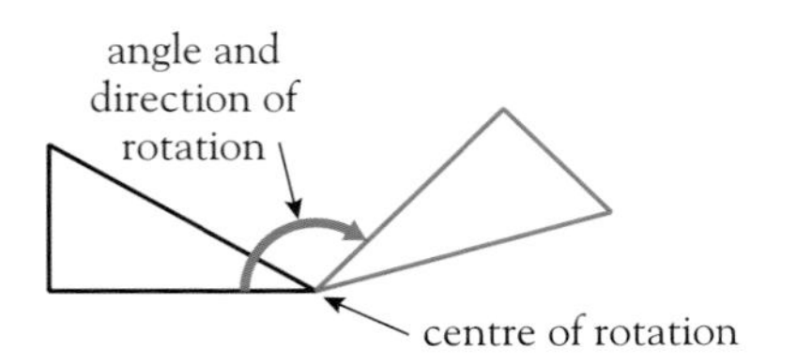

The triangle has been rotated in a clockwise direction through an angle about a centre of rotation on the triangle.

The **centre of rotation** can be inside, on, or outside the shape. The position of the image depends on the position of the centre of rotation.

It is the only point in a rotation that does not change its position.

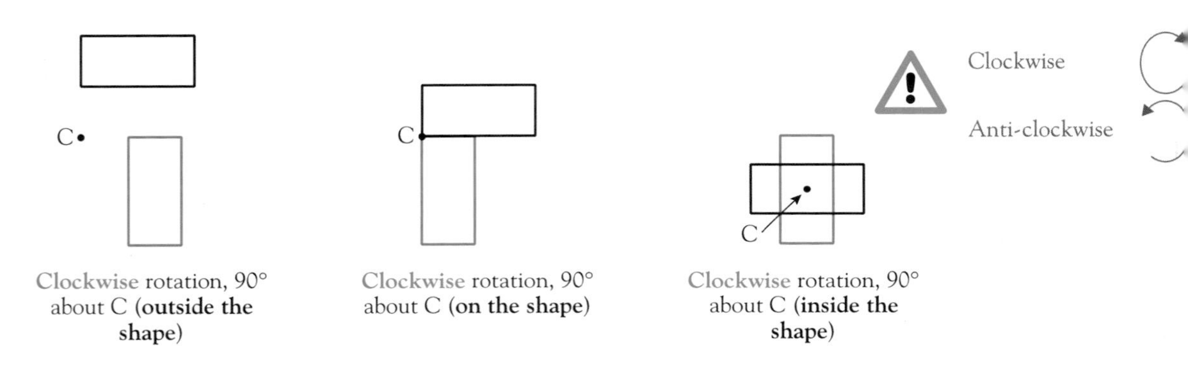

Clockwise rotation, 90° about C **(outside the shape)**

Clockwise rotation, 90° about C **(on the shape)**

Clockwise rotation, 90° about C **(inside the shape)**

The shape and size of the object remain unchanged under a rotation, only the position changes. The object and image are congruent.

To describe a rotation fully you must give

- the angle through which the object is to be rotated
- the direction (clockwise or anti-clockwise) that the object is to be rotated
- the position of the centre of rotation; on a grid, the centre of rotation is indicated by giving its co-ordinates.

It is often best to show a rotation using **tracing paper**. The object, axes and centre of rotation are drawn on the **tracing paper**. The object can then be easily rotated in the correct direction through the required angle.

You are allowed to use tracing paper in an exam. It should be available in the exam room. If you need it, **ask for it!**

ROTATION

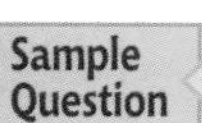

6 On the grid, rotate triangle ABC clockwise through 90° about the origin.

Label the rotated triangle A′B′C′.

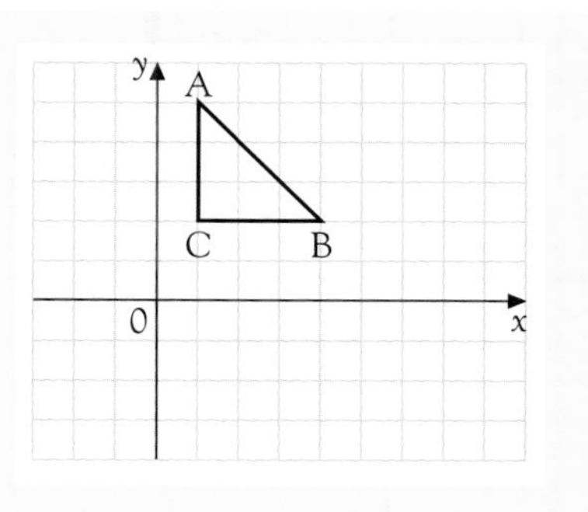

Answer (using tracing paper)

- Using tracing paper, trace the triangle ABC and axes (1).
- With the point of your compasses at the origin (0, 0), rotate the tracing paper clockwise (2) and (3) until your y axis is over the position of the x axis on the grid (4).
- Mark the new position of the triangle ABC and draw the triangle onto the grid. Label the triangle A′B′C′.

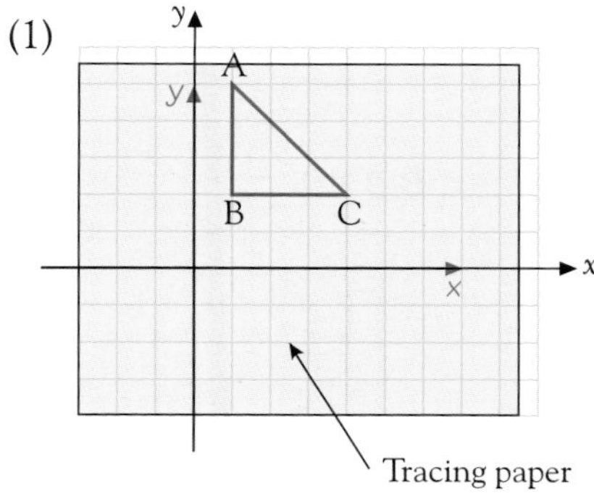

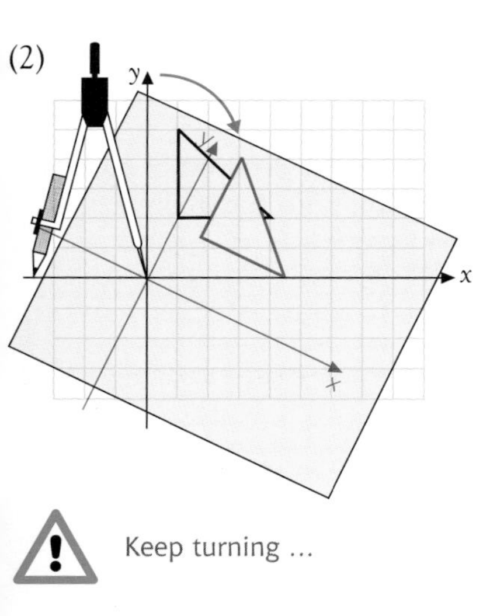

Keep turning ...

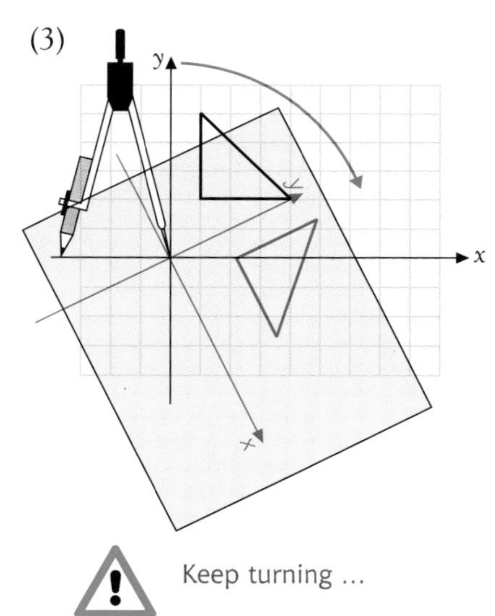

Keep turning ...

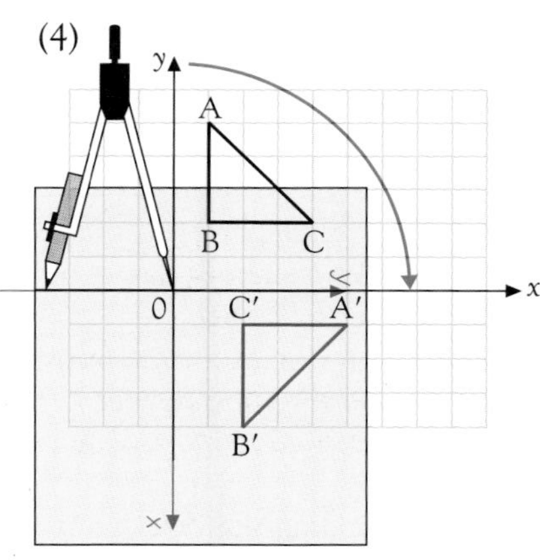

The triangle has now turned through 90°

Answer (without tracing paper)

- From the centre of rotation, draw a straight line to each corner of the triangle.
- Taking one line at a time, using a protractor or angle measurer, measure and draw a 90° angle with the centre of rotation as the corner of the angle.

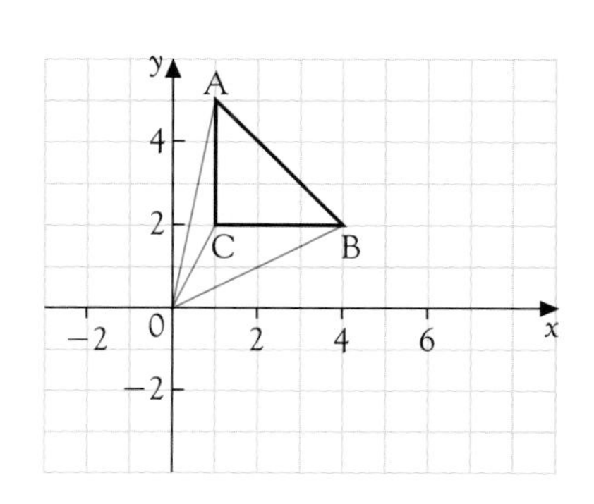

ROTATION

- Measure the distance from the centre of rotation to each corner and mark this distance on the corresponding angle lines that you have just drawn.
- Join the image points to form the triangle A′B′C′ and label the triangle.

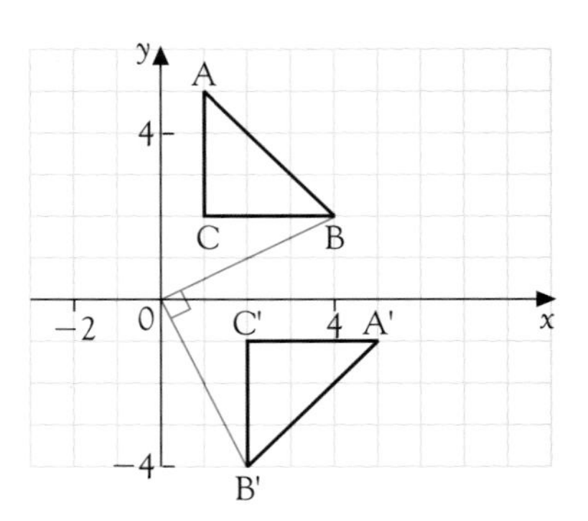

Exercise 8.3

1

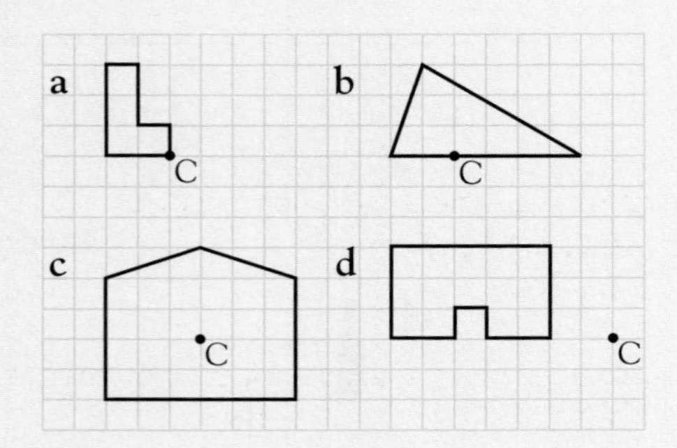

Copy each shape onto squared paper and rotate it 90° clockwise about the given centre of rotation, C. Shade the image.
(Do them one at a time and leave space around the shape for your answer.)

2 Copy the diagram below onto squared paper.

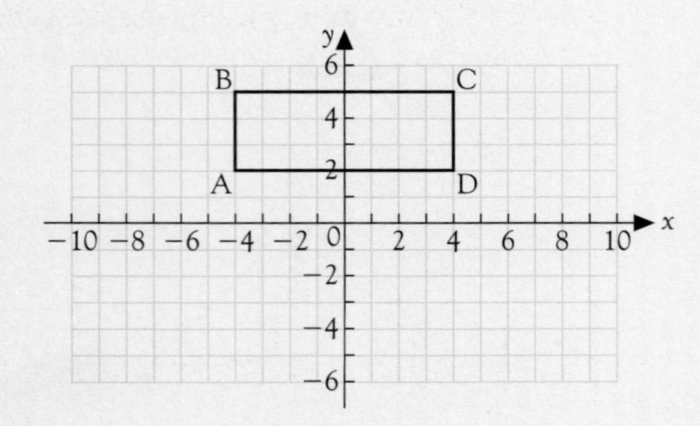

Rotate ABCD 180° with the origin as centre of rotation. Label the image A′B′C′D′.

3 On a grid, plot these points: P(3, 1), Q(1, 5), R(3, 6), S(4, 4). Join the points to form the shape PQRS.

Rotate PQRS 270° anticlockwise with centre of rotation (1, 1). Label the image P′Q′R′S′.

Describe an alternative rotation that would map PQRS onto P′Q′R′S′.

4 The diagram below shows the position of rectangle ABCD. Copy the diagram onto squared paper.

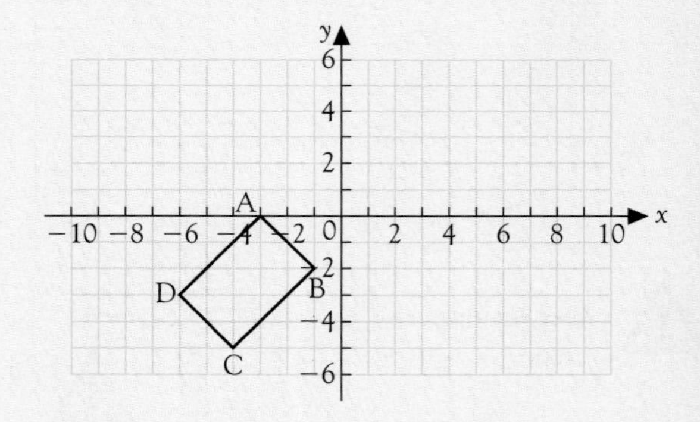

a Rotate rectangle ABCD clockwise 90° about the point (−1, −1). Label the image $A_1B_1C_1D_1$.

b Write down the co-ordinates of point D_1.

c Rotate $A_1B_1C_1D_1$ 180° about the origin. Label the image $A_2B_2C_2D_2$.

d Write down the co-ordinates of the point D_2.

e Rotate $A_2B_2C_2D_2$ anti-clockwise 90° about the point (1, −2). Label the image $A_3B_3C_3D_3$.

f Write down the co-ordinates of point D_3.

Enlargement

When the size of a shape is changed, but it is otherwise unaltered, the shape has been **enlarged**. This does not necessarily mean that the image is bigger; it might be smaller.

To define an enlargement you need to state a **scale factor**.

- A scale factor bigger than 1 makes the shape bigger.
- A fractional scale factor between 0 and 1 makes the shape smaller.

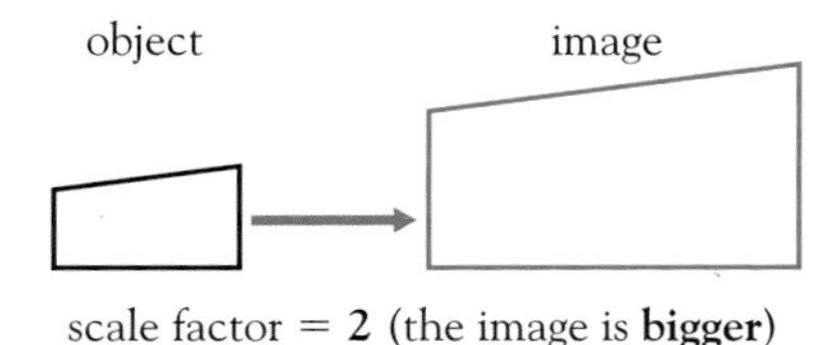

scale factor = **2** (the image is **bigger**)

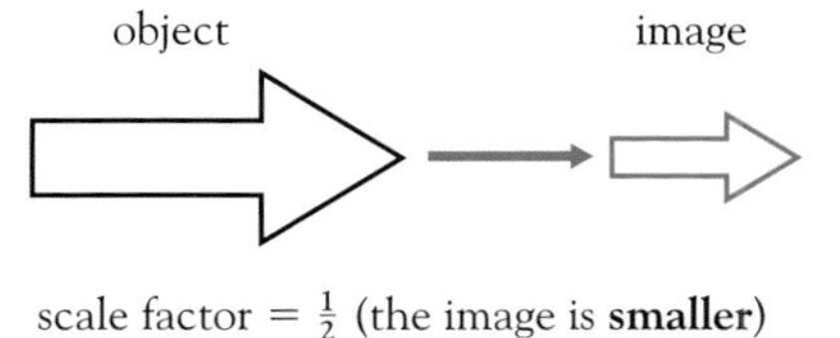

scale factor = $\frac{1}{2}$ (the image is **smaller**)

The shapes are not congruent. They are said to be **similar**.

In an enlargement all lines are enlarged by the same amount. This means that any pair of corresponding lengths in the image and the object are **in the same ratio**.
This ratio is the **scale factor** of the enlargement.

If you know that one shape is an enlargement of another, you can find the scale factor by dividing a length on the image by the corresponding length on the object.

For example:

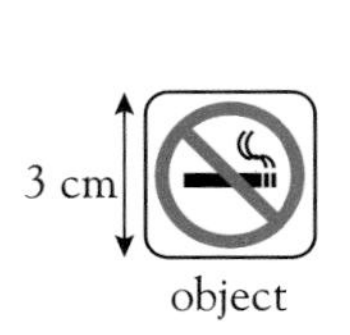

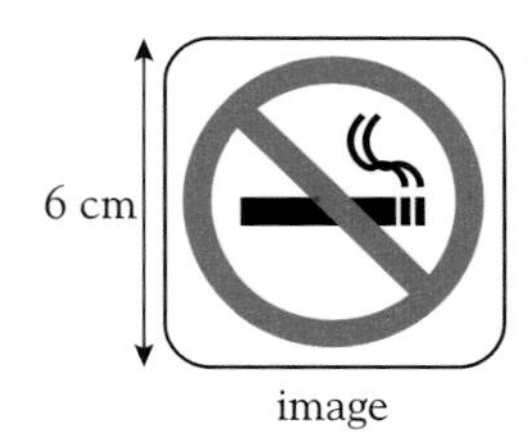

CHAPTER 10
Number II

To find the scale factor,
$6 \div 3 = 2$
Scale factor = 2

If you know the scale factor, you can draw the image of the object by multiplying each length on the object by the scale factor to find the corresponding length on the image.

For example:

To enlarge shape A by scale factor $1\frac{1}{2}$,
length on object × scale factor = length on image

$2 \times 1\frac{1}{2} = 3$

$4 \times 1\frac{1}{2} = 6$

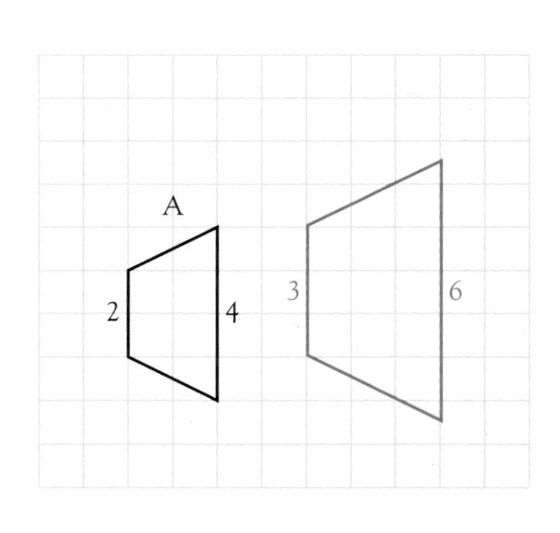

ENLARGEMENT

Sometimes you have to enlarge a shape using a **centre of enlargement** which can be inside, on, or outside the shape. The position of the image depends on the positioning of the centre of enlargement.

Centre of enlargement, C, outside the shape

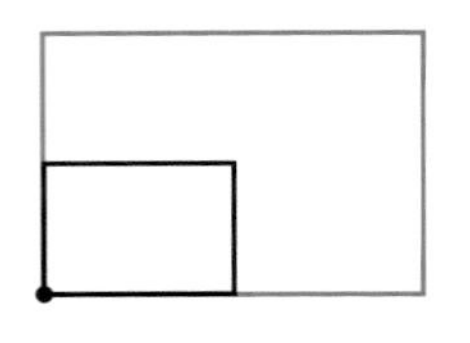

Centre of enlargement, C, on the shape

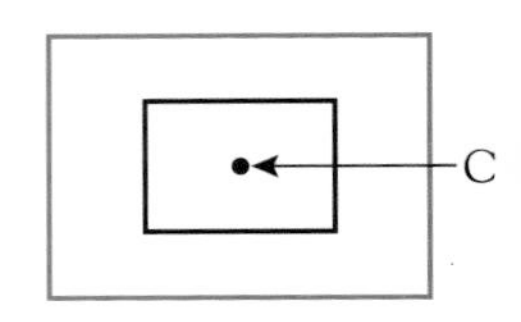

Centre of enlargement, C, inside the shape

To enlarge a shape from a given centre of enlargement it is useful to use the '**ray**' method, as in the following example.

Sample Question 7

Enlarge the shape ABCD by a scale factor of 2, with O as the centre of enlargement.

Label your enlargement A′B′C′D′.

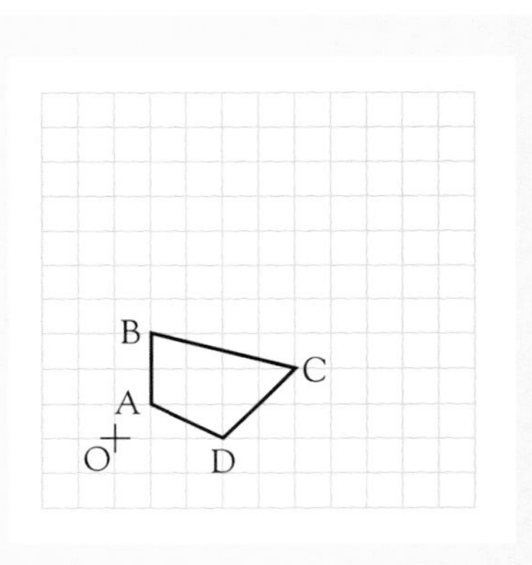

Answer

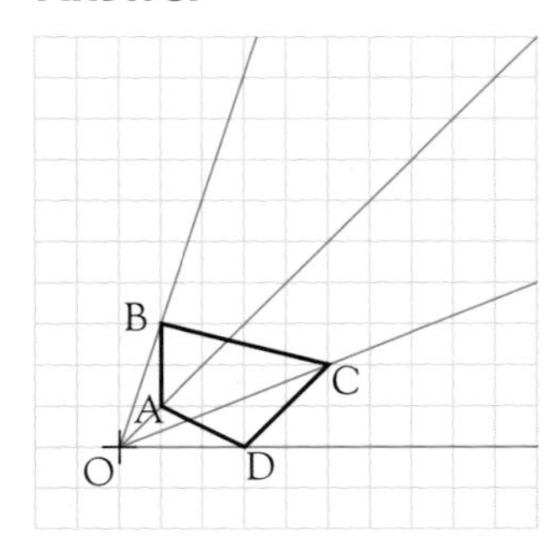

- From O, draw 'rays' through each point A, B, C and D.

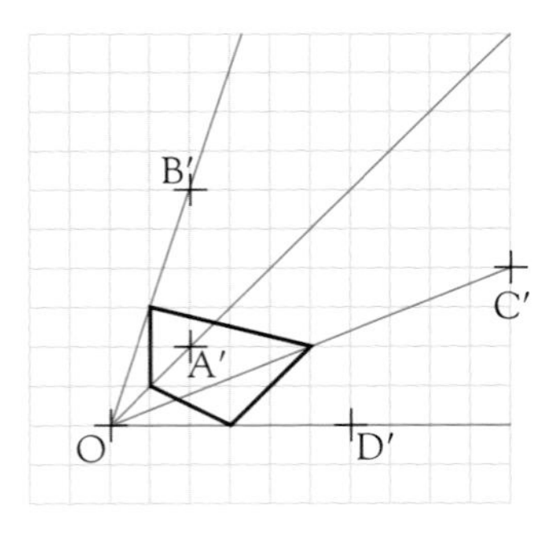

- Find the positions of A′, B′, C′ and D′ by using:
  OA′ = 2OA
  OB′ = 2OB
  OC′ = 2OC
  OD′ = 2OD

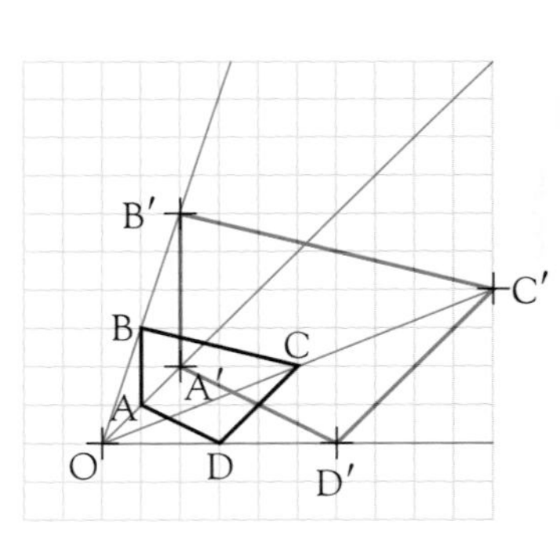

- Join A′, B′, C′ and D′ to form the image of ABCD.

ENLARGEMENT

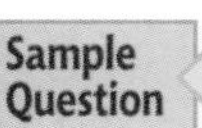

Sample Question 8

Enlarge the triangle ABC by scale factor $\frac{1}{2}$ with O as the centre of enlargement.

Label your enlargement A′B′C′.

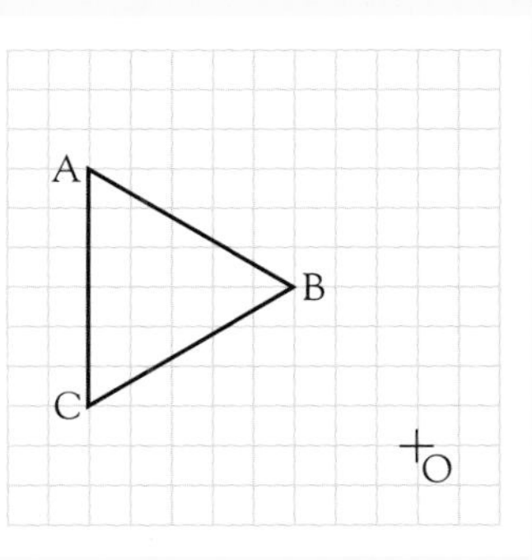

Answer

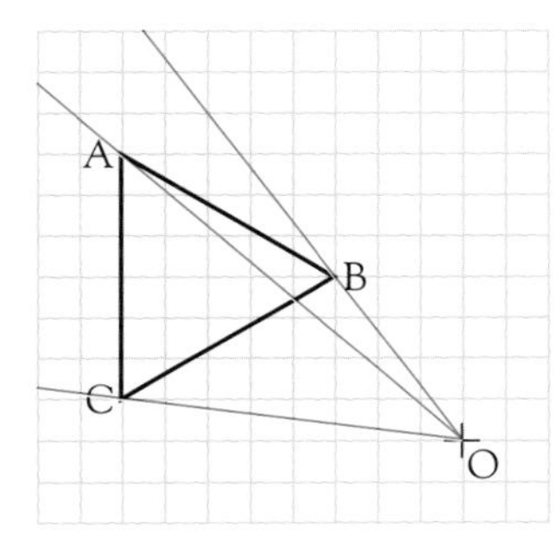

- Join O to each point A, B and C.

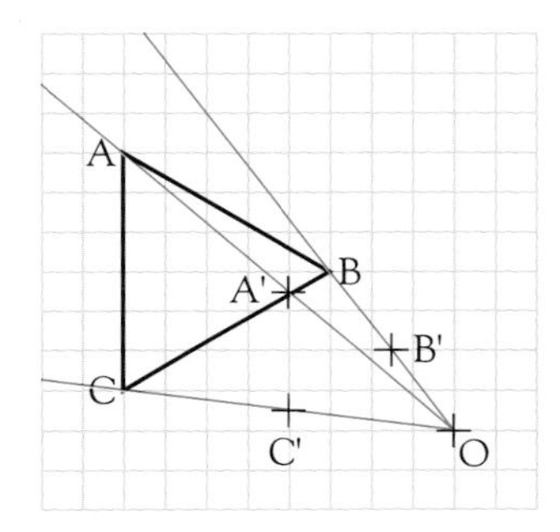

- Locate the positions of A', B' and C' using:

$OA' = \frac{1}{2}OA$
$OB' = \frac{1}{2}OB$
$OC' = \frac{1}{2}OC$

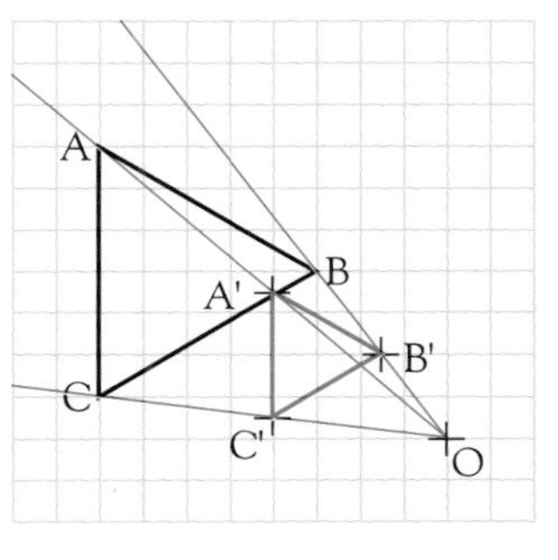

- Join A', B' and C' to form the image of ABC.

The image is smaller than the object when the scale factor is $\frac{1}{2}$

For an enlargement

- there is no effect on the shape of the object but the size and position of the object are both affected
- any pair of corresponding lengths on the image and the object are in the same ratio; this ratio is called the scale factor
- a fractional scale factor between 0 and 1 makes the image smaller than the object
- a scale factor bigger than 1 makes the image bigger than the object
- where appropriate you must state the position of the centre of enlargement.

ENLARGEMENT

Exercise 8.4

1 Copy each shape onto squared paper and enlarge it using the scale factor given.

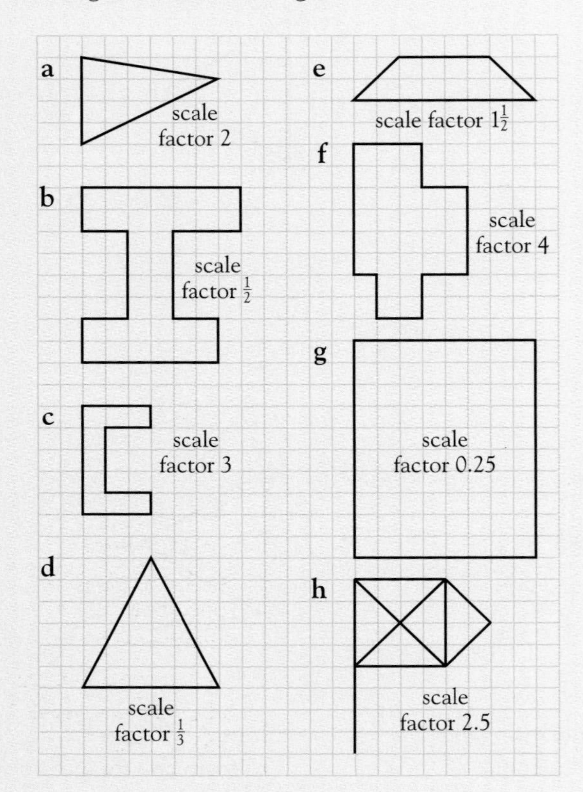

2 For each of the following enlargements:

a calculate the scale factor of the enlargement,

b calculate the length marked with a letter.

DO NOT take any measurements from the diagrams. They are not drawn to scale. You do not need to draw the diagrams.

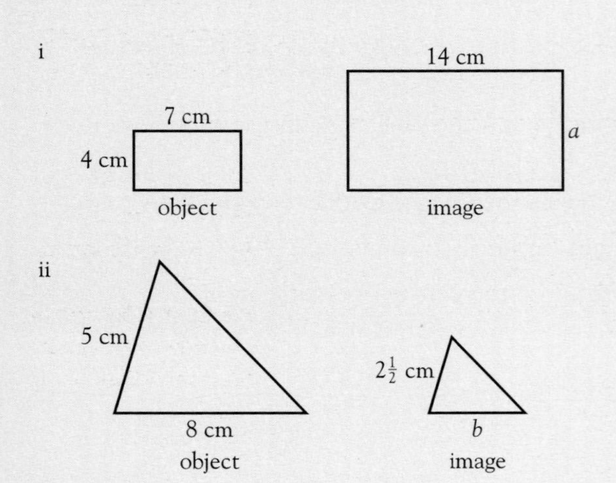

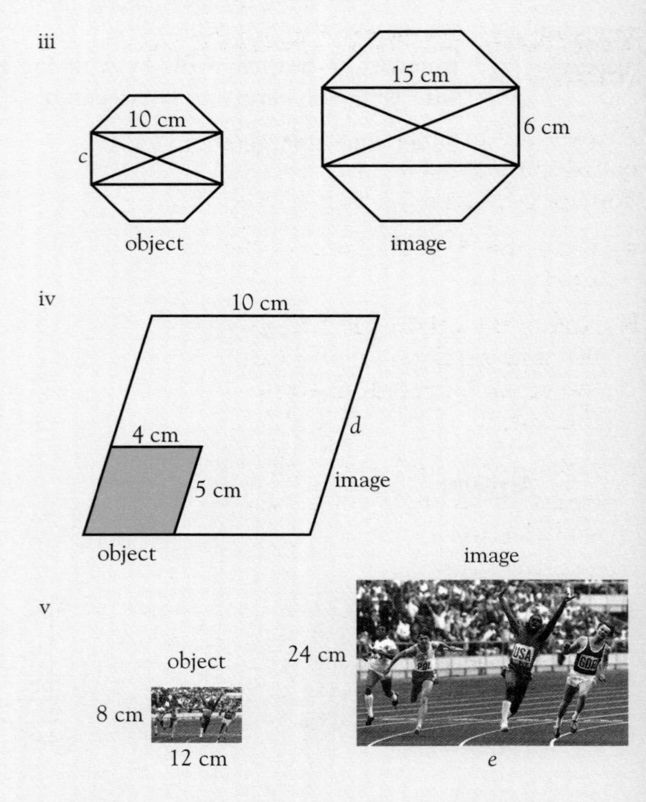

v

object

8 cm

12 cm

image

24 cm

e

3 Copy each of these shapes onto squared paper. Enlarge each shape using the given scale factor and centre of enlargement, C.

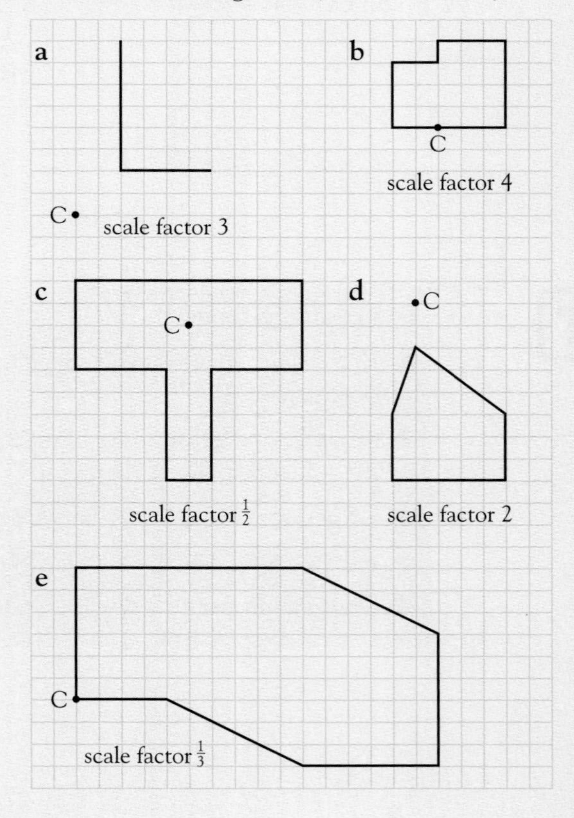

EXAM QUESTIONS

Worked Exam Question
[MEG]

Answer all three parts of this question on the diagram.
You are advised to use a pencil.

a Draw the reflection of F in the x axis.

b Rotate the original F through 90° anti-clockwise with O as the centre of rotation. Draw the image.

c Enlarge the original F, with centre of enlargement O and scale factor 2.

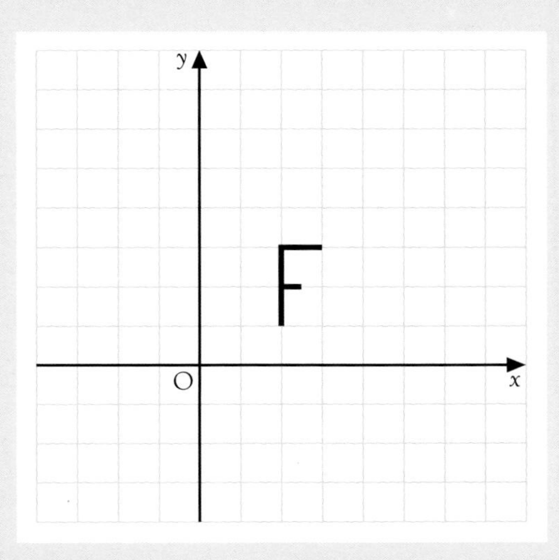

DO NOT use coloured pencils or pens – use an ordinary pencil as advised. Colour has only been used in this worked example to highlight the method and solutions.

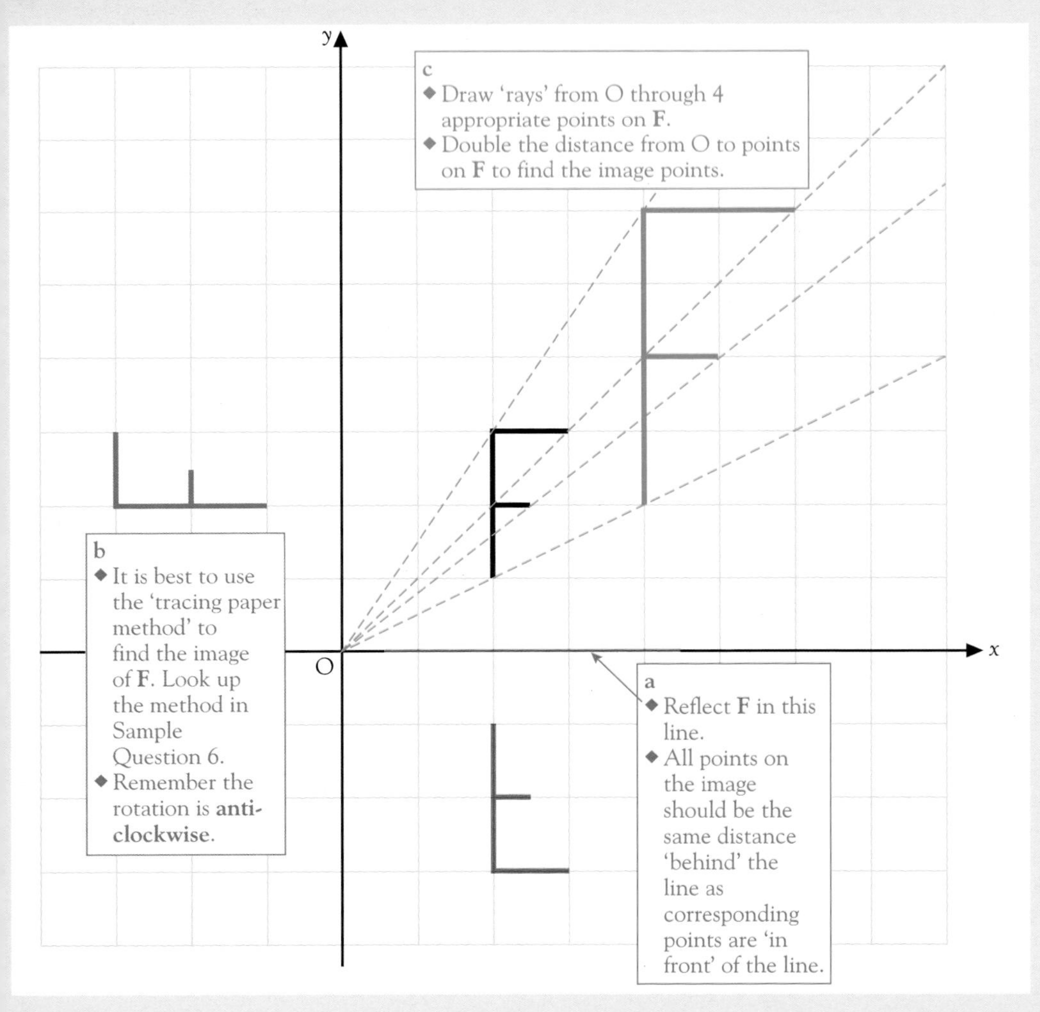

COMMENTS

Part c

A4 for correct enlargement with correct centre of enlargement

(**A2** for enlargement of the **F**, scale factor 2, any centre)

Part b

A3 for correct rotation, direction and angle

(**A2** for rotation of 90° clockwise about the origin,
A1 for any rotation clockwise or anticlockwise about any point)

Part a

A3 for correct reflection in x axis

(**A2** for reflection in the y axis,
A1 for a reflection in any other vertical or horizontal line)

10 marks

Exam Questions

1

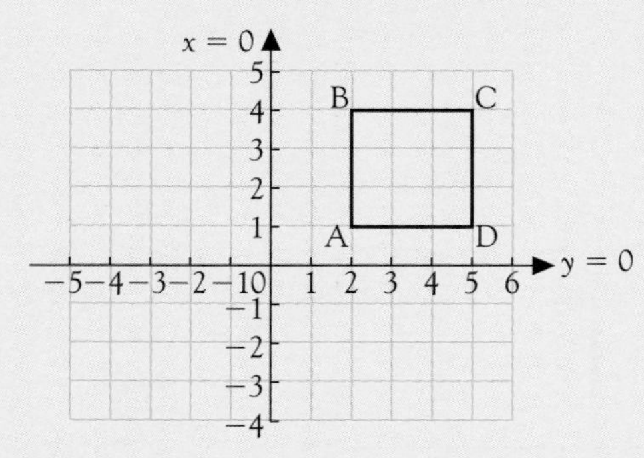

Copy the diagram above on squared paper.

a The square ABCD is reflected in the line $x = 1$. What are the new coordinates of C?

b The square ABCD is rotated through 180° about A. What are the new coordinates of C? [SEG]

2

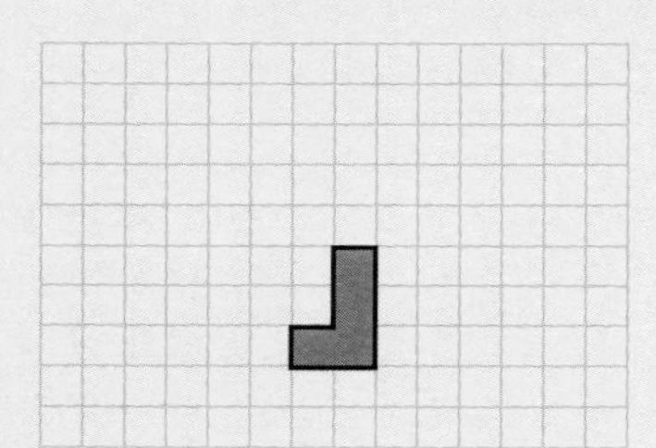

Copy this shape on squared paper.

a Draw the shape on the grid in the correct position after translating it with vector $\begin{pmatrix} -2 \\ 5 \end{pmatrix}$.

b What vector would translate the new shape back to the original shape? [MEG]

3 a In the following diagram, four triangles P, Q, R and S are marked. For each of the following mappings, state whether it is a reflection, a rotation, a translation or an enlargement.

i triangle P → triangle Q

ii triangle P → triangle R

iii triangle S → triangle P

b What is the angle of the rotation in part **a**?

c What is the scale factor of the enlargement in part **a**?

d Copy the diagram on the top half of a sheet of squared paper.

Draw the reflection of the whole diagram in the line XY.

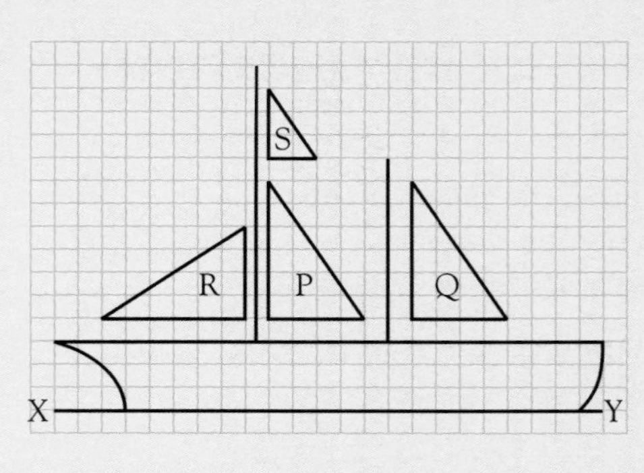

[MEG]

4 a Copy the diagram below on squared paper.

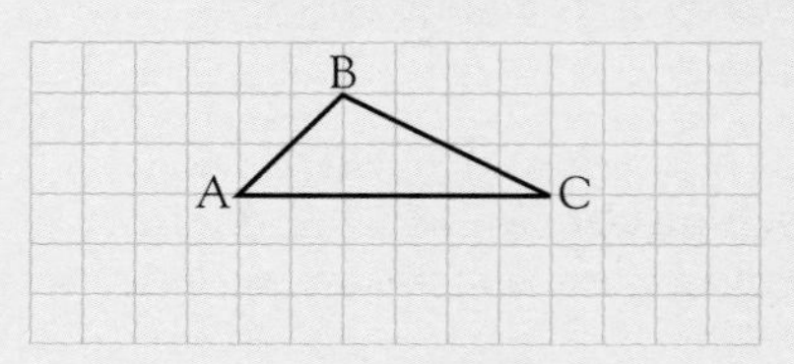

i Reflect triangle ABC in the line AC. Label as D the image of B.

ii Name the special quadrilateral ABCD.

b Copy the diagram below on squared paper.

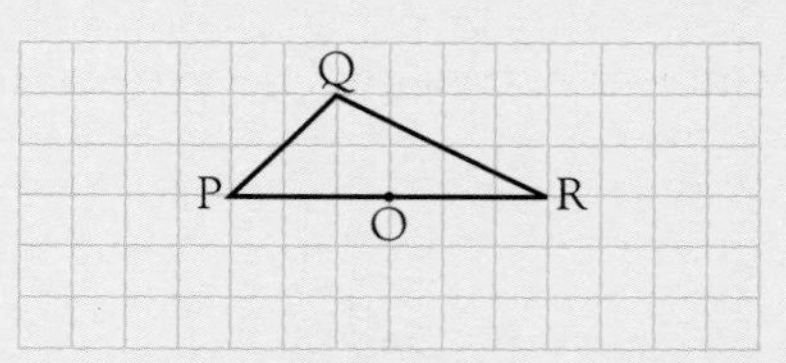

i Rotate triangle PQR through 180° about the point O. Label as S the image of Q.

ii Name the special quadrilateral PQRS.

[MEG, p]

5

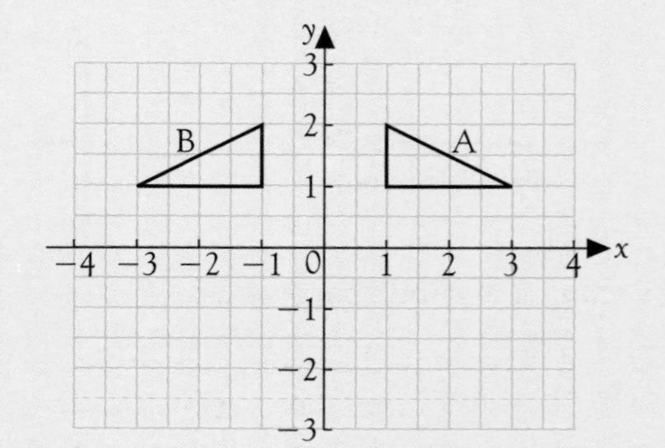

Copy the diagram on squared paper.
The diagram above shows two triangles A and B.

a Describe the single transformation that maps triangle A onto triangle B.

b Triangle A is mapped onto triangle C by a rotation of 90° clockwise about the origin, (0, 0).

Draw and label triangle C on your grid. [MEG]

EXAM QUESTIONS

6

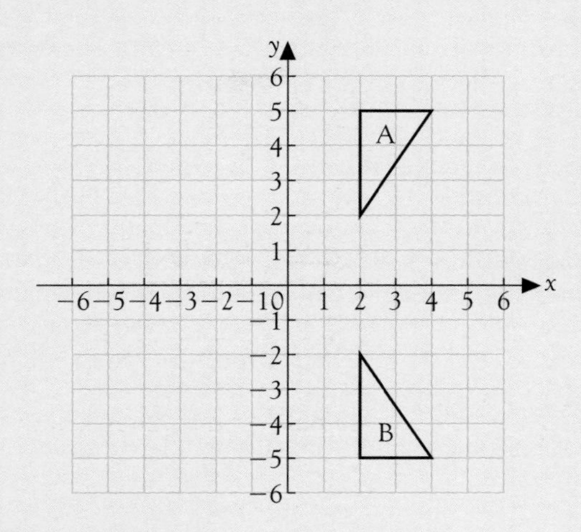

Copy the diagram.
The diagram above shows two triangles A and B.

a Describe fully the single transformation that maps triangle A onto triangle B.

b Triangle B is mapped onto triangle C by a rotation of 90° clockwise about the origin (0, 0).

Draw and label triangle C on your grid.

[MEG]

7

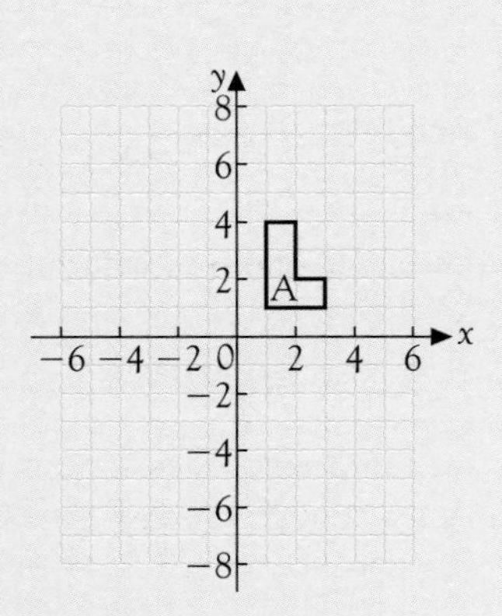

Copy these axes and shape A on graph paper.

a Shape A is rotated through 90° in an anti-clockwise direction about (0, 0). Draw its new position on your graph and label it B.

b Shape A is enlarged by scale factor 2, centre (0, 0). Draw the enlargement on your graph and label it C.

c Shape A is transformed by reflecting it in the mirror line $y = -1$. Draw the new position of shape A on your graph and label it D.

[SEG]

8

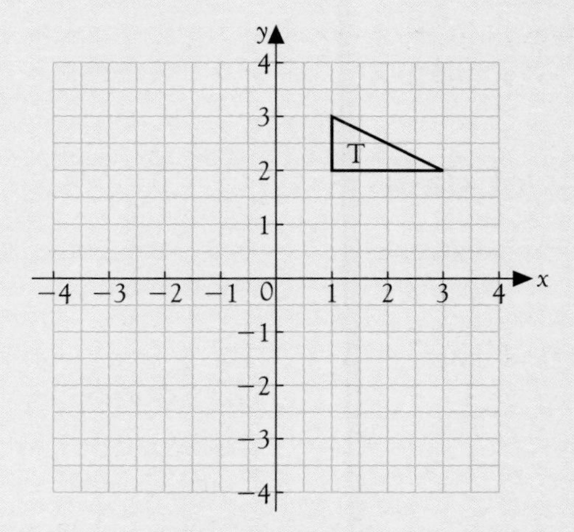

The diagram shows triangle T. Copy it on graph paper.

a Reflect triangle T in the y axis.
Label this triangle A.

b Rotate triangle T through 90° clockwise about the origin (0, 0).
Label this triangle B.

[MEG]

9 The sketch shows the position of a rectangle ABCD. Copy it on squared paper.

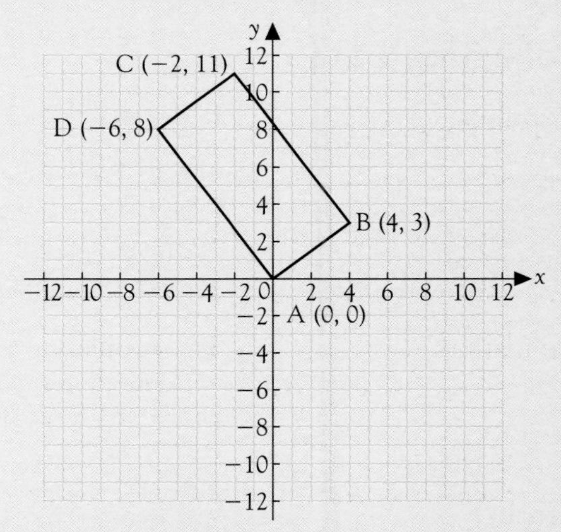

a The rectangle ABCD is reflected in the line $x = 4$ to give rectangle $A_1B_1C_1D_1$.
What are the co-ordinates of C_1?

b The rectangle ABCD is rotated about A anti-clockwise through 90° to give $A_2B_2C_2D_2$.
What are the co-ordinates of B_2?

c The rectangle ABCD is enlarged by scale factor 2, centre A.
What are the co-ordinates of the new position of B?

[SEG]

FACT SHEET 5: THE LANGUAGE OF DATA HANDLING

Data handling involves collecting, representing, analysing and interpreting information. This fact sheet will help you to remember some of the key ideas and words relating to collecting and representing data.

Types of data

Descriptive categories

This is sometimes known as **qualitative** data. **Categories**, usually described in words, are used, such as types of housing, modes of transport, favourite pop groups.

Numerical data

This is sometimes known as **quantitative** data in which **numbers** are used. Numerical data are either **discrete** or **continuous**:

Discrete data can take only particular **exact** values, usually whole numbers. Examples are

- the number of people in a car
- the number of goals in a match
- shoe sizes
- marks in an examination

Continuous data cannot take exact values. They are usually measurements and their accuracy depends on the accuracy of the measuring device. Examples are

- the temperature in different parts of the country
- the masses of bags of sweets
- the heights of police cadets

Collecting data

Planning

Plan carefully before you start to collect any data.

Decide what you want to know.

Decide how to collect your data, for example, you might conduct a survey by asking people to complete **questionnaires** or by carrying out an **observation** yourself; you might obtain your results by doing an **experiment**.

Population

The **population** is the name given to all the members of the group being considered, for example, everyone in your class, everyone in your school, all the people in Great Britain, all the snails in your garden, all the cars produced by a certain factory.

Census

In a **census** information is collected from every member of the population. The advantage of a census is that you have complete and accurate information. Disadvantages are that it is very time consuming to conduct a census and it may be very expensive.

You could destroy your population; for example, if you are testing the time that batteries operate before they fail, you would not have any working batteries left if you tested them all!

Sample

It is more usual to survey a **sample** from the population. Your sample must be **random**. This means that every member of the population has an equal chance of being chosen. You should be careful not to introduce **bias**, such as only asking your friends their opinions.

FACT SHEET 5: THE LANGUAGE OF DATA HANDLING

Questionnaires

Guidelines for writing a questionnaire

- Before you start designing a questionnaire be clear about what you want to know.
- When writing the questions think about how you are going to **represent and analyse** the answers.

DO NOT leave this until after you have obtained the answers.

CONSUMER SURVEY

Put a tick in the appropriate box or boxes.

1. Which daily newspaper did you buy last Monday?

Daily Telegraph ☐	Daily Mirror ☐	Sun ☐	Star ☐
Guardian ☐	Times ☐	Other ☐	None ☐

2. How often do you buy a Sunday newspaper?

- Allow for any possible answer. Providing a choice of answers with boxes ready to tick or complete is useful and makes the results easier to categorise.
- Ask short, simple questions that will not be misunderstood.
- Avoid descriptions that can be misinterpreted such as 'occasionally', 'sometimes', 'often', 'a lot'.
- Do not ask personal or embarrassing questions.
- Take care not to introduce bias in the way that you phrase the question, for example, if you are asking for an opinion do not suggest an answer by saying 'Sensible people think that this new road should not be built. Do you think the new road should be built?'
- Give instructions on how to complete the questionnaire.
- Before reproducing multiple copies of your questionnaire, **try out the questions on a small sample.**

It is better to find out the problems early on, before it is too late!

Observation sheet

Decide where to conduct the survey, when and for how long.

Prepare your **observation sheet** beforehand, deciding the categories you are looking out for. Do not try to look for too many things at one time.

Allow space to record your information. **Tally charts** are useful.

ROAD JUNCTION OBSERVATION SHEET

	Car	Bus	Lorry	Motor bike	Van	Bicycle
Turn Left	卌 卌 卌 卌 卌 卌 III	卌 卌 II	卌 I	卌 卌	卌 卌 卌 卌 卌 I	卌 III
	33	12	6	10	26	8
Turn Right	卌 卌 卌 卌 II	卌 卌	卌	III	卌 卌 I	I

FACT SHEET 5: THE LANGUAGE OF DATA HANDLING

Displaying data

Frequency tables

The word **frequency** is used for the number of times that something happened. Often the letter f is used as shorthand.

A **frequency table** (or **frequency distribution**) gives a summary of results.

a This is an example of data in **descriptive categories**.

Grade	A	B	C
Frequency, f	4	2	6

b This is an example of **ungrouped discrete data**.

Score on die	1	2	3	4	5	6
Frequency, f	3	4	6	5	4	2

c This is an example of **grouped continuous data**.

Age of teacher	Frequency, f
$20 \leqslant$ age < 30	10
$30 \leqslant$ age < 40	19
$40 \leqslant$ age < 50	23
$50 \leqslant$ age < 60	13

Anyone in the interval $20 \leqslant$ age < 30 has an age that is 20 or more, but less than 30

d This is an example of **grouped discrete data**.

Mark in test	1–10	11–20	21–30	31–40
Frequency, f	3	12	8	5

Bar charts

Data in descriptive categories and ungrouped discrete data can be represented in a **bar chart**.

The bars can be drawn horizontally or vertically and gaps can be left between the bars if preferred.

Bar chart to show number of students achieving grades A–C in GCSE Mathematics

Frequency

Grade

Bar chart to show scores on die

Frequency

Number on die

Categories are written in the middle of the bar.

FACT SHEET 5: THE LANGUAGE OF DATA HANDLING

Comparative bar charts

Comparative bar charts are useful for comparing sets of information quickly.

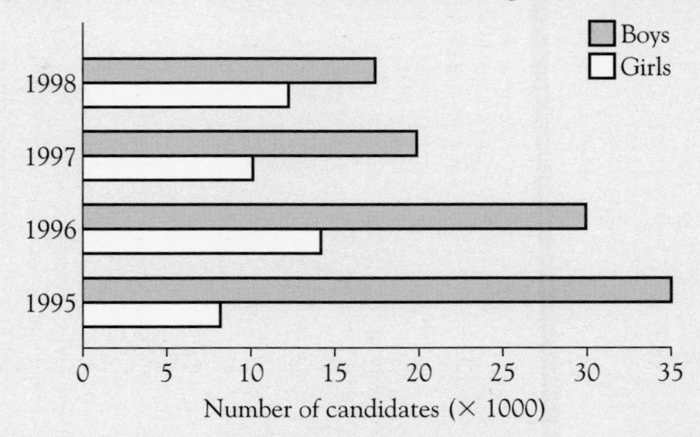

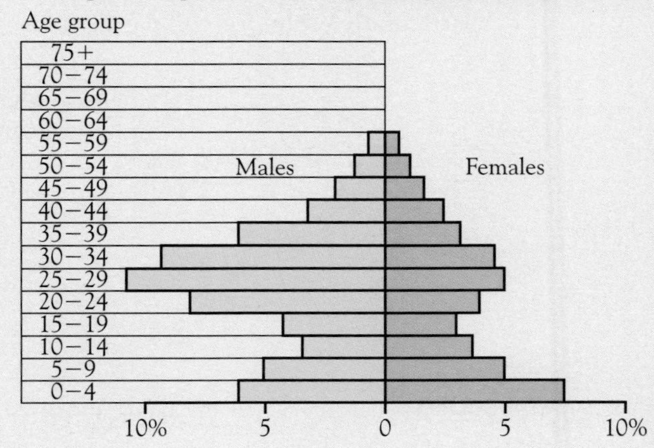

Histograms

A **histogram** can be used to represent **continuous data** or **grouped discrete data.**

In a histogram the **area of a bar represents the frequency**.

Provided that the intervals are all the same width the frequency can be represented by the height of the bar. There are no gaps between the bars.

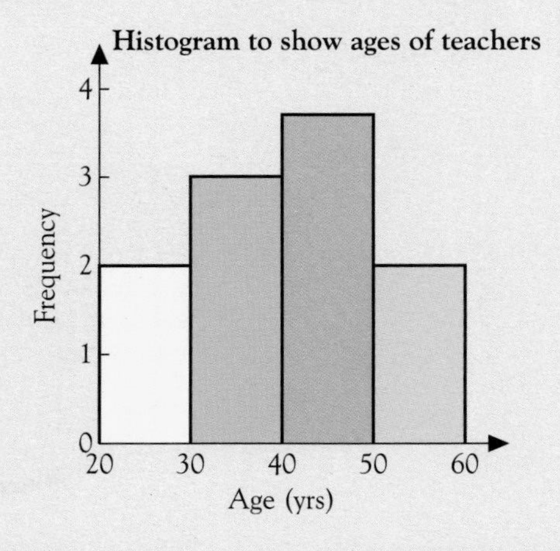

For continuous data the horizontal axis is labelled with the **class boundaries** 20, 30, 40, 50, 60.

FACT SHEET 5: THE LANGUAGE OF DATA HANDLING

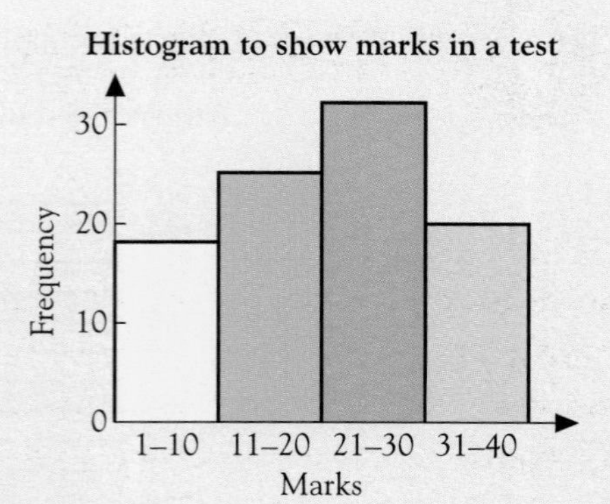

For grouped discrete data the horizontal axis is labelled with each interval.

Line graphs

Line graphs can be used to illustrate data, often over a period of time. They are particularly useful for comparing data.

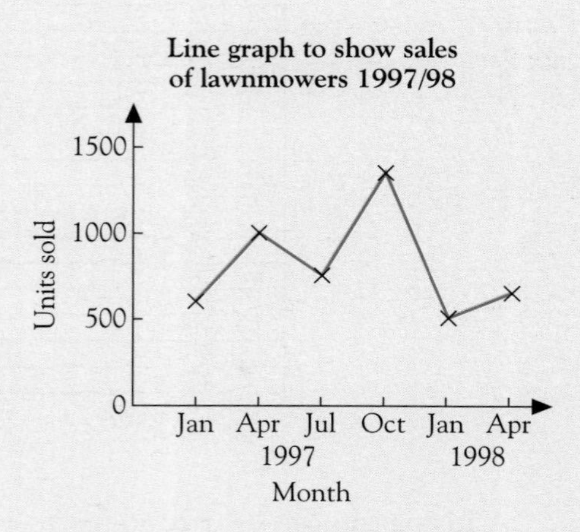

Pie charts

A **pie chart** is useful in showing how something has been divided, with each category having a 'slice' of the circle.

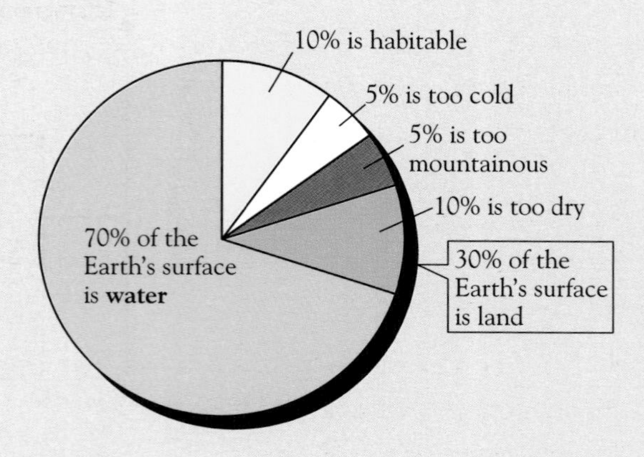

DATA HANDLING I

In this chapter you will
- **practise drawing and interpreting statistical diagrams including bar charts, histograms, pie charts and line graphs**
- **find and use the three averages: mean, mode and median of ungrouped data**
- **look for connections between two variables using scatter diagrams**

Drawing and interpreting statistical diagrams

Bar charts and histograms

Sample Question 1

Shoppers were asked to sample three cheeses, labelled A, B and C, and choose the tastiest. Their results are shown in the split bar chart.

a How many people were surveyed?

b What fraction chose B?

c What percentage of the women chose A?

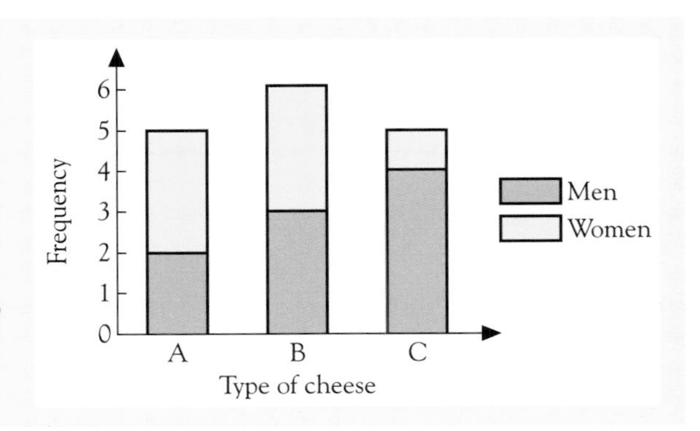

Answer

a

- Add the total frequencies by adding the heights of the bars.

  $5 + 6 + 5 = 16$

  16 people were surveyed.

b

- Write the number who chose B as a fraction of the total number.

  Number who chose B = 6

  Fraction who chose B $= \frac{6}{16} = \frac{3}{8}$

Give the fraction in its simplest form.

c

- Find how many women were surveyed.

  Number of women in a bar = total height of bar − height of men's section of bar

  Total number of women $= 3 + 3 + 1 = 7$

- Write the number of women who chose A as a fraction of 7, then change the fraction to a percentage.

  Fraction of women who chose A $= \frac{3}{7}$

  Percentage $= \frac{3}{7} \times 100\%$

  $= 42.85\ldots\%$

  43% of the women chose cheese A.

FACT SHEET 1
pages 13–15

Approximate your answer sensibly; 42.9 would also be acceptable.

STATISTICAL DIAGRAMS

Sample Question 2

A record was kept of the daily rainfall, in mm, at Sunnybrook Farm and the results are shown in the histogram.

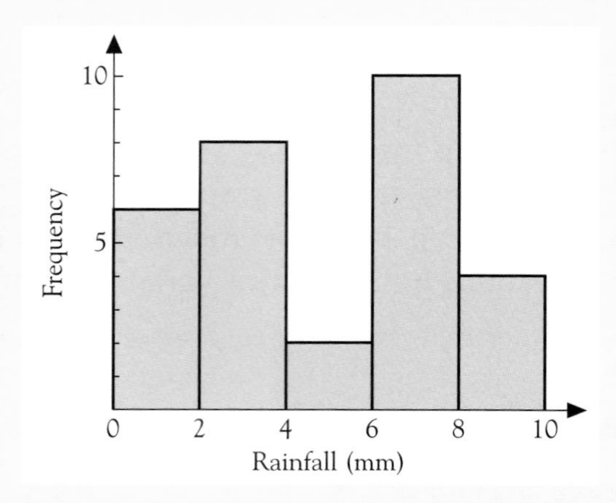

a Complete this frequency table and work out on how many days the survey was conducted.

Amount of rain, x mm	$0 \leqslant x < 2$	$2 \leqslant x < 4$	$4 \leqslant x < 6$	$6 \leqslant x < 8$	$8 \leqslant x < 10$
Frequency					

b The farmer said that 6 mm or more of rain fell on half the days. Was this correct?

Answer

a

- Find the number of days in each interval. Since all the intervals are equal in width, you can read off the heights of the bars.

$0 \leqslant x < 2$ means from 0 to just under 2. 0 would go in this interval, but 2 would go into the next interval.

Amount of rain, x mm	$0 \leqslant x < 2$	$2 \leqslant x < 4$	$4 \leqslant x < 6$	$6 \leqslant x < 8$	$8 \leqslant x < 10$
Frequency	6	8	2	10	4

- Add up all the frequencies to find the number of days.

$6 + 8 + 2 + 10 + 4 = 30$

The survey was conducted on 30 days.

b

- Add up the number of days on which 6 mm or more of rain fell by looking at the intervals from 6 to 8 and from 8 to 10.
- Compare this with half the total number of days.

Number of days with 6 mm or more of rain $= 10 + 4 = 14$
Half the total number of days $= 30 \div 2 = 15$

Since 14 is less than 15, the farmer was incorrect.

STATISTICAL DIAGRAMS

Pie charts

Sample Question 3

The table shows how the 150 pupils in Year 10 travelled to school on a particular day.

Walk	Bus	Car
83	22	45

Work out the angles needed to show the results in a pie chart.

Answer

Method 1

- Find the angle that represents one pupil first then work out the angle for each category.

150 pupils are represented by a complete circle of 360°.

One pupil is represented by $360 \div 150 = 2.4°$

Walk: 83 people $83 \times 2.4° = 199.2°$

Bus: 22 people $22 \times 2.4° = 52.8°$

Car: 45 people $45 \times 2.4° = 108°$

Angles needed are: Walk **199°**, Bus **53°**, Cycle 108°

Method 2

- Use fractional parts of a complete circle where a complete circle represents 150 pupils.

Walk: 83 pupils out of 150 walk to school, so find the angle for $\frac{83}{150}$ of a complete circle of 360°.

Walk: 83 people $\frac{83}{150} \times 360° = 199.2°$

Bus: 22 people $\frac{22}{150} \times 360° = 52.8°$

Car: 45 people $\frac{45}{150} \times 360° = 108°$

In practice, to draw the chart you would need to round the angles to the nearest whole number.

Always check that the angles add up to 360°

Sample Question 4

This question is about the way water is used in two Mozambique villages.

a In village A, 324 litres of water are used each day. The pie chart shows how the water is used.

i How much water (in litres) is used each day for cooking?

ii What fraction of the water used is given to animals?

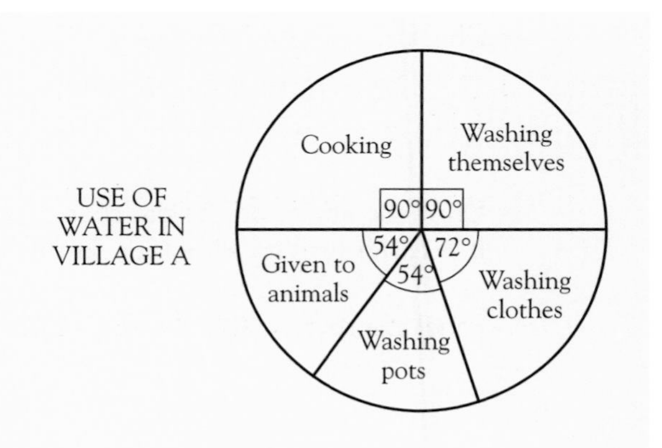

b In village B, the water is used as follows:

Cooking	20%
Washing themselves	50%
Washing clothes	20%
Washing pots	10%

Represent this information in a pie chart. [MEG]

STATISTICAL DIAGRAMS

Answer

a i

- Find the angle representing 'cooking' as a fraction of the complete angle of 360°.

  Fraction used for cooking $= \frac{90}{360} = \frac{1}{4}$

- Find $\frac{1}{4}$ of 324 litres.

  $324 \div 4 = 81$

  81 litres of water are used for cooking.

ii

- Write 54 as a fraction of 360 and simplify it.

  $\frac{54}{360} = \frac{3}{20}$

  $\frac{3}{20}$ is given to animals.

5 4 a b/c 3 6 0 =

Without a calculator, find numbers that divide into both the top and bottom of the fraction until you get to the simplest form.

b

- Work out the angle for each category. Since the amounts have been given in percentages, calculate the appropriate percentage of a complete circle of 360°.

FACT SHEET 1
pages 13–15

Angle for cooking = 20% of 360° = 72°

Angle for washing themselves = **50%** of 360° = 180°

$50\% = \frac{1}{2}$

Angle for washing clothes = 72° (same as cooking)

Angle for washing pots = **10%** of 360° = 36°

$10\% = \frac{1}{10}$, so just divide by 10

- Check that the total comes to 360°

  $72 + 180 + 72 + 36 = 360$

Check total:
$72 + 180 + 72 + 36 = 360$

- Draw the angles on the pie chart.
  It is easier to start with 180°, as this will be a diameter of the circle.

- Label the sectors, write in the angles and give the pie chart a title.

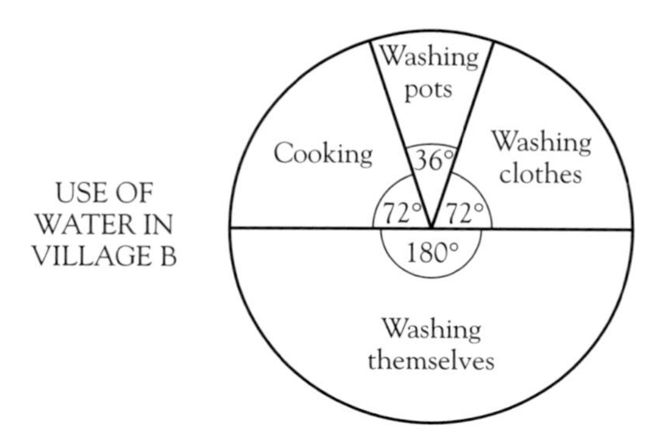

STATISTICAL DIAGRAMS

Exercise 9.1

1 Some children were asked what pets they had at home. Draw a bar chart to illustrate the results:

Dogs 9, Cats 13, Rabbits 4, Hamsters 5, Gerbils 4, Snakes 1

2 A garage noted the colours of cars ordered by men and women during August last year.

	Black	Red	Green	Yellow	Blue
Men	4	10	3	5	8
Women	4	5	7	1	7

a Draw a comparative bar chart to show the information.

b What percentage of men chose red?

c What fraction of the customers chose yellow?

d Draw a pie chart to illustrate the colours ordered by women.

3 Pupils in a science class tested soil samples, each weighing 120 g, to find the water, humus and mineral content.

a Simon drew this pie chart of his results.

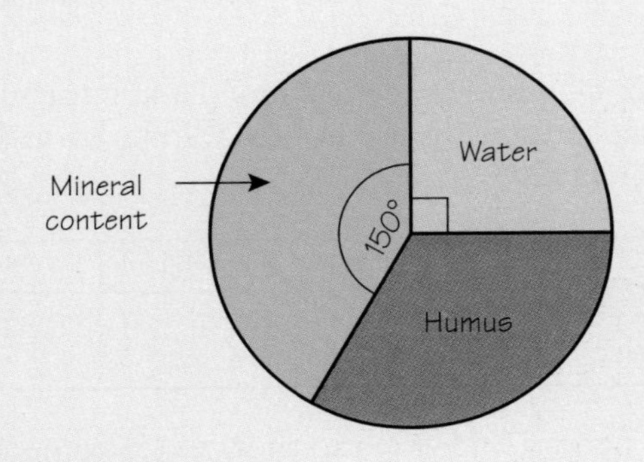

i What fraction of his sample was water?

ii What angle did he use to represent humus?

iii What was the weight of the mineral content?

b Helen's results were: water 15 g, humus 42 g, mineral content 63 g.

i What angle did she use for the water content?

ii Draw a pie chart to show Helen's results.

c Draw a comparative bar chart to show the results of both Simon and Helen.

4 This histogram shows the length of time, in minutes, that people waited in a supermarket queue.

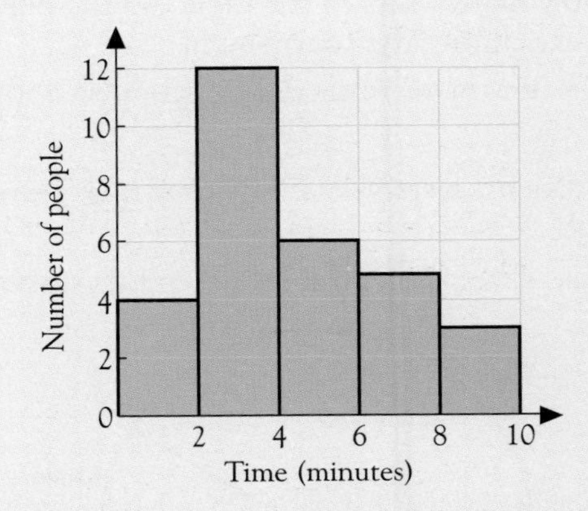

a How many people waited between 2 and 4 minutes?

b Use the histogram to complete the table of results:

Time spent waiting (in minutes)	Number of people
$0 \leqslant x < 2$	
$2 \leqslant x < 4$	
$4 \leqslant x < 6$	
$6 \leqslant x < 8$	
$8 \leqslant x < 10$	

c How many people were surveyed?

d How many people waited less than 4 minutes?

e A person was picked at random from the group to answer a questionnaire. What is the probability that the person chosen had waited between 6 and 10 minutes?

STATISTICAL DIAGRAMS

5 A snooker player recorded the number of points that he scored at each visit to the snooker table during a tournament. The results are shown in the frequency diagram.

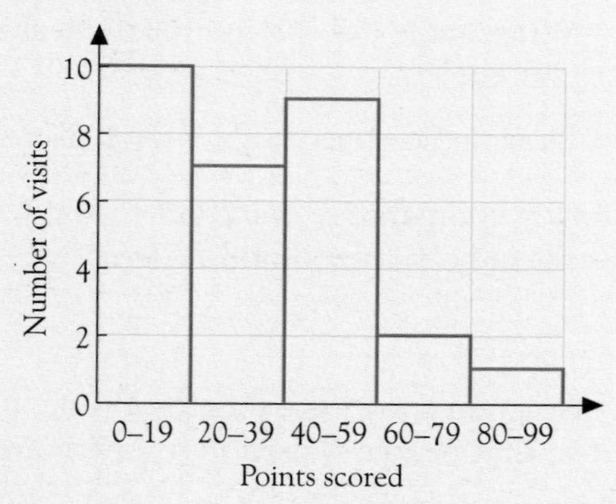

a How many visits to the table did the player make during the tournament?

b On how many visits did he score 60 or over?

6 720 students were asked how they travelled to school.

The pie chart shows the results of this survey.

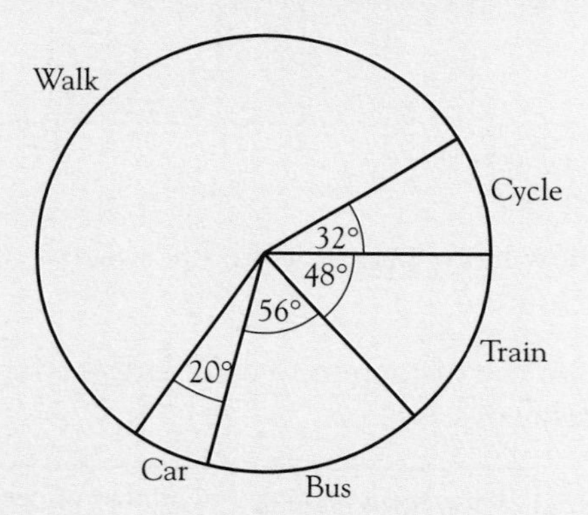

Work out how many of the students travelled to school by bus. [L, p]

7 In a school survey a record was made of all the reasons given by pupils for being absent. The results for one day are shown in the frequency diagram.

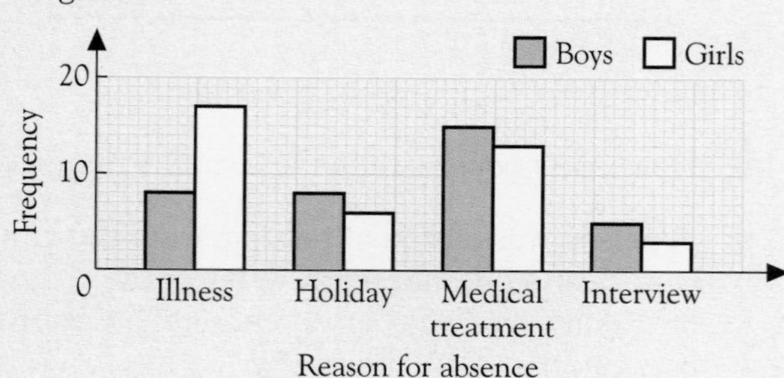

a Taking boys and girls together, which reason for absence is the mode?

b How many pupils were absent that day?

c What proportion of those absent for an interview were girls? [SEG]

8 A supermarket monitored the amounts spent at the "10 items or fewer" checkout. The results were as shown:

Amount in £ (x)	Frequency
$0 \leqslant x < 5$	19
$5 \leqslant x < 10$	43
$10 \leqslant x < 15$	57
$15 \leqslant x < 20$	10
$20 \leqslant x < 25$	1

Draw a histogram to illustrate the data.

9 Here is a frequency table of the animals on Mr McDonald's farm.

Animal	Hens	Sheep	Cows	Pigs	Geese
Frequency	30	80	104	20	6

a Draw a fully labelled pie chart to show this data.

An animal is chosen at random from Mr McDonald's farm.

b What is the probability that the animal:

i is a pig,

ii is a horse,

iii has four legs? [L]

10 The table shows the maximum and minimum temperatures (in °C) in Bristol from 1st to 10th March 1998.

Day	1st	2nd	3rd	4th	5th	6th	7th	8th	9th	10th
Maximum (°C)	8	10	12	10	10	10	12	9	10	9
Minimum (°C)	3	7	8	6	4	6	8	7	2	1

a Draw a line graph to show the maximum temperature each day.

b On the same diagram, draw a line graph to show the minimum temperature each day.

c On which day was the temperature difference least?

STATISTICAL DIAGRAMS

11

The diagram, which is not drawn to scale, shows how a council tackles the problem of what to do with domestic waste.

a Write down the fraction of the waste which is recycled.

b $\frac{2}{5}$ of the total waste is burnt.
Calculate the angle of the sector which represents the waste burnt.

c Calculate the percentage of the total waste which is used for landfill. [NEAB]

12 The following paragraph is taken from *The Night of The Big Wind* by Peter Carr.

"Trees, ten or twelve miles from the sea, were covered with salt brine. The surges of the sea, therefore, must have been whipped up, and whirled hundreds of miles inland."

a For the paragraph, copy and complete the grouped frequency table for the number of letters in a word.

Number of letters in a word	**Number of words**	
Class Interval	Tally (if required)	Frequency
1–3		11

b Using the values in the table, draw a frequency diagram on graph paper. [NI]

AVERAGES

Averages

It is often useful to describe data by giving a 'typical' or 'average' value that can be used to represent the data.

School report	
Science test result:	75%
Class average:	69%

There are three types of average in common use.

- The **mode** is the value that occurs the **most often**. It is the **most common** or **most popular** value.

To help you remember, notice that MOst and MOde begin with the same two letters.

- The **median** is the **middle value** when all the values are placed in order of size. If there is an odd number of values, there will be one middle one. If there is an even number of values, there will be two middle values and the median is taken as mid-way between them. To find the median in this case, add the two middle values and then divide by 2.

A **medium** size is a **middle** size. MEDIUM and MEDIAN sound similar when you say them.

- The **mean** is the average found by **sharing out equally** the total of all the values, so

$$\text{mean} = \frac{\text{total of all the values}}{\text{number of values}}$$

The mean is the average that is usually used in everyday descriptions.

When **sharing** out some sweets, it would be MEAN not to give everyone the same amount!

Sample Question 5

Annie recorded the number of goals scored by her hockey team in the 13 matches played last season. The results were

0, 3, 1, 0, 6, 2, 2, 0, 4, 0, 4, 4, 12

Find:

a the median number of goals per match,

b the mode,

c the mean number of goals per match.

Which average would Annie use to describe her team's results?

Answer

a

- Put the results in order of size and find the middle one.

0 0 0 0 1 2 (2) 3 4 4 4 6 12

<u>The median was 2 goals per match.</u>

Since there is an odd number of values, there is a middle one.

b

- Find the result that occurred most often.

  The mode was 0 goals per match.

c

- Find the total number of goals scored.

  Total $= 0 + 0 + 0 + 1 + 1 + 2 + 3 + 3 + 4 + 4 + 4 + 6 + 12 = 40$

- Share this total equally between the 13 matches by dividing by 13.

  Mean $= 40 \div 13 = \mathbf{3.1\ (1\ d.p.)}$

  The mean was 3.1 goals per match.

The mean is not necessarily a whole number. You may need to approximate your answer sensibly.

The mean gives the most favourable impression of the team, so Annie would probably use this average. She would be very unlikely to use the mode of 0 goals per match!

The extreme value of 12 helped to raise the mean average to Annie's advantage. Often, however, it is better to use the median as representative when there are extreme values. You will need to think carefully about this in relation to a particular situation.

Sometimes you will need to re-arrange the formula for the mean to find the total of all the values.

Since $\quad \text{mean} = \dfrac{\text{total of values}}{\text{number of values}}$

total of values = mean × number of values

Sample Question 6

The Spring Term consists of 10 weeks and each week Dave has a Spanish vocabulary test. He wants to get a mean mark for the term of at least 75%.

After 9 tests, his mean mark is 74%.

What is the lowest mark he can get in his last test in order to achieve his target?

Answer

- Find the total marks that Dave has obtained after 9 tests.

  $$\begin{aligned}\text{Total marks after 9 tests} &= \text{mean} \times \text{number of values}\\ &= 74 \times 9\\ &= 666\end{aligned}$$

- Find Dave's target total for 10 tests in order to achieve a mean of 75.

  Target total after 10 tests $= 75 \times 10 = 750$

- To find how many more marks he needs, find the difference between Dave's target total and actual total.

  Difference $= 750 - 666 = 84$

  Dave must get at least 84% in his last test.

AVERAGES

You need to take special care when the information is given in a **frequency table**.

Sample Question 7

Number of tickets	0	1	2	3	4	5	6
Number of teachers	2	7	5	2	0	3	1

Some teachers were asked how many National Lottery tickets they bought last week. The results are shown in the table.

a Which number of tickets is the mode?

b Work out the mean number of tickets.

c Find the median number of tickets. [L]

Answer

a

- Find the number of tickets that occurred most often. Look for the highest frequency.

The mode = 1 ticket

DO NOT write 7; this is the number of teachers who bought the modal number of tickets.

b

- Find the total number of tickets bought.

DO NOT add up the number of tickets column. This does not tell you how many tickets were bought altogether.

Work it out this way, using an extra column in the table to help you. In an examination the column is often drawn ready for you.

Number of tickets, x	Number of teachers, f	$f \times x$
0	2	$2 \times 0 = 0$
1	7	$7 \times 1 = 7$
2	5	$5 \times 2 = 10$
3	2	$2 \times 3 = 6$
4	0	$0 \times 4 = 0$
5	3	$3 \times 5 = 15$
6	1	$1 \times 6 = 6$
	20	Total = 44

How the total is built up:

2 teachers bought 0 tickets
7 teachers bought 1 ticket each, giving 7 tickets
5 teachers bought 2 tickets each, giving 10 tickets
2 teachers bought 3 tickets each, giving 6 tickets
0 teachers bought 4 tickets, giving 0 tickets
3 teachers bought 5 tickets each, giving 15 tickets
1 teacher bought 6 tickets, giving 6 tickets

- Divide the total number of tickets by the number of teachers.

You are putting **all** the tickets together and then sharing them out **equally** among the 20 teachers.

Mean = $44 \div 20$
= 2.2

The mean number of tickets = 2.2 per teacher

DO NOT round this answer. Leave it as 2.2 even though you cannot have 2.2 tickets.

c There are 20 teachers, so when they are put in order of how many tickets they bought, there will be 20 values. Since 20 is an even number there will be two middle values, the 10th and the 11th, so work out the number mid-way between them.

- Start to write the values in order. You need only go as far as the 11th value.

middle ones

0, 0, 1, 1, 1, 1, 1, 1, 1, (2, 2,)...

10th 11th

The median number of tickets = 2

The 10th and 11th values are both 2, so mid-way between them is also 2.

AVERAGES

Range

The **range** of a set of data gives an idea of how spread out the values are. It is the difference between the largest and the smallest values, so to find the range subtract the smallest value from the largest value.

Range = largest value − smallest value

For example, for the data

23, 25, 45, 31, 26, 37, 42

the largest value = 45, the smallest value is 23, so

range = 45 − 23 = 22

Do not write the range as 45 − 23 or 23–45 or 23 to 45. It is a single number, found by subtracting.

Comparing data using mean and range

To compare sets of data it is useful to give the typical value for each **and** also to indicate the range.

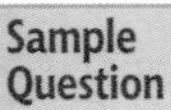

Sample Question 8

Twelve boys ran 100 metres. Their times, in seconds, are shown.

11.0	11.5	11.5	12.3	12.9	13.1
13.6	14.2	14.8	15.6	16.4	17.1

a **i** What is the range in the running times of these boys?

ii Calculate the mean running time.

b Twelve girls run 100 metres and their times are recorded.
The range in running times for the girls is 5.6 seconds and the mean is 14.3 seconds.
Comment on the differences in running times for boys and girls. [SEG]

Answer

a i

- Find the range by subtracting the smallest value from the largest value.

Largest value = 17.1, smallest value = 11.0

17.1 − 11.0 = 6.1

Range = 6.1 s

ii

- Find the total time by adding all the values then divide by 12.

Total time = 11.0 + 11.5 + … + 16.4 + 17.1
= 164 s

Mean = 164 ÷ 12
= 13.66 …

Mean running time = 13.7 s (1 d.p.)

b

- Use the **means and** the **ranges** to draw a conclusion.

Remember that the mean gives a typical value and the range tells you how spread out the data are.

The mean running time for the girls is higher, so the girls are slower on average than the boys. The times for the girls are, however, less spread out than the boys.

AVERAGES

Exercise 9.2

1 Over a two-week period a teacher recorded how many pupils were absent from her lesson out of a class of 30. The results were 2, 1, 3, 0, 2, 2, 1, 0, 9, 1, 2, 4, 2, 3.

Find:

a the median number of pupils absent per lesson,

b the mode,

c the mean.

2 Birdpool Park held its annual fishing competition in September.

The number of fish caught by each of 15 competitors in the time allowed was as follows:

1, 2, 2, 4, 4, 7, 7, 8,
9, 9, 9, 10, 10, 11, 11

a What was the modal number of fish caught?

b What was the median number of fish caught?

The numbers of fish caught by five more competitors were 2, 10, 7, 11, 7.

c For the 20 results now available calculate:
i the mode, ii the median. [SEG]

3 These are the prices of the Powerdriver cordless drill at six DIY stores.

£37, £38.50, £44, £38.50, £42, £43

Work out the mean cost of the Powerdriver drill. [MEG]

4 Write a list of five numbers that have a mean of 30. A sixth number, 42, is added to the list. What is the new mean?

5 Sophie sits nine module tests during her GCSE Science course.
Each paper has 18 multiple choice questions and her marks after eight tests are 8, 11, 9, 6, 8, 7, 10, 8.

a What is her median score?

b What is her mean score?

c If she needs a mean average score of 9 to get a Grade C, what score must she get in her final test to achieve this?

6

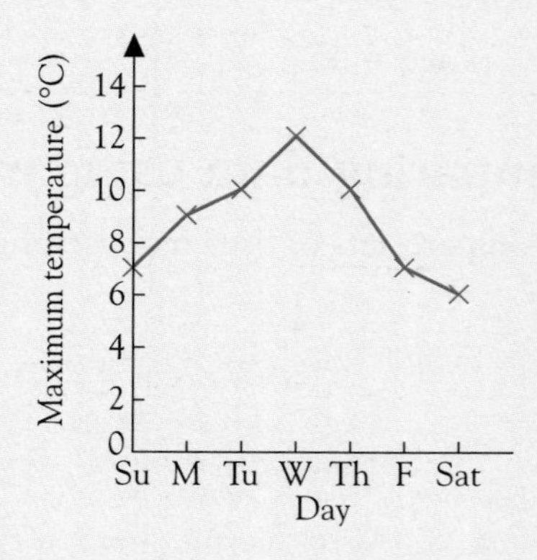

The line graph shows the maximum daily temperature over one week in January.
Find the mean maximum daily temperature for the week.

7 The temperatures at midnight in January 1995 in Shiverton were measured and recorded. The results were used to construct the frequency table.

Temperature in °C	0	1	2	3	4	5	6	7	8
Number of nights	4	5	5	3	3	7	3	0	1

a Work out the range of the temperatures.

b Work out the mean temperature. Give your answer correct to one decimal place. [L]

8 A quality control officer checks the number of blemished apples in a tray for 100 trays picked at random from the consignment sent to a supermarket. Each tray contains 36 apples. The table shows the results.

Number of blemished apples	0	1	2	3	4	5
Number of trays	26	41	18	8	6	1

AVERAGES

a Work out the mean number of blemished apples in a tray.

b What is the mean number of unblemished apples in a tray?

c What is the median number of blemished apples in a tray?

9 Pupils in a class needed to grow seeds ready for an experiment. They each planted 50 seeds and counted the number that germinated in their batch. The table shows the results.

Number germinating	40	41	42	43	44	45	46	47	48	49	50
Frequency	1	1	0	0	9	5	0	3	5	4	1

a How many pupils were there in the class?

b What is the median number of seeds germinating in a batch?

c Find the mean number of seeds germinating.

10 During a traffic survey, the number of people in each car passing a checkpoint was noted. The results are displayed in the bar chart.

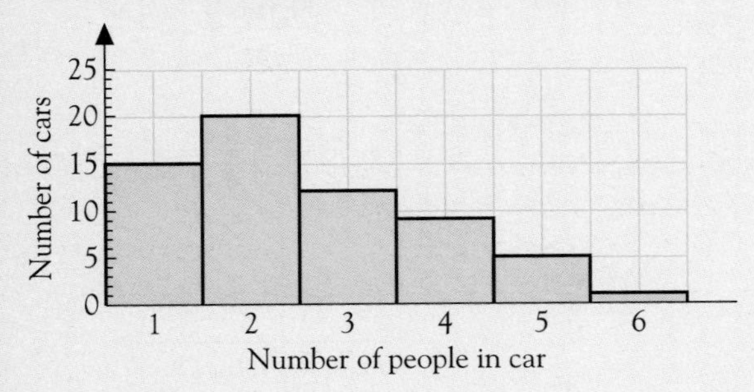

a How many cars were surveyed?

b Find the mean number of people in a car.

c An environmental group wanted to emphasise how few people cars carry. Would it be better to quote the mean or the mode?

d Would the group further its cause by publicising the median?

11 a The weekly wages of the 10 workers, in pounds, at the Bentgate factory are

170, 147, 170, 138, 162
163, 162, 149, 178, 561

i What is the range?

ii Calculate the mean wage.

b At the Penlight factory the range is £50 and the mean wage is £180. The employer at the Bentgate factory claims that his workers' wages are higher than those of the workers at the Penlight factory.
Give two reasons why this could be false. [MEG]

12 A mathematics tutor made this bar chart to show the examination grades obtained by his Advanced Level students.

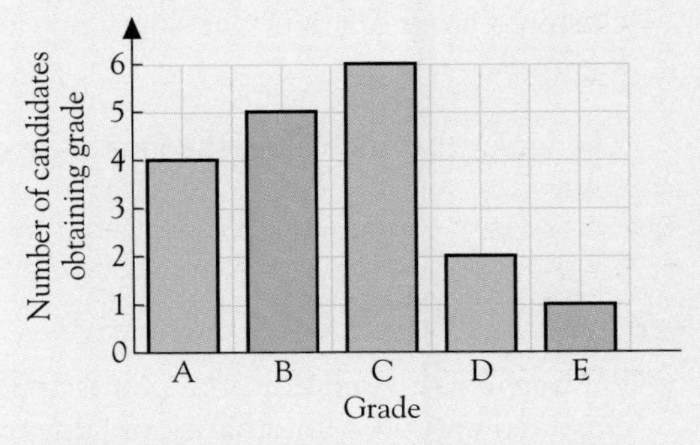

a What is the modal grade?

b At Advanced Level, each grade is worth points on the following scale:

A: 10, B: 8, C: 6, D: 4, E: 2

Find the mean points score of his candidates.

c Which average would he use to promote his results?

13 Eight judges each give a mark out of 6 in an ice-skating competition.
Oksana is given the following marks.

5.3, 5.7, 5.9, 5.4, 4.5, 5.7, 5.8, 5.7

The mean of these marks is 5.5, and the range is 1.4.
The rules say that the highest mark and the lowest mark are to be deleted.

5.3, 5.7, ~~5.9~~, 5.4, ~~4.5~~, 5.7, 5.8, 5.7

a i Find the mean of the six remaining marks.

ii Find the range of the six remaining marks.

b Do you think it is better to count all eight marks, or to count only the six remaining marks? Use the means and the ranges to explain your answer.

c The eight marks obtained by Tonya in the same competition have a mean of 5.2 and a range of 0.6.
Explain why none of her marks could be as high as 5.9. [MEG]

Scatter diagrams

A **scatter diagram** can be used to display two aspects of data, or two variables, at the same time, such as people's height and weight.

It helps to give an idea of whether there is any connection between the two variables. The word **correlation** is used to describe the relationship between the two variables.

For **positive correlation**, the points appear to be clustered around a line that slopes upwards, as in Figure A.

The larger the one value, the larger the other value.

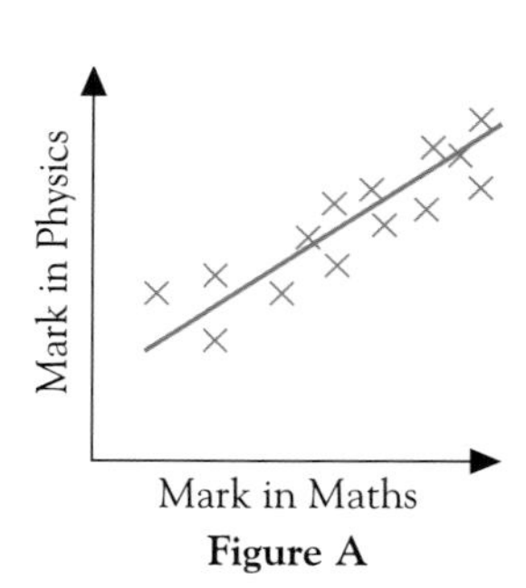

Figure A

For **negative correlation**, the points appear to be clustered around a line that slopes downwards, as in Figure B.

The larger the one value, the smaller the other value.

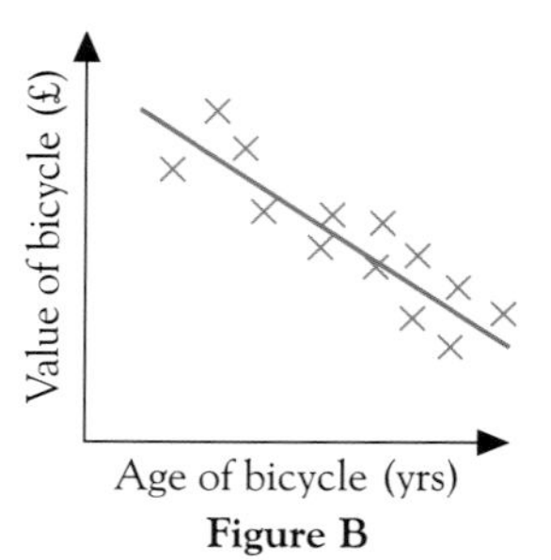

Figure B

If the points appear to be randomly scattered, with neither positive nor negative correlation, then the variables are described as having **no correlation**, as in Figure C.

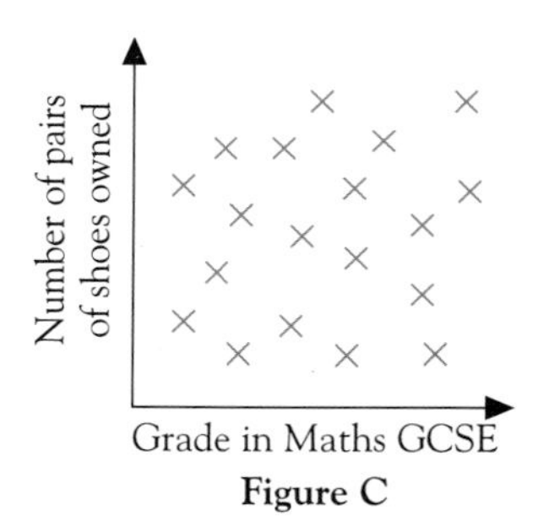

Figure C

Sample Question 9

After a junior golf tournament of 9 holes, the players were asked two questions.

1 What was your score?

2 How many hours did you practise in the month before the tournament?

Their replies are given below.

Score	41	41	42	42	43	45	45	46	47	47	48	48
Hours of practice	10	5	9	8	7	4	3	6	1	0	4	0

a Draw a scatter diagram to show this information.

b In golf, the player with the lowest score wins.

What does the scatter graph suggest about the relationship between the player's score and the practice time? [MEG]

a

- Draw a grid with the score on one axis (go from 40 to 48) and hours of practice on the other (go as far as 10).
- Plot each of the points in turn.

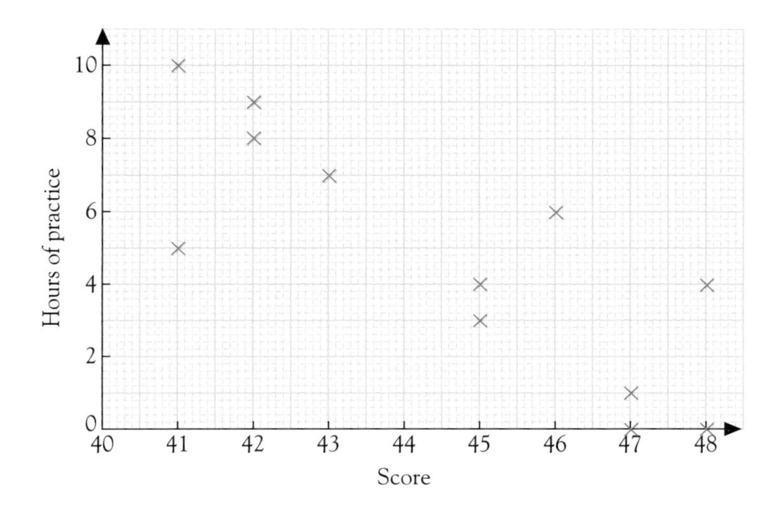

b

- Look for a general trend and describe it in terms of correlation.

  If you ignore (41, 5), the points generally indicate that a low score is related to a high practice time and a high score is related to a low practice time.

  There is negative correlation, suggesting the more you practice, the lower the score, so the better you will do.

Line of best fit

If the data show some correlation, you can draw a **line of best fit**. This is a line about which the points appear to be clustered.

To draw a line of best fit, try to draw the line so that there are the same number of points above it as below it.

It helps to place a clear ruler over the points so that you 'capture' most of them and then draw a line where the middle of the ruler is.

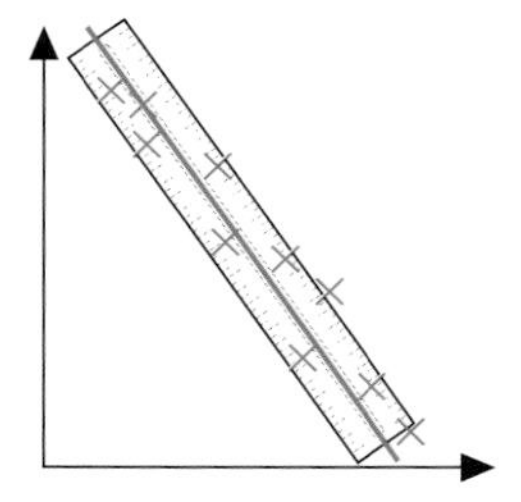

You can use the line of best fit to make predictions. If you know one value of a pair, you can estimate the other. The more the points are clustered about the line, the better the estimate.

SCATTER DIAGRAMS

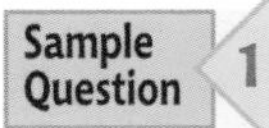

10 Ten people entered a craft competition.
Their displays of work were awarded marks by two different judges.

Competitor	A	B	C	D	E	F	G	H	I	J
First judge	90	35	60	15	95	25	5	100	70	45
Second judge	75	30	55	20	75	30	10	85	65	40

The table shows the marks that the two judges gave to each of the competitors.

a **i** On a grid, draw a scatter diagram to show this information.
ii Draw a line of best fit.

b A late entry was given 75 marks by the first judge.
Use your scatter diagram to estimate the mark that might have been given by the second judge. (Show how you found your answer.) [NEAB]

Answer

a **i**

- Plot the points in turn, being careful with the scales.
  On both axes on this grid, one small square represents two marks.

Always check the scales on **any** graph

ii

- Draw in a line so that the points are clustered around it.

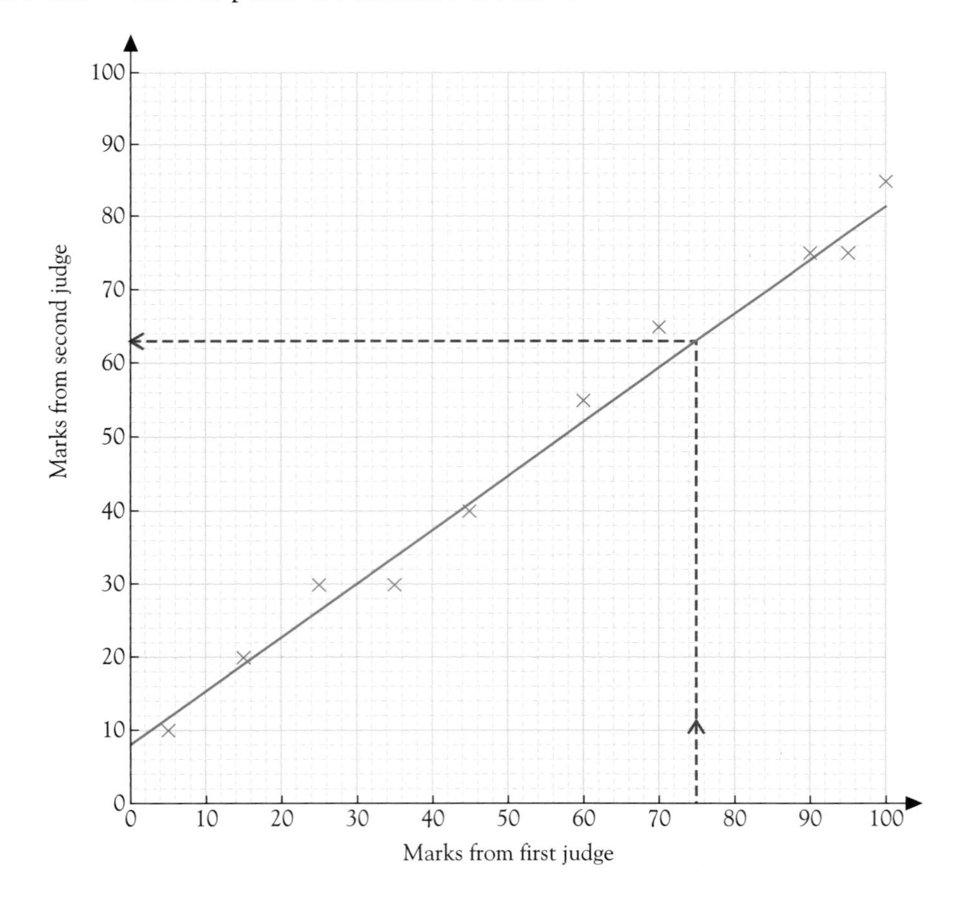

Notice that there is strong positive correlation. Although the first judge gave higher marks than the second, the judges generally agreed on the order.

b

- On the horizontal axis, find a mark of 75 from the first judge. Draw a line vertically up until you reach the line of best fit, then draw a line horizontally until you reach the vertical axis.

  Read off the mark from the second judge on the vertical axis.

  Remember to draw in the lines.
  This shows your working.

  Estimate of mark is 63

To help you draw a more accurate line of best fit, rather than just 'by eye', it is useful to know that the line should go through the point representing the **means of both sets of data.**

In Sample Question 10:

For the first judge, $\text{mean} = \dfrac{90 + 35 + 60 + \ldots + 45}{10} = 54$

For the second judge, $\text{mean} = \dfrac{75 + 30 + 55 + \ldots + 40}{10} = 48.5$

So the line should go through **(54, 48.5)**.

Plotting this point first and drawing the line through it will help you to fit the line.

TASK

Investigate whether there is any relationship between two variables related to the body, for example:

- height and mass
- length of hand and length of foot
- distance between eyes and circumference of head
- hand-span and arm-span

Collect data and display your results effectively.

Analyse them and use your findings to make conclusions.

SCATTER DIAGRAMS

Exercise 9.3

1 Brian has drawn some scatter diagrams.

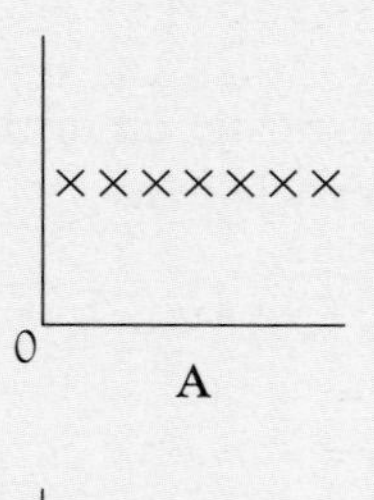

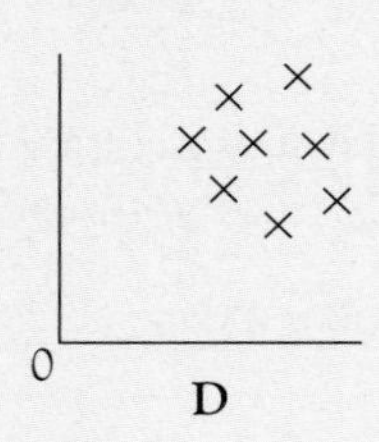

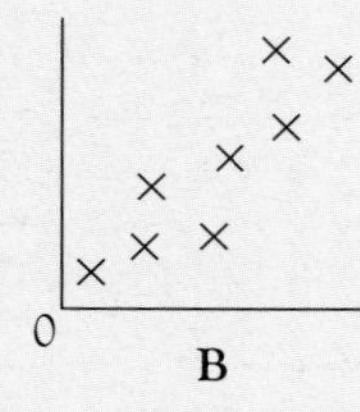

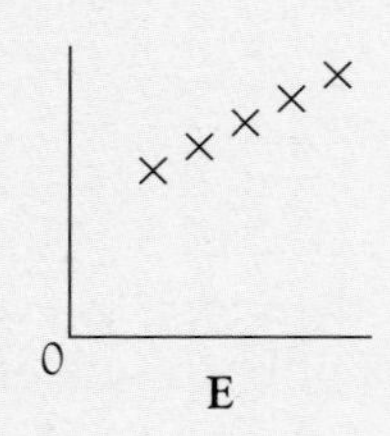

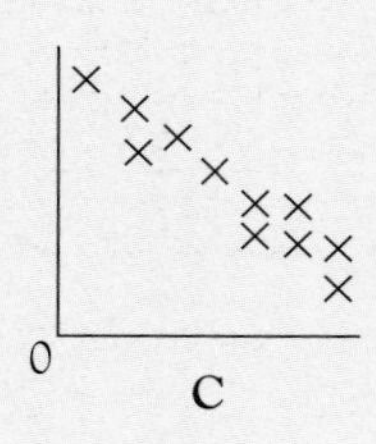

a One of these diagrams shows the value of *Aschico* 50cc mopeds plotted against their age.

Which diagram is most likely to show this?

b On another diagram he plotted the height of some pupils in his class against their marks in a Maths test.

Which diagram is most likely to show this? Explain your choice. [MEG]

2 On the planet Zarkon, there is a strange life form, small but intelligent. Zarkonians have various numbers of legs.

This one has three legs.

It is believed that a Zarkonian's intelligence is related to its number of legs.

Fifteen Zarkonians have had their intelligence tested.

This table gives the number (N) of legs and their intelligence score (I.S.).

N	I.S.	N	I.S.	N	I.S.
1	74	2	64	4	57
5	44	2	66	4	50
0	71	1	69	2	52
3	60	5	41	1	63
3	51	1	55	4	45

a Draw a scatter diagram for this information, with axes like these.

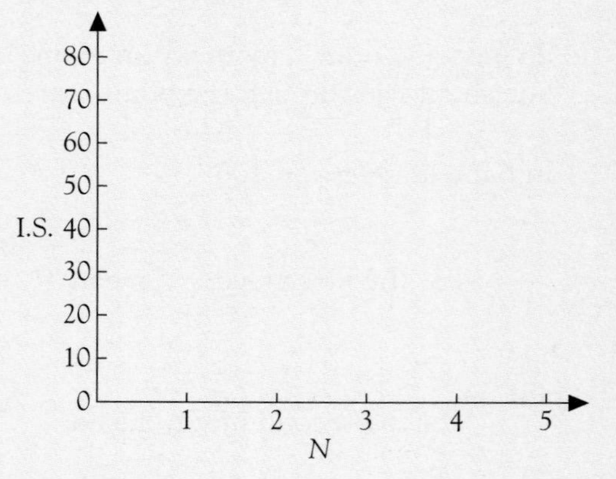

b Describe briefly what your diagram tells you. [MEG

3 This table gives the marks scored by pupils in a French and in a German test.

French	15	35	34	23	35	27	36	34	23	24	30	40	25	35
German	20	37	35	25	33	30	39	36	27	20	33	35	27	32

a Draw a scatter graph of the marks scored in th French and German tests.

b Describe the correlation between the marks scored in the two tests. [L

4

Mumtaz was studying a house plant for her Biology project. She measured the height, H cm, of each leaf up th stem of the plant together with the length L cm, of each leaf.

The table shows Mumtaz's measurements for 8 leaves.

Height (H cm)	10	15	22	25	28	32	35	40
Length (L cm)	23	19	17	18	16	15	16	13

a **i** On a grid with axes like those below, draw a scatter diagram to show this information.

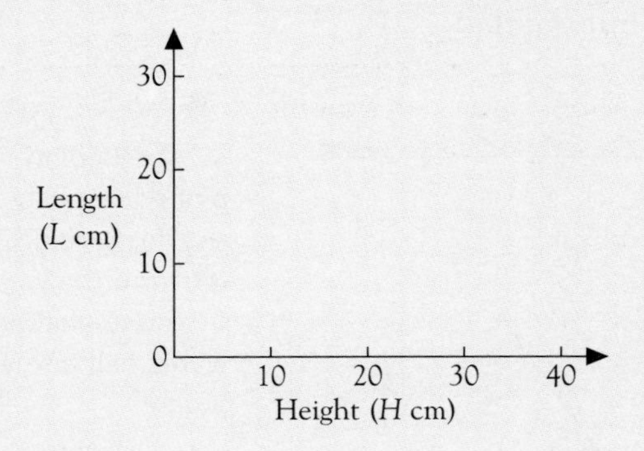

ii Draw a line of best fit.

b Mumtaz found a leaf which had dropped off the plant. Its length was 14 cm.

Use your scatter diagram to estimate the height of the leaf up the stem when it was on the plant. Show on the graph how you found your answer. [MEG]

5 The table gives information about the age and value of a number of cars of the same type.

Age (years)	1	3	$4\frac{1}{2}$	6	3	5	2	$5\frac{1}{2}$	4	7
Value (£)	8200	5900	4900	3800	6200	4500	7600	2200	5200	3200

a Use this information to draw a scatter graph, with axes like these.

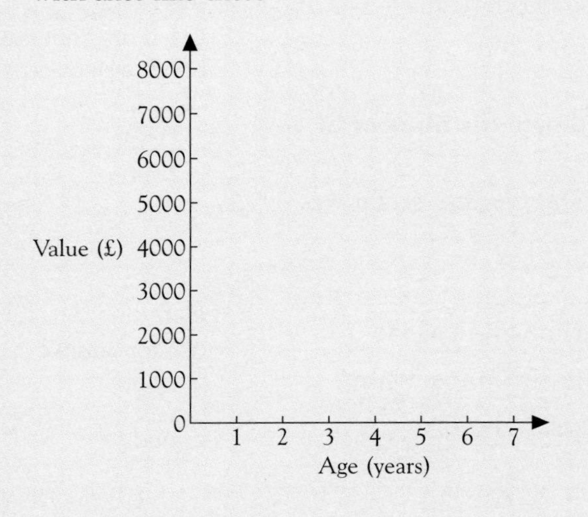

b What does the graph tell you about the value of these cars as they get older?

c The information is correct but the age and value of one of these cars looks out of place. Give a possible reason for this.

d Draw a line of best fit.

e John has a car of this type which is $3\frac{1}{2}$ years old and is in average condition.
Use the graph to estimate its value. [SEG]

6 Information about oil was recorded each year for 12 years. The table shows the amount of oil produced (in billions of barrels) and the average price of oil (in £ per barrel).

Amount of oil produced (billions of barrels)	Average price of oil (£ per barrel)
7.0	34
11.4	13
10.8	19
11.3	12
9.6	23
8.2	33
7.7	30
10.9	12.5
8.0	28.5
9.9	13.5
9.2	26.5
9.4	15.5

a Draw a scatter graph, using axes like these, to show the information in the table.

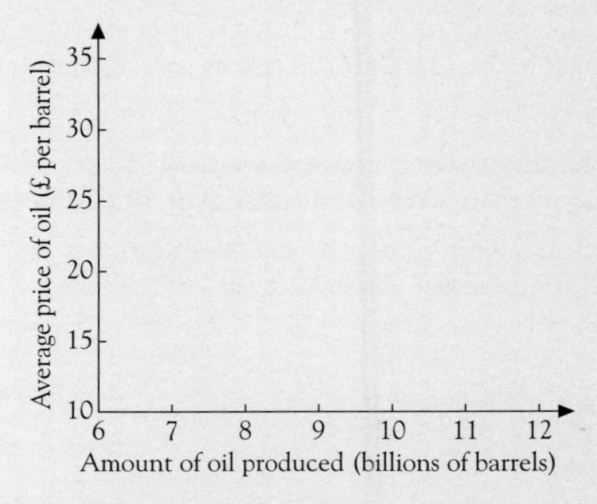

b Describe the correlation between the average price of oil and the amount of oil produced. [L]

Worked Exam Question

[MEG, p]

COMMENTS

An orchard contains nine young apple trees. The table shows the height of each tree and the number of apples growing on each.

Height (m)	1.5	1.9	1.6	2.2	2.1	1.3	2.6	2.1	1.4
Number of apples	12	15	20	17	20	8	26	22	10

a **On the axes below draw a scatter graph to illustrate this information.**

- Work out the scales on each axis.
  Height: 1 small square represents 0.1 m
  Number of apples: 1 small square represents 1 apple
- Plot the points in turn.

It is a good idea to tick them off in the table as you plot them.

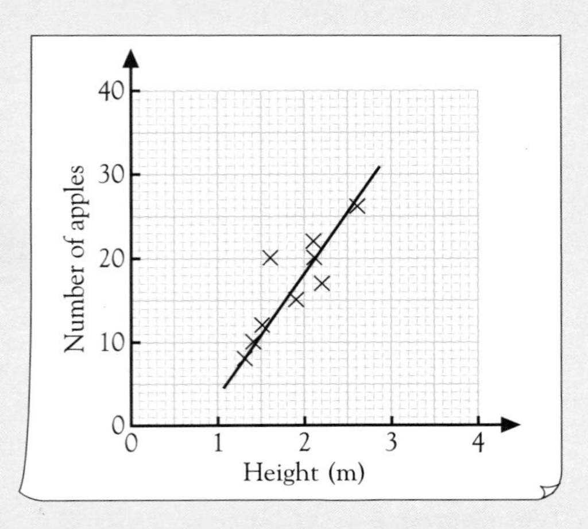

A3 for 9 points plotted correctly (**A2** for 8 points, **A1** for 6 or 7). (Accurate plotting to within half a small square)

3 marks

b **Comment briefly on the relationship between the height of the trees and the number of apples on the trees.**

- Look at the general trend. Say whether there is correlation or not and whether correlation is positive or negative.

You could say that the taller the tree, the more apples are growing on it.

There is good positive correlation.

B1 for positive correlation or equivalent statement.

1 mark

c **Add a line of best fit to your scatter graph.**

- Place your ruler so that there appears to be the same number of points above and below the line and the points are clustered around it.
- Draw the line, using a ruler.

B2 (**B1** if ruler not used or not well placed)

2 marks

d **Explain why it is not reasonable to use this line to estimate the number of apples on a tree of similar type but of height 4 m.**

- Relate your answer to the boundaries of the data used. The smallest height was 1.3 m and the largest 2.6 m.

It is unwise to make predictions for values outside the boundaries of your data.

4 m is much taller than the tallest tree in the data. The estimate might not be reliable for taller trees.

B2 (**B1** for sensible attempt)

2 marks

Exam Questions

1 500 people are asked to state their favourite snack. The pie chart shows their replies.

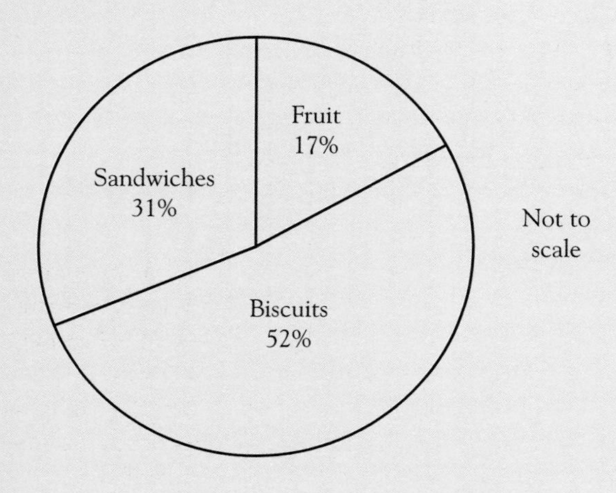

Of the 500 people, how many gave biscuits as their favourite snack? [NEAB]

2 In a school canteen the price of a banana depends on its weight.

The weights of 24 bananas for sale are recorded to the nearest gram as follows.

128 120 184 113 170 206 179 99

92 156 234 192 106 163 180 100

119 150 173 232 115 200 166 196

a The bananas are sold as small, medium or large. Copy and complete the grouped frequency table for these bananas.

Size	Weight (g)	Tally	Frequency
small	$80 \leq g < 120$		
medium	$120 \leq g < 180$		
large	$180 \leq g < 250$		

b Draw a pie chart to show the proportion of each size of banana. [SEG]

3 Julia plays in her school netball team. She will be selected to play for the county team next season if her mean score of goals for the school team after the first 7 games is at least 12.

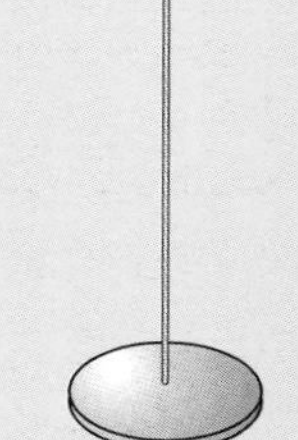

After 6 games her mean score of goals is 11.5.

a What is the total number of goals that Julia has scored in these 6 games?

b What is the least number of goals she must score in the next game in order to be chosen for the county team?

c In the seventh game she scores 13 goals.
Does her mean score of goals increase or decrease?
You must show all your working. [NEAB]

4 **a** The table shows the scores of 10 pupils in German and French tests.

German	18	25	20	38	50	15	44	20	25	42
French	22	22	15	33	42	21	38	15	32	33

Plot the data on a scatter diagram.

b Phil scored 30 in the German test. He was absent from the French test.
Use the scatter diagram to estimate his mark in the French test. [MEG, p]

5 11 people from different families were asked how many newspapers were usually bought by the family per week. The answers were

3, 10, 12, 8, 14, 10, 2, 6, 10, 4, 9

a What was the modal number of newspapers bought per week?

b What was the median number of newspapers bought per week?

c What was the mean number of newspapers bought per week? [NI]

EXAM QUESTIONS

6

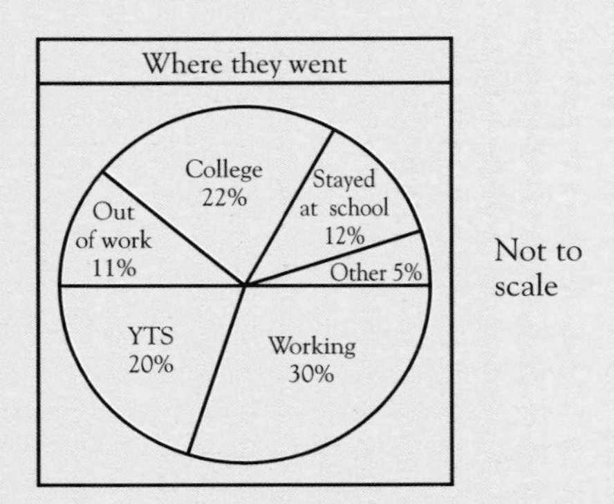

Not to scale

300 young people were asked what they did after completing Year 11 at school.

The pie chart shows the results of the survey.

a How many of the young people were working?

Gwen made an accurate drawing of the pie chart. She first drew the sector representing the young people out of work.

b Calculate the size of the angle of this sector. Give your answer correct to the nearest degree.

c Change to a decimal the percentage going to college.

d What fraction of the young people stayed at school? Give your answer in its simplest form. [L]

7 The table shows the best height cleared by 17 competitors in a high jump competition. Three competitors failed to clear any height at all.

Height (metres)	0	1.89	1.92	1.95	1.98	2.01	2.04
Frequency	3	3	3	2	1	4	1

a One of the 17 competitors is selected at random. Find the probability that this competitor jumped more than 2 metres.

b Write down the modal height.

c Find the median height.

d Find the mean height.

e Kerry is writing an article for an athletics magazine. Why would it not be sensible for her to use the mean height in her report?

[MEG]

8 The graph shows the number of people (in thousands) attending a holiday camp each month from March 1992 to February 1993.

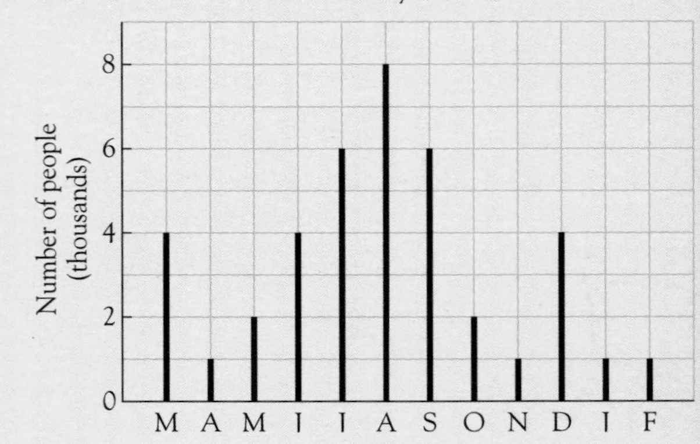

a How many people attended the holiday camp in April?

b Which was the modal month?

c What is the range of this data?

d Calculate the mean number of visitors per month.

e The holiday camp owners call March, April and May 'Spring'; June, July and August 'Summer'; September, October and November 'Autumn'; and December, January and February 'Winter'.

Complete this table to show the number of visitors (in thousands) to the holiday camp each season.

Spring	Summer	Autumn	Winter
7	18	9	

f Draw and label a pie chart to show the distribution of visitors by seasons.

You must show how you calculate the angles of your pie chart. [WJEC]

9 a

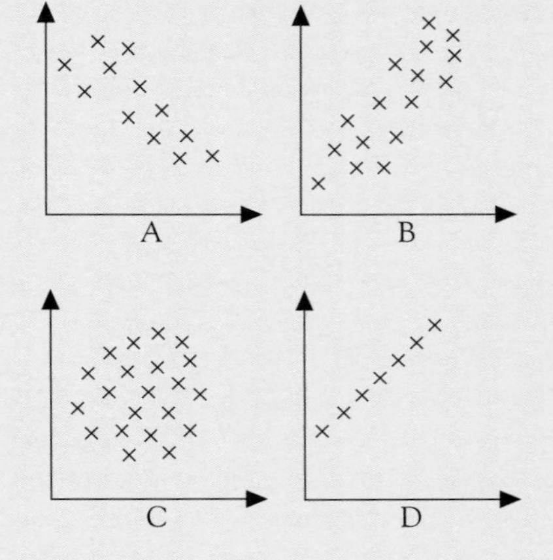

For each situation, choose the most appropriate from the four scatter diagrams.

i Boys' heights and their shoe sizes.

ii Men's weights and the times taken for the men to complete a crossword puzzle.

iii Ages of cars and their selling prices.

iv An example of negative correlation.

b The shoe sizes of 28 boys are recorded in the following table.

Shoe size	6	7	8	9	10
Frequency	9	6	6	5	2

i Write down the median shoe size.

ii Find the range of the shoe sizes.

iii The shoe sizes of seven girls have a median of 5 and a range of 6.

One of these girls takes size 11 shoes.

Give a possible list of the shoe sizes of the seven girls. [MEG]

REVISION CHECKLIST FOR STAGE 1

1. Perimeters

✓ Find the perimeter of a simple shape made up from rectangles

✓ Find the circumference of a circle

$C = \pi d$ 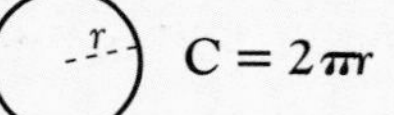 $C = 2\pi r$

✓ Find the lengths of arcs of circles

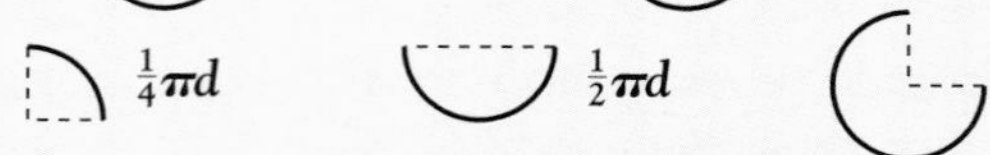

✓ Find the perimeter of shapes involving rectangles and circles

✓ Find the diameter or radius of a circle when you know the circumference $d = \frac{C}{\pi}$, $r = \frac{C}{2\pi}$

2. Number I

✓ Be able to interchange fractions, decimals and percentages

$\frac{3}{4} = 3 \div 4 = 0.75$ $\quad 35\% = \frac{35}{100} = \frac{7}{20}$ $\quad 0.42 = \frac{42}{100} = \frac{21}{50}$

$\frac{2}{5} = \frac{2}{5} \times 100\% = 40\%$ $\quad 80\% = \frac{80}{100} = 0.8$ $\quad 0.87 = 0.87 \times 100\% = 87\%$

✓ Find a fraction or percentage of something

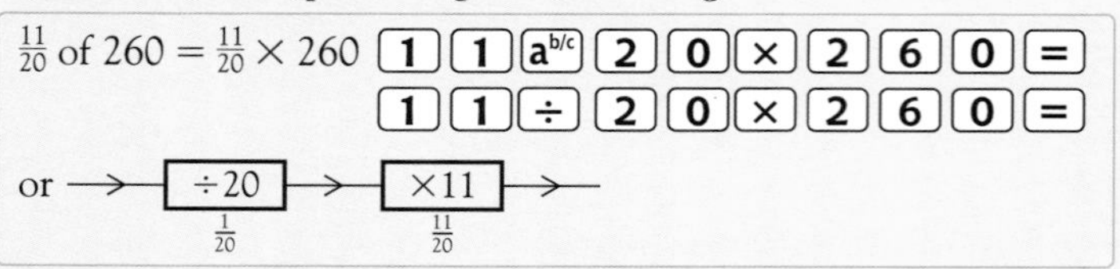

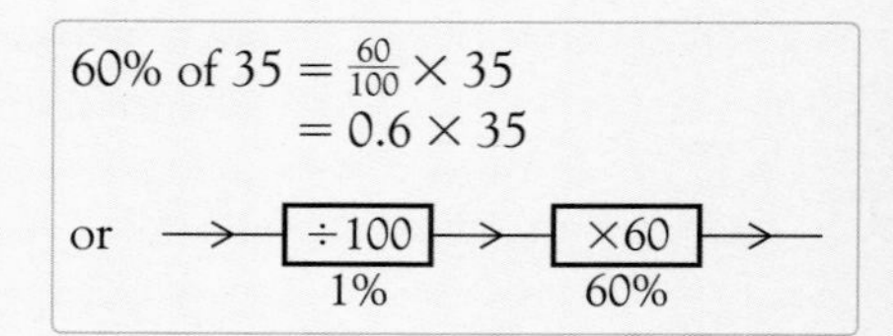

✓ Work out easy fractions and percentages without a calculator

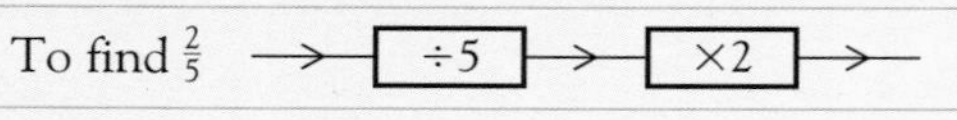

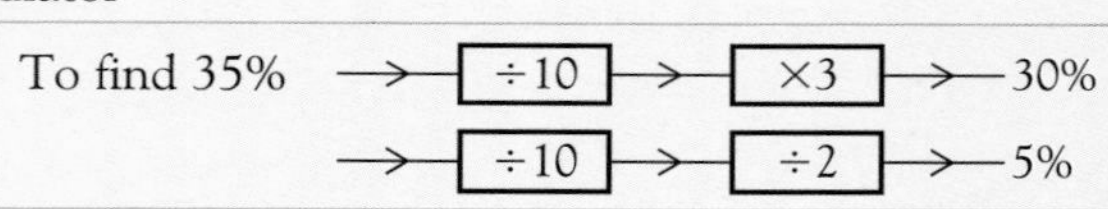

✓ Write a number to 1 significant figure (1 SF) and use it to make estimates

$27 \times 389 \approx 30 \times 400 = 12\,000$

3. Probability

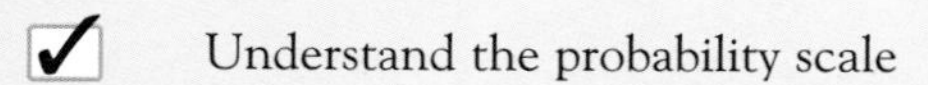

✓ Understand the probability scale

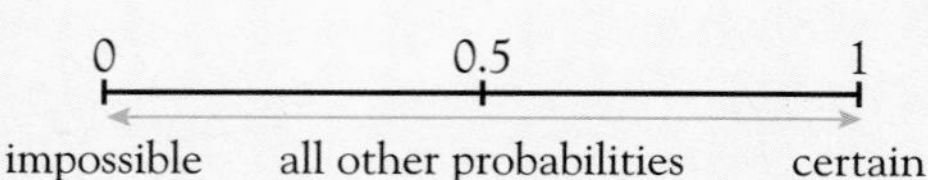

✓ Be able to list all the outcomes of an event systematically, including using a possibility space

✓ Know that, for equally likely events,

$$\text{P (event happens)} = \frac{\text{number of favourable outcomes}}{\text{total number of outcomes}}$$

✓ Know that mutually exclusive events cannot happen at the same time

✓ Know that, for mutually exclusive events,

P(A or B happens) = P(A happens) + P(B happens)

✓ Know that P(event does not happen) = 1 − P(event happens)

✓ Use relative frequency to estimate probability, where

$$\text{relative frequency} = \frac{\text{number of times event happened}}{\text{total number of trials}}$$

✓ Decide which method to use to estimate probability

equally likely outcomes | relative frequency | historical evidence

REVISION CHECKLIST FOR STAGE 1

4. Algebra I

✓ Be able to form an algebraic expression from a given situation

✓ Be able to substitute numbers into an expression to work out the value

If $x = 2$ and $y = 3$ then $7x^2 - 5y = 7 \times 2^2 - 5 \times 3 = 13$

✓ Know how to simplify an expression by collecting like terms

$12a + 3ab - 2a + 5ba + 3c = 10a + 8ab + 3c$

✓ Be able to write a formula and solve problems by substituting values into the formula

If $c = 12 + 8n$, when $n = 10$, $c = 12 + 8 \times 10 = 92$

✓ Be able to solve simple linear equations

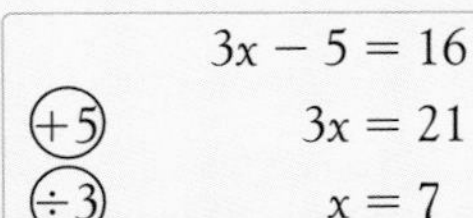

$3x - 5 = 16$

(+5) $3x = 21$

(÷3) $x = 7$

5. Geometry I

Know and use these angle facts about lines, triangles and quadrilaterals

✓ Angles on a straight line add up to 180°

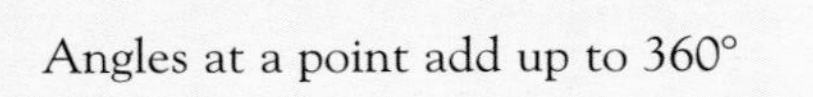

✓ Angles at a point add up to 360°

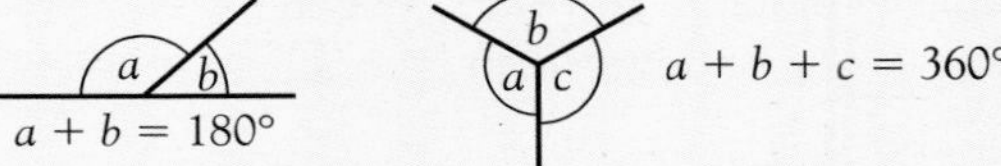

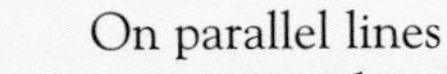

✓ On parallel lines
- corresponding (F) angles are equal ($a = b$)
- alternate (Z) angles are equal ($c = d$)
- interior($\sqsubset$) angles add up to 180° ($e + f = 180°$)

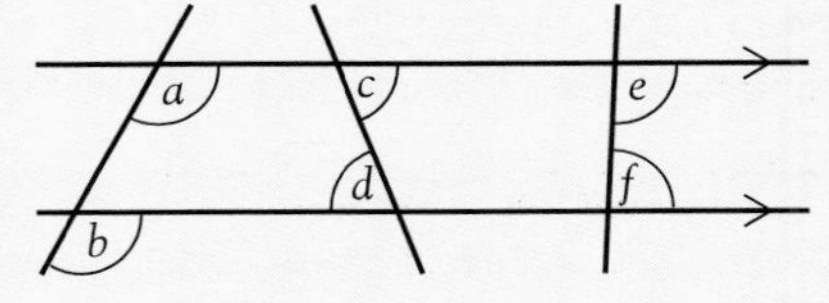

✓ The interior angles of a triangle add up to 180°

✓ An exterior angle of a triangle is equal to the sum of the interior opposite angles

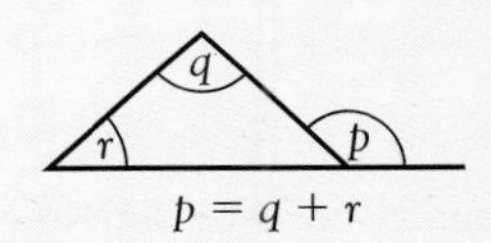

✓ The interior angles of a quadrilateral add up to 360°

✓ Be able to draw accurate diagrams

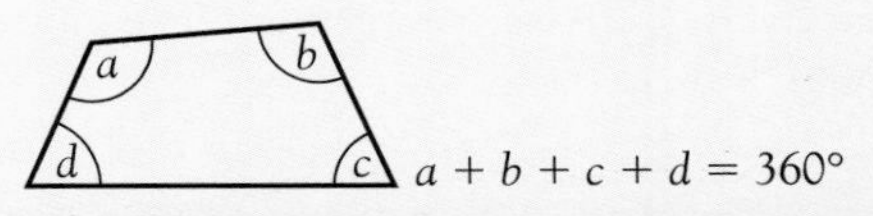

6. Linear Graphs I

✓ Make a table of values for a linear graph

✓ Be able to plot points to form a given straight line

$y = 2x + 1$

x	−1	0	1	2
y	−1	1	3	5

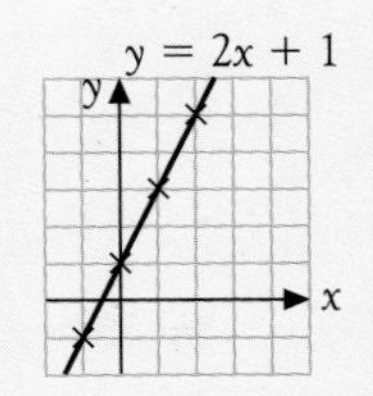

✓ Know how to plot special lines $x = 0$, $y = 0$, $y = x$, $y = -x$

✓ Know that $y = 0$ is the x axis and that $x = 0$ is the y axis

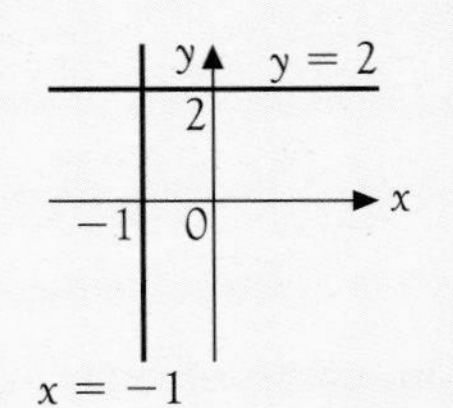

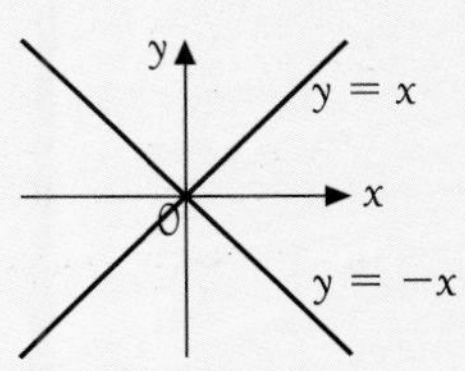

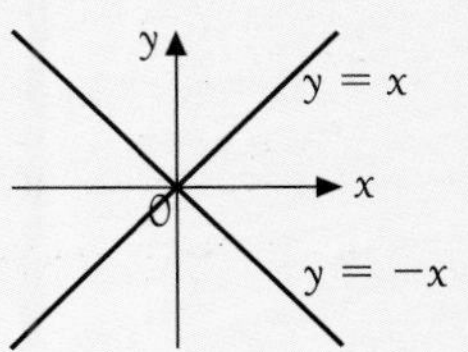

✓ Use graphs in practical situations

Conversion graphs

REVISION CHECKLIST FOR STAGE 1

7. Area and Volume I

- ✓ Work out the area of a rectangle

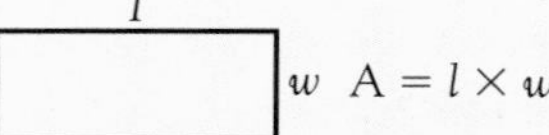

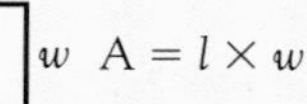

- ✓ Work out the area of a triangle

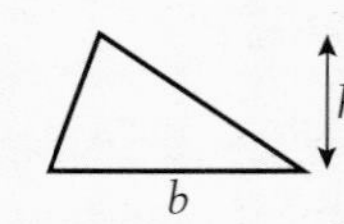

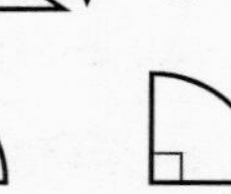

- ✓ Find the area of a circle and part of a circle
- ✓ Find the area of shapes involving rectangles and circles

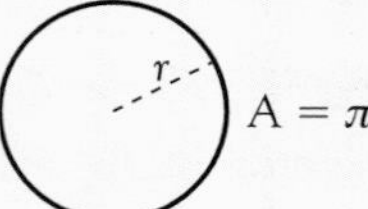

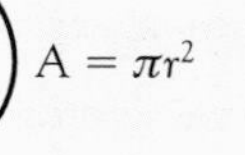

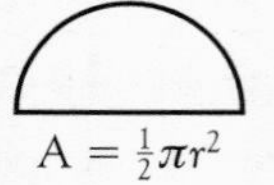

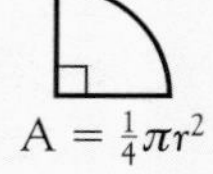

- ✓ Find the volume of a cuboid
- ✓ Draw the net of a cuboid

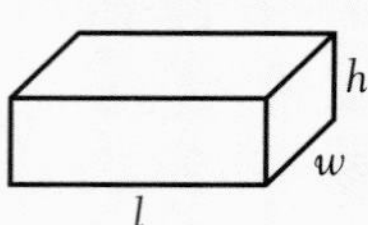

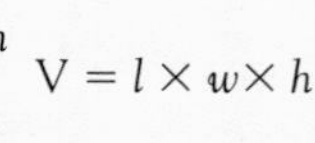

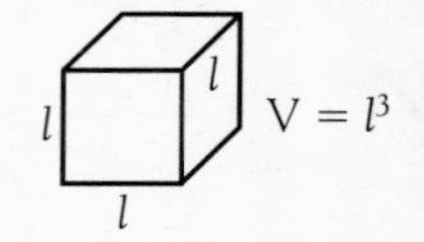

- ✓ Find the surface area of a cuboid by finding the area of each face

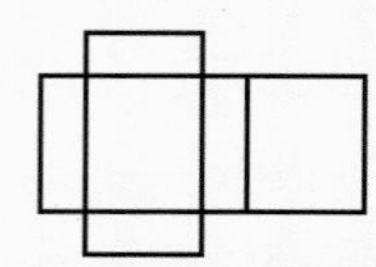

8. Transformations I

- ✓ Translate an object using a vector $\begin{pmatrix} a \\ b \end{pmatrix}$

  a tells how far along to go (→ +, ← −)
  b tells how far up or down to go (↑ +, ↓ −)

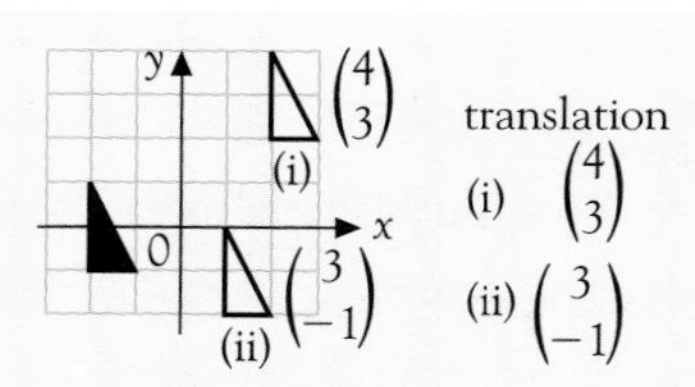

- ✓ Reflect a shape in a given mirror line

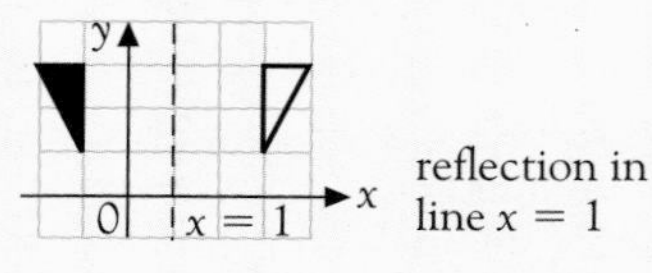

- ✓ Rotate a shape through given angle, in a given direction (clockwise or anti-clockwise) about a given centre

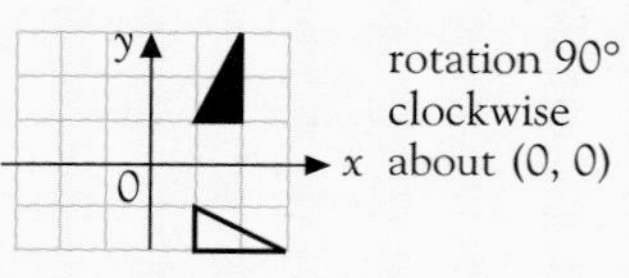

- ✓ Enlarge a shape by a given scale factor, using a given centre of enlargement if requested

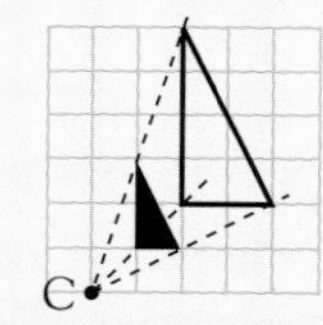

enlargement with scale factor 2, centre C

9. Data Handling I

- ✓ Draw and interpret bar charts, pie charts, histograms and line graphs
- ✓ Find the mean, mode and median of sets of data **mode:** most often **median:** middle when in order of size

$$\textbf{mean} = \frac{\text{total of all the values}}{\text{number of values}}, \qquad \textbf{mean} = \frac{\text{total of } (f \times x) \text{ column}}{\text{total of } f \text{ column}}$$

- ✓ Compare sets of data by looking at the means and ranges

  Range = highest value − lowest value

- ✓ Plot scatter diagrams to show the connection between two variables
- ✓ Draw a line of best fit on a scatter diagram
- ✓ Describe relationships in scatter diagrams using positive and negative correlation

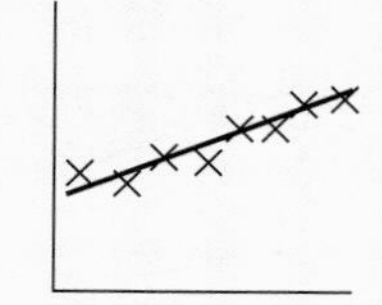

strong positive correlation

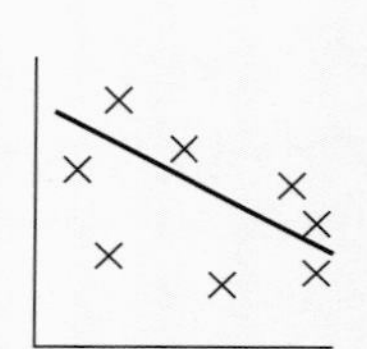

weak negative correlation

MIXED EXAM QUESTIONS: END OF STAGE 1

The same type of training shoe is advertised in three different shops.

Tom's Cheap Shoes	PETE'S MARKET STALL	City of Shoes
£25	£48	£40
plus VAT at $17\frac{1}{2}$%	WITH $\frac{1}{3}$ OFF	with 25% discount

Work out the cost of the training shoe in each of the three shops. [L]

2 **a** The number of new cars registered in August 1994 was 460 000.

i One quarter of the cars were red. Find the number of red cars.

ii 36% of the cars were blue. Find the number of blue cars.

iii There were 2400 black cars. What fraction of the cars were black?

b A car dealer sold 120 new cars in August. He made a note of the colours of the cars and decided to draw a pie chart to illustrate the popularity of different colours.

i Complete the table below to show the pie chart angles for each colour.

Car colour	Number of cars	Pie chart angle
Red	55	
Blue	32	
White	20	
Grey	13	
Total	120	

ii Represent this information on a pie chart. [MEG]

3 **a** The diagram shown has not been drawn accurately. When drawn accurately, POQ is a straight line, angle ROS is a right angle and $a + b + c = 218°$.

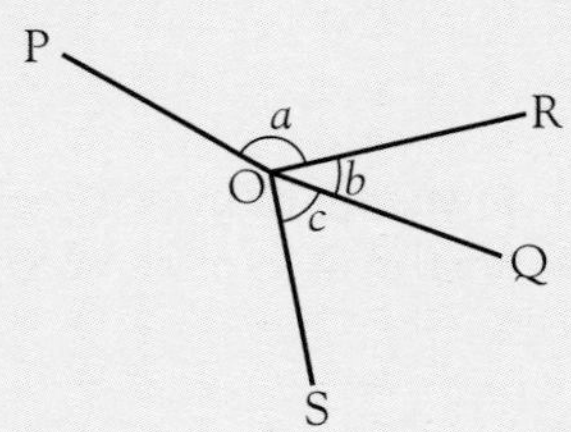

Work out the sizes of angle a and angle c.

b The diagram shows the frame of a rectangular gate with four parallel bars.

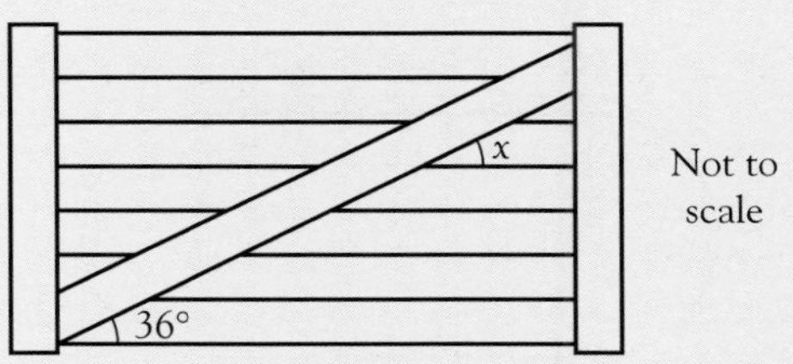

Write down the size of angle x giving a reason for your answer.

c ABCD is a kite.

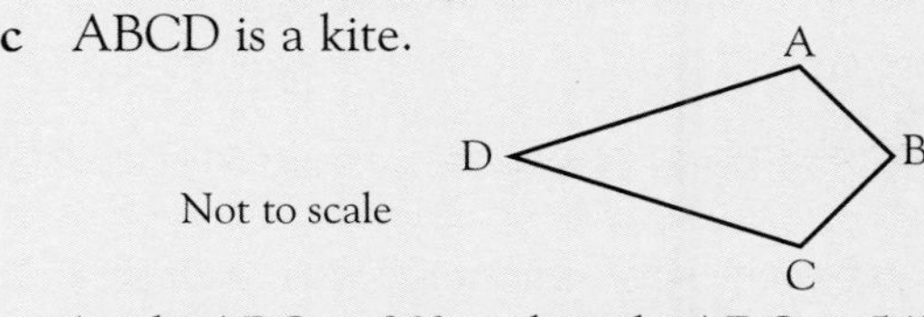

Angle ABC = 90° and angle ADC = 54°.

Calculate the size of angle BAD. [SEG]

4 A bag contains 20 sweets. There are 9 mints, 4 toffees and 7 sherbets.

Weishun picks a sweet at random from the bag.

a What is the probability that she picks:

i a mint,

ii a sherbet?

b The only sweets that Weishun does not like are toffees.
What is the probability that she picks a sweet she likes? [MEG]

5 A question in a survey about hunting was:

"Don't you agree that we should stop foxes being brutally killed by banning fox hunting?"

Another version of the same question was:

"Should fox hunting be banned?"

Which version would you use and why? Give two reasons. [MEG]

6 Copy these diagrams on squared paper and reflect the shapes in the mirror lines.

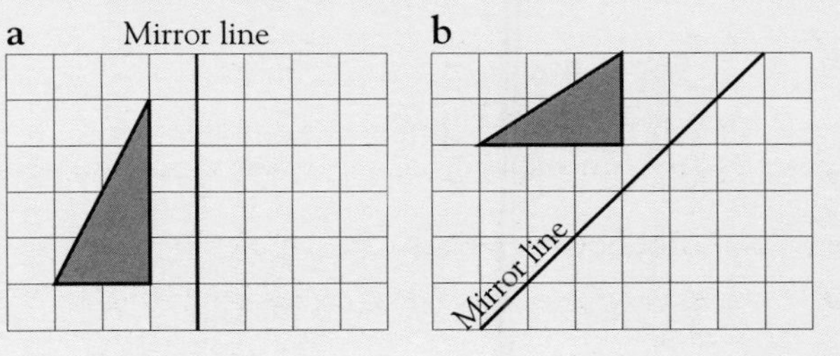

[L]

MIXED EXAM QUESTIONS: END OF STAGE 1

7 A sports arena is being planned. It is a rectangle and two semi-circles.

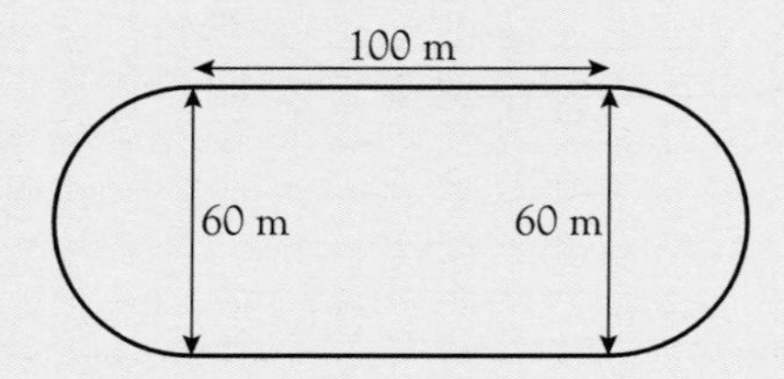

a Calculate the length of fencing required for the perimeter of the arena.

b Calculate the area of turf required to cover the arena. [MEG]

8 The table gives information about the age and value of nine bikes.

Age of bike (to the nearest year)	1	4	7	2	6	3	2	4	6
Value of bike (to the nearest £10)	270	160	40	200	60	190	230	110	100

a Use this information to draw a scatter graph.

b What does the scatter graph tell you about the connection between the age of bikes and their value?

c Use the scatter graph to estimate:

i the value of a bike which is 5 years old,

ii the age of a bike with a value of £250. [SEG]

9 Work out:

a **i** $\frac{5}{8}$ of £9.60,

ii 24% of 35 metres.

b Change $\frac{3}{8}$ into:

i a decimal fraction,

ii a percentage. [L]

10 **a** A bag contains just three balls.

Each ball is of a different colour: a white, a red and a black.

Anil chooses a ball.

Carl chooses one of the remaining two balls.

Mark has the ball that remains.

i Copy and complete the list of all the possible outcomes.

Anil	Carl	Mark
white	red	black

ii Write down the probability that Anil has a red ball.

b Another bag contains 7 green balls, 8 blue balls and 5 yellow balls.

A ball is chosen at random from this bag.

Write down the probability that this ball is:

i green,

ii not green. [MEG]

11 Here are the names of some quadrilaterals.

Square, Rectangle, Rhombus, Parallelogram, Trapezium, Kite

a Write down the names of the quadrilaterals which have two pairs of parallel sides.

b Write down the names of the quadrilaterals which must have two pairs of equal sides. [L]

12 Hemlata measured the lengths of 12 different leaves that she collected for a Science experiment.

Here are her results.

15 cm 12 cm 9 cm 10 cm 9 cm 7 cm
14 cm 10 cm 8 cm 9 cm 12 cm 6 cm

a For the leaves Hemlata collected, work out the:

i mean, **ii** mode,
iii range, **iv** median.

b Write down the probability that a leaf chosen at random from Hemlata's collection has a length:

i equal to 10 cm,

ii less than 10 cm. [L]

13 The diameter of a 10 franc coin is 23 millimetres.

a **i** Write down the radius of the coin.

ii Work out the area of the face of the coin.

Pierre rolled the coin across his desk for six complete turns of the coin.

b How far did it roll? [MEG]

MIXED EXAM QUESTIONS: END OF STAGE 1

14

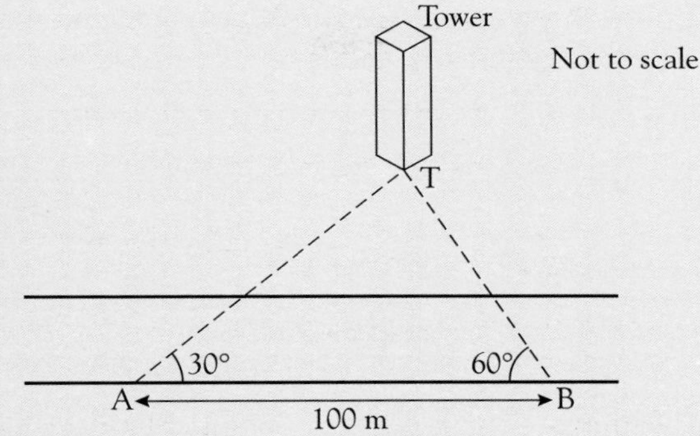

Ceri and Diane want to find how far away a tower T is on the other side of a river.

To do this they mark out a base line, AB, 100 metres long as shown on the diagram.

Next they measure the angles at the ends A and B between the base line and the lines of sight of the tower. These angles are 30° and 60°.

a Use ruler and compasses only to make a scale drawing of the situation.
Use a scale of 1 cm to represent 10 m.
Show clearly all your construction lines.

b Find the shortest distance of the tower, T, from the base line AB. [NEAB]

15 There are different ways of estimating probabilities:

Method 1: Use equally likely outcomes.
Method 2: Look at past records.
Method 3: Carry out a survey or do an experiment.

Look at the following situations. Say whether you would use Method 1, Method 2 or Method 3 to estimate the probability.

a The probability that it will rain in London on the next St. Swithin's Day.

b The probability that, when a stick of chalk is dropped, it will break into three pieces.

c The probability that, when an ordinary dice is thrown, the score will be 6.

d The probability that, when a drawing pin is thrown, it will land with its point upwards. [MEG]

16

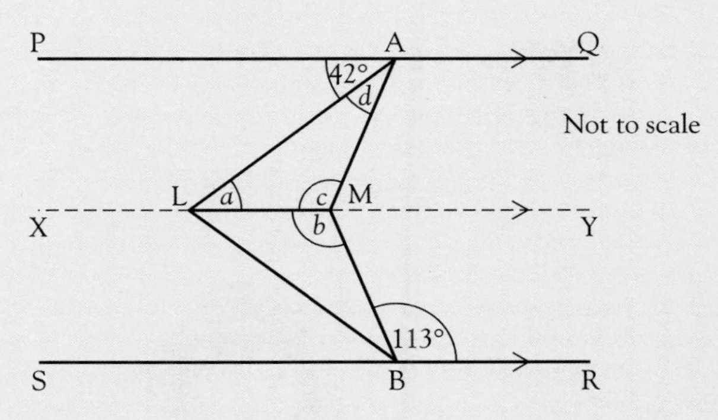

In the diagram, lines PQ, XY and SR are parallel.
Angle PAL = 42° and angle MBR = 113°.
Line XY is a line of symmetry of the figure.

a Find angle a, giving a reason for your answer.

b Find angle b, giving a reason for your answer.

c Explain why angle c is equal to angle b.

d Calculate angle d, explaining how you do this. [MEG]

17 Annie asked a group of teenagers to say how much time they spent doing homework one evening and how much time they spent watching TV.

Here is a scatter diagram to show the results:

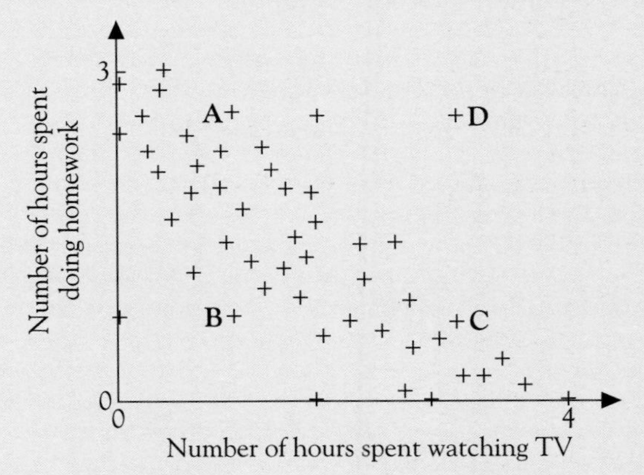

a Which of the four points **A**, **B**, **C** or **D** represents each of the statements shown below? Write one letter for each statement.

I watched a lot of TV last night and I also did a lot of homework.

This is represented by point ......

MIXED EXAM QUESTIONS: END OF STAGE 1

This is represented by point ……

This is represented by point ……

b Make up a statement which matches the fourth point.

c What does the graph tell you about the relationship between time spent watching TV and time spent doing homework?

d Annie also drew scatter diagrams which showed that:

> Older students tend to spend more time doing homework than younger students.
>
> There is no relationship between the time students spend watching TV and the time students spend sleeping.

On axes like those below, show what Annie's scatter diagrams may have looked like.

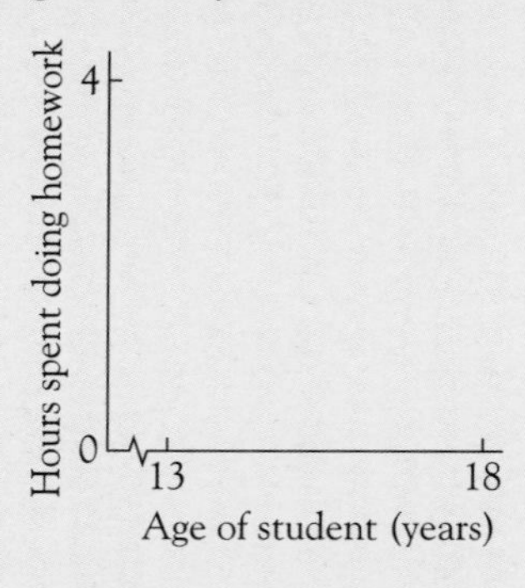

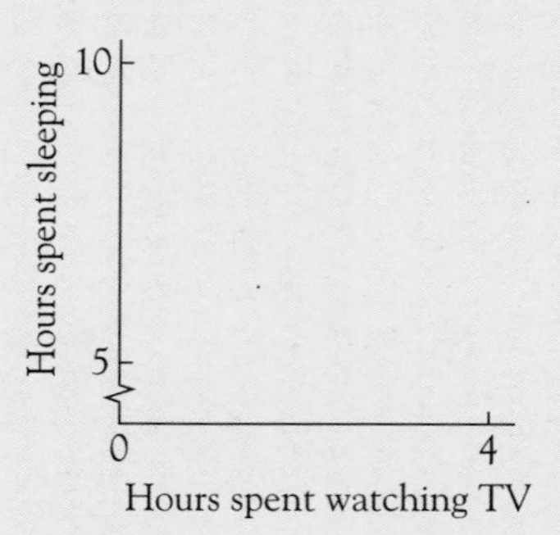

[NEAB]

18 Pierre throws a cricket bat into the air and makes a note of whether it lands on its front or its back. He does this 50 times. Here are his results.

	Tally	Frequency
Landed on front	𝍸 𝍸 𝍸 𝍸 |	
Landed on back	𝍸 𝍸 𝍸 𝍸 𝍸 ||||	

a Complete the frequency column in the table.

b From the table, work out the experimental probability of:

i the cricket bat landing on its front,

ii the cricket bat landing on its back.

c If you threw the bat in the air 100 times, about how many times would you expect it to land on its front? [MEG]

19 Write down an expression in terms of n and g for the total cost, in pence, of n buns at 18 pence each and 5 bread rolls at g pence each. [L]

20 You want to estimate the value of 21.2×31.2. Each number must be written to 1 significant figure.

a Write down a suitable calculation which could be used.

b State the value of this estimate. [L]

21

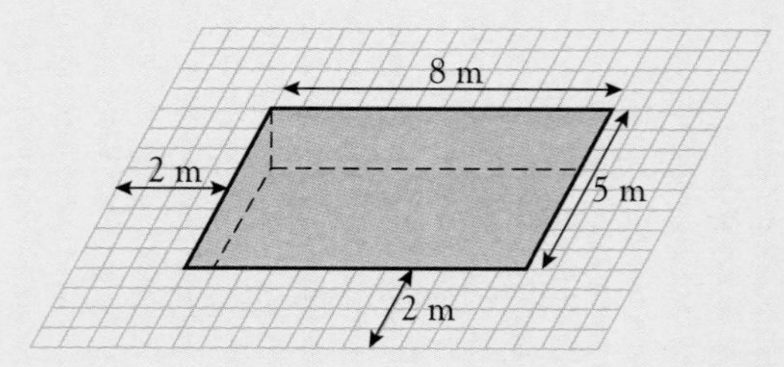

The diagram represents the babies' pool, with paving around, at a leisure centre.
The pool is rectangular, 8 m long by 5 m wide and has a depth of 0.6 m throughout.

a Work out the volume of the pool in m^3.

The paving around the pool is 2 m wide.

b Work out the area of the paving. [L]

MIXED EXAM QUESTIONS: END OF STAGE 1

22

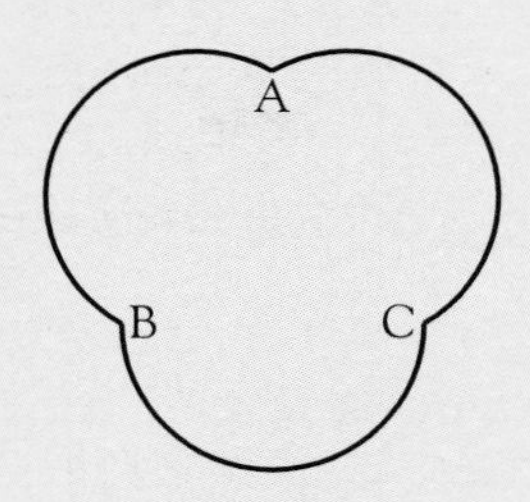

This shape consists of 3 semicircles.

The diameters of the semicircles, AB, BC and CA, are each 3 cm long.

a The points A, B, C are the corners of a triangle.
What special type of triangle is it?

b **i** Calculate the circumference of a circle with diameter 3 cm.

ii Calculate the perimeter of the shape drawn above. [MEG]

23 **a** Write down and simplify a formula for the total perimeter of this shape.

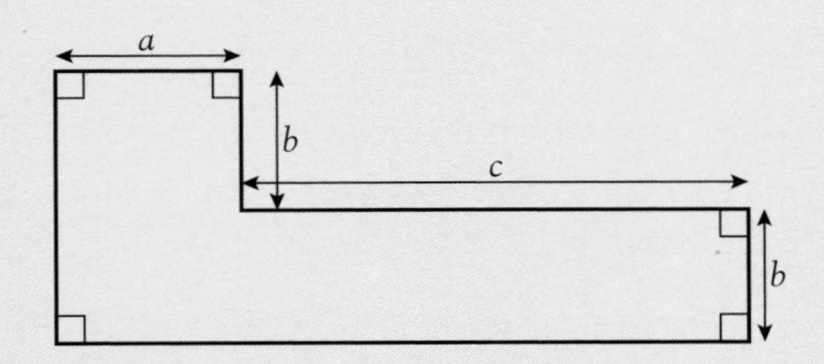

b Use your formula to work out the perimeter when $a = 3$, $b = 2$, $c = 4$. [L]

24 In 1990, a charity sold $2\frac{1}{4}$ million lottery tickets at 25p each.
80% of the money obtained was kept by the charity.

a Calculate the amount of money kept by the charity.

In 1991, the price of a lottery ticket fell by 20%.
Sales of lottery tickets increased by 20%.
80% of the money obtained was kept by the charity.

b Calculate the percentage change in the amount of money kept by the charity. [L]

25

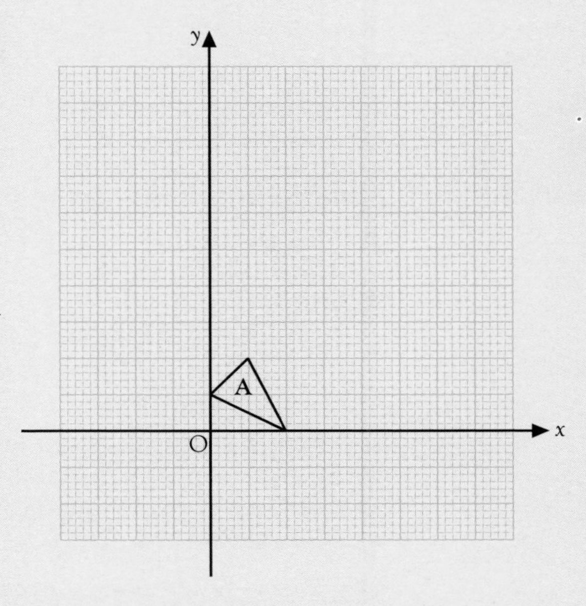

Copy this diagram onto graph paper.

a Reflect triangle **A** in the y axis. Label it **B**.

b Rotate triangle **A** through 90° clockwise, centre the origin, O. Label it **C**.

c Enlarge triangle **A** by scale factor 3, centre the origin, O. Label it **D**. [MEG, p]

26

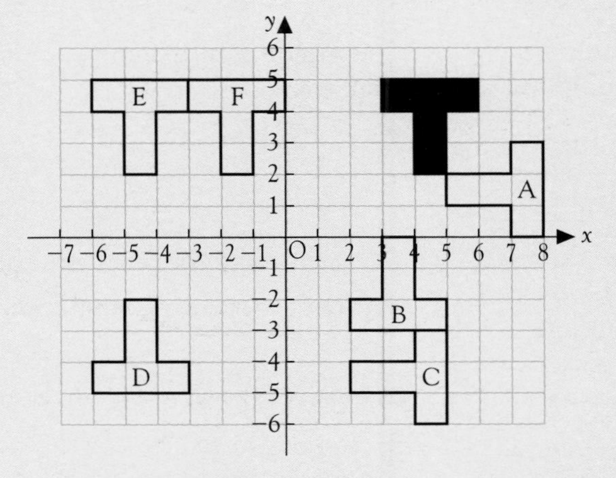

a Give the letter of the finishing position after:

i the shaded **T** shape has been reflected in the y axis,

ii the shaded **T** shape has been rotated $\frac{1}{4}$ turn clockwise about O,

iii the shaded **T** shape has been translated 6 units to the left.

b Describe the transformation which will map the shaded **T** shape onto shape D. [NEAB]

FACT SHEET 6: THE LANGUAGE OF NUMBER

Here are some key words about number that you should know.
Learn them and their meanings, using the examples to help you to understand.

Integers

Integers are whole numbers. They can be positive, negative or zero, for example -5, -1, 0, 4, 27, 345.

Multiples

Multiples of a number can be divided exactly by that number with no remainder.

- Multiples of 3:
  3, 6, 9, 12, 15, 18, 21, ...
- Multiples of 7:
  7, 14, 21, 28, 35, 42, ...
- Is 2594 a multiple of 6?
  $2594 \div 6 = 432.33$
  No, as there is a remainder
- Is 2104 a multiple of 8?
  $2104 \div 8 = 263$
  Yes, as it divides exactly

Multiples of 2 are called **even numbers**. Multiples of 2 end in 0, 2, 4, 6 or 8, for example 12, 34, 38, 2466, 5000.
Numbers which are not multiples of 2 are called **odd numbers**, for example 3, 23, 51, 279.

Quick ways of checking for multiples without a calculator:

- If a number is a **multiple of 3**, then its digits add up to 3, 6 or 9.

375 is a multiple of 3 since $375 \div 3 = 125$.
Check the digit sum of 375:
$3 + 7 + 5 = 15$
then again $1 + 5 = 6$

- If a number is a **multiple of 9**, then its digits add up to 9.

522 is a multiple of 9 since $522 \div 9 = 58$.
Check the digit sum of 522:
$5 + 2 + 2 = 9$

- If a number is a **multiple of 5**, then it ends in 0 or 5.

Examples are
35, 70, 195, 2005

Common multiples

A **common multiple** of two numbers is a number that can be divided exactly by both the numbers.

- $48 \div 4 = 12$, $48 \div 6 = 8$, so 48 can be divided exactly by both 4 and 6.
  48 is a common multiple of 4 and 6.
  Some other common multiples of 4 and 6 are 12, 24, 36, 60, ...

Lowest common multiple

The smallest of the common multiples is called the **lowest common multiple**.

- The lowest common multiple of 4 and 6 is 12

Factors

Factors of a number are whole numbers that divide into the number exactly.

- To find the factors of 12, find all the pairs of numbers that multiply to give 12:
  $12 = 1 \times 12$
  $12 = 2 \times 6$
  $12 = 3 \times 4$
  The factors of 12 are
  1, 2, 3, 4, 6 and 12
- To find the factors of 45, find all the pairs of numbers that multiply to give 45
  $45 = 1 \times 45$
  $45 = 3 \times 15$
  $45 = 5 \times 9$
  The factors of 45 are
  1, 3, 5, 9, 15, 45

Notice that 1 is a factor of **every** number.

Common factors

A **common factor** of two numbers is a factor of both numbers.

- 1 and 3 are the only numbers that are factors of both 12 and 45, so the common factors of 12 and 45 are 1 and 3

FACT SHEET 6: THE LANGUAGE OF NUMBER

Highest common factor

The largest of the common factors is called the **highest common factor**.
3 is the **highest common factor** of 12 and 45 as it is the highest number that divides exactly into both 12 and 45.

Prime numbers

A **prime number** is a number that has **exactly two factors**.
A prime number can be divided exactly only by 1 and itself.

◆ Is 7 a prime number?
$7 = 1 \times 7$
These are the only whole numbers that multiply to give 7,
so 7 has exactly two factors,
<u>7 is a prime number</u>

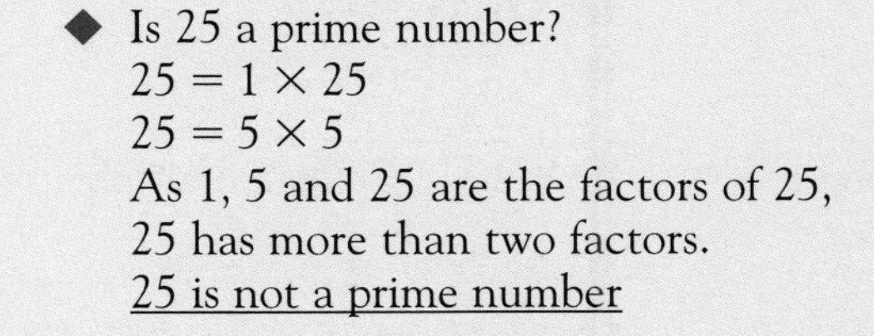

◆ Is 25 a prime number?
$25 = 1 \times 25$
$25 = 5 \times 5$
As 1, 5 and 25 are the factors of 25,
25 has more than two factors.
<u>25 is not a prime number</u>

The first prime number is 2.
The first few prime numbers are
2, 3, 5, 7, 11, 13, 17, 19, 23, 29, ...

Note that $1 = 1 \times 1$ so 1 has only one factor; it is not a prime number.
Also note that 2 is the only even prime number.

Prime factors

Prime factors are factors of a number that are also prime.

◆ Find the prime factors of 45.
The factors of 45 are 1, 3, 5, 9, 15, 45.
Of these, 3 and 5 are prime numbers, so the <u>prime factors of 45 are 3 and 5</u>.

Index numbers (powers)

Index numbers (indices) are small, raised numbers, sometimes called **powers**.

◆ $n^2 = n \times n$ This is 'n to the power of 2', usually read as 'n squared'.
◆ $n^3 = n \times n \times n$ This is 'n to the power of 3', usually read as 'n cubed'.
◆ $n^4 = n \times n \times n \times n$ This is 'n to the power of 4', sometimes shortened to 'n to the 4'. It is not correct to say 'n four'.

$2^5 = 2 \times 2 \times 2 \times 2 \times 2 = 32$
To find this using the calculator, use x^y button
Try it on your calculator.

2^5 is read as '2 to the power of 5' or '2 to the 5'

[2] [x^y] [5] [=] gives 32

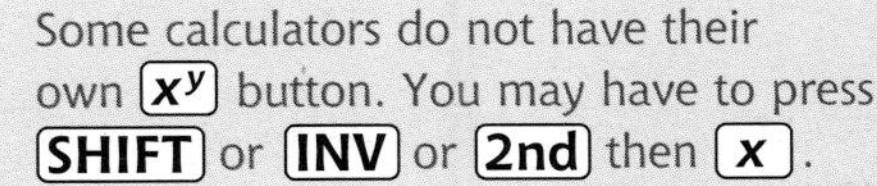

Some calculators do not have their own [x^y] button. You may have to press [SHIFT] or [INV] or [2nd] then [x].

To square a number

To **square** a number, multiply the number by itself.

◆ $3^2 = 3 \times 3 = 9$
◆ $1.4^2 = 1.4 \times 1.4 = 1.96$
◆ $(-8)^2 = (-8) \times (-8) = 64$

Remember that squaring a number is the same as 'raising it to the power of 2', so you could use the [x^y] button on the calculator – try it.

Make sure that you can use the [x^2] button on your calculator, especially when squaring a negative number.

Notice that the square of a number is **always positive**, whether the number being squared is positive or negative.

Square number

The square of a whole number is called a **perfect square** or a **square number**.
The first few square numbers are 1, 4, 9, 16, 25, 36
since $1^2 = 1$, $2^2 = 4$, $3^2 = 9$, $4^2 = 16$, $5^2 = 25$, $6^2 = 36$

FACT SHEET 6: THE LANGUAGE OF NUMBER

The square root of a number ($\sqrt{\ }$)

The **square root** of a number, when multiplied by itself, gives the number. The symbol for square root is $\sqrt{\ }$.

- $\sqrt{49} = 7$ because $7 \times 7 = 49$

Use the square root button [√] on your calculator to check these square roots.

- $\sqrt{121} = 11$,
- $\sqrt{0.0004} = 0.02$,
- $\sqrt{27.36} = 5.230 \ldots$

On some models [√] has to be keyed in before the number and on other models after the number. Make sure you know how to use yours.

To cube a number

To **cube** a number, multiply it by itself, then by itself again.

- $4^3 = 4 \times 4 \times 4 = 64$

On the calculator, key in [4] [×] [4] [×] [4] [=]

Alternatively, since cubing a number is the same as 'raising it to the power 3', use the [x^y] button. For example, to find 29^3 press

[2] [9] [x^y] [3] [=] giving $29^3 = 24389$

Cube numbers

The cube of a whole number is called a **perfect cube** or **cube number.** The first few cube numbers are 1, 8, 27, 64, 125, 216 since $1^3 = 1$, $2^3 = 8$, $3^3 = 27$, $4^3 = 64$, $5^3 = 125$, $6^3 = 216$.

Cube root of a number ($\sqrt[3]{\ }$)

The **cube root** of 343 is 7 because $7 \times 7 \times 7 = 343$. This is written $\sqrt[3]{343} = 7$

- $\sqrt[3]{27} = 3$
- $\sqrt[3]{0.008} = 0.2$,
- $\sqrt[3]{350} = 7.047 \ldots$

Try using the [∛] button on your calculator to check these. It is often on a SHIFT or 2nd function.

Writing a number as a product of primes

Every number can be written as a **product of prime numbers**, that is as a set of prime numbers multiplied together.

- Write 360 as a product of primes

$$\begin{aligned} 360 &= 36 \times 10 \\ &= 12 \times 3 \times 5 \times 2 \\ &= 4 \times 3 \times 3 \times 5 \times 2 \\ &= 2 \times 2 \times 3 \times 3 \times 5 \times 2 \\ &= 2 \times 2 \times 2 \times 3 \times 3 \times 5 \end{aligned}$$

Now write this in index form: $\underline{360 = 2^3 \times 3^2 \times 5}$

Break down the number into a pair of factors (36×10) then break down these numbers into other pairs ($36 = 12 \times 3$ and $10 = 5 \times 2$). Keep breaking them down into pairs until **all** factors are prime numbers.

Reciprocal

The **reciprocal of a number** n is $\frac{1}{n}$. The **reciprocal of a fraction** $\frac{a}{b}$ is $\frac{b}{a}$.

- the reciprocal of 2 is $\frac{1}{2}$; the reciprocal of 8 is $\frac{1}{8}$; the reciprocal of $\frac{2}{3}$ is $\frac{3}{2}$.

To find the reciprocal of a number, you can use the [1/x] button on the calculator. Sometimes this is [x^{-1}] on a [SHIFT] or [INV] or [2nd] button.

[2] [1/x] gives 0.5, [8] [1/x] gives 0.125

[2] [$a^{b/c}$] [3] [1/x] gives 1.5

Notice that the calculator gives these reciprocals as decimals. Check that they are the same as the fractions.

When you multiply a number by its reciprocal, the answer is 1.

$2 \times \frac{1}{2} = 1$ $\quad 8 \times \frac{1}{8} = 1$ $\quad \frac{2}{3} \times \frac{3}{2} = 1$

NUMBER II

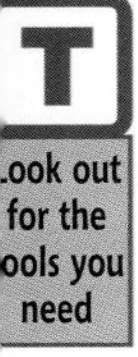

In this chapter you will learn how to
- **use different types of numbers**
- **approximate to a given number of significant figures**
- **find missing amounts in a ratio**
- **tackle problems involving direct proportion**
- **use ratios in enlargements and map scales**
- **share (or divide) something in a given ratio**

Types of number

For this section you need to know the words in FACT SHEET 6, The Language of Number (pages 174–176). Look them up if you are unsure.

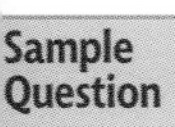

1 Mrs Hardy decided that her three children should help her occasionally with the tasks in the kitchen. She told them that John should help every second day, Katy every third day and David every fourth day, starting counting on January 1st.

a Who helped on January 16th?

b Which were the first three dates on which they all helped?

c Find the percentage of days in January when no-one helped Mrs Hardy.

Answer

This question is about **multiples**.

- Start to list the days when each child helped. This will show any patterns.

John
Multiples of 2
2, 4, 6, 8, 10, 12, 14, 16, …

Katy
Multiples of 3
3, 6, 9, 12, 15, 18, …

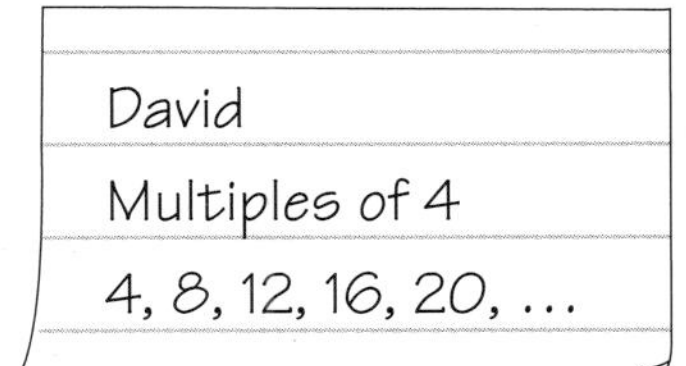

 Even numbers

 '3 times' table

 '4 times' table

a

- Look for 16 in the lists. Notice that 16 is a multiple of 2 and 4, but not of 3.
<u>John and David helped on January 16th.</u>

TYPES OF NUMBER

b

- Find the lowest number that is in all three lists.

  Lowest number = 12

This is the **lowest common multiple** (L.C.M.) of 2, 3 and 4.

- They all helped on days that were multiples of 12, so list the first three multiples of 12 and convert them to dates.

  12 ⟶ January 12th
  24 ⟶ January 24th
  36 ⟶ February 5th

Day 31 is January 31st, so day 32 is February 1st.

They all helped on January 12th, January 24th and February 5th.

c

- List the numbers up to 31 that are not multiples of 2, 3 or 4 and count how many there are.

First eliminate all the even numbers (multiples of 2) and then all the numbers not in the '3 times' table. The numbers in the '4 times' table have already been eliminated. Why?

1, 5, 7, 11, 13, 17, 19, 23, 25, 29, 31

Number of days = 11

- Write this number as a percentage of 31.

FACT SHEET 1 page 15

$$\text{Percentage of days} = \frac{11}{31} \times 100\%$$

Percentage of days = 35% (nearest whole number)

Sample Question 2

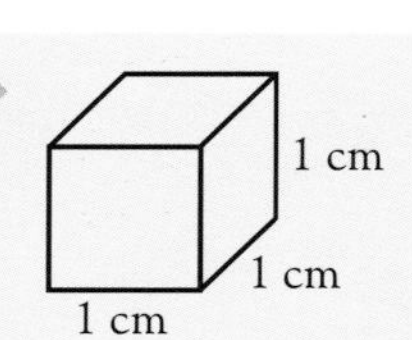

A large cube is made from 1728 small cubes like the one shown here.

a What is the length of a side of the large cube?

b What is the surface area of the large cube?

Answer

a

- Work out the volume of the large cube.

  Volume of small cube = $1 \times 1 \times 1 = 1\text{ cm}^3$

  Volume of large cube = $1728 \times 1 = 1728\text{ cm}^3$

- Write down the formula for a cube of side x cm.

$$V = x \times x \times x$$
$$= x^3$$

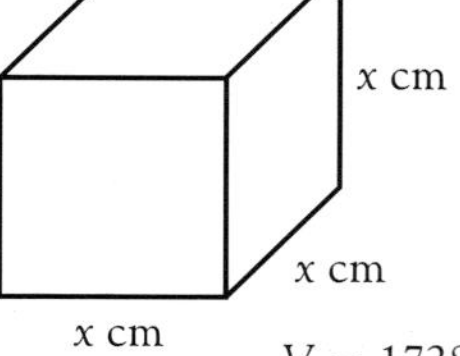

- Put the two volumes equal to each other

$$x^3 = 1728$$

- Find the cube root of 1728

$$x = \sqrt[3]{1728}$$
$$= 12$$

The large cube has sides of length 12 cm

Make sure you can do this on **your** calculator. On some, $\sqrt[3]{\ }$ is obtained by pressing **SHIFT** **+/–**

TYPES OF NUMBER

b

- Find the surface area of one face of the cube by working out the area of a square of side 12 cm then multiply by 6 as there are 6 identical faces.

12 cm

12 cm

Area of one face $= 12 \times 12 = 144\ \text{cm}^2$
Surface area of cube $= 6 \times 144$
<u>Surface area of cube $= 864\ \text{cm}^2$</u>

Sample Question 3

a Find all the factors of 72.

b Express 72 as a product of primes.

Answer

a

- Divide 72 by 1, 2, 3, … in turn, writing down the factors if the number divides exactly.

A **factor** divides exactly into the number.

		Factors
$72 \div 1 = 72$	$1 \times 72 = 72$	1, 72
$72 \div 2 = 36$	$2 \times 36 = 72$	2, 36
$72 \div 3 = 24$	$3 \times 24 = 72$	3, 24
$72 \div 4 = 18$	$4 \times 18 = 72$	4, 18
$72 \div 5 = 14.4$		
$72 \div 6 = 12$	$6 \times 12 = 72$	6, 12
$72 \div 7 = 10.28\ldots$		
$72 \div 8 = 9$	$8 \times 9 = 72$	8, 9

There is no need to try any more numbers, as you already know that 72 is divisible by 9.

As a general rule, to find all the factors, you need to test whether the number is divisible by all the integers up to the square root of the number.
$\sqrt{72} = 8.48\ldots$, so you need to try as far as 8.

<u>The factors of 72 are 1, 2, 3, 4, 6, 8, 9, 12, 18, 24, 36, 72</u>

b

- To express 72 as a product of primes, work systematically. Write down the pair of numbers with the lowest and highest factor (other than 1×72).

Remember that 1 is not a prime number.

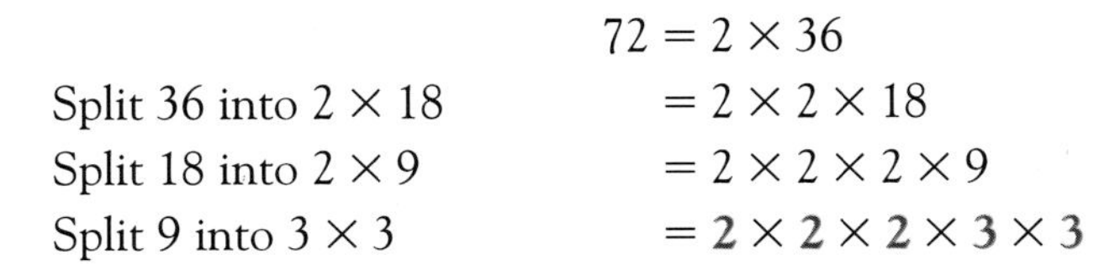

$72 = 2 \times 36$

Split 36 into 2×18 $\quad = 2 \times 2 \times 18$
Split 18 into 2×9 $\quad = 2 \times 2 \times 2 \times 9$
Split 9 into 3×3 $\quad = \mathbf{2 \times 2 \times 2 \times 3 \times 3}$

These are all prime numbers, so you cannot split them further.

- Write the numbers in index form.

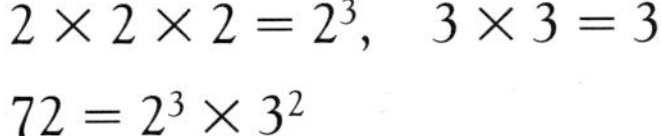

$2 \times 2 \times 2 = 2^3, \quad 3 \times 3 = 3^2$

<u>$72 = 2^3 \times 3^2$</u>

TYPES OF NUMBER

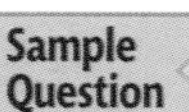

Sanjeet used her calculator to work out the volume, in cm^3, of a cuboid measuring 8000 cm by 7000 cm by 6000 cm.
The display showed

3.36 11

What answer should she write down?

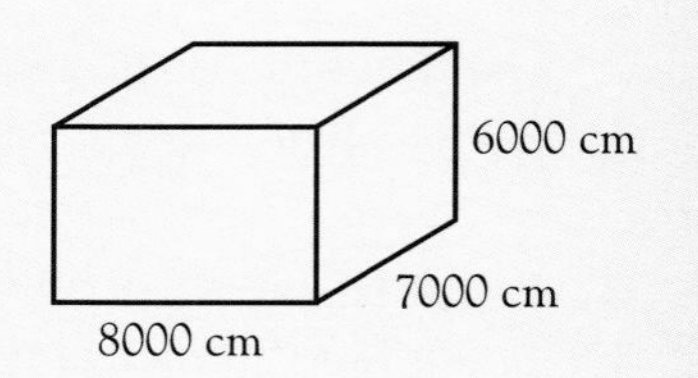

Answer

The number on the display is 'calculator shorthand' for 3.36×10^{11}

The format 3.36×10^{11} is called **standard form**. You will consider this topic further in chapter 23.

Some calculators show this properly 3.36×10^11
Check yours by trying 8000 × 7000 × 6000

◆ Work out 3.36×10^{11}

$$3.36 \times 10^{11} = 3.36 \times 10 \times 10 \times 10 \times 10 \times 10 \times 10 \times 10 \times 10 \times 10 \times 10 \times 10$$
$$= 336\,000\,000\,000$$

Move decimal point 11 places to the right.

Sanjeet could give the answer as 3.36×10^{11} cm^3 or 336 000 000 000 cm^3

Exercise 10.1

Answer questions 1 to 14 **without using a calculator.**

1 Which of the following numbers are multiples of 6?

32 24 18 6 56 60 3 27 42

2 Which of the following numbers are divisible by 9?

711 213 594 1224 829 7893 9460

3 Find the lowest common multiple of

a 8 and 12 **b** 15 and 20 **c** 4, 6 and 9

4 Write down all of the factors of

a 48 **b** 70 **c** 84

5 List all the prime numbers between 20 and 40.

6 In a game of bingo, the first ten numbers drawn are

27, 35, 5, 89, 50, 23, 16, 75, 3, 64

Which of these numbers are

a perfect squares
b perfect cubes
c multiples of 5
d prime numbers which are factors of 75?

7 From the following list of numbers write down those which are

a even numbers
b integers
c square numbers
d cube numbers
e prime numbers

36 2.4 8 19 64 $\frac{1}{2}$ 1 125

8 What are the values of

a 7^2 **b** 4^3 **c** 2^4 **d** $\sqrt[3]{27}$ **e** $\sqrt{25}$

9 Write down the reciprocal of each of the following numbers

a 5 **b** $\frac{3}{4}$ **c** $\frac{7}{3}$ **d** 0.5

TYPES OF NUMBER

10 a List all the factors of

i 24 ii 60 iii 75

b What is the highest common factor of

i 24 and 60

ii 60 and 75

iii 24, 60 and 75?

11 Express each of the following numbers as a product of primes

a 60 b 54 c 88

12 A prime number is less than 100 and its digits add up to 5. If it is 1 more than a multiple of 5, what is the number?

13 Write the following numbers in order of size, starting with the smallest.

2^5 $(-5)^2$ $\sqrt{400}$ $\sqrt[3]{27000}$

14 The National Lottery uses balls numbered from 1 to 49. Mark wishes to use one of the following methods to select the six numbers he needs to put on his lottery ticket. Which of the methods would give a possible selection of six numbers?

a multiples of 6

b multiples of 8

c perfect squares

d perfect cubes

e prime numbers above 25

For the rest of the exercise, use your calculator when necessary.

15 David has 1000 small cubes, each of side 1 cm. With these cubes he can build larger cubes with sides of 2 cm, 3 cm and so on, as shown in the sketch.

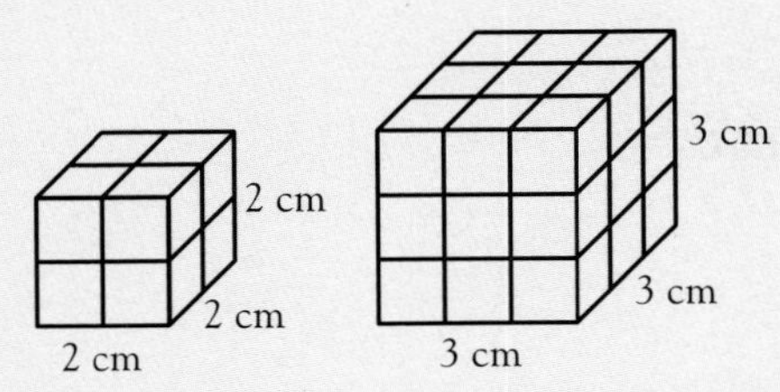

a The largest cube he can make uses all 1000 small cubes. What is the length of its sides?

b i How long is the side of the cube which he can make using 729 small cubes?

ii When he dismantles this cube he finds that he can lie all the small cubes on the floor in the shape of a square (as shown) with no small cubes left over. What is the length of the sides of the square?

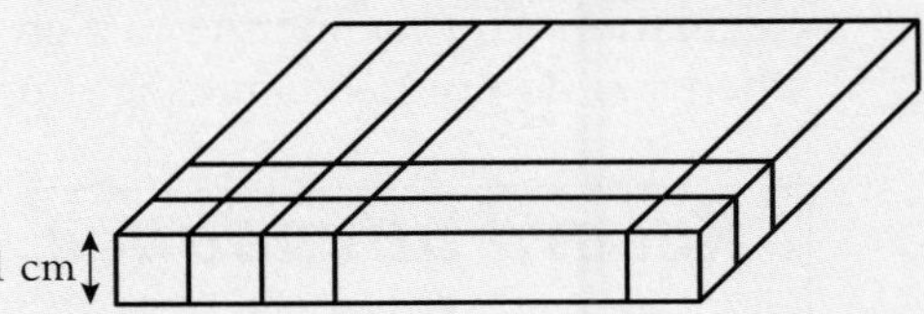

c There is only one other cube that he can build and then dismantle and make into a square 1 cm high without any small cubes left over. How many small cubes are used?

16 Find the value of x in each of the following equations:

a $x = 6^4$

b $2^x = 32$

c $x = \sqrt[3]{216}$

d $7 = \sqrt{x}$

17 a Write down all the prime numbers between 1 and 100.

b It is thought that all even numbers greater than 2 can be expressed as the sum of two prime numbers.
For example, $12 = 5 + 7$
and $26 = 3 + 23$ or $7 + 19$ or $13 + 13$

Write the following numbers as the sum of two primes in as many different ways as you can:

i 8 ii 22 iii 28 iv 54 v 100

18 Three buoys have lights attached which flash at regular intervals. The light on the first buoy flashes every 6 seconds, that on the second buoy flashes every 8 seconds and that on the third buoy every 10 seconds. Occasionally all three lights flash simultaneously. Find the length of the time interval between consecutive occasions on which this occurs.

19 Find the reciprocal of

a 10^3 b $\sqrt[3]{8}$ c $\dfrac{1}{5^2}$ d $\dfrac{3^2}{\sqrt{36}}$

20 a Write down all the square numbers from 1 to 100.

b Whole numbers can always be split into square numbers.
For example $54 = 49 + 4 + 1$ and $73 = 64 + 9$
Split the following numbers into square numbers, using as few as possible:

i 17 ii 35 iii 80 iv 92 v 104

SIGNIFICANT FIGURES

Significant figures

In chapter 2 you approximated numbers to 1 significant figure to work out an estimate for a calculation. Approximating to a given number of significant figures (s.f.) can be useful, especially when you do not need an exact number but just want a rough idea of its size.

Record attendance at rock concert

For example, if the number of people who went to the concert was 12 394, a newspaper would probably report this as 12 000 (which is an approximation to 2 s.f.) or 12 400 (approximating to 3 s.f.).

When approximating to a given number of significant figures, you can follow a rule to write down a 'lower' approximation and an 'upper' approximation, then decide which of these your number is closer to.

Approximating numbers bigger than 1

Sample Question 5 Approximate 28 752 to 3 significant figures.

Answer

- Give the lower approximation by writing down the first 3 digits of the number and replacing any other digits before a decimal point with zeros.

  For the number 28 752, lower approximation = 28 700

If you do not put in the zeros you lose an idea of the size of the number.

- Give the upper approximation by replacing the first 3 digits with the next 'number' up, i.e. replace 287 with 288 so upper approximation = 28 800
- Decide which approximation the number 28 752 is closer to by looking at the 4th digit of your original number using this rule:
  - if it is below 5, choose the lower approximation,
  - if it is 5 or above, choose the upper approximation.

  Since the 4th digit is 5, choose the upper approximation.

  <u>28 752 = 28 800 (3 s.f.)</u>

Approximating numbers between 0 and 1

Sample Question 6 Approximate 0.007 315 to 2 significant figures.

Answer

- Give the lower approximation by writing down the zeros until you get to the first non-zero digit. This is the first 'significant' figure. Write this down, then the next digit, even if the next digit is zero.

  For the number 0.007 315, lower approximation = 0.0073

DO NOT put in the zeros after these two digits.

SIGNIFICANT FIGURES

- Give the upper approximation by replacing the 2 digits with the next 'number' up, i.e. replace 73 with 74, so upper approximation = 0.0074
- Decide which approximation the number 0.007 315 is closer to by looking at the 3rd 'significant' digit of your original number, counting from the first digit that isn't zero:
  - if it is below 5, choose the lower approximation,
  - if it is 5 or above, choose the upper approximation.

  Since the 3rd 'significant' digit is 1, choose the lower approximation, so 0.007 315 = 0.0073 (2 s.f.)

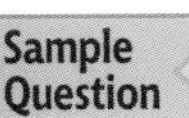

Write each of the numbers to the degree of accuracy specified.

a 6.0392 to 2 s.f.

b 491 093.2 to 4 s.f.

c 0.000 083 45 to 1 s.f.

d 0.503 72 to 3 s.f.

Answer

a

6.0392 to 2 s.f.
lower approximation = 6.0
upper approximation = 6.1
3rd digit is 3, so choose lower approximation
6.0392 = 6.0 (2 s.f.)

b

491 093.2 to 4 s.f.
lower approximation = 491 000
upper approximation = 491 100
5th digit is 9, so choose upper approximation
491 093.2 = 491 100 (4 s.f.)

c

0.000 083 45 to 1 s.f.
lower approximation = 0.000 08
upper approximation = 0.000 09
2nd 'significant' digit is 3,
so choose lower approximation
0.000 083 45 = 0.000 08 (1 s.f.)

d

0.503 72 to 3 s.f.
lower approximation = 0.503
upper approximation = 0.504
4th 'significant'digit is 7 so choose upper approximation
0.503 72 = 0.504 (3 s.f.)

Exercise 10.2

1 Write each of the numbers to the degree of accuracy specified.

a 7493 (2 s.f.)

b 4 296 321 (3 s.f.)

c 62.814 (2 s.f.)

d 200 943 (3 s.f.)

e 2.893 41 (4 s.f.)

f 0.006 935 (1 s.f.)

g 0.482 61 (3 s.f.)

h 0.092 83 (2 s.f.)

i 0.530 46 (3 s.f.)

j 0.000 102 04 (2 s.f.)

Do not use a calculator for this question.

2 A national daily newspaper claimed that it sold an average of 1 066 675 copies a day during the month of June and that the figure for a rival newspaper was 769 742.

a How many more copies than its rival did it sell on average each day in June?

b Find the difference between the numbers when they are approximated to

i 1 s.f. ii 2 s.f. iii 3 s.f.

Which approximation might the rival newspaper wish to use?

3 Find the area of a rectangle with width 3.24 m and length 6.59 m, giving your answer in square metres, correct to 3 s.f.

4 Change $\frac{2}{7}$ to a decimal, giving your answer to 2 s.f.

5 A circle has a circumference of 25 cm. Find its diameter, giving your answer to 2 s.f. (Use $\pi = 3.14$).

6 A number is approximated to 3 s.f. and its value is given as 41 300.

a What is the smallest number it could have been?

b If the number is an integer, what is the largest it could have been?

Ratio and Proportion

Ratios are used to compare amounts.

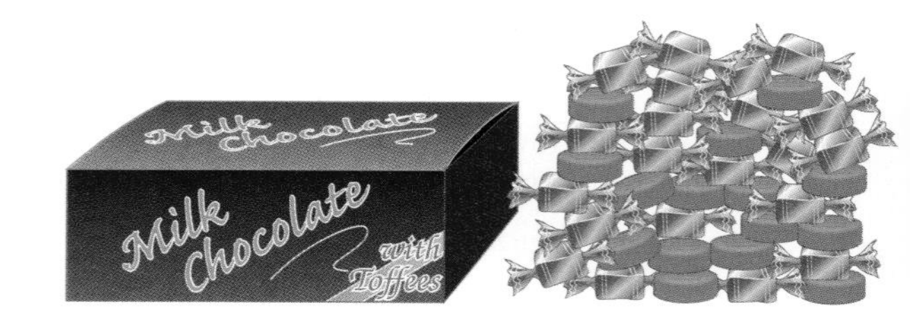

The contents of this box of chocolates and toffees are emptied out so that the number of each type can be compared.

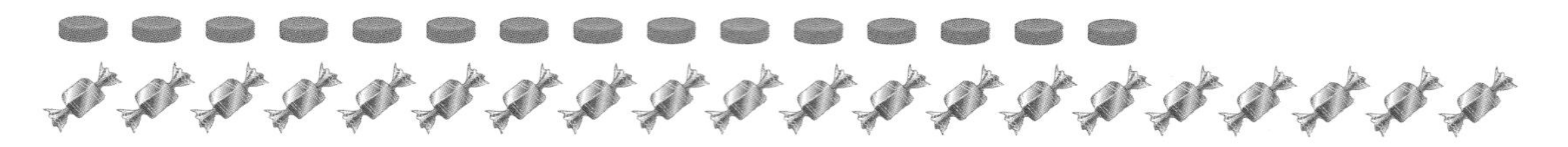

The ratio of chocolates to toffees is 15 to 20.
This is written chocolates : toffees = 15 : 20

The symbol : is written instead of 'to'.

The ratio can also be written as a fraction

$$\frac{\text{chocolates}}{\text{toffees}} = \frac{15}{20}$$

The sweets can be sorted into 5 equal groups. In each group, for every 3 chocolates there are 4 toffees, so chocolates : toffees = 3 : 4

Each number has been divided by 5.

The ratios 15 : 20 and 3 : 4 are said to be **equivalent** and 15 : 20 = 3 : 4

Compare equivalent fractions (page 13) where $\frac{15}{20} = \frac{3}{4}$

You can divide or multiply all the parts in a ratio by the same number to get an equivalent ratio.

A ratio is in its **simplest form** when it is written using whole numbers and there are no more numbers that will divide exactly into all the parts of the ratio.

If you are working with a ratio involving different units, such as 40 cm to 1.5 m, you must write them in a common unit, for example cm, before simplifying.

$$\begin{aligned} 40\text{ cm} : 1.5\text{ m} &= 40\text{ cm} : 150\text{ cm} \\ &= \mathbf{40 : 150} \\ &= 4 : 15 \end{aligned}$$

When the parts of the ratio are in a common unit, the unit can be left out.

RATIO AND PROPORTION

There are several ways of thinking about the ratio 3 : 4.

	chocolates	:	toffees	
	3	:	4	
÷ by 4	$\frac{3}{4}$	:	1	The number of chocolates is $\frac{3}{4}$ of the number of toffees
÷ by 3	1	:	$\frac{4}{3}$	$\frac{4}{3} = 1\frac{1}{3}$, so the number of toffees is $1\frac{1}{3}$ times the number of chocolates
÷ by (3 + 4) i.e. ÷ by 7	$\frac{3}{7}$	:	$\frac{4}{7}$	In every 7, 3 are chocolates and 4 are toffees, so $\frac{3}{7}$ are chocolates and $\frac{4}{7}$ are toffees

These ratios can be summarised.

$a : b = \frac{a}{b} : 1$ This is referred to as the form $n : 1$

$a : b = 1 : \frac{b}{a}$ This is referred to as the form $1 : n$

$a : b = \frac{a}{a+b} : \frac{b}{a+b}$

Remember $\frac{a}{b}$ is $a \div b$. It can be left as a fraction or worked out as a decimal.

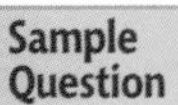

Sample Question 8

a Write these ratios in their simplest form.

i 24 : 54 **ii** 250 g : 3.5 kg **iii** $1\frac{1}{2} : 4$

b Write the ratio 5 : 9 in the form $1 : n$

c Write the ratio 8 : 3 in the form $n : 1$

Answer

a

i ◆ Find numbers that divide exactly into both parts.

24 : 54 = 12 : 27 (÷ by 2)

= 4 : 9 (÷ by 3)

You could do this in one step by dividing by 6

ii ◆ Change the ratio to the same unit, then simplify.

250 g : 3.5 kg = 250 g : 3500 g

= 250 : 3500

= 25 : 350 (÷ 10)

= 5 : 70 (÷ 5)

250 g : 3.5 kg = 1 : 14 (÷ 5)

Change kg to g, 1 kg = 1000 g

Check on calculator

[2] [5] [0] [a b/c] [3] [5] [0] [0] [=]

This gives the answer in fraction form.

iii ◆ Multiply both numbers by 2 to get a ratio with whole numbers.

$1\frac{1}{2} : 4 = 3 : 8$

On calculator

gives

RATIO AND PROPORTION

b

- Divide both parts by 5

$5:9 = 1:\frac{9}{5}$

$\underline{5:9 = 1:1.8}$

$\frac{9}{5} = 9 \div 5 = 1.8$
Also $\frac{9}{5} = 1\frac{4}{5}$, so you could write $1 : 1\frac{4}{5}$

c

- Divide both parts by 3

$8:3 = \frac{8}{3}:1$

$\underline{8:3 = 2\frac{2}{3}:1}$

$\frac{8}{3} = 8 \div 3 = 2.666...$
It is better to leave it as a fraction

8 $a^{b/c}$ 3 =

Can you do this without your calculator?
If not, find out how. FACT SHEET 1, Fractions, Decimals and Percentages (pages 13–15)

Sample Question 9 Squash and water are mixed to make a fruit drink. If 65% of the drink is water, write the ratio of squash to water in its simplest form.

Answer

- Find the percentage of the drink that is squash.

$100 - 65 = 35$

- Write the ratio of squash to water and simplify it.

squash : water = 35 : 65 (÷ by 5)

$\underline{\text{squash : water} = 7:13}$

Exercise 10.3

Do not use a calculator to do these questions.
When you have finished, use your calculator to check your answers.

1 Simplify these ratios

a 12 : 20
b 14 : 21
c 36 : 45
d 72 : 60
e 180 : 405
f 0.3 : 0.8
g 2500 : 7500
h 1 : 2.5
i 48 : 20

2 Write the following ratios in their simplest form.

a 240 cm : 3 m
b 2.5 m : 750 mm
c 1.6 kg : 320 g
d $\frac{1}{2}$ m : 300 cm
e 4 kg : 250 g
f 28 cl : 42 ml

3 Simplify these ratios

a $\frac{2}{5}:3$
b $\frac{3}{5}:\frac{7}{10}$
c $2\frac{2}{3}:\frac{4}{9}$

4 Write these ratios in the form $1:n$

a 5 : 4
b 4 : 5
c 2 : 7
d 3 : 7
e 100 : 248
f 1 cm : 1 km
g $\frac{1}{3}:3$
h $\frac{3}{5}:1$

5 Write these ratios in the form $n:1$

a 10 : 4
b 13 : 39
c 4 : 10
d 13 : 3
e 4 m : 2 cm
f $6^2:3^2$

6 A double decker bus has 36 seats upstairs and 42 seats downstairs. Find the ratio of seats upstairs to seats downstairs, giving your answer in its simplest form.

7 Fourteen of the thirty pupils in a class are girls. Find the simplest form for the ratio of girls to boys.

8 A shop sells brown and white loaves of bread. One day it sells 150 white loaves and 120 brown loaves. Write the ratio of white loaves to brown loaves

a in the form $n : 1$ b in the form $1 : n$

9 On a farm there are 54 cattle, 135 sheep and 18 goats.

a Write the ratio of cattle : sheep : goats in its simplest form.

b What is the ratio of cattle : goats in its simplest form?

10 Tessa has a box containing nails and screws. If $\frac{5}{7}$ of the contents of the box are nails, write the ratio of nails to screws in its simplest form.

Finding a missing amount in a ratio

In this large tin, there are 87 chocolates.
If chocolates and toffees are in the ratio 3 : 4, how many toffees are there?

You need to work out the missing number in the equivalent ratio $3 : 4 = 87 : \boxed{x}$.

There are several ways of doing this. Choose the method you prefer, or a different one provided that it **always** gives you the correct answer.

1 Using the 'multiplier'

- chocolate : toffee

  3 : 4

  87 : $\boxed{x}$

Write the ratios underneath each other, making sure they are both in the same order.

3 has been multiplied by something to get 87. This is called the 'multiplier' or 'multiplying factor'.

- To find what 3 has been multiplied by to get 87, divide 87 by 3.

  $87 \div 3 = 29$

This means that both parts of the ratio have been multiplied by 29, so to find x, multiply 4 by 29.

$4 \times 29 = 116$

<u>There are 116 toffees in the tin.</u>

Think whether your answer is sensible. For example, were you expecting more toffees than chocolates?

You can check your answer on the calculator.

[8][7][a b/c][1][1][6][=] gives [3⌟4] or [3⌐4]

showing that, in fractions, $\frac{87}{116} = \frac{3}{4}$ In ratios this means $87 : 116 = 3 : 4$

You could do all the working in one step, without working out the actual value of the multiplier, but leaving it as $87 \div 3$.

3 : 4

87 : $\boxed{x}$ $x = 4 \times 87 \div 3$

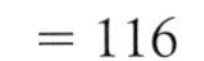

$= 116$

[4][×][8][7][÷][3][=]

Quick way: Notice that you multiply the two known 'diagonal numbers' (4×87) and divide by the 'diagonal' number opposite the unknown ($\div 3$).

2 Unitary method

This is really the same technique, but you explain the reasoning in a slightly different way.

$3 : 4$	3 parts represent 87
$87 : \boxed{x}$	so 1 part represents $87 \div 3 = 29$
	$\therefore$ 4 parts represents $4 \times 29 = 116$

Once again $x = 4 \times 87 \div 3 = 116$

When you work out 4 parts, you might write $87 \div 3 \times 4$ or $87 \times 4 \div 3$. The order in which you multiply or divide does not matter, but you must multiply by 4 and divide by 3.

3 Algebraic method

- Re-arrange the ratio, if necessary, so that the missing value is on the left.
- Write the ratios as fractions.
- Solve the equation.

$$\boxed{x} : 87 = 4 : 3$$

$$\frac{x}{87} = \frac{4}{3}$$

$$(\times 87) \qquad x = \frac{4}{3} \times 87$$

$$\underline{x = 116}$$

Note that $\frac{4}{3} \times 87$ is the same as $\frac{4 \times 87}{3}$

Sample Question 10 In a class of 30 pupils, 12 are girls. The ratio of boys to girls in the whole school is the same as in this class. There are 300 girls in the school. How many boys are there?

Answer

- Write the ratio of boys to girls in the class and simplify it.

Number of boys $= 30 - 12 = 18$

boys : girls $= 18 : 12 \qquad (\div 6)$

$= 3 : 2$

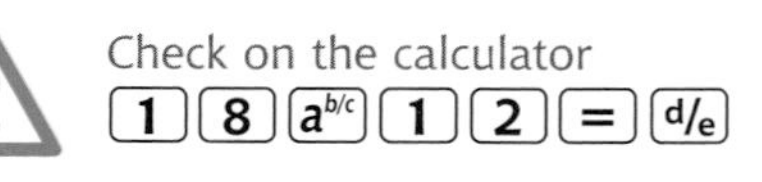

- Write the ratio of boys to girls in the school using x for the number of boys. Put this equal to $3 : 2$ and solve to find x.

$x : 300 = 3 : 2$

Multiplier method

$x : 300$

$3 : 2$

'multiplier' $= 300 \div 2$

$x = 3 \times 300 \div 2$

$\underline{x = 450}$

Unitary method

2 parts represent 300

1 part represents $300 \div 2$

$= 150$

3 parts represent 3×150

$\underline{x = 450}$

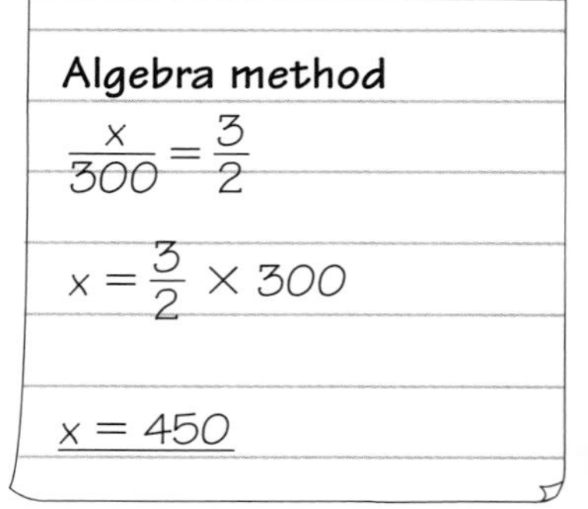

<u>There are 450 boys in the school.</u>

Sample Question 11 Uncle Bill left Jenny (aged 14) and Simon (aged 16) some money in his will with the instruction that the money should be divided in the ratio of their ages.
Jenny received £5600. How much did Simon receive?

Answer

- Write the ratio of the ages and put this equal to the ratio of money. Find the missing value.

	Jenny	:	Simon
Age ratio	14	:	16
Money ratio	5600	:	x

Multiplier method

14 : 16

5600 : x

'multiplier' = 5600 ÷ 14

$x = 16 \times 5600 \div 14$

$x = 6400$

Unitary method

14 parts represent 5600

1 part represents 5600 ÷ 14

16 parts represent

16 × 5600 ÷ 14

$x = 6400$

Algebra method

$x : 5600 = 16 : 14$

$\frac{x}{5600} = \frac{16}{14}$

$x = \frac{16}{14} \times 5600$

$x = 6400$

Simon received £6400

Direct proportion

If two quantities vary, but always remain in the same ratio, they are in **direct proportion**.

PETROL
70p per litre

The cost of petrol is in direct proportion to the amount you buy.

1 litre costs 70 p

25 litres cost 25 × 70 p = 1750 p = £17.50

$\frac{1}{2}$ litre costs 70 ÷ 2 = 35 p

The two quantities increase or decrease at the same rate.

You could draw a graph to show how the cost varies with the amount bought.

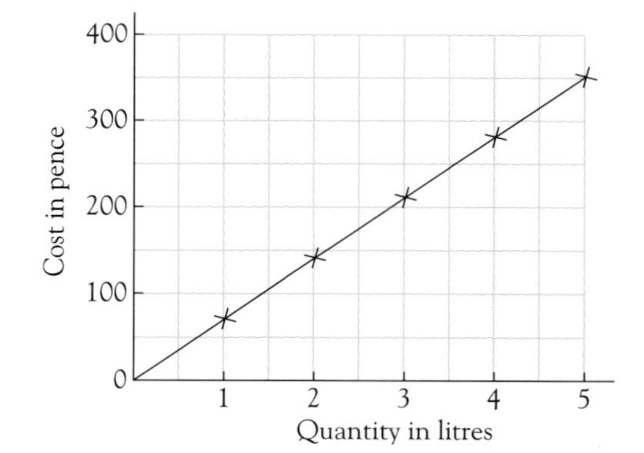

T When two quantities are in direct proportion, the graph is a straight line through (0, 0).

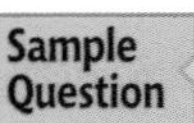

RATIO AND PROPORTION

Sample Question 12 A gardener wants to make up a solution of greenhouse shader. The instructions on the box say to mix 1.2 litres of water with 4 sachets of powder. He works out that he needs 9 sachets of powder to whiten his greenhouse. How much water should he use?

Answer

- The amount of water and the number of sachets are in direct proportion.

For 4 sachets, use 1.2 litres

For 1 sachet, use $1.2 \div 4 = 0.3$ litres

For 9 sachets, use $9 \times 0.3 = 2.7$ litres

<u>He should use 2.7 litres of water.</u>

This is the 'unitary method' for finding the missing amount in a ratio

In ratio form you could find x where

4 sachets : 1.2 litres = 9 sachets : $\boxed{x}$ litres

or

1.2 litres : $\boxed{x}$ litres = 4 sachets : 9 sachets.

These should both give the same answer – try them.

Common examples of the use of direct proportion are converting measurements, for example kilometres to miles, or converting money when you know the exchange rate.

Sample Question 13

a When going on holiday, Joe exchanged £200 for 520 000 Italian lira. Pat changed £300 into lira at the same exchange rate. How many lira did Pat receive?

b Samantha came to England from the USA for a holiday. She changed her dollars ($) into pounds (£) at the rate of $1 = £0.68. At Heathrow airport, Samantha bought a T-shirt for £12.99. What was the cost, in dollars, of the T-shirt? [MEG, p]

Answer

a

- Find how much Joe received for £100, then multiply by 3 to find the amount for £300.

For £200, he received 520 000 lira
For £100, he received $520\,000 \div 2 = 260\,000$
For £300, Pat received $260\,000 \times 3 = 780\,000$

<u>Pat received 780 000 lira.</u>

It is easier to work with £100, rather than £1, in this example

b

- Find the equivalent in dollars of £1, then multiply by 12.99.

£0.68 = $1

£1 = $\$\frac{1}{0.68} = \$1.470\ldots$

£12.99 = $12.99 \times \$1.470 = \$19.10$ (nearest cent)

<u>The T-shirt cost $19.10</u>

Using ratios

£0.68 : $1
£12.99 : $$x$

$x = 1 \times 12.99 \div 0.68$
$= \$19.10$

Exercise 10.4

1 The times taken by Amy and Laura to do their mathematics homework were in the ratio 2 : 3. Laura took half an hour. How long did Amy take?

2 In a mixed bed of tulips, the ratio of red to yellow is 2 : 5. There are 12 red tulips.

a How many are yellow?

b How many tulips are there in the bed?

3 The ratio of women to men in a sports club is 3 : 2. If there are 42 women, find the number of men.

4 I have some blackcurrant and redcurrant bushes in my garden in the ratio 1.5 : 1.
If I have four redcurrant bushes, how many blackcurrant bushes have I got?

5 Rick supports his local football team. He has attended home and away games in the ratio 9 : 2. If he has been to 126 home games, how many away matches has he been to?

6 Three brothers have ages in the ratio 2 : 4 : 5. The youngest brother is six years old. How old are the others?

7 When making mortar, sand and cement are mixed in the ratio 2 : 1. If half a tonne of sand is used to make the mortar for a wall, how many kg of cement are used? (Remember 1 tonne = 1000 kg)

8 A drink is made by mixing orange concentrate with water in the ratio 2 : 25. How many millilitres of concentrate are needed for each litre of water?

9 If 2 m of ribbon costs 96 pence, how much would it cost for

a 5 m b half a metre c 125 cm?

10 a Copy and complete the following table, giving values to 2 decimal places:

No of pints	2	4	6	8	10	12
No of litres				4.55		

b Draw a graph using pints on the horizontal axis and litres on the vertical axis.
Plot the points shown in the table and join them with a straight line.

c Use the graph to convert
i 5 pints to litres ii 7.2 pints to litres
iii 2 litres to pints iv 5.2 litres to pints

11 1 kg is equivalent to 2.2 lb. A farmer sells potatoes in sacks of 56 lb. How many kilograms is this?

12 Arshad is on holiday in Switzerland. A boat ride on one of the lakes costs 30 Swiss francs which is equivalent to £12. Find the price in pounds and pence of a train journey which costs 17 Swiss francs.

13 a When Sandra went on holiday to Greece she changed £500 into drachma. The travel agent gave her 478 drachma to the pound. How many drachma did she receive?

b She returned from holiday with 17 000 drachma. The travel agent will exchange this back to pounds at a rate of 492 drachma to the pound. Find, to the nearest pence, how much she will get.

14 The recipe shows the ingredients needed to make macaroni cheese for 4 people.
Work out how much of each ingredient I would need to make macaroni cheese for 7 people.

MACARONI CHEESE

Macaroni	120 g
Cheese	80 g
Cornflour	25 g
Milk	720 ml

15 Kieran knows that 8 km is approximately the same as 5 miles. The speed limits in France are:

Motorways	130 km/h
Dual Carriageways	110 km/h
Ordinary Roads	90 km/h
Built-up Areas	60 km/h

Calculate these speed limits in miles per hour, to the nearest m.p.h.

16 A 3 kg bag of washing powder is enough to do 20 normal machine washes.

a How many normal washes would you expect from a 5 kg bag?

b How much washing powder would be needed to do 4 normal washes? Give your answer in grams.

17 A yard is equivalent to 0.9144 metres. The distance between the wickets on a cricket field is 22 yards. How far would a batsman run if he scores 75 runs? Give your answer in km to 1 decimal place.

Ratio and enlargements

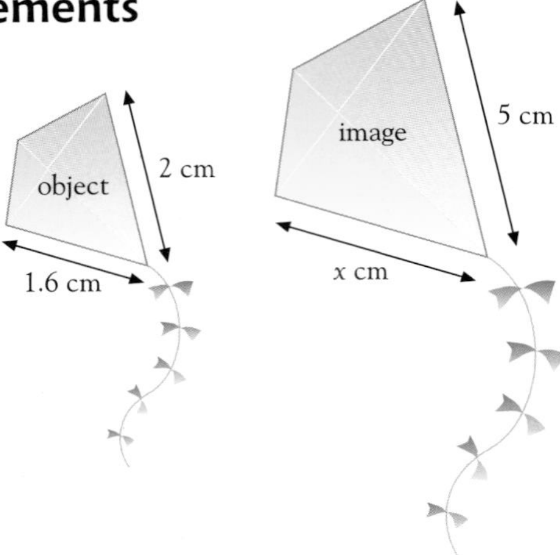

Look at the ratio of corresponding lengths on the image and the object.

$$\text{image length} : \text{object length} = 5 : 2$$
$$= \tfrac{5}{2} : 1$$
$$= 2\tfrac{1}{2} : 1$$

$2\frac{1}{2}$ is the **scale factor** of the enlargement.

This means that image length $= 2\frac{1}{2} \times$ object length

so $x = 2\frac{1}{2} \times 1.6$

$\underline{x = 4}$

Remember that if the scale factor is bigger than 1, the image is bigger than the object. If the scale factor is between 0 and 1, for example $\frac{1}{2}$ or $\frac{2}{3}$, the image is smaller than the object.

You could work out the missing length x without finding the scale factor first.

- Write down the two known corresponding lengths as a ratio.
- Write an equivalent ratio containing the unknown length.

image	:	object
5	:	2
$\boxed{x}$	:	1.6

so $x = 5 \times 1.6 \div 2 = 4$

Scale models and map scales

Scale 1:50

The scale for a model or for a drawing could be given as a scale factor, for example $\frac{1}{50}$, indicating that all lengths in the model or scale drawing are $\frac{1}{50}$ of the 'real life' length.

Sometimes the scale is given in ratio format using mixed units, for example 1 cm : 10 m or 4 cm : 1 km. If this is the case, it is often better to work in these mixed units when calculating missing lengths.

Sample Question 14

An architect draws a plan using a scale of 1 cm : 10 m

a How long is a wall which is 2.6 cm on the map?

b What is the distance on the plan if the real-life distance is 8 m?

Answer

a

$$1\text{ cm} : 10\text{ m}$$
$$2.6\text{ cm} : x\text{ m}$$
$$x = 2.6 \times 10$$
$$x = 26$$

The wall is 26 m long.

b

$$10\text{ m} : 1\text{ cm}$$
$$8\text{ m} : x\text{ cm}$$
$$x = 1 \times 8 \div 10$$
$$= 0.8$$

Distance on plan = 0.8 cm

Map scales are often given in the form $1 : n$, such as 1 : 25 000 or 1 : 50 000.

1 : 25 000 means that 1 cm is used to represent 25 000 cm. Similarly 1 inch represents 25 000 inches, 1 m represents 25 000 m, etc.

These are difficult to imagine, so change the ratio to mixed units, usually cm and m or cm and km.

$$1 : 25\,000 = 1\text{ cm} : 25\,000\text{ cm}$$
$$= 1\text{ cm} : 250\text{ m}$$
$$= 1\text{ cm} : 0.25\text{ km}$$

so 1 cm represents $\frac{1}{4}$ km

or 4 cm represents 1 km (multiplying by 4)

The map ratio 1 : 25 000 = 4 cm : 1 km

For example, a distance of 12 cm on the map represents 3 km.

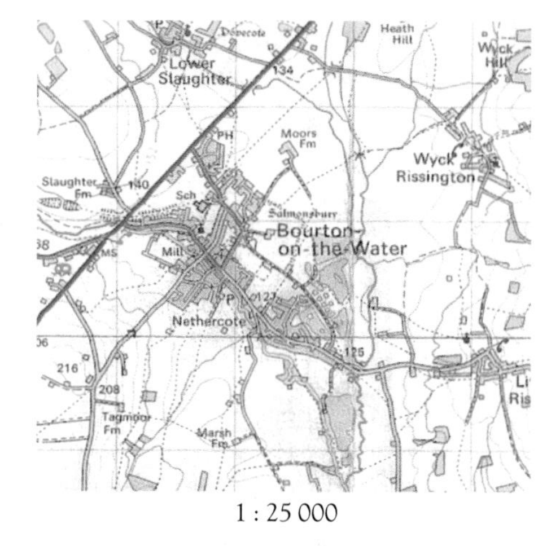

1 : 25 000

$$\text{Similarly } 1 : 50\,000 = 1\text{ cm} : 50\,000\text{ cm}$$
$$= 1\text{ cm} : 500\text{ m}$$
$$= 1\text{ cm} : \tfrac{1}{2}\text{ km}$$
$$= 2\text{ cm} : 1\text{ km}$$

The map ratio 1 : 50 000 = 2 cm : 1 km

For example, a distance of 12 cm on the map represents 6 km.

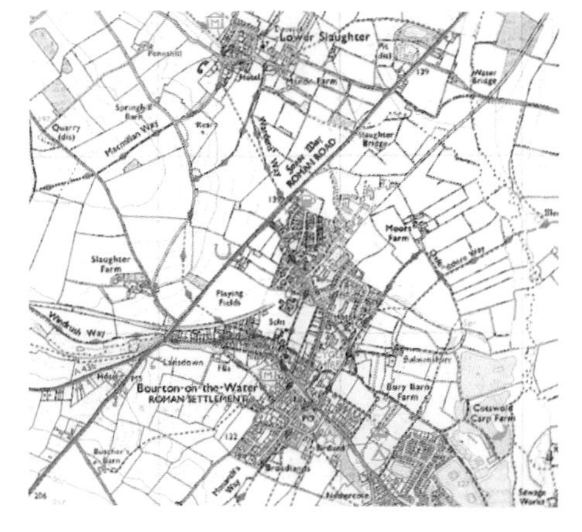

1 : 50 000

Notice that the scale 1 : 25 000 gives more detail than the scale 1 : 50 000

Exercise 10.5

1

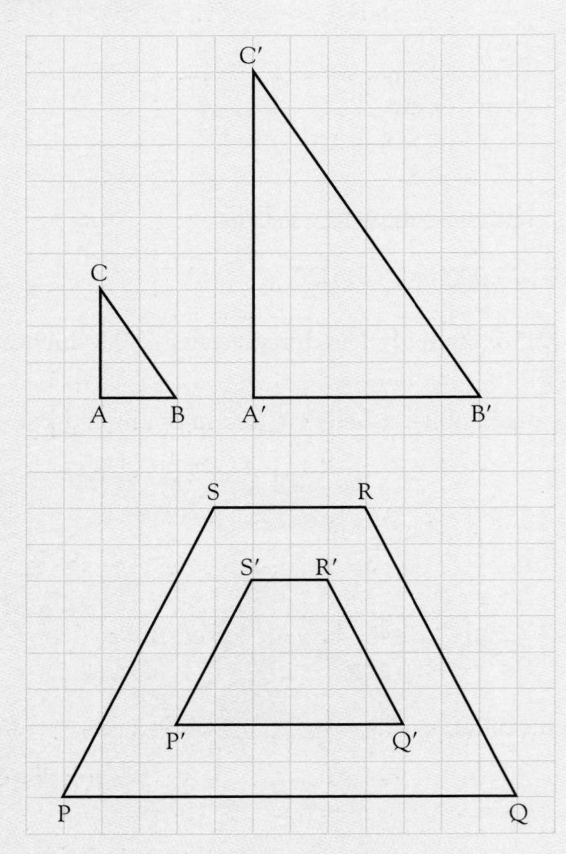

By measuring corresponding lengths, find the scale factor for

a the enlargement of triangle ABC to triangle A′B′C′,

b the enlargement of trapezium PQRS to P′Q′R′S′.

2 Triangle XYZ is enlarged to triangle X′Y′Z′ as shown in the sketch below (not drawn to scale)

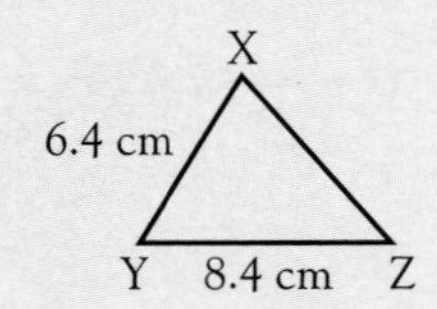

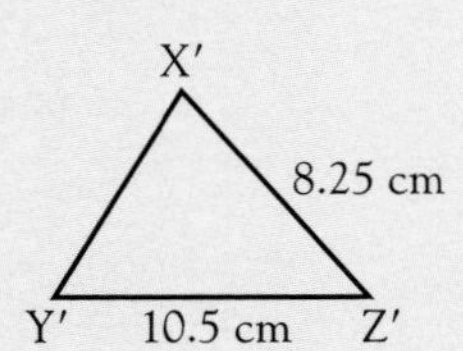

a Find the scale factor of the enlargement.

b Find the lengths of the sides X′Y′ and XZ.

3 Draw an enlargement of the kite shown in the diagram using a scale factor of

a 1.5 b 0.75

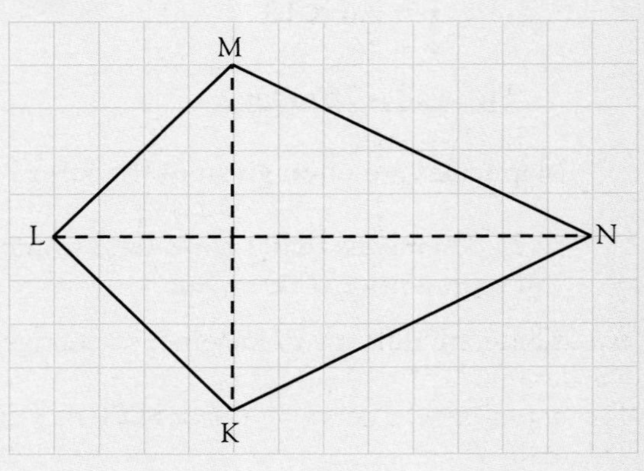

4 A plan of a garden is drawn using a scale of 1 : 50.

a Find the true dimensions of the following garden features whose dimensions on the plan are:

i lawn, 8 cm by 6 cm,

ii circular flower bed, diameter 2.4 cm.

b Find the dimensions on the plan of features whose true dimensions are given below.

i greenhouse base, 3.5 m by 2.25 m,

ii circular fish pond, diameter 1.2 m.

5 A model of a sports car is 19 cm long. If the real sports car is 3.8 m long, find

a the ratio of lengths in the model to the true lengths, giving your answer in the form $1 : n$,

b the width of the sports car if the model is 9 cm wide,

c the height of the model if the height of the real sports car is 1.2 m.

6 A photocopier has a number of settings so that it can change the size as well as copy items.

a A list of the scale factors is given below. For each of the settings state whether the copy would be larger or smaller than the original.

Scale factors 2 0.8 1.4 0.6 1

b The photocopier is set to enlarge a drawing by a scale factor of 1.4. If the dimensions of the drawing are 7.5 cm by 4.3 cm, find the dimensions of the enlargement.

7 Super-soup is sold in cans with diameter of 75 mm and height 110 mm. The manufacturer decides to introduce a new larger size. The new can is an enlargement of the old can with scale factor 1.26. Find, to the nearest mm, the diameter and height of the new can.

8 An architect builds a model of a building that she has designed. The height of the model is 84 mm. The actual height of the building is 21 m. Calculate

a the ratio of the height of the model to the height of the actual building in the form $1 : n$,

b the length of the building if the model is 120 mm long,

c the width of the model if the true width is 18 m.

9 An 'N' gauge model railway is made to a scale of $\frac{1}{150}$. If a model of a platform is 12 cm long, what would the actual length of the platform be in metres?

10 A model of an aircraft is made to a scale of 1 : 500 to be tested in a wind tunnel. Some of the dimensions of the model and the real aircraft are given in the table below. Complete the table by calculating the missing dimensions.

	Model	Real
Length	14 cm	
Wingspan		60 m
Height	1.2 cm	
Width of fuselage		4.5 m

11 A photograph which is 4 cm by 6 cm is enlarged using a scale factor of 1.5

a Find the dimensions of the enlargement.

b Find the ratio of the area of the original photograph to that of the enlargement, giving your answer in

i its simplest form,

ii the form $1 : n$.

12 A picture measuring 12 cm by 9 cm is mounted on a piece of card as shown. The border is 1.5 cm wide.

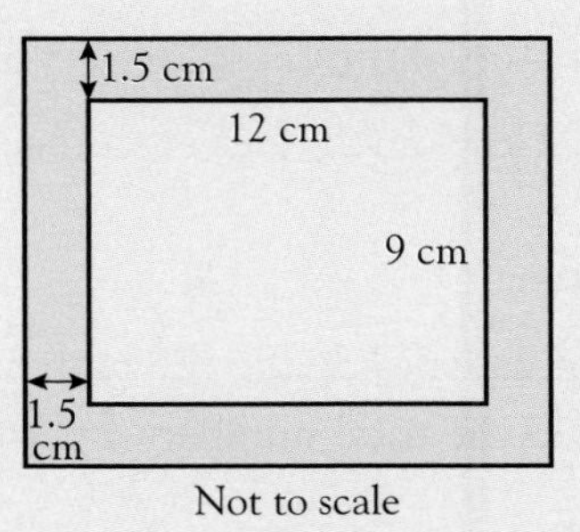

Not to scale

a Find the dimensions of the piece of card.

b Explain why the shape of the card is not an enlargement of the shape of the picture.

13 Alice is 4 ft 7 ins tall.

When she eats a cake marked 'Eat me', she is enlarged in the ratio 1 : 2.4.

How tall does she become? (Remember there are 12 inches in a foot.)

14 An ordnance survey map has a scale of 1 : 50 000.

a Find the actual distance between two villages which are 2.4 cm apart on the map.

b A farm track is 750 m long. How long is it on the map?

15 A plan is drawn of a bungalow using a scale of 1 cm to 2.5 m.

a Find the scale factor of the plan, giving your answer as a decimal.

b Find the dimensions on the plan of a rectangular room with length 4.2 m and width 3.5 m.

c Find the ratio of the area of the room on the plan to its true area. Give your answer in the form $1 : n$.

RATIO AND PROPORTION

Proportional division

This is when a quantity is 'shared' or divided in a given ratio.

Sample Question 15 Paul and John share £35 in the ratio 2 : 3.

How much does each receive?

Answer

- You have to divide the money into 2 parts and 3 parts.
  Find the total number of parts by adding the numbers in the ratio.

  $2 + 3 = 5$

- Find the amount in 1 part, by dividing £35 by 5

  $35 \div 5 = 7$

  1 part = £7
  Paul has 2 parts = 2 × £7 = £14
  John has 3 parts = 3 × £7 = £21

  <u>Paul has £14, John has £21</u>

Check 14 + 21 = 35

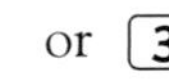

or 3 5 ÷ 5 × 2 =

Notice that $2 : 3 = \frac{2}{5} : \frac{3}{5}$
You have actually found $\frac{2}{5}$ of £35 and $\frac{3}{5}$ of £35.

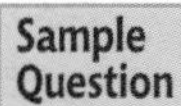

Sample Question 16 The three angles in this triangle are in the ratio 1 : 3 : 5.

Work out the size of each angle.

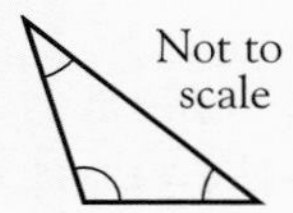

Answer

- Add the numbers in the ratio to find the total number of parts.

  $1 + 3 + 5 = 9$

- The sum of the angles in a triangle is 180°. Divide 180° by 9 to find the amount in 1 part.

  1 part = $180° \div 9 = 20°$

- Work out 3 parts and 5 parts.

Ratio	1	:	3	:	5
Angles	1 × 20° = 20°	:	3 × 20° = 60°	:	5 × 20° = 100°

<u>The angles are 20°, 60°, 100°</u>

Check
20 + 60 + 100 = 18

Sample Question 17 Black, red and green felt-tip pens are mixed in a drawer in the ratio 2 : 3 : 5. Danielle picks out a pen at random. What is the probability that it is red?

Answer

- Find the proportions of each colour in the drawer.

	black	:	red	:	green
	2	:	3	:	5
(÷ by 10)	$\frac{2}{10}$	:	$\frac{3}{10}$	:	$\frac{5}{10}$

So <u>P(red) = $\frac{3}{10}$</u>

2 + 3 + 5 = 10

Imagine the pens in bunches of 10 in which 2 are black, 3 are red and 5 are green.

:xercise 10.6

)o not use a calculator for this exercise.

)ivide each of these quantities in the given ratio.

1 20 in the ratio 2 : 3

2 15 in the ratio 1 : 2

3 42 in the ratio 3 : 4

4 80 in the ratio 5 : 3

5 84 in the ratio 5 : 9

6 7.5 in the ratio 2 : 1

7 54 in the ratio 1 : 2 : 3

8 600 in the ratio 3 : 5 : 4

9 450 in the ratio 2 : 2 : 5

10 165 in the ratio 2 : 3 : 6

:xercise 10.7

Jse a calculator only when necessary.

1 Sean has tapes and CDs of his favourite bands in the ratio 7 : 2. If he has a total of 54 recordings, how many of these are
  a tapes **b** CDs?

2 In an examination everyone gained credit or distinction. The ratio of candidates who gained credit to those who gained distinction was 7 : 3. What percentage gained distinction?

3 Each week Jason spends £3 on the National Lottery and Kelly spends £2. They divide any winnings in the ratio of how much they spent. This week they have won £64 for choosing 4 numbers correctly. How much should they each get?

4 A confectioner sells packets of mints. If the packets contain peppermints and spearmints in the ratio 3 : 2, find
  a the percentage of mints in a packet that are
    i peppermints **ii** spearmints
  b how many in a packet containing 35 mints are
    i peppermints **ii** spearmints

5 Daffodils, crocuses and tulips are planted in a border in the ratio of 5 : 3 : 2. If there are 150 bulbs, find the number of each type.

6 The angles in a quadrilateral are in the ratio 1 : 3 : 4 : 7. The sum of the angles is 360°. Find the size of each angle.

7 On her journey to school, Pat noted that the ratio of the time she spent on the bus to the time she spent walking was 2 : 5. If the total time was 35 minutes, how long did she spend walking?

8 A bag contains red and yellow counters in the ratio 5 : 3.
  a What percentage of the counters are
    i red **ii** yellow?
  b If a counter is taken from the bag at random, what is the probability that it is
    i red **ii** yellow?

9 Melanie and Peter decide to buy a computer costing £980 and agree to contribute towards the cost in the ratio of their annual salaries. If Melanie earns £15 000 per annum and Peter earns £13 000, how much should each contribute to the cost of the computer?

10 A bowl is made from bronze. Bronze is made from copper and zinc in the ratio 17 : 3.
  a What percentage of the bowl is copper?
  b If the bowl has mass 1.8 kg, find how much of this is copper.

11 Gemma, Ann and Kirsty are waitresses in a café. Each day they combine their tips and divide the total between them in the ratio of the hours they have worked. One day Gemma works seven hours, Ann works five hours, and Kirsty works four hours. If the total amount in tips is £23.20, how much does each waitress receive?

12 The ratio of boys to girls in a youth club is 14 : 11. If one of the members is chosen at random to represent the club, what is the probability that the person chosen will be
  a male
  b female?

13 Tom, Dick and Harry own shares in a company. Tom owns 50% of the shares, Dick owns 35% and Harry owns the remaining 15%. In the first year of trading the company makes a profit of £320 000. The partners decide to invest a quarter of this and divide the rest in the ratio of their share-holding. Calculate how much of the profits each of them receives.

EXAM QUESTIONS

Worked Exam Question
[MEG]

COMMENTS

Mrs Peters and Mr Wales invested in a new business.

a **They invested their money in the ratio of 5 : 3.**

i **Mrs Peters invested £75 000.**
How much did Mr Wales invest?

◆ Write out the equivalent ratios and solve to find x.

Remember that there are several ways of finding x. Use the method you prefer.

Mrs P : Mr W

5 : 3

75 000 : x

x = 3 × 75 000 ÷ 5

= 45 000

Answer £45 000

M1 for correct statements concerning ratios

A1 for correct answer

2 marks

ii **They agreed to share the profits in the same ratio as their investments.**
In the first year the business made £39 200 profit.
How much did Mr Wales receive?

◆ Find the total number of parts and then Mr Wales' share of them.

5 + 3 = 8

1 part = 39 200 ÷ 8

= 4900

3 parts = 3 × 4900 = 14 700

Answer £14 700

M1 for attempting to find the correct proportion

A1 for correct answer

2 marks

b **In the second year the profits increased from £39 200 to £41 944.**
Calculate the percentage increase in the profit.

◆ Find the amount of the increase.

◆ Write this as a fraction of the first year profits.

◆ Change this to a percentage.

Increase = £41 944 − £39 200

= £2744

% increase = $\frac{2744}{39\ 200}$ × 100%

= 7

Answer 7%

A1 for correct increase

M1 for correct denominator

A1 for correct answer

3 marks

Exam Questions

1 a Alwyn got 35% in his Science test. The total number of marks for the test was 120.

i What fraction of the marks did he get?

ii How many of the 120 marks did he obtain?

b Barbara obtained 33 marks out of 60 for French.
What percentage of the marks did she get?

c Colleen took two papers in Mathematics.
Her marks for these two papers were in the ratio 3 : 5.
The highest mark she got was 75.
Calculate the mark she got for the other paper. [NEAB]

2 Three people, Siad, Andrea and Duncan invested £90 000, £75 000, and £60 000 respectively in a business. They agreed to share any profit in the ratio of their investments.

a Write this ratio in the form $a : b : c$ as simply as possible.

b In 1997 they made £18 750 profit.
How much did Siad receive? [MEG]

3

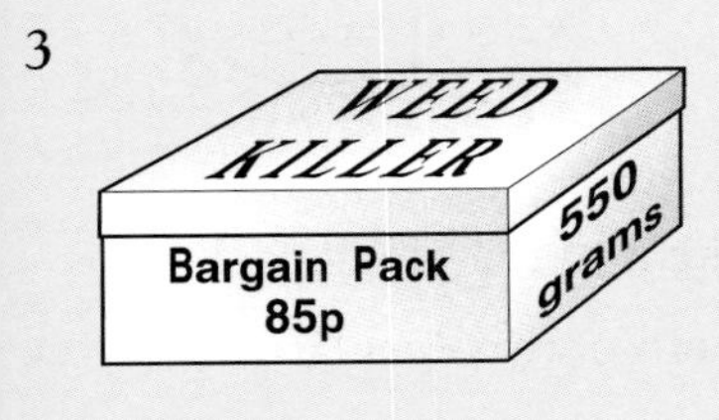

A manufacturer sells weed killer in two different packs, as shown in the diagrams above.

a i In the Bargain Pack, how many grams of weed killer do you get for 1 p?

ii In the Value Pack, how many grams of weed killer do you get for 1 p?

b Which of the two packs is the better value for money? [MEG]

4 Each week, John and Bev together place a bet of £4 on the football pools. John's contribution is £2.50 and Bev's is £1.50.

They agree to share any winnings in the same ratio as their contributions.

a Find the ratio of John's contribution to Bev's contribution in its simplest form.

b One week they win £1200.
Calculate John's share of this amount. [MEG]

5 Jane uses her slide projector to enlarge a rectangular slide which is 22 mm high and 34 mm wide. The height of the enlargement is 66 cm.

a Calculate the width of the enlargement in cm.

b Find the scale factor of the enlargement. [MEG]

6 a Express the following numbers as products of their prime factors.

i 72 ii 80

b Two cars go round a race track. The first car takes 1 minute 12 seconds to complete a circuit and the other car takes 1 minute 20 seconds. They start together on the starting line. Find the length of time, in minutes, before they are together again. [SEG]

7 The campsite has a children's pool and a main pool. The main pool is an enlargement of the children's pool.

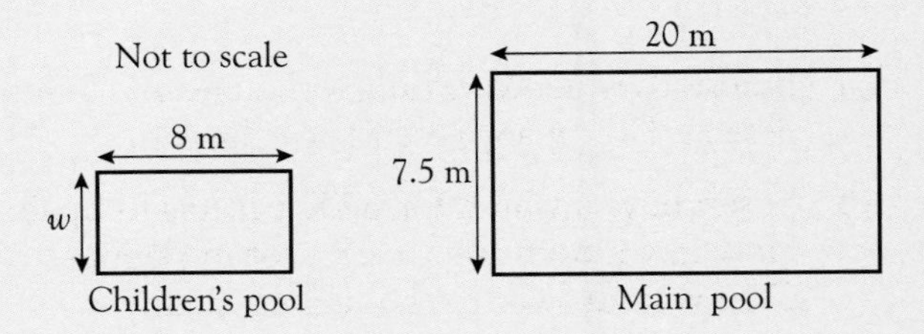

a Find the scale factor of the enlargement.

b Find the width, w, of the children's pool. [MEG, p]

8 On a motorway there are three lanes, an inside lane, a middle lane and an outside lane.
One day, at midday, the speed of the traffic on these three lanes was in the ratio 3 : 4 : 5. The speed on the outside lane was 70 miles per hour.
Calculate the speed on the inside lane. [NEAB]

9 It takes 100 g of flour to make 15 shortbread biscuits.

EXAM QUESTIONS

a How many shortbread biscuits can be made from 1 kg of flour?

b Calculate the weight of flour needed to make 24 biscuits. [MEG]

10 a Three members of the Smith family and five members of the Jones family have a meal in a restaurant.
They share the bill of £95.60 in the ratio of the number of people in each family.
How much did the Jones family pay?

b Mr Smith said, 'If the bill had been £82.40, I estimate we would pay about £30 as our share of the bill'. Write down how you would check Mr Smith's estimate **without** using a calculator. [MEG]

11 Citrola drink is made by mixing grapefruit juice and pineapple juice in the ratio 3 : 2.

a At a factory the drink is mixed in large drums. 40 litres of pineapple juice are used. How much grapefruit juice should be added?

b A bottle of Citrola holds 750 ml. How much of this is pineapple juice? [MEG]

12 Mr and Mrs Phillips and their daughter Susan go on holiday to Spain.
The rate of exchange between pounds and pesetas is £1 = 212.40 pesetas.

a Before they go, Mr Phillips changes £175 into pesetas. How many pesetas does he get?

b Susan buys a watch for 6900 pesetas in Spain. How much does the watch cost in pounds? Give your answer to the nearest penny.

c While on holiday in Spain, Mrs Phillips wants to change £290 into pesetas.

She estimates how many pesetas she will get. Write down a calculation she could do in her head. [MEG]

13 a In September 1997 there were 286 516 Turkish lira to one pound sterling. Write 286 516 correct to 3 significant figures.

b At the same time there were 1.6079 Argentinian pesos to one pound sterling. Write 1.6079 correct to 2 decimal places. [MEG]

14 A rectangular card is 15 cm wide and 28.5 cm long.

The card is enlarged to fit exactly a frame which is 85.5 cm long.

a What is the scale factor of the enlargement?

b How wide is the frame? [MEG]

15

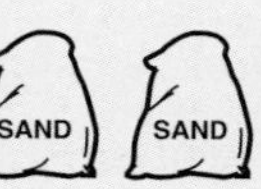

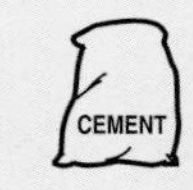

Mortar is made by mixing 5 parts by weight of sand with 1 part by weight of cement.
How much sand is needed to make 8400 kg of mortar? [L]

16 The width and depth of a Small size packet of pet food are 9 cm and 4.2 cm.

The Regular size is an enlargement of the Small size.

The width and height of the Regular size are 15 cm and 10 cm.

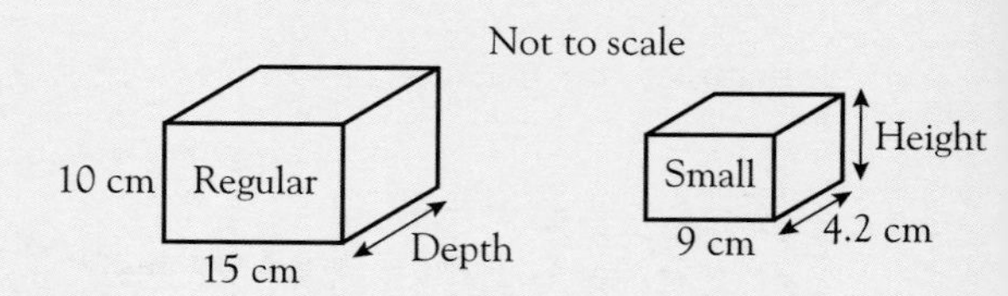

a Calculate the height of the Small packet.

b Calculate the depth of the Regular packet. [SEG]

17 By writing down the prime factors of 42 and 63, find the highest common factor of these numbers. You **must** show all your working. [SEG]

18 From the numbers 4, 12, 18, 20 and 23,

a which number is a prime number?

b which numbers are multiples of 4?

c which three numbers are factors of 60?

d which two numbers have a sum which is a square number?

e which two numbers have a sum which is a cube number? [NI]

EXAM QUESTIONS

19 A sketch plan of a shop is drawn on a scale of 1 : 50.
The shop is 650 centimetres wide.

How wide is the plan of the shop? [MEG]

20 Annabel and Henry are planning a party for 20 children. They have a shopping list of items for a party of 6 children. For the 6 children this includes:

12 balloons	9 jellies
24 sausage rolls	4 packets of biscuits

Rewrite this shopping list for a party of 20 children. [SEG]

21 This is a recipe for 24 scones.

600 g flour
250 g butter
100 g dried fruit
water to mix

a How much dried fruit is needed for 6 scones?

b How much flour would you need for 40 scones? [SEG]

22 Bronze is a mixture of copper and tin.
The bronze used for making nails is made from copper and tin in the ratio of 8 : 1 by weight.
A large bronze nail has a weight of 18 g.
What weight of copper does it contain? [MEG]

23

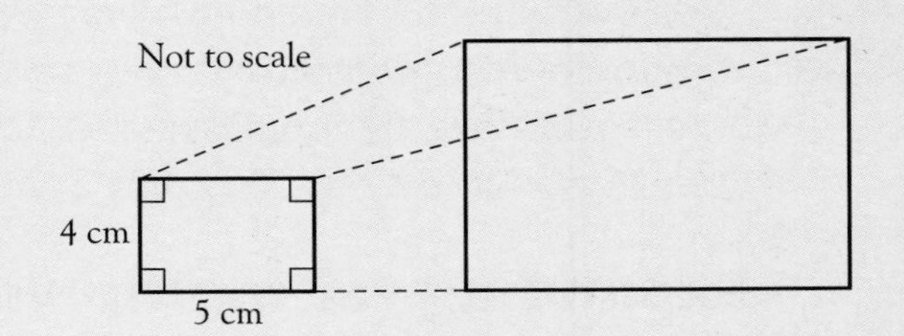

The diagram represents two photographs.

a Work out the area of the small photograph. State the units of your answer.

The photograph is enlarged by scale factor 3.

b Write down the measurements of the enlarged photograph.

c How many times bigger is the area of the enlarged photograph than the area of the small photograph? [L]

EXAMINATION TASK

Here are some numbers.

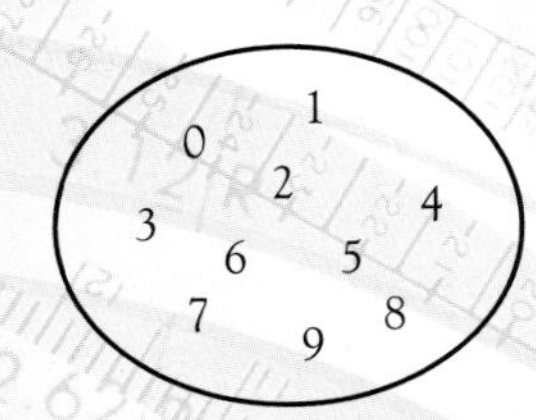

Step 1 Choose three of these numbers.

Step 2 Add them up. Call the answer **A**.

Step 3 Make all the possible two-figure numbers using the numbers in step 1. Add them up. Call the answer **B**.

Step 4 Find the ratio $\frac{\mathbf{B}}{\mathbf{A}}$

For example

Step 1 Choose 3, 5 and 9.

Step 2 $\mathbf{A} = 3 + 5 + 9 = 17$

Step 3 The possible two-figure numbers are 35, 39, 53, 59, 93 and 95.

$\mathbf{B} = 35 + 39 + 53 + 59 + 93 + 95 = 374$

Step 4 $\frac{\mathbf{B}}{\mathbf{A}} = \frac{374}{17} = 22$

Choose some other starting numbers.

Repeat the above steps.

Investigate the ratio of $\frac{\mathbf{B}}{\mathbf{A}}$ [SEG]

ALGEBRA II

In this chapter you will learn how to

- **expand (or multiply out) a bracket**
- **evaluate a formula involving brackets**
- **solve linear equations with unknowns on both sides and/or brackets**
- **solve linear inequalities and represent the result on a number line**

Expanding (or multiplying out) a bracket

The expression $2(x + 3)$ is a shorthand way of writing $2 \times (x + 3)$ or '2 lots of $(x + 3)$'.

You could write the expression as $(x + 3) + (x + 3)$ or $x + 3 + x + 3$ which you can simplify to $2x + 6$.

Comparing this with the original expression you can see that you have

$$2(x + 3) = (2 \times x) + (2 \times 3)$$
$$= 2x + 6$$

Each term inside the bracket is multiplied by 2. You have 'two lots of x plus two lots of 3'.

This is called '**expanding the bracket**' or '**multiplying out the bracket**'.

To expand a bracket, multiply each term inside the bracket, in turn, by the number or letter outside the bracket.

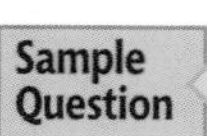

1 Simplify the following

a $4(x - 5)$ b $3(5x + 2y - 4z)$ c $x(x + 2)$

Answer

a

- Multiply each term inside the bracket by 4.

$$4(x - 5) = 4 \times x - 4 \times 5$$
$$4(x - 5) = 4x - 20$$

'4 lots of x minus 4 lots of 5'

b

- Multiply each term inside the bracket by 3.

$$3(5x + 2y - 4z) = 3 \times 5x + 3 \times 2y - 3 \times 4z$$
$$3(5x + 2y - 4z) = 15x + 6y - 12z$$

c

- Expand the bracket as before but remember that you are multiplying each term inside the bracket by x.

$$x(x + 2) = x \times x + x \times 2$$
$$x(x + 2) = x^2 + 2x$$

'x lots of x plus x lots of 2'
Remember that $x \times x = x^2$, $x \times 2 = 2x$

To simplify an expression containing brackets, first multiply out the brackets, then collect like terms.

EXPANDING BRACKETS

Sample Question 2 Simplify the following

a $12 + 3(x - 1)$ b $5(a + 2b - c) + 2a - 3c$

c $8x - 3(2x + 4)$ d $4 - (2x - 5)$

Answer

a

- The bracket part is $3(x - 1)$ so expand it first.

$$3(x - 1) = 3 \times x - 3 \times 1 = 3x - 3$$

12 is not part of the bracket. DO NOT add the 12 and 3

- Now simplify the complete expression.

$$12 + 3(x - 1) = 12 + 3x - 3$$

$$\underline{12 + 3(x - 1) = 9 + 3x}$$

b

- Expand the bracket and then simplify.

$$5(a + 2b - c) + 2a - 3c = 5 \times a + 5 \times 2b - 5 \times c + 2a - 3c$$

$$= 5a + 10b - \mathbf{5c} + 2a - \mathbf{3c}$$

$$\underline{5(a + 2b - c) + 2a - 3c = 7a + 10b - 8c}$$

$-5c - 3c = -8c$

c

- $$8x - 3(2x + 4) = 8x - 3 \times 2x - 3 \times (+4)$$

$$= 8x - 6x - 12$$

$$\underline{8x - 3(2x + 4) = 2x - 12}$$

You will need to take special care when simplifying expressions when a bracket term is being subtracted.

d

- There does not appear to be a number immediately in front of the bracket, but it really means $1(2x - 5)$. Write the expression as $4 - \mathbf{1}(2x - 5)$ and now expand the bracket.

$$4 - (2x - 5) = 4 - 1(2x - 5)$$

$$= 4 - 1 \times 2x - 1 \times (-5)$$

$$= 4 - 2x + 5$$

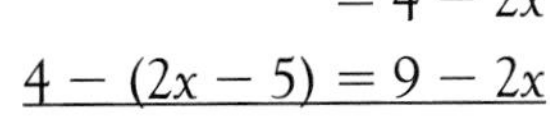

$$\underline{4 - (2x - 5) = 9 - 2x}$$

4 is not part of the bracket. DO NOT do 4 − 1

Remember $(-1) \times (-5) = (+5)$

Note that the **minus** sign **outside** the bracket has the effect of changing the signs **inside** the bracket when it is expanded.

Exercise 11.1

1 Expand the following brackets

a $3(x + 5)$

b $x(x - 4)$

c $2x(x + 7)$

d $6a(5 + 4x)$

e $3(2x + 4)$

f $\frac{1}{2}(6x + 10)$

g $10(a - 2c)$

h $6(3z - 4y + 5)$

i $\frac{1}{10}(100 - 40y)$

j $4(3a - \frac{1}{2}b)$

k $8(3x - 2y + z)$

l $\frac{1}{2}(\frac{1}{2}x + \frac{1}{4}y)$

EXPANDING BRACKETS

2 Simplify the following expressions

a $3x + 2(x + 5)$
b $12 + 2(x + 4)$
c $5 + 3(x - 1)$
d $12a + 4(a - b)$
e $x(x + 2) + 3(x + 2)$
f $3(x + 1) + 2(x + 4)$
g $4(2x - 3) + 6(x + 1)$
h $3(x + 1) + 6(x + 4)$
i $5(2a + 3b) + 4(c + 2b)$

3 Simplify the following expressions

a $3x - 2(x + 4)$
b $8a - 3(2a + 1)$
c $6y - 4(y - 2)$
d $10a - a(3 - 2a)$
e $8 + 3x - 4(x + 2)$
f $4(x + 3) - 2(x + 1)$
g $4x(x + 3) - 2x(1 - x)$
h $x(8 + x) - (x - 2)$
i $12 - 2(x - 1) + 4x$

Evaluating a formula containing brackets

Some formulae have brackets in them, for example $S = \frac{1}{2}(2a + b)$ and $P = 2(l + w)$

page 47

When **evaluating** formulae like this you can use one of these methods:

Either

- multiply out the bracket, simplify, if possible, then substitute the values.

or

- substitute the values into the formula straight away, work out the value inside the bracket then multiply by the value outside the bracket.

Sample Question 3

If $S = \frac{1}{2}(2a + b)$ find the value of S when $a = 5$ and $b = 0.8$.

Answer

Method 1

- Expand the bracket first

$$S = \frac{1}{2}(2a + b)$$
$$= \frac{1}{2} \times 2a + \frac{1}{2} \times b$$
$$= a + \frac{1}{2}b$$

- Now substitute $a = 5$ and $b = 0.8$

$$S = 5 + \frac{1}{2} \times 0.8$$
$$= 5 + 0.4$$
$$\underline{S = 5.4}$$

Method 2

- Substitute $a = 5$ and $b = 0.8$ into the formula and work out the value in the bracket first.

$$S = \frac{1}{2}(2a + b)$$
$$= \frac{1}{2}(2 \times 5 + 0.8)$$
$$= \frac{1}{2}(10 + 0.8)$$
$$= \frac{1}{2} \times 10.8$$
$$\underline{S = 5.4}$$

EXPANDING BRACKETS

Exercise 11.2

1 If $C = 4(3x - y)$, find the value of C when $x = 7$ and $y = 5$

2 Given that $M = 5(a + 4b)$, calculate the value of M when $a = 2$ and $b = 3$

3 Using $p = 3$, $q = 5$ and $r = 2$, find the value of L if $L = 2(6p - 2q + 7r)$

4 If $x = \frac{1}{3}(2m + 9n)$, find the value of x when $m = 12$ and $n = 4$

5 Use the formula $Q = r(4s + t)$ to calculate the value of Q when

a $r = 3$, $s = 1$ and $t = 2$

b $r = 6$, $s = 0$ and $t = -5$

c $r = -5$, $s = 4$ and $t = 3$

d $r = -2$, $s = -7$ and $t = 8$

6 The formula $C = \frac{5}{9}(F - 32)$ is used to convert temperatures from degrees Fahrenheit to degrees Centigrade where F is the temperature in °F and C is the temperature in °C.
Use the formula to find the value of C when

a $F = 50$

b $F = 65$

c $F = 23$

d $F = -13$

7 The average age of four children is calculated using the formula $Y = \frac{1}{4}(a + b + c + d)$ where a, b, c, and d are the ages in years of the four individual children. Calculate the average age of four children whose individual ages are 4 years, 6 years, 10 years and 12 years.

8 The area of the shaded part of the diagram is calculated using the formula $A = \pi(R^2 - r^2)$. Find the area, A, when $R = 68$ mm and $r = 36$ mm. Give your answer in mm² to 3 significant figures.
What is the area in cm²?

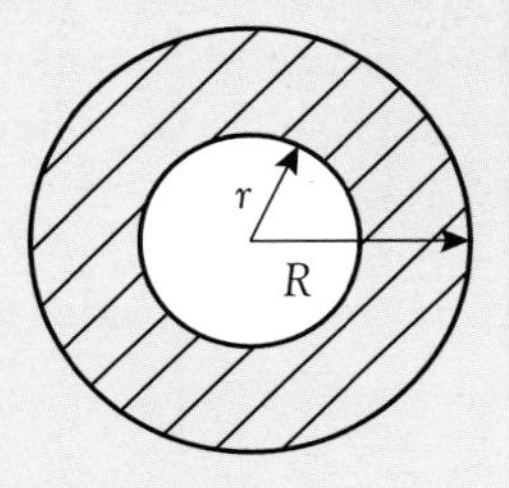

Take $\pi = 3.14$

9 The rectangular floor shown in the diagram has perimeter P and area A given by the formulae $P = 2x + 2(x - 2)$ and $A = x(x - 2)$

Work out the values of P and A when $x = 9.5$ m.

EQUATIONS

Solving linear equations

In chapter 4, you learned how to solve simple linear equations using a balance diagram, in the first instance, then by just writing down and carrying out the instructions.
You can use these methods to solve harder linear equations such as those with unknown quantities on **both** sides of the equation, e.g. $3x + 2 = 5x - 6$

Sample Question 4

Solve the equation $3x + 2 = 5x - 6$, shown in this balance diagram.

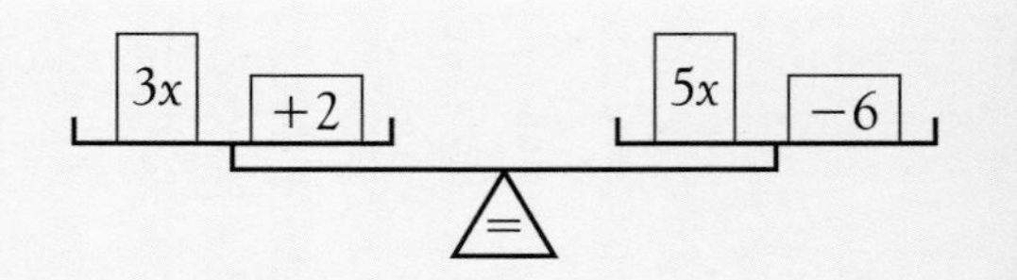

Answer

- Decide which side of the balance you want to have your unknowns (x) and which side the numbers. It is useful to collect the x terms on the side that has the largest x term.
  As $5x$ is larger than $3x$, collect the x terms on the right-hand side (RHS) and the numbers on the left-hand side (LHS).
- Subtract $3x$ from both sides of the balance.

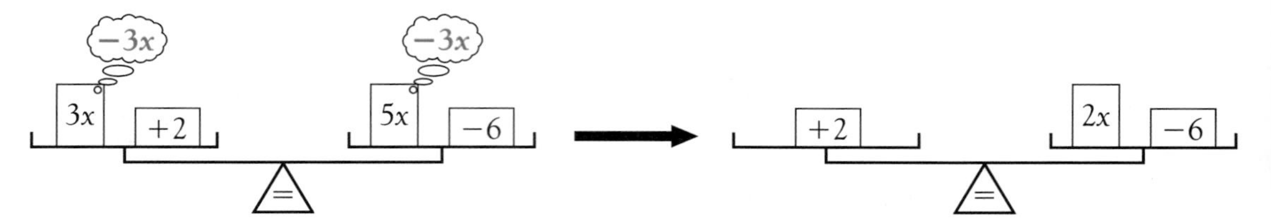

- To remove the -6 from the right-hand side you must add 6 to both sides of the balance.

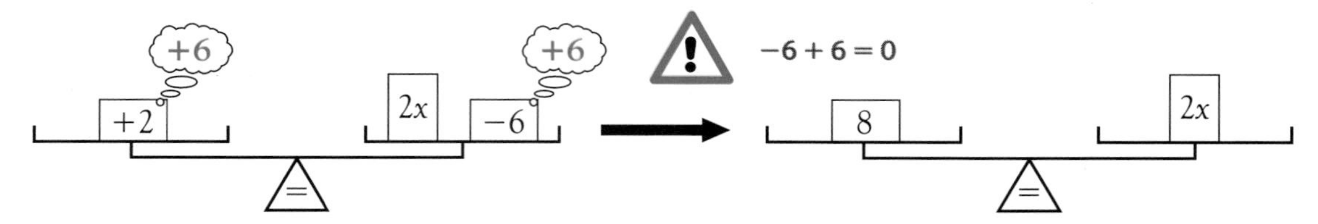

- 'Shrink' the $2x$ by dividing both sides by 2

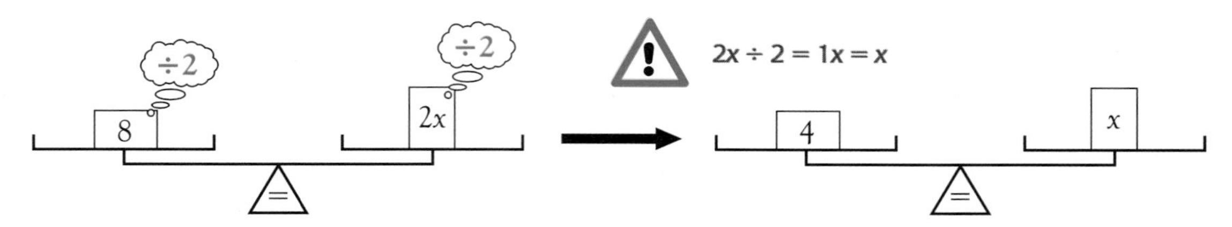

This leaves you with the required result:

So, if $3x + 2 = 5x - 6$ then $\underline{x = 4}$

Check your answer by substituting your value of x into the original equation and check whether the two sides 'balance'.

LHS	RHS
$(3 \times 4) + 2 = 14$	$(5 \times 4) - 6 = 14$

Your value works because LHS = RHS

EQUATIONS

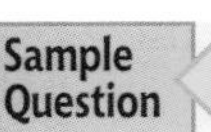

5 Solve the equation $7x - 3 = 2x - 13$.

Answer

- Draw the equation as a balance.
- Collect the x terms on the left-hand side as $7x$ is bigger than $2x$, and collect the numbers on the right-hand side.
- Subtract $2x$ from both sides of the balance.

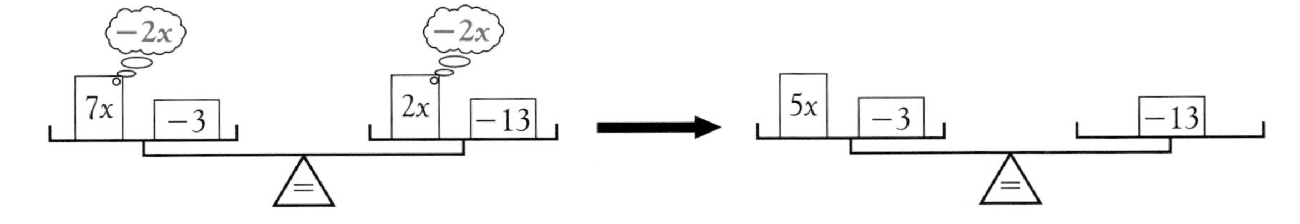

- Add 3 to both sides.

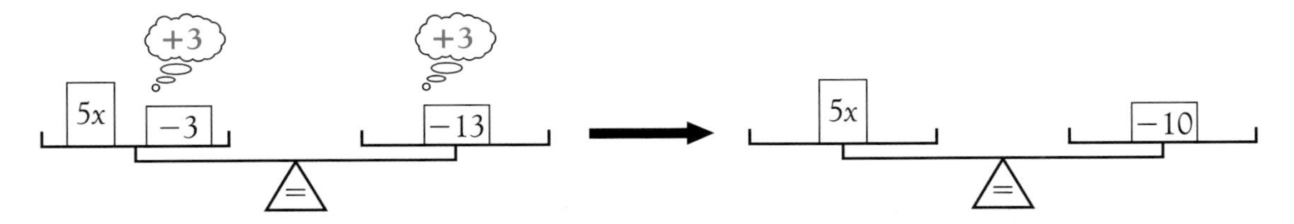

- 'Shrink' the x term by dividing both sides by 5.

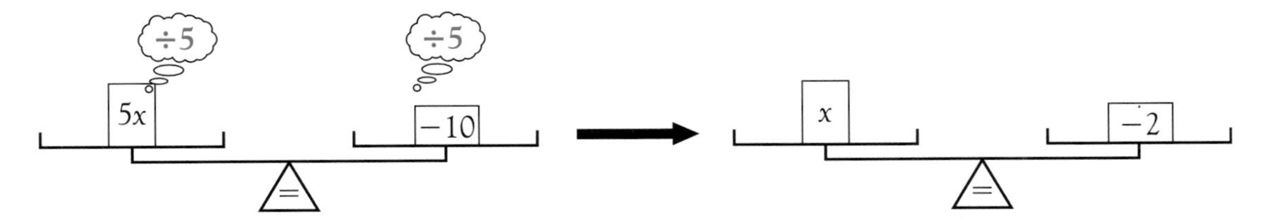

- This gives the required result.

So, if $7x - 3 = 2x - 13$, then $\underline{x = -2}$

Remember to check your answer.

Drawing the diagrams is very time consuming. You can solve these equations by leaving out the diagrams, just writing and carrying out the instructions as shown in the 'clouds' on the diagrams. The answers to Sample Questions 4 and 5 can be written like this:

4

	$3x + 2 = 5x - 6$
$-3x$	$2 = 2x - 6$
$+6$	$8 = 2x$
$\div 2$	$\underline{4 = x}$

5

	$7x - 3 = 2x - 13$
$-2x$	$5x - 3 = -13$
$+3$	$5x = -10$
$\div 5$	$\underline{x = -2}$

EQUATIONS

If your equation contains a bracket, expand the bracket and simplify **before** solving the equation.

Sample Question 6 Solve the equations

a $12 - 3x = 2(1 + x)$ b $12 - x = 3(6 - x)$

Answer

a

	$12 - 3x = 2(1 + x)$
Expand	$12 - \mathbf{3x} = 2 + \mathbf{2x}$
$+3x$	$12 = 2 + 5x$
-2	$10 = 5x$
$\div 5$	$\underline{2 = x}$

b

	$12 - x = 3(6 - x)$
Expand	$12 - \mathbf{x} = 18 - \mathbf{3x}$
$+3x$	$12 + 2x = 18$
-12	$2x = 6$
$\div 2$	$\underline{x = 3}$

In part **a**, collect the x terms on the right, as $2x$ is bigger than $-3x$.
In part **b**, collect the x terms on the left, as $-x$ is bigger than $-3x$.

Problems are often given in the form of a number puzzle as in the following example.

Sample Question 7 'I think of a number, add 5 to the number and then multiply the result by 2. The answer is the same as subtracting 1 from the number and multiplying the result by 3.' Write the number puzzle as an equation and solve your equation to find the number.

Answer

- Let the number be x.
- Look at the first part of the puzzle
  'I think of a number, add 5 ...' gives $x + 5$
  '... multiply the result by 2.' gives $2 \times (x + 5)$ or $2(x + 5)$
- Now look at the second part of the puzzle
  '... subtract 1 from the number ...' gives $x - 1$
  '... multiply the result by 3.' gives $3 \times (x - 1)$ or $3(x - 1)$
- Equate the two parts of the puzzle.
  $2(x + 5) = 3(x - 1)$

Equate means put them equal to each other.

- Solve the equation

	$2(x + 5) = 3(x - 1)$
Expand	$2x + 10 = 3x - 3$
$-2x$	$10 = x - 3$
Add 3	$\underline{13 = x}$

EQUATIONS

In this sort of question you will have to read the puzzle very carefully.
For example, notice the difference between these statements:

'... **subtract 3 from the number** ...' gives $x - 3$

but

'... **subtract my number from 3** ...' gives $3 - x$

Exercise 11.3

Solve the following equations.

Note that answers are not necessarily positive whole numbers.

You must show all the stages in your working.

1 $3x + 2 = x + 6$

2 $4x - 7 = 7x - 1$

3 $2x + 10 = 5x + 1$

4 $5x - 7 = 13x + 9$

5 $7x - 10 = 3x - 8$

6 $16x - 8 = 4 + 10x$

7 $5x + 1 = 7 - 2x$

8 $12x + 5 = 8x + 29$

9 $4(x + 2) = 12$

10 $3(2x - 5) = 9$

11 $2(x - 2) + 1 = 4$

12 $x + 3(x + 5) = 21$

13 $2(x + 1) = 3x - 2$

14 $2(x - 7) = 5(x - 3)$

15 $3x - (x + 2) = 8x$

16 $3(3x - 1) + 2(x + 3) = 14$

17 $3x + 2 = 18 - 3x$

18 $2(x + 3) = 18 - 6x$

19 $14 - 3x = 8 - x$

20 $6(x - 1) = 2(3 - x)$

21 Amy thinks of a number, multiplies it by 7 and then subtracts 12. The answer is the same as the number she started with. Write this puzzle as an equation and solve the equation to find the number that Amy thought of.

22 I start with a number, subtract 3 then multiply the result by 5. The answer is the same as adding 1 to the number. Write this puzzle as an equation and solve it to find the number I start with.

23 I think of a number, add 5 and multiply the result by 2. The answer is the same as subtracting 7 from my number and multiplying the result by 3. Write this puzzle as an equation and solve it to find the number I am thinking of.

24 David subtracts a number from 10 and then multiplies the result by 3. If he had added 2 to his number David would have got the same result. Write this number puzzle as an equation and solve it to find David's number.

25 When Jane multiplies a number by 8 she gets the same answer as when she subtracts 9 from the number and doubles the result. Write this number puzzle as an equation and solve it to find Jane's number.

Linear inequalities

This topic uses the symbols $>$, $<$, $\leqslant$, $\geqslant$. Here are some illustrations.

a $x > 3$ This can be read as 'x is greater than 3'.

If you are asked to give a value that x can take you could say 4 or 10 or 2500. Your value need not be a whole number, so you could say that it is 3.2 or $6\frac{1}{4}$ or 20.39. In fact you can give **any** number as long as it is greater than 3.

It could be very close to 3, for example 3.00000001, but it **cannot actually be 3**.

You can show possible values for x on a number line.

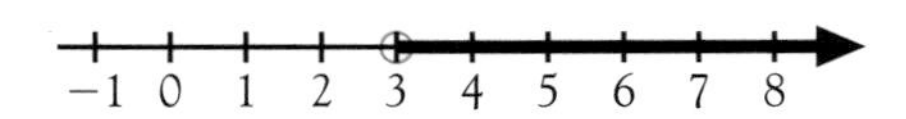

The arrow means that all numbers further to the right are included. The 'open' circle means that you cannot include 3.

b $x \geqslant 3$ This can be read as 'x is greater than or equal to 3'.

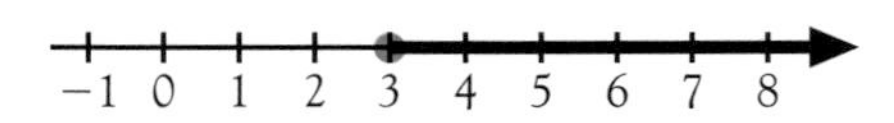

The 'filled' circle means that you **can** include 3.

In both of the inequalities $x > 3$ and $x \geqslant 3$, 3 is called the **lower boundary**.

c $x < 1$ This can be read as 'x is less than 1'.

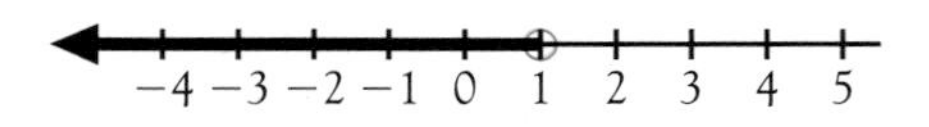

Note the 'open' circle showing that 1 is not included.

d $x \leqslant 1$ This can be read as 'x is less than or equal to 1'.

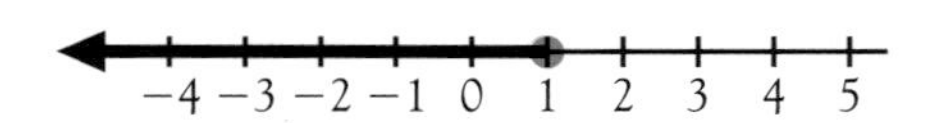

Note the 'filled' circle showing that 1 is included.

In both of the inequalities $x < 1$ and $x \leqslant 1$, 1 is called the **upper boundary**.

Notice that in an inequality, the bigger value is at the wider part of the symbol, the smaller value is at the point. This helps you to read inequalities from right to left as well as from left to right, for example,

	Inequality	Reading left to right	Reading right to left
i	$8 > 6$	8 is greater than 6	6 is less than 8
ii	$x \leqslant 5$	x is less than or equal to 5	5 is greater than or equal to x
iii	$2 \leqslant x$	2 is less than or equal to x	x is greater than or equal to 2

It is possible to combine **ii** and **iii** into one statement, as shown in **e**.

INEQUALITIES

e $2 \leq x \leq 5$

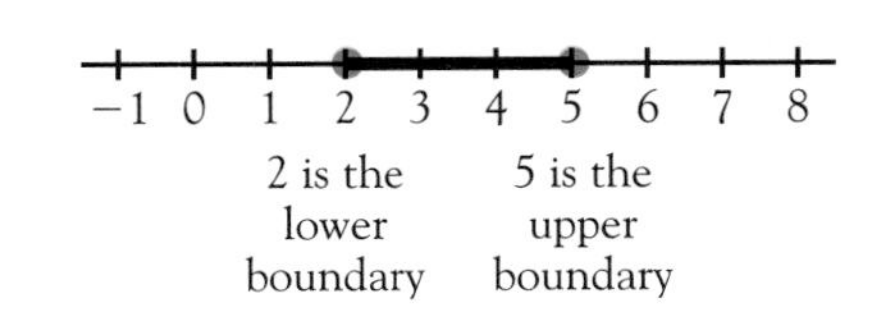

This means that x is greater than or equal to 2 **and** x is less than or equal to 5.
So x is 'sandwiched' between 2 and 5 inclusive.

Notice the 'filled in' circles: x could be 2, x could be 5.

f $-1 \leq x \leq 3$

'x is greater than or equal to -1 and x is less than or equal to 3'

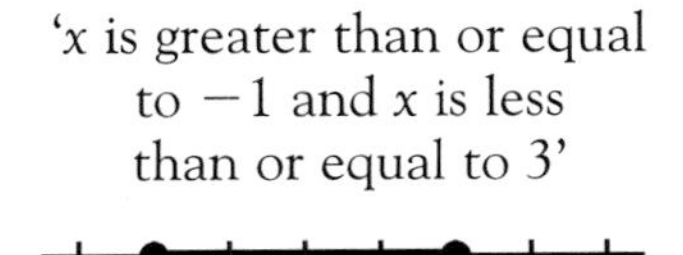

g $2 \leq x < 6$

'x is greater than or equal to 2 and x is less than 6'

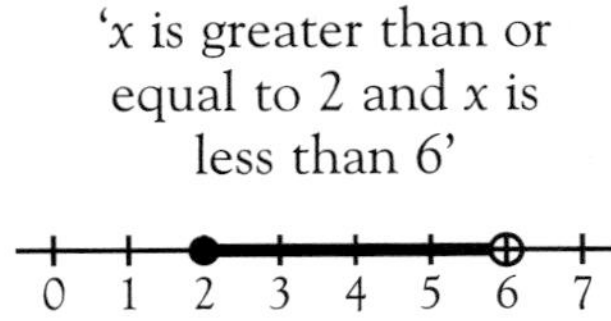

h $-3 < x < 0$

'x is greater than -3 and x is less than 0'

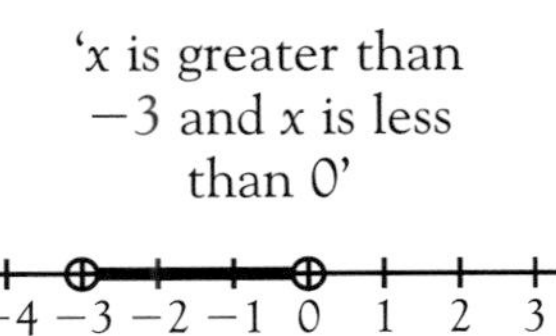

i $x < 2$ or $x > 6$

'x is less than 2 or x is greater than 6'

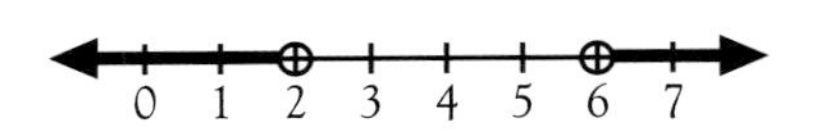

This is a special type.
DO NOT try to write the inequalities in one statement. They must be written separately.

In each example, notice the appropriate use of an 'open' or 'filled' circle on the number line

Remember that the inequality notation was used in Chapter 9 to show the upper and lower boundaries of a class interval, e.g. $10 \leq w < 20$. This meant that for the given class interval 10 is included in the interval but 20 is not included. The next interval starts with 20. You will use the notation again in Chapter 13 (Data Handling II).

Another use of this notation is to show the values that a number could have been before it was rounded off.

For example, the length of a pencil might be given as 9 cm (correct to the nearest cm).

This means that the actual length could be up to and including 0.5 cm shorter than 9 cm, but less than 0.5 cm longer than 9 cm.

This can be written as an inequality like this: $8.5 \leq \text{length} < 9.5$

The values 8.5 and 9.5 are the lower and upper boundaries respectively. The inequality covers all the values that the distance can be before it is rounded to the nearest centimetre.

The actual length of the pencil could have been any value between the boundaries, e.g. 8.5 cm, 8.75 cm, 8.9 cm, 9.2 cm **but not 9.5 cm.**

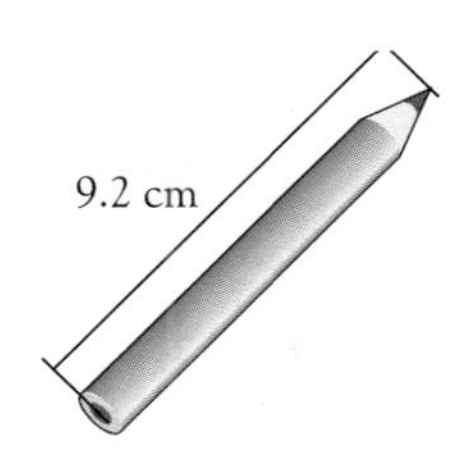

Writing the inequality is a lot quicker than trying to write down all the possible values!
Sometimes you are asked to state the **integer** values that satisfy an inequality, that is, say what **whole number values** the letter can take.

INEQUALITIES

Sample Question 8

a Show the following inequalities on a number line

i $x > -1$ ii $4 \leqslant x$ iii $-5 \leqslant x < 3$

b Write down the **integer** values of x that satisfy the following inequalities

i $3 < x < 8$ ii $-8 < x \leqslant -5$

Answer

a i

- x is **any** number greater than -1 **but not including** $\mathbf{-1}$, i.e. 0, 1, 2, 3, …
  This includes fractional parts as well.

−2 −1 0 1 2 3 4 5 $x > -1$

ii

- Reading from the letter, **right to left**, 'x is greater than or equal to 4', i.e. 4, 5, 6, 7, 8, …
  including fractional parts.

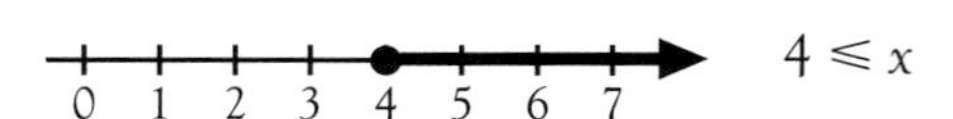

$4 \leqslant x$

iii

- Reading from the centre, **right to left**, 'x is greater than or equal to -5'.
  Reading from the centre, **left to right**, 'x is less than 3'.
  This means that x can take any value between -5 and 3, including -5, but **not including** 3.

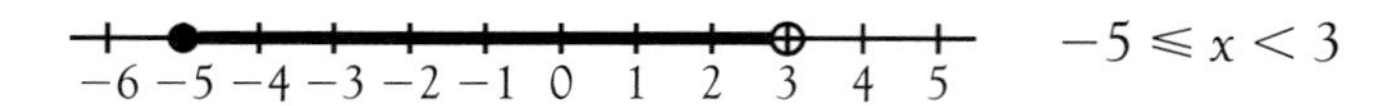

$-5 \leqslant x < 3$

b i

- For $3 < x < 8$, starting at the centre, read firstly from right to left then from left to right:
  'x is greater than 3, not including 3, and less than 8, not including 8'.
  As **only integer (whole number) values** are required,

<u>x can be either 4, 5, 6 or 7</u>

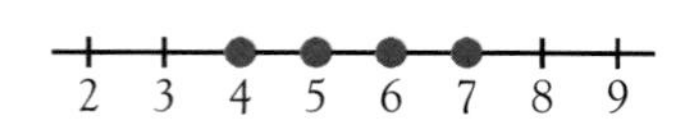

Notice that only integer values, shown by filled circles, are drawn on the number line.

ii

- 'x is greater than -8, not including -8, and less than or equal to -5'

<u>x can be either -7, -6 or -5</u>

−8 −7 −6 −5 −4

Exercise 11.4

Draw a number line to illustrate each of the inequalities 1–6:

1 $x < 4$

2 $x \geqslant 2$

3 $1 < x < 3$

4 $-2 \leqslant x \leqslant 1$

5 $x < -1$ or $x \geqslant 4$

6 $-2.5 < x \leqslant 4.5$

Write down the **integer** values of x that satisfy the inequalities 7–12:

7 $0 \leqslant x \leqslant 4$

8 $3 < x < 7$

9 $-2 < x \leqslant 5$

10 $-6 < x < -1$

11 $-7 \leqslant x < 0$

12 $-1.5 < x \leqslant 3.5$

Solving linear inequalities

Some inequalities need to be manipulated before you can state the values that satisfy the inequality. This is called **solving an inequality** and is very much like solving a simple equation obeying nearly all the same rules, with one exception explained below.

Solving an inequality involves finding the boundary, or boundaries if the inequality is a combined statement.

  Solve the inequality $2x + 1 > 7$

Answer

◆ Find the boundary by solving the inequality $2x + 1 > 7$ in the same way that you would solve the equation $2x + 1 = 7$

Solving the **inequality**

	$2x + 1 > 7$
-1	$2x > 6$
$\div 2$	$\underline{x > 3}$

Notice the similarity to solving the equation $2x + 1 = 7$

Solving the **equation**

	$2x + 1 = 7$
-1	$2x = 6$
$\div 2$	$\underline{x = 3}$

The lower boundary for the inequality is 3 but remember that 3 is **not** included.

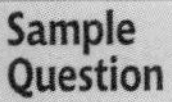  Solve these linear inequalities **a** $2(2x - 1) > 15$ **b** $7 - x \geqslant 9$

Answer

a

	$2(2x - 1) > 15$
Expand	$4x - 2 > 15$
$+2$	$4x > 17$
$\div 4$	$\underline{x > 4\frac{1}{4}}$

b

	$7 - x \geqslant 9$
$+x$	$7 \geqslant 9 + x$
-9	$-2 \geqslant x$
	$\underline{x \leqslant -2}$

It is better to have $+x$ than to have to multiply or divide by a negative number.

Part **b** illustrates an important difference between solving equations and inequalities.

In an inequality, multiplying or dividing by a negative number **reverses the inequality**,

for example $8 > 3$, but $-8 < -3$ (multiplying by -1)

AVOID HAVING TO DO THIS, IF POSSIBLE.

INEQUALITIES

Sometimes you are asked to solve an inequality that is a combination of two inequalities. In this case you will need to solve each 'half' as in Sample Question 11. You could do this in two stages, or in one.

Sample Question 11

a Find the solution of $-5 \leqslant 2x + 3 < 7$
b List all the integer solutions.

Answer

a

- In two stages

i $-5 \leqslant 2x + 3$
-3 $\quad -8 \leqslant 2x$
$\div 2$ $\quad -4 \leqslant x$

ii $2x + 3 < 7$
-3 $\quad 2x < 4$
$\div 2$ $\quad x < 2$

Combine these to give
$\underline{-4 \leqslant x < 2}$

- In one stage

$-5 \leqslant 2x + 3 < 7$
-3 $\quad -8 \leqslant 2x < 4$
$\div 2$ $\quad \underline{-4 \leqslant x < 2}$

b

- List the possible **integer** values i.e. all the integers that are smaller than 2 and bigger than or equal to -4.

$\underline{x = -4, -3, -2, -1, 0 \text{ or } 1}$

Exercise 11.5

1 Solve the following linear inequalities:

a $x + 5 \geqslant 11$
b $5x - 3 < 7$
c $3x + 7 > 19$
d $4x - 3 \leqslant 5$
e $9x - 1 \geqslant 17$
f $2x - 9 < 6$
g $8x - 21 < 5$
h $6x + 1 \geqslant 14$
i $3x + 9 > 0$
j $7x - 2 \leqslant 8$
k $5 < 2x - 6$
l $7 > 4x + 15$

2 Solve the following linear inequalities:

a $\dfrac{x}{3} + 1 \geqslant 4$
b $5 - 2x < 1$
c $3(x - 4) \leqslant 6$
d $\dfrac{2x - 1}{5} > 3$
e $5 < 11 - 3x$
f $\dfrac{3x}{2} - 4 \geqslant 5$
g $7(2x + 1) > 14$
h $\dfrac{3x + 1}{9} \leqslant 2$
i $3(1 - 4x) < 6$
j $13 > 5(2x + 7)$
k $9 \geqslant 15 - \dfrac{3x}{4}$
l $\dfrac{2(8 - x)}{3} \leqslant 4$

3 By solving the following inequalities, write down the largest and smallest integer values of x:

a $5 \leqslant x + 2 \leqslant 9$
b $3 \leqslant 2x - 1 < 11$
c $-1 < 5x - 6 < 14$
d $-5 < 4x + 3 < 11$
e $0 \leqslant 2(3x + 1) < 10$
f $-4 < 1 - 3x \leqslant 4$

4 a The maximum number of spectators allowed in a football stadium is given as 72 000 to the nearest 50. Using x to represent the number of spectators, copy and complete this inequality:
$\ldots \leqslant x < \ldots$

b The temperature in a freezer must be kept below -18°C. If t is the temperature in °C of the freezer, write an inequality to show this.

c I take at least a quarter of an hour to walk to the bus stop. Using x as the number of minutes it takes me to walk to the bus stop, write an inequality to show this.

d Ben collects foreign coins. He now has c coins in his collection. He has more than fifty but may have as many as 60 coins. Write an inequality to show this.

e A bus can seat 36 passengers downstairs and 28 passengers upstairs. Up to 10 more passengers are allowed to stand up on the bus. If all the seats are occupied and n is the total number of passengers on the bus, write down an inequality to show the minimum and maximum number of passengers on the bus.

f The length of a fence is given as 25 metres, correct to the nearest metre. If L is the length of the fence, write down an inequality to show the boundary values of the length of the fence.

5 a The mass of a tanker is 3 tonnes. There are x tankers in a convoy. Write down an expression for the total mass in tonnes of the x tankers.

b The convoy follows a route that involves passing over a bridge with a maximum load of 14 tonnes.

i Write down an inequality that links the total mass of the convoy and the maximum load that the bridge can carry.

ii Solve the inequality and state the maximum number of tankers that can be in the convoy.

6 a Winston buys an eraser costing 36 pence and x pencils costing 12 pence each.

Write down an expression for the total cost, in pence, of the eraser and the pencils.

b Winston only has £2 to spend. Write down and solve an inequality to find the maximum number of pencils that Winston can buy.

7 The mass of the packaging around a packet of biscuits is 45 g. Each biscuit in the packet has a mass of 20 g.

a If there are x biscuits in the packet, write down an expression for the total mass in grams of the biscuits.

b Given that the total mass of the packet of biscuits must not exceed 300 g, write down an inequality that links the total mass of the packet of biscuits, the mass of the packaging and the mass of the biscuits.

c Solve the inequality and state the maximum number of biscuits in the packet.

8 I think of a positive whole number, subtract 6 from it and then multiply the result by 7. The answer is less than 50.

a If x is the number I am thinking of, write down the puzzle as an inequality.

b Solve your inequality.

c Write down the largest number I could be thinking about.

9 A group of friends decide to make some extra pocket money by making and selling badges. They pay £5 to hire the badge-making equipment and buy badge blanks for 8 pence a blank. They decide to sell each badge for 20 pence.

a If x is the number of badges they make, write down an expression in pence for

i the total amount of money they spend,

ii the total amount of money they make from selling the badges.

b Assuming they sell all the badges that they make, use an inequality to find the minimum number of badges they need to make in order to make a profit.

10 Sam and her sister Rachel play a game of marbles. Rachel starts the game with 5 more marbles than Sam.

Rachel has x marbles.

a Write down an expression in x for the total number of marbles that the two sisters have.

b If together they have more than 30 marbles write down an inequality in x and solve your inequality.

c Write down the minimum number of marbles that Rachel has.

11 I think of a number, add 8 to it and divide the result by 3. The answer is greater than the number I first thought of.

a Write this number puzzle as an inequality, using x for the number I think of.

b Solve your inequality and illustrate your answer on a number line.

c If the number I first thought of was a positive integer, write down all the numbers it could have been.

Worked Exam Question

[NI]

a Solve the equation $\frac{1}{4}(x - 4) - \frac{1}{5}(x - 7) = 1$

- Multiply each term by 4
- Multiply each term by 5
- Expand the brackets
- Simplify
- Subtract 8 from both sides

Remember it is +28

$(x - 4) - \frac{4}{5}(x - 7) = 4$

$5(x - 4) - 4(x - 7) = 20$

$5x - 20 - 4x + 28 = 20$

$x + 8 = 20$

Answer ...$x = 12$......

You could do the first two steps in one by multiplying each term by 20.

COMMENTS

M1 attempting to multiply by 4 and by 5

A1 correct multiplication by 4, 5

A2 expanding each bracket correctly (A1 each bracket)

A1 correct answer

5 marks

b **n is a whole number less than 7.**
Solve the inequality $3n - 4 > 5$ and illustrate your answer on a number line

- Add 4 to both sides
- Divide both sides by 3
- The question tells you that n is a **whole number less than 7**.
  The solution to the inequality tells you that n is **greater than 3**.
  So, n can take the value **4, 5** or **6** as shown on the number line.

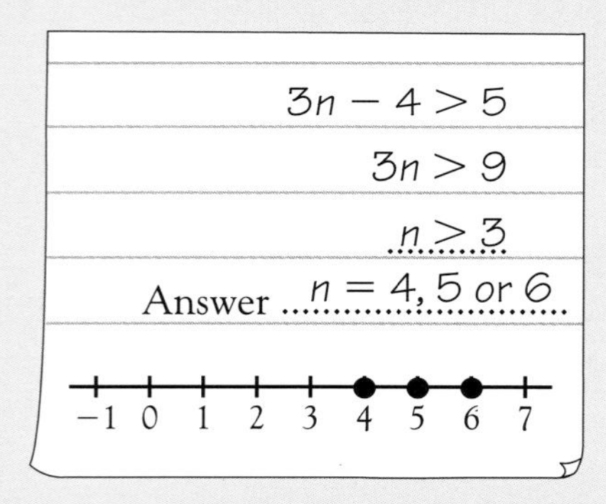

A1 adding 4 to both sides correctly

A1 correct inequality for n

A1 indicating correct integer values for n

3 marks

Exam Questions

1 Solve the equation $2(y + 3) = 7$ [MEG, p]

2 Solve $3(z - 4) = 30$ [L, p]

3 Solve the following equation
$5(q - 3) = 25$ [L, p]

4 Solve the equation $3(4x - 1) = 27$ [NEAB]

5 Solve the equation $3(x - 2.5) = 6$ [MEG, p]

6 Solve $4(z + 5) = 12$ [MEG, p]

7 Solve the equation $3(4y - 9) = 81$ [L]

8 The areas of two rectangles are the same.
By solving the equation $2(x + 2) = 2(4x - 1)$
find the area of one of the rectangles.

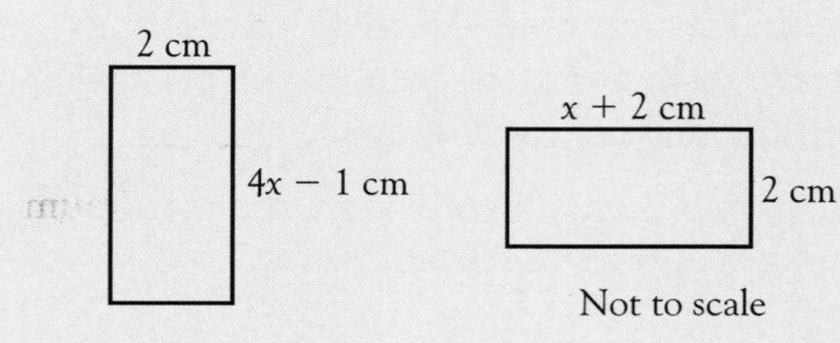

[SEG]

EXAM QUESTIONS

9 Solve this equation
$4(5 - x) + 3(3x + 4) = 25$ [MEG]

10 Solve $5(t + 2) - 3(t - 2) = 15$ [MEG, p]

11 Here is a formula for working out the perimeter of a rectangle $P = 2(l + w)$
Use the formula to work out the value of P when $l = 6$ and $w = 4$ [L]

12 $s = \frac{1}{2}(u + v)t$
Work out the value of s when $u = 10$, $v = -25$ and $t = 0.5$ [L]

13 The formula for the area, A, of a trapezium is
$A = \frac{1}{2}(a + b)h$
Work out the area A when $a = 7$ cm, $b = 12$ cm and $h = 4$ cm. [MEG]

14 The volume, V, of material inside a certain tube is given by the formula $V = 25\pi(R^2 - r^2)$
Calculate the value of V when $\pi = 3.1$, $R = 7.3$ and $r = 5.9$ [MEG]

15 Anna and Garry describe the same number. Anna says 'It is bigger than 4'. Garry says, 'It is smaller than 11'.
Write down all the possible values of the number. [NEAB]

16 x is an integer. Write down the greatest value of x for which $2x < 7$ [L]

17 List all integer values of n for which $-2 < n \leqslant 5$ [MEG, p]

18 Solve the inequality $3 - 2x > 8$ [MEG, p]

19 **a** Solve the inequality $\frac{1}{2}x + 1 < 3$

b Represent your answer to part **a** on a copy of this number line.

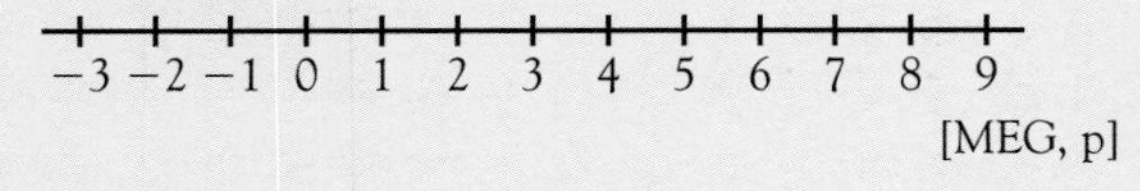

[MEG, p]

20 Solve the inequality $2x - 5 < 8$ [L, p]

21 **a** n is a whole number such that $-3 < n \leqslant 1$
List all the possible values of n.

b In a school race the winner's time was recorded as 9.7 seconds to the nearest 0.1 second.
Let the actual winning time be t seconds.

i On a copy of the number line, illustrate the range in which t could be.

9.7

ii Complete the inequality $\ldots \leqslant t < \ldots$ [L]

22 List all the possible integer values of n such that $-3 \leqslant n < 2$ [L]

23 **a** List all the possible values of x, where x is an integer, such that $-4 \leqslant x < 2$

b Solve $4x - 5 < -3$ [SEG, p]

24 Write down the values of n, where n is a whole number, such that $-6 < 2n \leqslant 10$ [MEG, p]

25 Solve the following inequalities for x

a $1 + 3x < 7$

b $4x - 3 > 3x - 2$ [NEAB]

26 Find the integer solutions of the inequality $-5 < a \leqslant 2$ [MEG]

27 Solve the inequality $7y > 2y - 3$ [L]

28 Solve the inequality $7n - 12 > 9$ [MEG, p]

29 **a** Solve the equation $5 - 2x = 11$

b Hence, or otherwise, solve the inequality $5 - 2x \leqslant 11$ [MEG, p]

30 Solve the inequality $3(x - 2) < x + 7$ [SEG, p]

31 **a** Solve the inequalities

i $5 + 3x \leqslant 20$

ii $5 + 3x \geqslant 2$

b Hence write down the range of values of x for which $2 \leqslant 5 + 3x \leqslant 20$ [MEG]

32 Solve the inequality $4x - 19 \leqslant 2x + 6$ [SEG]

33 **a** List all the integers which satisfy $-2 < n \leqslant 3$

b Ajaz said 'I thought of an integer, multiplied it by 3 then subtracted 2. The answer was between 47 and 62'.
List all the integers that Ajaz could have used. [MEG]

12 GEOMETRY II

In this chapter you will learn how to
- **measure and draw bearings**
- **make accurate drawings using drawing instruments**
- **identify and construct the path of a point (locus) that obeys a given rule**

Bearings

Directions can be given using **bearings**. There are two types – **compass bearings** (related to North, South, East and West) and **'three-figure' bearings**. A '**three-figure**' bearing is an angle that tells you where an object is relative to North on the compass. It gives the amount of turning in a **clockwise** direction and always starts from North. North is given the bearing 000°.

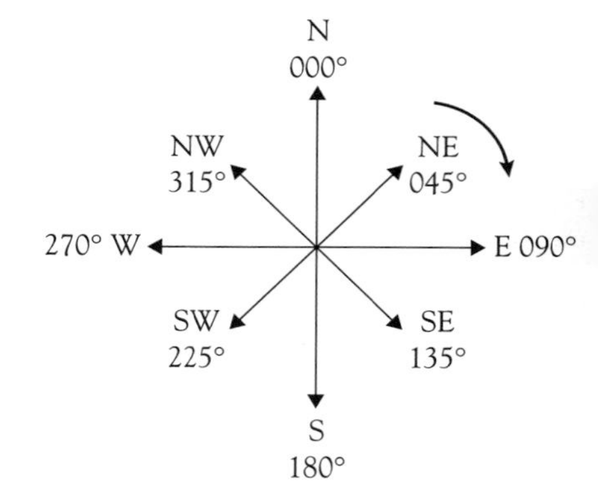

Note that there are three numbers in the bearing. All bearings must have three numbers.

A bearing less than 10° will start 00, e.g. 8° = 008°.
A bearing of 10° up to 99° will start with a 0, e.g. 47° = 047°.

To draw or measure a bearing you will need to use a 180° **protractor** or a 360° **angle measurer**.

Using a protractor if the bearing is less than 180°

- Place the protractor with the centre on the point that the bearing is taken from and with the curved edge facing to the **right**. 0° on the scale should be pointing **North**.
- Reading **clockwise**, read off the angle making sure you are using the 0, 10, 20, … scale **not** the 180, 170, 160, … scale.
- If the bearing is less than 100° make sure that you write down your answer with **3** figures.

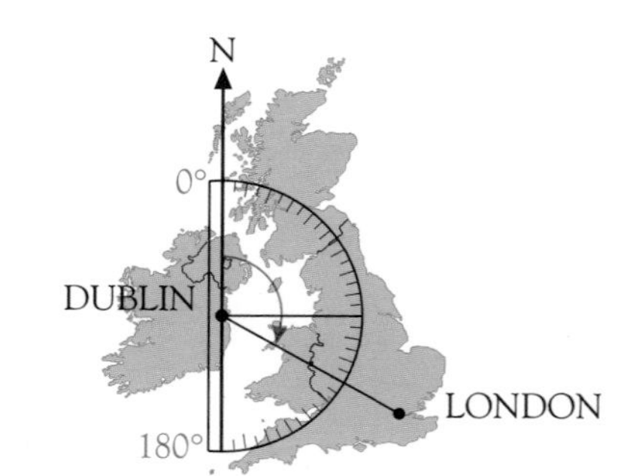

The bearing of London from Dublin is 120°

Using a protractor if the bearing is more than 180°

- Place the protractor with the centre on the point that the bearing is taken from and the curved edge facing to the **left**. 0° on the scale should be pointing **South**.
- Reading **clockwise**, read off the angle making sure you are using the 0, 10, 20, … scale **not** the 180, 170, 160, … scale.
- **Add this angle to 180° to find the bearing.**

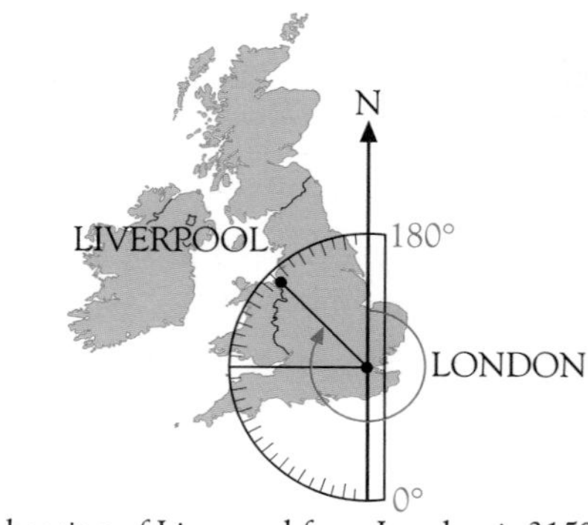

The bearing of Liverpool from London is 315°

BEARINGS

Using an angle measurer

- Place the angle measurer so that the 0° line is pointing **North** and the centre of the angle measurer is on the point from which the bearing is being taken.
- Read the bearing in a clockwise direction making sure that you use the scale that reads 0, 10, 20, … **not** the scale that reads 0, 350, 340, 330, … .
- Write down the bearing making sure that your answer has 3 figures.

The bearing of Glasgow from London is 330°

When you measure or draw a bearing it is important to place the centre of your protractor or angle measurer on the correct point! The correct starting point is usually the point that follows the word '**from**', e.g. in 'measure the bearing of Dublin **from** London' the starting point is London.

Often you are given the bearing from one point to another and are asked to give the bearing of the return journey. This is called the **back-bearing** or **reverse bearing**.

Sample Question 1

The bearing of Bidham from Rutworth is 132°.

Calculate the bearing of Rutworth from Bidham.

DO NOT measure the angle.
The diagram is not to scale.

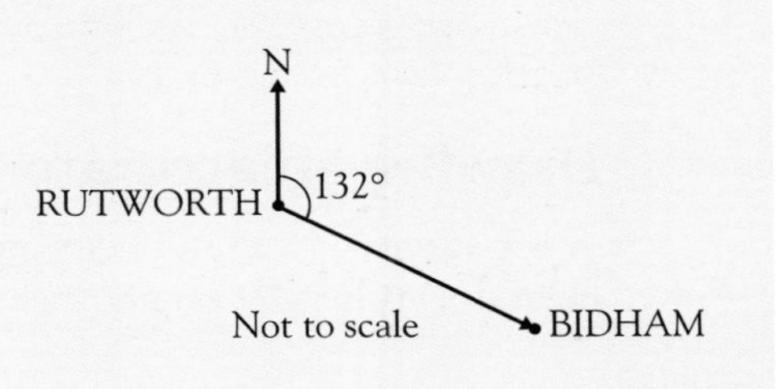

Answer

- Imagine walking from Rutworth to Bidham along a straight road. When you get to Bidham turn round in a clockwise direction to start your walk back to Rutworth. You turn through an angle of **180°**.
- **Add 180°** to the bearing of Bidham from Rutworth.

  132 + 180 = 312

  The bearing of Rutworth from Bidham is 312°

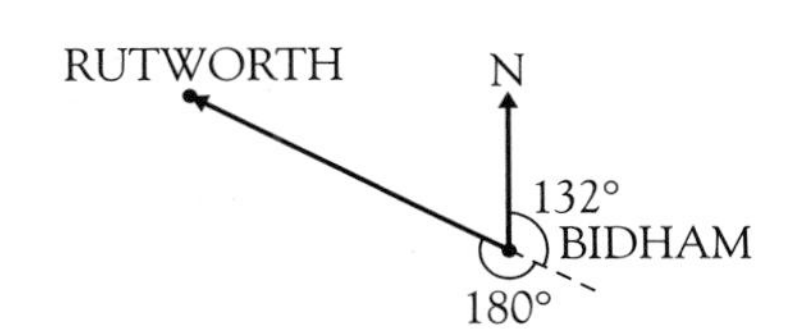

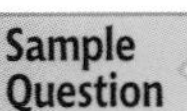

2

A helicopter is flown on a bearing of 220° from Wrexham to Swansea.
Calculate the bearing of the return journey from Swansea to Wrexham.

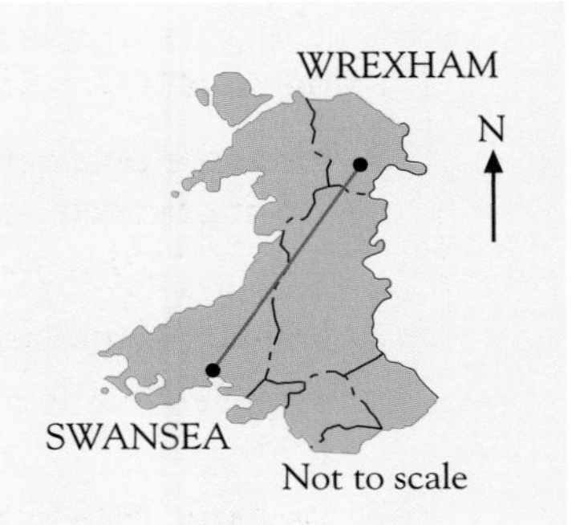

BEARINGS

Answer

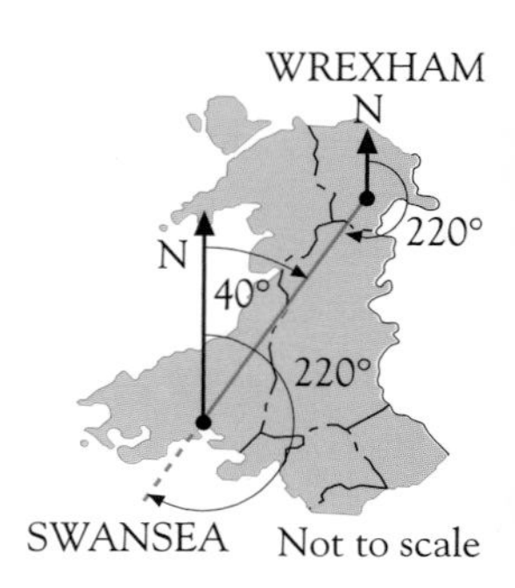

- As before, imagine the helicopter turning around at Swansea to go straight back to Wrexham but be careful with the calculation.
- If you **add** 180° to the original bearing you get an angle of **400°**! This is greater than a complete turn of 360° by 40°. So you can say that the bearing of Wrexham from Swansea is **040°**.
- Notice that the same result is obtained by **subtracting 180°** from the original bearing (220 − 180 = 40). So **if the original bearing is greater than 180° subtract 180° to calculate the reverse bearing.**

- A bearing is an angle that tells the direction a point must follow.
- All three-figure bearings start from North and are given in a clockwise direction.
- North has a bearing of 000°.
- To calculate a reverse bearing
  - a If the original bearing is **less than 180°**, add 180°.
  - b If the original bearing is **greater than 180°**, subtract 180°.

Drawing an accurate North line

A **set-square** is a useful instrument for helping to draw an accurate North line when a North line has been drawn for you on another part of the diagram.

Sample Question 3

A ship sails from A to a point B, 5 km away on a bearing of 120°. At B it changes direction and sails to a point C, 8 km away on a bearing of 030°.

a On the diagram, using a scale of 1 cm to 2 km, plot the course of the ship from A to B and then to C.

b Measure the bearing that the ship must take to sail directly back from C to A.

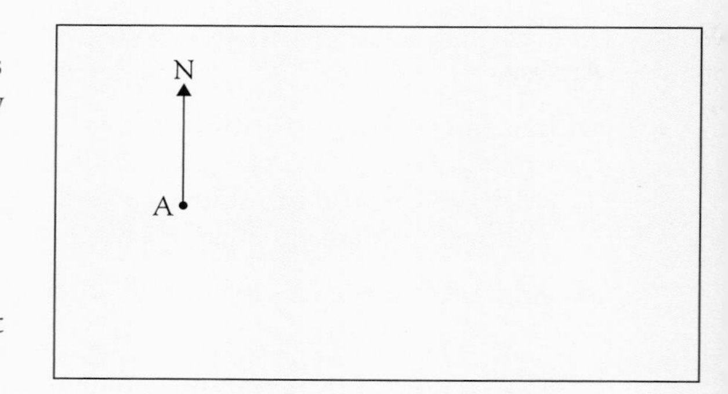

Answer

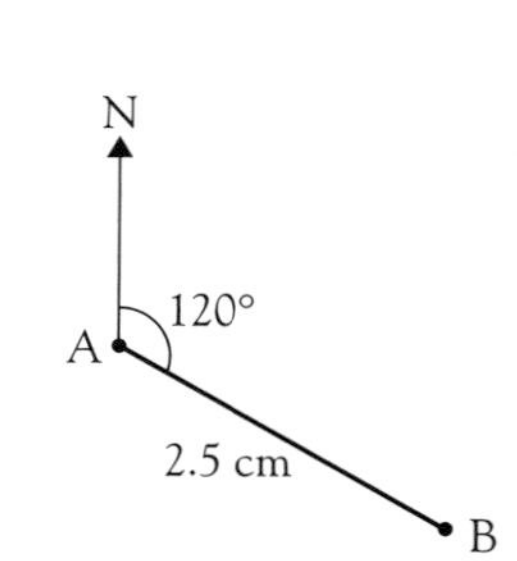

- Using your protractor or angle measurer, with the centre on A, and 0° pointing North, draw accurately an angle of 120°.
- Draw a straight line from A, on your measured bearing, a length of 2.5 cm (remember 1 cm = 2 km). Label the end of this line B.
- You now need to draw an accurate North line at B. Follow this procedure:

a Place a set-square so that part of the right angle is on the North line – see page 221 Fig. **i**.

b Place your ruler against the other part of the right angle so that you can slide the set-square along the ruler – Fig. **i**.

BEARINGS

c Slide the set-square along the ruler until the edge of the set-square is over the point B – Fig. **ii**.

d Keeping the set-square firmly in place draw a line upwards from B. This is your new North line – Fig. **iii**.

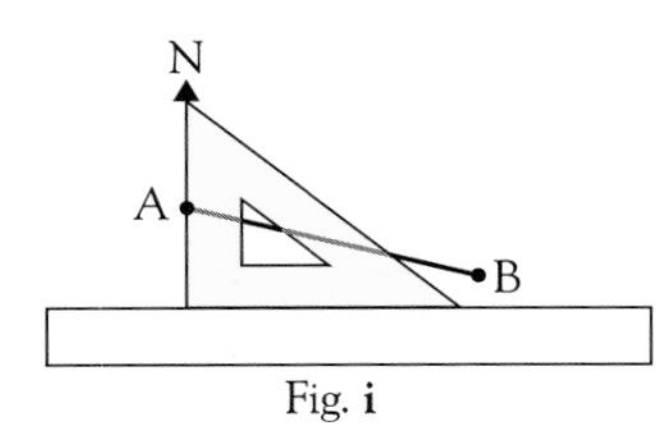

Fig. i

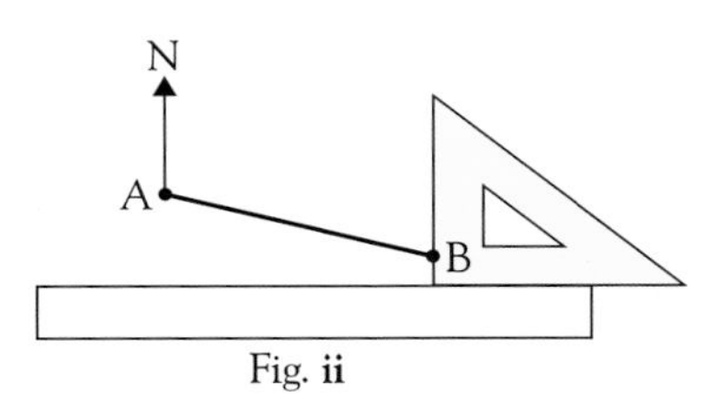

Fig. ii

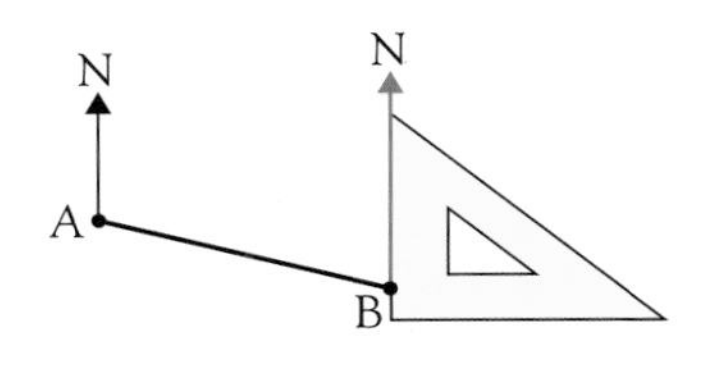

Fig. iii

- With the centre of your angle measurer or protractor on B, measure and mark an angle of 30° clockwise from the North line. Draw a straight line 4 cm long from B through your mark. Mark the end of this line C.
- Repeat the procedure for drawing a new North line at C.
- With the centre of your angle measurer or protractor at C measure the bearing of A from C.

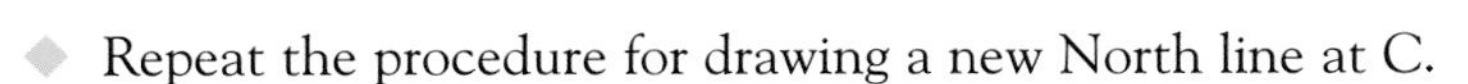
The bearing of A from C is 242°

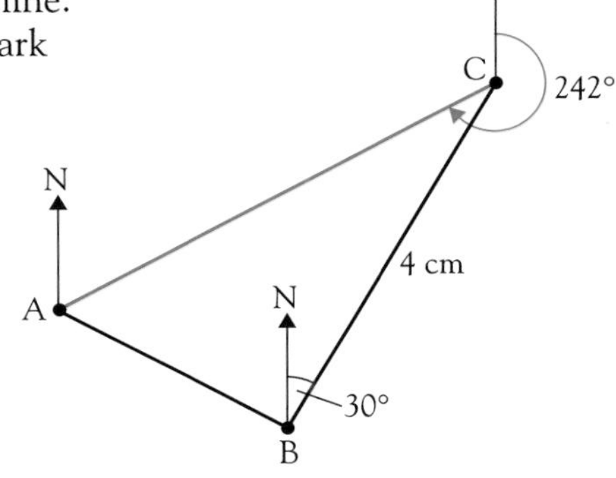

Exercise 12.1

Where necessary use tracing paper over the maps and other diagrams to enable you to measure the bearings requested.

1 From the top of Tegg's Nose, a hill in Cheshire, the directions of some other hills and places of interest are as shown in the diagram below. Give each of the directions as a bearing.

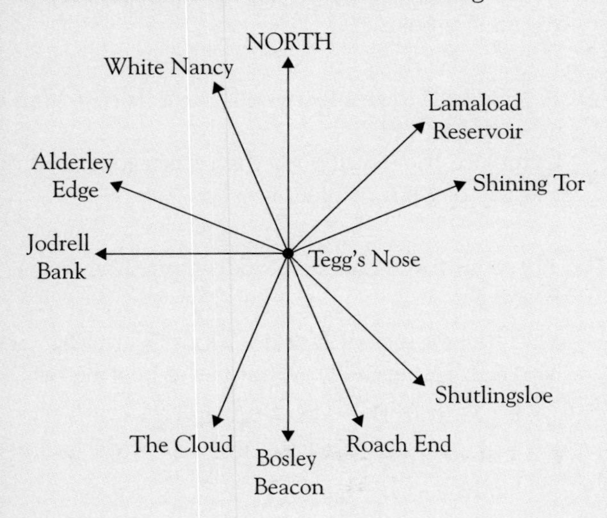

2 Measure the bearing of Paris from each of the other cities shown on the map of France below.

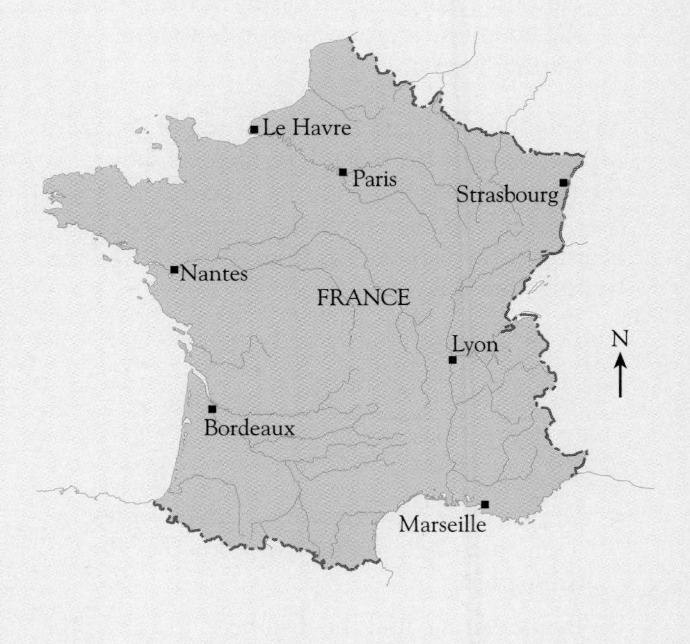

BEARINGS

3 The diagram shows a group of islands: Paros, Quintos, Remos and Scubados. Copy it onto tracing paper.

A weekly cruise is organised with the itinerary given below:

Day 1 Sail from Paros to Quintos
Day 2 Sightseeing on Quintos
Day 3 Sail from Quintos to Scubados
Day 4 Sightseeing on Scubados
Day 5 Sail from Scubados to Remos
Day 6 Sightseeing on Remos
Day 7 Return from Remos to Paros

Using the centre of each island, measure the bearings for the journeys on Days 1, 3, 5 and 7.

4 A boat travels 10.5 km due South and then 5.8 km on a bearing of 115°.

a Show this journey on a scale drawing.

b Find the distance and bearing of the boat at the end of its journey measured from its starting position.

5 Two coastguard stations A and B are 45 km apart on a coastline with B due East of A, as shown in the sketch. The coastguards at A and B both receive an SOS signal from a fishing boat. The bearing of the fishing boat from A is 164° and the bearing from B is 235°.

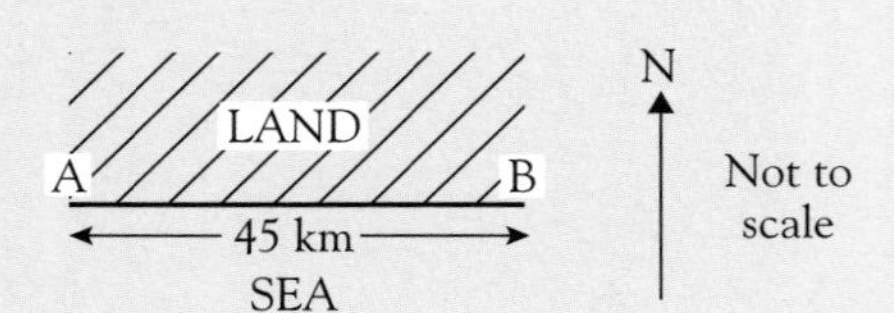

a Draw a scale drawing using 1 cm to represent 5 km, and use it to find the position of the fishing boat.

b By measuring, find the distances, in km, of the fishing boat from A and from B.

6 After leaving an airport, a plane flies 150 km on a bearing of 290°, followed by 108 km on a bearing of 032°, and then 240 km on a bearing of 105°.

Use a scale drawing to find the distance and bearing on which the plane would then need to fly to return straight back to the airport.

7 Sandy and Ken are joint owners of a racing pigeon. They have tagged a device to the pigeon's leg. The device transmits a signal, and they each have a receiver which they can use to find the bearing of the pigeon.

Sandy lives 12.5 km from Ken on a bearing of 036°.

During one race they each use their receiver from home.

a At one point in the race the pigeon is due West of Sandy's house and on a bearing of 318° from Ken's. Use a scale drawing to show the position of the pigeon and mark it with a letter P.

b Later in the race the pigeon is due North of Ken's house and on a bearing of 245° from Sandy's. Show the new position of the pigeon on your diagram and mark it with a letter Q.

c By joining PQ, find how far the pigeon has flown and the direction in which it has travelled.

8 A customs boat is trying to intercept a motor-boat whose occupants are suspected of smuggling. Initially the bearing of the motor-boat from the customs boat is 127°.

Calculate the bearing of the customs boat from the motor-boat.

9 A ferry sails from Fleetwood to the Isle of Man on a bearing of 286°.
Calculate the bearing for the return journey from the Isle of Man to Fleetwood.

10 Three air force bases are code-named as A, B and C.

a The bearings of B and C from A are 062° and 143°, respectively. What is the bearing of

i A from B ii A from C?

b The bearing of C from B is 195°. What is the bearing of B from C?

Constructions

In Chapter 5 (Geometry I) you learned how to make accurate drawings using a sharp pencil, a pair of compasses, a ruler, a set-square and an angle measurer (protractor). These were not classed as constructions but were accurate drawings.

Sometimes you are asked to make accurate drawings by **construction**. This means that you are only allowed to use your ruler and compass and you must show how you did the construction.

You are not allowed to use instruments such as protractors or set-squares.

Constructions should be done using a sharp HB pencil, **not** pens or coloured pencils. It is very important that **all** the **construction lines** are accurate and can be seen clearly, even if they do look a little untidy.

Construction 1 – The perpendicular bisector of a straight line

Perpendicular means 'at right angles to', i.e. 'at 90°'.

Bisect means to divide into two equal parts, i.e. to divide in half.

'Cuts it in half' is another way of saying 'divides it up in half'.

To construct the **perpendicular bisector** of a straight line you have to construct a line that is at right angles to the straight line and cuts it in half.

To bisect the line AB:

1 Open a pair of compasses to a setting that is more than half the length of AB.

2 With the point of your compass on point A, draw an arc above and below AB, as in Fig. **i**.

3 With the same compass setting put your compass point on point B. Draw another arc above and below the line AB and intersecting the first pair of arcs, as in Fig. **ii**.

4 Draw the straight line that joins the points where the two sets of arcs intersect, as in Fig. **iii**.

Now you have drawn a line perpendicular to AB cutting AB in half.

It is the perpendicular bisector of AB.

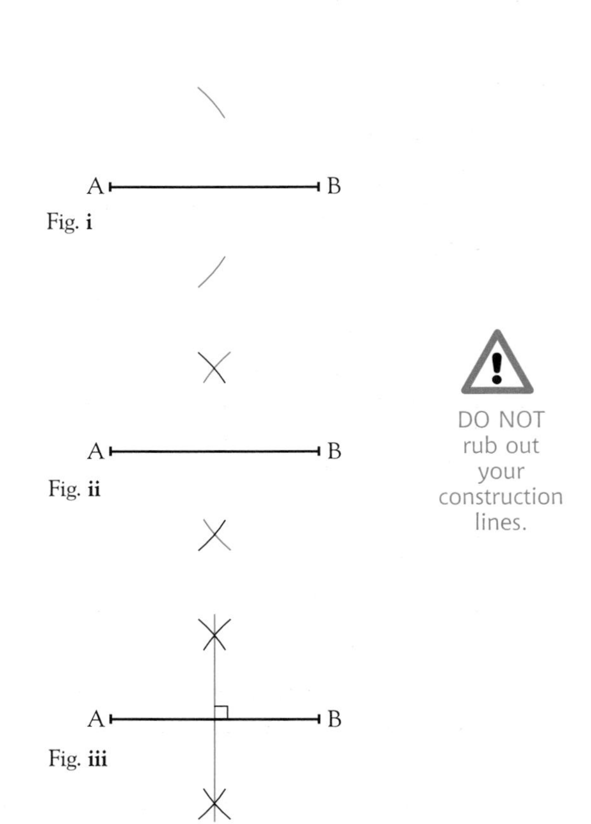

Construction 2 – The bisector of an angle

To bisect angle ABC:

1 Set your compasses to a sensible radius (not too large) and with the point on B draw an arc to cut AB at X and an arc to cut BC at Y, as in Fig. **iv**.

2 Keep the same radius and with the compass point on X and Y in turn, draw two arcs that intersect at Z, as in Fig. **v**.

3 Draw a straight line from B through Z, as in Fig. **vi**.

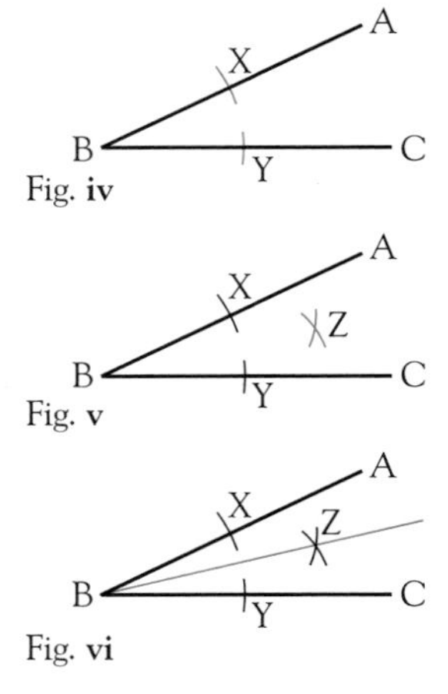

Fig. **iv**
Fig. **v**
Fig. **vi**

The line BZ cuts angle ABC in half so that $\angle ABZ = \angle CBZ$, i.e. BZ bisects angle ABC.

Construction 3 – A straight line parallel to another straight line, a given distance from it

1 Set your compasses to the required distance apart. Choose any point A on the line and draw an arc directly above the line.

2 Keep your compasses set to the same radius. Choose another point B on the line (not too close to the first) and with your compass point on this point draw a second arc directly above the line, as in Fig. **vii**.

3 Place your ruler so that its edge just touches the top of each arc. Draw a line with your ruler in this position, as in Fig. **viii**.

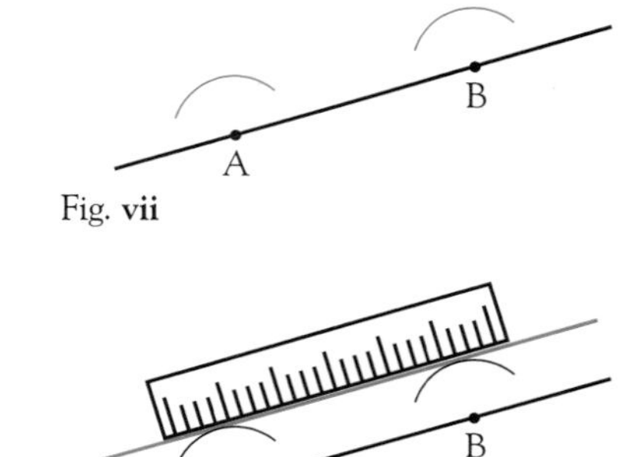

Fig. **vii**
Fig. **viii**

The line you have drawn is **parallel to the first line**, the **required distance away from it**.

You can construct another line that is parallel by drawing the arcs on the other side of the line.

To make an **accurate drawing** you may

- use any drawing equipment including angle measurers and set-squares

To make a **construction** you must

- use only a **sharp pencil**, **ruler** and **pair of compasses**,
- show all your construction lines even though your diagram might look untidy

CONSTRUCTIONS

Exercise 12.2

1 a On plain paper, draw accurate lines with the measurements given below:

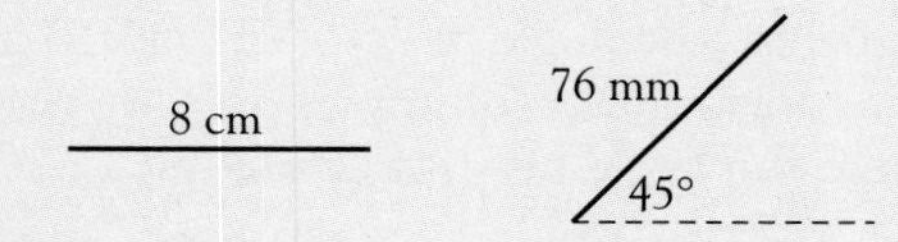

b Use compasses and ruler to draw the perpendicular bisectors of your lines.

c In each case use a ruler and protractor to check that the bisector passes through the mid-point of the line and is at right angles to it.

2 a Use a protractor and ruler to draw accurate angles of the sizes shown below:

i 68° ii 32° iii 110° iv 103°

b Use compasses and ruler to bisect your angles.

c Use a protractor to check.

3 Starting with diagrams like those shown below, construct the lines requested:

a Construct a line parallel to, and 4.6 cm to the right of **a**.

b Construct 2 lines parallel to and 12 mm from **b**.

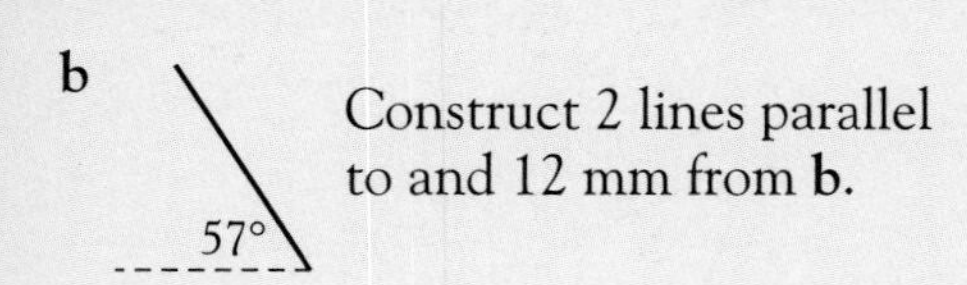

4 a Construct a triangle ABC with sides AB = 5.4 cm, BC = 6.2 cm and CA = 4.1 cm.

b Measure the angles of the triangle.

c Use compasses and ruler to bisect angles B and C.

d The bisectors meet at a point D. Measure the distances BD and CD.

5 a Construct an equilateral triangle with sides of length 8 cm.

b What is the size of each angle?

c Use a ruler and compasses to construct an angle of 30° on your diagram.

6 Starting with a line PN, 120 mm long, and using only compasses and ruler:

a construct the triangle PQR with dimensions as shown in the diagram

b construct the perpendicular bisectors of sides PQ and QR.

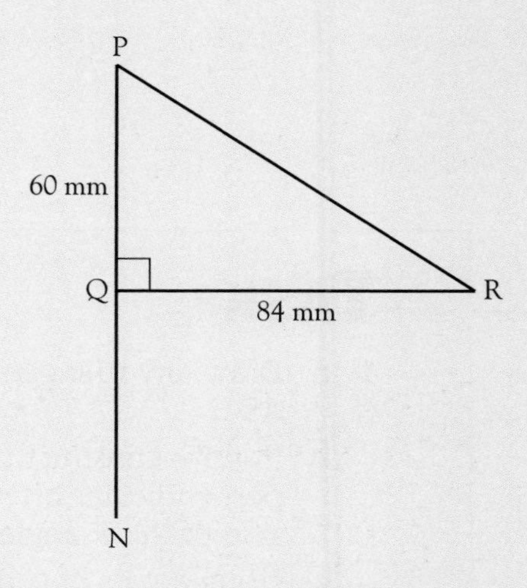

Where do the perpendicular bisectors meet?

7 The inner boundary of a running track has straight sides and semi-circular ends, with dimensions shown in the sketch.

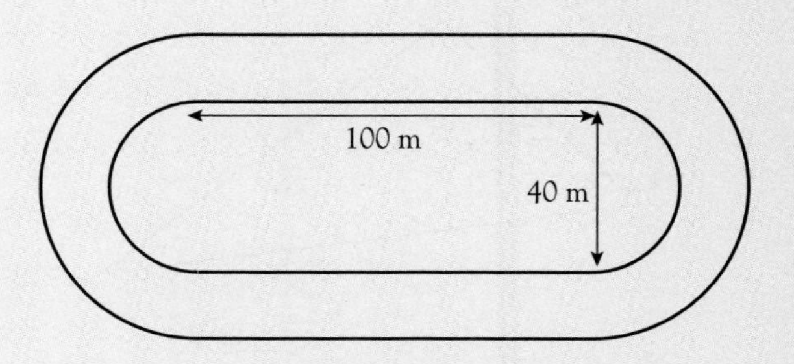

a Construct a scale drawing of this boundary using 1 cm to represent 10 m.

b The width of the track is 10 m.

Using ruler and compasses only, construct the outer boundary of the track.

CONSTRUCTIONS

8 On graph paper draw a set of axes with both x and y going from -3 to 3.
Plot the points P(-2, 1), Q(3, 3), R(2, -1) and S(-1, -2) and join them to form a quadrilateral PQRS.
Construct the perpendicular bisector of side RS and the bisector of angle S.
Through which point do they both pass?

9 a On graph paper draw a grid with both axes going from -3 to 8. On the grid draw and label triangle ABC with A (0, 3), B (5, 6) and C (8, 1).
b Construct the angle bisector of angle B and write down the co-ordinates of P, the point where the angle bisector crosses AC.
c Draw a circle with centre P and radius PB.
d The angle bisector of B crosses the circle at Q. Write down the co-ordinates of Q.
e What shape is ABCQ?

TASK

1 a Draw any triangle and label it ABC.
b In turn, construct the perpendicular bisector of each side of the triangle.
c Using a pair of compasses, draw a circle so that it passes through each of the points A, B and C. (Think! Where is the centre of the circle?)

2 a Draw another triangle, different from the first, and label it PQR.
b In turn, bisect each of the three angles of the triangle.
c Using a pair of compasses, draw a circle that is completely inside your triangle but **touches** each of the sides PQ, PR and QR only once. (Think! Where is the centre of this circle?)

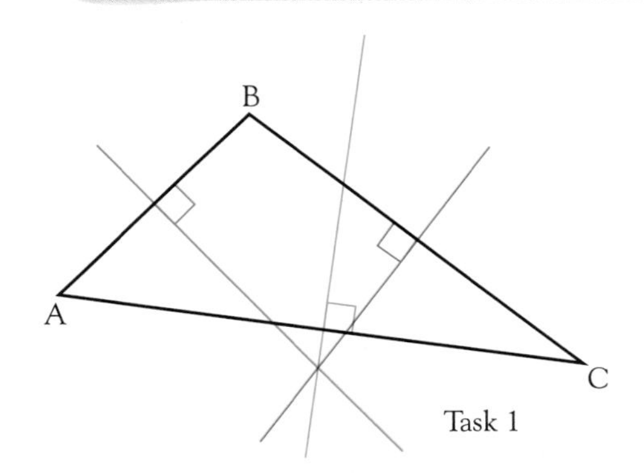

Task 1

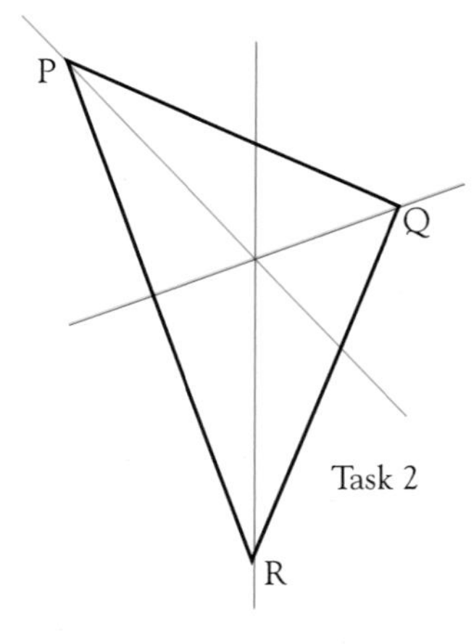

Task 2

In Task 1 you should have drawn the circle that passes through the three corners of a triangle. This circle is called the **circumcircle**. The intersection of the perpendicular bisectors of the sides is the centre of this circle.

In Task 2 you should have drawn the circle that is completely inside the triangle and whose circumference touches each of the three sides of the triangle. This circle is called the **inscribed circle**. The intersection of the angle bisectors is the centre of this circle.

Loci

Constructions are often used to trace out accurately the path of a point that obeys a given rule or law. This path is called the **locus** of the point. The plural of locus is **loci**.

The locus could be a straight line or a curve.
It could be a region with a boundary that might or might not be included.
Here are the most common loci that you will use.

1 The locus of a point that is a constant distance from a fixed point.

Imagine a special theatre stage where the audience sits around it. The distance of the front row of seats from the centre of the stage must be exactly 20 metres.

The picture shows all the possible positions of these seats. Notice that the seats form a **circle** of radius 20 m, with the centre of the circle being the centre of the stage.

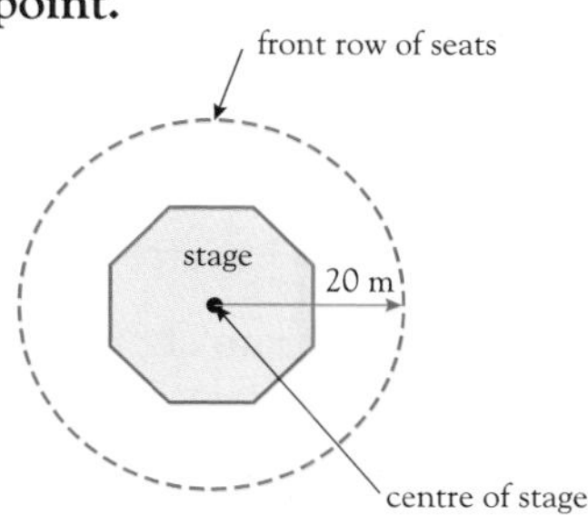

The locus of a point that is a constant distance from a fixed point is the circumference of a circle with centre at the fixed point and radius the distance from the fixed point.

2 The locus of a point that is equidistant from two fixed points.

Equidistant means an equal distance.

Imagine this situation. The island of Matland has been in a state of civil war for many years. The Royalist headquarters are in a town called Abar and the Republican headquarters are in Zuba.

The Royalists and Republicans eventually agree to share the island and the border is to be between the two main towns. A condition for the positioning of the border is that any point on the border must be exactly the same distance from Abar as it is from Zuba.

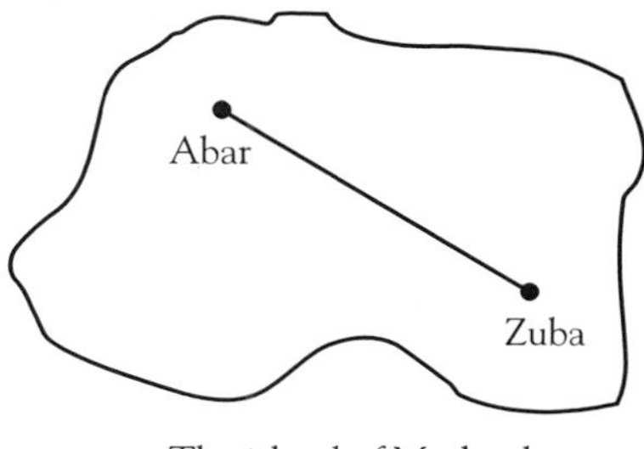

The island of Matland

To find the position of the border you would need to draw a straight line between Abar and Zuba. The most obvious point that meets the condition is the halfway point between the two towns. The other points lie on the **perpendicular bisector** of the line joining the towns.

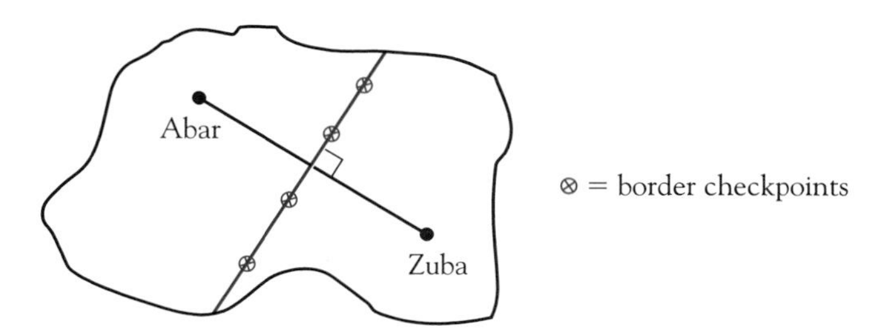

If you are at any of the border checkpoints indicated on the diagram, or at **any** point **on** the purple border line, you are the same distance from Abar as you are from Zuba.

If you are to the **left** of the purple line you are closer to Abar than to Zuba.
If you are to the **right** of the purple line you are closer to Zuba than to Abar.

The locus of a point that is equidistant from two fixed points is the perpendicular bisector of the straight line joining the two fixed points.

LOCI

Distance from a point to a line

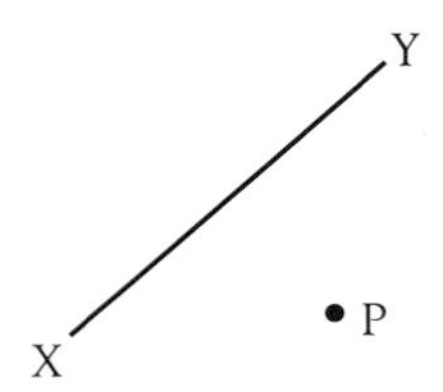

If you are asked to find the distance from P to the line XY, you would find the **perpendicular** distance.

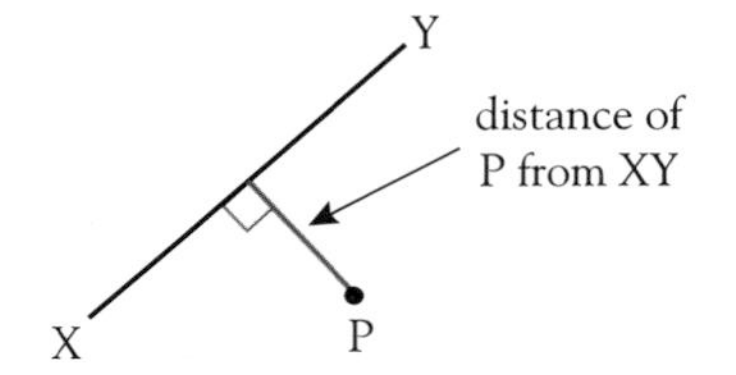

3 The locus of a point that is a fixed distance from a given line.

An underground electricity cable runs across my garden. I am told that the closest I can plant any shrubs is 1 metre from the cable.

The diagram shows all the points that lie 1 m from the cable. Notice that all these points form two straight lines, parallel to the cable, one on either side of the cable.

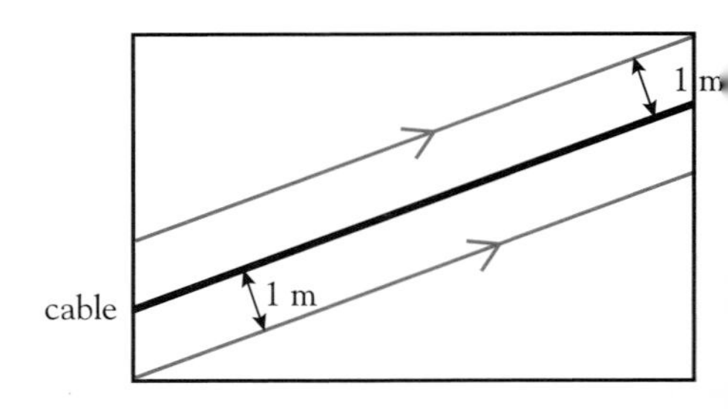

The locus of a point that is a fixed distance from a given line is a pair of lines, parallel to the fixed line, one on either side of it and at the fixed distance from it.

4 The locus of a point that is equidistant from two fixed intersecting straight lines.

The water pipe to a garden pond is to be situated so that it enters the garden at the intersection of two of the boundary walls and so that it is always the same distance from these two walls, i.e.

$$PX = PY$$

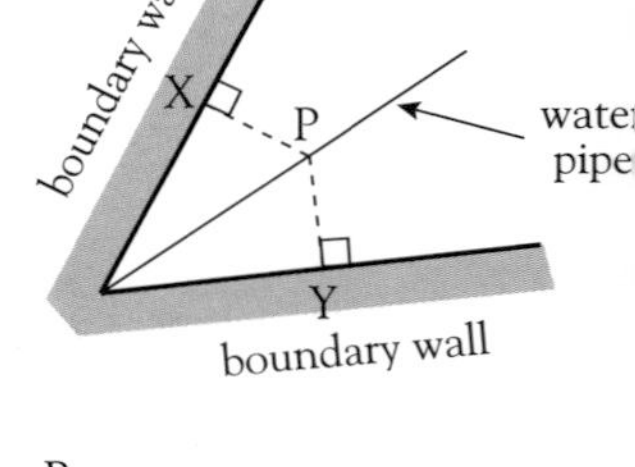

Remember the situation where you had to construct the locus of a point that is equidistant from two fixed points, i.e. the perpendicular bisector of the line joining the points.
You were in fact bisecting an angle of 180° (Fig. **i**).

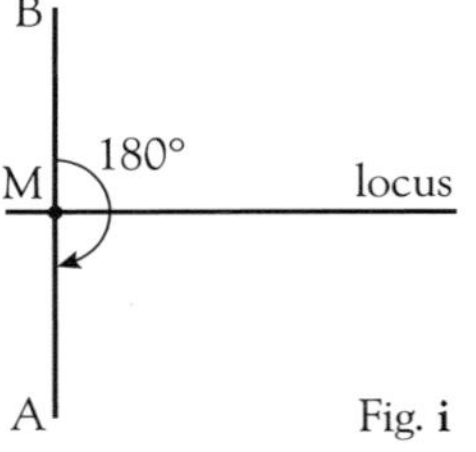

Fig. **i**

Now imagine an angle less than 180°, for example BMA in Fig. **ii**. You follow a similar procedure and the locus of a point that is equidistant from the two lines BM and AM is the bisector of the angle BMA.

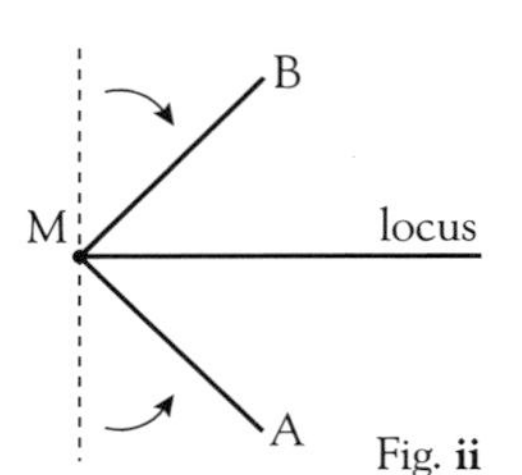

Fig. **ii**

The locus of the point that is equidistant from two fixed intersecting straight lines is the bisector of the angle between the lines.

Sample Question 4

An electrical firm is asked to fit an outdoor spotlight in a rectangular garden measuring 8 m by 5 m. The light must be the same distance from the two corners, A and B, of the back wall of the house but also has to be the same distance from the fence CD at the end of the garden and the side of the garden, BC. Using a scale of 1 cm to 1 m, draw a scale drawing of the garden and find the position of the spotlight. Mark its position L.

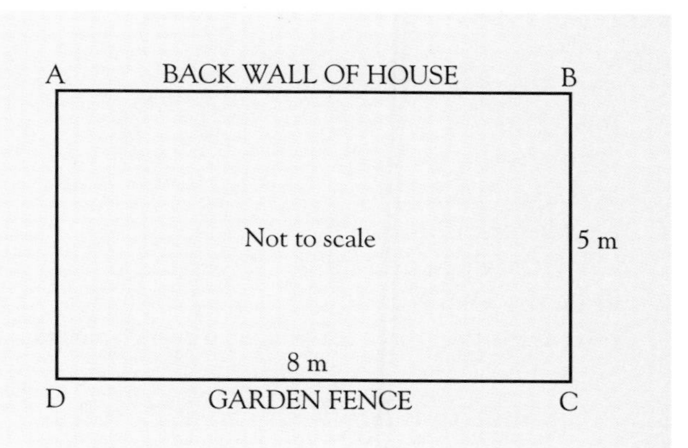

Answer

- Draw accurately a rectangle, ABCD, measuring 8 cm by 5 cm.
- If the spotlight is to be the same distance from A as from B it is on the **perpendicular bisector of the line AB**.
  Using a pair of compasses and a ruler construct this bisector. (See Fig. **i** for sketch.)

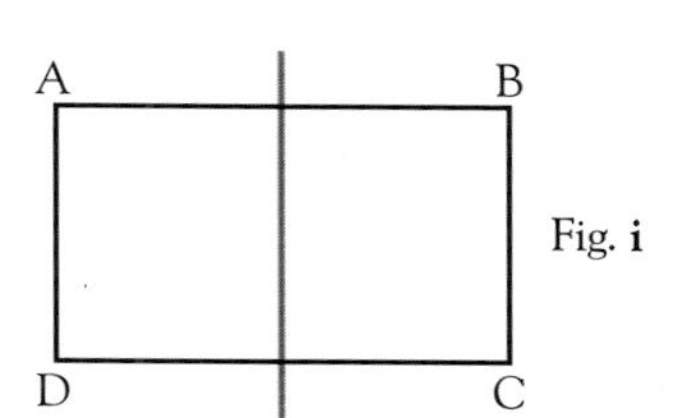

Fig. **i**

- If the spotlight is to be equidistant from the garden fence, CD, and the edge of the garden, BC, it is on the **bisector of the angle BCD**. Using a pair of compasses construct the angle bisector. (See Fig. **ii** for sketch.)

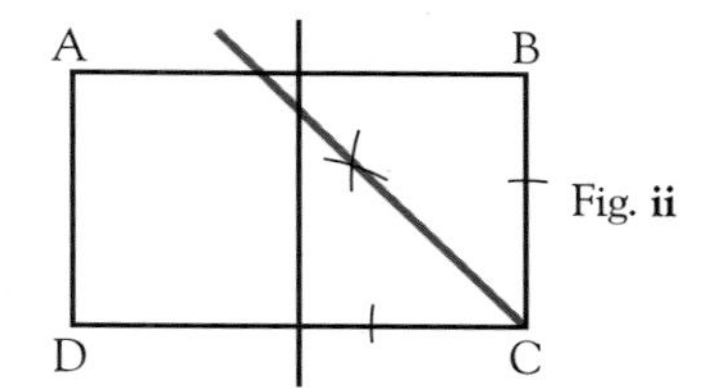

Fig. **ii**

- The point where the perpendicular bisector of the wall AB and the angle bisector of ∠BCD meet is the required position of the spotlight. Mark this position L. (See Fig. **iii** for sketch.)

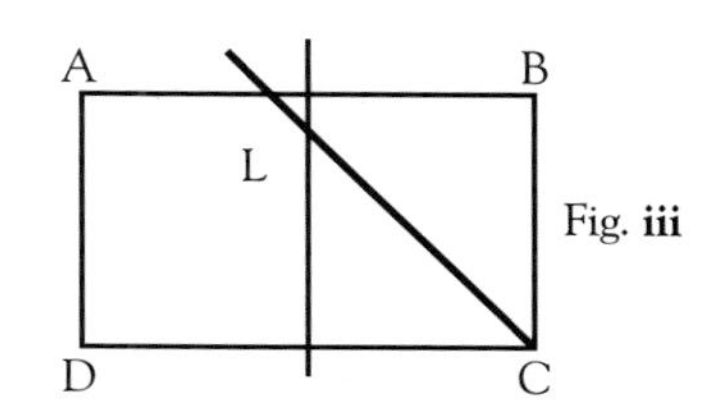

Fig. **iii**

Accurate diagram

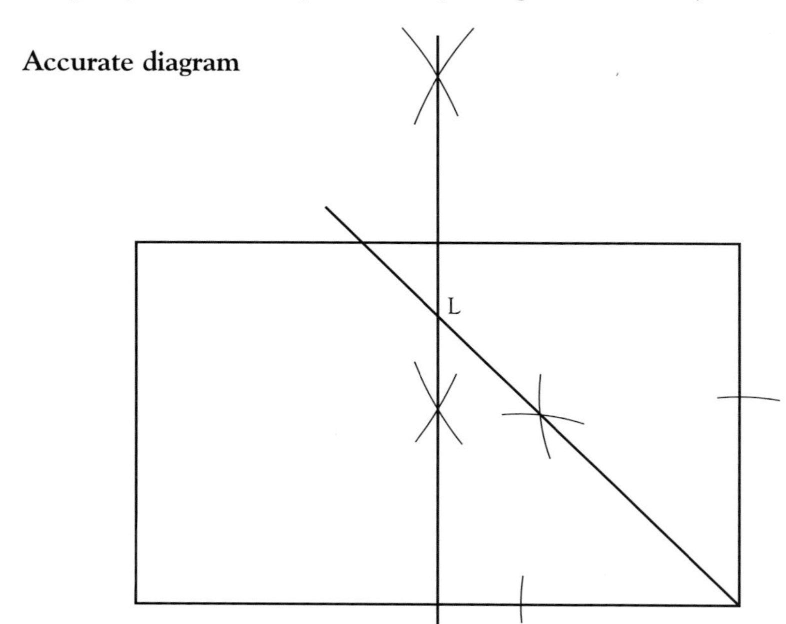

Remember to show the construction lines.

Exercise 12.3

1 Sam steers her rowing boat so that it passes between the rocks P and Q and remains equidistant from them. Copy the diagram and draw the locus of the boat's position as it travels between the rocks.

Q

P

2 A coastline runs North-East from A to B as shown in the sketch. The distance from A to B is 320 metres. A swimmer starts from a point in the sea 100 m from A and swims in the direction North-East, keeping 100 m from the shore. Show the locus of the swimmer's position on a scale drawing.

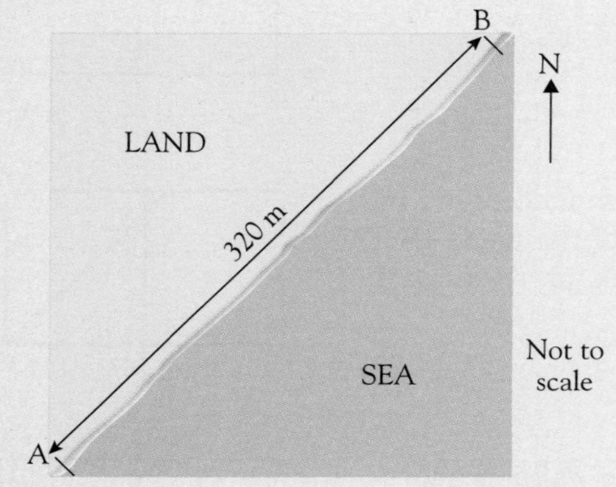

3 In one part of a putting competition all the competitors are instructed to position their golf balls 8 metres from the hole.

Draw a sketch to show the locus of the possible positions of the golf balls.

4 The sketch shows a rectangular lawn. James starts from the corner marked B and walks across the lawn so that he remains an equal distance from the edges AB and BC.

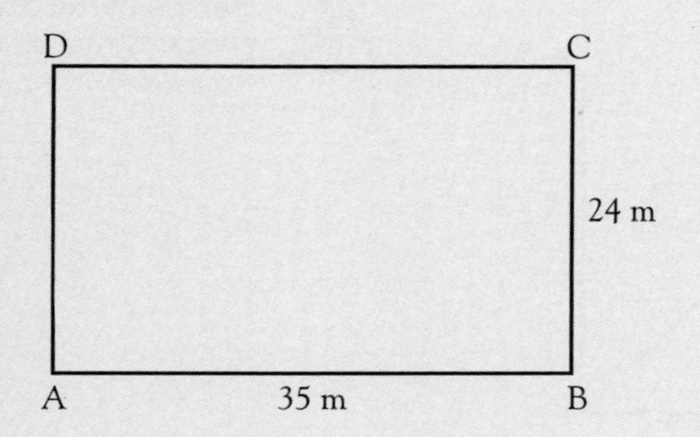

a Draw a scale diagram to show the locus of his position as he moves across the lawn.

b By measuring, find his distance from D when he reaches the edge CD.

5 Sketch the locus of a cyclist as she carries out each of the following tasks from a cycling proficiency test:

a Cycle between parallel lines AB and CD keeping an equal distance from each line.

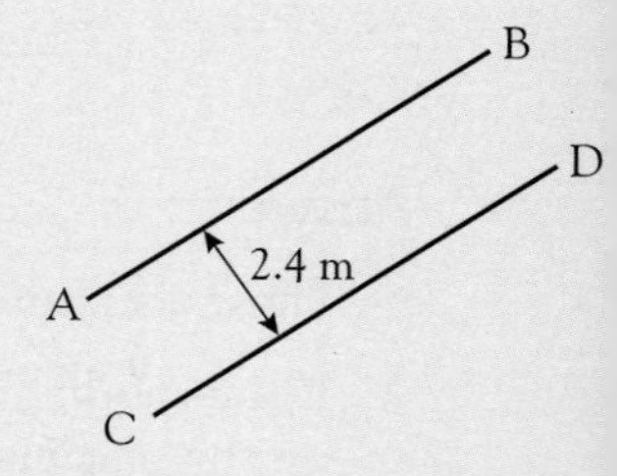

b Cycle around a traffic cone keeping a distance of 2 metres from it.

⊙ Traffic cone

c Starting from the point P, cycle between the lines PQ and PR keeping an equal distance from each line.

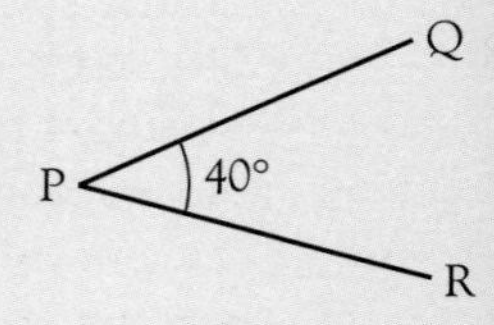

d Cycle between two traffic cones, keeping an equal distance from each.

⊙ 2.4 m ⊙

e Starting from P, cycle to Q between alternate cones, keeping at a distance of 1 m from each cone in turn.

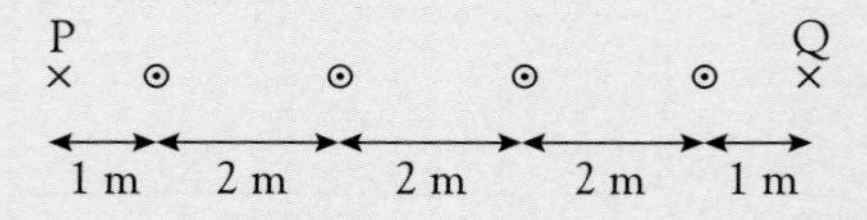

6 A ladder, AB, of length 5 metres, rests against a wall as shown in the diagram.

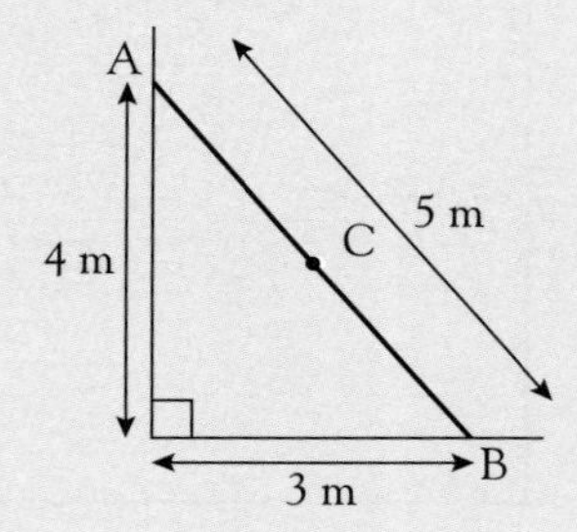

It slips down the wall until it lies on the ground. Sketch the locus of

a A

b B

c C, the mid-point of the ladder.

7 A cyclist travels along a straight horizontal road. Sketch the locus of

a the centre of the front wheel,

b a point on the circumference of the front wheel,

c the toe of one of the cyclist's shoes.

8 The sketch shows a bay with headlands, P and Q, which are 80 metres apart.

A jet-skier leaves a point, R, on the bay which is equidistant from P and Q and travels between P and Q, keeping an equal distance from them.

After he crosses the line PQ, he continues to jet-ski, keeping a constant distance of 40 metres from Q, until he reaches the beach.

Draw your own diagram, using a scale of 1 cm to 10 m.
By construction

a indicate the position of R,

b draw the locus of the jet skier's position.

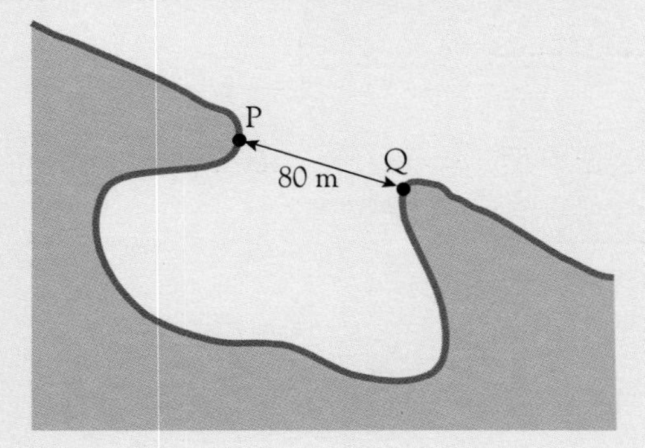

9 A triangular piece of land has vertices X, Y and Z. The lengths of the boundaries of the land are XY = 230 metres, XZ = 180 metres and YZ = 300 metres. Y is due North of Z and X is to the East of the line YZ.

a Draw a scale drawing of this piece of land and measure the bearings of Y and Z from X.

b There is a gas pipe which runs across this land. If the pipe is equidistant from X and Z at all points along its length, draw the position of the pipe on your diagram.

c The Water Board has decided that it also needs a pipe across this land.

It plans to lay the pipe from X across the land, so that the pipe is equidistant from the boundaries XY and XZ. Draw the position of the Water Board's pipe on your diagram.

d The pipes will cross at a point P. Mark P on your diagram and find its distance and bearing from X.

10 In a canoe race the circuit is marked by three buoys, A, B and C.

B is 200 metres North-West of A, and C is 350 metres from B on a bearing of 208°.

a Use a scale drawing to find the distance and bearing of the last section of the course from C to A.

b A referee is stationed in a boat which is an equal distance from the three buoys.

Find the position of the referee's boat and measure its distance from the buoys.

LOCI

A locus could be a region, e.g. **within** 10 m of a fixed point or **at least** 5 m from a wall. In these cases you could use shading to indicate the region but you must show the boundary for the region.

Sample Question 5

My dog Jasper likes to try to escape from the garden by digging under the fence. To stop him I put him on a lead that is attached to a 8 m long straight metal rail fixed horizontally to the ground so that the lead can slide easily along its length.

Jasper's lead is 2 m long.

Using a scale of 1 cm to 1 m, draw a scale drawing of the rail and indicate by shading the area of lawn that Jasper can play on.

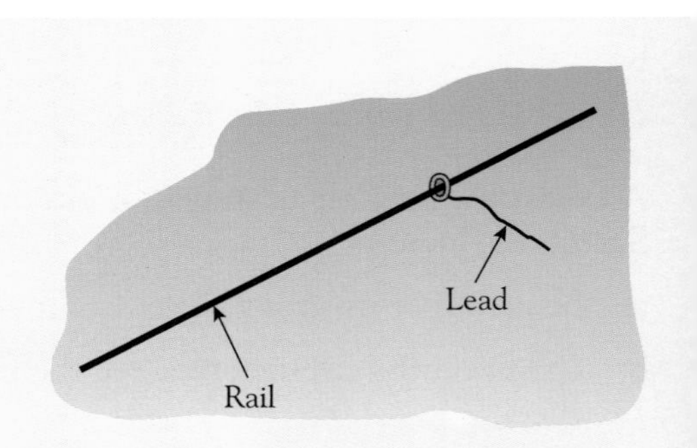

Answer

- Draw a line 8 cm long to represent the rail.
- There are **two** loci in this question. Firstly consider how far Jasper can reach to the sides of the rail. The maximum length of the lead is 2 m so he can reach 2 m either side of the rail along its length. This traces out a line parallel to the rail either side of it as in Fig. **i** so construct a line, 2 cm from the line representing the rail and parallel to it on either side of it.
- Secondly, when Jasper reaches the ends of the rail, the end becomes a fixed point so the path that will be traced now is a semi-circle, radius 2 m, centred on the end point of the rail (see Fig. **ii**), so construct two semi-circles each with radius 2 cm.
- These two loci form the boundary of the region in which Jasper can play. Shade the area inside this boundary to indicate the region.

Sketches

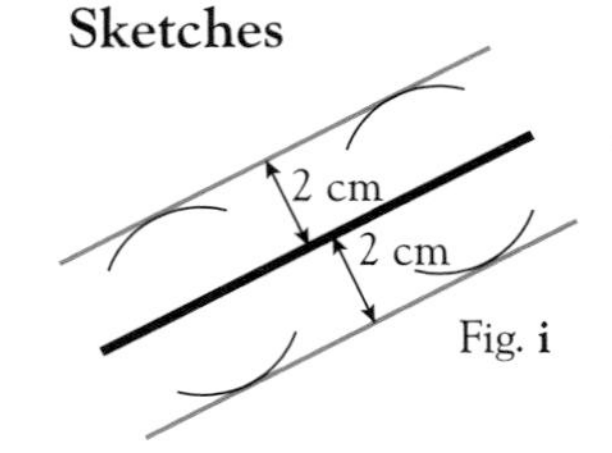

Fig. **i**

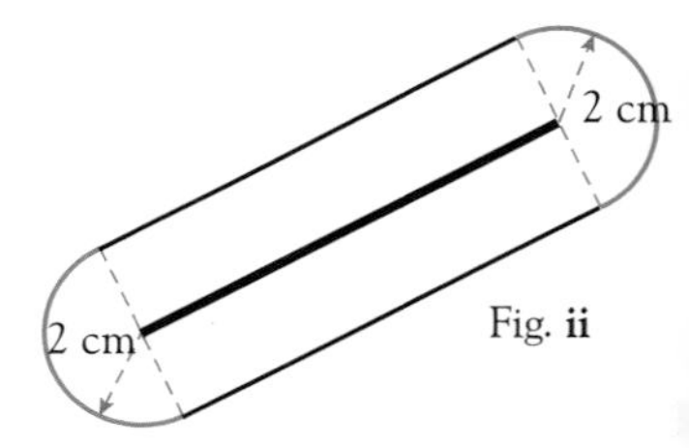

Fig. **ii**

Accurate diagram

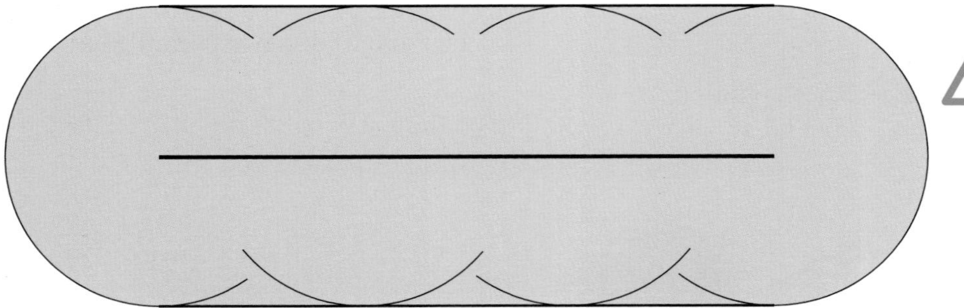

Remember to show all your construction lines.

Sample Question 6

The diagram shows two points P and Q.
On a diagram shade the region which contains all the points that satisfy both the following
– the distance from P is less than 3 cm
– the distance from P is greater than the distance from Q.

P × Q ×

[L]

Answer

- The boundary of the region that contains all points that are less than 3 cm from P is a **circle, radius 3 cm, centre P**. So all the points that are inside the circle are less than 3 cm from P.
- The boundary of the region that contains all the points that are a greater distance from P than from Q is the **perpendicular bisector of the straight line joining P and Q**. All the points to the right of this line are further from P than Q.
- The region required is the region to the right of the perpendicular bisector of the line PQ but only that part that is **inside** the circle with centre P as shown shaded in the diagram below.

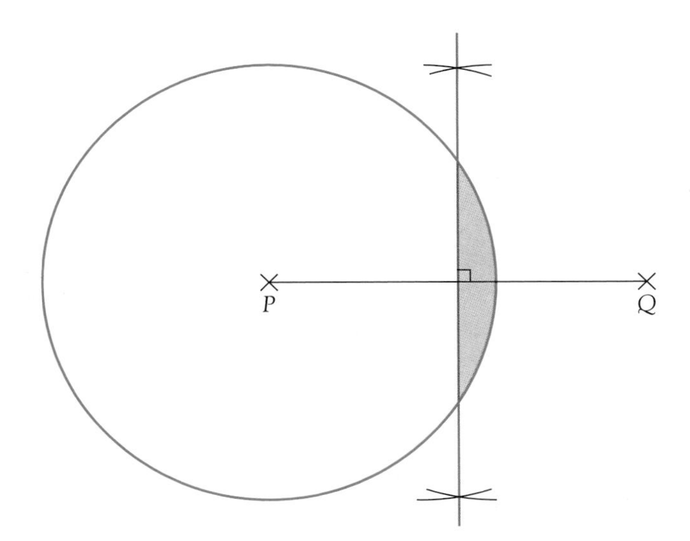

Exercise 12.4

1 A narrowboat is travelling along a straight canal which is 6 metres wide. There must be a gap of at least 1 metre between the boat and the sides of the canal.

Draw a diagram using a scale of 1 cm to 1 m and shade the part of the canal which may be used by the boat.

2

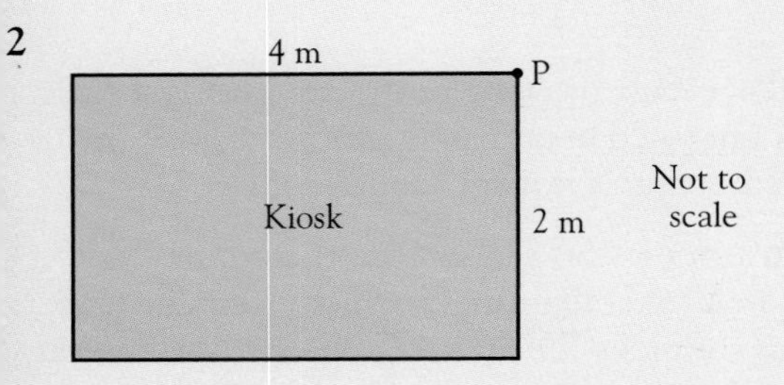

A dog is tied to a post P at the corner of a newsagent's kiosk whilst its owner buys a newspaper (see diagram).

The dog's lead can extend to a length of 3 m. Make an accurate scale diagram of the kiosk, using a scale of 1 cm to 1 m, and shade the region in which the dog can wait.

3 The distance between the centre stumps A and B of a wicket on a cricket field is 22 yards long, as shown.

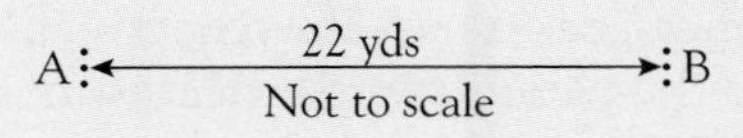

A fielder is told to stand between 4 yards and 12 yards from this line. The fielder must also be closer to wicket B than wicket A. Using a scale of 1 cm to 4 yds draw an accurate diagram to show the locus of all the possible places where the fielder could stand.

(Ignore the restrictions which would arise because of other players.)

4 A rectangular car park PQRS, with the dimensions shown, has security lights mounted at the corners P and Q.

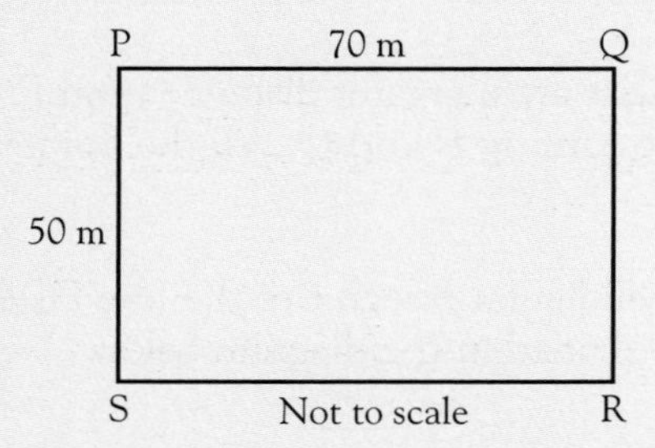

The lights can each illuminate the ground to a radius of 40 metres. Draw a diagram using a scale of 1 cm to 10 m and show the locus of points which are illuminated by the lights.

Also mark on your diagram a point X, on the boundary of the car park, where you would position a third identical light to give maximum light coverage.

5 A new waste disposal site is required for the villages Axton, Briarwood, Cambury and Daleby, whose positions are shown by A, B, C, D in the diagram. If the site is to be within the quadrilateral ABCD but at least 5 miles from each of the villages, copy the diagram on to tracing paper and show the locus of possible positions for the new site.

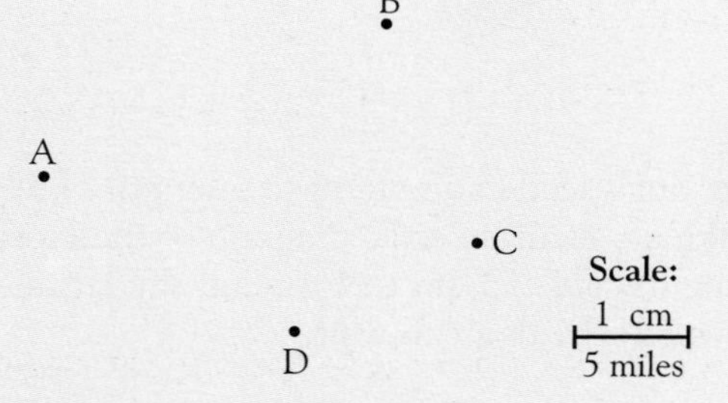

6 A city has three hospitals P, Q and R, whose positions are as shown in the diagram. When an ambulance goes on an emergency call-out it delivers the patients to the nearest hospital.

Copy the diagram and show the regions from which patients would be delivered to each of the hospitals.

7 Claire is going to plant a new plum tree in her garden.

The dimensions of the garden are as shown in the diagram.

The point **P** marks the position of an existing plum tree.

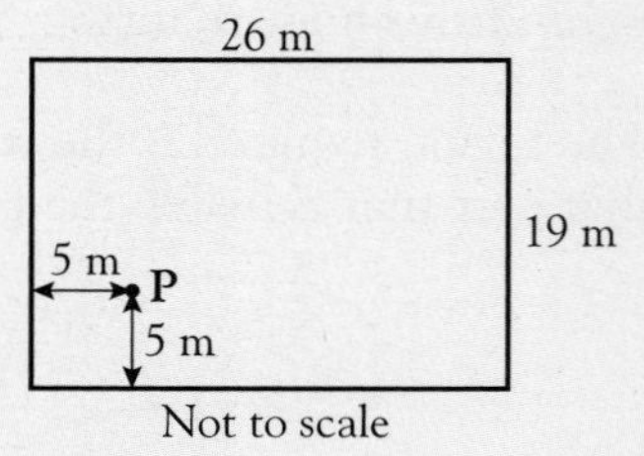

Claire decides that the new tree should be more than 2.5 metres from the boundaries of the garden and more than 4 metres from the existing tree.

The sales assistant at the garden centre advised her to plant it less than 10 metres away from the existing tree to encourage cross-pollination.

Using a scale of 1 cm to 2 m draw a scale diagram and show the region which satisfies all these requirements.

8 A sketch of the field used by a model aircraft club is shown in the diagram. New members are advised not to fly their planes within 25 metres of the wood, or 15 metres of the stream.

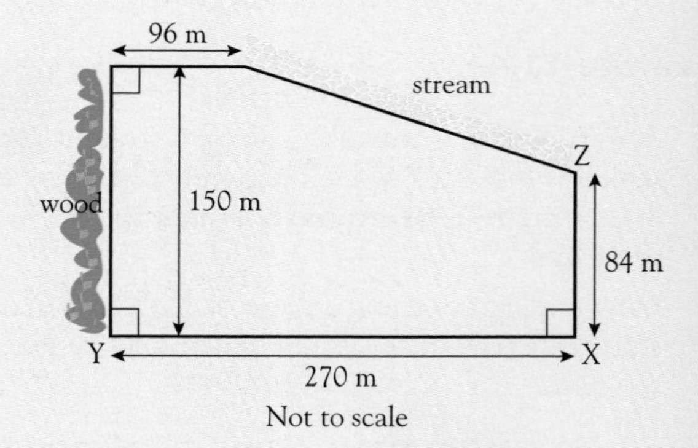

a Make a scale drawing of the field using a scale of 1 cm to 20 m and shade those regions in which flying is not recommended.

b Winston flies his plane from the corner X of the field on a flight path which is equidistant from the sides XY and XZ. Show this route on your drawing.

c How far can the plane fly before it enters the region which has been advised against?

9 Two radio beacons **A** and **C** are 100 km apart. The bearing of a third beacon **B** from **A** is 032° and the bearing of **B** from **C** is 317°.

The scale drawing below shows the positions of radio beacons **A** and **C**. (Scale 1 cm = 10 km)

a **i** Trace the diagram and complete the scale drawing to show the position of radio beacon **B**.

ii Use your scale drawing to find the bearing of **A** from **C**.

An aircraft flies so that it is equidistant from radio beacons **A** and **B**.

b Use your drawing to show the path of the aircraft's flight.

The radio beacon at **C** can transmit signals up to a distance of 60 km.

c **i** Show clearly on your diagram where the aircraft can receive signals from radio beacon **C**.

ii How far from **A** is the aircraft when it receives signals from radio beacon **C**?

[SEG]

N

C

TASK

1 Draw a circle with radius 5 cm.

2 Draw any diameter and label its end points A and B.

3 Mark any point on the circumference of the circle and label the point C.

4 Join A → C and B → C to form a triangle ABC.

5 Measure the angle ACB.

6 Choose another 5 points, C, on the circumference of the circle and repeat steps 4 and 5 above. (C does not have to be in the same half of the circle as your first choice.)

7 What do you notice about the size of ∠ACB?

C

A B

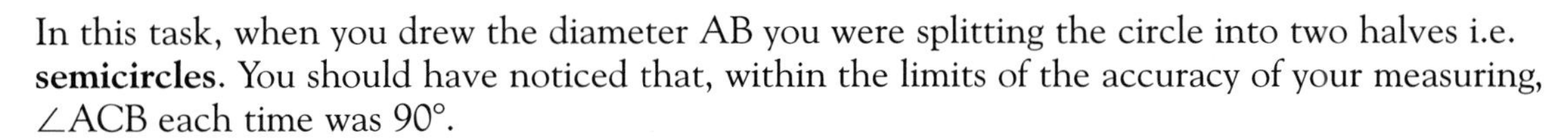

In this task, when you drew the diameter AB you were splitting the circle into two halves i.e. **semicircles**. You should have noticed that, within the limits of the accuracy of your measuring, ∠ACB each time was 90°.

It does not matter where the point C is as long as it is not in the same position as A or B!

∠ACB will **always** be a **right angle (90°)**. ∠ACB is called **the angle in a semicircle**.

The angle in a semicircle is always 90°.

EXAM QUESTIONS

Worked Exam Question
[SEG]

COMMENTS

Two oil rigs, A and C, are in the sea near the port of *D*. The diagram is drawn to a scale of 1 cm to 10 km.

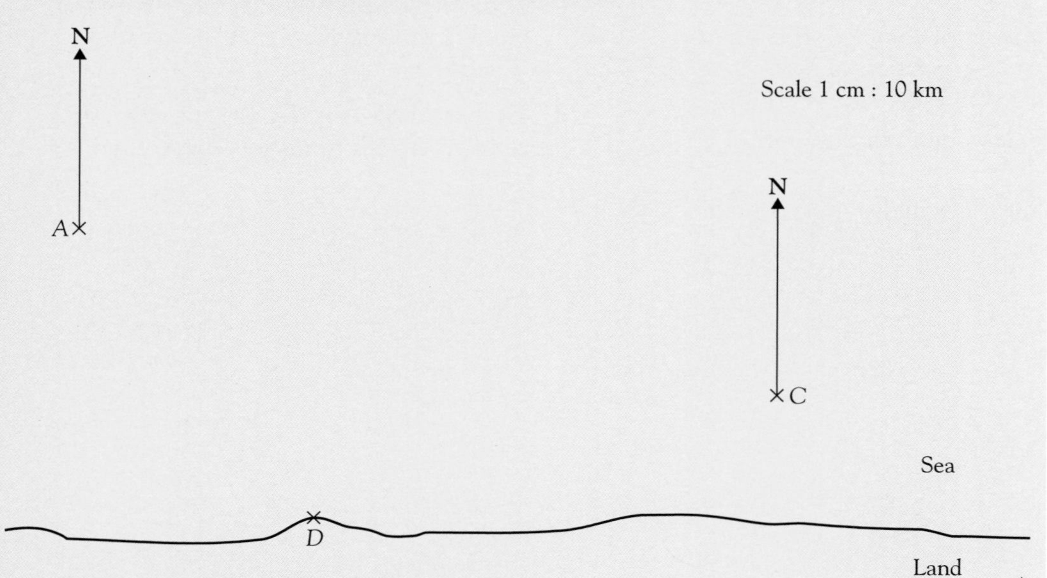

a Find the distance and bearing of A from *D*.

- Draw a straight line joining *A* and *D*.
- Using your set-square and ruler draw a North line at *D*.
- Measure the length of *AD* and convert it to km.
- Measure the angle marked in red on the diagram.

Answer Distance ...50... km

Bearing ...322... degrees

A1 correct distance ± 1 km.
A1 correct bearing ± 1°
2 marks

b A third oil rig, B, is on a bearing of 117° from A, and a bearing of 267° from C. Show the position of B on the diagram.

c i A fourth oil rig is the same distance from A and from C. Construct the locus of all possible positions of the fourth oil rig on the diagram.

ii The fourth oil rig, *E*, is the same distance from A and from C and is 65 km from *D*. Mark the position of *E* on the diagram with a cross.

b
- Measure and mark 117° from North with centre of protractor on *A*. Draw a line from *A* through the mark.
- Measure and mark 267° from North, centre C. Draw a line from C through the mark.
- Mark point *B* at the intersection of lines.

c i
- Join A to C.
- Construct the perpendicular bisector of AC.

c ii
- Set your compasses to 6.5 cm and draw an arc, centre *D*, cutting the perpendicular bisector of AC.
- Mark the point where they intersect with a '×'.

This is point *E*.

A1 117° drawn accurately ± 1°
A1 267° drawn accurately ± 1°
2 marks

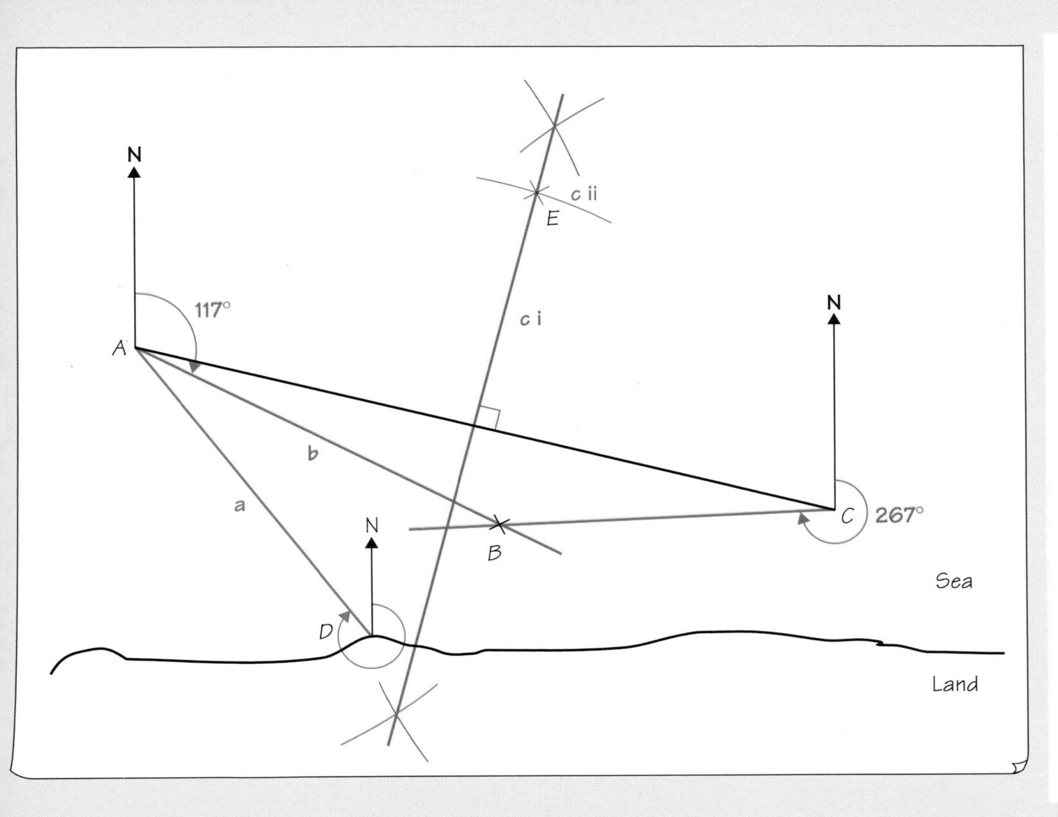

M1 perpendicular bisector of AC constructed accurately with construction lines shown (No construction lines shown – M0)

1 mark

M1 arc with radius 6.5 cm accurately drawn, intersecting with perpendicular bisector of AC. Intersection indicated with a '×'

1 mark

Exam Questions

1 ABCD is a square in which A is north of C.

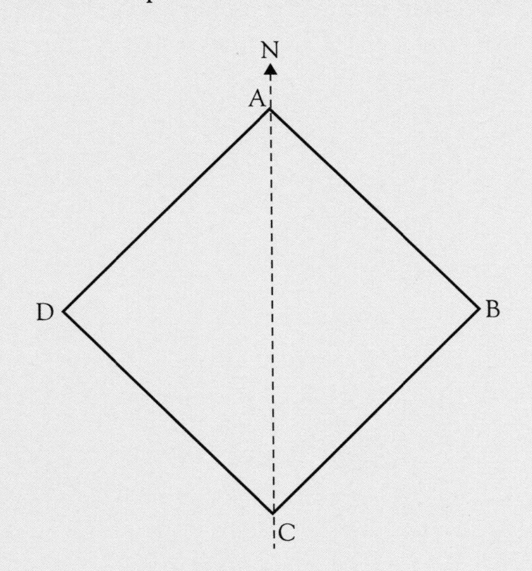

Write down the three-figure bearing of:

a B from C

b D from B

c D from C [NI]

2

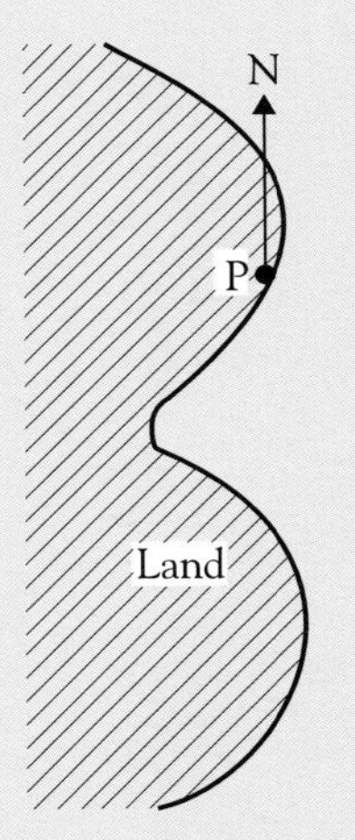

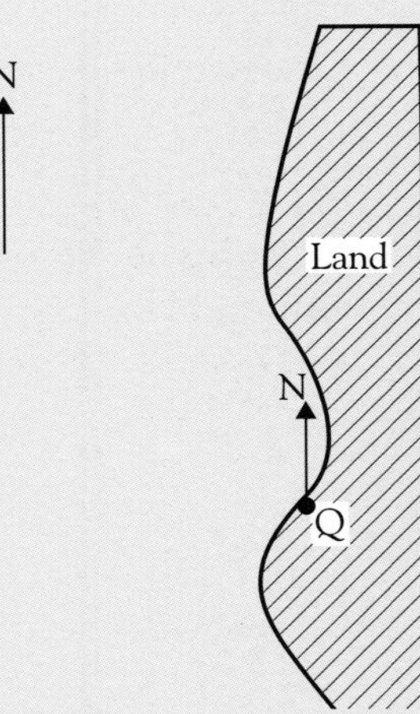

Two ports, P and Q, are shown on the map. Measure and write down the bearing of P from Q. [L]

3 The map at the top of the next page shows an island.

The scale of the map is 1 : 50 000.

EXAM QUESTIONS

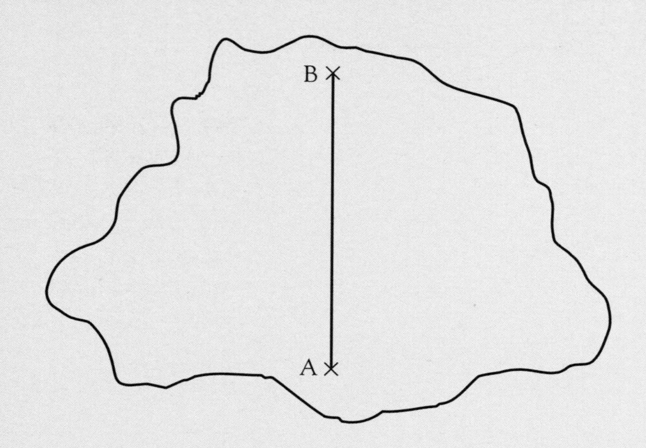

a What is the distance in metres between A and B?

b B is due north of A.

C is the highest point on the island.

The bearing of C from A is 048°.

The bearing of C from B is 127°.

Trace the map and mark the position of C on the map. [MEG]

4 The diagram shows a map of part of the North Devon coast.

The bearing of a ship from Hartland Point is 070°.

Its bearing from Appledore is 320°.

Trace the map.

Showing your construction lines, mark the position of the ship on the map. Label the position with the letter *S*.

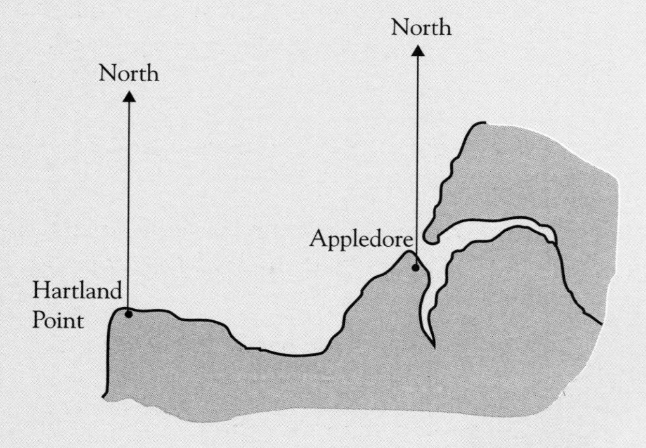

[MEG]

5 Byron and Shelley are two dogs.

Byron's lead is 1 m long. One end of the lead can slide along a rail *PQ*, which is fixed to the wall of the house.

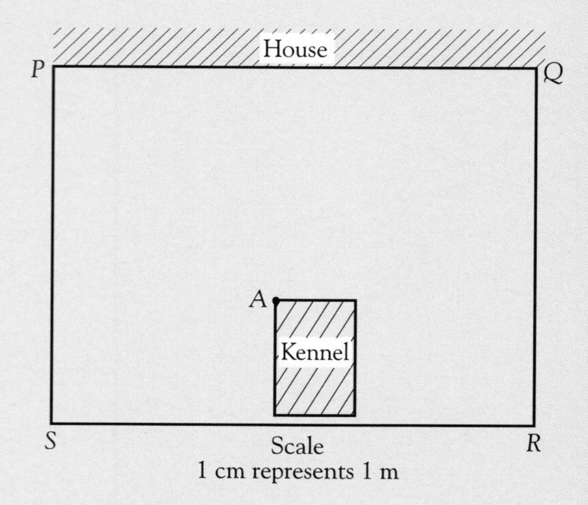

Shelley's lead is 1.5 m long. One end of this lead is attached to a post *A*, at the corner of his kennel.

The scale diagram above represents the fenced garden *PQRS* where the dogs live.

Trace the diagram and show all the possible positions of each dog if the leads remain tight.

[MEG, p]

6 The diagram below is the plan of a campsite drawn to a scale of 1 centimetre to 10 metres.

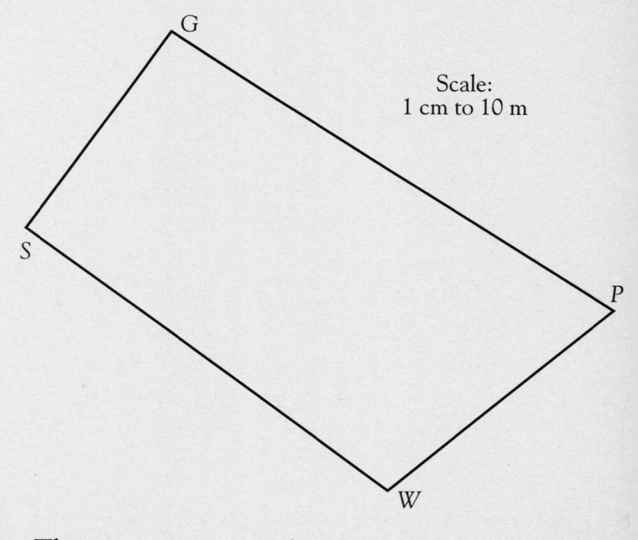

The campsite owner decides to put a new water tap on the campsite.

The tap must be the same distance from the shop, *S*, as it is from the gate, *G*.

a Trace the diagram and draw accurately the locus of all points equidistant from G and S.

The water system is such that the tap must be no more than 50 metres from the washblock, *W*.

b Construct accurately the locus of points which are exactly 50 m from *W*.

c Show on the diagram all possible positions for the tap. [MEG, p]

7 Two straight roads are shown in the diagram.

A new gas pipe is to be laid from Bere equidistant from the two roads.

The diagram is drawn to a scale of 1 cm to 2 km.

a Copy the diagram and construct, on it, the path of the gas pipe.

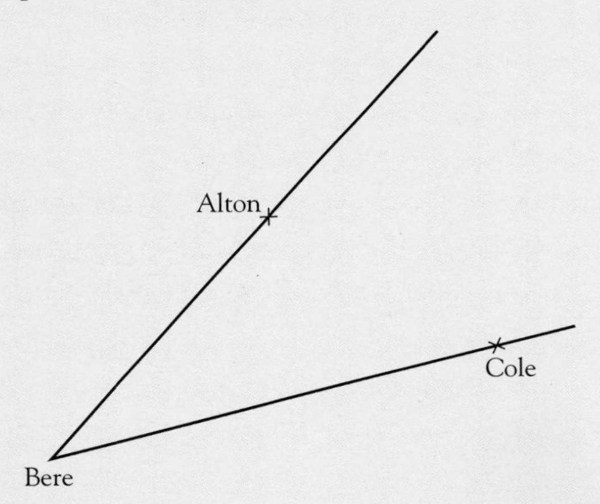

b The gas board needs a construction site depot.

The depot must be equidistant from Bere and Cole.

The depot must be less than 4 km from Alton.

i On your diagram draw loci to represent this information.

The depot must be nearer the road through Cole than the road through Alton.

ii Mark on your diagram with a cross a possible position for the site depot which satisfies all of these conditions. [SEG]

8 A scale drawing of a display area for fireworks is a rectangle 8.5 cm by 5 cm. The scale used is 1 cm to 2 m.

Mr Ward places a rope outside the area to keep the audience a safe distance away.

The rope is always 4 m from the edge of the display area.

Do a scale drawing (in the middle of your page) showing the display area. Draw accurately the position of the rope on the scale drawing. [SEG]

9

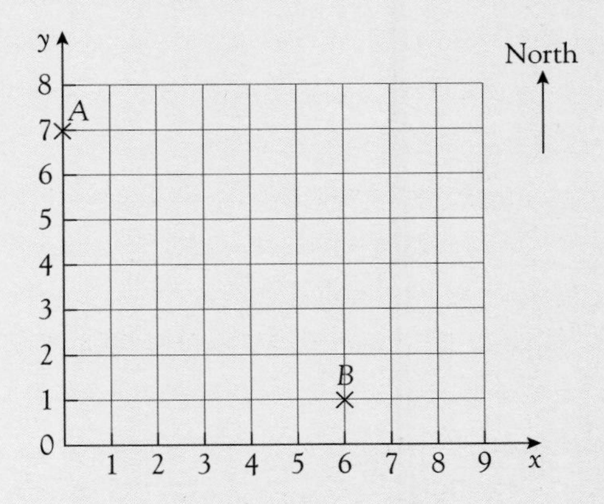

The grid above is used to locate towns on a map.

a **i** Copy the grid onto squared paper using a scale of 1 cm to 1 unit on both axes.

On your diagram, 1 cm represents 1 km.

Draw the line *AB*.

Measure and write down the bearing of town *B* from town *A*.

ii Town C is 3 km on a bearing of 270° from town *B*.

Write down the co-ordinates of town C.

b **i** Draw the locus of all points on the grid which are 5 km from town *A*.

ii Draw the locus of all points on the grid which are equidistant from towns *A* and *B*.

iii Mark and label the position, M, of a monument which is north of the line *AB*, 5 km from town *A* and equidistant from towns *A* and *B*. [MEG]

10 Sandtown council wants to build a pavilion on a playing field.

They decided to site it at an equal distance from both gates and not more than 250 m from the toilet block.

Trace the scale drawing below, and on it make suitable constructions to locate the site.

Leave in your construction lines.

Show clearly where the pavilion can be built.

EXAM QUESTIONS

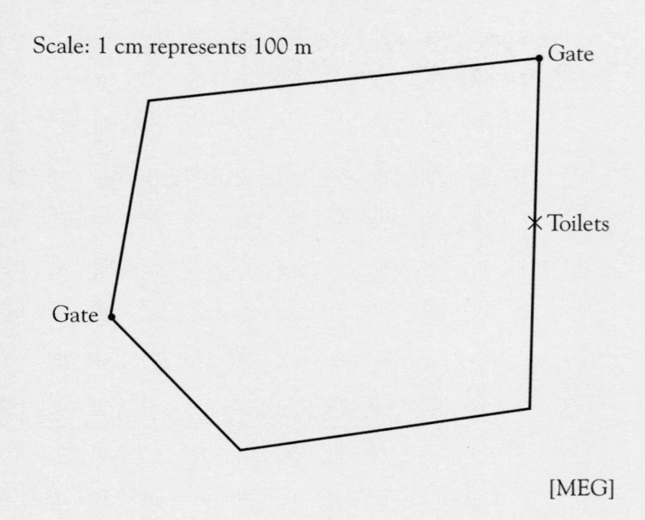

[MEG]

11 Two ships A and B both hear a distress signal from a fishing boat.

The positions of A and B are shown on the map below.

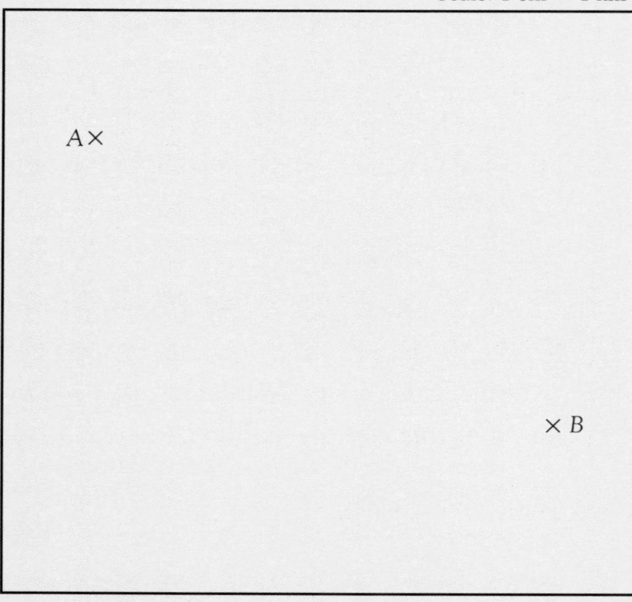

The map is drawn using a scale of 1 cm to represent 1 km.

The fishing boat is less than 4 km from ship A and is less than 4.5 km from ship B.

A helicopter pilot sees that the fishing boat is nearer to ship A than to ship B.

Copy the diagram and use accurate construction to show the region which contains the fishing boat. Shade this region. [NEAB]

12 The map shows part of a coastline and a coastguard station.

1 cm on the map represents 2 km.

A ship is 12 km from the coastguard station on a bearing of 160°.

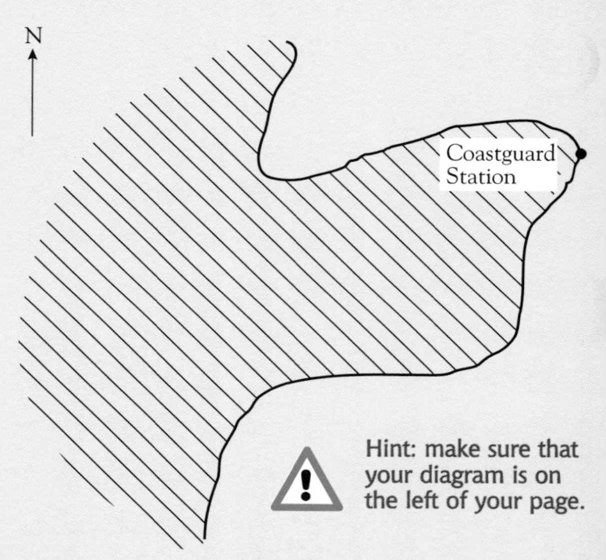

Hint: make sure that your diagram is on the left of your page.

a Trace the map and plot the position of the ship from the coastguard station, using a scale of 1 cm to represent 2 km.

It is not safe for ships to come within 6 km of the coastguard station.

b Shade the area on the map which is less than 6 km from the coastguard station. [L, p]

DATA HANDLING II

In this chapter you will
- **revise how to find the mean of raw data**
- **learn how to estimate the mean of grouped data**
- **learn how to draw frequency polygons**
- **compare data using frequency polygons**

Formula for the mean of raw data

Raw data are data that have not been grouped into intervals

page 150

You already know that the **mean** of a set of data is the average found by dividing the total of all the values by the number of values.

The formula is $\text{mean} = \dfrac{\text{total of all the values}}{\text{number of values}}$

This can be written in mathematical shorthand as follows:

$\bar{x} = \dfrac{\Sigma x}{n}$

$\bar{x}$, pronounced 'x bar' is the symbol for the mean,

Σx, pronounced 'sigma x' is the symbol for 'the total of all the values',

n is the number of values.

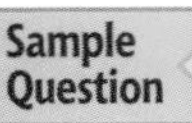

Five friends obtained these marks out of 50 in a test: 25, 29, 41, 36, 43

Find their mean mark.

Answer

- Find the total of all the marks, then use it in the formula for the mean.

$$\Sigma x = 25 + 29 + 41 + 36 + 43 = 174$$

$$\bar{x} = \frac{\Sigma x}{n} = \frac{174}{5} = 34.8$$

$n = 5$, since there are 5 values.

Their mean mark was 34.8

You do not have to use the letter x. If, for example, the variable is the time taken, you might decide to use the letter t. In this case you would write $\bar{t} = \dfrac{\Sigma t}{n}$

MEANS

Sample Question 2

An experiment was carried out to investigate the time taken to complete a simple task. The twelve male volunteers had a mean time of 1.6 minutes and the eight female volunteers had a mean time of 2.1 minutes.

a Find the mean time, in minutes, of all the volunteers.

b Write your answer in seconds.

Answer

a

- Find the total time taken by the males, using

$$\overline{m} = \frac{\Sigma m}{n} \qquad \text{with } \overline{m} = 1.6 \text{ and } n = 12$$

$$1.6 = \frac{\Sigma m}{12} \qquad \text{so } \Sigma m = 12 \times 1.6 = 19.2 \text{ minutes}$$

- Find the total time taken by the females, using

$$\bar{f} = \frac{\Sigma f}{n} \qquad \text{with } \bar{f} = 2.1 \text{ and } n = 8$$

$$2.1 = \frac{\Sigma f}{8} \qquad \text{so } \Sigma f = 8 \times 2.1 = 16.8 \text{ minutes}$$

- Work out the time taken by all 20 volunteers and find the mean for all 20.

$$\begin{aligned} \text{total time} &= \Sigma m + \Sigma f \\ &= 19.2 + 16.8 \\ &= 36 \text{ minutes} \\ \text{mean time} &= \frac{36}{20} \end{aligned}$$

$\underline{\text{mean time} = 1.8 \text{ minutes}}$

b

- Change minutes to seconds by multiplying by 60

$1.8 \times 60 = 108$

$\underline{\text{mean time} = 108 \text{ seconds}}$

1 min = 60 secs

Formula for the mean when data are in a frequency distribution

x	f	$f \times x$
⋮	⋮	⋮
	$\Sigma f =$	$\Sigma fx =$

To find the mean

- multiply each value x by its frequency f and complete the $f \times x$ column
- add up the numbers in the $f \times x$ column. This gives Σfx, the total of all the values
- add up the number in the f column. This gives Σf, the number of values.

$$\bar{x} = \frac{\Sigma fx}{\Sigma f}$$

Σfx is $\Sigma(f \times x)$ but the bracket and multiplication sign are usually left out.

MEANS

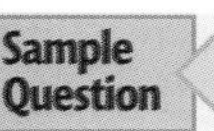

The Under 14 Hockey Team's performance in 20 matches was summarised in the following table, showing the number of goals scored by the team.

Work out the mean number of goals per match.

Number of goals (x)	0	1	2	3	4
Number of matches (f)	4	6	7	2	1

Answer

- Write the table in column form and work out $f \times x$
- Add up the numbers in the $f \times x$ column to find the total number of goals, Σfx
- Divide by the number of matches, Σf

x	f	$f \times x$
0	4	0
1	6	6
2	7	14
3	2	6
4	1	4
	$\Sigma f = 20$	$\Sigma fx = 30$

Show all your working. This will enable you to get method marks, even if you make a slip in the arithmetic.

$$\bar{x} = \frac{\Sigma fx}{\Sigma f}$$

$$= \frac{30}{20}$$

$$= 1.5$$

It is useful to quote the formula. This helps you to be clear about what to do.

The mean number of goals per match was 1.5

Mean of grouped data

When data have been grouped into intervals, the mean cannot be calculated exactly, as you do not know what the individual values are, only that they lie in a particular interval. It is, however, possible to work out an **estimate** of the mean.

The table shows a summary of the results when 100 apples were picked at random from an orchard and weighed.

Mass (in g)	$40 \leqslant x < 50$	$50 \leqslant x < 60$	$60 \leqslant x < 70$	$70 \leqslant x < 80$	$80 \leqslant x < 90$
Frequency	6	15	32	28	19

To estimate the total mass of the six apples in the 40 g to 50 g interval, take the mid-interval value of 45 g as representative and assume that they were all 45 g. Estimate the total mass of these six apples as $6 \times 45\text{ g} = 270\text{ g}$.

This might be an overestimate, it might be an underestimate, but without the original data it is the best that you can do!

MEANS

If the mid-interval value for each interval is called x, you will need to find $f \times x$ for each interval. This is best done in a table in column form. In an examination the columns are often drawn ready for you to complete.

Mass (in g)	Frequency, f	Mid-interval, x	$f \times x$
$40 \leqslant x < 50$	6	45	270
$50 \leqslant x < 60$	15	55	825
$60 \leqslant x < 70$	32	65	2080
$70 \leqslant x < 80$	28	75	2100
$80 \leqslant x < 90$	19	85	1615
	$\Sigma f = 100$		$\Sigma fx = 6890$

Do not add up this column.

$$\bar{x} = \frac{\Sigma fx}{\Sigma f} = \frac{6890}{100} = 68.9$$

The mean mass of an apple is 68.9 g

You can add up the numbers as you go along, as well as writing them in the $f \times x$ column, by using M+ on your calculator and watching the display.

First make sure that the memory is clear 0 Min

Then key in

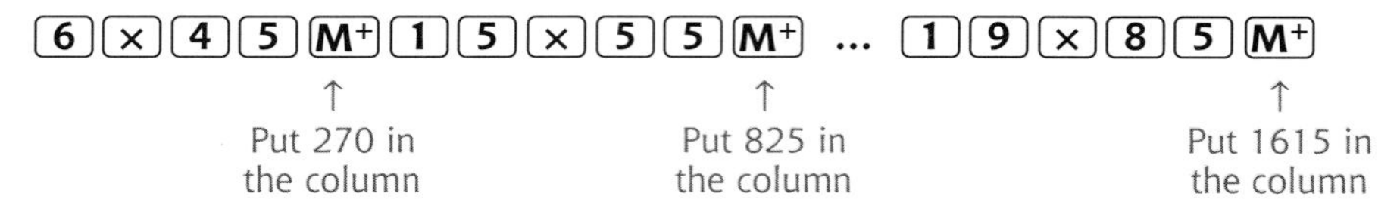

To find the total, press MR

Finding mid-interval values

For **discrete data**, add the smallest and largest possible values that the interval could contain, then divide by 2, for example

Number of flowers on a plant	Mid-interval value
0–4	2
5–9	7
10–14	12
15–19	17
20–24	22

In the 5–9 interval, smallest possible value = 5, largest possible value = 9. The mid-interval value is

$$\frac{5+9}{2} = \frac{14}{2} = 7$$

For **continuous data**, add the two boundary values of the interval, then divide by 2, for example

a

Length of leaf (in cm)	Mid-interval value
$0 \leqslant x < 5$	2.5
$5 \leqslant x < 10$	7.5
$10 \leqslant x < 15$	12.5
$15 \leqslant x < 20$	17.5

In the interval $5 \leqslant x < 10$ the boundary values are 5 and 10. The mid-interval value is

$$\frac{5+10}{2} = \frac{15}{2} = 7.5$$

MEANS

b

Length of telephone calls (in mins)	Mid-interval value
0 –	1.5
3 –	**4.5**
6 –	7.5
9 –	10.5

In the interval 3– the boundary values are 3 and 6. The mid-interval value is

$$\frac{3+6}{2} = \frac{9}{2} = 4.5$$

Assume that the upper boundary of the last interval is 12.

c

Temperature (to nearest degree)	Mid-interval value
10 – 14	12
15 – 19	**17**
20 – 24	22
25 – 29	27

The interval 15–19 contains values from 14.5 to 19.5, since temperatures were measured to the nearest degree, for example 14.7° would be in this interval. The mid-interval value is

$$\frac{14.5+19.5}{2} = \frac{34}{2} = 17$$

d

Age (in completed years)	Mid-interval value
5 – 9	7.5
10 – 14	**12.5**
15 – 19	17.5
20 – 24	22.5

Look at the 10–14 interval. Someone who will be 15 tomorrow is in this interval. The boundary points are 10 and 15.

$$\frac{10+15}{2} = \frac{25}{2} = 12.5$$

Sample Question 4

For a statistics project, Duncan was investigating the lengths of sentences used by different authors. He counted the number of words in each of the first 100 sentences in Alice in Wonderland by Lewis Carroll, and summarised the results in this table.

Number of words in sentence	1–10	11–20	21–30	31–40	41–50	51–60	61–70	71–80	81–90	91–100	101–110
Number of sentences	25	24	16	11	10	8	2	2	0	0	2

Find the mean number of words in a sentence.

Answer

- Write the data in a table and find the mid-interval values, x
- Complete the $f \times x$ column and work out the total Σfx
- Use the formula $\bar{x} = \dfrac{\Sigma fx}{\Sigma f}$ to find the mean

MEANS

Number of words in a sentence	Frequency, f	Mid-interval value, x	$f \times x$
1–10	25	**5.5**	137.5
11–20	24	15.5	372
21–30	16	25.5	408
31–40	11	35.5	390.5
41–50	10	45.5	455
51–60	8	55.5	444
61–70	2	65.5	131
71–80	2	75.5	151
81–90	**0**	**85.5**	**0**
91–100	0	95.5	0
101–110	2	105.5	211
	$\Sigma f = 100$		$\Sigma fx = 2700$

For 1–10 interval

$\frac{1 + 10}{2} = \frac{11}{2} = 5.5$

Remember

$0 \times 85.5 = 0$

$$\bar{x} = \frac{\Sigma fx}{\Sigma f}$$

$$= \frac{2700}{100}$$

$$= 27$$

The mean number of words in a sentence was 27

Sample Question 5 The speed of 100 cars on a motorway was recorded. This is the distribution:

Speed, s (miles/hour)	$40 \leqslant s < 50$	$50 \leqslant s < 60$	$60 \leqslant s < 70$	$70 \leqslant s < 80$	$80 \leqslant s < 90$	$90 \leqslant s < 100$	
Number of cars	2	15	34	37	8	4	Total 100

a Calculate an estimate of the mean speed of these cars.

The speed limit is 70 miles/hour.

b Francis uses the information in the table to estimate the probability that the next car to pass is travelling at less than 70 miles/hour. What is this probability? [MEG]

Answer

a

- Find the mid-interval value for each interval by adding the boundary points then dividing by 2. Call it x.
- Work out $f \times x$ for each interval and add them up to find Σfx
- Find the mean by dividing this total by Σf, the number of cars in the survey.

MEANS

Speed	Frequency, f	Mid-interval value, x	$f \times x$
$40 \leqslant s < 50$	2	45	90
$50 \leqslant s < 60$	15	55	825
$60 \leqslant s < 70$	34	65	2210
$70 \leqslant s < 80$	37	75	2775
$80 \leqslant s < 90$	8	85	680
$90 \leqslant s < 100$	4	95	380
	$\Sigma f = 100$		$\Sigma fx = 6960$

Do not add up the mid-interval column.

$$\bar{x} = \frac{\Sigma fx}{\Sigma f}$$

$$= \frac{6960}{100}$$

$$= 69.6$$

The mean speed is 69.6 miles/hour.

b

◆ Work out how many cars in the sample were travelling at less than 70 miles/hour, by adding the numbers in the intervals $40 \leqslant s < 50$, $50 \leqslant s < 60$ and $60 \leqslant s < 70$

$2 + 15 + 34 = 51$

page 34

◆ Use **probability** = $\dfrac{\textbf{number of favourable outcomes}}{\textbf{total number of outcomes}}$

$$\text{Probability} = \frac{51}{100}$$

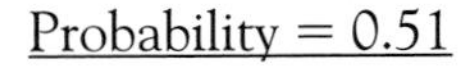

Probability = 0.51

Look up probability ideas in chapter 3 if you are unsure.

Exercise 13.1

1 A cold virus swept through St. Benedict's School. The number of absences in each class was recorded:

13 5 10 8 11 7 3 4 5 6

8 5 4 1 0 2 3 2 1 2

Find the mean number of absences per class.

2 A class took a test. The mean mark of the 20 boys in the class was 17.4. The mean mark of the 10 girls in the class was 13.8.

a Calculate the mean mark for the whole class.

5 pupils in another class took the test.
Their marks, written in order, were 1, 2, 3, 4 and x.

The mean of these 5 marks is equal to twice the median of these 5 marks.

b Calculate the value of x. [L]

3 As part of their revision for SATs, some Year 9 students were given a mental test. The results were as follows

Mark x	0	1	2	3	4	5	6	7	8	9	10
Frequency, f	0	1	3	6	8	3	7	4	5	2	1

Find the mean, median and modal test scores.

MEANS

4 A youth worker was concerned that the members of his youth club were showing signs of tiredness. He conducted a survey to find out the number of hours a week they were studying outside school hours.

Hours worked, h	Frequency, f
$0 \leqslant h < 4$	6
$4 \leqslant h < 8$	11
$8 \leqslant h < 12$	4
$12 \leqslant h < 16$	3
$16 \leqslant h < 20$	0
$20 \leqslant h < 24$	1
	Total 25

Find an estimate of the mean number of hours in a week spent studying.

5 a The headteacher of an infant school carried out a survey of the new pupils in the reception classes to find out how many brothers and sisters each pupil had. The results are summarised in the table.

Number of brothers and sisters, x	0	1	2	3	4	5
Frequency, f	12	19	11	6	1	1

i How many pupils had more than two brothers and sisters?

ii How many pupils were in the reception classes?

iii A pupil was chosen at random from the reception classes. What is the probability that the pupil had less than three brothers and sisters?

iv Find the mean number of brothers and sisters per pupil in the reception classes.

b The previous year's survey results were smudged when coffee was spilt on the piece of paper.

Number of brothers and sisters, x	0	1	2	3	4	5
Frequency, f	15	25	[illegible]	3	0	2

There were 55 pupils in the reception classes in the previous year.
Was the current year's mean higher or lower than the previous year's, and by how much?

6 A horticulturist collected data on a new variety of rose. For each plant in the sample, she recorded the amount of time it was in flower. Her results are shown in the histogram.

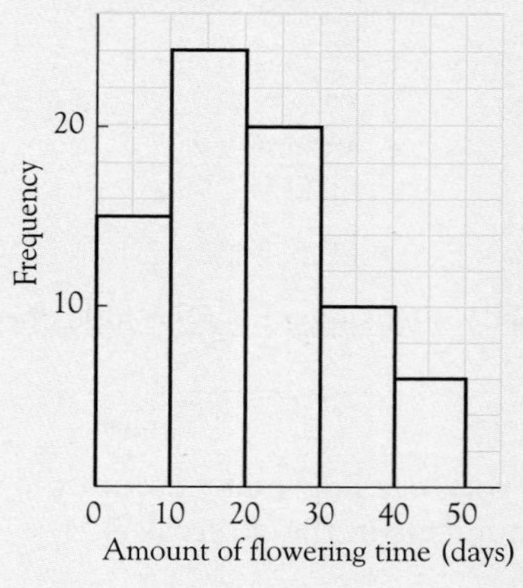

a How many plants were there in the sample?

b Copy and complete the frequency table for this sample.

Days flowering, d	Number of plants, f
$0 \leqslant d < 10$	15
$10 \leqslant d < 20$	
$20 \leqslant d < 30$	
$30 \leqslant d < 40$	
$40 \leqslant d < 50$	

c Estimate the mean time that this new variety of rose was in flower. (You will need to put more columns in your table.)

7 A team of 30 volunteers collected money for a charity outside a number of supermarkets one Saturday. The amounts they collected, rounded to the nearest £, are given in the table.

MEANS

Amount of money, £	Mid-interval value, x	Frequency, f	$f \times x$
1–10		1	
11–20		4	
21–30		3	
31–40		6	
41–50		5	
51–60		8	
61–70		2	
71–80		1	
		$\Sigma f =$	$\Sigma fx =$

a Copy and complete the table.

b Find an estimate of the mean amount of money collected per person.

8 A survey is carried out to find the neck size of a sample of 600 men.
The results of the survey are shown in the grouped frequency table below.

Neck size (cm)	Frequency
20–25	4
25–30	21
30–35	180
35–40	234
40–45	116
45–50	36
50–55	9
	600

Use mid-interval values to calculate an estimate of the mean neck size. [MEG]

9 Local residents were very concerned about the speed of cars travelling along their road at the time that young children were coming home from school. Police from the traffic unit recorded the speed of all vehicles passing a checkpoint between 2.30 pm and 5.00 pm one weekday.

Speed in mph m	Frequency, f
$0 \leqslant m < 10$	1
$10 \leqslant m < 20$	14
$20 \leqslant m < 30$	28
$30 \leqslant m < 40$	24
$40 \leqslant m < 50$	10
$50 \leqslant m < 60$	5

a How many vehicles travelled along that road between 2.30 pm and 5.00 pm on the survey day?

b Estimate the mean speed.

c The speed limit along the road is 30 mph. How many vehicles were travelling less than 30 mph?

d What is the probability that a vehicle will be travelling at 30 mph or over?

e Should traffic calming measures be introduced?
Give reasons for your answer.

10 A water polo team coach noted the number of goals scored by the under 16 team over a season.

Goals scored

22	1	3	18	13
15	6	15	20	17
12	7	10	2	10
10	21	9	8	14

a Calculate the mean number of goals scored.

b Copy and complete the grouped frequency table and hence find an estimate of the mean number of goals scored.

Goals scored	Frequency, f	Mid-interval value, x	$f \times x$
1–5			
6–10			
11–15			
16–20			
21–25			
	$\Sigma f =$		$\Sigma fx =$

c Is your estimate of the mean a good estimate?

Frequency polygons

If you put a cross at the middle of the top of each bar in a histogram and then join the points with straight lines, you have drawn what is called a **frequency polygon**.

In the diagram a histogram has been drawn to show the heights of pupils given in the table. A frequency polygon has been drawn on it.

Height (cm)	$130 \leq x < 140$	$140 \leq x < 150$	$150 \leq x < 160$	$160 \leq x < 170$	$170 \leq x < 180$
Frequency	4	13	40	25	10

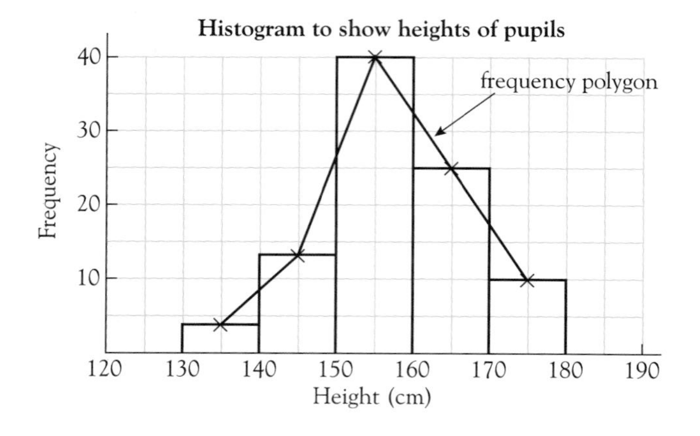

In all the examples in this book, the intervals are of equal width. If you choose to group data into unequal intervals, for example when doing your coursework, ask your teacher how to adjust the heights of the bars on the histogram or the points to plot on the frequency polygon.

The frequency polygon can be drawn on its own.

To draw the frequency polygon without drawing the histogram first:

Provided you have **equal width intervals**

- work out the mid-interval value for each interval
- plot this against the frequency
- join up the points with straight lines
- extend the lines to the horizontal axis if you wish, as shown in the diagram on page 251

Height	Mid-interval value	Frequency
$130 \leq x < 140$	135	4
$140 \leq x < 150$	145	13
$150 \leq x < 160$	155	40
$160 \leq x < 170$	165	25
$170 \leq x < 180$	175	10

In the table above, the mid-points of the intervals are 135, 145, 155, 165 and 175.

To draw the frequency polygon without drawing the histogram first, plot the following points and join them with straight lines.

(135, 4), (145, 13), (155, 40), (165, 25), (175, 10)

If you wish to extend your polygon to the horizontal axis, plot (125, 0) and (185, 0) and extend your lines to them.

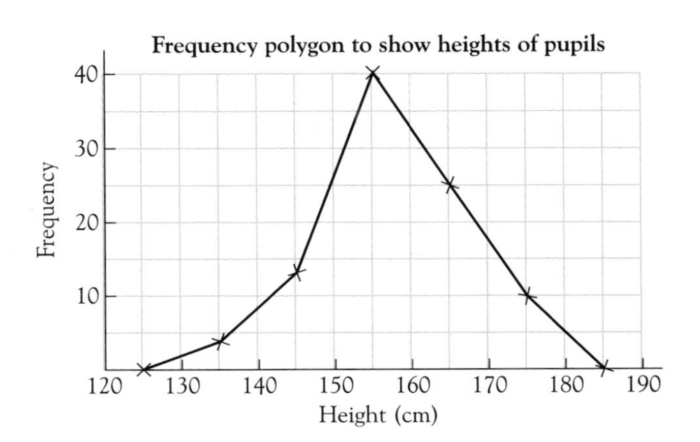

Modal class

When data have been grouped into intervals, it is not possible to say what the mode is as you do not know all the individual values. You can, though, describe the **modal class**.

The modal class is the interval with the tallest bar in the histogram

When the intervals are of equal width, the modal class is the interval with the greatest frequency. The highest point on the frequency polygon shows the mid-interval value of the modal class.

The modal class for the heights of pupils described above is the interval $150 \leqslant x < 160$

Comparing sets of data

Frequency polygons are very useful when you want to compare sets of data. This is because you can draw more than one frequency polygon on a diagram.

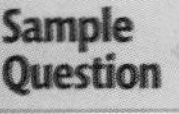

A gardener tests a fertiliser.
He grows some tomatoes with the fertiliser and some without.
He records the weights of all the tomatoes grown.

Weight (grams)	Frequency	
	With fertiliser	Without fertiliser
$50 < W \leqslant 100$	10	2
$100 < W \leqslant 150$	15	42
$150 < W \leqslant 200$	55	46
$200 < W \leqslant 250$	53	41
$250 < W \leqslant 300$	17	34
$300 < W \leqslant 350$	8	1

a Draw a frequency polygon for each distribution on a grid, clearly indicating which is **with fertiliser**, and which is **without fertiliser**.

b Use the frequency polygons to compare the effects of the fertiliser. [NEAB]

FREQUENCY POLYGONS

Answer

a

- Work out the mid-interval value for each interval.

  For the interval $50 < W \leqslant 100$, mid-interval value $= \dfrac{50 + 100}{2} = 75$

  For the interval $100 < W \leqslant 150$, mid-interval value $= \dfrac{100 + 150}{2} = 125$

  The mid-interval values are 75, 125, 175, 225, 275, 325.

- For data 'with fertiliser' plot (75, 10), (125, 15), (175, 55), (225, 53), (275, 17), (325, 8)
- Join points with straight lines and extend back to (25, 0) and (375, 0)
- Label the polygon.
- Repeat for 'without fertiliser' data, plotting (25, 0), (75, 2), (125, 42), (175, 46), (225, 41), (275, 34), (325, 1), (375, 0).

In an examination, make an extra column on the table given on the question paper and jot down the mid-interval values for each interval.

Frequency polygon to show weights of tomatoes

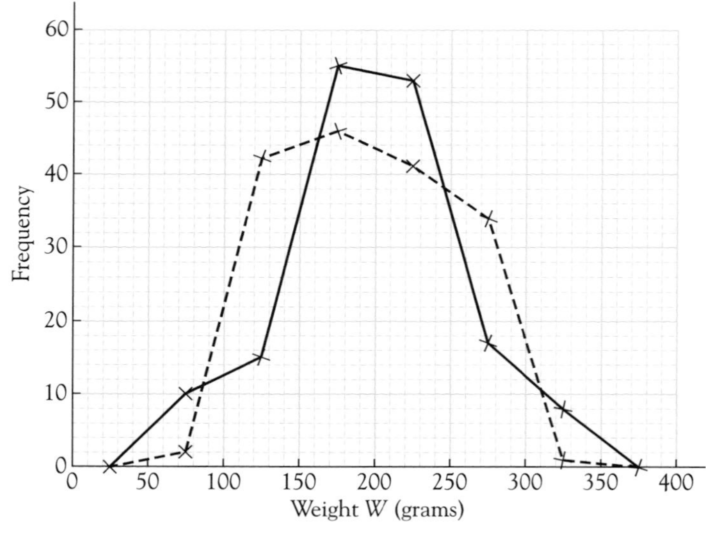

Key:
—— with fertiliser
- - - - without fertiliser

Remember to indicate which polygon is which.

Take care with the scales.
On the horizontal axis, one square represents 10 grams.
On the vertical axis, one square represents 2 tomatoes.

b

- Notice that the tomatoes grown with fertiliser have weights that are more clustered and less spread out than those without. You could describe this by saying:

  Tomatoes grown with fertiliser are of a more uniform weight.
  There are not as many light or heavy ones as those grown without fertiliser.

FREQUENCY POLYGONS

Exercise 13.2

Pupils attending the school computer club were asked to record the time that they spent using the computer. Mrs Disk put the results in a frequency table.

Time spent, t (mins)	Frequency, f
$0 \leqslant t < 10$	4
$10 \leqslant t < 20$	1
$20 \leqslant t < 30$	3
$30 \leqslant t < 40$	6
$40 \leqslant t < 50$	17
$50 \leqslant t < 60$	9

a Draw a frequency polygon to illustrate the results.

b Find an estimate of the mean time spent using the computer.

A video rental shop took a random sample of 30 customers and recorded the number of videos each rented over a 6 month period.

Number of videos	Frequency, f
1–5	5
6–10	13
11–15	6
16–20	3
21–25	2
26–30	0
31–35	1

a Find an estimate of the mean number of videos rented by a customer.

b Draw a frequency polygon to illustrate the results of the survey.

A jelly manufacturer tested two setting agents, X and Y, to find out which caused the jelly to set more quickly without loss of flavour.

A sample of 100 jellies using Agent X and a sample of 100 jellies using Agent Y were set in the same conditions with the following results.

Agent X Frequency	Setting time h (hours)	Agent Y Frequency
1	$0 \leqslant h < 2$	22
3	$2 \leqslant h < 4$	35
5	$4 \leqslant h < 6$	21
17	$6 \leqslant h < 8$	15
27	$8 \leqslant h < 10$	5
33	$10 \leqslant h < 12$	1
14	$12 \leqslant h < 14$	1
100		100

a Find the mean setting time for each setting agent.

b On the same axes, show a frequency polygon for each setting agent, indicating clearly which is which.

c Assuming there was no loss of flavour, which setting agent should the manufacturer use? Give a reason for your answer.

4

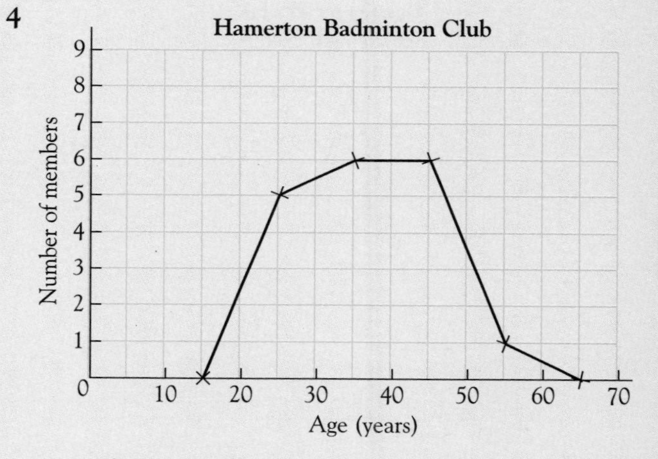

The frequency polygon shows the age distribution of the women members of the Hamerton badminton club.

Copy the frequency polygon for the women. The age distribution for the 23 men in the club is given in the table.

Age (years)	Number	Mid-point
10–19	2	15
20–29	6	
30–39	7	
40–49	5	
50–59	2	
60–69	1	

FREQUENCY POLYGONS

a Draw a frequency polygon for the men on the same graph as for the women.

b Which is the modal group for the men?

c Calculate an estimate for the mean age of the men.

The mean age of the women is 36.7 years.

d Can you draw any conclusions about the two distributions?
Explain your answer. [MEG]

5 A fish farmer trialled a new brand of fish food, separating a batch of fish, all approximately the same size, into two pools with equal numbers in each pool.

The fish in Pool A were fed on the new brand and those in Pool B were fed on the usual brand.

A month later the farmer measured the lengths of the fish in each pool.

The frequency polygon shows the results from Pool A.

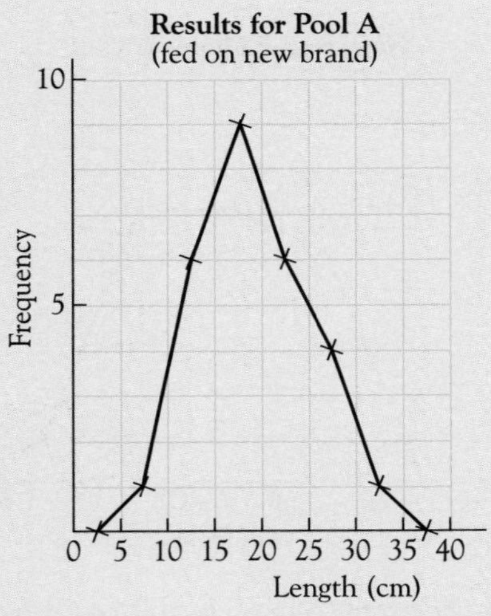

a Copy and complete the frequency table for fish in Pool A.

Length, l (cm)	Frequency, f
$0 \leqslant l < 5$	0
$5 \leqslant l < 10$	1
$10 \leqslant l < 15$	6
$15 \leqslant l < 20$	
$20 \leqslant l < 25$	
$25 \leqslant l < 30$	
$30 \leqslant l < 35$	
$35 \leqslant l < 40$	

b How many fish were in Pool A?

c Estimate the mean length of fish fed on the new brand.

The results for the fish in Pool B (fed on the usual fish food) were

Length, l (cm)	Frequency, f
$0 \leqslant l < 5$	0
$5 \leqslant l < 10$	2
$10 \leqslant l < 15$	3
$15 \leqslant l < 20$	4
$20 \leqslant l < 25$	6
$25 \leqslant l < 30$	7
$30 \leqslant l < 35$	5
$35 \leqslant l < 40$	0

d Copy the frequency polygon for Pool A and draw the frequency polygon for Pool B on the same diagram.

e By looking at the two polygons, decide which food you would recommend the farmer to use in future. Give your reasons.

f The mean length of fish fed on the usual brand is 22.7 cm. Does this value support your recommendation? Explain your answer.

6 The table gives information about the weight of the potato crop produced by 100 potato plants of two different types.

Weight of potatoes per plant (w kg)	Number of plants Type X	Number of plants Type Y
$0 \leqslant w < 0.5$	0	0
$0.5 \leqslant w < 1$	3	0
$1 \leqslant w < 1.5$	12	6
$1.5 \leqslant w < 2$	55	39
$2 \leqslant w < 2.5$	23	32
$2.5 \leqslant w < 3$	7	23
$3 \leqslant w < 3.5$	0	0

a Draw a frequency polygon for each type of potato.
Use a scale of 2 cm to 0.5 kg on the horizontal axis and 2 cm to 10 plants on the vertical axis.

b Which type of potato produces the heavier crop?

c i Which type of potato has more variation in the weight of the crop?

ii Give a reason for your answer. [SEG]

Worked Exam Question
[NEAB]

50 children take part in a sponsored spell to raise money for 'Comic Relief'. They are given 40 words to learn and are then tested on them. The results are shown in the table.

Correct spellings	Frequency
1 to 10	1
11 to 20	7
21 to 30	26
31 to 40	16

a Calculate the estimate of the mean number of spellings each child got correct.

- Extend the table on the examination paper by putting in a column for the mid-interval value and a column for $f \times x$
- Complete these columns and find Σf and Σfx

For the interval 1 to 10, mid-interval value $= \frac{1+10}{2} = 5.5$

Correct spellings	Frequency, f	Mid-interval value, x	f × x
1 to 10	1	5.5	5.5
11 to 20	7	15.5	108.5
21 to 30	26	25.5	663
31 to 40	16	35.5	568
	$\Sigma f = 50$		$\Sigma fx = 1345$

- Show the working for the mean

$$\bar{x} = \frac{\Sigma fx}{\Sigma f} = \frac{1345}{50} = 26.9$$

Answer ...26.9...

b For each word a child spells correctly they receive 5 p from their sponsor. Altogether £592 was raised by this sponsored spell. No person sponsored more than one child. Use your answer to part a to estimate the total number of people who sponsored the children.

- On average, each child spells 26.9 words correctly, so on average each child raises 26.9 × 5 p = 134.5 p
- Find how many lots of 134.5 p there are in £592.

Change £592 to pence.

Average amount raised per child = 26.9 × 5 = 134.5 p

59 200 ÷ 134.5 = 440.14 ...

Answer ...440 sponsors...

COMMENTS

M1 for attempting to use mid-interval values

M1

A1 for correct answer

3 marks

M1 Marks awarded for using your answer to **a** even if it was wrong

A1 440, 441 or 440.14... would score the mark

2 marks

Exam Questions

1 The values of 30 cars for sale in a showroom are given in the table.

Value (£)	Frequency	Mid-interval value	
0 and less than 2000	4	1000	
2000 and less than 4000	10		
4000 and less than 6000	7		
6000 and less than 8000	5		
8000 and less than 10000	4		
	30		

Use mid-interval values to calculate an estimate of the mean value of the cars. [MEG]

2 The graph shows the number of hours a sample of people spent viewing television one week during the summer.

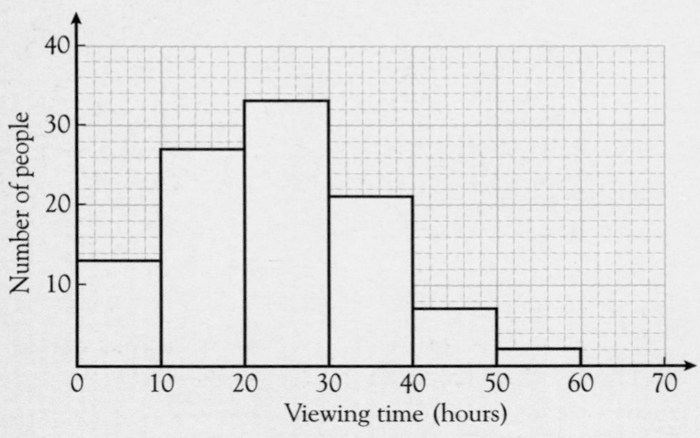

a Copy and complete the frequency table for this sample.

Viewing time h hours	Number of people
$0 \leqslant h < 10$	13
$10 \leqslant h < 20$	27
$20 \leqslant h < 30$	33
$30 \leqslant h < 40$	
$40 \leqslant h < 50$	
$50 \leqslant h < 60$	

b Another survey is carried out during the winter. State **one** difference you would expect to see in the data.

c Use the mid-points of the class intervals to calculate the mean viewing time for these people. [SEG]

3 The students at Loovilla College decided to have a biscuit eating competition.

A random sample of 25 students was taken.

The table shows the numbers of students eating different numbers of biscuits in four minutes.

Number of Biscuits eaten in 4 minutes	Mid-point	Frequency (Number of students)	
1–5		2	
6–10		8	
11–15		7	
16–20		5	
21–25		2	
26–30		1	
		25	

a Calculate an estimate of the mean number of biscuits eaten in 4 minutes.

b Write down the modal class interval.

c 250 students entered the competition. Estimate how many of them will eat more thar 20 biscuits in the four minutes. [L

4 The table below gives information about the expected lifetimes, in hours, of 200 light bulbs.

Lifetime (t)	Frequency
$0 < t \leqslant 400$	32
$400 < t \leqslant 800$	56
$800 < t \leqslant 1200$	90
$1200 < t \leqslant 1600$	16
$1600 < t \leqslant 2000$	6

a Mr Jones buys one of the light bulbs.
  i What is the probability that it will not last more than 400 hours?
  ii What is the probability that it will last at least 800 hours but not more than 1600 hours?

b Draw a frequency polygon to illustrate the information in the table. [MEC

5 Mrs Wilson wants to sell her herd of dairy cows. A buyer will need to know the herd's average dail yield of milk.

The daily milk yield, p litres, is monitored over 5 weeks.

The table below shows the results of this survey.

EXAM QUESTIONS

Milk yield (p litres)	Frequency (f)
$140 \leqslant p < 145$	3
$145 \leqslant p < 150$	5
$150 \leqslant p < 155$	9
$155 \leqslant p < 160$	6
$160 \leqslant p < 165$	8
$165 \leqslant p < 170$	4
Total	35

a Mrs Wilson finds the modal class for the daily yield. What is the value?

b Calculate the estimated mean daily milk yield.

c Which is the more suitable average for Mrs Wilson to use?
Give a reason for your answer. [NEAB]

One hundred people, selected at random, are weighed. The results are shown in the table.

Weight (w kg)	Frequency
$50 \leqslant w < 60$	7
$60 \leqslant w < 70$	49
$70 \leqslant w < 80$	29
$80 \leqslant w < 90$	15

Of the people weighed, 40 were females and the mean of their weights is 65.9 kg.
Calculate the mean weight of the males. [SEG]

A survey was carried out to find out how much time was needed by a group of pupils to complete homework set on a particular Monday evening.

Time, t hours, spent on homework	Number of pupils		
0	3		
$0 \leqslant t < 1$	14		
$1 \leqslant t < 2$	17		
$2 \leqslant t < 3$	5		
$3 \leqslant t < 4$	1		

Calculate an estimate for the mean time spent on homework by the pupils in the group. [L]

8 The weights, in kilograms, of 50 girls in year 11 at a school have been grouped into intervals of 10 kilograms as shown below.

Weight (w kg)	Frequency	Mid-interval	
$40 \leqslant w < 50$	8	45	
$50 \leqslant w < 60$	23		
$60 \leqslant w < 70$	10		
$70 \leqslant w < 80$	7		
$80 \leqslant w < 90$	2		

a Using the mid-interval values, calculate an estimate of the mean weight of these girls.

b Write down the maximum possible range of these weights.

c The mean weight of 50 boys in year 11 at the school is 74 kg with a range of 18 kg.
Is the person with the lowest weight of these 100 pupils a boy or a girl?
Explain your answer. [MEG]

9 Cole's sells furniture and will deliver up to a distance of 20 miles.

The diagram shows the delivery charges made by Cole's.

The table shows the information in the diagram and also the number of deliveries made in the first week of May 1994.

Distance (d) from Cole's in miles	Delivery charge in pounds	Number of deliveries			
$0 < d \leqslant 5$	15	27			
$5 < d \leqslant 10$	20	11			
$10 < d \leqslant 15$	25	8			
$15 < d \leqslant 20$	30	4			

a Calculate the mean charge per delivery for these deliveries.

b Calculate an estimate for the mean distance of the customers' homes from Cole's. [L]

In this chapter you will
- **learn how to factorise an expression by taking out common factors**
- **practise substituting into a formula**
- **learn how to re-arrange (transpose or change the subject of) a formula**
- **learn how to identify linear sequences and find the formula for the *n*th term**

Factorising expressions

Highest common factors

Work through these examples which remind you about **factors**.

page 174

a $15 = 3 \times 5$ 3 is a factor of 15 because it divides exactly into 15.
$21 = 3 \times 7$ 3 is also a factor of 21, so 3 is a **common factor** of 15 and 21.

In fact 3 is the **highest common factor** (H.C.F.) of 15 and 21 as it is the highest number that divides into both 15 and 21.

Writing each number as a **product of prime numbers** makes it easier to find the H.C.F. as you can then pair off the numbers when they appear in both.

b $12 = 2 \times 6 = 2 \times 2 \times 3$
$30 = 2 \times 15 = 2 \times 3 \times 5$

Only one of the twos for 12 has been paired off.

You can see that one 2 and one 3 have been paired off, so the **highest common factor** (H.C.F.) of 12 and 30 is $2 \times 3 = 6$.

c $24 = 2 \times 12 = 2 \times 2 \times 6 = 2 \times 2 \times 2 \times 3$
$40 = 2 \times 20 = 2 \times 2 \times 10 = 2 \times 2 \times 2 \times 5$

Notice that all the twos have been paired off this time.

The highest common factor of 24 and 40 is $2 \times 2 \times 2 = 8$.
This means that 8 is the highest number that divides exactly into both 24 and 40.

d $3p = 3 \times p$
$3q = 3 \times q$
$3r = 3 \times r$

The highest common factor of $3p$, $3q$ and $3r$ is 3.

e $4ab^2c = 2 \times 2 \times a \times b \times b \times c$
$6ca^2 = 2 \times 3 \times c \times a \times a$

The highest common factor of $4ab^2c$ and $6ca^2$ is $2 \times a \times c = 2ac$

FACTORISING

Exercise 14.1

Find the highest common factor of each of the following

1 60, 72
2 $4a$, $12c$
3 $3ab$, $12ac$
4 $15x$, 5
5 ca^2, $5ac$
6 $10a$, $12b$
7 $8a$, $12b$, $16ab$
8 $3x^2$, $4x$
9 $4xy$, $6x^2$
10 pq^2, $2pq$
11 x^3, $4x^2$
12 $8ab^2c$, $12a^2bc^2$

'Taking out' common factors

Reminder about multiplying out brackets:

i $4(c + 3d) = 4 \times c + 4 \times 3d = 4c + 12d$
Notice that 4 is a common factor of $4c$ and of $12d$

ii $x(x - 3) = x \times x - x \times 3 = x^2 - 3x$
x is a common factor of x^2 and of $3x$

If you are asked to **factorise** $4c + 2d$ you can do this by 'taking out' the common factor 4, writing it before the brackets:

$$4c + 12d = 4(c + 3d)$$

To factorise $x^2 - 3x$ 'take out' the common factor x

$$x^2 - 3x = x(x - 3)$$

You are reversing the process of expanding a bracket.

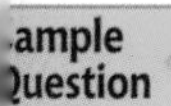
Example Question 1

Cliona and Lulu each did this question for homework:

'Factorise $4ab + 6bc$'

Cliona wrote $b(4a + 6c)$, Lulu wrote $2(2ab + 3bc)$
Who was correct? Explain your answer.

Answer

- Check each answer by expanding the brackets to see whether you obtain the original expression.

Cliona: $b(4a + 6c) = b \times 4a + b \times 6c$
$= 4ab + 6bc$, Cliona is correct.

Lulu: $2(2ab + 3bc) = 2 \times 2ab + 2 \times 3bc$
$= 4ab + 6bc$, Lulu is correct.

Both expressions have been factorised correctly, but neither is factorised completely, or **fully factorised**.

Look again at Cliona's answer.

The part in brackets can be factorised further, since $4a + 6c = 2(2a + 3c)$

So she could finish off her answer by writing

$$4ab + 6bc = b(4a + 6c)$$
$$= b \times 2(2a + 3b)$$
$$= 2b(2a + 3b)$$

Always look for more factors in the bracket. If you find any, take them out until there are no more. How could Lulu finish off her answer?

FACTORISING

This is now fully factorised as the terms in the bracket do not have a common factor.

One way of being sure of factorising completely is to take out the highest common factor of all the terms.

Sample Question 2

Factorise fully

a $pqr + r$

b $5xy^2 - 10x^2yz^2$

Answer

a

- Take out the common factor r

$$pqr + r = r(pq + 1)$$

Notice that you must put 1 in the bracket as $r \times 1 = r$. Do not leave it blank.

b

- Find the highest common factor of each term

$$5xy^2z = 5 \times x \times y \times y \times z$$

$$10x^2yz^2 = 2 \times 5 \times x \times x \times y \times z \times z$$

$$\text{H.C.F.} = 5 \times x \times y \times z = 5xyz$$

Take this common factor outside the bracket

$$5xy^2z - 10x^2yz^2 = 5xyz\,(\ldots - \ldots)$$

$$\underline{5xy^2z - 10x^2yz^2 = 5xyz\,(y - 2xz)}$$

To decide what goes inside the bracket, look at the factors left in each term when the common factors have been taken out.

Sometimes factors can be useful in working out numerical expressions without a calculator.

Sample Question 3

Re-write this expression using a bracket and work out the value **without using a calculator.**

$$15 \times 48 + 15 \times 12$$

Answer

- Take out the common factor 15

$$15 \times 48 + 15 \times 12 = 15(48 + 12)$$

$$= 15 \times 60$$

$$\underline{15 \times 48 + 15 \times 12 = 900}$$

$48 + 12 = 60$
$15 \times 60 = 150 \times 6 = 900$

FACTORISING

Exercise 14.2

Factorise the following expressions, by taking out the common factors:

1 $3x + 15$
2 $2x + 6$
3 $4x - 12$
4 $5a - 35$
5 $6x - 4$
6 $14y + 2$
7 $8x - 4$
8 $6m - 30$
9 $7a + 21$
10 $8x - 20$
11 $15p + 5$
12 $12x - 8$
13 $16 - 12a$
14 $9 + 12p$
15 $3a + 3b$
16 $10p - 12q$
17 $20x + 12y$
18 $8b - 6a$
19 $6m - 10n$
20 $18r + 12s$
21 $xy - 2x$
22 $pq + pr$
23 $ab + 4b$
24 $5a + ax$
25 $2pq + 3qr$
26 $2st - 4s$
27 $6xy - 3x$
28 $4ap + 6bp$
29 $10xy + 15xz$
30 $14pq + 7rs$
31 $x^2 + 2x$
32 $3x^2 - 3x$
33 $a^2 + 4a$
34 $y^2 - 18y$
35 $x^2 + xy$
36 $2x^2 + 4x$
37 $5p^2 - 10p$
38 $6a^2 + 8a$
39 $12y - 4y^2$
40 $3x - 9x^2$
41 $2x^2 + 6xy$
42 $8x^2 - 4ax$
43 $10a^2 - 4ab$
44 $15pq + 3q^2$
45 $18st - 2t^2$
46 $xy - 8y^2$
47 $7pq + 2q^2$
48 $p^2q + 4pq$
49 $3yz + xy^2$
50 $ab - a^2b$
51 $ac^2 + 2ac$
52 $x^2y - xy^2$
53 $a^2b^2 + 3ab$
54 $4p^2q + 2pq^2$
55 $12xy^2 - 6x^2y$
56 $3r^2s + 6s^2t$
57 $6ay^2 + 3by$
58 $5s^2t - st$
59 $16np + 2np^2$
60 $x^3 + x^2$
61 $x^3 - 7x$
62 $a^3 + 8a^2$
63 $4p^3 + 2pq$
64 $6x^3 - 3x^2$
65 $3ab - ac + 5ad$
66 $x^3 - 2x^2 + 5x$
67 $p^3 - p^2 - p$
68 $b^3 + b^2 - ab$

Rewrite each of the following expressions using a bracket and work out their values **without using a calculator.**

69 $19 \times 7 - 9 \times 7$
70 $13 \times 27 - 13 \times 17$
71 $36 \times 8 + 36 \times 12$
72 $58^2 - 58 \times 56$
73 $6.5^2 + 6.5 \times 3.5$
74 $4.2 \times 2.9 - 3.2 \times 2.9$
75 $76 \times 39 + 39 \times 24$
76 $236^2 - 36 \times 236$
77 $23.4 \times 16.6 + 23.4^2$
78 $5\frac{1}{2} \times 9 - 3\frac{1}{2} \times 9$
79 $6\frac{3}{7} \times 17 + 3\frac{4}{7} \times 17$
80 $9.65 \times 87.2 + 12.8 \times 9.65$

Factorise the right-hand side of the formulae given in questions 81 to 86.

81 $P = 2L + 2W$
82 $I = mv - mu$
83 $L = l + l\alpha t$
84 $s = \frac{1}{2}ut + \frac{1}{2}vt$
85 $A = \pi R^2 - \pi r^2$
86 $A = \frac{1}{2}ha + \frac{1}{2}hb$

87 The sketch shows the dimensions of the inside lane of a running track. The length of the track, L, and the area, A, inside the track are given by the following formulae:

$L = 2\pi r + 200 \qquad A = \pi r^2 + 200r$

Factorise each of these formulae.

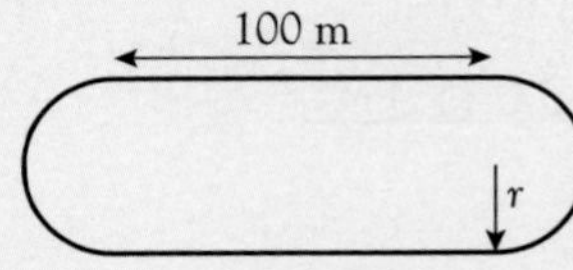

88 Factorise the formula for the total surface area of a cylinder:

$S = 2\pi r^2 + 2\pi rh$

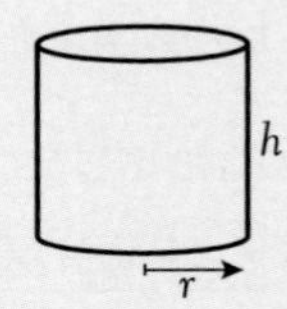

FORMULAE

Working with formulae

A formula shows the relationship between two or more variables. Remember that a formula must contain an equals sign.

Often when solving problems you need to work out the value of something (i.e. evaluate it) by using a formula. The following sample questions and exercise give you an opportunity to practise this technique.

Sample Question 4

$P = \dfrac{a^2 + b^2}{a + b}$ $\qquad Q = \sqrt{(b^2 - 4ac)}$

Use your calculator to find the values of P and Q when $a = 2.7$, $b = 8.2$, $c = -3.1$ [MEG]

Answer

- Substitute the values for a and b into the formula for P, then use your calculator to work out P

$$P = \frac{2.7^2 + 8.2^2}{2.7 + 8.2}$$

Write down the values you are using in the formula.

$$= \frac{74.53}{10.9}$$

$$= 6.8376\ldots$$

$$= \underline{6.8\ (1\text{ d.p.})}$$

Approximate sensibly, stating your degree of approximation and method. You could use decimal places or significant figures.

Calculator note

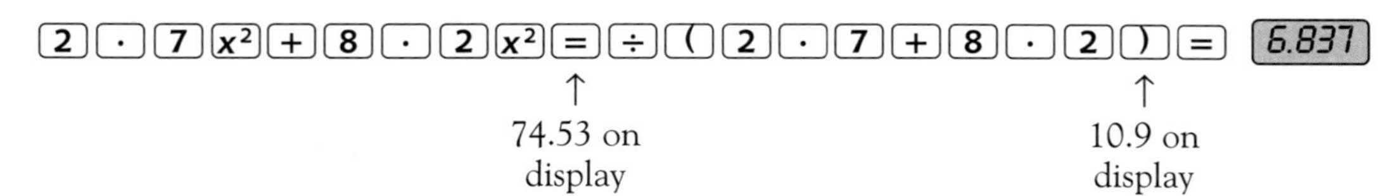

Alternatively you could find 2.7 + 8.2 first and store it in your calculator memory.

- Now substitute for a, b and c into the formula for Q

$$Q = \sqrt{b^2 - 4ac}$$

$$= \sqrt{8.2^2 - 4 \times 2.7 \times (-3.1)}$$

$$= \sqrt{100.72}$$

$$= 10.035\ldots$$

$$\underline{Q = 10.0\ (1\text{ d.p.})}$$

Calculator note

The order for keying in depends on your calculator model. On some you key in [√] first, on others at the end. Make sure you know what to do on **your** calculator. It could be one of these sequences.

[√] [(] [8] [·] [2] [x²] [−] [4] [×] [2] [·] [7] [×] [3] [·] [1] [+/−] [)] [=]

or [8] [·] [2] [x²] [−] [4] [×] [2] [·] [7] [×] [3] [·] [1] [+/−] [=] [√]

FORMULAE

Do not use a calculator for this question.

Sample Question **5** The formula used to find the focal length (f) of a lens is

$$\frac{1}{f} = \frac{1}{u} + \frac{1}{v}$$

where u is the distance of the object from the lens,
and v is the distance of the image from the lens.
Find the **exact** value of f when $u = 4$ and $v = 5$.

Answer

You cannot get f straight away from the formula so DO NOT start by writing $f = \dots$

- Substitute your known values into the formula.

$$\frac{1}{f} = \frac{1}{4} + \frac{1}{5}$$

- Add the fractions by changing them to equivalent fractions with the same denominator. The smallest number that can be divided by both 4 and 5 is 20.

$$\frac{1}{4} = \frac{1 \times 5}{4 \times 5} = \frac{5}{20}, \quad \frac{1}{5} = \frac{1 \times 4}{5 \times 4} = \frac{4}{20}$$

Chapter 2 page 23

$$\frac{1}{4} + \frac{1}{5} = \frac{5}{20} + \frac{4}{20} = \frac{9}{20}$$

DO NOT write 40, remember 5 twentieths and 4 twentieths is 9 twentieths.

$$\frac{1}{f} = \frac{9}{20}$$

- The reciprocal of $\frac{1}{f}$ is $\frac{f}{1}$, and the reciprocal of $\frac{9}{20}$ is $\frac{20}{9}$
Put these equal:

FACT SHEET 6: The Language of Number pages 174–176

Reciprocal of $\frac{a}{b}$ is $\frac{b}{a}$

$$\frac{f}{1} = \frac{20}{9}$$

$$f = \frac{20}{9} \times 1 = \frac{20}{9} = 2\frac{2}{9}$$

$2\frac{2}{9}$ is **exact.** If you had used your calculator it would give the answer as 2.222... which is not exact.

FORMULAE

Exercise 14.3

Do questions 1–8 without a calculator.

1 The cost of hiring a motor-boat consists of a fixed charge of £5 plus a variable charge of £4 per hour. The formula for the total cost in pounds is $C = 4h + 5$ where h is the number of hours for which the motor-boat is borrowed.

Find C when

a $h = 2$ **b** $h = 6$ **c** $h = 1\frac{1}{2}$ **d** $h = \frac{3}{4}$

2 A farmer keeps free-range hens and ducks. The number of eggs he expects to collect each week is given by the formula $E = 5h + 2d$ where h is the number of hens on the farm and d is the number of ducks.

Find E when $h = 10$ and $d = 8$

3 If an object is thrown up into the air with a velocity of 36 metres per second, its velocity (v m/s) after t seconds is given by the formula $v = 36 - 10t$

Find v when

a $t = 3$ **b** $t = 4$ **c** $t = 3.6$

4 The price charged by a removal firm depends partly on the total time taken to load and unload the van (t hours), and partly on the distance travelled between loading and unloading (d miles). The formula giving the total price, P in pounds is $P = 80t + 3d$

Calculate the value of P when

a $t = 2$ and $d = 50$ **b** $t = 3.5$ and $d = 12$

5 $s = \frac{1}{2}(u + v)t$

Work out the value of s when $u = 6$, $v = 8$ and $t = 10$

6 $v^2 = u^2 + 2as$

Work out the value of v when

a $u = 4$, $a = 3$ and $s = 8$

b $u = 10$, $a = -8$ and $s = 4$

7 The length, l cm, of the diagonal of this rectangle is given by the formula $l = \sqrt{a^2 + b^2}$
Find l when $a = 6$ cm and $b = 8$ cm

8 Using the formula $\frac{1}{f} = \frac{1}{u} + \frac{1}{v}$, find the exact value of f when

a $u = 2, v = 3$ **b** $u = 4, v = 12$

c $u = 3, v = -4$.

You may use a calculator *when necessary* in the following questions.

9 When goods are bought on credit, the monthly payment in pounds is $P = \frac{C - D}{n}$ where £C is the cost of the item (including interest), £D is the deposit paid and n is the number of months over which the payments are made.
Evaluate P when $C = 152$, $D = 32$ and $n = 12$

10 Jenny has a Saturday job at kennels. Her first task of the day is to feed the cats, dogs and rabbits. The time taken in minutes to fill the food bowls is given by the formula $T = 2c + 3d + r$ where c is the number of cats, d the number of dogs and r the number of rabbits. Find T when

a $c = 8$, $d = 9$ and $r = 6$

b $c = 15$, $d = 20$ and $r = 7$

11 The time taken to cook a chicken is $t = 22(2m + 1)$ minutes where m is the mass of the chicken in kilograms.
How long would it take (in hours and minutes) to cook chickens whose masses are given by:

a $m = 2$ **b** $m = 1.5$

12 Using the formula $y = x^2 + 3x - 2$ find the value of y when

a $x = 4$ **b** $x = 3.9$ **c** $x = -2.6$

13 The formula for the volume of a sphere is $\frac{4}{3}\pi r^3$ where r is the radius of the sphere. Calculate the volume of a sphere of radius 2.6 cm, giving your answer to 3 s.f. Use $\pi = 3.14$.

14 The area of the curved surface of this cone is given by the formula $A = \pi r\sqrt{h^2 + r^2}$
Find A when $r = 4$ cm and $h = 6$ cm
Use $\pi = 3.14$ and give your answer to 3 s.f.

15 $s = ut + \frac{1}{2}at^2$

Work out the value of s when

a $u = 3$, $t = 40$, $a = 4$

b $u = 6$, $t = 9$, $a = -\frac{1}{4}$

Re-arranging a formula

In chapter 1 you learnt that the formula for finding the circumference C of a circle with diameter d is $C = \pi d$. For a particular value of d, you can work out C.
C is known as the **subject** of the formula $C = \pi d$

If you know the circumference and want to work out the diameter you can re-arrange the formula, writing it 'backwards' as $d = \frac{C}{\pi}$. In this format, d is the subject.

The relationship can be shown in these flow diagrams.

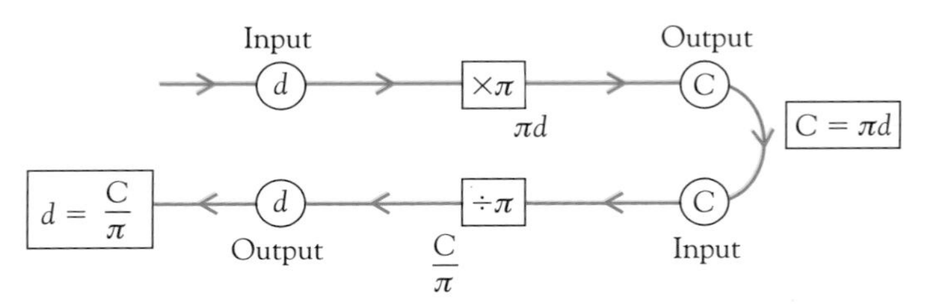

Re-arranging a formula is also referred to as **changing the subject** of the formula or **transposing** the formula.

Sample Question 6 The pupils in Steve's class went to the theatre. The total cost P, in £, of the visit depended on n, the number of children who went, according to the formula $P = 5n + 12$

a If 17 children went, what was the total cost?

b If the total cost was £152 how many children went?

Answer

a

- Substitute $n = 17$ into the formula

$$\begin{aligned} P &= 5n + 12 \\ &= 5 \times 17 + 12 \\ &= 97 \end{aligned}$$

<u>The total cost was £97</u>

You want to find P which is the subject of the formula.

b

- Substitute $P = 152$ into the formula

$$152 = 5n + 12$$

You want to find n but it is not the subject of the formula.

- Solve the equation to find n.

$$5n + 12 = 152$$

Subtract 12 $\quad 5n = 152 - 12$

$$5n = 140$$

Divide by 5 $\quad n = \frac{140}{5}$

$$n = 28$$

<u>28 pupils went to the theatre.</u>

FORMULAE

Alternatively, you can re-arrange the formula first so that n is the subject, then substitute the value for P.

$$5n + 12 = P$$

Subtract 12 $\quad 5n = P - 12$

Divide by 5 $\quad n = \frac{P - 12}{5}$

Now substitute $P = 152$ $\quad n = \frac{152 - 12}{5}$

$\underline{n = 28}$

It helps to re-write the equation so that the side that contains what you want is on the left.

Notice that you have followed exactly the same steps algebraically in the two methods. The difference is in when you substitute and when you can find particular numerical values.

The re-arrangement of the formula can be shown in these diagrams:

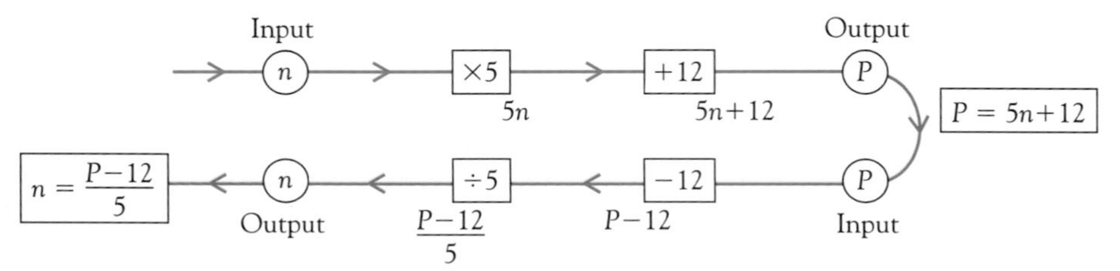

Sample Question 7

$y = \frac{\sqrt{A}}{3}$

a Find y when A = 36.

b Re-arrange the formula to **express A in terms of y.**

c Use your re-arranged formula to find A when y = 4.

This means make A the subject.

Answer

a

- Substitute A = 36 into the formula for y

$y = \frac{\sqrt{36}}{3}$

$y = \frac{6}{3}$

$\underline{y = 2}$

b

- Show on a flow diagram what is done to A to get y

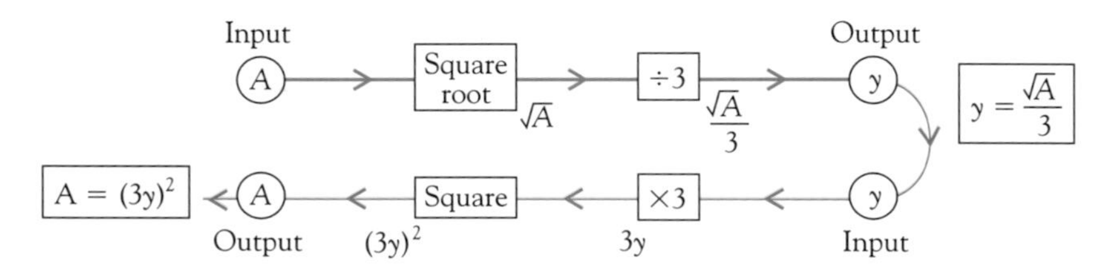

$A = (3y)^2$
$A = 3y \times 3y$
$\underline{A = 9y^2}$

Remember that when you square $3y$ you must find $3y \times 3y$

c Substitute y = 4 into the formula for A

$A = 9 \times 4^2$
$\underline{A = 144}$

FORMULAE

Sample Question 8 Re-arrange these formulae to express y in terms of x.

a $2x + 5y = 10$ **b** $x = 12 - 3y$ **c** $x = \frac{5}{y} + 3$

Answer

As neither letter is the subject, it is not helpful to use a flow diagram method.

a

- Re-arrange the equation as though you are solving it to find y, trying to get y on its own.

$$2x + 5y = 10$$

Subtract $2x$ $\quad 5y = 10 - 2x$

Divide by 5 $\quad y = \frac{10 - 2x}{5}$

DO NOT try to take this any further. This is as far as you can go.

b

- $x = 12 - 3y$

Add $3y$ $\quad 3y + x = 12$

Subtract x $\quad 3y = 12 - x$

Divide by 3 $\quad y = \frac{12 - x}{3}$

When the letter you want is in a negative term ($-3y$) then add it to both sides so that it will occur as a positive term in the equation.

Flow diagram method for b

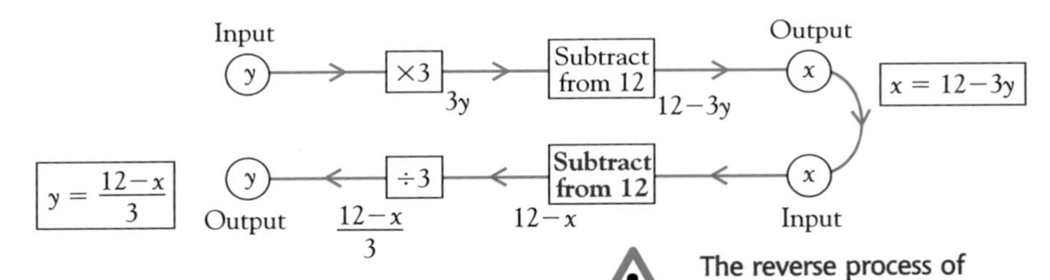

The reverse process of 'subtract from 12' is 'subtract from 12'.

c

- Get the term containing y on its own

$$x = \frac{5}{y} + 3$$

Subtract 3 $\quad x - 3 = \frac{5}{y}$

Multiply by y $\quad y(x - 3) = 5$

Divide by $(x - 3)$ $\quad y = \frac{5}{x - 3}$

You need to re-arrange so that y is not the denominator of a fraction.

Flow diagram method for c

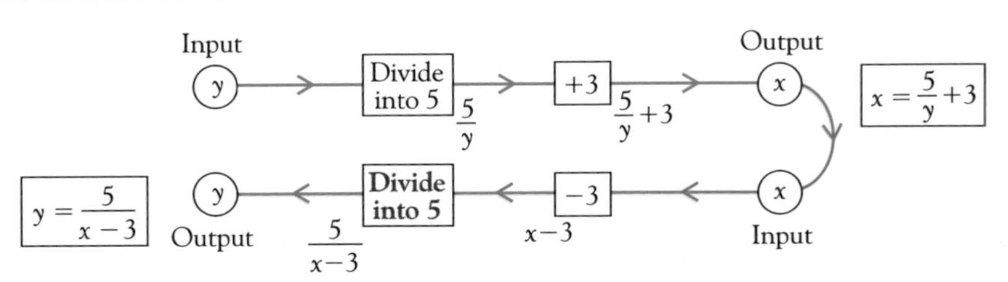

b and **c** are examples of special cases. Try to remember them

The reverse process of 'divide into 5' is 'divide into 5'.

FORMULAE

Sample Question 9 A rectangle has sides of length a and b. The perimeter P is given by the formula

$$P = 2(a + b)$$

Make a the subject of the formula.

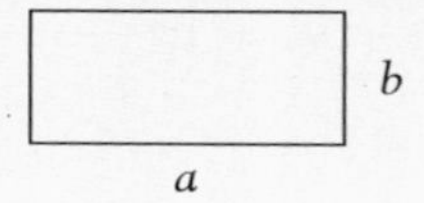

Answer

Method 1

$P = 2(a + b)$

Divide by 2 $\quad \dfrac{P}{2} = a + b$

Subtract b $\quad \dfrac{P}{2} - b = a$

$a = \dfrac{P}{2} - b$

Method 2

$P = 2(a + b)$

Expand brackets $\quad P = 2a + 2b$

Subtract $2b$ $\quad P - 2b = 2a$

Divide by 2 $\quad \dfrac{P - 2b}{2} = a$

$a = \dfrac{P - 2b}{2}$

The two answers are the same algebraically. To test this, substitute values for P and b to see whether you get the same value for a.

Exercise 14.4

1 A salesman earns a basic wage of £120 per week plus £7 for each order he takes.
His total wage in pounds is given by
$W = 7n + 120$ where n is the number of orders.

a Calculate W when
  i $n = 8$ ii $n = 20$

b Find the number of orders taken in a week when his total wage is:
  i £134
  ii £204

2 Imran buys a box of dirty garden ornaments for £9. He cleans them up and then sells them for £2 each. If he sells n ornaments, the profit in pounds is $P = 2n - 9$

a Find P if
  i $n = 15$
  ii $n = 4$

b How many ornaments would he need to sell for a profit of
  i £1
  ii £25

3 The amount of paint needed to cover a surface of area A m² with n coats is

$$V = \frac{nA}{13} \text{ litres}$$

a Find the amount needed to cover an area of 32.5 m² with 3 coats of paint.

b How many coats of paint would you be able to give a wall of area 16 m², using a 5-litre tin of paint?

4 Carl usually walks to work. Most of the route is along a road where he is often picked up by a friend who travels by car. The time taken for the journey, t minutes, depends on the distance he walks along the road, w km, before he is picked up.

The formula relating t and w is $t = 9w + 8$

a Re-arrange the formula to make w the subject.

b Find how far Carl walks along the road on days when his journey time is
  i 35 minutes
  ii three-quarters of an hour.

FORMULAE

5 When a plumber is called out on an emergency she charges a call-out fee of £50 and an additional £20 per hour. The total charge in pounds is given by the formula $C = 10(2t + 5)$ where t is the time in hours she spends on the job.

a Calculate the charges for emergency call-outs lasting

i 2 hours

ii 45 minutes

b Re-arrange the formula to make the subject t

c How long would the call-out time be for a job which cost £110?

6 A travel agent makes a charge of £15 for exchanging currencies. When he gives a customer K Norwegian kroner in exchange for £P he uses the formula $K = 12(P - 15)$

a Calculate K when $P = 350$

b By re-arranging the formula, calculate P when $K = 1800$.

7 Make x the subject of the formula in each of the following:

a $y = 4x - 5$

b $y = \frac{x}{4}$

c $y = 3(x + a)$

d $y = 4x^2$

e $y = 16 - 4x$

f $3y + 8x = 24$

g $y = \frac{12}{x}$

h $y^2 = x^2 - 4$

i $y = abx$

j $2y - x^2 = 20$

k $y = \frac{3x}{4}$

l $y = \sqrt{\frac{5x}{2}}$

8 David uses his car to deliver boxes of apples for a local farmer to the supermarket where they are sold. The farmer gives him £2 per box delivered and each return journey costs David £4 in petrol.

David's profit in pounds is $P = 2(B - 2J)$

Calculate
a P when $B = 32$ and $J = 4$
b B when $P = 20$ and $J = 2$
c J when $P = 34$ and $B = 23$

9 In motorway driving, my car travels 10 miles for each litre of petrol used. When I fill the tank at the service station it contains 60 litres of petrol.

The amount left, p litres, after travelling another m miles, is given by $p = 60 - \frac{m}{10}$

a How much petrol will be left 250 miles after the tank was filled?

b When there are 20 litres of petrol left in the tank, how many miles have been travelled since the tank was filled with petrol?

10 The total surface area, S, of a cuboid with a square base of side x and height h is given by the formula $S = 2x^2 + 4xh$

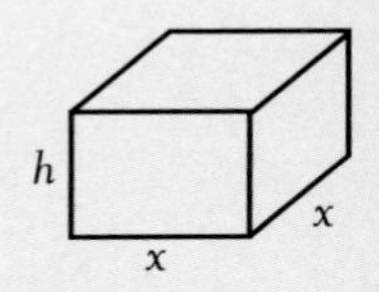

a Factorise $2x^2 + 4xh$ completely.

b Make h the subject of the formula $S = 2x^2 + 4xh$

11 Jason has £5 to spend and decides to buy some coloured pencils which cost 24 pence each. If he buys n pencils the amount he has left in pence is $p = 500 - 24n$

a Find how much he has left if he buys

i 10 pencils ii 18 pencils

b Re-arrange the formula to make n the subject.

c Find the number of pencils he buys if the amount left is £3.08.

12 In a tax system, the tax payable in pounds on a gross income of £G is $T = \frac{G - A}{5}$ where £A are allowances against tax.

a Calculate T when $G = 12\,000$ and $A = 4800$

b Find the formula for G

c Calculate G when $T = 1860$ and $A = 5250$

13 The formula for converting temperatures from F degrees Fahrenheit to C degrees Centigrade is:

$$C = \frac{5(F - 32)}{9}$$

a Evaluate C when

i $F = 59$ ii $F = 212$ iii $F = 14$

b By transposing the formula, calculate F when

i $C = 59$ ii $C = 95$ iii $C = -80$

Sequences

A **sequence** is a list of numbers, called **terms**, formed by using a rule. When trying to find the rule it is useful to look at the **difference**, *d*, between **consecutive terms**, i.e. terms that are next to each other.

a 3, 7, 11, 15, 19, …
(differences: +4 +4 +4 +4)

d is always 4: it is said to be **constant** as it does not change. The rule is 'add 4 to the term before', so the next term is 23.

b 1, 3, 6, 10, 15, …
(differences: +2 +3 +4 +5)

d is not constant, it is increasing by 1 each time, so the next term is $15 + 6 = 21$.

c 1, 3, 7, 15, 31, …
(differences: +2 +4 +8 +16)

d is not constant, it is doubling each time, so the next term is $31 + 32 = 63$.

d 99, 94, 89, 84, 79, …
(differences: −5 −5 −5 −5)

d is constant and is always -5. The rule is 'add -5 to the term before' which is the same as saying 'subtract 5 from the term before' so the next term is $79 + (-5) = 79 - 5 = 74$.

Linear sequences

In this chapter you will be investigating linear sequences.

In a **linear sequence** the difference, *d*, between consecutive terms is constant.

Examples **a** and **d** above are linear sequences, the others are not.

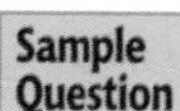

Sample Question 10 Find the 5th term of this linear sequence: 2, 8, 14, …

Answer

- Look at the difference between consecutive terms

  2, 8, 14, …
  (differences: +6 +6)

  $d = 6$, so you add 6 to the previous term to get the next term.

- Use the rule 'add 6'

  4th term $= 14 + 6 = 20$, 5th term $= 20 + 6 = \underline{26}$

There is a shorthand way of writing which term in the sequence you are referring to.

Write u_1, for the first term, u_2 for the second term and so on. For example, for the 25th term you would write u_{25}.

You can write a **general term**, called the **nth term**, u_n, where *n* is any number you want.

If you know the value of *d* it is easy to find the next term from the one before. It would, however, be very time consuming to work out, for example, u_{100} (the hundredth term). There is a way of finding **any** term in a sequence without knowing the previous term.

Calculating any term of a linear sequence

Look at this illustration. Consider a gardener erecting a fence using posts and identical panels.

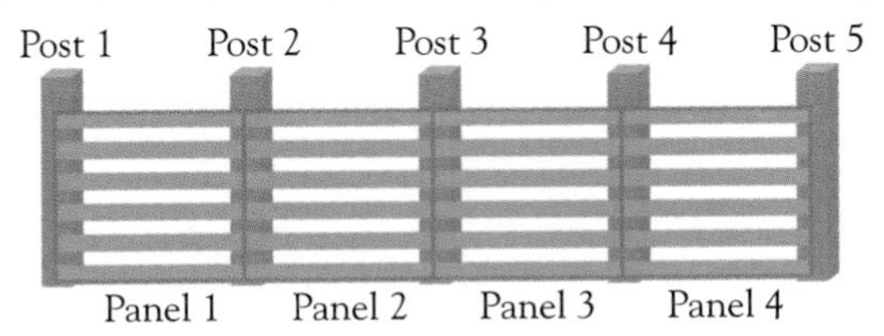

So far the gardener has used 5 posts and 4 fence panels. If he uses 8 posts he will need 7 panels, for 10 posts he will need 9 panels and so on. He always needs one less fence panel than the number of posts used.

If you think of the posts as being the terms and the panels as the constant difference, then you can work out any term in the sequence if you know the first term, u, and the difference, d.

For the sequence, 2, 8, 14, ... with first term $u_1 = 2$ and $d = 6$.

- to find the 8th term, u_8, add 7 lots of d to the first term

$$u_8 = u_1 + 7d$$
$$= 2 + 7 \times 6$$
$$= 44$$

$u_1 = 2$ u_2 u_3 u_4 u_5 u_6 u_7 u_8

+6 +6 +6 +6 +6 +6 +6

You can write a general formula to find any term, called the nth term:

- to find the nth term, add $(n - 1)$ lots of d to the first term

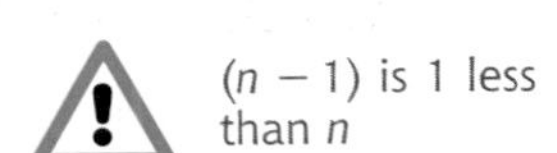
$(n - 1)$ is 1 less than n

T

In words: nth term = 1st term + $(n - 1) \times$ constant difference

In algebra: $u_n = u_1 + (n - 1)d$

This looks complicated, but is easy to use if you remember the posts and the fence panels! For example, in these linear sequences

a 3, 8, 13, 18, ... $u_1 = 3$ and $d = 5$

25th term, $u_{25} = 3 + 24 \times 5$

$u_{25} = 123$

24 is one less than 25

b 87, 83, 79, 75, ... $u_1 = 87$ and $d = -4$

10th term, $u_{10} = 87 + 9 \times (-4)$

$u_{10} = 51$

9 is one less than 10

Alternative way of writing a formula for the nth term, u_n of a linear sequence

Consider this linear sequence 7, 9, 11, 13, 15, ... with first term $u_1 = 7$ and $d = 2$.

Using
$$u_n = u_1 + (n - 1)d$$
$$u_n = 7 + (n - 1) \times 2$$
$$= 7 + 2(n - 1)$$
$$= 7 + 2n - 2$$
$$= 2n + 5$$

The formula for the nth term is $u_n = 2n + 5$

You can check that this is correct by substituting $n = 1$, $n = 2$, $n = 3$, etc. into the formula to see whether you get the first term, second term, third term, etc.
When $n = 1$, $u_1 = 2 \times 1 + 5 = 7$
When $n = 2$, $u_2 = 2 \times 2 + 5 = 9$
When $n = 3$, $u_3 = 2 \times 3 + 3 = 11$ and so on.

It is easy to write down the formula **straight away** if you notice two things:

- The difference $d = 2$ and the formula contains $2n$
- The $+5$ in the formula comes from $7 - 2$ which is the first term − difference i.e. $u_1 - d$

So another way of writing the formula for the nth term is

In words: $u_n = \text{difference} \times n + (\text{first term} - \text{difference})$
In algebra: $u_n = d \times n + (u_1 - d)$

Sample Question 11

a Find a formula for the nth term of the sequence 2, 8, 14, 20, …

b Use your formula to find the 10th term

c 104 is a number in the sequence. Which term is it?

Answer

a

- Find the constant difference d

2, 8, 14, 20, …
+6 +6 +6

$d = 6$

- Use the formula

$$u_n = \text{difference} \times n + (\text{first term} - \text{difference})$$

Wrting this all in algebra, with $u_1 = 2$ and $d = 6$

$u_n = d \times n + (u_1 - d)$
$u_n = 6 \times n + (2 - 6)$
$u_n = 6n + (-4)$
$\underline{u_n = 6n - 4}$

- Work out

first term − difference $= 2 - 6 = -4$

so $u_n = 6n + (-4)$
$\underline{u_n = 6n - 4}$

Check:

When $n = 1$, $u_1 = 6 \times 1 - 4 = 2$ correct
When $n = 2$, $u_2 = 6 \times 2 - 4 = 8$ correct

If you prefer to use the format for the nth term obtained by thinking about the 'posts' and 'fence panels' you would write

$u_n = u_1 + (n - 1)d$
$u_n = 2 + (n - 1) \times 6$
$u_n = 2 + 6n - 6$
$u_n = 6n - 4$

The resulting formula would be the same.

b

- Substitute $n = 10$ into the formula $u_n = 6n - 4$

$u_{10} = 6 \times 10 - 4 = 56$

<u>The tenth term is 56</u>

c

- Substitute 104 for u_n in the formula $u_n = 6n - 4$
- Solve the equation formed

$$104 = 6n - 4$$

Add 4 $\quad 108 = 6n$

Divide by 6 $\quad 18 = n$

104 is the 18th term of the sequence.

If you remember that the nth term = $d \times n$ + 'something' but forget how to find what to add, just try out the formula with, for example, $n = 1$.

Exercise 14.5

1 For each part of the question:

i Decide whether the sequence is linear.

ii If it is linear, find the next two terms and the 20th term.

a 12, 15, 18, 21, 24, …

b 60, 50, 40, 30, 20, …

c 5, 6, 8, 11, 14, …

d 1, 1.5, 2, 2.5, 3, …

e 20, 26, 32, 38, 46, …

f −5, −1, 3, 7, 11, …

g 2, 4, 8, 16, 32, …

h 1.4, 1.6, 1.8, 2, 2.2, …

i 2, 3, 5, 8, 13, …

j 800, 750, 700, 650, 600, …

2 For each part of the question, find the next two terms and a formula for the nth term:

a 2, 7, 12, 17, …

b 3, 7, 11, 15, …

c 22, 19, 16, 13, …

d 11, 4, −3, −10, …

e 20, 29, 38, 47, …

f −1, −5, −9, −13, …

g 4.5, 8, 11.5, 15, …

h $1, 1\frac{1}{4}, 1\frac{1}{2}, 1\frac{3}{4}, \ldots$

i 7, 6.8, 6.6, 6.4, …

3 The nth term of a sequence is $3n + 5$. Write down the first 4 terms.

4 The nth term of a sequence is $120 - 2n$. Write down the first 3 terms and the 10th term.

5 a i Write down the first five even numbers.

ii Find a formula for the nth even number.

b Find a formula for the nth odd number.

6 In each part of the question find a formula for the nth term of the linear sequence, then use your formula to find the 20th term and the 50th term.

a 1, 5, 9, 13, …

b 8, 11, 14, 17, …

c 17, 15, 13, 11, …

d 1, 1.4, 1.8, 2.2, …

e −5.5, −4, −2.5, …

f $-5\frac{1}{2}, -5\frac{1}{4}, -5, -4\frac{3}{4}, \ldots$

7 a Find the nth term of the sequence 1, 5, 9, 13, 17, …

b Which term of the sequence is 85?

8 a Find the nth term of the sequence 5, 2, −1, −4, −7, …

b Use your rule to find the 50th term.

c Which term of the sequence is −37?

9 a Write down the next two terms of the sequence which begins $\frac{1}{3}, \frac{2}{5}, \frac{3}{7}, \frac{4}{9}, \ldots$

b Write down a formula for the nth term of the denominator.

c Write down a formula for the nth term of the sequence.

Sample Question 12

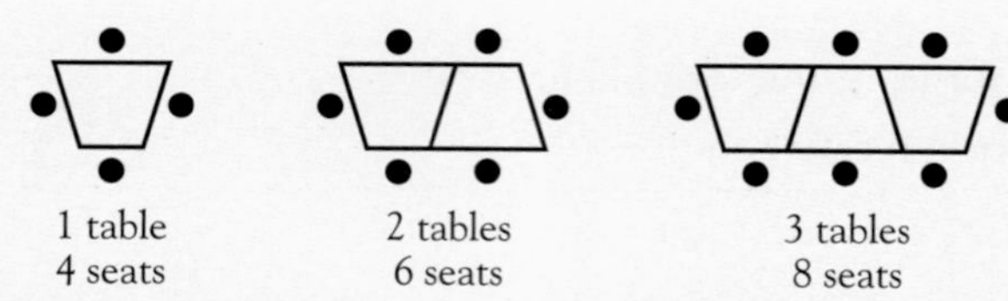

A special 4-sided table is used in the canteen. The diagrams show the number of students who can sit around the tables when they are joined in a line.

a Draw a diagram to show the seats for 4 tables placed in a line.

b If ten tables are put together, what is the maximum number of students that can sit around them?

c How many students can sit around n tables arranged in this way?

d 38 students are sitting at a row of tables. What is the smallest number of tables there could be?

e Write a formula connecting s, the number of seats and t, the number of tables

Answer

a

◆ Follow the pattern and draw the diagram.

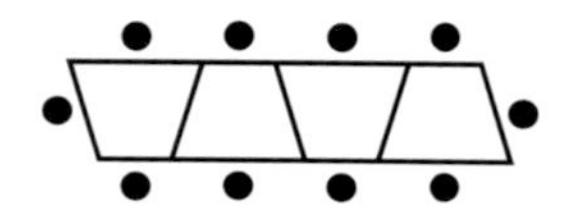

4 tables
10 seats

b

◆ Summarise the results in a table.

Number of tables (t)	1	2	3	4
Number of seats (s)	4	6	8	10
	+2	+2	+2	

$d = 2$

The number of seats forms a linear sequence 4, 6, 8, 10, …
The number of tables is the number of the term in the sequence.

◆ For 10 tables, find the 10th term in the sequence where

$$\begin{aligned} \text{10th term} &= \text{1st term} + 9 \times d \\ &= 4 + 9 \times 2 = 22 \end{aligned}$$

A maximum of 22 students can sit around 10 tables.

c

◆ Find the nth term, u_n, of the sequence using

$$\begin{aligned} u_n &= d \times n + (\text{first term} - \text{difference}) \\ &= 2n + (4 - 2) \\ &= 2n + 2 \end{aligned}$$

$2n + 2$ students can sit around n tables.

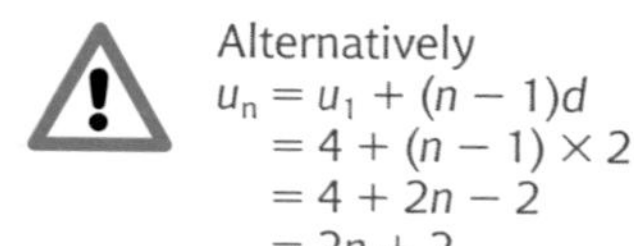

Alternatively
$u_n = u_1 + (n - 1)d$
$= 4 + (n - 1) \times 2$
$= 4 + 2n - 2$
$= 2n + 2$

d

◆ Substitute $u_n = 38$ into the formula and solve the equation

$38 = 2n + 2$

Subtract 2 $\quad 36 = 2n$

Divide by 2 $\quad 18 = n$

The smallest number of tables is 18

e

◆ The formula $\quad u_n = 2n + 2$
is the same as saying

number of seats = 2 × number of tables + 2

i.e. $\quad s = 2t + 2$

LINEAR PATTERNS

Sample Question 13 When Mrs Brown invites friends to tea, the number of sandwiches and cakes she makes depends on the number of guests she has invited, as indicated in the table below.

Number of guests (n)	1	2	3	4	5	and so on
Number of sandwiches (s)	7	10	13	16	19	
Number of cakes (c)	5	7	9	11	13	

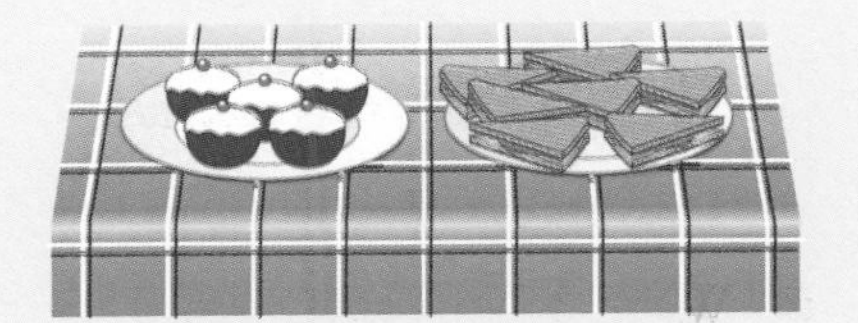

a Find a formula for s in terms of n.

b Find a formula for c in terms of n.

c Use your two formulae to calculate the number of sandwiches and cakes she made when she invited 12 friends.

Answer

a

- The number of sandwiches forms a linear sequence

7, 10, 13, 16, …

+3 +3 +3 $d = 3$

- You can use the relationship

number of sandwiches = 3 × (number of guests) + (first term − difference)

$s = 3n + (7 - 3)$

$\underline{s = 3n + 4}$

Check: When $n = 1$, $s = 3 \times 1 + 4 = 7$, which is correct.

b

- The number of cakes forms a linear sequence

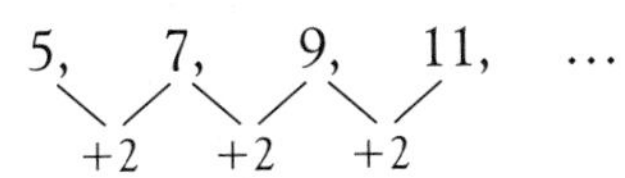

$d = 2$

- You can use the relationship

number of cakes = 2 × (number of guests) + (first term − difference)

$c = 2n + (5 - 2)$

$\underline{c = 2n + 3}$

Check: When $n = 1$, $c = 2 \times 1 + 3 = 5$, which is correct.

c

- Use $n = 12$ in both formulae

$s = 3n + 4$
$= 3 \times 12 + 4$
$= 40$

She makes 40 sandwiches.

$c = 2n + 3$
$= 2 \times 12 + 3$
$= 27$

She makes 27 cakes.

Note that to use this method

- the numbers in the top row of the table, which represent the number of the terms in the sequence, must start from 1 and increase consecutively, i.e. 1, 2, 3, 4, 5, …
- the numbers in the bottom row, which represent the values of the terms, must have a constant difference.

LINEAR PATTERNS

Exercise 14.6

1 Hexagons are used to make a sequence of patterns as shown.

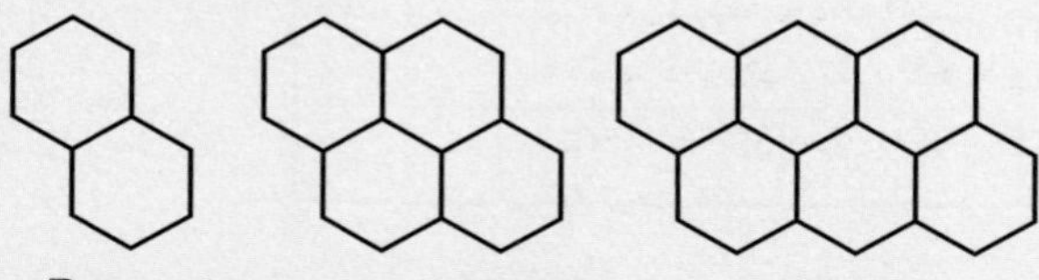

a How many outside edges has Pattern 8?

b Write down a rule to find the number of outside edges for Pattern n. [SEG]

2 Jen is planning a fruit garden. She intends to fill one area with apple trees and gooseberry bushes. Some of the possible combinations are:

Number of apple trees (n)	1	2	3	4	5
Number of gooseberry bushes (g)	14	12	10	8	6

a Find a formula for g in terms of n.

b How many gooseberry bushes would there be room for, if there were 7 apple trees?

c What is the maximum number of

i gooseberry bushes

ii apple trees?

3

Pattern 1

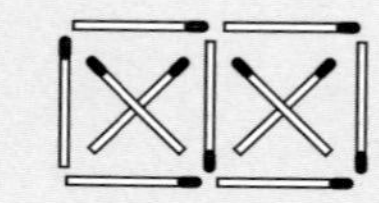
Pattern 2

Pattern 3

Matchsticks are laid out to make the patterns shown above.

a Copy and complete this table

Pattern	1	2	3	4	5
Matchsticks		11			

b How many matchsticks will there be in Pattern 10?

c Write down a formula for the number of matchsticks in the nth pattern.

d If 361 matchsticks are used, which pattern number is it?

e Using no more than 8 matchsticks for Pattern 1, design a new arrangement of matches. Continue the pattern and write a formula for the number of matches in the nth term of **your** sequence.

4 The cost of a taxi depends on the length of the journey as shown below:

No of miles (n)	1	2	3	4	5	6
Cost of taxi (£C)	2.40	3.20	4.00	4.80	5.60	6.40

a Find the linear relationship between C and n.

b Calculate the cost of journeys of length

i 9 miles

ii 3.5 miles

iii 5.7 miles

c Find the distance you could travel by taxi for

i £4.20

ii £7

5 When testing the boiler of a central heating system, the temperature of the water, T °C, is measured at minute intervals. The results are as follows:

Heating time (n minutes)	1	2	3	4	5	6
Temperature (T° C)	25	33	41	49	57	65

a Assuming a linear relationship, find a formula for T in terms of n.

b Use the formula to predict

i the temperature of the water after it is heated for $8\frac{1}{2}$ minutes

ii the time taken for the water to reach boiling point (100 °C).

6 The cost of camping in a farmer's field depends on the time spent there as shown in the table below:

Length of stay (n days)	1	2	3	4	5	6	7
Cost (£C)	9	13	17	21	25	29	33

a Assuming a linear relationship, find a formula for C in terms of n.

b Use the formula to find the cost of camping for

i 10 days

ii a fortnight.

Worked Exam Question

[SEG]

a **A sequence begins 2, 9, 16, 23, …**
i **What must be added to 23 to give the next number in the sequence?**
ii **Explain how you could use the first term in the sequence to find the 10th term without writing down all the terms.**

◆ Find the difference between the terms.

$d = 7$, the sequence is linear.

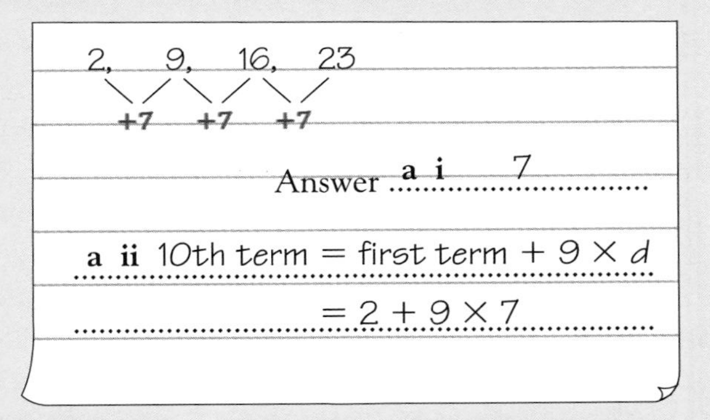

2, 9, 16, 23

+7 +7 +7

Answer **a i** 7

a ii 10th term = first term + 9 × *d*

= 2 + 9 × 7

COMMENTS

B1 for correct answer

B1 for correct statement

2 marks

b i **A new sequence begins 4, 10, 16, 22, …**
What is the 20th term in the sequence?
ii **Write an expression for the *n*th term in this sequence.**

◆ **i** Find the differences. If they are constant use the formula connecting first term and difference.

$d = 6$

20th term = first term + 19d

$= 4 + 19 \times 6$

Answer **b i** 118

M1

A1 for 118

◆ **ii** Use $u_n = d \times n +$ (first term − difference)

$u_n = 6n + (4 - 6)$

Answer **b ii** $u_n = 6n - 2$

B2 (1 mark for $6n$ or $6n$ + something or 'add 6 to previous term')

4 marks

Do not start by writing $n = \ldots$
Write $u_n = \ldots$ or nth term = …

Exam Questions

1 a Consider the sequence

$1, 4, 7, 10, 13, \ldots$

i Write down the 6th term.

ii Write down the nth term.

b i $s = ut + \dfrac{at^2}{2}$

Work out the value of s when $t = 0.4$, $u = 15$ and $a = -9.8$

Show all your working.

ii $x = \sqrt{p + q^2 - r^3}$

Work out the value of x when $p = 5, q = -7$ and $r = -3$

Show all your working. [MEG]

2 a The nth term of this sequence is $3n + 1$

Term	1	2	3	4	5
Sequence	4	7			

Write down the next three terms of the sequence.

b Another sequence starts

$6, 10, 14, 18, \ldots$

For this sequence write down

i the next three terms,

ii the nth term. [MEG]

3 Use the formula $A = \frac{1}{4}c\sqrt{4a^2 - c^2}$ to calculate the value of A given that $c = 7.23$ and $a = 8.76$

Give your answer correct to 1 decimal place. [L]

4 a Factorise $x^2 + 3x$

b Multiply out and simplify $3(2x - y) - 2y$

c Re-arrange the formula $F = 2C + 30$ to give C in terms of F. [SEG, p]

5 The price of a handtool of size S cm is P pence. The formula connecting P and S is $P = 20 + 12S^2$

a Calculate the price of a handtool of size 3 cm.

b Calculate the size of a handtool whose price is 95 p.

c Re-arrange the formula $P = 20 + 12\,S^2$ to express S in terms of P. [MEG]

6 Trevor rears turkeys to sell at Christmas. He estimates that the cost, £C, of rearing each turkey is given by the formula

$C = 0.1W + 0.05\,W^2$

where W kilograms is the weight of a turkey.

a What is the estimated cost of rearing a 9-kilogram turkey?

b He advises his customers that they should cook their turkeys using the formula

$T = 40\,W + 20$

where T minutes is the cooking time for each turkey.

i Re-arrange the formula to make W the subject.

ii Anne uses this formula and cooks her turkey for 7 hours.

What is the weight of the turkey which Anne cooks? [NEAB]

7 Sheep enclosures are built using fences and posts. The enclosures are always built in a row.

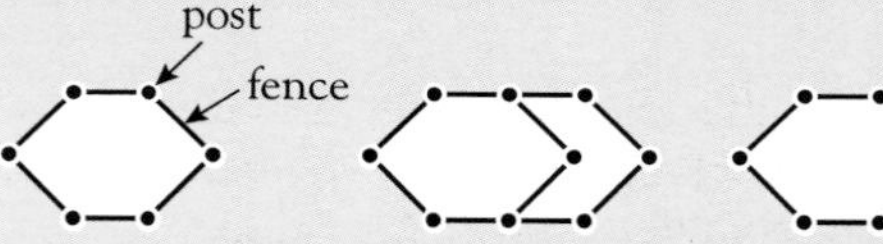

a i Sketch four enclosures in a row.

ii Sketch five enclosures in a row.

b Copy and complete the table below.

Number of enclosures	1	2	3	4	5	6	7	8
Number of posts	6	9	12					

c Work out the number of posts needed for 20 enclosures in a row.

d Write down an expression to find the number of posts needed for n enclosures in a row. [L]

8 The formula $f = \dfrac{uv}{u + v}$ is used in the study of light.

a Calculate f when $u = 14.9$ and $v = -10.2$. Give your answer correct to 3 significant figures.

b By rounding the values of u and v in part **a** to 2 significant figures, check whether your answer to part **a** is reasonable.
Show your working. [MEG]

EXAM QUESTIONS

9 a Factorise completely $12p^2q - 15pq^2$

b The cost, C pence, of printing n party invitations is given by $C = 120 + 40n$. Find a formula for n in terms of C. [MEG, p]

10 Factorise completely $4x^2 - 6x$ [L, p]

11 The length of a man's forearm (f cm) and his height (h cm) are approximately related by the formula $h = 3f + 90$

a Part of the skeleton of a man is found and the forearm is 20 cm long. Use the formula to estimate the man's height.

b A man's height is 162 cm. Use the formula to estimate the length of his forearm.

c George is 1 year old and he is 70 cm tall. Find the value the formula gives for the length of his forearm and state why this value is impossible.

d Use the formula to find an expression for f in terms of h. [MEG]

12 Joe is making a pattern by surrounding black equilateral triangles with white equilateral triangles.

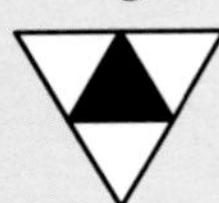
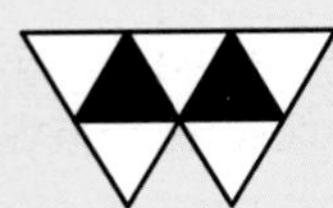
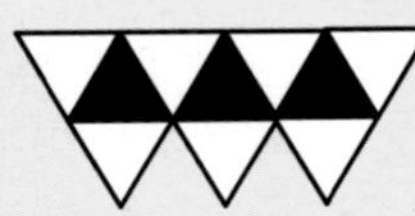

a Write down a general rule for the number of white triangles, w, in terms of the number of black triangles, b, in the pattern.

Joe has 92 white triangles.

b What is the greatest number of black triangles that he can surround in the pattern? [L]

13 The air temperature, T °C, outside an aircraft flying at a height of h feet is given by the formula

$$T = 26 - \frac{h}{500}$$

An aircraft is flying at a height of 27 000 feet.

a Use the formula to calculate the air temperature outside the aircraft.

The air temperature outside an aircraft is −52 °C.

b Calculate the height of the aircraft. [L]

14 a A 4x B; 2x + 7; D C — Not to scale

$ABCD$ is a rectangle in which $AB = 4x$ cm and $BC = 2x + 7$ cm

i Express the area of the rectangle in terms of x.

ii Express the perimeter of the rectangle in terms of x, giving your answer in its simplest form.

b The length of the perimeter, P cm, of an equilateral triangle is given by the formula

$$P = 6y - 9$$

i Find the value of P when $y = 3.5$

ii Express y in terms of P.

iii Factorise $6y - 9$

iv Express the length of one side of the equilateral triangle in terms of y. [MEG]

15 A rocket is fired vertically upwards with velocity u metres per second.
After t seconds the rocket's velocity, v metres per second, is given by the formula

$$v = u + gt$$

a Calculate v when $u = 100$, $g = -9.8$ and $t = 5$

b Re-arrange the formula to express t in terms of v, u and g.

c Calculate t when $u = 93.5$, $g = -9.8$ and $v = 20$ [NEAB]

16 A teacher read out to her class:

'P equals four point five one times nine Y plus three cubed'.

Sue wrote down:

$$P = 4.51 \times 9Y + 3^3$$

a If $Y = -2$, work out Sue's value for P.

John wrote down:

$$P = 4.51 \times (9Y + 3)^3$$

b If $Y = -2$, work out John's value for P. [MEG]

AREA AND VOLUME II

In this chapter you will
- **learn how to find the area of a parallelogram and trapezium**
- **practise finding areas in mixed problems**
- **learn how to find the volume of a prism**
- **learn how to find the surface area of a solid from its net**

FACT SHEET 3 page 69

Area of a parallelogram

A **parallelogram** is a quadrilateral with two pairs of parallel sides. To find the area of a parallelogram you need to know the length of one of the sides (call it the 'base') and the perpendicular distance between it and the side parallel to it (call this the 'height').

You can use any side as the base, but remember that the height must be the **perpendicular** distance between that side and the side parallel to it.

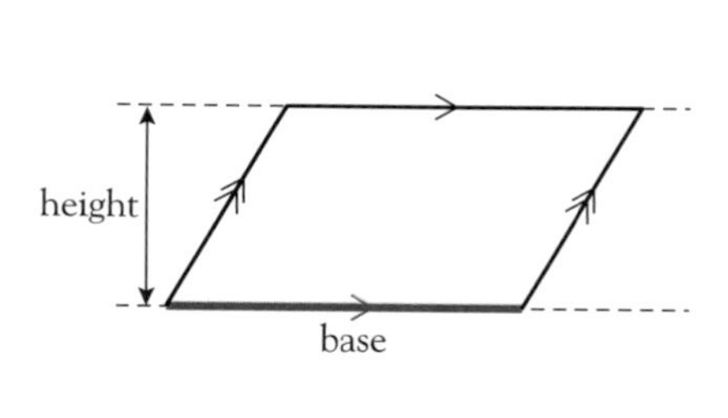

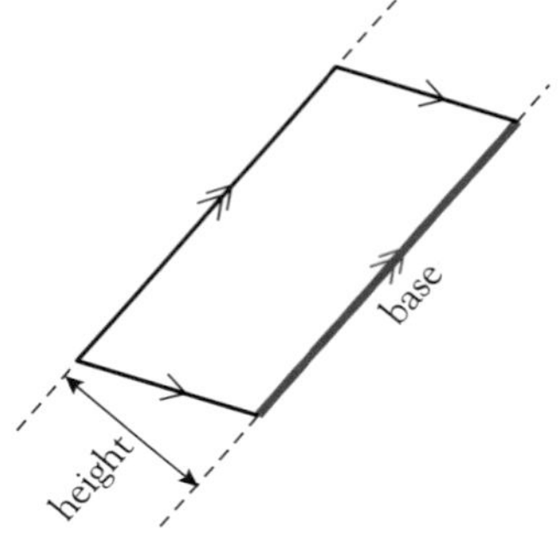

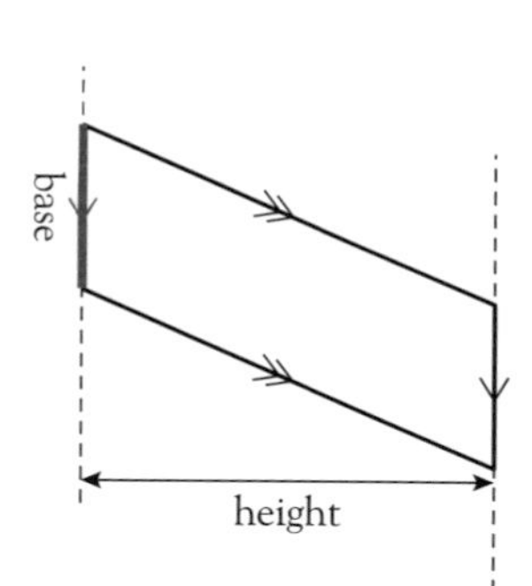

To work out the formula for the area of a parallelogram, draw in a diagonal. This splits the parallelogram into two equal triangles.

page 104

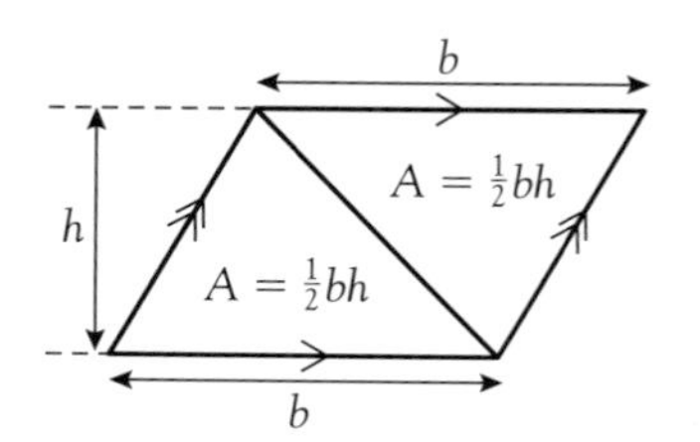

Area of one triangle $= \frac{1}{2}bh$

Area of two triangles $= \frac{1}{2}bh + \frac{1}{2}bh = bh$

Area of parallelogram $= bh$

Area of a parallelogram = base × perpendicular height
$= b \times h$

MORE AREAS

Area of a trapezium

A trapezium is a quadrilateral with just one pair of parallel sides.
To find the area of a trapezium you need to know the lengths of the parallels, call them a and b, and the **perpendicular distance** between the parallels, call it h.

FACT
-IEET 3
age 69

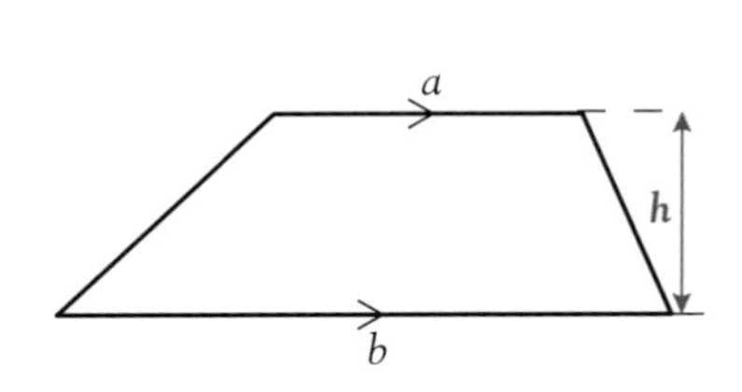

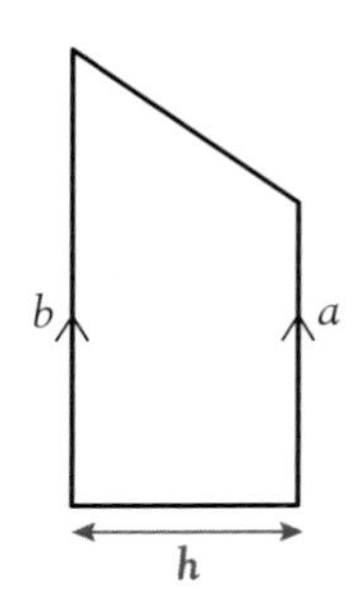

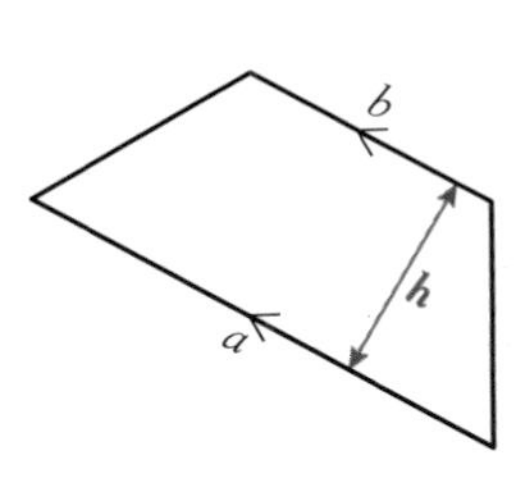

To work out the formula for the area of a trapezium, draw in a diagonal. This splits the trapezium into two different sized triangles.

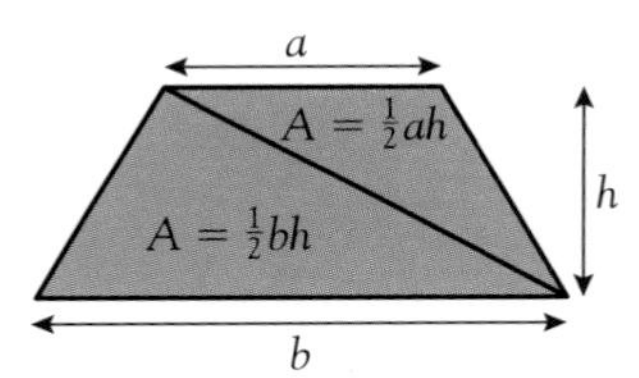

$$\text{Area} = \tfrac{1}{2}ah + \tfrac{1}{2}bh$$
$$= \tfrac{1}{2}\mathbf{h}(a + b)$$
$$= \tfrac{1}{2}(\mathbf{a} + \mathbf{b})\mathbf{h}$$

Take out $\frac{1}{2}h$ as a factor.

The formula is often written in this order to make it easier to remember.

Area of a trapezium
$= \frac{1}{2}$(sum of parallel sides × perpendicular distance between them)
$= \frac{1}{2}(a + b)h$

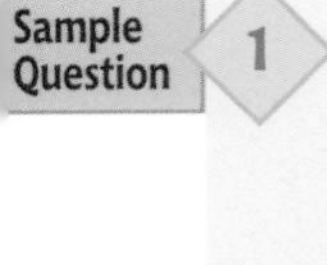

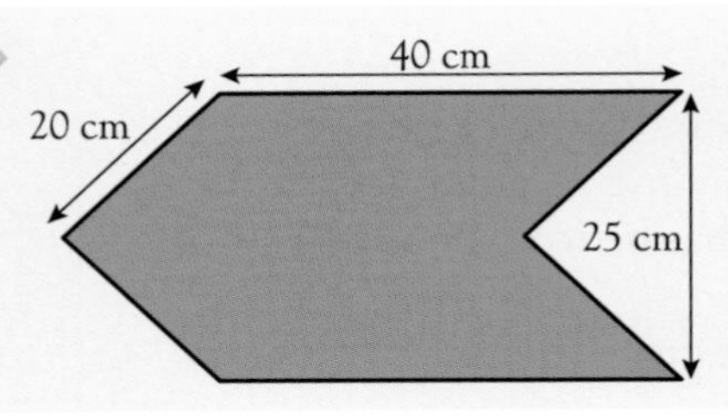

This arrow head is formed from two identical parallelograms.

Find the area of material used to make it.

Answer

- Sketch one of the parallelograms and put in the base and height.

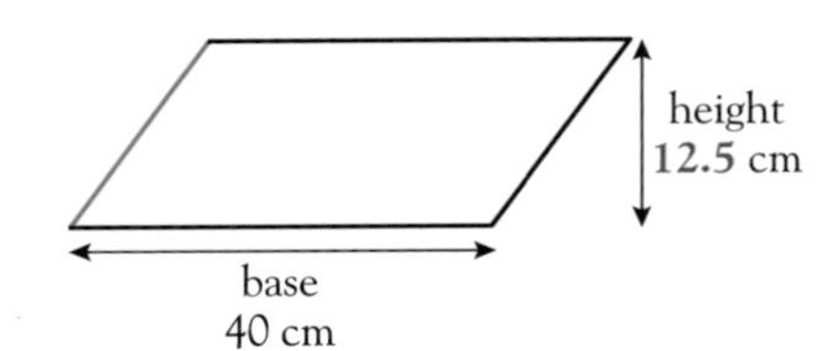

DO NOT use the slant height of 20 cm.

$25 \div 2 = 12.5$

MORE AREAS

- Find the area of the parallelogram, then double it to find the area of the arrowhead.

$$\text{Area of parallelogram} = 6 \times h = 40 \times 12.5 = 500\ \text{cm}^2$$

$$\text{Area of arrowhead} = 2 \times 500$$

$$\underline{\text{Area} = 1000\ \text{cm}^2}$$

Sample Question 2 The trapezium has the same area as the parallelogram. Find the length of the side marked x.

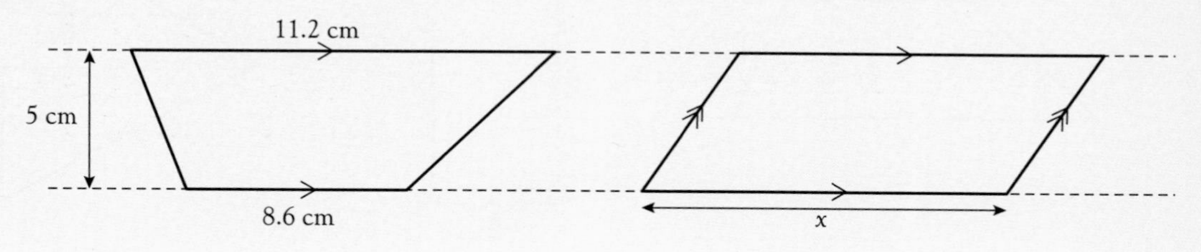

Answer

- Find the area of the trapezium

$$A = \tfrac{1}{2}(a + b)h = \tfrac{1}{2}(11.2 + 8.6) \times 5 = 49.5\ \text{cm}^2$$

There are several ways of doing this on the calculator, including

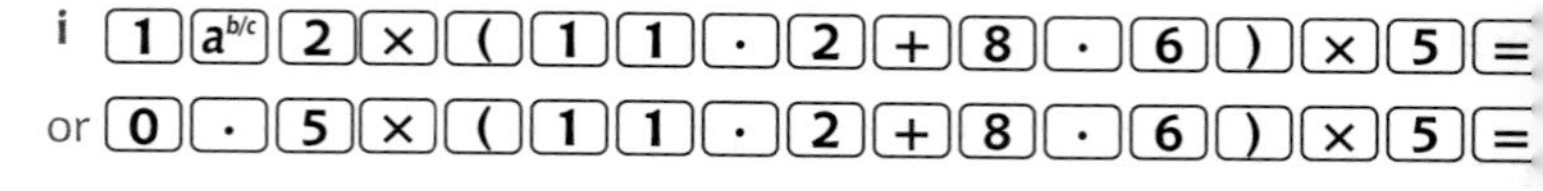

- Put this answer equal to the area of a parallelogram with base x and height 5

$$\text{Area} = \text{base} \times \text{height}$$
$$49.5 = x \times 5$$
$$x = 49.5 \div 5$$
$$\underline{x = 9.9\ \text{cm}}$$

Divide the area by 5 to find x.

Sample Question 3 Mr Wilkins had a circular lawn which he wanted to treat with fertilizer. The instructions on the packet stated 'use 25 grams for every square metre of lawn'. He worked out that he would need 1 kg. What was the radius of his lawn?
Give your answer, in m, correct to the nearest cm.
Take $\pi = 3.14$.

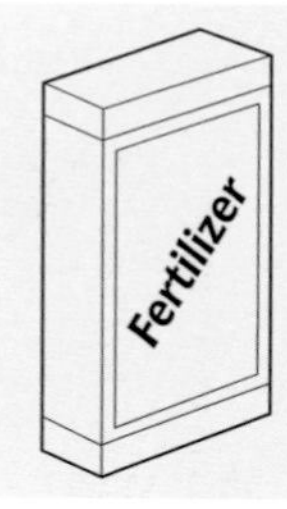

Answer

- Work out how many lots of 25 g there are in 1 kg.
This will give the number of square metres that can be treated for 1 kg and so will tell you the area of the lawn.

$$1000 \div 25 = 40$$

Area of lawn $= 40\ \text{m}^2$

1 kg = 1000 g

MORE AREAS

- Put this equal to the area of a circle with radius r, then solve the equation to find r.

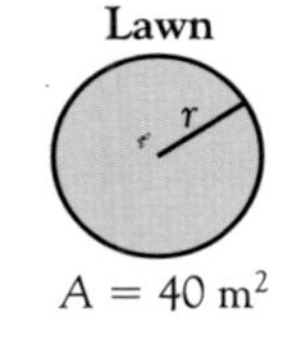

$$\pi r^2 = 40$$

$$r^2 = \frac{40}{\pi}$$

$$r = \sqrt{\frac{40}{3.14}}$$

$$= 3.569\ldots$$

Radius = **3.57 m** (to the nearest cm)

Make sure you can do this on **your** calculator.
Models differ. You may have to do this

[4] [0] [÷] [3] [·] [1] [4] [=] [√]

or [4] [0] [÷] [3] [·] [1] [4] [=] [√] [=]

or [√] [(] [4] [0] [÷] [3] [·] [1] [4] [)] [=]

r = 3 m 57 cm (to the nearest cm)

xercise 15.1

1 this exercise, use $\pi = 3.14$.

Find the shaded areas.

a

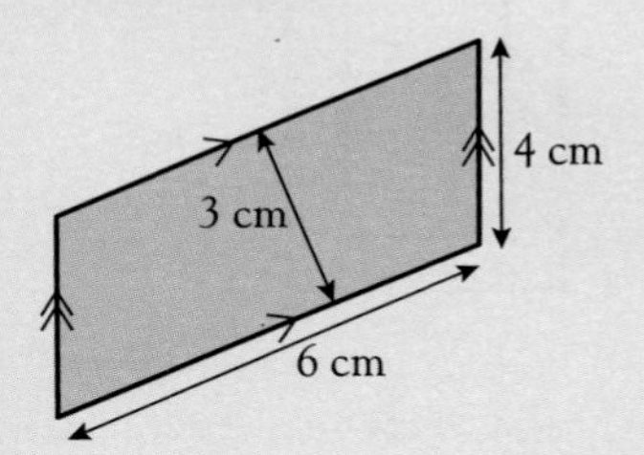

d

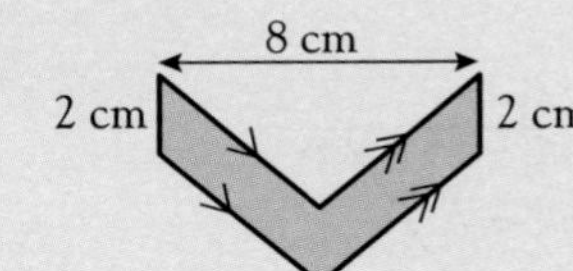

b

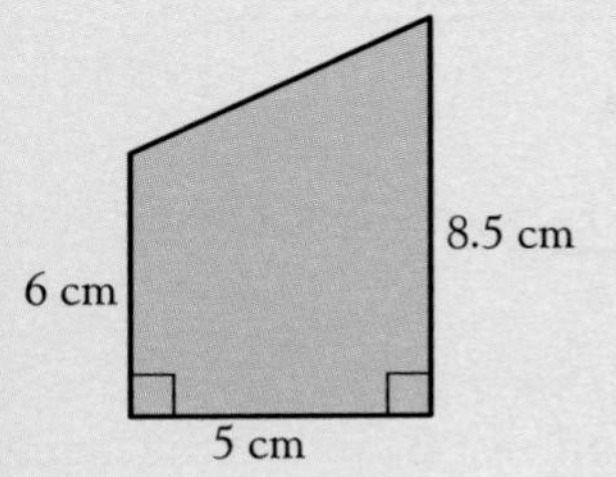

e

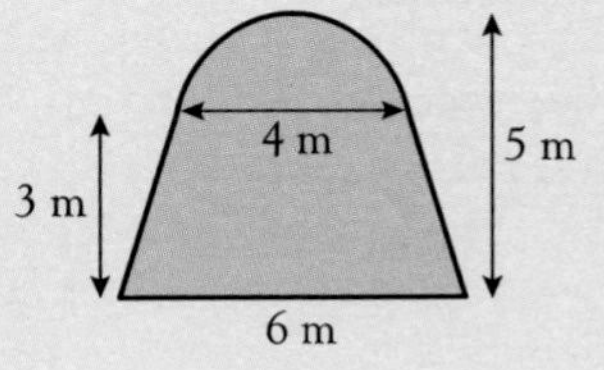

c

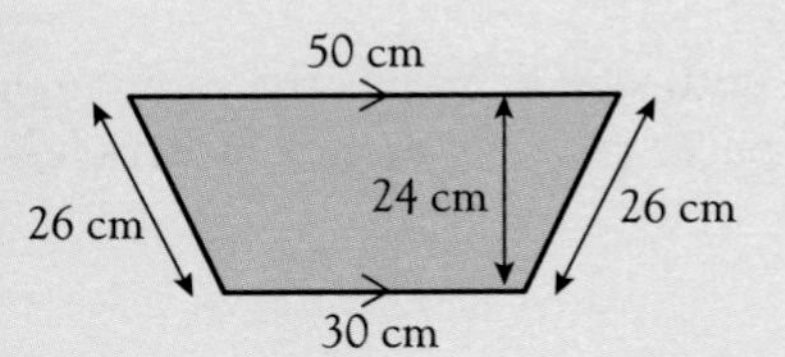

f

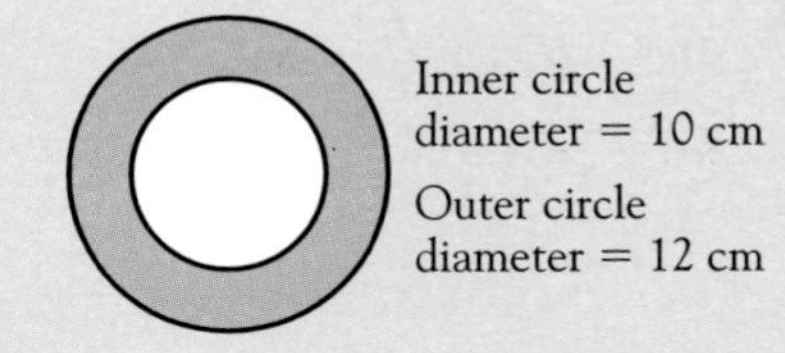

MORE AREAS

2 On graph paper, draw a grid using a scale of 1 cm to 1 unit from -6 to 6 on both axes.

a Plot each of the following shapes on the grid by joining the points with straight lines.

Shape 1: $(-2, -3)$, $(2, 1)$, $(-2, 5)$, $(-2, -3)$

Shape 2: $(2, -2)$, $(6, -2)$, $(5, 3)$, $(3, 3)$, $(2, -2)$

Shape 3: $(-5, 1)$, $(-3, 2)$, $(-3, -1)$, $(-5, -2)$, $(-5, 1)$

Shape 4: $(2, 4)$, $(3, 6)$, $(-2, 6)$, $(1, 4)$, $(2, 4)$

Shape 5: $(-6, -6)$, $(-2, -6)$, $(4, -4)$, $(1, -4)$, $(-6, -6)$

b Write down the mathematical name of each shape.

c Find the area of each shape.

3 The perimeter of this trapezium is 36 cm.

a Find x.

b Find the area of the trapezium.

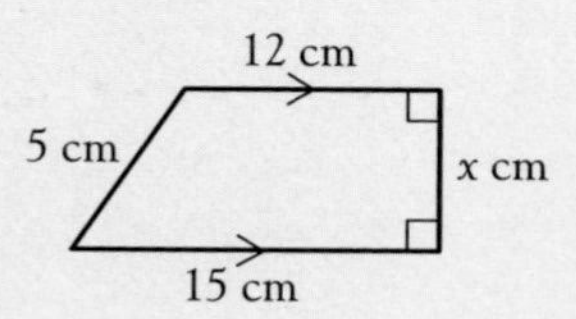

4 a Find h.

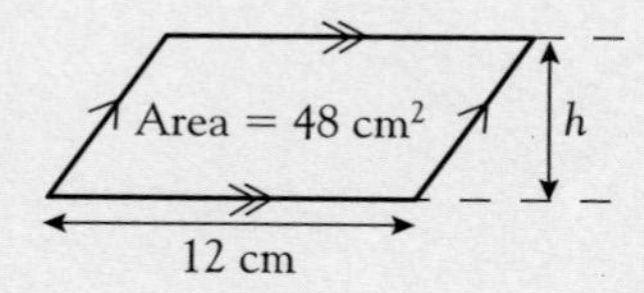

b Find x.

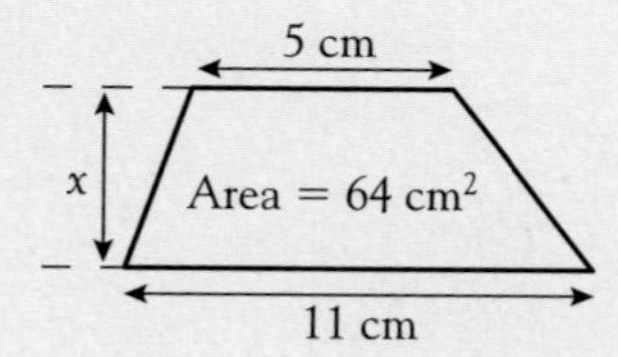

c Find a.

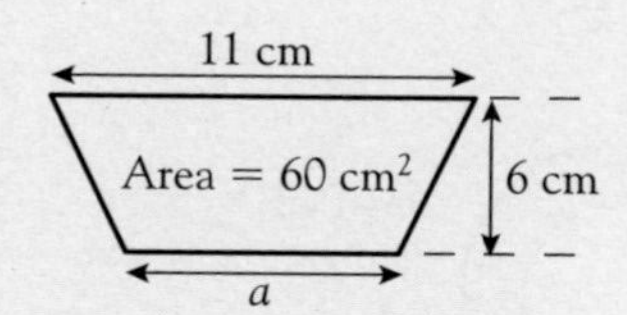

d Find the radius of the circle.

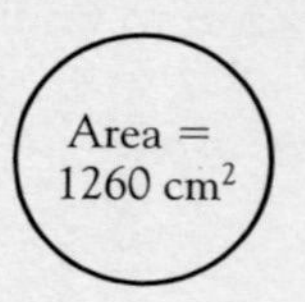

e Find y.

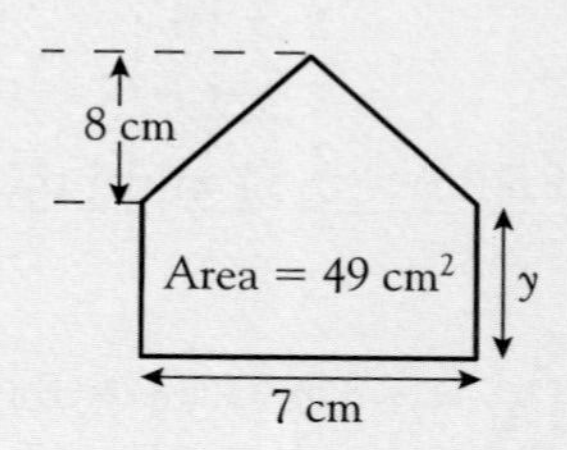

5 Find the area of these mirror frames.

a

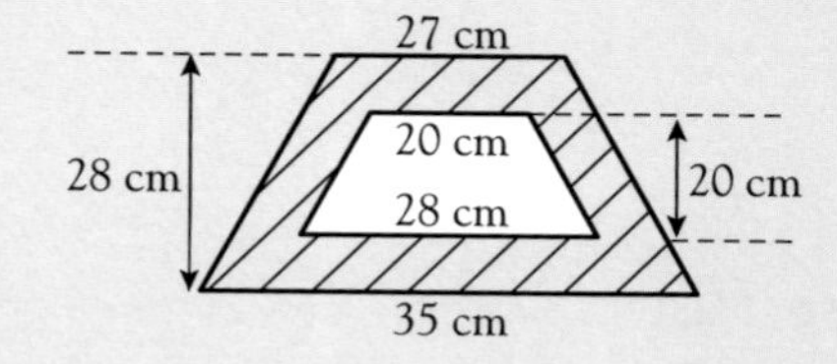

b

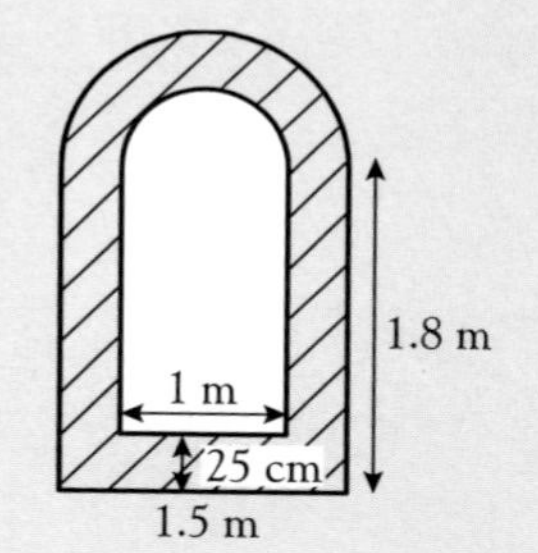

6 This is the shape of one of the walls of a swimming pool. Find its area.

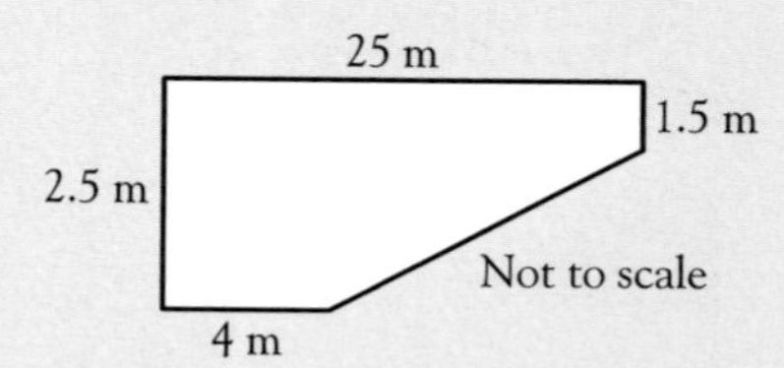

7 a A circle with diameter 8 cm has the same area as a square. What is the length of the side of the square?

b A square with side 8 cm has the same area as a circle. What is the diameter of the circle?

VOLUMES OF PRISMS

Volume of a prism

A **prism** is a 3-D shape (called a solid) that has the **same cross-section** running through it. You can see the cross-section by looking at one of the two identical 'ends' of the prism.

In these examples of prisms, the cross-section has been shaded.

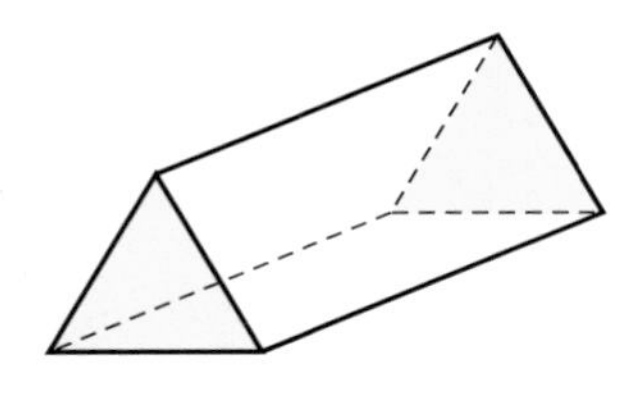
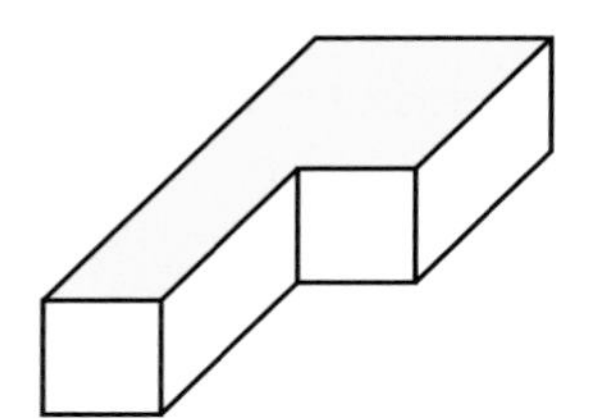
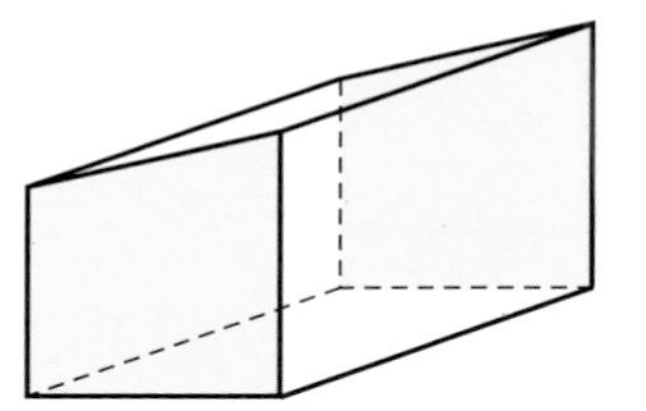
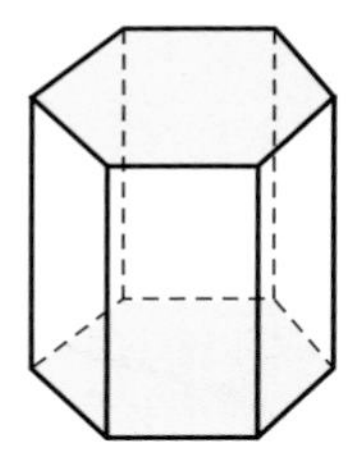

You could say that each prism has a 'uniform' or 'constant' cross-section.

The prism is not necessarily 'standing' on its cross-section.

The distance between the two end cross-sections is usually called the **length** of the prism, though it is sometimes referred to as the **height** or **thickness** of the prism.

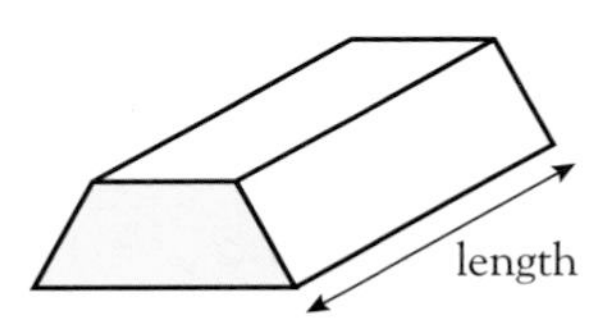

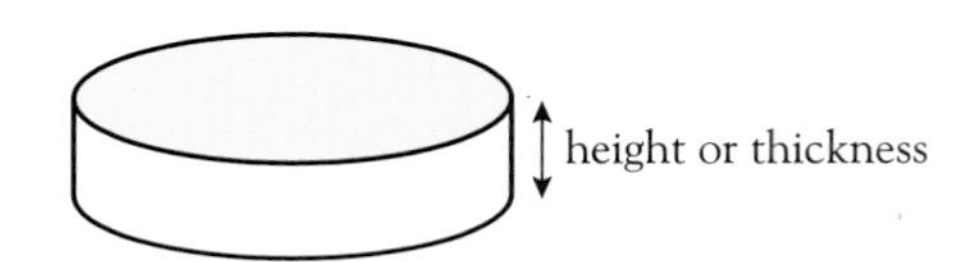

The volume of a prism = area of the cross-section × length

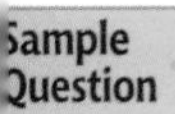

Sample Question 4

The diagram shows a lean-to shed in the shape of a prism, standing on a horizontal base 2.5 m by 5 m. The walls of the shed are vertical and the cross-section of ABCD is a trapezium.

AB = 3 m, BC = 2.5 m and CD = 2 m

Calculate

a the area, in m^2, of the cross-section ABCD,

b the volume of the shed.

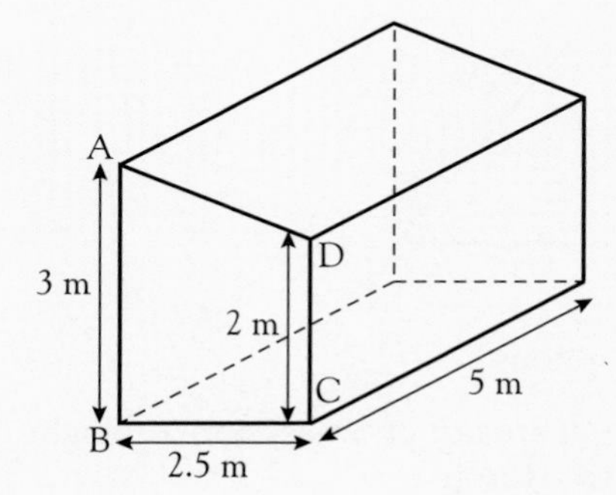

VOLUMES OF PRISMS

Answer

a

- The cross-section is a trapezium. Sketch the cross-section and work out its area.

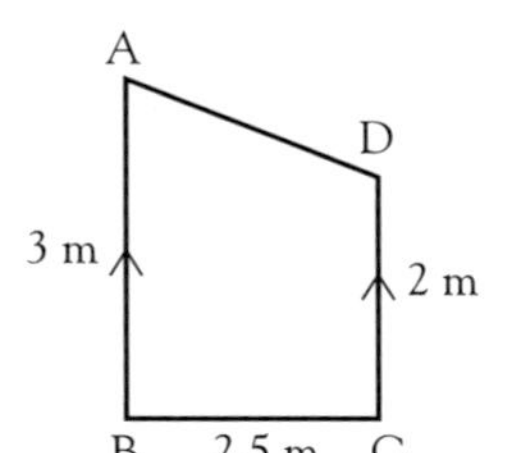

$$\text{Area} = \tfrac{1}{2}(a + b)h$$
$$= \tfrac{1}{2} \times (3 + 2) \times 2.5$$
$$\underline{\text{Area} = 6.25\ \text{m}^2}$$

You could split this trapezium into a rectangle and a triangle, but it is quicker to work out the area in one step.

b

- Find the volume using the formula for the volume of a prism.

$$\text{Volume} = \text{area of cross-section} \times \text{length}$$
$$= 6.25 \times 5$$
$$= 31.25$$
$$\underline{\text{Volume of shed} = 31.25\ \mathbf{m^3}}$$

Remember that volume is given in cubic units.

Sample Question 5 Find the volume of this cardboard packaging sleeve.

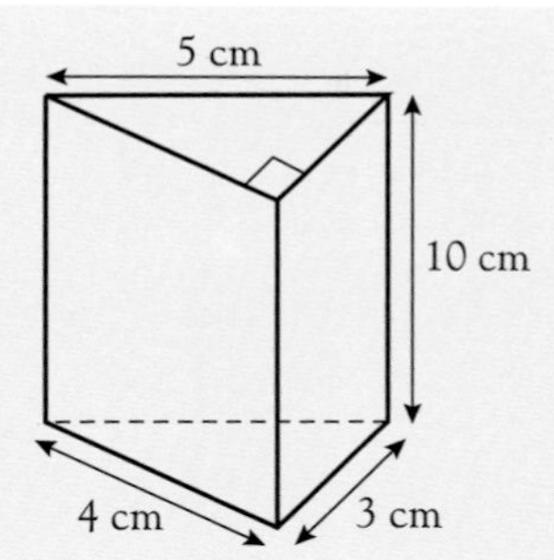

Answer

- The cross-section is a right-angled triangle. Sketch it and work out its area.

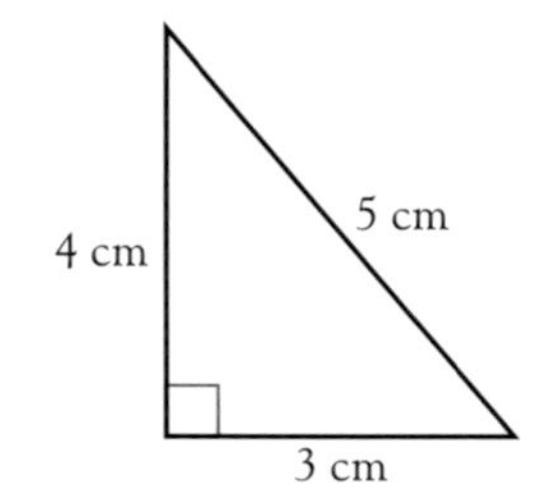

$$\text{Area} = \frac{b \times h}{2}$$
$$= \frac{3 \times 4}{2}$$
$$\text{Area} = 6\ \text{cm}^2$$

When you sketch the triangle, draw the 90° angle properly. This helps you to see the base and height of the triangle.

- Work out the volume.

$$\text{Volume} = \text{area of cross-section} \times \text{height of prism}$$
$$= 6 \times 10$$
$$\underline{\text{Volume} = 60\ \text{cm}^3}$$

Volume of a cylinder

A **cylinder** is a prism with a circle as cross-section.

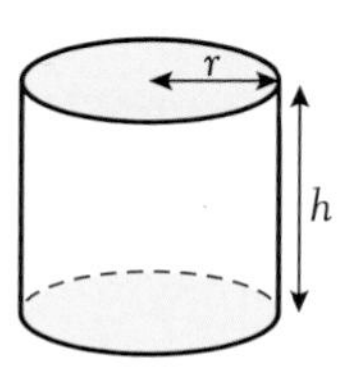

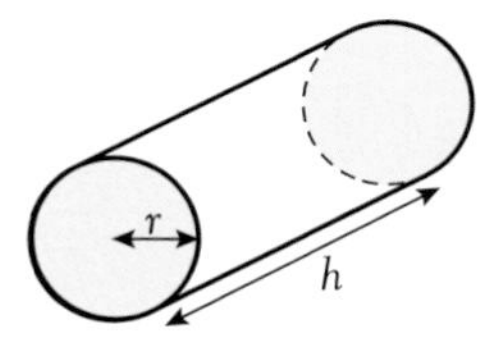

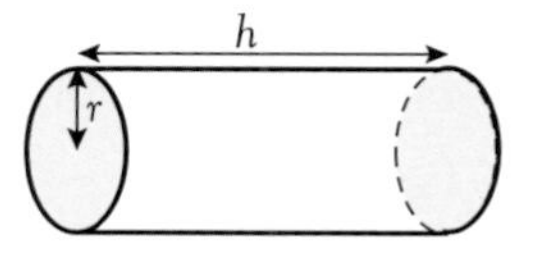

age 107

Area of cross-section $= \text{area of a circle}$
$= \pi r^2$

Volume of a cylinder $= \text{area of cross-section} \times \text{height}$
$= \pi r^2 \times h$
$= \pi r^2 h$

Cross-section

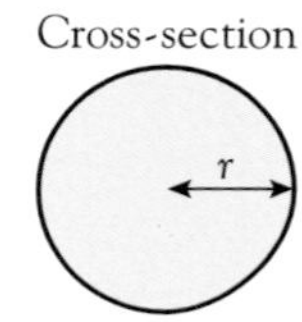

Volume, V, of a cylinder with radius r and height h is

$$V = \pi r^2 h$$

You can use this formula to find the volume of a cylinder in one step.

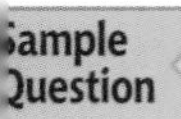

6 'Bradley's Soup' is canned by a small family business. Each morning they make 200 litres of soup. This is put in cylindrical tins, each of which is 8.4 cm high and has a diameter of 7 cm.

How many of these tins can be filled from the 200 litres of soup?

[NEAB]

Answer

- Find the radius of the circular cross-section by dividing the diameter by 2.

  $r = 7 \div 2 = 3.5$ cm

- Find the volume, in cm^3, of 1 tin.

  $V = \pi r^2 h$
  $= \pi \times 3.5^2 \times 8.4$
  $= \mathbf{323.26} \ldots \text{ cm}^3$

Use $\boxed{\pi}$ on your calculator or use 3.14 or 3.142

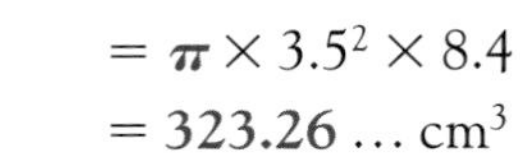

Store 323.26... in the memory of your calculator. There is no need to approximate it.

- Change 200 litres to cm^3

  $200 \times \mathbf{1000} = 200\,000$

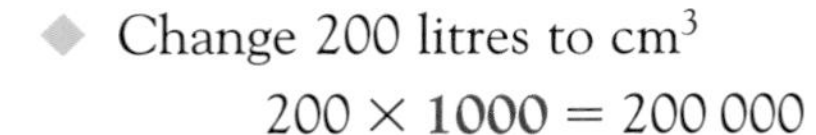

1 litre = 1000 cm^3
Learn this.

- Divide 200 000 by the volume of 1 tin to find how many tins can be filled.

  $200\,000 \div 323.26 \ldots = 618.67 \ldots$

- Round this number **down** to find the number of complete tins that can be filled.

  618 tins can be filled.

VOLUMES OF PRISMS

Sample Question 7

A child's paddling pool has a circular cross-section and vertical sides.

The radius of the pool is 1.5 m and it holds 4 m^3 of water when full.

Find the depth of water in the pool when it is full.

Use $\pi = 3.14$.

[MEG]

Answer

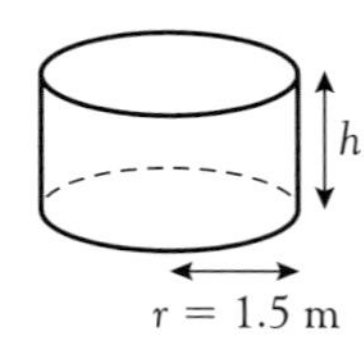

The pool is a cylinder with $r = 1.5$ m, $V = 4\text{ m}^3$ and depth h.

- Write out the formula for the volume of a cylinder, then substitute in the known values.

$$V = \pi r^2 h$$
$$4 = \mathbf{3.14 \times 1.5^2} \times h$$
$$4 = 7.065 \times h$$

Work out $3.14 \times 1.5^2 = 7.065$

- Solve the equation to find h.

$$h = \frac{4}{7.065}$$
$$= 0.5661 \ldots \text{ m}$$
$$= 56.61 \ldots \text{ cm}$$
$$= \mathbf{57 \text{ cm (nearest cm)}}$$

It is sensible to give the answer in cm, to the nearest cm.

The depth of water is 57 cm.

Sample Question 8

Jelly beans are sold in tubes which are cylinders of radius 3 cm and height 12 cm.

a Calculate the volume of one of the tubes.

Tubes of jelly beans are packed into a carton in the shape of a cuboid measuring 60 cm by 30 cm by 12 cm.

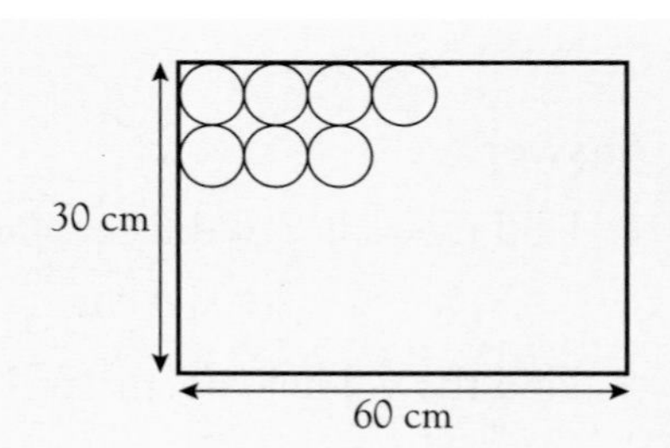

b i How many tubes may be packed into a carton?

ii Find the volume of empty space in a carton filled with tubes of jelly beans. [MEG]

Answer

a

- Find the volume of a cylinder with $r = 3$, $h = 12$

$$V = \pi r^2 h$$
$$= 3.14 \times 3^2 \times 12$$

Volume = 339.12 cm^3

VOLUMES OF PRISMS

b i

- The diameter of a tube is 6 cm.
  Find how many tubes fit along the 60 cm length and how many along the 30 cm length

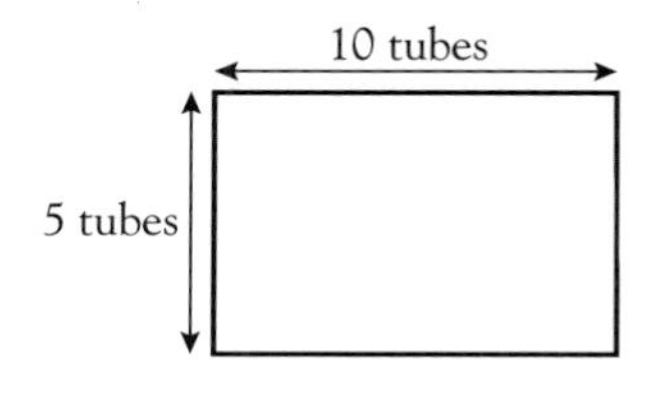

$60 \div 6 = 10, \quad 30 \div 6 = 5$

Number of tubes $= 10 \times 5 = \underline{50}$

ii

- Subtract the volume of 50 tubes from the volume of the box.

Volume of box $= 60 \times 30 \times 12 = 21\,600\ \text{cm}^3$

Volume of 50 tubes $= 50 \times 339.12 = 16\,956\ \text{cm}^3$

Volume of empty space $= 21\,600 - 16\,956$

$\underline{\text{Volume} = 4644\ \text{cm}^3}$

Exercise 15.2

Use $\pi = 3.14$.

1 Find the volume of each of these prisms

a

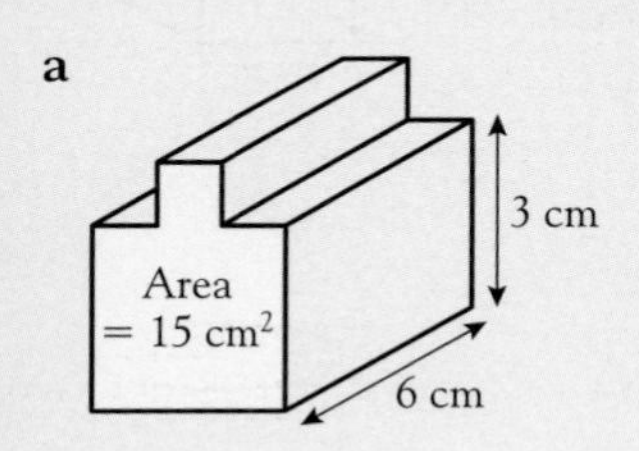

b

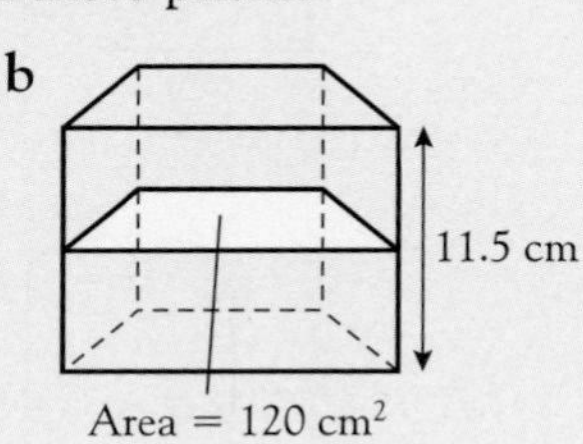

2 Find the volume of these prisms. It is usually helpful to draw the cross-section first and find its area.

a

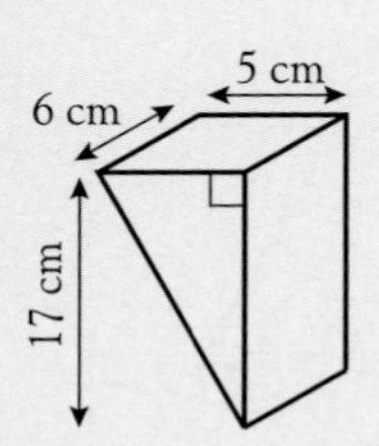

b

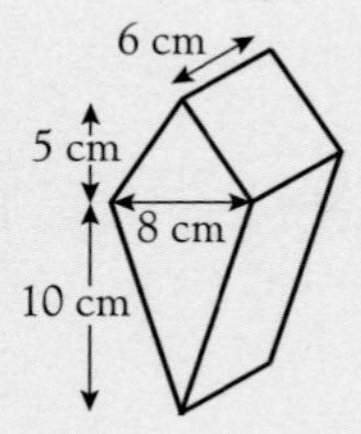

c

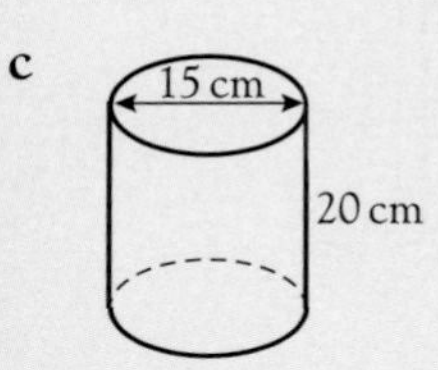

d

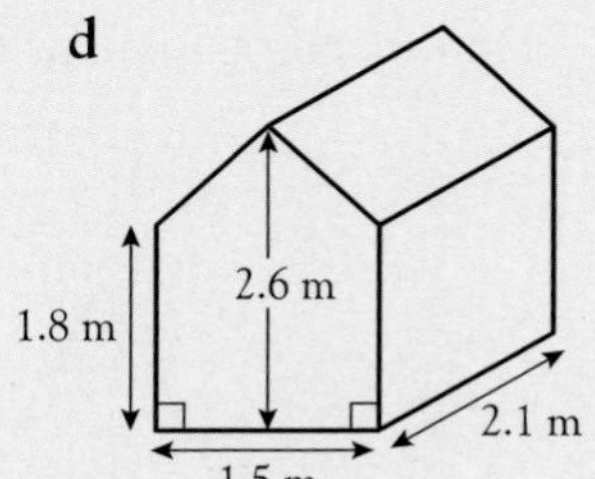

e

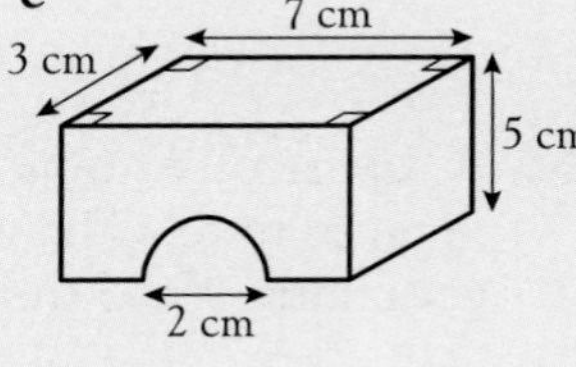
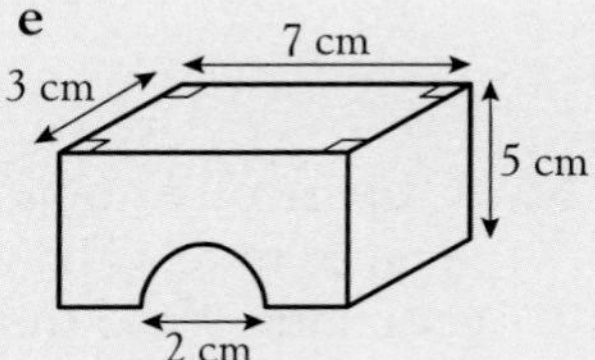

3 Find the volume of this child's building block.

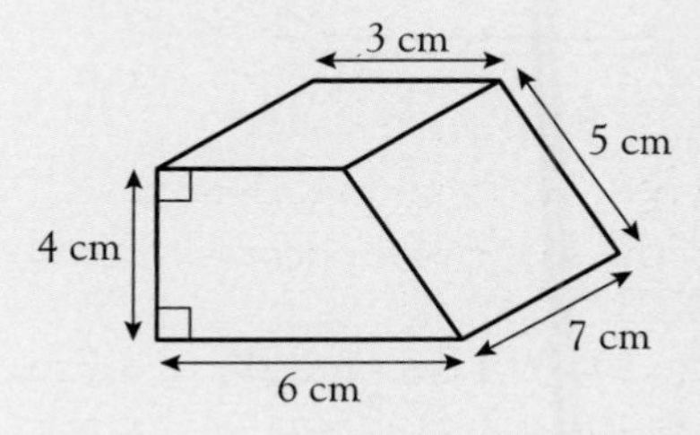

4 These two solid blocks of chocolate have been produced for sale at Christmas. Which block has the greater volume? Show your working.

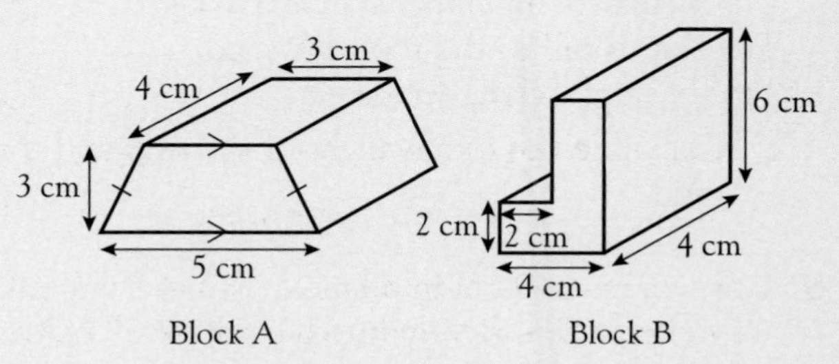

5 The photograph shows the cross-section of a £2 coin. The coin is 28 mm in diameter and 2.5 mm thick. The diameter of the central silver section is 20 mm.

a What is the area of the cross-section of the coin?

b What is the area of the silver part of the cross-section?

c One million coins were minted. What volume of each metal was needed?

VOLUMES OF PRISMS

6 A section of semi-circular guttering runs for 11.25 m along the bottom of a steep barn roof. The farmer finds that it regularly overflows in heavy rain so decides to replace the 9 cm diameter guttering with 12 cm diameter guttering.

What extra volume of water will the new guttering hold before overflowing?

7

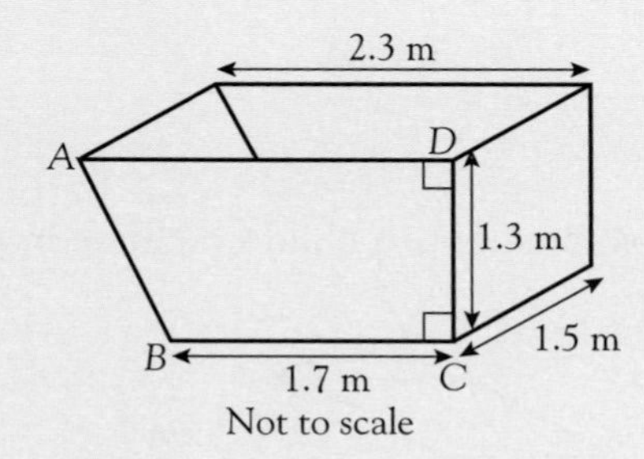

A skip is in the shape of a prism with cross-section $ABCD$.
$AD = 2.3$ m, $DC = 1.3$ m and $BC = 1.7$ m.
The width of the skip is 1.5 m.

a Calculate the area of the shape $ABCD$.

b Calculate the volume of the skip.

The weight of an empty skip is 650 kg.
The skip is full to the top with sand.
1 m^3 of sand weighs 4300 kg.

c Calculate the total weight of the skip and the sand. [L]

8 Tins which are 7 cm in diameter and 4.5 cm tall are to be packed tightly into a cardboard box in quantities of 48.

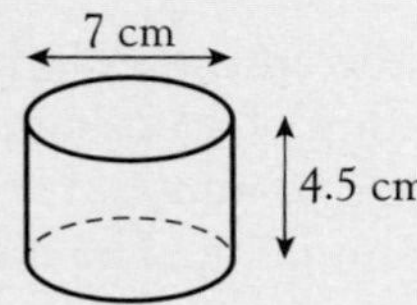

a Suggest internal measurements for four differently shaped boxes which would be suitable.

b Would all the boxes have the same volume of wasted space? Show your working.

9 A restaurateur serves diners with spring water from 1 litre bottles. The glasses are cylindrical with internal measurements diameter 6 cm, height 8 cm.

The glasses are filled three-quarters full.

How many servings can be made from 1 bottle?

10 Each pair of solids has the same volume. Find the missing lengths.

a

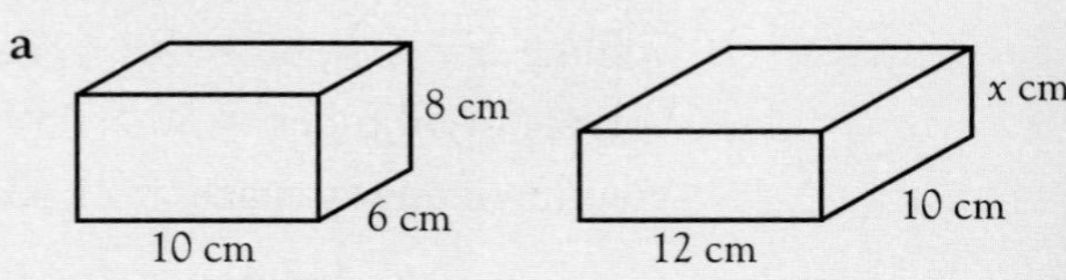

b

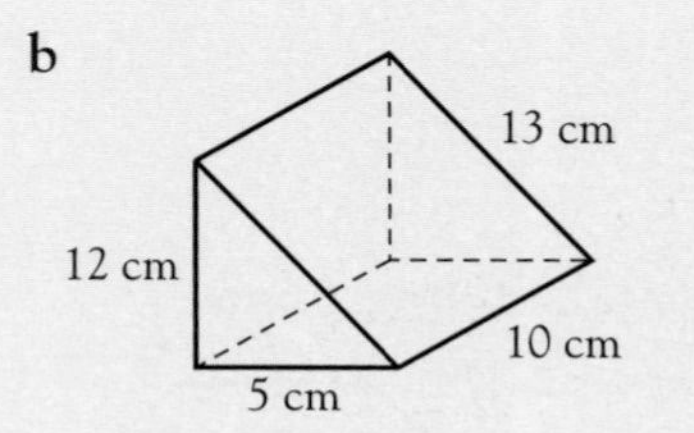

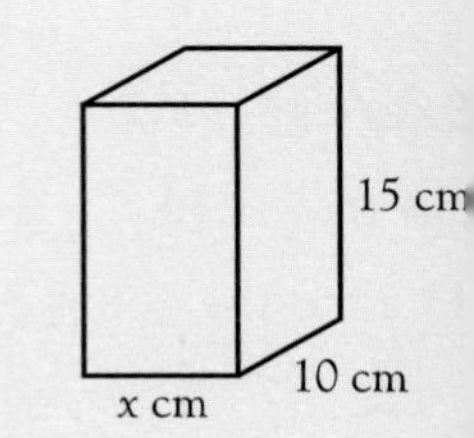

c

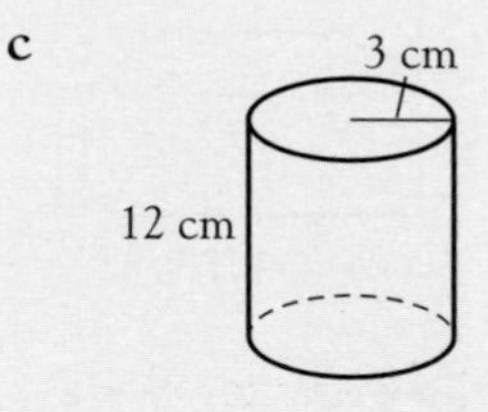

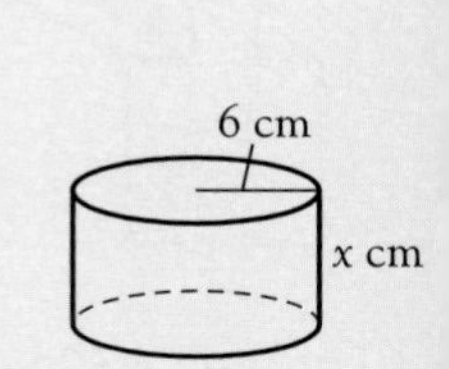

11 This tent has a volume of 105 cubic feet. Find its perpendicular height.

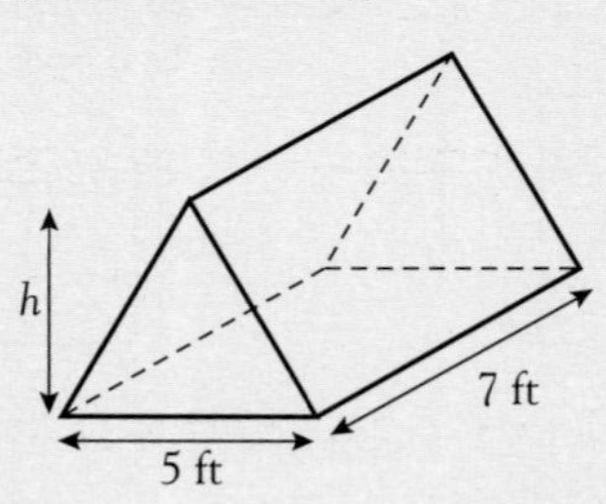

12 A square based tank with height 2 m takes 12 minutes to fill. The water flows at a rate of 1.5 m^3 per minute. Find the length x.

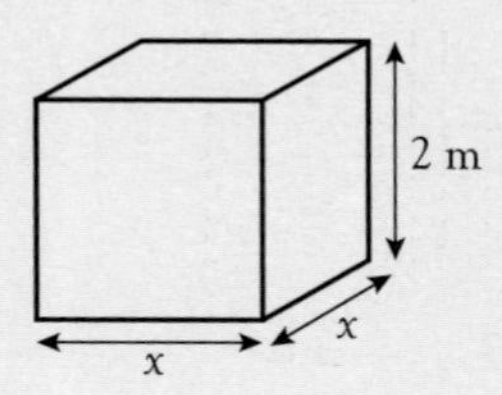

Surface areas of solids

To find the surface area of a solid, find the area of each of its faces.

Drawing the **net** of the solid sometimes makes it easier to work out the areas.

Sample Question 9

a

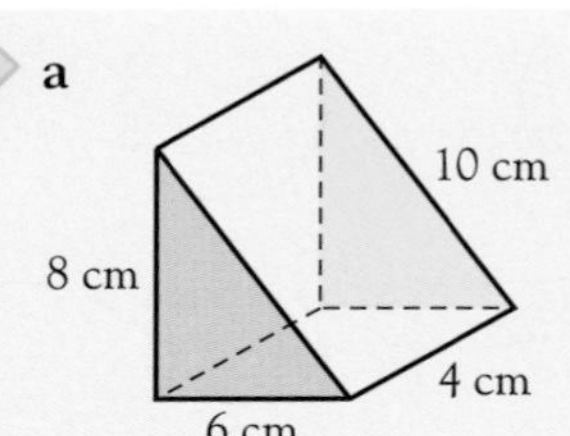

b

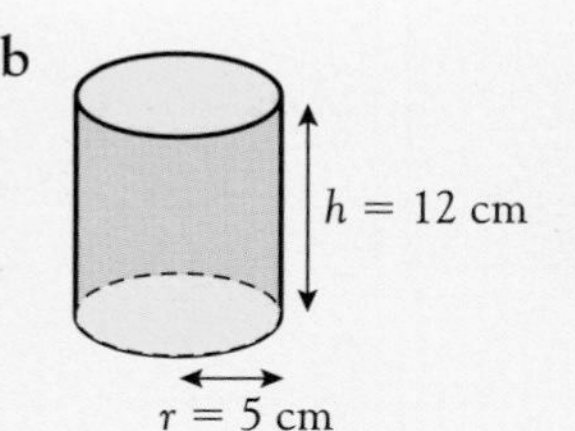

i For each prism sketch a net, showing all the measurements.

ii Find the surface area of the prism by finding the area of the net.

Answer

a i

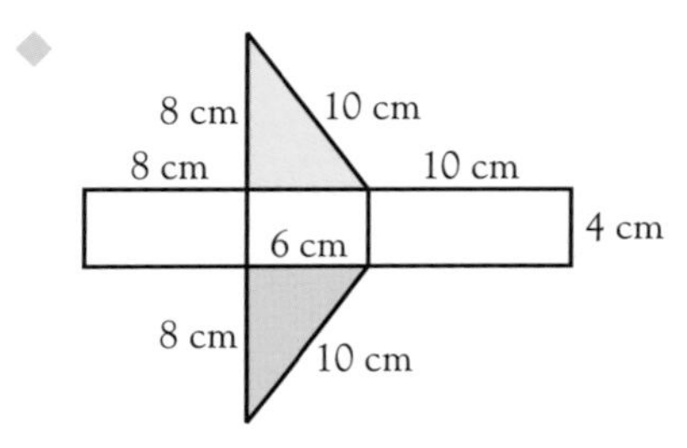

ii

- Add the areas of the two triangles and three rectangles.

$$\text{Area of one triangle} = \tfrac{1}{2} \times 6 \times 8 = 24\,\mathbf{cm^2}$$

$$\text{Total area} = 24 + 24 + 10 \times 4 + 6 \times 4 + 8 \times 4 = 144\,\mathbf{cm^2}$$

Remember that area is measured in square units.

The surface area of the prism is 144 **cm²**

b

- Draw the net in three parts – the top, the bottom and the curved section.
- Sketch the curved section by thinking of the cylinder being cut open along the line shown and placed flat to form a rectangle.

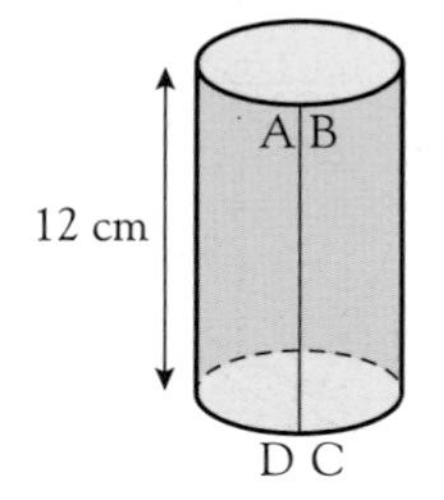

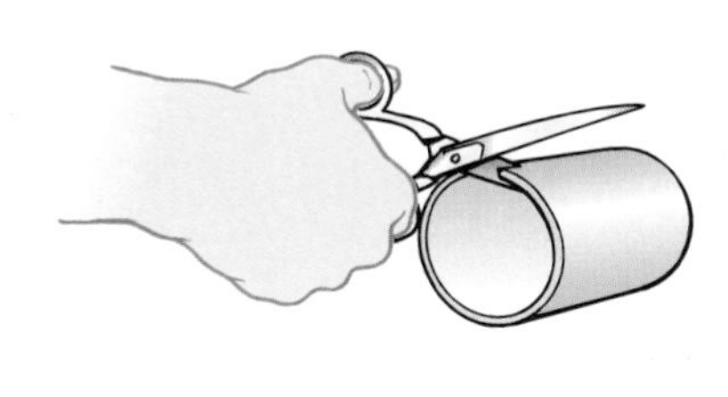

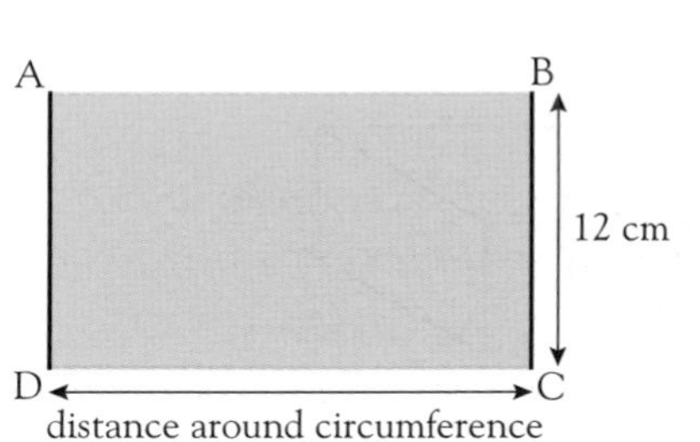

SURFACE AREAS

- The length of the rectangle is the distance around the circumference of a circle with radius 5 cm.

$C = 2\pi r$
$= 2 \times 3.14 \times 5$
$= 31.4$ cm

Net:

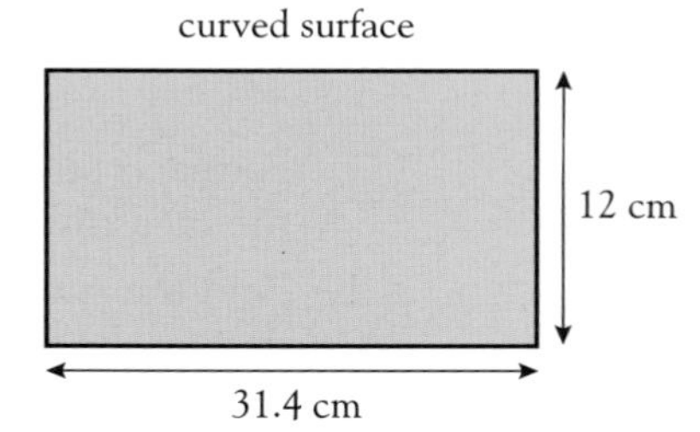

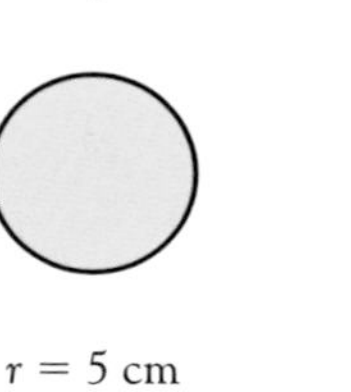

Area of curved surface
$= 31.4 \times 12$
$= 376.8\ \text{cm}^2$

Area of one circle $= \pi r^2$
$= 3.14 \times 5^2$
$= 78.5\ \text{cm}^2$

Area of net $= 376.8 + 78.5 + 78.5 = 533.8\ \text{cm}^2$

Total surface area $= 533.8\ \text{cm}^2$

You can write a formula for the curved surface area and for the total surface area of a closed cylinder.

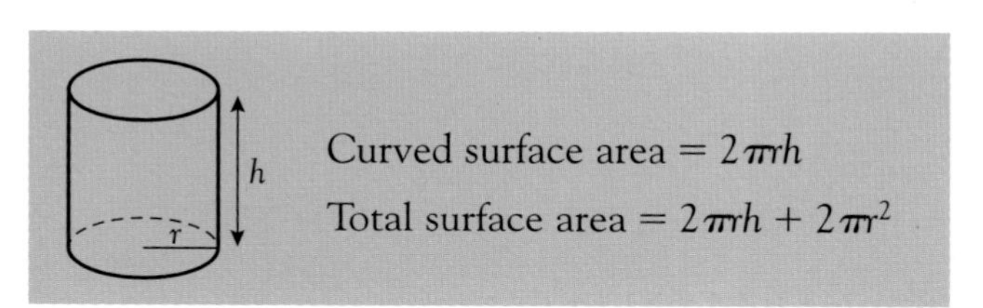

Curved surface area $= 2\pi rh$

Total surface area $= 2\pi rh + 2\pi r^2$

Remember to add the area of two circles if the cylinder is closed.

Exercise 15.3

1 Sketch the net of each prism and calculate the surface area.

a

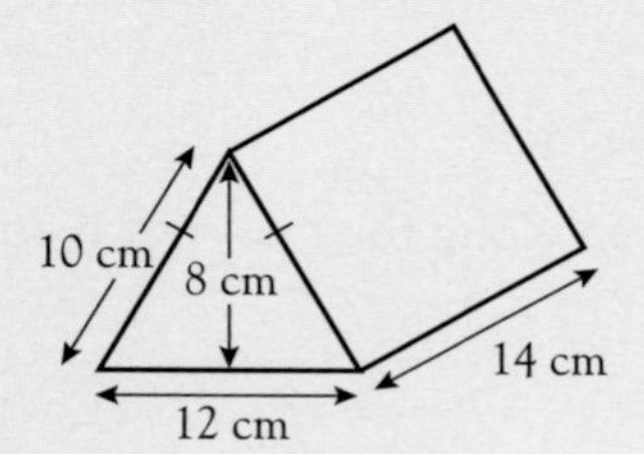

b

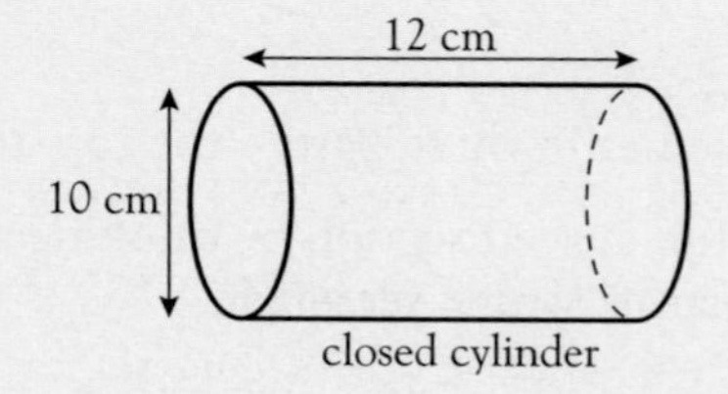

c

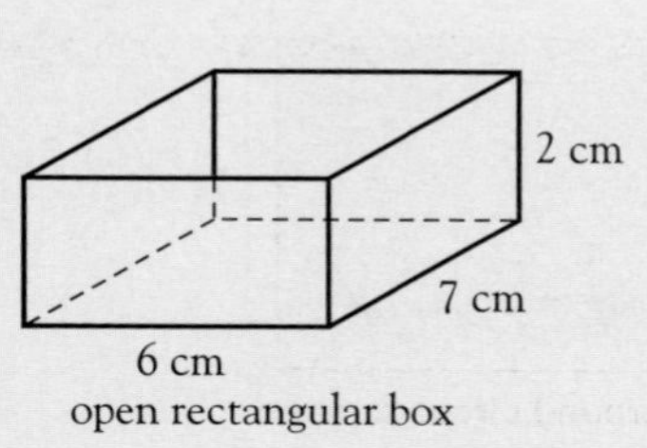

d

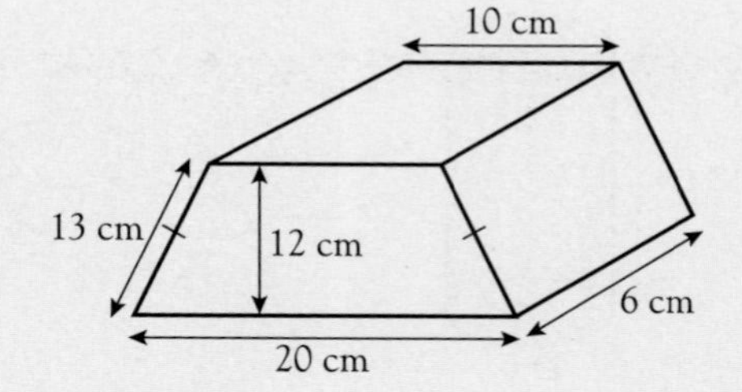

SURFACE AREAS

2 A child's toy is made from cubes with sides of length 3 cm, 5 cm and 7 cm.

  a Find the surface area of each cube.

  b Write down the ratio of the surface areas in the form $x : y : z$ in its simplest form.

  c The cubes are stacked as shown.

  Find the visible surface area of each cube.

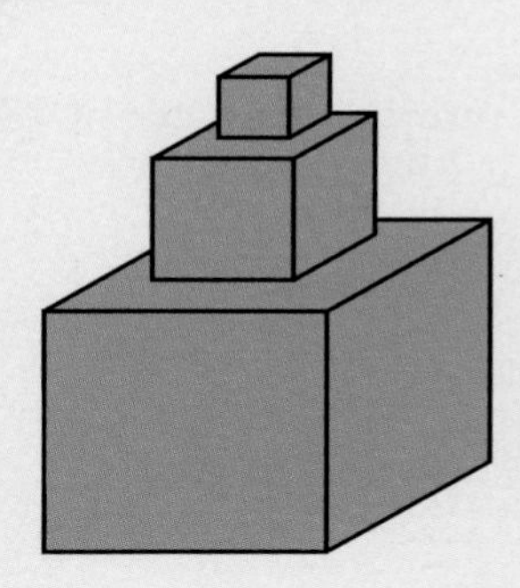

3 These two cylinders have the same **curved** surface area.

  a Find h.

  b Which cylinder has the greater volume?

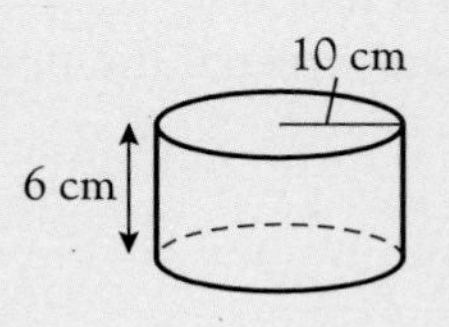

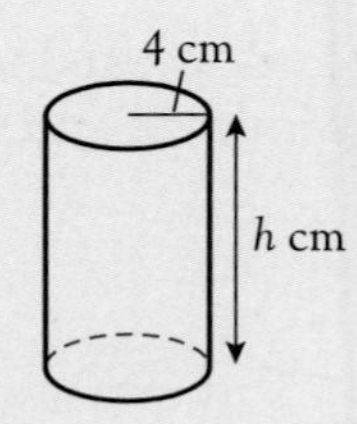

4 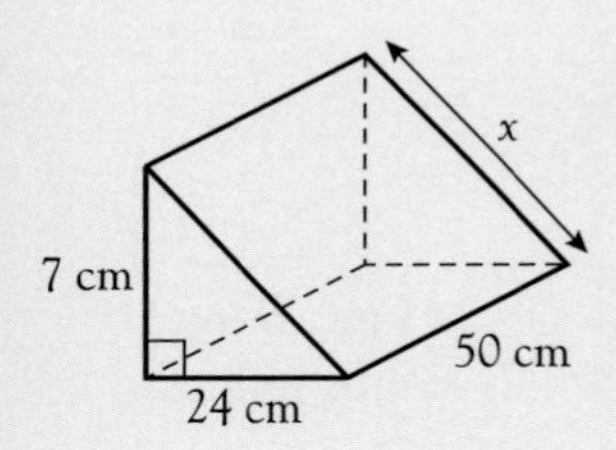

This cube has a surface area of 600 cm^2. Find the length of a side of the cube.

5 This triangular-based prism has a surface area of 2968 cm^2. Find the length x.

TASK

- Design and make a selection of prisms each with a volume of 72 cm^3.
- For each prism, work out the surface area.
- Which prism has the smallest surface area?

Worked Exam Question
[SEG]

A cylindrical water tank has an internal diameter of 78 cm.
The depth of water in the tank is 65 cm.

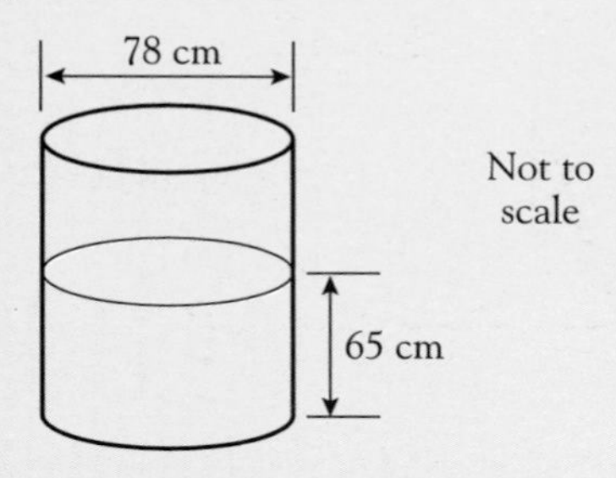

All of the water in the tank is used to fill a paddling pool.
The paddling pool is cylindrical and has an internal diameter of 2.1 m.

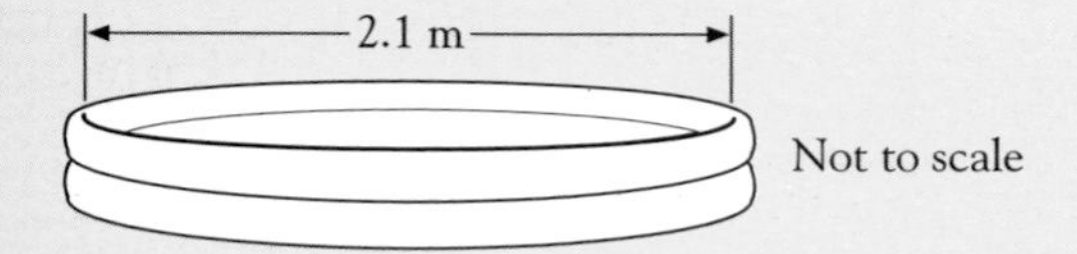

Calculate the depth of water in the paddling pool, stating your units.

Answer

- Work out the volume of the water in the tank.

  Find the radius first.

$r = 78 \div 2 = 39$

$V = \pi r^2 h$

$= 3.14 \times 39^2 \times 65$

$= 310\,436.1 \text{ cm}^3$

- Work out the radius of the pool, in cm.
- Write down the formula for the volume of the pool and put this equal to the volume of the water.

Remember to put in the units as requested.

$r = 210 \div 2 = 105 \text{ cm}$

$V = \pi \times r^2 \times h$

$310\,436.1 = 3.14 \times 105^2 \times h$

$= 34\,618.5h$

$h = \dfrac{310\,436.1}{34\,618.5}$

$= 8.967\ldots$

Answer 9.0 **cm** (1 d.p.)

COMMENTS

M1 for correct volume formula

M1 for equating the volumes

M1 for calculation for h

A1 for correct answer

M1 for answer in correct units

5 marks

Exam Questions

1 Kai has two cake tins. Both tins are prisms.

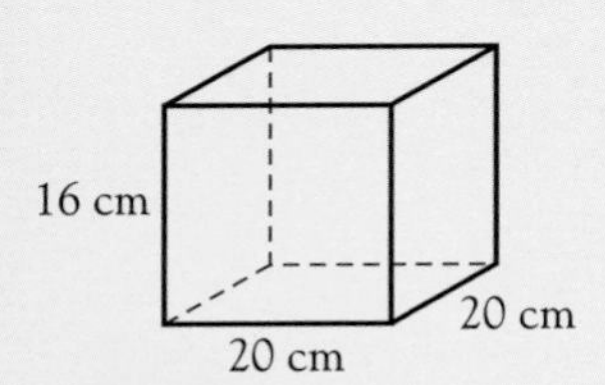

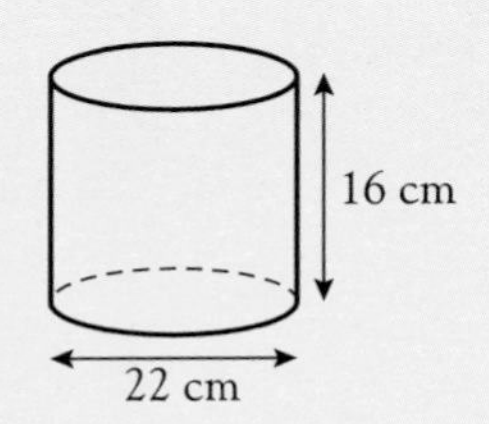

One has a square base, the other has a circular base.

The diagrams show the sizes of these tins.

One tin will hold more cake mixture than the other.

Calculate the **difference** between the amounts of mixture when both tins are completely filled. Give your answer to the nearest cm^3. [MEG]

2 This is a triangular prism.

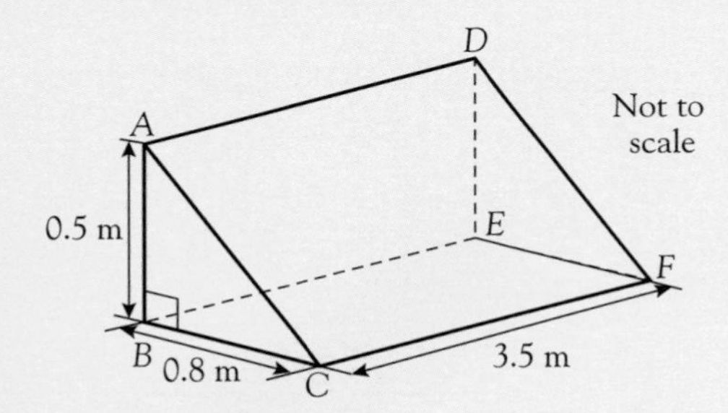

$AB = 0.5$ m, $BC = 0.8$ m, $CF = 3.5$ m

Angle $ABC = 90°$

a Calculate the area of the triangular end ABC.

b Calculate the volume of the prism. [SEG, p]

3 A box for Easter Eggs is in the shape of a prism.

The diagram shows the cross-section ABCD of the box.

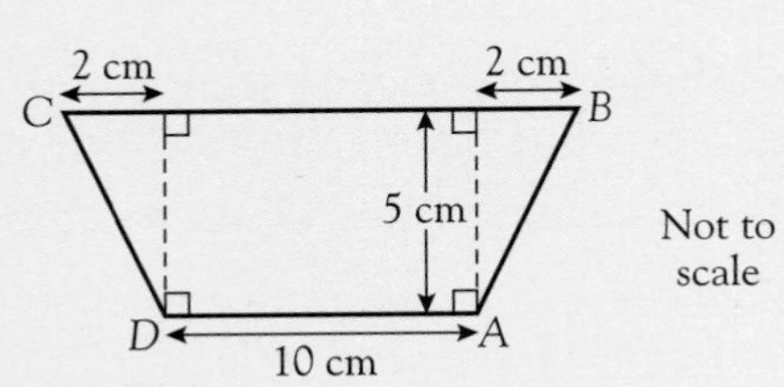

Not to scale

a Calculate the area of ABCD.

b The box is 8 cm wide.

Calculate the volume of the box. [MEG]

4 ABCD is an isosceles trapezium where AD = 10 cm, BC = 5 cm and the perpendicular height is 7.1 cm.

All four sides are tangents to the circle.

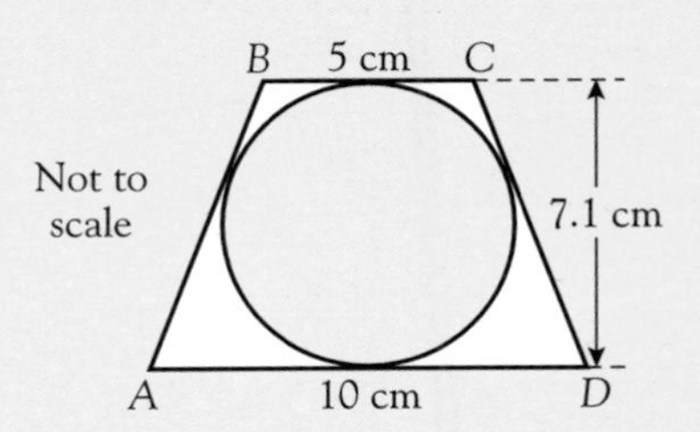

a Calculate the area of the circle.

b Calculate the total area shown white on the diagram. [MEG]

5 A cylindrical can has a radius of 6 centimetres.

a Calculate the area of the circular end of the can.

(Use the π button on your calculator or $\pi = 3.14$.)

The capacity of the can is 2000 cm^3.

b Calculate the height of the can.

Give your answer correct to 1 decimal place. [L]

6 A tray is circular, with a rim as shown in the diagram. The tray is horizontal.

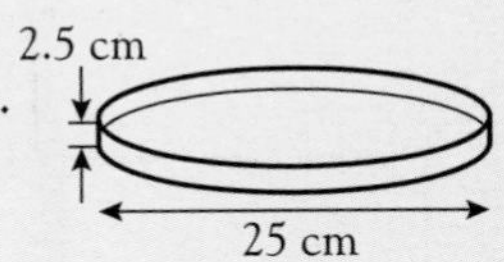

a What is the area of the base of the tray?

John spilt half a litre of orange juice onto the tray.

b What was the depth of the liquid on the tray? Give your answer to the nearest millimetre. [MEG]

7

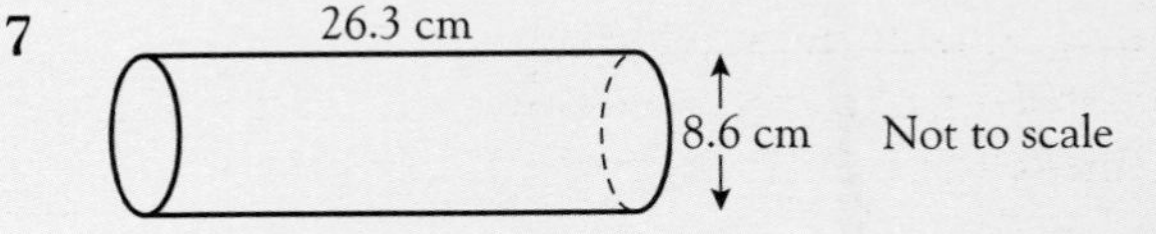

The diagram shows a cylinder.
The height of the cylinder is 26.3 cm.
The diameter of the base of the cylinder is 8.6 cm.

Calculate the volume of the cylinder.
Give your answer correct to 3 significant figures. [L]

EXAM QUESTIONS

8 Fred's compost bin is in the shape of a cylinder, diameter 80 cm and height 150 cm.

a Calculate

i the area of the base,

ii the volume of the bin.

This is Fred's wheelbarrow.
Its shape is a prism.

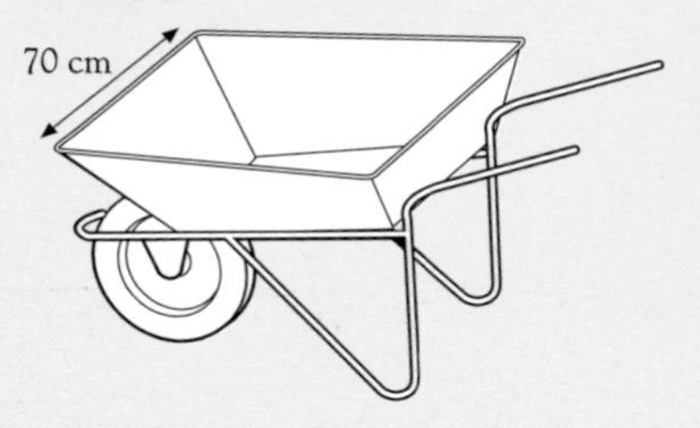

The cross-section is this trapezium.
The dimensions are in centimetres.

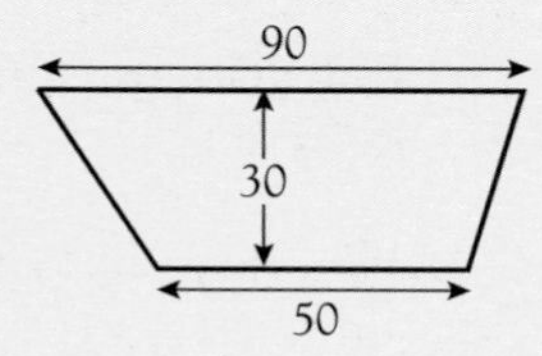

b Calculate

i the area of the cross-section,

ii the volume of the wheelbarrow.

c Based on your calculations, find how many wheelbarrow loads it will take to fill the bin.

[MEG]

9 This loaf of bread is cut into 30 slices, as shown.

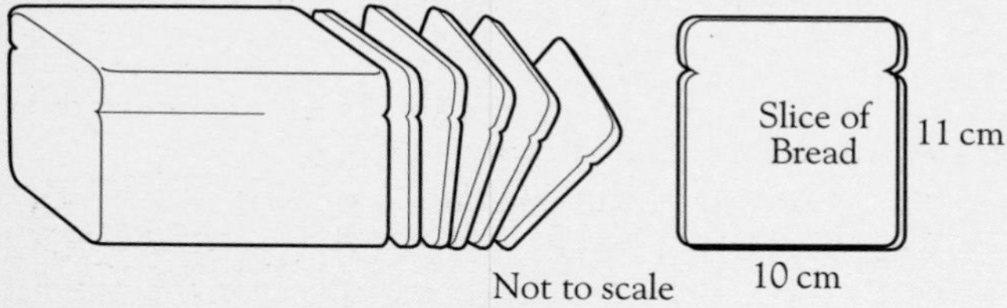

Each slice is approximately the shape of a cuboid of width 10 cm, length 11 cm and depth 8 mm.

a Calculate the volume of the loaf in cubic centimetres.

The same volume of bread is used to make a round loaf of length 24 cm.

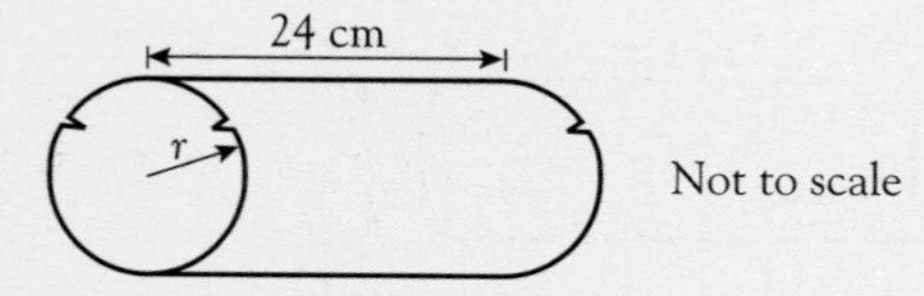

b What is the radius of the round loaf?

Take π to be 3.14 or use the π key on your calculator.

[SEG]

10

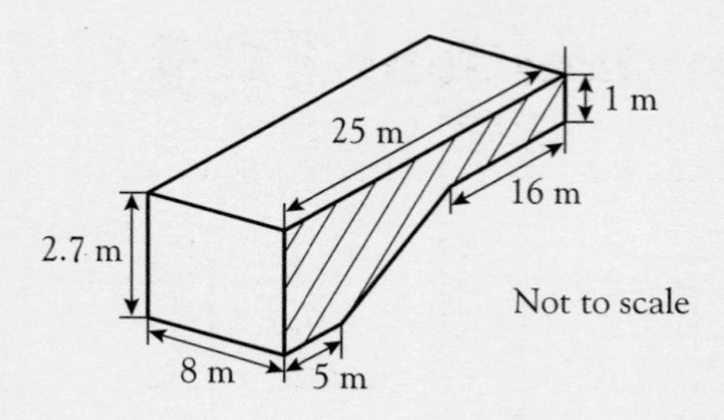

The diagram represents a swimming pool.

The pool has vertical sides.

The pool is 8 m wide.

a Calculate the area of the shaded cross-section.

The swimming pool is completely filled with water.

b Calculate the volume of water in the pool.

64 m^3 of water leaks out of the pool.

c Calculate the distance by which the water level falls.

[L]

11 Mr Maragh wanted to make a garden pond.

He dug out 5.2 m^3 of soil.

He hired a skip to take away the soil.

The skip is a prism and has the dimensions marked on the sketch below.

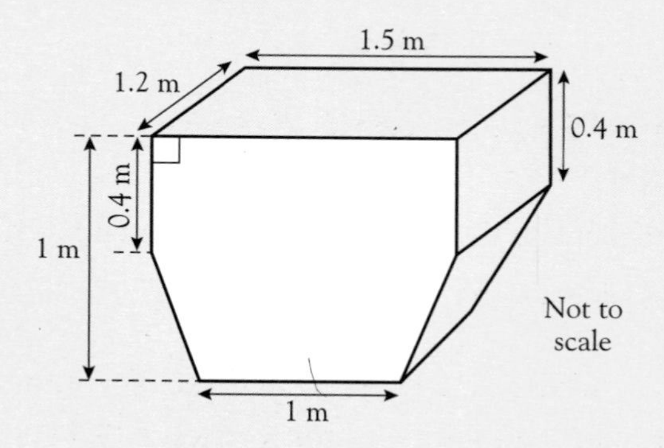

a Calculate

i the area of the cross-section of the skip (the white part),

ii the volume of the skip.

b How many skips are needed to remove all the soil? (They can only be filled up level with the top.)

[MEG]

16 TRANSFORMATIONS II

In this chapter you will learn

- **how to find the angle and centre of a rotation**
- **how to describe a single transformation that has the same effect as a combination of transformations**

Finding the centre and angle of rotation

In Chapter 8 you learned that when an object is **rotated** it turns through an angle about a **fixed point** called the **centre of rotation.**

page 128

Sometimes, you are told that a shape has been rotated and you are asked to identify the centre of rotation, the angle of rotation or both from the positions of the object and image.

Quite often this can be done **'by eye'**, just by looking at the diagrams, especially if the centre of rotation is on the perimeter of the shape.

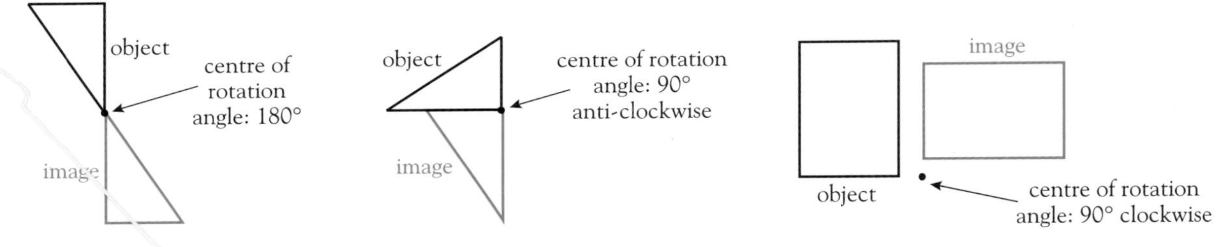

Remember that the centre of rotation is the only point that does not change its position. Look for a point that does not move.

To find the centre of rotation 'by eye'

- Look for a point on the perimeter of the shape that does not move. If there is one, then this is the centre of rotation.
- If you cannot find a point that does not move, look at the relative positions of the object and image. Look for a point that you think might be the centre of rotation.
- Test out your point using tracing paper and a pair of compasses or pencil point to hold down the tracing paper at your suspected point for the centre.
- Decide the direction and angle of rotation. Usually the rotation will be 90°, 180° or 270°.

A clockwise rotation of 270° is the same as an anti-clockwise rotation of 90°.
It does not matter which direction the rotation is if the angle is 180°.

When it is not possible to identify the centre of rotation 'by eye', you can use a drawing method. This uses the fact that any point on the image is the same distance away from the centre of rotation as the corresponding point on the object.

ROTATIONS

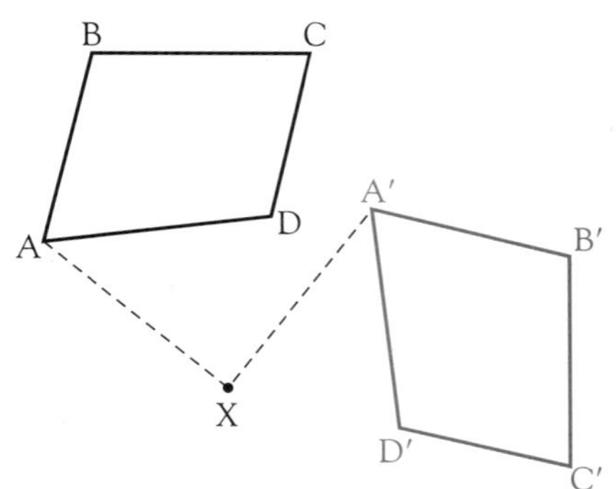

If ABCD is rotated to A′B′C′D′ about the centre of rotation X, then

$$AX = A'X$$

This means that A and A′ are equidistant from X.

page 227

From Chapter 12, in Constructions and Locus, you know that the locus of a point that is equidistant from two fixed points is the **perpendicular bisector** of the line joining the two fixed points.

So X lies on the perpendicular bisector of AA′. Similarly, it also lies on the perpendicular bisectors of BB′, CC′ and DD′.

If the position of the centre of rotation is unknown, it can be found by drawing any two of these bisectors and finding where they cross.

Sample Question 1

The diagram shows the image A′B′C′D′ of the object ABCD after a rotation.

a Identify the centre of rotation. Label it X.

b Measure the angle of rotation and state the direction.

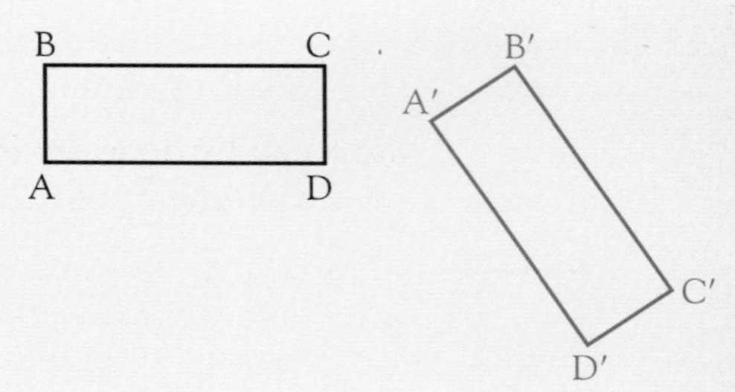

Answer

a

To find the centre of rotation:

Step 1

- Join A to A′.
- Construct the perpendicular **bisector** of AA′.

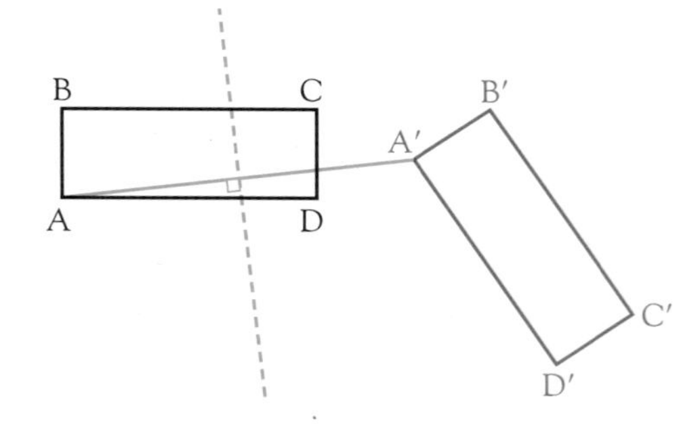

Alternatively, you can find the halfway point of AA′ and draw, using an angle measurer, a 90° angle above and below this point. Construction is more accurate, though.

Step 2

- Join **D to D′**.
- Construct the perpendicular bisector of DD′. Make the line long enough to meet the bisector of AA′.
- Where the two bisectors meet is the centre of rotation. Label it X.

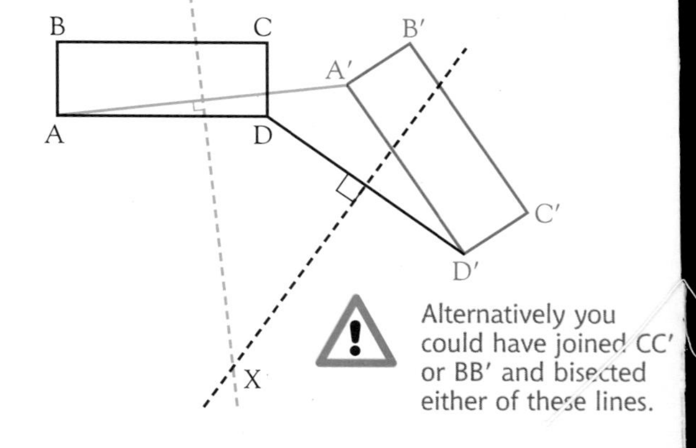

Alternatively you could have joined CC′ or BB′ and bisected either of these lines.

ROTATIONS

b

To find the angle of rotation:

- Draw a straight line from A to X and from X to A′.
- Measure the angle between AX and A′X. This is the angle of rotation
- State the direction.

<u>The rotation is clockwise through 56°</u>

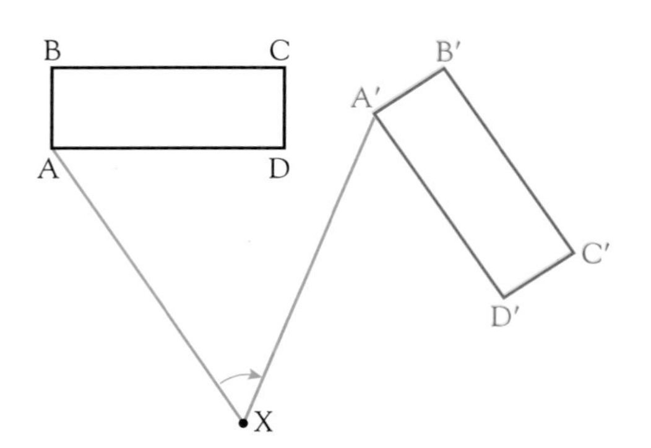

The angle of rotation is NOT the angle between the perpendicular bisectors!

In an examination tracing paper should be available for you to check that you have found the centre of rotation correctly. You will probably have to ask for the tracing paper!

To find the centre and angle of rotation

- Join two pairs of corresponding points on the object and image and construct the perpendicular bisector of each.
  Where the two bisectors meet is the centre of rotation.
- Draw two lines from the centre of rotation, one to a point on the object, the other to the corresponding point on the image.
  Measure the angle between these lines. This is the angle of rotation.
- Note the direction of the rotation.

Exercise 16.1

(You may need tracing paper to help you to copy some of the diagrams.)

1 Copy the following diagrams on to squared paper. (Solid lines show the original shape, broken lines show the image after a rotation.)

For each diagram

i locate the centre of rotation 'by eye'.

ii write down the direction and angle of rotation.

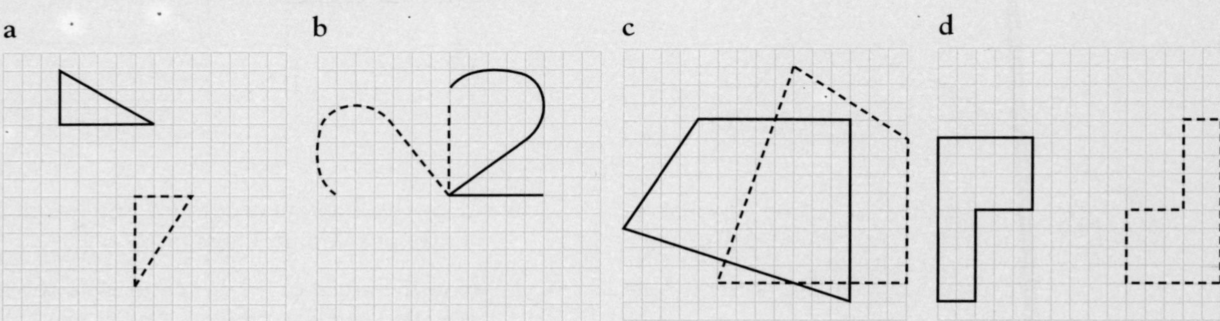

ROTATIONS

2 The diagram shows a regular hexagon (solid lines) and its image (broken lines) after a rotation.

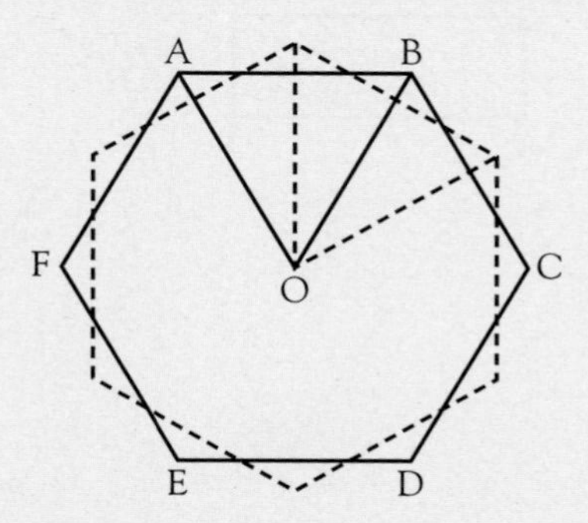

a Where is the centre of rotation?

b Measure and write down the angle of rotation and its direction.

3 The diagram shows a regular pentagon (solid lines) and its image (broken lines) after a rotation.

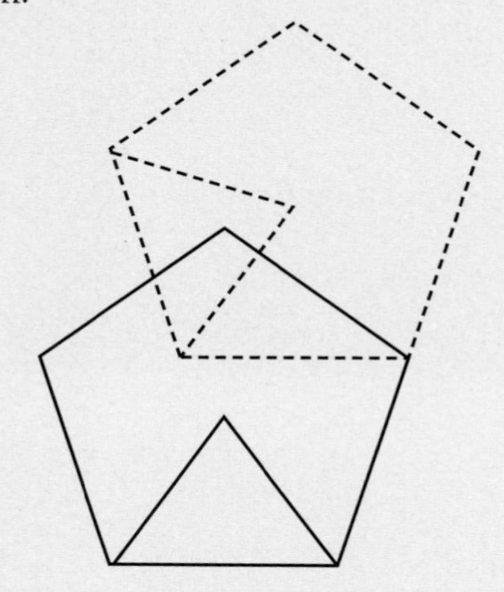

a Copy the diagram accurately and mark on your diagram the centre of rotation.

b Measure and write down the angle of rotation.

c State the direction of the rotation.

4

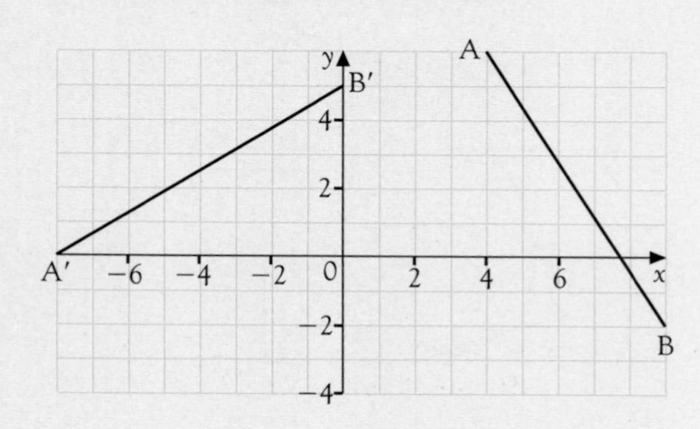

The diagram shows the line AB and its image A′B′ following a rotation. Copy the diagram accurately onto squared paper.

a Construct the perpendicular bisector of the line joining A to A′.

b Construct the perpendicular bisector of the line joining B to B′.

c Write down the co-ordinates of the centre of rotation.

d Find the angle and direction of the rotation.

5 The diagram shows the image, A′B′C′D′, of the diamond, ABCD, after a rotation.

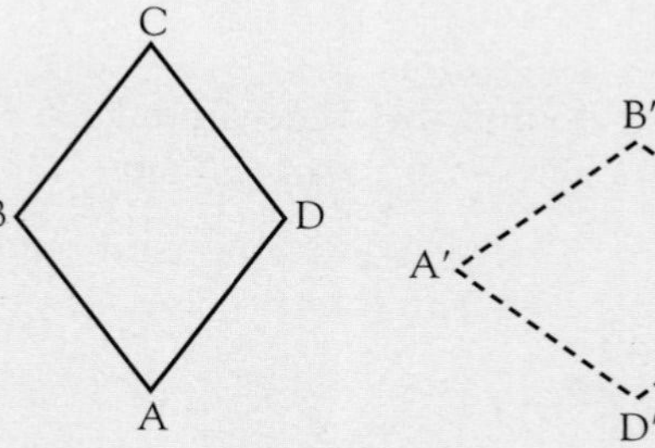

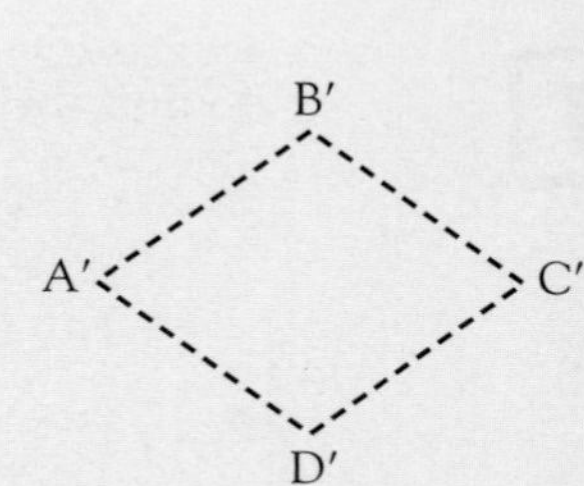

a Copy the diagram accurately and, using a geometrical construction, find the centre of rotation. Label it X.

b Find the angle of rotation and state the direction of the rotation.

6 Triangle ABC is rotated onto A′B′C′.

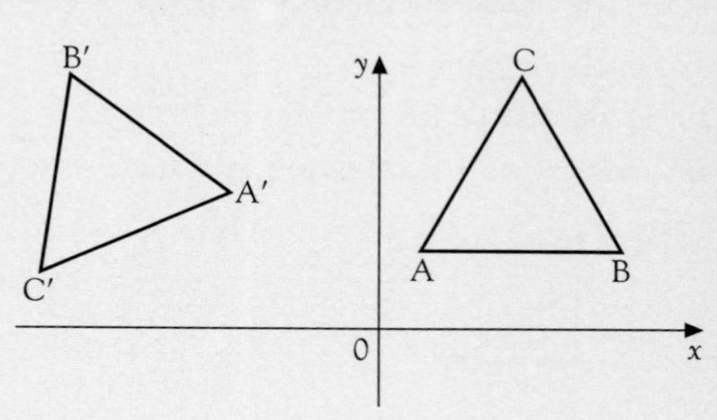

a Copy the diagram accurately and locate the centre of rotation. Label it X.

b Find the angle of rotation and state the direction of the rotation.

Combined transformations

A combined transformation is one that is formed by one transformation followed by another. Repeated transformations may produce a pattern.

This pattern is formed from repeated reflections.

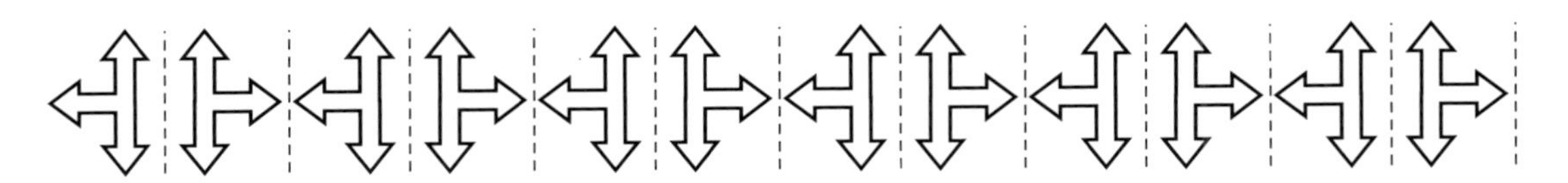

The transformations used need not be the same. This pattern is a combination of a rotation followed by a reflection.

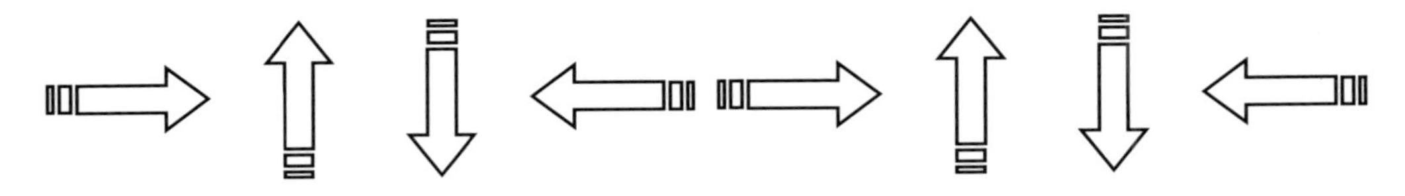

Two transformations can often be replaced by a single transformation. When it does not matter in which order the transformations take place, the transformations are said to be **commutative**. Be careful. This is not always the case.

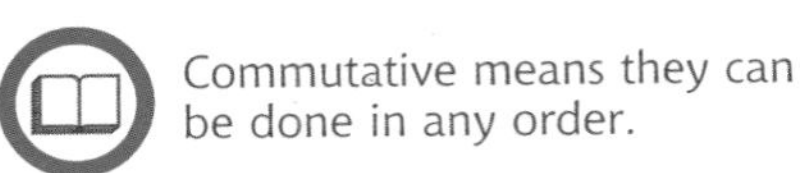

Commutative means they can be done in any order.

Sample Question 2

a On the grid, reflect the triangle ABC in the y axis. Label the reflected triangle A′B′C′.

b Reflect the triangle A′B′C′ in the x axis. Label the reflection A″B″C″.

c Describe fully the single transformations that would replace the two reflections.

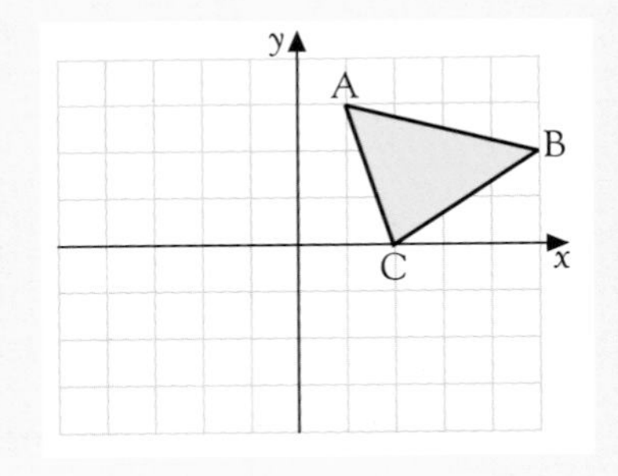

Answer

a

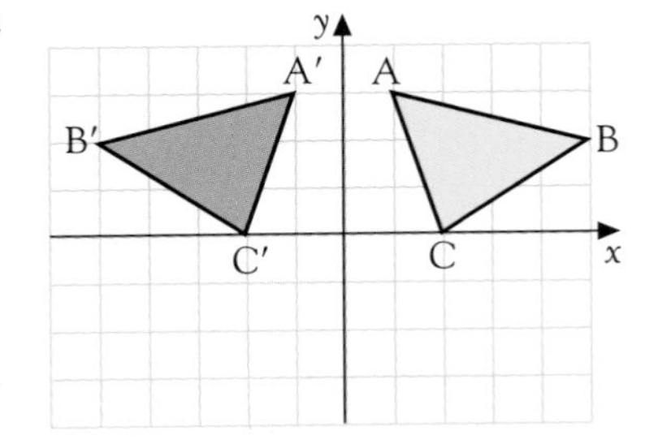

◆ Reflect triangle ABC in the y axis and label it A′B′C′.

b

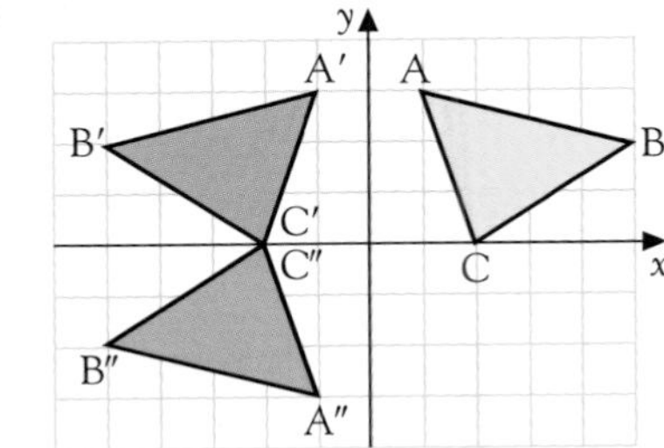

◆ Reflect triangle A′B′C′ in the x axis and label it A″B″C″.

COMBINING TRANSFORMATIONS

c

By joining corresponding pairs of points you can see that the triangle ABC has been **rotated 180° about the origin (0, 0)** onto A″B″C″.

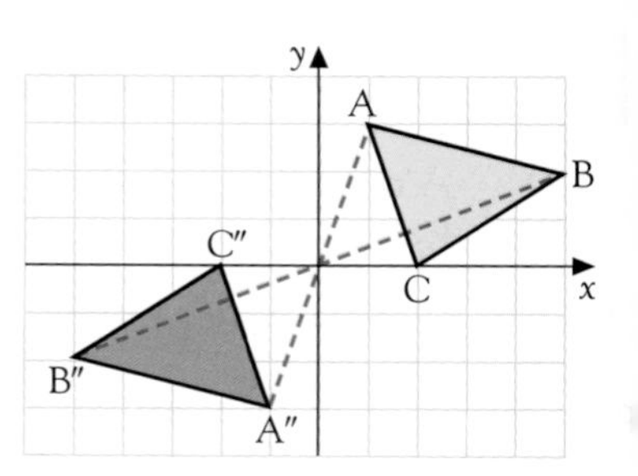

Sample Question 3

a Reflect the letter 'T' in the line $y = 1$.

b Enlarge the reflected 'T' with scale factor 2 and centre of enlargement at the point (3, 1).

c If the order of the transformation is reversed, is the end result the same?

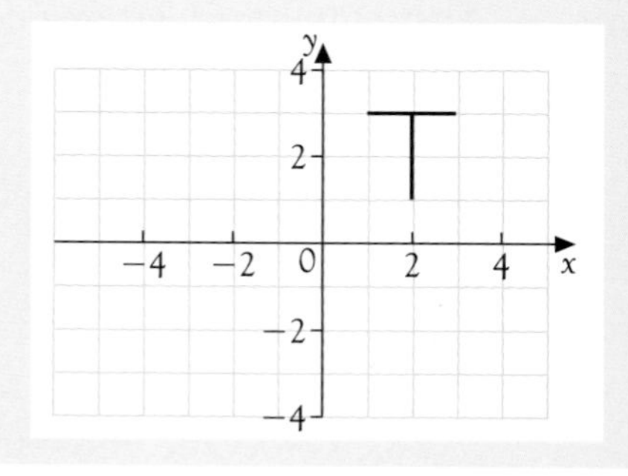

Answer

a

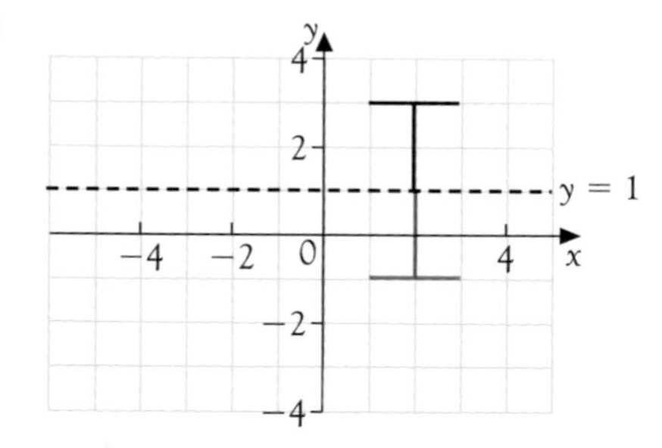

- Identify and draw the line $y = 1$
- Reflect 'T' in this line

b

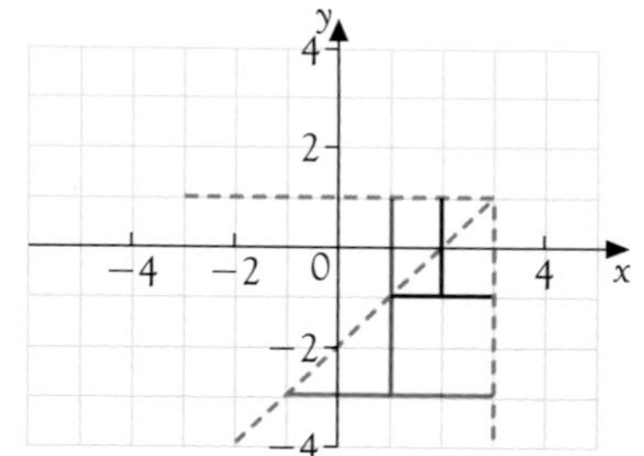

page 131

- Find the point (3, 1)
- Enlarge the reflected 'T' by scale factor 2

c

- Enlarge the original 'T' first then reflect the enlargement in the line $y = 1$
- Compare the resulting image

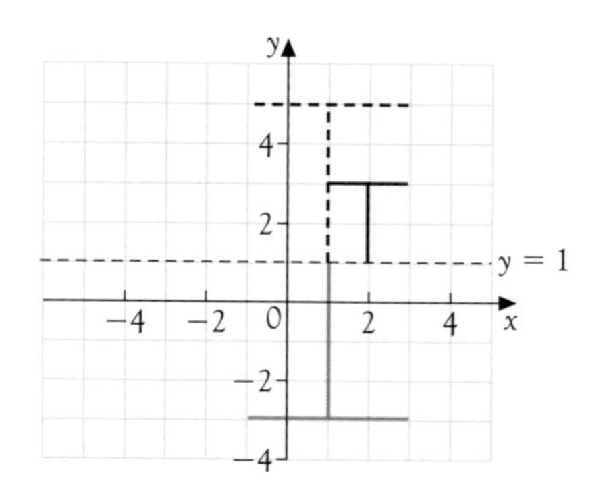

By comparing the two final images of the original 'T' you can see that the order in which the transformations are carried out does not matter in this example.
The end result is the same.

COMBINING TRANSFORMATIONS

Sample Question 4

a Rotate the rectangle A 90° clockwise about the point (0, 0). Label the image B.

b **Translate** the rectangle B using the column vector $\begin{pmatrix}2\\-2\end{pmatrix}$. Label the image C.

c If rectangle A is translated first, then the image rotated, will the position of rectangle C be in the same position as before? Show how you reach your decision.

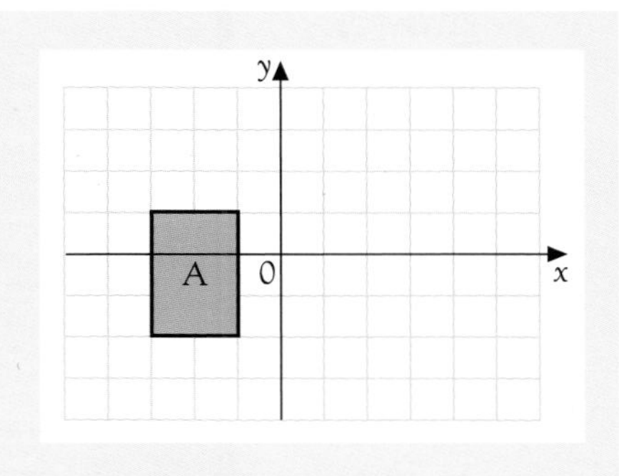

page 121

Answer

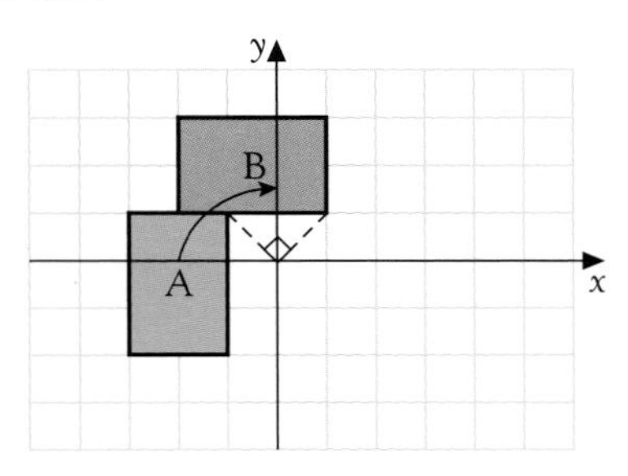

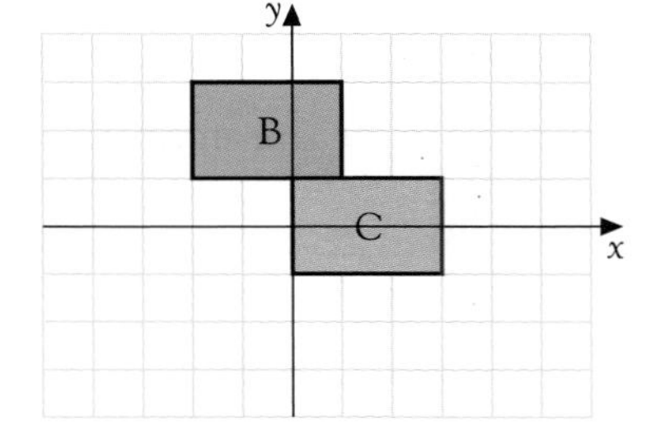

Use tracing paper.

a

- **Rotate** each point of A 90° clockwise about (0, 0).
- Label the image B.

b

- Translate each point 2 to the right and 2 down (−2).
- Label the image C.

c

- On a new grid **translate** A using $\begin{pmatrix}2\\-2\end{pmatrix}$. Label the image B.
- **Rotate** B 90° clockwise about (0, 0). Label the image C.
- Compare the two positions of C.

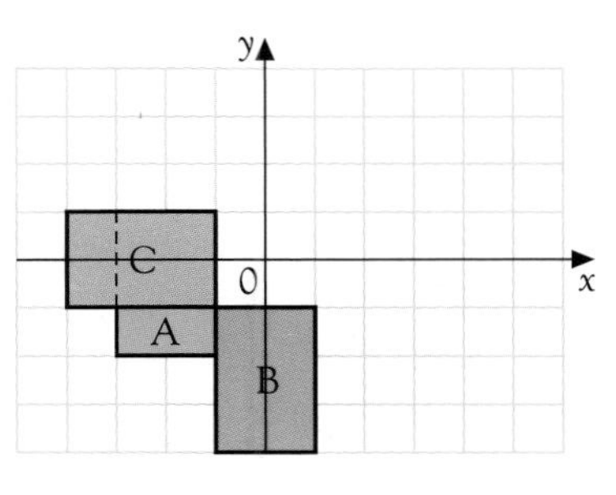

You can see that the positions of C are different so in this case **the order in which the transformations are performed gives different positions for the final image.**

Sometimes, two transformations can be replaced by a single transformation.

Sample Question 5

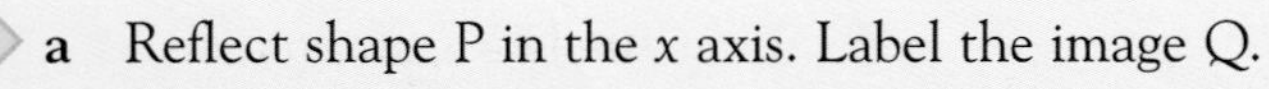

a Reflect shape P in the *x* axis. Label the image Q.

b Reflect shape Q in the *y* axis. Label the image R.

c Describe fully the single transformation that will transform shape P on to shape R.

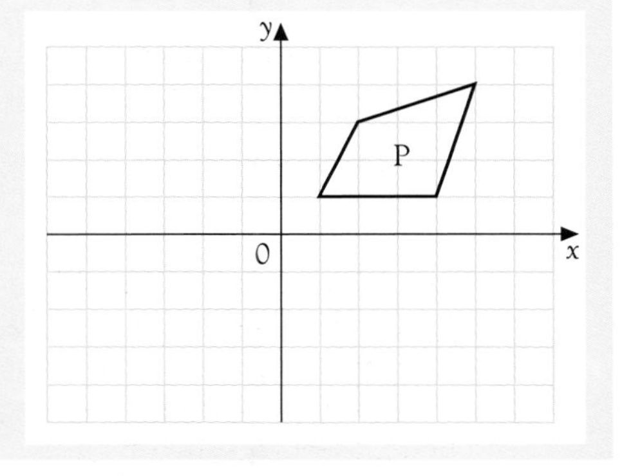

COMBINING TRANSFORMATIONS

Answer

a

- Reflect P in the x axis and label the image Q.

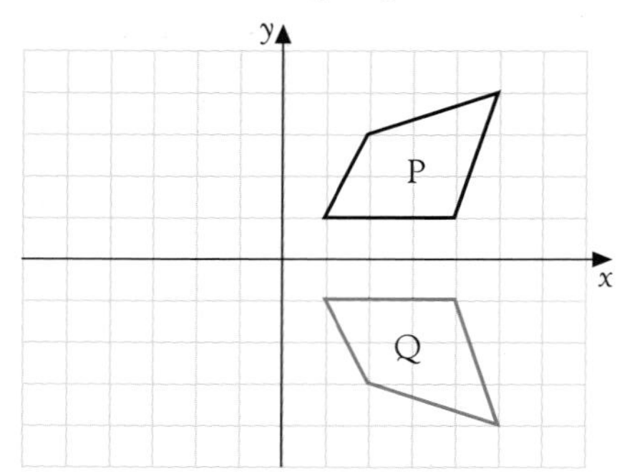

b

- Reflect Q in the y axis and label the image R.

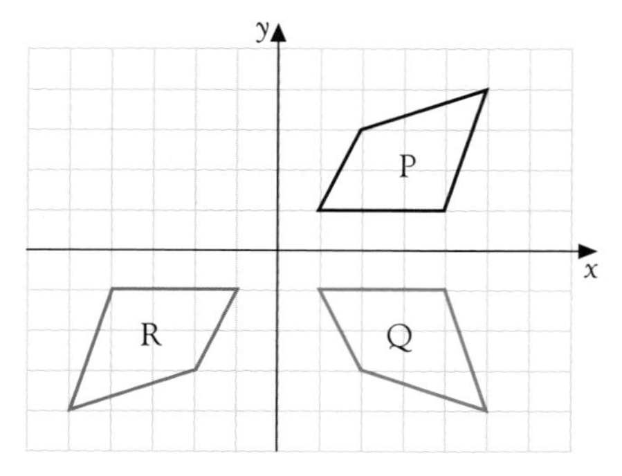

- To find the single transformation decide whether R is a reflection of P. Join corresponding points on P and R. R is not a **reflection** of P.
- Is R a translation of P? R is not a translation because it is not the same way up as P.
- R is not an enlargement of P as R is the same size as P.
- By joining corresponding points on P and R you can see that R is a rotation of P through an angle of 180° about (0, 0)

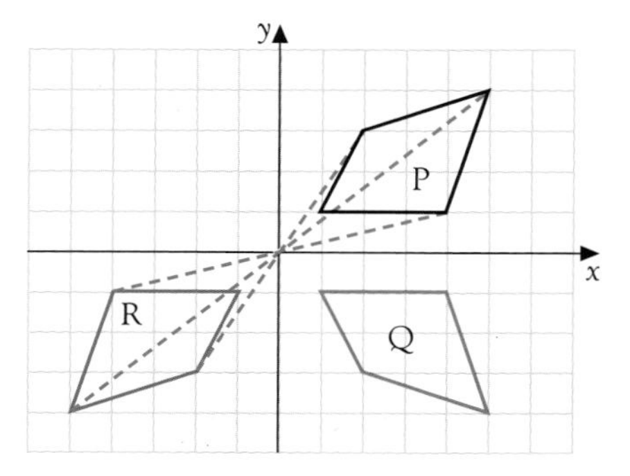

If R was a reflection of P the lines joining corresponding points on P and R would have been parallel.

Exercise 16.2

1 The pattern is formed by repeating a pair of transformations, in order. Copy the diagram onto squared paper and

a identify the two transformations,

b draw the next two figures in the pattern.

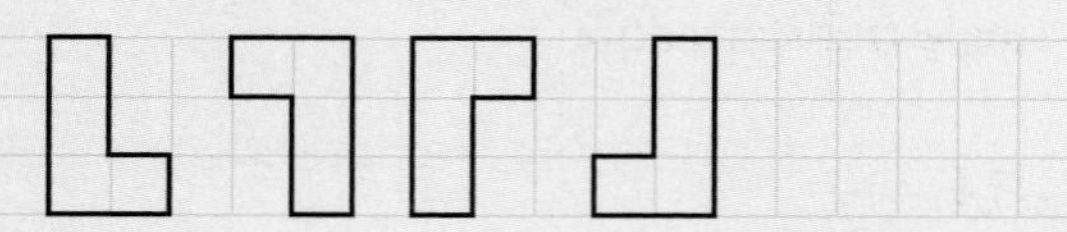

2 Copy this diagram onto squared paper.

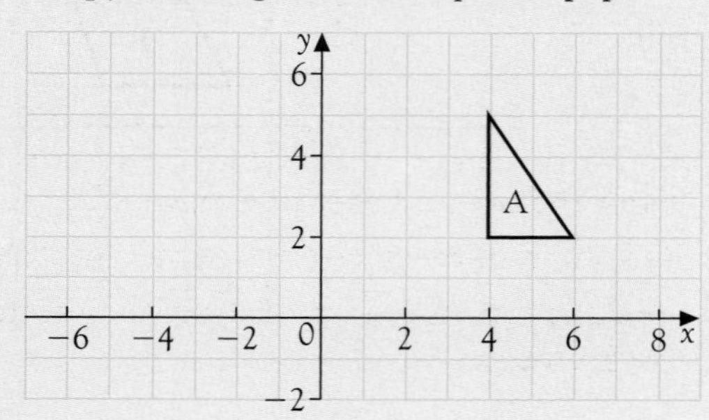

a On your diagram, reflect triangle A in the y axis. Label the image B.

b Draw the image of B after a reflection in the line $y = 2$. Label it C.

c Transform C under the translation given by $\begin{pmatrix} 12 \\ 0 \end{pmatrix}$. Label the image D.

d What **single** transformation will map A to D?

3 Draw a grid with both axes going from −8 to 8. On the grid, plot the points (0, 5), (0, 1), (−3, 1). Join the points to form a triangle. Label the triangle X. On the same grid:

a Reflect X in the line $x = 2$. Label the image Y.

b Reflect Y in the line $y = -1$. Label the image Z.

c Describe fully the single transformation that maps X on to Z.

COMBINING TRANSFORMATIONS

4 The letter F in the diagram is to be reflected in the x axis and the image is to be reflected in the y axis.

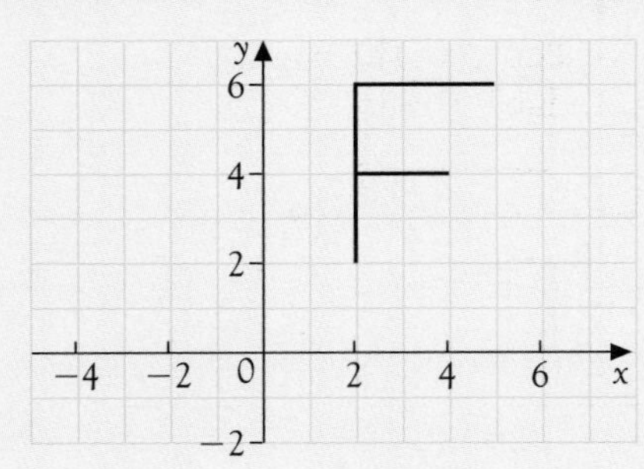

a Copy the diagram drawing both axes from -6 to 6. Carry out the two required transformations.

b Describe fully the single transformation that would have the same result as transforming the letter F using the two reflections.

5

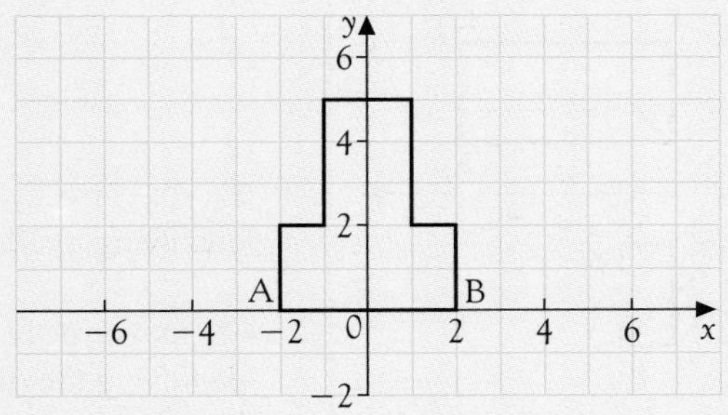

Copy the diagram onto squared paper drawing both axes from -6 to 6.

a Rotate the 'upside-down' T shape through 90° anti-clockwise with centre of rotation A. Label the image P.

b On the same diagram, rotate the original T shape through 90° clockwise about the point B. Label the image Q.

c Describe fully the single transformation that would map P onto Q.

6 Copy this diagram onto squared paper drawing both axes from -6 to 6.

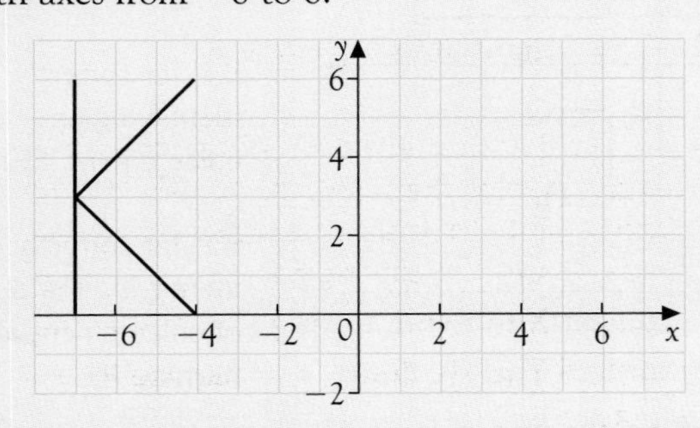

a Reflect the letter K in the line $x = -1$. Draw the image in red.

b Reflect your red K in the line $x = 1$. Draw the image in green.

c Describe fully the single transformation that maps the original K onto the green K.

7 Copy this diagram onto squared paper drawing both axes from -6 to 6.

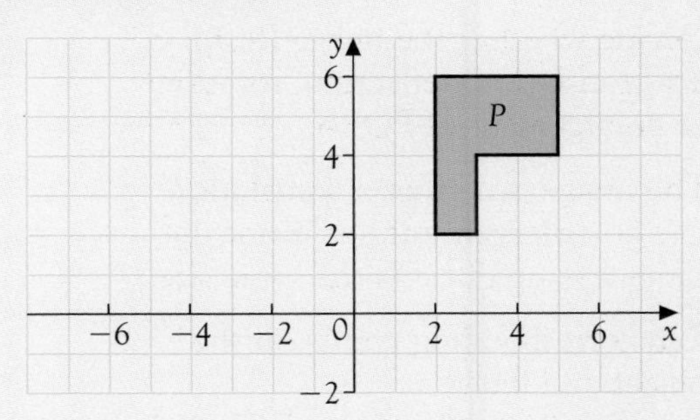

a Rotate the shape P through 90° clockwise about the origin. Label the image Q.

b Reflect Q in the line $y = x$. Label the image R.

c Describe fully the single transformation that maps P onto R.

8 Draw a grid going from -8 to 8 on both axes.

a On the grid, draw the quadrilateral ABCD where A = (2, 2), B = (5, 3), C = (5, 5) and D = (2, 5).

b Reflect ABCD in the line $y = -x$. Label the image A′B′C′D′.

c Reflect A′B′C′D′ in the x axis. Label the image A″B″C″D″.

d Describe fully the single transformation that maps ABCD on to A″B″C″D″.

9 Copy this diagram onto squared paper drawing both axes from -6 to 6.

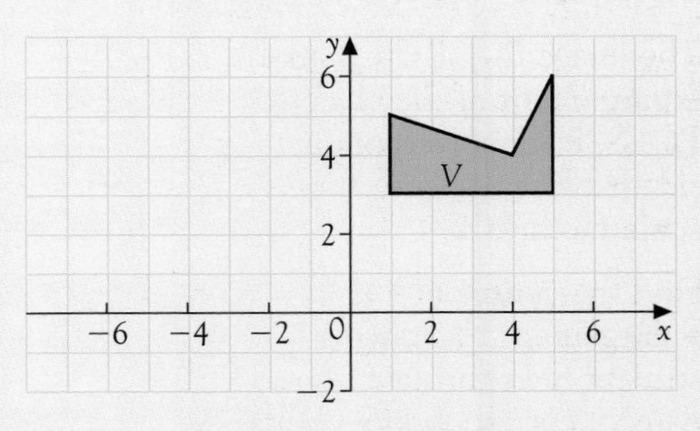

a Rotate the shape V through 90° clockwise about the origin. Label the image W.

b Reflect W in the y axis. Label the image X.

c Describe fully the single transformation that maps V on to X.

Worked Exam Question

[SEG]

Describe fully the single transformation which maps:

a $PQRS$ **onto** $P_1Q_1R_1S_1$

- Decide whether the image $P_1Q_1R_1S_1$ is a translation, reflection, rotation or enlargement of PQRS.
- The image is neither a **translation** nor an **enlargement** – it is not the same way up but it is the same size.
- Join corresponding points on the object and image.
- The lines joining P to P_1, Q to Q_1, R to R_1 and S to S_1 are **parallel**, so the transformation is a **reflection.** (By using tracing paper you can see that the image is not a rotation.)
- Decide what the mirror line is: find the mid-points of the lines joining corresponding points on the object and image. Join these mid-points with a straight line.
- By inspection you can see that the green line on the diagram is a special line, $y = x$.

Chapter 8 page 121

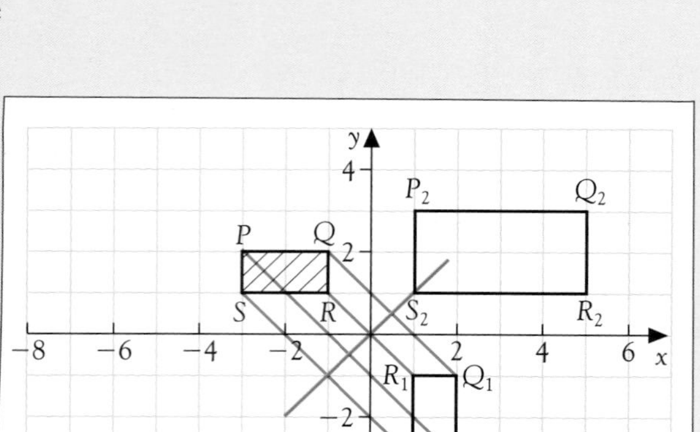

$P_1Q_1R_1S_1$ is a reflection of $PQRS$ in the line $y = x$

COMMENTS

B1 for correctly identifying a reflection

B1 for correctly identifying the mirror line $y = x$

2 marks

b $PQRS$ **onto** $P_2Q_2R_2S_2$.

- The image $P_2Q_2R_2S_2$, is an **enlargement** of the object, $PQRS$. Each corresponding side is twice as big so the **scale factor** is 2.
- Find the **centre of enlargement** by drawing straight lines through two pairs of corresponding points on the object and image, making the lines long enough to cross.

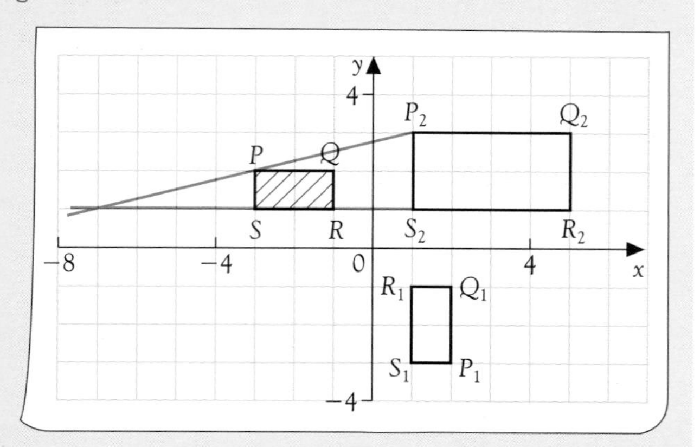

$P_2Q_2R_2S_2$ is an enlargement of $PQRS$, scale factor 2, with centre of enlargement $(-7, 1)$

B1 for correctly identifying an enlargement, SF2

B1 for correctly identifying the centre of enlargement **and naming it**

2 marks

Exam Questions

1

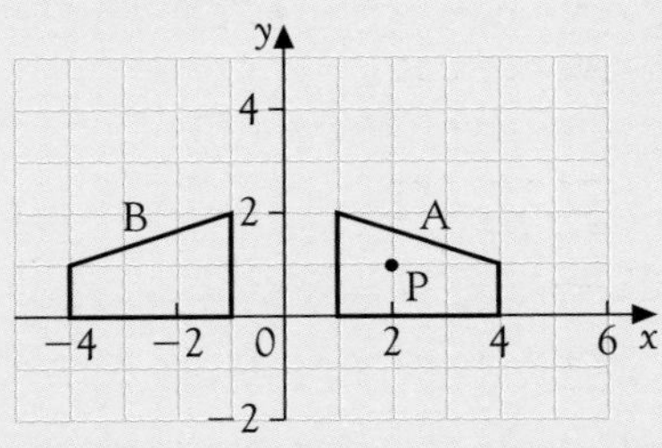

The diagram shows shapes A and B.

Describe the **single** transformation that maps shape A onto shape B. [MEG, p]

2

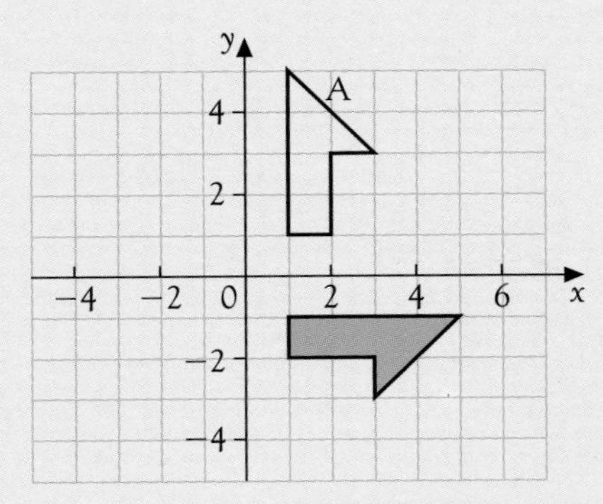

Describe fully the single transformation that will transform the shape labelled A to the shaded shape. [NEAB, p]

3

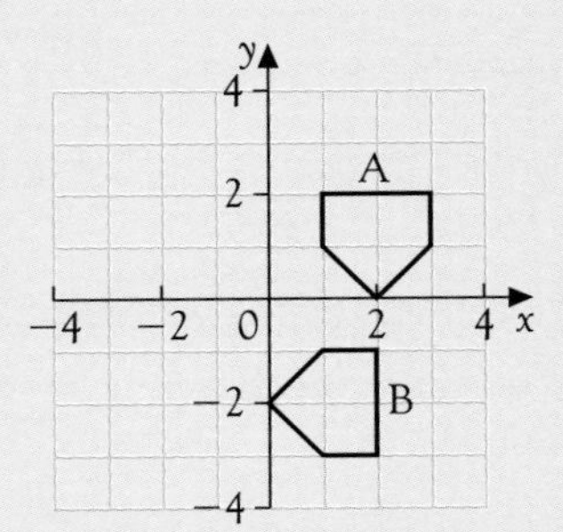

The diagram shows shapes A and B.

Describe fully the **single** transformation that maps shape A onto shape B. [MEG, p]

4

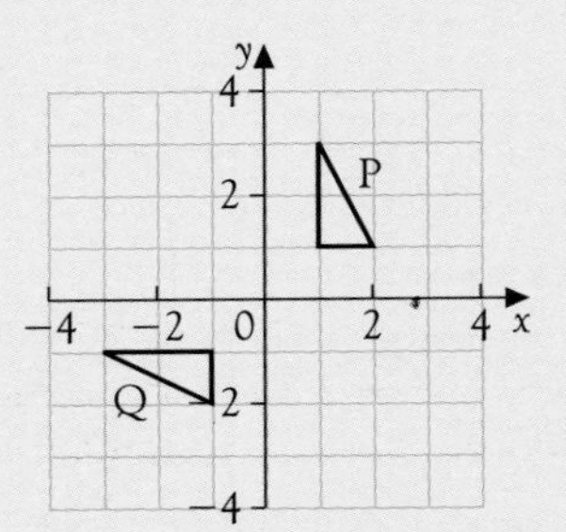

Describe the transformation that maps triangle P onto triangle Q. [MEG, p]

5

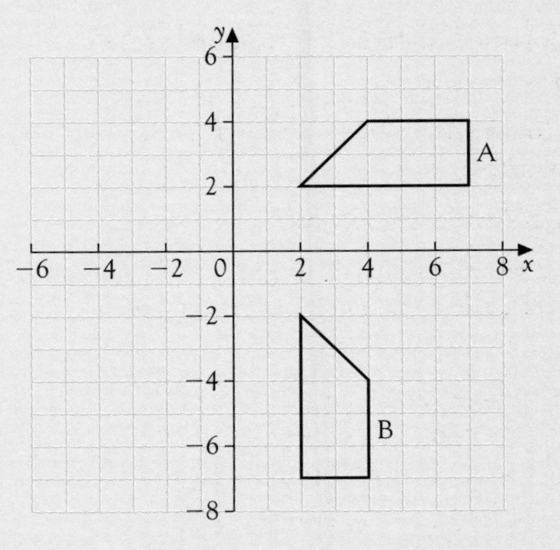

The diagram shows two shapes A and B.

Describe the single transformation that maps shape A onto shape B. [MEG, p]

6

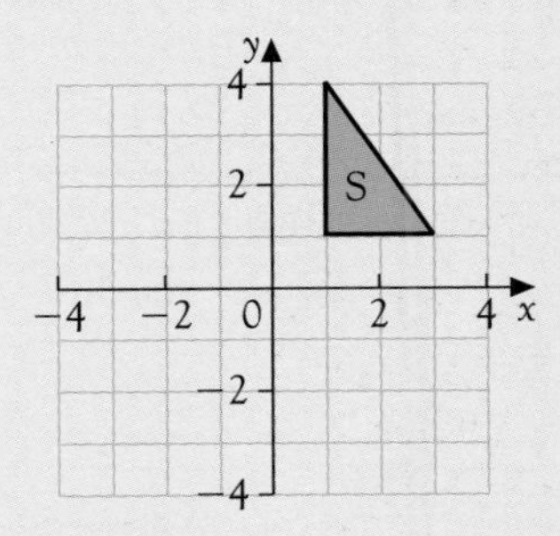

Copy the diagram onto squared paper.

a Reflect shape S in the line $x = 0$. Label the new shape R.

b Rotate the new shape R through an angle of 90° anti-clockwise using (0, 0) as a centre of rotation. Label the new shape T.

c Describe fully the single transformation that will move shape T back onto shape S. [L]

7

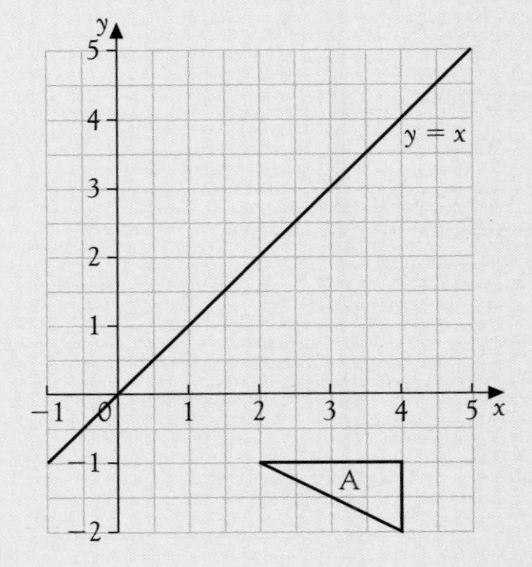

EXAM QUESTIONS

Copy the diagram onto squared paper.

a Reflect the triangle A in the x axis.
Label the reflection B.

b Reflect the triangle B in the line $y = x$.
Label the reflection C.

c Describe fully the single transformation which maps the triangle A onto the triangle C. [L]

8

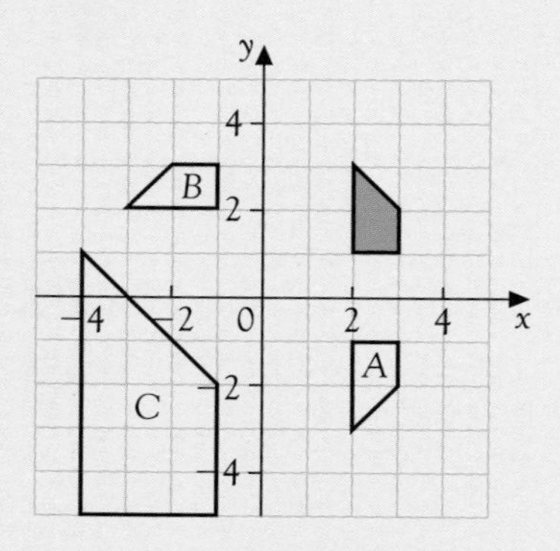

Describe fully a single transformation that would map the shaded shape on to

a shape A,

b shape B,

c shape C. [NEAB]

9

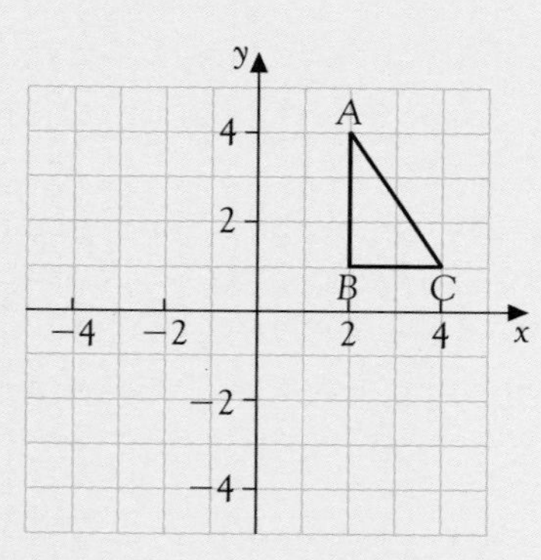

Copy the grid onto squared paper.

a Rotate triangle ABC 90° clockwise about (0, 0).
Label the new triangle LMN.

b Write down the co-ordinates of point L.

c Reflect triangle LMN in the x axis.
Label the new triangle PQR.

d Which single transformation maps triangle ABC onto triangle PQR? [WJEC]

10

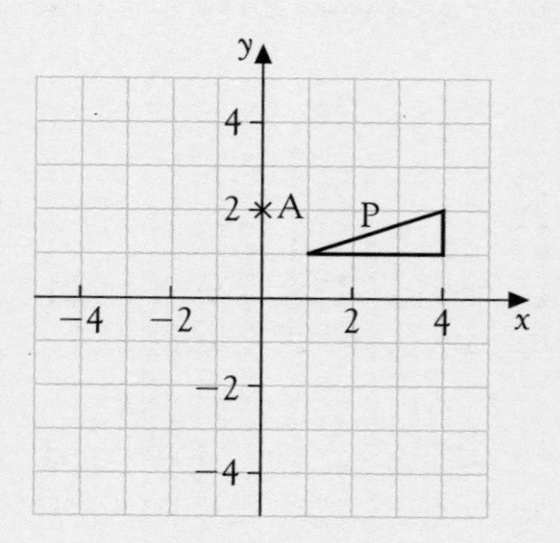

Draw a grid with both axes going from −6 to 6 and copy the triangle P. Mark the point A.

a Draw the image of triangle P after an anti-clockwise rotation of 90° about point A (0, 2). Label the image Q.

b Now reflect Q in the y axis. Label the new image R.

c A reflection maps triangle P directly onto R.

i Draw the line of reflection on the diagram.

ii Write down the equation of the line of reflection. [MEG]

17 PYTHAGORAS' RULE

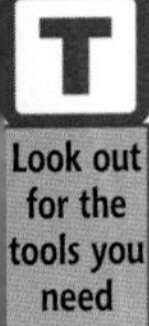

In this chapter you will

- **learn about Pythagoras' rule for right-angled triangles**
- **use it to find the length of the hypotenuse**
- **use it to find the length of one of the shorter sides**
- **solve problems using Pythagoras' rule**

TASK

In this task you will be investigating the relationship between the areas of squares that are fitted onto the sides of triangles.

Step 1

- Draw a triangle, using ruler and compasses, with sides $a = 1$ cm, $b = 2$ cm, $c = 3$ cm (see page 81). Position your triangle so that there is some space around it and label the sides carefully.

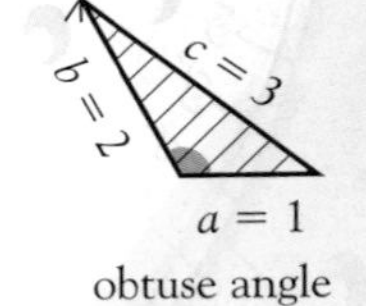

Step 2

- Shade in the angle between a and b and decide whether it is acute (smaller than 90°), a right angle (90°) or obtuse (between 90° and 180°).

Step 3

- Using your set-square, draw a square to fit on each side of the triangle and work out the area of each square.
  The area of the square that is fitted onto side a is a^2.

Step 4

- Draw out a table like the one below and fill in the columns for question 1.

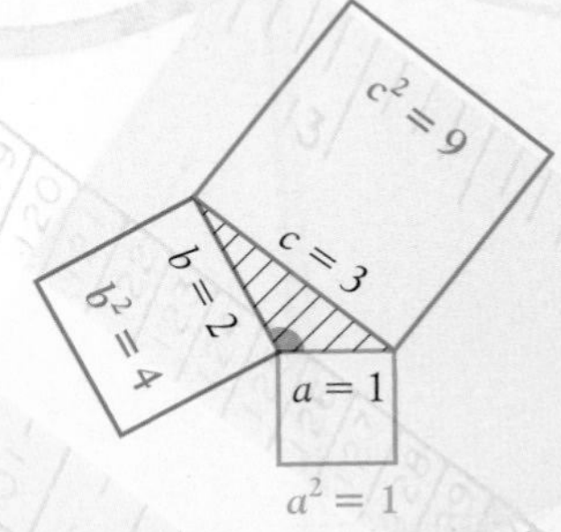

Step 5

- Repeat Steps 1 to 3 for each of the other triangles, complete the table and comment on your results.

Remember that $c^2 > a^2 + b^2$ means c^2 **is bigger than** $a^2 + b^2$, and $c^2 < a^2 + b^2$ means c^2 **is smaller than** $a^2 + b^2$.

	a	b	c	a^2	b^2	c^2	$a^2 + b^2$	Is $c^2 > a^2 + b^2$ or $c^2 = a^2 + b^2$ or $c^2 < a^2 + b^2$?	Acute, right or obtuse angle between a and b?
i	1	2	3						
ii	2.5	2	3						
iii	3	4	5						
iv	2	2	5						
v	1.5	3.5	4						
vi	2.5	6	6.5						
vii	2	3.5	2.5						
viii	2	3	3.5						
ix	1.5	2	2.5						

PYTHAGORAS' RULE

In your investigation you should have found the following:

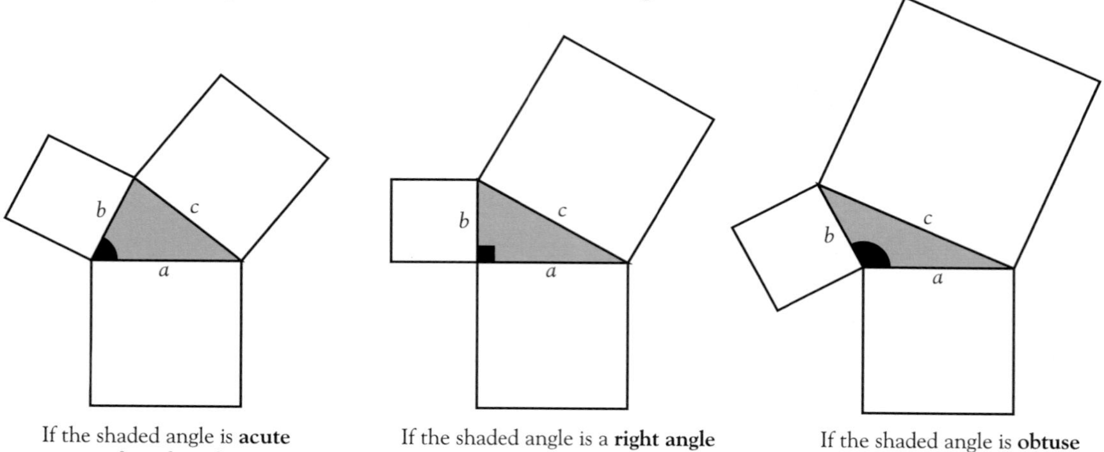

If the shaded angle is **acute** $c^2 < a^2 + b^2$ | If the shaded angle is a **right angle** $c^2 = a^2 + b^2$ | If the shaded angle is **obtuse** $c^2 > a^2 + b^2$

Right-angled triangles

You will now look in more detail at **right-angled triangles**, using the result $\mathbf{c^2 = a^2 + b^2}$. This is a very useful formula. Although it connects squares, it can be used to find missing lengths in triangles.

Sample Question 1 Calculate the missing areas of the squares and lengths of the sides in these right-angled triangles.

a

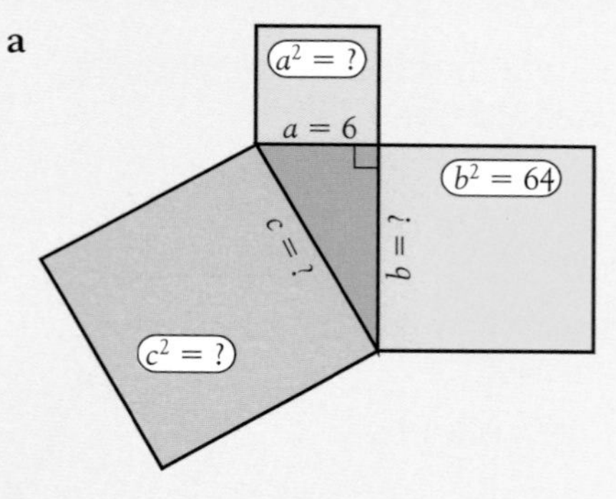

b

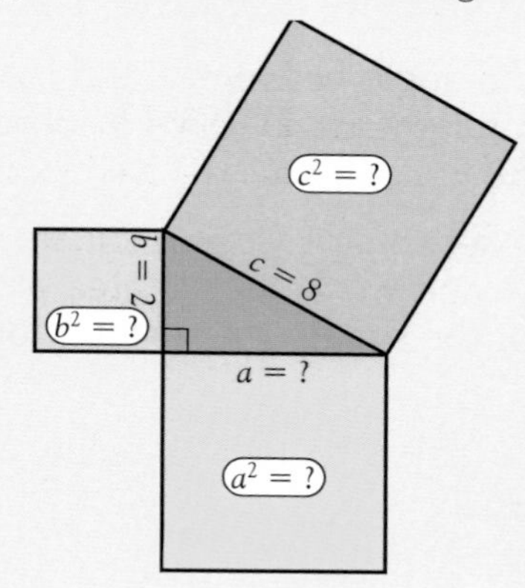

Answer

a

- Use $b^2 = 64$ to find b

  $b = \sqrt{64} = \underline{8}$

- Use $a = 6$ to find a^2

  $a^2 = 6^2 = \underline{36}$

- Find $a^2 + b^2$ and put this equal to c^2

  $a^2 + b^2 = 36 + 64 = 100$

  $c^2 = 100$

  $c = \sqrt{100}$

  $\underline{c = 10}$

b

- Use $b = 2$ to find b^2

  $b^2 = 2^2 = \underline{4}$

- Use $c = 8$ to find c^2

  $c^2 = 8^2 = \underline{64}$

- Use $a^2 = c^2 - b^2$

  $a^2 = 64 - 4 = \underline{60}$

  $a = \sqrt{60}$

  $= 7.745\ldots$

  $\underline{a = 7.7 \text{ (1 d.p.)}}$

PYTHAGORAS' RULE

Exercise 17.1

1 Calculate the missing areas of the squares and lengths of the sides in each of the following. Notice that **all** the triangles are right-angled.

a

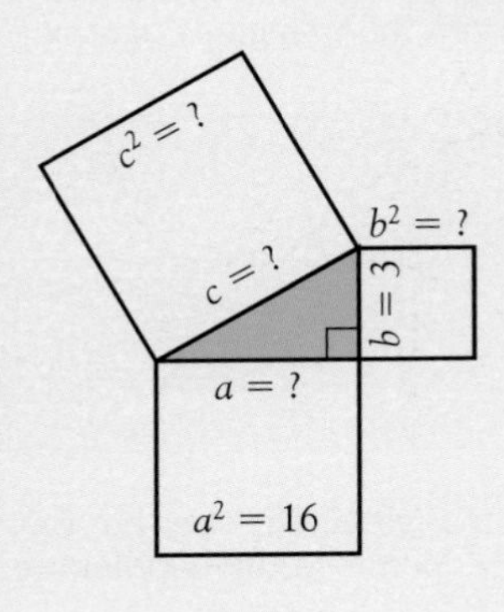

b

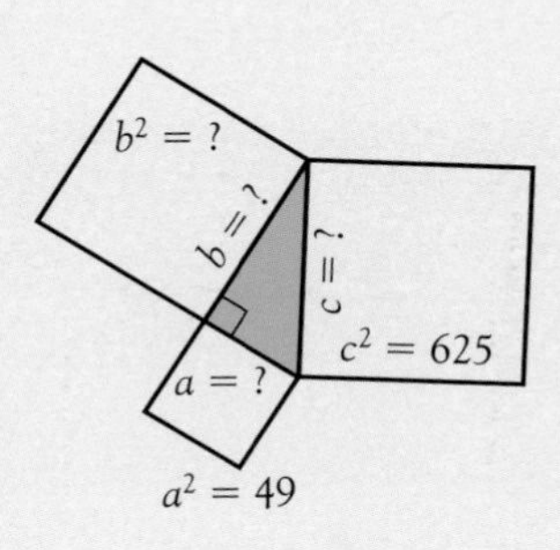

c

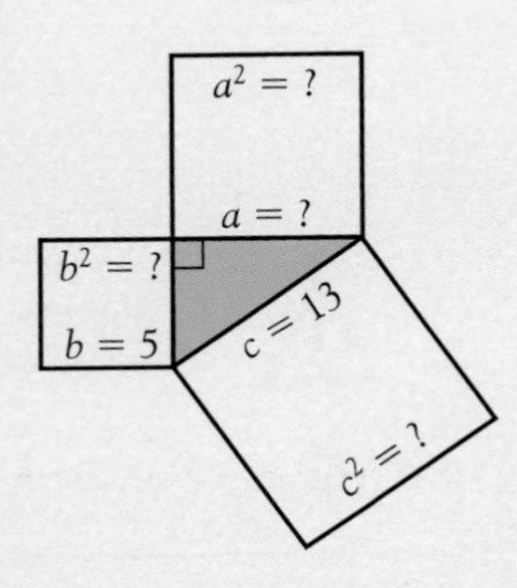

d

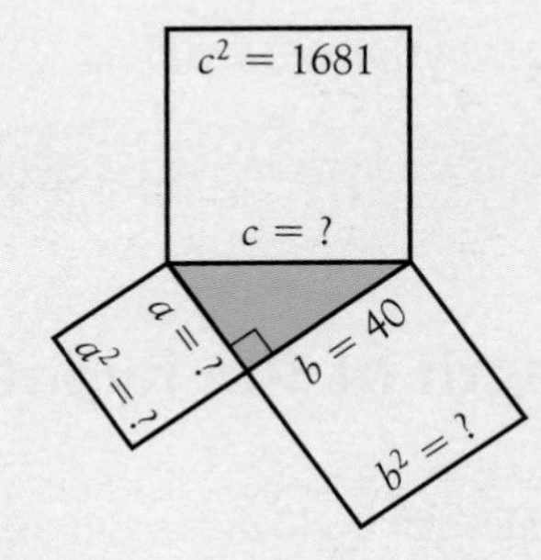

2 A triangle with sides 3 cm, 4 cm and 5 cm has a right angle because $3^2 + 4^2 = 5^2$. Which of the following are right-angled triangles?

a

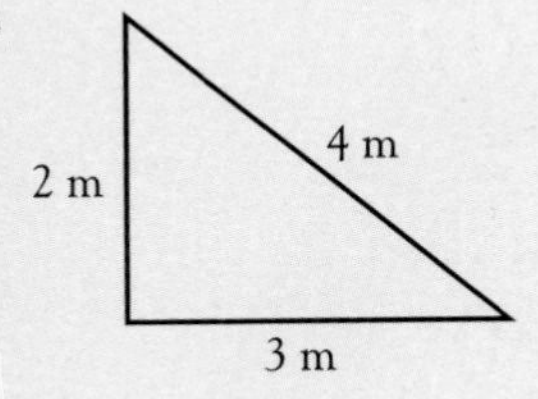

d

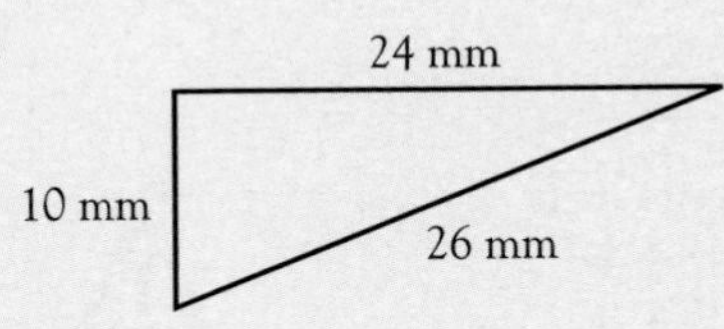

b

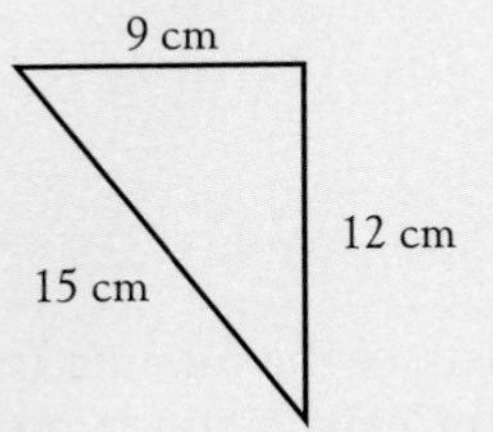

e

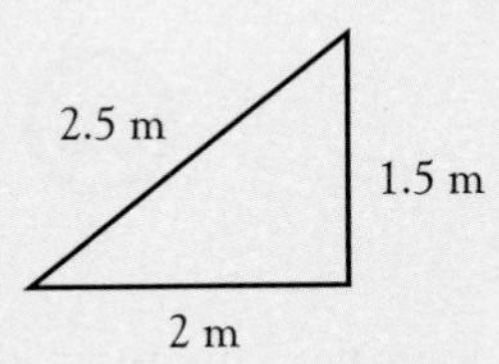

c

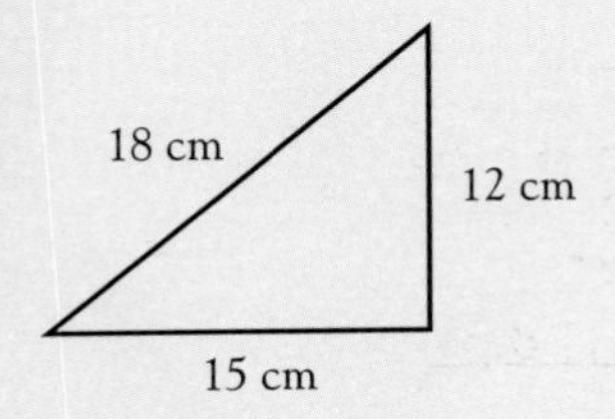

f

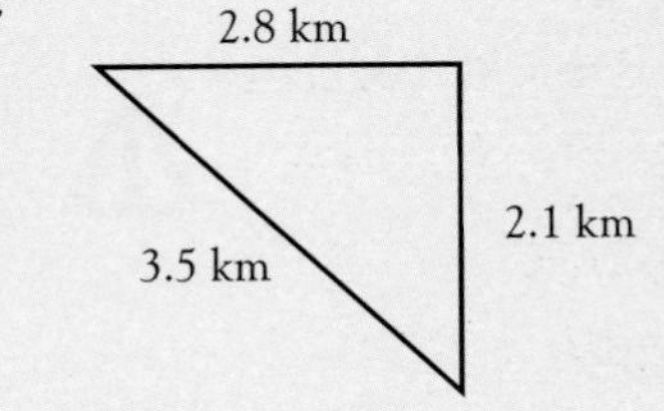

PYTHAGORAS' RULE

Pythagoras' Rule

You have been using a very important rule in mathematics, called Pythagoras' rule (sometimes called Pythagoras' theorem).

You must remember that you can only use this rule when the triangle is right-angled.

Pythagoras' rule states that in a right-angled triangle,

$$c^2 = a^2 + b^2$$

where c is the hypotenuse and a and b are the other two sides.

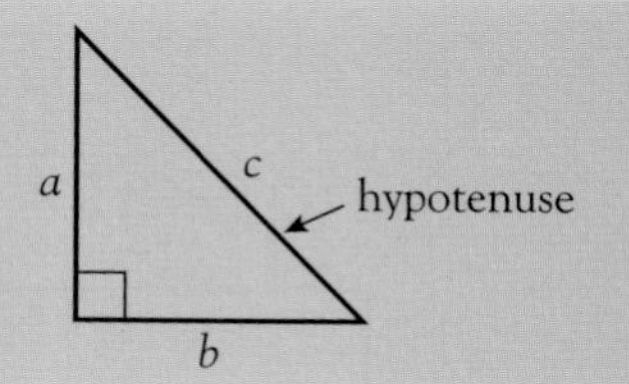

The hypotenuse is the longest side of the triangle and it does not form part of the right angle. The right angle mark points to the hypotenuse.

Sometimes people learn Pythagoras' rule as

'The square on the hypotenuse is equal to the sum of the squares on the other two sides.'

To find the sum, you add.

Finding the length of the hypotenuse

Sample Question 2

In this right-angled triangle $a = 4$ cm and $b = 7$ cm.
Find the length c, giving your answer to 2 s.f.

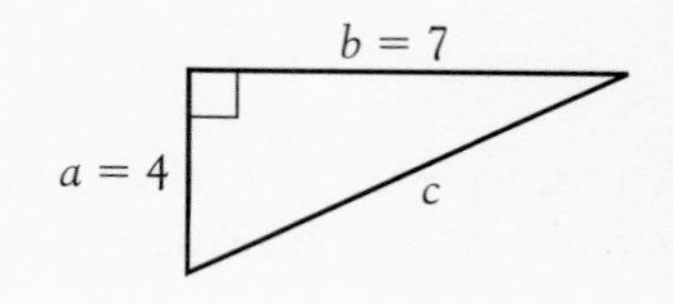

Answer

- Use Pythagoras' rule

$$c^2 = a^2 + b^2$$
$$c^2 = 4^2 + 7^2$$
$$= 65$$

You do not have to draw out the squares on the sides of the triangle. You can just picture them and write down the areas as a^2, b^2 and c^2.

- Find the square root

$$c = \sqrt{65}$$
$$= 8.062 \ldots$$

$\underline{c = 8.1 \text{ cm (2 s.f.)}}$

Significant figures, page 182

If you wish you can do your working more efficiently by making c the subject of the formula (rather than c^2), then substituting the values for a and b and finally working out the calculation in one go on your calculator:

$$c^2 = a^2 + b^2$$
$$c = \sqrt{a^2 + b^2}$$
$$= \sqrt{4^2 + 7^2}$$
$$= \sqrt{65}$$
$$= 8.062 \ldots$$

$\underline{c = 8.1 \text{ cm (2 s.f.)}}$

It is helpful to write down the value of $4^2 + 7^2$ as it makes it easier to check your work.

PYTHAGORAS' RULE

Check how to do this on **your** calculator. It could be

65 shows on display

You do not need to press = at the end on some calculators.

or

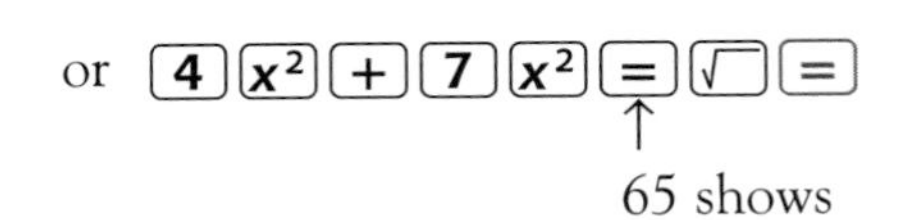

65 shows

You should always check your answer to make sure that it is sensible using the following facts:

- The hypotenuse is the longest side.
- The sum of the lengths of the two shorter sides is greater than the length of the hypotenuse.

A note about labelling

The sides of the triangle are not always labelled a, b and c.
You could use any letters, usually written in lower case, for example:

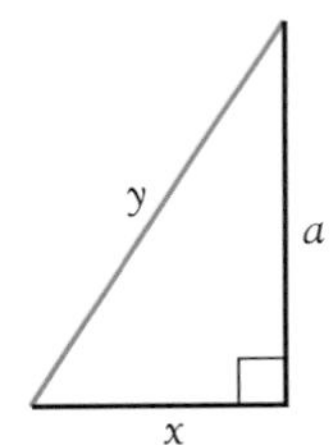

$y^2 = x^2 + a^2$

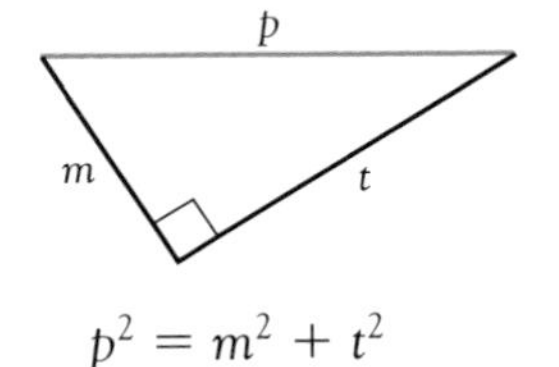

$p^2 = m^2 + t^2$

The hypotenuse is shown in red.

You could use capital letters to label the points of the triangle.

The distance between A and B can be written as AB or BA.

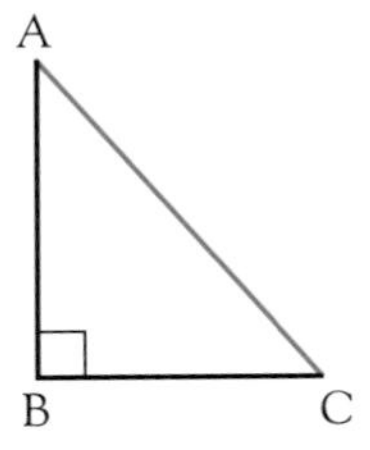

$AC^2 = AB^2 + BC^2$

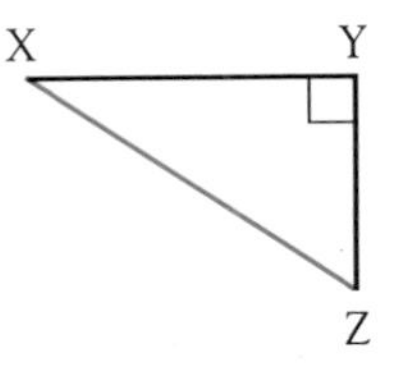

$XZ^2 = XY^2 + YZ^2$

Sample Question 3

A boat sailed due north from its mooring M to a marker buoy P, 2.5 km away. It then changed course and sailed 4.3 km east to a second buoy Q. How far was it then from its mooring? Give your answer in km, to 1 d.p.

Answer

- Draw a sketch, marking in all the details given in the question.

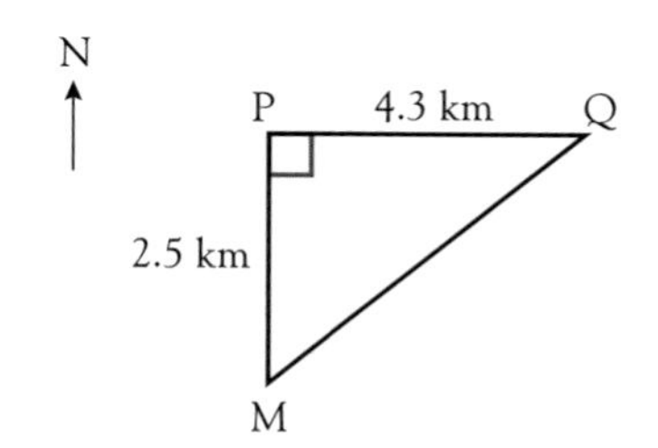

In a sketch you do not use accurate measurements, but you should use a ruler to draw the lines.

East is 90° clockwise from North, so there is a right angle at P. Mark it on your diagram.

- Find the hypotenuse, MQ, using Pythagoras' rule.

$$\begin{aligned} MQ^2 &= MP^2 + PQ^2 \\ &= 2.5^2 + 4.3^2 \\ &= 24.74 \\ MQ &= \sqrt{24.74} \\ &= 4.973 \ldots \\ &= 5.0 \text{ (1 d.p.)} \end{aligned}$$

<u>The marker buoy Q is 5.0 km from M.</u>

If you want to write your working making MQ the subject of the formula first:

$$\begin{aligned} MQ &= \sqrt{MP^2 + PQ^2} \\ &= \sqrt{2.5^2 + 4.3^2} \\ &= \sqrt{24.74} \\ &= 4.973 \ldots \\ &= 5.0 \text{ (1 d.p.)} \end{aligned}$$

- Check your answer
  - Is the hypotenuse the longest side? Yes, since 5.0 is bigger than 2.5 and 4.3.
  - Is MP + PQ longer than the hypotenuse?

$$MP + PQ = 2.5 + 4.3 = 6.8$$
$$MQ = 5.0$$

Yes, MP + PQ **is** longer than MQ

Exercise 17.2

1 Find the length of the hypotenuse in each of the following triangles:
(Give your answers correct to 3 significant figures.)

a

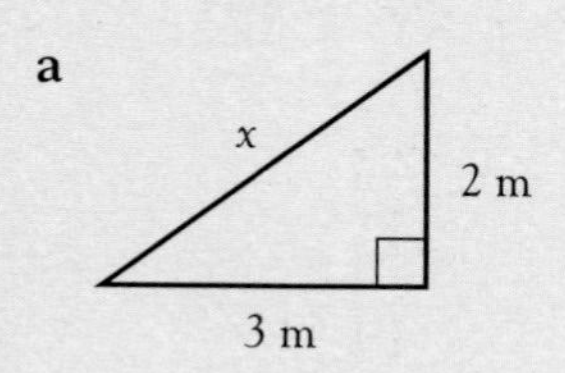

b

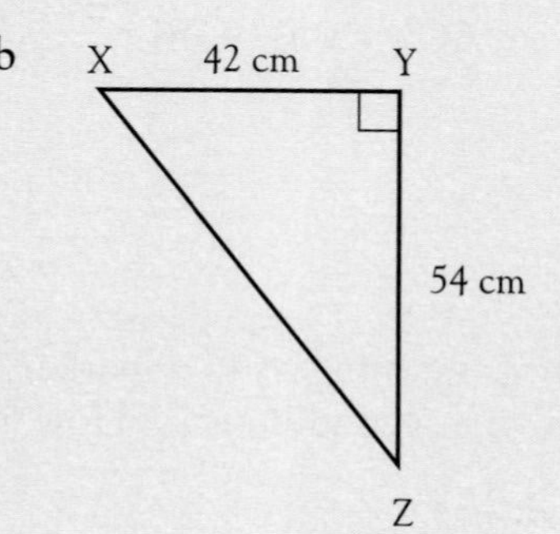

c

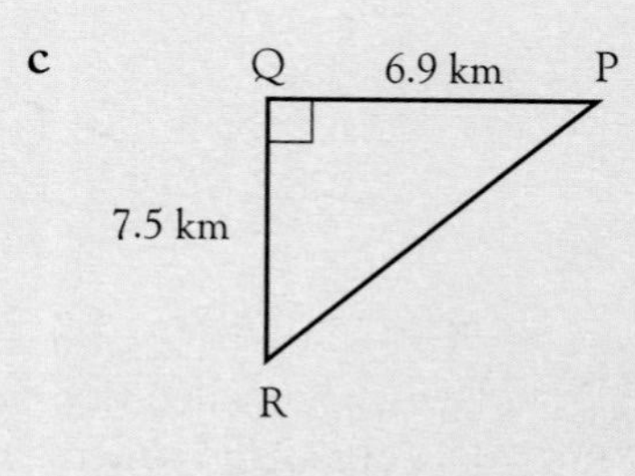

PYTHAGORAS' RULE

2 The sketch shows the end of a ramp.
Calculate the length, l, in metres correct to 2 decimal places.

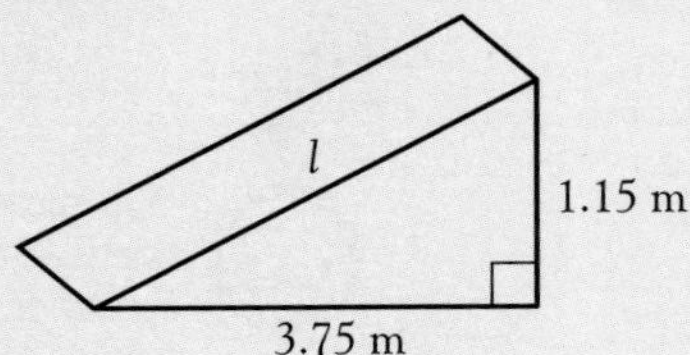

3 A rectangular gate ABCD is strengthened by adding a piece of wood along the diagonal DB as shown.
Calculate the length of this piece of wood, correct to 3 significant figures.

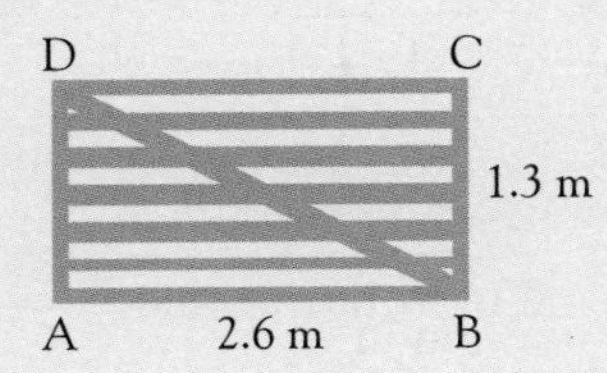

4 The diagram shows a shelf bracket ABC.
Calculate the length of AB, correct to the nearest mm.

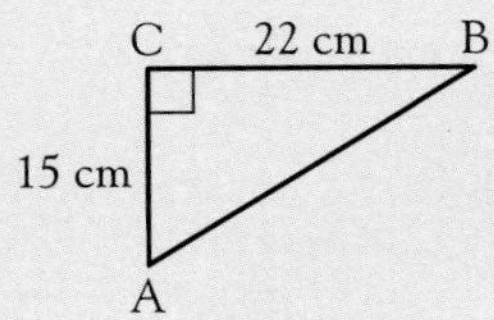

5 Ashton is 25 km due south of Woodfield and Canford is 31 km due west of Woodfield.

a Draw a diagram to show the relative positions of Woodfield, Ashton and Canford.

b Calculate the distance of Canford from Ashton. Give your answer correct to 3 significant figures.

6

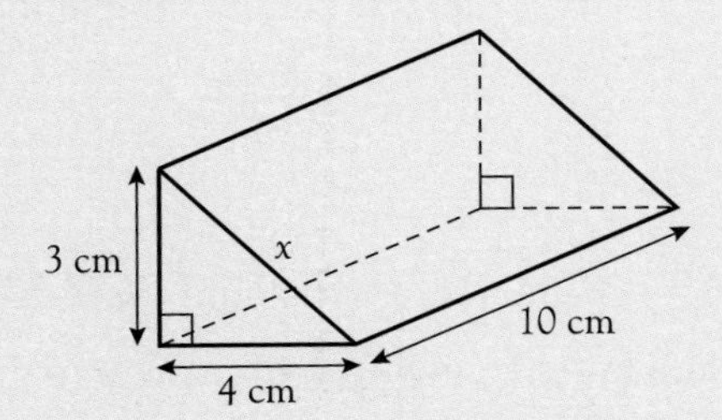

a Find the area of the triangular cross-section of this prism.

b Calculate the value of x.

c Calculate the surface area of the prism.

7

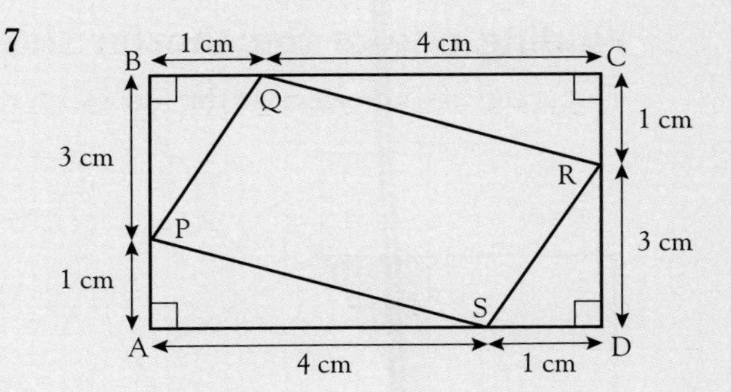

a Calculate the length PQ.

b Calculate the length QR.

c Calculate the perimeter of the quadrilateral PQRS.

8 A ship leaves port P and sails North for 12 km. It then changes course and travels East for 14 km before anchoring.

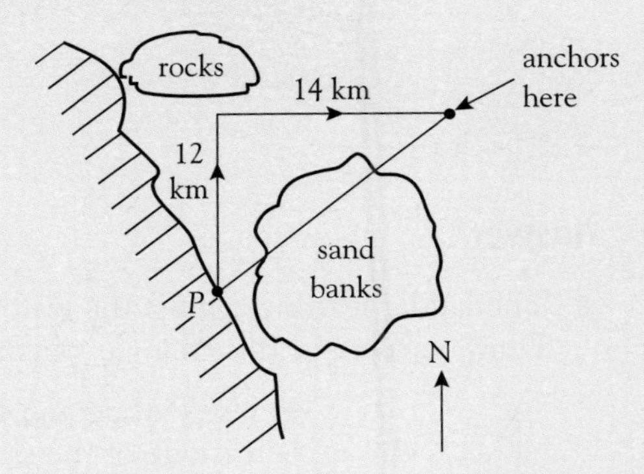

Calculate how far away from P the ship is anchored. [MEG]

9 Calculate the length of the diagonal of a square of side 10 cm.

10 The diagram shows the end of a lean-to shed.
Find the length of the slant edge.

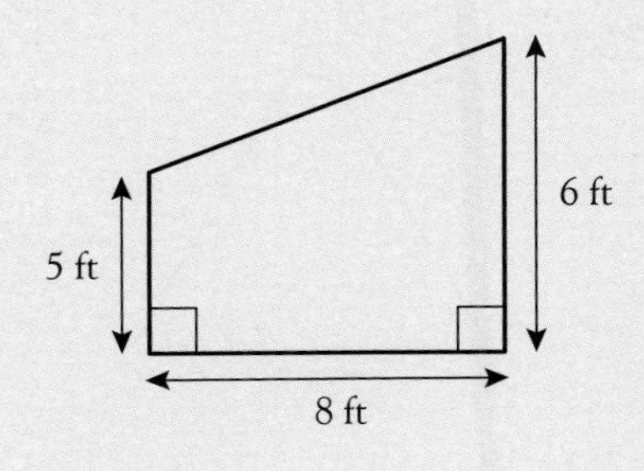

PYTHAGORAS' RULE

Finding one of the shorter sides

This is easy if you remember the squares on the sides of the triangle

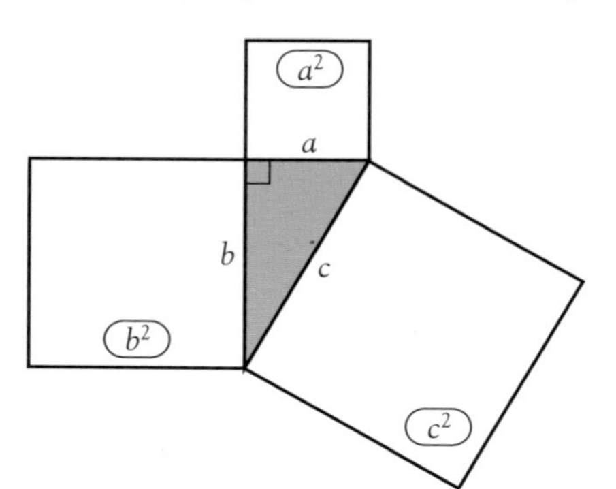

$c^2 = a^2 + b^2$

Therefore $a^2 = c^2 - b^2$

$a = \sqrt{c^2 - b^2}$

Also $b^2 = c^2 - a^2$

$b = \sqrt{c^2 - a^2}$

You have re-arranged the formula to make a or b the subject (page 265).

Notice that to find one of the shorter sides you have to subtract something. Remember that if c is the hypotenuse, c^2 is the 'most', so you must not add more!

Sample Question 4

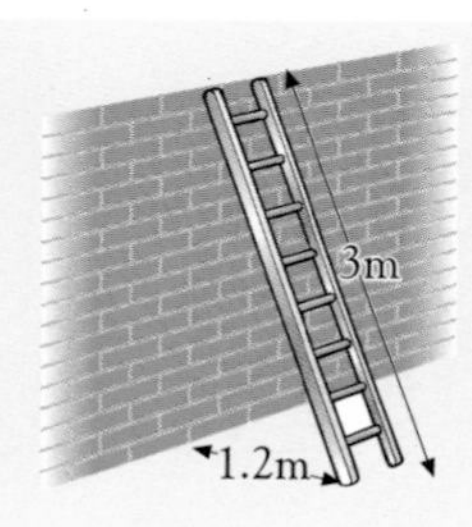

A ladder 3 m long is leaning against a wall.

The foot of the ladder is 1.2 m from the wall.

How far up the ladder does the wall reach?

Give your answer to the nearest cm.

Answer

- The ladder leaning against the wall forms a right-angled triangle. Draw a sketch of this triangle, putting in the measurements.

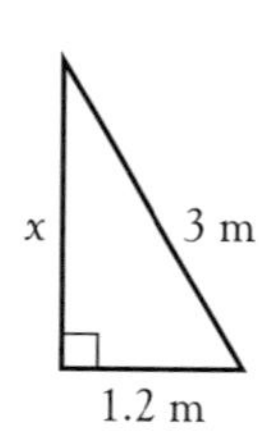

- Use Pythagoras' rule:
  x is one of the shorter sides, so the formula with x^2 as the subject contains a minus sign.

$$x^2 = 3^2 - 1.2^2$$
$$= 7.56$$
$$x = \sqrt{7.56}$$
$$= 2.749\ldots$$

- x is in metres, so to give your answer to the nearest cm, you need to approximate to 2 d.p.

$x = 2.75$ m (nearest cm)

The ladder reaches 2.75 m up the wall.

Sample Question 5

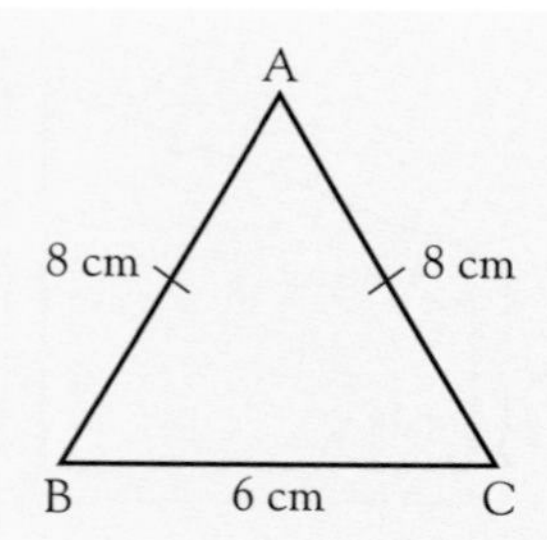

In triangle ABC, AB = 8 cm, AC = 8 cm and BC = 6 cm.

Find the area of triangle ABC, giving your answer to 3 significant figures.

This question does not appear to have anything to do with Pythagoras' rule as the triangle does not have a right angle.

PYTHAGORAS' RULE

Answer

- Write down the formula for the **area of a triangle**.

page 104

$$\text{Area} = \frac{\text{base} \times \text{height}}{2}$$

- Draw a sketch taking BC as the base and draw in the perpendicular height. You will need to find this height.

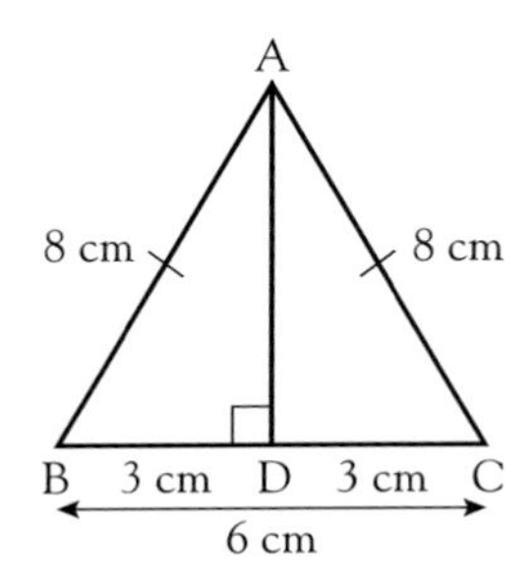

△ABC is isosceles, so AD is a line of symmetry. It is at 90° to BC and D is the mid-point of BC. You have now formed two congruent right-angled triangles.

- Sketch one of the right-angled triangles – it does not matter which one.

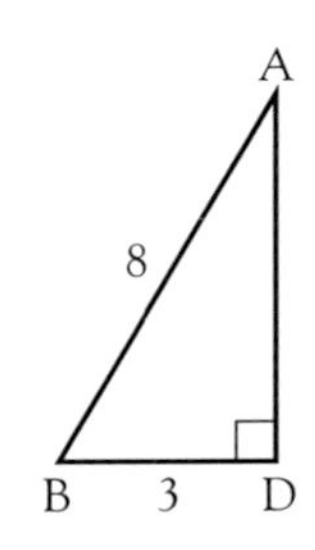

- Use Pythagoras' rule to find AD.
  Notice that it is one of the shorter sides.

$$\begin{aligned} AD^2 &= AB^2 - BD^2 \\ &= 8^2 - 3^2 \\ &= 55 \\ AD &= \sqrt{55} \\ &= 7.416\ldots \end{aligned}$$

Check:
hypotenuse (AB) longest side ✓
AD + BD > AB ✓

DO NOT round this answer as you are going to need it to find the area. Put it into your calculator memory **Min**.

- Find the area of △ABC with base BC = 6 cm, height AD = 7.416 … cm

$$\text{Area} = 6 \times \frac{\mathbf{7.416\ldots}}{2}$$

$$= 22.24\ldots$$

<u>Area = 22.2 cm² (**3 s.f.**)</u>

If 7.416 … is still on your calculator display, just multiply by 6 and divide by 2. If it is not on your display, recall it from the memory

Remember to approximate your final answer to 3 s.f. as requested in the question.

PYTHAGORAS' RULE

Exercise 17.3

1 Calculate the length of the sides marked by letters, giving your answers to 3 significant figures.

a
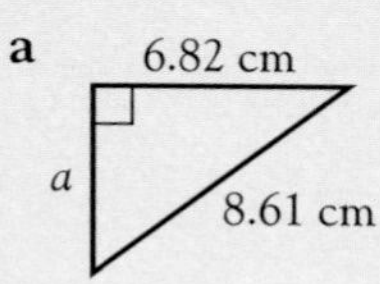

b
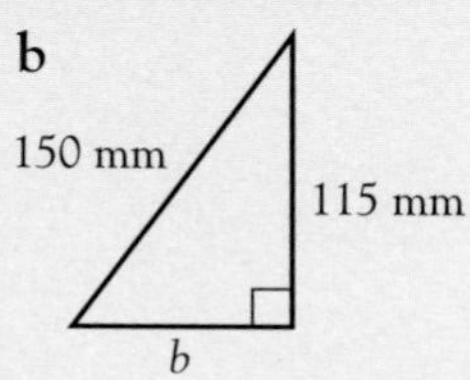

c
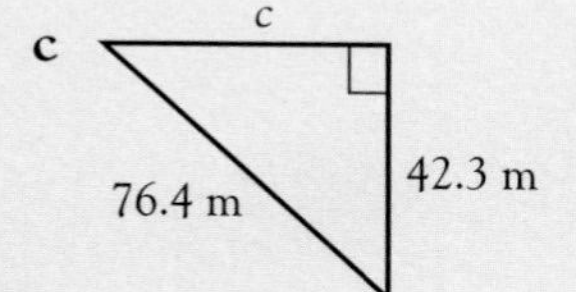

d
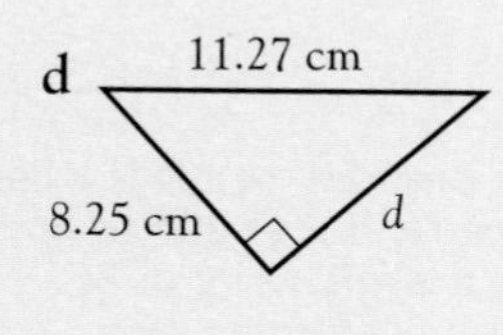

2 A piece of cheese is in the shape of a triangular prism as shown.

Without using a calculator, find

a the height, h

b the volume of cheese.

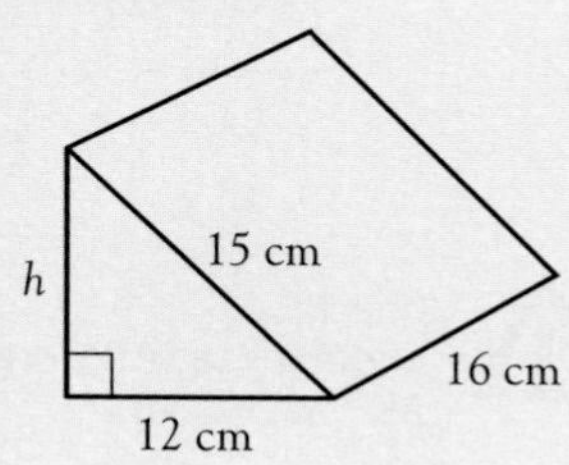

3 An isosceles triangle ABC has equal sides AB and AC of length 84 mm. The other side BC is 96 mm long. Determine:

a the height, h

b the area of the triangle.

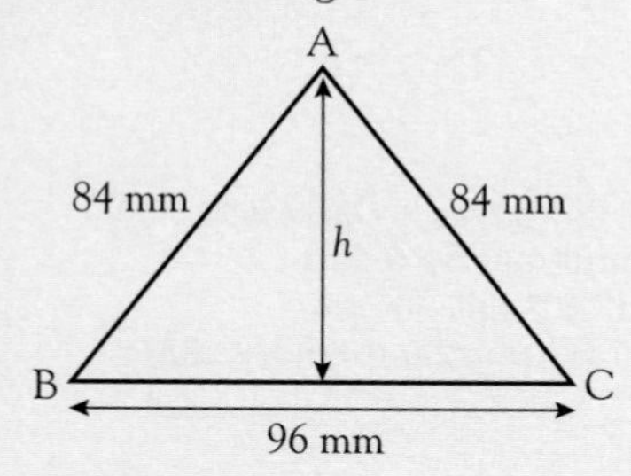

4 a A ladder of length 7.5 metres rests against the wall of a house as shown in the sketch. The foot of the ladder is 1.45 metres from the bottom of the wall. Find, correct to 2 decimal places, the height in metres of the top of the ladder above the ground.

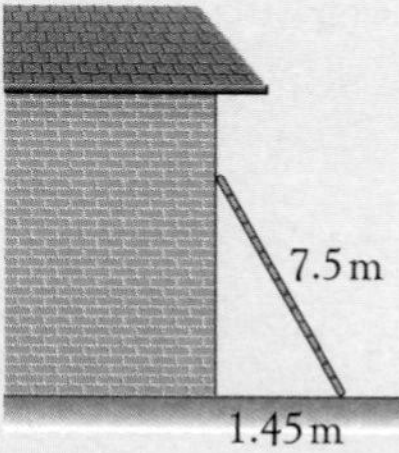

b The bottom of the ladder is moved nearer to the wall so that the top rests on the wall at a height of 7.45 metres above the ground.

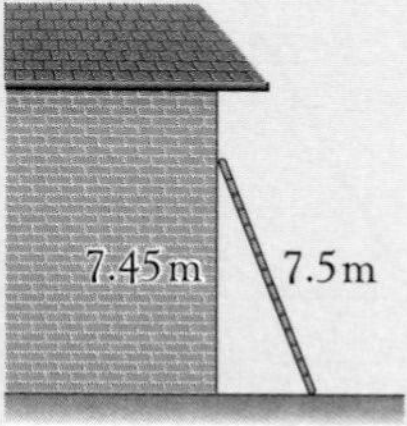

Calculate, to 2 decimal places, the distance in metres between the foot of the ladder and the bottom of the wall.

5

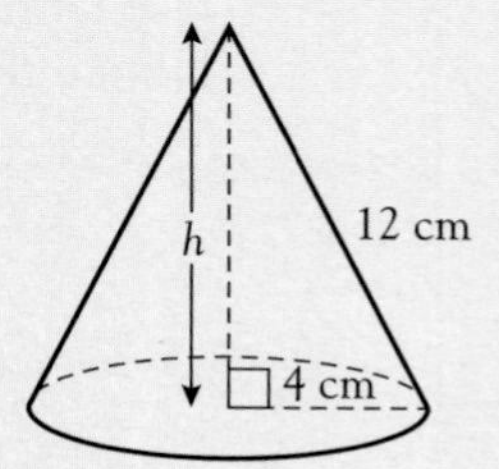

The slant height of this cone is 12 cm and the base radius is 4 cm.

a Find the perpendicular height, h, of the cone. Give your answer to 3 s.f.

b The formula for the volume of this cone is $V = \frac{1}{3}\pi r^2 h$ where r is the base radius and h is the perpendicular height.

Work out the volume of the cone giving your answer to 2 s.f.

6 Calculate the perpendicular height of an equilateral triangle with side 8 cm.

7 Find the sides marked x and y and hence work out the perimeter of triangle ABC.

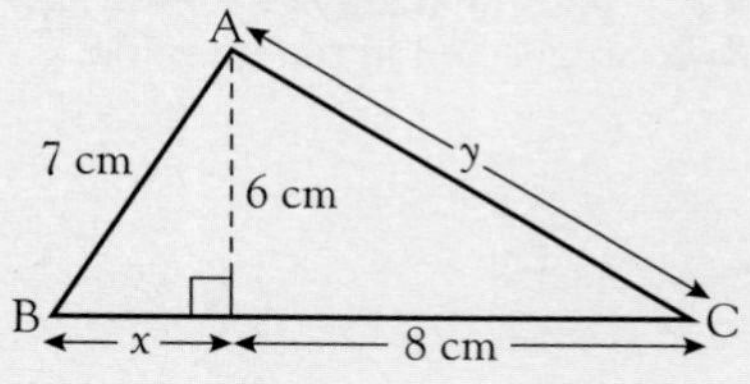

PYTHAGORAS' RULE

Mixed problems

When you do not appear to have a right-angled triangle, look out for situations in which you can add a line or use a geometry fact to get one.

Triangles (isosceles and equilateral)

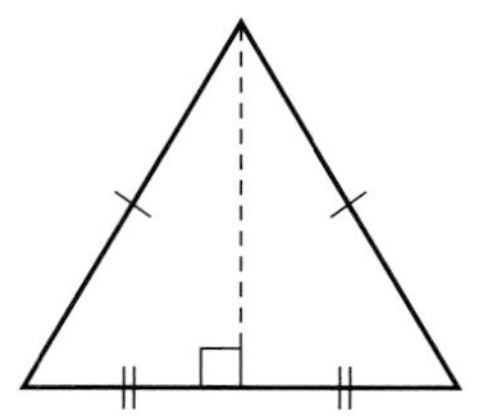

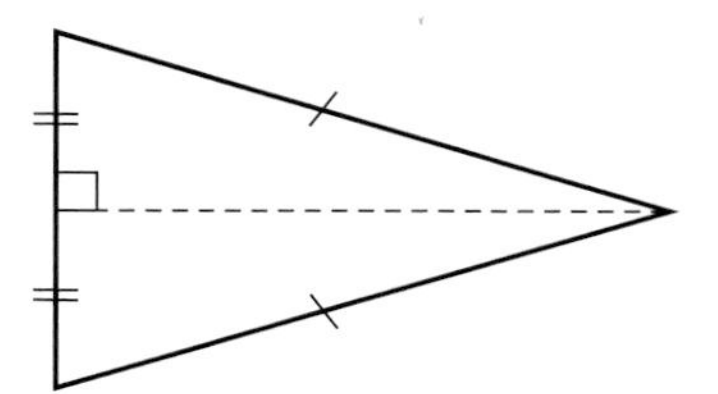

Trapezia

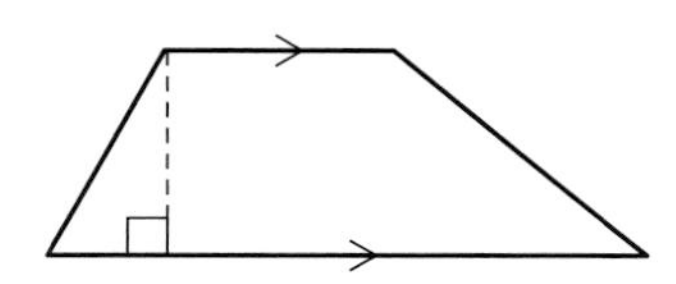

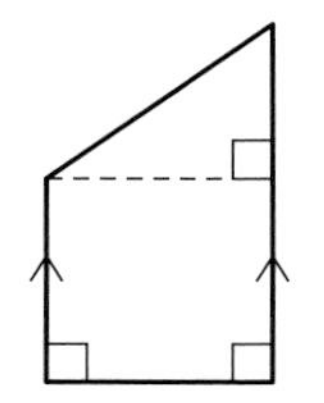

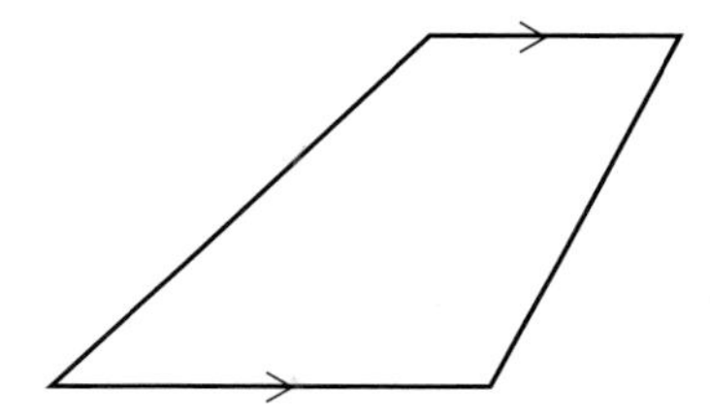

Triangles inside circles

Chapter 12 page 235

O

Angle in a semicircle is 90°

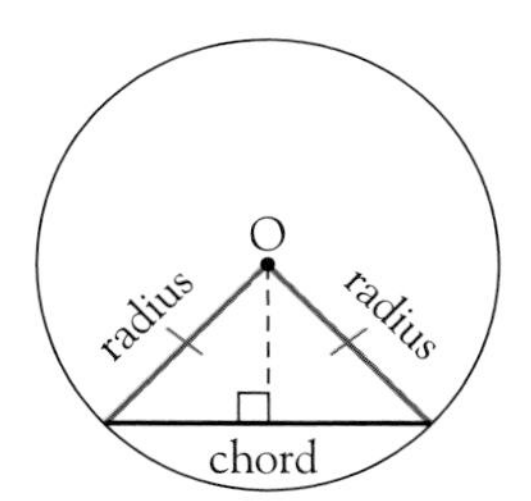

Finding the distance from a chord to the centre of a circle

Look out for the isosceles triangle with the radii as equal lengths.

Special quadrilaterals

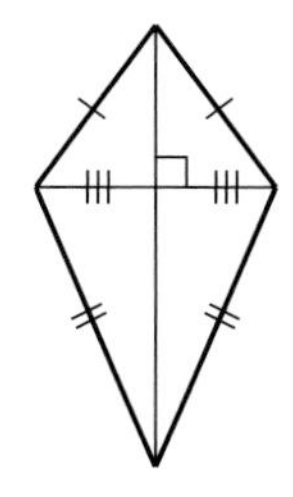

Diagonals of a kite intersect at 90°

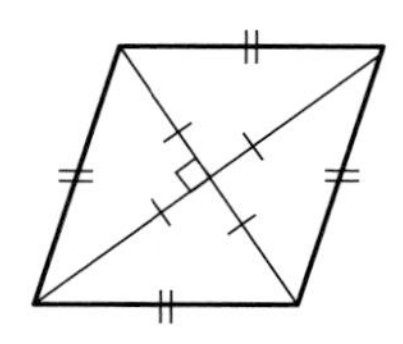

Diagonals of a rhombus cut each other in half at 90°

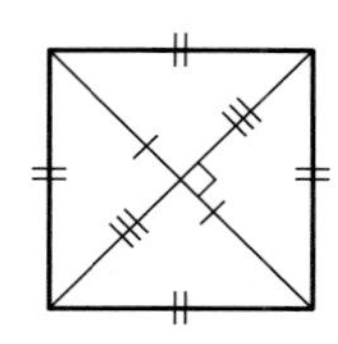

A square is a special rhombus

PYTHAGORAS' RULE

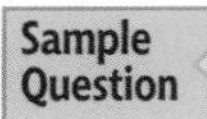

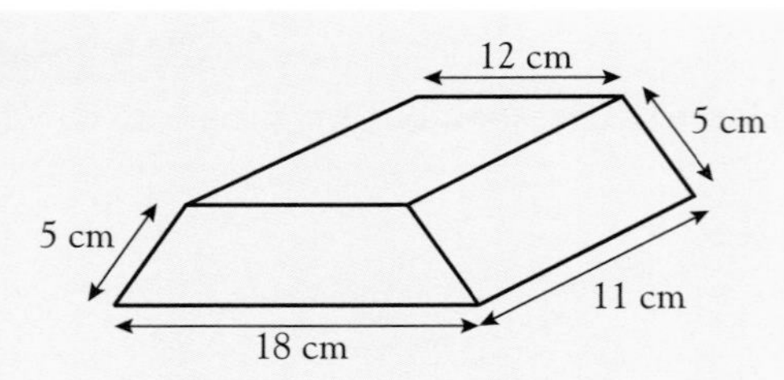

Find the volume of this trapezoidal prism.

Answer

- Think what you will need to find to calculate the volume.

  Volume of a prism = area of cross-section × length

  You know the length; it is 11 cm, but you will have to work out the area of the cross-section.

- The cross-section is a trapezium. Sketch it and put in all the measurements.

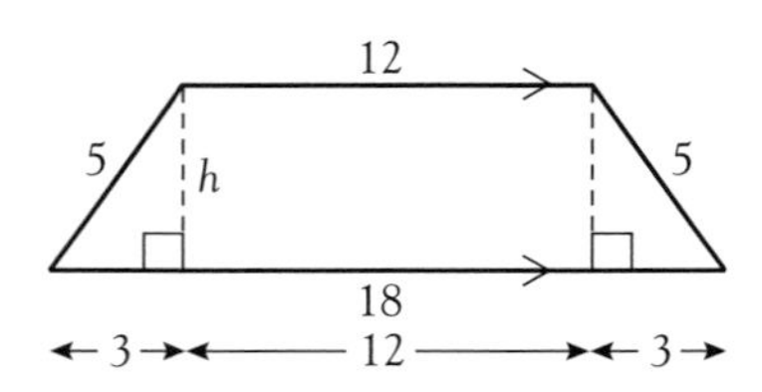

- To find the area you will need the distance between the parallel lines, i.e. the perpendicular height, h. Draw this on the diagram.

- Work out how the bottom length of 18 cm is split.

- Sketch one of the triangles and use Pythagoras' rule to find h.

$$\begin{aligned} h^2 &= 5^2 - 3^2 \\ &= 16 \\ h &= \sqrt{16} \\ h &= 4 \end{aligned}$$

- Calculate the area of the trapezium (see page 281).

$$\begin{aligned} A &= \tfrac{1}{2}(a + b)h \\ &= \tfrac{1}{2} \times (18 + 12) \times 4 \\ &= 60 \text{ cm}^2 \end{aligned}$$

$a + b$ is the sum of the parallel sides.

- Calculate the volume

$$\begin{aligned} V &= \text{Area of cross-section} \times \text{length} \\ &= 60 \times 11 \\ &= 660 \end{aligned}$$

Remember to put in the correct units for volume at the end.

Volume of the prism = 660 cm^3

PYTHAGORAS' RULE

Sample Question 

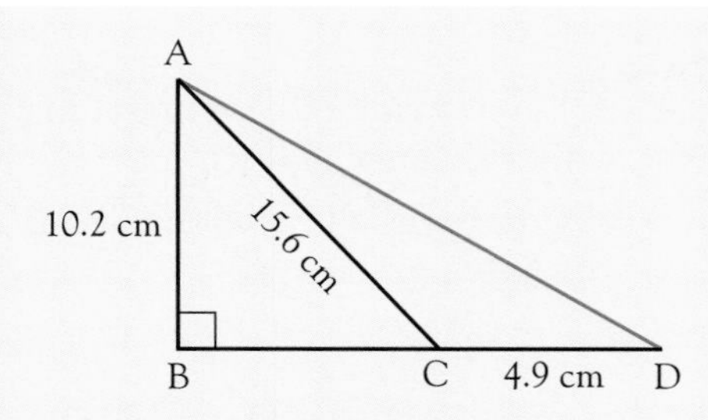

Calculate the length AD, giving your answer to 2 s.f.

Answer

AD is the third side of △ACD but DO NOT try to use Pythagoras' rule in △ACD as it is not right-angled.

- Sketch △ABC and calculate BC (shorter side) using Pythagoras' rule.

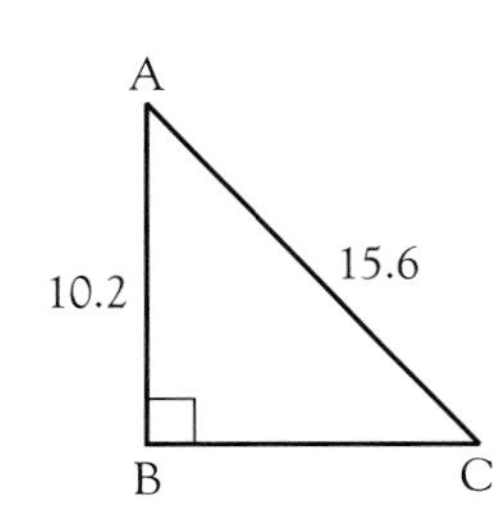

$$\begin{aligned} BC^2 &= AC^2 - AB^2 \\ &= 15.6^2 - 10.2^2 \\ &= 139.32 \\ BC &= \sqrt{139.32} \\ &= \mathbf{11.80\ldots} \end{aligned}$$

DO NOT approximate 11.80 ... Leave it on your calculator display and write down the first four figures followed by ...

- Work out the length of BD.

$$\begin{aligned} BD &= BC + CD \\ &= 11.80\ldots + 4.9 \\ &= \mathbf{16.70\ldots} \end{aligned}$$

Put 16.70 ... into your calculator memory.

- Look at △ABD and calculate AD (hypotenuse) by Pythagoras' rule.

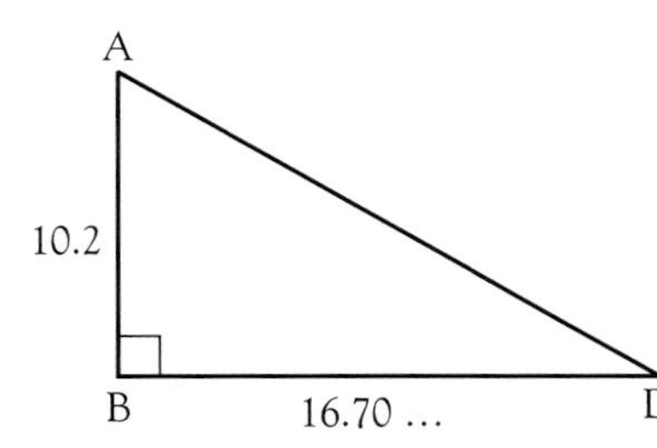

$$\begin{aligned} AD^2 &= AB^2 + BD^2 \\ &= 10.2^2 + \mathbf{16.70\ldots}^2 \\ &= 383.04\ldots \\ AD &= \sqrt{383.04\ldots} \\ &= 19.57\ldots \\ AD &= \mathbf{20\ cm\ (2\ s.f.)} \end{aligned}$$

Use 16.70 ... from your display or recall it from the memory.

Approximate your final answer – not before.

Do not clear your calculator at any stage until you are sure that you will not need the number again. It is wise to store it in your calculator memory rather than write down the whole display each time.

PYTHAGORAS' RULE

Exercise 17.4

1 A telephone cable stretches from the top of a post to the end of a block of flats, as shown.

Calculate the length of the cable.

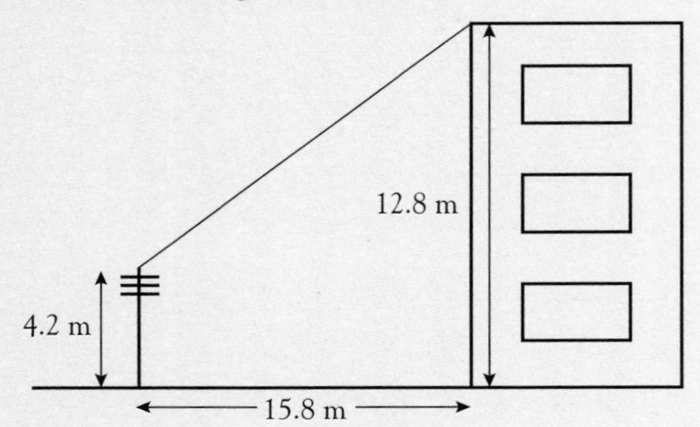

2 For the triangle shown below, some of the possible values of a, b and c are shown in the table.

a Use Pythagoras' theorem to find the missing values.

a	b	c
	4	5
5	12	
7		25
	40	41
11	60	
13		85
	112	113

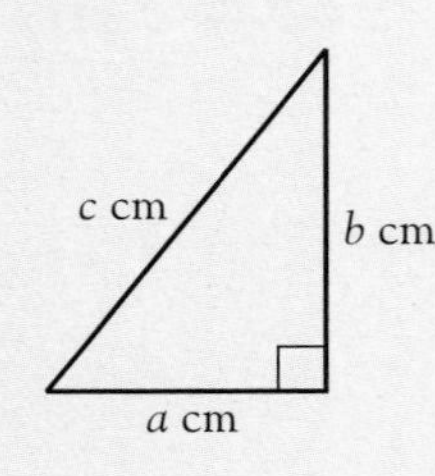

b What type of numbers are in column a?

c What is the relationship between the values of b and c?

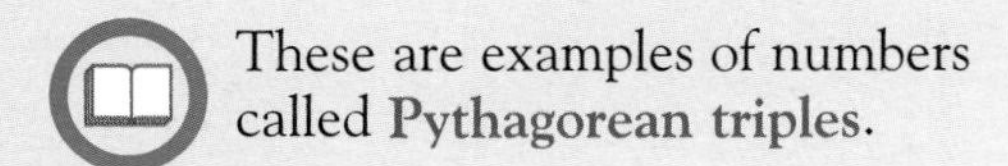

These are examples of numbers called **Pythagorean triples**.

3 Quintown is 16 km north-west of Pitbury and Redbridge is 27 km south-west of Pitbury, as shown.

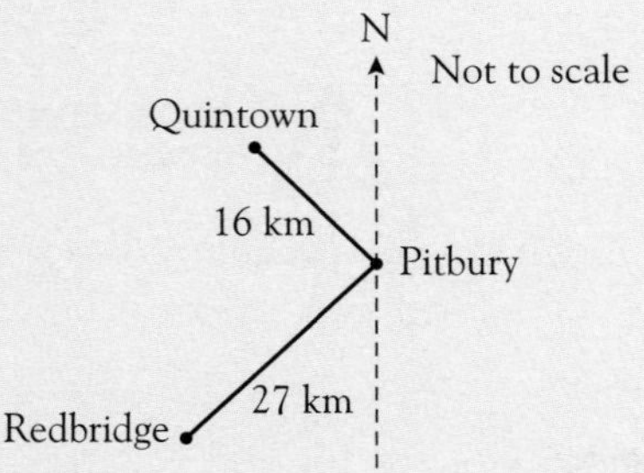

Calculate, to the nearest kilometre, the distance between Quintown and Redbridge.

4 A kite is made by covering a framework of rods with a sheet of plastic.

The rod AB is 15 cm long, BC is 25 cm long and BD is 18 cm long.

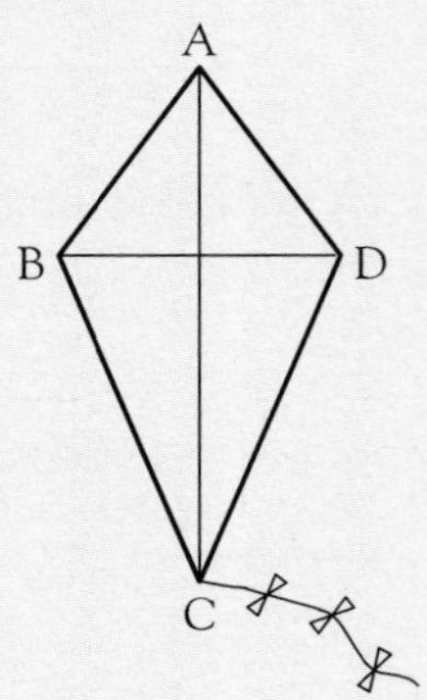

Find, giving your answers to 2 significant figures,

a the length of AC

b the total length of the rods

c the total area of plastic sheet needed to cover the kite.

5 A ridge tent has the dimensions shown in the sketch. Find

a h, the height of the tent

b the total area of the canvas.

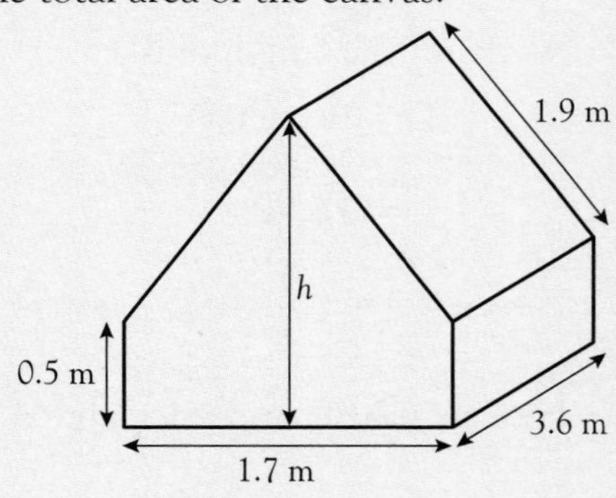

6 Giving your answers to 2 significant figures, for each part find

i the value of x, **ii** the shaded area.

a

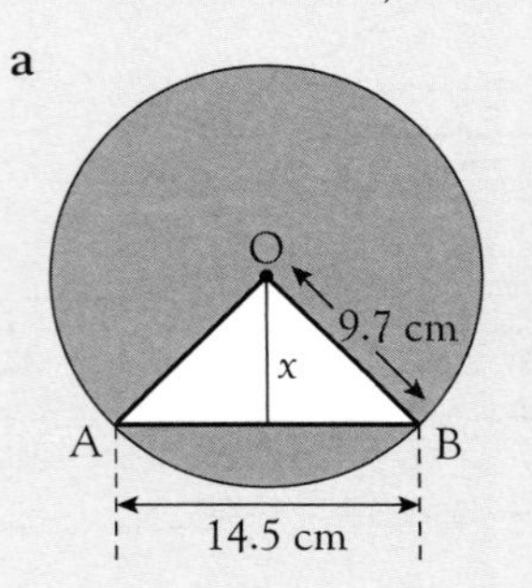

b

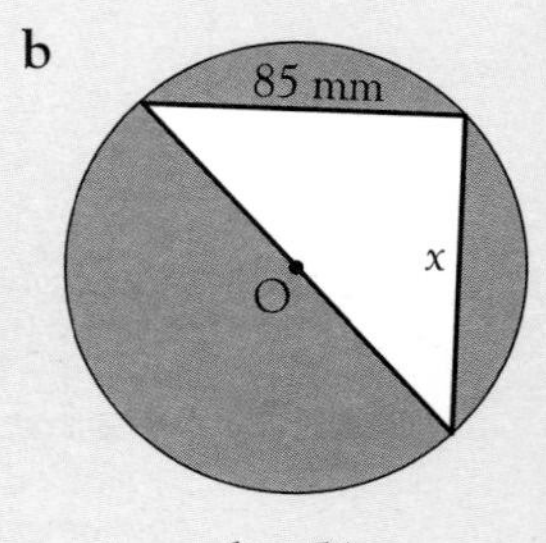

radius 54 mm

PYTHAGORAS' RULE

7 PQR is a right-angled triangle. PQ is of length 3 m and QR is of length 2 m. Calculate the length of PR.

Not to scale

R, P, Q, 2 m, 3 m

[L]

8

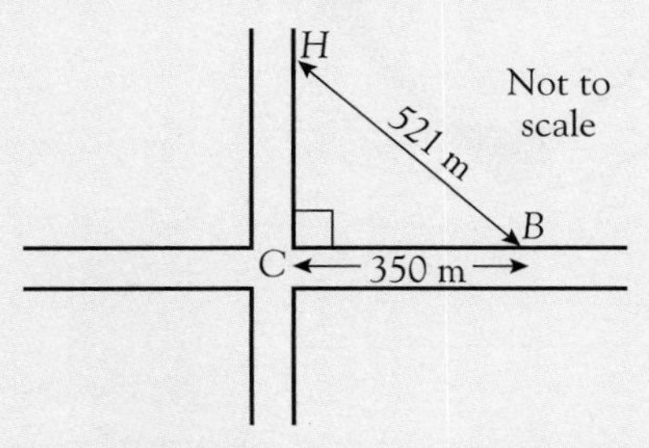

Mohamed takes a short cut from his home (*H*) to the bus stop (*B*) along a footpath *HB*.

How much further would it be for Mohamed to walk to the bus stop by going from *H* to the corner (*C*) and then from *C* to *B*?

Give your answer to the nearest metre. [MEG]

9 In a game two umpires stand on a pitch as shown.

How far apart are the two umpires?

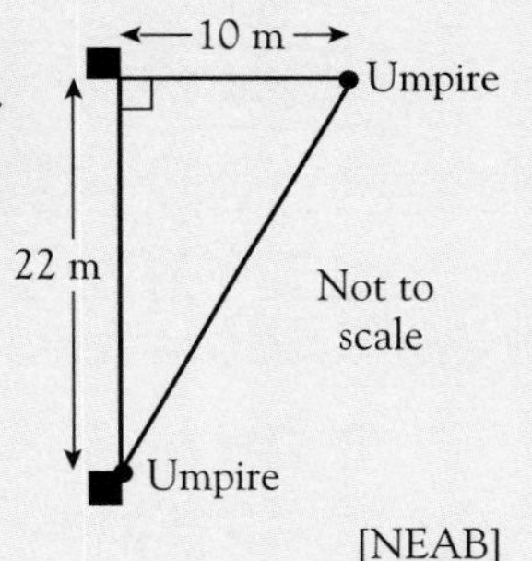

[NEAB]

10 The cross-section of a road tunnel is part of a circle of radius 5.0 metres, as shown.

The width of the road is 8.2 metres.

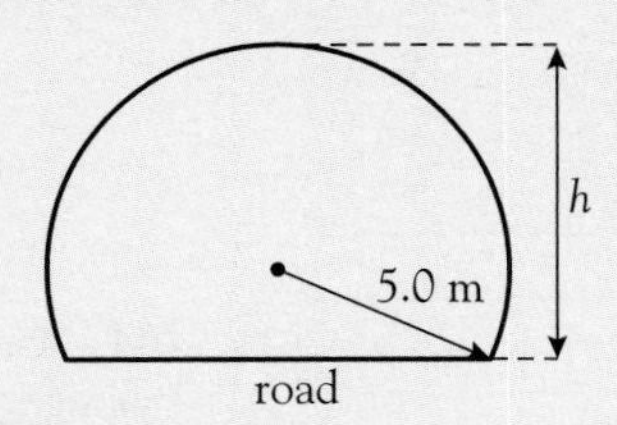

Find, correct to 2 significant figures, *h*, the height of the tunnel.

11 In this rhombus AB = 5 cm and BD = 6 cm.

a Find the length AC.

b Find the area of the rhombus.

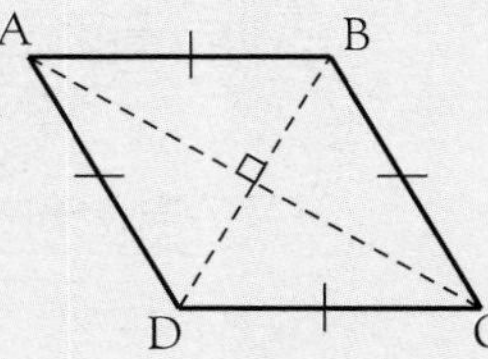

12 A gold ingot is a prism, with trapezium cross-section, with dimensions as shown in the sketch.

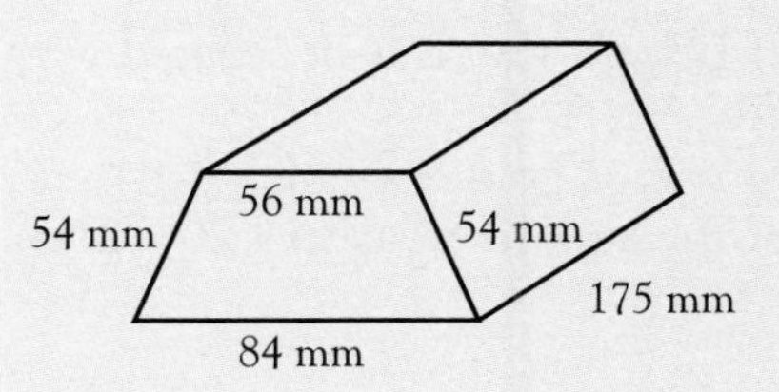

Find, to 2 significant figures

a the height of the trapezium cross-section,

b the area of the trapezium,

c the volume of the ingot.

13

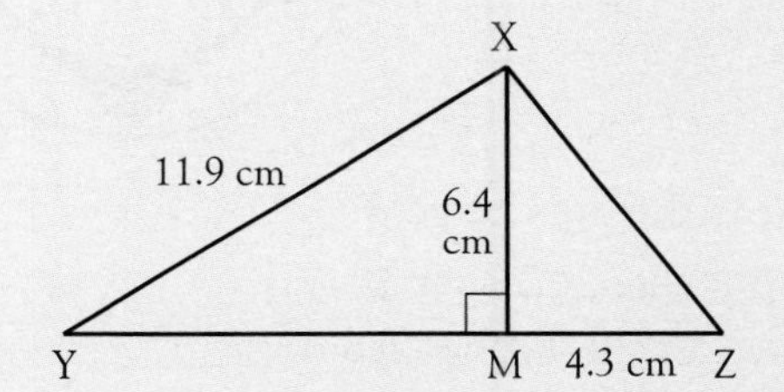

In the diagram XM is perpendicular to YZ, YX = 11.9 cm, MZ = 4.3 cm and XM = 6.4 cm.

a Find the length of XZ.

b Find the length of YM.

c Find the area of triangle XYZ.

14 A square-based pyramid has the net shown.

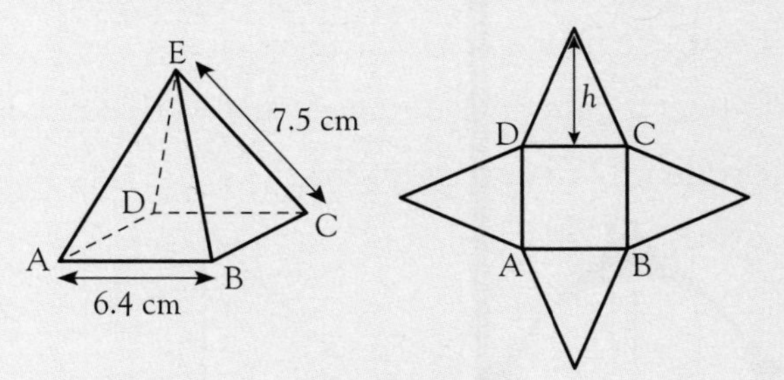

The base has sides of length 6.4 cm and the slant edges are 7.5 cm long.

Calculate

a the length of the diagonal of the base, AC,

b the height of the triangular sides, *h*,

c the total surface area of the pyramid.

EXAM QUESTIONS

Worked Exam Question
[MEG]

The face of Brian's watch is decorated with two circles and a square.
The shaded part is gold.
One side of the square measures 20.0 mm.

a What is the radius of the small circle?

◆ Write the measurements for the square onto the diagram and put in a diameter of the smaller circle. It is then easy to work out the radius.

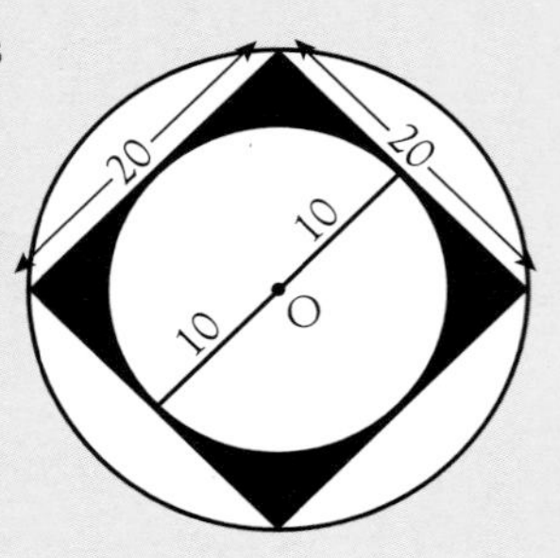

Answer 10 mm

b What is the area of gold?

◆ Find the area of the circle and subtract this from the area of the square.

You can use the calculator value for π. If you do, your answer will be 85.840 ...
Approximate this sensibly.

Area of circle $= \pi r^2$
$= 3.14 \times 10^2$
$= 314$
Area of square $= 20 \times 20$
$= 400$
Area of gold $= 400 - 314$
$= 86$

Answer 86 mm^2

c Calculate the radius of the large circle.

◆ Draw in a diameter of the larger circle.

◆ Extract a right-angled triangle and work out d using Pythagoras' rule.

◆ Divide by 2 to find the radius.

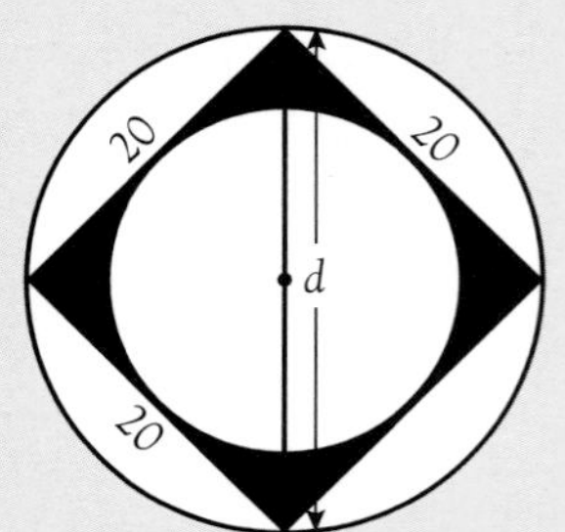

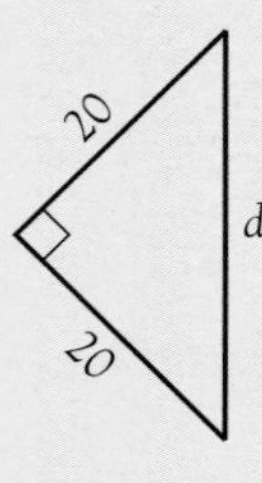

$d^2 = 20^2 + 20^2$
$= 800$
$d = \sqrt{800}$
$= 28.28 \ldots$
$r = 14.14 \ldots$

Answer 14.1 mm

At first sight this question does not appear to be about Pythagoras' rule. Watch out for questions that mix topics.

COMMENTS

A1 for correct answer
1 mark

M1 for area of circle
M1 for area of square
A1 for correct answer
3 marks

M1 for using Pythagoras' rule correctly
M1 for finding the square root
A1 for correct answer
3 marks

Exam Questions

1 Pauline is building a greenhouse.

The base, $PQRS$, of the greenhouse is a rectangle measuring 2.6 m by 1.4 m.

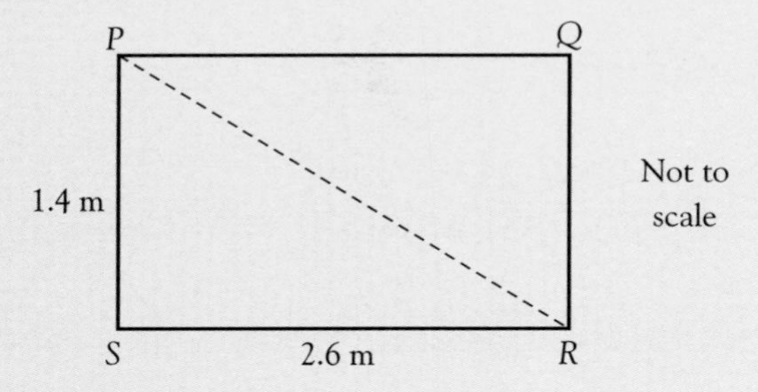

Calculate the length of PR when the base is rectangular.

You **must** show all your working. [SEG, p]

2 The diagram shows a symmetrical trapezium.

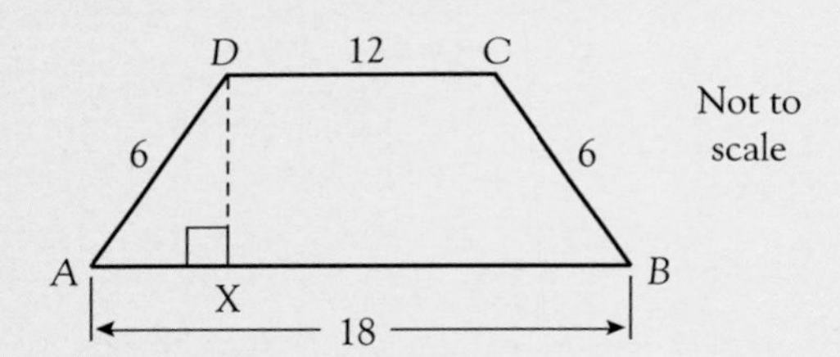

$AB = 18$ cm, $DC = 12$ cm, $AD = CB = 6$ cm
Angle $AXD = 90°$

a Write down the length of AX.

b Calculate the length of DX.

c Calculate the area of the trapezium. [SEG]

3 A rubbish skip has the shape of a prism.

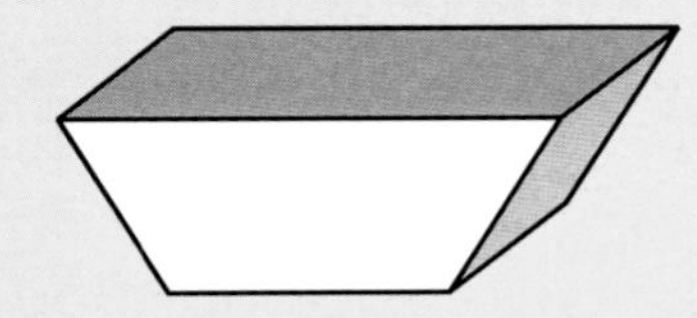

The cross-section of the skip is an isosceles trapezium ABCD, where AB = 3.2 m, DC = 2 m and AD = BC = 1.5 m.

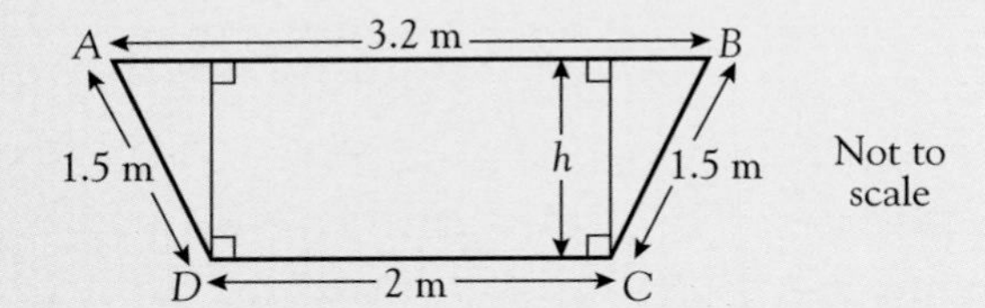

a Calculate the height, h, of the side of the skip.

b The skip is 1.7 m wide. Calculate the volume of the skip. [MEG]

4 A tent is 4 m wide and 2.5 m high.

Use Pythagoras' rule to work out the length of the sloping side s.

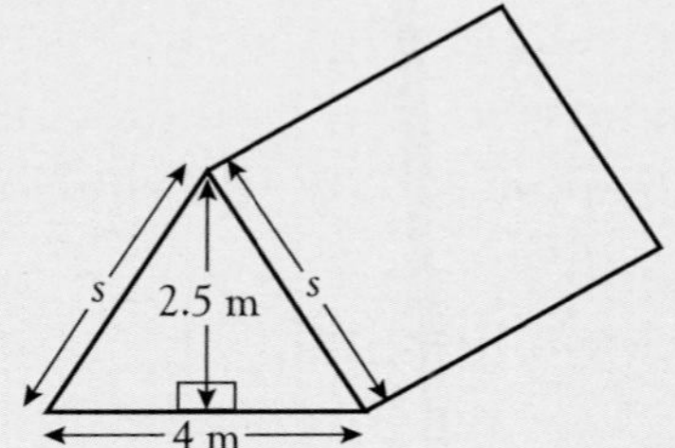

[MEG]

5 The diagram shows a large tent.

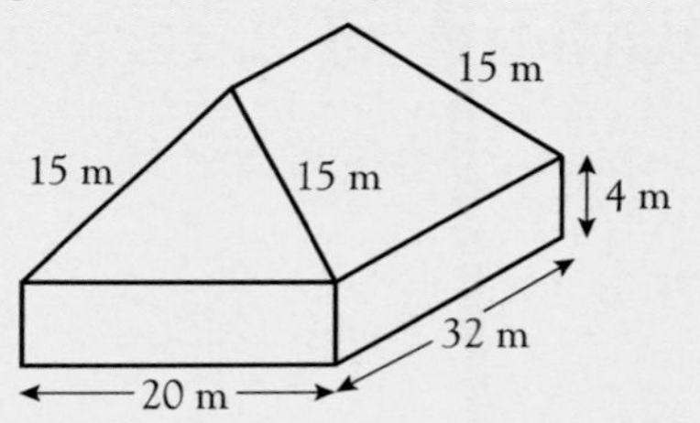

a The four vertical walls are rectangles.

Find the total area of the 4 vertical walls.

b One of the triangles used in the roof is shown below.

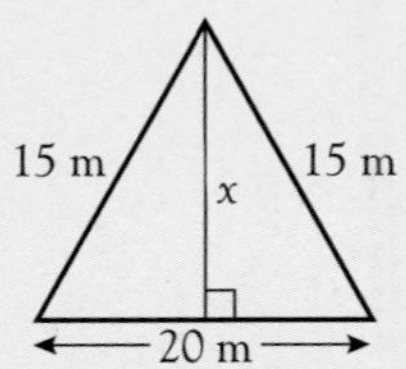

i Calculate the height, x, of the triangle.

ii Find the area of the triangle.

c One of the trapeziums used in the roof is shown below.

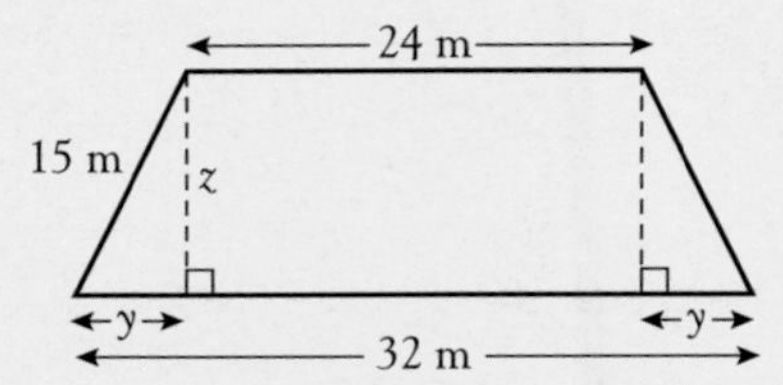

i Find the length marked y.

ii Calculate the height, z, of the trapezium.

iii Find the area of the trapezium.

d Find the total surface area of the tent. [MEG]

18 LINEAR GRAPHS II

In this chapter you will learn

- **how to work out the gradient of a straight line**
- **about practical applications of gradients**
- **about the 'gradient–intercept' form of the equation of a line**
- **how to find the equation connecting two variables, when the relationship is linear**

Gradient of a straight line

These diagrams show slopes in two different directions. They are known as positive and negative slopes.

Positive slopes

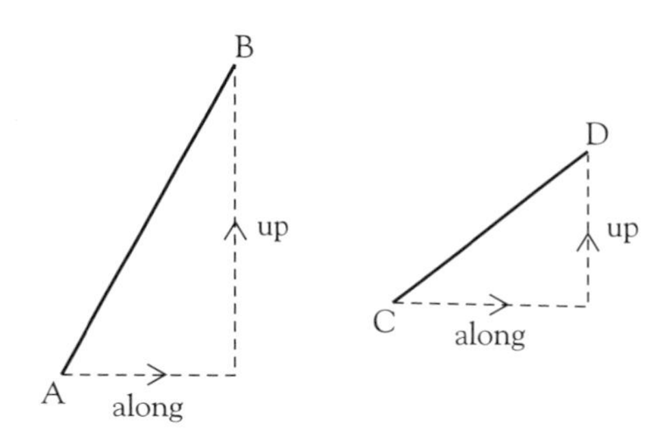

Negative slopes

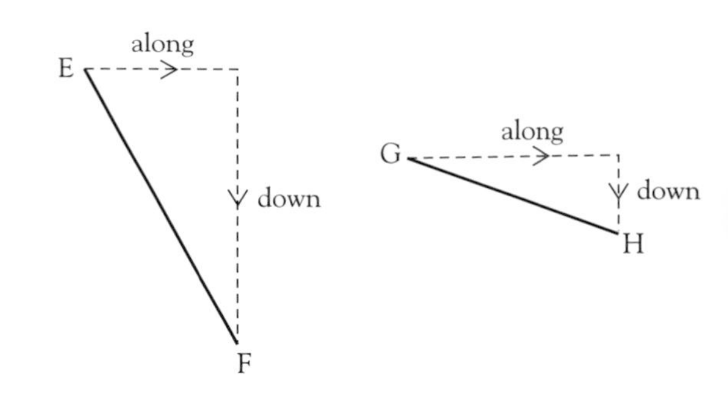

As you go along horizontally, in the positive direction
the line goes **up** for positive slopes and the line goes **down** for negative slopes.

Notice that AB is steeper than CD and that EF is steeper than GH.

The **gradient** of a line is a number that describes the slope. The gradient tells you which way the line is sloping +/ or −\, and how steep it is.

The gradient compares the **vertical** change with the **horizontal** change in the positive direction.

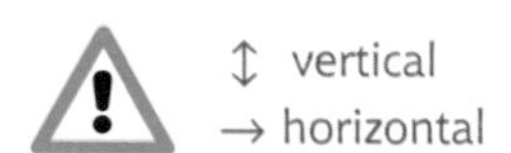

$$\text{Gradient} = \frac{\text{vertical change}}{\text{horizontal change}}$$

GRADIENT

For the line AB:

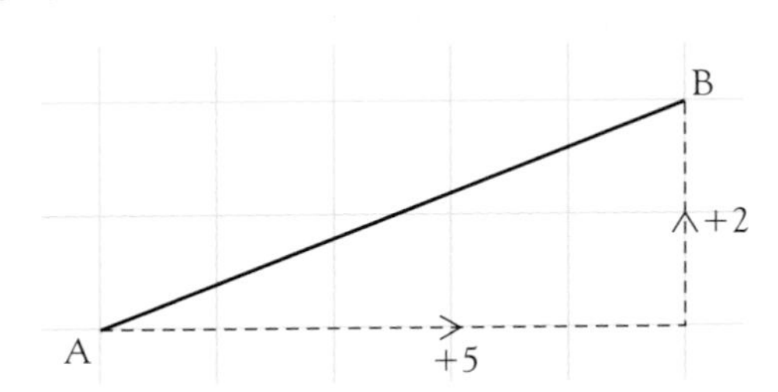

vertical change $= 2$

horizontal change $= 5$

$$\text{gradient} = \frac{\text{vertical change}}{\text{horizontal change}} = \frac{2}{5}$$

For the line CD:

vertical change $= -3$

horizontal change $= 1$

$$\text{gradient} = \frac{\text{vertical change}}{\textbf{horizontal change}} = \frac{-3}{1} = -3$$

You go down, so the vertical change is negative.

Always write out the horizontal change from left to right so that it always increases.

The gradient can be written as a fraction or a decimal, for example the gradient of AB can be written as 0.4. Sometimes it is a whole number, as in the gradient of CD.

Parallel lines

Here are two sets of parallel lines, one with positive gradients and the other with negative gradients.

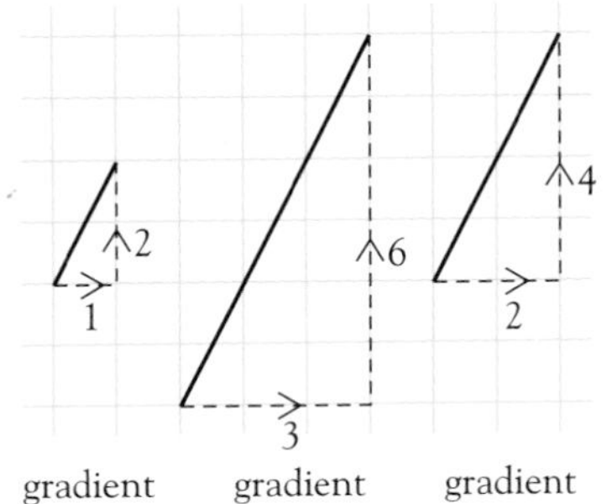

gradient $\frac{2}{1} = 2$ gradient $\frac{6}{3} = 2$ gradient $\frac{4}{2} = 2$

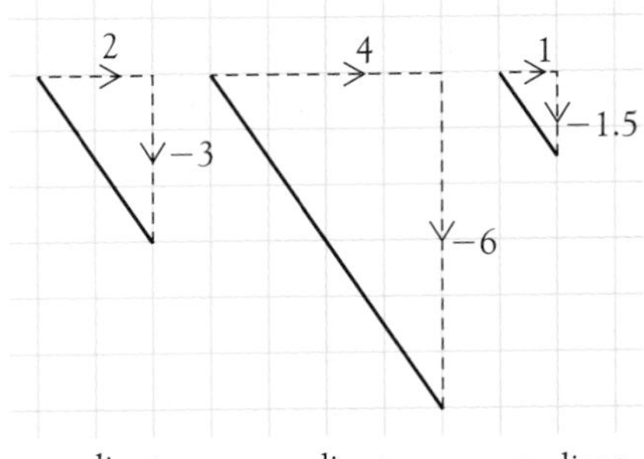

gradient $\frac{-3}{2} = -1.5$ gradient $\frac{-6}{4} = -1.5$ gradient $\frac{-1.5}{1} = -1.5$

Parallel lines have the same gradient.

Sample Question 1

a On a grid, plot these points in order and join them up with straight lines to make the pentagon ABCDE.
A(1, 2), B(3, 6), C(5, 6), D(7, 2), E(7, 1)

b Work out the gradients of the following lines

i AB **ii** CD **iii** BC

c Describe the slope of DE. What happens when you try to calculate the gradient?

GRADIENT

Answer

a

- Draw a grid and plot the shape.

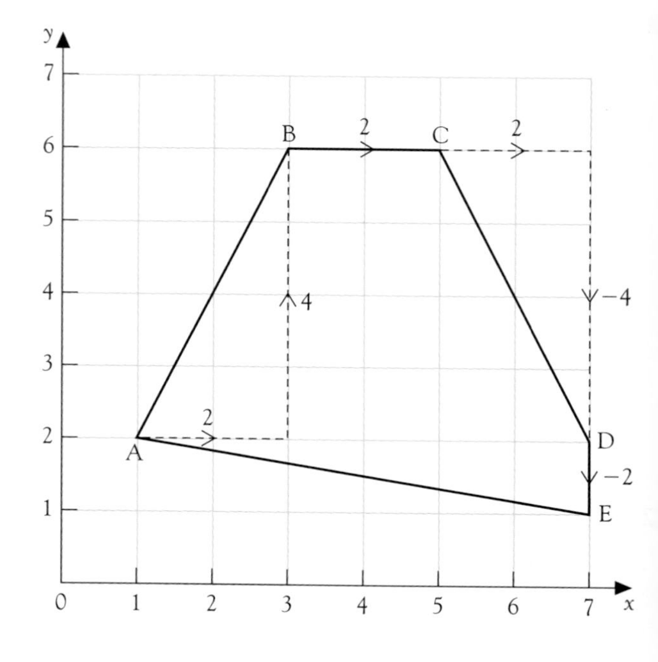

b

- Draw in dotted lines to help you find the vertical and horizontal changes.

 i gradient of AB $= \frac{4}{2} = 2$

 ii gradient of CD $= \frac{-4}{2} = -2$

 iii gradient of BC $= \frac{0}{2} = 0$

c

- DE is vertical.

 $\frac{\text{vertical change}}{\text{horizontal change}} = \frac{-1}{0}$

> You cannot divide by zero. If you try dividing by a very very small number close to zero such as 0.00001, you get 100000 which is a very large number. You could describe the gradient of DE as 'infinite'.

The gradient of a straight line is the same all the way along the line. To find the gradient you do not have to use the end points. You can choose **any two points** on the line and find the gradient between them.

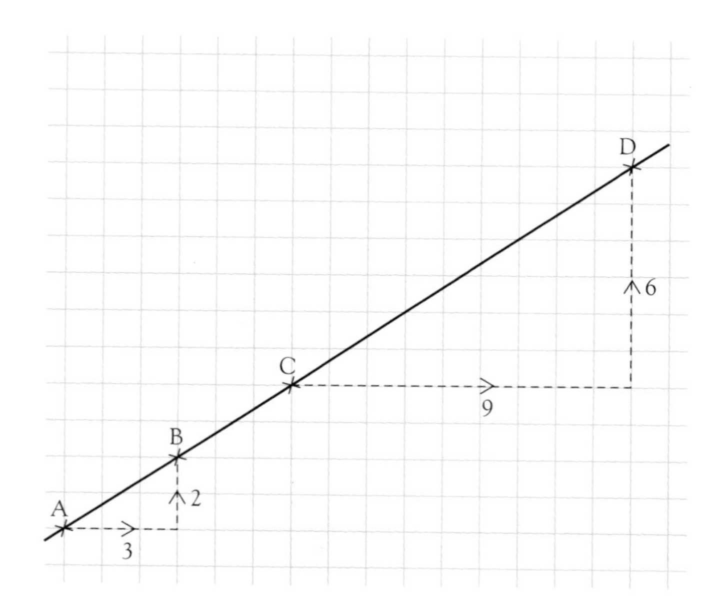

If you choose A and B:

$$\text{gradient} = \frac{2}{3}$$

If you choose C and D

$$\text{gradient} = \frac{6}{9}$$

Simplify

$$= \frac{2}{3}$$

Gradient of the line $= \frac{2}{3}$

It does not matter which two points you choose, but it is better to use points that go exactly through the corners of squares on the grid, rather than those that lie in between squares.

GRADIENT

Exercise 18.1

1 For each of the lines in the diagrams, state whether the gradient is positive (+), negative (−) or zero.

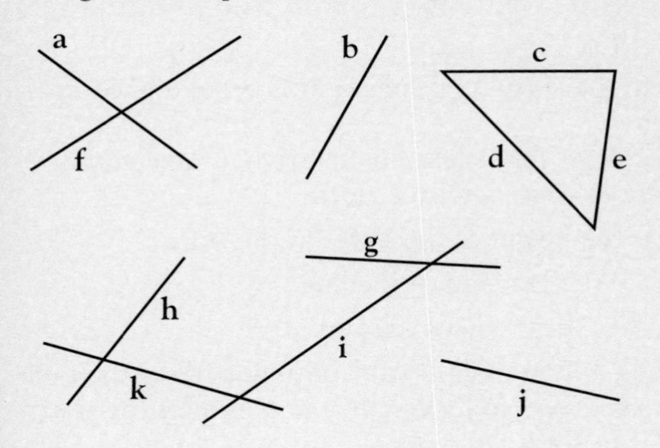

2 Find the gradient of each line.

Give each answer as a whole number or a fraction in its lowest terms.

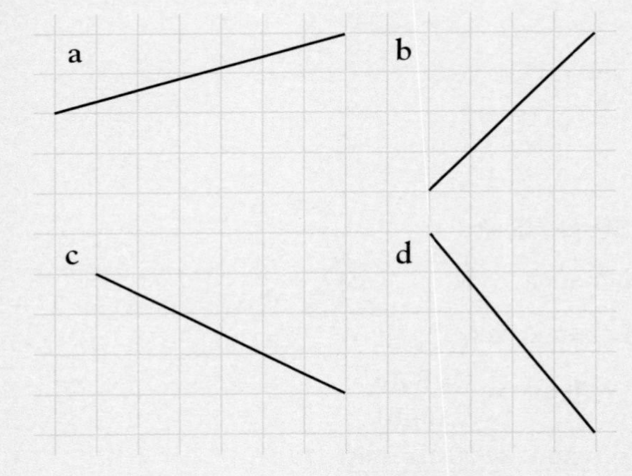

3 Calculate the gradient of each line, giving your answers to 2 significant figures. The sketches are not to scale.

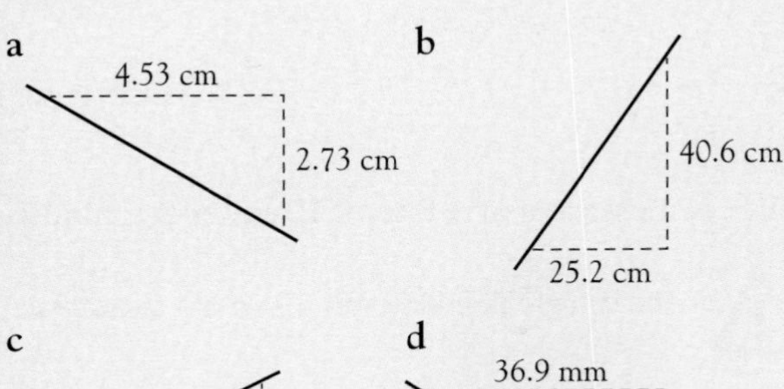

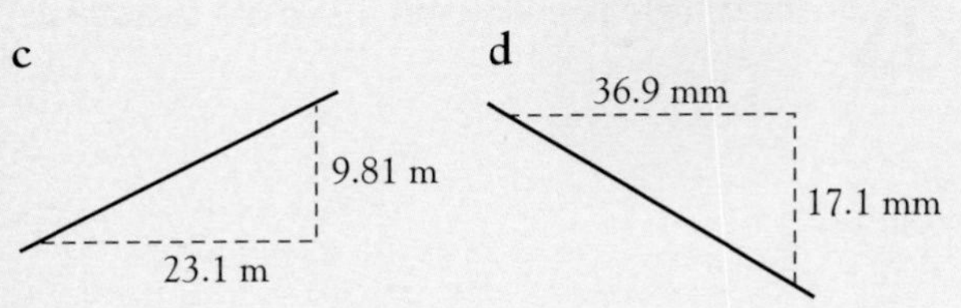

Which line is the steepest?

4 Find the gradients of each of the three lines.

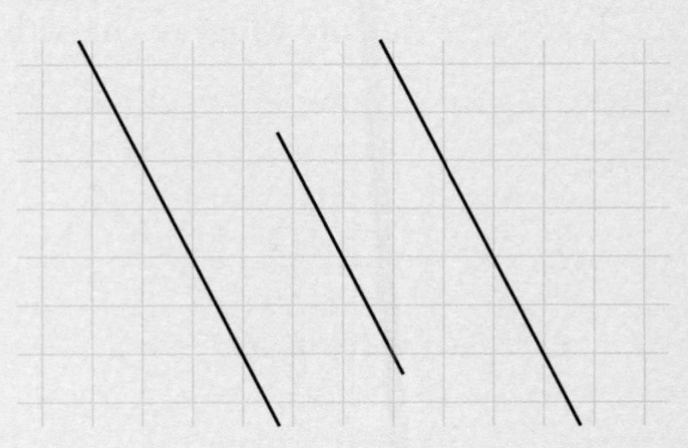

What do your answers tell you about the lines?

5 a Plot the following points on a grid then join them to make the shape ABCDA.
A(1, 8), B(4, 4), C(7, 8), D(4, 1)

b Calculate the gradient of each side.

6 a Plot the following points on a grid then join them to make the shape ABCDEFA.
A(0, 4), B(2, 7), C(5, 7), D(7, 4), E(5, 1), F(2, 1)

b Calculate the gradient of each side.

c Name three pairs of parallel lines.

7 a Plot the following points on a grid, joining them to make the shape ABCDEA.
A(0, 0), B(1, 6), C(4, 7), D(7, 7), E(7, 3)

b Calculate the gradients of the sides AB, BC, CD and EA.

c What can you say about the slope of DE?

8 Use three **different** pairs of points to calculate the gradient of the line. What do your answers tell you about the slope of a straight line?

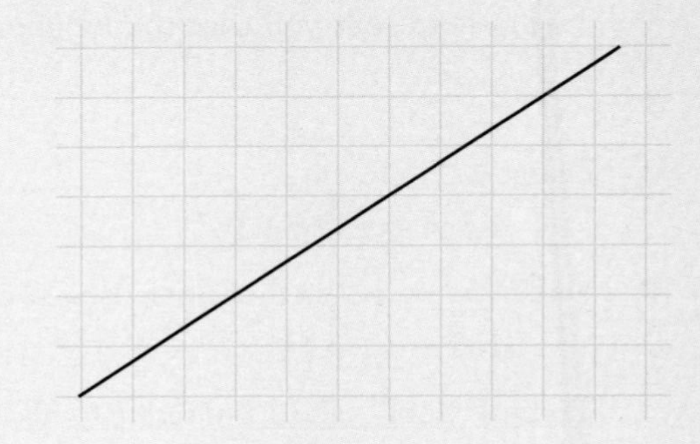

GRADIENT

Practical applications of gradients

The gradient tells you how one quantity is varying relative to the other.
It gives the **rate of change** of one with the other.

Sample Question 2

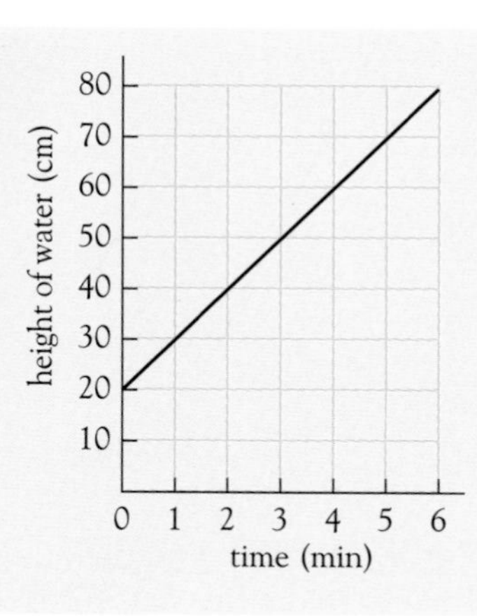

A tank is in the shape of a cuboid. It is being topped up with water.
The graph shows the height, in cm, of the water in the tank at various times after the start.

a What is the height of the water at the start?

b Calculate the gradient of the line.

c What does the gradient tell you?

d The tank is 1.5 m tall. If you continued filling the tank at the same rate, in how many minutes after the start would it be full?

Answer

a

- At the start, $t = 0$, so read off the height when the time is zero.

  Height at start = 20 cm

b

- Draw suitable lines on your graph to work out the gradient

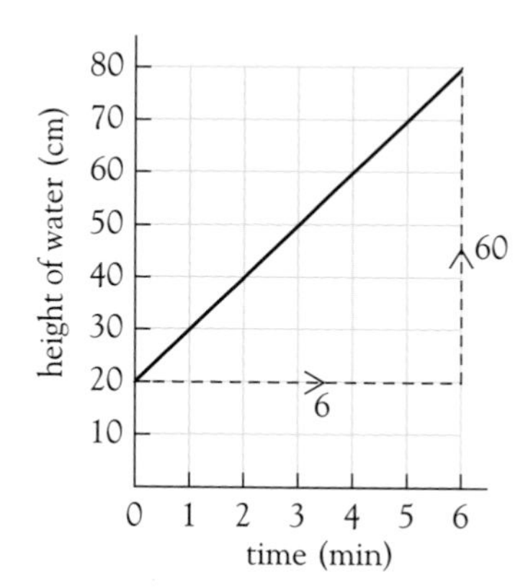

vertical change = 60

horizontal change = 6

$$\text{gradient} = \frac{60 \text{ (cm)}}{6 \text{ (min)}}$$

gradient = **10 cm/min**

DO NOT just count squares. Read the scale carefully.

If you write in the units, you can see that you are dividing cm by minutes. In this case the gradient has a unit cm/min or cm per minute.

c

- The gradient tells you that the height of the water is increasing at a rate of 10 cm every minute.

The straight line indicates that it is a steady rate.

d

- Change 1.5 m to cm.

  1.5 m = 150 cm

- Divide 150 by 10, as the water rises 10 cm every minute

  $150 \div 10 = 15$

  The tank would take 15 minutes to fill.

GRADIENT

Sample Question 3

A new freezer is at room temperature of 20 °C when it is switched on. The graph shows the temperature in the freezer at various times during the next hour.

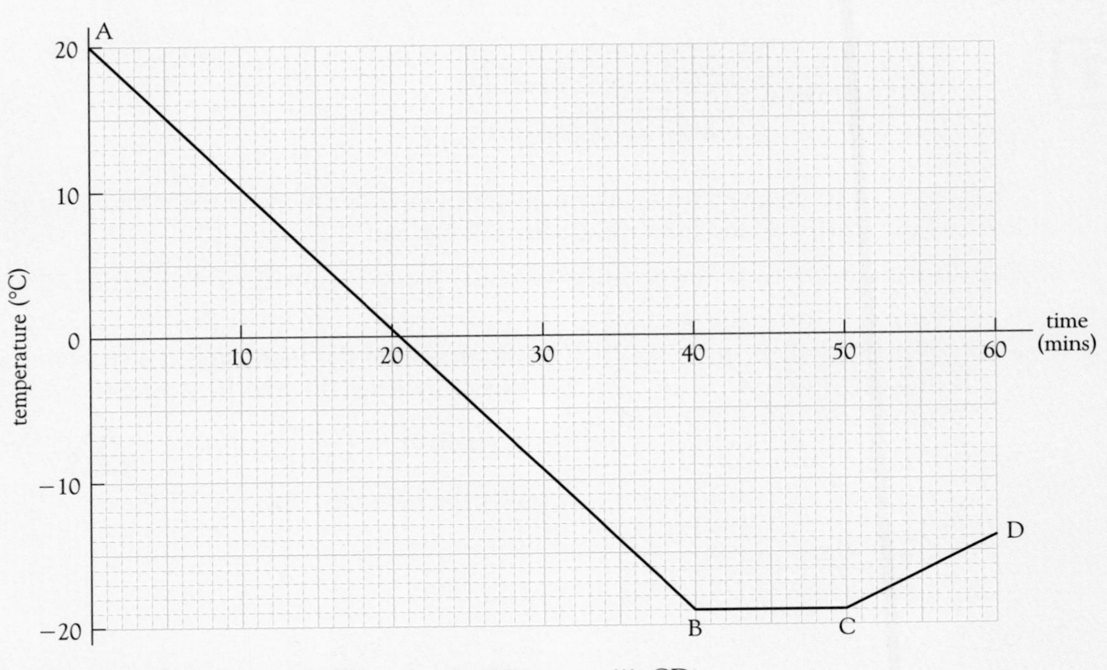

a Find the gradient of **i** AB **ii** BD **iii** CD

b Describe what happens during the hour.

Answer

a

i

A 40 (mins)
−38 (°C)
B

$$\text{gradient} = \frac{\text{vertical change}}{\text{horizontal change}}$$

$$= \frac{-39\ (°C)}{40\ (\text{mins})}$$

$$= \mathbf{-0.975}\ °C/\text{min}$$

The negative gradient indicates the temperature is falling.

ii

B —— C

gradient = 0 °C/min

iii

D
5 (°C)
C
10 (mins)

$$\text{gradient} = \frac{5 (°C)}{10\ (\text{mins})}$$

$$= 0.5\ °C/\text{min}$$

b

- For the first 40 minutes the temperature is decreasing at a steady rate of 0.975 °C/min. The temperature then remains constant at −19°C for 10 minutes. It then rises steadily, at a rate of 0.5 °C/min for 10 minutes.

GRADIENT

Speed, distance and time

An important application of gradients is in finding the speed of something.

Speed tells you how distance is varying with time.

If you walk at a constant speed of 6 km/hour then in 1 hour you cover 6 km, in 2 hours 12 km, in 3 hours 18 km and so on.

$$\text{Average speed} = \frac{\text{distance travelled}}{\text{time taken}}$$

The speed can be obtained from a distance–time graph. If the speed is constant, the graph is a straight line and the speed is found from the gradient of the line.

The diagrams below show the distances of three objects from position P at various times.

a

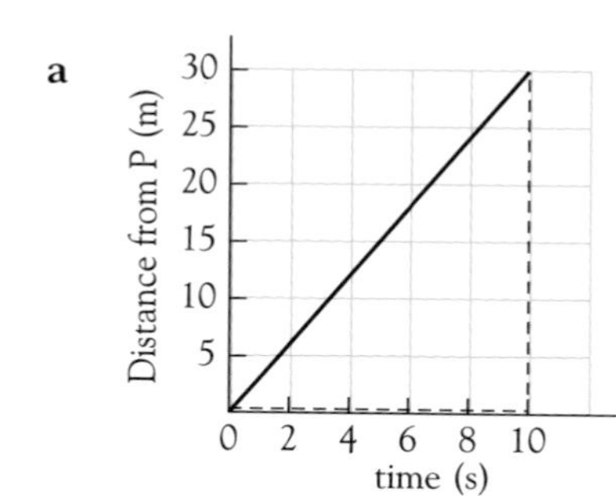

The distance from P is increasing at a constant rate.
The object travelled 30 m in 10 seconds.

$$\textbf{gradient} = \frac{30\text{ (m)}}{10\text{ (s)}} = 3\text{ m/s}$$

This gives the speed.

<u>Speed of the object = 3 m/s</u>

b

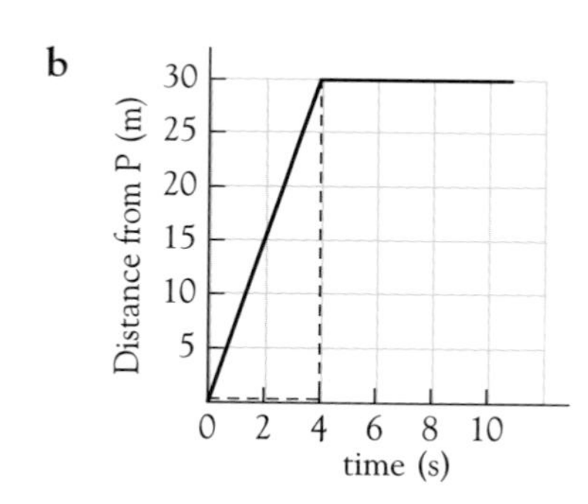

For the first 4 seconds

$$\text{speed} = \frac{30}{4} = 7.5\text{ m/s}$$

Note that the steeper the gradient, the faster the speed.

For the next 6 seconds, the distance from P remained at 30 m.
The object did not move. Its speed was zero.

c

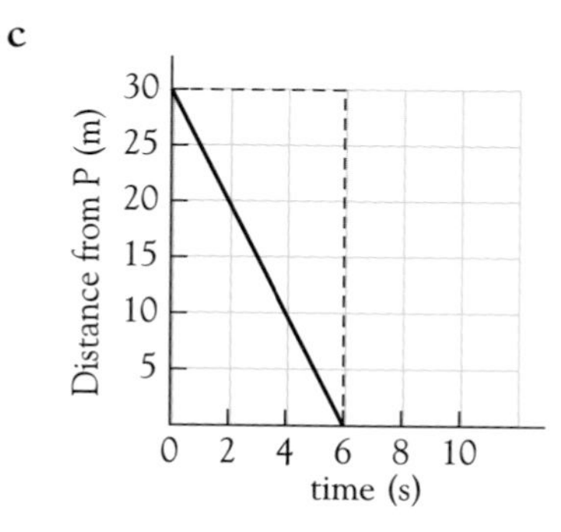

From 0 to 6 seconds

$$\text{gradient} = \frac{-30}{6} = -5\text{ m/s}$$

The negative gradient indicates that the object moved towards P.

<u>Speed = 5 m/s</u>

GRADIENT

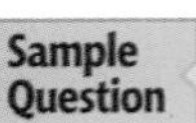

Sample Question 4 Mrs Hutchins set out from home at 10.00 a.m. After jogging at a steady rate for 15 minutes she reached the post box 3 km away. She rested for 5 minutes and then walked home at a steady rate of 5 km/h.

a Show the first 20 minutes of her journey on a distance–time graph.

b At what speed, in km/h, did she jog?

c At what time did she arrive home? Show this last part of her journey on the graph.

Answer

a

◆ Draw a grid and label the axes. Choose a scale on each axis that is easy to read.

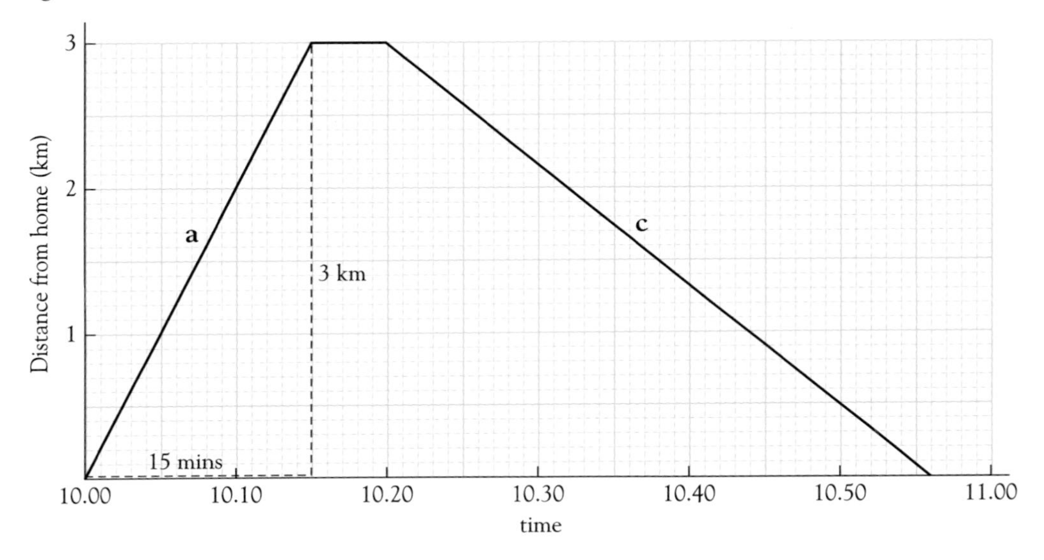

b

◆ For the first 15 minutes

$$\text{speed} = \frac{\text{distance travelled}}{\text{time taken}} = \frac{3 \text{ (km)}}{15 \text{ (mins)}} = 0.2 \textbf{ km/min}$$

You need to change this speed from km/min to km/h. To do this, multiply by 60.

◆ You need to change this to km/h.

In 1 minute she travels 0.2 km

In 1 hour (i.e. 60 minutes) she travels **60** × 0.2 km = 12 km

Mrs Hutchins jogged at a speed of 12 km/h

c

◆ She walked 3 km at a speed of 5 km/h,

i.e. 5 km in 60 mins
1 km in 60 ÷ 5 = 12 mins
3 km in 3 × 12 mins = 36 mins

◆ Add 36 minutes to 10.20 a.m.

She arrived home at 10.56 a.m.

Check: The gradient of this section is $\frac{-3}{36} = -\frac{1}{12}$ km/min

$\frac{1}{12}$ km/min $= \frac{1}{12} \times 60$ km/h $= 5$ km/h

The negative indicates she is travelling **towards** home.

GRADIENT

The D-S-T triangle

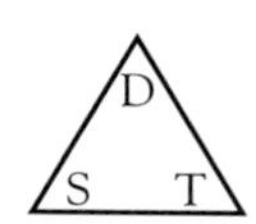

When working with distance (D), speed (S) and time (T) it is useful to remember the connection between them with the aid of the D-S-T triangle.

It works like this:

a To find S, cover up S on the triangle

The D is above the T, so

$$S = \frac{D}{T}$$

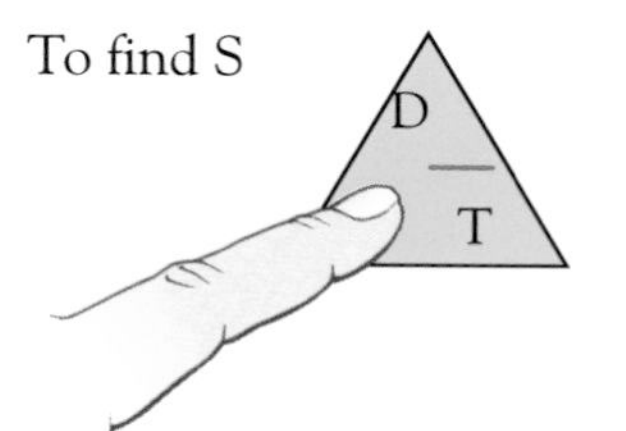

b To find T, cover up T on the triangle

The D is above the S, so

$$T = \frac{D}{S}$$

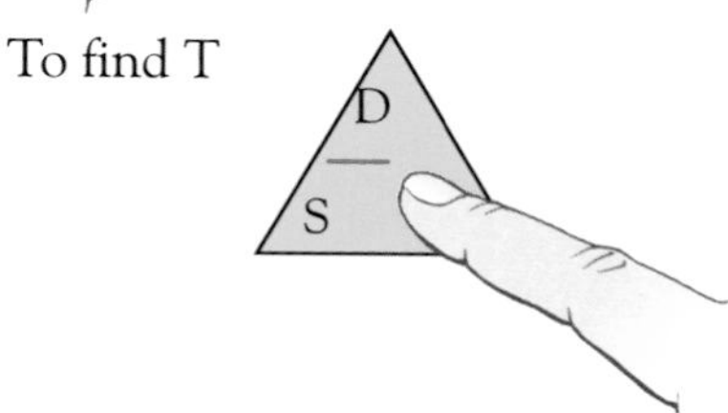

c To find D, cover up D on the triangle

The S and the T are next to each other so

$$D = S \times T$$

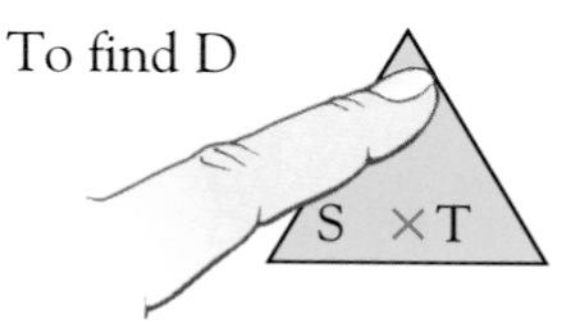

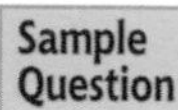

5 **a** A train travels at an average speed of 50 miles per hour for a distance of 70 miles. How long does it take?

b Joe walks at an average speed of 6.5 km/h for 45 minutes. How far does he walk?

Answer

a

- You are given S = 50 (miles per hour), D = 70 (miles)
- Use  $T = \frac{D}{S} = \frac{70}{50} = 1.2$ hours
- Change 0.2 h to minutes by multiplying by 60.

  $0.2 \text{ h} = 0.2 \times 60 \text{ mins} = 12 \text{ mins}$

  <u>The train takes 1 h 12 mins</u>

b

- The speed is given in miles per hour, so change 45 minutes to hours.

  $45 \text{ mins} = \frac{45}{60} \text{ h} = 0.75 \text{ h}$

- You are given S = 6.5 (km/h), T = 0.75 h, use

  $D = S \times T = 6.5 \times 0.75 = 4.875 \text{ km}$

  <u>Joe walks 4.875 km</u>

GRADIENT

Exercise 18.2

1 When Diane was going on holiday to Spain she was given a conversion graph relating the number of pesetas to pounds sterling.

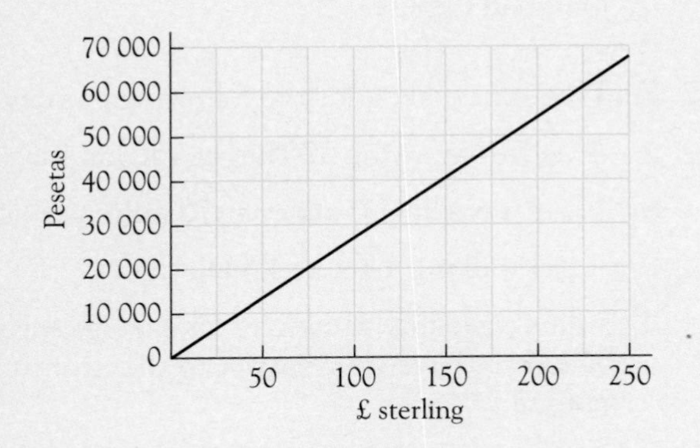

a Use the graph to find how many pesetas Diane got for £200.

b What is the gradient of the line? (Take care with units.)

c What is the practical meaning of the gradient?

2 Some scouts wish to hire a campsite for an international camp. The cost, depending on the number of campers, is shown on the graph.

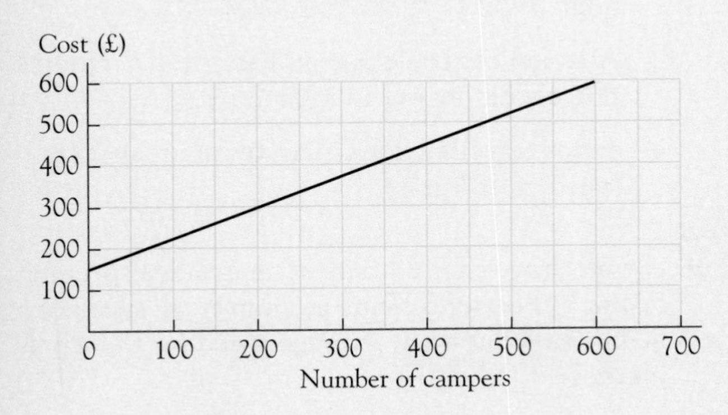

a What is the total cost for 400 campers?

b What is the gradient of the line?

c What is the extra cost per camper?

d What would happen if no one turned up for the camp?

3 A washing machine heats the water at a steady rate as shown by the line AB.

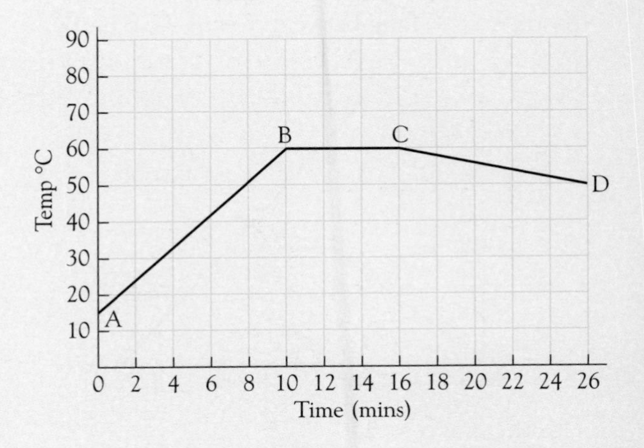

a What is the temperature of the water at the beginning?

b What is the gradient of the line AB?

c What does this tell you?

d Describe what the lines BC and CD represent.

4 The distance from Upton to Dorchester is 20 miles.
The diagram shows the distance–time graph of a cyclist travelling from Upton to Dorchester.

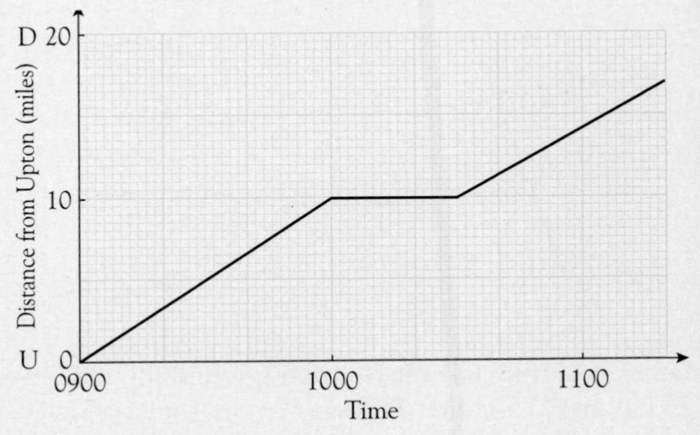

D = Dorchester
U = Upton

At 0940 a motorist leaves **Dorchester** to travel to Upton along the same road as the cyclist. The motorist travels at an average speed of 40 mph.

a Copy the diagram and draw on it the distance–time graph of the motorist.

b At what time did the motorist pass the cyclist? [SEG]

GRADIENT

5 The travel graph shows the journeys of two trains X and Y travelling between towns A, B and C.

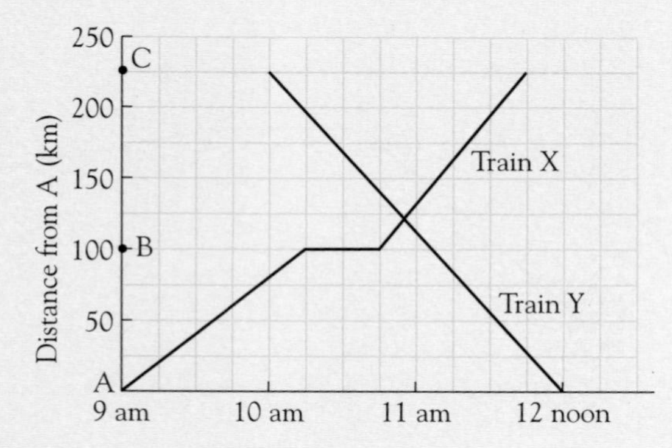

Use the graph to find

a The average speed of train X for each of the three sections of the journey.

b The average speed of train Y. Explain the negative gradient.

c The average speed of train X between A and C.

d At what time and at what distance from A do the trains cross?
What happens to train X at town B?

6

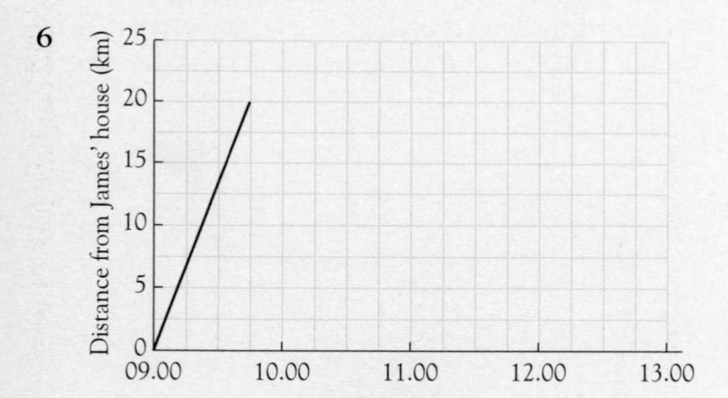

James is cycling to visit his friend, Jonathan, who lives 25 km away. The first 20 km of his journey is shown on the graph. At this point he gets a puncture and walks the rest of the way at a steady rate of 10 km/h. He stays at Jonathan's house whilst he repairs the puncture and has a drink, and leaves at 11.45 a.m. He then travels at a steady speed of 30 km/h for 30 minutes, has a break of 15 minutes and finishes the journey at a steady speed of 20 km/h.

a Copy and complete the graph using the information given.

b At what time did James reach Jonathan's house and how long did he stay there?

c How far was James from home when he had his break on the return journey?

d When did he arrive home?

e What was the average speed of the journey *to* Jonathan's house?

7 Find the times taken for the following journeys.

a A helicopter flying 300 km at 150 km/h.

b A car travelling 35 miles at 70 miles per hour.

c A girl walking 3 km at 4.5 km/h.

d A bus travelling 516 km at an average speed of 50 km/h. (Give your answer to the nearest minute.)

e A car travelling at 70 miles per hour for 23 miles. (Give your answer to the nearest minute.)

f A train travelling at an average speed of 105 mph for 150 miles. (Give your answer to the nearest minute.)

8 Find the distance travelled on the journeys.

a A $2\frac{1}{2}$ hour journey at an average speed of 30 km/h.

b A car travelling for 20 minutes at 60 miles per hour.

c A stolen car travelling at 40 m/s for 2 minutes. (Give your answer in kilometres.)

d A bus travelling for 7 minutes at an average speed of 43 km/h.

9 A train travels for 2 hours at an average speed of 50 km/h then for 3 hours at an average speed of 60 km/h. What is the average speed for the whole journey?

10 A coach travels for 6 hours between two towns A and B which are 180 km apart. On the return journey from B to A the average speed is reduced by 3 km/h. Calculate the time for the return journey.

11 For the first $1\frac{1}{2}$ hours of a 210 mile journey the average speed was 30 miles per hour. If the average speed for the remainder of the journey was 45 miles per hour, calculate the average speed for the whole journey.

EQUATION OF A STRAIGHT LINE

Straight line graphs

In Chapter 6 you used a table of values to draw a **straight line** from its equation, lines such as

$$y = 4x + 1, \quad y = 3x - 2, \quad y = -x + 5$$

page 89

These equations are written in a general format

$y = mx + c$ where m and c are numbers.

- In the line $y = 4x + 1$, $m = 4$ and $c = 1$
- In the line $y = 3x - 2$, $m = 3$ and $c = -2$

$3x - 2$ is the same as $3x + (-2)$

- In the line $y = -x + 5$, $m = -1$ and $c = 5$

$-x$ means $(-1) \times x$

TASK

You are going to investigate linear graphs using the general formula

$$y = mx + c$$

a

- Choose a particular value for c.
- Make up the equations of 6 lines each with your chosen value of c but **different values** of m.

Choose some positive and some negative values for m!

- Plot all the lines on the same grid labelling each line carefully. You will need to do all your tables of values first so that you know how big to draw the grid.
- Write down anything you notice.

b

- Now choose a particular value for m.
- Make up the equations of 6 lines, each with your chosen value of m but different values of c.
- Do a second grid and plot these 6 lines on it.
- Write down anything you notice.

EQUATION OF A STRAIGHT LINE

The meaning of *m* and *c* in the straight line equation $y = mx + c$

Sample Question 6

a On separate grids draw the lines $y = 2x + 1$ and $y = -\frac{1}{2}x - 1$

b Find the gradient of each line.

c Where does each line cross the y axis?

d Comment on your answers.

Answer

a

◆ $y = 2x + 1$

x	−2	−1	0	1	2
y	−3	−1	1	3	5

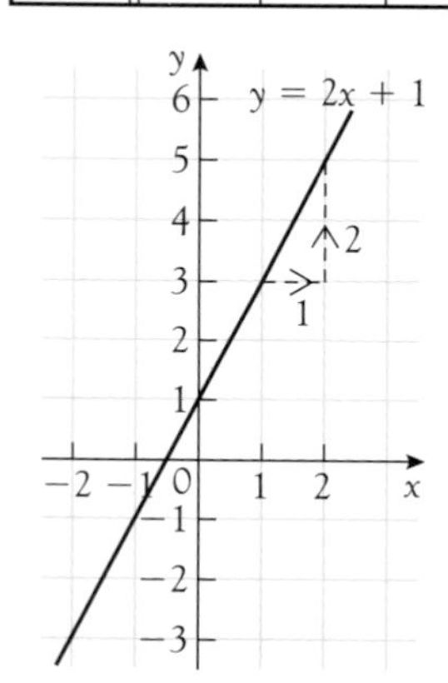

b

◆ gradient $= \frac{2}{1} = 2$

c

◆ The line crosses the y axis at 1

d

◆ The equation is $y = 2x + 1$

↑ The gradient is 2

↑ The line crosses the y axis at 1

a

◆ $y = -\frac{1}{2}x - 1$

x	−4	−2	0	2
y	1	0	−1	−2

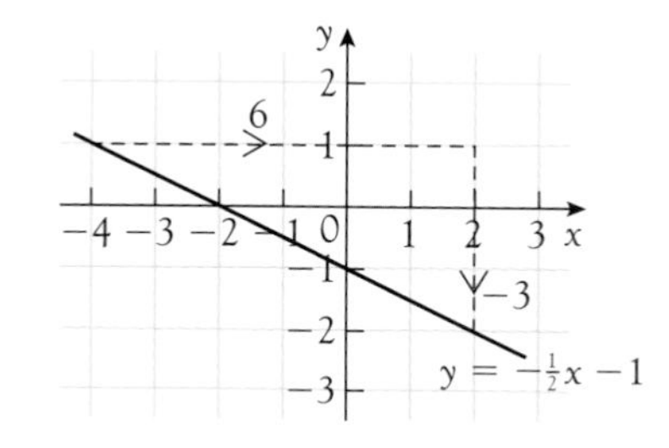

b

◆ gradient $= \frac{-3}{6} = -\frac{1}{2}$

c

◆ The line crosses the y axis at −1

d

◆ The equation is $y = -\frac{1}{2}x - 1$

↑ The gradient is $-\frac{1}{2}$

↑ The line crosses the y axis at −1

When the equation of a straight line is written in the form

$$y = mx + c$$

m is the gradient and
c is the value where the line crosses the y axis.

c is known as the intercept on the y axis, and this form of the equation is called the 'gradient–intercept' form.

EQUATION OF A STRAIGHT LINE

It is possible to draw a sketch of a line from its equation without plotting all the points accurately, as in these examples:

a $y = 4x - 2$
gradient, $m = 4$
intercept on y axis, $c = -2$

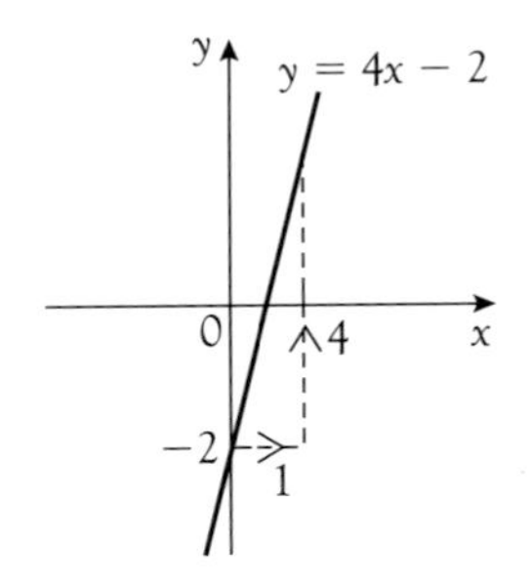

b $y = -2x + 1$
gradient, $m = -2$
intercept on y axis, $c = +1$

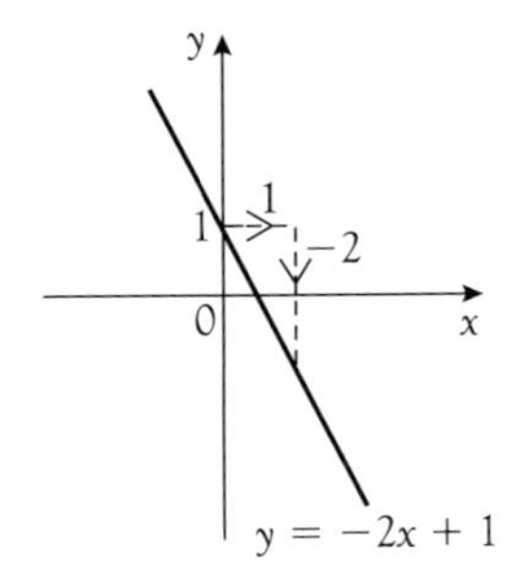

This method can also be used to draw the lines accurately by plotting the intercept on the y axis, then counting squares, according to the gradient, to find another point on the line.

Sample Question 7

Write down the equation of the line through (0, 3) that is parallel to the line with equation

$$y = \tfrac{2}{3}x - 1$$

Draw a sketch showing both lines.

Answer

- Write down the gradient of $y = \frac{2}{3}x - 1$

  gradient $= \frac{2}{3}$

- Write the equation of the new line as $y = mx + c$ and find m and c.

  As the new line is parallel to $y = \frac{2}{3}x - 1$, it has the same gradient, so $m = \frac{2}{3}$.

  The new line goes through (0, 3). This means that when $x = 0$, $y = 3$ so it crosses the y axis at 3, i.e. $c = 3$

  Equation is $y = \frac{2}{3}x + 3$

- Sketch parallel lines, both with gradient $\frac{2}{3}$, one crossing the y axis at -1 and the other crossing the y axis at 3.

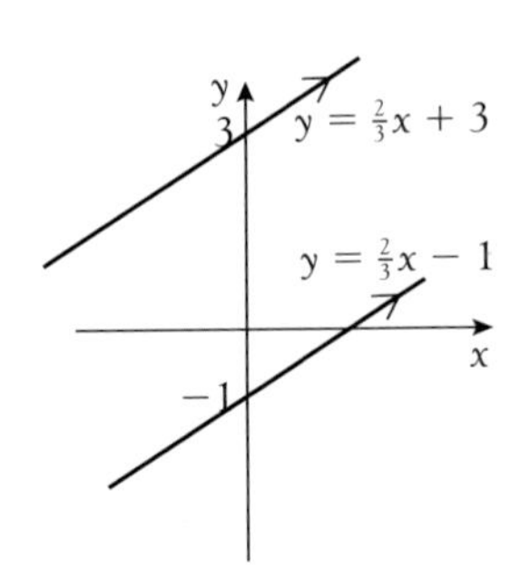

EQUATION OF A STRAIGHT LINE

Special case

When the line goes through the origin (0, 0), the line crosses the y axis at 0.

This means that in the equation $y = mx + c$, $c = 0$ and you just have $y = mx$, as in these examples:

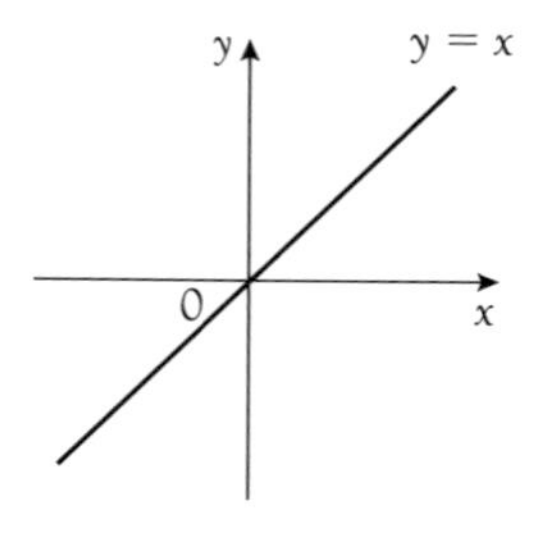

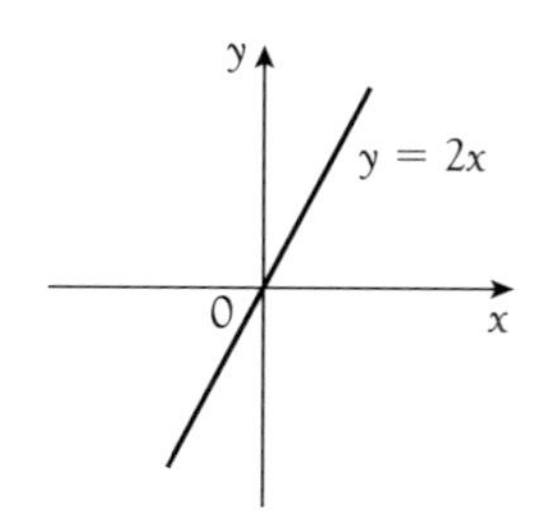

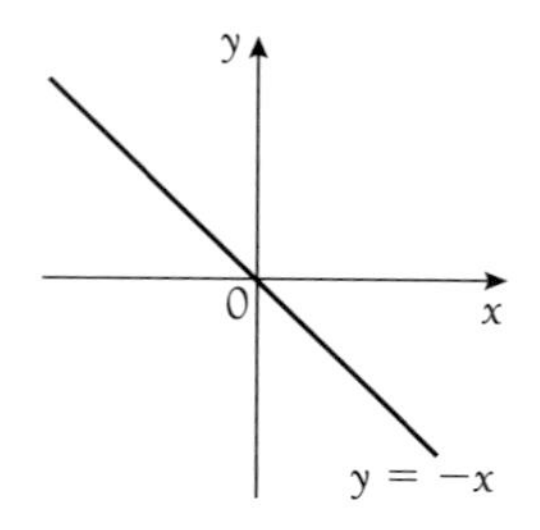

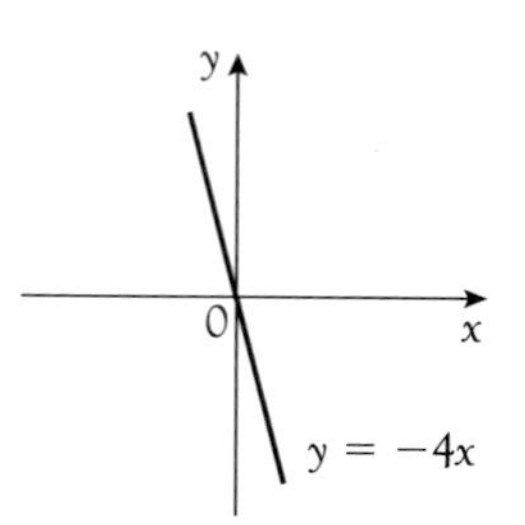

A straight line through the origin has equation of the form $y = mx$, where m is the gradient.

Lines that are not in the form $y = mx + c$

If the equation of the line is not written in the form $y = mx + c$, and you want to work out the gradient, m, and the y-intercept, c, from the equation, you will need to **re-arrange** it so that y is the subject.

page 265

For example

a $\quad 2y - x = 6$

$+x \quad 2y = 6 + x$

$\div 2 \quad y = 3 + \frac{x}{2}$

 $\frac{x}{2} = \frac{1}{2}x$

$y = \frac{1}{2}x + 3$

Write it as $y = mx + c$ putting the x term first.

$\uparrow \quad \uparrow$

$m = \frac{1}{2} \quad c = 3$

b $\quad 3x + 4y = 12$

$-3x \quad 4y = 12 - 3x$

$\div 4 \quad y = 3 - \frac{3x}{4}$

$\frac{3x}{4} = \frac{3}{4}x$

$y = -\frac{3}{4}x + 3$

$\uparrow \quad \uparrow$

$m = -\frac{3}{4} \quad c = 3$

 Put the x term first.

EQUATION OF A STRAIGHT LINE

'Cover up' method for plotting a line

Use the 'cover-up' method when you have a positive x term and a positive y term, both on the same side of the equation.

If you are asked to plot a graph accurately, from a table of values, and the equation is written like example **b** on the previous page, $3x + 4y = 12$, it is a good idea to use the 'cover up' method to find where the line cuts each of the axes.

Consider the line with equation $3x + 4y = 12$

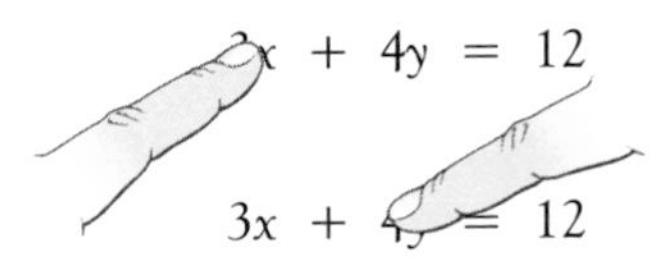

- Cover up the $3x$ part of the equation.
  In doing this you are making $x = 0$.
  This leaves $4y = 12$, so $y = 3$.

  The line cuts through the y axis at the point (0, 3).

- Next cover up the $4y$ part of the equation.
  In doing this you are making $y = 0$.
  This leaves $3x = 12$, giving $x = 4$.

  The line cuts the x axis at the point (4, 0).

Table for $3x + 4y = 12$

x	0	4
y	3	0

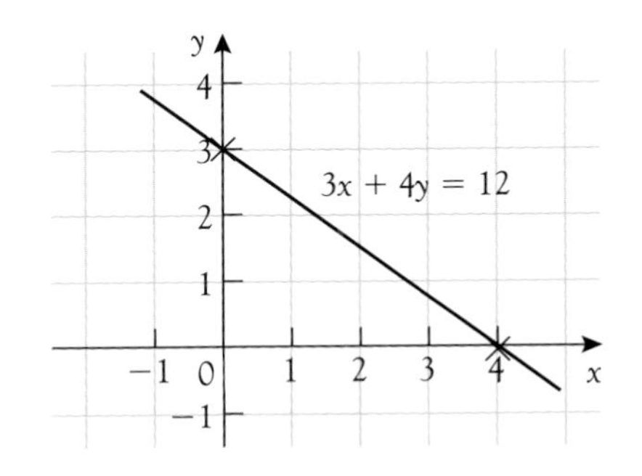

Just to be sure, you should try another point as a check – perhaps $x = 2$

When $x = 2$ $\quad 3 \times 2 + 4y = 12$

$$6 + 4y = 12$$
$$4y = 6$$
$$y = 1.5$$

The line should go through (2, 1.5).

Sample Question 8

The table shows the largest quantity of salt, w grams, which can be dissolved in a beaker of water at temperature t °C.

t °C	10	20	25	30	40	50	60
w grams	54	58	60	62	66	70	74

a Plot the points and draw a graph to illustrate this information.

b Use your graph to find

  i the lowest temperature at which 63 g of salt will dissolve in the water.

  ii the largest amount of salt that will dissolve in the water at 44 °C.

c **i** The equation of the graph is of the form

$$w = at + b$$

  Use your graph to estimate the values of the constants a and b.

  ii Use the equation to calculate the largest amount of salt which will dissolve in the water at 95 °C.

[NEAB]

EQUATION OF A STRAIGHT LINE

Answer

a

- Plot the points, taking care with the scales.
- Join the points with a ruler to form a straight line.

b

i

- Find 63 on the 'quantity of salt' (vertical) axis, go horizontally right to meet your line then vertically down to temperature axis.

  Temperature = 33 °C

ii

- Find 44 on the 'temperature' axis, go up to the line, then left to the 'quantity of salt' axis.

  Amount of salt = 67.5 g

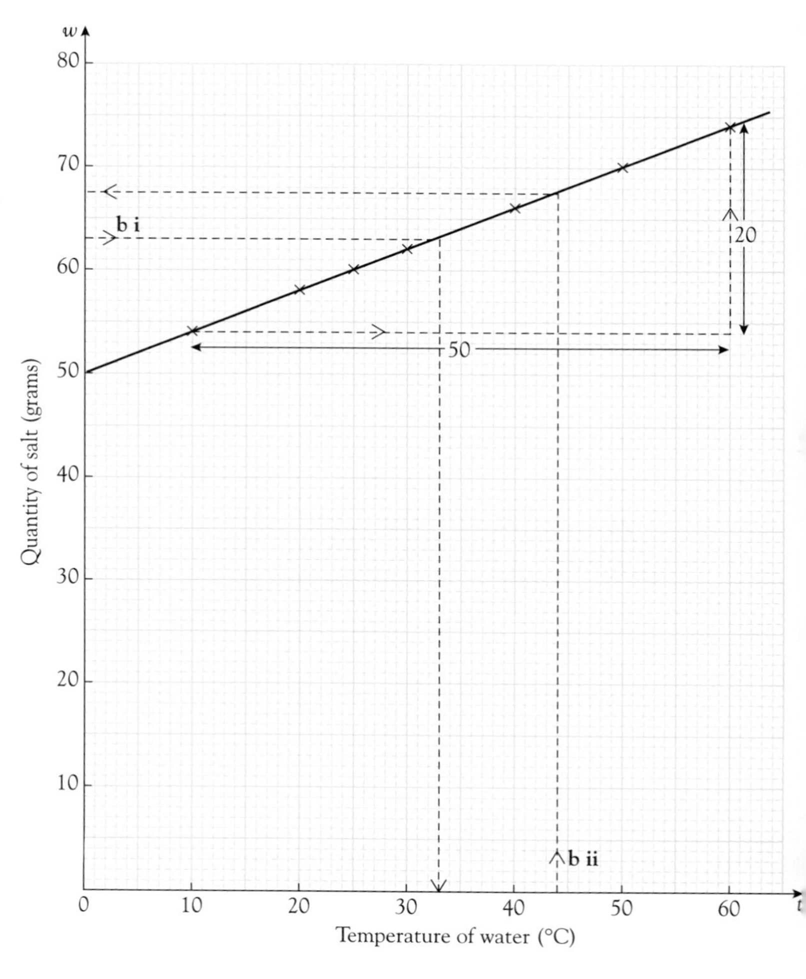

c

i

- Compare $w = at + b$ with $y = mx + c$ ($a \to m$, $b \to c$)

  w compares with the y axis and t compares with the x axis so in the equation

  $$w = at + b$$

  a gives the gradient, and b the intercept on the w axis

- Find the gradient by looking at two points, for example (10, 54) and (60, 74)

  $$\text{gradient} = \frac{\text{vertical change}}{\text{horizontal change}} = \frac{20}{50} = 0.4, \text{ so } a = 0.4$$

- Take your straight line back so that it crosses the w axis and read off the value where it crosses this axis. This is the value b.

  $b = 50$

ii From the graph the equation of the line is $w = 0.4t + 50$

- Substitute $t = 95$ into the equation and work out w.

  $$w = 0.4 \times 95 + 50$$
  $$= 88$$

  Largest amount = 88 g

You are assuming that the same relationship applies at much higher temperatures. As a general rule you should take care when using your line outside the range of your data.

EQUATION OF A STRAIGHT LINE

Exercise 18.3

1 Complete the tables of values for the straight lines given and in each case draw the line on a set of axes using values of x and y from -5 to $+5$.

Draw a separate grid for each line.

a

x	-1	1	3
$y = 2x - 3$			

b

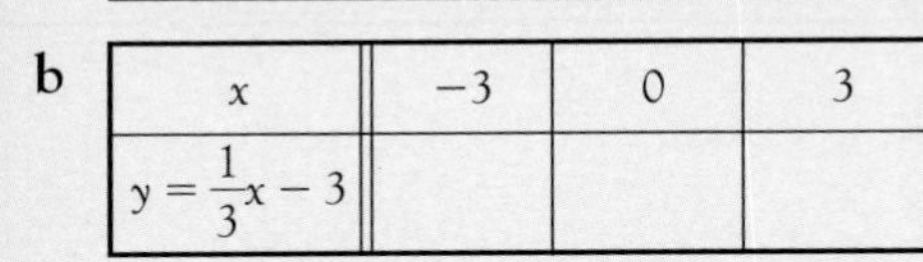

x	-3	0	3
$y = \frac{1}{3}x - 3$			

i Find the gradient of each line.

ii Where does each line cross the y axis?

iii Use the equations of the lines to comment on the answers to **i** and **ii**.

2 Which of the following lines are parallel?

a $y = -2x + 5$ **d** $y = 2x - 2$

b $y = x - 2$ **e** $y = 3 + x$

c $y = 9 - 2x$ **f** $y - 2x = 5$

3 Which of the following lines have the same intercept on the y axis?

a $y = 3x - 1$ **b** $y = 1 - 3x$

c $y = x + 1$ **d** $y = x - 3$

Which of the lines are parallel?

4 Write down the gradient and intercept on the y axis of each of the following lines and use your answer to sketch each line.

a $y = 3x + 5$ **f** $y = 5x + 1$

b $y = x - 6$ **g** $y = -x + 3$

c $y = 10 - 4x$ **h** $y = 6 - \frac{1}{4}x$

d $y = \frac{1}{3}x$ **i** $y = 3$

e $y = 2x - 4$ **j** $y = 1 - x$

5 Find the equations of the lines described.

a The line passes through the point $(0, -3)$ and has a gradient of 4.

b The line passes through the point $(0, 5)$ and has a gradient of -3.

c The line passes through the origin and has a gradient of $-\frac{2}{3}$.

d The line is parallel to the line $y = \frac{1}{2}x - 1$ and passes through the point $(0, 2)$.

e The line is parallel to the x axis and passes through $(1, 2)$.

6 For each of the lines find

i the gradient

ii the intercept on the y axis

iii the equation of the line.

a

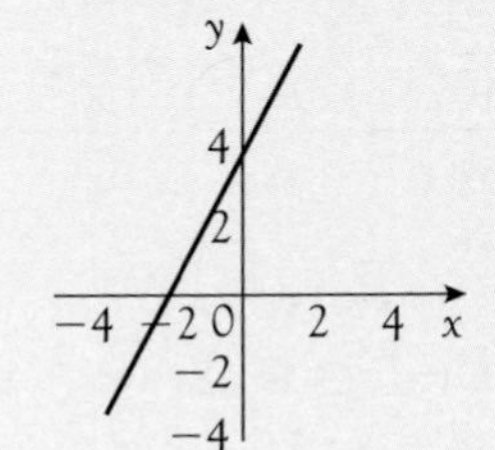

d

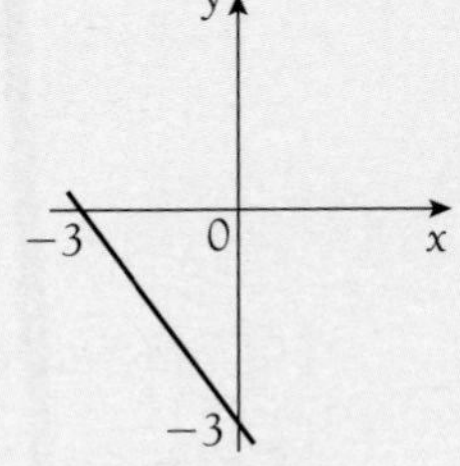

b

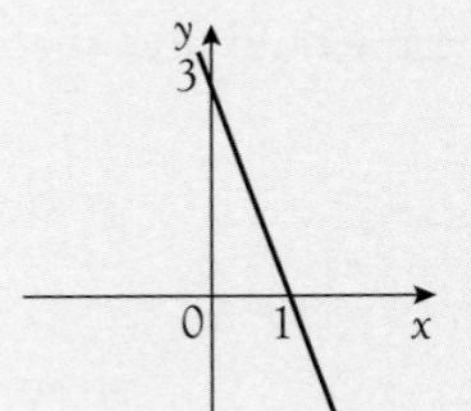

e

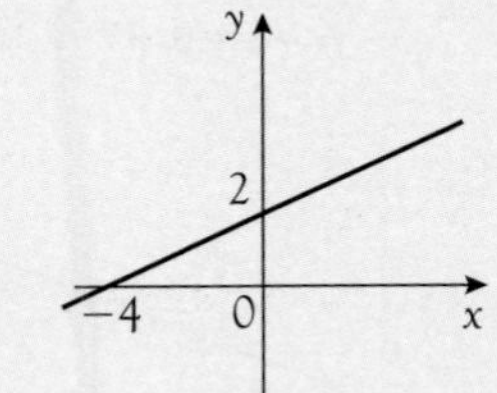

c

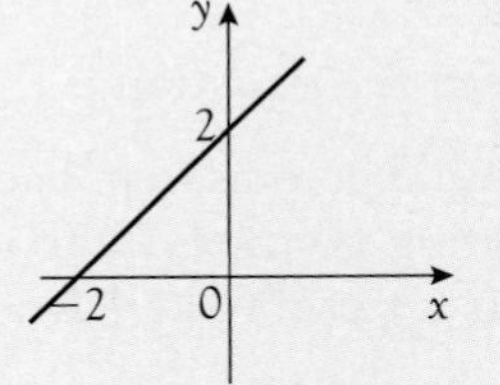

f

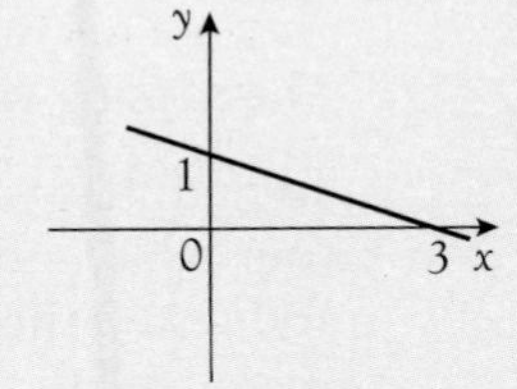

7 Re-arrange the equations to make the form $y = mx + c$. Hence write down the values of the gradient (m) and the intercept (c).

a $7x + y = 2$ **b** $y - x = 5$

c $2x - y = 4$ **d** $3x + 2y = 6$

8 Using the 'cover up' method find the intercepts on the x and y axes of the lines $3x + 2y = 6$ and $x + 2y = 2$. Hence sketch the lines.

9 The line $y = mx + c$ is parallel to the line $y = 3x$ and passes through the point $(1, 4)$. Find the values of m and c.

10 Find the equation of the line which is parallel to $y = -2x + 3$ and passes through the point $(-2, 3)$.

EQUATION OF A STRAIGHT LINE

Linear relationships in investigations

When you are doing an investigation you often have a table of values relating one variable to another. If the relationship is linear, it may be possible to spot this and work out the equation from the table.

Consider the table given in Sample Question 8.

You need to consider t values which have a constant difference between them. This would be the case if you ignored $t = 25$. Cross it out.

t	10	20	~~25~~	30	40	50
w	54	58	~~60~~	62	66	70

Now work out the differences in t values and the differences in w values.

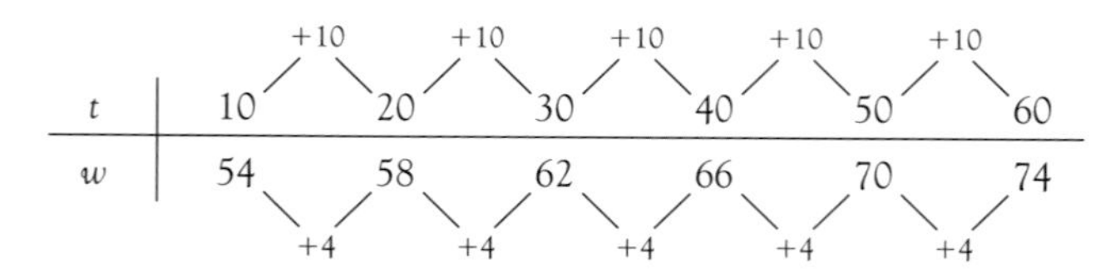

The difference in w values is always $+4$ for a difference in t values of $+10$

This means that w increases by 4 for every increase of 10 in t.

If you draw a line, you would find that

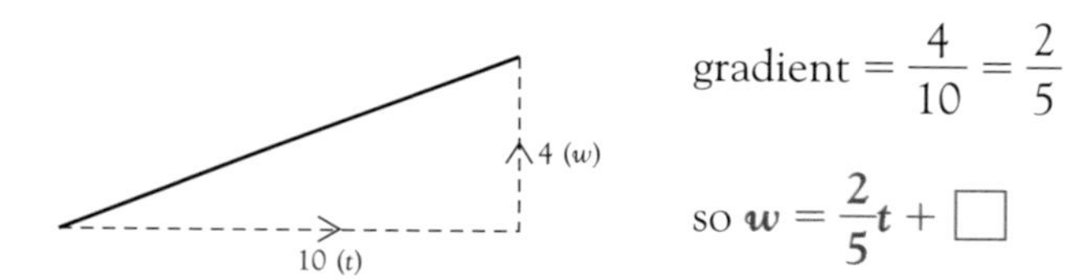

$$\text{gradient} = \frac{4}{10} = \frac{2}{5}$$

$$\text{so } w = \frac{2}{5}t + \square$$

Using gradient–intercept form '$y = mx + c$', writing t instead of x and w instead of y.

The missing value in the equation is the intercept on the w axis, i.e. the value of w when $t = 0$.

To find this, you could go one step 'backwards' in your table, subtracting 10 from 10 to give $t = 0$. You would then need to subtract **4 from 54** to give $w = 50$

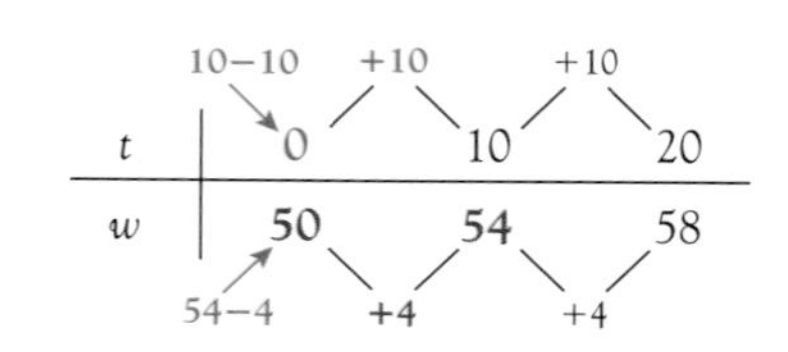

$$\text{Equation is } w = \frac{2}{5}t + 50$$

Alternatively, once you know $w = \frac{2}{5}t + \square$ you can find the missing value by choosing a value of t from the table, perhaps $t = 10$ and work backwards in the formula.

When $t = 10$, the formula gives $w = \frac{2}{5} \times 10 + \square$

$$w = 4 + \square$$

When $t = 10$, the table gives $w = 54$

$$\therefore\ 4 + \square = 54$$

$$\square = 50$$

$$\text{Equation is } w = \frac{2}{5}t + 50$$

EQUATION OF A STRAIGHT LINE

Sample Question 9

Freddie is investigating the relationship between the number of fences, f, and the number of posts, p, needed to build square enclosures in a line and has drawn these diagrams.

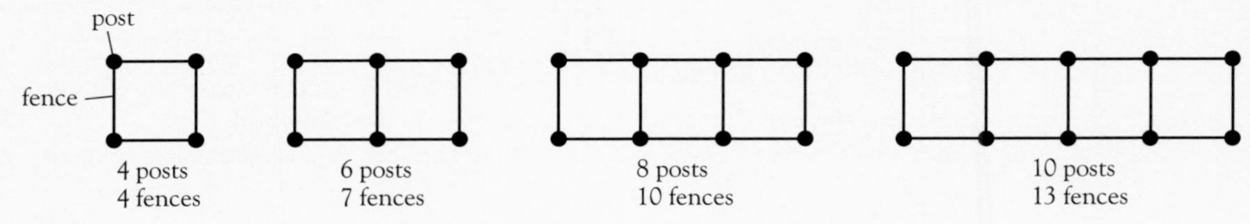

a Complete a table of values

p	4	6	8	10
f				

and use it to explain the type of relationship between f and p.

b Work out the equation in the form $f = \ldots$

c Illustrate the information on a graph.

Answer

a

- Complete the table and look at differences.

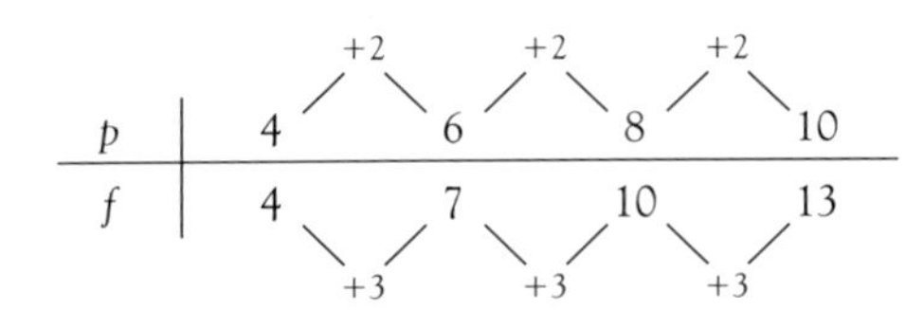

As the differences for p are constant and the differences for f are constant, the relationship between f and p is linear.

b

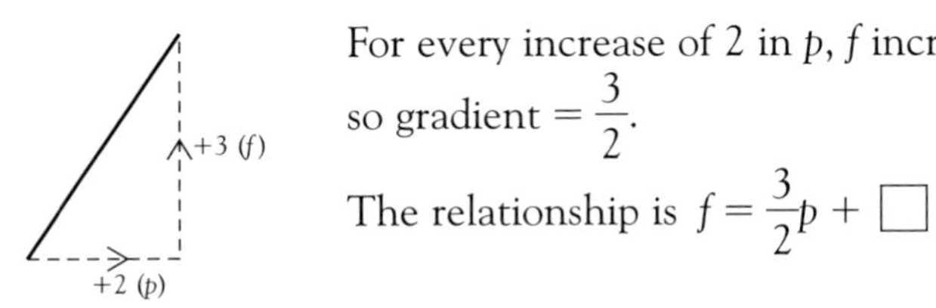

For every increase of 2 in p, f increases by 3,

so gradient $= \frac{3}{2}$.

The relationship is $f = \frac{3}{2}p + \square$

- To find the missing values, you could use one of the known results in the table, e.g. when $p = 4$, $f = 4$

When $p = 4$, the formula gives $f = \frac{3}{2} \times 4 + \square$

$f = 6 + \square$

When $p = 4$, the table gives $f = 4$

$\therefore \quad 6 + \square = 4$

$\square = 4 - 6 = -2$

The relationship is $f = \frac{3}{2}p - 2$

You could write $\frac{3}{2}$ as a decimal, so $f = 1.5p - 2$. It is, however, often better to work in fractions, especially in investigations, as it makes it easier to spot general patterns.

EQUATION OF A STRAIGHT LINE

c

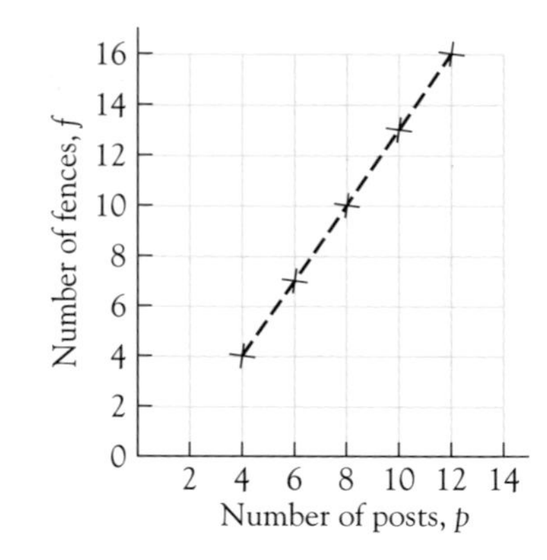

The line is shown dotted because the formula only holds at points when p is an even number. For example you could not have 5 posts as this wouldn't form a square enclosure, nor would it be possible to have 5.8 posts.

Note also that the lowest possible value of p is 4.

TASK

Continue the investigation on posts and fences, looking at enclosures in a line, a double line, a triple line, etc.

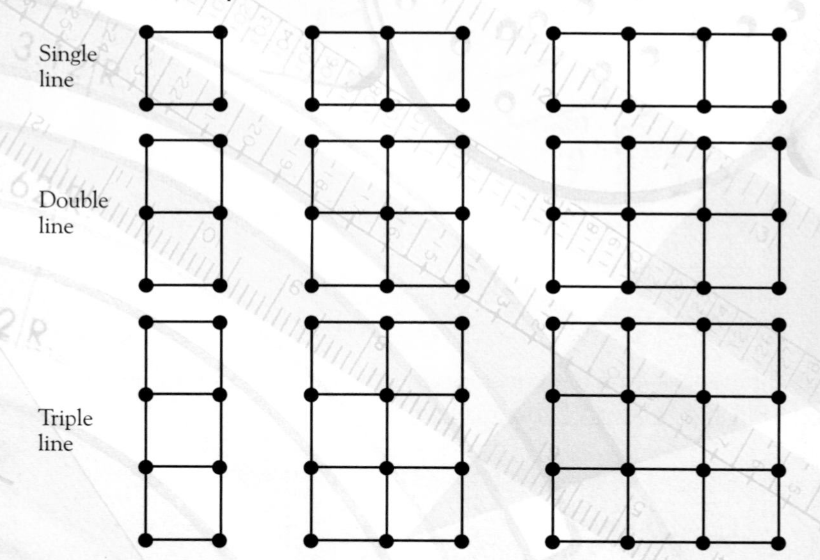

Find general relationships between the number of posts and/or the number of fences and/or the number of enclosures.

Worked Exam Question
[SEG]

Television repair charges depend on the length of time taken for the repair, as shown on the graph.

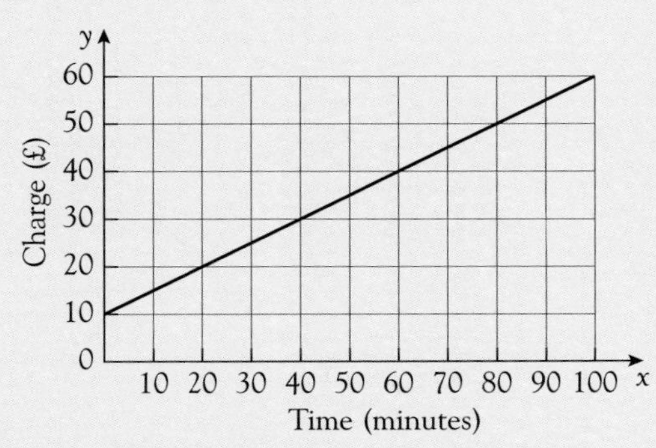

The charge is made up of a fixed amount plus an extra amount which depends on the time.

a Calculate the gradient of the line.

◆ Choose two points and draw lines on the graph to work out the vertical and horizontal change.

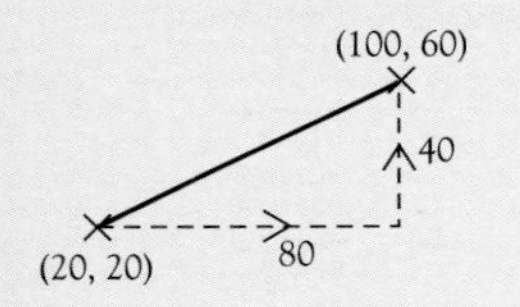

$$\text{gradient} = \frac{40}{80}$$
$$= 0.5$$

Answer ......0.5......

COMMENTS

M1 for attempting to find $\frac{\text{vertical change}}{\text{horizontal change}}$

A1 for correct answer

2 marks

b Write down the equation of the line.

◆ Use $y = mx + c$ where m is the gradient and c is the intercept on the y axis, so $m = 0.5$ and $c = 10$

Note the possible 'follow through' from **a**.
Use your answer even if you think it might be wrong.

Answer ....$y = 0.5x + 10$....

B1 for $0.5x$ or your answer to **a** $\times x$.

B1 for $+10$

2 marks

c Mr Swann hopes that the repair to his television will cost £84 or less. Write down an inequality to represent this information and use it to calculate the maximum time which could be spent on the repair.

◆ In the equation $y = 0.5x + 10$, y is the charge and x is the time taken.

You want the charge to be less than or equal to 84.
Write this as an inequality and solve it.

charge $\leqslant 84$

$\therefore\ 0.5x + 10 \leqslant 84$

(−10) $0.5x \leqslant 74$

(÷0.5) $x \leqslant 148$

Answer ....Max. time = 148 mins....

M1 for inequality (mark given if you use your answer to *b*

A1 for correct answer

2 marks

Exam Questions

1 The following graph can be used to convert between pounds (p) and dollars (d).

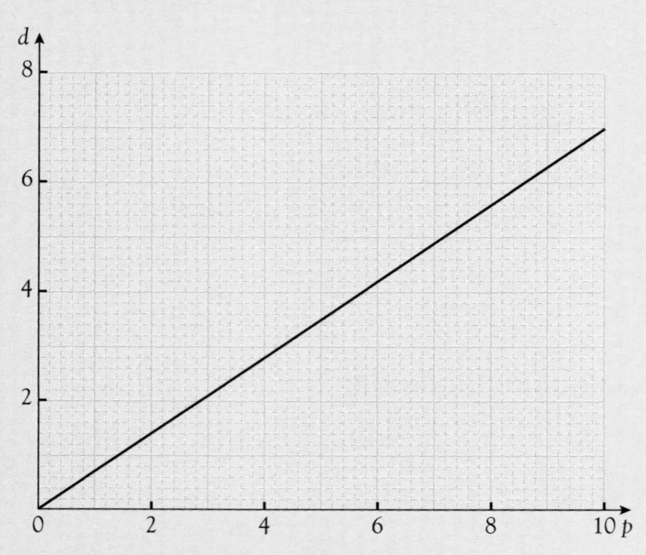

a Find the gradient of the line.

b What does this gradient represent? [MEG]

2 Louise is cycling in the countryside. The sketch is a distance–time graph for part of her journey.

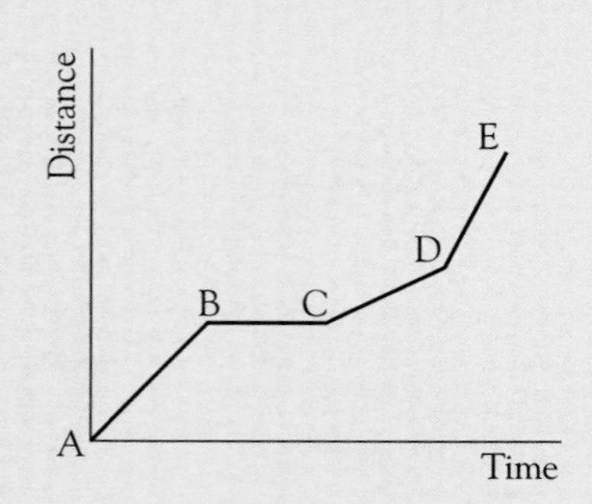

Copy and complete the following sentences with different parts of the graph.

Louise is not moving from .... to ....

Louise is cycling the quickest from .... to ....

Louise is cycling the slowest from .... to ....

[MEG]

3 This graph shows the volume of water in a bath as it is being filled.

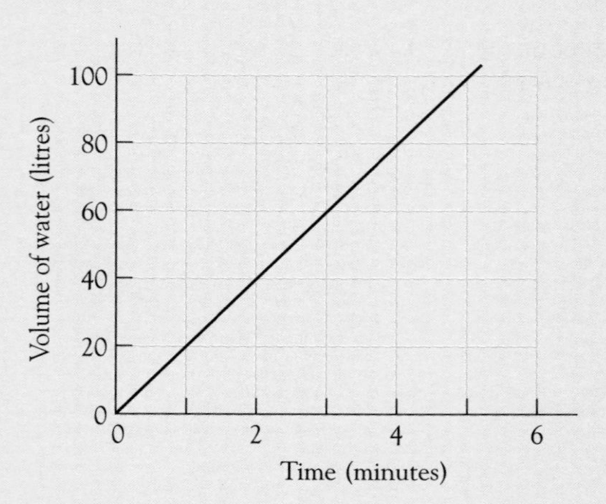

a At what rate is the bath filling? (Make sure you include units in your answer.)

b V is the volume of water in the bath in litres and t is the time in minutes since the tap was turned on.

Write down a formula connecting V and t.

[MEG]

4

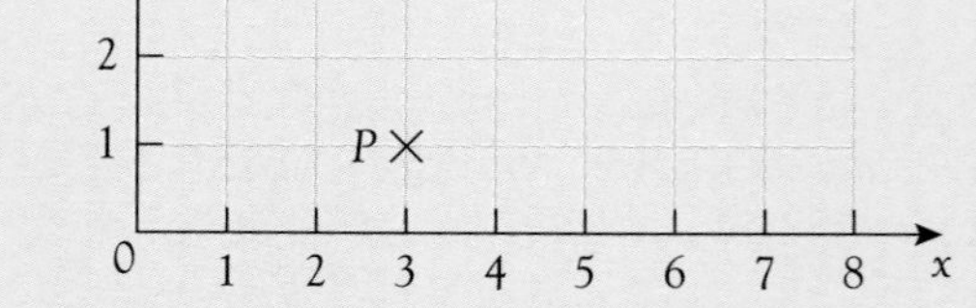

a Work out the gradient of the line joining P and Q.

b Calculate the length PQ.

c P is fixed but Q moves so that PQ remains a constant length. Describe fully the locus of Q.

[MEG]

EXAM QUESTIONS

6 Jennifer walks from Corfe Castle to Wareham Forest and then returns to Corfe Castle. The travel graph of her journey is shown.

a At what time did Jennifer leave Corfe Castle?

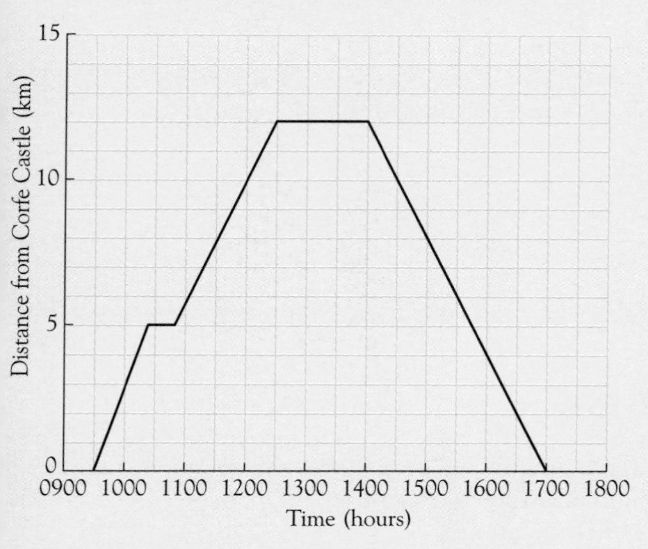

b How far from Wareham Forest did Jennifer make her first stop?

c Jennifer had lunch at Wareham Forest. How many minutes did she stop for lunch?

d At what average speed did Jennifer walk back from Wareham Forest to Corfe Castle?

[SEG]

6 The line $y = x$ has been drawn on the grid.

a Copy the diagram and on your grid
 i draw the graph of $y = 2x - 1$ for values of x from -1 to 2.

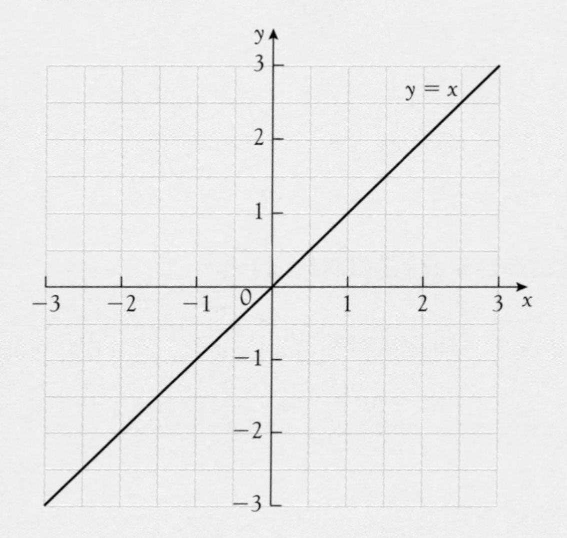

 ii draw the image of $y = 2x - 1$ when it is reflected in the line $y = x$. Label your image line L.

b **i** Calculate the gradient of line L.
 ii For the line L write down the value of y when $x = 0$.
 iii Write down the equation of the line L.

[NEAB]

7 This sketch graph shows what happened to the volume of water in a bath.

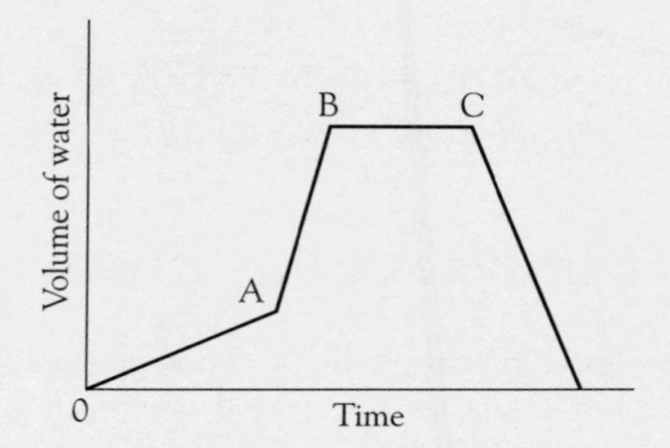

At the start the plug was put in and the hot tap was turned on full.

a Describe what you think happened at A.

b What happened at B?

c What happened at C?

[MEG]

8 Steve travelled from home to school by walking to a bus stop and then catching a school bus.

a Use the information below to construct a travel graph showing Steve's journey. Use time as x axis (0800–0840 using a scale of 2 cm for 5 mins) and distance travelled as y axis (0–10 km using a scale of 1 cm for 1 km).

Steve left home at 0800.

He walked at 6 km/h for 10 minutes.

He then waited for 5 minutes before catching the bus.

The bus took him a further 8 km to school at a steady speed of 32 km/h.

b How far is Steve from home at 0820?

c **i** How long would it take Steve to cycle from home to school at an average speed of 15 km/h? Give your answer in minutes.
 ii Steve cycles at 15 km/h and wants to arrive at the same time as the bus in part **a**. At what time must he leave home?

[MEG]

SIMULTANEOUS LINEAR EQUATIONS

In this chapter you will learn how to solve simultaneous linear equations
- **graphically**
- **by elimination**
- **by substitution**
- **in a problem solving situation**

Simultaneous linear equations

The equation $x + 3 = 5$ has **one** solution, $x = 2$. No other value for x makes the equation true.
The equation $y = 2x + 1$ has **many** solutions that make the equation true, for example, $x = 2$, $y = 5$; or $x = -1$, $y = -1$. These are just two of the possible solutions.

In fact there is an **infinite** number of solutions to the equation $y = 2x + 1$

Notice that in the equation $y = 2x + 1$ there are two unknowns or **variables**, x and y, whereas in the equation $x + 3 = 5$ there is only one unknown, x, hence the **unique** solution.

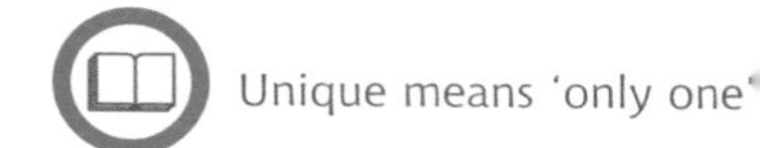

Now consider the equation $y = -x + 4$.
This too has many solutions, such as $x = 4$, $y = 0$ or $x = -2$, $y = 6$.

Plotting the graphs of $y = -x + 4$ and $y = 2x + 1$ on a grid, you can see that the two lines cross in one place. This is called the **point of intersection** of the two lines. At this point, the values of x and y are identical for both equations. These values of x and y make **both** equations true.

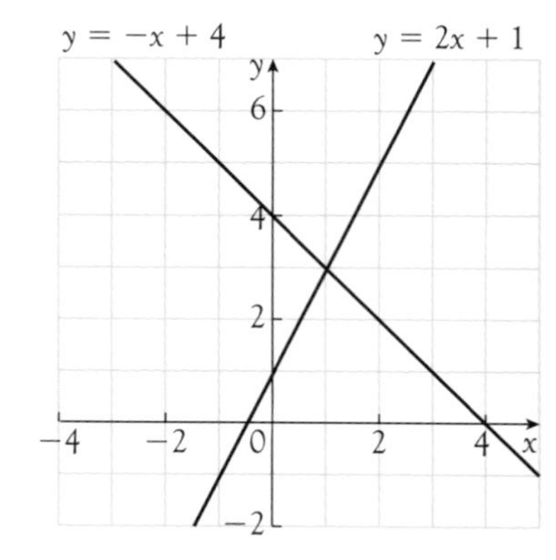

In finding them you are **solving the equations simultaneously**.

$y = 2x + 1$ and $y = -x + 4$ are called **simultaneous linear equations**.

At the intersection, $x = 1$ and $y = 3$.
This is the **solution** of the simultaneous equations.

Solving simultaneous equations graphically

To solve a pair of simultaneous linear equations graphically:
- On a grid, draw both lines accurately.
- Read off the values of the variables at the point of intersection of the lines.

Unless the lines are parallel they will intersect somewhere.

GRAPHICAL METHOD

Sample Question 1

Solve graphically the simultaneous equations $y = 2x + 3$

$y = -x + 9$

Answer

- Choose 3 values for x and find the corresponding values for y for the line $y = 2x + 3$
- On a grid, draw the line $y = 2x + 3$ and label the line with its equation.

x	-2	1	4
y	-1	5	11

- Repeat the process to draw the line $y = -x + 9$

x	-2	0	2
y	11	9	7

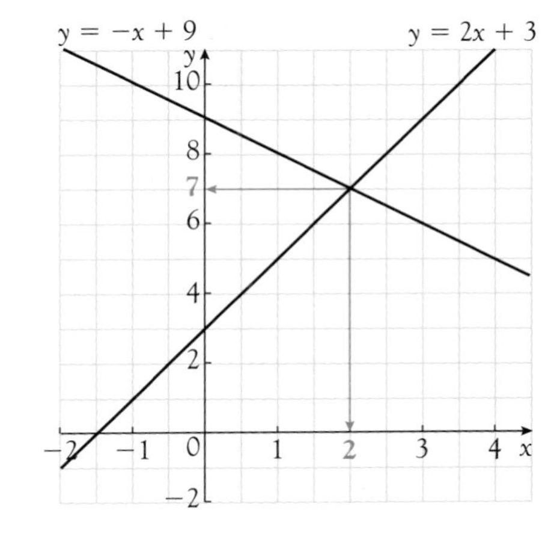

- From the point of intersection draw horizontal and vertical lines to the axes and read off the values of x and y from the axes.
- At the point of intersection $x = 2$, $y = 7$

The solution is $x = 2, y = 7$

You must give **both** values.

Sample Question 2

The graph of $3x + y = 9$ has been drawn on a grid.

a On the same grid, draw the line with equation $y = x - 1$

b Write down the value of x and y that **satisfies both equations.**

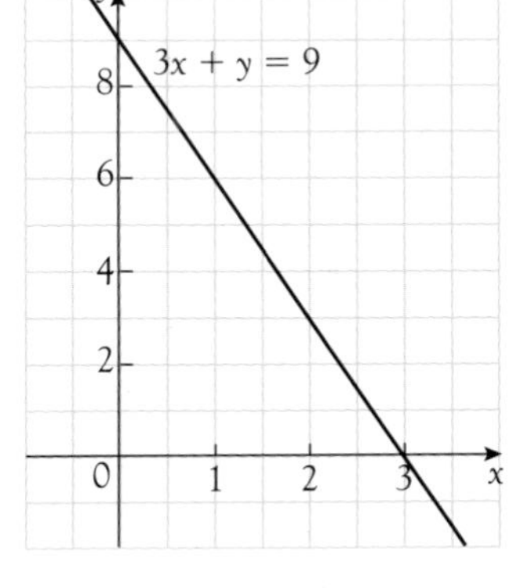

'Satisfies both equations' is another way of saying 'makes both equations true'.

Answer

- On the grid, draw the line $y = x - 1$

x	0	1	2
y	-1	0	1

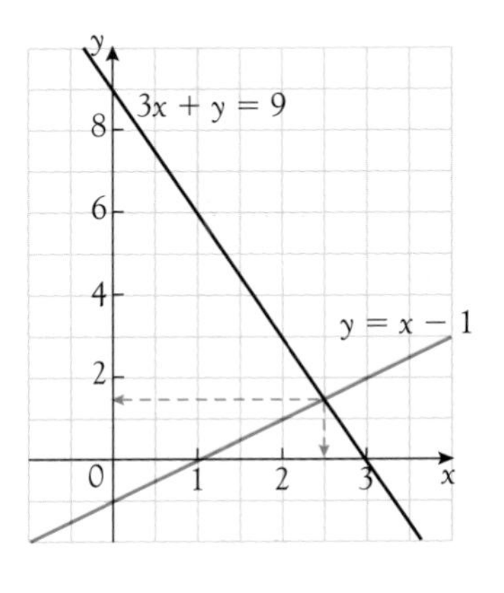

- **Draw vertical and horizontal lines** from the point of intersection of the lines to the x and y axes.
- Read off the values of x and y at this intersection.

The value of x and y that satisfies both equations is $x = 2\frac{1}{2}$, $y = 1\frac{1}{2}$

GRAPHICAL METHOD

If the equation of the line that you are asked to draw is of the form $ax + by = c$ with $a > 0$ and $b > 0$ use the 'cover-up' method. This is illustrated in Sample Question 3.

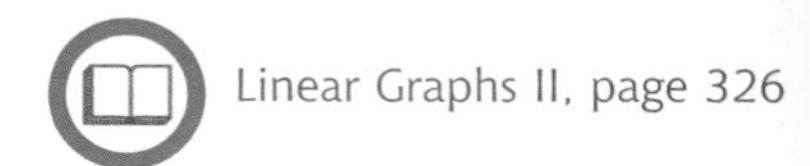
Linear Graphs II, page 326

Sample Question 3

A pizza shop delivers to Tim and Jenny regularly. Jenny lives 7 miles further from the shop than Tim does. In one week the shop delivered 3 times to Tim and twice to Jenny. This was a total distance of 24 miles.

Tim lives x miles from the shop and Jenny lives y miles from the shop.

a Using the information given, write down two equations that involve x and y.

b On a suitable grid, draw the graphs of your equations.

c Using your diagram, write down the distance that Tim and Jenny each live from the pizza shop.

Answer

a

- Write an equation that uses the information about how far from the shop Tim and Jenny live.
- Write an equation that uses the information about the total number of miles travelled.
  Total distance to Tim's $= 3 \times x$
  Total distance to Jenny's $= 2 \times y$ } Add

a

Distance from shop:

Jenny's distance = Tim's distance +7

$y = x + 7$

Number of miles travelled:

$3x + 2y = 24$

b

- For $y = x + 7$ choose three values for x and calculate the corresponding values for y.
- Use the 'cover-up' method to find out where $3x + 2y = 24$ cuts the x and y axes.
- You can see that the maximum value for x is 8 and the minimum value for y is 12. Draw a grid using only **positive x and y** with x going from 0 to 8 and y from 0 to 12.

Only positive x and y are needed as they cannot live a negative distance from the shop.

- Draw the two lines on your grid. Label them.
- Read off the values for x and y where the lines intersect: $x = 2$, $y = 9$.

  Tim lives 2 miles and Jenny lives 9 miles from the shop.

b

For $y = x + 7$:

x	0	2	4
y	7	9	11

For $3x + 2y = 24$:

'Cover-up' x (make $x = 0$)
$2y = 24$
$y = 12$
Cuts y axis at $(0, 12)$

'Cover-up' y (make $y = 0$)
$3x = 24$
$x = 8$
Cuts x axis at $(8, 0)$

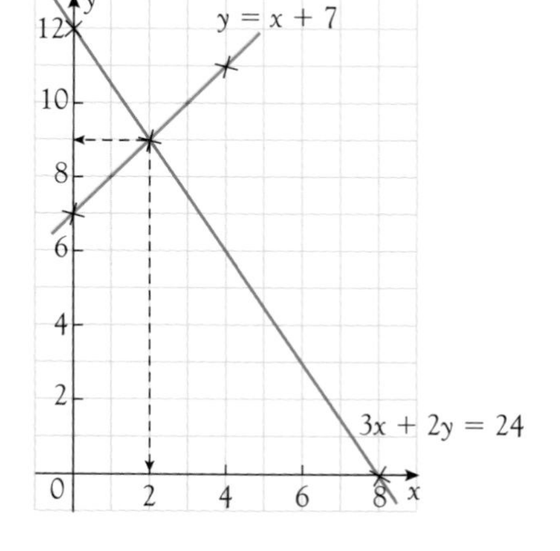

GRAPHICAL METHOD

Exercise 19.1

1 The graph of $2x + 3y = 6$ is shown on the grid.

a Copy the grid and draw on it this line and the line whose equation is $y = 2x - 6$

b Write down the value of x and y that satisfies both equations.

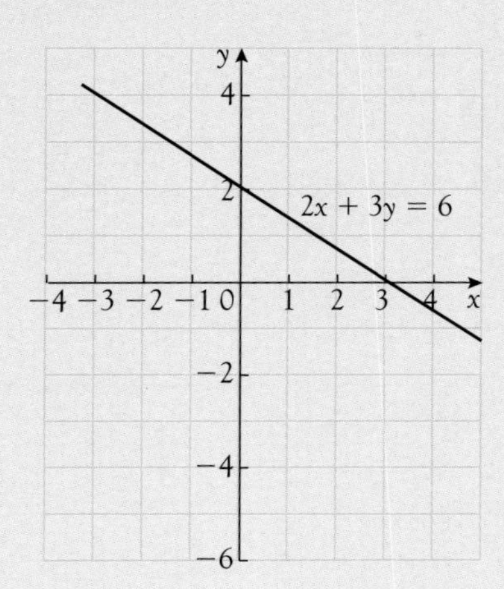

2 a Copy and complete the tables of values for the straight lines given.

i $y = 2x - 3$

x	-1	2	5
y			

ii $y = \frac{1}{3}x + 2$

x	-3	0	3
y			

b On the same set of axes, draw the graph of each line.

c Use your graphs to solve the simultaneous equations

$$y = 2x - 3$$
$$y = \frac{1}{3}x + 2$$

3 Draw a grid with x values from 0 to 4 and y values from -6 to 6

On the grid, draw the lines $y = x - 4$ and $2x + y = 5$

Use your diagram to solve the simultaneous equations $y = x - 4$ and $2x + y = 5$

4 Use a graphical method to find estimates for the value of x and y that satisfies the simultaneous equations.

$$2x + 5y = 10$$
$$y = x + 3$$

Give your answers correct to 1 d.p.

5 A coach trip for two adults and three children costs a total of £66.

a Using £x to represent the cost for an adult and £y the cost for a child, explain why $2x + 3y = 66$

b The same trip for one adult and two children costs a total of £38. Write down another equation involving x and y.

c Use a graph to solve the two equations simultaneously and hence write down the cost for an adult and the cost for a child.

6 Claire adds together her marks for maths and English. Her total mark is 150.

She got 30 more marks in maths than in English.

Using x to represent her maths mark and y to represent her English mark, write down two equations connecting x and y.

Using a graphical method, find her mark for maths and her mark for English.

7 A customer in a café buys two cups of tea and three cups of coffee. The bill comes to £3.70. Another customer in the same café buys three cups of tea and five cups of coffee at a total cost of £5.95.

a Using x pence to represent the cost of a cup of tea and y pence to represent the cost of a cup of coffee, write down two equations connecting x and y.

b Use a graphical method to solve the two equations simultaneously and hence write down the cost of a cup of tea and the cost of a cup of coffee.

8 The sum of the ages of a man and his son is 38 years. If the son multiplies his age by 4 the result is three years less than his father's age. Use a graphical method to find the ages of the man and his son.

ELIMINATION

Solving by elimination

The graphical method of solving simultaneous equations is not always very accurate especially if the solutions are not whole numbers.

To be sure of accuracy in these cases you need an algebraic method.

Elimination is one such method and it is so called because you try to eliminate ('get rid of') one of the variables from the two equations by adding or subtracting the equations.

Sample Question 4 Solve **a** $3x + y = 13$, $x + y = 5$ **b** $4x - 2y = 5$, $3x + 2y = 9$

Answer

a

- Label the equations ① and ② so that you can identify the original equations.

$3x + y = 13$ ①
$x + y = 5$ ②

- Look at the left hand side of the equations. Notice that the y term in each equation is the same size ($1y$) and they are both positive.

- **Eliminate** the y from the equations by **subtracting** one equation from the other. Try to make it so that the resulting x term is **positive**. In this case **subtract equation ② from equation ①**. ($3x - x = 2x$, $y - y = 0$, $13 - 5 = 8$) Solve the resulting linear equation in x.

This is not always possible.

Subtract equ^n ② from equ^n ①

equ^n is short for 'equation'

$2x = 8$

$x = 4$

- Substitute your value for x into one of the original equations – it does not matter which one – and solve the resulting equation to find the value of y.

Substitute $x = 4$ into equ^n ②

$4 + y = 5$

$y = 1$

- Check that the values you have found for x and y satisfy the equation that you have not used by checking for a balance between the LHS and the RHS when the values of x and y are substituted in.

[Check using equ^n ①

LHS: $3 \times 4 + 1 = 13$

RHS: 13

LHS = RHS]

You could substitute into equ^n ① but equ^n ② is easier to solve

$x = 4, y = 1$

ELIMINATION

b

FACT SHEET 2 pages 47–50

- Notice this time that the **coefficient** of y is the same but that one is negative and the other is positive. If you subtract the equations you will not eliminate the y. To eliminate y **add** the two equations, then follow the same procedure as in part (a).

$4x - 2y = 5$ ①
$3x + 2y = 9$ ②

Add equ^n ① to equ^n ②

$7x = 14$

$x = 2$

Substitute for x in **equ^n** ②

$3 \times 2 + 2y = 9$

$6 + 2y = 9$

$2y = 3$

$y = 1\frac{1}{2}$

[**Check** using equ^n ①

LHS: $4 \times 2 - 2 \times 1\frac{1}{2} = 8 - 3 = 5$

RHS: 5

LHS = RHS]

$x = 2, y = 1\frac{1}{2}$

When you substitute your value for x into one of the equations use equation ② because you will get a positive y. If you use the other equation you will have to cope with negative numbers. Try to avoid this. Unfortunately it is not always possible to do so.

The 'check' can often be done quickly in your head and does not have to be written down. It is still very important that the check **IS** done whether mentally or otherwise.

It is not necessary to check the values in both equations as you know it works in one of them already. **You used it to get the values!**

- If the signs of the letter that you are eliminating are both the **same**, i.e. both positive or both negative, then **subtract** the two equations.
- If the signs of the letter that you are eliminating are **different**, i.e. one is positive and the other is negative, then **add** the two equations.

Exercise 19.2

1 Solve each pair of equations simultaneously by elimination.

a $x + y = 12$
$x - y = 4$

b $3s + t = 10$
$s + t = 8$

c $4x - 3y = 2$
$2x + 3y = 28$

d $2x + 5y = 14$
$-4x + 5y = 2$

e $x - 3y = 1$
$x - y = 5$

f $2x + 5y = 46$
$-2x + 3y = 18$

g $p + 3q = 10$
$4p - 3q = -5$

h $2x + 4y = 11$
$4x + 4y = 16$

2 I think of two numbers a and b. The sum of the two numbers is 29. If I subtract b from a the result is 7. Find the value of a and b.

3 A taxi firm charges a fixed amount, £A, for a journey plus an extra amount, £M, for every extra mile (or part of a mile) travelled. John travels 7 miles and pays £3.95 and Sara pays £2.55 for a journey of 3 miles. Calculate the values of A and M.

ELIMINATION

4 5 bread rolls and 2 doughnuts cost £1.60.
8 bread rolls and 2 doughnuts cost £2.20.

a Find the cost of 1 doughnut.

b How much did Susie pay for 6 bread rolls and 4 doughnuts?

5 Find the value of x and of y if

a $2y = 3x + 4$
$4y = 8 - 3x$

b $18 - 3x = 2y$
$10 + 3x = 5y$

What if the coefficients of the letters are not the same?

Sometimes the coefficients of neither of the letters are the same. In this case you are allowed to 'adjust' the coefficients of an equation by multiplying all the terms by an appropriate number.

It could be that you only have to multiply one of the equations but it may be necessary to multiply both equations so that one of the letters has the same coefficients.

Multiplying one equation

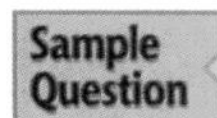

5 Solve the simultaneous equations $3x - y = 1$
$x + 2y = 12$

Answer

- Label the equations ① and ②.
- The x coefficients are different and so are the y coefficients. Decide which letter you want to eliminate.

  If you choose x you will have to subtract the two equations because both the x coefficients are positive so they have the same sign.

  If you choose y you can add the two equations since the y coefficients have different signs.

  Choose to eliminate y.

  It is easier to add than subtract!
- Multiply all the terms in equation ① by 2. Relabel it ③.
- Add equations ③ and ② to eliminate y.
- Solve the resulting equation in x.
- Substitute your value of x into one of the **original** equations. Choose the one with the **positive y** (equation ②).
- Check your values of x and y using equation ①.

$3x - y = 1$ ①
$x + 2y = 12$ ②

× equⁿ ① by 2

$6x - 2y = 2$ ③

$x + 2y = 12$ ②

Write equⁿ ③ below equⁿ ②.

Add equⁿ ③ and equⁿ ②

$7x = 14$

$x = 2$

Substitute for x in equⁿ ②

$2 + 2y = 12$

$2y = 10$

$y = 5$

[Check in equⁿ ①:

LHS: $3 \times 2 - 5 = 6 - 5 = 1$

RHS: 1

LHS = RHS]

$x = 2, y = 5$

ELIMINATION

Sample Question 6 Solve the simultaneous equations $3x + 2y = 2$

$x + 3y = 10$

Answer

- Label the equations ① and ② so that you can identify them.
- Multiply equⁿ ② by 3 to make the coefficients of x the same. Re-label it equⁿ ③.
- Subtract equⁿ ① from equⁿ ③. This gives positive y.
- Solve the resulting equation in y.
- Substitute your value of y into one of the original equations and find the resulting value of x.
- Check your values for x and y using the other original equation.

$3x + 2y = 2$ ①
$x + 3y = 10$ ②

× equⁿ ② by 3
$3x + 9y = 30$ ③
$3x + 2y = 2$ ①
(Write equⁿ ① below equⁿ ③.)
Subtract equⁿ ① from equⁿ ③
$7y = 28$
$y = 4$

Substitute for y in equⁿ ①
$3x + 2 \times 4 = 2$
$3x + 8 = 2$
$3x = -6$
$x = -2$

[Check in equⁿ ②:
LHS: $(-2) + 3\times4 = -2 + 12 = 10$
RHS: 10
LHS = RHS]
$x = -2, y = 4$

Multiplying both equations

Sample Question 7 Solve the following pairs of simultaneous equations:

a $4x + 3y = -5$
$3x + 5y = -12$

b $2x - 3y = -5$
$3x + 2y = 5\frac{1}{2}$

Answer

a

- Label the equations ① and ②.
- You can choose to make the coefficients of x or y the same as all the coefficients are positive. Make the x coefficients equal.
- Multiply equation ① by 3. Re-label it equⁿ ③.
- Multiply equation ② by 4. Re-label it equⁿ ④.
- Subtract equⁿ ③ from equⁿ ④. Watch out for negative numbers.
- Solve the resulting equation in y.
- Substitute your value of y into equⁿ ① (one of the original equations) and solve the equation to find the value for x.
- Check your values using equⁿ ②.

$4x + 3y = -5$ ①
$3x + 5y = -12$ ②

$12x + 9y = -15$ ③
$12x + 20y = -48$ ④
Subtract equⁿ ③ from equⁿ ④
$11y = -33$
($-48 - (-15) = -48 + 15 = -33$)
$y = -3$
Substitute for y in equⁿ ①
$4x + 3\times-3 = -5$
$4x + (-9) = -5$
$4x = 4$
$x = 1$

[Check in equⁿ ②:
LHS: $3\times1 + 5\times(-3) = 3 - 15 = -12$
RHS: -12
LHS = RHS]
$x = 1, y = -3$

ELIMINATION

b

- Label the equations ① and ②.
- In this case the coefficients of y are different signs so adjust these to make them the same so that you can add.
- Multiply equation ① by 2. Label it ③.
- Multiply equation ② by 3. Label it ④.
- Add equation ③ to equation ④.
- Solve the resulting equation in x.
- Substitute your value of x into equation ② and solve it to find the value of y.
- Check your values using equation ①.

$2x - 3y = -5$ ①
$3x + 2y = 5\frac{1}{2}$ ②

$4x - 6y = -10$ ③
$9x + 6y = 16\frac{1}{2}$ ④

Add equ^n ③ to equ^n ④

$13x = 6\frac{1}{2}$

$x = \frac{1}{2}$

Substitute for x in equ^n ②

$3 \times \frac{1}{2} + 2y = 5\frac{1}{2}$

$1\frac{1}{2} + 2y = 5\frac{1}{2}$

$2y = 4$

$y = 2$

[Check in equ^n ①:

LHS: $2 \times \frac{1}{2} - 3 \times 2 = 1 - 6 = -5$

RHS: -5

LHS = RHS]

$x = \frac{1}{2}, y = 2$

When solving simultaneous equations by elimination

- Label both equations so that you can identify them easily.
- Compare the coefficients of 'like' letters in both equations. If one of the 'like' letters has the same coefficient then eliminate the letter by either adding or subtracting the two equations.
- If the 'like' letters do not have the same coefficients
  - 'adjust' **one** of the equations by multiplying by an appropriate number or 'adjust' **both** equations by multiplying each equation by an appropriate number (a different number for each equation),
  - re-label the equation(s),
  - eliminate one of the letters by adding or subtracting the equations.
- Solve the resulting linear equation to find the value of the letter that is left.
- Substitute this letter into one of the **original** equations (not one of the re-labelled equations) to find the value of the other letter.
- Check that the values of the letters satisfy the original equation that you have not used. This can be done on paper or mentally.

ELIMINATION

If you want to solve a pair of simultaneous equations by **elimination** the letters should be on the same side of the equations and in the same order. If they are not you need to re-arrange the equations so that they are.

For example, if you are asked to solve the two equations $\left.\begin{aligned} 2x &= 3y - 5 \\ 2y + 3x &= 5\tfrac{1}{2} \end{aligned}\right\}$ you could re-arrange

them as $\left.\begin{aligned} 2x - 3y &= -5 \\ 3x + 2y &= 5\tfrac{1}{2} \end{aligned}\right\}$ and proceed as in Sample Question 7b above.

You will need to take great care when re-arranging.

Exercise 19.3

1 By adjusting the coefficients of one of the equations solve the following pairs of simultaneous equations. (You may have to re-arrange one or both of the equations as well.)

a $2x + 3y = 23$
$3x + y = 24$

b $2p + 5q = 7$
$8p - 3q = 5$

c $3m - 5n - 11 = 0$
$m - 2n - 4 = 0$

d $4a - 3b = 12$
$2a - b = 6$

e $2x = 3y + 6$
$x + y = 8$

f $3a + 4b = 18$
$3a = 2b$

g $5x - 3y = 6$
$3x + y = 5$

h $2b + 4c = 17$
$3c = 11 - b$

2 By adjusting the coefficients of both equations solve these simultaneous equations:

a $2x + 5y = 16$
$3x + 2y = 13$

b $5a - 3b = 26$
$4a + 2b = 34$

c $3x - 14 = 2y$
$10x - 3y = 65$

d $2x + 7y = -3$
$5x + 3y = 7$

e $2x - 3y = -2.5$
$3x = 2y$

f $3a + 2b = 75$
$4a + 5b = 128$

g $2m - 3n = -7$
$5m + 2n = 1.5$

h $5x - 2y = 7.1$
$2x + 5y = 6.9$

3 A full price ticket for a Saturday match at Silkton football club costs £x. The reduced price for senior citizens (aged 60 and over) is £y.
One Saturday Dai took his father and uncle (aged 63 and 70 years, respectively) to a match. He paid a total of £27 for the three tickets.
The next week he took his father and a 30-year-old friend. He paid a total of £31.50 for these tickets.

a Calculate the price of a full price ticket.

b Calculate the cost of a reduced price ticket.

4 Jane is 24 years older than her daughter Kate.

a Using the letters x and y to represent Kate's and Jane's ages, write down an equation that represents this information.

In two years' time, if Jane multiples her age by 2 and Kate multiples her age by 3 the result will be the same.

b Write down a different equation using this new information.

c Use your equations to find Kate's and Jane's ages now.

5 Alison was selling tickets for a concert in aid of the local hospice. She sold x tickets at £4 each and y tickets at £7.50 receiving a total of £203. She sold a total of 35 tickets. Calculate how many of each value of ticket were sold.

6 Lucy pays a restaurant bill of £95 using £5 and £10 notes.

a If the number of £5 notes she uses is f and the number of £10 notes is t, write down an equation connecting f and t.

b Lucy uses 14 notes altogether. Write down a different equation connecting f and t.

c Solve the equations simultaneously to find how many notes of each value she uses.

Substitution

Rather than re-arranging the equations it may be better to use a substitution method.

Look at this pair of simultaneous equations

$y = 2x + 1$ ①

$y = x + 4$ ②

In solving them together you are finding the value of x and y that satisfy both equations. You know that the value of y in equation ① and the value of y in equation ② are the same.
This means that the quantities on the RHS of each equation must also be the same as each other.
So $2x + 1 = x + 4$.
You can now go on to solve this equation to find the value of x and hence the value of y. This method of solving a pair of simultaneous equations is called **substitution**.

Sample Question 8 Solve the following simultaneous equations

a $y = 2x + 5$, $y = 5x + 1$

b $x + y = 8$, $y = x + 2$

Answer

a

- Label the equations ① and ②.
- Because the y has the same value in both equations the right hand sides of each equation are also equal so put them equal to each other.
- Solve the resulting equation to find the value of x.
- Substitute for x in equation ① to find the value of y.
- Check your values for x and y by substituting them into equation ②.

Try to avoid using decimals especially if you are asked for the EXACT answer. Make use of the $a^{b/c}$ button on you calculator to check if you're not sure about working with fractions.

$y = 2x + 5$ ①
$y = 5x + 1$ ②

$2x + 5 = 5x + 1$
$5 = 3x + 1$
$4 = 3x$
$x = \frac{4}{3} = 1\frac{1}{3}$

Substitute for x in equ^n ①
$y = 2 \times 1\frac{1}{3} + 5$
$y = 7\frac{2}{3}$

[Check in equ^n ②
LHS: $7\frac{2}{3}$
RHS: $5 \times 1\frac{1}{3} + 1 = 7\frac{2}{3}$
LHS = RHS]

$x = 1\frac{1}{3}, y = 7\frac{2}{3}$

b

- Label the equations ① and ②.
- Put a bracket around the expression on the RHS of equ^n ② and use this expression as a substitute for y in equ^n ①.
- Simplify and solve the resulting equation in x.
- Substitute for x in equ^n ②.
- Check your values for x and y using equ^n ①.

You must take care when using this method when minus signs are involved. It is important to put a bracket around the expression that you are going to use as a substitution. If you do not you may miss the fact that there may be two negatives multiplied together.

$x + y = 8$ ①
$y = x + 2$ ②

$x + (x + 2) = 8$
$2x + 2 = 8$
$2x = 6$
$x = 3$

Substitute for x in equ^n ②
$y = 3 + 2$
$y = 5$

[Check in equ^n ①
LHS: $3 + 5 = 8$
RHS: 8
LHS = RHS]

$x = 3, y = 5$

SUBSTITUTION

Sample Question 9 Solve the simultaneous equations $4x - y = 4$
$y = 3x - 1$

Answer

- Label the equations ① and ②.
- Bracket the expression $3x - 1$ from equuⁿ ② and substitute it for y into equⁿ ①.
- Expand the bracket **taking care with the negative signs**.
- Simplify and solve the equation in x.
- Substitute your value of x into equⁿ ② to find the value of y.
- Check your values for x and y by substituting them into equⁿ ①.

$4x - y = 4$ ①
$y = 3x - 1$ ②
Substitute $(3x - 1)$ into equⁿ ①
$4x - \mathbf{(3x - 1)} = 4$
$4x - 3x + 1 = 4$
$x + 1 = 4$
$x = 3$
Substitute for x in equⁿ ②
$y = 3 \times 3 - 1$
$y = 8$
[Check in equⁿ ①
LHS: $4 \times 3 - 8 = 12 - 8 = 4$
RHS: 4
LHS = RHS]
$x = 3, y = 8$

When solving simultaneous equations by substitution:
If each equation starts with '*same letter* ='
- put the right hand sides of the equations equal to each other,
- simplify and solve the resulting equation,
- substitute the value of the letter into one of the original equations to find the value of the other letter,
- check your answer by substituting into the equation that has not been used,

If one of the equations starts '*letter* = *expression in a different letter*' and the other equation has both letters on the same side of the equation,
- put a bracket around the '*expression in a different letter*' and substitute it in the other equation.
- simplify and solve the resulting equation,
- substitute this value into the equation starting '*letter* =' to find the value of the other letter,
- check your answer by substituting into the unused equation.

Exercise 19.4

Solve the following pairs of simultaneous equations using a substitution method. (You may need to re-arrange one of the equations into a suitable form first.)

1. $y = x + 4$
   $x + y = 18$
2. $x + y = 4$
   $2x + y = 5$
3. $2x + y = -3$
   $x = y - 3$
4. $y = x - 1$
   $3x - y = 5$
5. $9x + 2y = 28$
   $3x + y = 11$
6. $p = 8 - q$
   $3p - 2 = q$
7. $a = b - 3$
   $3b - a = 7$
8. $y = 5 + 2x$
   $2x - 3y = -27$

EXAM QUESTIONS

Worked Exam Question
[MEG]

a

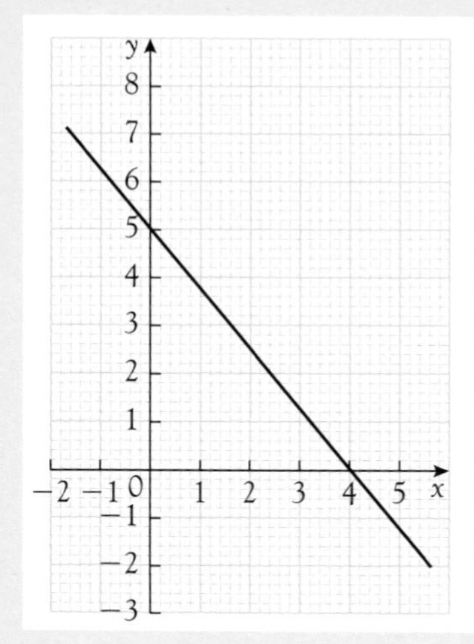

i The graph of $5x + 4y = 20$ is shown in the diagram on the left.
On the diagram, draw the graph of $y = 2x$

ii Use the graphs to find the solution of the simultaneous equations

$$5x + 4y = 20$$
$$y = 2x$$

Give the value of x and the value of y to one decimal place.

i

- Do a table of values

x	0	2	4
y	0	4	8

- Plot the values in the table and draw a **straight line** through the points.
- Label the line.

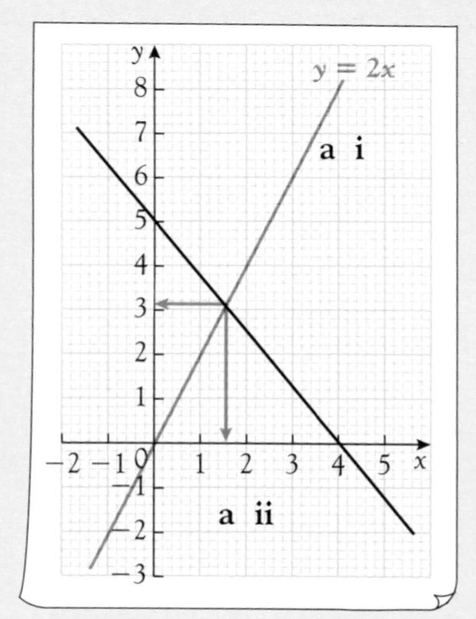

ii

- Draw **horizontal and vertical lines** to the axes from the point of intersection.
- Read off the value of x and the value of y as accurately as possible bearing in mind the instruction to give the values to 1 d.p.

Answer **a ii** $x =$ 1.6 $y =$ 3.1

b **Calculate the exact solution of the simultaneous equations**

$$5x + 4y = 20 \quad \text{—①}$$
$$2x - y = 0 \quad \text{—②}$$

× ② by 4

$8x - 4y = 0$ ③

$5x + 4y = 20$ ①

Add $13x = 20$

$x = \frac{20}{13}$

Substitute into ②

$2 \times \frac{20}{13} - y = 0 \quad \frac{40}{13} = y$

Answer **b** $x = \frac{20}{13}$ $y = \frac{40}{13}$

Write on the question.

Do not use decimals. Fractions are exact.

COMMENTS

a i

A1 graph through (0, 0) and (2, 4)

A1 straight line – not parallel to x or y axes

A1 correct straight line

3 marks

a ii

A1 $x = 1.5$ to 1.6

A1 $y = 3.0$ to 3.1
[For both answers f.t. is allowed if 2 lines are seen intersecting.]

2 marks

b

M2 correct method used to find x or y
[This may be implied by $13x = k$ or $13y = k$]

A1 $x = \frac{20}{13}$ or $1\frac{7}{13}$

A1 $y = \frac{40}{13}$ or $3\frac{1}{13}$

If no working shown

B2 for $x = \frac{20}{13}$

B2 for $y = \frac{40}{13}$
After A0, a mark is given for **both** $x = 1.54$ **and** $y = 3.08$ (or better)

4 marks

Exam Questions

1 A shop sells pens costing x pence each and pencils costing y pence each.
Sanjit pays 67 pence for 3 pens and 2 pencils.
Fiona pays 86 pence for 5 pens and 1 pencil.

a Form two equations in x and y.

b Use your equations to find the cost of a pen and the cost of a pencil.
You must show all your working. [SEG]

2 A discount shop is selling tapes and CDs.
The tapes are all at one price.
The CDs are all at another price.
Jane buys three CDs and a tape and pays £8.92.
Paul buys two CDs and five tapes and pays £12.23.
How much does one CD cost? Show your equations and working clearly. [MEG]

3 Solve the simultaneous equations

$$2x + 3y = 23$$
$$x - y = 4$$

[L]

4 Solve the simultaneous equations

$$x + 2y = 12$$
$$3x - 10y = 0$$

[SEG, p]

5 Mrs Rogers bought 3 blouses and 2 scarves.
She paid £26.
Miss Summers bought 4 blouses and 1 scarf.
She paid £28.
The cost of a blouse was x pounds.
The cost of a scarf was y pounds.

a Use the information to write down two equations in x and y.

b Solve these equations to find the cost of one blouse. [L]

6 The cost, w pounds, of a chest of drawers may be calculated using the formula $w = k + md$
k and m are constants and d is the number of drawers.

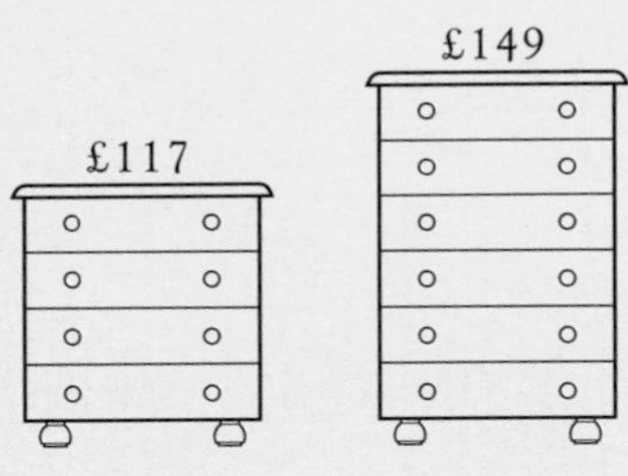

The cost of a chest of drawers with 4 drawers is £117.
The cost of a chest of drawers with 6 drawers is £149.

a Use the information to write down two equations in k and m.

b Solve the equations to find the value of m.

[L, p]

7 Solve the simultaneous equations

$$3x + y = 13$$
$$2x - 3y = 16$$

[L]

8 Use a graphical method to solve the simultaneous equations

$$y = 2x - 3$$
$$y = 10 - x$$

[MEG]

9 The picture shows a pattern of cards.

The number of cards, C, needed to make a house of cards with S storeys, is given by the formula

$$C = aS^2 + bS$$

where a and b represent numbers.

a Use some of the information given at the beginning of the question to show that
$2 = a + b$ and $7 = 4a + 2b$

b Solve these two equations simultaneously to find the numbers a and b. [NEAB, p]

10 Use a graphical method to solve the simultaneous equations

$$7x + 4y = 28$$
$$4x + 5y = 20$$

[MEG]

EXAM QUESTIONS

11 At an indoor market in Blackpool there is a stall where they sell articles at either 50p or £1 each. On 3 October they sold 229 articles and took £140.
Let the number of articles sold at 50p be x and the number of articles sold at £1 be y.

a Explain why $x + 2y = 280$

b Write down another equation involving x and y.

c Solve the two equations algebraically and find how many articles were sold at each price.

[MEG]

12

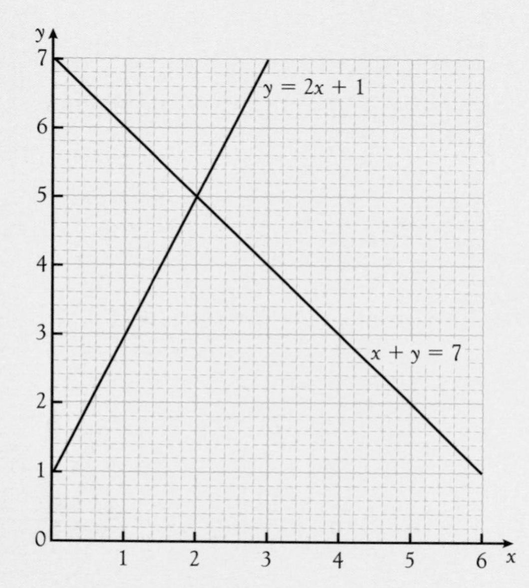

The diagram shows the graphs of the equations

$$y = 2x + 1 \text{ and } x + y = 7$$

Use the diagram to solve the simultaneous equations

$$y = 2x + 1$$
$$x + y = 7$$

[L]

13 L-number patterns are formed by subtracting the corner number from the top number to give the right hand number. For example:

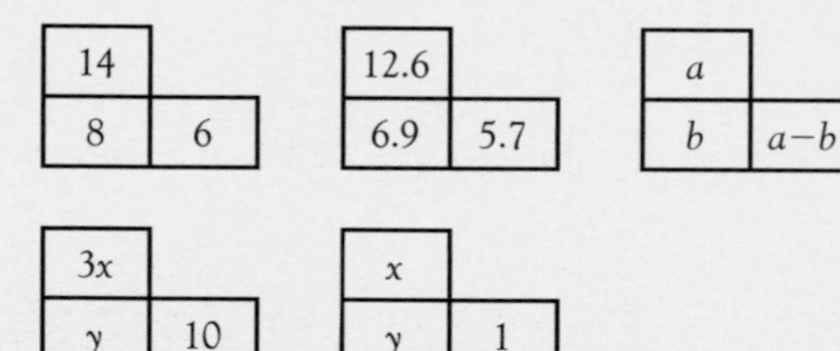

These L-number patterns give the equations

$$3x - y = 10$$
$$x - y = 1$$

Solve these equations to find the value of x and the value of y.

[MEG]

14 a i For the equation $y = 3x + 2$ copy and complete this table.

x	0	3	6
y			

ii On a grid with x axis from 0–6 and y axis from 0–20, draw the graph of $y = 3x + 2$ for values of x between 0 and 6.

b On the same axes draw the graph of $y = 18 - 2x$ for values of x between 0 and 6.

c Write down the values of x and y where the two lines cross.

[MEG]

15 The diagram shows the graph of $y = x + 3$

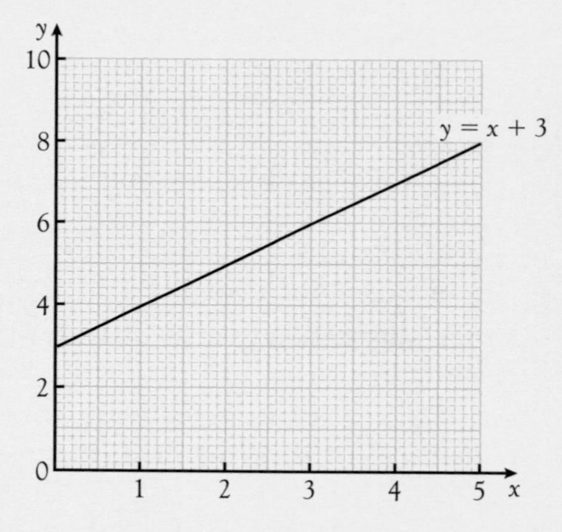

a Copy the diagram, and draw the lines

i $y = 2x - 1$

ii $y = 10 - 2x$

b Use the graphs to solve the simultaneous equations

i $y = x + 3$
$y = 10 - 2x$

ii $y = x + 3$
$y = 2x - 1$

[MEG]

20 TRIGONOMETRY

In this chapter you will
- **learn about the trigonometric ratios of sin, cos and tan**
- **use trig ratios to find missing lengths in right-angled triangles**
- **use trig ratios to find missing angles in right-angled triangles**
- **solve problems involving missing lengths and angles, using trigonometry and Pythagoras' rule**

TASK

In this task you will be investigating the relationships between lengths and angles in right-angled triangles.

You will need a protractor or angle measurer, a ruler, a set-square and a sharp HB pencil.

Step 1

- Draw six different triangles. The sides can be any length you like, but each triangle must have a 90° and a 35° angle in it.
- Label each triangle so that angle A = 35°, angle B = 90° and angle C is the third angle.

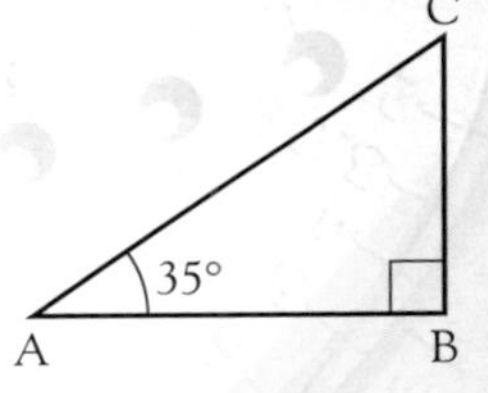

Step 2

- Copy this table

	BC	AB	AC	$\frac{BC}{AC}$	$\frac{AB}{AC}$	$\frac{BC}{AB}$
Triangle 1						
Triangle 2						
Triangle 3						
Triangle 4						
Triangle 5						
Triangle 6						

Step 3

- For each triangle, measure the lengths of BC, AB and AC, in mm. Write each length, to the nearest mm, in the appropriate spaces in the table.

Step 4

- Use your calculator to work out the values in the remaining three columns.

  For example, $\frac{BC}{AC}$ means work out BC ÷ AC.

  If the answer is not exact, give it correct to 3 decimal places.

Step 5

- Write down anything that you notice. Look at others' results and add any further comments.

TRIG RATIOS

In your investigation you should have found that the six values in the column $\frac{BC}{AC}$ were all around the same number, about 0.57. In the column $\frac{AB}{AC}$ all the numbers should have been around 0.82 and in the column $\frac{BC}{AB}$ around 0.7.

Instead of drawing triangles accurately, then measuring and calculating, it is possible to obtain the values of the ratios from your calculator.

Trigonometric definitions

You will need some definitions concerning the sides of a right-angled triangle.

In this triangle one of the angles is marked θ (this is the Greek letter theta, pronounced 'theeta').

This side is **opposite angle θ**.

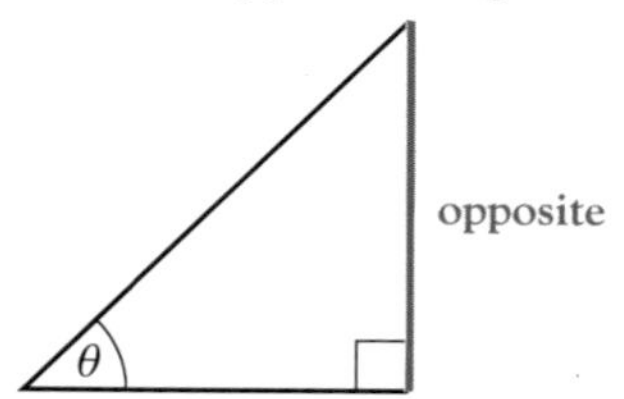

Remember

- the side opposite θ does not form part of angle θ,
- it does not touch θ at all.

This side is **adjacent to angle θ**.

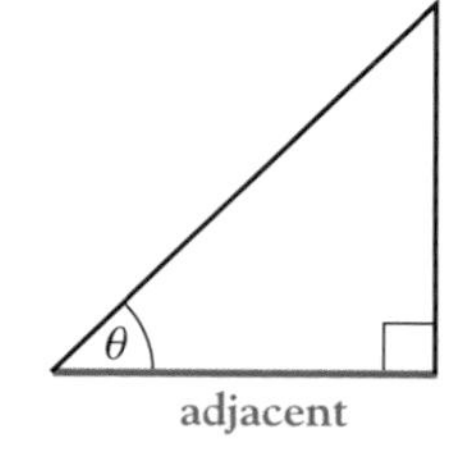

Remember

- the side adjacent to θ is the side 'next to' it,
- the adjacent, together with the hypotenuse, forms angle θ.

This side is the **hypotenuse**.

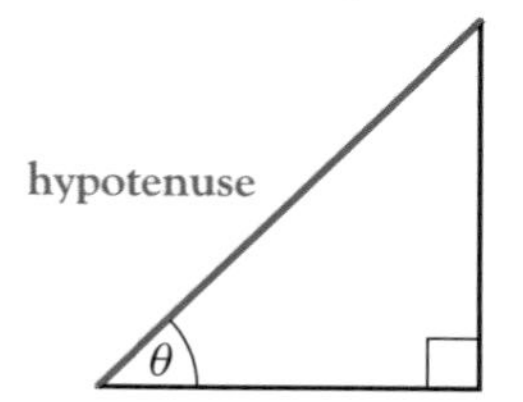

Remember

- it is the longest side,
- the right angle symbol 'points' towards it,
- it does not form part of the right angle,

The names of the sides are often shortened to opp, adj and hyp.

Trigonometric ratios

The adjacent, opposite and hypotenuse are combined in three different ratios, called trigonometric ratios and each has a special name. They are

- the sine ratio (abbreviated to sin – although you still read it as sine, to rhyme with wine, not sin to rhyme with win!)
- the cosine ratio (abbreviated to cos and pronounced 'coz'),
- the tangent ratio (abbreviated to tan).

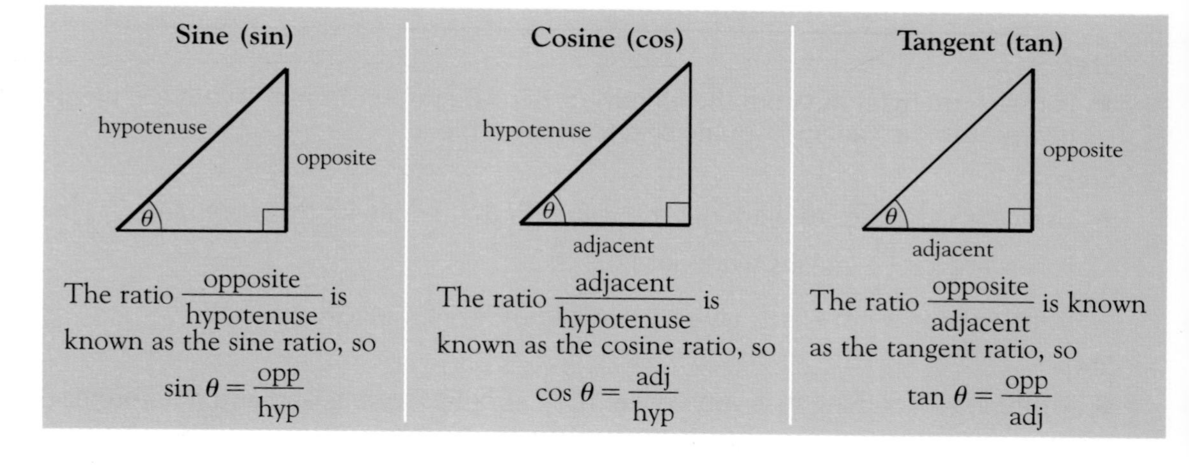

TRIG RATIOS

Using a calculator to find the values of trigonometric ratios

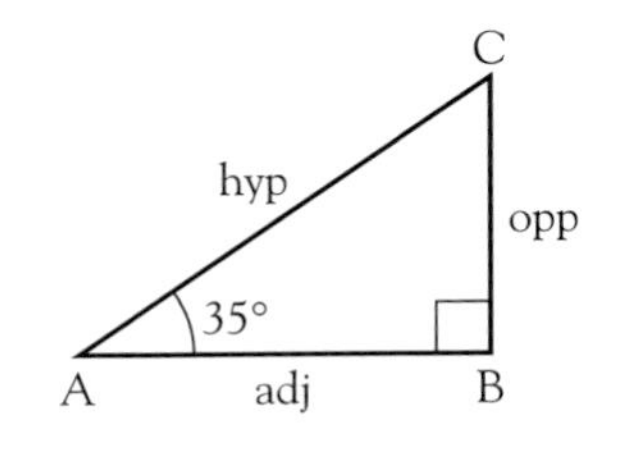

Look again at the results of your task.

The ratios that you found, $\frac{BC}{AC}$, $\frac{AB}{AC}$ and $\frac{BC}{AB}$ are trigonometric ratios.

In your triangle, angle θ was 35°,

so $\frac{BC}{AC} = \frac{\text{opp}}{\text{hyp}} = \sin 35°$, $\frac{AB}{AC} = \frac{\text{adj}}{\text{hyp}} = \cos 35°$, and $\frac{BC}{AB} = \frac{\text{opp}}{\text{adj}} = \tan 35°$.

All scientific calculators are programmed to give you the values of the sin, cos or tan of angles, so you can check how accurate your answers were by using your calculator.

IMPORTANT: You must always make sure that your calculator is working in **degrees**. Look for D or DEG on your calculator display. This is known as **degree mode**.

Do not use your calculator if you see RAD or R, or GRA or G. You **must** reset the mode as these do not work with angles in degrees.

Look up in your calculator manual how to do this.

There are two main types of calculator that differ in how to use them for trigonometry. This book will refer to them as Type 1 and Type 2. Find out which type you have and follow the instructions for that type throughout this chapter.

Type 1: key in the word first, then the angle

sin 3 5 = 0.573576...

cos 3 5 = 0.819152...

tan 3 5 = 0.700207...

Type 2: key in the angle first, then the word.

3 5 sin 0.573576...

3 5 cos 0.819152...

3 5 tan 0.700207...

There is no need to press = if you use Type 2.

How close were the values in your table to these values given by the calculator?

The number of figures given depends on the number of digits on your calculator display. In the following examples, the first four figures have been written down, then ... to indicate that there are more figures.

Try these and other angles between 0 and 90°. You should find that for angles between 0 and 90°, sin and cos always lie between 0 and 1, but tan takes values bigger than 0.

$\sin 32° = 0.5299...$ $\cos 27° = 0.8910...$ $\tan 75° = 3.732...$

$\sin 61° = 0.8746...$ $\cos 86° = 0.0697...$ $\tan 53° = 1.327...$

There are just three 'nice' angles where the value does not go on and on, one for cos, one for sin and one for tan. Can you find them?

Using trigonometric ratios to find missing lengths

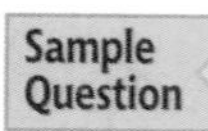

1

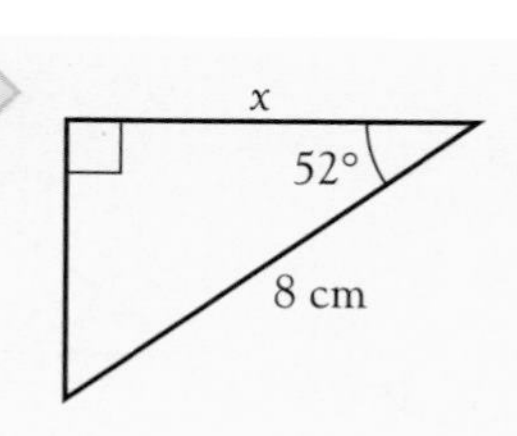

Find the length x.

Give your answer correct to 2 significant figures.

Answer

- The angle you are given is 52°.

  Decide, in relation to 52°, which side of the triangle you want to find and which side you know.

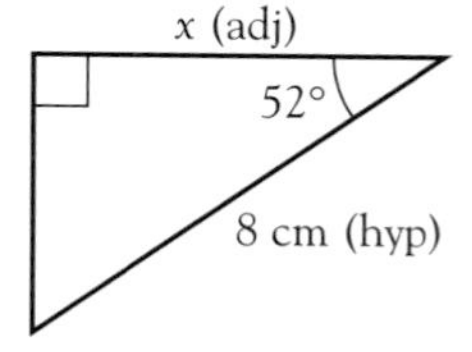

- You want 'adj', you know 'hyp'.
  Draw a sketch and write these on it.
- **Decide which** of the three trig ratios to use and write it down (see page 366).

You are nearly always given credit for choosing the correct ratio in an examination even if you go on to make a mistake.

$$\cos \theta = \frac{\text{adj}}{\text{hyp}}$$

- Substitute the known values

$$\cos 52° = \frac{x}{8}$$

- Use your calculator to find cos 52°.

Type 1 [cos] [5] [2] [=] 0.6156...

Type 2 [5] [2] [cos] 0.6156...

$$\cos 52° = \mathbf{0.6156...}$$

so $0.6156... = \frac{x}{8}$

- Solve the equation to find x by multiplying both sides by 8.

$$x = 8 \times 0.6156...$$
$$= 4.925...$$
$$\underline{x = \mathbf{4.9\ cm\ (2\ s.f.)}}$$

0.6156... [×] [8] [=]

This number should be on display.

Put in the units at the end and approximate as requested.

Note: If you wish, you can do the working in one go, without writing down the value of cos 52°, as follows:

$$\cos 52° = \frac{x}{8}$$

×8 $\quad x = 8 \times \cos 52°$

$$= 4.925...$$
$$\underline{x = 4.9\text{ cm (2 s.f.)}}$$

Type 1 [8] [×] [cos] [5] [2] [=]

Type 2 [8] [×] [5] [2] [cos] [=]

0.6156... shows on display

Sample Question 2

An aircraft flies from X on a bearing of 050° for 9.3 km to Y.
How far east is it now from its starting position?

FINDING LENGTHS

Answer

- Draw a sketch.
- Work out angle XYZ.

  $X\hat{Y}Z = 50°$
  (alternate angles on parallel lines)

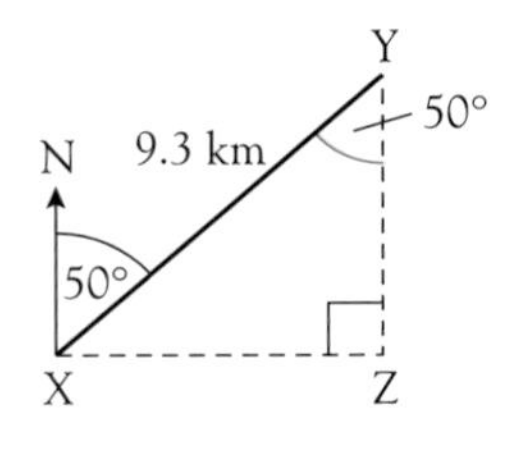

Remember that bearings are measured clockwise from North.

- You need to find XZ, so use △XYZ.
- You want opp, you know hyp, use sin

  $$\sin\theta = \frac{\text{opp}}{\text{hyp}}$$

XZ tells you how far east the aircraft is from its starting position.

- Substitute the known values.

  $$\sin 50° = \frac{XZ}{9.3}$$

- Find the value of sin 50° on your calculator.

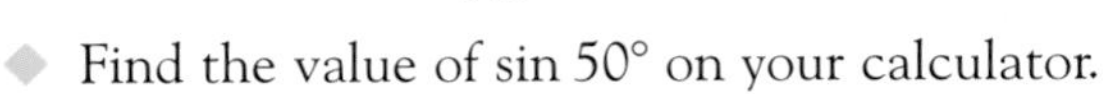

  $\sin 50° = 0.7660\ldots$

Type 1	sin 5 0 =	0.7660...
Type 2	5 0 sin	0.7660...

so $$0.7660 = \frac{XZ}{9.3}$$

- Solve the equation.

  $XZ = 9.3 \times 0.7660\ldots$
  $= 7.124\ldots$
  $= 7.12$ km (3 s.f.)

0.7660... × 9 · 3 =
↑
This should still be on your display

The aircraft is now 7.12 km east of its starting position.

Doing the working in one go:

$$\sin 50° = \frac{XZ}{9.3}$$

$XZ = 9.3 \times \sin 50°$
$= 7.124\ldots$
$= 7.12$ km (3 s.f.)

Type 1	9 · 3 × sin 5 0 =
Type 2	9 · 3 × 5 0 sin =

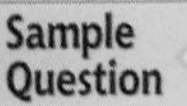

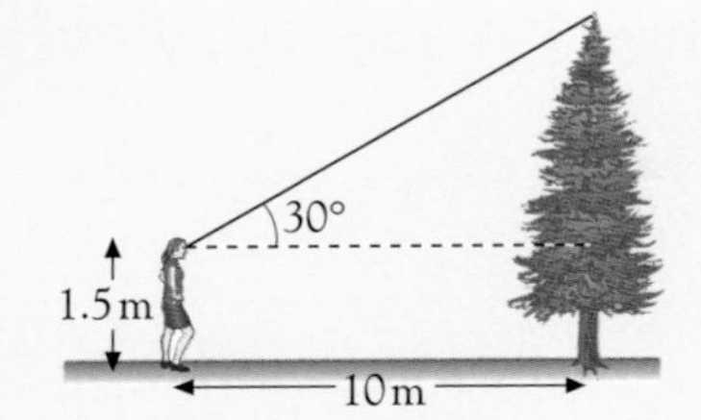

Brenda stands 10 m from the base of the conifer tree in her garden and measures the angle of elevation of the top of the tree to be 30°.

Brenda's eye level is 1.5 m above the ground. What is the height of the tree? Give your answer to the nearest 10 cm.

FINDING LENGTHS

Answer

Do not draw Brenda or the tree!

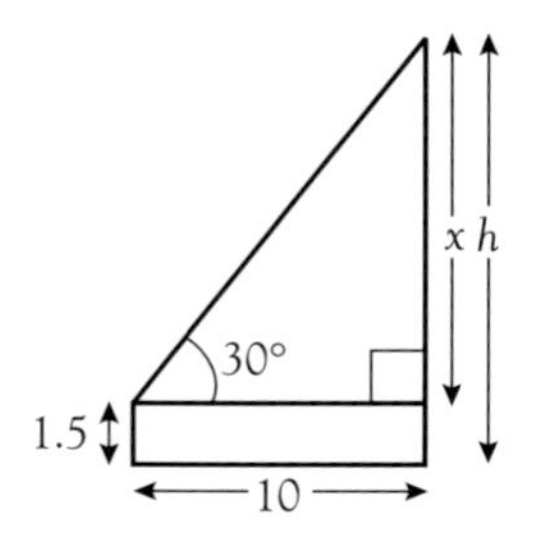

- Draw a sketch showing the lines and angles.
- Write in the values you know and decide what you need to find.
- Write letters for unknown lengths.
  You want to find h but you will need to find x first.

- Look at the right-angled triangle and sketch it.

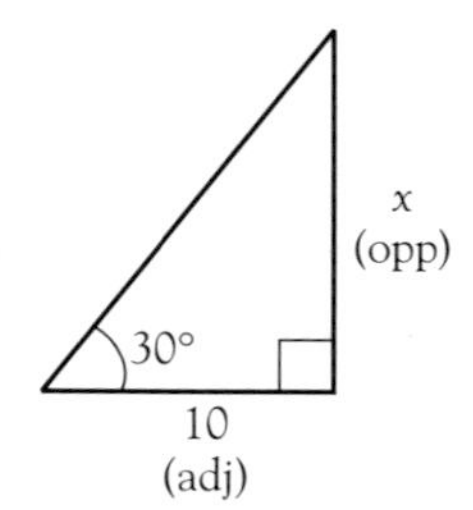

You want opp, you know adj, use tan

$$\tan \theta = \frac{\text{opp}}{\text{adj}}$$

$$\tan 30° = \frac{x}{10}$$

$$x = 10 \times \tan 30°$$

$$= 5.773\ldots$$

You can of course write down the value of tan 30° if you wish.

Do not approximate at this stage as you need this answer to find the height h.
There is no need to write down all the numbers on the display, perhaps just the first four.

- Add 1.5 to find the height

$$h = 5.773\ldots + 1.5$$

$$= 7.273\ldots$$

$$= 7.30\text{ m (to nearest 10 cm)}$$

Do not clear the value of x from your calculator. Just add 1.5 to it.

The height of the tree = 7.30 m

On calculator

Type 1 [1] [0] [×] [tan] [3] [0] [=] [+] [1] [·] [5] [=]

Type 2 [1] [0] [×] [3] [0] [tan] [=] [+] [1] [·] [5] [=]

The first [=] is in fact not necessary if you think ahead and know that you are going to add 1.5. Try the calculator without pressing [=] and watch the calculator display.

Remember when using trig ratios:

- You **must** have a right-angled triangle. If one is not obvious, look for it. You may have to add a line to create one.
- Decide which of the two angles other than 90° you are considering.
- Decide which length is opposite it, which is adjacent to it and which is the hypotenuse.
- Try to **learn the ratios**, but check them with your formula sheet in an examination.

To learn the ratios make up a sentence, using the first letters of the ratios, for example

Spotty **O**llie **h**as	**c**aught **a** **h**orrible	**t**ype **o**f **a**cne!
$s = \frac{o}{h}$	$c = \frac{a}{h}$	$t = \frac{o}{a}$

Some people learn

SOH CAH TOA

Make up **your own sentence** that **you** can remember.

FINDING LENGTHS

Exercise 20.1

1 Calculate, to 2 significant figures, the lengths of the sides marked with letters.
Think carefully whether to use sin, cos or tan. The diagrams are not to scale.

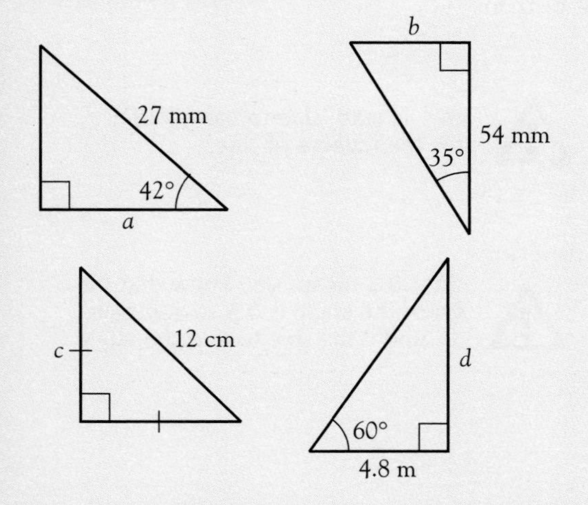

2 A see-saw is 3.6 metres long and when one end is touching the ground the see-saw is at an angle of 30° to the horizontal as shown.
What is the height above the ground of the other end?

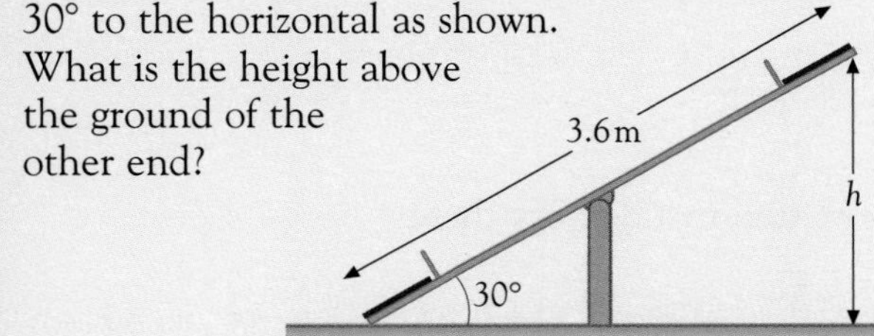

3 A water-skier uses a ramp which is inclined at 72° to the vertical as shown in the sketch. If the length of the ramp is 6.2 metres, find the height, h, of the end of the ramp above the water.

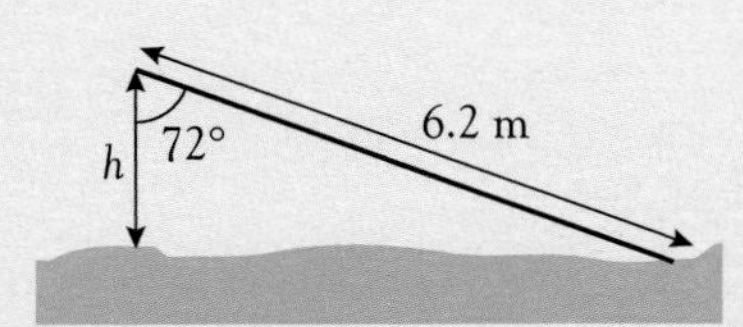

4 When an aircraft takes off it climbs at an angle of 32° to the horizontal.
Find to the nearest metre, the altitude it will reach when it has travelled 1 kilometre from take-off.

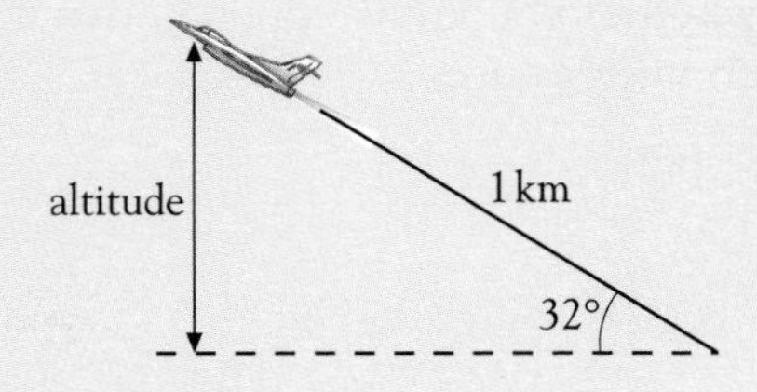

5 The angle of elevation of the top of a tree from a point on the ground 150 metres from its base is 25°. Calculate the height of the tree, giving your answer to the nearest metre.

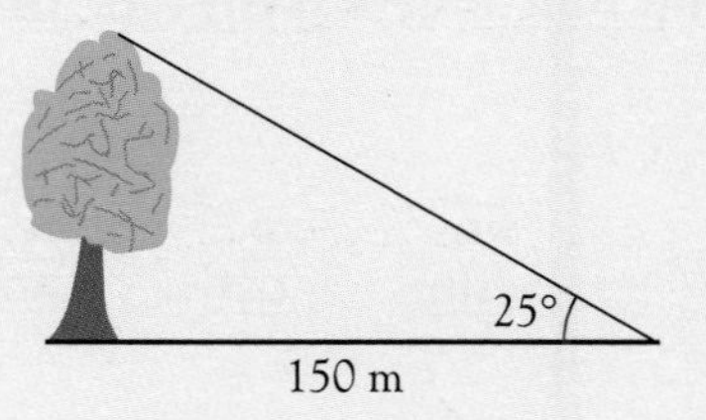

6 A vertical radio mast, AB, of height 27 metres stands on horizontal ground. It is held in place by two cables AC and AD which make angles of 52° and 49° with the vertical (as shown).
Calculate the distance DC.

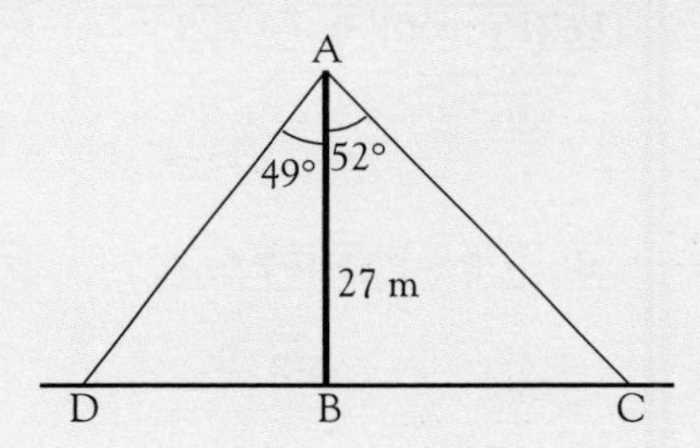

7 A shelf of width 20 cm is held in a horizontal position by a support of length 19 cm which meets the shelf at an angle of 48° as shown in the sketch.
Find the distance, x, by which the shelf overhangs the support. Give your answer in cm to 2 s.f.

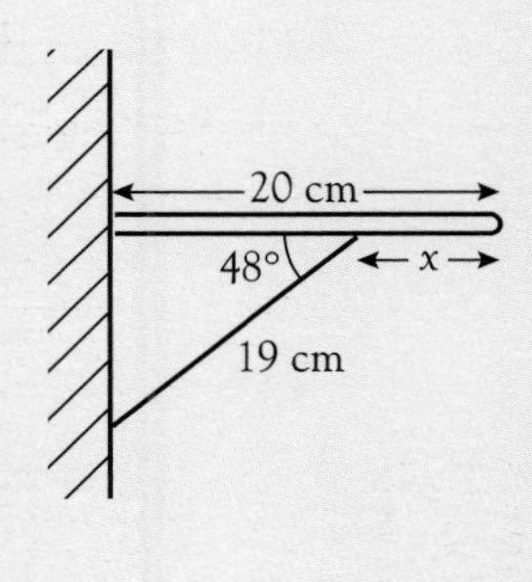

8 An isosceles triangle has equal sides.
AB = AC = 7.5 cm and angle B = angle C = 65° as shown in the sketch.

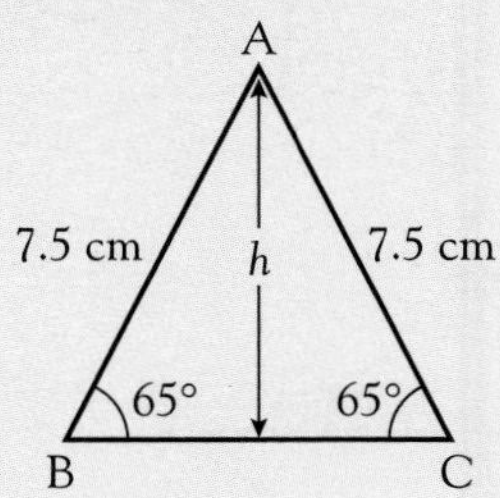

a Calculate the height, h.

b Calculate the length BC.

c Calculate the area of triangle ABC.

Using trigonometric ratios to find missing angles

If you know two lengths in a right-angled triangle, you can work out the angles.

To do this you will need to use the **inverse trig functions** on your calculator. These are usually on the [2nd function] or [SHIFT] or [INV] operation of the trig buttons.

Look for

$\sin^{-1}$ [sin] $\cos^{-1}$ [cos] $\tan^{-1}$ [tan]

$\sin^{-1}$ is read 'sine to the minus 1'. It is the **inverse** of sine.

If you know that $\sin \theta = 0.5$, to find θ use the inverse function

$\theta = \sin^{-1} 0.5$

$= 30°$

$\sin^{-1} 0.5$ means you know that the sin of the angle is 0.5 so you have to 'undo' the sine to find the angle.

Type 1 [SHIFT] [sin] [0] [·] [5] [=] → 30 ([SHIFT] [sin] = [$\sin^{-1}$])

Type 2 [0] [·] [5] [SHIFT] [sin] → 30 ([SHIFT] [sin] = [$\sin^{-1}$])

When the answer is not exact, the angle is usually written to 1 d.p. Try these:

$\cos \theta = 0.8$	$\tan \theta = 2.413$
$\theta = \cos^{-1} 0.8$	$\theta = \tan^{-1} 2.413$
$= 36.86\ldots$	$= 67.48\ldots$
$= \underline{36.9° \text{ (1 d.p.)}}$	$= \underline{67.5° \text{ (1 d.p.)}}$

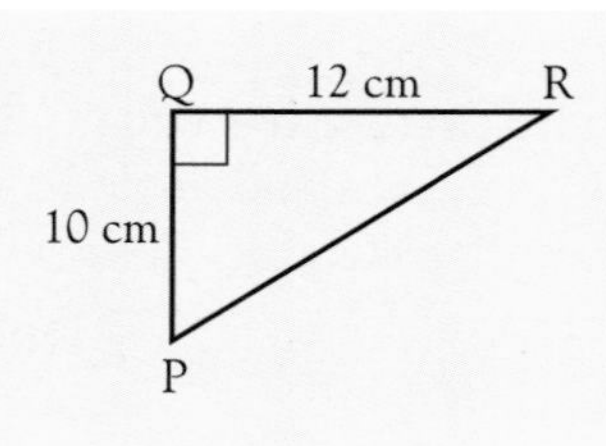

Calculate angle P, giving your answer in degrees to 1 decimal place.

Answer

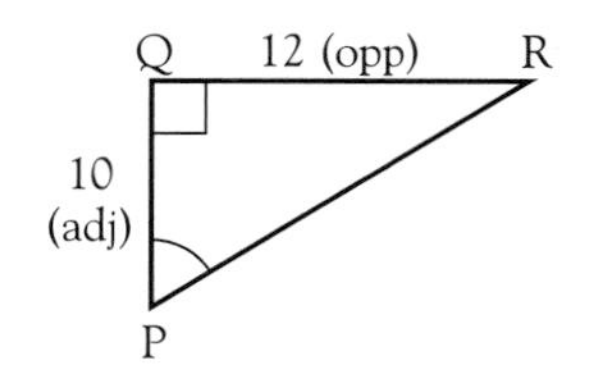

- Draw a sketch and write in whether you know opp, adj, hyp in relation to angle P. Then decide which ratio to use.

  You know opp and adj, so use tan.

  $$\tan \theta = \frac{\text{opp}}{\text{adj}}$$

FINDING ANGLES

- Substitute the known values.

$$\tan P = \frac{12}{10}$$
$$= 1.2$$
$$P = \tan^{-1} 1.2$$
$$= 50.19\ldots$$
$$\underline{P = 50.2° \text{ (1 d.p.)}}$$

Type 1 [1] [2] [÷] [1] [0] [=] [SHIFT] [tan] [=] (SHIFT tan = $\tan^{-1}$)

Type 2 [1] [2] [÷] [1] [0] [=] [SHIFT] [tan] (SHIFT tan = $\tan^{-1}$)

If you prefer, you need not write down the value of $\frac{12}{10}$.

Watch the display.

$$\tan P = \frac{12}{10}$$
$$P = \tan^{-1}\left(\frac{12}{10}\right)$$
$$= 50.19\ldots$$
$$\underline{P = 50.2° \text{ (1 d.p.)}}$$

Type 1 [SHIFT] [tan] [(] [1] [2] [÷] [1] [0] [)] [=] (SHIFT tan = $\tan^{-1}$; after [)] display shows 1.2)

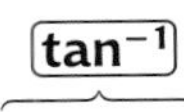

Type 2 [1] [2] [÷] [1] [0] [=] [SHIFT] [tan] (SHIFT tan = $\tan^{-1}$)

For type 2 calculators, you need to work out 12 ÷ 10 first.

Sample Question 5

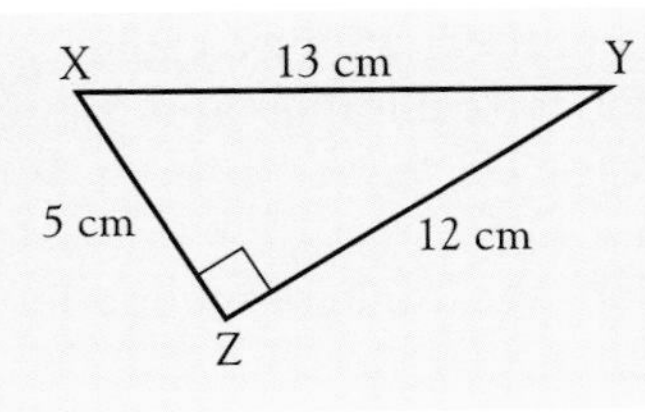

Find the smallest angle in this triangle.

Answer

- Decide which is the **smallest angle.**
- Draw a sketch.

The smallest angle is opposite the smallest side, so angle Y is needed.

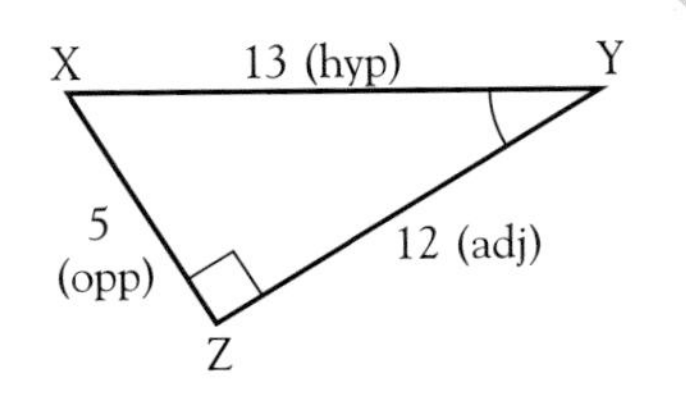

- As you know all three lengths, you can choose which ratio to use. If you choose sin

$$\sin \theta = \frac{\text{opp}}{\text{hyp}}$$
$$\sin Y = \frac{5}{13}$$
$$= 0.384\ldots$$

Do not approximate here. Leave all the numbers on your display.

$$Y = \sin^{-1} 0.384\ldots$$
$$= 22.61\ldots$$
$$\underline{Y = \mathbf{22.6°} \text{ (1 d.p.)}}$$

Type 1 [5] [÷] [1] [3] [=] [SHIFT] [sin] [=] (SHIFT sin = $\sin^{-1}$)

Type 2 [5] [÷] [1] [3] [=] [SHIFT] [sin] (SHIFT sin = $\sin^{-1}$)

In one step:

$$\sin Y = \frac{5}{13}$$
$$Y = \sin^{-1}\left(\frac{5}{13}\right)$$
$$= 22.61\ldots$$
$$= \underline{22.6° \text{ (1 d.p.)}}$$

Remember that the answer is an angle and the units are degrees.

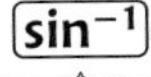

Type 1 [SHIFT] [sin] [(] [5] [÷] [1] [3] [)] [=] (SHIFT sin = $\sin^{-1}$)

Type 2 As type 2 above.

FINDING ANGLES

If you choose cos

$$\cos Y = \frac{\text{adj}}{\text{hyp}}$$
$$= \frac{12}{13}$$
$$Y = \cos^{-1}\left(\frac{12}{13}\right)$$
$$= 22.61\ldots$$
$$\underline{Y = 22.6° \text{ (1 d.p.)}}$$

If you choose tan

$$\tan Y = \frac{\text{opp}}{\text{adj}}$$
$$= \frac{5}{12}$$
$$Y = \tan^{-1}\left(\frac{5}{12}\right)$$
$$= 22.61\ldots$$
$$\underline{Y = 22.6° \text{ (1 d.p.)}}$$

Exercise 20.2

1 Calculate, to the nearest degree, the angles shaded in the following triangles:

a

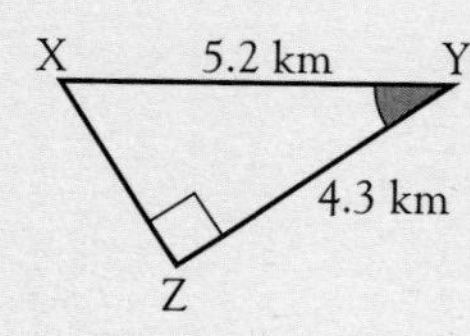

b

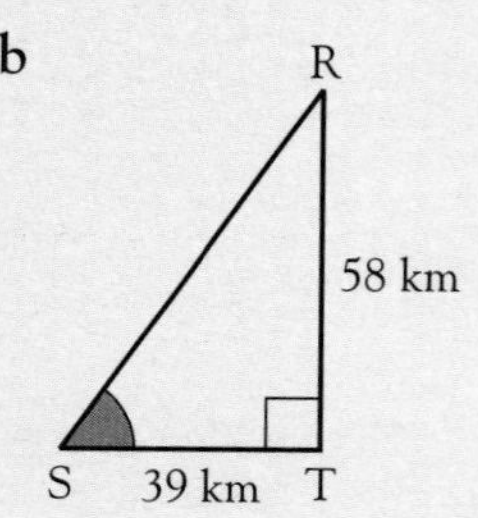

c

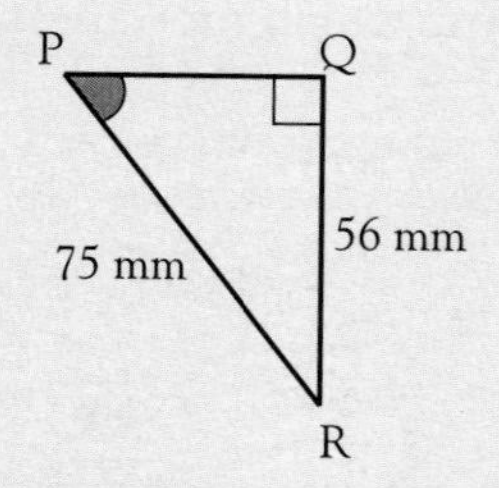

d

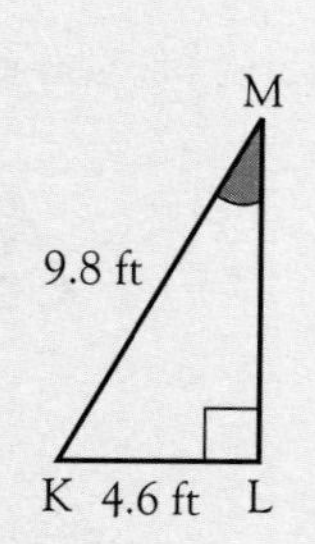

e

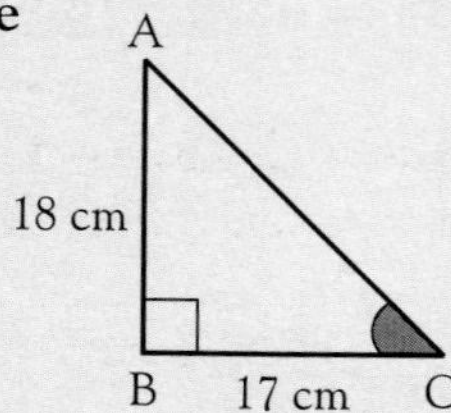

f

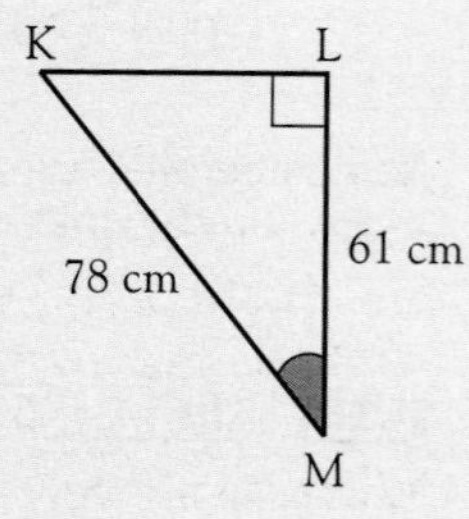

2 A ramp is to be built to allow wheelchairs to go along a path which has a step.
The step is 16 cm high and the ramp is to extend for 1.5 metres in front of the step as shown.
Calculate the angle of inclination of the ramp, θ.

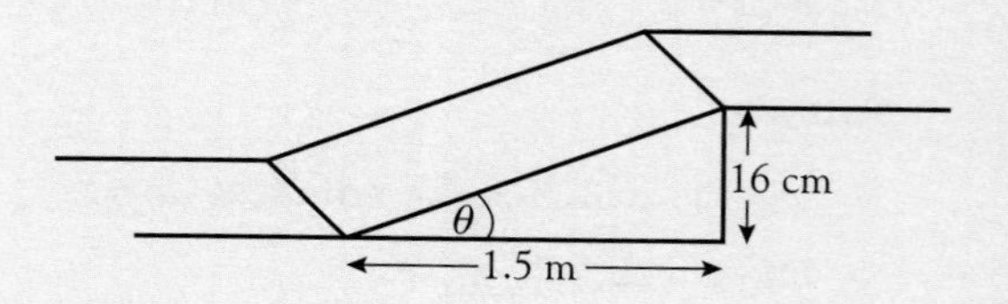

3 A ladder of length 4 metres rests against a window sill which is at a height of 3.75 metres above the ground.
Find the angle the ladder makes with the horizontal.

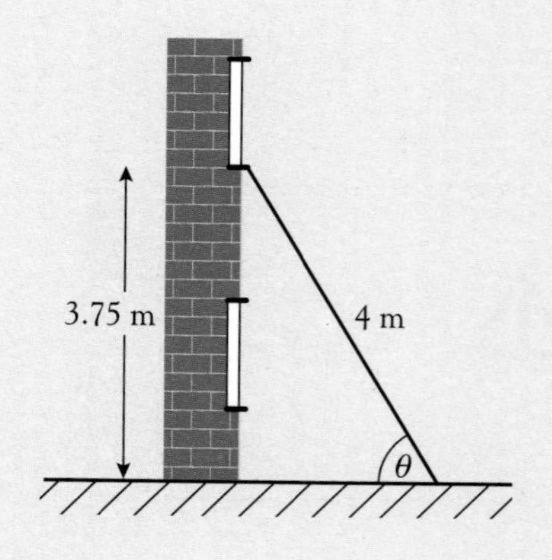

FINDING ANGLES

4 In a 'Roll a Penny' game at a fairground, 1 p coins are released down a chute onto a horizontal board. The player wins if the coin comes to rest completely within any square on the board.
If the chute is 8 cm long and projects 5 cm onto the board, find the angle the chute makes with the horizontal.

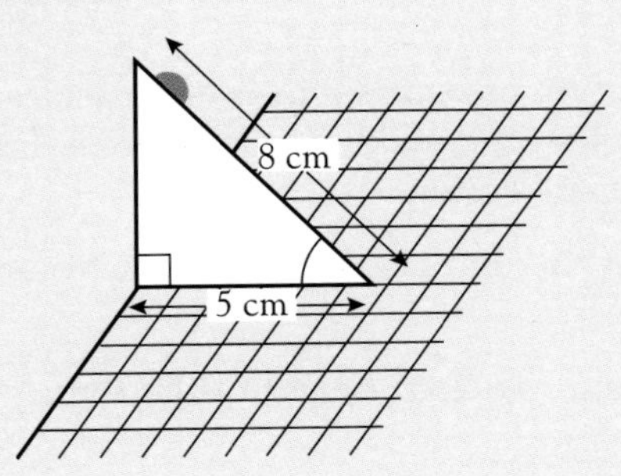

5 The North Yorkshire Moors Railway includes one section on an incline of 1 in 49. For every 49 metres of track, the height increases by 1 metre. Find, to 2 significant figures, the angle between the track and the horizontal.

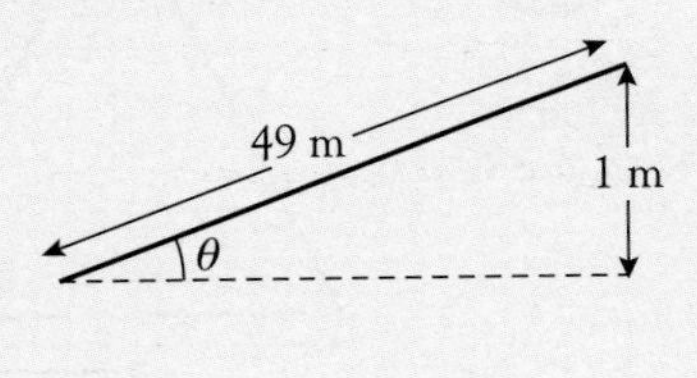

6

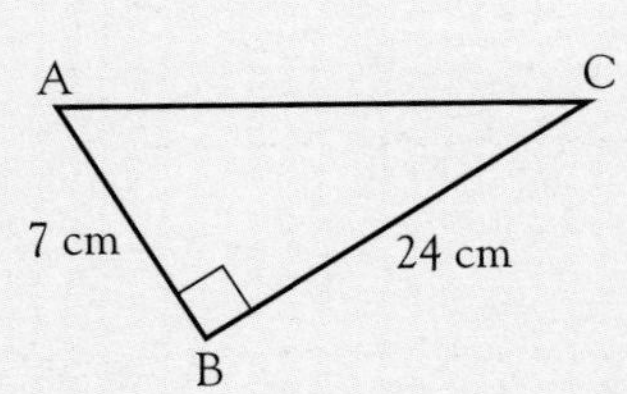

Calculate

a the length of AC.

b the smallest angle in $\triangle ABC$.

7 Find the angle of elevation of the top of a flagpole of height 3.2 metres from a point on the ground 12 metres from its base (see diagram).

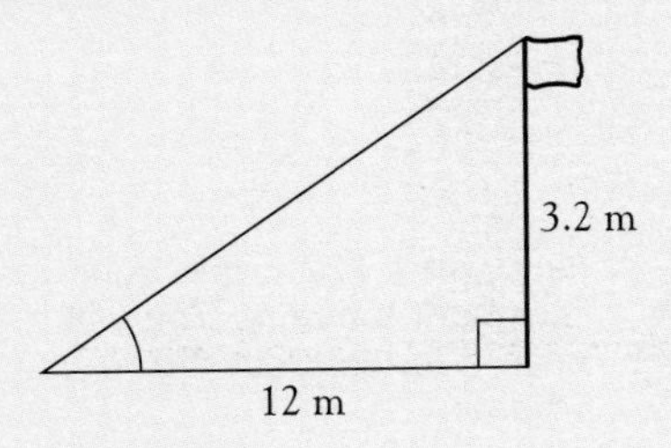

8 Gavin is flying his kite. The string is 36 metres long and he holds the lower end at a distance of 1.3 metres above the ground.
Find the angle at which the string is inclined to the horizontal when the kite is flying at a height of 32.2 m.

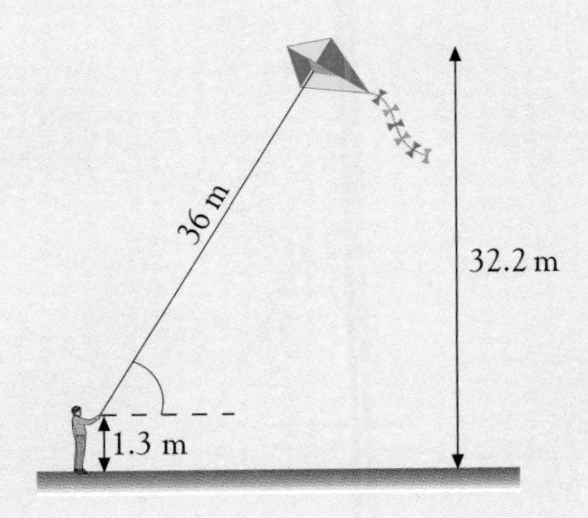

9 A boat sets off from the end of a pier, P. It travels 24 km East and then 17 km South. Find its bearing and distance from P after this journey.

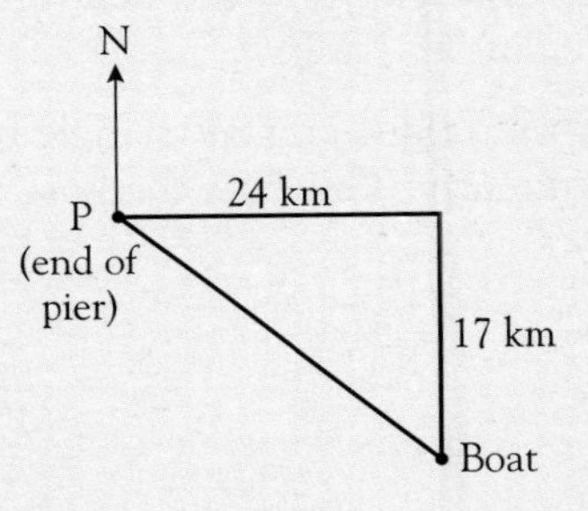

10

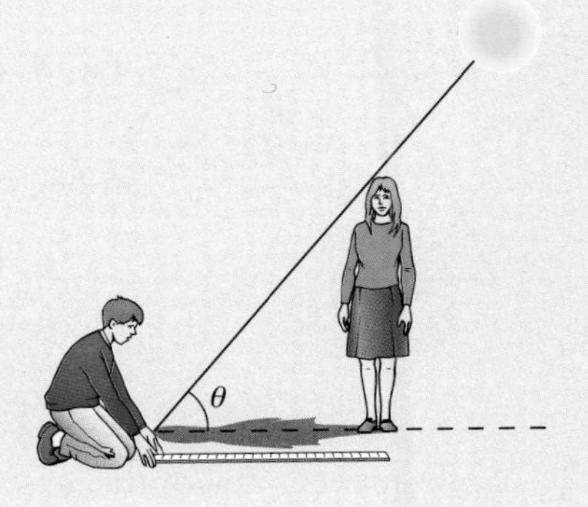

Jack is calculating θ, the angle of elevation of the sun, by measuring Rachel's height and then the length of her shadow.
He finds that Rachel's height is 158 cm but the length of her shadow is only 106 cm.
What is the angle of elevation of the sun?

Mixed problems

Sometimes you need to add a line to your diagram to form a right angle, or you need to find one length first in order to find another length.

Sample Question 6

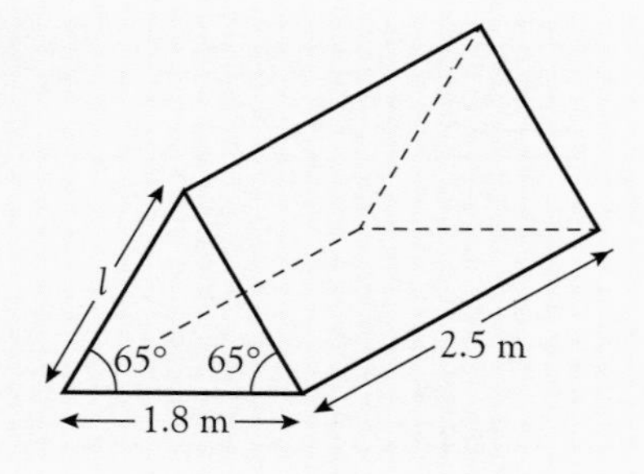

a The cross section of Amy's tent is an isosceles triangle with base angles 65° and base length 1.8 m. Find the area of the cross-section.

b The length of the tent is 2.5 m. Find the volume of the tent.

c Find l, the slant (sloping) height of the tent.

Answer

a

- Draw a sketch of the **cross-section**. To find the area of the triangle you will need to find the perpendicular height, h.
- Form a right-angled triangle by drawing the line of symmetry. This cuts the base in half at 90°.

Area and Volume II page 280

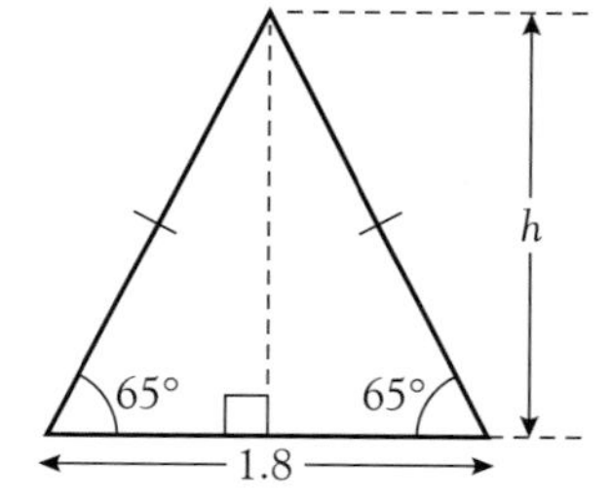

- Sketch one of the right-angled triangles and use trigonometry to find h.

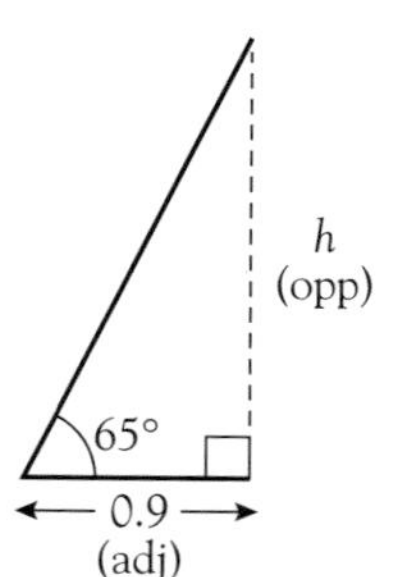

You want opp, you know adj, use tan

$$\tan\,\theta = \frac{\text{opp}}{\text{adj}}$$

$$\tan 65° = \frac{h}{0.9}$$

$$h = 0.9 \times \tan 65°$$

$$h = 1.930\ldots$$

Do not clear your display. Leave all the figures on it. You will need them to work out the area.

- Work out the area of the isosceles triangle that forms the cross-section.

$$\text{Area} = \frac{b \times \text{h}}{2}$$

$$= \frac{1.8 \times 1.930\ldots}{2}$$

$$= \mathbf{1.737\ldots}$$

Area = 1.73 cm^2 (3 s.f.)

Now put 1.737... into your calculator memory. You will need it to work out the volume.

b

- Volume = area of cross-section × length

  = **1.737...** × 2.5

  = 4.342...

  Volume = 4.34 cm^2 (3 s.f.)

Recall 1.737... from your memory if it is not still on your display.

Remember that if you use a value calculated earlier in the question, you must not use the approximated value. Use all the figures if possible, but at least 4 figures if your final answer is needed to 3 s.f.

c

- To find the slant height, l, you can use the value found in part **a** for the perpendicular height of the triangle.

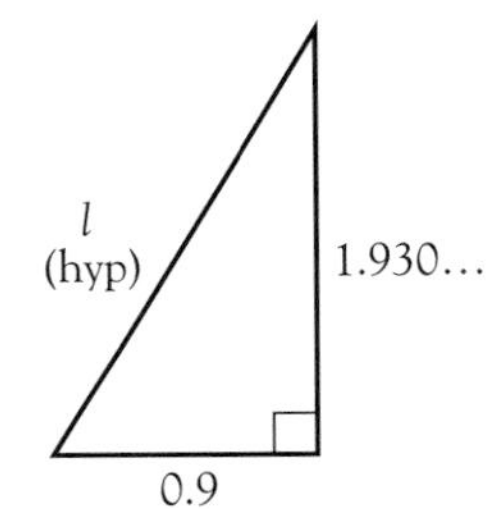

- You know two lengths in the right-angled triangle so you can use Pythagoras' rule to find the hypotenuse, l.

$l^2 = \mathbf{1.903^2...} + 0.9^2$

$= 4.535...$

$l = \sqrt{4.535...}$

$= 2.129...$

$= 2.1$ m (1 d.p.)

If you have lost the full figure 1.903... calculate it again or use the four figures 1.903.

Slant height of tent = 2.1 m

Alternative method using trigonometry

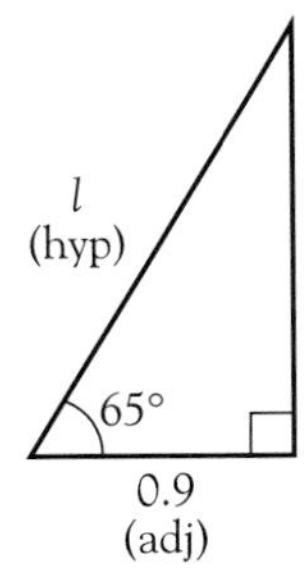

- You want hyp, you know adj, use cos.

$$\cos\theta = \frac{\text{adj}}{\textbf{hyp}}$$

$$\cos 65° = \frac{0.9}{l}$$

Notice this is different from the usual equation as the missing length is on the bottom of the fraction.

- Re-arrange the equation to make l the subject.

Multiply by l $\quad l \times \cos 65° = 0.9$

Divide by cos 65° $\quad l = \dfrac{0.9}{\mathbf{\cos 65°}}$

$l = 2.129...$

$l = 2.1$ m (1 d.p.)

Notice that l and cos 65° have changed places.

Type 1 [0] [·] [9] [÷] [cos] [6] [5] [=]

Type 2 [0] [·] [9] [÷] [6] [5] [cos] [=]

LENGTHS AND ANGLES

Exercise 20.3

1 An isosceles triangle PQR has sides PQ = PR = 82 mm and QR = 74 mm. Calculate the angles of this triangle.

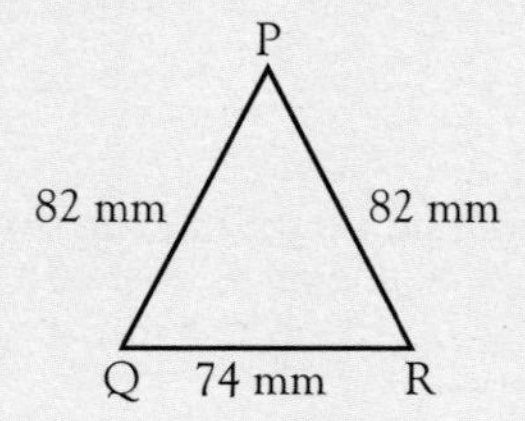

2 A rhombus has equal sides of length 84 mm. Its longer diagonal is 150 mm.

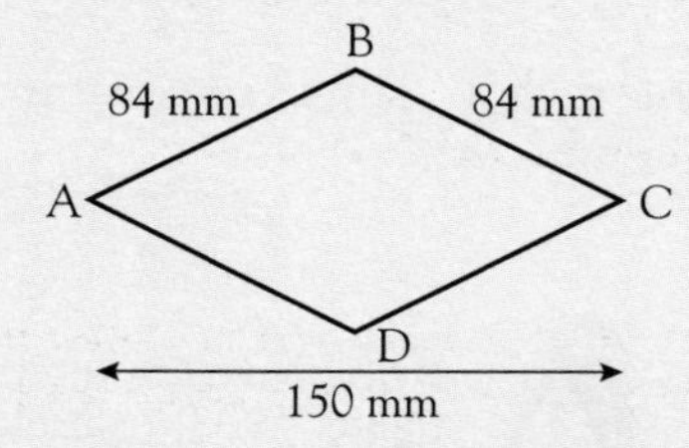

Find the angles of the rhombus, giving your answers to the nearest degree, and calculate the length of the shorter diagonal.

3 In the diagram, O is the centre of the circle. Find length a and angle θ, giving your answers correct to 3 significant figures.

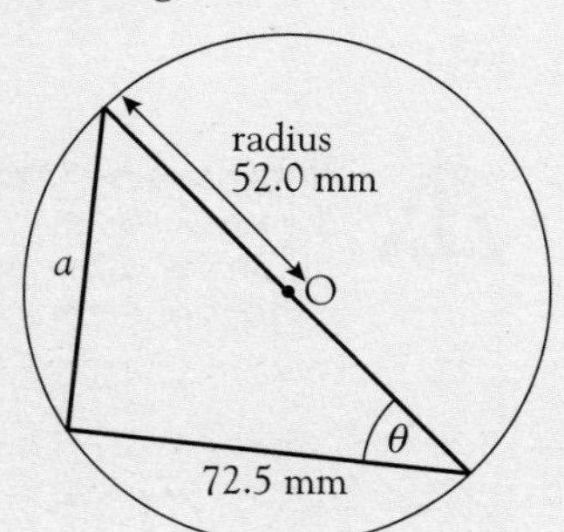

4 An aerial is mounted on top of a block of flats as shown in the sketch. From a point on the ground 45 metres away from the flats, the angles of elevation of the top and bottom of the aerial are 32° and 29°, respectively.

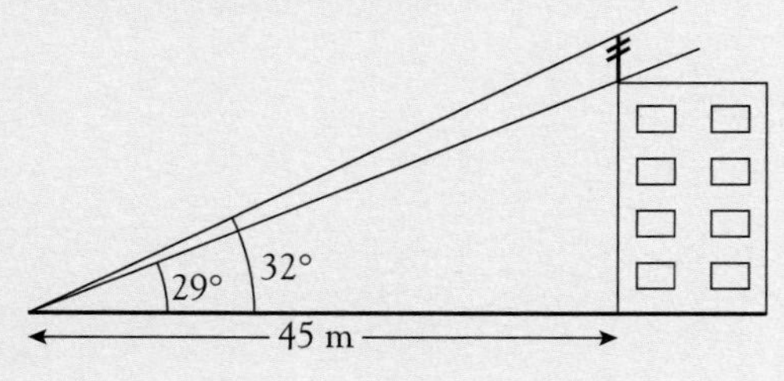

Calculate

a the height of the block of flats

b the height of the aerial.

5 The sketch shows the end view of a house. The house is 22 feet wide and the roof starts at a height of 21 feet and rises to a height of 28 feet.

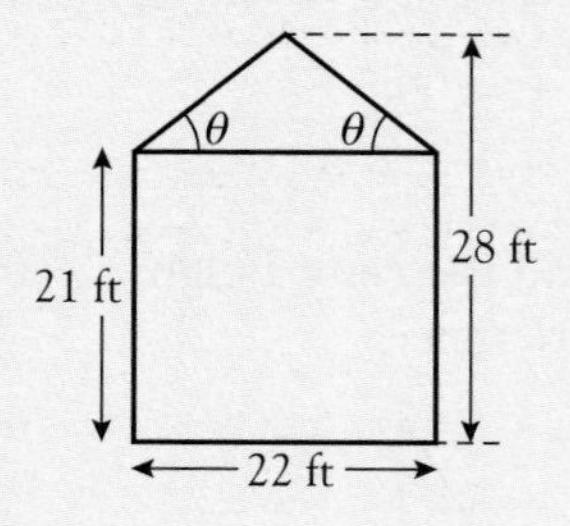

Calculate the angle, θ, at which the sides of the roof are inclined to the horizontal.

6 A conservatory is attached to the side of a house. The front of the conservatory is 2.5 metres high and its roof panels are 2.9 metres long and inclined at 72° to the vertical, as shown in the sketch.

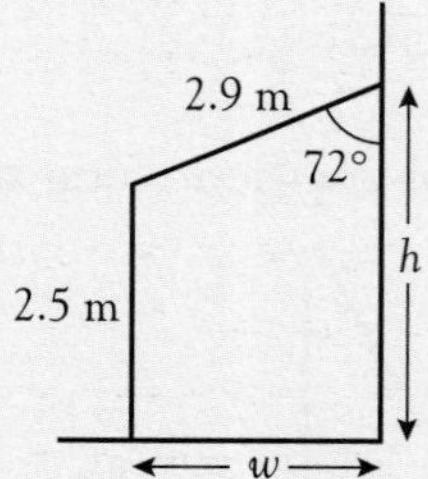

Find

a the width of the conservatory, w

b the height, h, at which the roof meets the house.

7 The sketch shows a chord AB drawn in a circle whose centre is O. The point C is the mid-point of AB.

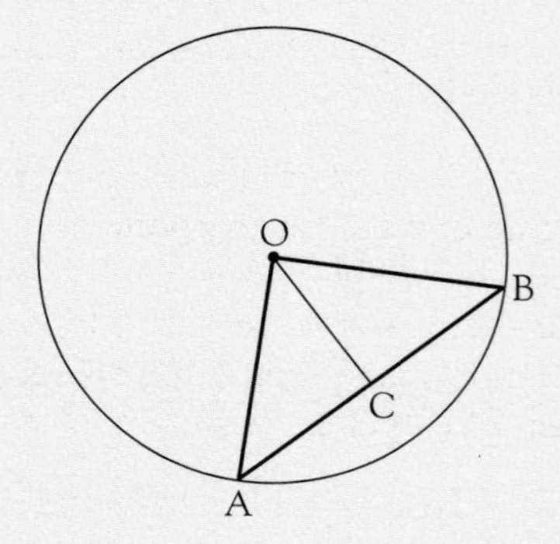

a Find angle OAB, if OC = 6 cm and the radius of the circle is 8 cm.

b Find the length of AB if angle AOB = 114° and the diameter of the circle is 14 cm.

c If AB = 2.6 metres and OC = 1.2 metres, find

i angle AOB

ii the radius of the circle.

8 Each side of a swing consists of an A-frame as shown in the diagram.

The sides AB and AC are each inclined at 68.5° to the horizontal and are 326 cm long.

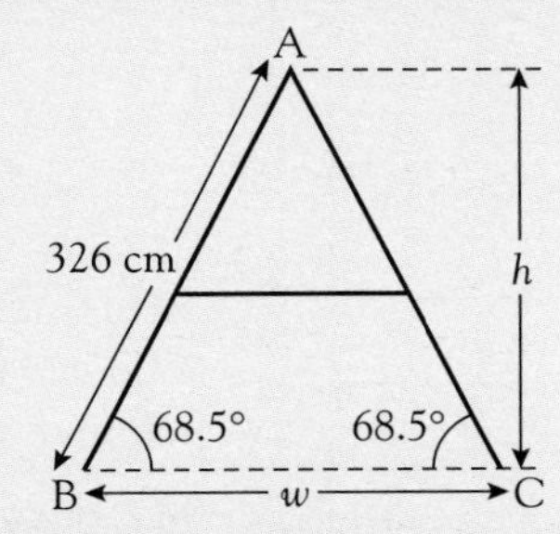

a Calculate the height, h, and width, w, of the A-frame.

b The horizontal bar joins the mid-points of AB and AC. Find the length of this bar.

9 The water in a swimming pool is 1 m deep at the shallow end and 3 m deep at the deep end. The pool is 25 m long and 12.5 m wide.

The water forms a prism the cross-section of which is a trapezium, as shown in the sketch.

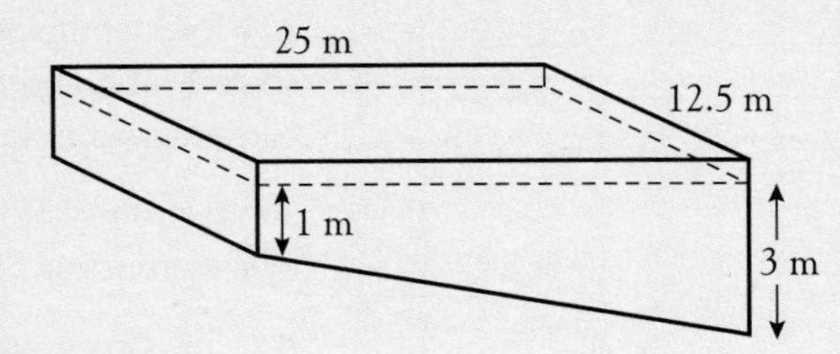

Calculate

a the angle the sloping floor makes with the horizontal

b the area of the cross-section of the water

c the volume of water in the pool.

10 A cable car ride is in two sections, PQ and QR. The height of Q and R above P and the length of each section of cable are as shown in the sketch.

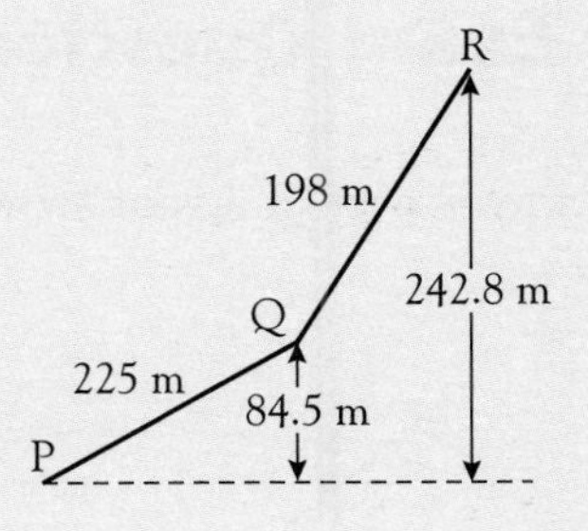

a Find the angle between each section of cable and the horizontal.

b If the cable sections are replaced by a single cable joining P directly to R, find

i the angle of inclination of the new cable to the horizontal,

ii the length of the new cable.

11 An observer is located in a lighthouse at a height of 43.6 m above the sea. She observes two yachts, X and Y, in line with the foot of the lighthouse at angles of depression of 24.7° and 16.4° respectively.

Calculate the distance between the yachts.

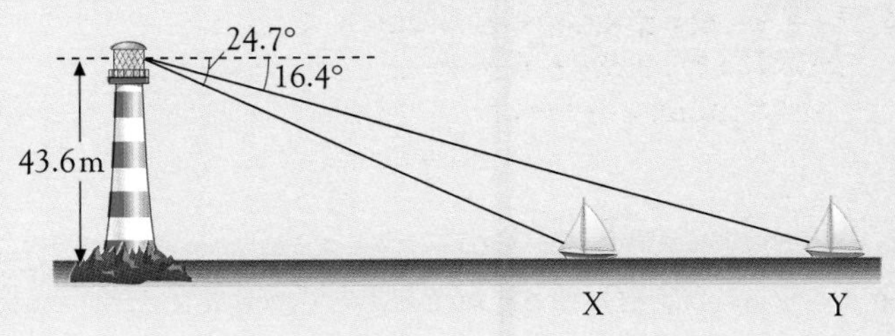

12 The sketch shows a section of trellis fencing.

In the rhombus ABCD, the distance AC = 144 mm and BD = 108 mm.

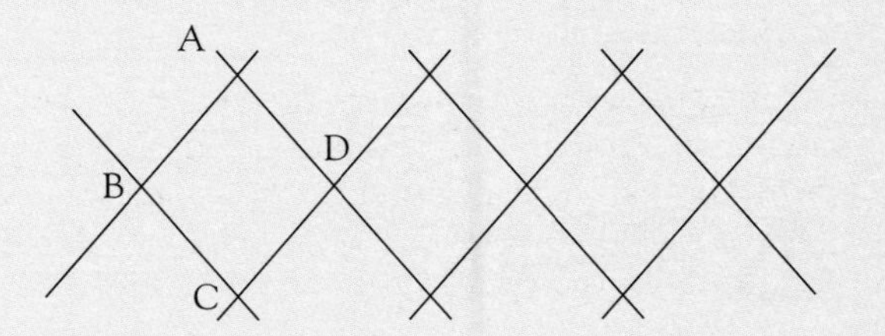

Calculate

a the length AB,

b angle ABC.

EXAM QUESTIONS

Worked Exam Question
[SEG]

The diagram shows part of a roof structure.

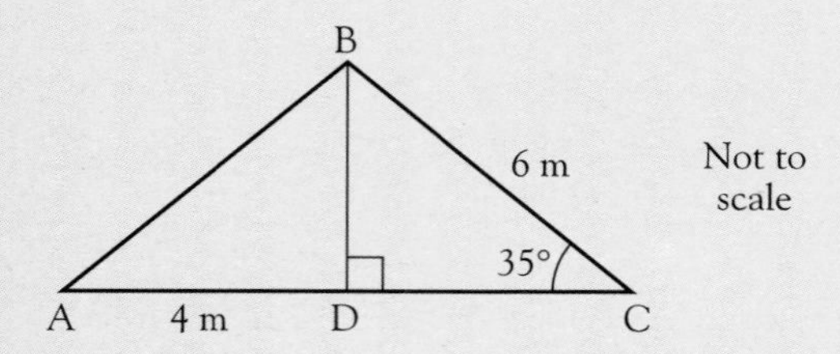

***AD* = 4 m, *BC* = 6 m and angle *BCD* = 35°.**
***BD* is perpendicular to *AC*.**

a Calculate the height of *BD*.

◆ Look at △*BCD* and use the sin ratio

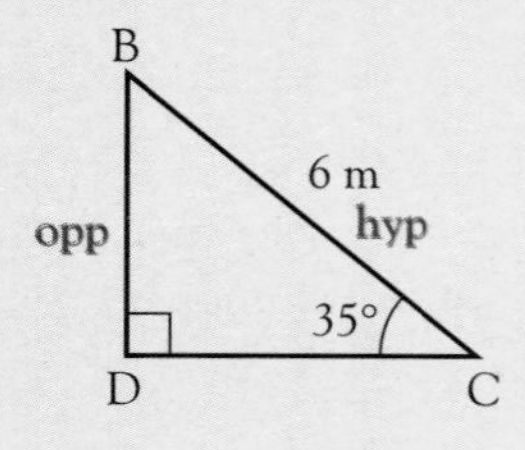

Write opp and hyp on the diagram drawn on the question paper.

$$\sin 35° = \frac{\text{opp}}{\text{hyp}}$$

$$\sin 35° = \frac{BD}{6}$$

$$BD = 6 \times \sin 35°$$

$$= 3.441\ldots$$

Answer 3.44 m

M1 for identifying sin ratio and for correct substitution

M1 for correct re-arranging

A1 for correct answer

3 marks

b Calculate angle BAC.

◆ Look at △*BAD*, use your answer from **a** for *BD* and use the tan ratio for angle *A*.

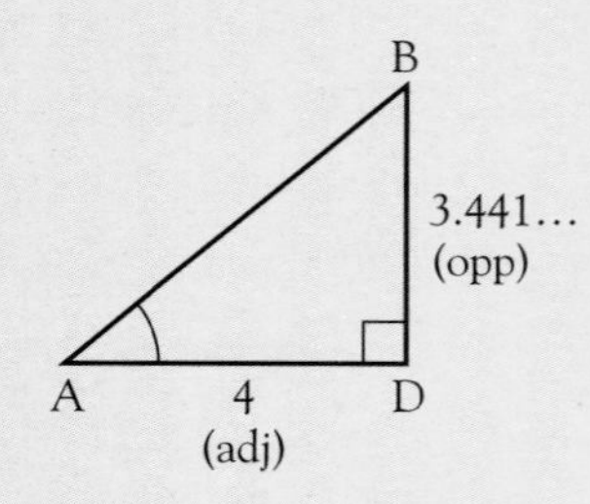

Note that angle BAC = angle BAD

$$\tan A = \frac{\text{opp}}{\text{adj}}$$

$$= \frac{3.441\ldots}{4}$$

$$= 0.860\ldots$$

$$A = \tan^{-1} 0.860\ldots$$

$$= 40.70\ldots$$

Answer 40.7°

M1 for identifying correct ratio and for correct substitution

A1 for using the inverse function

A1 for correct answer

3 marks

You will gain the marks if you use your answer to BD, even if it was wrong. This is known as 'follow through'.

Exam Questions

1 A wire 18 m long runs from the top of a pole to the ground as shown in the diagram. The wire makes an angle of 35° with the ground.

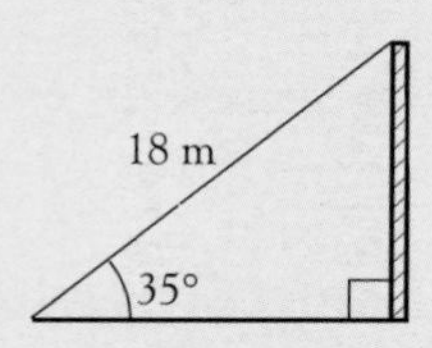

Calculate the height of the pole.
Give your answer to a suitable degree of accuracy.

[NEAB]

2 The diagram shows the end view of the framework for a sports arena stand.

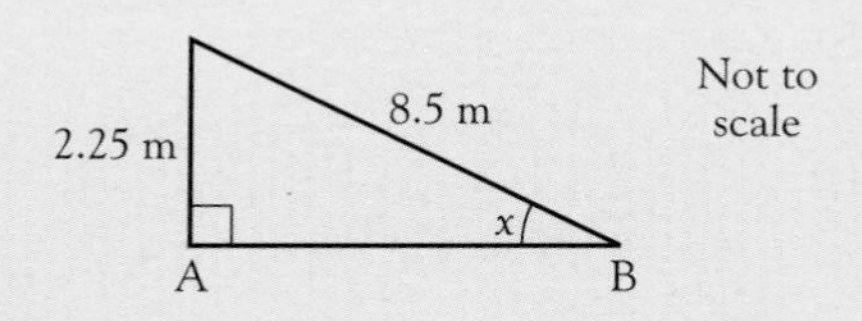

a Calculate the distance AB.

b Calculate the angle x. [NEAB]

3

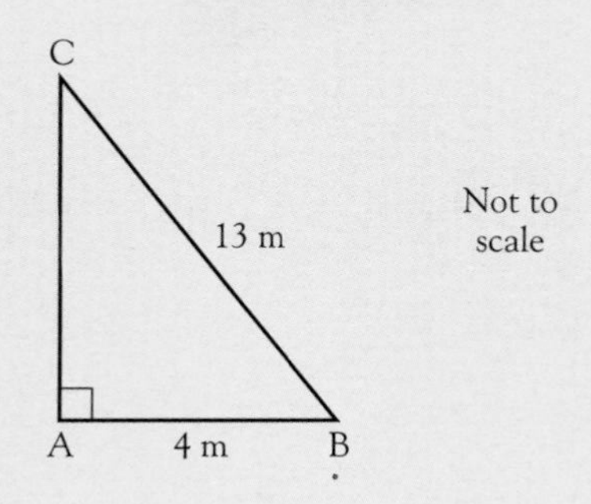

ABC is a right-angled triangle.
AB is of length 4 m and BC is of length 13 m.

a Calculate the length of AC.

b Calculate the size of angle ABC. [L]

4 The diagram shows an 8 m ladder leaning against a wall.

Instructions for finding the safest position for the foot of the ladder are also given.

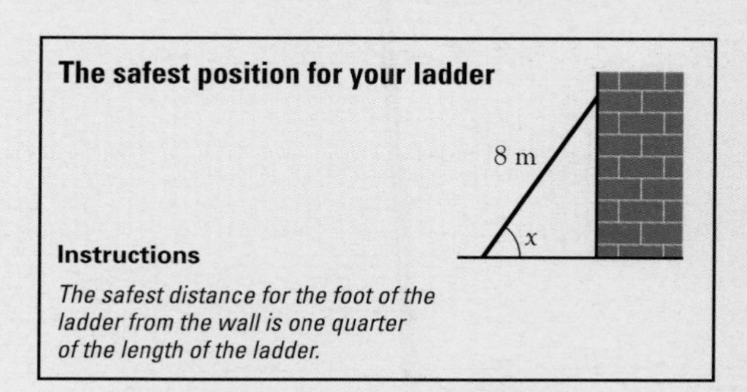

a An 8 m ladder is placed against a wall using the safety instructions above.
Calculate the size of the angle marked x.

b Use trigonometry to calculate how far up the wall this ladder will reach when it is in the safest position. [NEAB]

5

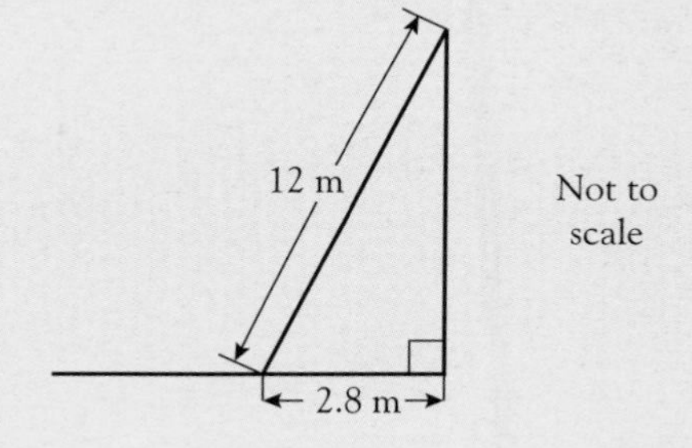

A ladder 12 m long leans against the vertical wall of a house. The foot of the ladder is 2.8 m from the wall on horizontal ground.

a Calculate the height of the top of the ladder above the ground.

b Calculate the size of the angle which the ladder makes with the ground. [MEG]

6

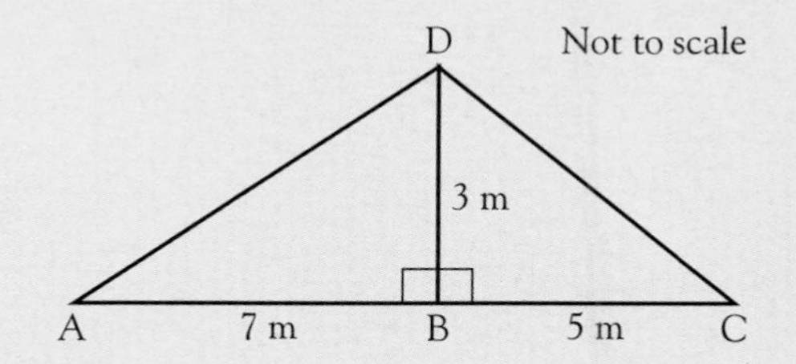

The diagram shows a roofing frame ABCD.
AB = 7 m, BC = 5 m, DB = 3 m,
angle ABD = angle DBC = 90°

a Calculate the length of AD.

b Calculate the size of angle DCB. [MEG]

EXAM QUESTIONS

7

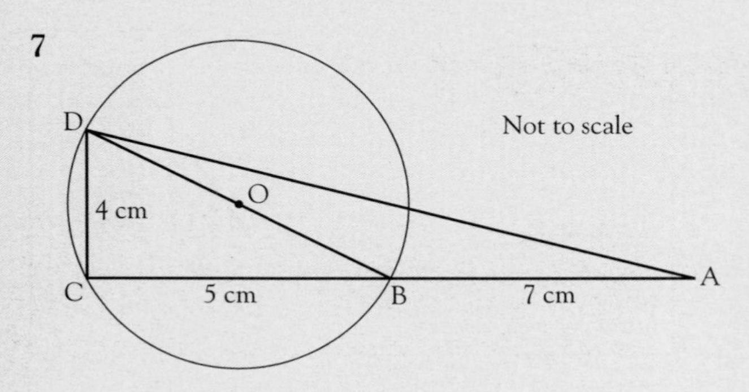

In the diagram, ABC is a straight line and O is the centre of the circle.

AB = 7 cm, BC = 5 cm and CD = 4 cm

a Explain why angle ACD = 90°

b Use trigonometry to calculate angle CAD.

c Calculate

i the length of BD

ii the radius of the circle. [MEG]

8

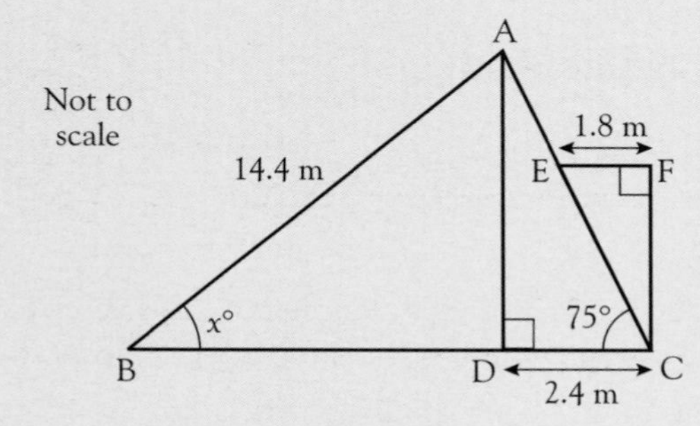

The diagram represents a triangular roof frame ABC with a window frame EFC.
BDC and EF are horizontal and AD and FC are vertical.

a Calculate the height AD.

b Calculate the size of the angle marked $x°$ in the diagram.

c Calculate FC. [MEG]

9

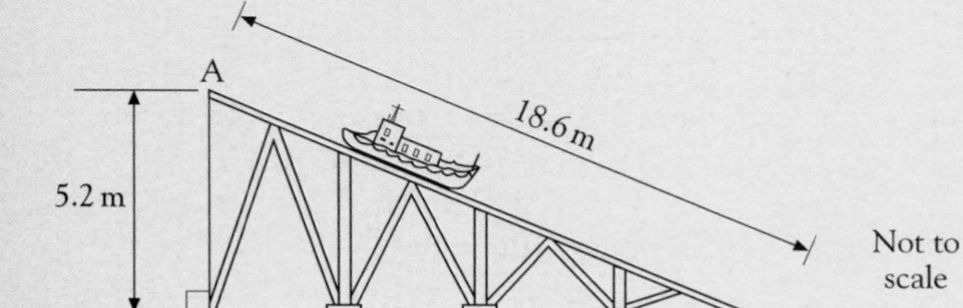

The diagram represents a ramp AB for a lifeboat. AC is vertical and CB is horizontal.

a Using a scale of 1 cm to represent 2 m, draw an accurate scale diagram of triangle ABC.

b Use your scale diagram to find

i the size of angle ABC

ii the actual length, in metres, of CB.

c The results obtained from the scale drawing may not be very accurate. More accurate results can be obtained by calculation.

i Use trigonometry to **calculate** the size of angle ABC correct to three significant figures.

ii Use Pythagoras' theorem to **calculate** the length of CB correct to three significant figures. [MEG

10

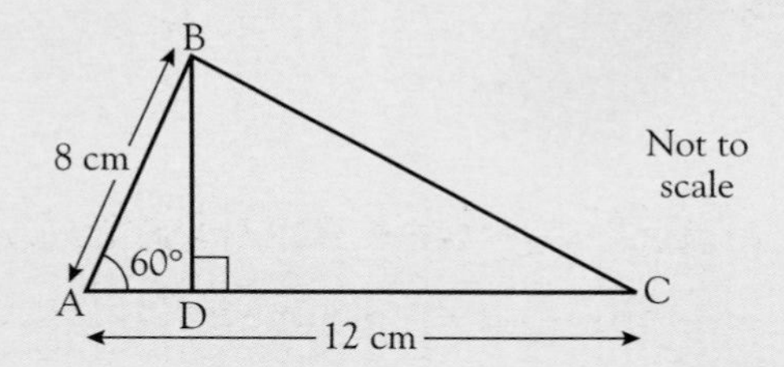

a Calculate the area of the triangle ABC. Show your working.

b Calculate the length DC. [MEG

11 In the diagram, ABCDEF represents the wall of a bedroom.

The vertical edges AB and FE are each 1.25 m high.

The edges AF = 5.10 m and CD = 3.20 m are both horizontal.

The sloping edges BC and DE are each 1.50 m long.

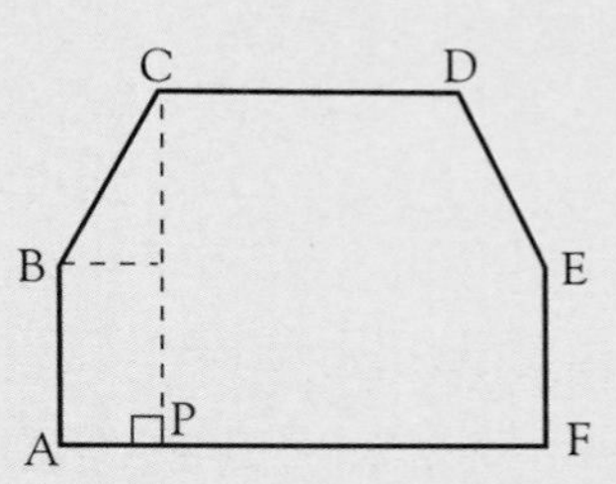

Giving each answer correct to three significant figures, calculate

a CP, the vertical height of the wall

b the angle at which BC is inclined to the horizontal. [N]

21 QUADRATICS I

In this chapter you will learn how to
- multiply out two brackets
- draw quadratic graphs and use them to solve quadratic equations
- solve quadratic equations by trial and improvement

pter 14
ge 258

In Chapter 14 you learnt how to expand or 'multiply out' a bracket, for example

a $5(x + 3) = 5 \times x + 5 \times 3$
$= 5x + 15$

You multiply, in turn, each term inside the bracket by the numbers or letters outside it.

b $x(x + 3) = x \times x + x \times 3$
$= x^2 + 3x$

Multiplying out two brackets

Sometimes you will need to multiply out two brackets, for example to find the area of this rectangle:

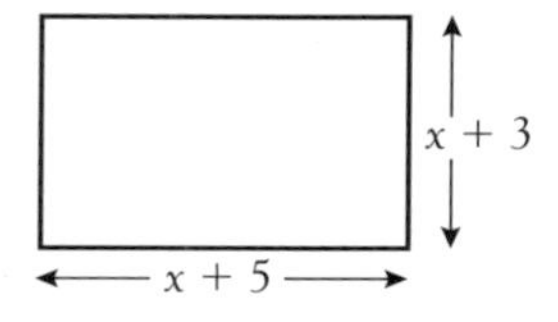

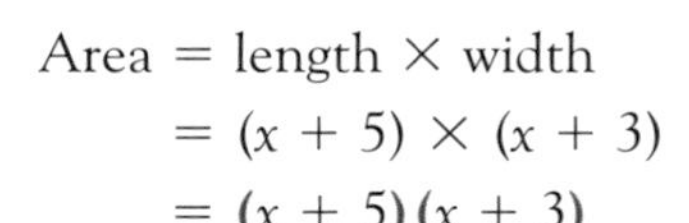

Area = length × width
$= (x + 5) \times (x + 3)$
$= (x + 5)(x + 3)$

It is usual to leave out the multiplication sign between the two brackets.
It avoids confusion with the letter x.

To multiply out these two brackets, think of it as $(x + 5)$ lots of $(x + 3)$.
This is the same as x lots of $(x + 3)$ plus 5 lots of $(x + 3)$.

$$(x + 5)(x + 3) = x(x + 3) + 5(x + 3)$$
$$= x \times x + x \times 3 + 5 \times x + 5 \times 3$$
$$= x^2 + \mathbf{3x + 5x} + 15$$
$$= x^2 + 8x + 15$$

Simplify $3x + 5x = 8x$

If you substitute a particular value of x into both formats, they should come to the same answer, for example when $x = 4$

$$(x + 5)(x + 3) = (4 + 5)(4 + 3)$$
$$= 9 \times 7$$
$$= 63$$

$$x^2 + 8x + 15 = 4^2 + 8 \times 4 + 15$$
$$= 16 + 32 + 15$$
$$= 63$$

They are the same.

Note that although you could write the expression $x^2 + 8x + 15$ in any order, for example $15 + x^2 + 8x$ or $8x + 15 + x^2$, it is usual, and helpful for further work, to write the x^2 term first, then the x term, then the number.

ample uestion **1** Expand and simplify $(x + 2)(x + 6)$

EXPANDING BRACKETS

Answer

$$
\begin{aligned}
(x + 2)(x + 6) &= x \text{ lots of } (x + 6) + 2 \text{ lots of } (x + 6)\\
&= x(x + 6) + 2(x + 6)\\
&= x^2 + \mathbf{6x} + \mathbf{2x} + 12\\
&= \underline{x^2 + 8x + 12}
\end{aligned}
$$

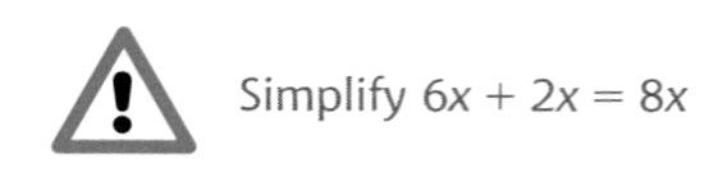

With practice you should be able to do the first two lines of writing shown above in your head. You could put marks on your question to remind you of the steps, like this:

Step 1

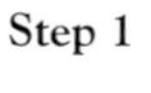

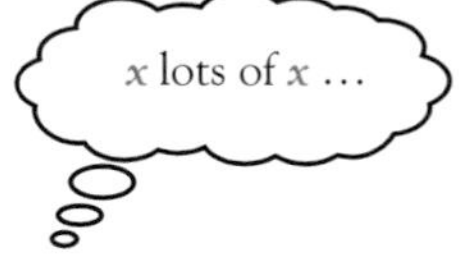

$(x + 2)(x + 6) = x^2 \ldots$

Step 2

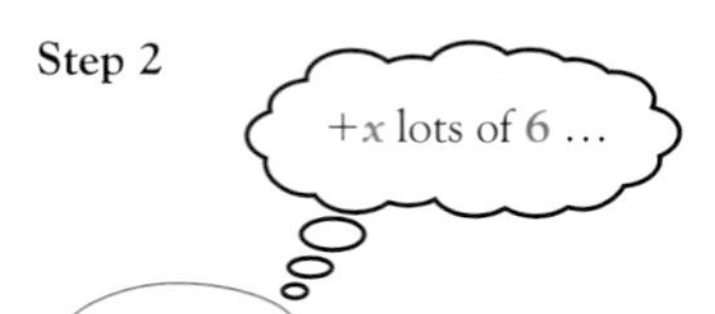

$(x + 2)(x + 6) = x^2 + 6x \ldots$

Step 3

+2 lots of x …

$(x + 2)(x + 6) = x^2 + 6x + 2x \ldots$

Step 4

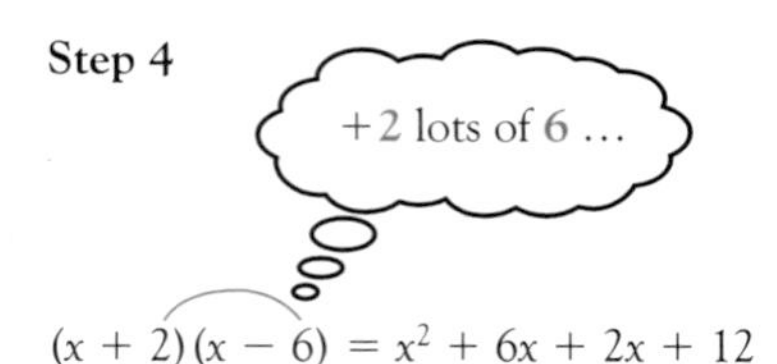

$(x + 2)(x - 6) = x^2 + 6x + 2x + 12$

You could just write

$$
\begin{aligned}
(x + 2)(x + 6) &= x^2 + 6x + 2x + 12\\
&= x^2 + 8x + 12
\end{aligned}
$$

Another good way of showing your working is in a multiplication square

$\times$	x	$+6$
x	x^2	$+6x$
$+2$	$+2x$	$+12$

$(x + 2)(x + 6) = x^2 + 8x + 12$

Sample Question 2 Expand and simplify **a** $(2x + 3)(x + 5)$ **b** $(3x + 4y)(2x + 5y)$

Answer

a

$$
\begin{aligned}
(2x + 3)(x + 5) &= \mathbf{2x \times x} + 2x \times 5 + 3 \times x + 3 \times 5\\
&= 2x^2 + 10x + 3x + 15\\
&= \underline{2x^2 + 13x + 15}
\end{aligned}
$$

$2x \times x$
$= 2 \times x \times x$
$= 2x^2$

b

$$
\begin{aligned}
(3x + 4y)(2x + 5y) &= 3x \times 2x + 3x \times 5y + 4y \times 2x + 4y \times 5y\\
&= 6x^2 + 15xy + \mathbf{8yx} + 20y^2\\
&= \underline{6x^2 + 23xy + 20y^2}
\end{aligned}
$$

$8yx = 8xy$

EXPANDING BRACKETS

Using multiplication squares to show your working

a

$\times$	x	$+5$
$2x$	$2x^2$	$+10x$
$+3$	$+3x$	$+15$

$(2x + 3)(x + 5) = 2x^2 + 13x + 15$

b

$\times$	$2x$	$+5y$
$3x$	$6x^2$	$+15xy$
$+4y$	$+8xy$	$+20y^2$

$(3x + 4y)(2x + 5y) = 6x^2 + 23xy + 20y^2$

You need to take special care when dealing with minus signs in the brackets:

i $(x - 4)(x + 1) = x(x + 1) - 4(x + 1)$
$= x^2 + x - 4x - 4 \times 1$
$= x^2 - 3x - 4$

$+x - 4x = -3x$

ii $(x + 3)(x - 5) = x(x - 5) + 3(x - 5)$
$= x^2 - 5x + 3x + 3 \times (-5)$
$= x^2 - 2x - 15$

$+3 \times (-5) = -15$

$-5x + 3x = -2x$

iii $(x - 3)(x - 4) = (x - 4) - 3(x - 4)$
$= x^2 - 4x - 3x - 3 \times (-4)$
$= x^2 - 7x + 12$

$-3 \times (-4) = +12$

$-4x - 3x = -7x$

Using multiplication squares to show your working

i

$\times$	x	$+1$
x	x^2	$+x$
-4	$-4x$	-4

$(x - 4)(x + 1) = x^2 - 3x - 4$

ii

$\times$	x	-5
x	x^2	$-5x$
$+3$	$-3x$	-15

$(x + 3)(x - 5) = x^2 - 2x - 15$

iii

$\times$	x	-4
x	x^2	$-4x$
-3	$-3x$	$+12$

$(x - 3)(x - 4) = x^2 - 7x + 12$

Sample Question 3 Expand and simplify **a** $(4x + 3y)(2x - 5y)$ **b** $(x - 2y)(x + 2y)$

Answer

a $(4x + 3y)(2x - 5y) = 4x \times 2x + 4x(-5y) + 3y \times 2x + 3y(-5y)$
$= 8x^2 - 20xy + 6yx - 15y^2$
$= \underline{8x^2 - 14xy - 15y^2}$

b $(x - 2y)(x + 2y) = x^2 + x \times 2y - 2y \times x - 2y \times 2y$
$= x^2 + 2xy - 2yx - 4y^2$
$= \underline{x^2 - 4y^2}$

$2xy$ is the same as $2yx$, so $2xy - 2yx = 0$

EXPANDING BRACKETS

There are two special cases to note:

Special case a

i $(x-4)(x+4) = x^2 + \mathbf{4x} - \mathbf{4x} - 16$
$= x^2 - 16$

$+4x - 4x = 0$
There is no x term in the expansion.

ii $(2p-3q)(2p+3q) = 4p^2 + \mathbf{6pq} - \mathbf{6qp} - 9q^2$
$= 4p^2 - 9q^2$

$6pq = 6qp$
so $6pq - 6qp = 0$

In general

$(a-b)(a+b) = a^2 - b^2$
where a and b can be numbers or letters

This is a very useful result and is called the 'difference of two squares'. Learn it.

Special case b Squaring an expression

i $(2x+3y)^2 = (2x+3y)(2x+3y)$
$= 4x^2 + 6xy + 6yx + 9y^2$
$= 4x^2 + 12xy + 9y^2$

Do not be tempted just to square $2x$ and square $3y$. Always write out the two brackets before expanding.

ii $(x-5)^2 = (x-5)(x-5)$
$= x^2 - 5x - 5x + 25$
$= x^2 - 10x + 25$

Likewise here, write out the two brackets.

In general, if a and b are any numbers or letters

$(a+b)^2 = a^2 + 2ab + b^2$
$(a-b)^2 = a^2 - 2ab + b^2$

Although you could learn this general formula, it is probably better to work it out each time, writing out the two brackets and then multiplying.

Exercise 21.1

Expand and simplify each set of brackets.

1 $(x+2)(x+5)$
2 $(x+1)(x+7)$
3 $(a+1)(a+2)$
4 $(p-2)(p+3)$
5 $(x+4)(x-5)$
6 $(x-7)(x+3)$
7 $(a-7)(a-2)$
8 $(2+a)(6+a)$
9 $(3-x)(4-x)$
10 $(3p+2)(2p-3)$
11 $(2x+7)(3x+2)$
12 $(7a-3)(8a-1)$
13 $(3x-2)(6x+1)$
14 $(a-3)(a+3)$
15 $(2x+1)(3x-2)$
16 $(3p-1)(3p+1)$
17 $(2x+1)(x-2)$
18 $(3t-1)(7t-1)$
19 $(4x+5)(4x-5)$
20 $(7-2x)(7-3x)$
21 $(7x-3)(7x+2)$
22 $(5a-7)(3a+4)$
23 $(3y-2x)(4y-3x)$
24 $(6p+q)(5p-4q)$
25 $(m+1)^2$
26 $(3-2x)(2x+1)$
27 $(5m-2n)(m+n)$
28 $(3t+2)^2$
29 $(6a-b)(2a+3b)$
30 $(p-q)^2$
31 $(3x-2)(7x+2)$
32 $(5y+x)(5y-x)$
33 $(5x+2y)^2$
34 $(5x-3)^2$
35 $(7-2x)(2x+5)$
36 $(a+5)(5-a)$
37 $(2x-3y)(2x+3y)$
38 $(6p-q)(p-7q)$
39 $(11-7a)(2a+9)$
40 $(3t+13)(2t-17)$

Quadratic graphs

Quadratic graphs are graphs whose equations are of the form $y = ax^2 + bx + c$

a, b and c are numbers, b and c can be any numbers, but **a cannot be zero.**

If $a = 0$, then you would have $0x^2$, so there would not be an x^2 term and the equation would become $y = bx + c$.
This gives a linear graph not a quadratic one.

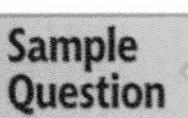

a Complete the table of values and draw the graph of $y = x^2 + x - 2$

x	−3	−2	−1	0	1	2	3	4
y								

b Draw in the line of symmetry on the graph.
What is the equation of this line of symmetry?

Answer

a

- Substitute each of the x values in turn into the equation and work out the y value. Complete the table.

When $x = -3$, $y = (-3)^2 + (-3) - 2 = 9 - 3 - 2 = 4$

You should not need to use a calculator when $x = -3$, but try it on your calculator so that you can work out the value if x is not a 'nice' number!

[3] [+/−] [x²] [+] [3] [+/−] [−] [2] [=]

When $x = -2$, $y = (-2)^2 + (-2) - 2 = 4 - 2 - 2 = 0$

When $x = -1$, $y = (-1)^2 + (-1) - 2 = 1 - 1 - 2 = -2$

When $x = 0$, $y = 0^2 + 0 - 2 = -2$

When $x = 1$, $y = 1^2 + 1 - 2 = 0$

When $x = 2$, $y = 2^2 + 2 - 2 = 4$

When $x = 3$, $y = 3^2 + 3 - 2 = 10$

x	−3	−2	−1	0	1	2	3
y	4	0	−2	−2	0	4	10

- Plot the points and join them up to form a **smooth** curve.

Do not use a ruler to join the points.

QUADRATIC GRAPHS

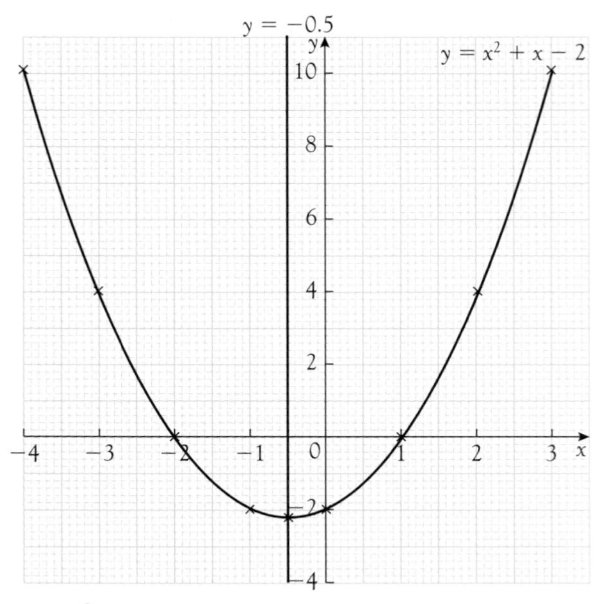

Remember to write the equation beside the curve. This is called 'labelling the curve'.

Do not draw a straight line between $(-1, 2)$ and $(0, -2)$. The curve dips to its lowest value half-way between, when $x = -0.5$. To find the turning point accurately, work out y when $x = -0.5$.

$y = (-0.5)^2 + (-0.5) - 2 = -2.25$

Comparing the equation $y = x^2 + x - 2$ with $y = ax^2 + bx + c$, $a = 1$, $b = 1$ and $c = -2$

Linear Graphs I page 89

b

- The line of symmetry is the vertical line through $x = -0.5$. Draw it on your graph.

  The equation of this line is $\underline{x = -0.5}$

Note that the graph continues on the left and right beyond the x value chosen in the table.

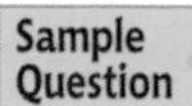

Sample Question 5 Draw the graph of $y = -x^2 + 2x$ for $-1 \leqslant x \leqslant 3$

Answer

- The instruction to use $-1 \leqslant x \leqslant 3$ means that you should draw your grid with x axis going from -1 to 3, so work out your table of values in this range.

  When $x = -1$, $y = -(-1)^2 + 2 \times (-1) = -1 - 2 = -3$ On calculator

  When $x = 0$, $y = -0^2 + 2 \times 0 = 0$

Do not use a calculator for this one – it is too easy!

  When $x = 1$, $y = \mathbf{-(1^2)} + 2 \times 1 = -1 + 2 = 1$

Watch out here, it is $-(1^2)$ which is -1.

  When $x = 2$, $y = -(2^2) + 2 \times 2 = -4 + 4 = 0$

  When $x = 3$, $y = -(3^2) + 2 \times 3 = -9 + 6 = -3$

x	−1	0	1	2	3
y	−3	0	1	0	−3

QUADRATIC GRAPHS

- Now draw your grid. You cannot draw it before you have worked out the y values as you do not know how far you will need to go on the y axis.

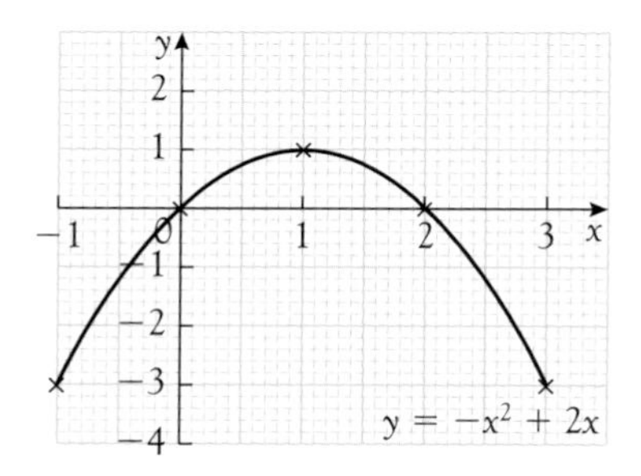

Notice that the line of symmetry is a vertical line through 1 on the x axis. This has equation $x = 1$.

Comparing the equation $y = 2x^2 + 2x$ with $y = ax^2 + bx + c$, $a = -1$, $b = 2$ and $c = 0$.

You can tell the general shape of the curve from its equation if you compare it with the general formula

$$y = ax^2 + bx + c$$

If a is positive, i.e. $a > 0$ the curve is like this:

It has a 'minimum turning point'

If a is negative, i.e. $a < 0$ the curve is like this:

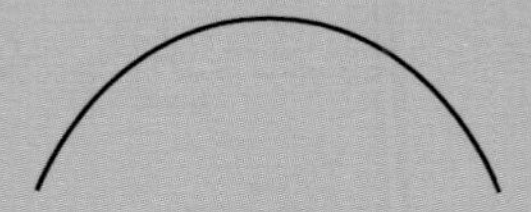

It has a 'maximum turning point'

Remember positive: 'smiley' curve.

Remember negative: 'sad' curve.

TASK

You are going to investigate graphs of the form $y = ax^2 + bx + c$ by trying different values for a, b and c.

Either

- do a table of values and plot the graph

or

- use a graphics calculator or computer graph-plotting software to obtain the curve.

a Investigate $y = ax^2$

b Investigate $y = x^2 + c$

c Investigate $y = x^2 + bx$

d Investigate $y = ax^2 + bx + c$

For each section, write down anything you notice, together with any general rules you find.

QUADRATIC GRAPHS

Quadratic equations

These are examples of quadratic equations:

$$x^2 = 12, \quad x^2 = 2x - 1, \quad 3x^2 = 6x, \quad 2x^2 + 5x - 3 = 0$$

In a quadratic equation, the highest power of x is x^2. A quadratic equation might contain an x term, it might contain a number term, it **must** contain an x^2 term.

A general way of defining a **quadratic equation** is to say that it can be written in the form

$$ax^2 + bx + c = 0$$

where a, b and c are numbers and a cannot be zero.

In the examples above
$x^2 = 12$ can be re-arranged as $x^2 - 12 = 0$
$x^2 = 2x - 1$ as $x^2 - 2x + 1 = 0$
$3x^2 = 6x$ as $3x^2 - 6x = 0$

To **solve** a quadratic equation, you have to find the values of x that fit in the equation so that the left hand side equals the right hand side. Sometimes there are two values, sometimes one value and sometimes there are **no values**.

This is different from linear equations when there is always a solution.

There are several ways of attempting to solve a quadratic equation. Two are shown in this chapter (by drawing graphs and by trial and improvement) and a third way (by factorising) is shown in chapter 25.

Solving quadratic equations using graphs

Sample Question 6

a On one grid, draw the graphs of $y = x^2$, $y = 12$ and $y = 2x - 1$

b Use your graphs to solve the quadratic equations
i $x^2 = 12$ **ii** $x^2 = 2x - 1$

Answer

a

- Do tables of values and draw the graphs.
- Label the graphs carefully.

$y = x^2$

x	−4	−3	−2	−1	0	1	2	3	4
y	16	9	4	1	0	1	4	9	16

$y = 2x - 1$

x	−2	0	2	4
y	−5	−1	3	7

QUADRATIC GRAPHS

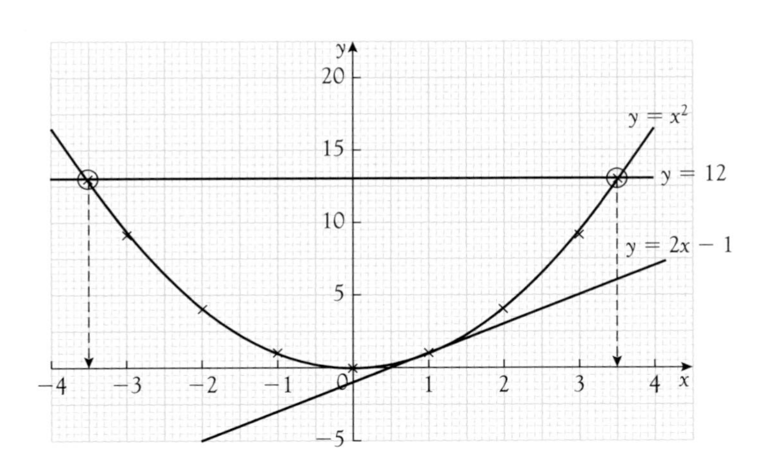

b

i

- To solve $x^2 = 12$, find the points of intersection of the graphs $y = x^2$ and $y = 12$.
- Read off the x values of the points of intersection. It is helpful to draw in a dotted line to the x axis to show your working.

<u>Solutions are $x = \mathbf{3.5}$, $x = \mathbf{-3.5}$</u>

> Try these values.
> When $x = 3.5$, $x^2 = 12.25$ (check)
> When $x = 23.5$, $x^2 = 12.25$
> These answers are not exact, as x^2 should come to 12, but they are as accurate as you can get from the graph.

ii

- To solve $x^2 = 2x - 1$, find the intersection of the graphs of $y = x^2$ and $y = 2x - 1$ and read off the x value of this point.

<u>Solution is $x = 1$</u>

> Notice that equation **i** had two solutions and equation **ii** had one solution.

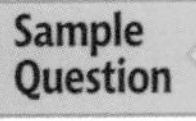

a Complete the table of values for $y = x^2 - 2x + 3$ and draw the graph.

x	−1	0	1	2	3
y					

b Use your graph to find the number of solutions of the equation $x^2 - 2x + 3 = 0$

Answer

a

x	−1	0	1	2	3
y	6	3	2	3	6

When $x = -1$, $y = (-1)^2 - 2(-1) + 3$
$= 1 + 2 + 3 = 6$

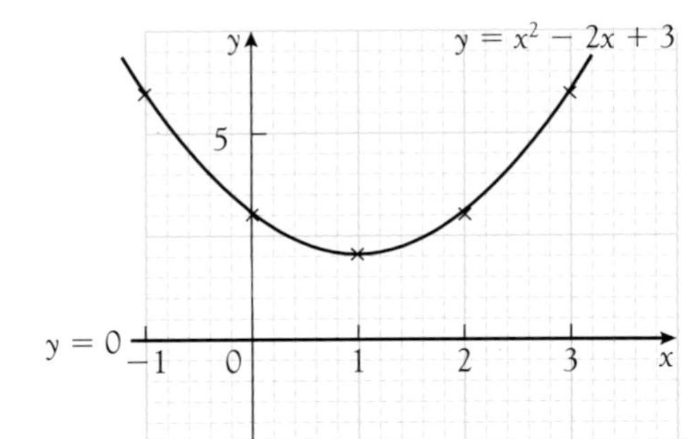

b

- Draw the line $y = 0$.

This is the x axis.

- Find the number of intersections of the curve and line.
They do not cross

<u>The equation $x^2 - 2x + 3 = 0$ has no solutions.</u>

QUADRATIC GRAPHS

Sample Question 8

An L-shaped work top has dimensions as shown in the diagram. All lengths are in feet.

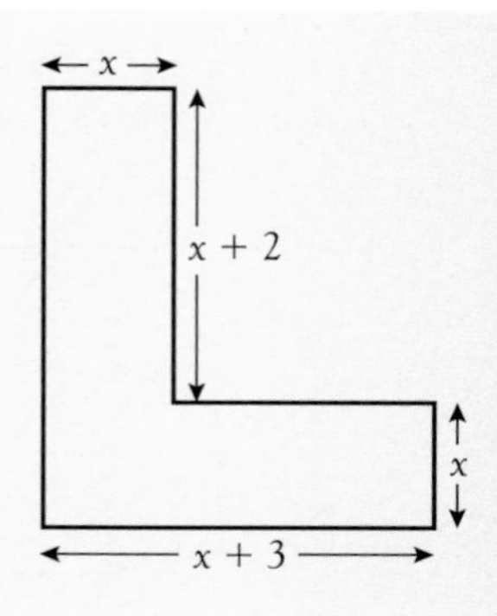

a Show that the area A of the work top is given by $A = 2x^2 + 5x$

b Draw a graph of $A = 2x^2 + 5x$, taking x values from 0 to 4.

c The work top has an area of 30 square feet. Find the value of x from your graph.

Answer

a

- Split the work top into two rectangles and work out the area of each one

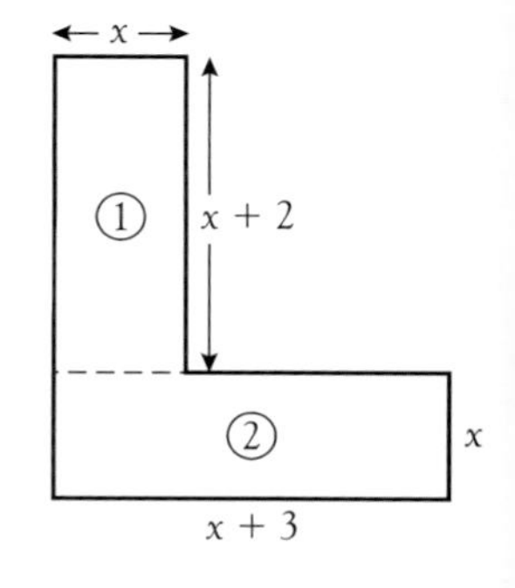

Area 1 = $x(x + 2)$

Area 2 = $x(x + 3)$

Total area A = Area 1 + Area 2

$A = x(x + 2) + x(x + 3)$

$A = x^2 + 2x + x^2 + 3x$

$A = 2x^2 + 5x$

When you are asked to show a given answer, do the working yourself and hope that it comes to the answer given.

b

- Do a table of values and draw the graph

Label the vertical axis *A* (not *y*).

x	0	1	2	3	4
A	0	7	18	33	52

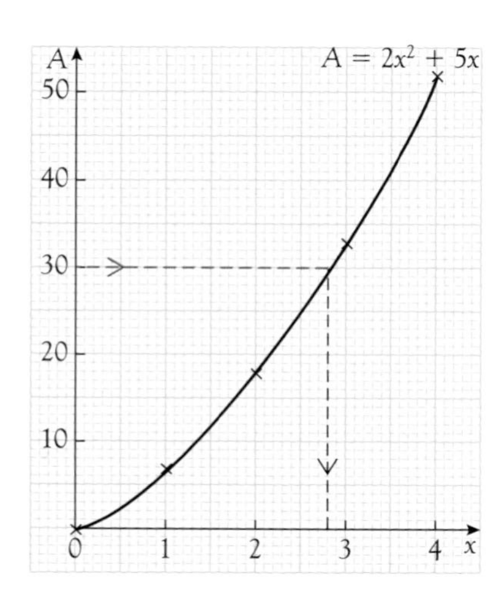

c

- Find A = 30 on the vertical axis.
- Draw a **horizontal line** to meet the curve.

This is the line $A = 30$.

- Read off the x value at this point.

$\underline{x = 2.8}$

The equation $2x^2 + 5x = 30$ does in fact have two solutions; the second solution is negative and so would not be relevant in this question as x is a length and has to be positive.

QUADRATIC GRAPHS

Exercise 21.2

1 Copy and complete the table of values for the graph of $y = 3x^2 - 10$

x	-3	-2	-1	0	1	2	3
$y = 3x^2 - 10$			-7			2	

a Draw the graph of $y = 3x^2 - 10$

b Use your graph to find the value of y when $x = 2.5$
Check using your calculator.

c Use your graph to find the values of x when $y = -5$

2 a Draw the graph of $y = x^2 - 3x$ for values of x between -3 and $+5$.

b Use your graph to find the value of y when $x = 3.8$ (Check your answer by calculation.)

c Use your graph to find the values of x when $y = -1.5$

d What is the equation of the line of symmetry?

3 a Draw the graph of $y = 3x - x^2$ for values of x between -3 and $+5$.

b Explain any differences between this graph and the graph in question 2

c Use your graph to find the value of y when $x = -2.5$

d Use your graph to find the values of x when $y = 6.5$

4 Copy and complete the table of values for the graph of $y = 3x^2 - 2x - 1$

x	-2	-1	0	1	2	3
$y = 3x^2 - 2x - 1$	15					20

a Draw the graph of $y = 3x^2 - 2x - 1$

b Use your graph to find the value of y when $x = 2.4$

c Use your graph to find the values of x when $y = 12$

d Use your graph to solve the equation

$3x^2 - 2x - 1 = 0$

e What is the equation of the line of symmetry?

5 a Draw the graph of $y = 4x^2 - 8x - 5$ for values of x from -2 to 4.

b Use your graph to estimate the value of y when $x = -1.7$ (Check your answer by calculation.)

c Use your graph to solve the equations

i $4x^2 - 8x - 5 = 0$

ii $4x^2 - 8x - 5 = -4$

6 On one set of axes, draw the graphs of $y = x^2 + 3x - 1$ and $y = \frac{1}{2}x + 1$ for values of x from -4 to 2.

a Use your diagram to solve the equation

$x^2 + 3x - 1 = 0$

b Use your diagram to solve the equation

$x^2 + 3x - 1 = \frac{1}{2}x + 1$

Simplify this equation so that it is in its simplest form.

7 a By putting $y = 0$ in the equation $y = x^2 - 9$, show that the graph of $y = x^2 - 9$ crosses the x axis at the points $(-3, 0)$ and $(3, 0)$.

b Find where the graph crosses the y axis.

c Using these points, draw a rough sketch of the graph of $y = x^2 - 9$

d Use a similar method to draw a rough sketch of the graph of $y = 16 - x^2$

8 A rectangular field has the dimensions shown in the diagram. All dimensions are in metres.

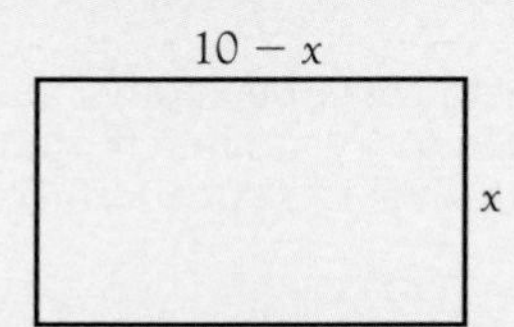

If A is the area of the field, show that $A = 10x - x^2$

a Taking values of x from 0 to 10, draw the graph of $A = 10x - x^2$

b Use your graph to estimate the dimensions of the rectangle when $A = 20\text{ m}^2$

c For what value of x is the area a maximum?

d What is the maximum area and what shape will the field be in this case?

Solving quadratic equations by 'trial and improvement'

In this method, which you could use for other types of equations as well as quadratic ones, you make a guess at a solution and then try it in the equation to see whether it works.

For example, to find a solution of $x^2 - 3x = 7$

try $x = 3$, $x^2 - 3x = 3^2 - 3 \times 3 = 0$ (too small)

You are trying to get 7.

Now 'improve' your guess:

try $x = 4$, $x^2 - 3x = 4^2 - 3 \times 4 = 4$ (too small)
try $x = 5$, $x^2 - 3x = 25 - 3 \times 5 = 10$ (too big)

You now know that there is a solution between $x = 4$ and $x = 5$.

You can go on improving your guess and trying it in the equation, until you have a solution to a particular degree of accuracy.

If you are asked to find a solution correct to 1 decimal place, you need to find two x values, within one decimal place of each other, one of which gives a number below 7 and the other a number above 7 when you try it in the equation.

try $x = 4.5$, $x^2 - 3x = 4.5^2 - 3 \times 4.5 = 6.75$ (too small)
try $x = 4.6$, $x^2 - 3x = 4.6^2 - 3 \times 4.6 = 7.36$ (too big)

So the solution is between 4.5 and 4.6. You need to find out whether it is closer to 4.5 or to 4.6
To do this, try half-way between 4.5 and 4.6, i.e. 4.55

When $x = 4.55$, $4.55^2 - 3 \times 4.5 = 7.0525$ (too big)

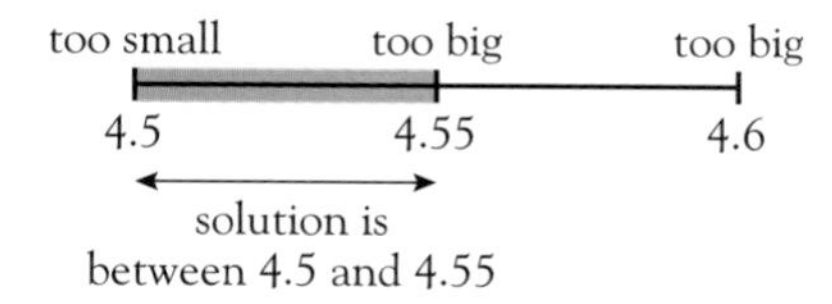

You know that the solution is between 4.5 and 4.55. This means that it is closer to 4.5 than to 4.6, so the solution, to 1 decimal place, is 4.5

A solution of $x^2 - 3x = 7$ is 4.5 (to 1 d.p.)

There could be another solution. You would need to try a different starting point (perhaps negative) to search further.

You need to show all your trials and give evidence that you have found a solution to the required accuracy.

It is useful to summarise your trials in a table as in the next example.

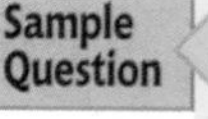

9 The equation $x^2 + x - 28 = 0$ has a solution between $x = 4$ and $x = 5$. Use the method of trial and improvement to find this solution, correct to 2 decimal places.

Show all your working.

Answer

- Use your calculator to work out $x^2 + x - 28$ when $x = 4$ and when $x = 5$ and write the values in a table.

$4^2 + 4 - 28 = -8$
$5^2 + 5 - 28 = 2$

- Refine x, trying different values and filling in the table as you go, deciding whether your value for $x^2 + x - 28$ is too big or too small. Remember you are trying to get 0.

x	Value of $x^2 + x - 28$	Outcome (trying to get 0)
4	−8	too small
5	2	too big
4.5	−3.25	too small
4.8	−0.16	too small
4.9	0.91	too big
4.85	0.3725	too big
4.82	0.0524	too big
4.81	−0.0539	too small

Decisions:
x is between 4 and 5,
try 4.5 (half-way between)
between 4.5 and 5, try 4.8
between 4.8 and 5, try 4.9
between 4.8 and 4.9, try 4.85
go smaller
go smaller
x is between 4.81 and 4.82

As you have to give your x value correct to 2 d.p. you need to find two x numbers within consecutive digits in the **second decimal place**, one giving an outcome too big and the other too small.

- You know that x is between 4.81 and 4.82.

  To find out which it is closer to, look at x half-way between these, i.e. $x = 4.815$

  $$4.815^2 + 4.815 - 28 = -0.000\,775 \text{ (too small)}$$

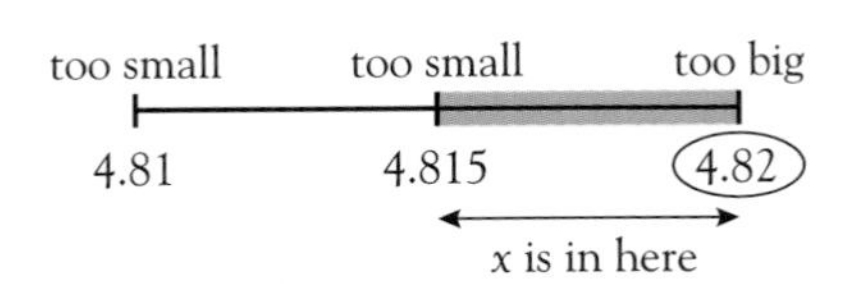

The x value is closer to 4.82 than it is to 4.81

$x = 4.82$ (2 d.p.)

Sometimes you have to solve a quadratic equation as part of a problem and use your value to answer the question, as in the next sample question.

Sample Question 10

All the lengths in this question are in metres.
A rectangular room has length $2x - 1$ and width $x + 1$

a Multiply out and simplify $(2x - 1)(x + 1)$

b The area of the room is 26 m^2. Show that $2x^2 + x - 27 = 0$

c The equation $2x^2 + x - 27 = 0$ has a solution between 3 and 4.
Using trial and improvement, or otherwise, find this solution correct to 2 decimal places and hence find the length and width of the room.
You must show all your trials. [MEG]

Answer

a

- $(2x - 1)(x + 1) = 2x \times x + 2x \times 1 - 1 \times x - 1 \times 1$

  $= 2x^2 + 2x - x - 1$

  $= \underline{2x^2 + x - 1}$

b

- Draw a sketch of the room and work out the area.

$x + 1$

$2x - 1$

Area = length × width

$= (2x - 1) \times (x + 1)$

- Use part **a** to expand and simplify.
  Area $= 2x^2 + x - 1$

- Use the fact that the area is 26

  $2x^2 + x - 1 = 26$

- Re-arrange the equation by subtracting 26 from both sides

  $\underline{2x^2 + x - \mathbf{27} = 0}$

$-1 - 26 = -27$

c

- Try values between 3 and 4 until you get two x values, within consecutive digits in the second decimal place, one giving an outcome too big and the other too small.

Trial x	$2x^2 + x - 27$	Outcome (trying to get to 0)
3	−13	too small
4	9	too big
3.5	1	too big
3.4	−0.48	too small
3.45	0.255	too big
3.44	0.1072	too big
3.43	−0.0402	too small

Between 3 and 4

Between 3.4 and 3.5

The x value is between 3.43 and 3.44

- To check which it is closer to, try half-way between these, $x = 3.435$

  When $x = 3.435$, $2x^2 + x - 27 = 0.03345$ (too big)

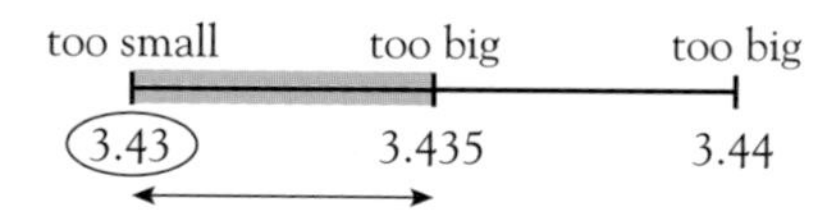

x value is closer to 3.43 than to 3.44

$\underline{x = 3.43 \text{ (2 d.p.)}}$

- Put this value of x into the expressions for the length and the width.

  length $= 2x - 1 = 2 \times 3.43 - 1 = 5.86$

  width $= x + 1 = 3.43 + 1 = 4.43$

<u>length = 5.86 m, width = 4.43 m</u>

Check your answer by working out the area.
Area = 5.86 × 4.43 = 25.9598 (**very** close to 26)

Exercise 21.3

1 A solution of the equation

$$2x^2 + x = 8$$

lies between $x = 1$ and $x = 2$.

Use the method of trial and improvement to find this solution of the equation

$$2x^2 + x = 8$$

Give your answer to 1 decimal place.

2 Show that the equation

$$x^2 - 10 = 0$$

has a solution between 3 and 4.

Use trial and improvement to find this solution to 2 decimal places.

3 Use the method of trial and improvement to find the positive solution of the equation

$$5x^2 + 2x - 11 = 0$$

4 Starting with the value $x = 2$, use trial and improvement to find the positive solution of the equation

$$2x^2 - 3x - 1 = 0$$

Give your answer to 2 decimal places.

5 Show that the equation

$$7x^2 + 3x = 2$$

has a solution between 0.3 and 0.4.

Use trial and improvement to find this solution to 2 decimal places.

6 A right-angled triangle has dimensions as shown in the diagram (x in cm).

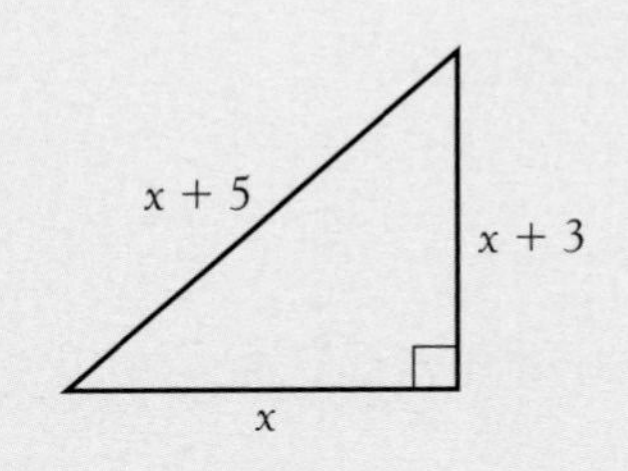

Use Pythagoras' theorem to write down an equation for x.

By multiplying out the brackets and simplifying, show that

$$x^2 - 4x - 16 = 0$$

Use trial and improvement to find the value of x correct to 2 decimal places.

Write down the lengths of the three sides of the triangle, each correct to 2 decimal places.

7 A rectangular piece of cardboard of length 6 cm and width 5 cm has a piece cut out with dimensions x and $(x + 2)$.

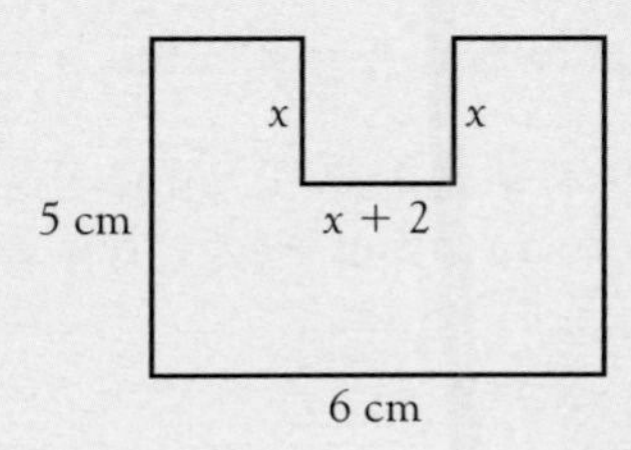

a What is the largest possible value of x?

b Show that the area, A, of the new shape of cardboard, can be written as

$$A = 30 - 2x - x^2$$

c Draw the graph of $A = 30 - 2x - x^2$, taking values of x from 0 to the value given in **a**.

d Use your graph to estimate the value of x when $A = 20\text{ cm}^2$.

e Use trial and improvement to estimate the value of x to 2 decimal places.

8 Amy is four years older than William. Let William's age be x.

a How old is Amy?

Let y be the result when you multiply their ages.

b i Write down an equation connecting y and x.

ii If $y = 38$, estimate Amy's age using a trial and improvement method. Give your answer in years and months.

EXAM QUESTIONS

Worked Exam Question
[MEG]

COMMENTS

Kevin uses a 10-metre length of wire netting to make a rectangular enclosure for his pet rabbit.

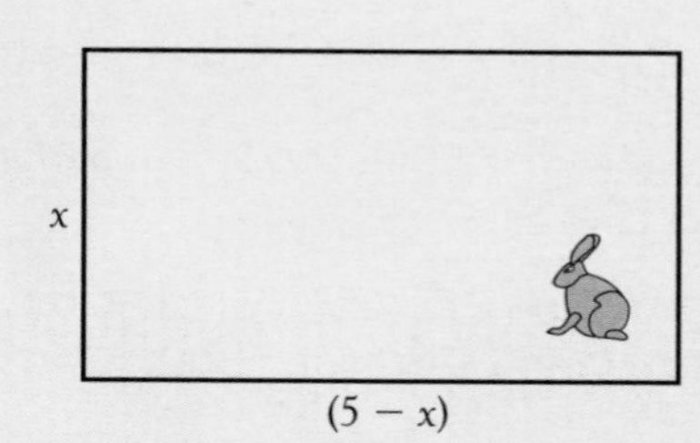

When the width of the enclosure is x metres, the area, A, is given by

$$A = 5x - x^2$$

a i Show that, when A = 2, x has a value between 0.4 and 0.5

◆ Substitute $x = 0.4$ and $x = 0.5$ into the equation; one value of A should be smaller than 2, the other bigger than 2.

Answer $x = 0.4,\ A = 5 \times 0.4 - 0.4^2 = 1.84$
$x = 0.5,\ A = 5 \times 0.5 - 0.5^2 = 2.25$

M1 for testing a value which gives $A < 2$
M1 for testing a value which gives $A > 2$

2 marks

ii Use trial and improvement to find this value of x, correct to 2 decimal places.

x	$5x - x^2$	Outcome
0.45	2.0475	too big
0.44	2.0064	too big
0.43	1.9651	too small
0.435	1.985 …	too small

too small too small too big
0.43 0.435 (0.44)

Answer 0.44 (2 d.p.)

You should do at least five trials altogether

M1 for testing one value in between 0.4 and 0.45
A1 for correct answer

2 marks

b The graph below shows the values of A for values of x from 0 to 5.

i Use the graph to find the values of x for which A = 3.

ii Find the maximum area of the enclosure.

◆ Draw the line A = 3 and read off the x values of the points of intersection of the line with the curve.

- Read off the highest value of A on the graph.

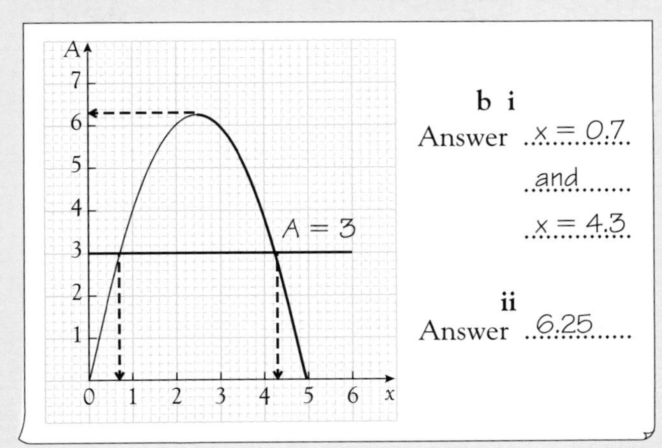

B1 for x value of 0.6 to 0.8
B1 for x value of 4.2 to 4.4
2 marks

B1 for answer of 6.1 to 6.3
1 mark

iii Give a geometrical description of the shape of the enclosure when the area is maximum.

- Read off the x value when A is maximum.
- Work out the length $5 - x$ and make your conclusion.

$x = 2.5$
$5 - x = 5 - 2.5 = 2.5$
Answer square with side 2.5 m

B2 for stating 'square' **and** 'side 2.5 m'
B1 for stating 'square' only
2 marks

Exam Questions

1 a Copy and complete the table of values for the equation $y = x^2 - 5$

x	-3	-2	-1	0	1	2	3
x^2	9						
$y = x^2 - 5$	4						

b Draw the graph of $y = x^2 - 5$, for values of x from -3 to 3.

c Use your graph to solve the equation $x^2 - 5 = 0$ [MEG]

2 Use the method of trial and improvement, or otherwise, to find the positive solution of

$$x^2 + x = 177$$

Give your answer correct to 3 significant figures. [L]

3 a Factorise completely $12x^2 - 6x$

b Remove brackets and simplify $(5x + 4)(3x - 7)$ [MEG]

4 a Copy and complete the table for $y = 4 - x^2$

x	-3	-2	-1	0	1	2	3
y	-5			4	3		

b On a grid, draw the graph of $y = 4 - x^2$

c Use the graph to solve the equation

$$4 - x^2 = 1$$

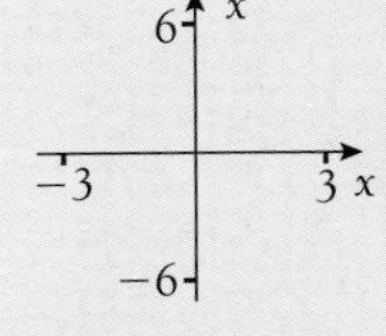

2 cm = 1 on x axis
1 cm = 1 on y axis

[MEG]

5 a Copy and complete the table below and draw the graph of the mapping $y = x^2 - 3$

x	-3	-2	-1	0	1	2	3
$y = x^2 - 3$							

b Use your graph to write down the two values of x for which $y = 3$. [SEG]

EXAM QUESTIONS

6 a i Multiply out $4x(x + 3)$

ii Multiply out and simplify $(2x + 3)(2x + 3)$

b

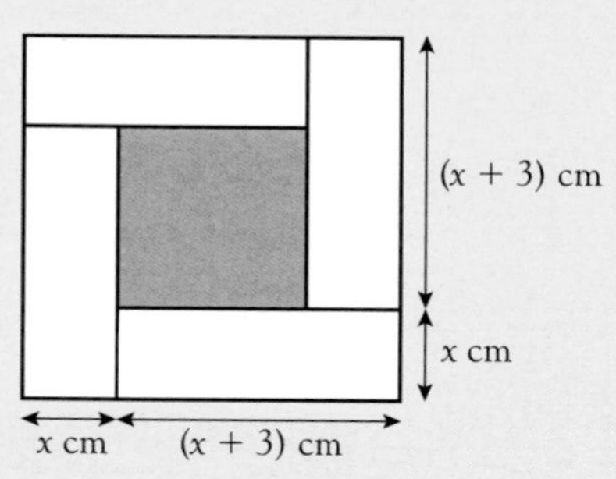

Four identical rectangular tiles are placed around a square tile as shown in the diagram.

Using your answers to **a**, or otherwise, find the area of the square tile. [NEAB]

7 a Draw the graphs of $y = 2x - 4$ and $y = -x^2$ for values of x from -4 to 4.

b Use the graphs to find the smallest value of x which gives the same value of y in both equations. [SEG]

8 A rocket is launched vertically upwards at a speed of 30 m/s.

The height, h metres, of the rocket above the ground after t seconds is given by the formula

$$h = 30t - 5t^2$$

a Copy the table and use the formula to complete it.

t	0	1	2	3	4	5	6
h	0	25	40		40	25	0

b On a grid with time t as horizontal axis (from 0 to 6 s) and height h as vertical axis (from 0 to 50 m),

i plot the points from the table,

ii join the points up with a smooth curve.

c Use your graph to find when the rocket is 30 m above the ground. [MEG]

9 Use a 'trial and improvement' method, starting at $x = 3$, to work out a solution of the equation $x^2 - 2x = 6$

Give your answer correct to 1 decimal place.

Show **all** the stages of working. [L]

10 A firm makes a range of cylindrical water tanks.

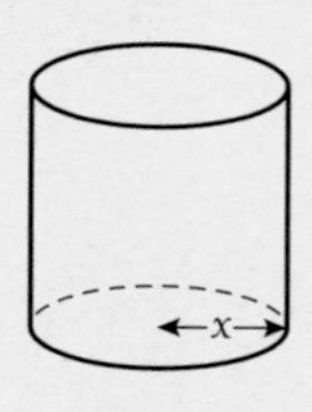

Each tank has the same height.

The volume, V m^3, of a tank with radius x metres is given by the formula

$$V = 6x^2$$

The table shows the values of V for the given values of x.

x	0	1	2	3	4
V	0	6	24	54	96

a On a grid, plot points to represent these values.

Join your points up with a smooth curve.

b Use your graph to estimate

i the volume of a tank with a radius of 2.5 m,

ii the radius of a tank with a volume of 20 m^3. [MEG]

11 The equation $x^2 - x = 7$ has a solution between $x = 3$ and $x = 4$. Use trial and improvement to find this solution correct to one decimal place.

Show all your working clearly. [MEG, p]

12 a Simplify $3x - 2xy + 7y + 4x + yx$

b Factorise completely $6p^2 - 8p$

c Expand and simplify $(2z - 3)(z + 5)$ [MEG]

13 The length of a rectangle is 2 cm more than its width.

a Calculate the area of the rectangle

i when the width is 4 cm,

ii when the width is 5 cm.

b Using a trial and improvement method, find the width of the rectangle given that its area is 32 cm^2.

Give your answer in centimetres correct to one decimal place. (You must show all your trials.) [MEG]

GEOMETRY III

In this chapter you will learn how to
- **calculate the sum of the interior angles of a polygon**
- **calculate the sum of the exterior angles of a polygon**
- **calculate an interior angle and an exterior angle of a regular polygon**
- **check whether shapes tessellate.**

page 68

Interior angles of a polygon

A **polygon** is a flat shape with three or more straight sides.

Look at the following methods for finding the sum of the interior angles of a polygon and the size of each interior angle of a **regular** polygon.

The interior angles are shaded

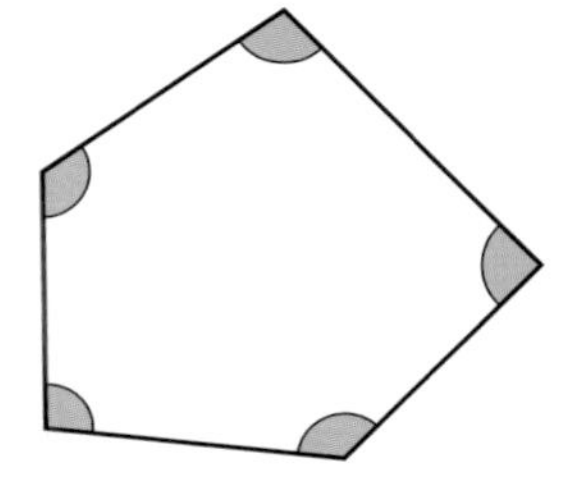

In a **regular** polygon, the sides are all equal in length and the interior angles are equal.

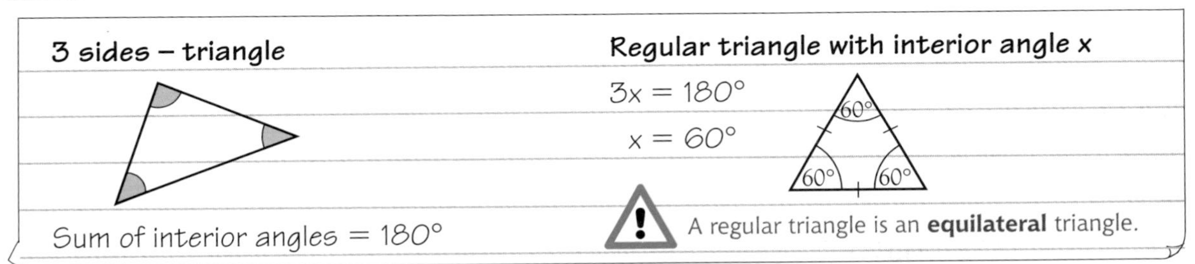

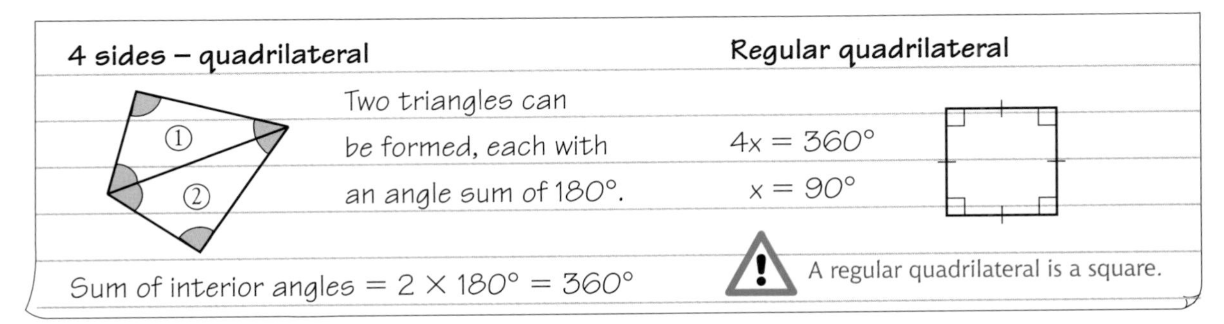

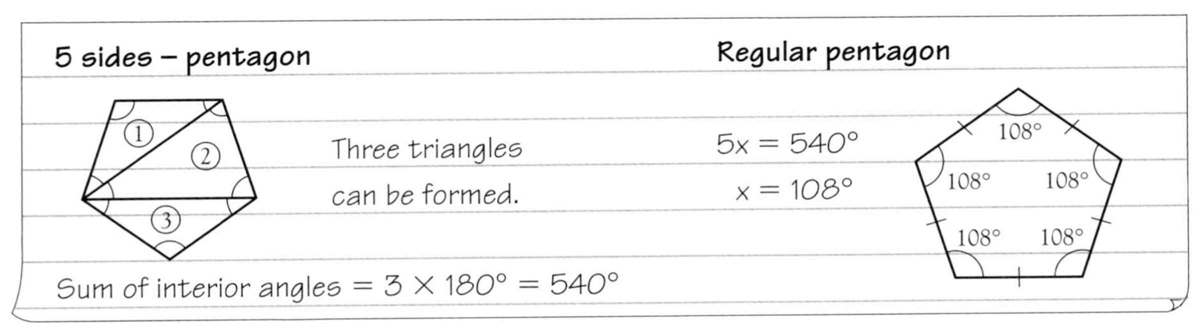

POLYGONS

TASK

Continue for polygons with 6, 7, 8, 9, 10, 11 and 12 sides. Write a rule for finding the sum of the interior angles of any polygon.

In the task you should have found that when a polygon has n sides, then the number of triangles that can be formed is two less than n. This is written $(n - 2)$.

In 1 triangle, sum of interior angles = 180°

So in $(n - 2)$ triangles, sum of interior angles = $(n - 2) \times 180°$

All the triangle angles form the interior angles of the polygon.

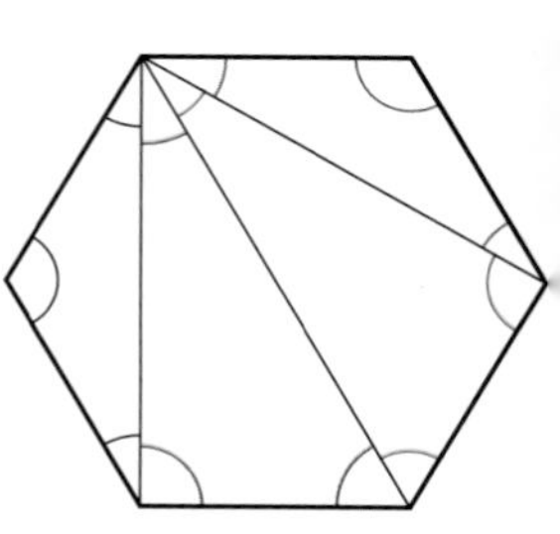

In a **regular polygon**, each interior angle is the same, so to find the size of an interior angle you have to divide the sum of the interior angles by the number of angles.
If the polygon has n sides, there are n angles, so divide by n.

In an n-sided polygon
- Sum of the interior angles = $(n - 2) \times 180°$

In a **regular** n-sided polygon
- each interior angle = $\frac{(n-2) \times 180°}{n}$

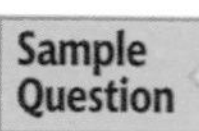

1
a Find the sum of the interior angles of a 20-sided polygon.
b Find the size of each interior angle in a regular 20-sided polygon.

Answer

a

- It is difficult to draw a polygon with 20 sides so just imagine it.
  Starting from one point on the polygon,
  you could form 18 triangles by drawing lines to the other points.

$n = 20$, so $(n - 2) = 18$

Sum of interior angles $= (n - 2) \times 180°$
$= \mathbf{18} \times 180°$
$= \underline{3240°}$

b

- Share the total equally between the 20 angles

  $3240 \div 20 = 162°$

  Each interior angle of a regular 20-sided polygon is 162°

Sample Question 2

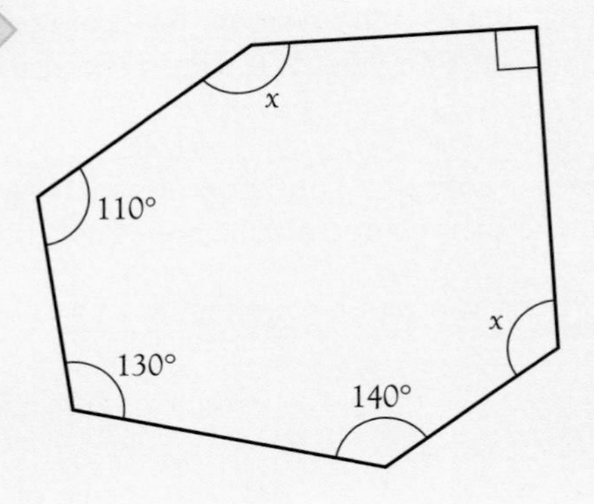

a What shape is this?

b Calculate the value of x.

Answer

a

- Count the number of sides.

  There are 6 sides, so the shape is a hexagon.

b

- Find the sum of the interior angles of a hexagon using sum $= (n - 2) \times 180°$, with $n = 6$

  Sum of interior angles $= 4 \times 180°$

  $= 720°$

$n - 2 = 6 - 2 = 4$

- Add up all the angles in the diagram and put the sum equal to 720°.

  $x + 110 + 130 + 140 + x + 90 = 720°$

- Simplify and solve the equation

  $2x + 470 = 720$

  -470 $\quad 2x = 250$

  $\div 2$ $\quad x = 125°$

POLYGONS

Exercise 22.1

1 Calculate the sum of the interior angles of polygons with

a 7 sides **b** 15 sides

c 22 sides **d** 100 sides

2 Find the number of sides of the polygon if the sum of the interior angles is:

a 1260° **b** 1980°

c 2520° **d** 3420°

3 Find the number of sides and the size of each interior angle of a regular polygon with interior sum of

a 1800° **b** 1440°

c 3240° **d** 3960°

4 Calculate the size of each interior angle of a regular polygon with

a 6 sides **b** 9 sides

c 18 sides **d** 30 sides

5 Find the unknown angles a, b, c, d.

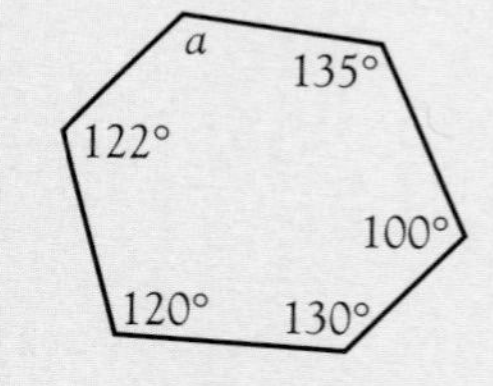

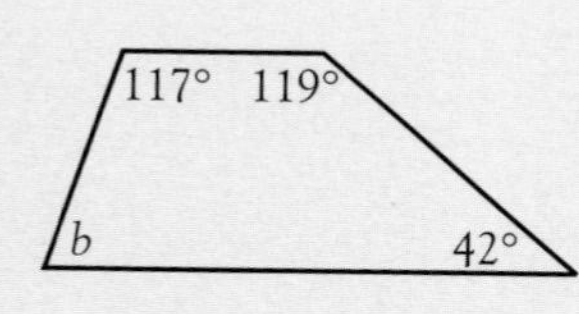

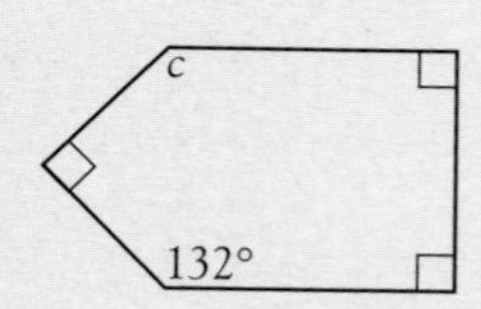

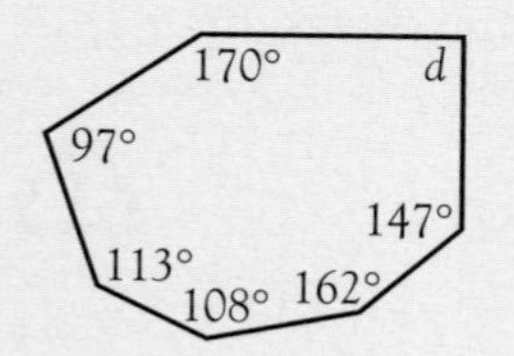

6 Find the unknown angle x in each diagram.

a

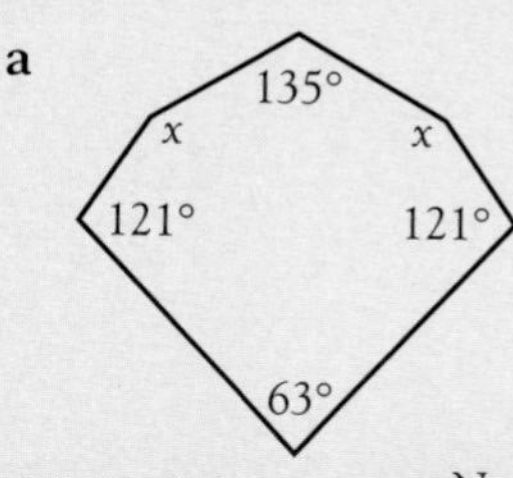

b

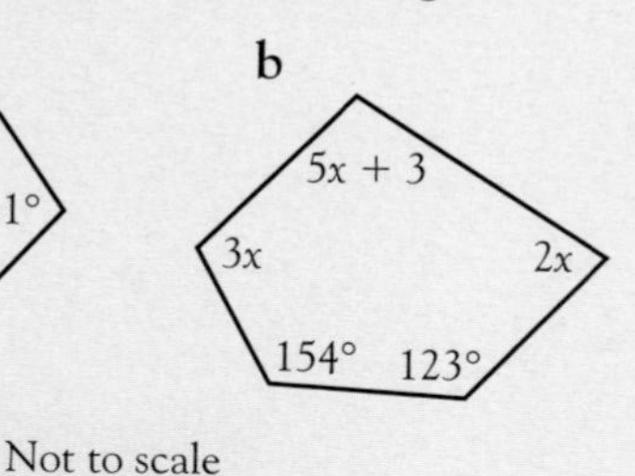

Not to scale

c

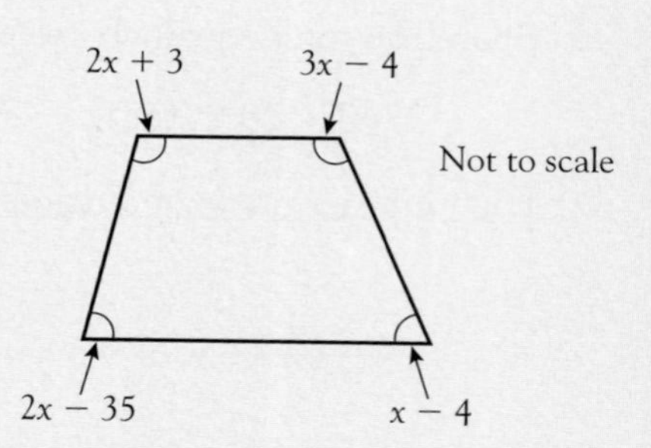

7 Find the size of each interior angle of a regular octagon.

8 Five angles of a hexagon are equal to each other. The sixth angle is 100°. Find the size of the other angles.

9 Two angles of a pentagon are each equal to 102°. If the other three angles are equal, find their size.

10 ABCD is a regular pentagon

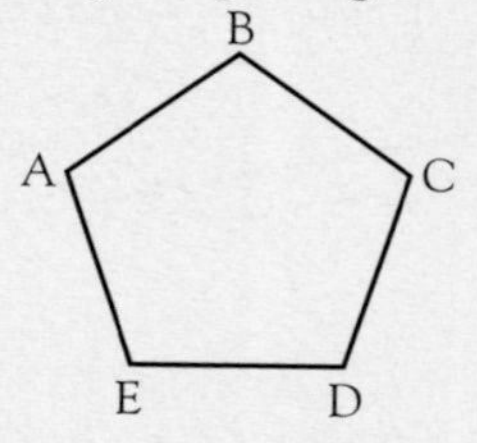

a Calculate **i** ∠BCD **ii** ∠BCA **iii** ∠ACD

b What can you say about the lines AC and ED?

c Give a reason for your answer.

11 ABCDEFGH is a regular octagon.

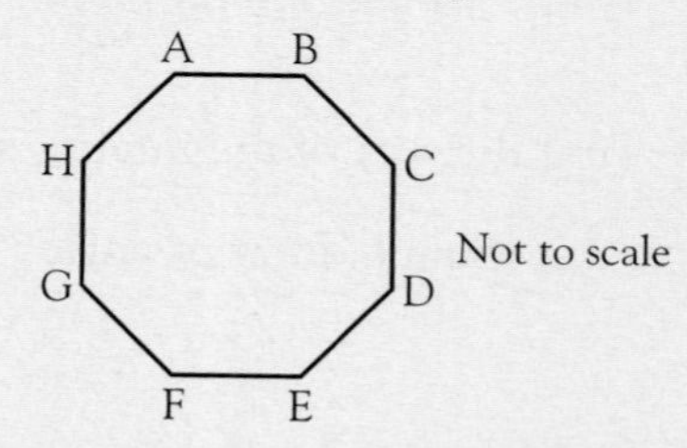

Calculate the angles HAB, BAC, AHD, HGD and DGF.

12 The shape shown has been drawn by extending the sides of a regular pentagon.
Find the angles a, b and c.

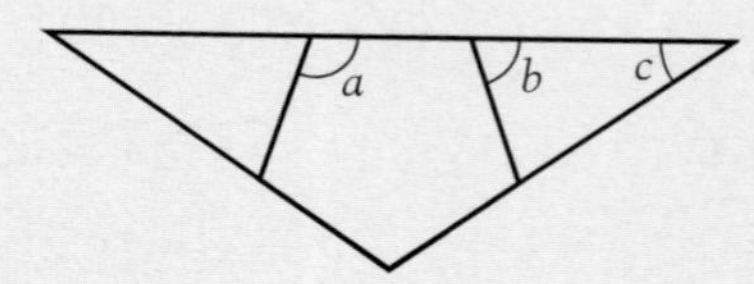

POLYGONS

Exterior angles of a polygon

page 69

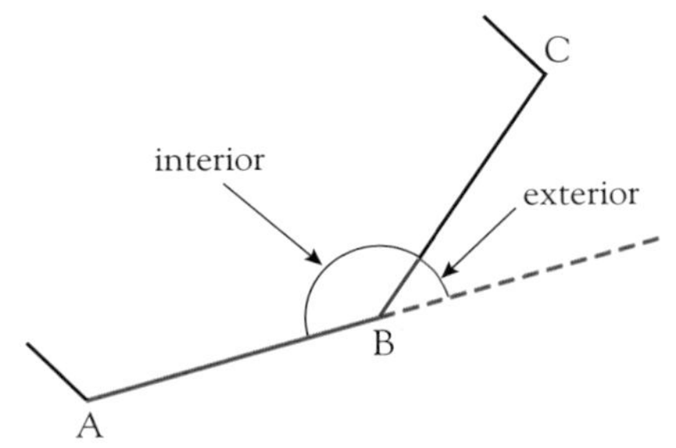

AB and BC are two sides of a polygon with interior angle ABC.

Angle CBD is an **exterior angle**.

Notice that the two angles form a straight line, so interior angle + exterior angle = 180°

In this polygon, imagine taking your pencil for a walk around the shape. Lay it flat, turn it through each exterior angle along the shape and note the total angle turned through from start to finish.

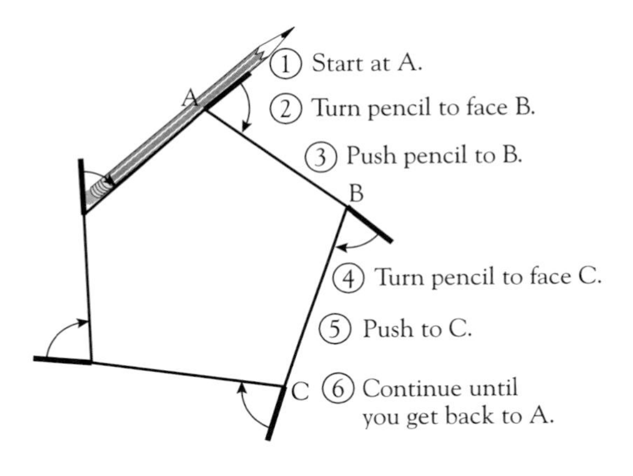

The pencil makes one complete turn, i.e. it turns through 360°.

This result is true for any polygon. It does not depend on the number of sides.

The sum of the exterior angles of **any** polygon is 360°.

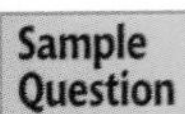

Sample Question 3

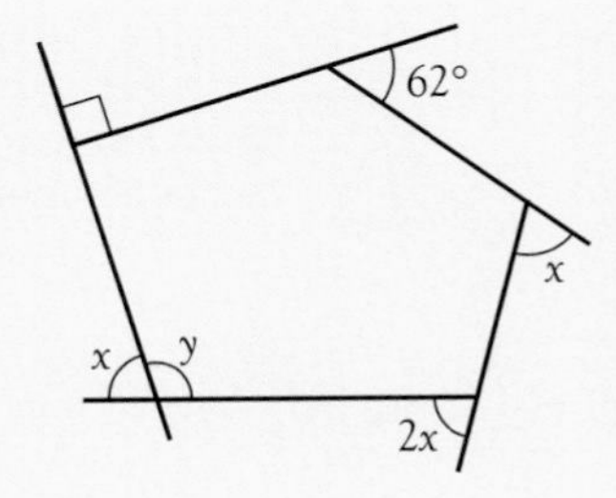

a Calculate x.

b Calculate y.

c A regular polygon is drawn with the same number of sides. What is the size of an exterior angle of this regular polygon?

Answer

a

- Add the exterior angles, put the total equal to 360° and solve the equation.

$$90 + 62 + x + 2x + x = 360$$

Simplify $\quad 152 + 4x = 360$

$-152 \quad 4x = 208$

$\div 4 \quad \underline{x = 52°}$

Do not include y. It is not an exterior angle.

POLYGONS

b

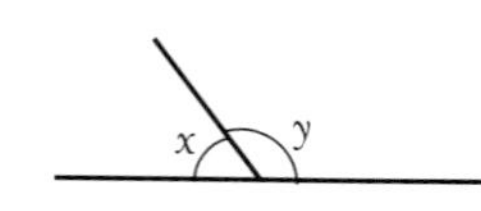

- Use the fact that angles in a straight line add up to 180°.

$y = 180 - 52$

$\underline{y = 128°}$

c

- The polygon has 5 sides, so divide 360° by 5.

$360 \div 5 = 72$

<u>Each exterior angle = 72°</u>

You can write a general rule

> In a regular polygon with n sides, each exterior angle $= \left(\dfrac{360}{n}\right)^{\circ}$

If you know an exterior angle of a regular polygon, you can calculate the sum of the interior angles, using the fact that

$$\text{interior angle} + \text{exterior angle} = 180°,$$

exterior angle

interior angle

For example:

a In a regular hexagon (6 sides)

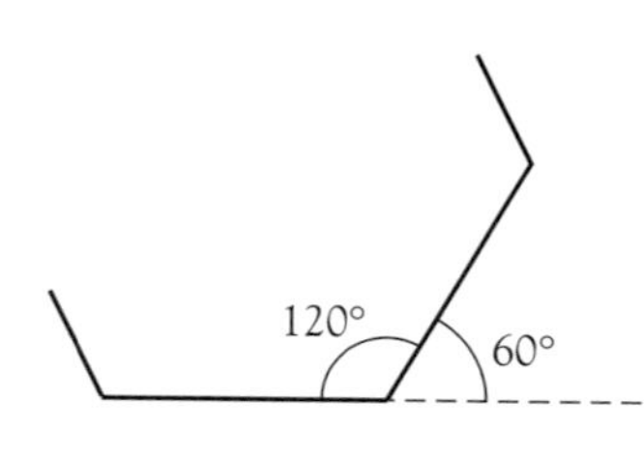

Each exterior angle $= \dfrac{360}{6} = 60°$

Each interior angle $= 180° - 60° = 120°$

Sum of interior angles $= 6 \times 120° = 720°$

There are 6 interior angles in a hexagon.

b In a regular decagon (10 sides)

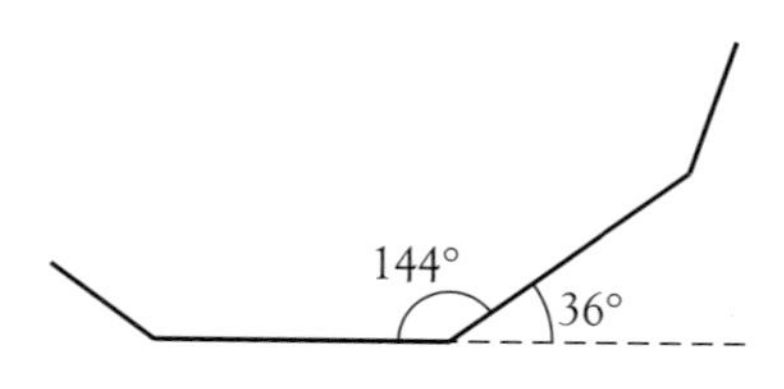

Each exterior angle $= \dfrac{360}{10} = 36°$

Each interior angle $= 180° - 36° = 144°$

Sum of interior angles $= 10 \times 144° = 1440°$

Sample Question 4 In a regular polygon, each exterior angle is 24°.

a How many sides has the polygon?

b What is the size of each interior angle?

c What is the sum of the interior angles of the polygon?

POLYGONS

Answer

a

- Use sum of exterior angles = 360°
- Find how many angles of size 24° there are in 360°.

  $360 \div 24 = 15$

  The polygon has 15 sides.

b

- Use interior angle + exterior angle = 180°

$$\begin{aligned} x + 24° &= 180° \\ x &= 180° - 24° \\ &= \underline{156°} \end{aligned}$$

c

- There are 15 interior angles, each 156°

  Sum $= 15 \times 156° = \underline{2340°}$

Compare this with the formula:

$$\begin{aligned} \text{Sum of interior angles} &= (n - 2) \times 180° \\ &= \mathbf{13} \times 180° \\ &= 2340° \end{aligned}$$

$n = 15$
so $n - 2 = 13$

Sample Question 5 In a regular polygon, each interior angle is 160°. How many sides has the polygon?

Answer

- Work out each **exterior** angle using interior angle + exterior angle = 180°
  interior angle = 180° − 160° = 20°
- Divide this into 360° to find how many exterior angles there are

  $360 \div 20 = 18$

  The polygon has 18 sides.

Symmetries of regular polygons

a **Line symmetry**

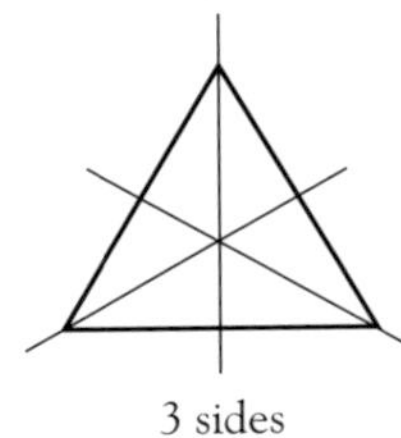

3 sides
3 lines of symmetry

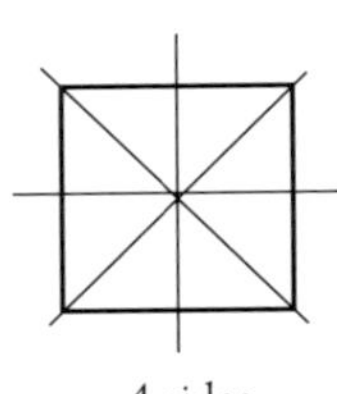

4 sides
4 lines of symmetry

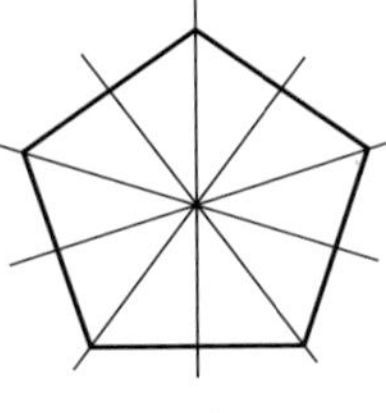

5 sides
5 lines of symmetry

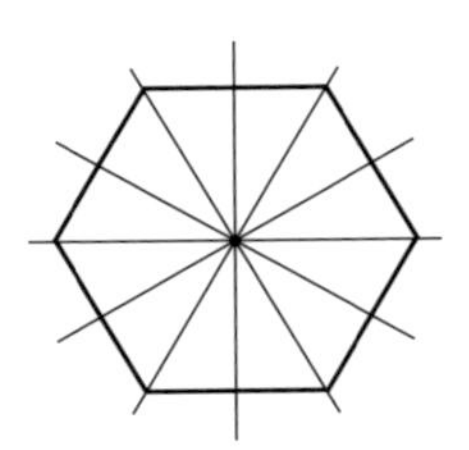

6 sides
6 lines of symmetry

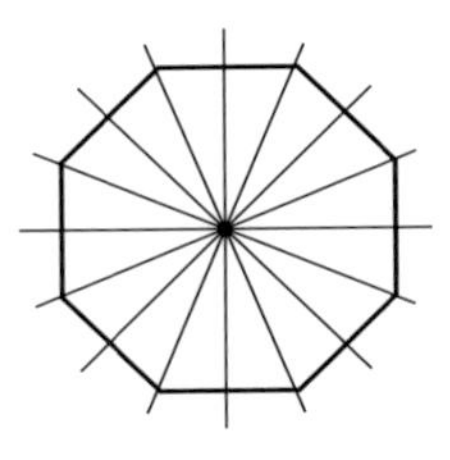

8 sides
8 lines of symmetry

An n-sided **regular** polygon has n lines of symmetry.

POLYGONS

b Rotational symmetry

The order of rotational symmetry is the number of times a tracing of the shape fits exactly on top of it when the tracing is rotated through one complete turn about the centre of the shape.

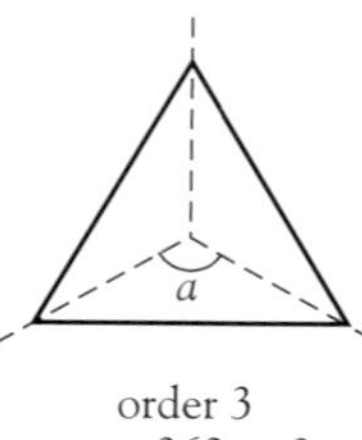

order 3
$a = 360 \div 3$
$= 120°$

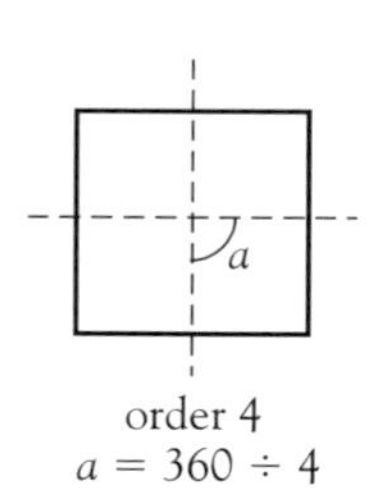

order 4
$a = 360 \div 4$
$= 90°$

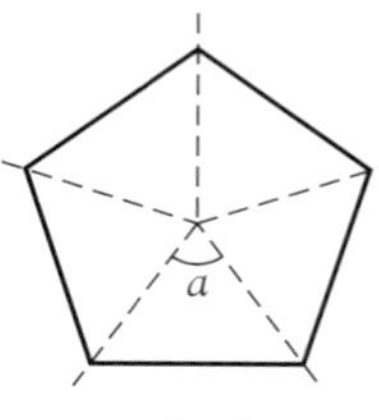

order 5
$a = 360 \div 5$
$= 72°$

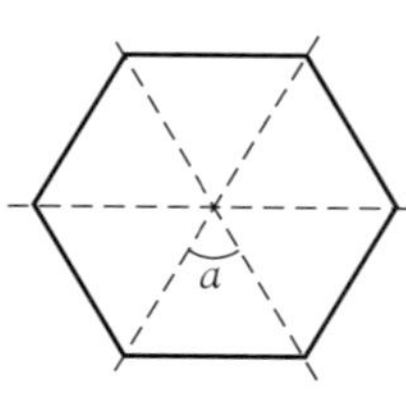

order 6
$a = 360 \div 6$
$= 6°$

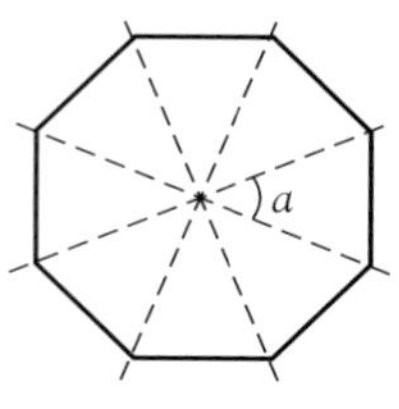

order 8
$a = 360 \div 8$
$= 45°$

An n-sided **regular** polygon has order of rotational symmetry n.

You can use this fact to draw a regular polygon accurately using the 'spokes in a circle' method.

Sample Question 6

a Draw a regular pentagon inside a circle of radius 3 cm.

b **Measure** the length of a side of the pentagon.

c **Calculate** the length of a side of the pentagon.

d Comment on the symmetries of the pentagon.

Answer

a

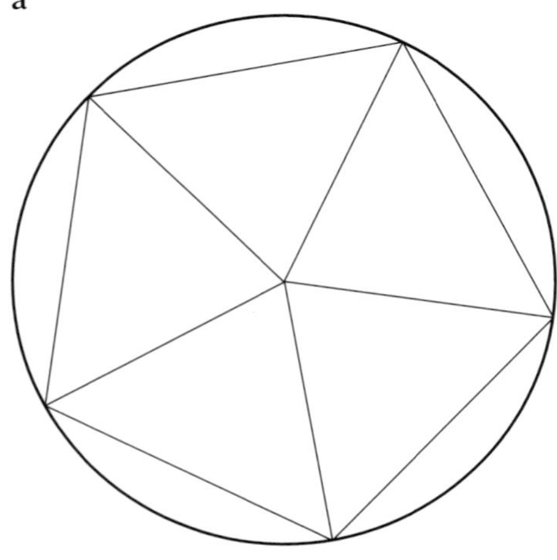

- Draw a circle with radius 3 cm.
- Draw 5 equally spaced 'spokes' from the centre to the circumference, measuring each one accurately so that the angle is

  $360 \div 5 = 72°$

- Join the points on the circumference with straight lines. These form a regular pentagon.

b

- Measure the length of a side

  length = 3.5 cm

c

- Sketch one of the inside triangles
  $A\hat{O}D = 72°$ and OA = OB = 3 cm (radii)
  The triangle is isosceles

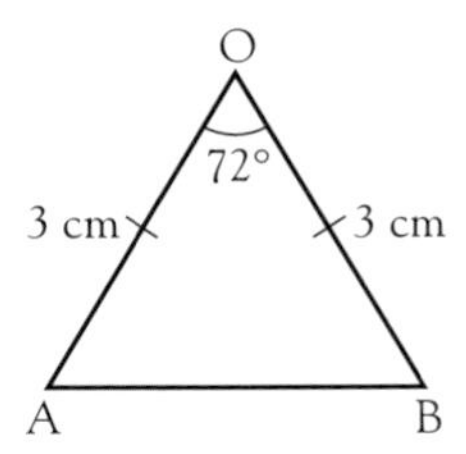

- Form a right-angled triangle by drawing a line of symmetry.
  Sketch this triangle and work out $A\hat{O}D$.
  $A\hat{O}D = 72° \div 2 = 36°$

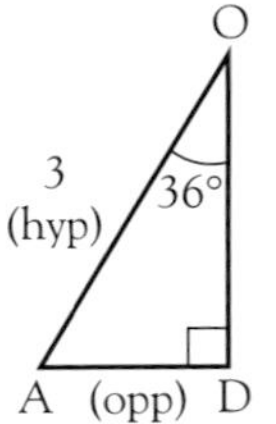

page 368

- Calculate AD using trigonometry.
  You want opp, you know hyp, use sin.

$$\sin \theta = \frac{\text{opp}}{\text{hyp}}$$

$$\sin 36° = \frac{\text{AD}}{3}$$

$$\text{AD} = 3 \times \sin 36°$$

$$= 1.763\ldots$$

There is no need to write down all the numbers on the calculator. Do not clear the display as you are about to use it again.

- Double this to find AB

  AB = 3.526…

  AB = **3.5 cm** (1 d.p.)

This agrees well with the measurement from the accurate drawing.

d

- Note that the pentagon is regular and comment on line and rotational symmetry.

  The pentagon has 5 lines of symmetry and also rotational symmetry of order 5.

POLYGONS

Exercise 22.2

1 Find the number of sides of the regular polygon with an exterior angle of

a 60° b 40°

c 20° d 12°

2 Find the exterior angle and hence the number of sides of a regular polygon when each interior angle is

a 170° b 135°

c 165° d 150°

3 Find **a** the size of the exterior angle and **b** the size of the interior angle of the following regular polygons

i pentagon (5 sides) iv nonagon (9 sides)

ii hexagon (6 sides) v decagon (10 sides)

iii octagon (8 sides)

4 The size of an interior angle of a regular polygon is twice the size of the exterior angle. Find the number of sides and name the polygon.

5 The interior angle of a regular polygon is three times the size of the exterior angle. Find the number of sides and name the polygon.

6 The interior angle of a regular polygon is four times the size of the exterior angle. Find the number of sides and name the polygon.

7 Find the value of x in each of the polygons.

a

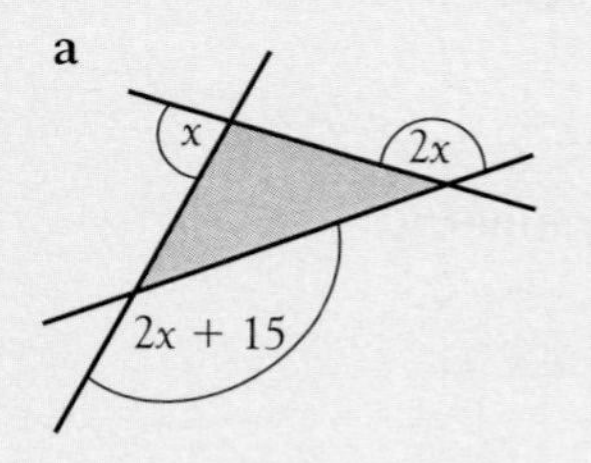

b

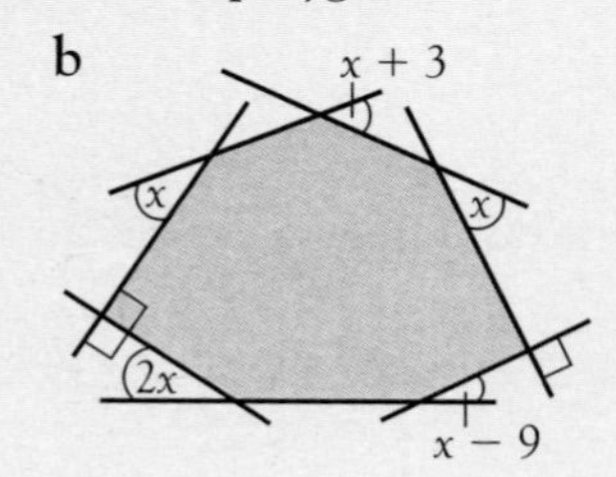

8 The diagram shows a regular octagon with a square drawn on one side and a regular pentagon drawn on another side.

a Calculate the angle marked a.

b Calculate the angle marked b.

9 Three triangles are joined together to form ABCDE.

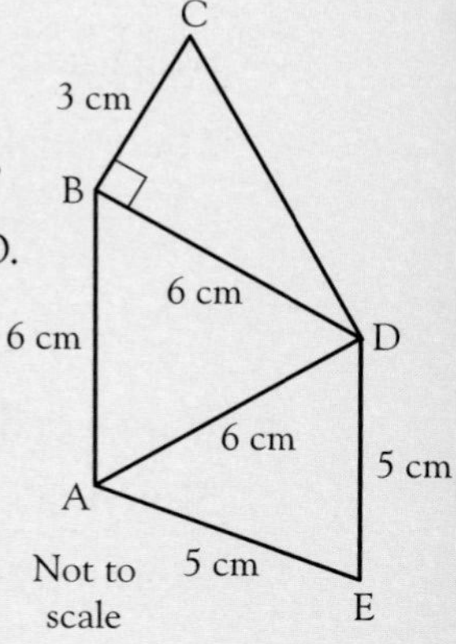

a What shape is ABCDE?

b What is the size of $A\hat{B}D$.

c Calculate $B\hat{C}D$.

d Calculate $D\hat{A}E$.

e Calculate $A\hat{E}D$.

f Find the sum of all the angles of polygon ABCDE.

10 a Draw an accurate diagram of a regular octagon inside a circle of radius 4 cm.

b Measure the length of a side of the octagon.

c Calculate the length of a side using trigonometry.

11 Draw a circle of radius 3 cm. Inside the circle, draw an equilateral triangle.

a Measure the length of one of the sides and the perpendicular height of the triangle.

b Use trigonometry to calculate the length of a side and Pythagoras' theorem to calculate the height of the triangle.

c Calculate the area of the triangle correct to one decimal place.

12 The diagram shows a plan view of a children's roundabout comprising a regular hexagon inside a circle, of radius 1.5 m.

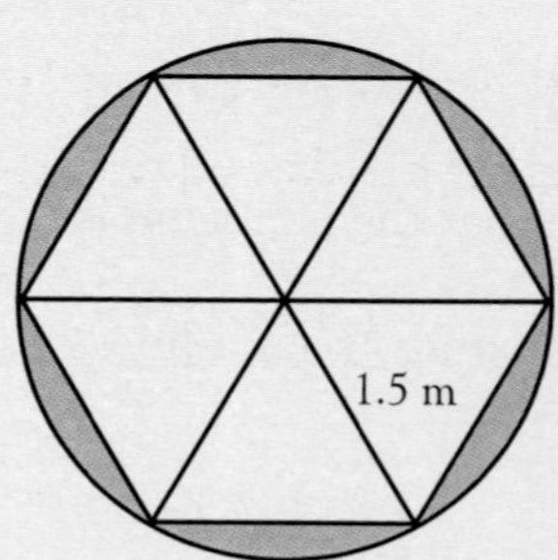

a Calculate the perpendicular height of one of the triangles in the diagram.

b Calculate the area of the hexagon

c Calculate the area of the circle lying outside the roundabout, shown shaded.

d This shaded area is to be filled with sand to a depth of 25 cm. Find the volume of sand required.

Tessellations

If shapes are translated, reflected or rotated to form a pattern with no gaps or overlaps between the shapes, the pattern is called a **tessellation**. If the pattern is continued it would go on and on. Sometimes in a tessellation all the shapes are **congruent**. This means that they are all the same shape and size. In other tessellations, a combination of shapes may be used.

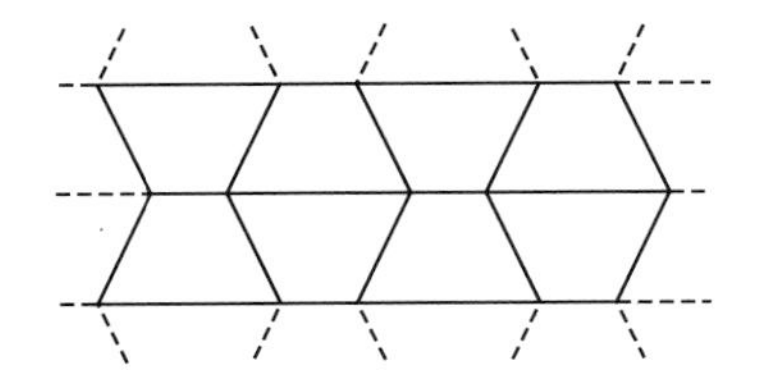

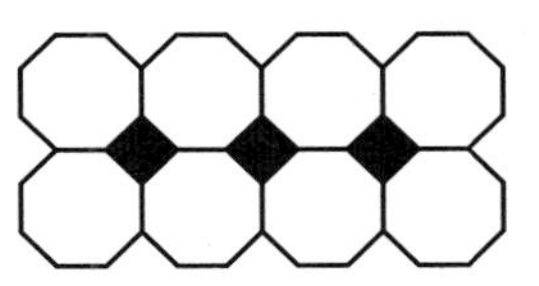

A tessellation using octagons and squares

In a tessellation made from polygons, at any vertex (point where shapes meet) all the interior angles must add up to 360°.

Some regular polygons tessellate on their own:

Equilateral triangle

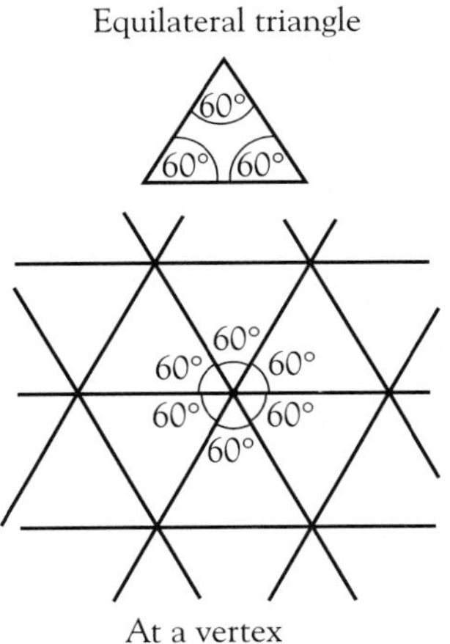

At a vertex
$6 \times 60 = 360°$

Square

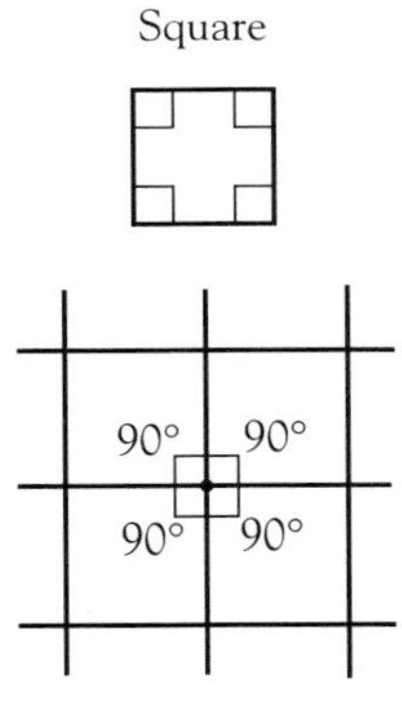

At a vertex
$4 \times 90 = 360°$

Regular hexagon

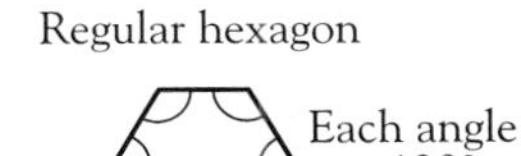

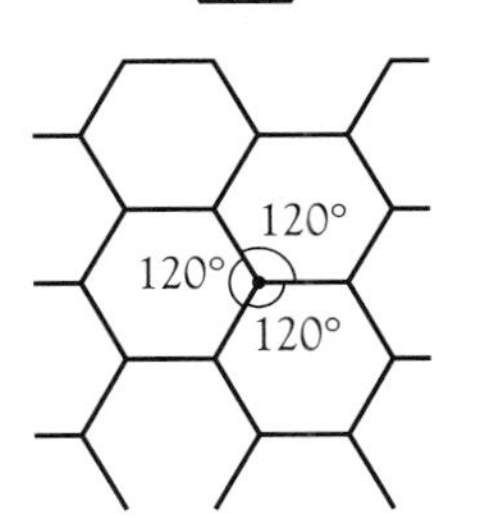

At a vertex
$3 \times 120 = 360°$

Using two different regular polygons:

Squares and equilateral triangles

At a vertex

$$90° + 90° + 60° + 60° + 60° = 360°$$

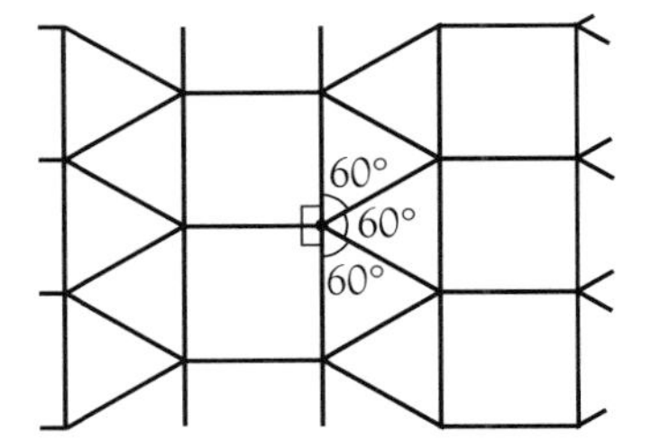

TASK

Investigate tessellations that can be made from polygons.

Explain how you can use interior angles to find out whether shapes tessellate **before** you draw them.

Worked Exam Question
[WJEC]

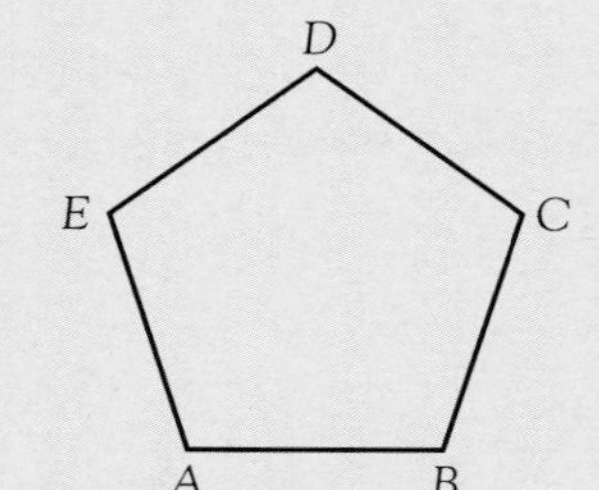

Not to scale

In a sailboard race contestants follow a course which is a regular pentagon *ABCDE*.

B is due east of A.

***D* is north of the line *AB*.**

a What is the bearing of C from B?

◆ Mark the angle needed on the diagram – stand at *B*, face north, turn clockwise to face C.

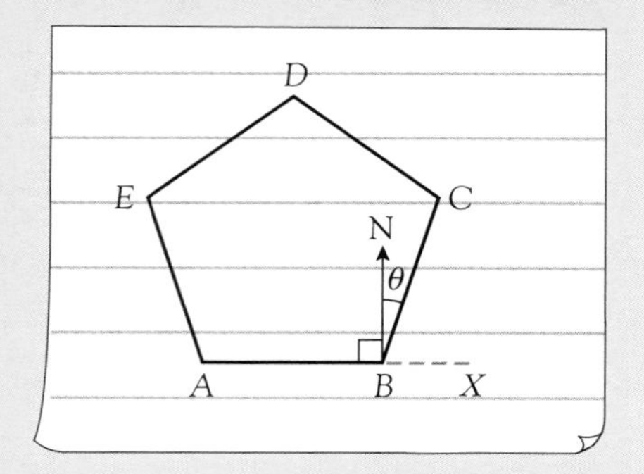

◆ Work out angle *ABC* by finding the exterior angle of a regular pentagon (*CBX*) then subtracting this from 180° to find the interior angle.

◆ Find angle *NBC* by subtracting 90°.

◆ Write the angle as a three-figure bearing.

Exterior angle $CBX = \frac{360}{5} = 72°$

Interior angle $ABC = 180° - 72°$

$ABC = 108°$

angle $NBC = 108° - 90°$

$= 18°$

Answer 018°

COMMENTS

M1 for finding exterior angle

M1 for subtracting 90°

A1 for correct bearing written correctly with three figures

3 marks

b What is the bearing of *D* from C?

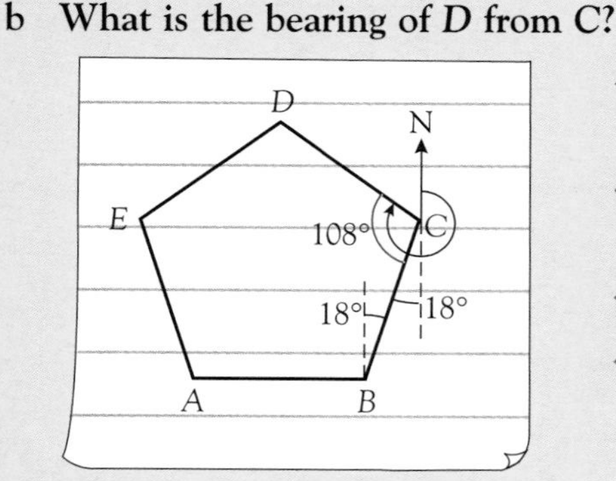

◆ Mark the angle needed on the diagram, putting in the north line at C.

◆ Use alternate angles on parallel lines to put in 18° on diagram.

◆ Put in **interior angle *BCD*** = 108°

◆ Add the angles at C.

M1 for indication of angle required for the bearing

$180 + 18 + 108 = 306$

Answer 306°

A1 for correct addition of all angles

2 marks

Exam Questions

a The diagram shows a quadrilateral.

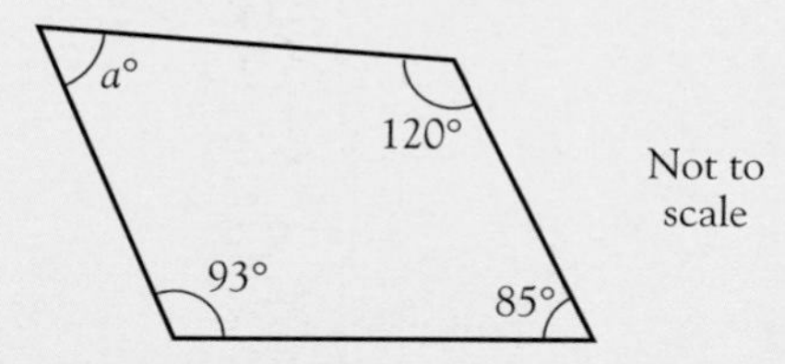

Work out the size of the angle marked $a°$.

b The diagram shows a regular hexagon.

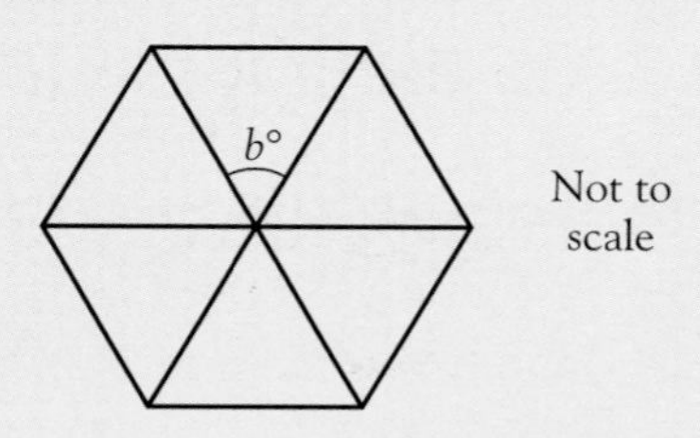

Work out the size of the angle marked $b°$.

c The diagram shows a regular octagon.

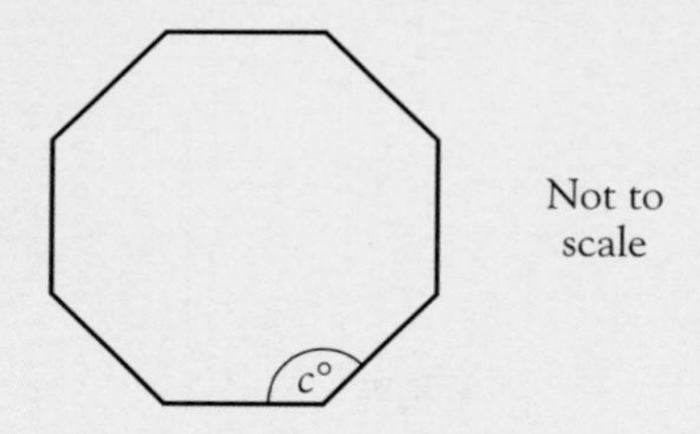

Work out the size of the angle marked $c°$. [L]

This diagram shows a regular hexagon and a regular pentagon with equal length sides.
The diagram is not drawn accurately.

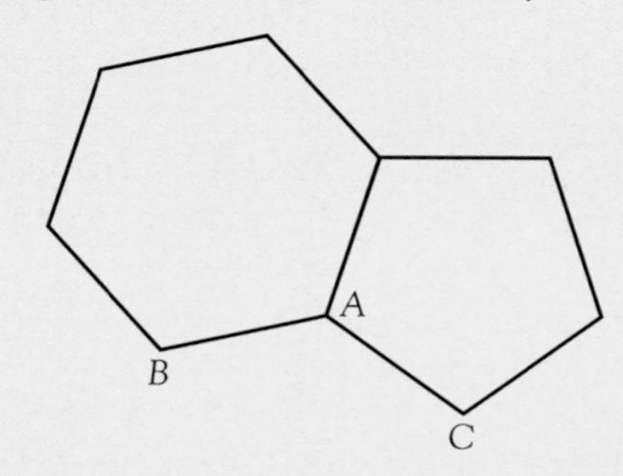

a Work out the size of one interior angle of the pentagon.

Each interior angle of the hexagon is 120°.

b Work out the sizes of the angles of triangle ABC. [MEG]

3 Karen has joined some of the vertices in a regular pentagon.

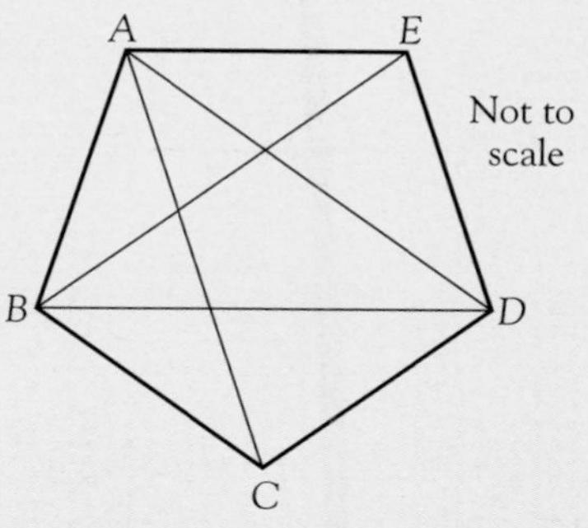

a Calculate the size of angle ABC.

b Triangles have been formed by the lines drawn in the pentagon.
Name a triangle which is congruent to triangle BED. [SEG, p]

4 The diagram represents a regular pentagon with two of its lines of symmetry shown.

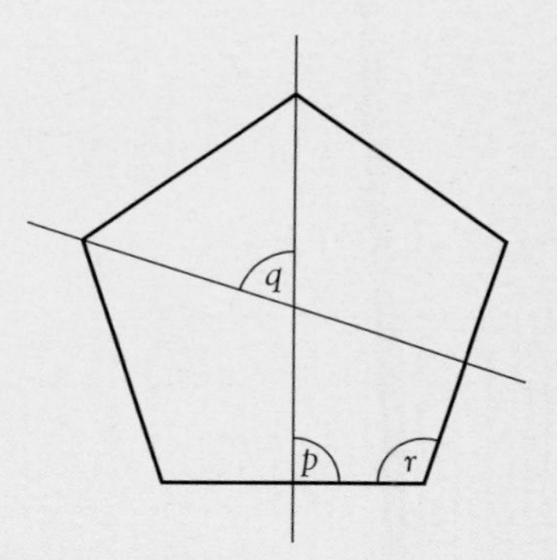

a Write down the value of angle p.

b Calculate the value of
i angle q,.
ii angle r. [NEAB]

5 The diagram shows a regular pentagon $ABCDE$.
BD is parallel to XY. Angle $CBD = 36°$.

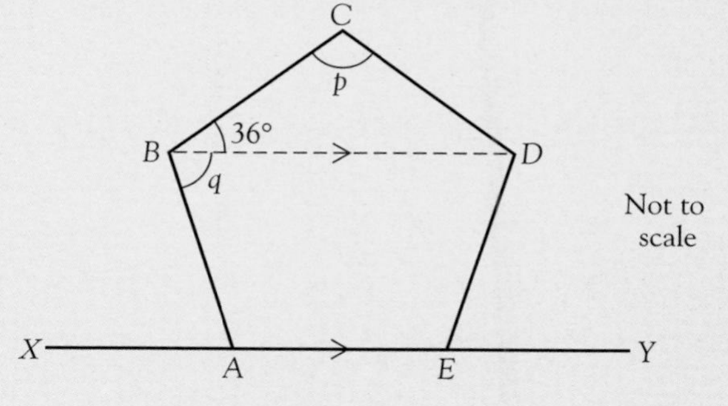

Work out the sizes of angles p and q. [SEG, p]

EXAM QUESTIONS

6 The diagram shows three identical rhombuses, P, Q and T.

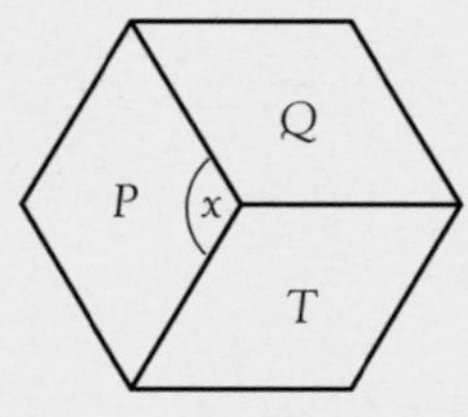

a Explain why angle x is 120°.

b Rhombus Q can be rotated onto rhombus T.

i Copy the diagram and mark a centre of rotation on it.

ii State the angle of rotation.

c Write down the order of rotational symmetry of

i a rhombus

ii a regular hexagon

d The given shape could also represent a three-dimensional shape. What is this shape?

[NEAB]

7 a When the sides of a regular polygon are produced in order, what is the sum of all the exterior angles in degrees?

b Find, for a regular hexagon

i the size of each exterior angle

ii the size of each interior angle.

c Three regular polygons A, B, C fit together at point 0.

A is a square and B is a regular hexagon.

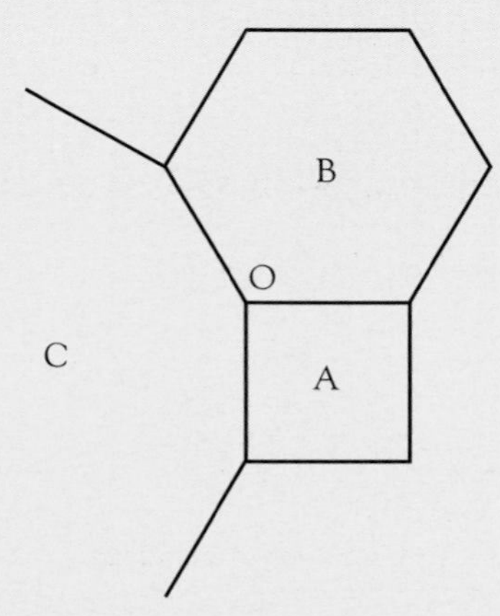

How many sides has C? [NI]

8

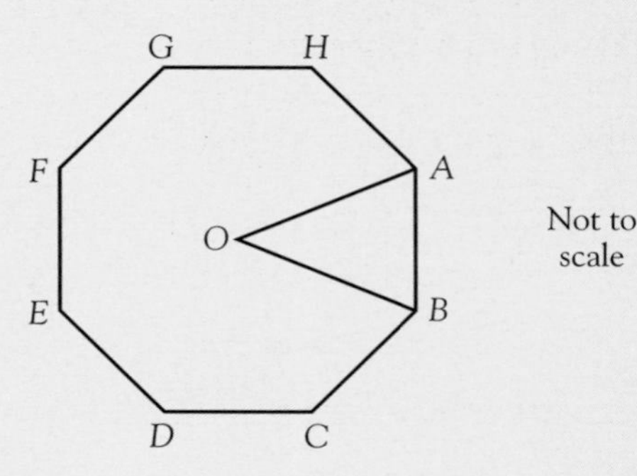

a $ABCDEFGH$ is a regular octagon with centre O.

i Calculate angle AOB. Show your working clearly.

ii Calculate angle ABC. Show your working clearly.

b The diagram shows the top of a table which is covered by four identical regular octagonal tiles and sixteen identical triangular tiles.

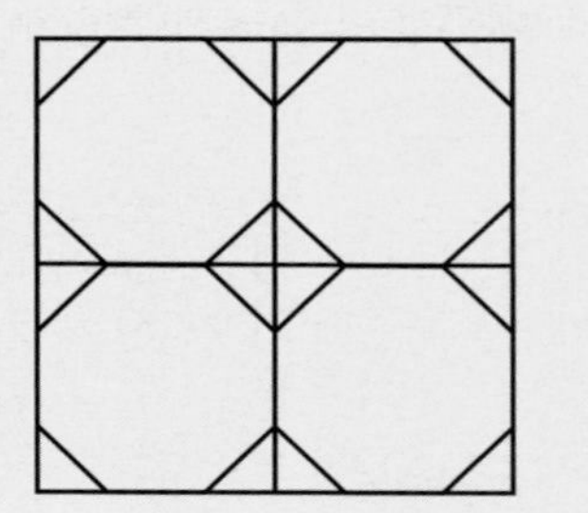

Explain why the quadrilateral formed in the centre of the table top is a square. [MEG]

9 Kensak was investigating regular polygons.

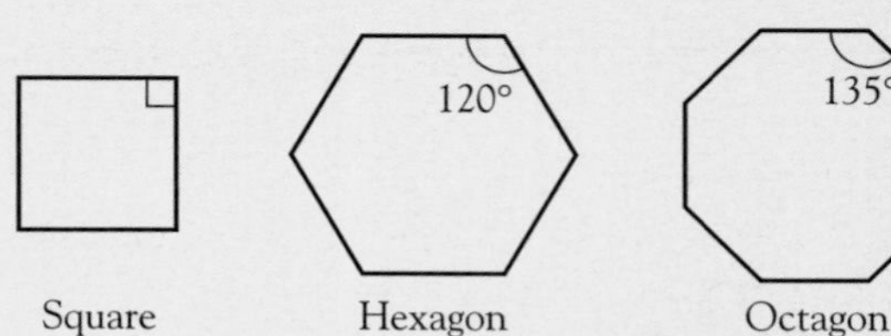

He said 'If a regular figure has an even number of sides then the internal angles are always a whole number of degrees'.

Investigate whether he is correct. [MEG]

REVISION CHECKLIST FOR STAGE 2

10. Number II

✓ Know about different types of numbers: factors, primes, prime factors, multiples, index numbers, squares, square roots, cubes, cube roots, reciprocals

✓ Know how to approximate to a given number of significant figures

$12.496 = 12.5$ (3 s.f.) | $0.004\,216 = 0.0042$ (2 s.f.) | $43\,781 = 43\,800$ (3 s.f.)

✓ Know about equivalent ratios and how to write a ratio in the form $n : 1$ and $1 : n$

$12 : 15 = 4 : 5$ (÷ by 3) | $4 : 9 = 1 : \frac{9}{4} = 1 : 2\frac{1}{4}$ | $2 : 3 = \frac{2}{3} : 1$

✓ Find the missing amount in a ratio, for example find x if $5 : 2 = x : 12$

Multiplier method
$5 : 2$ multiplier
$x : 12$ $= 12 \div 2$
$x = 5 \times 12 \div 2 = 30$

Unitary method
2 parts represent 12
1 part represents 6
5 parts represent $5 \times 6 = 30$

Algebraic method
$\frac{5}{2} = \frac{x}{12}$
(×12) $\frac{5}{2} \times 12 = x$
$x = 30$

✓ Know about direct proportion, enlargements and map scales

straight line through (0, 0)

In an enlargement
scale factor $= \dfrac{\text{image length}}{\text{object length}}$

Map scale 1 : 5000
= 1 cm : 5000 cm
= 1 cm : 50 m

✓ Be able to share in a given ratio

Share £20 in the ratio 1 : 3
1 + 3 = 4, so divide into 4 equal parts, £20 ÷ 4 = £5
1 share = £5, 3 shares = 3 × £5 = £15

11. Algebra II

✓ Be able to expand, or multiply out a bracket

$4y(3x + 5y) = 4y \times 3x + 4y \times 5y$
$= 12yx + 20y^2$

✓ Solve linear equations with brackets and with x terms both sides

	$4(2x - 3) = 3(x + 1)$
Expand brackets	$8x - 12 = 3x + 3$
$+12$	$8x = 3x + 15$
$-3x$	$5x = 15$
$\div 5$	$\underline{x = 3}$

✓ Show linear inequalities on a number line and solve inequalities

$-1 \leqslant x < 3$
−2 −1 0 1 2 3 4

$3x + 4 > 13$
$3x > 9$
$x > 3$

$-6 \leqslant 2x - 4 \leqslant 10$
(+4) $-2 \leqslant 2x \leqslant 14$
(÷2) $1 \leqslant x \leqslant 7$

12. Geometry II

✓ Be able to draw accurately and use three-figure bearings

To find the bearing of A from B
Stand at B,
face North,
turn clockwise to face A

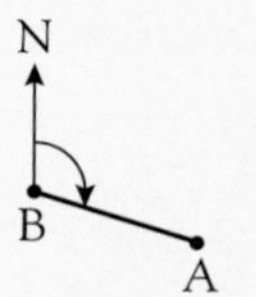

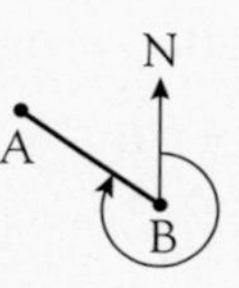

✓ Be able to construct a perpendicular bisector and angle bisector and know and use locus ideas

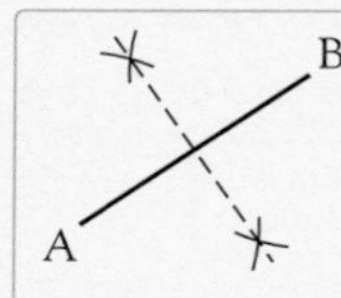

The locus of points equidistant from A and B is the perpendicular bisector of AB.

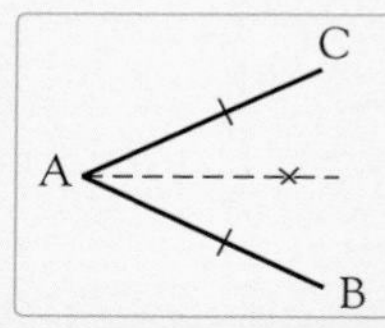

The locus of points equidistant from lines AB and AC is the angle bisector of angle CAB.

REVISION CHECKLIST FOR STAGE 2

13. Data Handling II

✓ Be able to find the mean of raw data and data in a frequency distribution

Raw data $\bar{x} = \frac{\Sigma x}{n}$

Frequency distribution $\bar{x} = \frac{\Sigma fx}{\Sigma f}$

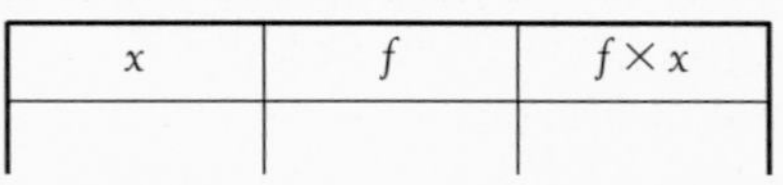

x	f	$f \times x$

✓ Be able to find the mid-interval value of an interval

Add the two boundary points and divide by 2

✓ Use mid-interval values to calculate the mean of grouped data

✓ Be able to draw and interpret frequency polygons (for data with equal class intervals)

Plot frequency against mid-interval value
Join points with a straight line

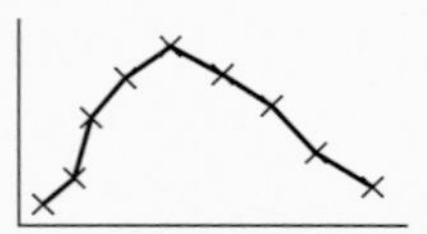

14. Algebra III

✓ Be able to find the highest common factor (H.C.F.) of expressions

H.C.F. of $4x^2y$ and $6xy^2$ is $2xy$

✓ Be able to take out common factors in an expression $4x^2y + 6xy^2 = 2xy(2x + 3y)$

✓ Be able to re-arrange a formula (or change the subject of a formula)

Make x the subject of $y = 3x^2 - 2$

$+2 \qquad y + 2 = 3x^2$

$\div 3 \qquad x^2 = \frac{y+2}{3}$

$\sqrt{\ } \qquad x = \sqrt{\frac{y+2}{3}}$

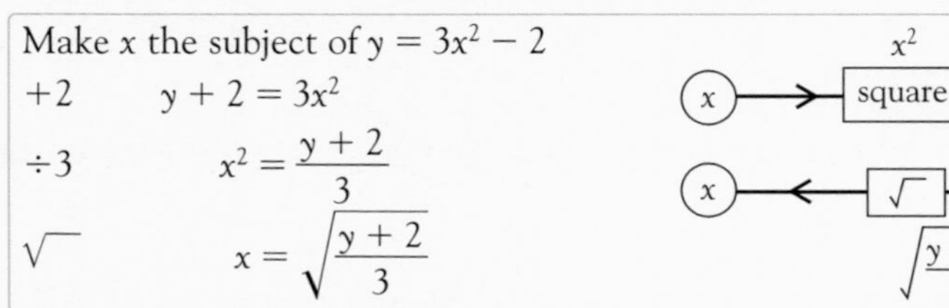

✓ Know when a sequence is linear, continue the sequence and find the nth term

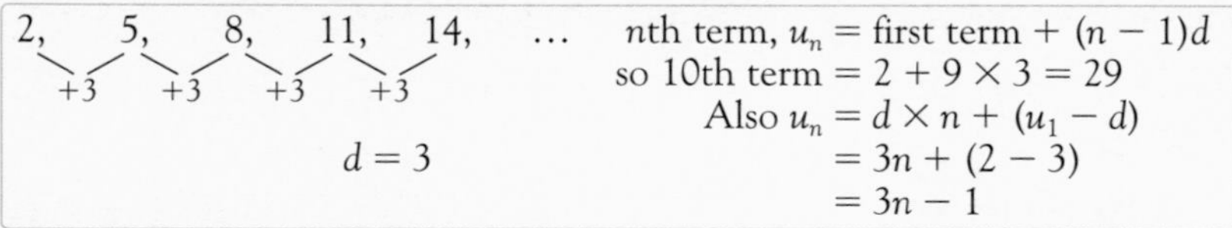

2, 5, 8, 11, 14, ...
+3 +3 +3 +3

$d = 3$

nth term, u_n = first term + $(n - 1)d$
so 10th term $= 2 + 9 \times 3 = 29$
Also $u_n = d \times n + (u_1 - d)$
$= 3n + (2 - 3)$
$= 3n - 1$

15. Area and Volume II

✓ Use the formula for the area of a parallelogram and for the area of a trapezium

Parallelogram

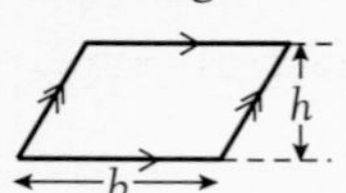

$A = b \times h$

Trapezium

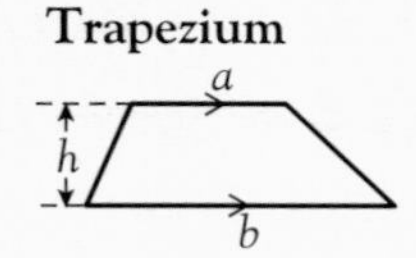

$A = \frac{1}{2}(a + b)h$

✓ Be able to find the volume of a prism

Volume of a prism = area of cross-section × length

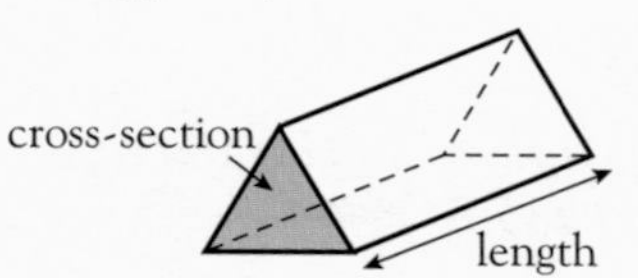

✓ Be able to find the volume of a cylinder

Volume of a cylinder $= \pi r^2 h$

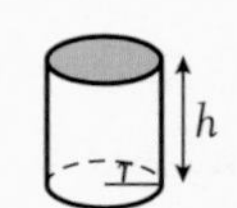

Area of cross-section $= \pi r^2$

✓ Find the surface area of a solid by working out the area of each face

✓ Know that the curved surface area of a cylinder is $2\pi rh$ and that the total surface area is $2\pi rh + 2\pi r^2$

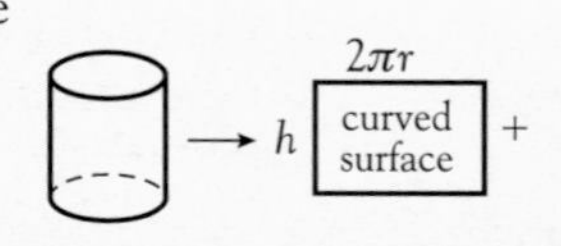

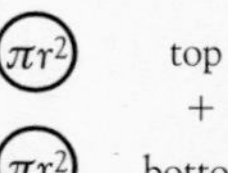

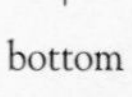

top + bottom

REVISION CHECKLIST FOR STAGE 2

16. Transformations II

✓ Be able to construct the centre of rotation when its position is not obvious by guesswork

- Join two corresponding points, one on the object, one on the image
- Construct the perpendicular bisector of the line joining the points
- Repeat process with two more points
- The centre of rotation is the intersection of the perpendicular bisectors

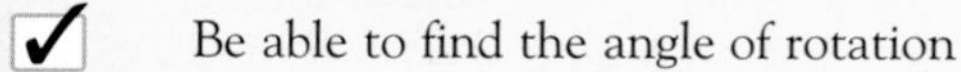

✓ Be able to find the angle of rotation

- Join a point on the object to the centre of rotation, then to the corresponding point on the image
- Measure the angle at the centre of rotation between these lines
- This is the angle of rotation

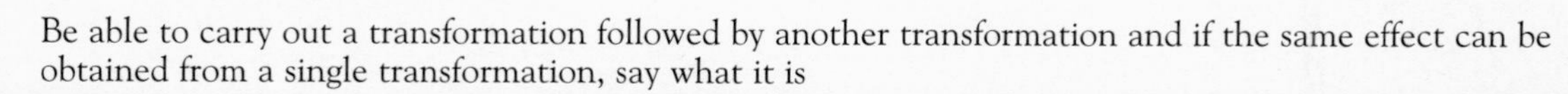

✓ Be able to carry out a transformation followed by another transformation and if the same effect can be obtained from a single transformation, say what it is

17 and 20. Pythagoras' Rule and Trigonometry

✓ Know and be able to use Pythagoras' rule

'In a right-angled triangle, the square on the hypotenuse is equal to the sum of the squares on the other two sides'

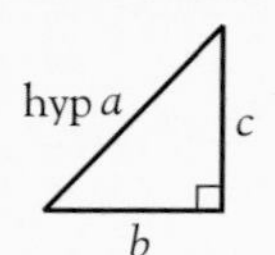

To find the hypotenuse:

$a^2 = b^2 + c^2$

$a = \sqrt{b^2 + c^2}$

To find a shorter side:

$b^2 = a^2 - c^2$

$b = \sqrt{a^2 - c^2}$

✓ Recognise the three trigonometric ratios

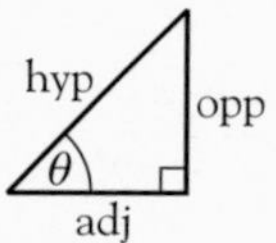

$\sin\theta = \frac{\text{opp}}{\text{hyp}}$ $\quad \cos\theta = \frac{\text{adj}}{\text{hyp}}$ $\quad \tan\theta = \frac{\text{opp}}{\text{adj}}$

✓ Use trig ratios to find missing lengths and angles

To find a missing length

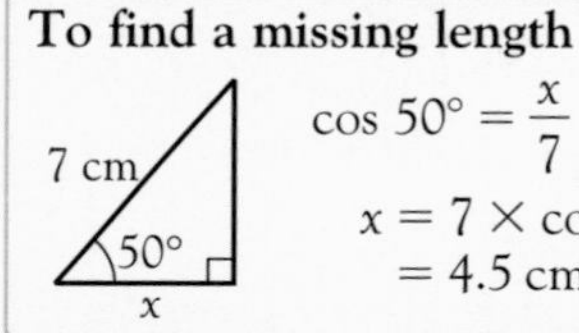

$\cos 50° = \frac{x}{7}$

$x = 7 \times \cos 50°$
$= 4.5$ cm (2 s.f.)

To find a missing angle

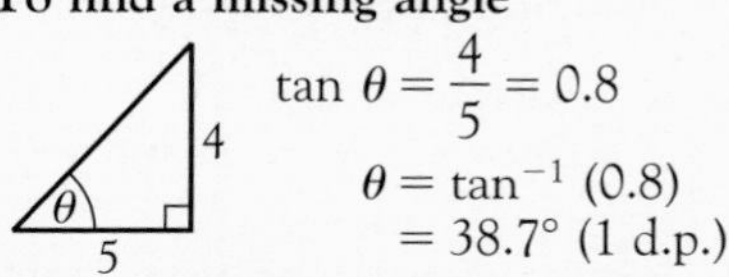

$\tan\theta = \frac{4}{5} = 0.8$

$\theta = \tan^{-1}(0.8)$
$= 38.7°$ (1 d.p.)

✓ Be able to spot right-angled triangles in diagrams:

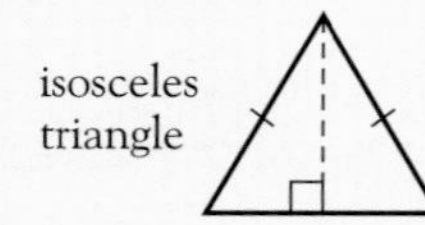

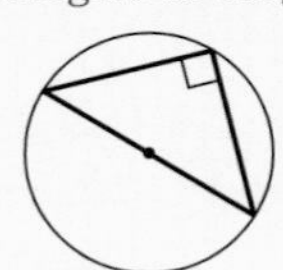

angle in semi-circle

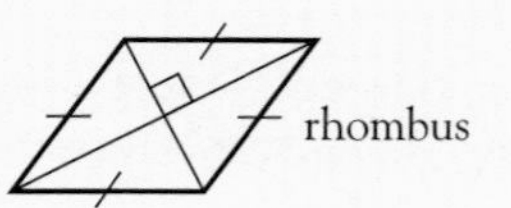

18. Linear Graphs II

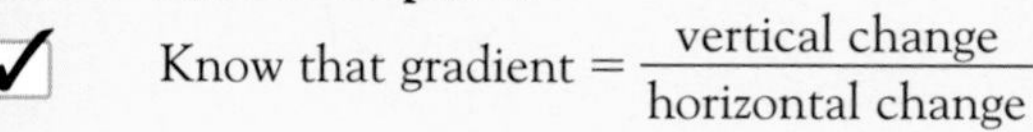

✓ Know that gradient $= \frac{\text{vertical change}}{\text{horizontal change}}$

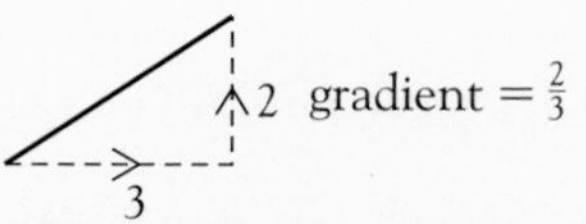

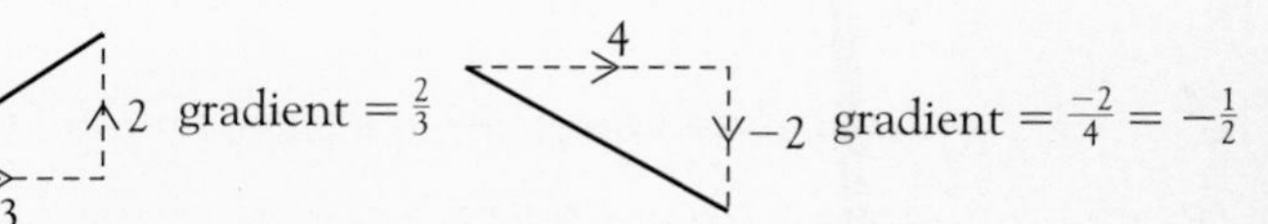

✓ Use the gradient of a graph to find the rate of change of one variable with another, for example in a distance–time graph, the gradient gives the speed

✓ Solve problems involving distance, speed and time

D / S T $\quad D = S \times T \quad S = \frac{D}{T} \quad T = \frac{D}{S}$

✓ Be able to sketch and interpret equations of straight lines written in 'gradient-intercept' form

$y = mx + c$ $\quad$ m is gradient
c is intercept on y axis

✓ Be able to recognise a linear relationship from a table and find its equation

REVISION CHECKLIST FOR STAGE 2

19. Simultaneous Linear Equations

Be able to solve simultaneous linear equations

✔ by drawing lines on a grid and finding where they intersect

Solve $2x + 3y = 12$
$y = 2x - 4$

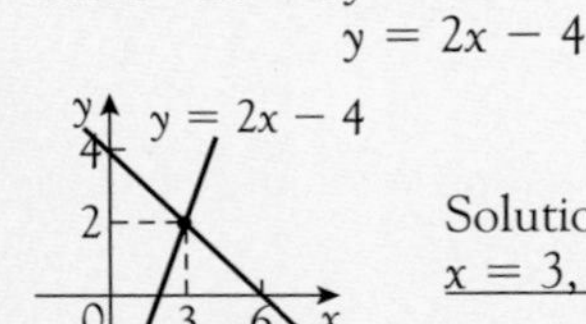

Solution is
$\underline{x = 3,\ y = 2}$

✔ by elimination

Solve $2x + 3y = 12$ ①
$5x - 3y = 2$ ②

① + ② $7x = 14$
$x = 2$

Substitute in ①
$4 + 3y = 12$
$3y = 8$
$y = \frac{8}{3} = 2\frac{2}{3}$

$\underline{x = 2,\ y = 2\frac{2}{3}}$

✔ by substitution

Solve $2x + 3y = 12$ ①
$y = 2x - 4$ ②

Substitute ② into ①
$2x + 3(2x - 4) = 12$
$2x + 6x - 12 = 12$
$8x = 24$
$x = 3$

Substitute into ②
$y = 6 - 4 = 2$

$\underline{x = 3,\ y = 2}$

21. Quadratics 1

✔ Be able to multiply out two brackets

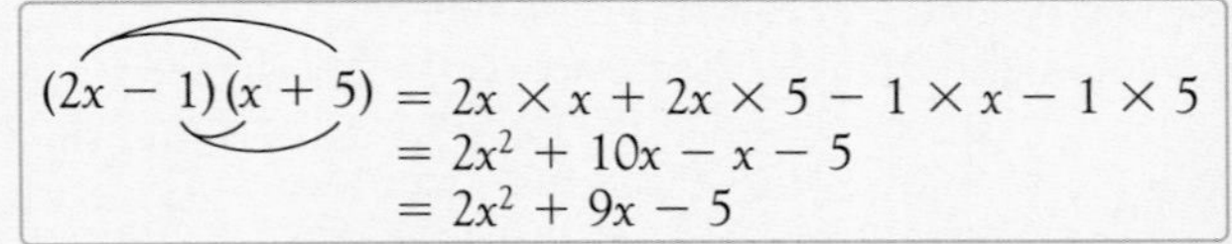
$$(2x - 1)(x + 5) = 2x \times x + 2x \times 5 - 1 \times x - 1 \times 5$$
$$= 2x^2 + 10x - x - 5$$
$$= 2x^2 + 9x - 5$$

✔ Remember special cases

Difference of two squares
$(a - b)(a + b) = a^2 - b^2$

Squaring
$(a + b)^2 = a^2 + 2ab + b^2$
$(a - b)^2 = a^2 - 2ab + b^2$

✔ Be able to draw quadratic graphs of $y = ax^2 + bx + c$

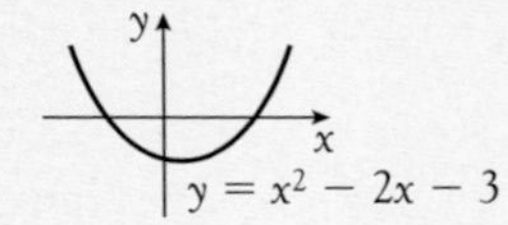

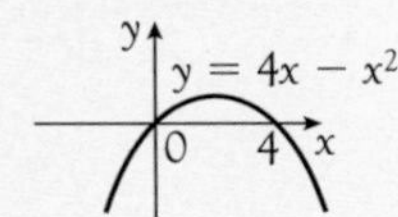

Remember

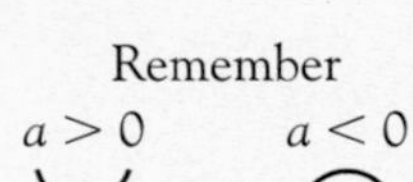

✔ Be able to solve quadratic equations using graphs

Solve $x^2 + 2x = 5$
Draw $y = x^2 + 2x$ and $y = 5$

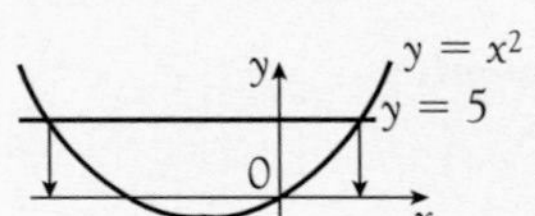

Read off x values where they cross.

✔ Be able to solve quadratic equations by trial and improvement

Solve $x^2 + 3x - 1 = 0$

x	$x^2 + 3x - 1$	outcome
0	−1	too small
1	3	too big etc.

✔ Know when you have done sufficient trials to obtain your answer to a given degree of accuracy

22. Geometry III

✔ Know that the sum of the interior angles of an n-sided polygon is $(n - 2) \times 180°$

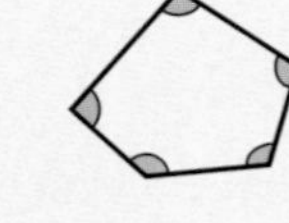

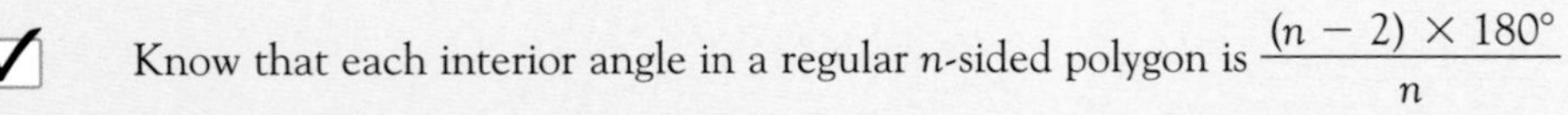

✔ Know that each interior angle in a regular n-sided polygon is $\frac{(n - 2) \times 180°}{n}$

✔ Know that the sum of the exterior angles of any polygon is 360°

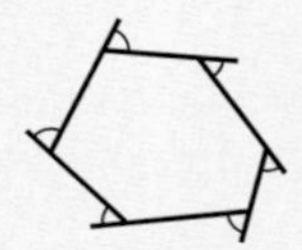

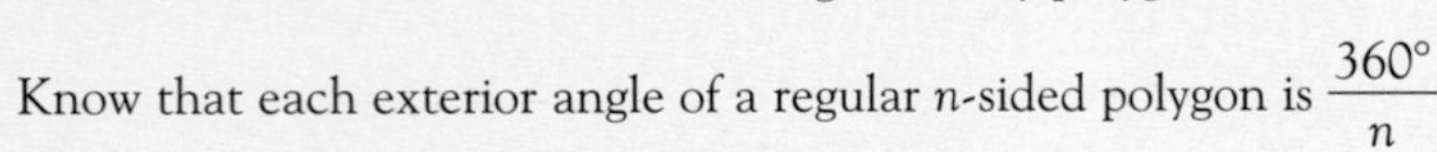

✔ Know that each exterior angle of a regular n-sided polygon is $\frac{360°}{n}$

✔ Know that in a tessellation made of polygons at any vertex the sum of the interior angles is 360°

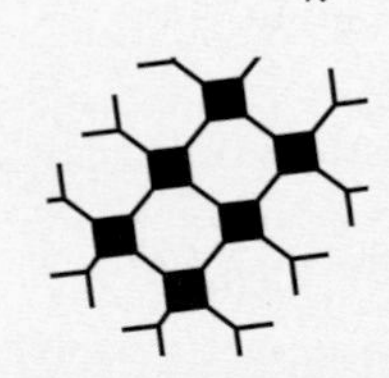

MIXED EXAM QUESTIONS: END OF STAGE 2

1 Three friends start up a stationery shop. Anne invests £11 000, Sangita £7000 and Tony £8000. The shop makes a profit of £12 480 in the first year and they agree to divide this in the ratio of their investments. How much does Tony receive? [SEG]

2 Solve the equation $3(x + 2) = 18 - 5x$ [WJEC]

3 The sketch below shows an equilateral triangle of side 7.5 cm.

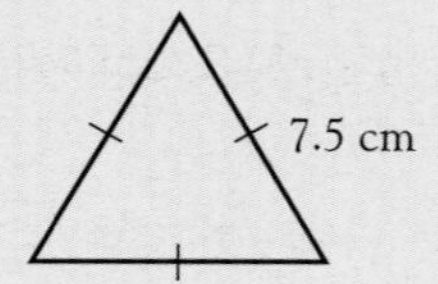

A formula for calculating the area of this triangle is

$A = \frac{\sqrt{3}}{4}s^2$, where s is the length of the side.

a Calculate the area of the triangle using this formula.

b Sinead wants to draw an equilateral triangle that has an area of 100 cm². How long should each side be? [NEAB]

4 A cylindrical ice hockey puck has a radius of 3.8 cm. The height of the puck is 2.5 cm.

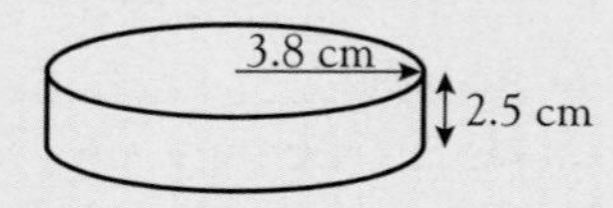

a Find the area of the cross-section of the puck.

b Find the volume of the puck. [MEG]

5 a i Draw the graph of $y = 2x^2$ for values of x from 0 to 3

ii Use your graph to find a value of x when $y = 12$

b Solve the simultaneous equations

$$3x - 2y = 8$$
$$x + 4y = 5$$

[SEG]

6 a Triangle T is rotated through 180° about the point (0, 0). Its image is triangle U.

Draw and label triangle U.

b Triangle U is now rotated through 180° about the point (1, −2). Its image is triangle V.

Draw and label triangle V.

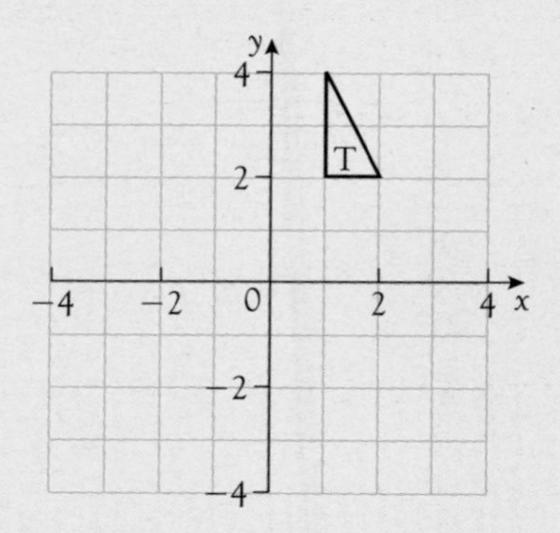

c Describe fully the single transformation which maps triangle T onto triangle V. [MEG]

7 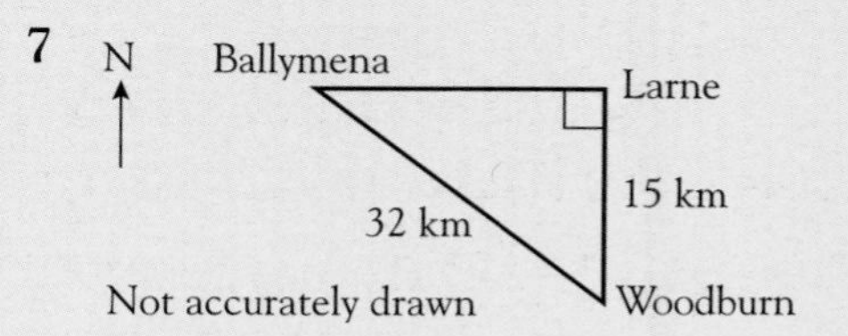

Ballymena is due West of Larne.
Woodburn is 15 km due South of Larne.
Ballymena is 32 km from Woodburn.

a Calculate the distance of Larne from Ballymena. Give your answer in kilometres, correct to 1 decimal place.

b Calculate the bearing of Ballymena from Woodburn. [L]

8 Mr Chan plans to leave £6000 in his will to be divided between his grandchildren, Seng and Wai, in the ratio of their ages.

a At present, Seng is 14 years old and Wai is 16 years old.
How much will Seng receive if Mr Chan dies now?

b How much will Seng receive if Mr Chan dies in 10 years' time? [MEG]

9 Factorise completely $2p^3q^2 - 4p^2q^3$ [L]

10 Bronwen is given a pendant in the shape of a regular polygon. The interior angle between two of the sides is 150°. How many sides has the polygon? [MEG]

11 Class 11A conducted an experiment to see whether boys or girls in their year could hold their breath longer.

MIXED EXAM QUESTIONS: END OF STAGE 2

The information collected is shown below.

Time (T seconds)	Mid-point	BOYS frequency	GIRLS frequency
$10 \leqslant T < 20$	15	0	1
$20 \leqslant T < 30$	25	2	8
$30 \leqslant T < 40$	35	17	16
$40 \leqslant T < 50$	45	21	24
$50 \leqslant T < 60$	55	16	9
$60 \leqslant T < 70$	65	7	3
$70 \leqslant T < 80$	75	4	2
$80 \leqslant T < 90$	85	0	1

a A boy and a girl are picked at random from the year.
What is the probability that

i the boy can hold his breath for at least 60 s,

ii the girl can hold her breath for at least 60 s?

b Calculate an estimate for the mean time for the girls.

c The mean time for the boys is 48.1 s.
Is it fair to say that boys can hold their breath longer than girls?
Give arguments for and against.
Give a total of three arguments. [MEG]

12 **a** Simplify $2x + 3y + 5x - 4y$

b Remove the brackets $2(x - 2y)$

c Remove the brackets giving your answer in the simplest form $(x - 4)^2$

d Simplify $(4x^3)^2$ [NI]

13 L-shapes are pinned on a notice board as shown below.

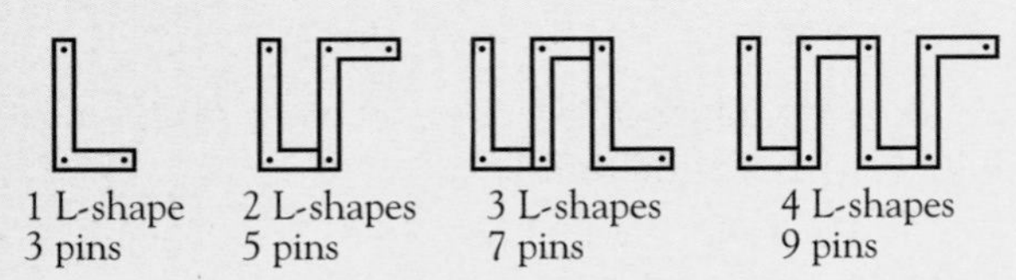

There is a rule to calculate the number of pins needed, if you know the number of L-shapes.

a Without drawing, calculate how many pins are used when there are 30 L-shapes pinned in this way.

b Write down, in words, the rule for finding the number of pins from the number of L-shapes. [WJEC]

14 The diagram shows a side view of two buildings.

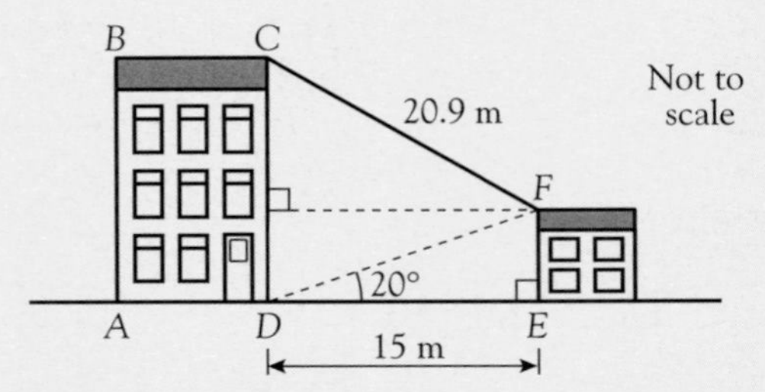

The length $DE = 15$ m. Angle $FDE = 20°$.

a Calculate the height EF.

A telephone wire stretches from C to F.
The length $CF = 20.9$ m.

b Calculate the size of angle CFD. [SEG]

15 **a** On a grid, draw the graph of $y = x^2 - x - 4$
Use values of x between -2 and $+3$

b Use your graph to write down an estimate for

i the minimum value of y

ii the solutions of the equation $x^2 - x - 4 = 0$ [L]

16 Sophie takes part in a sponsored walk each year.
The money she raises is divided between two local charities, A and B, in the ratio 5 : 3.

a In 1992 she raised a total of £48.
How much did she give to charity A?

b In 1993 she gave £21 to charity B.
How much did she raise altogether? [MEG]

17

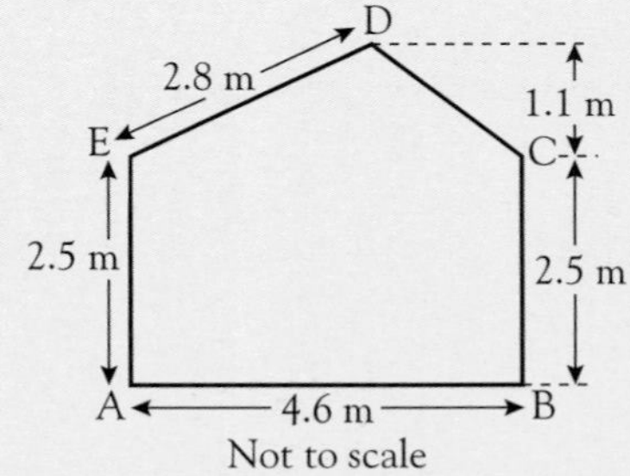

The diagram shows the cross-section ABCDE of a shed which has the shape of a prism. The shed is 7 metres long. AB is horizontal. AE and BC are vertical.

a Calculate the area of the cross-section ABCDE.

b Calculate the volume of the shed.

c Calculate the angle which DE makes with the horizontal. [MEG]

MIXED EXAM QUESTIONS: END OF STAGE 2

18 Graph paper must be used for this question.

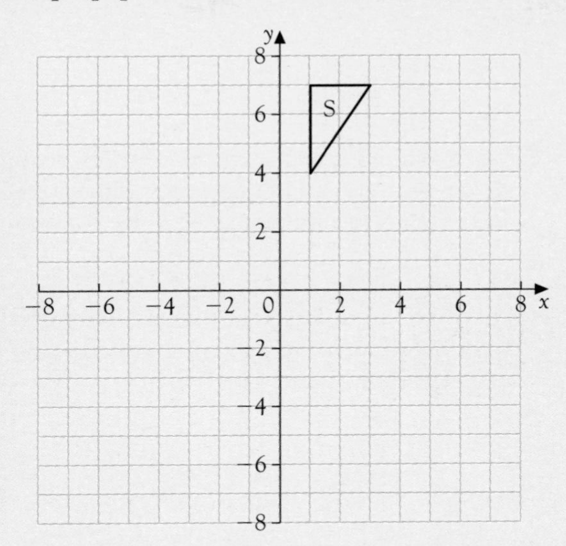

a Copy the diagram above on to graph paper, using a scale of 1 centimetre to represent 1 unit on each axis. The vertices of the triangle S are at the points $(1, 4)$, $(1, 7)$ and $(3, 7)$.

b i Draw the image of triangle S after reflection in the line $x = 4$. Label the image T.

ii Draw the image of the triangle T after reflection in the x axis. Label the image V.

c Draw the image of the triangle S after a rotation of 180° about the origin. Label the image W.

d Describe fully the single transformation which would map triangle W onto triangle V. [MEG]

19 Tomato soup is sold in cylindrical tins.
Each tin has a base radius of 3.5 cm and a height of 12 cm.

a Calculate the volume of soup in a full tin.
Take π to be 3.14 or use the π key on your calculator.

b Mark has a full tin of tomato soup for dinner. He pours the soup into a cylindrical bowl of radius 7 cm.

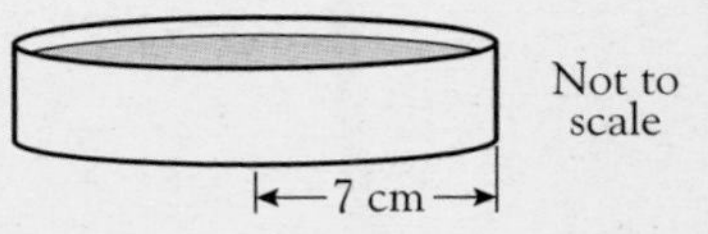

What is the depth of the soup in the bowl? [SEG]

20 The table below shows some corresponding values of x and y, where $y = x^2 - 2x$

x	−1	0	1	2	3
y	3	0	−1	0	3

a Using these values, draw on a grid the graph of $y = x^2 - 2x$ for values of x from −1 to 3.

b i On the grid, draw the line of symmetry of the graph. Label it LS.

ii Write down the equation of the line of symmetry.

c The value of $x^2 - 2x$ when $x = 5$ is 15.
Write down another value of x for which $x^2 - 2x = 15$ [MEG]

21

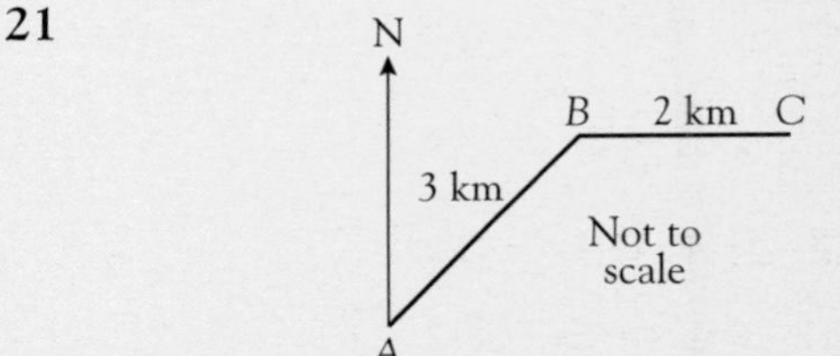

The diagram shows the positions of three small islands A, B, C.
B is 3 km North East of A. C is 2 km due East of B.

a i Using a scale of 2 cm to represent 1 km, draw the triangle ABC accurately.

ii Use your diagram to find the distance and bearing of island C from island A.

b A yacht Y sails so that it is always 1.5 km from C.
In your diagram, draw and label the path of Y.

c A speedboat S moves so that it is always the same distance from A as it is from B.
In your diagram, draw and label the path of S.

d i Mark clearly in your diagram the point Y_1 on the path of Y and the point S_1 on the path of S such that Y_1S_1 is the shortest distance between the two paths.

ii Hence find, in km, the shortest distance between the two paths. [MEG]

MIXED EXAM QUESTIONS: END OF STAGE 2

22 **a** Solve the simultaneous equations

$$3y = 2x - 5$$
$$y = x - 4$$

b Use the formula $y = \dfrac{(x - c)}{\sqrt{(m^2 + 6)}}$ to calculate the value of y when $x = 20$, $c = 4.7$ and $m = \frac{1}{2}$

c Solve the inequality $2x - 5 < 8$ [L]

23

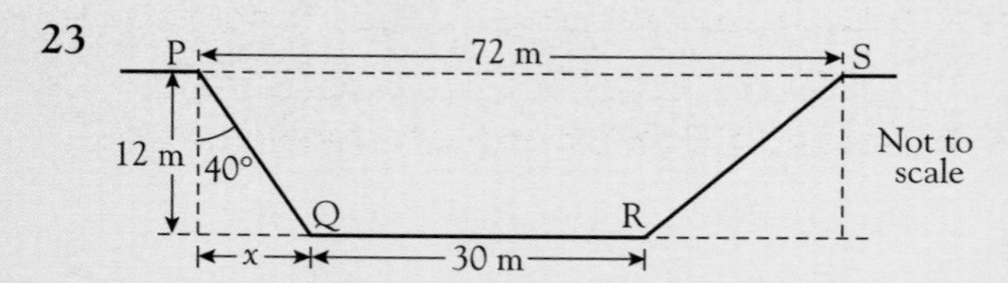

The diagram shows the cross-section $PQRS$ of a motorway cutting. PS and QR are horizontal. $PS = 72$ m, $QR = 30$ m and the vertical height of the cutting is 12 m.

a Calculate the area of the trapezium $PQRS$.

b The bank PQ makes an angle of 40° with the vertical. Calculate, correct to three significant figures, the horizontal distance between P and Q (marked x in the diagram). [MEG]

24 **a** Expand $(2x + 1)(x + 4)$

b Factorise completely $4x^2 - 6x$ [L]

25 George has to find a solution to the equation $x^2 + 2x = 10$, correct to one decimal place.

First he tries $x = 3.0$ and finds that the value of $x^2 + 2x$ is 15.

By trying other values of x find a solution of the equation $x^2 + 2x = 10$, correct to one decimal place. You **must** show all your working. [SEG]

26 The maximum speed, s km/h, at which a train can travel round a circular arc of railway track with a radius of r metres is given by

$$s = k\sqrt{r}$$

When the radius is 100 m, the maximum speed is 35 km/h. What is the maximum speed when the radius is 500 m? Hint: find k first. [MEG]

27 In a road accident the police use a formula of $s = \dfrac{v^2}{200}$ to calculate the length of skid, s metres, for a vehicle travelling at a speed of v kilometres per hour.

a A van skids while travelling at 120 km per hour. Find the length of the skid.

b Two cars, travelling towards each other, start to skid at the same instant. One is travelling at 40 km per hour, the other at 80 km per hour. Find the minimum distance they must be apart if they are not to collide.

c Re-arrange the above formula to make v the subject.

d Using your new formula, or otherwise, find the speed of a vehicle which skidded for 40 metres. [MEG]

23 NUMBER III

In this chapter you will learn how to
- **calculate using index numbers and with numbers written in standard form**
- **calculate percentage profit, percentage loss and percentage error**
- **find amounts after repeated percentage change**
- **find the original quantity when you are given the result of a percentage change**
- **multiply and divide fractions**

Index numbers (or indices)

FACT SHEET 6

Consider the number 3^4.
3 is known as the **base** and 4 is the **index** or **power**.
You would say '3 to the power of 4' and to calculate it you would multiply 3 by itself 4 times, so

$$3^4 = 3 \times 3 \times 3 \times 3 = 81$$

On your calculator, key in 3 x^y 4 =

Note that the plural of index is indices, pronounced 'indeeseas'.

When numbers are written in index form with the same base there are quick ways that can be used when calculating.

a $3^2 \times 3^4 = (3 \times 3) \times (3 \times 3 \times 3 \times 3)$
$= 3 \times 3 \times 3 \times 3 \times 3 \times 3$
$= 3^6$

Look at the indices
$2 + 4 = 6$

You add the indices, so $3^2 \times 3^4 = 3^{2+4} = 3^6$

b $2^7 \div 2^3 = \dfrac{\overset{1}{\cancel{2}} \times \overset{1}{\cancel{2}} \times \overset{1}{\cancel{2}} \times 2 \times 2 \times 2 \times 2}{\underset{1}{\cancel{2}} \times \underset{1}{\cancel{2}} \times \underset{1}{\cancel{2}}}$

Cancel down

$= \dfrac{2 \times 2 \times 2 \times 2}{1}$

$= 2^4$

Look at the indices
$7 - 3 = 4$

You subtract the indices, so $2^7 \div 2^3 = 2^{7-3} = 2^4$

If the letter a is used for the base, and n and m for the indices, you can write a general rule:

When numbers are written in index form, with the same base,
- to multiply the numbers, add the indices
  $a^m \times a^n = a^{m+n}$
- to divide the numbers, subtract the indices
  $a^m \div a^n = a^{m-n}$

INDEX NUMBERS

Look carefully at the following examples.

a When the index number is zero

If you work out $2^3 \div 2^3$ according to the index rule for division, you get $2^3 \div 2^3 = 2^{3-3} = 2^0$

But you know that $2^3 = 8$ and $8 \div 8 = 1$

Try it on your calculator. [2] [x^y] [0] [=]

This means that $2^0 = 1$

This holds for **any** base, not just for a base of 2, so you can write a general rule:

$$a^0 = 1$$

b When the index number is 1

$2^4 \div 2^3 = 2^{4-3} = 2^1$, but $2^4 \div 2^3 = \dfrac{\cancel{2}^1 \times \cancel{2}^1 \times \cancel{2}^1 \times 2}{\cancel{2}_1 \times \cancel{2}_1 \times \cancel{2}_1} = 2$, so $2^1 = 2$

You can write a general rule

$$a^1 = a$$

c When the index number is negative

$2^4 \div 2^5 = 2^{4-5} = 2^{-1}$, but $2^4 \div 2^5 = \dfrac{\cancel{2}^1 \times \cancel{2}^1 \times \cancel{2}^1 \times \cancel{2}^1}{\cancel{2}_1 \times \cancel{2}_1 \times \cancel{2}_1 \times \cancel{2}_1 \times 2} = \dfrac{1}{2}$ so $2^{-1} = \dfrac{1}{2}$

On calculator [2] [x^y] [1] [+/-] [=] gives 0.5

The calculator gives the number in decimal format.

Now consider

$5^4 \div 5^6 = 5^{4-6} = 5^{-2}$, but $5^4 \div 5^6 = \dfrac{\cancel{5}^1 \times \cancel{5}^1 \times \cancel{5}^1 \times \cancel{5}^1}{\cancel{5}_1 \times \cancel{5}_1 \times \cancel{5}_1 \times \cancel{5}_1 \times 5 \times 5} = \dfrac{1}{5^2}$ so $5^{-2} = \dfrac{1}{5^2}$

On your calculator, check that [5] [x^y] [2] [+/-] [=] gives the same number as [1] [÷] [5] [x^2] [=]

You can write a general rule

$$a^{-n} = \frac{1}{a^n}$$

DO NOT think that a negative index means that the ordinary format is a negative number. It means $\frac{1}{\square}$.

INDEX NUMBERS

Sample Question 1

Simplify, leaving your answer in index form,

a $2^3 \times 3^4 \times 3 \times 2^2 \times 2^5$ b $\dfrac{5^3 \times 2^5}{2^3 \times 5^4}$ c $(5^3)^2$

Answer

a

- Combine numbers with the same base using the index rule for multiplication

$$2^3 \times 3^4 \times 3 \times 2^2 \times 2^5 = (2^3 \times 2^2 \times 2^5) \times (3^4 \times 3^1)$$
$$= 2^{3+2+5} \times 3^{4+1}$$

$3 = 3^1$

$$= \underline{2^{10} \times 3^5}$$

You cannot combine these any further as they have different bases.

b

- Write in index form and simplify the terms with the same base

$$\frac{5^3 \times 2^5}{2^3 \times 5^4} = 5^3 \times 2^5 \div 2^3 \div 5^4$$
$$= 5^3 \div 5^4 \times 2^5 \div 2^3$$

You are dividing by 2^3 and by 5^4.

$$= 5^{3-4} \times 2^{5-3}$$
$$= \underline{5^{-1} \times 2^2}$$

c

$$(5^3)^2 = 5^3 \times 5^3$$
$$= 5^{3+3}$$
$$\underline{(5^3)^2 = 5^6}$$

In general

$$(a^m)^n = a^{m \times n}$$

Sample Question 2

Simplify $\dfrac{3x^2 \times 4x^6}{6x^3}$

Answer

- Simplify the number terms and the x terms separately.

$$\frac{3x^2 \times 4x^6}{6x^3} = \frac{\overset{2}{\cancel{12}} \times x^2 \times x^6}{\cancel{6} \times x^3}$$

Cancel down the numbers.

$$= \frac{2x^8}{x^3}$$

$x^2 \times x^6 = x^8$

$$= 2 \times x^8 \div x^3$$
$$= \underline{2x^5}$$

$x^8 \div x^3 = x^5$

INDEX NUMBERS

Exercise 23.1

1 Find the values of

a 3^4
b 4^5
c 6^3
d 2^{10}
e 9^4
f 7^5
g 1360^1
h 17.63^0
i 1.5^3
j 0.5^4
k 18^5
l 100^6
m $4^3 \times 2^3$
n $2^7 \div 4^2$
o $(-3)^3 \times 2^2$
p $3^3 \times (-2)^5$
q $-3^2 \times 7^4$
r $-4^2 \times -2^3$

2 Write each of the following as a fraction

a 2^{-1}
b 3^{-2}
c 4^{-3}
d 10^{-5}
e 5^{-4}
f 10^{-1}

3 Write each of the following in a form involving a negative index

a $\frac{1}{5}$
b $\frac{1}{6^3}$
c $\frac{1}{100}$
d $\frac{1}{1000}$
e $\frac{1}{t^2}$
f $\frac{2}{x^3}$
g $\frac{1}{3y^4}$
h $\frac{3}{4t^{10}}$

4 Write each expression without a negative index

a x^{-3}
b t^{-4}
c $3y^{-2}$
d $4x^{-5}$
e $2a^{-1}$
f $\frac{1}{2}x^{-4}$
g $\frac{1}{4}m^{-3}$
h $\frac{3}{4}t^{-4}$

5 Find the value of

a $(-1)^0$
b $(-1)^1$
c $(-1)^2$
d $(-1)^3$
e $(-1)^4$
f $(-1)^5$
g $(-1)^6$

What do you notice about your answers?
Use this knowledge to write down the values of

h $(-1)^{76}$
i $(-1)^{89}$
j $(-1)^{100}$
k $(-1)^{101}$

6 a What power of 2 is 32?
b What power of 3 is 27?
c What power of 4 is 64?

7 Find the values of the following expressions when $x = 3$, $y = 2$ and $z = -5$.

a x^{-1}
b z^2
c y^{-2}
d $\frac{x^2y^3}{z}$
e $\frac{x^3 + z^2}{y^2}$
f $\frac{x^4 - 2z^2 - y}{y^2 + z^2}$

8 Work out 7^3. If I work out 3^7 by mistake, what would be the difference in the answer?

9 Simplify

a $5^4 \times 5^6$
b $3^6 \div 3^4$
c $2^2 \times 2^{-3}$
d $7^{-3} \div 7^4$
e $7^3 \div 7^4$
f $6^3 \div 6^{-5}$
g $\frac{10^4 \times 10^7}{10^5}$
h $\frac{(2^2)^3 \times 2^5}{2^7}$
i $\frac{x^4 \times x^2}{x^6}$
j $\frac{y^6}{y^2 \times y^{-3}}$
k $\frac{3a^2 \times 6a}{2a^5}$
l $\frac{12t^5}{4t^2 \times 3t^3}$

10 Simplify leaving your answer in index form

a $\frac{3^3 \times 2^7 \times 3^2 \times 2^6}{2^4 \times 3}$
b $\frac{(5^2)^3 \times 3^5}{3^3 \times 5^7}$
c $6a^4b^{-3}c \times 5a^2b^{-1}c^3$
d $3x^2y^3 \times 4x^{-3}y^{-4}$
e $\frac{-4ab^2}{8b}$
f $(-x^3) \div (-x^2)$
g $\frac{18a^2b}{3ab^3}$
h $\frac{-xyz}{zyx}$
i $\frac{2s^2 \times 4t^3}{3st \times 8st^2}$

STANDARD FORM

Standard Form

All numbers can be written in a special format, called **standard form** (sometimes known as standard index form or scientific notation).

You have probably seen this already when your calculator has converted an answer to standard form possibly because it was a very large number, such as 27 000 000 000 000 or a number close to zero, such as 0.000 000 0014

Depending on your calculator model, the display may have shown

2.7×10^{13}	or	2.7^{13}	or	2.7 13
1.4×10^{-09}		1.4^{-09}		1.4 −09

This first display is the closest to being correct, the others are written in 'calculator shorthand'.

In standard form, the two numbers are as follows:

$$27\,000\,000\,000\,000 = 2.7 \times 10^{13}$$

$$0.000\,000\,0014 = 1.4 \times 10^{-9}$$

Each has been written as a number between 1 and 10 multiplied by a power of 10.

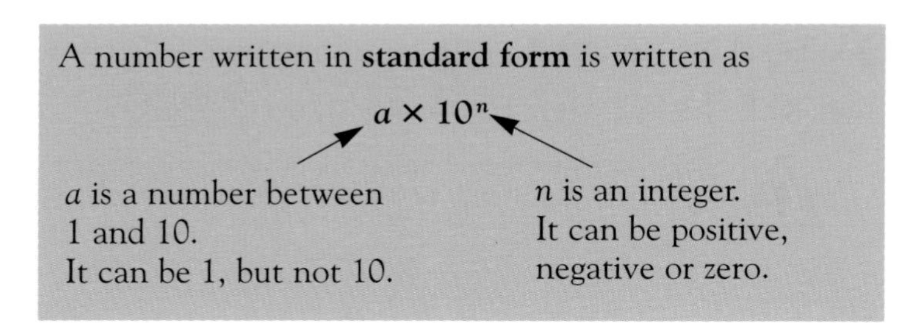

Converting standard form numbers to ordinary numbers

When n is positive or zero

a $6.9 \times 10^2 = 6.9 \times 100 = 690$

Move decimal point 2 places to right.

b $4 \times 10^5 = 4 \times 100\,000 = 400\,000$

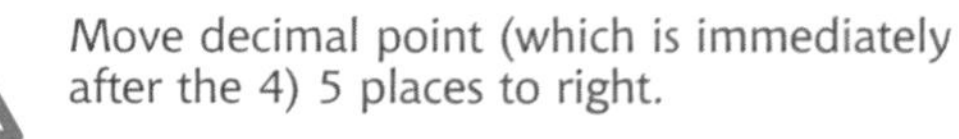

c $3.92 \times 10^1 = 3.92 \times 10 = 39.2$

Move decimal point 1 place to right.

d $8.5 \times 10^0 = 8.5 \times 1 = 8.5$

$10^0 = 1$
Do not move the decimal point at all.

STANDARD FORM

When n is negative

e $7.32 \times 10^{-1} = 7.32 \times \frac{1}{10}$
$= 7.32 \div 10$
$= 0.732$

$10^{-1} = \frac{1}{10}$
To find $\frac{1}{10}$ of 7.32, divide by 10.
Move the decimal point 1 place to the left.

f $3.1 \times 10^{-2} = 3.1 \times \frac{1}{100}$
$= 3.1 \div 100$
$= 0.031$

$10^{-2} = \frac{1}{10^2} = \frac{1}{100}$
To find $\frac{1}{100}$ of 3.1, divide by 100.
Move the decimal point 2 places to the left.

g $5.93 \times 10^{-5} = 0.000\,059\,3$

Following the pattern, move the decimal point 5 places to the left. You need to write in the zeros.

Notice that when written in standard form
- numbers that are 10 or above have a positive power of 10
- numbers from 1 but below 10 have a zero power of 10
- numbers between 0 and 1 have a negative power of 10.

Using the calculator

You can key in numbers written in standard form using the [EXP] button. On some types of calculator it will be labelled [EE].

Keying in [a] [EXP] [b], where a and b are numbers, has the effect of entering $a \times 10^b$

a To enter 5.9×10^3
- key in [5] [·] [9] [EXP] [3]. The display shows the calculator shorthand.
- show this as an ordinary number by pressing [=]. The display shows [5900].

Take note – there will be occasions when the display still shows the calculator shorthand, or it loses some of the end digits. Watch out!

b To enter 2.35×10^{-2}
- key in [2] [·] [3] [5] [EXP] [2] [+/−]
- press [=]. The display shows [0.0235]

c To enter 2.35×10^{-4}
- key in [2] [·] [3] [5] [EXP] [4] [+/−]
- press [=]. Most calculators show [0.000 235], but older models may show the calculator shorthand. Check yours.

You must only use [EXP] when numbers are in standard form. For example, to enter the number 10^3, you must write it as 1×10^3 first and key in [1] [EXP] [3].

IMPORTANT CALCULATOR NOTE

Numbers written in standard form can also be entered using [x^y] as follows:

a 5.9×10^3 [5] [·] [9] [×] [1] [0] [x^y] [3] [=]

b 2.35×10^{-2} [2] [·] [3] [5] [×] [1] [0] [x^y] [2] [+/−] [=]

DO NOT confuse the two methods. Decide which you want to use and stick to it.

STANDARD FORM

Converting ordinary numbers to standard form

a $4852.6 = a \times 10^n$

n will be positive as 4852.6 is bigger than 10.

To find *a*, move the decimal point until you get a number between 1 and 10; *n* is the number of places you have to move the decimal point to achieve this.

4 8 5 2 . 6 (3 places, so $n = 3$)

$4852.6 = 4.8526 \times 10^3$

DO NOT leave out any digits unless you are asked to approximate the number.

b $416\,000 = a \times 10^n$

n will be positive.

To find *a*, put in the decimal point, then move it until you get a number between 1 and 10.

4 1 6 0 0 0 . (5 places, so $n = 5$)

$416\,000 = 4.16 \times 10^5$

Do not put 4.16000×10^5

c $0.002\,58 = a \times 10^n$

n will be negative as the number is less than 1.

To find *a*, move the decimal point until you get a number between 1 and 10.

0 . 0 0 2 5 8 (3 places, so $n = -3$)

$0.002\,58 = 2.58 \times 10^{-3}$

Calculating when numbers are in standard form

You need to be able to add, subtract, multiply or divide numbers in standard form, **with or without** a calculator!

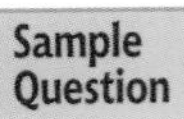

Sample Question 3

Work out, giving your answer in standard form

a $(2.1 \times 10^5) + (3 \times 10^3)$
b $(6.23 \times 10^{-3}) - (2.5 \times 10^{-4})$
c $(3 \times 10^{-2}) \times (4.8 \times 10^5)$
d $(5 \times 10^4) \div (2 \times 10^2)$

Answer (without, then with a calculator)

a

- Change each number to ordinary form, then add.

$$2.1 \times 10^5 = 210\,000$$
$$3 \times 10^3 = 3000 \; +$$
$$(2.1 \times 10^5) + (3 \times 10^3) = 213\,000$$
$$(2.1 \times 10^5) + (3 \times 10^3) = \mathbf{2.13 \times 10^5}$$

Change back to standard form.

On calculator:

- Key in
  [2] [.] [1] [EXP] [5] [+]
  [3] [EXP] [3] [=] 213 000
- Convert to standard form

STANDARD FORM

b

- Change to ordinary numbers, then subtract.

$$6.23 \times 10^{-3} = 0.006\,23$$
$$2.5 \times 10^{-4} = 0.000\,25 \; -$$
$$(6.23 \times 10^{-3}) - (2.5 \times 10^{-4}) = 0.005\,98$$
$$\underline{(6.23 \times 10^{-3}) - (2.5 \times 10^{-4}) = \mathbf{5.98 \times 10^{-3}}}$$

On calculator:

- Key in
  6 · 2 3 EXP 3 +/– –
  2 · 5 EXP 4 +/– = 0.00598
- Convert to standard form

Now change back to standard form.

If your calculator display gives the answer in calculator shorthand for standard form such as 5.98⁻⁰³ do not write 5.98^{-3}. Write it properly as 5.98×10^{-3}.

Note that, when adding or subtracting using the x^y method on the calculator, you do not need to key in the brackets.

c

- Multiply the numbers and multiply the powers of 10.

$$(3 \times 10^{-2}) \times (4.8 \times 10^5) = 3 \times 4.8 \times 10^{-2} \times 10^5$$
$$= 14.4 \times 10^3$$
$$= 1.44 \times 10^1 \times 10^3$$
$$\underline{(3 \times 10^{-2}) \times (4.8 \times 10^5) = 1.44 \times 10^4}$$

On calculator:

- Key in
  3 EXP 2 +/– ×
  4 · 8 EXP 5 = 14400
- Convert to standard form

Note that, if you use x^y method when multiplying, you do not need to key in brackets around 3×10^{-2} or 4.8×10^5.

d

- Deal with the numbers and the powers of 10 separately.

$$(5 \times 10^4) \div (2 \times 10^2) = \frac{5 \times 10^4}{2 \times 10^2}$$
$$= \frac{5}{2} \times \frac{10^4}{10^2}$$
$$= 5 \div 2 \times 10^4 \div 10^2$$
$$\underline{(5 \times 10^4) \div (2 \times 10^2) = 2.5 \times 10^2}$$

On calculator:

- Key in
  5 EXP 4 ÷ 2 EXP 2 = 250
- Convert to standard form

Note that if you use x^y method when dividing, you **must** put a bracket around (2×10^2) when keying in on the calculator. Can you figure out why?

STANDARD FORM

Sample Question 4

The population of Brazil is 1.15×10^8. The population of Peru is 1.71×10^7.

a How many more people live in Brazil than in Peru? Give your answer in standard form.

The population density of a country is calculated as

$$\text{Population density} = \frac{\text{Population}}{\text{Area}}$$

The area of Brazil is 3.29×10^6 square miles.

b Calculate the population density of Brazil.

The population density of Peru is 34.5.

c Calculate the area of Peru. Give your answer in standard form. [MEG]

Answer

a

- Find

Population of Brazil − Population of Peru

$$1.15 \times 10^8 - 1.71 \times 10^7 = 97\,900\,000$$
$$= \underline{9.79 \times 10^7}$$

Remember to give the answer in standard form.

On calculator:

- Key in
  1 · 1 5 EXP 8 −
  1 · 7 1 EXP 7 =
- Convert to standard form

b

- Substitute the values into the formula.

$$\text{Population density} = \frac{1.15 \times 10^8}{3.29 \times 10^6}$$
$$= 34.954\ldots$$

On calculator:

- Key in
  1 · 1 5 EXP 8 ÷
  3 · 2 9 EXP 6 =

- Approximate this sensibly

<u>Population density = 35</u>

Note that the units are people/square mile.

c

- Substitute into the formula writing A for area.

$$34.5 = \frac{1.71 \times 10^7}{A}$$

- Re-arrange the formula.

$$34.5 \times A = 1.71 \times 10^7 \quad \text{(multiply by A)}$$
$$A = \frac{1.71 \times 10^7}{34.5} \quad \text{(divide by 34.5)}$$
$$= 495\,652.17$$

On calculator:

- Key in
  1 · 7 1 EXP 7 ÷
  3 4 · 5 =
- Convert to standard form

- Approximate sensibly and write the answer in standard form.

<u>Area = 4.96×10^5 square miles (3 s.f.)</u>

STANDARD FORM

Exercise 23.2

1 Write as an ordinary decimal number

a 3.6×10^5
b 4.75×10^4
c 5.4×10^{-1}
d 5.84×10^7
e 1.4×10^{-4}
f 7.236×10^2
g 9.8×10^{-8}
h 2×10^6
i 8.25×10^{-5}
j 9.63×10^2
k 4.7×10^{-2}
l 6.6×10^{-6}

2 The time between collisions of oxygen molecules at 0 °C is about 2×10^{-10} seconds. Write this as an ordinary number.

3 The mass of the sun is 1.99×10^{30} kg. Write this mass in tonnes, in standard form.
(1 tonne = 1000 kg)

4 Write in standard index form

a 346
b 7260
c 34
d 0.567
e 1 000 000
f 0.017
g 965 700
h 1.6
i 0.0007
j 0.2
k 0.0064
l 144.78

5 One gram is 0.035 27 ounces. Write this in standard form.

6 The sun has a diameter of 1 392 000 km. Write its diameter in standard form.

7 For each pair of numbers, state which is the larger.

a 2.079×10^{10} or 6.99×10^8
b 4.75×10^{-6} or 4.75×10^5
c 8.36×10^{-7} or 5.965×10^{-6}

8 Calculate each of the following leaving your answer in standard form (correct to 3 significant figures if it cannot be given exactly).

a $(3.56 \times 10^{10}) + (7.49 \times 10^9)$
b $(7.63 \times 10^{-8}) + (7.86 \times 10^{-10})$
c $(4.86 \times 10^9) - (3.65 \times 10^8)$
d $(2.43 \times 10^{-12}) - (4.92 \times 10^{-10})$
e $(4.2 \times 10^7) \times (3.4 \times 10^9)$
f $(1.7 \times 10^{11}) \times (1.9 \times 10^{-3})$
g $(6.7 \times 10^{-4}) \times (2.66 \times 10^{-3})$
h $(4.9 \times 10^6) \div (3.5 \times 10^{10})$
i $(6.5 \times 10^{-6}) \div (7.3 \times 10^{-4})$
j $(2.6 \times 10^5) \div (6.5 \times 10^{-8})$
k $(9.3 \times 10^{-2}) \times (8.5 \times 10^{-3})$
l $(5.3 \times 10^{-23}) \times (9.45 \times 10^{20})$

9 The mass of the sun is 1.99×10^{30} kg and the mass of the earth is 5.976×10^{24} kg. Write down the ratio of the mass of the earth to the mass of the sun. Find this ratio in the form $1 : n$.

10 A light-year is the distance travelled by light in one year. If the speed of light is $2.997\,925 \times 10^8$ metres per second, how far is one light-year in metres?

11 The table shows populations in the UK for the year 1995.

United Kingdom (total)	$5.860\,58 \times 10^7$
England	$4.682\,08 \times 10^7$
Wales	2.8135×10^6
Scotland	5.1802×10^6
Northern Ireland	1.5377×10^6

a How many more people were living in England than in Wales?

b Find the ratio of the population of Northern Ireland to that of the UK as a whole in the form $1 : n$.

12 The masses of an electron, a neutron and a proton are 9.1083×10^{-28}, 1.6747×10^{-24} and $1.672\,39 \times 10^{-24}$ grams, respectively. List the masses in decreasing order of mass.

13 The expenditure on defence personnel in the United Kingdom fell from 10 504 (£ million) in 1992/93 to an estimated 8279 (£ million) for 1997/98.

a Express these two amounts in standard form.

b What was the fall in expenditure?

c What was the percentage decrease in expenditure?

14 The distance of a particular star from the earth is 8.547×10^{15} km. If light travels at approximately 2.98×10^5 km/s, find how long light from the star will take to reach the earth. Give your answer to the nearest year.

Percentages

Percentage profit and loss

If you buy something, then sell it at a higher price, you make a profit. If you sell it at a lower price, you make a loss. You can calculate the percentage profit or loss:

$$\text{Percentage profit} = \frac{\text{Amount of profit}}{\text{Original price}} \times 100\%$$

$$\text{Percentage loss} = \frac{\text{Amount of loss}}{\text{Original price}} \times 100\%$$

Notice that the original price is always on the bottom of the fraction.

Sample Question 5

The school mini-enterprise company bought 40 toy rabbits at £5 each.

a They sold 30 at £7 each. Find the percentage profit on each of these rabbits.

b Unfortunately they were unable to sell the remaining 10, so reduced the price to £3.50 each. Find the percentage loss on these.

c What was the overall percentage profit or loss?

Answer

a

- Find the profit made on each rabbit when sold for £7.

  Profit = £7 − £5 = £2

- Express this as a percentage of the amount the company paid for a rabbit.

$$\text{Percentage profit} = \frac{\text{Profit}}{\text{Original price}} \times 100\%$$

$$= \frac{2}{5} \times 100\%$$

<u>Percentage profit = 40%</u>

b

- Find the loss on each rabbit when sold for £3.50.

  Loss = £5 − £3.50 = £1.50

- Express this as a percentage of £5.

$$\text{Percentage loss} = \frac{\text{Loss}}{\text{Original price}} \times 100\%$$

$$= \frac{1.5}{5} \times 100\%$$

<u>Percentage loss = 30%</u>

c

- Find the total amount paid by the company for the rabbits.

  40 × £5 = £200

- Find the total amount raised by the sale.

  30 × £7 + 10 × £3.50 = £245

- Find the profit and express this as a percentage of £200.

  Profit = £245 − £200 = £45

$$\text{Percentage profit} = \frac{45}{200} \times 100\% = 22.5\%$$

Percentage error

If you are measuring something and you make an error in your measurement (over or under the true amount) then you can work out the percentage error as follows.

$$\text{Percentage error} = \frac{\text{Amount of error}}{\text{True value}} \times 100\%$$

Sample Question 6 When measuring the length of a line that was 20 cm long, Neil wrote down 19.6 cm. What was his percentage error?

Answer

- Find the amount of error.

  Error = 20 − 19.6 = 0.4 cm

- Express this as a percentage of the true length.

$$\text{Percentage error} = \frac{\text{Error}}{\text{True length}} \times 100\%$$

$$= \frac{0.4}{20} \times 100\%$$

$$= 2\%$$

<u>He made a 2% error in his measurement.</u>

Increase by a given percentage

Sample Question 7 Sharon's salary was £12 000 a year. She was awarded a pay rise of 5%. Calculate her new salary.

Answer

Method 1: In two stages

- 1 Find the increase by finding 5% of 12 000.

$$\frac{5}{100} \times 12\,000 = 600$$

  or **0.05** × 12 000 = 600

$5\% = \frac{5}{100} = 0.05$
This decimal format method is very useful.

- 2 Add the increase to the original salary.

  £12 000 + £600 = <u>£12 600</u>

Method 2: In one stage

- Call the original amount 100%.
- If you increase this by 5% you get (100 + 5)%, i.e. 105% of the original amount. To find the new salary, find 105% of 12 000.
- Remember $105\% = \frac{105}{100} = \mathbf{1.05}$

1.05 is known as the multiplying factor.

- Calculate 1.05 × 12 000 = 12 600

  <u>New salary = £12 600</u>

	Further examples	Multiplying factor	Calculation
a	Increase 270 by 10%	$(100 + 10)\% = 110\% = 1.1$	$1.1 \times 270 = 297$
b	Increase 35 by 20%	$(100 + 20)\% = 120\% = 1.2$	$1.2 \times 35 = 42$
c	Increase 4000 by 25%	$(100 + 25)\% = 125\% = 1.25$	$1.25 \times 4000 = 5000$
d	Increase 50 by $17\frac{1}{2}\%$	$(100 + 17\frac{1}{2})\% = 117\frac{1}{2}\% = 1.175$	$1.175 \times 50 = 58.75$

Decrease by a given percentage

Sample Question  8

SALE!
All prices reduced by 15%

Calculate the sale price of a television set which originally cost £260.

Answer

Method 1: In two stages

- 1 Find the decrease by finding 15% of £260.

$$\frac{15}{100} \times 260 = 0.15 \times 260 = 39$$

- 2 Subtract this from the original amount.

$$260 - 39 = 221$$

Sale price = £221

Method 2: In one stage

- Call the original amount 100%.
- If you decrease this by 15% you get (100 − 15)%, i.e. 85% of the original amount. To find the sale price, find 85% of £260.
- Remember $85\% = \frac{85}{100} = 0.85$

The multiplying factor is 0.85

- Calculate $0.85 \times 260 = 221$

Sale price = £221

	Further examples	Multiplying factor	Calculation
a	Decrease 20 by 5%	$(100 - 5)\% = 95\% = 0.95$	$0.95 \times 20 = 19$
b	Decrease 650 by 12%	$(100 - 12)\% = 88\% = 0.88$	$0.88 \times 650 = 572$
c	Decrease 125 by 30%	$(100 - 30)\% = 70\% = 0.7$	$0.7 \times 125 = 87.5$
d	Decrease 5000 by 20%	$(100 - 20)\% = 80\% = 0.8$	$0.8 \times 5000 = 4000$

Sample Question 9

In 1996 there were 400 pupils at St. Mark's School. In 1997 the number dropped by 5%. In 1998 there was a 5% increase on the 1997 figure. How many pupils were there in 1998?

DO NOT jump in and say that the answer is 400. It is not 400.

Answer

- Find the number in 1997 by decreasing 400 by 5%.

  To decrease by 5%, multiply by (100 − 5)%, i.e. 95% = 0.95

  $0.95 \times 400 = 380$

- Find the number in 1998 by increasing 380 by 5%.

  Multiply by (100 + 5)%, i.e. 105% = 1.05

  $1.05 \times 380 = 399$

  There were 399 pupils in 1998.

Notice that when you decrease by 5%, then increase by 5%, you do not get back to the original number. Can you figure out why not?

Non-calculator note:

page 21

Calculating the final amount in one stage by using a multiplying factor is a useful method, especially if you are using a calculator. If however you are working without a calculator it is probably easier and quicker to find the separate amounts, for example

a Increase 270 by 10%

$10\% \text{ of } 270 = 270 \div 10$

$= 27$

$270 + 27 = \underline{297}$

b Decrease 20 by 5%

$10\% \text{ of } 20 = 2$

$\therefore \quad 5\% \text{ of } 20 = 1$

$20 - 1 = \underline{19}$

c Increase 640 by $17\frac{1}{2}\%$

$17\frac{1}{2} = 10\% + 5\% + 2\frac{1}{2}\%$

$10\% \text{ of } 640 = 64$

$5\% \text{ of } 640 = 32$

$2\frac{1}{2}\% \text{ of } 640 = 16$

$64 + 32 + 16 = 112$

$640 + 112 = \underline{752}$

Repeated percentage change

You can use multiplying factors to work out the final amount when a quantity is repeatedly increased or decreased by a given percentage.

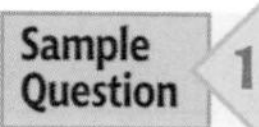

Sample Question 10

SAVINGS ACCOUNT

Guaranteed compound interest rate of 7% a year

You invest £250 and at the end of each year the interest is added to your account. The interest for the year is calculated on the amount in the account at the start of each year.

What is the account worth

a after 1 year,

b after 2 years,

c after 10 years?

Answer

a

- After 1 year, the interest (7% of £250) is added to the £250, so you have (100 + 7)% of 250, i.e. 107% of £250.

$$\mathbf{1.07} \times £250 = £267.50$$

Multiplying factor is 1.07

After 1 year, the account is worth £267.50

b

- At the end of the second year, the interest (7% of £267.50) is added to £267.50, so you have 107% of £267.50.

$$1.07 \times £267.50 = £286.225$$

After 2 years, the account is worth £286.23

c

- For the amount after 2 years, you actually calculated

$$1.07 \times 1.07 \times £250 = (1.07)^2 \times £250$$

After 3 years, the account is worth $(1.07)^3 \times £250$.

After 4 years, the account is worth $(1.07)^4 \times £250$ and so on.

After 10 years the account is worth $(1.07)^{10} \times £250$.

$$(1.07)^{10} \times £250 = £491.787 \ldots$$

[1] [.] [0] [7] [x^y] [1] [0] [×] [2] [5] [0] [=]

After 10 years, the account is worth £491.79 (nearest p)

Sample Question 11 A car **depreciates** by 12% of its value each year. It is worth £9000 now. How much will it be worth in four years' time?

Depreciates means that it decreases in value.

Answer

- Find the multiplying factor for a 12% decrease.

$$(100 - 12)\% = 88\% = 0.88$$

- Apply this 4 times by multiplying by $(0.88)^4$

$$(0.88)^4 \times 9000 = 5397.25 \ldots$$

[0] [.] [8] [8] [x^y] [4] [×] [9] [0] [0] [0] [=]

In four years' time, the car will be worth £5397 (nearest £)

Exercise 23.3

Section A

This section should be completed without using a calculator.

1 Write down the multiplying factors when **increasing** by the following percentages

a 15% b $12\frac{1}{2}$% c 8%

d $3\frac{1}{2}$% e 50% f $2\frac{1}{4}$%

2 Write down the multiplying factors when **decreasing** by the following percentages

a 13% b $8\frac{1}{2}$% c 28%

d $14\frac{1}{2}$% e 25% f $7\frac{1}{4}$%

3 a Increase £100 by 20%
b Increase £70 by 10%
c Decrease £100 by 15%
d Decrease £30 by 5%
e Increase £20 by 30%
f Decrease £600 by 7%
g Increase £700 by 10%
h Decrease £900 by 20%

4 Sophie was earning £200 a week and received an increase of 10%. Two weeks later her salary was reduced by 10%. What was her new weekly wage after the reduction?

5 When selling a house, an estate agent charges 2% commission on the first £60 000 plus 1% on any amount over this.
What would be the charge on a house selling for

a £60 000 b £90 000?

6 VAT is charged at $17\frac{1}{2}$%.
Use $17\frac{1}{2}\% = 10\% + 5\% + 2\frac{1}{2}\%$ to find the VAT payable on goods with a pre-VAT price of

a £200 b £720 c £92 d £174

7 The price of a meal in a restaurant is £9.60 plus $17\frac{1}{2}$% VAT. What is the full cost of the meal?

8 A 1 kg bag of flour was found to weigh only 988 g. What was the percentage error?

9 A garage advertises a second hand car for £4000 with 2% discount for a cash payment. If I pay cash, how much will the car cost me?

10 A shop offers a discount of 8p in the pound. What percentage discount is this and how much would an article advertised for £30 cost?

11 A market stall holder buys articles for 20p each and sells them at 50% profit. How much does he sell them for?

Section B

A calculator may be used in this section.

1 Use a multiplying factor to
a Increase £67 by 35%
b Increase £735 by $14\frac{1}{2}$%
c Decrease £123 by 12%
d Decrease £563 by $22\frac{1}{2}$%

2 A bank pays 7% compound interest each year on the money in a savings account. If Amy has £343 in the bank initally, how much will she have after

a 1 year b 2 years c 5 years?

3 Jane's watch gave the time for her completing a test as 13 minutes. Her actual time was 13 minutes and 58 seconds. What was the percentage error?

4 Jack bought a piece of rope advertised as 6 m long. When he measured it, he found it was 5.96 metres. What was the percentage error in the length?

5 The speedometer of Steve's car shows 70 mph when he is, in fact, travelling at 73 mph? What is the percentage error?

6 A car loses 15% of its value every year. If the car is worth £7500 at the beginning of the year, how much will it be worth

a at the end of the year

b after 2 years

c after 6 years?

7 Mrs Williams bought some scarves at £4 each.

a She sold 50 at £7 each. What was her percentage profit on each scarf?

b In her autumn sale she advertised the remaining scarves for £3 each. Find the percentage loss on each of these scarves.

c If she bought 65 scarves altogether, find the overall percentage profit or loss.

8 A group of three people visits a restaurant and each person has the set meal costing £7.50. They also have a bottle of wine between them costing £6.95 plus VAT at $17\frac{1}{2}$%.
What was the total bill if they also pay a 10% service charge?

Finding the original quantity, knowing a percentage of it

Sometimes you need to find the original amount when you are told the value of a certain percentage of it, for example, find x if 60% of x is 24.

Some answers are easy to work out without using a calculator, especially if you remember the **fraction equivalents of percentages** such as

$10\% = \frac{1}{10}$, $25\% = \frac{1}{4}$, $50\% = \frac{1}{2}$, $75\% = \frac{3}{4}$, $20\% = \frac{1}{5}$

Work through these examples, **without a calculator**, trying to follow the reasoning.

a 50% of x is 6

$50\% = \frac{1}{2}$

$\therefore \quad \frac{1}{2}x = 6$

$x = 6 \times 2$

$\underline{x = 12}$

FACT SHEET 1, page 14
Writing percentages as fractions or decimals

b 10% of x is 8

$10\% = \frac{1}{10}$

$\therefore \quad \frac{1}{10}x = 8$

$x = 8 \times 10$

$\underline{x = 80}$

c 60% of x is 24

10% of $x = 24 \div 6 = 4$

100% of $x = 4 \times 10$

$\underline{x = 40}$

d 75% of x is 36

$75\% = \frac{3}{4}$

$\therefore \quad \frac{3}{4}x = 36$

$\frac{1}{4}x = 36 \div 3 = 12$

$x = 12 \times 4$

$\underline{x = 48}$

e 5% of x is 12

10% of $x = 12 \times 2 = 24$

100% of $x = 24 \times 10$

$\underline{x = 240}$

f x is increased by 10% to give 66

110% of $x = 66$

10% of $x = 66 \div 11 = 6$

100% of $x = 6 \times 10$

$\underline{x = 60}$

g x is decreased by 10% to give 27

90% of $x = 27$

10% of $x = 27 \div 9 = 3$

100% of $x = 3 \times 10$

$\underline{x = 30}$

When the numbers are not as easy to work with, it may be necessary to use a calculator. The following sample questions illustrate two methods that are often used.

Sample Question 12 Mr and Mrs Davies paid a deposit of £9750 for their new house. This was 15% of the purchase price of the house. What was the total cost of the house?

Answer

- Let x be the cost of the house, so 15% of x is £9750.

Method 1

- Find 1%, then 100%.

$$15\% \text{ of } x = 9750$$
$$1\% \text{ of } x = \frac{9750}{15}$$
$$100\% \text{ of } x = \frac{9750}{15} \times 100$$
$$= 65\,000$$

Method 2

- Use multiplying factor

$$15\% \text{ of } x = 9750$$
$$0.15 \times x = 9750$$
$$(\div 0.15) \quad x = \frac{9750}{0.15}$$
$$= 65\,000$$

The total cost of the house was £65 000

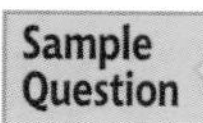

13

CHURCH FÊTE
£1395 raised
10% up on last year!

How much was raised at the church fête last year?

Answer

- Let x be the amount raised last year.
  An increase of 10% on x is 110% of x, so 110% of x = £1395

Method 1

$$110\% \text{ of } x = 1395$$
$$1\% \text{ of } x = \frac{1395}{110}$$
$$100\% \text{ of } x = \frac{1395}{110} \times 100$$
$$= 1250$$

Method 2

$$110\% \text{ of } x = 1395$$
$$1.1 \times x = 1395$$
$$(\div 1.1) \quad x = \frac{1395}{1.1}$$
$$= 1250$$

The amount raised last year was £1250

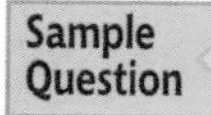

14

SALE
Prices cut by **30%**
TV now **£209.30**

What is the pre-sale price of the television set?

Answer

- Let x be the pre-sale price.
  If x is reduced by 30%, you have $(100 - 30)\%$, i.e. 70% of x

Method 1

$$70\% \text{ of } x = 209.30$$
$$1\% \text{ of } x = \frac{209.30}{70}$$
$$100\% \text{ of } x = \frac{209.30}{70} \times 100$$
$$= 299$$

Method 2

$$70\% \text{ of } x = 209.30$$
$$0.7x = 209.30$$
$$(\div 0.7) \quad x = \frac{209.30}{0.7}$$
$$= 299$$

The pre-sale price was £299

Exercise 23.4

Section A

This section should be completed without using a calculator.

1 Find the original amount (100%) in each case.

a 10% is £36
b 20% is 40 m
c 75% is 24 kg
d 5% is £8
e 50% is £2400
f 25% is 20 cm
g 40% is 12 g
h $33\frac{1}{3}$% is 90 m
i 60% is £12
j 70% is 210 kg

2 If the selling price of a skirt is £22 and the profit is 10%, what was the cost price?

3 Carl bought a car then sold it for £3300 at a loss of 25%. How much did he pay for it?

4 You are paid 20% commission on all articles that you sell. If you are paid £180 commission, what is the value of the articles you sold?

Section B

A calculator may be used in this section.

1 Joyce is paid a basic weekly wage of £150 and 5% commission on her sales. If, in one week she receives a total wage of £210, find the value of her sales for that week.

2 A car lost 22% of its value in the first year. If its value at the end of the year was £12 600, what was it worth at the beginning of the year?

3 Tim sold his watch for £73.50, making a profit of 5%. How much had he paid for it?

4 The prices including VAT of $17\frac{1}{2}$% of articles A, B and C were £24.50, £89.70 and £3450. Find the prices of the articles without the VAT.

5 Find the VAT at $17\frac{1}{2}$% paid on articles which, with VAT, cost

a £66 b £79 c £1100 d £24 000

6 82% of the people eligible voted in a club election. How many people were eligible to vote if 72 people did not vote?

7 A college has 720 female students. 55% of the students are male. What is the total number of students in the college?

8 Gerald spends 12% of his net monthly earnings on gas and electricity. What are his net earnings for a year if he spends £75 a month on gas and electricity?

9 Aunt Gertrude leaves a sum of money to her three grandchildren. Tom is to receive 25%, Ian 35% and Jane 40%.

a If Jane receives £1200 what was the total amount left by aunt Gertrude?

b How much did Tom and Ian receive?

10 When selling houses, Tamil charges 2% commission on the first £35 000 plus $1\frac{1}{2}$% on any amount over this.

a What would be the charge on a house selling for i £40 000 ii £85 000?

b Tamil is paid £925 for selling a house. How much does the house sell for?

11

SALE
30% off all prices
FURTHER REDUCTION OF 10% OF SALE PRICE ON LAST DAY

a If I buy a coat originally costing £120 on the last day of the sale, how much will I pay for it?

b What total percentage reduction will I receive?

12 Alex's insurance company gives him a 60% reduction on his premium for no claims. He then gets a 10% reduction on the discounted amount for paying the first £200 of any claim. How much will he pay on a premium of £820? What is his percentage saving?

13 An oil tank is 35% full. If it takes a further 325 litres to fill it up, what is the capacity of the tank?

PERCENTAGES/FRACTIONS

More calculations with fractions

Multiplying fractions

a

$\frac{3}{5}$ shaded

b 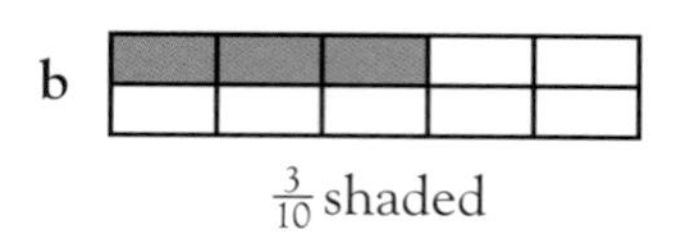

$\frac{3}{10}$ shaded

The amount shaded in **b** is half the amount shaded in **a**.

You can see from the diagrams that $\frac{1}{2}$ of $\frac{3}{5}$ is $\frac{3}{10}$. To find $\frac{1}{2}$ of $\frac{3}{5}$, multiply $\frac{1}{2}$ by $\frac{3}{5}$.

$$\frac{1}{2} \times \frac{3}{5} = \frac{1 \times 3}{2 \times 5} = \frac{3}{10}$$

On a calculator:

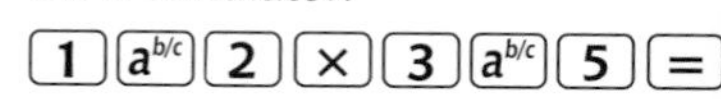

To multiply fractions, multiply the numerators (top numbers) and multiply the denominators (bottom numbers)

so $\quad \frac{a}{b} \times \frac{c}{d} = \frac{a \times c}{b \times d}$

Examples

a $\frac{2}{7} \times \frac{3}{5} = \frac{2 \times 3}{7 \times 5} = \frac{6}{35}$

2 $a^{b/c}$ 7 × 3 $a^{b/c}$ 5 =

b $\frac{3}{4} \times \frac{2}{3} = \frac{3 \times 2}{4 \times 3} = \frac{\overset{1}{\cancel{6}}}{\underset{2}{\cancel{12}}} = \frac{1}{2}$

Simplify $\frac{6}{12}$ by dividing top and bottom by 6.

It is often easier to **simplify before you multiply**. Look for a number that divides exactly into a number on the top and a number on the bottom.

$$\frac{3}{4} \times \frac{2}{3} = \frac{\overset{1}{\cancel{3}}}{4} \times \frac{2}{\underset{1}{\cancel{3}}}$$

3 goes into the 3 on the top and the 3 on the bottom.

$$= \frac{\overset{1}{\cancel{3}}}{\underset{2}{\cancel{4}}} \times \frac{\overset{1}{\cancel{2}}}{\underset{1}{\cancel{3}}}$$

2 goes into the 2 on the top and the 4 on the bottom.

$$= \frac{1 \times 1}{2 \times 1} = \frac{1}{2}$$

You would probably do all the working in one stage: $\frac{3}{4} \times \frac{2}{3} = \frac{\overset{1}{\cancel{3}}}{\underset{2}{\cancel{4}}} \times \frac{\overset{1}{\cancel{2}}}{\underset{1}{\cancel{3}}} = \frac{1}{2}$

If you are working without a calculator and you are multiplying mixed numbers, you must change them to 'top-heavy' fractions first.

$$1\tfrac{1}{2} \times 2\tfrac{1}{4} = \frac{3}{2} \times \frac{9}{4}$$

⚠ $1\tfrac{1}{2} = \tfrac{3}{2}$, $2\tfrac{1}{4} = \tfrac{9}{4}$

FACT SHEET 6 pages 174–176

$$= \frac{3 \times 9}{2 \times 4}$$

$$= \frac{27}{8}$$

$$= 3\tfrac{3}{8}$$

Change back to a mixed number.

If you are using a calculator, the fractions can be left as mixed numbers:

[1] [a b/c] [1] [a b/c] [2] [×] [2] [a b/c] [1] [a b/c] [4] [=]

Multiplying a number by its reciprocal

The reciprocal of 4 is $\frac{1}{4}$ $\qquad 4 \times \frac{1}{4} = \frac{4}{1} \times \frac{1}{4} = \frac{4}{4} = 1$

The reciprocal of $\frac{2}{3}$ is $\frac{3}{2}$ $\qquad \frac{2}{3} \times \frac{3}{2} = \frac{6}{6} = 1$

The reciprocal of $10\tfrac{1}{2}$ is $\frac{2}{21}$ $\qquad 10\tfrac{1}{2} \times \frac{2}{21} = \frac{\cancel{21}^{1}}{\cancel{2}_{1}} \times \frac{\cancel{2}^{1}}{\cancel{21}_{1}} = 1$

$10\tfrac{1}{2} = \frac{21}{2}$

When you multiply a number by its reciprocal, the answer is 1,

so $\qquad \frac{a}{b} \times \frac{b}{a} = 1$

Dividing fractions

$3 \div \frac{1}{4}$ means 'how many quarters in 3 whole ones?'

There are 4 quarters in 1 whole one, so in 3 whole ones there are 12 quarters, i.e. $3 \div \frac{1}{4} = 12$

You can see that dividing by $\frac{1}{4}$ has the same effect as multiplying by 4.

On calculator: 

4 is the reciprocal of $\frac{1}{4}$

Consider $\frac{3}{5} \div \frac{3}{4} = \dfrac{\frac{3}{5}}{\frac{3}{4}}$

$= \dfrac{\frac{3}{5} \times \frac{4}{3}}{\frac{3}{4} \times \frac{4}{3}}$

⚠ Multiply top and bottom by $\frac{4}{3}$, the reciprocal of $\frac{3}{4}$

$= \dfrac{\frac{3}{5} \times \frac{4}{3}}{1}$

⚠ $\frac{3}{4} \times \frac{4}{3} = 1$

$= \dfrac{3}{5} \times \dfrac{4}{3}$

So $\dfrac{3}{5} \div \dfrac{3}{4} = \dfrac{\cancel{3}^{1}}{5} \times \dfrac{4}{\cancel{3}_{1}} = \dfrac{4}{5}$

⚠ Instead of dividing by $\frac{3}{4}$, you multiply by $\frac{4}{3}$

To divide on the calculator straight away, key in [3] [a b/c] [5] [÷] [3] [a b/c] [4] [=]

If you are working without a calculator, you must change mixed numbers to 'top heavy' fractions first, for example

$\frac{7}{8} \div 1\frac{1}{4} = \dfrac{7}{8} \div \dfrac{5}{4}$

⚠ $1\frac{1}{4} = \frac{5}{4}$

$= \dfrac{7}{\cancel{8}_{2}} \times \dfrac{\cancel{4}^{1}}{5}$

⚠ Multiply by the reciprocal of $\frac{5}{4}$

$= \dfrac{7}{10}$

On the calculator you can do the division straight away by keying in

[7] [a b/c] [8] [÷] [1] [a b/c] [1] [a b/c] [4] [=]

Sample Question 15

$10\frac{1}{2}$ kg of tea is bagged into packets, each containing $\frac{3}{8}$ kg. How many packets are there? You must show your working. Do not use a calculator.

Answer

- Find how many lots of $\frac{3}{8}$ there are in $10\frac{1}{2}$ by calculating $10\frac{1}{2} \div \frac{3}{8}$

$10\frac{1}{2} \div \frac{3}{8} = \dfrac{21}{2} \div \dfrac{3}{8}$

⚠ Change $10\frac{1}{2}$ to 'top-heavy' format

$= \dfrac{\cancel{21}^{7}}{\cancel{2}_{1}} \times \dfrac{\cancel{8}^{4}}{\cancel{3}_{1}}$

⚠ Multiply by the reciprocal of $\frac{3}{8}$

$= \dfrac{28}{1}$

$= 28$

There are 28 packets.

Exercise 23.5

Do not use a calculator in this exercise.

1 Simplify

a $\frac{2}{3} \times \frac{1}{2}$

b $\frac{4}{5} \times \frac{15}{16}$

c $\frac{2}{7} \times \frac{14}{15}$

d $\frac{3}{8} \times \frac{4}{9}$

e $\frac{6}{10} \times \frac{5}{18}$

f $\frac{7}{6} \times \frac{9}{28} \times \frac{8}{30}$

g $\frac{11}{12} \times \frac{36}{55} \times \frac{5}{3}$

h $\frac{8xy^3}{9} \times \frac{3}{4y^2}$

i $\frac{-3a}{2b} \times \frac{-b}{6a^2}$

j $\frac{4ab}{9a^2} \times \frac{6}{b^2} \times \frac{3b}{2}$

2 Write down the reciprocals of

a 7

b $\frac{1}{6}$

c $\frac{4}{5}$

d $1\frac{3}{8}$

3 Simplify

a $\frac{8}{15} \div \frac{2}{5}$

b $\frac{7}{16} \times \frac{3}{2} \div \frac{21}{24}$

c $\frac{7}{12} \div 1\frac{1}{6}$

d $\frac{4xy^2}{5} \div \frac{2x^2y}{15}$

4 Calculate, giving your answer in its lowest form and as a mixed fraction if appropriate.

a $\frac{8}{9} \times 1\frac{4}{5}$

b $2\frac{3}{4} \times 3\frac{1}{5}$

c $3\frac{1}{2} \div 1\frac{1}{2}$

d $6\frac{7}{8} \div 2\frac{1}{5}$

5 How many $\frac{2}{3}$ litre bottles can be filled from a container holding 100 litres?

6 How many $\frac{3}{4}$ kg drums of cornflour can be filled from a tub containing $55\frac{1}{2}$ kg?

7 a A rectangle has length $1\frac{2}{5}$ m and width $\frac{5}{8}$ m.

i Find its area, in m^2.

ii Find its area, in cm^2.

b A square has side $2\frac{1}{2}$ cm. Work out its area.

TASK

$\frac{4}{5}$	$\frac{2}{3}$	$2\frac{2}{5}$
$1\frac{1}{2}$	$2\frac{1}{4}$	$\frac{3}{4}$

- Choose two fractions from the box and add them.
- Choose another two fractions and add them.
- Which two fractions should you choose to obtain the smallest possible total when you add them, using fractions from the box?
- Investigate further to find which two to choose to get the smallest answer when
  - **a** you subtract them,
  - **b** you multiply them,
  - **c** you divide them.
- What about when you want to obtain the largest possible answer?

Worked Exam Question

[NEAB]

a **Light takes about 12 minutes and 40 seconds to reach the planet Mars from the Sun.**
Light travels at approximately 299 800 kilometres per second.

Calculate the approximate distance of the Sun from Mars.
Give your answer in standard form correct to 2 significant figures.

◆ Use 

D = S × T

◆ The speed is given in km/s, so time is needed in seconds.

◆ Change your answer to standard form and correct it to 2 s.f.

$T = 12 \times 60 + 40 = 760$ secs
$D = 299\,800 \times 760$
$= 227\,848\,000$

Answer 2.3×10^8 km

COMMENTS

M1 for changing to seconds
M1 for multiplying speed by time

A1 for correct answer

3 marks

b **The distance from the Earth to the Moon is approximately 384 400 km.**
The distance from the Earth to the Sun is approximately 1.496×10^8 km.
Use these approximations to express the ratio

distance of Earth to Moon : distance of Earth to Sun

in the form 1 : *n* where *n* is a whole number.

◆ Write the ratio, then divide to get the form 1 : *n*

Give *n* to the nearest whole number.

$384\,400 : 1.496 \times 10^8$

$= 1 : \dfrac{1.496 \times 10^8}{384\,400} = 1 : 389.177\ldots$

Answer 1 : 389

M1

A1 (390 accepted)

2 marks

c **Light travels at the rate of approximately 186 000 miles per second.**
Light takes 12 years to reach Earth from a particular star.

Find the approximate distance, in miles, of this star from the Earth.
Give your answer in standard form correct to 3 significant figures.

◆ Multiply 186 000 by the number of seconds in 12 years.

◆ Write your answer in standard form to 3 s.f.

$D = 186\,000 \times 60 \times 60 \times 24 \times 365 \times 12$
$= 7.038\ldots \times 10^{13}$

Answer 7.04×10^{13} miles

B1 for correct numbers in standard form
B1 for 3 s.f.

2 marks

Exam Questions

1 a Eric wanted to calculate $\left(\frac{4.8 + 1.27}{1.2}\right)^2$.

He pressed these keys to get his answer.

(4 . 8 + 1 . 2 7
÷ 1 . 2) x^2

He got the answer 34.320 069
The answer is wrong.
Show a correct calculator sequence and calculate the correct answer.

b Hannah does the following calculation.

$$\frac{8.7 \times 102}{(3.1)^2}$$

i **Without** using your calculator, work out an approximate answer to this calculation.
You **must** show all your working.

ii Hannah's answer is 923.4.
Explain whether Hannah's answer is of the right order of magnitude. [SEG]

2 In a sale, a teapot was reduced by 20% from its original price of £9.50. I bought it on the last day of the sale, when there was 30% off the sale prices.

a How much did I pay for the teapot?

b What percentage reduction did I get altogether? [MEG]

3 £500 is invested for 2 years at 6% per annum compound interest.

a Work out the total interest earned over the 2 years.

£250 is invested for 3 years at 7% per annum compound interest.

b By what single number must £250 be multiplied to obtain the total amount at the end of the 3 years? [L]

4 The value of an oil painting was £20 000 at the end of 1990.
Its value increased by 5% each year.

Find the value at the end of

a 1991,

b 1992,

c 1996. [MEG]

5 $F = \frac{GmM}{r^2}$

Calculate F when $G = 6.67 \times 10^{-11}$, $m = 65$, $M = 6.02 \times 10^{24}$, $r = 6.40 \times 10^6$. [NI]

6 Flyway Tours undertake short distance, medium distance and long distance flights.
Last year $\frac{1}{4}$ of their flights were short distance and $\frac{2}{3}$ were medium distance.

a What fraction of Flyway Tours' flights were long distance?
Write your answer as a fraction in its simplest form.

Flyway Tours flew 630 long distance flights. The destination of $\frac{2}{5}$ of the long distance flights was Canada and $\frac{1}{3}$ of these were to the city of Vancouver in Canada.

b Calculate the number of flights to Vancouver last year. [SEG]

7 The following information about vitamin content is printed on the side of a breakfast cereal packet.

	100 g serving	30 g serving
Vitamin C	… mg	16.2 mg
Vitamin B6	1.7 mg	… mg

a Calculate and fill in the missing values in a copy of the table.

b The 16.2 mg of vitamin C is 24% of the recommended daily amount of vitamin C.
What is the recommended daily amount of vitamin C?

c The recommended daily amount of vitamin B6 is 2 mg.
What percentage of the recommended daily amount is provided by the 100 g serving? [SEG]

8 The mass of one atom of oxygen is given as 2.66×10^{-23} grams.
The mass of one atom of hydrogen is given as 1.67×10^{-24} grams.

a Find the difference in mass between one atom of oxygen and one atom of hydrogen.

b A molecule of water contains two atoms of hydrogen and one atom of oxygen.

i Calculate the mass of one molecule of water.

ii Calculate the number of molecules in 1 gram of water. [NEAB]

EXAM QUESTIONS

9 **a** The approximate population of the United Kingdom is given in standard form as 5.2×10^7. Write this as an ordinary number.

b The thickness of grade A paper is 6.0×10^{-2} cm. Grade B paper is twice as thick as grade A. Calculate, in centimetres, the thickness of grade B paper.
Write your answer in standard form. [SEG]

10 Calculate, to the nearest penny, the compound interest on £900 invested at 7% per annum for three years. [WJEC]

11 **a** Work out $\dfrac{3^4 \times 3^5}{3^7 \times 3^2}$

b Find the difference between 5×10^{-3} and 5^{-3}

c Simplify $4a^2c^6 \times (2ab)^3$ [MEG]

12 During the 1970s the value of houses increased by 8% per year.
In 1975 the value of a house was £60 000.

a What was the value of the house in 1976?

b In which year was the value of the house first more than £72 000?

c What was the value of the house in 1974? [MEG]

13 In 1997 a store added VAT at $17\frac{1}{2}$% to the basic price of furniture.

a A sofa had a basic price of £650.
What was its selling price inclusive of VAT?

b During a sale the store said 'we pay the VAT'. What percentage discount on the normal selling price was this store giving? [MEG]

14 **a** The distance between Earth and the Moon is 384 000 km correct to 3 significant figures.
Write this distance

i as an interval approximation,

ii in standard form.

b The distance between the Sun and Earth is 1.5×10^8 km.
Light travels at 3.0×10^5 km per second.
How long does it take light to travel from the Sun to the Earth?
Give your answer in seconds.

c Mercury is 5.8×10^7 km from the Sun.
On a certain day the Sun, Mercury and Earth are in a straight line with Mercury between the Sun and Earth.
How far is Mercury from Earth?
Give your answer in standard form. [MEG]

15 Some students are using calculators to work out four questions.

Question 1 $\dfrac{2.34 + 1.76}{3.22 + 1.85}$

Question 2 $\dfrac{2.34 + 1.76}{3.22} + 1.85$

Question 3 $2.34 + \dfrac{1.76}{3.22} + 1.85$

Question 4 $2.34 + \dfrac{1.76}{3.22 + 1.85}$

a Tom presses keys as follows.

[2] [.] [3] [4] [+] [1] [.] [7] [6]
[÷] [3] [.] [2] [2] [+] [1] [.] [8] [5] [=]

For which of the four questions is this the correct method?

b Jayne presses keys as follows.

[2] [.] [3] [4] [+] [1] [.] [7] [6] [=]
[÷] [3] [.] [2] [2] [+] [1] [.] [8] [5] [=]

For which of the four questions is this the correct method? [SEG]

16 A crystal has rectangular cross-section of length 5.2×10^{-3} mm and width 8.7×10^{-4} mm.

a **i** Calculate the area of cross-section. Give your answer in standard form.

ii Calculate the perimeter of the cross-section. Give your answer in standard form.

The thickness of the crystal is 0.041 mm.

b Fiona estimated that fifty thousand crystals would be needed for the total thickness to be about 2 cm.

i **Without using a calculator**, show that she was incorrect in her estimate.

ii What estimate should Fiona have given for the total thickness of fifty thousand crystals? [MEG]

DATA HANDLING III

In this chapter you will learn how to
- **find the probabilities of combined events including the use of the '*and*' and '*or*' rules**
- **draw and use tree diagrams to calculate the probabilities of combined events**
- **create a cumulative frequency distribution and draw a cumulative frequency curve**
- **find estimates for the median, quartiles and inter-quartile range from a cumulative frequency curve**
- **compare two cumulative frequency distributions**

Addition law

In Chapter 3 (Probability) you learned that for equally likely outcomes

$$\text{the probability of an event happening} = \frac{\text{number of times event can happen}}{\text{total number of outcomes}}$$

You also learned that for two **mutually exclusive events** A and B

Remember, mutually exclusive means that the events cannot happen at the same time.

P(A or B) = P(A) + P(B)

Chapter 3 page 30

This only works for mutually exclusive events and is called the **addition law** for probabilities. This is sometimes called the **'or'** law

This law extends to more than two mutually exclusive events, for example,

P(A or B or C or D) = P(A) + P(B) + P(C) + P(D)

So to be able to apply the addition law you have to decide whether the events are mutually exclusive.

Because you are adding fractions the result will always be bigger than any of the fractions you are using but will never exceed 1.

If A, B, C, … are mutually exclusive events
- The probability of one event or another event happening is found by **adding** together the probabilities of the individual events.
- P(A or B or C or …) = P(A) + P(B) + P(C) + … This is called the **addition law for probabilities**, sometimes known as the **'or'** law.

Sample Question 1

Decide whether each of the following situations is mutually exclusive and hence say whether you are able to apply the addition law for probability:

a Michael is colouring in a pattern and in his pencil case he has a selection of coloured pencils. He has 1 HB brown, 1 H brown, 2 HB red, 2 HB yellow, 1 HB blue, 1 2H blue and 1 HB black pencil.
What is the probability that he will select, at random, a pencil that is blue or HB?

b Louis is making a soft toy for his Technology project. In a tin there are blue eyes, brown eyes, green eyes and black eyes. Louis chooses an eye at random from the tin. What is the probability that he chooses a blue or green eye?

Answer

a

- In the pencil case there is a mixture of colours and a mixture of types of pencil. Decide whether there is a blue pencil in the case that is also HB.

  Yes, there is a blue HB pencil so the events 'HB pencil' and 'blue pencil' are **not mutually exclusive** (they can happen at the same time). The addition law **cannot** be applied, so P(blue or HB) $\neq$ P(blue) + P(HB).

  $\neq$ means 'does not equal'.

b

- In one selection can Louis choose more than one eye? Obviously he cannot so he can only select one colour eye. This means that in one selection he cannot choose an eye that is blue **and** green at the same time.

  Therefore the events 'choose a blue eye' and 'choose a green eye' **are mutually exclusive** and the addition law **can** be applied, so P(blue or green) = P(blue) + P(green).

Exercise 24.1

1 Decide whether the following pairs of events are mutually exclusive

a i Drawing an ace from a pack of cards
 ii Drawing a spade from a pack of cards

b i Throwing a six on a die
 ii Throwing a four on a die

c i Travelling to work by train
 ii Travelling to work by public transport

d i Chris travels by bus
 ii Chris does not travel by bus

e i Throwing a double on a pair of dice
 ii Getting a total score of 10 on a pair of dice

f i Throwing a double on a pair of dice
 ii Getting a total score of 9 on a pair of dice

2 A bag contains 4 blue, 5 red and 3 green counters. One is drawn at random. What is the probability that the counter is

a blue b red

c blue or red d not red?

3 A card is drawn at random from a pack of 52 playing cards. Find the probability of

a drawing an ace

b drawing a picture card (jack, queen or king)

Are events **a** and **b** mutually exclusive?

c What is the probability of drawing an ace or a picture card?

4 On any day, the probability that I drive to work is $\frac{2}{3}$ and the probability that I go by bus is $\frac{1}{6}$. If I do not drive or go by bus, I cycle.

What is the probability that tomorrow I will

a drive or go by bus b cycle?

5 Keith is trying to decide where to go on holiday. He is going to choose between Cyprus, Tenerife, Malta and Crete. If the probability of choosing Cyprus is $\frac{1}{3}$, the probability of choosing Tenerife is $\frac{2}{5}$ and the probability of choosing Malta is $\frac{1}{5}$, find

a the probability of him choosing Cyprus or Tenerife,

b the probability of him choosing Cyprus or Malta,

c the probability of him choosing Cyprus, Tenerife or Malta,

d the probability of him choosing Crete.

6 In a survey to find the number of cars at each house in a road of 30 houses, the following results were obtained.

Number of cars	0	1	2	3	4
Number of houses	3	12	12	2	1

If a house is chosen at random, find the probability of the number of cars being

a 2 b 3 or 4 c more than 1

PROBABILITY LAWS

Multiplication law

If you pick a card at random from an ordinary pack of 52 playing cards, tear up the card and throw it away, then pick a second card from the pack, will the first pick have any effect on what is picked the second time?

Yes, because on the second pick there are less cards to choose from. The result of the second pick depends on what was selected the first time.

If you select a card at random from an ordinary pack of 52 playing cards, then spin a coin in the air, does what card is picked affect the way in which the coin will land?

No. The events are totally unrelated. The one event has no effect on the result of the other event. The events are said to be **independent**.

Consider the events 'choosing a letter from the word MATHS' and 'spinning a coin in the air'.

You can draw a **possibility space** to show all the outcomes.

Chapter 3 page 30

	M	A	T	H	S
Head	M, Head	A, Head	T, Head	H, Head	S, Head
Tail	M, Tail	A, Tail	T, Tail	H, Tail	S, Tail

What is the probability of picking the letter 'A' and the coin landing on a Tail?
It does not matter whether you spin the coin first or pick the letter first. The events are independent. From the table you can see that there is only one favourable outcome out of a total of 10 possible outcomes. So

$$P(\text{A,Tail}) = \frac{\text{number of ways of selecting A,Tail}}{\text{total number of outcomes}} = \frac{1}{10}$$

Looking at the probabilities of the separate events, P(picking A) $= \frac{1}{5}$, P(Tail) $= \frac{1}{2}$

Multiplying the two probabilities you have $\frac{1}{5} \times \frac{1}{2} = \frac{1}{10}$. Notice that this is the same result as before.

Multiplying the separate probabilities gives the probability of both the events happening.

This is called the **multiplication law for probabilities** and can only be applied to independent events. The law can be applied to any number of independent events, not just two. This is sometimes called the **'and'** law. It gives the probability of one event happening **and** another event happening.

Because you are multiplying two or more fractions together the result will always be smaller than any of the fractions that you are using. This also applies if you are using decimals.

If A, B, C ... are independent events

- The probability of an event followed by another event is found by **multiplying** together the individual probabilities for each event.
- P(A **and** B **and** C **and** ...) = P(A) ×P(B) × P(C) × ... This is called the **multiplication law for probabilities,** sometimes known as the **'and'** law.

PROBABILITY LAWS

Sample Question 2 In my sock drawer I have 10 identically styled pairs of socks of different colours. There are 2 brown pairs, 3 grey pairs, 4 green pairs and 1 black pair. I also have four pairs of trousers in my wardrobe, 2 black pairs and 2 brown pairs. If I select a pair of socks at random from the drawer and a pair of trousers at random from the wardrobe, what is the probability that I will select a black pair of trousers and a black pair of socks?

Answer

- The events are independent as the colour of socks selected has no effect on the colour of the trousers selected.
- Use the multiplication law. Multiply the probability of selecting a black pair of socks by the probability of selecting a black pair of trousers.

Cancel fractions to their simplest form and avoid decimals.

P(pair of black socks) $= \frac{1}{10}$

P(pair of black trousers) $= \frac{2}{4} = \frac{1}{2}$

P(black socks and black trousers) $= \frac{1}{10} \times \frac{1}{2}$

$= \frac{1}{20}$

P(selecting black socks and black trousers) $= \frac{1}{20}$

Exercise 24.2

1 A coin and a die are thrown together. What is the probability of getting

a a head and a six?

b a tail and a score less than 3?

2 Newtown Youth Club needs one boy and one girl to present gifts to visiting speakers. There are 20 boys and 15 girls to choose from. If a boy and a girl are chosen at random, what is the probability that brother and sister, Tim and Jane, are chosen?

3 In a multiple choice exam all questions have four possible answers. If a pupil guesses all the answers, what is the probability that

a the first answer is correct?

b the first four answers are correct?

If there are 36 questions on the paper how many would you expect to get right by merely guessing?

4 A coin is tossed three times. What is the probability of getting

a three heads

b three tails

c one head and two tails in that order?

5 A biased spinner is coloured as indicated in the diagram.

The probabilities of the spinner stopping on red, yellow or green are 0.4, 0.5 and 0.1, respectively.

If the spinner is spun twice, find the probability that the spinner will stop on

a red both times,

b red neither time,

c green first time and yellow second time.

6 A die is thrown and a card picked from a pack of 52.

What is the probability of getting

a a six on the die and an ace on the card?

b a six on both the die and the card?

7 Stuart, Stephen and Paul take part in a shooting competition. The probabilities of them hitting the target are 0.8, 0.6 and 0.7, respectively.

Find the probability that

a they all hit the target,

b Stuart and Paul hit the target, but Stephen misses.

Tree diagrams

Listing outcomes can be a very hazardous job and you have already met one way of making life easier, namely a possibility space. A **tree diagram** is another way of listing outcomes and helps to simplify the calculation of probabilities when combined events are concerned.

Each 'branch' of the tree indicates the outcome at each stage. Each route along the tree leads to a combined event. The tree is usually created across the page but can be drawn down the page.

Sample Question 3

In John's sock drawer there are 2 grey, 4 white and 3 black pairs of identically styled socks. In his wardrobe he has 3 pairs of grey and 2 pairs of black trousers.

a Draw a tree diagram to show the outcomes of John selecting, at random, a pair of socks from the drawer and a pair of trousers from the wardrobe.

b What is the probability that John will select

i a white pair of socks and a black pair of trousers?

ii a pair of socks and a pair of trousers that match (i.e. are the same colour)?

Answer

a

- Choose a letter to represent each choice of colour. In this case black = B, grey = G, white = W.
- Start the tree from a point and list the three choices of socks.
- Draw a branch from each sock choice to indicate the choice of trousers.
- At the end of the tree write the outcome of each 'route' along the tree.

Socks	Trousers	Outcome
B	B	B, B
	G	B, G
W	B	W, B
	G	W, G
G	B	G, B
	G	G, G

b

- As it stands this diagram cannot be used to calculate the probability of an outcome as the outcomes are **not equally likely**.

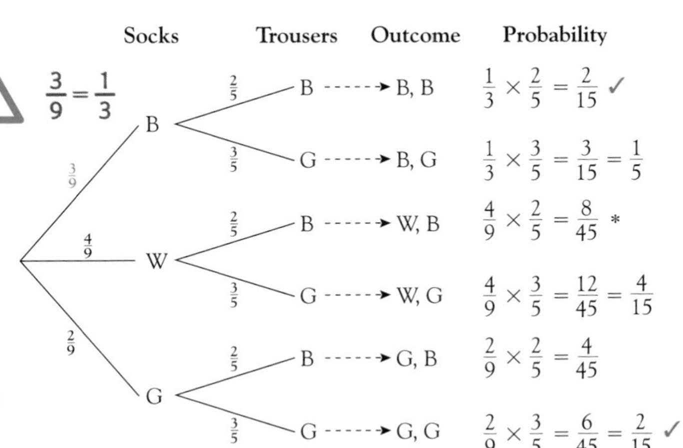

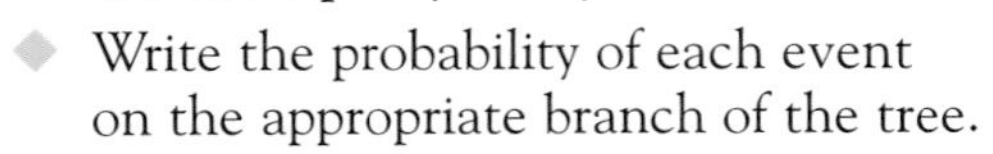

- Write the probability of each event on the appropriate branch of the tree.
- Using the **multiplication law** work out the probability of each outcome. Put this at the end of each set of branches.

Multiply as you go 'across' the diagram from branch to branch. ('Across' sounds like a '×' which means multiply.)

i

- Look at the outcome that gives a 'white pair of socks **and** a black pair of trousers', marked with *.

ii

- Tick (✓) the outcomes that give the same colour socks and trousers.
- As the final outcomes are **mutually exclusive** P(matching colour) = P(B,B) + P(G,G)

i $P(W, B) = \frac{8}{45}$

ii $P(\text{matching colour}) = P(B,B) + P(G,G)$

$= \frac{2}{15} + \frac{2}{15} = \frac{4}{15}$

TREE DIAGRAMS

Sample Question 4

Lemmy and Marlene play a game of tennis and a game of table-tennis.
The probability that Marlene will win the tennis is 0.65. The probability that Lemmy will win the table-tennis is 0.8. There are no draws.

a Draw a tree diagram to illustrate the possible outcomes of the two games.

b Use your tree diagram to find the probability that Marlene wins

i exactly one game

ii at least one game.

Answer

a

- There are two outcomes to a game, Lenny wins or Marlene wins. Work out the probabilities of Lemmy winning the tennis and of Marlene winning the table-tennis.
- Draw two branches to represent the two outcomes of the **tennis game**.
- From each of these outcomes draw two branches to represent the two outcomes of the **table-tennis game**.
- Put the probabilities on the branches and work out, using the multiplication law, the probabilities of each combined event, e.g. P(L wins both games).

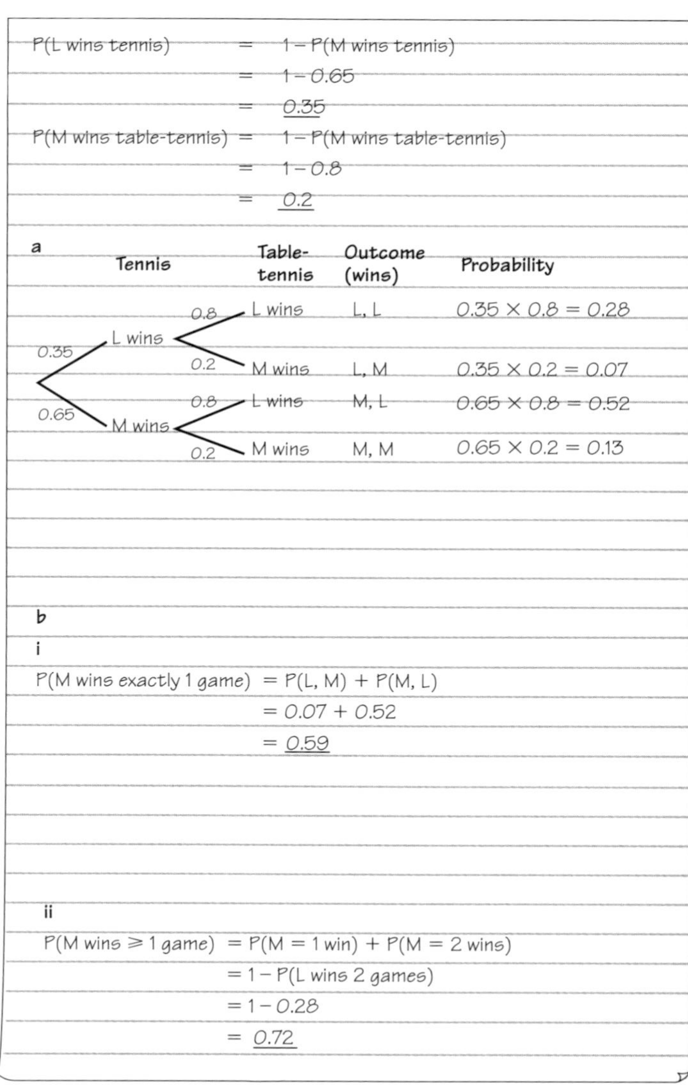

b

i

- For Marlene to win exactly one game she must win the tennis or the table-tennis but not both.
- Using the addition law calculate the probability that Marlene wins one game.

ii

- 'Marlene wins at least one game' means that Marlene wins 1 game or 2 games. This leaves only one option – Lemmy wins both games. So, instead of having to work out the 3 probabilities and adding them, you can say that $P(M \geqslant \mathbf{1} \text{ win}) = 1 - P(L \text{ wins } 2 \text{ games})$.

$\geqslant 1$ is shorthand for '1 or more'.

TREE DIAGRAMS

Sometimes, care needs to be taken when deciding what the outcome of a branch should be. Here is one such occasion.

Sample Question 5

The probability that Amy will pass her first aid exam is 0.8.
The probability that Bob will pass his first aid exam is 0.7.
They are both taking their exam next Monday.

a Draw a tree diagram to show the possible outcomes of their exam.

b Use your tree diagram to work out the probability that

i only one of them passes,

ii at least one of them passes.

Answer

a

- In answering this question it is best to consider the outcomes of each person's first aid exam separately. Draw a tree, writing P for pass and F for fail.

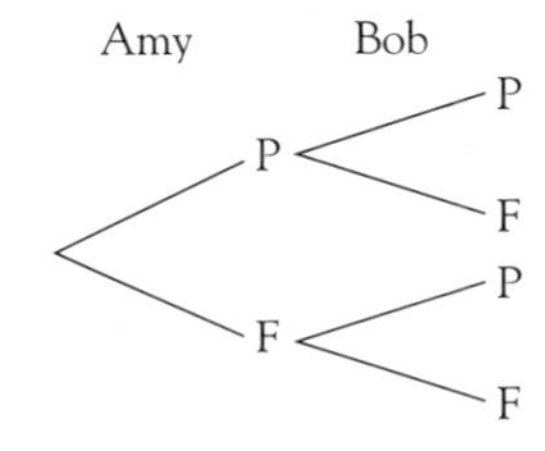

- Work out the probability of Amy failing the exam and of Bob failing the exam.

P(Amy fails) $= 1 - $ P(Amy passes)
$= 1 - 0.8$
$= \underline{0.2}$

P(Bob fails) $= 1 - $ P(Bob passes)
$= 1 - 0.7$
$= \underline{0.3}$

- Put the probabilities on to the branches, write down the outcomes and work out the probability of each combined outcome (use the multiplication law for independent events).

Amy	Bob	Outcome	Probability
P (0.8)	P (0.7)	P, P	$0.8 \times 0.7 = 0.56$
P (0.8)	F (0.3)	P, F	$0.8 \times 0.3 = 0.24$ ✓
F (0.2)	P (0.7)	F, P	$0.2 \times 0.7 = 0.14$ ✓
F (0.2)	F (0.3)	F, F	$0.2 \times 0.3 = 0.06$

b

i

- Put a tick (✓) by the outcomes that show the one person passing and the other person failing. There are two such cases.
- Using the addition law (as these outcomes are mutually exclusive), add up the probabilities of these events happening.

b
i P(only one passes) $= 0.24 + 0.14$
$= \underline{0.38}$

ii

- 'At least one of them passes' means that either Amy passes but not Bob, or Bob passes but not Amy or both of them pass. This only leaves one other option, i.e. they both fail. Use

P(at least one passes) $= 1 - $ P(both fail)

ii P(at least one passes) $= 1 - $ P(both fail)
$= 1 - 0.06$
$= \underline{0.94}$

TREE DIAGRAMS

Exercise 24.3

1 In a town, the probability that a family will have a freezer is $\frac{2}{3}$ and the probability that a family will have a computer is $\frac{1}{4}$.

These probabilities are independent.

a Copy and complete the probability tree diagram below.

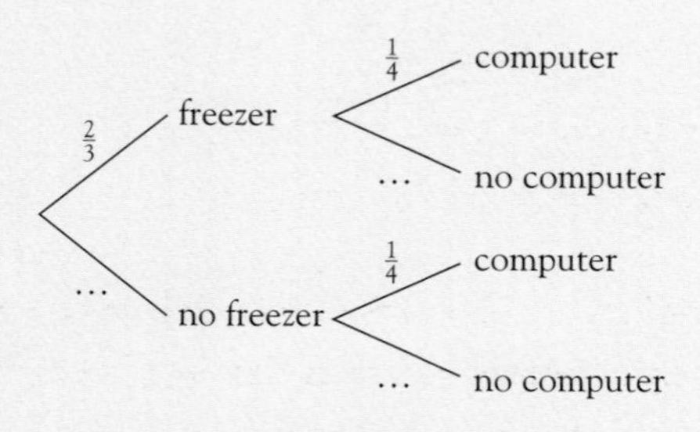

b Calculate the probability that a family in the town will have a freezer but not a computer. [L]

2 I take a ball, at random, from a bag containing five blue and seven red balls. I then take a ball, at random, from a second bag containing three blue and six red balls.

Copy and complete the tree diagram showing the possible outcomes and the probabilities.

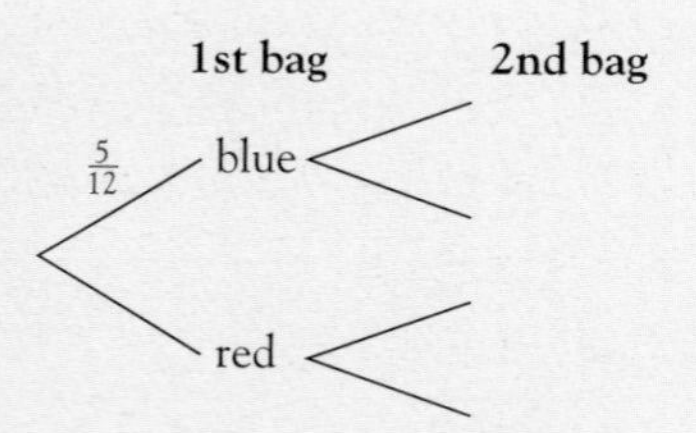

Use the tree diagram to find the probability that I take

a two blue balls,

b balls of different colours,

c at least one red ball,

d balls of the same colour.

3 Ahmed and Kate play a game of tennis. The probability that Ahmed will win is $\frac{5}{8}$.

Ahmed and Kate play a game of snooker. The probability that Kate will win is $\frac{4}{7}$.

a Copy and complete the probability tree diagram below.

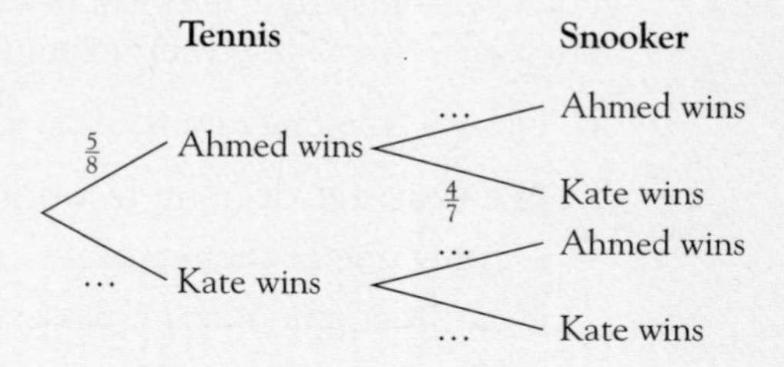

b Calculate the probability that Kate will win both games. [L]

4 When driving to work, Kirsty can be delayed at two roundabouts. The probability of a delay at roundabout A is $\frac{3}{7}$ and at roundabout B, $\frac{1}{3}$. Using a tree diagram, work out the probability that

a she will have a clear run to work,

b she will be delayed at one roundabout only.

5 A letter has a first-class stamp on it.

The probability that it will be delivered on the next working day is 0.86.

a What is the probability that the letter will **not** be delivered on the next working day?

Sam posts two letters with first-class stamps.

b Copy and complete the tree diagram.

Write all the missing probabilities on the appropriate branches.

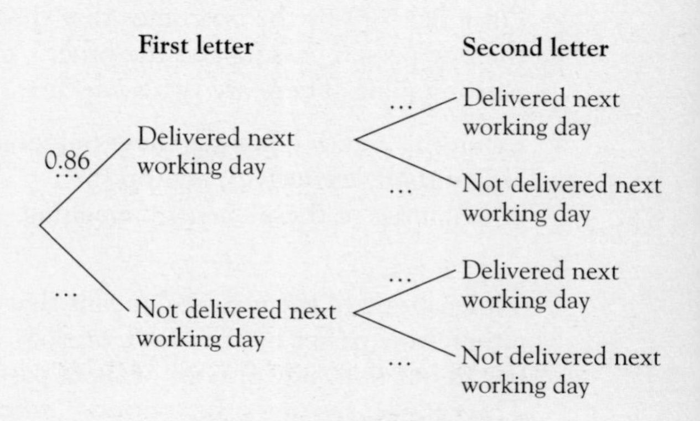

c Calculate the probability that both letters will be delivered on the next working day. [L]

TREE DIAGRAMS

6 Karen has a black jacket and a navy jacket in her wardrobe, together with a black skirt, a navy skirt and a grey skirt. She gets up early one morning and chooses a skirt and a jacket without turning the light on.

a Draw a tree diagram to show all her choices.

b What is the probability that the skirt and jacket will be the same colour?

c What is the probability that the black jacket will be chosen with the navy or grey skirt?

7 A coin is tossed three times.
Complete the tree diagram to show all the possible outcomes.

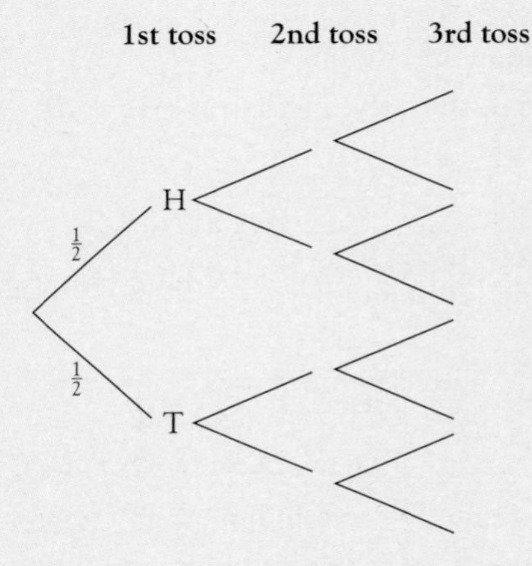

a Find the probability of tossing three heads.

b Find the probability of tossing two heads and a tail, in any order.

c Find the probability of tossing at least one head.

8 The probability that Kevin drives to work is 0.6, the probability that he walks is 0.3 and the probability that he cycles is 0.1.
Draw a probability tree to show the possible outcomes for Kevin's journeys to work on Monday and Tuesday in the first week of March.
Find the probability that

a he walks on both days,

b he drives on Monday, but not on Tuesday,

c he travels by different methods on the two days,

d he cycles on at least one day.

9 Helen has a box of 15 chocolates consisting of six with soft centres, five with nut centres and four with toffee centres. She has another box of 15 chocolates with equal numbers of soft, nut and toffee centres.
Draw a tree diagram showing the possible outcomes when Helen chooses a chocolate, at random, from each box.
What is the probability that she chooses

a two nut centres,

b a soft centre and a toffee centre,

c chocolates with different centres?

10 The probability of a person having brown eyes is $\frac{1}{4}$.
The probability of a person having blue eyes is $\frac{1}{3}$.
Two people are chosen at random.
Work out the probability that

a both people will have brown eyes,

b one person will have blue eyes and the other person will have brown eyes. [L]

11 To go from bus stop A to bus stop B a bus has to pass through 2 sets of traffic lights.
The traffic lights work independently of each other.
The probability of the first set of lights being green is $\frac{1}{3}$ and for the second set of lights it is $\frac{1}{4}$.
By drawing a tree diagram, or otherwise, find the probability that a bus is stopped by:

a both sets of lights,

b just one set of lights. [L]

12 The probability that a washing machine will break down in the first 5 years of use is 0.27.
The probability that a television will break down in the first 5 years of use is 0.17.
Mr. Khan buys a washing machine and a television on the same day.

a Calculate the probability that, in the five years after that day, the television will NOT break down.

b Calculate the probability that, in the five years after that day, both the washing machine and the television will break down. [L]

Median and quartiles

When analysing data it is useful to find the **median** (the middle value when the data is arranged in order of size) and also consider how spread out the data are.

There are three situations being considered:

a Discrete raw data
b Data in an ungrouped frequency distribution
c Data in a grouped frequency distribution

a Discrete raw data

For **ungrouped discrete data** you can find the middle value using this rule

If there are n observations, arranged in order of size, the middle value is the $\frac{1}{2}(n + 1)$th observation.

You find that:

if n is odd, there is **one middle value** and this is the median,

if n is even, there are **two middle values**. If these are c and d, then the median is $\frac{1}{2}(c + d)$.

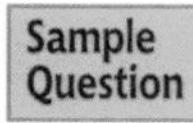

Sample Question 6 Find the median of each set of data:

a 4, 8, 7, 8, 5, 9, 3, 2, 6 b 32, 35, 31, 39, 21, 34, 56, 23

Answer

a

- Arrange the nine numbers in order of size.
- As the number of observations is odd there is **one** middle value.
- Find the $\frac{1}{2}(9 + 1)$th value, i.e. the 5th value.

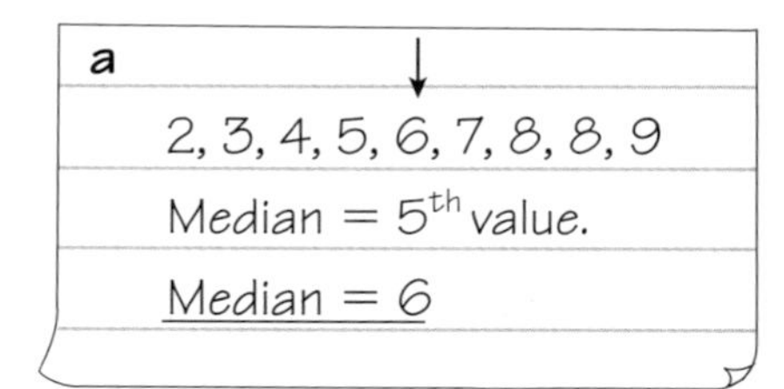

a
2, 3, 4, 5, 6, 7, 8, 8, 9
Median = 5th value.
Median = 6

b

- Arrange the 8 numbers in order of size.
- As there is an even number of observations there are **two** middle values and the median is halfway between them.
- Find the $\frac{1}{2}(8 + 1)$th value, i.e. halfway between the 4th and 5th value.

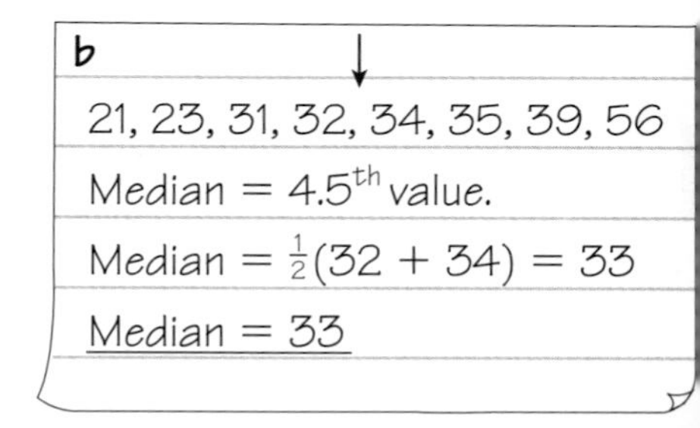

b
21, 23, 31, 32, 34, 35, 39, 56
Median = 4.5th value.
Median = $\frac{1}{2}(32 + 34) = 33$
Median = 33

The **range** of the data tells you how far apart the highest and lowest values are. This gives you a good idea of how spread out the data are but it is affected by extreme values.

By dividing the data into 4 equal parts you are able to consider the range of the **middle 50%** of the observations, thus ignoring any extreme values.

The **quartiles** divide in half the data either side of the median. When the data are written in ascending order

the quartile to the left of the median is called the **lower quartile**.

the quartile to the right of the median is called the **upper quartile**.

The difference between these quartiles is called the **inter-quartile range**.

Sample Question 7 Find the inter-quartile range of these sets of data:

a 4, 4, 4, 5, 6, 8, 8, 9, 11, 15, 26 b 125, 128, 136, 137, 138, 140, 159, 160

Answer

a

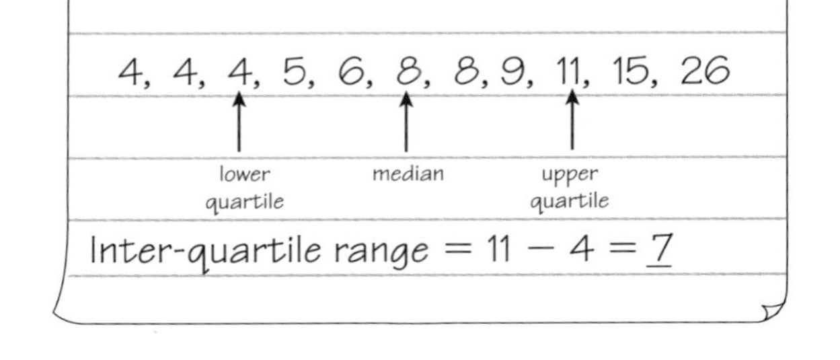

- Find the median first.
- Find the middle value of the observations to the left of the median. This is the lower quartile.
- Find the middle value of the observations to the right of the median. This is the upper quartile.
- Use inter-quartile range = upper quartile − lower quartile

b

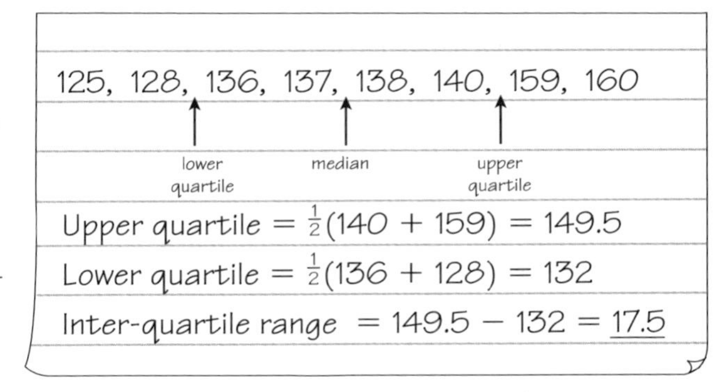

- Find the median first.
- As there is an even number of values to the left and to the right of the median the quartiles will be between two values. Find the quartiles.
- Use inter-quartile range = upper quartile − lower quartile

To find the quartiles for discrete data you can use this method.

Sometimes the following rules are used for n observations, arranged in order of size: lower quartile is the $\frac{1}{4}(n+1)$th observation and the upper quartile is the $\frac{3}{4}(n+1)$th observation.

Inter-quartile range = upper quartile − lower quartile
It gives the range of the middle 50% of the values.

Exercise 24.4

1 Find the median values of the following sets of data.

a 4, 8, 1, 9, 11, 6, 3

b 13 cm, 20 cm, 3 cm, 14 cm, 22 cm, 35 cm, 2 cm, 38 cm

c 8, 7, 18, 46, 2, 4, 16, 39, 25

d 135 kg, 165 kg, 169 kg, 155 kg, 160 kg, 153 kg

2 Find the median, quartiles and inter-quartile range of these sets of data.

a 48, 29, 12, 18, 25, 32, 13, 16, 15

b 5, 3, 2, 2, 3, 6, 2, 5

c 20 m, 14 m, 22 m, 15 m, 18 m, 14 m, 18 m

d 98.2, 96.3, 99.8, 97.5, 89.1, 91.9, 96.7, 94.8

3 20 torch batteries were tested and the running times were measured to the nearest hour. The results are shown in the table:

39	36	50	48	46	37	53	41	38	56
43	32	46	49	36	48	47	42	29	49

Calculate the median and inter-quartile range of the lifetime of the batteries.

If a battery is used in a clock, it will run for 120 times as long. What would be the median lifetime of a battery used in a clock? Give your answer in days.

4 Representatives of youth associations who had travelled to an annual conference were asked to record their mileage, d, and time taken, t (to the nearest minute), to get to the conference. The data is given in the table.

d	116	15	99	134	126	125	140	57	75
t	135	29	186	160	175	129	200	56	95

Find the median and inter-quartile range of both sets of data.

5 The two sets of data show the marks (out of 60) gained by two sets of students in a mathematics test.

Set A 33, 39, 36, 48, 45, 36, 36, 48, 39, 36, 48, 42

Set B 48, 24, 21, 48, 21, 36, 45, 24, 18, 54, 60

Find the median mark and inter-quartile range for each set of marks.

Comment on any differences in the performance of the two sets of students.

Cumulative frequency

b Data in an ungrouped frequency distribution

To find the median and quartiles of data in the form of an **ungrouped frequency distribution**, it is useful to find the **cumulative frequency**.

This is the total frequency up to a particular item.

8 The table shows the number of pets owned by the 29 children in Class 10H.

a Find the median number of pets.

b Find the quartiles and inter-quartile range.

Number of pets	0	1	2	3	4	5
Frequency	6	8	6	4	3	2

Answer

a

- Construct a cumulative frequency distribution.

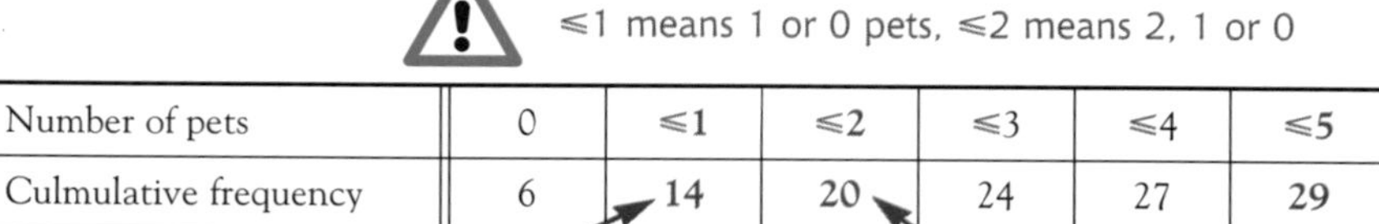

Number of pets	0	≤1	≤2	≤3	≤4	≤5
Culmulative frequency	6	14	20	24	27	29

- Since there are 29 values the median is the $\frac{1}{2}(29 + 1)$th, i.e. the 15th value.
- There are 14 children with 0 or 1 pet and 20 children with 0, 1 or 2 pets, so the 15th child must have 2 pets.

The median = 2 pets per child

b

- lower quartile $= \frac{1}{4}(29 + 1)$th value $=$ **7.5th** value $= 1$
- upper quartile $= \frac{3}{4}(29 + 1)$th value $=$ 22.5th value $= 3$
- inter-quartile range $=$ upper quartile $-$ lower quartile $= 3 - 1 = 2$

Inter-quartile range = 2 pets per child

6 children have no pets,
14 children have 0 or 1 pet
so the 7.5th value is 1.

c Data in a grouped frequency distribution

For grouped continuous data

- the median is the $\frac{n}{2}$th value,
- the lower quartile is the $\frac{n}{4}$th value,
- the upper quartile is the $\frac{3n}{4}$th value.

If the frequency distribution is grouped then the cumulative frequency table uses the **upper class boundary** of the group. The cumulative frequency is the 'running total' up to a particular upper class boundary. Approximate values of the median and quartiles can be found from the table.

Sample Question 9 The table shows the speed, s, of vehicles as they join a motorway.

a Construct a cumulative frequency table.

b In which class intervals are the median, upper and lower quartiles?

Speed (mph)	$30 < s \leqslant 40$	$40 < s \leqslant 50$	$50 < s \leqslant 60$	$60 < s \leqslant 70$	$70 < s \leqslant 80$
Frequency	4	15	33	11	1

Answer

a

- The speed columns in the cumulative frequency table should read $\leqslant 40$, $\leqslant 50$, $\leqslant 60$, $\leqslant 70$ and $\leqslant 80$ as you are creating a 'running total' of vehicles with speeds up to the upper boundary of each class interval. For example, the cumulative total for $\leqslant 60$ means the **total** number of vehicles travelling 60 mph or less.
- The cumulative total in the last column ($\leqslant 80$) should be the same as the sum of the frequencies, 64 in this case.

Speed (mph)	$\leqslant 40$	$\leqslant 50$	$\leqslant 60$	$\leqslant 70$	$\leqslant 80$
Cumulative frequency	4	19	52	63	64

b

For grouped continuous data

- the median is the $\frac{n}{2}$th value. This is the 32nd value which is in the interval $50 < s \leqslant 60$
- the lower quartile is the $\frac{n}{4}$th value. This is the 16th value which is in the interval $40 < s \leqslant 50$
- the upper quartile is the $\frac{3n}{4}$th value. This is the 48th value which is in the interval $50 < s \leqslant 60$

Cumulative frequency curve

As you can see from Sample Question 9, it was only possible to put the median and quartiles into specific class intervals. In fact, as you do not have all the data it is not possible to give an exact value for these measures, only an estimate.

To find an estimate of the median, you can draw a **cumulative frequency curve** using points that you obtained in your cumulative frequency table.

So for the data in Sample Question 9 you would draw a grid and plot the points (**40**, 4), (**50**, 19), (**60**, 52), (**70**, 63) and (**80**, 64). Since no speeds were less than 30, you could also plot (30, 0).

The '*x*' co-ordinate must be the upper boundary of the class interval.

As you cannot assume that the values within a class interval are evenly distributed, join the points with a smooth curve, not straight lines.

From this curve you are able to give a better estimate for the median and quartiles of the distribution.

Sample Question 10

a Draw a cumulative frequency curve for the data in Sample Question 9.

b Use your curve to find an estimate for
  i the median,
  ii the upper and lower quartiles,
  iii the inter-quartile range.

c What is the probability that a vehicle joining the motorway, selected at random, will be travelling faster than 65 mph?

Answer

Often, on an examination paper, the grid and axes are drawn for you.

a

- On squared paper, draw a set of axes. Label the horizontal axis 'speed' and the vertical axis 'cumulative frequency'. The horizontal axis should be from 30 to 80 and the vertical axis from 0 to 64.
- Plot the points (30, 0) (40, 4), (50, 19), (60, 52), (70, 63) and (80, 64). Join the points with a smooth curve. The line does not have to go to the origin, (0, 0), but it stops at (80, 64).

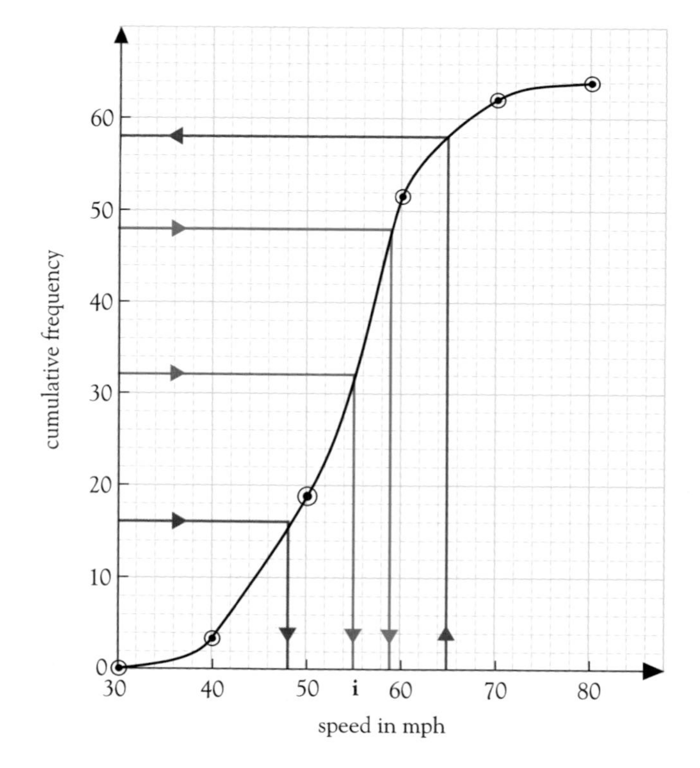

b i

- You know that the median is the 32nd value so find 32 on the 'cumulative frequency' axis. Draw a horizontal line to meet the curve and, from this point draw a vertical line to the 'speed' axis.
- Read off the speed from the axis. This is an estimate for the median speed.

ii

- The **lower quartile** is the 16th value so repeat the procedure as for the median but from 16 on the 'cumulative frequency' axis.
- Repeat for the 48th value to find the **upper quartile**.

iii

- Use inter-quartile range = upper quartile − lower quartile.

c

- From 65 on the 'speed' axis draw a **vertical line** to meet your curve and then draw a **horizontal line** to meet the 'cumulative frequency' axis.
- Read off the value from the 'cumulative frequency' axis. This gives you an estimate for the number of vehicles that were travelling at 65 mph or less. Subtract this from 64 to find the number travelling at more than 60 mph. Then use

P(speed > 65 mph) is $\frac{64 - \text{your reading}}{64}$

b Reading from the graph

i Median = 55 mph

ii Upper quartile = 59 mph

Lower quartile = 48 mph

iii Inter-quartile range = 59 − 48 = 11 mph

c Number of vehicles travelling ≤ 65 mph = 58

P(speed > 65 mph) = $\frac{64-58}{64} = \frac{6}{64}$

P(speed > 65 mph) = $\frac{3}{32}$

Note that, in an examination, whilst any estimate for the median or quartiles will fall into a particular range, the answers that **you** give must come from **your graph. The 'working' lines are very important and you should not leave them out.** As long as your answer falls into the examiner's range you will get the marks. Don't worry if your answer is not exactly the same as somebody else's. Your graph will probably be slightly different to theirs!

Sample Question 11

The graph shows the cumulative frequency curve for the serving times at Costsave.

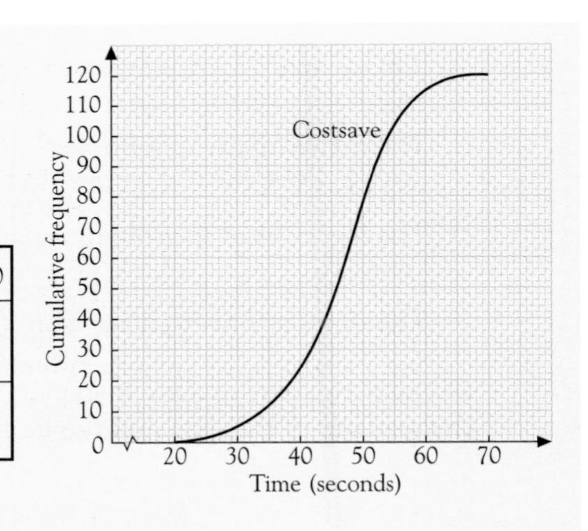

a On the same axes, show the cumulative frequency curve for the serving times at Pricewell. The table is given below.

Time (sec)	20–30	30–40	40–50	50–60	60–70
Number of customers at Pricewell	4	17	48	16	35
Number of customers at Costsave	5	20	54	36	5

b Copy and complete the table below.

	Pricewell	Costsave
lower quartile		41
upper quartile		53
inter-quartile range		12

c Use the information in the table to comment on the difference in the distributions of the serving times at Pricewell and Costsave. [MEG, p]

Answer

a

- Work out the cumulative frequency for Pricewell using the upper boundary of each class interval as a column heading.

Time (sec)	⩽30	⩽40	⩽50	⩽60	⩽70
Cumulative frequency	4	21	69	85	120

It does not matter whether you use ⩽ or <, as the boundary will still be the same number.

- Plot the cumulative frequency against the end of each interval.
- Draw a smooth curve through the points, plotting (20, 0) also.

b

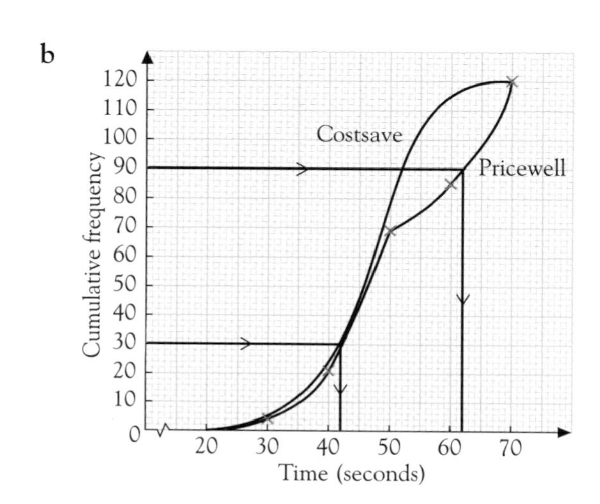

- As the data are continuous

  lower quartile = $\frac{n}{4}$th value, i.e. 30th value

  upper quartile = $\frac{3n}{4}$th value, i.e. 90th value

- Draw your 'method' lines on the graph and read off the values of the quartiles from the 'Time' axis.
- Put your results into the table.

	Pricewell	Costsave
lower quartile	42	41
upper quartile	62	53
inter-quartile range	20	12

c The higher upper quartile for Pricewell shows that more customers have to wait longer to be served at Pricewell than at Costsave.

The bigger inter-quartile range at Pricewell shows that the serving times are more spread than at Costsave.

The times at Pricewell are more dispersed at the upper end.

In comparing sets of data, if you are not told what to compare or comment on, you should compare the medians of the data and the inter-quartile ranges.

Exercise 24.5

1 The table shows the number of television sets owned by 50 families chosen at random from a school of 350 pupils.

Number of TVs	0	1	2	3	4	5
Number of families	1	10	13	20	4	2

Construct a cumulative frequency distribution for the data.

a Find the median number of televisions per family.

b Find the quartiles and the inter-quartile range.

CUMULATIVE FREQUENCY

2 The marks gained out of ten in a test given to 29 students were as follows

Mark	0	1	2	3	4	5	6	7	8	9	10
Students	0	0	0	1	3	2	4	4	5	6	4

Construct a cumulative frequency distribution for the data.

a Find the median mark.

b Find the quartiles and the inter-quartile range.

3 The cumulative frequency graph shows the monthly salaries of 120 members of a sports club.

Cumulative frequency curve for monthly salaries

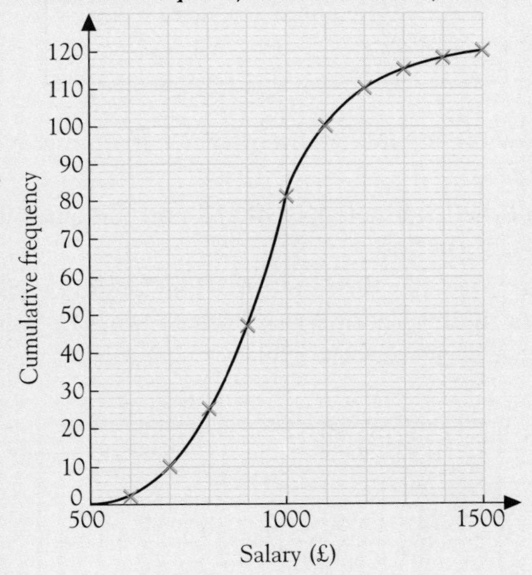

a Copy the following table and use the graph to complete it.

Monthly salary (£x)	Cumulative frequency
$\leqslant$ £500	
$\leqslant$ £600	
$\leqslant$ £700	
$\leqslant$ £800	
$\leqslant$ £900	
$\leqslant$ £1000	
$\leqslant$ £1100	
$\leqslant$ £1200	
$\leqslant$ £1300	
$\leqslant$ £1400	
$\leqslant$ £1500	

b How many members earned between £800 and £900?

c From the graph, estimate the median salary.

d Use the graph to estimate the inter-quartile range.

4 The times spent at the checkout by 200 customers at a supermarket are represented by the cumulative frequency graph.

Cumulative frequency curve for time spent at the checkout

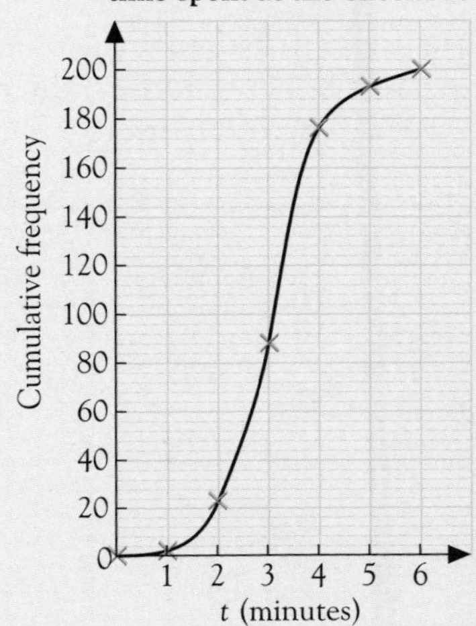

a Use the graph to estimate the median time spent by the customers.

b Use the graph to estimate the inter-quartile range of the times.

c How many customers managed to get through in less than $3\frac{1}{2}$ minutes?

d How many took longer than $4\frac{1}{2}$ minutes?

5 The exam marks of 200 boys are shown in the table.

Mark	Frequency	Cumulative frequency	
$0 < m \leqslant 10$	4	$\leqslant 10$	4
$10 < m \leqslant 20$	16	$\leqslant 20$	20
$20 < m \leqslant 30$	23	$\leqslant 30$	
$30 < m \leqslant 40$	62	$\leqslant 40$	
$40 < m \leqslant 50$	53	$\leqslant 50$	
$50 < m \leqslant 60$	28	$\leqslant 60$	
$60 < m \leqslant 70$	9	$\leqslant 70$	
$70 < m \leqslant 80$	3	$\leqslant 80$	
$80 < m \leqslant 90$	1	$\leqslant 90$	
$90 < m \leqslant 100$	1	$\leqslant 100$	

a Copy and complete the table to give the cumulative frequencies.

b Draw a cumulative frequency graph for the data.

c Use the graph to estimate the median mark and the inter-quartile range.

d If the pass mark is 45, how many boys pass?

e What percentage of the boys gain more than 53 marks?

6 150 people were asked to estimate the number of hours (to the nearest hour) that they watched television in a week. The results were

Number of hours	Frequency	Upper boundary	Cumulative frequency
0–5	2	5.5	
6–10	18	10.5	
11–15	22		
16–20	44		
21–25	41		
26–30	13		
31–35	7		
36–40	3		

a Copy and complete the table and use it to draw the cumulative frequency curve.

b Estimate the inter-quartile range.

c What percentage of the people watch more than 23 hours of television in a week?

d 16% of the people watch for less than x hours a week. Find the value of x.

7 The weights of 50 women were recorded as follows.

Weight (kg)	Frequency	Cumulative frequency
$45 < w \leq 50$	2	
$50 < w \leq 55$	5	
$55 < w \leq 60$	15	
$60 < w \leq 65$	14	
$65 < w \leq 70$	9	
$70 < w \leq 75$	3	
$75 < w \leq 80$	2	

a Compile a cumulative frequency table for the data.

b Draw a cumulative frequency graph.

c Use the graph to estimate the median weight and the inter-quartile range.

d What percentage of the women weigh more than 68 kg?

e Another group of women have a median weight of 68 kg and an inter-quartile range of 10 kg. Compare the two groups.

8 The cumulative frequency table shows the duration of 80 calls from the telephone kiosk on the corner of North Street.

Length of call (minutes)	Cumulative frequency
0	0
<3	4
<6	18
<9	43
<12	63
<18	78
<21	80

Draw the cumulative frequency curve on graph paper.

Find the median length of call and the quartiles of the distribution.

The frequency distribution of the length of calls from the telephone kiosk on the corner of South Street is given in the table.

Length of call	Frequency
$0 \leq t < 3$	22
$3 \leq t < 6$	28
$6 \leq t < 9$	14
$9 \leq t < 12$	10
$12 \leq t < 15$	4
$15 \leq t < 18$	1
$18 \leq t < 21$	1

On the same set of axes as the graph for North Street, draw the cumulative frequency curve for the length of calls from the South Street kiosk. Use your graphs to complete the following table.

	North Street	South Street
median		
lower quartile		
upper quartile		
inter-quartile range		

Use the table to compare the lengths of calls at the two kiosks.

CUMULATIVE FREQUENCY

9 The manager of a motorway service area conducted a survey one lunchtime to find how long 120 customers spent in the restaurant. His results are shown in the table below.

Time (t minutes)	$0 < t \leqslant 10$	$10 < t \leqslant 20$	$20 < t \leqslant 30$	$30 < t \leqslant 40$	$40 < t \leqslant 50$
Number of customers	16	60	28	10	6

a Copy and complete the table below.

Time (t minutes)	$t \leqslant 10$	$t \leqslant 20$	$t \leqslant 30$	$t \leqslant 40$	$t \leqslant 50$
Number of customers					

b Draw a cumulative frequency curve for the time spent in the restaurant.

c Use the cumulative frequency curve to find the median of this distribution.

d The manager said that one third of the customers spent more than 25 minutes in the restaurant. Does the cumulative frequency curve support the manager's claim? Give numerical values to support your conclusion. [MEG]

10 A sample of 80 eggs from a farm were weighed. The results are shown in the table below.

Mass (m grams)	$40 < m \leqslant 45$	$45 < m \leqslant 50$	$50 < m \leqslant 55$
Number of eggs	2	19	24

Mass (m grams)	$55 < m \leqslant 60$	$60 < m \leqslant 65$	$65 < m \leqslant 70$
Number of eggs	21	13	1

a An egg is classified as large if it weighs more than 60 g.
  i One egg is picked at random from the sample.
  What is the probability that it is large?
  ii A shopkeeper buys 2400 eggs from the farmer.
  How many would he expect to be large?

b Copy and complete the cumulative frequency curve for the data.

m (not more than)	45	50	55	60	65	70
Number of eggs	2					

c Draw a cumulative frequency curve for the data.

d Find for this data
  i the median,
  ii the inter-quartile range. [MEG]

11 A survey is made of all 120 houses on an estate. The floor area, in m², of each house is recorded.

The results are shown in the cumulative frequency table.

Floor area (x) in m²	Cumulative frequency
$0 < x \leqslant 100$	4
$0 < x \leqslant 150$	20
$0 < x \leqslant 200$	49
$0 < x \leqslant 250$	97
$0 < x \leqslant 300$	114
$0 < x \leqslant 350$	118
$0 < x \leqslant 400$	120

a Draw a cumulative frequency graph for the table.

b Use your cumulative frequency graph to estimate the inter-quartile range of the floor areas of the houses.

The houses on the estate with the greatest floor areas are called luxury houses. 10% of the houses are luxury houses.

c Use your graph to estimate the minimum floor area for a luxury house. [L]

12 The prices, P, in pounds of 120 holidays offered by a travel firm are shown in the table.

Price (£P)	$400 \leqslant P < 500$	$500 \leqslant P < 600$	$600 \leqslant P < 700$
Number	14	39	33

Price (£P)	$700 \leqslant P < 800$	$800 \leqslant P < 900$	$900 \leqslant P < 1000$
Number	20	10	4

a Copy and complete the cumulative frequency table.

Price (£P)	$P < 400$	$P < 500$	$P < 600$	$P < 700$	$P < 800$	$P < 900$	$P < 1000$
Cumulative frequency	0						

b On a grid draw the cumulative frequency graph.

c Use your graph to estimate
  i the median cost,
  ii the inter-quartile range of the cost,
  iii the percentage of holidays that cost more than £850. [MEG]

Worked Exam Question
[NEAB]

Two archers shoot at a target.
The probability that Marion hits is $\frac{2}{5}$.
The probability that Robin hits is $\frac{1}{3}$.

a i Marion shoots at the target once. What is the probability that she misses it?

- P(Marion misses) = 1 − P(Marion hits)

ii Marion shoots at the target 3 times. Calculate the probability that she misses the target each time.

- Assuming that the events are independent, you can use the multiplication law.
- P(miss each time) = P(miss) × P(miss) × P(miss)

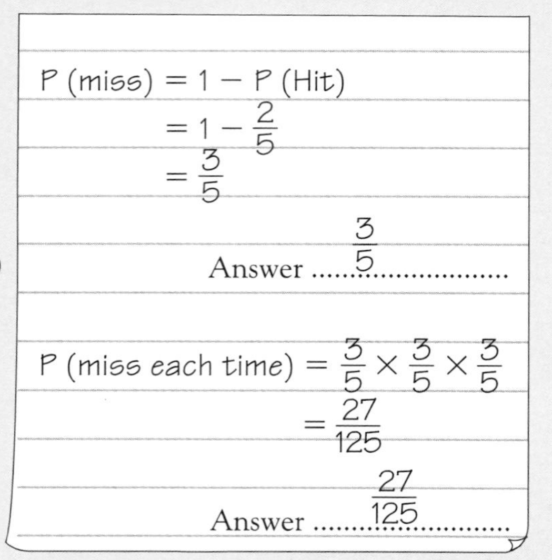

COMMENTS

M1 correct method

A1 correct answer

M1 use of multiplication law

A1 correct answer (f.t. from student's answer to **a i**)

4 marks

b Marion and Robin each shoot once. Complete the tree diagram.

- You already know the probability of Marion hitting and you have worked out the probability of her missing. Put these on the diagram.
- Work out P(Robin misses) as you did for Marion. Put this on the diagram.
- Write the outcomes at the end of each branch.

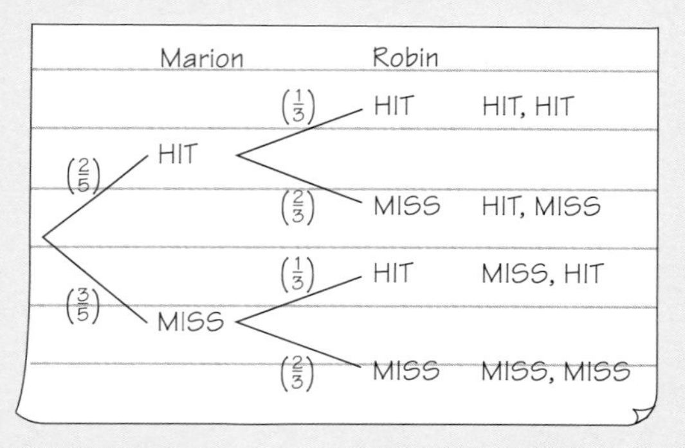

B1 correct placement of Marion's probabilities.

A1 correctly working out P(Robin misses) and correct placement

2 marks

c Calculate the probability that

i they both hit the target,

- Find the outcome that gives (HIT, HIT)
- Use the multiplication law to find P(H,H)

$$P(H, H) = \frac{2}{5} \times \frac{1}{3} = \frac{2}{15}$$

Answer ..$\frac{2}{15}$..

M1 use of multiplication law

A1 correct answer (f.t. from tree diagram)

2 marks

ii at least one of them hits the target.

- 'At least one hits' means Marion hits or Robin hits or both hit. This leaves the outcome 'both miss'.
- Use P(≥1 hit) = 1 − P(both miss)

$$P(\geq 1 \text{ hit}) = 1 - P(M, M)$$
$$= 1 - \left(\frac{3}{5} \times \frac{2}{3}\right)$$
$$= 1 - \frac{2}{5} = \frac{3}{5}$$

Answer ..$\frac{3}{5}$..

M1 any acceptable method including calculation of P(H,M) & P(M,H) then adding to P(H,H)

A1 correct answer (f.t. from tree diagram)

2 marks

Exam Questions

1 A target has a bull's-eye worth 10 points, and an outer ring worth 3 points.

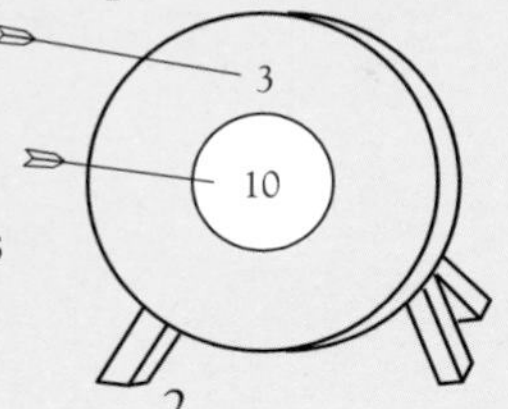

When Siobhan fires an arrow at the target:
the probability that she misses the target completely is $\frac{1}{9}$;
the probability that she scores 10, is $\frac{2}{9}$;
the probability that she scores 3, is $\frac{2}{3}$.

Siobhan fires two arrows.

a Calculate the probability that Siobhan fails to score.

b Calculate the probability that she scores a total of 13 points.

c Calculate the probability that she scores **either** no points **or** 6 points. [SEG]

2 Alison, Brenda, Claire and Donna are the only runners in a race.
The probabilities of Alison, Brenda and Claire winning the race are shown below.

Alison	Brenda	Claire	Donna
0.31	0.28	0.24	

a Work out the probability that Donna will win the race.

b Calculate the probability that either Alison or Claire will win the race.

Hannah and Tracy play each other in a tennis match.
The probability of Hannah winning the tennis match is 0.47.

c Copy and complete the probability tree diagram.

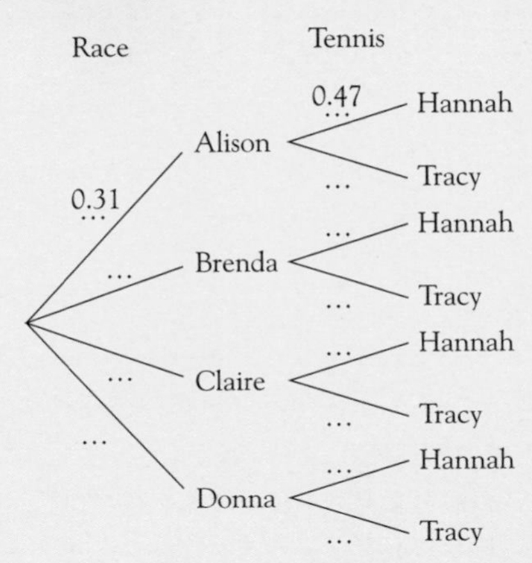

d Calculate the probability that Brenda will win the race and Tracy will win the tennis match. [L]

3 When I answer the telephone the call is never for me.
Half the calls are for my daughter Janette.
One-third of them are for my son Glen.
The rest are for my wife Barbara.

a I answer the telephone twice this evening.
Calculate the probability that
i the first call will be for Barbara,
ii both calls will be for Barbara.

b The probability that both these two telephone calls are for Janette is $\frac{1}{4}$.
The probability that they are both for Glen is $\frac{1}{9}$.
Calculate the probability that either they are both for Janette or both for Glen. [NEAB]

4 Tim is blowing up balloons for a party. He takes a balloon out of a packet.
The probability that the balloon is red is 0.4.
The probability that it is blue is 0.5.
The probability that a balloon bursts while he is blowing it up is 0.02, independently of its colour.

a Copy and complete this tree diagram to show the different probabilities.

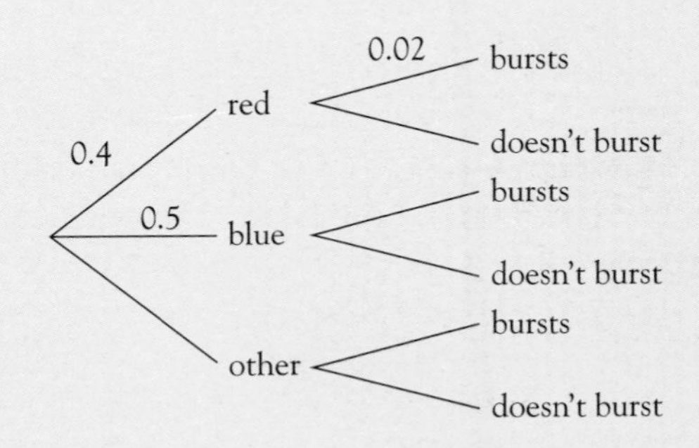

b What is the probability that Tim takes out a red balloon and it bursts while he is blowing it up?

c What is the probability that either the balloon is red or it bursts, but not both? [MEG]

5 Jane's mother says she will let her choose a pudding on her birthday and on the next day.
The probability that Jane will choose ice-cream on her birthday is $\frac{7}{10}$.

EXAM QUESTIONS

If Jane chooses ice-cream on her birthday, there is a probability of $\frac{2}{5}$ that she will choose it on the next day.

If Jane does not choose ice-cream on her birthday, there is a probability of $\frac{9}{10}$ that she will choose it on the next day.

a Copy and complete the tree diagram to show all the probabilities for each day.

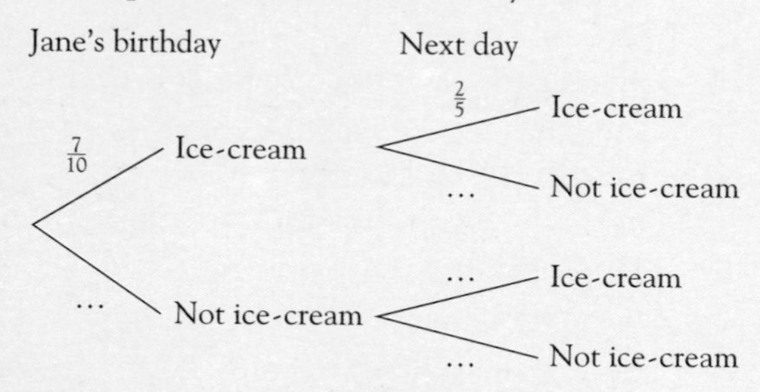

b What is the probability that Jane will choose ice-cream on the next day and not on her birthday?

c What is the probability that Jane will choose ice-cream on at least one of these two days?

[MEG]

6 A spinner, with its edges numbered one to four, is biased.

For one spin, the probability of scoring 1 is 0.2, the probability of scoring 3 is 0.3, and the probability of scoring 4 is 0.15.

a Calculate the probability of scoring 2 with one spin.

b The spinner is used in a board game called 'Steeplechase'. In the game, a player's counter is moved forwards at each turn by the score shown on the spinner.

If the player's counter lands on one of the two squares numbered 27 and 28 (labelled 'WATER JUMP'), the player is out of the game.

23	24	25	26	WATER JUMP 27	28	29	30	31

i Ann's counter is on square 26. Find the probability that she will **not** be out of the game after one more turn.

ii Peter's counter is on square 25. Find the probability that, after two more turns, his counter will be on square 29.

[MEG]

7 Ahmed recorded how long it had taken him to finish each of his last 100 homeworks then constructed a cumulative frequency curve to show the results.

Use the graph below to answer the following questions.

a How many homeworks took 20 minutes or less?

b Estimate the median time taken to finish a homework.

c Estimate the upper quartile time.

d Estimate the lower quartile time.

e Calculate the inter-quartile time.

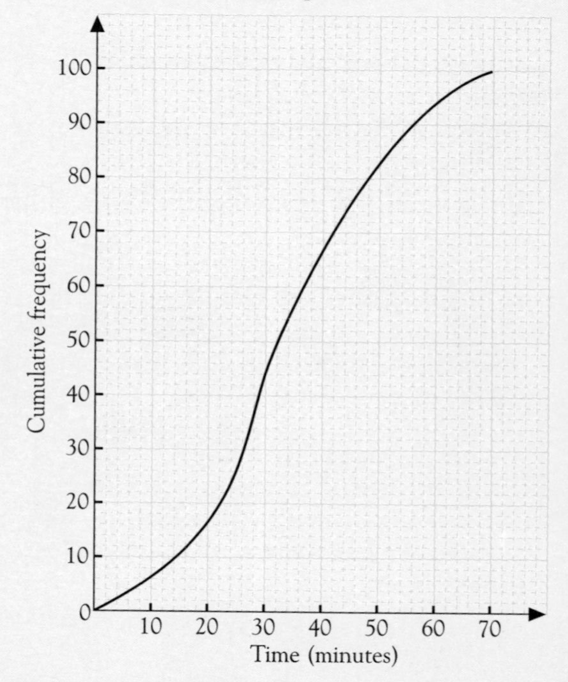

[L]

8 Twenty-five people took part in a competition.

The points scored are grouped in the frequency table below.

Points scored	Number of people		
1 to 5	1		
6 to 10	2		
11 to 15	5		
16 to 20	7		
21 to 25	8		
26 to 30	2		

a Work out the class interval which contains the median.

b Work out an estimate for the mean number of points scored.

c Complete the table below to show the cumulative frequency for this data.

Points scored	Cumulative frequency
1 to 5	
1 to 10	
1 to 15	
1 to 20	
1 to 25	
1 to 30	

d Draw a cumulative frequency graph for this data. [L]

9 The 110 pupils in Year 11 at Grange Valley School were asked how many KMP tasks they had completed in the previous school term. The table below shows the results obtained.

Number of KMP tasks completed	Number of pupils (frequency)	Cumulative frequency
1–5	2	2
6–10	6	8
11–15	15	23
16–20	21	
21–25	29	
26–30	23	
31–35	11	
36–40	3	110

a Complete the cumulative frequency column in the table.

b On graph paper, draw the cumulative frequency curve (ogive).

c Use your graph paper to estimate the median number of KMP tasks completed.

The Mathematics department at Grange Valley awards certificates to those Year 11 pupils who complete 29 tasks or more in a term.

d Use your graph to estimate the number of pupils who should be awarded certificates for their work last term.

e Use your graph to work out an estimate for the median for this data. [L]

10 The lengths of time, t minutes, of the tracks on my compact discs are summarised in the table below.

Time (t mins)	$0 < t \leq 1$	$1 < t \leq 2$	$2 < t \leq 3$
Number of tracks	2	7	13

Time (t mins)	$3 < t \leq 4$	$4 < t \leq 5$	$5 < t \leq 6$
Number of tracks	23	18	5

a Do a cumulative frequency table.

b Draw a cumulative frequency curve for the information.

c Find
i the median time,
ii the inter-quartile range.

d What percentage of the tracks last for more than $4\frac{1}{2}$ minutes? [MEG]

11 The mass of each of 60 apples was recorded to the nearest gram.

Mass	80–	85–	90–	95–	100–	105–	110–115
Frequency	3	7	13	15	12	8	2

a Calculate the values of the cumulative frequencies and copy and complete the table.

Mass	<80	<85	<90	<95	<100	<105	<110	<115
Cumulative frequency								

b Draw the cumulative frequency curve.

c Use your graph to estimate the inter-quartile range of the mass of the apples.

d Use your graph to estimate the number of apples that have a mass of less than 106 g.

e Seventeen of the apples are rejected as they are too heavy. What is the maximum weight of the apples that are accepted? [SEG]

25 QUADRATICS II

In this chapter you will learn how to
- **factorise quadratic expressions of the type $x^2 + bx + c$**
- **solve quadratic equations of the type $x^2 + bx + c = 0$**
- **find the *n*th term of a quadratic expression**

Factorising quadratic expressions of the type $x^2 + bx + c$

In chapter 21 you learned how to multiply out two brackets, for example

$$(x + 3)(x + 5) = x^2 + 5x + 3x + 15$$
$$= x^2 + 8x + 15$$

Remember how each of the terms is obtained:

$$(x + 3)(x + 5) = x^2 \quad + \quad 5x \quad + \quad 3x \quad + \quad 15$$
$$(x \quad)(x \quad) \qquad (x \quad)(\quad +5) \qquad (\quad +3)(x \quad) \qquad (\quad +3)(\quad +5)$$

If you prefer to use a multiplication square, look at these diagrams

First term

×	x	$+5$
x	x^2	
$+3$		

Middle terms

×	x	$+5$
x		$+5x$
$+3$	$+3x$	

Last term

×	x	$+5$
x		
$+3$		15

If you start with $x^2 + 8x + 15$ and re-write it in the form $(x + 3)(x + 5)$, this process is known as **factorising**. You have written the expression as two terms multiplied by each other and these terms, $(x + 3)$ and $(x + 5)$, are **factors**. Look at this example.

Factorise $x^2 + 7x + 12$
You need to complete the two brackets so that

$$x^2 + 7x + 12 = (\qquad)(\qquad)$$

Step 1: To get x^2, you know that the first term in each bracket must be x

$$x^2 + 7x + 12 = (x \qquad)(x \qquad)$$

Step 2: Look at the last term. When you multiply the second term in each bracket, you need to get $+12$

$$x^2 + 7x + 12 = (x \ \square)(x \ \square)$$

There are several pairs of numbers which give $+12$ when when you multiply them. They are 12×1, $(-12) \times (-1)$, 6×2, $(-6) \times (-2)$, 3×4, $(-3) \times (-4)$.
You have to decide which pair it is.

FACTORISING

Step 3: To decide which pair it is, look at the middle term.

$x^2 + 7x + 12 = (x\ \square)(x\ \square)$ When you **add** these you need to get $+7x$

Ask yourself: What two numbers multiply to give $+12$ and add to give $+7$?

The two numbers are 3 and 4 since $3 \times 4 = 12$ and $3 + 4 = 7$
Check whether this works

×	x	$+4$
x	x^2	$+4x$
$+3$	$+3x$	$+12$

$$(x + 3)(x + 4) = x^2 + 4x + 3x + 12$$
$$= x^2 + 7x + 12$$

so this **is** correct

$x^2 + 7x + 12 = \mathbf{(x + 3)(x + 4)}$

$(x + 3)(x + 4)$ is the factorised form of the expression $x^2 + 7x + 12$

You can see from these multiplication squares why the other pairs of numbers that multiply to give 12 would not work.

×	x	$+12$
x		$+12x$
$+1$	$+x$	

Middle term $= +13x$

×	x	$+12$
x		$+12x$
-1	$-x$	

Middle term $= -13x$

×	x	$+6$
x		$+6x$
$+2$	$+2x$	

Middle term $= +8x$

×	x	-6
x		$-6x$
-2	$-2x$	

Middle term $= -8x$

×	x	-3
x		$-3x$
-4	$-4x$	

Middle term $= -7x$

With practice you should be able to decide quite quickly which numbers to use.
Each time you will need to ask yourself:
What two numbers multiply to give the number term **and** add to give the number in the x term?

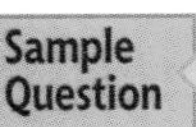

Factorise

a $x^2 + 6x - 16$ **b** $x^2 - 7x - 30$ **c** $x^2 - 5x + 6$

Answer

a

- Ask yourself: What two numbers multiply to give -16 (the number term) and add to give $+6$ (the number in the x term). Write it in shorthand like this:

  × to give -16
  \+ to give 6

  Clue: one number must be positive, the other negative, as $(+) \times (-) = (-)$

  The numbers are 8 and -2 since $8 \times (-2) = -16$ and $8 + (-2) = 6$

- Write x in each bracket, then fit in the numbers.

  $x^2 + 6x - 16 = (x + 8)(x - 2)$

Check that your answer is correct by multiplying it out to see whether you get $x^2 + 6x - 16$.

You could write $x^2 + 6x - 16 = (x - 2)(x + 8)$
It does not matter which bracket is written first.

b

- Ask yourself: What two numbers multiply to give −30 and add to give −7?

  i.e. × to give **−30**
  + to give −7

One must be positive, the other negative, since (+) × (−) = (−)

The numbers are **−10 and 3**

$(-10) \times 3 = -30$
$(-10) + 3 = -7$

$\underline{x^2 - 7x - 30 = (x - 10)(x + 3)}$

Check by multiplying out.

c

- Ask yourself: What two numbers multiply to give +6 and add to give −5?

  i.e. × to give **+6**
  + to give **−5**

If they multiply to give a positive number, the numbers are both positive or both negative. But if they are both positive, they would add to give a positive number. So they must both be negative.

The numbers are **−3** and **−2**

$(-3) \times (-2) = 6$
$(-3) + (-2) = -5$

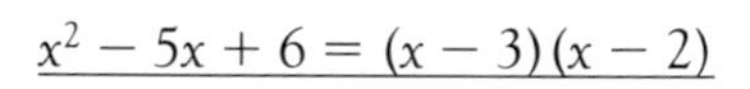
$\underline{x^2 - 5x + 6 = (x - 3)(x - 2)}$

Check.

Here are some clues to help you decide whether you are looking for two positive numbers, two negative numbers or one positive and one negative number.

If the numbers × to give (+) + to give (+) they are both positive	If the numbers × to give (+) + to give (−) they are both negative
If the numbers × to give (−) + to give (+) One is positive, one is negative and the larger 'number' is positive	If the numbers × to give (−) + to give (−) One is positive, one is negative and the larger 'number' is negative

Sample Question 2 Factorise $x^2 - 25$

Answer

- There is no x term. You could think of this as $0x$. So you want two numbers that multiply to give −25 and add to give 0. The numbers are 5 and −5.

  $\underline{x^2 - 25 = (x + 5)(x - 5)}$

FACTORISING

When you subtract, you find the difference. In the expression $x^2 - 25$, a square number, 5^2, is being subtracted from another square term, x^2. Factorising this type has a special name: **factorising the difference between two squares.**

Another example would be $x^2 - 64 = (x - 8)(x + 8)$

It is useful to learn a general rule

$$x^2 - a^2 = (x - a)(x + a)$$

You must have a minus between the two squared terms.

Exercise 25.1

1 **a** Find numbers a and b so that each pair of equations is satisfied

i $a \times b = 24$
$a + b = 10$

ii $a \times b = 12$
$a + b = 7$

iii $a \times b = -10$
$a + b = 3$

iv $a \times b = -12$
$a + b = 4$

v $a \times b = 36$
$a + b = -15$

vi $a \times b = -63$
$a + b = -2$

b Write down two numbers which

i multiply to give 81 and add to give -18
ii multiply to give -56 and add to give -1
iii multiply to give 48 and add to give -14
iv multiply to give -28 and add to give 12
v multiply to give -48 and add to give 13
vi multiply to give -36 and add to give 0

2 Factorise the following

a $x^2 + 5x + 6$
b $x^2 + 7x + 10$
c $x^2 + 12x + 35$
d $x^2 - 8x + 12$
e $x^2 - 4x + 3$
f $x^2 - 11x + 24$
g $x^2 - 7x + 12$
h $x^2 + 5x - 6$
i $x^2 + 4x - 5$
j $x^2 - 4x - 12$
k $x^2 - 2x - 8$
l $x^2 - 9x - 22$
m $x^2 + 11x - 12$
n $x^2 - x - 2$
o $x^2 - 8x + 16$
p $a^2 - 3a - 18$
q $p^2 + 20p + 100$
r $b^2 - 12b + 36$
s $q^2 - 10q + 16$
t $w^2 + 22w + 21$
u $t^2 + t - 56$
v $m^2 - 2m - 63$
w $n^2 - n - 42$
x $y^2 + 6y + 9$
y $c^2 - 3c - 10$
z $d^2 - d - 72$

3 Factorise the following (hint: look for a common factor)

a $x^2 + 5x$
b $x^2 - 3x$
c $y^2 - 20y$
d $a^2 + 2a$
e $5x^2 - 25x$
f $7a^2 - 21a$

4 Factorise the following

a $x^2 - 9$
b $a^2 - 16$
c $b^2 - 81$
d $36 - w^2$
e $x^2 - 121$
f $4 - x^2$

5 Factorise the following completely (hint: look for a common factor first)

a $2x^2 + 6x + 4$
b $3x^2 - 27x + 42$
c $3x^2 - 9x + 6$
d $3x^2 - 12$
e $7x^2 + 7x$
f $4x^2 - 16$

6 Factorise the following completely

a $x^2 + x - 6$
b $3a^2 + 12a$
c $n^2 - 10n + 25$
d $t^2 - 144$
e $b^2 + 12b + 36$
f $2y^2 + 16y + 32$
g $25w^2 - 100$
h $x^2 - x - 20$
i $5m^2 + 25m + 20$
j $x^2 - 25x$
k $2s^2 - 4s + 2$
l $9u^2 - 81$

QUADRATIC EQUATIONS

Solving quadratic equations of the type $x^2 + bx + c = 0$

If you are asked to 'solve' $x^2 + 3x - 10 = 0$ you have to find values of x that give nought when you substitute them into $x^2 + 3x - 10$. In chapter 21 you looked at a graphical method and a trial and improvement method for finding the solution.

If the quadratic expression factorises, there is an easy method which works like this:

- Factorise the expression

  $$x^2 + 3x - 10 = (x - 2)(x + 5)$$

  This is really $(x - 2) \times (x + 5)$

- Re-write the equation $x^2 + 3x - 10 = 0$ using the factors.

  $$(x - 2) \times (x + 5) = 0$$

- Remember that when you multiply something by nought, the answer is nought, so put each of the factors equal to nought, in turn:

  $(x - 2) \times (x + 5) = 0$ or $(x - 2) \times (x + 5) = 0$

  i.e. $0 \times (x + 5) = 0$ or $(x - 2) \times 0 = 0$

  so $\mathbf{x - 2 = 0}$ or $\mathbf{x + 5 = 0}$

  $x = 2$ $\quad$ $x = -5$

Solve the two simple equations.

  The solutions are $x = 2$ and $x = -5$

- Check that they are correct by substituting into $x^2 + 3x - 10$. The value obtained should be nought.

  When $x = 2$, $\quad x^2 + 3x - 10 = 2^2 + 3 \times 2 - 10 = 4 + 6 - 10 = 0$

  When $x = -5$, $\quad x^2 + 3x - 10 = (-5)^2 + 3 \times (-5) - 10 = 25 - 15 - 10 = 0$

Sample Question 3 Solve these quadratic equations

a $x^2 + 6x + 8 = 0$

b $x^2 = 11x - 24$

c $x^2 - 100 = 0$

d $x^2 - 6x + 9$

e $x^2 - 6x = 0$

Answer

a

- Write the expression in factorised form, then put each factor equal to nought and solve the simple equation.

  $$x^2 + 6x + 8 = 0$$

  $$(x + 2)(x + 4) = 0$$

  Either $x + 2 = 0$ or $x + 4 = 0$

  $\underline{\mathbf{x = -2}}$ $\quad$ $\underline{\mathbf{x = -4}}$

Check that these work.

b

- Re-write the equation so that it is in the form $x^2 + bx + c = 0$, then solve by factorising.

  $$x^2 = 11x - 24$$

  $$x^2 - 11x + 24 = 0$$

  $$(x - 3)(x - 8) = 0$$

  Either $x - 3 = 0$ or $x - 8 = 0$

  $\underline{\mathbf{x = 3}}$ $\quad$ $\underline{\mathbf{x = 8}}$

Check.

QUADRATIC EQUATIONS

c

$$x^2 - 100 = 0$$
$$(x - 10)(x + 10) = 0$$

Either $x - 10 = 0$ or $x + 10 = 0$

$\underline{x = 10}$ $\underline{x = -10}$

Remember this special type: difference between two squares.

d

$$x^2 + 6x + 9 = 0$$
$$(x + 3)(x + 3) = 0$$
$$x + 3 = 0$$
$$\underline{x = -3}$$

Both brackets are the same, so just write $x + 3 = 0$. There is no need to write it twice!

Note that there is only one value of x in this example.

e

$$x^2 - 6x = 0$$
$$x(x - 6) = 0$$

Either $\underline{x = 0}$ or $x - 6 = 0$

$\underline{x = 6}$

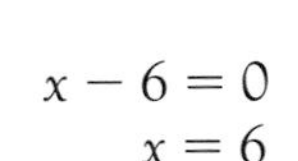

In this example there is no constant term. It is possible to take out a **common factor**.

Sometimes you need to solve a quadratic equation in order to find the answer to a problem.

In this rectangle, the length is $(x + 5)$ cm and the width is $(x + 2)$ cm. The area of the rectangle is 54 cm^2. Find x and hence find the dimensions of the rectangle.

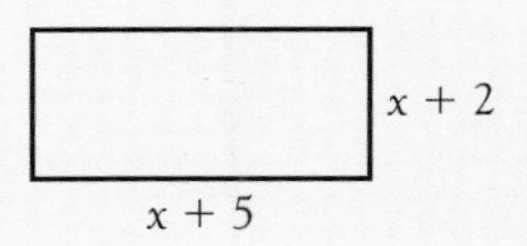

Answer

- Write the expression for the area

$$(x + 5)(x + 2) = 54$$

DO NOT write $x + 5 = 54$ or $x + 2 = 54$. This is nonsense!

- Multiply out the brackets and re-arrange the terms to get an equation of the form $x^2 + bx + c = 0$.

$$x^2 + 2x + 5x + 10 = 54$$
$$x^2 + 7x + 10 = 54$$

Subtract 54 $\quad x^2 + 7x - 44 = 0$

- Factorise and solve

$$(x - 4)(x + 11) = 0$$

Either $x - 4 = 0$ or $x + 11 = 0$

$x = 4$ $\quad x = -11$

You cannot have $x = -11$ as this would give a length of $-11 + 5 = -6$ which is impossible. This solution to the equation is not applicable to this problem.

$\underline{x = 4}$

Length $= x + 5 = 4 + 5 = 9$ cm

Width $= x + 2 = 4 + 2 = 6$ cm

<u>The rectangle has dimensions 9 cm by 6 cm</u>

QUADRATIC EQUATIONS

Exercise 25.2

1 Solve the equations

a $(x-3)(x-1)=0$
b $(x-5)(x+6)=0$
c $x(x-7)=0$
d $(x-3)(x+3)=0$
e $(x+7)(x-7)=0$
f $6(x+5)(x-2)=0$
g $(x+12)(x+13)=0$
h $(x-9)^2=0$
i $12x(x-10)=0$
j $(10-x)(x-5)=0$
k $(7+x)(8-x)=0$
l $7(x+10)^2=0$
m $(x+13)(x-12)=0$

2 Solve the equations

a $x^2+5x+4=0$
b $x^2-10x+21=0$
c $x^2-4x+4=0$
d $x^2-10x+24=0$
e $y^2-7y-8=0$
f $y^2+10y+16=0$
g $x^2-16=0$
h $x^2+14x=0$
i $x^2+10x+25=0$
j $y^2+y-20=0$
k $y^2-14y+49=0$
l $2x^2+12x-32=0$
m $y^2+6y=0$
n $x^2+5x-24=0$
o $8x^2+24x=0$
p $5y^2-10y=0$
q $x^2+2x-8=0$
r $27-3y^2=0$

3 Re-arrange the equations then solve them

a $x^2+1=2x$
b $x^2=x+30$
c $x^2=9x$
d $5x^2=30x$
e $11x=x^2+28$
f $17=x^2+16x$
g $18-3x=x^2$
h $49=x^2$

4 I think of a number greater than zero, add on seven and multiply the result by the number I first thought of. My answer is sixty.

a Write my calculation as an algebraic equation.
b Re-arrange your equation to get a quadratic equation.
c Solve this equation to find the number I first thought of.

5 Ted is three years older than Shamshir and the product of their ages is seventy. Letting Ted's age be x years, form an equation in x.

Use the equation to find the ages of the two boys.

6 The length of a rectangle is 2 cm greater than its width. The area of the rectangle is 120 cm^2. If the width is x cm, form an equation in x. Use this equation to find the width and the length of the rectangle.

7 The dimensions of a rectangle are as shown in the diagram (all lengths in inches).

$x+5$
Area = 48 in^2
$x-3$

Form an equation in x.

Find the lengths of the sides of the rectangle.

8 An open box is formed by cutting out squares of side x cm from each corner of a cardboard rectangle of length 30 cm and width 20 cm. If the area of the base of the completed box is 416 cm^2, show that

$x^2-25x+46=0$

What is the length x?

Find the volume of the completed box.

Quadratic sequences

Look at this sequence

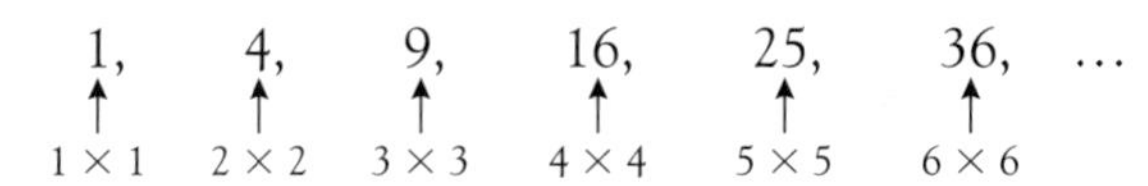

It is easy to spot that the terms are squares of consecutive numbers, starting at 1.

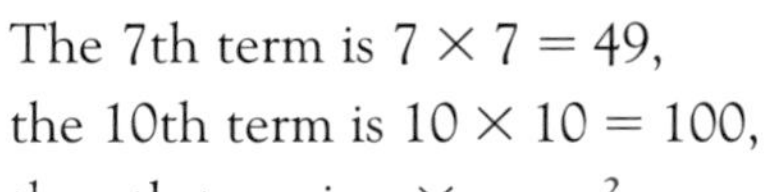

The 7th term is $7 \times 7 = 49$,
the 10th term is $10 \times 10 = 100$,
the nth term is $n \times n = \mathbf{n^2}$

n^2 is a quadratic expression.

Investigate the differences between the terms:

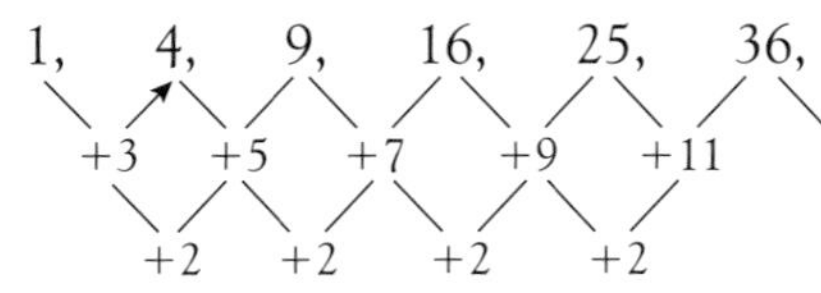

The differences increase by 2 each time.

The 'difference of the differences' is always 2. It is constant. This is a feature of quadratic sequences.

In a quadratic sequence, the difference of the differences (or second difference) is constant.

Sample Question 5 Generate the first five terms of the sequence with nth term $2n^2 - n$ and check that the second difference is constant.

Answer

- Put $n = 1, 2, 3, 4$ and 5, in turn, into the expression $2n^2 - n$ to generate the first five terms.
  When $n = 1$, $2n^2 - n = 2 \times 1^2 - 1 = 1$
  When $n = 2$, $2n^2 - n = 2 \times 2^2 - 2 = 6$
  When $n = 3$, $2n^2 - n = 2 \times 3^2 - 3 = 15$
  When $n = 4$, $2n^2 - n = 2 \times 4^2 - 4 = 28$
  When $n = 5$, $2n^2 - n = 2 \times 5^2 - 5 = 45$

The sequence is 1, 6, 15, 28, 45, ...

- Look at differences +5 +9 +13 +17
- Look at second difference +4 +4 +4

  The second difference is always 4, so it is constant.

Sample Question 6 Write down the next three terms of the sequence 100, 99, 96, 91, 84, ...

Answer

- Look at the first and second differences and continue the pattern

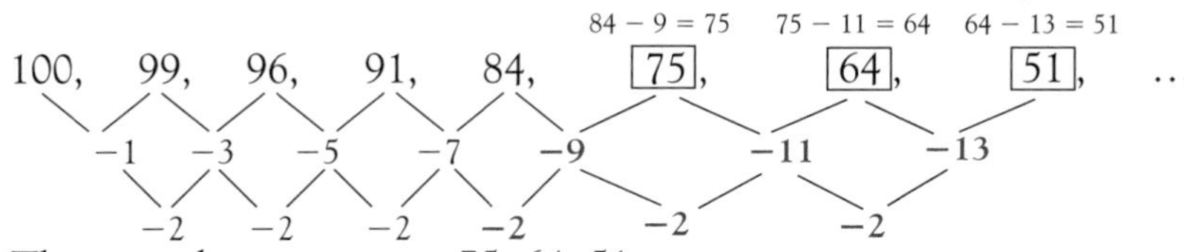

The next three terms are 75, 64, 51

Finding the *n*th term of a quadratic sequence

Sometimes the *n*th term can be spotted easily when you put the information in a table:

Consider the sequence 2, 6, 12, 20, 30, 42, 56, …

Term Number	1	2	3	4	5	6	7
Value of term	2	6	12	20	30	42	56

Notice that $2 \times 3 = 6$ (2nd term $= 2 \times 3$) and $5 \times 6 = 30$ (5th term $= 5 \times 6$)

Look for patterns formed by **multiplying** the term numbers.

You can see that, for example

6th term $= 6 \times 7 = 42$

7th term $= 7 \times 8 = 56$

So *n*th term $= n \times (\mathbf{n + 1})$

$= n^2 + n$

$(n + 1)$ is one more than n.

Check one of the values, for example $n = 4$.
When $n = 4$
$n^2 + n = 16 + 4 = 20$, which is correct.

Sample Question 7 Find the *n*th term of the quadratic sequence 0, 1, 3, 6, 10, 15, …

Answer

- Put the values in a table and look for a pattern between the term numbers and the value.

Term number	1	2	3	4	5	6
Value of term	0	1	3	6	10	15

- Look at the 5th term, for example. The value is 10. You can get this as follows.

  $5 \times 4 = 20, \quad 20 \div 2 = 10$

- Try this for the 4th term; the value is 6.

  $4 \times 3 = 12, \quad 12 \div 2 = 6$

The pattern is that for a particular term, you multiply the term number by one less than it, and divide by 2.

So *n*th term $= n \times (\mathbf{n - 1}) \div 2$

$$= \frac{n(n-1)}{2}$$

One less than n is $(n - 1)$.

If you prefer, you can write it as *n*th term $= \frac{1}{2}n(n - 1)$

or *n*th term $= \frac{1}{2}n^2 - \frac{1}{2}n$

Check with $n = 6$

6th term $= \frac{1}{2} \times 6 \times 5 = 15$ or

6th term $= \frac{1}{2} \times 6^2 - \frac{1}{2} \times 6 = 18 - 3 = 15$

Algebraic method for finding the *n*th term

When the *n*th term cannot be spotted easily, it may be quicker to use an algebraic method.
To see how it works look at this table which gives values of y for various values of n in the relationship $y = an^2 + bn + c$. The table starts with $n = 0$.

When $n = 0$, $y = a \times 0^2 + b \times 0 + c = c$

When $n = 1$, $y = a \times 1^2 + b \times 1 + c = a + b + c$

When $n = 2$, $y = a \times 2^2 + b \times 2 + c = 4a + 2b + c$ and so on

n	0	1	2	3	4
y	c	$a + b + c$	$4a + 2b + c$	$9a + 3b + c$	$16a + 4b + c$

First difference $a + b$ $3a + b$ $5a + b$ $7a + b$

Second difference $2a$ $2a$ $2a$

- The 'second difference' is constant and gives the value of $2a$. To find a, halve this value.
- The first 'first difference' is $a + b$. Once you have worked out a, you can substitute it into $a + b$ to find b.
- The value of y when $n = 0$ gives c.

You can use this theory to find the values of a, b and c in a particular case.

There is, however, a difficulty with this method. You do not usually know the value when $n = 0$; you do not have a '0th term' in your sequence.

For example, for the sequence 6, 11, 20, 33, 50, … do the table and leave a gap for $n = 0$. When you have found the differences, work backwards, following the pattern, to find the value when $n = 0$.

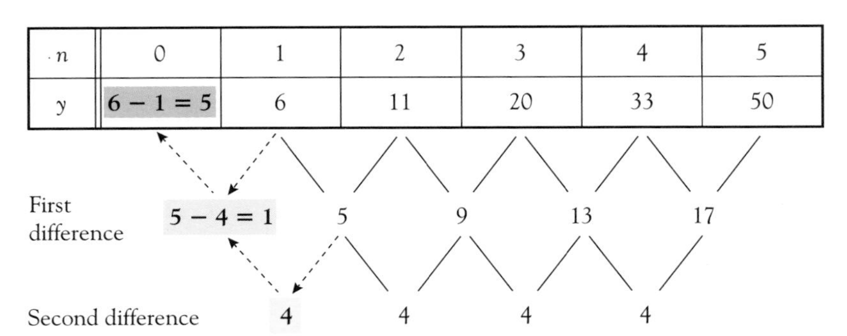

n	0	1	2	3	4	5
y	**6 − 1 = 5**	6	11	20	33	50

Find the second difference then work backwards.

For $y = an^2 + bn + c$

$2a = 4$
$a = 2$

$a + b = 1$
$2 + b = 1$
$b = 1 - 2$
$b = -1$

$c = 5$

Substitute the values found for a, b and c into the general equation $y = an^2 + bn + c$

The relationship is $y = 2n^2 - n + 5$. This gives the *n*th term.

So the *n*th term is $\underline{2n^2 - n + 5}$.

Check with a particular value for *n*.

Sample Question 8

Arifa is investigating the number of squares needed to form these ‘growing crosses’. She uses d for the diagram number and s for the number of squares.

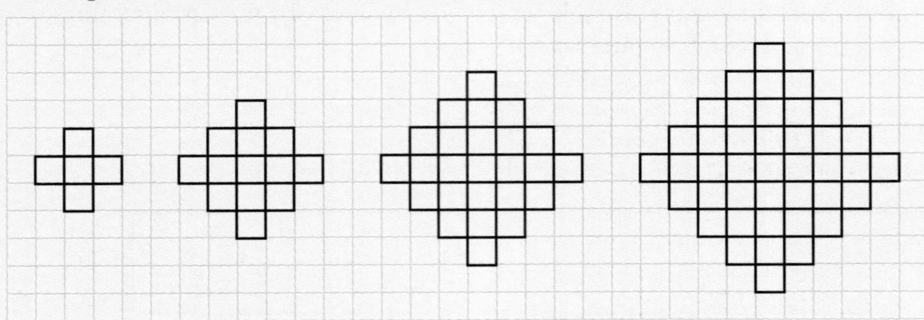

$d = 1$	$d = 2$	$d = 3$	$d = 4$
$s = 5$	$s = 13$	$s = 25$	$s = 41$

Work out the formula connecting s and d in the form $s = \ldots$

Use your formula to find the number of squares needed for the eighth diagram in the sequence.

Answer

- Put the results into a table, leaving a gap for $d = 0$.
- Work out the first and second differences and then work backwards to find s when $d = 0$.

d	0	1	2	3	4
s	1	5	13	25	41

$5 - 4 = 1$

First differences: 4, 8, 12, 16

$8 - 4 = 4$

Second differences: 4, 4, 4, 4

The relationship is quadratic as the second difference is constant.

- For the quadratic relationship $s = ad^2 + bd + c$,

  the ‘second difference’ is $2a$, so $2a = 4$
  $\underline{a = 2}$

  the first ‘first difference’ is $a + b$, i.e. $2 + b$ so $2 + b = 4$
  $\underline{b = 2}$

  the value of s when $d = 0$ gives c, so $\underline{c = 1}$

- Substitute $a = 2$, $b = 2$ and $c = 1$ into $s = ad^2 + bd + c$

  The relationship is $\underline{s = 2d^2 + 2d + 1}$

Remember to check that the formula works for a pair of numbers in your table.

- To find the number of squares needed for the 8th diagram, substitute $d = 8$

  $$s = 2 \times 8^2 + 2 \times 8 + 1 = 145$$

  <u>145 squares are needed.</u>

xercise 25.3

Generate the first five terms of the sequence with nth term $n^2 - 2n$. Check that the second difference is constant and find its value.

Generate the first five terms of the sequence with nth term $3n^2 + 1$. Check that the second difference is constant and find its value.

Generate the first five terms of the sequence with nth term $2n^2 - n + 4$. Check that the second difference is constant and find its value.

Write down the next three terms of the sequence

3, 8, 15, 24, 35, ...

Show that the second difference is constant and find its value.

Write down the next three terms of the sequence

59, 52, 39, 20, −5, ...

Show that the second difference is constant and find its value.

Write down the nth term for each sequence.

a −1, 2, 7, 14, 23, ...

b 4, 9, 16, 25, 36, ...

c 1, 6, 13, 22, 33, ...

Find, by using first and second differences, the nth term of each sequence.

a 1, 12, 29, 52, 81, ...

b −6, 0, 10, 24, 42, ...

c 5, 21, 47, 83, 129, ...

d −1, −1, 3, 11, 23, ...

e 2, 11, 28, 53, 86, ...

f 1, 3, 7, 13, 21, ...

8 A sequence is formed from the number of dots in the following diagrams:

Diagram 1 Diagram 2 Diagram 3

Draw the next two diagrams.

Copy and complete the table for the number of dots.

Diagram	1	2	3	4	5	6
Number of dots						

Find the nth term of the sequence.

(Note that these numbers are called triangular numbers because of the way they arise.)

9 Alan is playing at making triangles with matchsticks as illustrated in the diagrams.

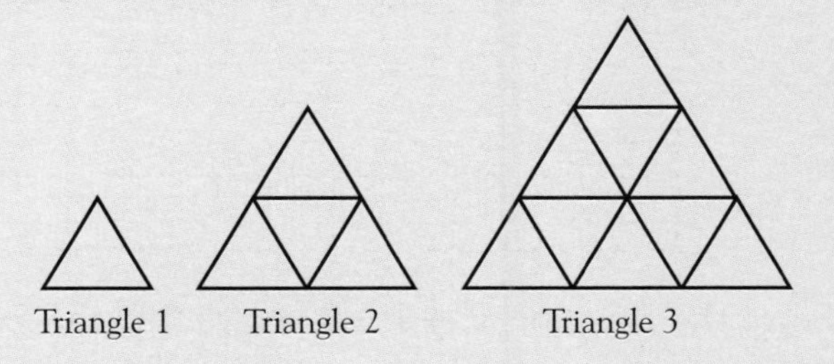

Triangle 1 Triangle 2 Triangle 3

How many matches will he require to make the next large triangle?

Copy and complete the table for the number of matchsticks.

Triangle	1	2	3	4	5	6
Number of matchsticks						

Find the nth term of this sequence.

Worked Exam Question
[SEG]

COMMENTS

These two rectangles have the same area.

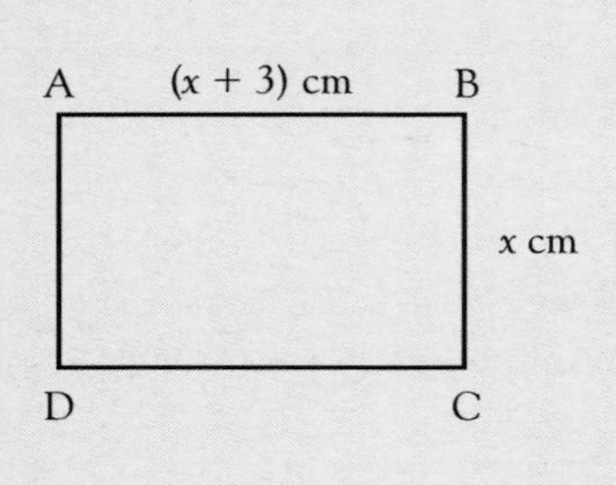

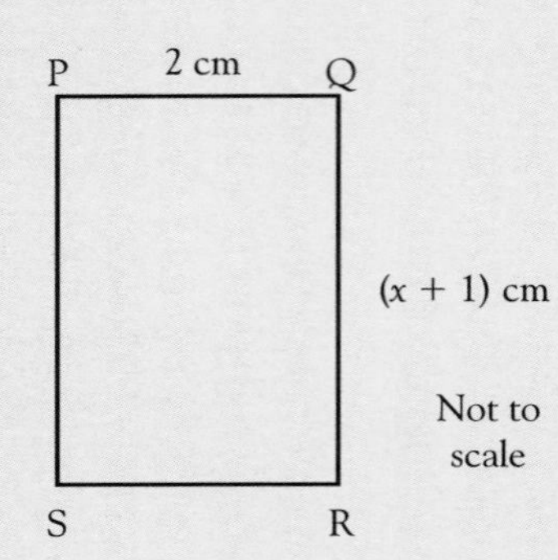

a Form an equation in x and show that it can be simplified to $x^2 + x - 2 = 0$

- Find expressions for the area of each rectangle and put them equal to each other.
- Simplify the equation and write it in the format requested.

Area ABCD $= x(x + 3)$

Area PQRS $= 2(x + 1)$

$x(x + 3) = 2(x + 1)$

$x^2 + 3x = 2x + 2$

$x^2 + x - 2 = 0$

B1 for one correct expression

B1 for putting the areas equal

B1 for expanding and rearranging correctly

3 marks

b Solve the equation $x^2 + x - 2 = 0$ to find the length of BC.

- Try to factorise the expression $x^2 + x - 2$ then use the factors to solve the equation.

To factorise, ask what two numbers multiply to give −2 and add to give 1.

Note that $BC = x$

$(x + 2)(x - 1) = 0$

Either $x + 2 = 0$ or $x - 1 = 0$

$x = -2$ $x = 1$

Ignore $x = -2$ (negative length)

so $x = 1$

Answer ....BC = 1 cm....

M1 for factorising $x^2 + x - 2$ correctly

A1 for correct answer

2 marks

Exam Questions

1 Maria has two brothers Aaron and Jason.
Maria is x years old.
Aaron is three years older than Maria.
Jason is two years younger than Maria

a Given that the product of the two brothers' ages is 126 show that

$$x^2 + x - 132 = 0$$

b Use the equation to find the ages of the three children. [MEG]

2 a Factorise $x^2 - 3x - 10$

b Solve the equation $x^2 + 3x = 0$

c Solve the inequality $-1 \leq 3x + 2 < 5$ [SEG]

3

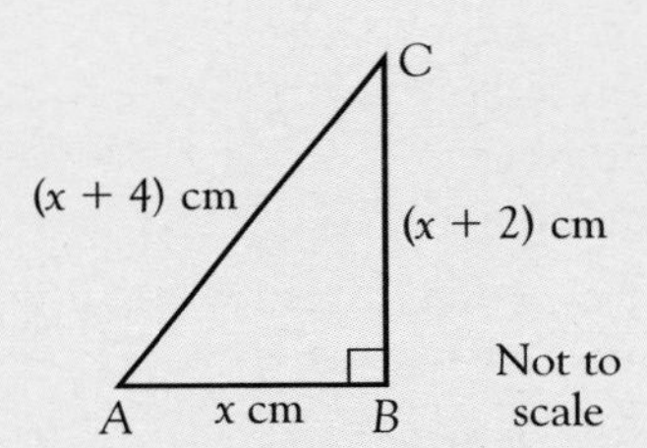

ABC is a triangle with a right angle at B.
The length of its three sides are x cm, $(x + 2)$ cm and $(x + 4)$ cm.

a Use Pythagoras' Theorem to show that x satisfies the equation

$$x^2 - 4x - 12 = 0$$

b Solve the equation

$$x^2 - 4x - 12 = 0$$

c Use your solutions in **b** to write down the lengths of the sides of the triangle. [WJEC]

4 Look at this sequence

2 8 18 32

a Write down the next term of the sequence.

b Write down the 25th term of the sequence.

c Write down the nth term of the sequence. [MEG]

5 These shapes are made with matches.
Shape 1 uses 4 matches
Shape 2 uses 12 matches
Shape 3 uses 24 matches.

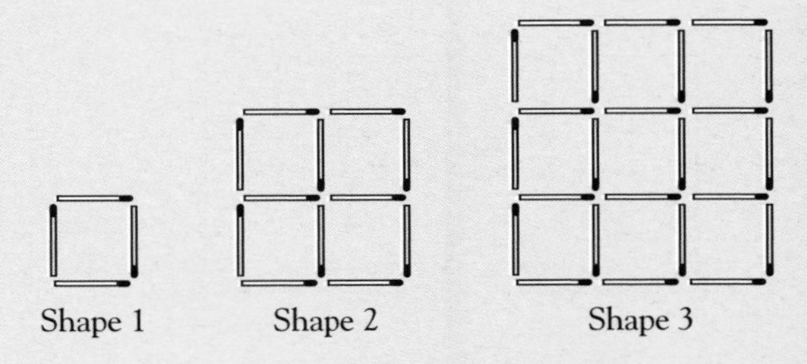

a Draw the next shape in the sequence.

b Complete the table.

Shape Number	1	2	3	4	5
Number of matches	4	12	24		

c Explain how you can work out the number of matches in the 6th shape without drawing the shape. [MEG]

6 Find a formula, in terms of n, for the area of the nth rectangle in this sequence.

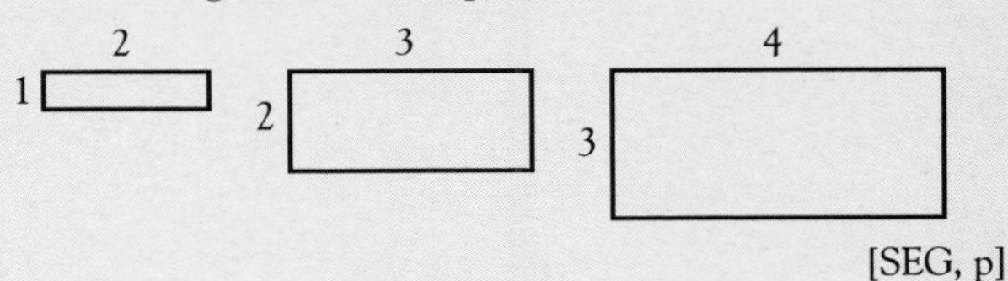

[SEG, p]

7 Equilateral triangles are combined together to form shapes.
The number of triangles in the shape form a sequence.

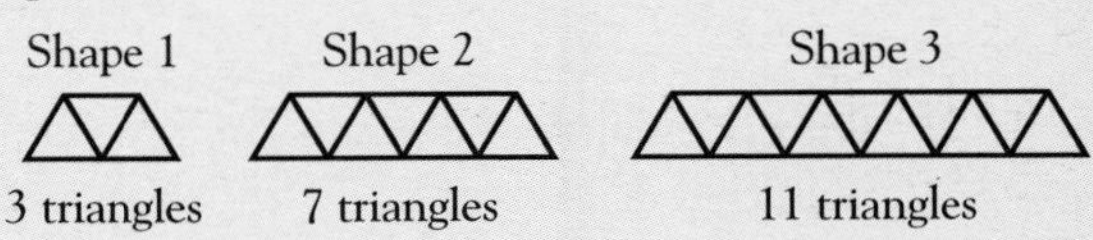

a Find a rule in terms of n for the number of triangles in the nth shape in the sequence.

b The triangles are now grouped together differently as shown below.

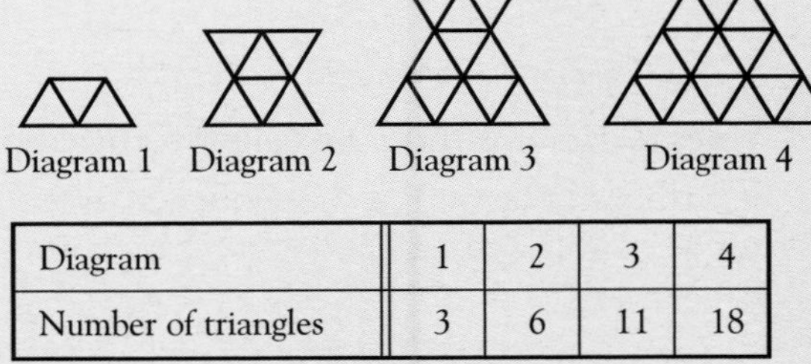

Diagram	1	2	3	4
Number of triangles	3	6	11	18

Find a rule in terms of n for the number of triangles in the nth diagram in the sequence. [SEG]

EXAM QUESTIONS

8

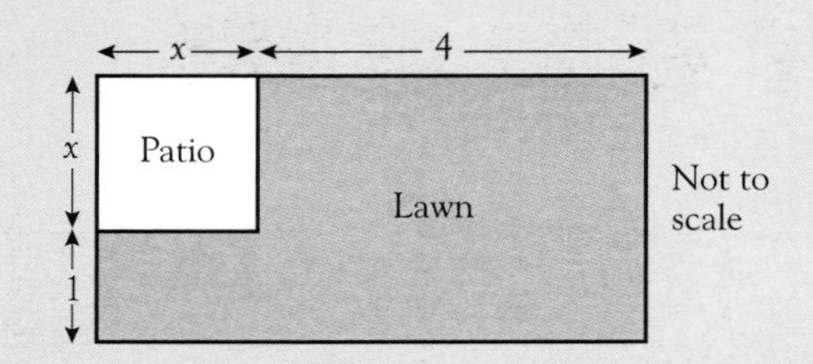

All lengths in this question are in metres.

A rectangular garden has a square patio of side x metres in one corner. The remainder of the garden is a lawn.

a The **total** area of the rectangular garden is 54 m^2. Show that x is a solution of the equation

$$x^2 + 5x - 50 = 0$$

b **i** Solve the equation $x^2 + 5x - 50 = 0$

ii Hence find the area of the patio. [MEG]

9 A rectangle has a length of $(x + 5)$ cm and a width of $(x - 2)$ cm.

$(x + 5)$ cm

$(x - 2)$ cm

a If the perimeter of the rectangle is 24 cm, what is the value of x?

b **i** If the area of the rectangle is 60 cm^2, show that

$$x^2 + 3x - 70 = 0$$

ii Find the value of x when the area of the rectangle is 60 cm^2. [NEAB]

10 The following diagrams show the beginning of a sequence.

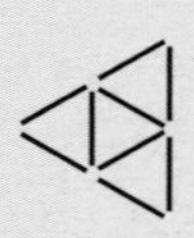

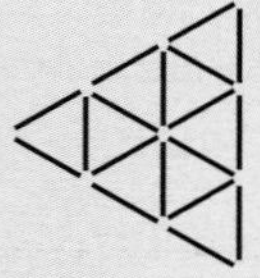

Diagram 1 Diagram 2 Diagram 3

Lines drawn like this | are called vertical lines.

For the nth diagram in the sequence, write, in terms of n, the number of non-vertical lines. [SEG]

11 **a** Factorise $x^2 + 4x - 12$

Hence or otherwise,

b solve $x^2 + 4x - 12 = 0$ [L, p

12 Javid and Anita try to find different ways of exploring the sequence

4, 10, 18, 28, 40, …

a Javid writes

1st number $4 = 1 \times 4 = 1 \times (1 + 3)$
2nd number $10 = 2 \times 5 = 2 \times (2 + 3)$
3rd number $18 = 3 \times 6 = 3 \times (3 + 3)$
4th number $28 = 4 \times 7 = 4 \times (4 + 3)$

i How would Javid write down the 5th number?

ii How would Javid write down the nth number?

b Anita writes

1st number $4 = 2 \times 3 - 2$
2nd number $10 = 3 \times 4 - 2$
3rd number $18 = 4 \times 5 - 2$
4th number $28 = 5 \times 6 - 2$

i How would Anita write down the 5th number?

ii How would Anita write down the nth number?

c Show how you would prove that Javid's expression and Anita's expression for the nth number are the same. [MEG

13 **a** Multiply out the brackets and simplify

$$(2x + 3)(2x - 3)$$

b Factorise

i $x^2 + 6x$

ii $x^2 + 6x + 8$

c Solve the equation $(x - 3)(x + 5) = 0$

[MEG

14 The first five terms of a sequence are shown below.

Number of term (n)	1	2	3	4	5
Term	2	5	10	17	26

Find, in terms of n, an expression for the nth term of the sequence. [MEG, p

26 FURTHER SHAPE AND SPACE

In this chapter you will
- learn how to distinguish formulae for perimeters, areas and volumes
- learn how to find missing lengths and angles in similar shapes
- practise finding missing lengths and angles in harder questions involving Pythagoras' rule, trigonometry and similar shapes

Dimensions

Perimeter

A perimeter is measured in a length unit such as cm, m or km. It is said to have a **length dimension**, or a **dimension of 1**. This can be written $[L^1]$.

To find the perimeter of this triangle, you add the lengths a, b and c.

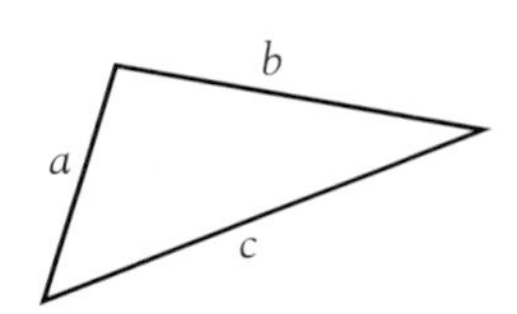

$$a + b + c = \text{perimeter of triangle}$$

In dimensions: (length) + (length) + (length) = (length)

Using shorthand: $[L^1] + [L^1] + [L^1] = [L^1]$

You can **add** lengths to lengths, and the answer is a length.
The same rule applies to subtraction.

To find the perimeter of this circle (i.e. the circumference), you find $\pi \times d$ where π is a number and d is a length. π is said to have **no** dimension, or dimension 0. This can be written $[L^0]$.

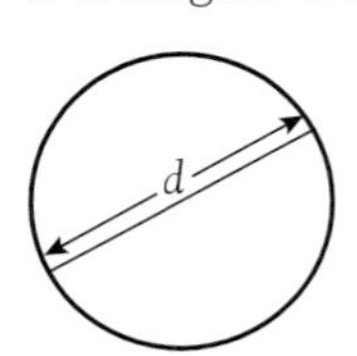

Consider $\pi \times d = \text{perimeter of circle}$

In dimensions: (number) × (length) = (length)

In shorthand: $[L^0] \times [L^1] = [L^1]$

Notice $0 + 1 = 1$.
When **multiplying**, add the dimensions (index numbers of L) to find the dimension of the answer.

Area

Area is measured in cm^2, m^2, km^2: i.e. a **length unit squared**. It is said to have a **dimension of 2**, written $[L^2]$.

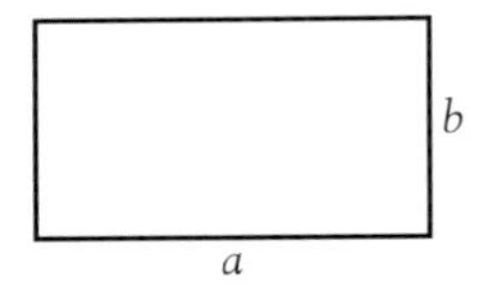

To find the area of this rectangle, you multiply the lengths a and b.

$$a \times b = \text{area of rectangle}$$

In dimensions: (length) × (length) = (area)

In shorthand : $[L^1] \times [L^1] = [L^2]$

Since you are multiplying, add the dimensions: $1 + 1 = 2$

DIMENSIONS

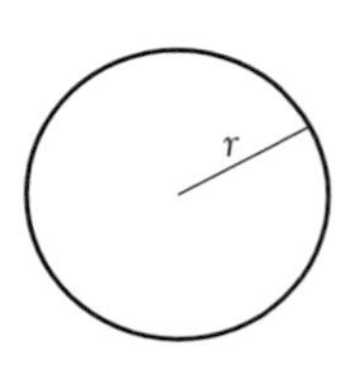

To find the area of this circle, you multiply π by r^2.

Consider $\pi \times r^2 =$ area of circle

i.e. $\pi \times r \times r =$ area

In dimensions: (number) $\times$ (length) $\times$ (length) $=$ (area)

$$[L^0] \times [L^1] \times [L^1] = [L^2]$$

$0 + 1 + 1 = 2$

Notice r^2 has dimension $[L^2]$.

Consider the dimensions for the total surface area of a cylinder with base radius r and height h.

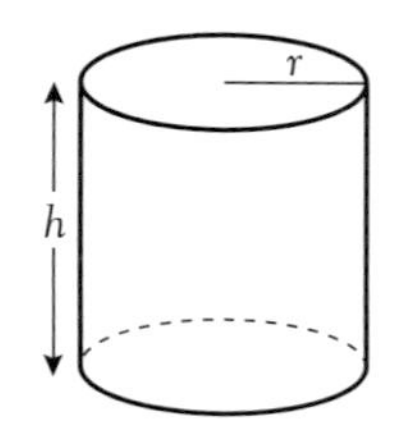

Total surface area $= 2\pi rh + 2\pi r^2$

$$\text{Dimensions} = [L^0] \times [L^0] \times [L^1] \times [L^1] + [L^0] \times [L^0] \times [L^2]$$

$$= [L^2] + [L^2]$$

$$= [L^2]$$

DO NOT add these index numbers. Remember that when you add an area to an area, the answer is also an area.

Note that it is not possible to add unless the dimensions are the same. For example you would not be able to add a length $[L^1]$ to an area $[L^2]$. In this case you would be trying to add cm to cm^2, which is nonsense.

Volumes

Volume is measured in cm^3, m^3, km^3: i.e. a **length unit cubed**. It is said to have a **dimension of 3**, written $[L^3]$.

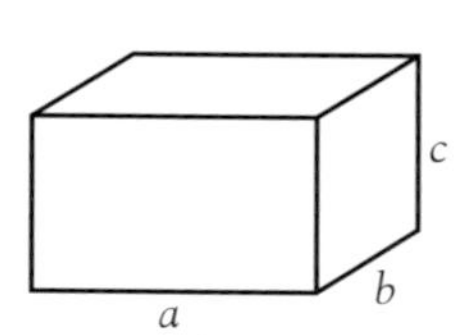

To find the volume of this cuboid, you multiply the three lengths a, b and c.

$a \times b \times c =$ volume of cuboid

In dimensions: (length) $\times$ (length) $\times$ (length) $=$ (volume)

$$[L^1] \times [L^1] \times [L^1] = [L^3]$$

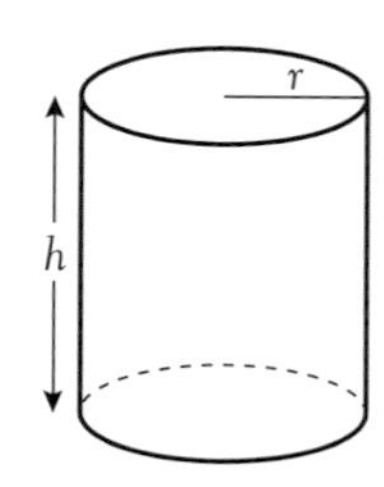

To find the volume of this cylinder, use $\pi r^2 h =$ volume of cylinder

In dimensions: (number) $\times$ (length^2) $\times$ (length) $=$ volume

$$[L^0] \times [L^2] \times [L^1] = [L^3]$$

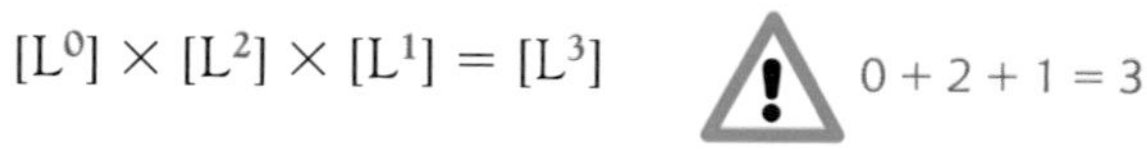

Remember that you can only add volumes to volumes, and the answer is also a volume, i.e. $[L^3] + [L^3] = [L^3]$

Summary of dimensions

- A constant or number has no dimension, written $[L^0]$
- A perimeter has dimension 1, written $[L^1]$
- An area has dimension 2, written $[L^2]$
- A volume has dimension 3, written $[L^3]$
- 3 is the maximum dimension
- You can only add or subtract quantities with the same dimension. The answer is also that dimension
- When you multiply quantities, add their dimensions to find the dimension of the answer

DIMENSIONS

Sample Question 1

In the following expressions, a and b both represent lengths. By considering the dimensions of each expression, write down whether it represents a perimeter, an area, a volume or none of these.

a $\pi a + 2b$ **b** $\frac{1}{3}\pi a^2b^2$ **c** a^2b **d** $a(a + b)$

Answer

a

- Look at the dimensions of each part of the expression.

Consider $\pi a + 2b$

$$\text{Dimension} = [L^0] \times [L^1] + [L^0] \times [L^1]$$
$$= [L^1] + [L^1]$$
$$= [L^1]$$

(length) + (length) = (length)

$\pi a + 2b$ is a perimeter.

b

- Consider $\frac{1}{3}\pi a^2b^2$

$$\text{Dimension} = [L^0] \times [L^0] \times [L^2] \times [L^2]$$
$$= [L^4]$$

Dimension of $a^2 = [L^2]$

The maximum value for a dimension is 3, so this is not a valid expression.
$\frac{1}{3}\pi a^2b^2$ is not a perimeter, area or volume.

c

- Consider a^2b

$$\text{Dimension} = [L^2] \times [L^1]$$
$$= [L^3]$$

$2 + 1 = 3$

a^2b is a volume.

d

- Consider $a(a + b)$

Method 1

- Expand the bracket first, then look at the dimensions

$$a(a + b) = a^2 + ab$$

$$\text{Dimension} = [L^2] + [L^1] \times [L^1]$$
$$= \underset{\text{(area)}}{[L^2]} + \underset{\text{(area)}}{[L^2]}$$
$$= \underset{\text{(area)}}{[L^2]}$$

$a(a + b)$ is an area.

Method 2

- Find the dimension of the bracket first

Consider $a + b$

$$\text{Dimension} = \underset{\text{length}}{[L^1]} + \underset{\text{length}}{[L^1]} = \underset{\text{length}}{[L^1]}$$

- Now consider $a \times (a + b)$

$$\text{Dimension} = [L^1] \times [L^1] = \underset{\text{(area)}}{[L^2]}$$

DIMENSIONS

Sample Question 2

This toy consists of a hemisphere attached to a cone.
Using dimensions, work out which of these expressions could be

a the total volume of the toy,

b the total surface area of the toy.

i $\frac{2}{3}\pi r^3 + \frac{1}{3}\pi r^2 h$ **ii** $\frac{2}{3}\pi r^2 + \pi r^2 h$ **iii** $\pi r(2r + l)$

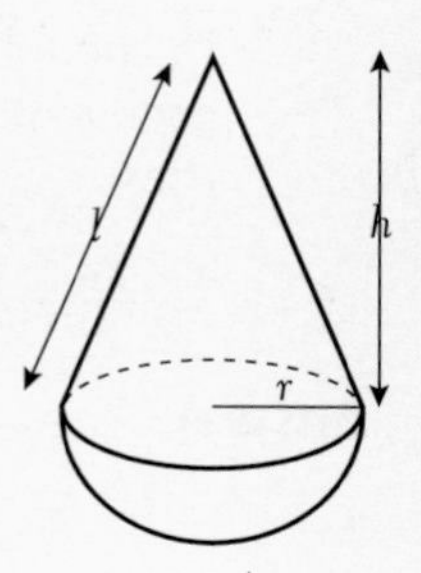

Answer

i

- Work out the dimension of $\frac{2}{3}\pi r^3$ and of $\frac{1}{3}\pi r^2 h$

  Consider $\frac{2}{3}\pi r^3$

  Dimension $= [L^0] \times [L^0] \times [L^3] = [L^3]$

  Consider $\frac{1}{3}\pi r^2 h$

  Dimension $= [L^0] \times [L^0] \times [L^2] \times [L^1] = [L^3]$

- Now consider the full expression $\frac{2}{3}\pi r^3 + \frac{1}{3}\pi r^2 h$

  Dimension $=$ $[L^3]$ (volume) $+$ $[L^3]$ (volume) $=$ $[L^3]$ (volume)

  Since the dimensions are the same, the quantities can be added.

  $\frac{2}{3}\pi r^3 + \frac{1}{3}\pi r^2 h$ could be the formula for the <u>total volume of the toy</u>.

ii

- Work out the dimension of $\frac{2}{3}\pi r^2$ and of $\pi r^2 h$

  Consider $\frac{2}{3}\pi r^2$

  Dimension $= [L^0] \times [L^0] \times [L^2] = [L^2]$

  Consider $\pi r^2 h$

  Dimension $= [L^0] \times [L^2] \times [L^1] = [L^3]$

- Now consider the full expression $\frac{2}{3}\pi r^2 + \pi r^2 h$

  Dimension $= [L^2] + [L^3]$

  It is not possible to add quantities with different dimensions, so this could not be the formula for the volume or the surface area.

  $\frac{2}{3}\pi r^2 + \pi r^2 h$ is <u>not a valid formula</u>.

iii

- Look at the bracket.

  Consider $2r + l$

  Dimension $= [L^0] \times [L^1] + [L^1] = [L^1] + [L^1] = [L^1]$

  You can add, as the dimensions are the same. The answer is also $[L^1]$.

- Look at the complete expression, using $[L^1]$ for the bracket.

  Consider $\pi r(2r + l)$, i.e. $\pi r \times (2r + l)$

  Dimension $= [L^0] \times [L^1] \times [L^1] = [L^2]$

  $\pi r(2r + l)$ could be the formula for the <u>total surface area of the toy</u>.

DIMENSIONS

Dividing quantities

page 285

The volume of a prism is given by the formula

$$\text{volume} = \text{area of cross-section} \times \text{length}$$

If you re-arrange this formula, you get

$$\text{length} = \frac{\text{volume}}{\text{area of cross-section}}$$

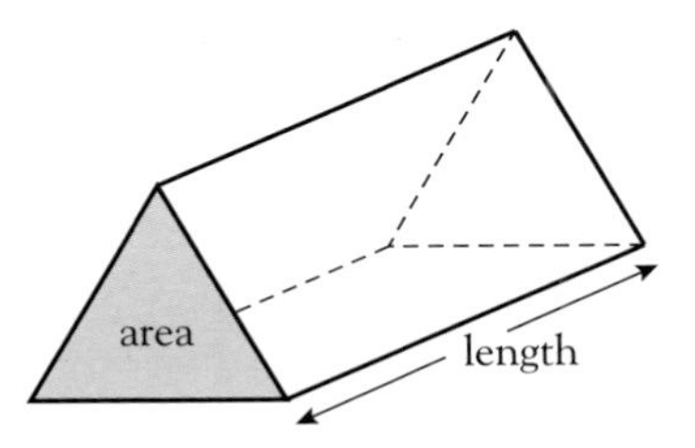

In dimensions:

$$[L^1] = \frac{[L^3]}{[L^2]}$$

Notice that 3 − 2 = 1

When dividing quantities, subtract the dimensions to find the dimension of the answer.

Sample Question 3

In the expression $\frac{x^3 + y^3}{2y}$, x and y are lengths.
Work out the dimension and indicate what it might be finding.

Answer

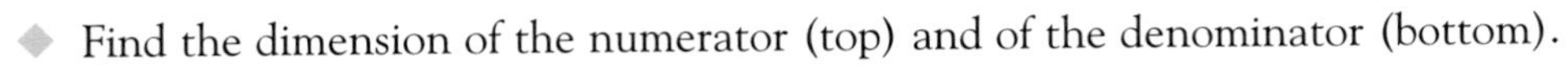

- Find the dimension of the numerator (top) and of the denominator (bottom).

Consider $x^3 + y^3$

Dimension $= [L^3] + [L^3] = [L^3]$

Consider $2y$

Dimension $= [L^0] \times [L^1] = [L^1]$

- Combine the results

Consider $\frac{x^3 + y^3}{2y}$

Dimension $= \frac{[L^3]}{[L^1]} = [L^2]$

You are dividing quantities, so subtract the dimensions to find the dimension of the answer.

Since the dimension is 2, the expression could be finding an area.

Exercise 26.1

In this exercise all numbers and π do not have dimensions.

1 The letters a and b both represent lengths. By considering the dimensions of each of the expressions, say whether the expression represents a length, an area, a volume or none of these.

a $\pi a^2 b + \pi ab$

b $4\pi a^2$

c $\frac{1}{3}\pi a^2 b$

d $2\pi a^2 + 2\pi ab$

e $a^3 + 2a^2 b$

f πab

g $\frac{1}{3}\pi b^2 + \pi a$

h $\pi(a + b) - b$

i $\frac{(a + b)^2}{b}$

j $\frac{a^3 b}{a + b}$

DIMENSIONS

2 A hot water tank consists of a hemisphere on top of a cylinder.

a By considering dimensions decide which formula represents the surface area of the tank

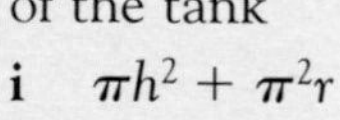

i $\pi h^2 + \pi^2 r$

ii $3\pi r^2 + 2\pi rh$

iii $2\pi r^2 + 2\pi r$

iv $2\pi r + \pi r^2 h$

b By considering dimensions decide which formula represents the volume of the tank.

i $\frac{4}{3}\pi r(r^2 + h)$

ii $\frac{4}{3}\pi r^2 h + \pi r^2$

iii $4\pi r^2 + \pi r^2 h$

iv $\frac{2}{3}\pi r^3 + \pi r^2 h$

3 The diagram represents a plan of an athletics stadium, consisting of a rectangle with semi-circular ends.

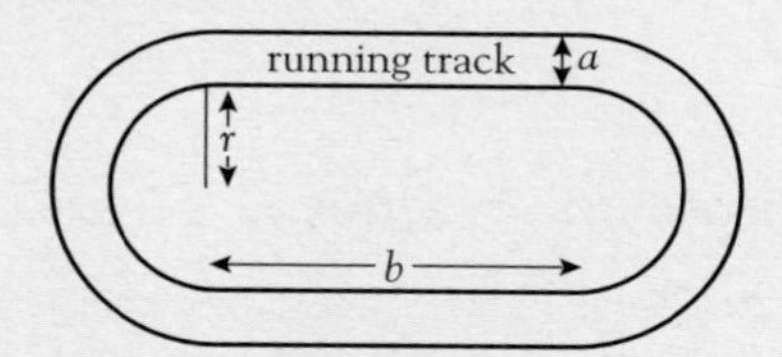

By considering dimensions, decide which expression will give

a the perimeter of the inside of the running track

b the area of the running track.

i $ab^2 + \pi^2(2r + a)$	iv $2ab + \pi a(2r + a)$
ii $\pi + 2a + b^2$	v $ab + \pi^2(r + a)$
iii $\pi(r + ab)$	vi $2\pi r + 2b$

4 In the following expressions, r and h each represent a length.

a For each expression, state whether it represents a length, an area, a volume or none of these.

i $2\pi r^2$

ii $r^2(r + h)$

iii $\sqrt{r^2 + h^2 + rh}$

iv $\dfrac{r^3}{h}$

v $3r + rh$

b Explain your answer to part **a v**. [MEG]

5 The expressions shown below can be used to calculate lengths, areas or volumes of various shapes.

The letters r and h represent lengths. π, 2, 3, 4, 5 and 10 are numbers which have no dimensions.

$r(\pi + 2)$ $\quad \dfrac{4r^2\pi}{h}$ $\quad r(r + 4h)$ $\quad \dfrac{rh}{4}$ $\quad \dfrac{4r^3}{5}$

$10r^3\pi$ $\quad \pi(r + 2h)$ $\quad \dfrac{3r^3}{h}$ $\quad r^2(h + \pi r)$

Write down each of the expressions which can be used to calculate an area. [L]

6

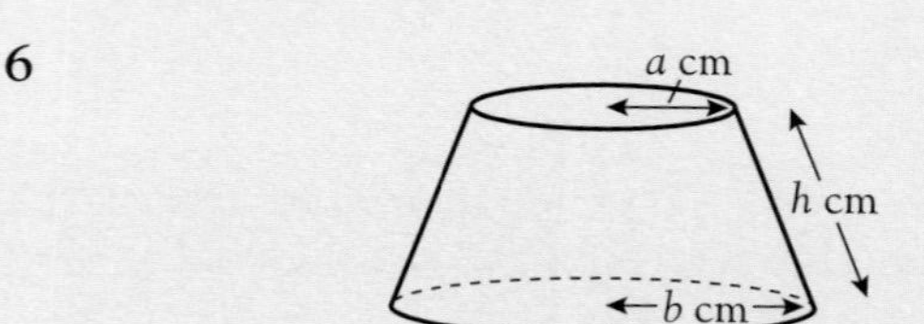

One of the expressions in the list can be used to calculate the area of material needed to make the curved surface of the lampshade in the diagram.

a $\pi h(a + b)^2$ b $\pi h^2(a + b)$

c $\pi h(a + b)$ d $\pi h^2(a + b)^2$

State which expression is correct. Give a reason for your answer. [MEG]

7 Here are some expressions.

$\pi r^2 l$	$2\pi r^2$	$4\pi r^3$	$abrl$	$\dfrac{abl}{r}$	$3(a^2 + b^2)r$	πrl

The letters r, l, a and b represent lengths. π, 2, 3, and 4 are numbers that have no dimensions.

Three of the expressions represent volumes. Which are they? [L]

8 A waste-paper bin is a prism on a hexagonal base. If the length of each side of the hexagon is x and the height is h, which expression could give

a the surface area of the bin

b the volume of the bin?

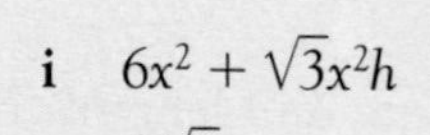

i $6x^2 + \sqrt{3}x^2 h$	iv $6xh + \dfrac{3\sqrt{3}}{2}x^2$
ii $\dfrac{3\sqrt{3}}{2}x^2 h$	v $\dfrac{\sqrt{3}}{2}xh + x^2 h$
iii $\dfrac{3\sqrt{3}}{2}x^2 h^2$	vi $\dfrac{\sqrt{3}}{2}xh + \pi h$

SIMILAR SHAPES

Similar shapes

A′B′C′D′ is an **enlargement** of ABCD, with scale factor 2.

ansformations I page 121

The quadrilaterals are the same shape, but one is larger than the other.

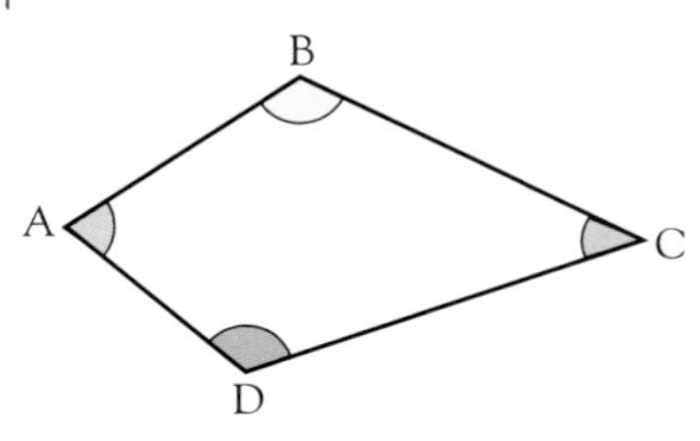

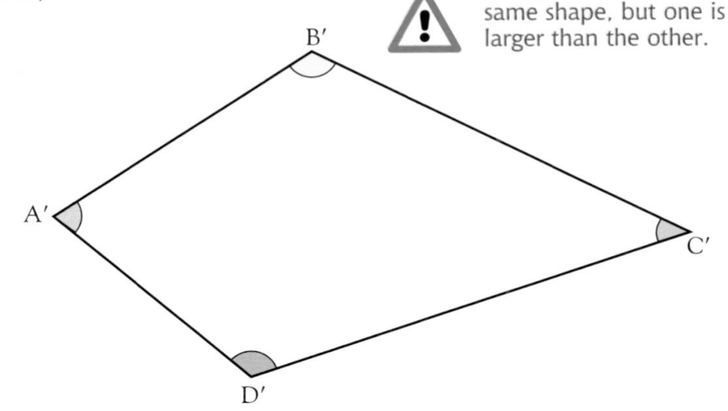

Number II page 192

The scale factor tells you the **ratio of corresponding lengths** so A′B′ : AB = 2, i.e. $\frac{A'B'}{AB} = 2$

In fact $\frac{A'B'}{AB} = 2$, $\frac{B'C'}{BC} = 2$, $\frac{D'C'}{DC} = 2$, $\frac{A'D'}{AD} = 2$

The corresponding angles on the two shapes are the same

$$\hat{A} = \hat{A}', \quad \hat{B} = \hat{B}', \quad \hat{C} = \hat{C}', \quad \hat{D} = \hat{D}'$$

If corresponding angles on the shapes are equal, the shapes are said to be equiangular.

The two shapes are said to be **similar**.

Two shapes are similar if
- corresponding angles on the shapes are equal (i.e. they are equiangular) **and**
- the ratio of corresponding sides is the same for all lengths.

In mathematics, similar does not just mean that they look alike. The above conditions must be satisfied.

Sample Question 4

a Two circles have radii 2 cm and 3 cm. Are they similar?

b Two rectangles have dimensions 2 cm by 5 cm and 5 cm by 12 cm. Are they similar?

Answer

a
- There are no angles to compare.
- Compare lengths on the two circles (radii, diameters, circumferences).

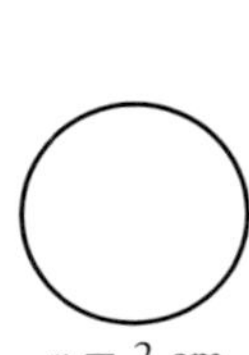

$r = 2$ cm

$C = 2 \times \pi \times r$
$= 2 \times 3.14 \times 2$
$= 12.56$

$r = 3$ cm

$C = 2 \times \pi \times r$
$= 2 \times 3.14 \times 3$
$= 18.84$

Ratio of radii $= 3 : 2 = \frac{3}{2} = 1.5$

Ratio of diameters $= 6 : 4 = \frac{6}{4} = 1.5$

Ratio of circumferences $= \frac{18.84}{12.56} = 1.5$

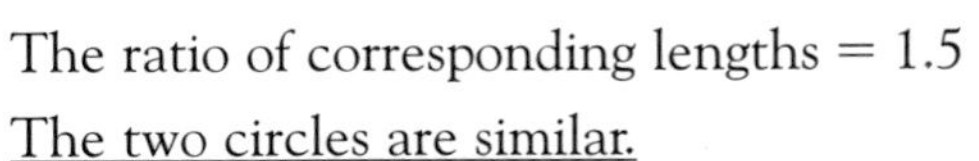

The ratio of corresponding lengths = 1.5

The two circles are similar.

In fact all circles are similar to each other.

b

- Sketch the rectangles

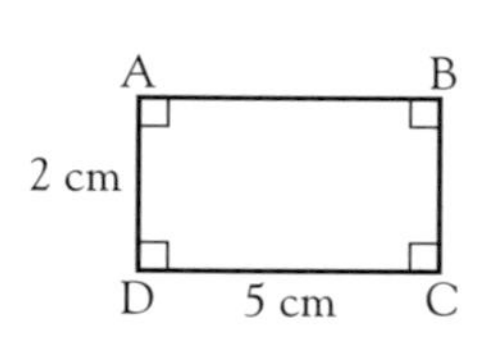

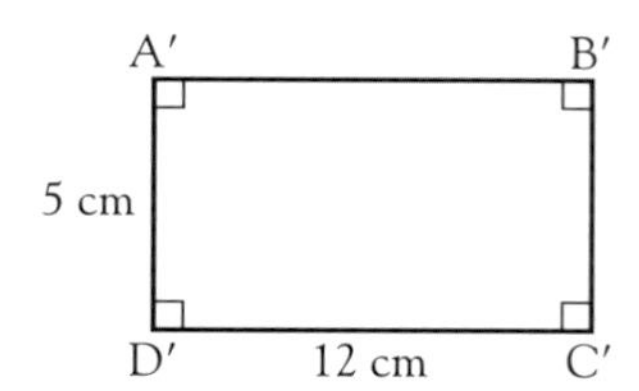

- Check corresponding angles.
  All the angles are 90° on both diagrams, so corresponding angles are equal.

- Check ratio of widths and ratio of lengths to see whether they are the same.

$$\text{Ratio of widths} = 5 : 2 = \frac{5}{2} = 2.5$$

$$\text{Ratio of lengths} = 12 : 5 = \frac{12}{5} = 2.4$$

- Make your decision.
  The ratios of corresponding lengths are not the same.

  The two rectangles are not similar.

Similar triangles

When comparing two **triangles**, you only need to show **one** of the conditions for similarity. This is because if one happens, the other happens also. (You would need advanced trigonometry techniques to prove this)

> If the ratios of corresponding sides in two triangles are equal, then the triangles are **similar**.

Consider triangles ABC and XYZ, which are not drawn to scale.

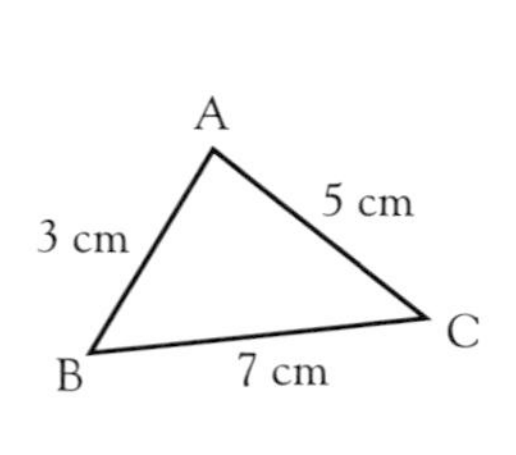

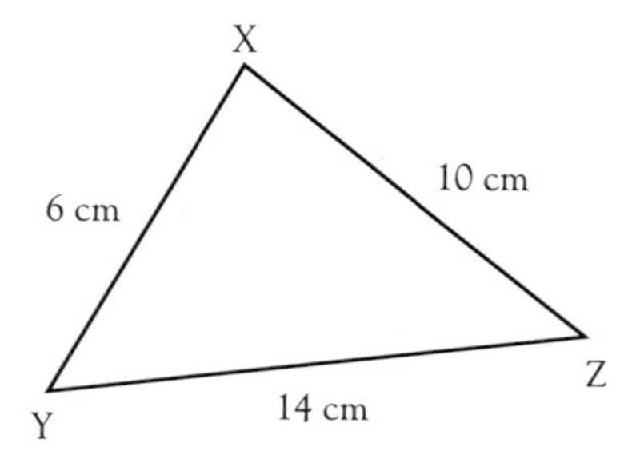

$$XY : AB = \frac{XY}{AB} = \frac{6}{3} = 2 \qquad XZ : AC = \frac{XZ}{AC} = \frac{10}{5} = 2 \qquad YZ : BC = \frac{YZ}{BC} = \frac{14}{7} = 2$$

Since the ratios of corresponding sides are equal, triangles XYZ and ABC are similar.

It is possible, using advanced trigonometry, to show that $\hat{A} = \hat{X}$, $\hat{B} = \hat{Y}$ and $\hat{C} = \hat{Z}$. Therefore corresponding angles on the triangles are equal and the triangles are equiangular.

SIMILAR SHAPES

If two triangles are **equiangular**, they are **similar**.

Consider triangles DEF and PQR.

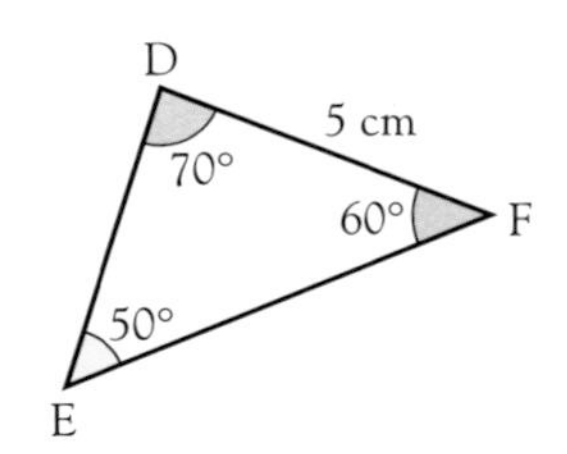

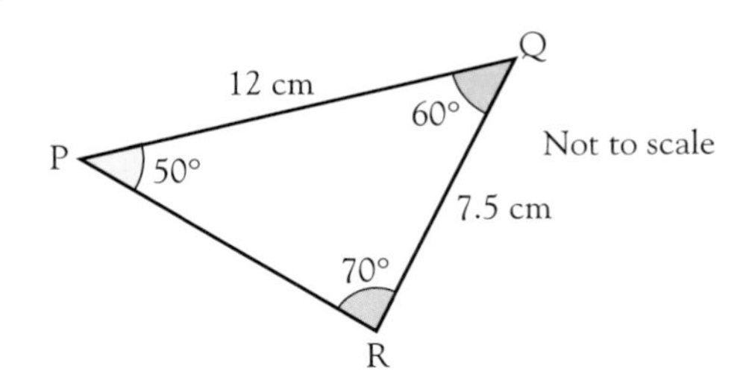

$\hat{D} = \hat{R}$
$\hat{E} = \hat{P}$
$\hat{F} = \hat{Q}$

Corresponding angles are equal, i.e. the triangles are equiangular, so they are similar.

It can be shown, using advanced trigonometry, that the ratios of corresponding sides are equal. You can use this fact to work out missing lengths.

Sometimes it is obvious which lengths correspond on the two triangles, but when it is not easy to decide, a good way to work it out is to write the corresponding angles underneath each other like this:

Triangles $\frac{\textbf{DEF}}{\textbf{RPQ}}$ are similar.

$\hat{D}$ and $\hat{R}$ are both 70°.
$\hat{E}$ and $\hat{P}$ are both 50°.
$\hat{F}$ and $\hat{Q}$ are both 60°.

Then pair off the letters as ratios

$$\frac{\textbf{DE}}{\textbf{RP}} = \frac{\textbf{EF}}{\textbf{PQ}} = \frac{\textbf{DF}}{\textbf{RQ}}$$

(first two letters of each) (last two letters of each) (first and last letters of each)

This shows that DE and RP correspond, EF and PQ correspond, DF and RQ correspond

This format also gives the ratios that have to be equal.

Writing in the lengths given in the triangles:

$$\frac{DE}{RP} = \frac{EF}{12} = \frac{5}{7.5}$$

You need to know both lengths in one complete ratio.

$$\frac{EF}{12} = \frac{5}{7.5}$$

Multiply by 12 $\quad EF = \frac{5}{7.5} \times 12$

$= 8$

$\underline{EF = 8 \text{ cm}}$

To find DE you would need to know RP and vice-versa.

Note that it would not matter if you had written the initial statement about the similar triangles with the letters in a different order, provided that the correct letters are below each other

for example $\frac{DFE}{RQP}$ or $\frac{EDF}{PRQ}$.

Sometimes you will need to prove that the triangles are equiangular first, as in the following sample question.

SIMILAR SHAPES

Sample Question 5

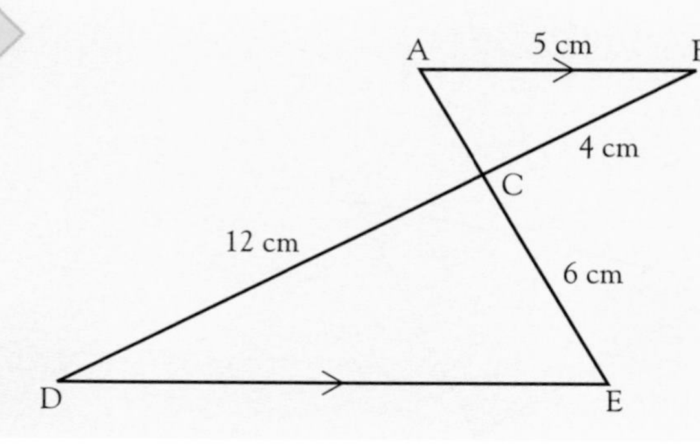

In the diagram, AB is parallel to DE.

AB = 5 cm, BC = 4 cm, DC = 12 cm and CE = 6 cm.

Calculate

a AC

b DE

Answer

Geometry I page 70

- Not enough information has been given to solve this using trigonometry or Pythagoras' rule, so look for similar triangles. In this example you need **parallel line facts.**

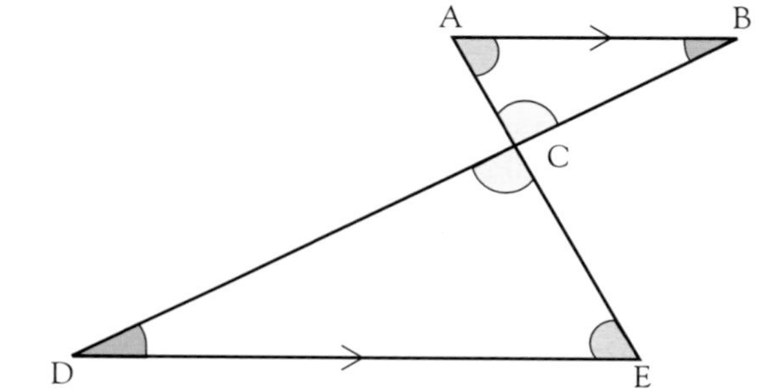

$\hat{D} = \hat{B}$ (alternate angles on parallel lines)

$\hat{A} = \hat{E}$ (alternate angles on parallel lines)

$A\hat{C}B = D\hat{C}E$ (vertically opposite angles at a point are equal)

So corresponding angles in the two triangles are equal.

The triangles are equiangular, therefore they are **similar**.

- Write the corresponding letters underneath each other

Triangles $\begin{matrix}ABC\\EDC\end{matrix}$ are similar

- Pair off the ratios

$$\frac{AB}{ED} = \frac{BC}{DC} = \frac{AC}{EC}$$

- Fill in any values that you know

$$\frac{5}{ED} = \frac{4}{12} = \frac{AC}{6}$$

Notice that you know one complete ratio $\frac{4}{12}$.

- To find AC look at

$$\frac{\mathbf{4}}{\mathbf{12}} = \frac{\mathbf{AC}}{\mathbf{6}}$$

Multiply by 6 $\quad 6 \times \frac{4}{12} = AC$

$\underline{AC = 2\text{ cm}}$

If you cancel you can see the answer straight away.

$\frac{\not{4}^{1}}{\not{12}_{3}} = \frac{AC}{6}$

$\frac{1}{3} = \frac{2}{6}$, so AC = 2

SIMILAR SHAPES

Sample Question 6

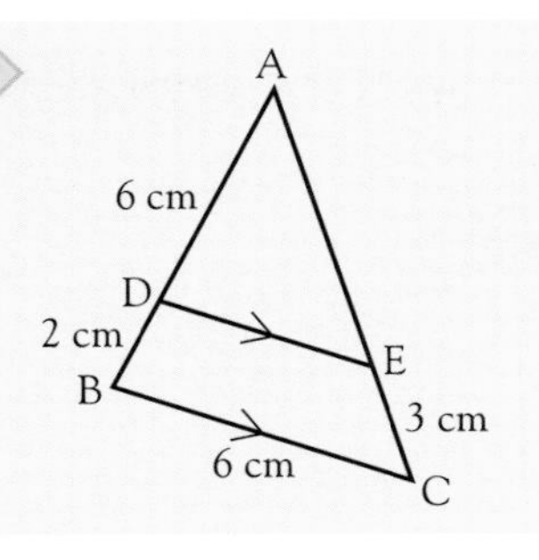

Not to scale

The diagram shows a triangle ABC.

DE is parallel to BC.
Calculate the lengths

a DE,

b AC. [SEG, p]

Answer

- Mark on the diagram angles that are equal. Use parallel line facts.

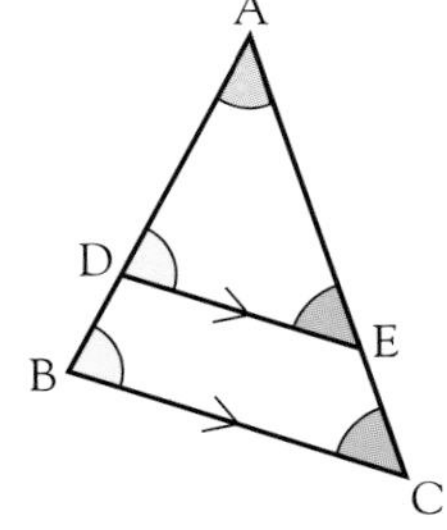

Corresponding angles on parallel lines are equal,

so $\quad A\hat{D}E = A\hat{B}C$

and $\quad A\hat{E}D = A\hat{C}B$

- **Split** the diagram into **two triangles** and put in any lengths that you know. Look for similar triangles.

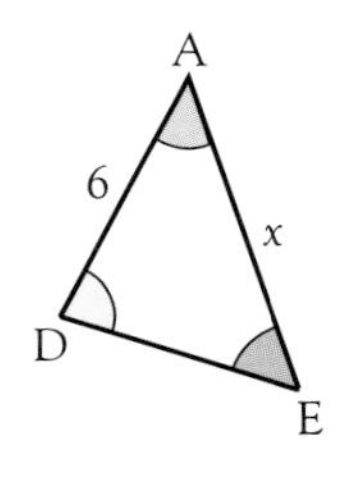

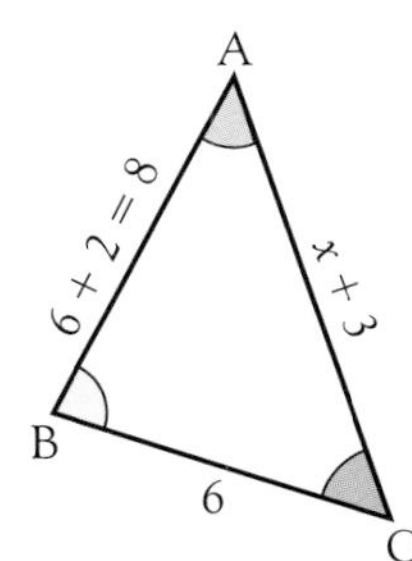

You will see the information more clearly if you draw the two separate triangles.

$\hat{A}$ is common to both triangles, so the triangles are equiangular.

Therefore the two triangles are similar.

- Put corresponding angles underneath each other and pair off the ratios. Then write in any values that you know.

Triangles $\frac{\text{ADE}}{\text{ABC}}$ are similar, so

$$\frac{\text{AD}}{\text{AB}} = \frac{\text{DE}}{\text{BC}} = \frac{\text{AE}}{\text{AC}}$$

$$\frac{6}{8} = \frac{\text{DE}}{6} = \frac{x}{x+3}$$

a To find DE, use

$$\frac{6}{8} = \frac{\text{DE}}{6}$$

$$6 \times \frac{6}{8} = \text{DE}$$

$$\underline{\text{DE} = 4.5 \text{ cm}}$$

b To find AC, first find AE where AE $= x$

$$\frac{6}{8} = \frac{x}{x+3}$$

$$6(x+3) = 8x$$

$$6x + 18 = 8x$$

$$18 = 2x$$

$$x = 9$$

$$\therefore \; \underline{\text{AC} = 9 + 3 = 12 \text{ cm}}$$

SIMILAR SHAPES

Sample Question 7

a Are these two triangles similar? Show your working to support your answer.

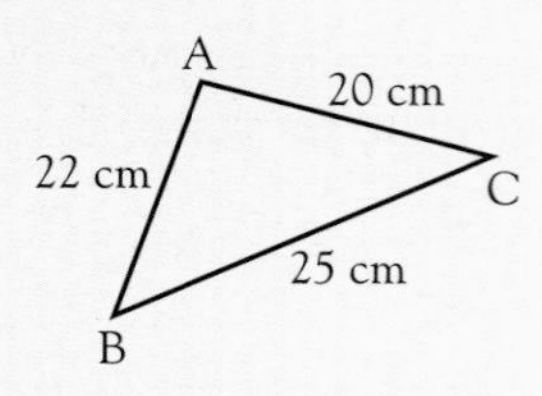

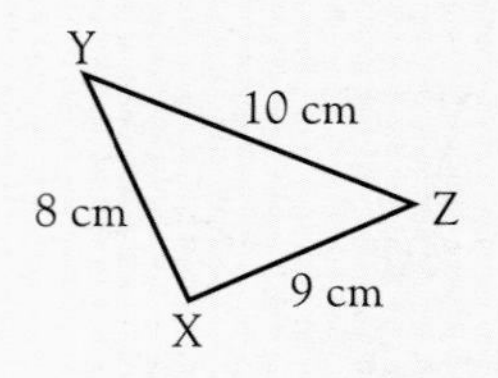

b Use similar triangles to find the length of SR.

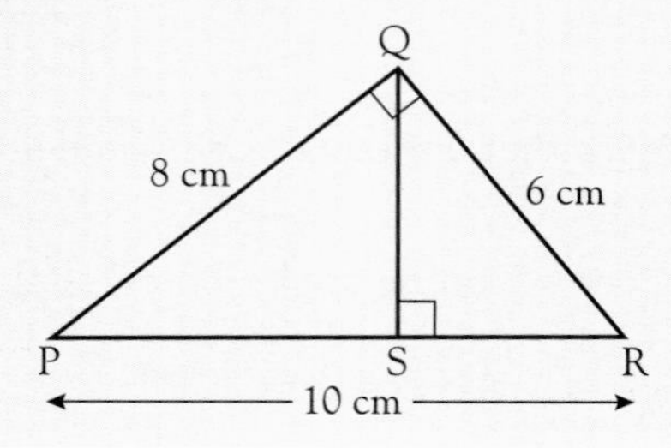

Answer

a

- Compare ratios of corresponding sides

longest sides $25 : 10 = \frac{25}{10} = 2.5$

shortest sides $20 : 8 = \frac{20}{8} = 2.5$

'middle' sides $22 : 9 = \frac{22}{9} = 2.44\ldots$

The ratios of corresponding sides are not equal, so the triangles are not similar.

b

- Find two similar triangles by splitting the diagram

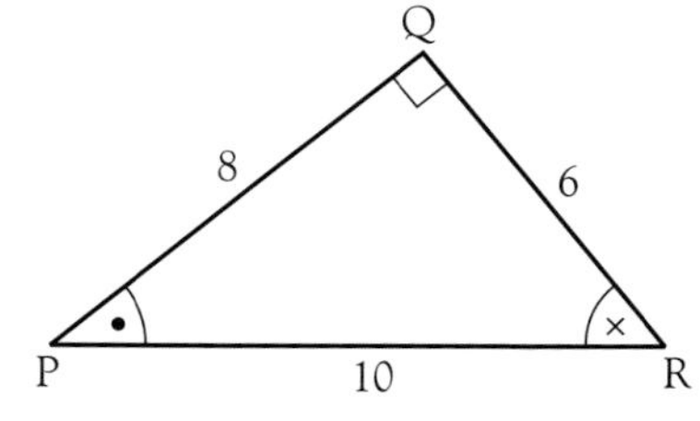

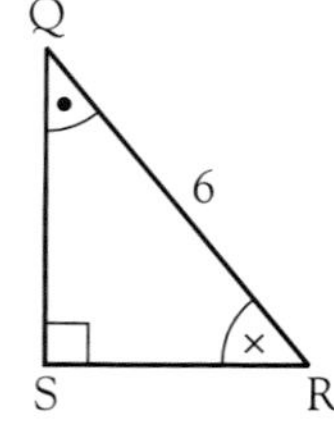

(i) $P\hat{Q}R = Q\hat{S}R = 90°$
(ii) $\hat{R}$ is common to both
(iii) $Q\hat{P}R = S\hat{Q}R$
(third angle in triangle with 90° and ×)

Triangles $\begin{matrix}PQR\\QSR\end{matrix}$ are similar, because they are equiangular.

- Pair off the ratios and put in known values

$$\frac{PQ}{QS} = \frac{QR}{SR} = \frac{PR}{QR}$$

$$\frac{8}{QS} = \frac{6}{SR} = \frac{10}{6}$$

- Find SR

$\frac{6}{SR} = \frac{10}{6}$ is the same as $\frac{SR}{6} = \frac{6}{10}$

Multiply by 6 $SR = \frac{6}{10} \times 6$

SR = 3.6 cm

SIMILAR SHAPES

Exercise 26.2

1 Test the following pairs of triangles for similarity. Give reasons for your decision.

a

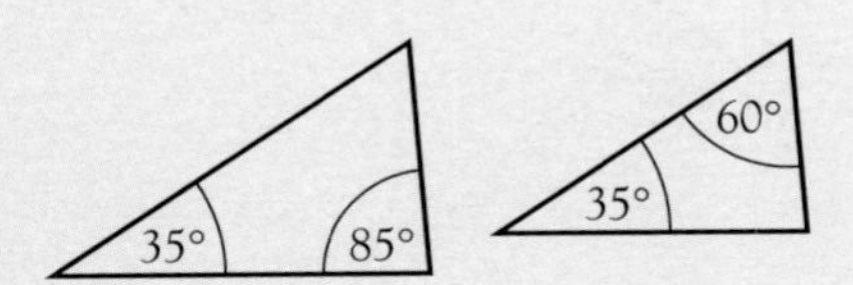

b

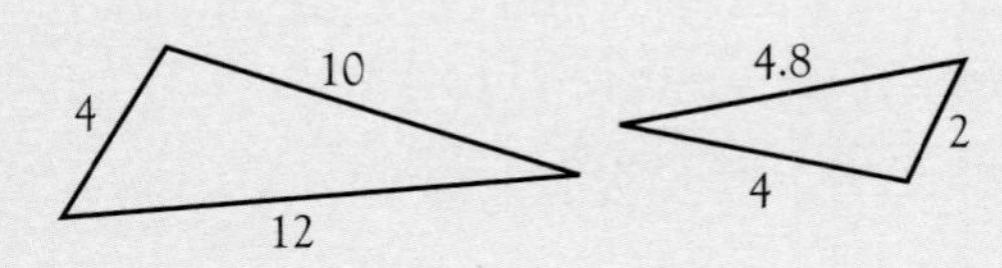

2 A square cloth is made by joining four squares together as shown in the diagram.

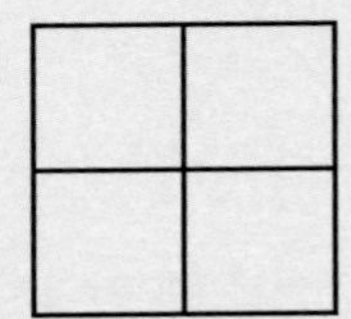

Is the large square similar to each of the other squares?

3 Jane wants to put a photograph, measuring 20 cm by 15 cm, into a rectangular frame so that there is a border of 3 cm all the way around. Will the framed photograph be similar to the photograph?

4 A5 paper is the same size as a sheet of A4 paper cut in half as in the diagram. Measure a sheet of A4 paper, then cut or fold it in half, and measure the A5 paper. Are the two pieces similar?

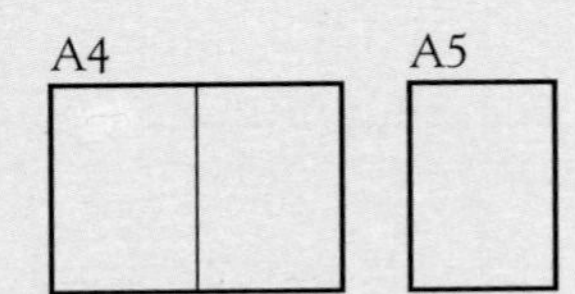

5 In the diagram $\hat{A} = \hat{C}$

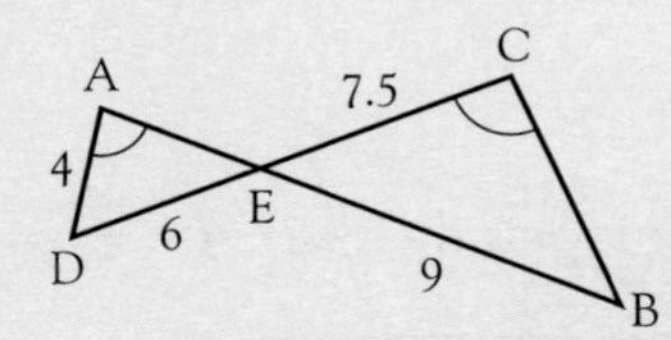

a Explain why triangles AED and CEB are similar.

b Find the length of BC and AE.

6 In the following pairs of similar figures (not drawn to scale), find the lengths represented by letters.

a

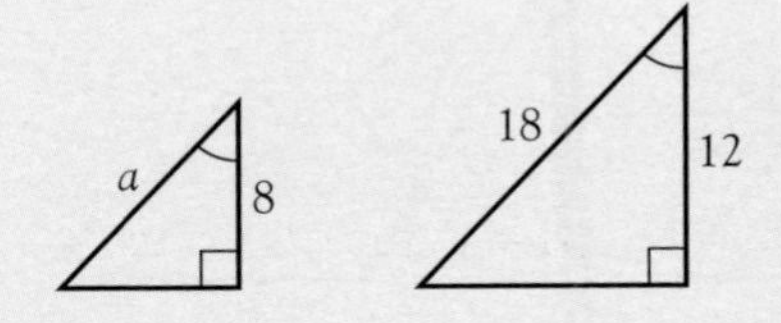

b

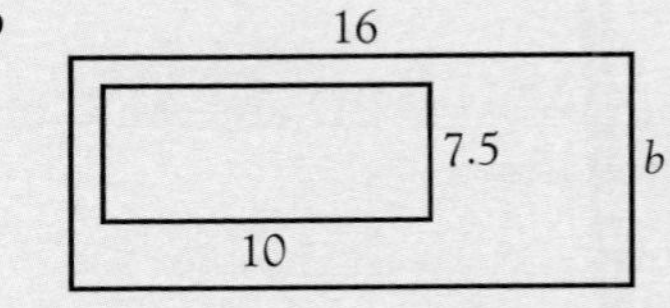

c

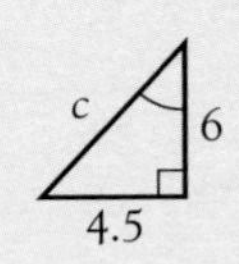

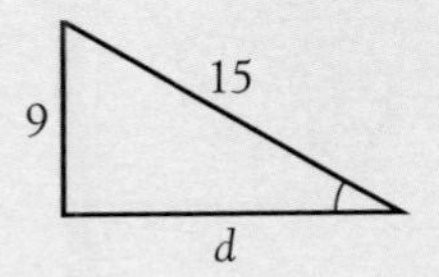

7 By considering pairs of similar triangles, find the lengths represented by letters.

a

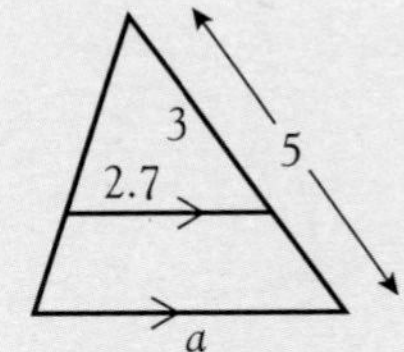

b

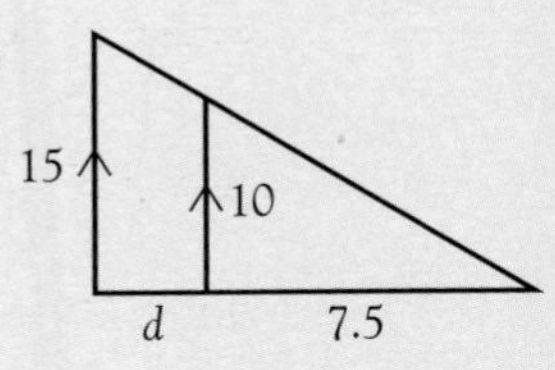

c

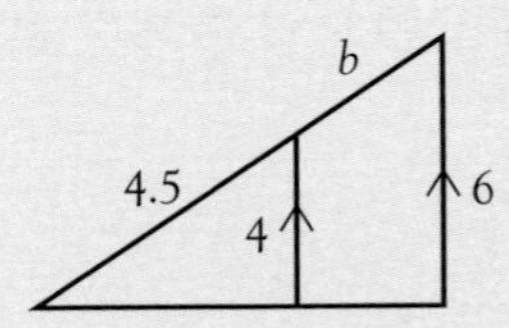

d

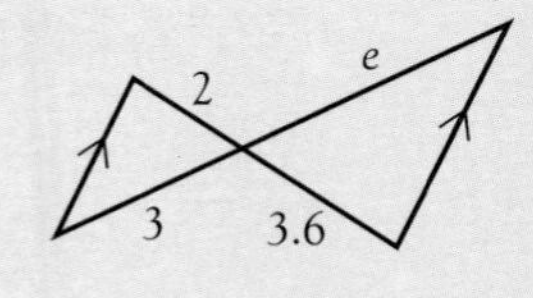

8

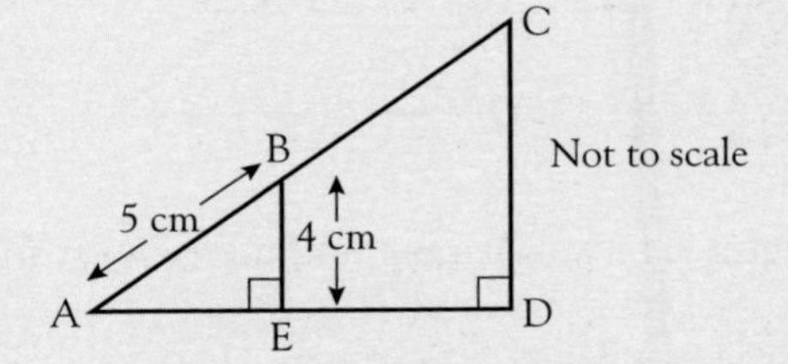

AB : AC = 1 : 3

a Work out the length of CD.

b Work out the length of BC. [L]

SIMILAR SHAPES

9 For each part, find a pair of similar triangles and calculate the lengths marked x and y.

a **b**

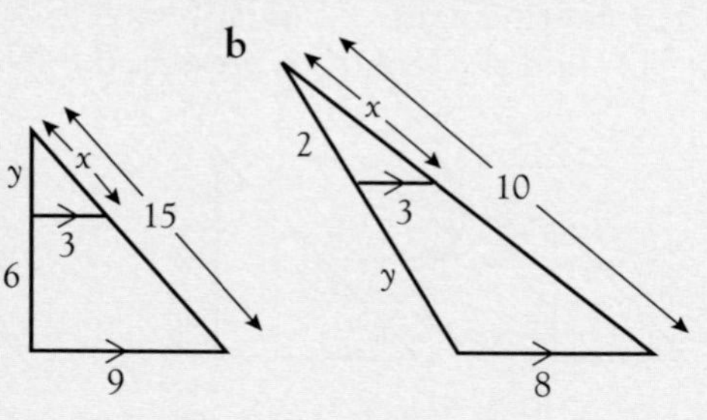

10 Show that triangles ABC, ABD and ADC are similar. Find the lengths x and y.

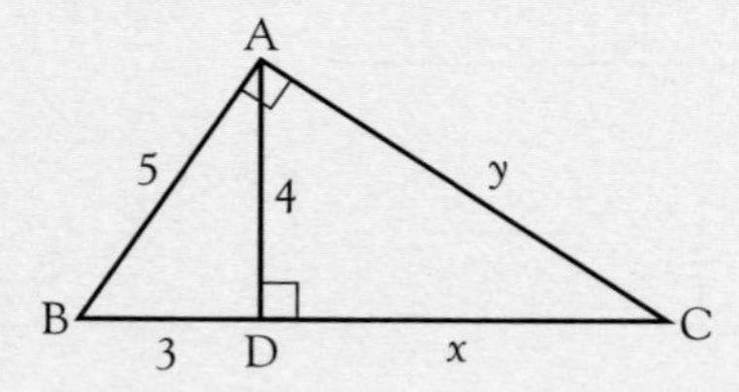

11 **a** Explain why triangles PTS and RTQ are similar.

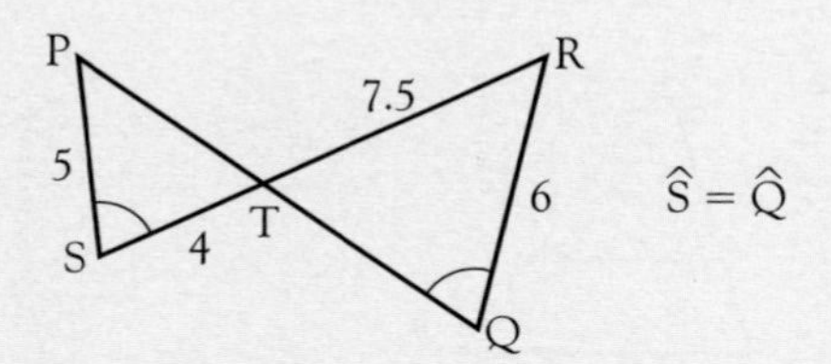

b The lengths given are in cm. Find the lengths of PT and TQ.

12

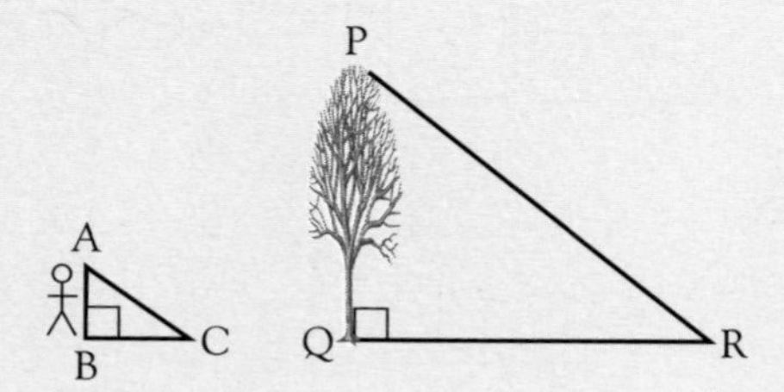

The diagram shows Mr Hall and a tree with their shadows cast by the sun at 12 noon.

The triangles ABC and PQR are similar.

Mr Hall knows he is 180 cm tall.

He measures his own shadow, BC, as 75 cm.
He measures the tree's shadow, QR as 220 cm.

Calculate the height, PQ, of the tree. [NEAB]

13

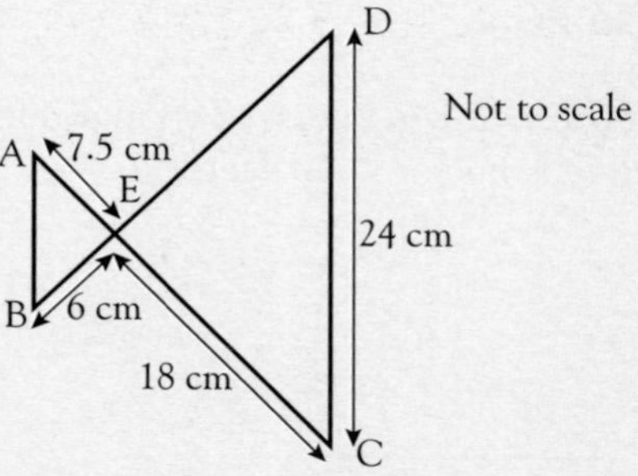

The triangles ABE and CDE are similar.

Line AB is parallel to line DC.

Calculate the length of AB. [SEG]

14

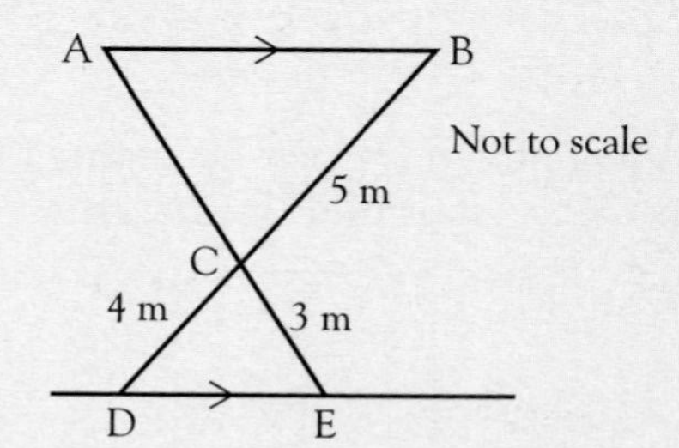

In the diagram CD = 4 metres, CE = 3 metres and BC = 5 metres.

AB is parallel to DE.

ACE and BCD are straight lines.

a Explain why triangle ABC is similar to triangle EDC.

b Calculate the length of AC. [L]

15

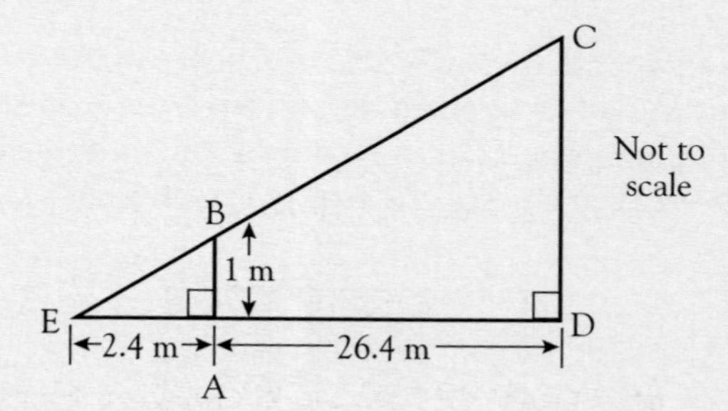

Using the measurements in the diagram, calculate the length of CD. [L]

16

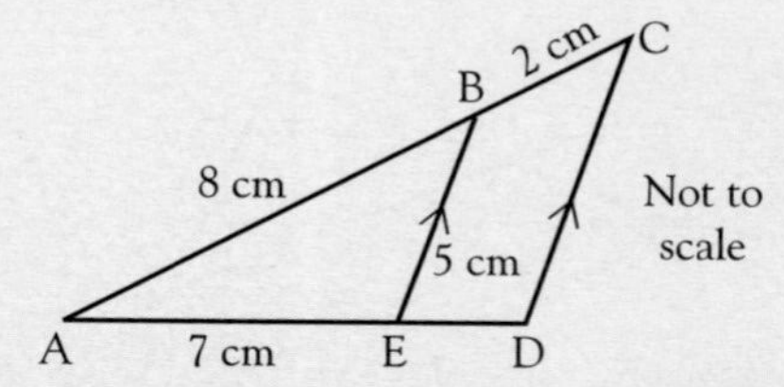

BE is parallel to CD, AB = 8 cm, BC = 2 cm, AE = 7 cm and EB = 5 cm. Find DC and ED.

MIXED QUESTIONS

Further trigonometry

a Finding the hypotenuse

Chapter 17 page 312

If you know the lengths of the two shorter sides in a right-angled triangle, you would find the hypotenuse using **Pythagoras' rule** where $AC^2 = AB^2 + BC^2$.

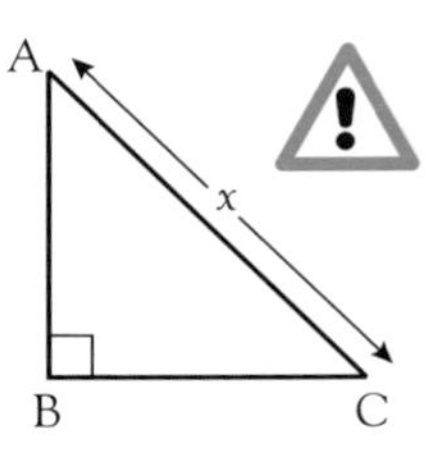

Use Pythagoras' rule when both AB and BC are known.

Chapter 20 page 377

If, however, you just know one of the shorter sides but you also know an angle, then you could find the hypotenuse using **trigonometry**. The method is shown in the following sample question.

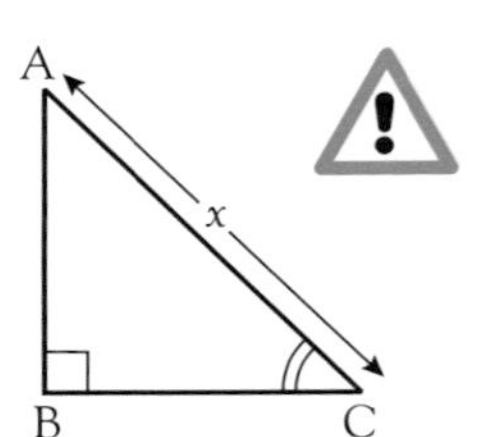

Use trigonometry when $\hat{C}$ is known and one of AB or BC is known.

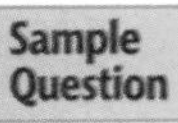

8 Triangle ABC is right angled at B, with $\hat{C} = 30°$ and AB = 5 cm. Find the length of AC.

Answer

- Draw a sketch and identify which trig ratio is needed. You know opp, you want hyp so use sin.

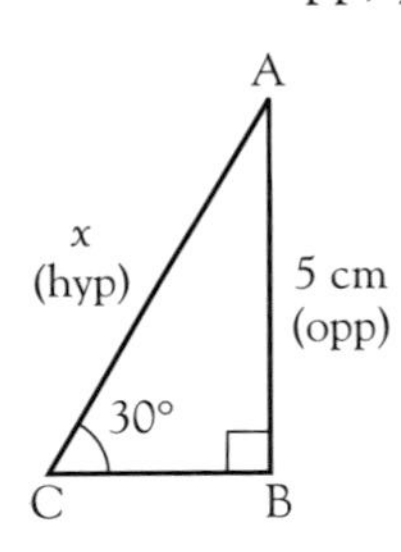

$$\sin 30° = \frac{\text{opp}}{\text{hyp}}$$

$$\sin 30° = \frac{5}{x}$$

Notice that x is on the bottom of the fraction.

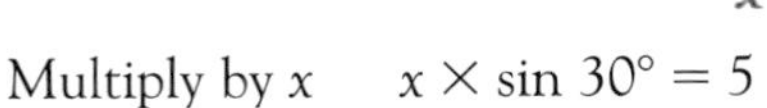

Multiply by x $\quad x \times \sin 30° = 5$

Divide by sin 30° $\quad x = \frac{5}{\sin 30°}$

Notice that x and sin 30 have changed places.

$\underline{x = 10 \text{ cm}}$

On calculator: Type 1: [5] [÷] [sin] [3] [0] [=]
Type 2: [5] [÷] [3] [0] [sin] [=]

Do not forget to press [=].

b Finding the adjacent side, using tan

A similar method can be used for solving the trig equation when you want to find the adjacent side and you know the opposite.

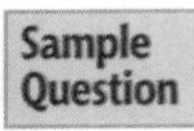

9 Find x, giving your answer to 2 s.f.

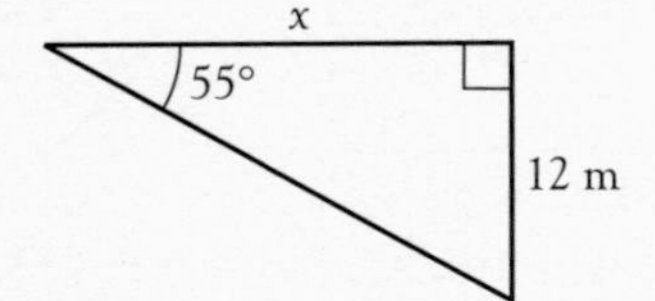

MIXED QUESTIONS

Answer

- Identify which trig ratio is needed, set up the equation and solve it.

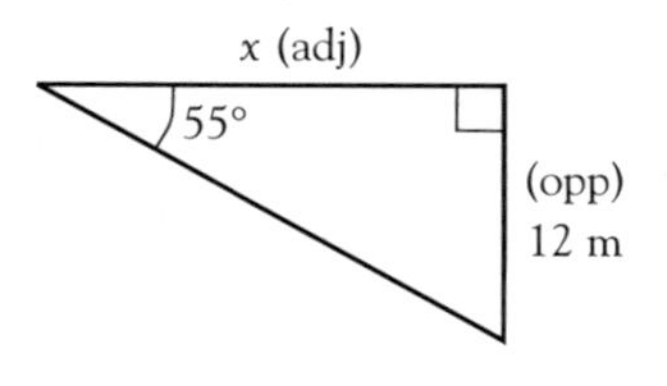

You know opp, you want adj, use tan.

$$\tan 55° = \frac{\text{opp}}{\text{adj}}$$

$$\tan 55° = \frac{12}{x}$$

Multiply by x $\quad x \times \tan 55° = 12$

Divide by $\tan 55°$ $\quad x = \dfrac{12}{\tan 55°}$

$$x = 8.402\ldots$$

$$\underline{x = 8.4 \text{ m (2 s.f.)}}$$

Note: It is also possible, in this type of question, to find x, using the tan ratio, but working with the other angle.
The third angle in the triangle is $180 - (55 + 90) = 35°$.

In relation to 35°, x is now opposite and 12 is adjacent.

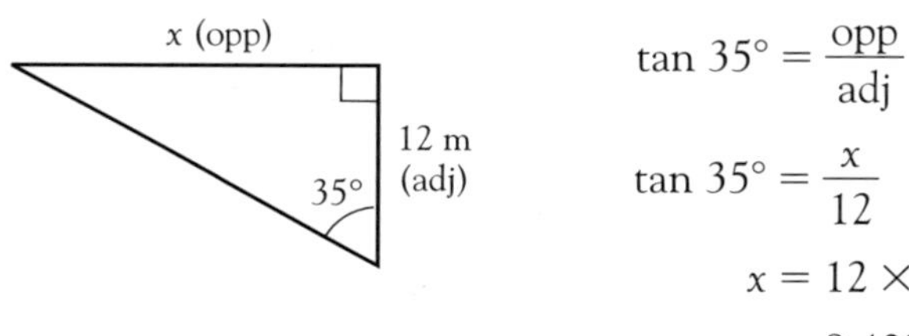

$$\tan 35° = \frac{\text{opp}}{\text{adj}}$$

$$\tan 35° = \frac{x}{12}$$

$$x = 12 \times \tan 35°$$

$$x = 8.402\ldots$$

$$\underline{x = 8.4 \text{ m (2 s.f.)}}$$

In the following exercise you will practise finding missing lengths using trigonometry and Pythagoras' rule, and finding missing angles using trigonometry. Look up the methods in chapters 17 and 20 if you are unsure about them.

Exercise 26.3

1 Calculate the length of the sides marked by letters. Give your answers to 3 significant figures.

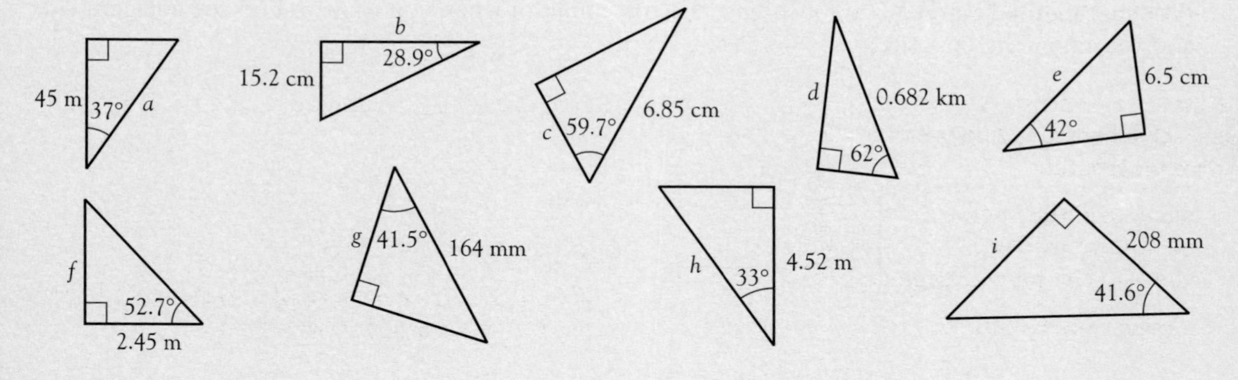

MIXED QUESTIONS

2 A conveyor belt is used to carry components from one level of a factory to another.

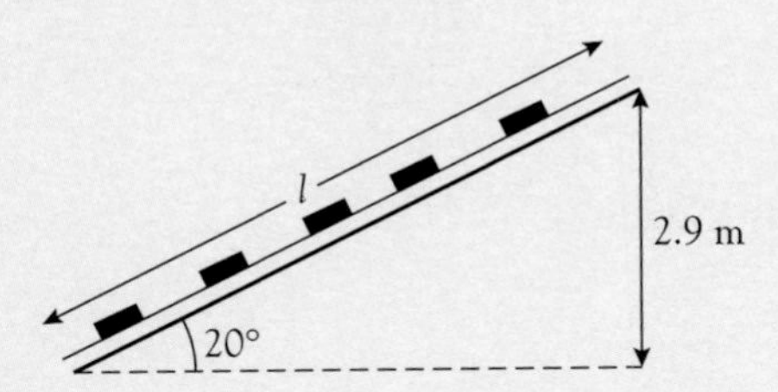

The belt rises by 2.9 metres at an angle of 20° as shown in the sketch.
Calculate, to 2 significant figures, the length of the conveyor belt, l, using trigonometry.

3 In triangle PQR, $\hat{Q} = 90°$, $\hat{R} = 40°$ and QR = 6 cm.

a Find PR, using trigonometry only.

b **i** Find PQ using trigonometry.

ii Using your value for PQ, find PR using Pythagoras' rule.

Comment on your answers to **a** and **b**.

4 The sketch below shows the cross section of a hill of height 110 metres which has artificial ski slopes built on each side, and a horizontal platform on the top.

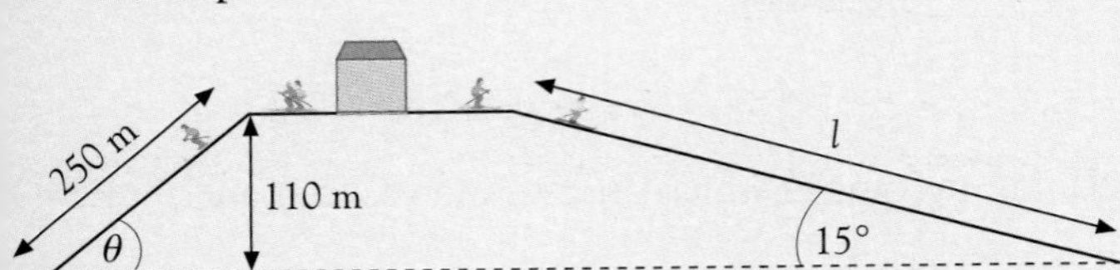

a The advanced slope is 250 metres long.
Find the angle of this slope to the horizontal (θ).

b The beginners' slope is constructed at 15° to the horizontal.
Find the length of this slope (l).

5 William is 1.78 m tall and he casts a shadow 1.5 m long on horizontal ground.

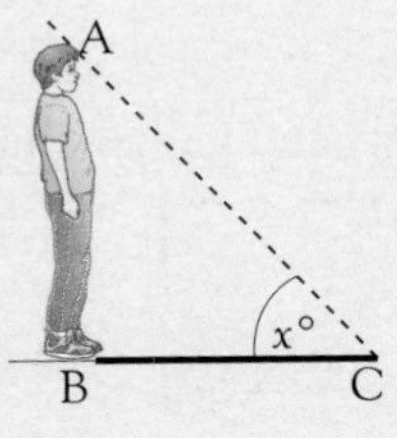

a Calculate x, the angle of elevation of the sun, correct to the nearest degree.

b Calculate the length AC.

6

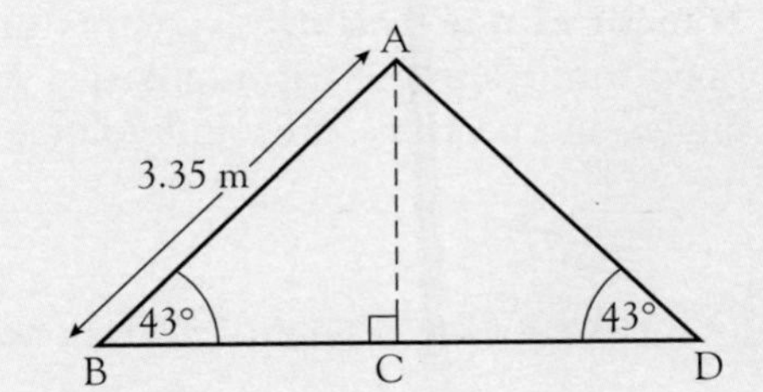

The diagram shows the cross-section of a loft. The length AB is 3.35 m. The roof slopes at 43° to the horizontal.

a Calculate the height AC.

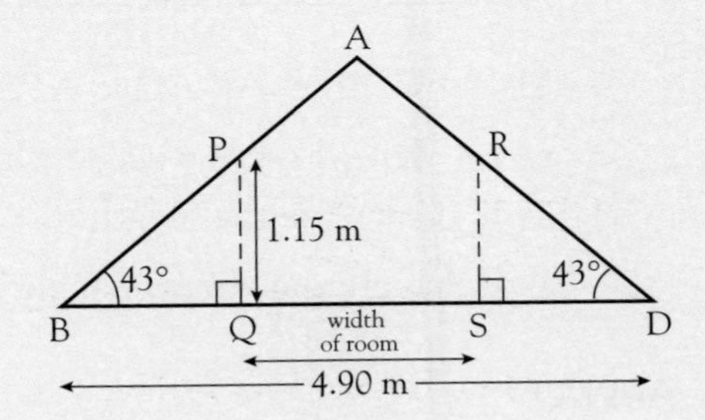

The loft is 4.90 m wide.
The lines PQ and RS show where walls, 1.15 m high, are to be built to make a room in the loft.

b **i** Calculate the length BQ.

ii Calculate the width of this room. [MEG]

7 The diameter of a Big Wheel is 15 metres and it carries 8 cars equally spaced around its circumference, as shown.

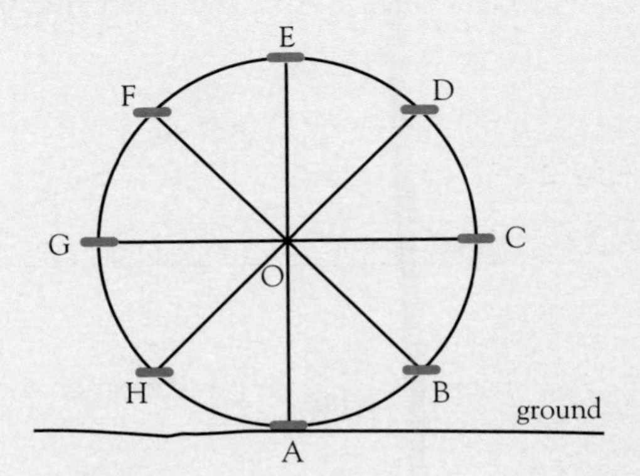

Find

a angle AOB

b the height of each of the cars above the ground, in the positions shown.

MIXED QUESTIONS

Sometimes problems involve a mixture of methods, so you need to be on the look-out for opportunities to use trigonometry, Pythagoras' rule or similar triangles. You may also need work on angles and bearings, area and volume.

Sample Question 10 A roof has a symmetrical frame, with dimensions as shown.

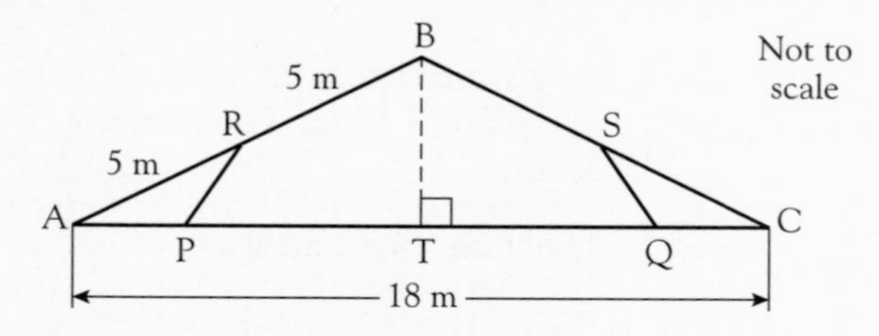

AB = BC, PR = AP, Angle ATB = 90°

a i Write down a triangle which is similar to triangle ABC.

ii Calculate the length PR.

b Calculate the value of angle BAT. [SEG]

Answer

a

- Look at the angles and mark in anything that you know about them. Look for angles that are equal.

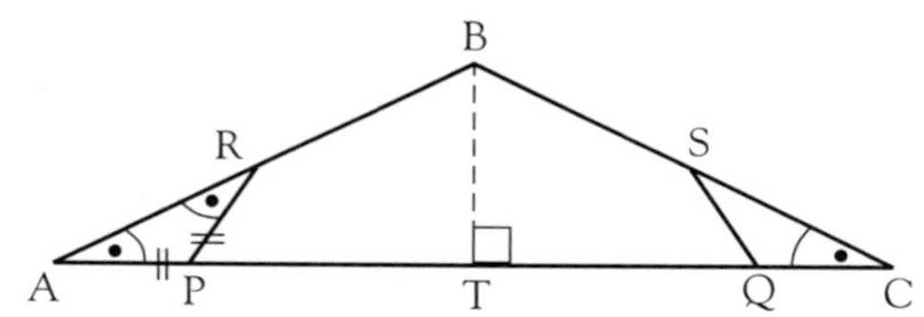

$\hat{RAP} = \hat{SCQ}$ ($\triangle$ABC is isosceles)

$\hat{RAP} = \hat{ARP}$ ($\triangle$ARP is isosceles)

- Draw triangle ABC and notice that triangle ARP has two angles which are the same as in triangle ABC.

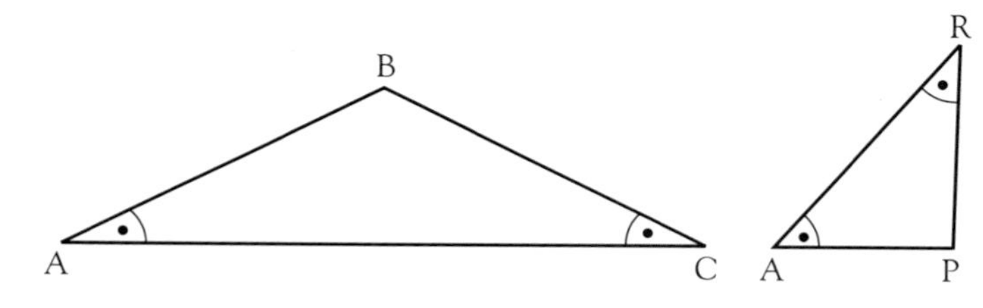

$\therefore$ $\hat{B} = \hat{P}$ (third angle in triangles) and the triangles are equiangular.

- Triangles $\frac{\text{ABC}}{\text{APR}}$ are similar, since they are equiangular.

- Write out the ratios, pairing them off. Write in any values that you know.

$$\frac{AB}{AP} = \frac{BC}{PR} = \frac{AC}{AR}$$

$$\frac{10}{AP} = \frac{10}{PR} = \frac{18}{5}$$

MIXED QUESTIONS

- Use $\dfrac{10}{PR} = \dfrac{18}{5}$

$$\frac{PR}{10} = \frac{5}{18}$$
$$PR = \frac{5}{18} \times 10$$
$$= 2.77\ldots$$
$$\underline{PR = 2.8\text{ m (2 s.f.)}}$$

b

- Draw △ABT and note that AT = 9 m, since triangle ABC is isosceles
- You know adj and hyp, so use cos.

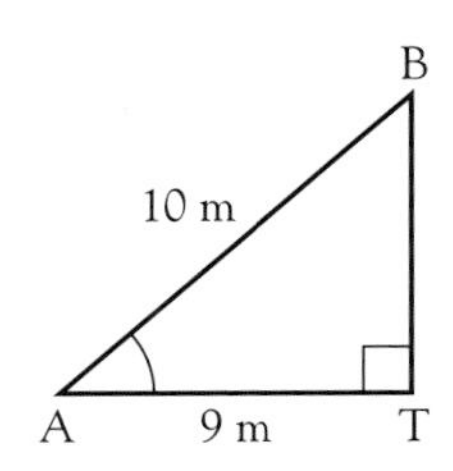

$$\cos \hat{A} = \frac{9}{10}$$
$$= 0.9$$
$$\hat{A} = \cos^{-1}(0.9)$$
$$= 25.84\ldots$$
$$\underline{\hat{A} = 25.8° \text{ (1 d.p.)}}$$

Sample Question **11** The diagram shows a tent which is 3 m long and 2.6 m wide. The tent has a volume of 8.58 m³. Find θ, the angle between the sloping side of the tent and the ground.

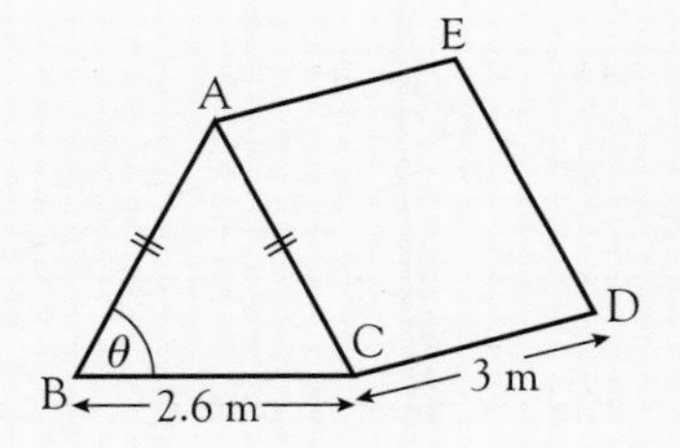

Answer

- Look at the triangle containing θ. You know that it is isosceles and that the base length is 2.6 m. In order to find θ, you will need more information.
- You are told that the volume is 8.58 m³. Think what information you can get from this. It is a **prism**.

Volume of a prism = **area of cross-section** × length

8.58 = area of ABC × 3

Area of ABC = 8.58 ÷ 3 = 2.86 m²

⚠ The cross-section of the prism is the isosceles triangle ABC.

- Now look again at triangle ABC.
Draw in the perpendicular height AX. Since the triangle is isosceles, this cuts BC in half.

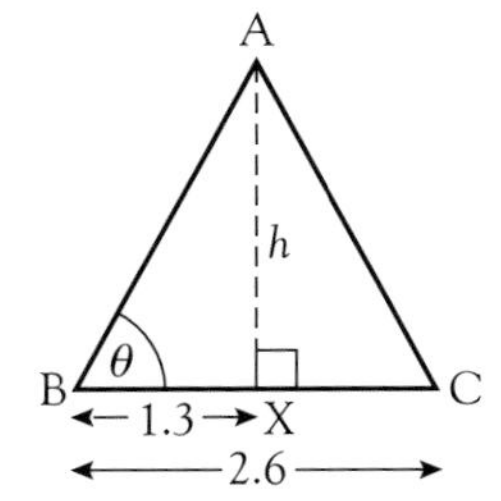

$$\frac{\text{base} \times \text{height}}{2} = \text{Area of triangle}$$
$$\frac{2.6 \times h}{2} = 2.86$$

(×2) $\quad 2.6h = 5.72$

(÷2.6) $\quad h = 2.2$

MIXED QUESTIONS

◆ Now look at triangle ABX and use trigonometry to find θ.

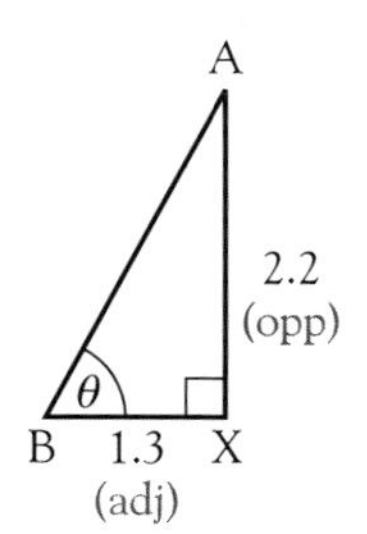

$$\tan \theta = \frac{2.2}{1.3}$$

$$\theta = \tan^{-1}\left(\frac{2.2}{1.3}\right)$$

$$= 59.42 \ldots$$

$$\underline{\theta = 59.4^\circ \text{ (1 d.p.)}}$$

You know opp and adj so use tan.

This is an example of an unstructured question in which you have to decide the steps that will lead you to a solution.

Exercise 26.4

1 The model of the cross-section of a roof is illustrated below.

BC = 6 cm
CD = 9 cm
Angle CDE = 19.5°

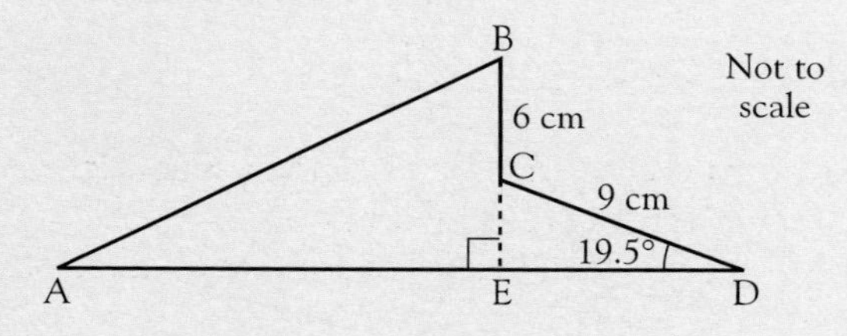

a Calculate the length of CE.

b Triangles ABE and DCE are similar triangles with angle BAE = angle CDE.
Calculate the length of AB. [SEG]

2 The sketch shows a roundabout, viewed from above. It is in the shape of a regular pentagon ABCDE. The sides are all of length 1.54 m, and the mid-point of side CD is denoted by F.

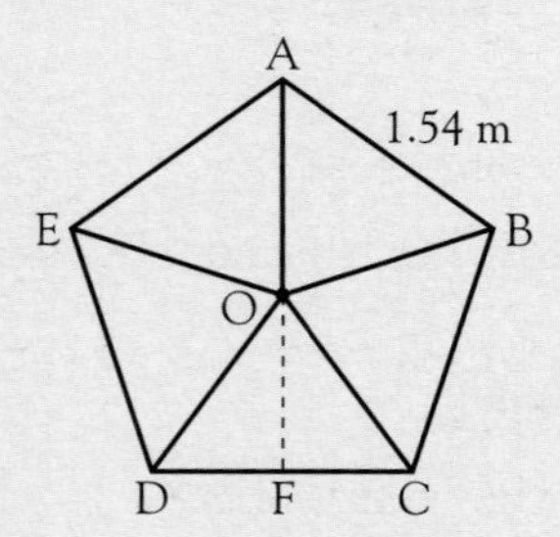

Find

a angle AOB,

b the distance OA,

c the distance OF,

d the area of the pentagon.

3

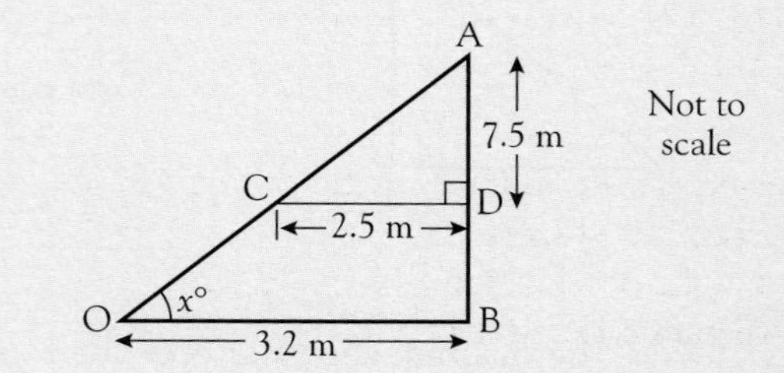

a Using the measurements shown in the diagram, calculate the length of the line AB.

Angle AOB = x°

b Write down the value of $\tan x$°. [L]

4 A water trough is in the shape of a prism whose cross-section is an isosceles trapezium ABCD, as shown.
The trough is 40 cm wide at the top, 18 cm wide at the bottom and 1.8 metres long. Its sloping sides are inclined at 58° to the horizontal.
Calculate

a the depth of the trough, d,

b the area of the trapezium ABCD,

c the volume of the trough in litres.

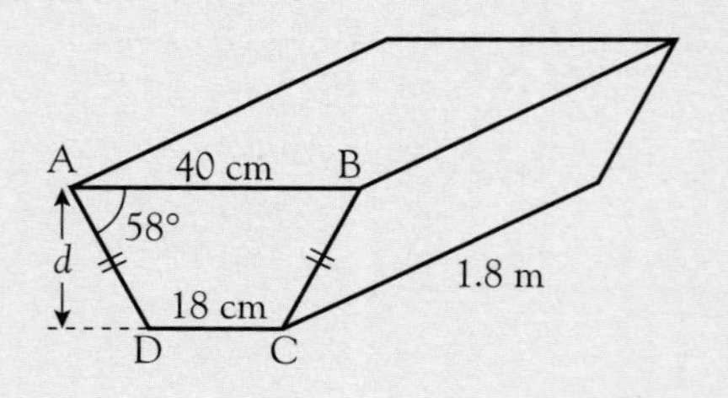

MIXED QUESTIONS

5 A gift box is in the shape of a prism and its cross-section is a pentagon, shown in the diagram.

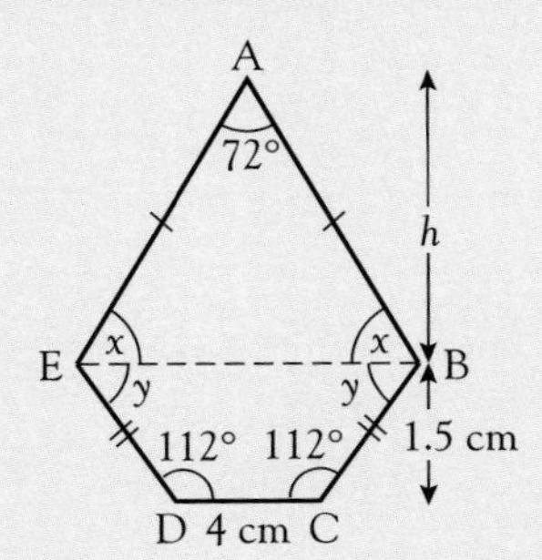

a Work out the values of x and y.

b Calculate the length of EB.

c Calculate the height, h, of $\triangle$ABE.

d The length of the box is 10 cm.
Find its volume.

6 A design for a children's slide, ABCD, for a playground is shown.
The height of the slide BD is 136 cm.
The distance DC is 195 cm.

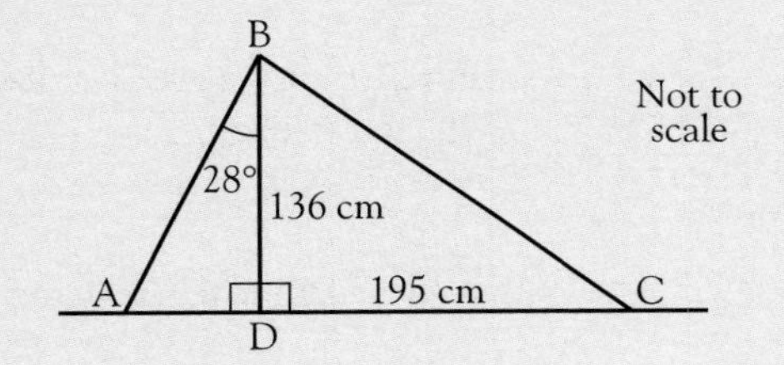

a i Calculate the angle BCD.

ii Angle ABD is 28°.
Calculate the distance AB.

b Another design of slide, EFGH, has angle FEH = 57°.
Triangle EFH is similar to triangle FGH but the two triangles are not the same size.

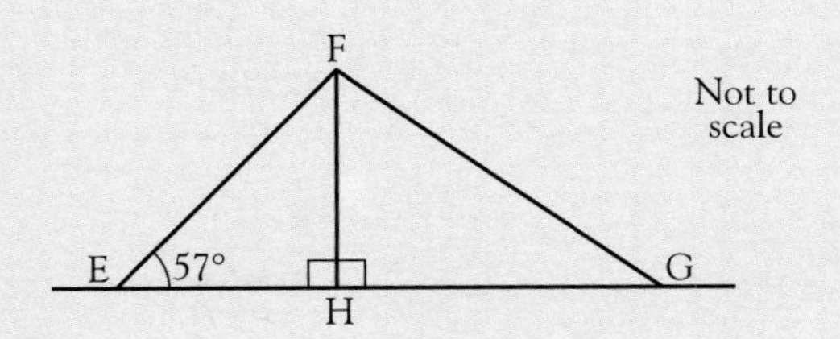

Find the size of angle FGH. [SEG]

7 The diagram shows a regular pentagon inside a circle with centre O and radius 5 cm. X is the mid-point of AE.

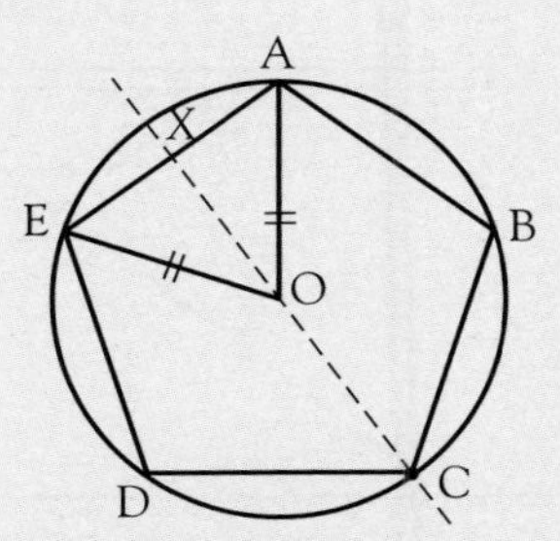

a Calculate the angle AOE.

b Calculate the length CX.

c Calculate the length of a side of the pentagon.

d Calculate the area of the part of the circle outside the pentagon.

8

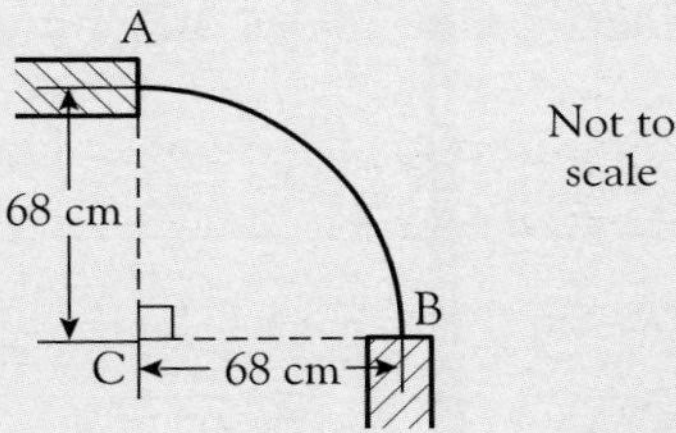

The diagram represents the plan of a window frame.
The arc AB is a quarter of a circle.
The centre of the circle is at C and the radius of the circle is 68 cm.

a Calculate the length of the arc AB.
(Use $\pi = 3.14$ or use the π button on your calculator.)
Give your answer correct to 3 significant figures.

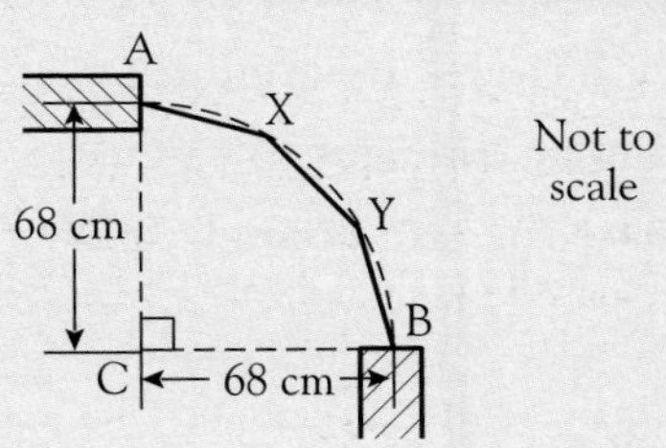

The window frame in **a** is replaced by double glazed panels.
These panels are made only in straight lengths.
The arc AB is replaced by three identical panels AX, XY and YB.

b Calculate the length of AX.
Give your answer correct to 2 significant figures.

[L]

Worked Exam Question
[MEG]

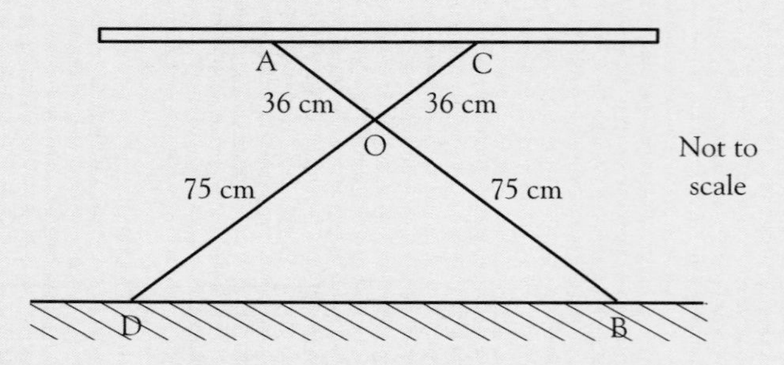

The diagram shows how an ironing board is supported by two legs AOB and COD. The legs are hinged at O, and C is hinged to the ironing board.

The distance between A and C can be varied.

a i When angle BOD = 80°, work out the size of angle OBD.

ii What facts about angles did you use?

◆ Use angle facts in isosceles triangle. Draw a sketch.

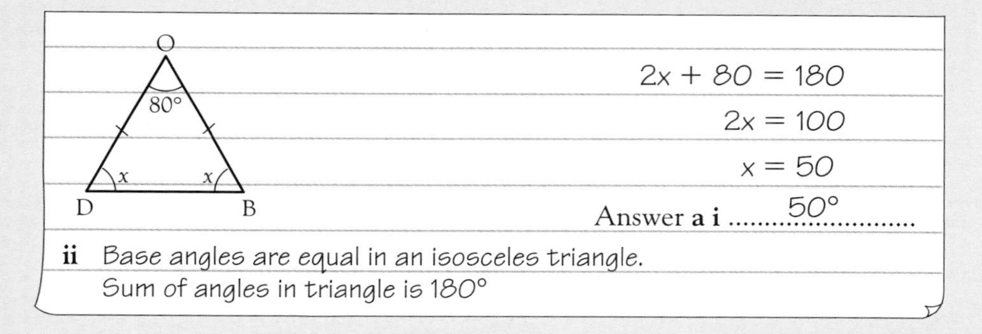

b The ironing board is placed on a horizontal floor and adjusted so that BD = 92 cm.

i Use Pythagoras' theorem to calculate the height of O above the floor.

ii Use similar triangles to calculate the distance AC.

iii Calculate the size of angle OBD.

i

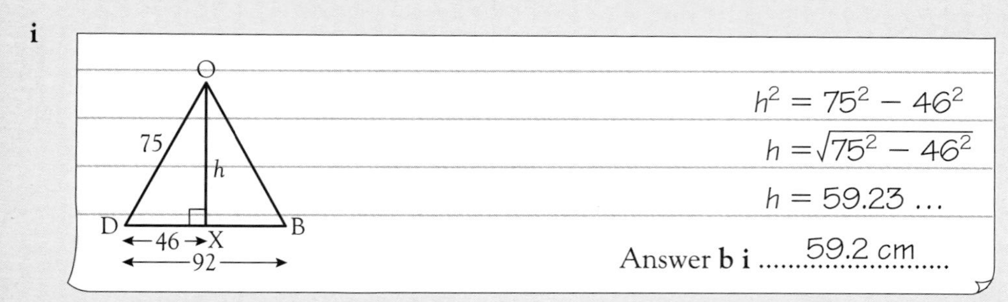

COMMENTS

A1 for 50°

A1 for $\hat{B} = \hat{D}$ stated

A1 for sum

3 marks

M2 (**M1** for $h^2 + 46^2 = 75^2$)

A1

3 marks

EXAM QUESTIONS

i

- Show equal angles on a diagram (use alternate angles on parallel lines are equal)
- State similar triangles and write out the equal ratios

Triangles $\frac{AOC}{BOD}$ are similar (because they are equiangular)

$$\frac{AO}{BO} = \frac{OC}{OD} = \frac{AC}{92}$$

$$\frac{36}{75} = \frac{OC}{OD} = \frac{AC}{92} \qquad AC = 92 \times \frac{36}{75} = 44.16$$

Answer ii 44.16 cm

COMMENTS

M2 for similar triangles and correct ratio for AC

A1 for correct answer

3 marks

ii

- Note that $O\hat{B}D = O\hat{B}X$ and look at △OBX.
- Use trigonometry to find $\hat{B}$. You know adj and hyp, so use cos.

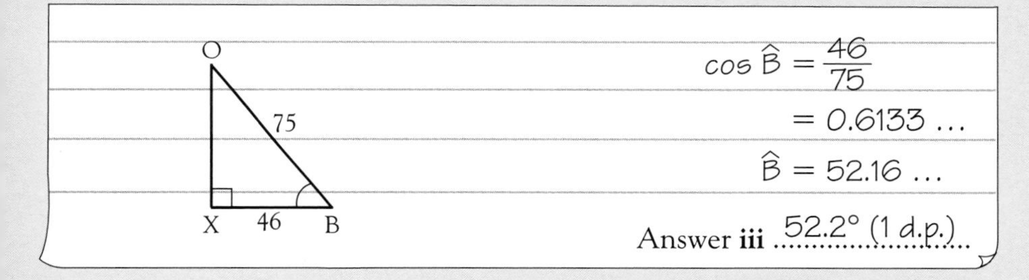

M2 for cos ratio used correctly

A1 for correct angle

3 marks

The ironing board is adjusted again so that AC is 90 cm above the floor. Calculate the height of O above the floor.

- Look at the diagram like this.

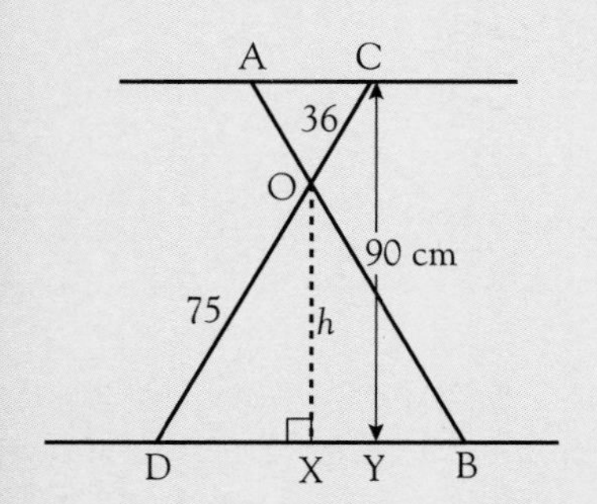

- Pick out △ODX and △CDY. They are equiangular (using parallel line facts) so they are similar.

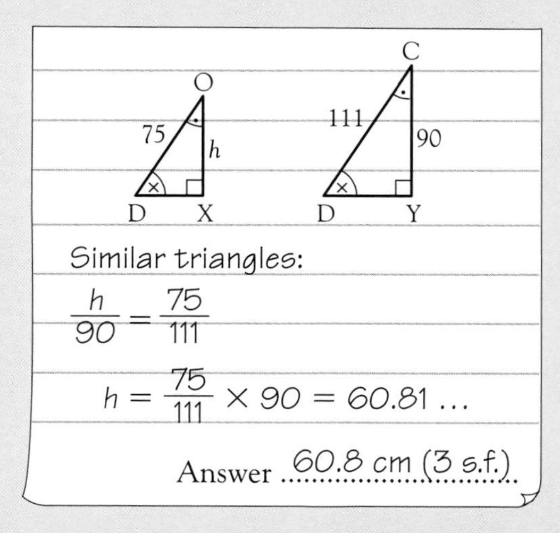

M2 for similar triangles and correct ratio for h

A1 for correct answer

3 marks

Exam Questions

1 For this question, no marks will be given for answers obtained by scale drawing.

a

The diagram shows how Ravi intends to hang a picture on a wall. He will fasten a cord to the points A and B on the picture frame. The cord will pass over two hooks P and Q on the wall. APQB is a trapezium. AB and PQ are horizontal.

i Calculate the length of AP.

ii Calculate the total length of the cord.

iii Calculate, correct to the nearest degree, the size of the angle marked x in the diagram.

b

As an alternative, Ravi considers using just one hook R.
Given that angle RAB = angle RBA = 50°, calculate the length of the cord. [MEG]

2

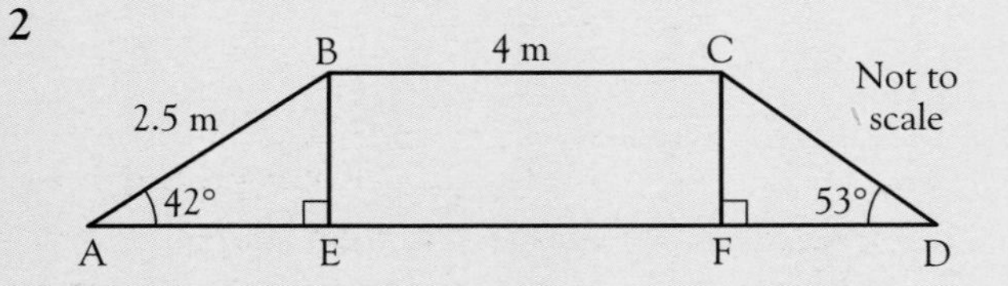

In a dog obedience test, a dog has to climb a ladder from A to B, walk along a horizontal plank from B to C, then run down a slope from C to D. The line AEFD is horizontal.

The ladder is 2.5 m long and makes an angle of 42° with the horizontal ground.

a Calculate the distance AE.

b Calculate the height, BE, of the horizontal plank above the ground.

The plank is 4 m long and the slope, CD, makes an angle of 53° with the ground.

c Calculate the total **horizontal** distance between A and D. [MEG]

3 The diagram shows two vertical lines AB and CD of lengths 4 cm and 12 cm.
BFD is a horizontal line.
E is the intersection of AD and BC.

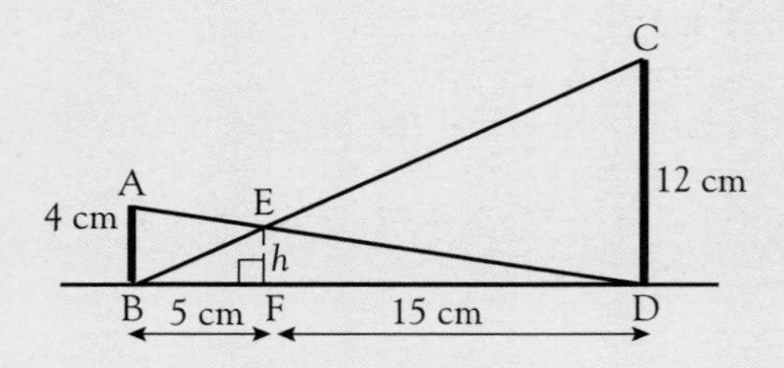

a i Write down a triangle that is similar to triangle BEF.

ii Use these similar triangles to calculate the height marked h.

b CD is gradually moved closer to AB.
Describe what happens to the height h. [NEAB]

4 State whether or not the triangles ABC and XYZ below are similar.
Show working to support your answer.

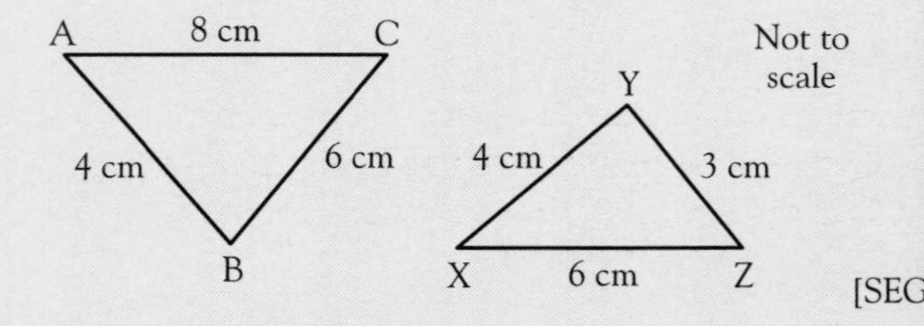

[SEG]

5 A ship sails on a two-stage journey from A to B to C.
The first stage of the journey from A to B is shown.
A to B is a journey of 90 km on a bearing of 032°.

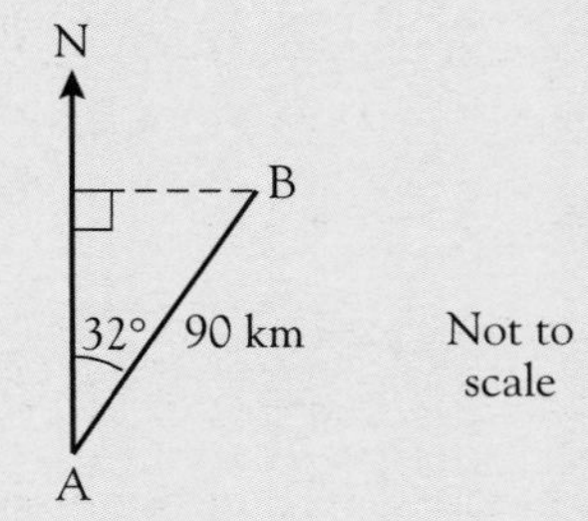

EXAM QUESTIONS

a Calculate the distance travelled east during the first stage of this journey.

The second stage of the journey from B to C is a distance 150 km on a bearing of 090°.

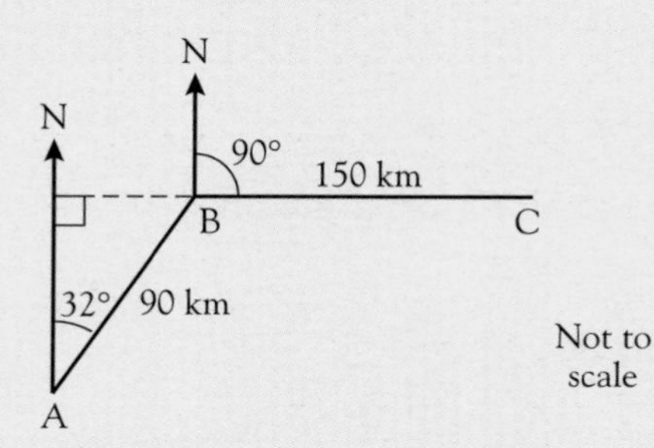

b Find the total distance travelled east on the journey from A to C.
Hence calculate the bearing of C from A. [SEG]

In the following expressions, a and b both represent lengths.

$3a + 3b$	a^2b^2
ab	$ab + b$
$\frac{a+b}{b}$	a^2b
a^3b	a^3b^3

By considering the dimensions of each expression, write down one which could represent

a a perimeter, **b** an area, **c** a volume.

[MEG]

From a point P on level ground, a surveyor measures the angle of elevation of R, the top of a building, as 32°.
He walks 12 metres towards the building to point Q and measures the angle of elevation of R as 57°.
X is the point on PR such that angle PXQ = 90°.

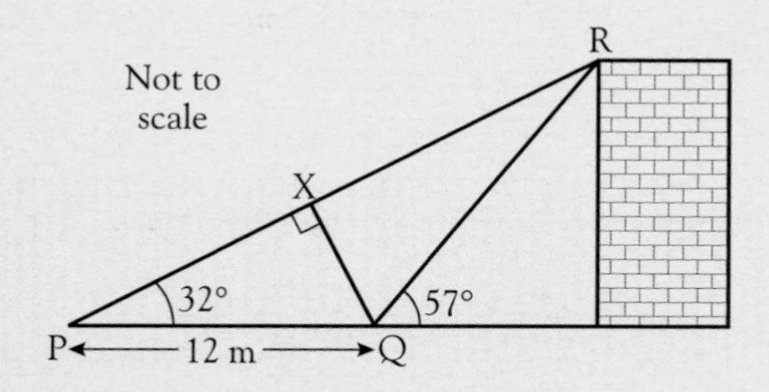

a Show that QX = 6.36 m.

b i Explain why angle PRQ is 25°.

ii Hence find the length of QR.

c Calculate the height of the building. [MEG]

8 a The British Rail logo shown below is to be enlarged for a new poster.

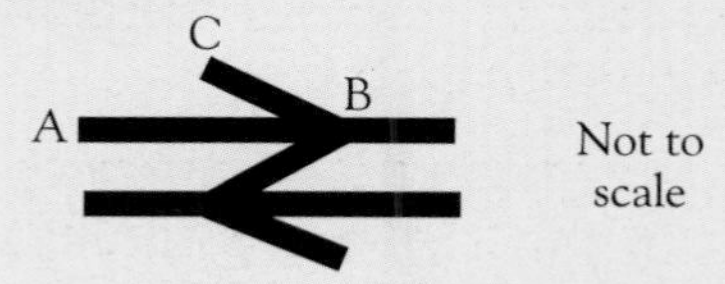

The distance AB is 3 cm on the original shape.
On the enlarged shape AB is 13.5 cm and BC is 7.8 cm.
What is the distance BC on the original shape?

b In the diagram AB = 3 cm, BC = 2 cm and AE = 2 cm.
Angles AEB and ACD are equal.

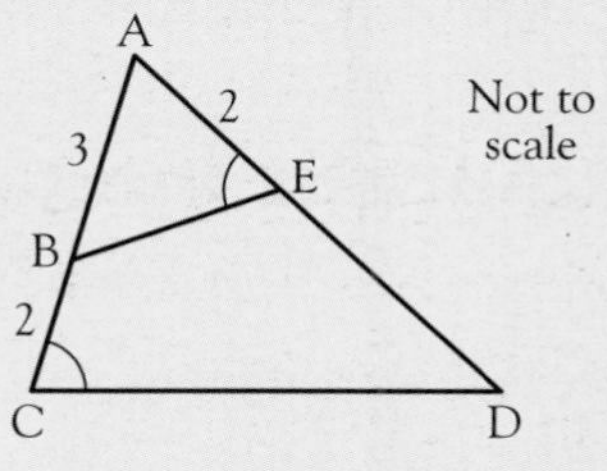

i Explain why triangle ABE is similar to triangle ADC.

ii Calculate the length of ED. [SEG]

9

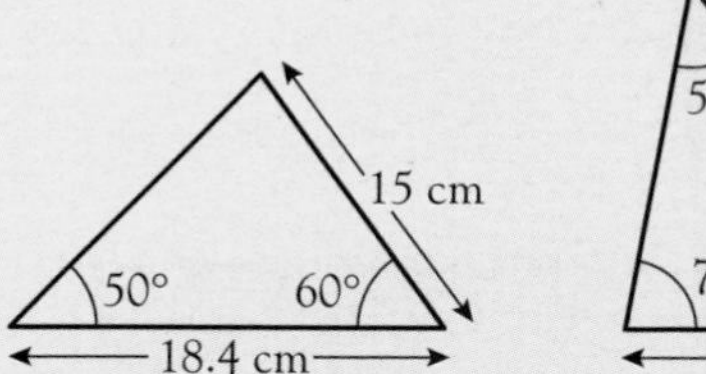

a Show clearly that these two triangles are similar.

b Calculate the value of x. [NEAB]

10

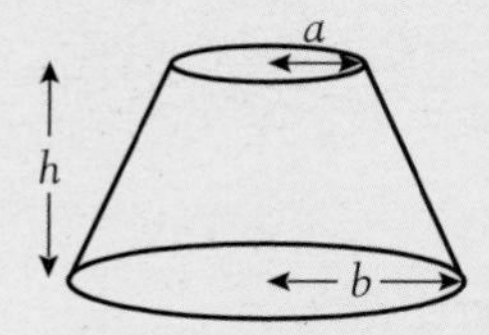

The diagram represents a solid shape.
From the expressions below, choose the one that represents the volume of the solid shape.
π and $\frac{1}{3}$ are numbers which have no dimensions.
a, b and h are lengths.

$\frac{1}{3}\pi(b^2 - ab + a^2)$	$\frac{1}{3}\pi(a^2 + b^2)$
$\frac{1}{3}\pi h(b^2 + ab + a^2)$	$\frac{1}{3}\pi h^2(b^2 - a^2)$
$\frac{1}{3}\pi h^2(b^2 - ab + a^2)$	

Write down the correct expression. [L]

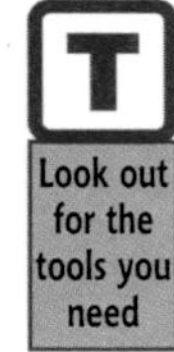

Look out for the tools you need

In this chapter you will learn how to
- **solve quadratic inequalities**
- **represent inequalities in two variables as regions on a graph**
- **solve cubic equations by trial and improvement**
- **draw cubic and reciprocal graphs**
- **sketch standard graphs**
- **use graphs to model real life situations.**

Quadratic inequalities

Consider first the quadratic equation $x^2 = 9$

This has two solutions, $x = 3$ or $x = -3$

Remember $(-3)^2 = 9$

Now consider these **quadratic inequalities**

a $x^2 \leqslant 9$ **b** $x^2 < 9$ **c** $x^2 \geqslant 9$ **d** $x^2 > 9$

For each of these, -3 and 3 are the **boundary points**, where 3 and -3 are the solutions of $x^2 = 9$.

The diagrams below show the values of x that satisfy each inequality.

a $x^2 \leqslant 9$

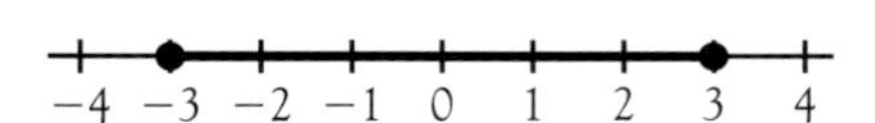

The values are $-3 \leqslant x \leqslant 3$

Any number sandwiched between -3 and 3 including the boundary values of 3 and -3, satisfies the inequality $x^2 \leqslant 9$.
Note the closed or 'filled-in' circles.

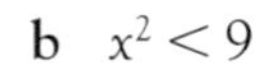

b $x^2 < 9$

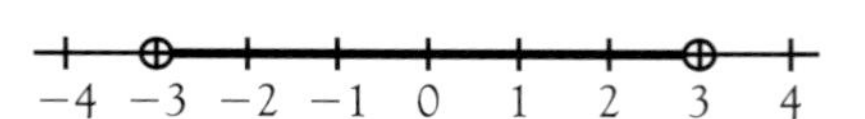

The values are $-3 < x < 3$

Any number sandwiched between -3 and 3 not including 3 and -3, satisfies the inequality $x^2 < 9$.
Note the open circles.

c $x^2 \geqslant 9$

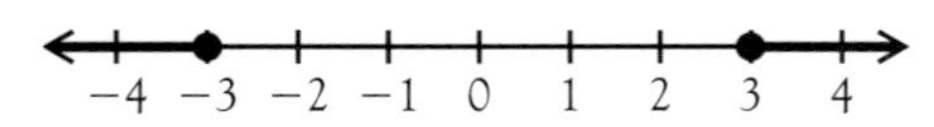

The values are $x \leqslant -3$ or $x \geqslant 3$

Any number at the extremes of the number line, including 3 and -3, satisfies the inequality $x^2 \geqslant 9$.
Note the filled-in circles.

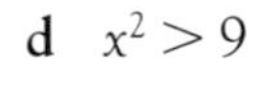

d $x^2 > 9$

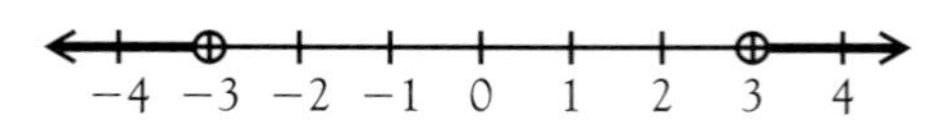

The values are $x < -3$ or $x > 3$

Any number at the extremes of the number line, not including -3 or 3, satisfies the inequality $x^2 > 9$.
Note the open circles.

Notice that in parts **a** and **b** the answer is given in a combined statement, whereas in parts **c** and **d** the answer is given in two statements.

QUADRATIC INEQUALITIES

Sample Question 1 Find the solution set in each of the following and show it on a number line.

a $12 - 3x^2 \leqslant 0$ b $x^2 - 3x < 0$

Answer

a

- Re-arrange $12 - 3x^2 \leqslant 0$ to make $3x^2$ the subject.

  $12 \leqslant 3x^2$ i.e. $3x^2 \geqslant 12$

- Simplify the inequality by dividing both sides by 3.

  $x^2 \geqslant 4$

- Find the boundary points by solving $x^2 = 4$ and decide whether to include them in your answer.

  $x^2 = 4$ gives $x = 2, x = -2$ (include the boundary points because inequality contains $\geqslant$)

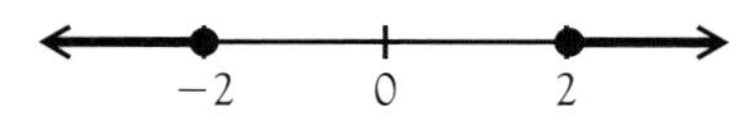

The solution set is $x \leqslant -2, x \geqslant 2$

When you square a number bigger than 2, or smaller than −2, the answer is bigger than 4.

b

- Find the boundary points by solving $x^2 - 3x = 0$

  $x(x - 3) = 0$

  so $x = 0$ or $x - 3 = 0$

  $x = 3$

Quadratic equations page 477

- Decide whether to include the boundary points (do not include since inequality contains $<$ sign).
- Put 0 and 3 on the number line and decide whether the required x values lie between them or at the extremes.

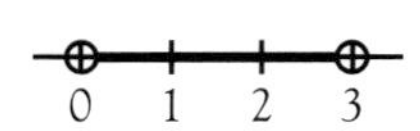

Try $x = 2$
$x^2 - 3x = 4 - 6 = -2 < 0$
so $x = 2$ is in the required part of the number line.

The solution set is $0 < x < 3$

Exercise 27.1

1 For each of the following quadratic inequalities, find the solution set and show it on a number line. Remember to draw a 'filled' circle if the boundary number is included and an 'open' circle if it is not.

a $x^2 \leqslant 4$
b $x^2 < 25$
c $x^2 \geqslant 16$
d $x^2 > 100$
e $4x^2 < 36$
f $3x^2 \geqslant 3$
g $x^2 > 0$
h $x^2 - 49 < 0$
i $2.25 - x^2 < 0$
j $x^2 - 2x > 0$
k $x^2 + 4x \leqslant 0$
l $(x - 2)(x - 4) \leqslant 0$

2 Write down the integer values of x that satisfy each of the following quadratic inequalities.

a $x^2 + 5 \leqslant 6$
b $x^2 < 16$
c $3x^2 < 12$
d $9 - x^2 > 0$
e $x^2 - 3x \leqslant 0$
f $5x - x^2 > 0$

3 Danielle thought of an integer, squared it and then added 3. She said that her answer was smaller than 15. Write down all the possible numbers she could have thought of.

4 What is the smallest integer such that $x^2 \leqslant 32$?

REGIONS

Inequalities in two variables

Think about this problem:

Jellybeans are 2p each.
Chewies are 3p each.
You want to spend **exactly** 30p.
How many of each can you buy?

You could solve this problem by trial and improvement, but unless you are systematic it is difficult to make sure that you have found all the combinations.

A good method to use is to form an equation and draw a graph.

Imagine buying x jellybeans; these would cost $2x$ pence.
Imagine buying y chewies; these would cost $3y$ pence.
The total cost is $2x + 3y$ pence.
You spend 30 pence, so $2x + 3y = 30$.

Draw the line $2x + 3y = 30$ on a grid.

x	0	15
y	10	0

Remember the 'cover up' method for the table (page 341)

You can only buy whole numbers of sweets, so mark the points on the line with whole number co-ordinates. These show all the possible combinations of jellybeans and chewies that you can buy, paying exactly 30p.

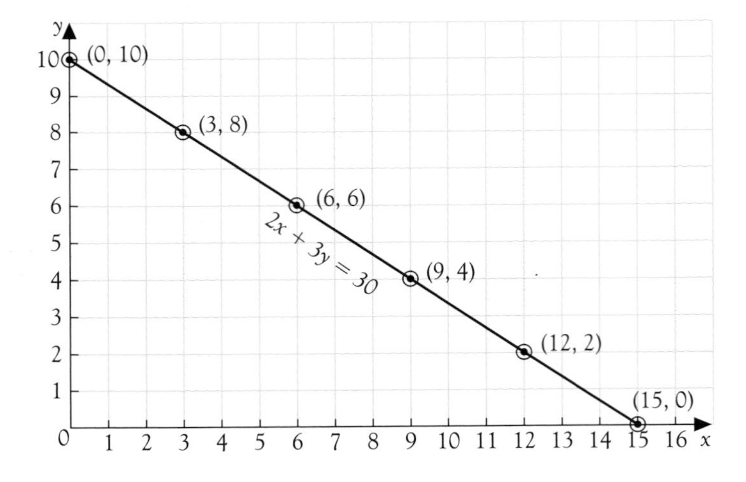

You can buy the following combinations:

0 jellybeans, 10 chewies
cost $= 2 \times 0 + 3 \times 10 = 30$p
3 jellybeans, 8 chewies
cost $= 2 \times 3 + 3 \times 8 = 30$p
6 jellybeans, 6 chewies
cost $= 2 \times 6 + 3 \times 6 = 30$p
9 jellybeans, 4 chewies
cost $= 2 \times 9 + 3 \times 4 = 30$p
12 jellybeans, 2 chewies
cost $= 2 \times 12 + 3 \times 2 = 30$p
15 jellybeans, 0 chewies
cost $= 2 \times 15 + 3 \times 0 = 30$p

At any point on the line, $2x + 3y = 30$ but only those with whole number co-ordinates are needed in this example!

The line splits the grid into two **regions**.

REGIONS

a Look at some points in the region below the line (left unshaded).

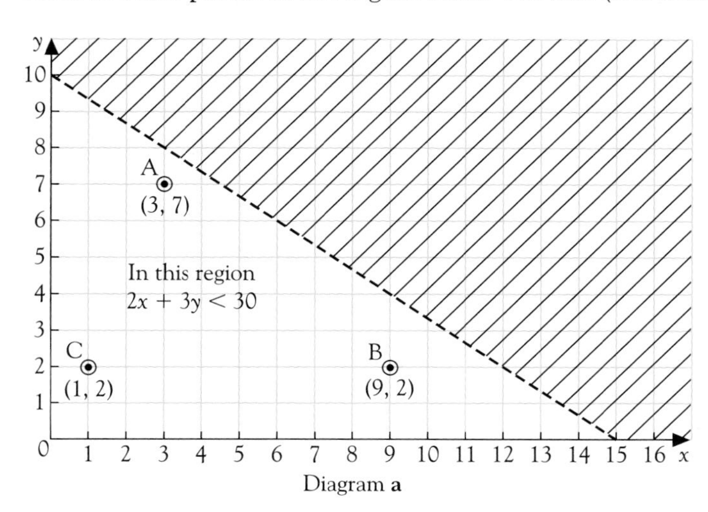

Diagram **a**

At A, $x = 3$ and $y = 7$, so you would buy 3 jellybeans and 7 chewies

Cost $= 2 \times 3 + 3 \times 7 = 27$p

At B, $x = 9$ and $y = 2$, so you would buy 9 jellybeans and 2 chewies

Cost $= 2 \times 9 + 3 \times 2 = 24$p

At C, $x = 1$ and $y = 2$, so you would buy 1 jellybean and 2 chewies

Cost $= 2 \times 1 + 3 \times 2 = 8$p

For **all** points in the unshaded region, you would spend **less than 30p**.

This is the region where $2x + 3y < 30$.

This is an inequality.

Note that the line $2x + 3y = 30$ has been drawn as a broken line, not a solid one. This is to show that it is not included in the region where $2x + 3y < 30$.

b Now look at points in the region above the line (now left unshaded).

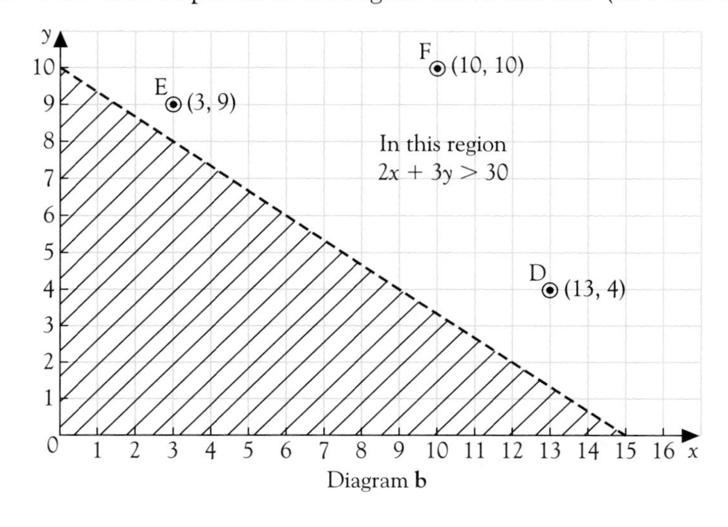

Diagram **b**

At D, $x = 13$ and $y = 4$

Cost $= 2 \times 13 + 3 \times 4 = 38$p

At E, $x = 3$ and $y = 9$

Cost $= 2 \times 3 + 3 \times 9 = 33$p

At F, $x = 10$ and $y = 10$

Cost $= 2 \times 10 + 3 \times 10 = 50$p

For **all** points in the unshaded region, the cost is **more than 30p**.

This is the region where $2x + 3y > 30$.

Notice the three situations

$2x + 3y < 30$ (You spend less than 30p)
$2x + 3y > 30$ (You spend more than 30p)
$2x + 3y = 30$ (You spend exactly 30p)

REGIONS

If you want to **include** points on the line, then you would draw it as a **solid** line, not a broken one, as in the diagrams below.

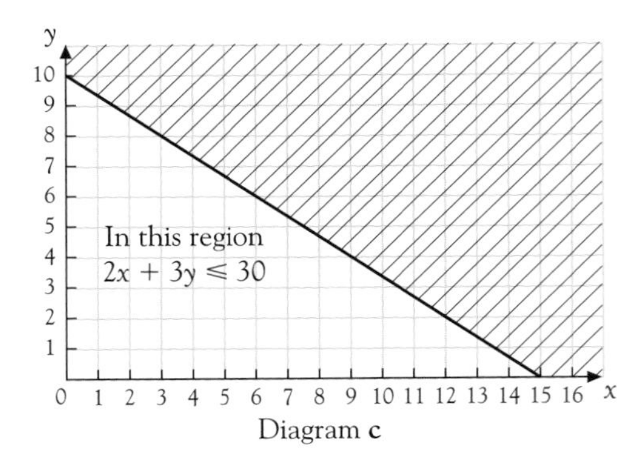

Diagram **c**

This unshaded region **includes** the line, so it shows where you spend **30p or less**.

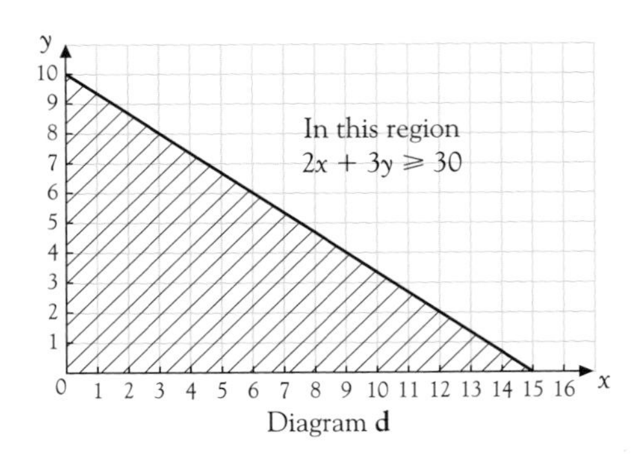

Diagram **d**

This unshaded region **includes** the line, so it shows where you spend **30p or more**.

Consider now the following situations which you want to be satisfied.

1 You want to spend no more than 30p

$$2x + 3y \leqslant 30$$

This is shown unshaded on diagram **c** above.

2 You want to buy at least two jellybeans

$$x \geqslant 2$$

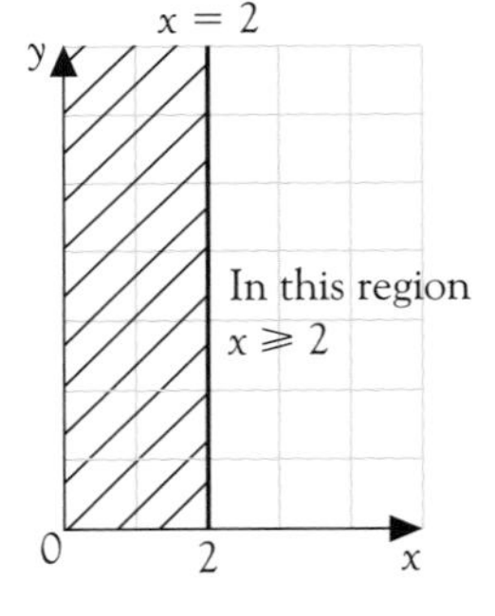

- To show this on the diagram draw the boundary line $x = 2$. It is to be included in the region, so draw a solid line.
- Identify where $x \geqslant 2$.

Notice that any point to the right of the line $x = 2$ has an x co-ordinate greater than 2 so the region required is to the right of the line.

- Leave the region $x \geqslant 2$ unshaded, so shade to the left of the line $x = 2$.

3 You want to buy more than twice as many chewies as jellybeans. The inequality is

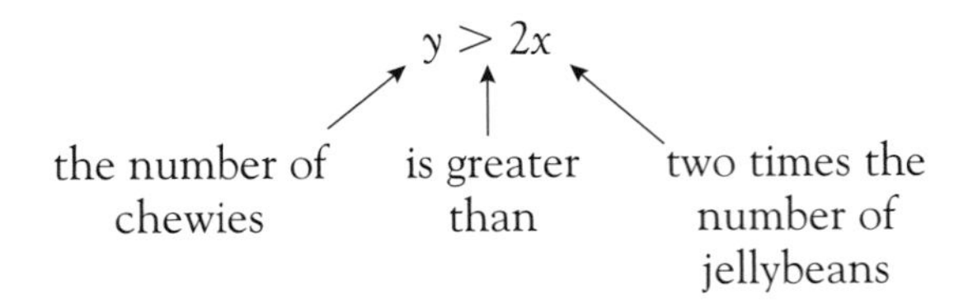

REGIONS

- Draw the boundary line $y = 2x$ and identify where $y > 2x$.

Since you want $y > 2x$, the line is not included in the region, so draw it as a broken line.

- To decide on the region, choose **any** point e.g. (2, 10) and test it. The y co-ordinate is 10, $2 \times x$ co-ordinate is 4.

  Since $10 > 4$, $y > 2x$ so (2, 10) lies in the region $y > 2x$.

(2, 10)
In this region $y > 2x$
$y = 2x$

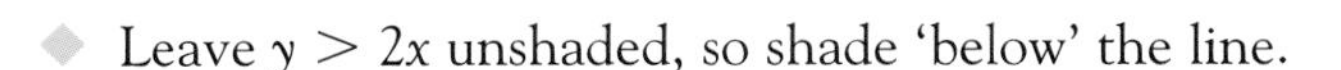

- Leave $y > 2x$ unshaded, so shade 'below' the line.

In practice you would show all three conditions on one diagram: the region left unshaded is where all three conditions are satisfied.

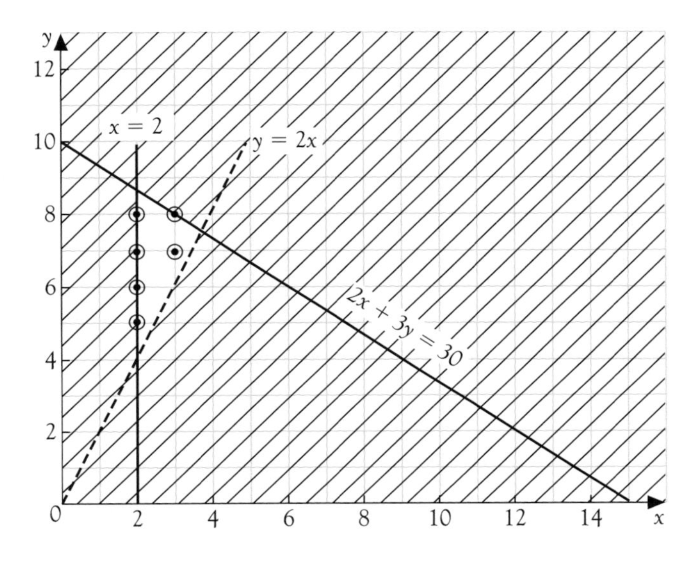

There are only 6 possible combinations (shown as ⊙ on diagram):

2 jellybeans, 5 chewies
2 jellybeans, 6 chewies
2 jellybeans, 7 chewies
2 jellybeans, 8 chewies
3 jellybeans, 7 chewies
3 jellybeans, 8 chewies

In only the last combination would you spend exactly 30p.

It is difficult to label all the regions on a diagram, so you should usually just label the lines, as in Sample Question 1. Remember that there is an equals sign in the equation of the line.

Sample Question 2

The ingredients for one Empire Cake include 200 g of self-raising flour and one egg.

The ingredients for one Fruit Cake include 100 g of self-raising flour and three eggs.

Alphonso has 1200 g of self-raising flour and 12 eggs.

He makes x Empire Cakes and y Fruit Cakes.

Using the weight of self-raising flour available means that x and y must satisfy the inequality $2x + y \leq 12$.

a Write down another inequality which x and y must satisfy, other than $x \geq 0$ and $y \geq 0$.

b By drawing straight lines on a grid and shading, indicate the region within which x and y must lie to satisfy all the inequalities.
The line $2x + y = 12$ has been drawn for you (see diagram on next page). [L]

REGIONS

Answer

a

- Look at the condition regarding eggs:

  An Empire Cake uses one egg, so x Empire Cakes use x eggs.

  A Fruit Cake uses three eggs, so y Fruit Cakes use $3y$ eggs.

  The total number of eggs used is $x + 3y$.

  Since Alphonso has 12 eggs, he cannot use more than 12 eggs; he must either use all the 12 eggs or fewer than 12 eggs. This gives you the inequality

  $\underline{x + 3y \leqslant 12}$

b

- Look at the line already drawn. It is $2x + y = 12$, and decide which side of the line shows where $2x + y \leqslant 12$ by choosing a point, say (2, 5), and checking it.

  At (2, 5), $2x + y = 2 \times 2 + 5 = 9$. This is less than 12, so (2, 5) lies in the region $2x + y \leqslant 12$. Shade the other side of the line. You could think of this as 'shading out' the region not required.

- Draw the line $x + 3y = 12$. Do a table using the 'cover-up' method and remember to label the line.

x	0	12
y	4	0

- Decide which side shows $x + 3y \leqslant 12$ by checking a point, say (8, 3).

  At (8, 3), $x + 3y = 8 + 9 = 17$. This is not less than 12, so (8, 3) is in the region **not required**. It is in the region to be shaded.

- Indicate clearly, by writing R in it, that the final region left unshaded is the one required for the answer.

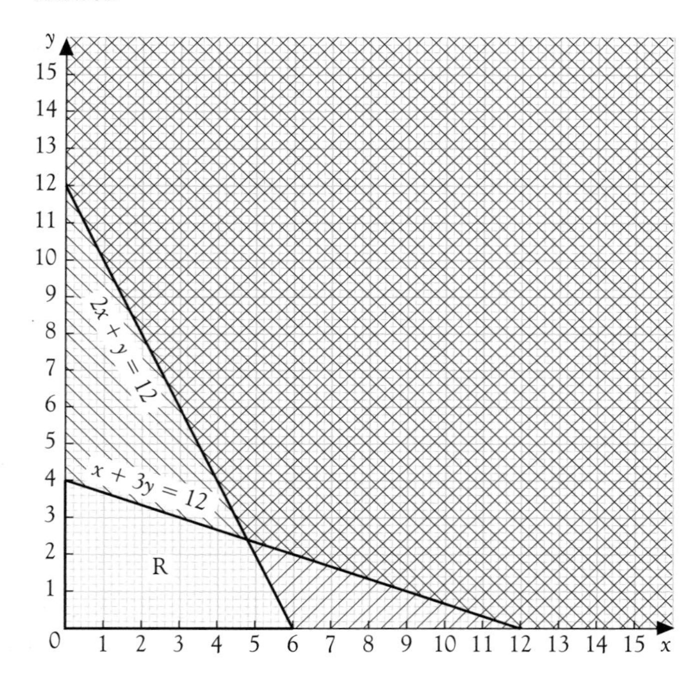

If you are not told which region to shade, it is better to leave the required region unshaded.

Sometimes, however, you are **told** to **shade the required region**, not leave it unshaded. In this case it is a good idea to shade what is not needed lightly in pencil, identify the answer region, and then shade it boldly in a colour – putting in a label or a key to explain that it represents your answer.

REGIONS

Exercise 27.2

Remember that when you are drawing or describing regions
- a solid line means that the line is included as part of the region and you must use either $\leqslant$ or $\geqslant$
- a dotted line means that the line is not included as part of the region and you must use either $<$ or $>$

1 Write inequalities to describe the regions that are **not shaded** in each of the following diagrams:

a

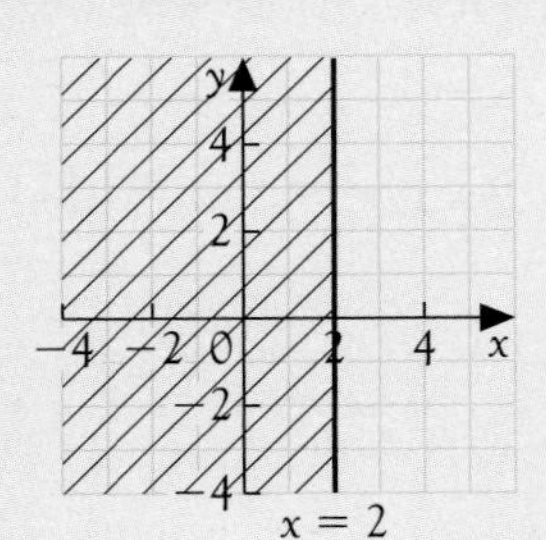

b

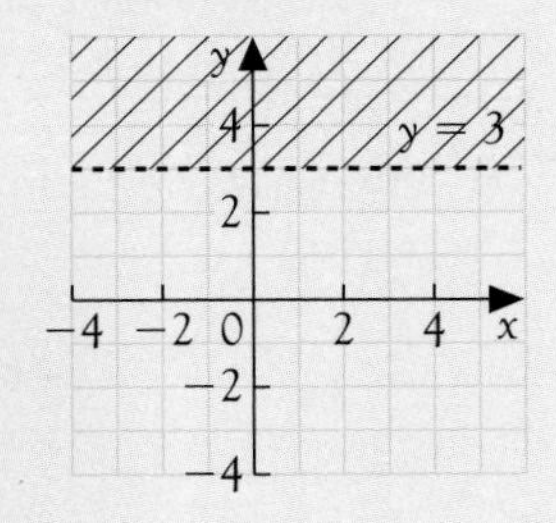

c

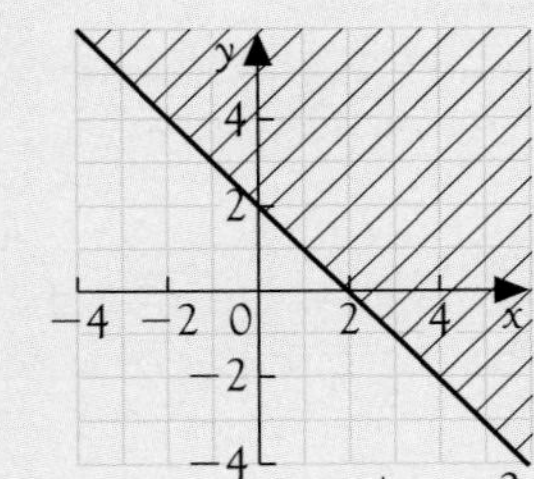

d

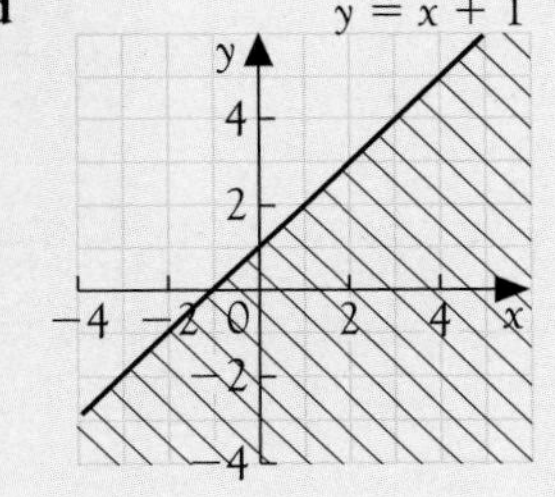

2 **a** Draw a grid with x axis going from -4 to 4 and y axis from -2 to 7.

b Copy and complete the table for the line $y = x + 3$

x	−4	−2	0	2	4
y					

c Draw and label the line $y = x + 3$

d Show, by leaving it unshaded, the region defined by its inequality $y \leqslant x + 3$. Label the region R.

3 For each part, draw a grid with both axes going from -4 to 6. Draw and label a suitable boundary line and then show, by leaving it unshaded, the region defined by the given inequality. Label the required region R.

a $y \geqslant \frac{1}{2}x - 1$

b $y < 3x + 1$

c $x + y < 5$

d $y \geqslant 1$

e $y \leqslant x$

f $x > -2$

g $2x + y < 6$

h $y \geqslant 3 - x$

i $y > x - 2$

4 The unshaded region in the diagram is described by the inequality $-2 \leqslant x \leqslant 3$.

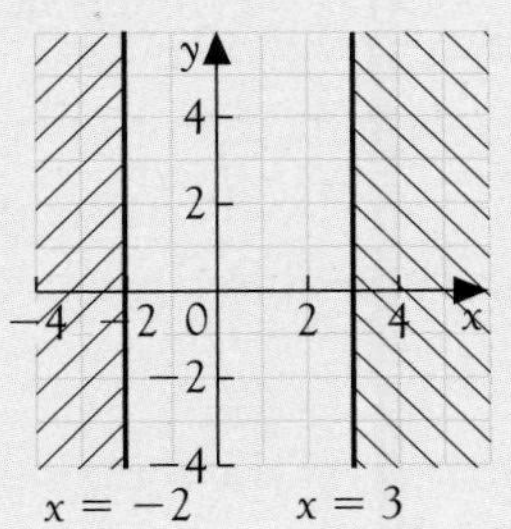

Drawing a separate grid for each part, show, by leaving it unshaded, the region where

a $1 \leqslant x \leqslant 4$

b $-2 \leqslant y \leqslant 4$

c $-3 < x < -1$

d $0 \leqslant x < 3$

e $1 \leqslant x \leqslant 4$ and $2 \leqslant y \leqslant 4$

f $0 \leqslant x \leqslant 3$ and $0 \leqslant y \leqslant 2$

5 For each part

a draw and label a grid with both axes going from -4 to 6

b show, by leaving it unshaded, the region satisfied by all three inequalities. Remember to draw solid or dotted boundary lines as required.

i $y \geqslant x - 1,\ x \geqslant -1,\ 2y + x \leqslant 4$

ii $x \leqslant 2,\ x + y \leqslant 3,\ y \leqslant x + 3$

iii $x + y < 2,\ y < x,\ y \geqslant -2$

iv $y \geqslant x - 2,\ y \geqslant -x - 2,\ y < 2$

v $x \geqslant 0,\ y \geqslant 0,\ 2x + 3y \leqslant 12,\ 3x + 2y \leqslant 12$

vi $x + y \leqslant 5,\ x + y > 1,\ x > 2$

6 **a** Draw and label both axes from 0 to 6 and leave unshaded the region given by the following inequalities.

i $x + y \leqslant 4,\ x > 1,\ y \geqslant 0$

ii $y < x + 3,\ y > x,\ x > 0,\ x < 3$

b Write down all possible points in the required region that have integer values of x and y.

REGIONS

7 A film producer is choosing children for a crowd scene. She chooses x boys and y girls.

Write each of the following statements as an inequality.

a She does not want more than 50 children in the crowd.

b She wants at least 10 boys.

c She wants fewer than 30 girls.

d She wants more girls than boys.

e She wants at least twenty more girls than boys.

f She wants the crowd to contain at least two boys for every girl.

8 At each performance of the school play the number of people in the audience must satisfy the following conditions.

i The number of children in the audience must be less than 250.

ii The maximum size of the audience must be 300.

iii There must be at least twice as many children as adults.

On any one evening there are x children and y adults in the audience.

a Write down the three inequalities that x and y must satisfy other than $x \geqslant 0$ and $y \geqslant 0$.

b By drawing straight lines and shading, indicate the region within which x and y must lie to satisfy all the inequalities.

(Draw both axes from 0 to 400 and use a scale of 1 cm to 50.) [L]

9 A contractor hiring earth moving equipment has a choice of two machines.

Type A which costs £50 per day to hire.
Type B which costs £20 per day to hire.

Let x denote the number of **Type A** machines hired and y the number of **Type B** machines hired.

a The contractor can spend up to £1400 per day on hiring machines.
Explain why $5x + 2y \leqslant 140$.

A **Type A** machine needs one person to operate it.
A **Type B** machine needs 4 people to operate it.

b The contractor has a labour force of 64 people. Write down an inequality which fits these conditions.

c Represent the inequalities $x \geqslant 0$, $y \geqslant 0$, $5x + 2y \leqslant 140$ and your inequality in **b** on a diagram. By shading the regions **not** required, find the region that satisfies all these conditions and label it R. (Draw both axes from 0 to 80.) [SEG]

10 By using inequalities, describe the unshaded region:

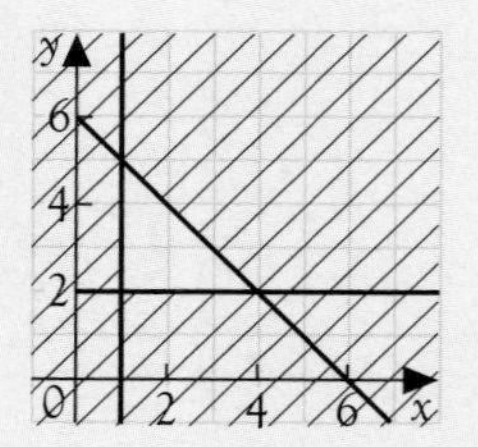

11 Susie is ordering food for a party. Let x be the number of pizzas and y be the number of quiches that she orders.

a She wants to buy at least two pizzas and at least one quiche.
Write down the two inequalities which represent these conditions.

b Each pizza is divided into 12 portions and each quiche is divided in 6 portions.
The total number of portions Susie needs is more than 36 and less than 72.
Write down two more inequalities and show that they simplify to $2x + y > 6$ and $2x + y < 12$.

c Draw a grid with both axes going from 0 to 12. The point (x, y) represents x pizzas and y quiches. On your grid, indicate clearly, by leaving it unshaded, the region in which (x, y) must lie.

d Mark with a cross on your diagram each point which represents a possible combination of pizzas and quiches that satisfies all the conditions.

e Pizzas cost £3 each and quiches cost £2 each. Which of the possible combinations would cost the most money?

Solving cubic equations using trial and improvement

A cubic equation in x is an equation with x^3 as the highest power of x.

Examples of cubic expressions are $x^3 = 64$, $2x^3 - 4x - 6 = 0$, $x^3 + x^2 = 25$.

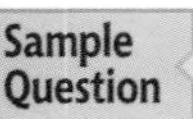

The following sample question revises the method used to find solutions of equations by **trial and improvement.**

age 394

Sample Question 3

The equation $x^3 - 4x = 25$ has a solution between 3 and 4. Find this solution correct to two decimal places. You must show all your trials.

Answer

- Draw up a table showing your trials. Start with $x = 3$ and $x = 4$, substituting into $x^3 - 4x$. Then refine x so that the outcome is nearer 25.

x	$x^3 - 4x$	Outcome
3	15	Too small
4	48	Too big
3.5	28.875	Too big
3.4	25.704	Too big
3.3	22.737	Too small
3.39	25.398	Too big
3.38	25.094	**Too big**
3.37	24.792	**Too small**
3.375	24.943	Too small

You are trying to get 25.

x appears to be closer to 3.4.

x is between 3.38 and 3.37. Now try x half-way between these.

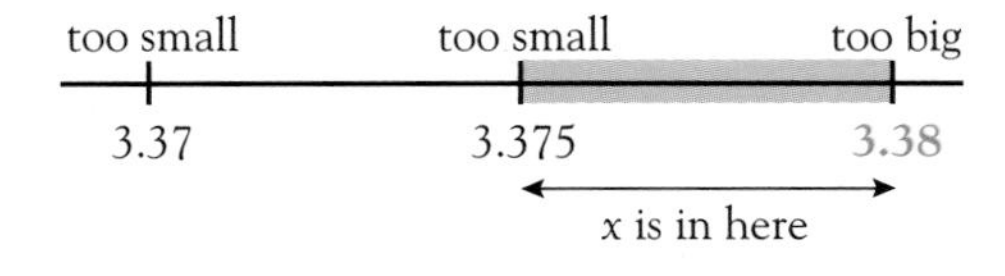

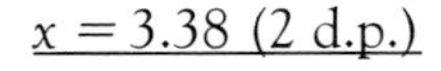

The x value is closer to 3.38 than it is to 3.37.

$x = 3.38$ (2 d.p.)

Cubic curves

It is possible to solve cubic equations using graphs.

Sample Question 4

a Complete the table of values for $y = x^3 + x^2 - 2x$.

x	−2.5	−2	−1.5	−1	−0.5	0	0.5	1	1.5	2
y										

b Draw the graph of $y = x^3 + x^2 - 2x$ for x values from -2.5 to 2.

c Draw the line $y = 1$ on the same grid.

d Use your graphs to solve the equation $x^3 + x^2 - 2x = 1$, giving values to 1 decimal place.

Answer

a

◆ Work out the y values for each x value in turn.

When $x = -2.5$, $y = (-2.5)^3 + (-2.5)^2 - 2 \times (-2.5) = -4.375$

On calculator:

2 . 5 +/- x^y 3 + 2 . 5 +/- x^2 − 2 × 2 . 5 +/- =

The completed table should look like this.

x	−2.5	−2	−1.5	−1	−0.5	0	0.5	1	1.5	2
y	−4.375	0	1.875	2	1.125	0	−0.625	0	2.625	8

b

◆ Plot the points choosing a suitable scale.

◆ Join the points with a smooth curve.

c

◆ Draw the line $y = 1$.

Remember this is a horizontal line through 1 on y axis.

d

◆ To solve the equation, read off the x values where the curve crosses the line.

$x = -1.7, -0.3, 1.3$

These are approximate answers. Their accuracy depends on how accurately you have drawn the graph, and can read the scale.

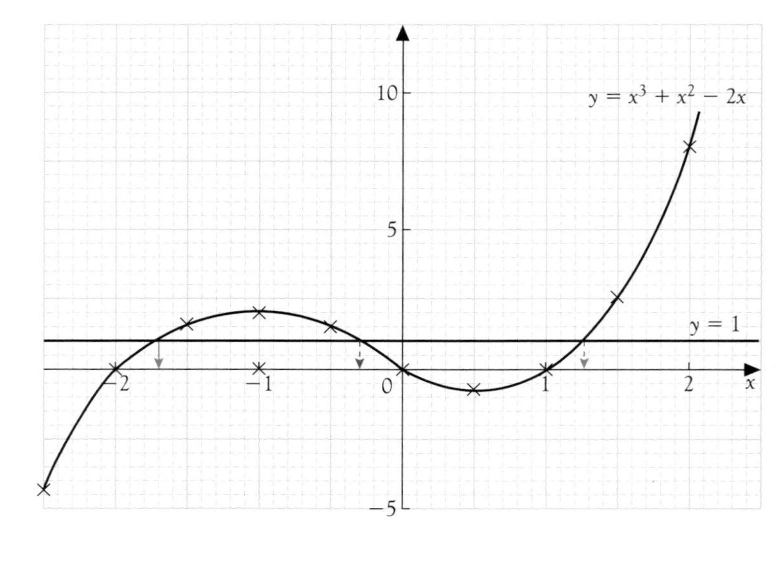

The following sketches show the general shapes of cubic curves:

a **If the x^3 term is positive.**

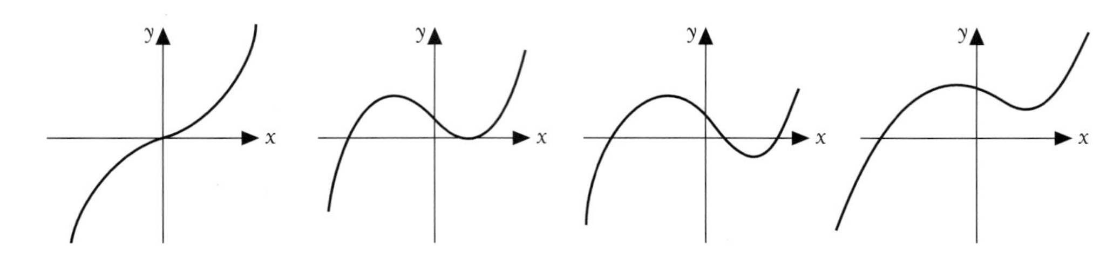

If the x^3 term is negative.

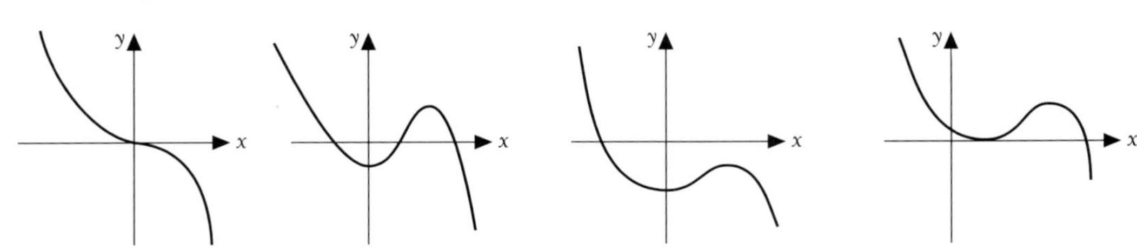

The cubic curve can meet the x axis at one, two or three points.

Reciprocal curves

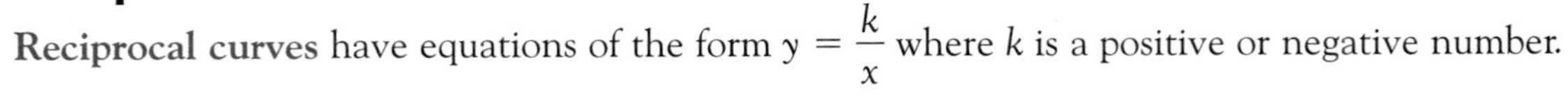

Reciprocal curves have equations of the form $y = \dfrac{k}{x}$ where k is a positive or negative number.

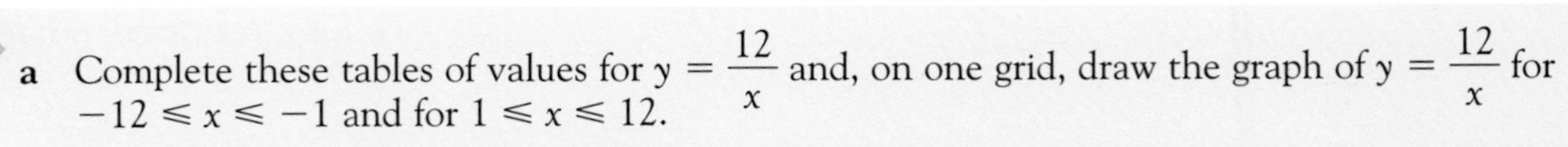

Sample Question 5

a Complete these tables of values for $y = \dfrac{12}{x}$ and, on one grid, draw the graph of $y = \dfrac{12}{x}$ for $-12 \leqslant x \leqslant -1$ and for $1 \leqslant x \leqslant 12$.

x	−12	−8	−6	−4	−3	−2	−1.5	−1
y								

x	1	1.5	2	3	4	6	8	12
y								

b What is the y value when $x = 0.1$?

c What is the y value when $x = -0.1$?

Answer

- Work out each y value in turn

when $x = -12$, $y = \dfrac{12}{-12} = -1$

when $x = -8$, $y = \dfrac{12}{-8} = -1.5$

when $x = -6$, $y = \dfrac{12}{-6} = -2$ and so on.

Notice that $x \times y = 12$ for each of these pairs of values.
$xy = 12$ is another way of writing the equation of the curve.

The completed tables should look like this.

x	−12	−8	−6	−4	−3	−2	−1.5	−1
y	−1	−1.5	−2	−3	−4	−6	−8	−12

x	1	1.5	2	3	4	6	8	12
y	12	8	6	4	3	2	1.5	1

◆ Draw the two parts on a grid, choosing a suitable scale on each axis.

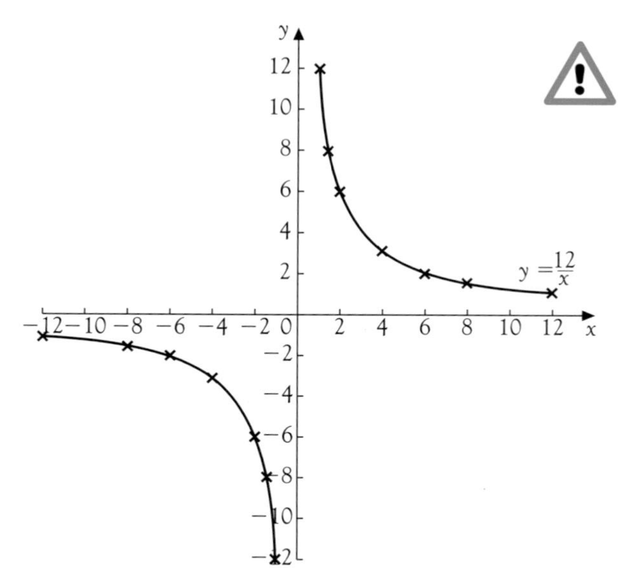

This is a special curve. Notice that it has two separate branches.

b When $x = 0.1$, $y = \frac{12}{0.1} = 120$.
This is a very large number.

c When $x = -0.1$, $y = \frac{12}{-0.1} = -120$.
This is a very small number.

The graph does not exist for $x = 0$. You cannot divide by zero. The lines $x = 0$ and $y = 0$ are called asymptotes for this curve.

Sketch graphs

page 88

You should be able to recognise and sketch the following graphs. Examples are shown of each type.

1 $y = ax$

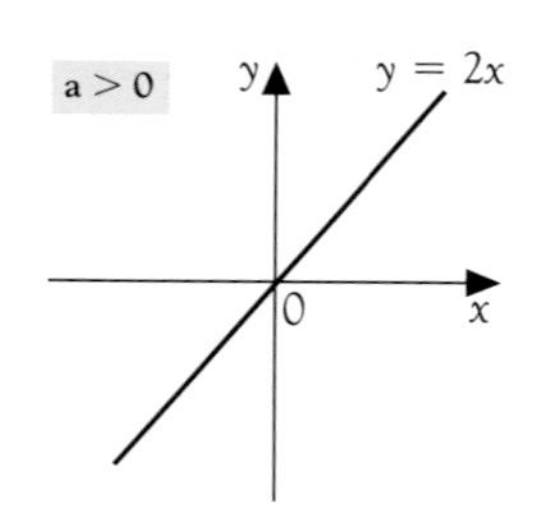

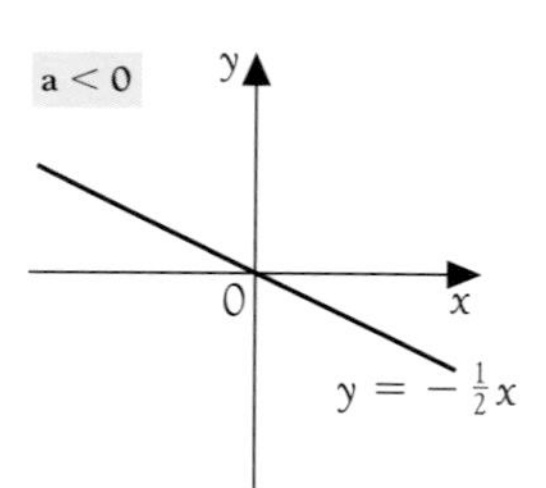

This is a straight line through the origin.

If $a > 0$, the gradient is positive.

If $a < 0$, the gradient is negative.

2 $y = ax + b$

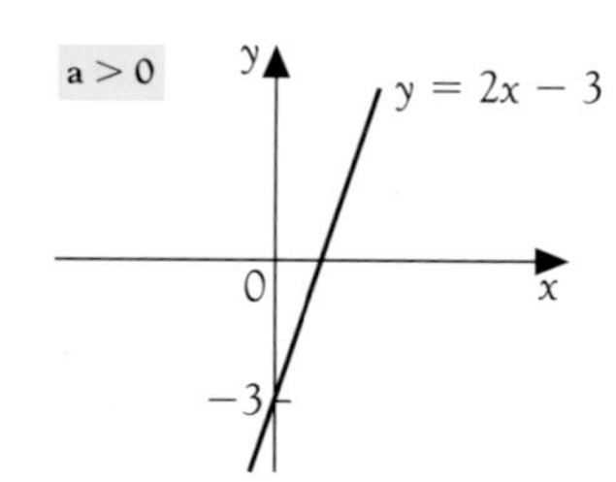

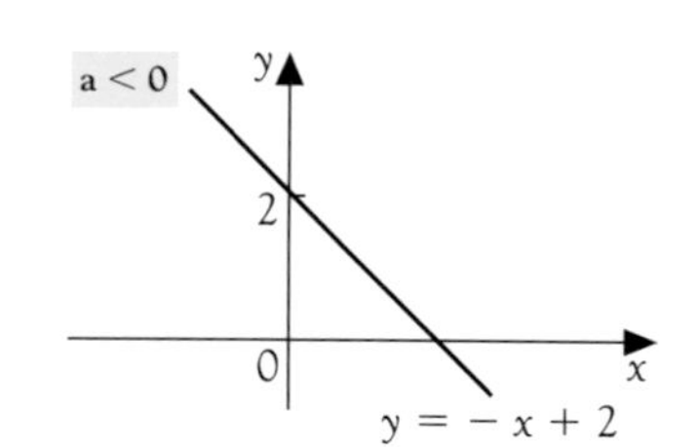

This is a straight line, with gradient a. It goes through $(0, b)$.

page 389

3 $y = ax^2$

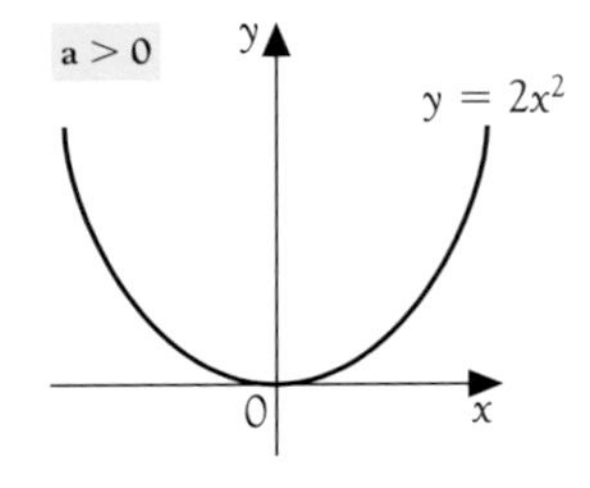

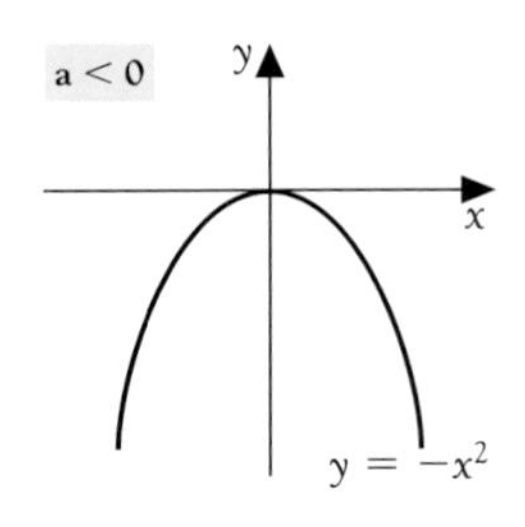

This is a quadratic curve, through (0, 0).

4 $y = ax^3$

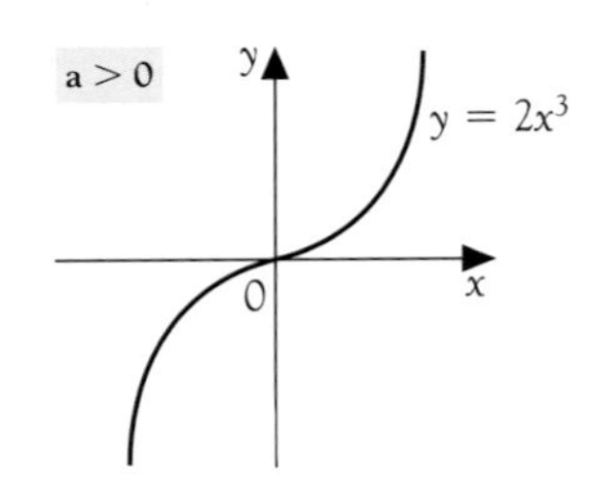

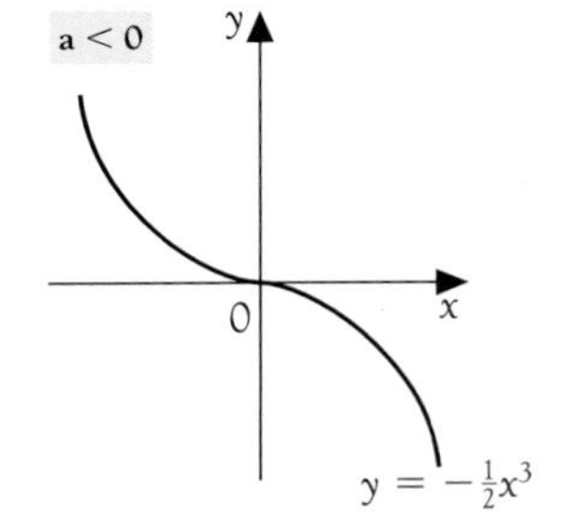

This is a cubic curve, through (0, 0).

5 $y = \frac{k}{x}$

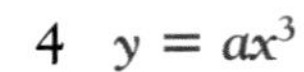

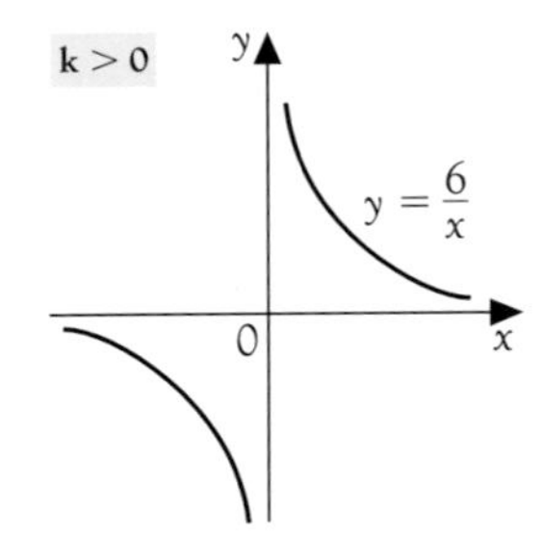

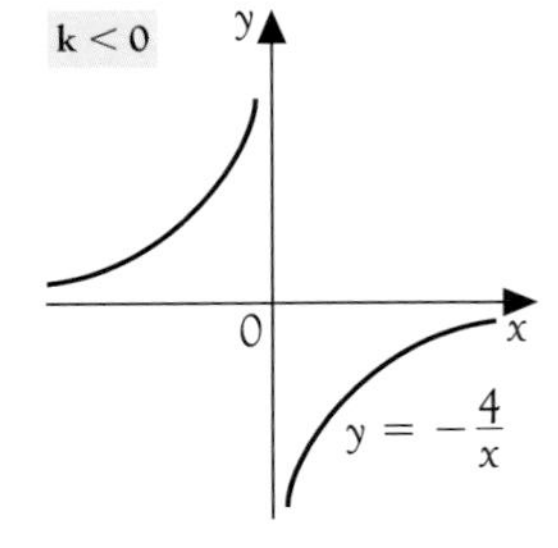

This is a reciprocal curve. It does not exist for $x = 0$.

Exercise 27.3

1 The equation $x^3 - 6x = 120$ has a solution between 5 and 6. Using trial and improvement, find this solution, correct to 2 decimal places. You must shows all your trials.

2 Show that the equation $2x^3 + x^2 - 10 = 0$ has a solution between 1 and 2. Find this solution, correct to 2 decimal places. You must show all your working.

3 The equation $x^3 + 3x^2 = 3$ has a solution between -2 and -3.

a Find this solution correct to one decimal place.

b There are two more solutions to this equation, one between -1 and -2 and the other greater than 0. Find these two solutions, each correct to 1 decimal place.

EQUATIONS AND GRAPHS

4 For the curve $y = x^3 + 2$

a Copy and complete this table of values.

x	-2	-1	0	1	2	3
y	-6					29

b Draw the graph for values of x from $x = -2$ to $x = 3$. Use a scale of 2 cm to 1 unit on the x axis and 2 cm to 10 units on the y axis.

c Read off the value of x for which $y = 20$ and use it to write down the cube root of 18.

5 Draw the graph of $y = x^3 - x$ for values of x from -2 to 2.

6 a Copy and complete the table of values for the graph of $y = \frac{6}{x}$. Remember that the graph does not exist when $x = 0$.

x	-12	-6	-4	-3	-2	-1	-0.5
y	-0.5					-6	

x	0.5	1	2	3	4	6	12
y			3		1.5		

b Using a scale of 1 cm to 2 units on both axes, draw the graph of $y = \frac{6}{x}$.

c Your graph should have two lines of symmetry. Draw in these lines and write down their equations.

d Your graph should have rotational symmetry about (0, 0). What is the order of the rotational symmetry?

7 a Draw a grid with both axes going from -8 to 8.

b On your grid, draw the graph of $y = \frac{-4}{x}$.

8 The diagrams show sketches of the following graphs.

a $y = x^3 - 4x$

b $y = \frac{10}{x}$

c $y = 4x - x^2$

d $y = -x^3 + 2$

e $y = 2x^2 + 1$

f $y = 2x + 1$

Identify which graph is which.

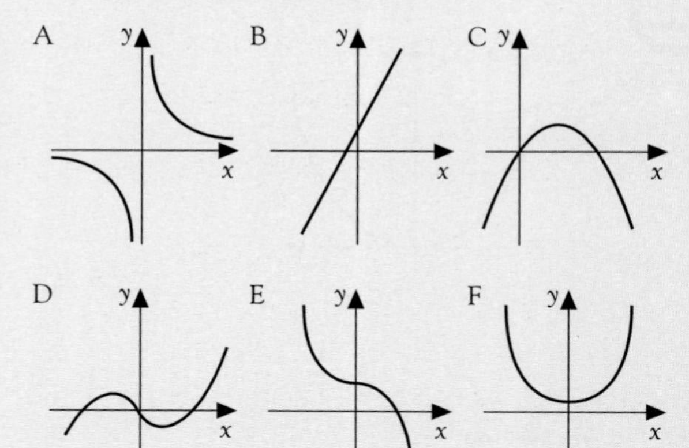

9 A cuboid has dimensions x cm by $(10 - x)$ cm by $(10 - x)$ cm.
Its volume, V cm^3, is given by the formula

$$V = x^3 - 20x^2 + 100x$$

a Copy and complete this table of values.

x	0	1	2	3	4	5	6	7	8	9	10
V			128			125			32		

b Draw a graph of V against x for values of x from 0 to 10.

c Draw the line $V = 100$ on your grid.

d Use your diagram to find the two values of x for which the volume of the cuboid is 100 cm^3.

10 Sketch these graphs.

a $y = 2x^2$

b $y = -3x$

c $y = -x^3$

d $y = \frac{-3}{x}$

e $y = 2x - 5$

f $y = -5x^2$

g $y = \frac{5}{x}$

h $y = \frac{1}{2}x^3$

11 a Copy and complete the table of values for the curve $y = 10 - 2x^3$.

x	-2	-1.5	-1	-0.5	0	0.5	1	1.5	2
y		16.75		10.25				3.25	

b Draw a grid with x values going from -2 to 2 and y values from -10 to 30.

c On your grid, draw the graph of $y = 10 - 2x^3$.

d By drawing a suitable straight line on your diagram, solve the equation $10 - 2x^3 = 5$.

Practical graphs

A graph shows the relationship between two variables. Here are some practical examples.

a In a chemical reaction, you might measure how much product is formed in a certain time, and draw a graph.
The slope (or gradient) tells you how quickly the reaction is going at that time.

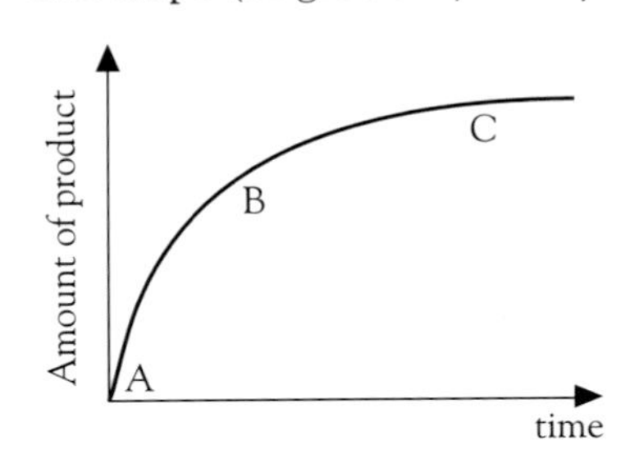

At A the curve is steepest. The reaction is fastest here.

At B the curve is less steep. The reaction is slowing down.

At C the slope of the curve is zero. The reaction has finished and no more product is formed.

b The graph shows the radioactive decay curve for the ^{14}C isotope.

The curve is steepest at A, showing that the rate of decay is greatest at the beginning.

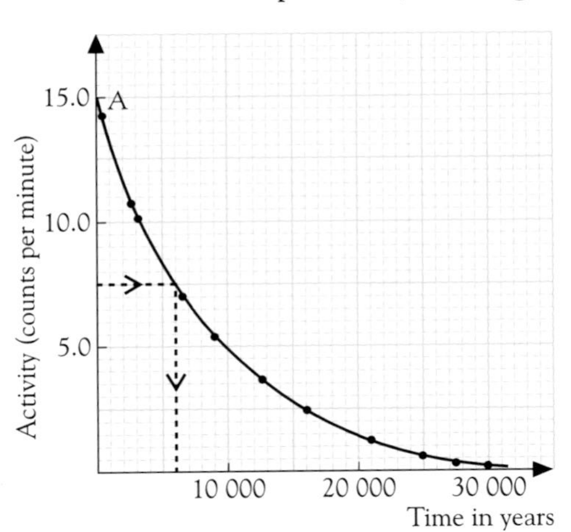

The time taken for the activity count to halve is called the **half-life**.

To work out the half-life, find the value on the vertical axis where the number is half what it was at the start (i.e. 7.5). Draw a horizontal line until it meets the curve, then a vertical line to the 'time' axis. Read off the value.

<u>The half-life of ^{14}C is 6000 years.</u>

c This graph shows the water uptake by the roots and the water loss from the leaves of a plant over a 24-hour period.

Notice that the water loss is greatest during the early afternoon when it is warmest.

During this time the water uptake is less than the water loss, so the plant is likely to wilt.

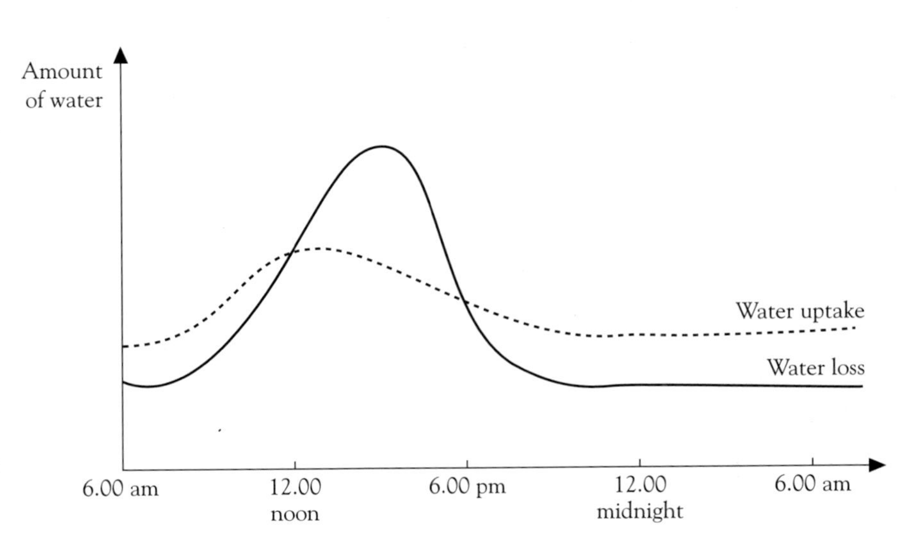

EQUATIONS AND GRAPHS

Sample Question 6 Look at these graphs.

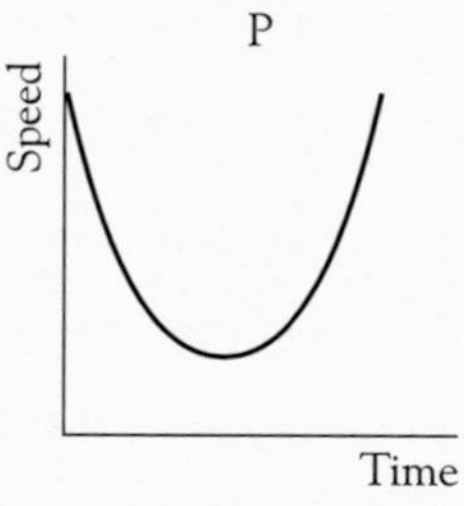

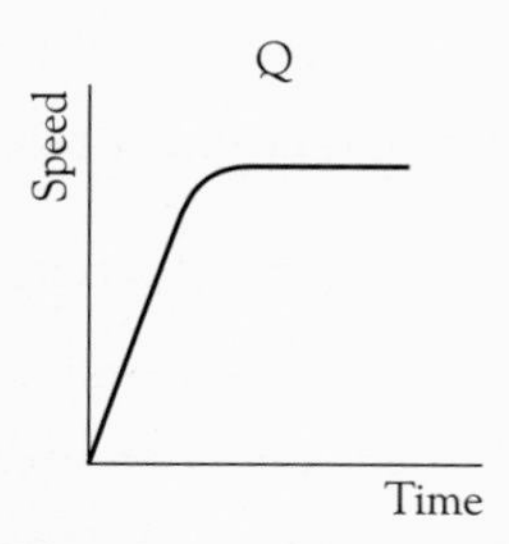

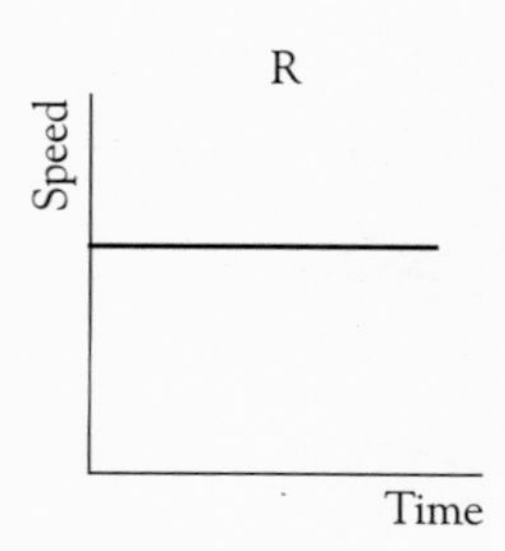

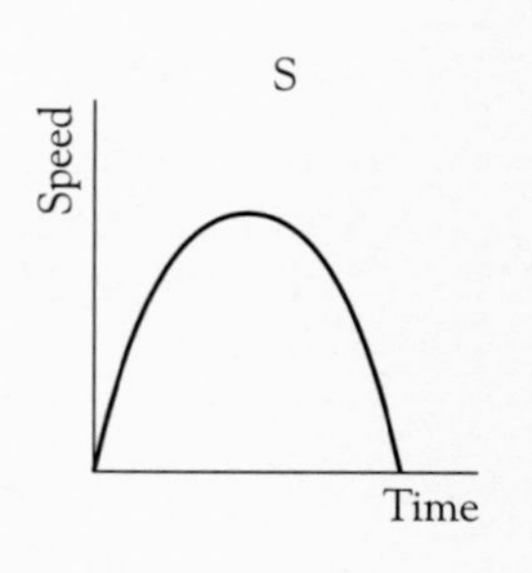

a Which graph shows a cyclist travelling at constant speed?

b Which graph shows a cyclist accelerating and then travelling at constant speed?

c Which graph shows a cyclist slowing down as she goes up a hill then speeding up as she goes down the other side? [MEG]

Answer

a

- If the speed is constant, it does not change. This is shown by <u>graph R</u>.

b

- If the cyclist is accelerating, the speed is increasing [graph Q or S]. Since the speed is then constant, this is shown by <u>graph Q</u>.

c

- The speed is decreasing, then increasing. This is shown by <u>graph P</u>.

Do not put graph S because it looks like a hill!

Exercise 27.4

1 Decide which graph matches each relationship.

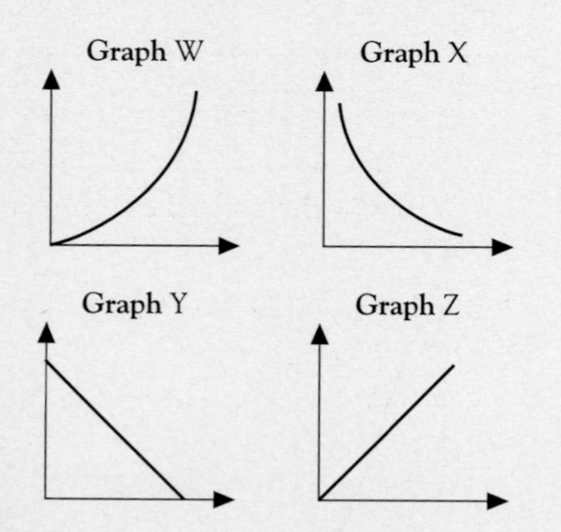

Relationships

A The area of a circle plotted against its radius.

B The circumference of a circle plotted against its radius.

C The length of a rectangle of area 24 cm^2 plotted against its width. [SEG]

2 The cone is being filled with liquid at a steady rate.

At time t, the height of the liquid is h and the radius of the surface of the liquid is r.

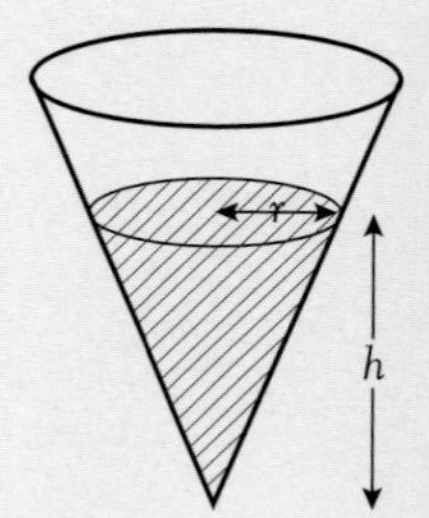

a Draw a sketch to show how the height varies with time.

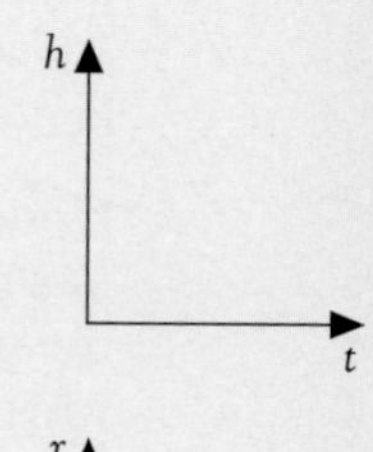

b Draw a sketch to show how the radius varies with the height.

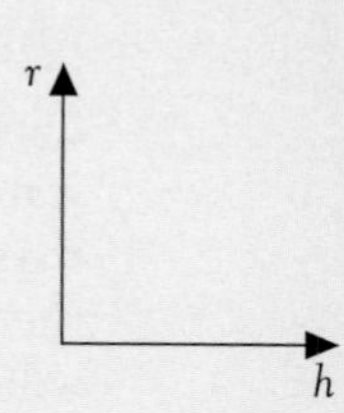

3 Look at these graphs of speed against time.

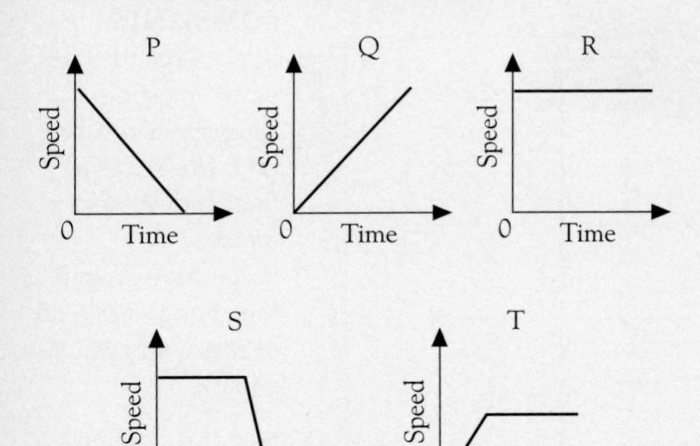

a Which of them represents someone walking steadily?

b Which of them represents someone running to catch a bus but missing it?

c Sketch a graph to show the speed of a train starting from a station, then going at a constant speed and then stopping at the next station. [MEG]

4

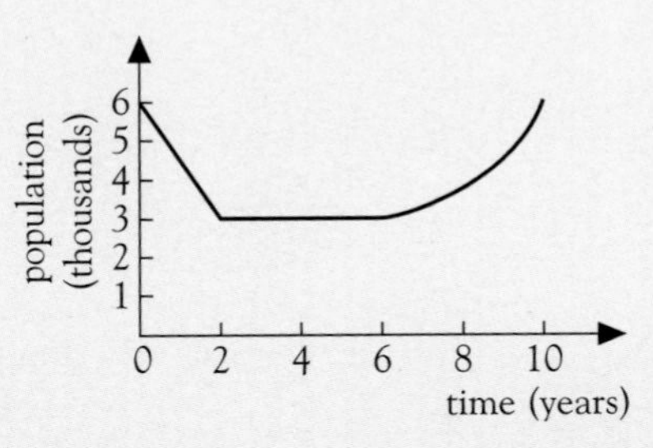

The graph shows how the population of a town changed over a period of ten years.

a At what rate was the population decreasing during the first two years?

b When was the population increasing most rapidly?

c For how long was the population constant?

5

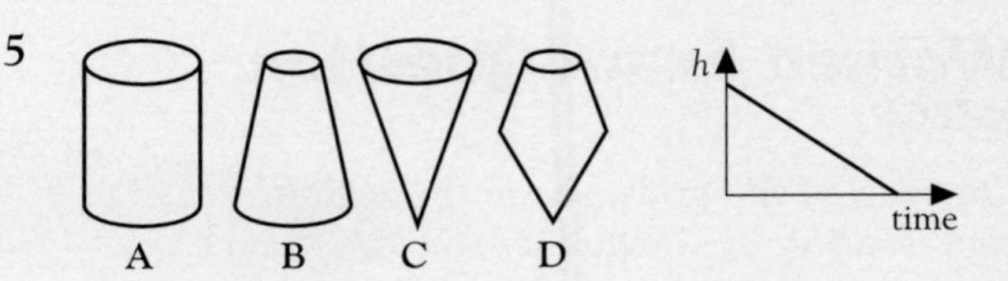

Water is leaking at a steady rate from one of these four containers labelled A to D.

The graph shows how the height, h, of the water in the container changes with time.

Write down the letter of the container which matches the graph. [L]

6 Match the section of the graph with the situation.

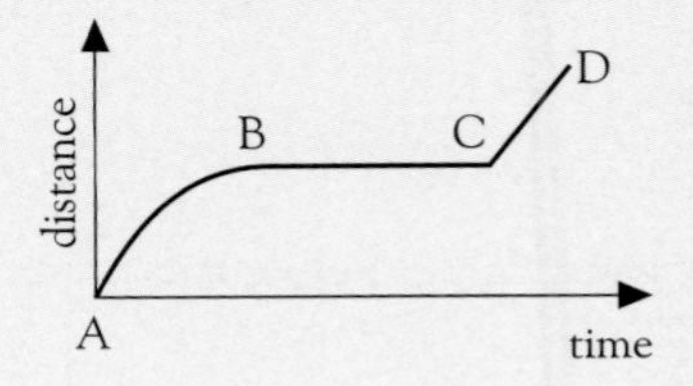

a The speed was zero.

b The car was travelling with a steady speed.

c The car was slowing down.

7 When a ball was thrown in the air, its height, h metres, above the ground varied with the time, t seconds, such that $h = 20t - 5t^2$.

a Copy and complete this table and draw the graph of h against t.

t	0	0.5	1	1.5	2	2.5	3	3.5	4
h									

b How long did the ball take to reach its greatest height?

c During the first two seconds, when was the speed of the ball greatest?

d Describe the speed of the ball from $t = 2$ to $t = 4$ seconds.

EXAM QUESTIONS

Worked Exam Question

[MEG]

On each of the grids, shade the region representing the inequality stated.

a $x \geqslant 2$

Notice that you are asked to shade the required region.

The boundary lines for each region are solid lines.

- Draw and label the line $x = 2$.
- Shade where $x \geqslant 2$, i.e. to the right.

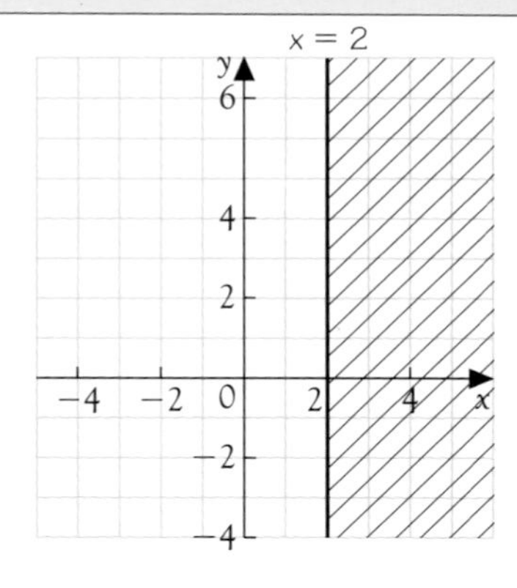

b $y \leqslant -1$

- Draw and label $y = -1$.
- Shade where $y \leqslant -1$, i.e. below the line.

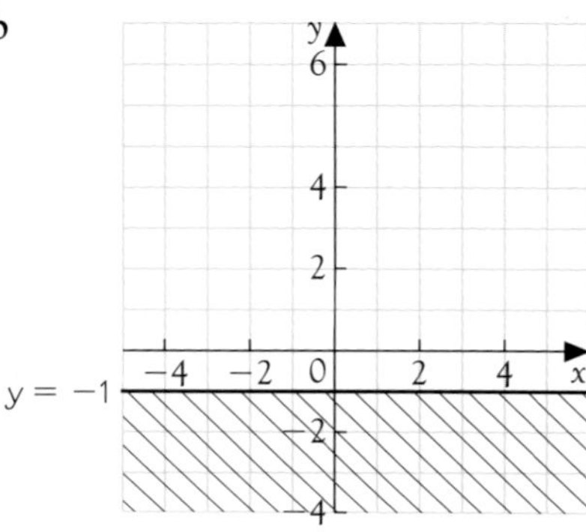

c $x + y \leqslant 3$

- Draw $x + y = 3$ and label it.

x	0	3
y	3	0

- Decide which side shows $x + y \leqslant 3$ and shade it.

Try (0, 0)
$x + y = 0 < 3$ so (0, 0) is in required region.

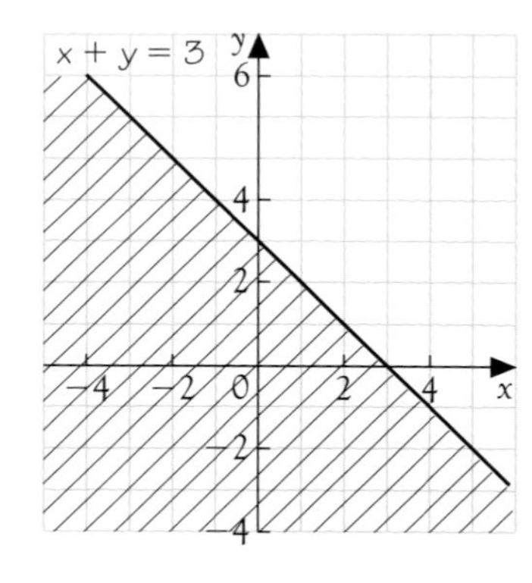

d $y \leqslant 2x - 2$

- Draw $y = 2x - 2$ and label it.

x	0	1	2
y	−2	0	2

- Decide which side shows $y \leqslant 2x - 2$.

Try (0, 0)
$y = 0$, $2x - 2 = -2$, $y > -2$
(0, 0) is **not** in required region.
Shade the other region.

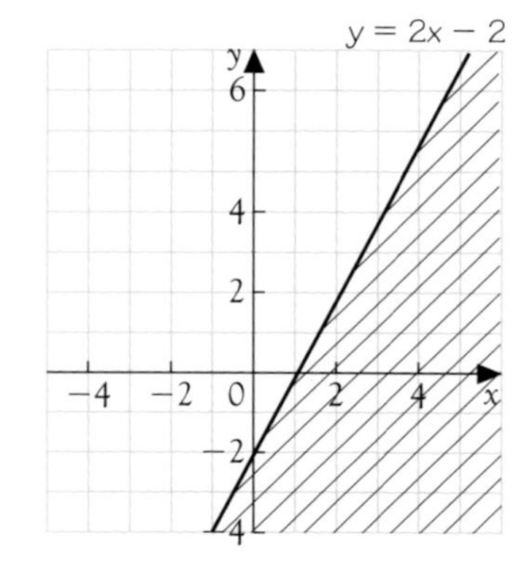

COMMENTS

In all parts, the lines drawn must be complete across the grid. Lines drawn freehand will lose a mark.

If you have shaded the wrong side in **all** of them, you will lose a mark.

a Correct region
1 mark

b Correct region
1 mark

c **A1** Line $x + y = 3$ drawn
A1 Correct region
2 marks

d **A1** Line $y = 2x - 2$ drawn
A1 Correct region
2 marks

Exam Questions

1 a Solve the inequality

$$7x + 3 > 13x + 5$$

b Copy the diagram and label with the letter R the single region which satisfies all of these inequalities.

$$y < \frac{1}{2}x + 1, \quad x > 6, \quad y > 3$$

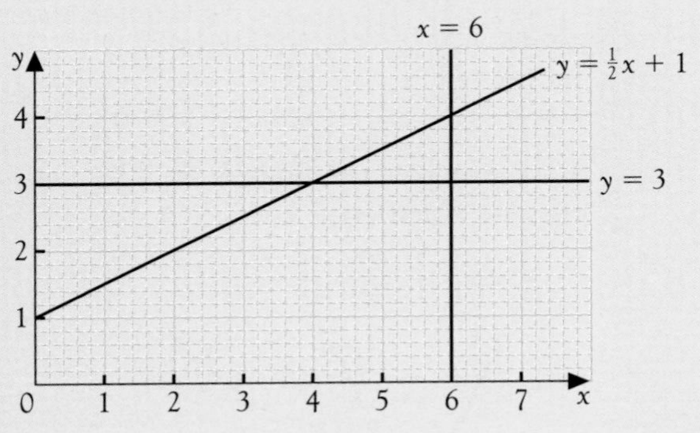

[SEG]

2 a Find the integer values of x for which $0 < 3 - x < 5$.

b Copy the diagram below and indicate clearly on it the region for which $y \geqslant 0$, $y \leqslant 3 - x$ and $1 \leqslant x \leqslant 2$.

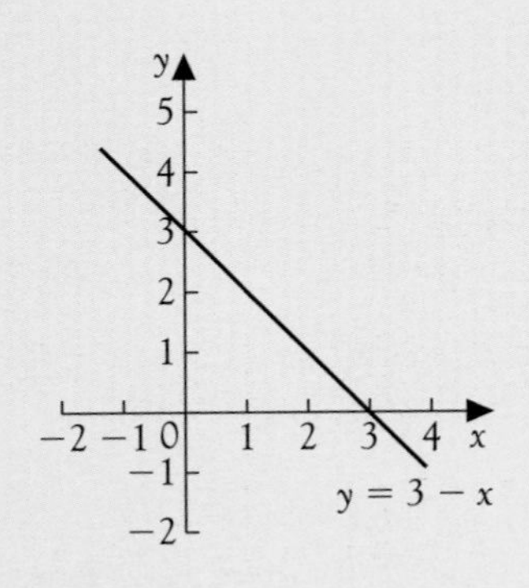

[MEG]

3 The equation $x^3 - x = 45$ has an answer which is a positive number.

a Show that, correct to the nearest whole number, this answer is $x = 4$.

b Using a trial and improvement method, find this answer correct to one decimal place. (You must show all your trials.) [MEG]

4 Some Year 11 pupils are organising a disco.

END OF TERM DISCO

Tickets: £2 each at the door

Special Offer:

There are a limited number of tickets for £1 each if you pay in advance.

The sale of tickets has to raise at least £100. The inequality for this is

$$a + 2d \geqslant 100$$

where a is the number of tickets sold in advance and d is the number of tickets sold at the door.

a 53 tickets are sold in advance.
What is the minimum number of tickets they need to sell at the door?

b A sketch of the line $a + 2d = 100$ is drawn below.
Draw this sketch and shade in the region which represents $a + 2d \geqslant 100$.

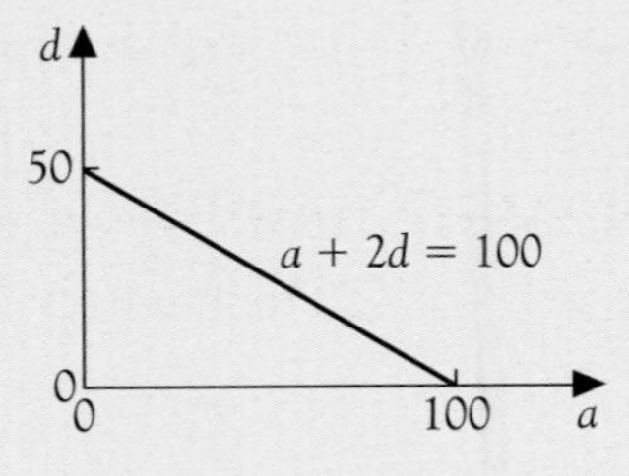

c They will use the school hall for the disco.
Fire regulations state that a maximum of 85 people can use the hall.

i Use this condition to write down an inequality involving a and d.

ii 53 tickets have been sold in advance. The Fire regulations must not be broken. What is the maximum number of tickets that can be sold at the door?

d Use your answers to parts **a** and **c** to write down the range of possible values for d.

[NEAB]

5 For the curve $y = x^3 - 3x + 8$

a complete the table of values,

x	-3	-2	-1	0	1	2	3
y	-10		10	8		10	

b copy the grid below and draw the graph for values from $x = -3$ to $x = 3$.

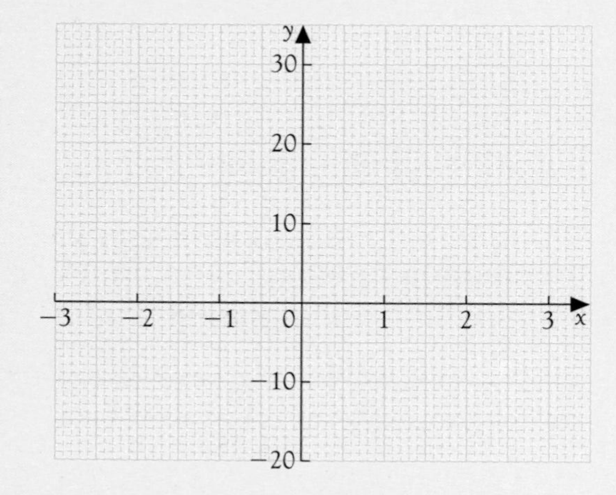

c Read from your graph all the values of x for which $y = 7$.

Give your answer to one decimal place. [MEG]

6 Use the method of trial and improvement to solve the equation

$$x^3 - 2x = 37$$

Give your answer correct to two decimal places. You must show **all** your working. [L]

7 The diagram shows a rough sketch of the graphs $y = 2x + 1$ and $y = 5$. The lines intersect at Q.

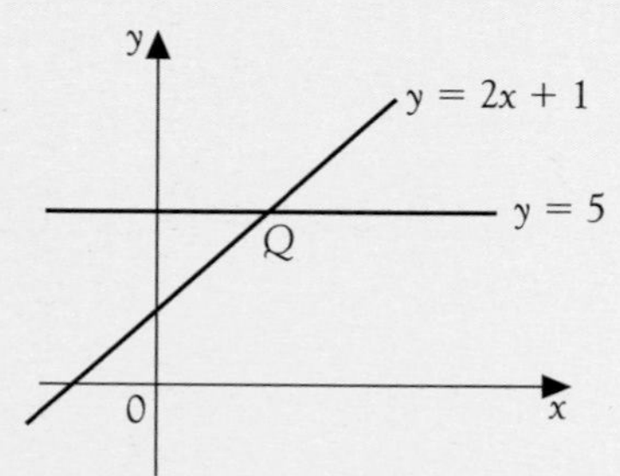

a What are the coordinates of Q?

b A line parallel to $y = 2x + 1$ crosses the y axis at $(0, -2)$. What is the equation of this line?

c Draw a sketch of the diagram and label with the letter R the single region which satisfies all of these inequalities.

$x \geqslant 0$ $\quad 0 \leqslant y \leqslant 5$ $\quad y \leqslant 2x + 1$ [SEG]

8 a Solve the inequality $7n - 12 > 9$.

b Find the integer values of x for which $x^2 \leqslant 4$. [MEG]

9 The equation $x^3 + x^2 = 95$ has a solution between 4 and 5.

Using trial and improvement, find this solution correct to **two** decimal places. You must show all your trials. [MEG]

10 The lines shown have equations

i $3x - y = 1$ iii $2x + y = 5$

ii $y = x$ iv $2y - x = 6$

a State which of the lines A, B, C or D fits each of these equations.

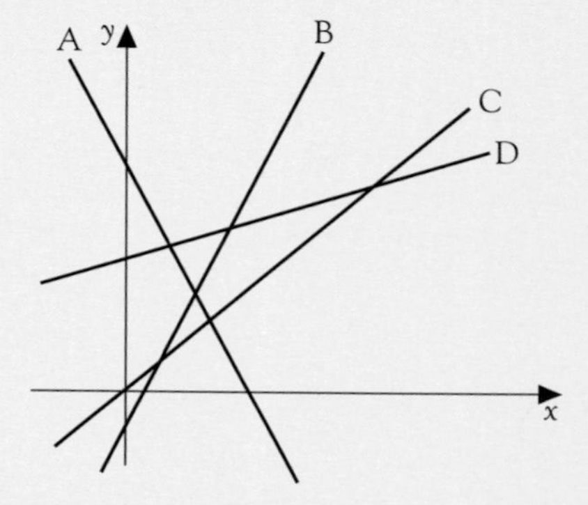

b Copy the diagram and show clearly the region which satisfies all the following conditions:

$3x - y \geqslant 1$

$2x + y \geqslant 5$

$y \geqslant x$

$2y - x \leqslant 6$ [MEG]

11 a Water is poured at a constant rate into each of the four containers shown.

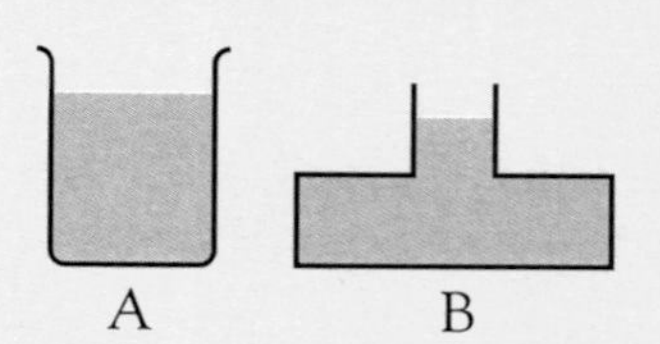

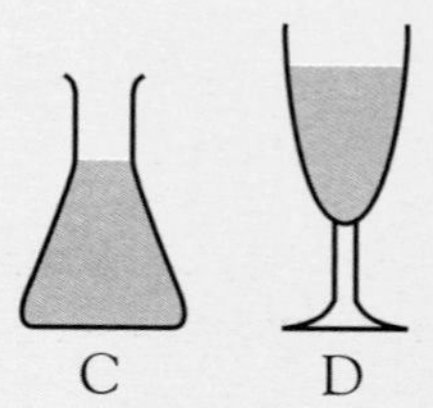

The graphs below show the height of water in the containers as they are being filled.

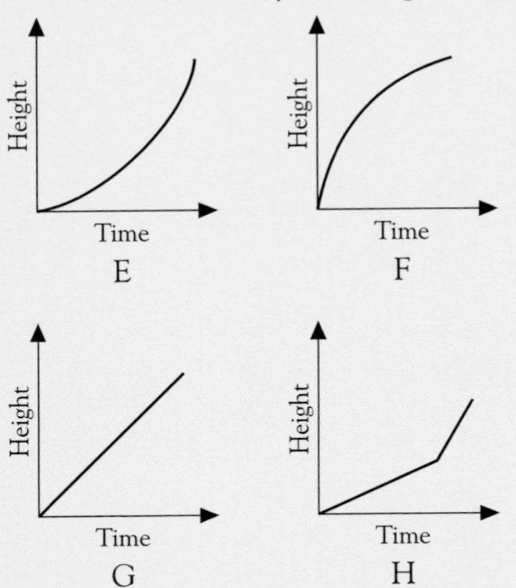

Write down the letter of the graph that matches each container.

b Draw similar graphs for each of the containers below.

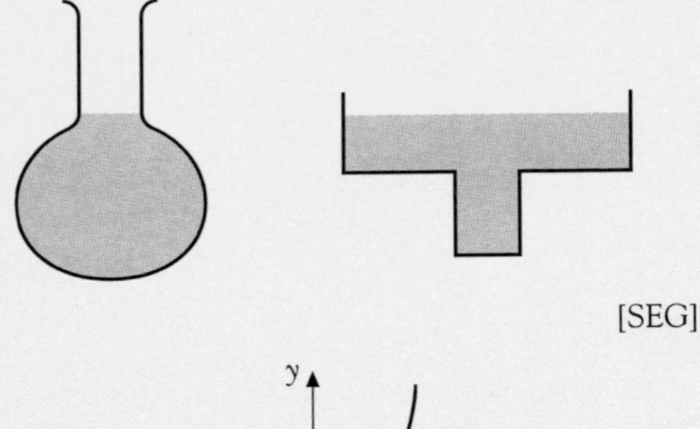

[SEG]

12

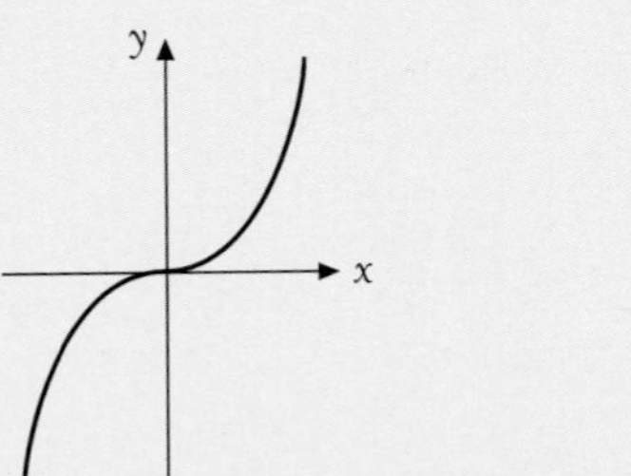

The diagram shows a sketch of the graph of $y = x^n$, where n is a whole number.

a Explain how you can tell that n is not equal to 4.

b State a possible value of n. [MEG]

13 Mr Watt wants to tile his kitchen.
Plain tiles cost 20p each.
Patterned tiles cost £1 each.
Mr Watt can spend a maximum of £300.
He buys x plain tiles and y patterned tiles.

a Show that $x + 5y \leqslant 1500$.

He will use at least one patterned tile to every 5 plain tiles.

b Express this condition as an inequality which x and y must satisfy.

c By drawing straight lines and shading, indicate the region within which x and y must lie to satisfy both the inequalities.
[Draw a grid with both axes going from 0 to 1600, using a scale of 2 cm to represent 200.]

[L]

14 a Each of the four graphs below represents *one* of the following *five* situations.

Situation A A ball rolling from rest down a smooth slope for 4 seconds.

Situation B A ball thrown vertically upwards and caught by the thrower 4 seconds later.

Situation C A train travelling at a constant speed of 15 m/s taking 4 seconds to pass through a tunnel.

Situation D The first 4 seconds of a sprinter's 100 m race.

Situation E A lift setting off from one floor and stopping 4 seconds later at the next floor.

Write down the letter of the situation which matches each graph.

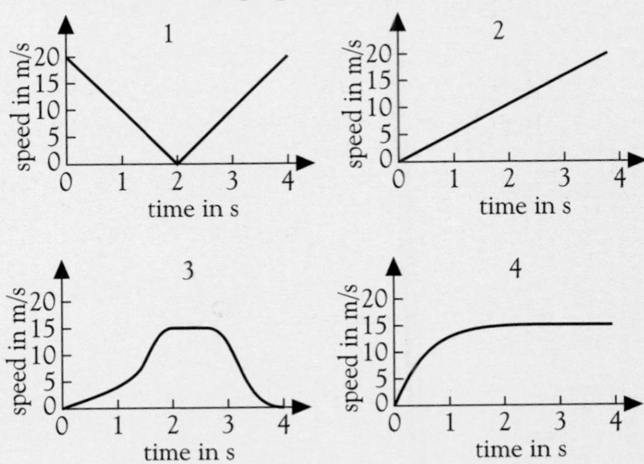

b One of the situations, A, B, C, D or E, does not match any of the four graphs drawn in **a**. On a diagram like this, sketch the speed-time graph for this situation.

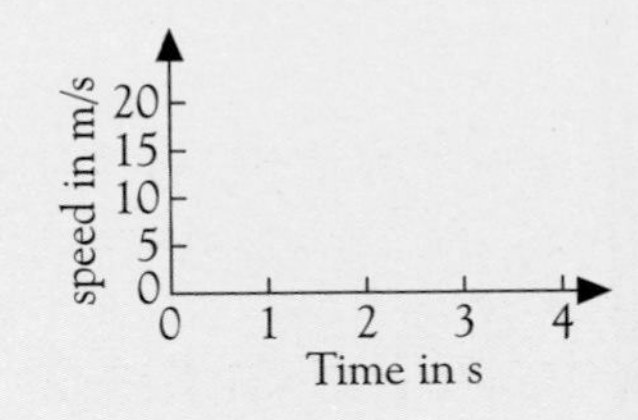

[NEAB]

REVISION CHECKLIST FOR STAGE 3

23. Number III

✓ Know how to multiply and divide numbers written in index form with the same base using

$a^m \times a^n = a^{m+n}, \quad a^m \div a^n = a^{m-n}$

✓ Know that $a^0 = 1, \quad a^1 = a, \quad a^{-1} = \frac{1}{a}, \quad a^{-2} = \frac{1}{a^2}, \quad a^{-n} = \frac{1}{a^n}$

✓ Write ordinary numbers in standard form and vice versa

$6.54 \times 10^5 = 654\,000, \quad 5.3 \times 10^{-2} = 0.053$
$429.3 = 4.293 \times 10^2, \quad 0.000\,002\,1 = 2.1 \times 10^{-6}$

✓ Calculate using standard form, including efficient use of a calculator

✓ Find percentage profit and loss and percentage error

✓ Increase or decrease by a given percentage and perform repeated proportional change

To increase by 15%, multiply by 1.15

To decrease by 10%, multiply by 0.9

✓ Be able to find the original quantity, given a percentage of it

✓ Be able to multiply and divide fractions without a calculator

$\frac{a}{b} \times \frac{c}{d} = \frac{a \times c}{b \times d}, \qquad \frac{a}{b} \div \frac{c}{d} = \frac{a}{b} \times \frac{d}{c} = \frac{a \times d}{b \times c}$

Change mixed numbers to 'top-heavy' fractions.

24. Data Handling III

✓ Use the addition law for mutually exclusive events

P(A or B) = P(A) + P(B)

✓ Use the multiplication law for independent events

P(A and B) = P(A) × P(B)

✓ Use tree diagrams to find probabilities of combined events

If $P(A) = \frac{1}{4}$, $P(B) = \frac{2}{7}$ and A and B are independent events

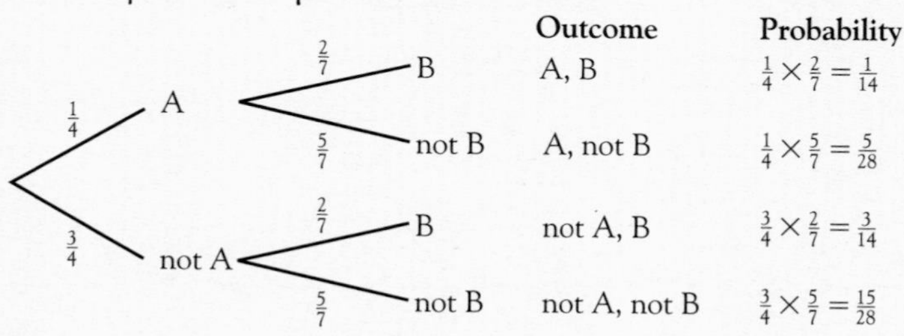

Outcome	Probability
A, B	$\frac{1}{4} \times \frac{2}{7} = \frac{1}{14}$
A, not B	$\frac{1}{4} \times \frac{5}{7} = \frac{5}{28}$
not A, B	$\frac{3}{4} \times \frac{2}{7} = \frac{3}{14}$
not A, not B	$\frac{3}{4} \times \frac{5}{7} = \frac{15}{28}$

✓ Also P(at least one of A or B happens) = 1 − P(neither happens) = $1 - \frac{15}{28} = \frac{13}{28}$

✓ Be able to compile cumulative frequency tables and draw and interpret cumulative frequency curves

✓ Be able to find the median and the quartiles, and calculate the inter-quartile range

Inter-quartile range = upper quartile − lower quartile

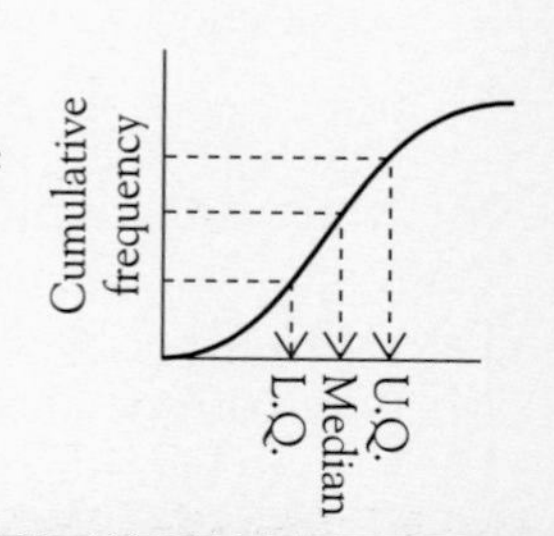

REVISION CHECKLIST FOR STAGE 3

25. Quadratics II

✓ Be able to factorise quadratics of the type $x^2 + bx + c$ and $x^2 - a^2$

i $x^2 + 10x + 21 = (x + 3)(x + 7)$
ii $x^2 - 2x - 8 = (x - 4)(x + 2)$
iii $x^2 + 4x - 5 = (x + 5)(x - 1)$
iv $x^2 - 9x + 20 = (x - 4)(x - 5)$
v $x^2 - 9 = (x - 3)(x + 3)$
vi $x^2 - 100 = (x - 10)(x + 10)$

✓ Be able to solve quadratic equations by factorising

i $x^2 + 4x - 5 = 0$
$(x + 5)(x - 1) = 0$
Either $x + 5 = 0$ or $x - 1 = 0$
$\underline{x = -5}$ $\underline{x = 1}$

ii $2x^2 - 50 = 0$
$2(x^2 - 25) = 0$
$2(x - 5)(x + 5) = 0$
Either $x - 5 = 0$
$\underline{x = 5}$
or $x + 5 = 0$
$\underline{x = -5}$

✓ Be able to recognise a quadratic sequence (constant second difference) and work out the nth term

e.g. for 1, 4, 8, 13, 19, …

n	0	1	2	3	4	5
value	−1	1	4	8	13	19

c (= −1)
a + b: +2 +3 +4 +5 +6
2a: +1 +1 +1 +1

nth term $= an^2 + bn + c$
$2a = 1, \quad a = \frac{1}{2}$
$a + b = 2, \quad \frac{1}{2} + b = 2 \quad b = 1\frac{1}{2}$
$c = -1$
nth term $= \frac{1}{2}n^2 + 1\frac{1}{2}n - 1$

Check 4th term $= \frac{1}{2} \times 4^2 + 1\frac{1}{2} \times 4 - 1 = 8 + 6 - 1 = 13$

26. Further Shape and Space

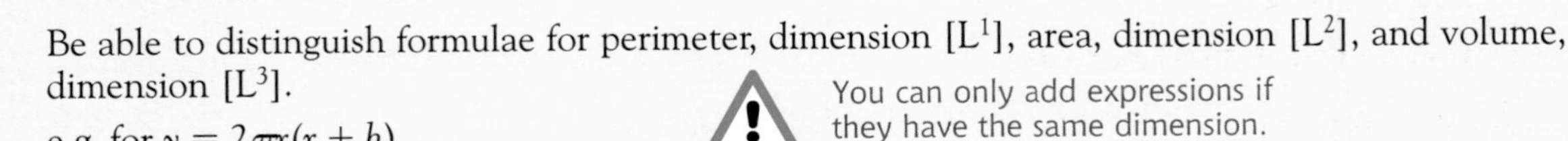

✓ Be able to distinguish formulae for perimeter, dimension $[L^1]$, area, dimension $[L^2]$, and volume, dimension $[L^3]$.

e.g. for $y = 2\pi r(r + h)$
$r + h$ has dimension $[L^1] + [L^1] = [L^1]$
so y has dimension $[L^0] \times [L^0] \times [L^1] \times [L^1] = [L^2]$

⚠ You can only add expressions if they have the same dimension.

When multiplying, add the dimensions $0 + 0 + 1 + 1 = 2$

✓ Be able to find missing lengths and angles in similar figures

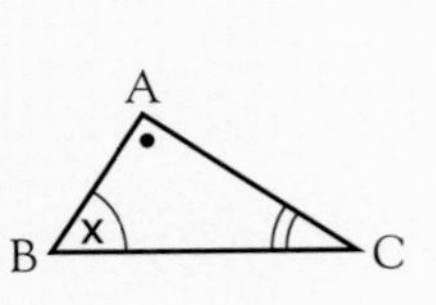

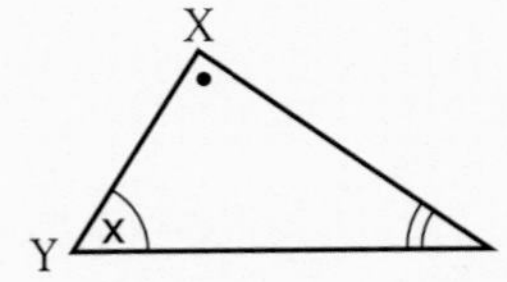

Triangles $\frac{ABC}{XYZ}$ are similar

so $\dfrac{AB}{XY} = \dfrac{BC}{YZ} = \dfrac{AC}{XZ}$

✓ Solve more complicated trigonometry problems including finding the hypotenuse using sine or cosine and finding the adjacent using tan

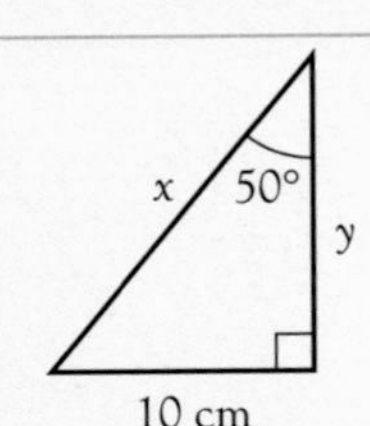

$\sin 50° = \dfrac{10}{x}$

$x = \dfrac{10}{\sin 50°}$

$= 13.1$ cm (3 s.f.)

$\tan 50° = \dfrac{10}{y}$

$y = \dfrac{10}{\tan 50°}$

$= 8.39$ cm (3 s.f.)

REVISION CHECKLIST FOR STAGE 3

27. Algebra 4

☑ Be able to solve quadratic inequalities

a $x^2 \leqslant 49$

−7 0 7

$\underline{-7 \leqslant x \leqslant 7}$

b $32 - 2x^2 < 0$

$32 < 2x^2$

$16 < x^2$

$\therefore \quad x^2 > 16$

−4 0 4

$\underline{x < -4 \text{ or } x > 4}$

☑ Be able to draw and identify regions, using a solid line if the line is included and a broken line if it is not

Shade the region **not** required unless requested otherwise.

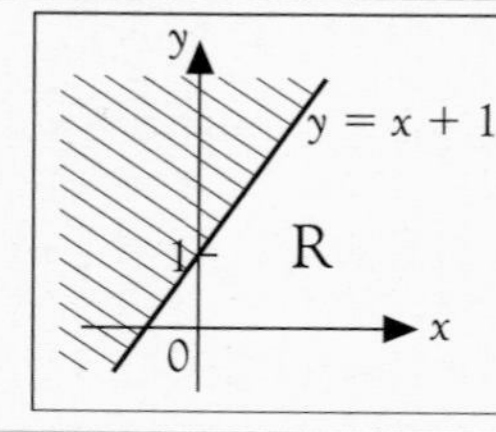

R is where $y \leqslant x + 1$

☑ Be able to solve cubic equations by trial and improvement

☑ Be able to draw cubic graphs and use them to solve equations

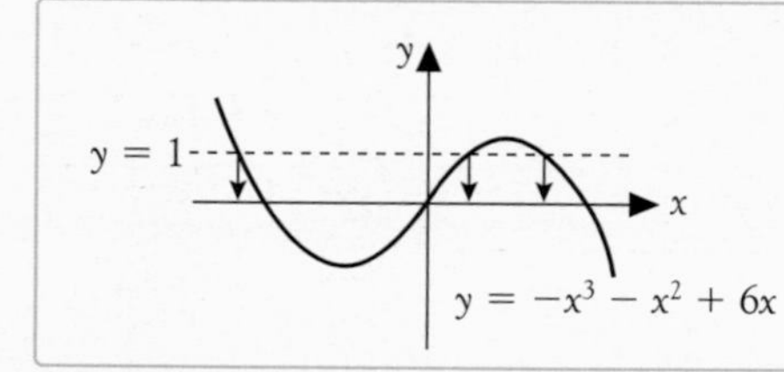

To solve $-x^3 - x^2 + 6x = 1$, draw $y = 1$ also and find where line and curve cross.

Read off the x values.

☑ Recognise and be able to sketch these standard curves

$$y = ax, \quad y = ax + b, \quad y = ax^2, \quad y = ax^3, \quad y = \frac{a}{x}$$

☑ Use graphs in real life practical situations

MIXED EXAM QUESTIONS: END OF STAGE 3

1 A thunderstorm is taking place 60 km away.
Light travels 3×10^5 km in one second.

a How long does it take for the light from the lightning to travel 60 km?

Sound travels 1.226×10^3 km in one hour.

b **i** How many kilometres does sound travel in one second?

ii How many seconds does it take for the sound of the thunder to travel 60 km? Write your answer in standard form.

[MEG]

2

1 km = $\frac{5}{8}$ mile. 1 gallon = 4.54 litre.
Change 9 litres per 100 km into miles per gallon.

[L]

3 **a** Patricia bought a new video.
The price of the video was £370 plus $17\frac{1}{2}\%$ VAT.
Calculate how much she paid for the video.

b A television was advertised as £399.50 including $17\frac{1}{2}\%$ VAT.
Calculate the VAT charged on the television.

[MEG]

4 A dice is biased as follows:
The probability of scoring a 6 is 0.4.
The probability of scoring a 5 is 0.2.

a Julia throws the dice once. Calculate the probability that the score will be 5 or 6.

b Jeff throws the dice twice. Calculate the probability that both scores will be 6's.

[SEG]

5 **a** Factorise completely $14n - 4n^2$

b Find the integer values of n for which $14n - 4n^2 > 0$.

[MEG, p]

6 **a** Draw a grid, using a scale of 1 cm to 1 unit, with both axes going from 0 to 10.

i On the grid, draw the lines

$$x = 2, \quad y = 4, \quad y = 8 - x$$

ii Shade and label, with the letter R, the region for which points (x, y) satisfy the three inequalities

$$x \geqslant 2, \quad y \leqslant 4, \quad y \leqslant 8 - x$$

b **i** Solve the equation $5 - 2x = 11$

ii Hence, or otherwise, solve the inequality $5 - 2x \leqslant 11$.

[MEG]

7 Taking l, b, h, and r to be lengths, copy and complete the table to show which of the expressions 1, 2, 3, 4 or 5 denotes length, area, volume.

	Expression
Length	
Area	
Volume	

Expression

1 $h^3(l + b)$

2 $\pi r(l + r)$

3 $4(l + b)^2$

4 $\pi\sqrt{(h^2 + b^2)}$

5 $b^2\left(\frac{h}{3} + l\right)$

[NI]

8

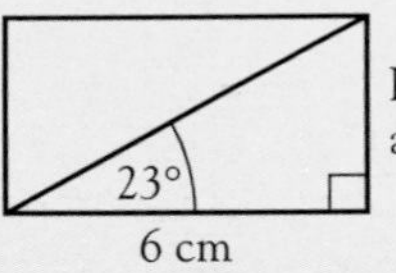

Diagram NOT accurately drawn

The diagram represents a rectangle which is 6 cm long.
A diagonal makes an angle of 23° with a 6 cm side.
Calculate the length of a diagonal.
Give your answer correct to 3 significant figures.

[L]

9 The diagram shows a child's play brick in the shape of a prism.

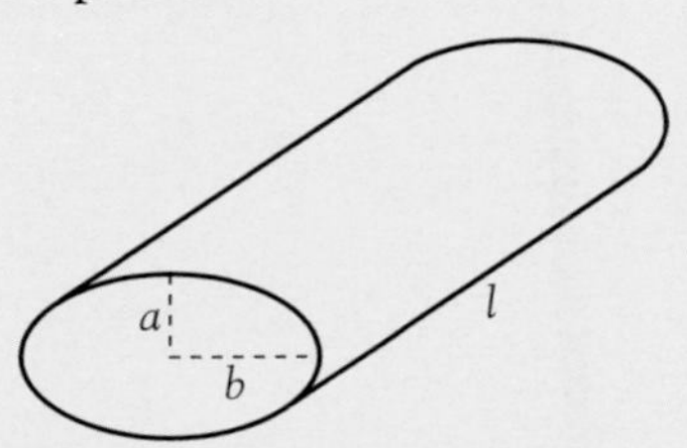

The following expressions represent certain quantities connected with this prism.

$$\pi ab, \quad \pi(a + b), \quad \pi abl, \quad \pi(a + b)l$$

Which of these expressions represent areas? [SEG]

MIXED EXAM QUESTIONS: END OF STAGE 3

10

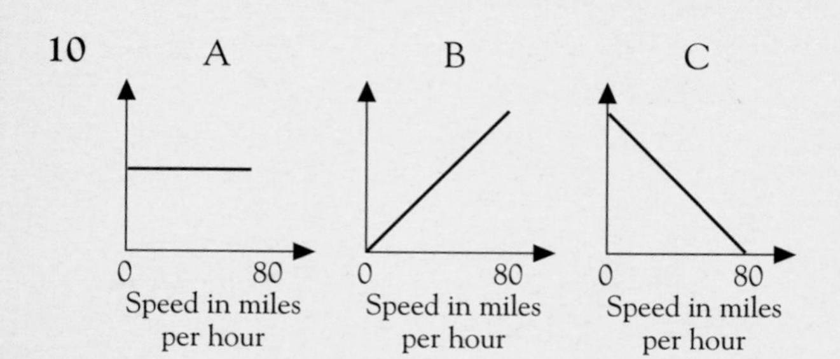

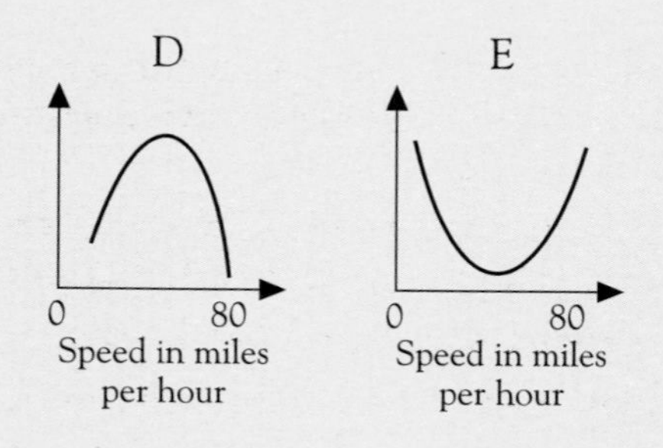

The diagrams show the shapes of five graphs A, B, C, D, and E.

The vertical axes have not been labelled.

On one of the graphs, the missing label is 'Speed in km per hour'.

a Write down the letter of this graph.

On one of the graphs the missing label is 'Petrol consumption in miles per gallon'.

It shows that the car travels furthest on 1 gallon of petrol when it is travelling at 56 miles per hour.

b Write down the letter of this graph. [L]

11 A **US** Centillion is the number 10^{303}

A **UK** Centillion is the number 10^{600}

a How many **US** Centillions are there in a **UK** Centillion? Give your answer in standard form.

b Write the number 40 **US** Centillions in standard form. [L]

12 An aeroplane is flying from Leeds (L) to London Heathrow (H).

It flies 150 miles on a bearing of 136° to a point A. It then turns through 90° and flies the final 80 miles to H.

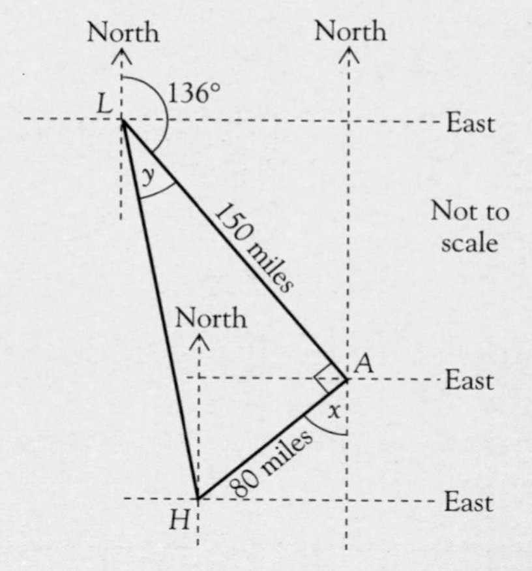

a **i** Show clearly why the angle marked x is equal to 46°.

ii Give the bearing of H from A.

b Use Pythagoras' Theorem to calculate the distance LH.

c **i** Calculate the size of the angle marked y.

ii Work out the bearing of L from H.

[NEAB]

13 **a** On a grid, draw the graph of $y = x^2 - 3x$

Take x values from -2 to 4 and use a scale of 2 cm to 1 unit. Take y values from -3 to 10 and use a scale of 1 cm to 1 unit.

b Use your graph to find the two solutions of the equation $x^2 - 3x = -1$

c Using your graph, or otherwise, solve the inequality $x^2 - 3x < 0$. [MEG]

14

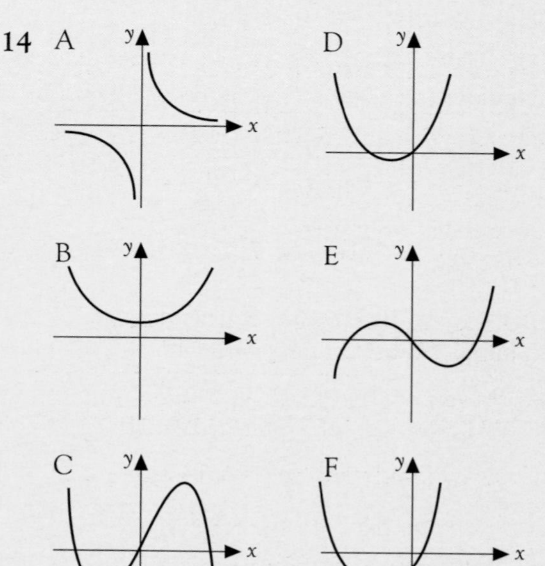

MIXED EXAM QUESTIONS: END OF STAGE 3

Each of the equations in the table represents one of the graphs A to F.

Copy the table and write the letter of each graph in the correct place in the table.

Equation	Graph
$y = x^2 + 3x$	
$y = x - x^3$	
$y = x^3 - 2x$	
$y = x^2 + 2x - 4$	
$y = \frac{4}{x}$	
$y = x^2 + 3$	

[L]

15 A shop is having a sale. Each day, prices are reduced by 20% of the price on the previous day.

Before the start of the sale, the price of a television is £450.

On the first day of the sale, the price is reduced by 20%.

a Work out the price of the television on
  i the first day of the sale,
  ii the third day of the sale.

On the first day of the sale, the price of a cooker is £300.

b Work out the price of the cooker before the start of the sale. [L]

16 At *Fred's Flea Circus* the fleas all entered the long jump competition.

The results of 90 jumps are given in the table.

Length of jump (cm)	0–10	10–20	20–30	30–40	40–50	50–60
Number of jumps	3	7	24	22	21	13

a Copy and complete the cumulative frequency table.

Length not more than (cm)	10	20	30	40	50	60
Number of jumps	3					

b Draw a cumulative frequency graph for the data.

c Find, for this set of data,
  i the median,
  ii the inter-quartile range. [MEG]

17 The diagram shows a polygon with six sides and nine diagonals.

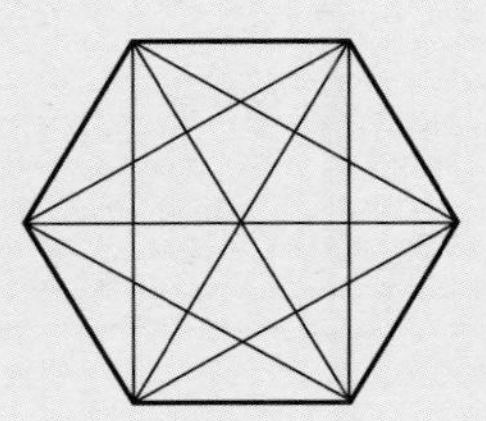

The formula $2d = n^2 - 3n$ can be used to find the number of diagonals, d, in a polygon with n sides.

a Use the formula to find the number of diagonals in a polygon with 8 sides.

b Another polygon has 54 diagonals.
  How many sides has this polygon?

c Explain why a polygon cannot have seven diagonals. [SEG]

18 **a** Find the value of $x^3 + 5x$ when $x = 3.7$

b Use trial and improvement to solve

$$x^3 + 5x = 60$$

Give the value of x correct to two decimal places. [MEG]

19 In the formula $R^3 = KT^2$,

T is the time, in years, that it takes a planet to make one orbit round the Sun,

R is the distance, in miles, of the planet from the Sun,

K is a constant number.

a The planet Earth is 9.3×10^7 miles from the Sun.
  It takes 1 year to make one orbit round the Sun.
  Use the formula to calculate the value of K.
  Give your answer in standard form.

b Re-arrange the formula to express T in terms of R and K.

c The planet Venus is 6.7×10^7 miles from the Sun.
  Using your answers for **a** and **b**, or otherwise, calculate the time it takes Venus to make one orbit round the Sun. [NEAB]

MIXED EXAM QUESTIONS: END OF STAGE 3

20

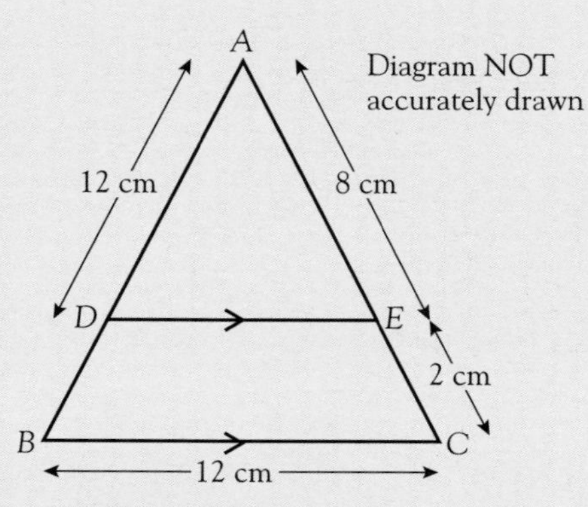

DE is parallel to *BC*.

ADB and *AEC* are straight lines.

$AD = 12$ cm. $BC = 12$ cm.

$AE = 8$ cm. $EC = 2$ cm.

Calculate the length of

a *DE*

b *DB* [L]

21

Continent	Population	Area (m^2)
Europe	6.82×10^8	1.05×10^{10}
Asia	2.96×10^9	4.35×10^{10}

$$\text{Population density} = \frac{\text{Population}}{\text{Area}}$$

Which of these two continents has the larger population density?

You **must** show all your working. [SEG]

22 Fred imports cars from the USA. He sells them in the UK.

In May 1996, he bought a car in the USA for $24 000.

The exchange rate was £1 = $1.50.

It cost him £900 to import the car.

He sold the car in the UK. He made a profit of 20% on his total costs.

a Calculate the selling price of the car. Give your answer in pounds.

Fred's selling price in May 1996 for a car was 12.5% more than his selling price for the same model of car in May 1995.

In May 1996, the selling price of a car was £22 680.

b Calculate the selling price of the same model of car in May 1995. [L]

23 A sample of forty trout is taken at a fish farm.

Mass (g)	40–	50–	60–	70–	80–	90–100
Frequency	2	3	8	9	13	5

a Complete the cumulative frequency table.

Mass (g)	<50	<60	<70	<80	<90	<100
Cumulative frequency						

b Draw the cumulative frequency graph.

c Find the median mass of the trout.

d Find the inter-quartile range of the mass of the trout.

e A second sample of trout has a median mass of 75 g, an upper quartile of 93 g and a lower quartile of 55 g.

Compare and comment on the spread of the data in these two samples. [SEG]

24 The number 10^{100} is called a googol.

a Write the number 50 googols in standard index form.

A nanometre is 10^{-9} metres.

b Write 50 nanometres, in metres.
Give your answer in standard index form. [L]

25 The line $x = 5$ is drawn on the axes below.

Copy the graph. On the same axes draw the appropriate boundary lines and carefully shade the area that contains all the points satisfying the inequalities $x > 5$, $y > 8$ and $y < 30 - 2x$.

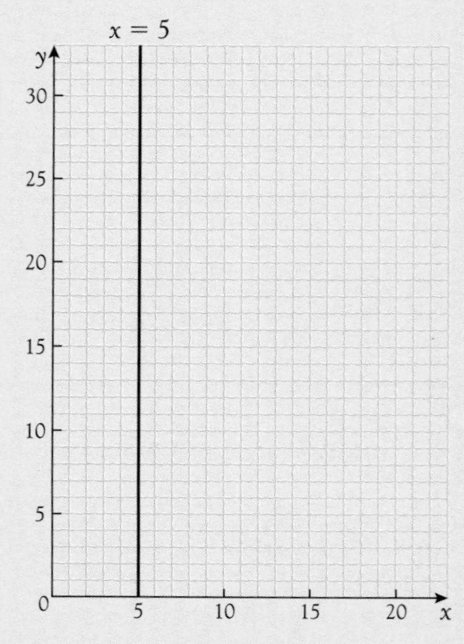

[SEG]

Exercise 1.1 (p. 3)

1 a 22 cm b 34 cm
c 32 cm d 32 cm
2 119 mm
3 a 40 m b 39 m
c 64 cm d 14 mm
4 23 cm
5 a 5 m b £2.50
6 2050 m
7 10 cm
8 4 cm
9 a 27 m b 2.4 cm

Exercise 1.2 (p. 6)

1 a 25.12 cm b 47.1 cm
c 72.22 cm d 5.7 cm (1 d.p.)
2 a 10.8 cm (1 d.p.) b 50.24 m
c 40.5 cm (1 d.p.) d 70.65 mm
3 a 15.7 m b 20.1 m (1 d.p.)
4 7.2 m (1 d.p.)
5 78.5 cm
6 132.52 cm
7 7
8 45.42 cm
9 22 m
10 37.68 cm

Exercise 1.3 (p. 9)

1 a 23.9 m (1 d.p.) b 27.9 m (1 d.p.)
c 4.0 m (1 d.p.)
2 5.7 cm (1 d.p.)
3 9.6 m (1 d.p.)
4 a $d = \frac{c}{\pi}$
b i 5.4 in (1 d.p.)
ii 4.9 in (1 d.p.)
iii 10.2 cm (1 d.p.)
5 50 yds
6 16 cm
7 a 40 m b 111.4 m
8 8.5 m

Exam Questions: Chp 1 (p. 11)

1 a 19 m b 9 c £67.86
2 dinner plate
3 a 125.6 cm b 150.72 cm
c 280 cm d £3.36
4 79.8 m
5 2
6 a i 1.256 m ii 6.456 m
7 a 398 ft b 100.48 ft
8 a 94.26 cm b 32 cm
9 a 157 cm b 50
10 a 471 cm b 3

Exercise 2.1 (p. 20)

1 72
2 $\frac{3}{10}$, $\frac{2}{5}$, 0.44, 45%
3 a Tony
b Jane $66\frac{2}{3}$%, Tony 75%
4 a $\frac{1}{10}$ b 25%
5 3.3 m
6 a £18 b £22.40
7 a 440 b No
8 £150
9 a £1,200 b £375
c £925 d 37%
10

Fraction		$\frac{17}{20}$	$\frac{1}{5}$			$\frac{3}{25}$	$\frac{51}{100}$
Decimal	0.6	0.85		$0.\dot{3}$	0.45	0.12	
%	60		20	$33\frac{1}{3}$	45		51

11 30%, $\frac{1}{3}$, 39%, $\frac{2}{5}$, 0.405, $\frac{3}{7}$
12 a £2.50 b £2.25 c 45%
13 a 130 b Yes
14 $1\frac{3}{4}$ cm
15 1000 g
16 a £61.25 b £411.25

Exercise 2.2 (p. 21)

1 8
2 8
3 22
4 41
5 8
6 18
7 5
8 4
9 150
10 12
11 80
12 12
13 120
14 30
15 21
16 16

Exercise 2.3 (p. 21)

1 2.7
2 340
3 0.631
4 0.0324
5 54
6 0.004
7 27.5
8 0.3
9 12.03
10 0.64
11 26
12 0.0000351
13 6.3
14 63
15 1800
16 1.5

Exercise 2.4 (p. 22)

1 16
2 6
3 7
4 150
5 15
6 24
7 28
8 24
9 7
10 280
11 8
12 200
13 9
14 99
15 104
16 60

Exercise 2.5 (p. 24)

1 a $\frac{3}{4}$ b $\frac{11}{12}$ c $\frac{31}{35}$ d $\frac{19}{24}$
e $\frac{11}{12}$ f $\frac{1}{4}$ g $\frac{1}{2}$ h $\frac{3}{14}$
i $\frac{9}{40}$ j $\frac{1}{10}$
2 a $\frac{1}{4}$, $\frac{3}{10}$, $\frac{7}{20}$, $\frac{2}{5}$ b $\frac{4}{9}$, $\frac{1}{2}$, $\frac{11}{18}$, $\frac{2}{3}$, $\frac{3}{4}$
3 $\frac{1}{4}$

Exercise 2.6 (p. 24)

1 $\frac{11}{15}$
2 500 ml
3 4
4 a 0.4 b 0.73 c 0.6
d 0.35 e 0.12 f 0.26
5 a 28p b £27.50
6 45, 30, 20, 54, 36, 24
7 60%
8 £10
9 28p
10 £5
11 $\frac{1}{12}$
12 a £60 b $\frac{3}{5}$
13 20
14 a 70p b £4.70

Exercise 2.7 (p. 26)

1 a 450 d 320 000 g 150
b 1800 e 10 h 10
c 140 f 10

In question 2 onwards, several answers are possible.

2 £300, underestimate
3 £30
4 £60
5 27 g
6 a ii b iv, v
7 5000
8 180 cm
9 240 cm

Exam Questions A: Chp 2 (p. 28)

1 300 × 20
2 83
3 a £1,800 (18 × 100) or
  20 ≯ 100 = £2,000
  b £1,872   c 7
4 Not the right size
5 3 × 50p + 5 × £2 = £11.50
6 a £2,025   b £32

Exam Questions B: Chp 2 (p. 28)

1 a £32,000   b 20%
2 a i 0.375   ii 0.38
  b 62.5%
  c 25%, 0.3, $\frac{3}{8}$, $\frac{1}{2}$, 0.6, 67%
3 a £159.60   b £17.75   c £167.58
4 3.5%; Yes
5 15%
6 a 99 g   b 80 g
7 a 0.8
  b 0.096, $\frac{4}{5}$, 0.805, 0.85
8 a 115 000   b 165 600   c $\frac{3}{575}$
9 a 48   b 45%
10 £1,205
11 37.5%
12 a 20%   b 5400 million
13 £3.60
14 16%
15 a £70   b £280
16 a $\frac{14}{100}$, $\frac{7}{50}$ b $\frac{17}{20}$ c 0.97
17 a 0.647   b $\frac{3}{5}$, 0.63, $\frac{11}{17}$, 65%

Exercise 3.1 (p. 31)

2 2, 5, 4, 1, 3
3 College, Don't know, Job, Nothing, Year 12
4 a Very likely   b Impossible
  c Very unlikely   d Very likely

Exercise 3.2 (p. 33)

1

Laura	Jamie	Amy
P	Y	G
P	G	Y
Y	G	P
Y	P	G
G	Y	P
G	P	Y

2 1 and 2, 1 and 3, 1 and 4
  2 and 3, 2 and 4, 3 and 4

3

Original	Original
Original	Chicken
Original	Beef
Original	Lamb
Chicken	Chicken
Chicken	Beef
Chicken	Lamb
Beef	Lamb
Lamb	Lamb

4 In pence: 11, 12, 15, 21, 22, 25, 51, 52, 55

5

First die \ Second die	1	2	3	4	5	6
1	1	2	3	4	5	6
2	2	4	6	8	10	12
3	3	6	9	12	15	18
4	4	8	12	16	20	24
5	5	10	15	20	25	30
6	6	12	18	24	30	36

6

Design	Outer part
red	blue
red	green
blue	red
blue	green
white	red
white	blue
white	green

7

Marble from first bag \ Marble from second bag	R	B	B	Y
R	RR	RB	RB	RY
R	RR	RB	RB	RY
B	BR	BB	BB	BY
Y	YR	YB	YB	YY

8

Anil	Carl	Mark
white	red	black
white	black	red
red	white	black
red	black	white
black	red	white
black	white	red

Exercise 3.3 (p. 38)

1 a $\frac{7}{20}$ b $\frac{3}{5}$
2 a $\frac{1}{5}$ b $\frac{2}{5}$
3 $\frac{1}{60}$
4 0.15
5 a $\frac{22}{81}$ b $\frac{1}{3}$ c $\frac{1}{9}$ d $\frac{1}{2}$
6 a $\frac{3}{7}$ b $\frac{1}{7}$ c $\frac{2}{7}$ d $\frac{3}{7}$ e $\frac{4}{7}$
7 a

Question 3 \ Question 8	A	C	D
A	AA	AC	AD
B	BA	BC	BD
C	CA	CC	CD
D	DA	DC	DD
E	EA	EC	ED

  b $\frac{1}{15}$ c $\frac{2}{5}$ d $\frac{7}{15}$

8 a

Eight-faced die \ Six-faced die	1	2	3	4	5	6
1	1	2	3	4	5	6
2	2	4	6	8	10	12
3	3	6	9	12	15	18
4	4	8	12	16	20	24
5	5	10	15	20	25	30
6	6	12	18	24	30	36
7	7	14	21	28	35	42
8	8	16	24	32	40	48

  b i $\frac{7}{24}$ ii $\frac{19}{48}$ iii $\frac{7}{48}$ iv $\frac{1}{4}$
9 a 5
  b i 0.4   ii 0.6   iii 0.3   iv 0.25
10 a $\frac{4}{7}$ b $\frac{17}{28}$ c $\frac{19}{28}$ d $\frac{3}{16}$

Exercise 3.4 (p. 40)

1 Longalife 0.025, Superbat 0.034 …, Cellendure 0.02
2 Cellendure
3 0.975

Exercise 3.5 (p. 41)

1 a Biased   b 76
2 742
3 527
4 1030
5 a 0.42   b 72
6 9
7 570
8 3000
9 132
10 a 38   b 17

Exercise 3.6 (p. 43)

1 a 1   d 3 or 2   g 3   i 3
  b 3   e 2 or 3   h 2   j 1
  c 3   f 2

Exam Questions: Chp 3 (p. 45)

1 a i Near zero   ii 1
  b (5, 6) (6, 5) (5, 5) (6, 6)
  c A = middle, B = Very close to zero
2 $\frac{3}{8}$
3 a $\frac{6}{25}$ b $\frac{3}{25}$ c $\frac{12}{25}$
4 a $\frac{1}{6}$ b see page 32 table
  c i $\frac{1}{12}$ ii $\frac{1}{4}$
5 a 0.352   b 35
6 a 0.3   b Approach 0.25
7 a i $\frac{1}{8}$ ii $\frac{3}{8}$
  b

Ferrari	Ferrari
Ferrari	Benetton
Ferrari	Tyrell
Ferrari	McLaren
Benetton	Benetton
Benetton	Tyrell
Benetton	McLaren
Tyrell	Tyrell
Tyrell	McLaren

  c Not equally likely outcomes.
  To use equally likely outcomes you would need to list, for example Benetton, Ferrari as well as Ferrari, Benetton.
  d 0. He hasn't got two McLaren cars.
8 a i 0.31   ii 0.77
  b Not mutually exclusive events as someone could have seen both.

9 a $\frac{3}{20}$ b $\frac{2}{5}$

10 a
7	8	9	10	11	12
6	7	8	9	10	11
	6	7	8	9	10
		6	7	8	9
		5	6	7	8

b $\frac{1}{6}$ c 45 d £4.50

Exercise 4.1 (p. 52)

1 a i 176 ii $22n$
b i bn ii $\frac{bn}{100}$
2 a 180 b $20c$ c $\frac{180 + 20c}{100}$
3 $25n + 10m$
4 a i 135 ii $50w + 35$
b $\frac{50w + 35}{60}$
5 $15f + 35s$
6 $20 - s$
7 $30 + 12h$
8 a £400 b $5N$ c $5N - 45$
9 a $5n$ b $\frac{5n}{12}$ c $\frac{2500n}{12}$
10 a $30E$ b $\frac{30E - B}{50}$

Exercise 4.2 (p. 54)

1 $5a$
2 $5b$
3 $9c$
4 Not possible
5 e
6 $7a + 5b$
7 $8a + 2c$
8 $6f + 2g + 2$
9 $10ab$
10 $5d + 3e$
11 $2x^2 + 6x$
12 Not possible
13 $3c^2 + 2c + 10$
14 0
15 $10a$
16 $24b$
17 $8ac$
18 a^3
19 $8e^2c^2$
20 $12a^2$
21 $-15ab$
22 $60bcd$
23 $4e^2d$
24 $56h^2k^5$

Exercise 4.3 (p. 55)

1 a 7 d 1 g 0
b 26 e 24 h 18
c 10 f 12 i −20
2 a 11 d 17 g 1080
b 33 e 168 h −648
c −11 f 54 i 315

Exercise 4.4 (p. 57)

1 a $B = T - G$ b $G = T - B$
c $T = B + G$
2 a $T = J + F$ b $F = T - J$
c $J = T - F$
3 a $P = T \div B$ b $B = T \div P$
c $T = B \times P$
4 $C = AX + BY$
5 $T = 3BP$
6 $T = R + W$, $R = T - W$
7 $S = B + 2A$, $B = S - 2A$
8 $T = bM$, $b = T \div M$
9 $T = CP$, $P = T \div C$
10 $t = 5f$, $f = t \div 5$

Exercise 4.5 (p. 58)

1 a 115p b 325p
2 £725
3 2000 volts
4 24°F
5 1962.5 cm^3
6 $4\frac{4}{9}$ g/cm^3
7 4 m/s^2
8 62.5 watts
9 224 m
10 37.5

Exercise 4.6 (p. 63)

1 a 6 b 7 c 30 d 3 e 16
2 a 6 b 15 c 3 d 3
e 25 f 75 g 90
3 a 1.75 b 5 c 58 d 9
4 a $10x$ b $40x$ c $50x + 50 = 300$; 5
5 a $x - 30$ b $4x - 60$ c 115 m, 85 m
6 $3x = 63$; 42

Exam Questions: Chp 4 (p. 65)

1 a £3.30
b i $20n + 30 = 250$ ii 11
2 a £76 b i −20 ii Loss of £20
3 a £480.44 b $d + 24m$
4 a $P = 3x + 5$
b i $3x + 5 = 38$
ii 11 cm, 13 cm, 14 cm
5 a 11 b 9 c 4
6 a $x + 3$ b $P = 4x + 6$
c $4x + 6 = 32$; 6.5
7 a 5 b 24 c 6
8 a £74 b 6
9 a £4.60 b $60 + 50W$
c $C = 60 + 50W$ d 9
10 a £22 b 9; $C = 7.5d + 12$ c £42

11 a $x + y$ b $5x + 5y$
12 a 13 b 26
13 a $15n$
b i $68 + 15n$
ii $68 + 15n = 188$ iii 8
14 a 6 metres b $10 - x$ metres
c $y - z$ metres
15 a $3a + 6$ b $10b + 4$

Exercise 5.1 (p. 71)

1 $a = 48°$
2 $x = 100°$
3 $x = 33°$, $y = 147°$, $p = 147°$
4 $b = 20°$, $c = 150°$
5 $x = 75°$
6 $y = 60°$, $z = 60°$
7 $c = 142°$, $d = 38°$
8 $m = 45°$, $n = 110°$
9 $a = 30°$, $2a = 60°$
10 $x = 135°$
11 $x = 320°$
12 $x = 45°$, $y = 45°$

Exercise 5.2 (p. 74)

1 a $a = 53°$
b $x = 120°$
c $c = 100°$
d $r = 45°$, $p = 45°$
e $a = 64°$, $b = 116°$
f $c = 70°$, $b = 110°$, $a = 70°$, $y = 70°$, $x = 110°$
g $b = 76°$, $a = 59°$, $c = 76°$
h $a = 54°$, $b = 62°$
i $a = 140°$, $b = 40°$, $c = 40°$
2 a 43° b 137° c 118° d 137°

Exercise 5.3 (p. 79)

1 a 133° b 45° c 60° d 88°
e 30°, 60° f $x = 60°$, $y = 120°$
g 73° h 75°, 150°
i $a = c = 65°$, $b = d = 115°$
2 a 55° b 40°
3 $a = 35°$, $b = 35°$, $c = 55°$
4 $m = 63°$, $n = 63°$, $p = 63°$, $q = 117°$
5 42°, 72°, 66°
6 a $d = 30°$, $a = 30°$, $c = 160°$, $b = 20°$
b $a = 55°$, $b = 45°$
c $a = 58°$, $b = 58°$, $c = 64°$
7 $x = 60°$, $y = 30°$, $z = 60°$

Exercise 5.4 (p. 82)

(Answers are approximate)

1 a 4.9 cm d 14.2 cm
b 104° e 2.4 cm, 6.2 cm
c 10 cm, 53° f 11.7 cm
2 b 120 m
3 b 97 km c 118 km
4 Equilateral triangle with side 7 cm

Exam Questions: Chp 5 (p. 84)

1 a $p = 56°$, $q = 124°$
  b $r = 60°$, $s = 120°$ c $t = 102°$
2 $a = 120°$, $b = 40°$
3 $a = 110°$, $b = 65°$
4 $A\hat{B}C$, $B\hat{C}D$; 43°, 148°
5 $a = 56°$, $b = 56°$, $c = 68°$, $d = 56°$
6 $a = 76°$, $b = 111°$
7 $x = 104°$, $y = 34°$, $z = 104°$
8 $a = 58°$, $b = 32°$, $c = 47°$
9 20°
10 a i 40°   ii isosceles
  b i 102°   ii 40°
11 a 360°   b i 67°   ii 65°
12 b 7.2 cm
13 b 14.5 m   c Approx 16.3 m

Exercise 6.1 (p. 93)

1

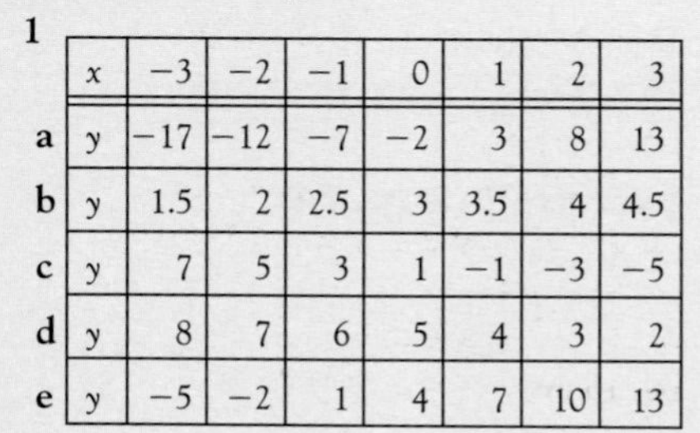

	x	−3	−2	−1	0	1	2	3
a	y	−17	−12	−7	−2	3	8	13
b	y	1.5	2	2.5	3	3.5	4	4.5
c	y	7	5	3	1	−1	−3	−5
d	y	8	7	6	5	4	3	2
e	y	−5	−2	1	4	7	10	13

2 a no   b yes   c yes
  d yes   e yes
3 a 4, 5, 7, 8
  b

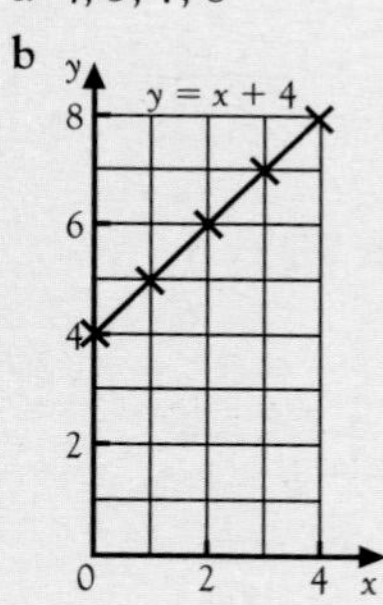

4 a 4, 3, 1, 0, −1
  b, c

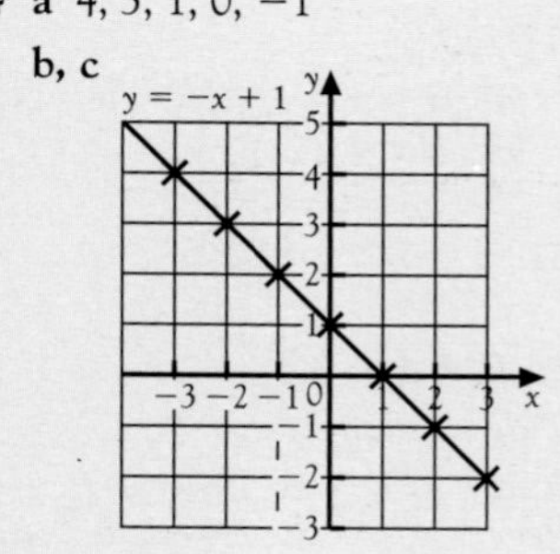

5 a

x	−2	−1	0	1	2	3	4	5	6
y	−3	−2.5	−2	−1.5	−1	−0.5	0	0.5	1

  b

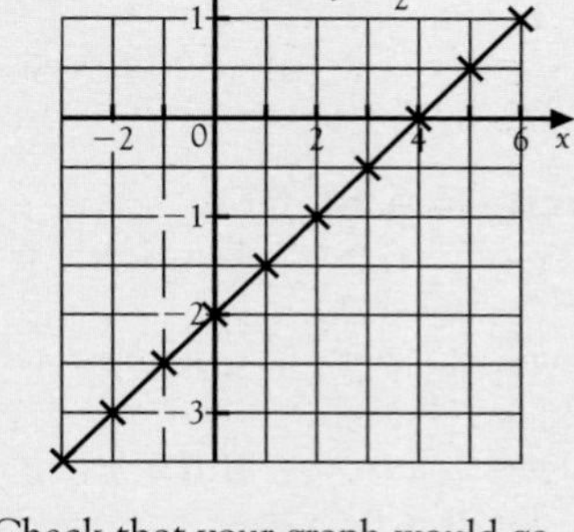

6 Check that your graph would go through these points:
  a (0, −5), (1, −2), (2, 1)
  b (−1, 1), (0, 3), (1, 5)
  c (−2, 8), (0, 4), (2, 0)
  d (−1, −5), (0, 1), (1, 7)
  e (−4, −1), (0, 0), (4, 1)
  f (−2, 0), (0, −2), (2, −4)
  g (−1, 10), (0, 3), (1, −4)
  h (−1, 2), (0, 6), (1, 10)

Exercise 6.2 (p. 95)

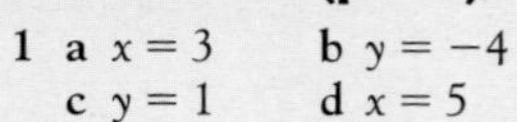

1 a $x = 3$ b $y = -4$
  c $y = 1$ d $x = 5$
2

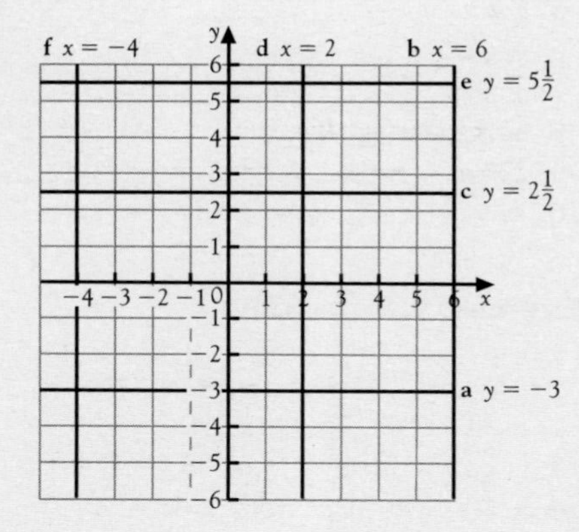

Exercise 6.3 (p. 98)

1 a 103.5   b 19
2 7.5, 27.5, 52.5

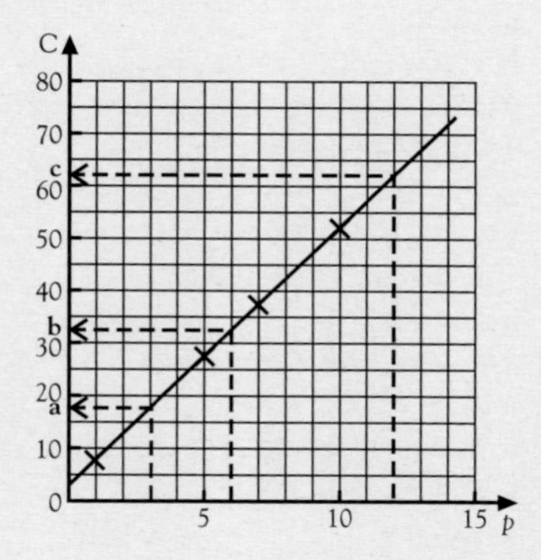

  a £17.50   b £32.50   c £62.50
3 40, 115   a £74   b 216 miles

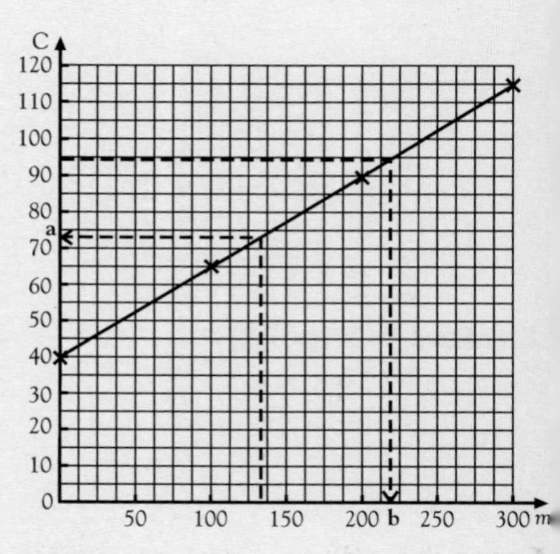

4

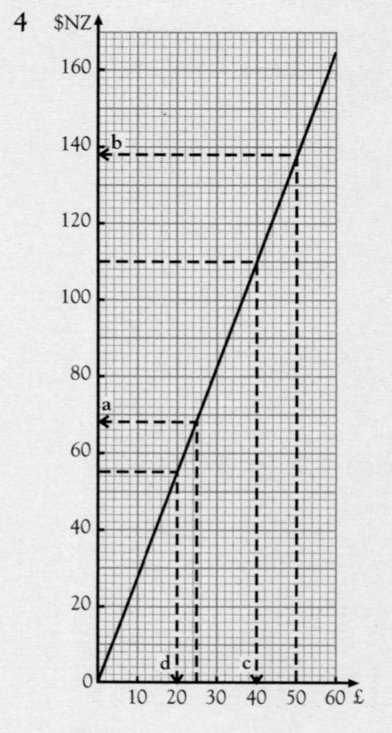

  a \$NZ 68   b £40
  c \$NZ 138   d £20
5 a 17   b 105 pence

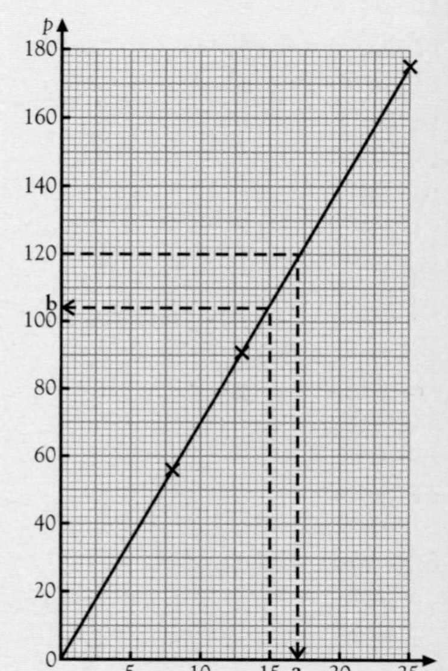

6 a i 200 km   ii 144 km

b i 2 hrs 6 mins   ii 48 mins

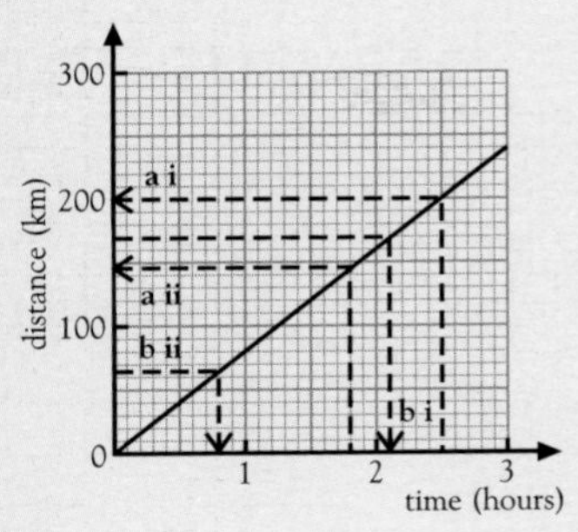

Exam Questions: Chp 6 (p. 100)

1 b Draw line horizontally from 15 pounds, then vertically to kg axis. 6.8 kg

2 a

x	−2	0	2	4	6
y	−2	−1	0	1	2

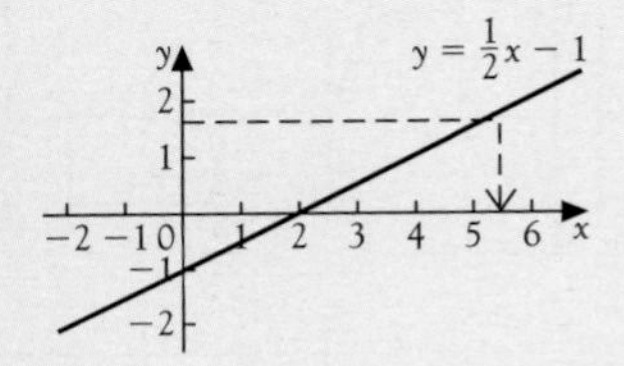

b 5.4

3 a

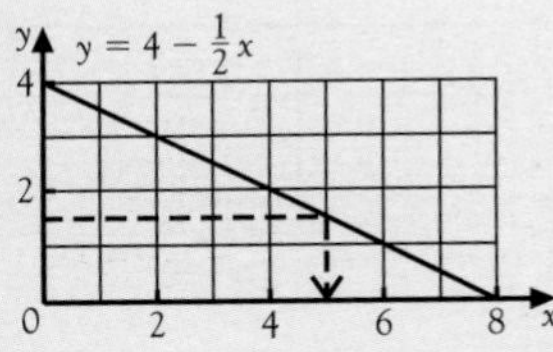

b i After 5 minutes

ii The tank is empty

4 a

x	0	30	60
y	4	10	16

b

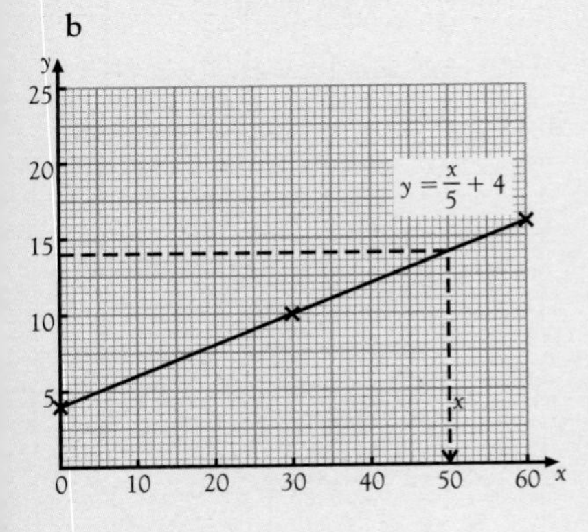

c 50 inches

5 a 21 DM

b £6.50

6 a −7, −4, 2, 5

b

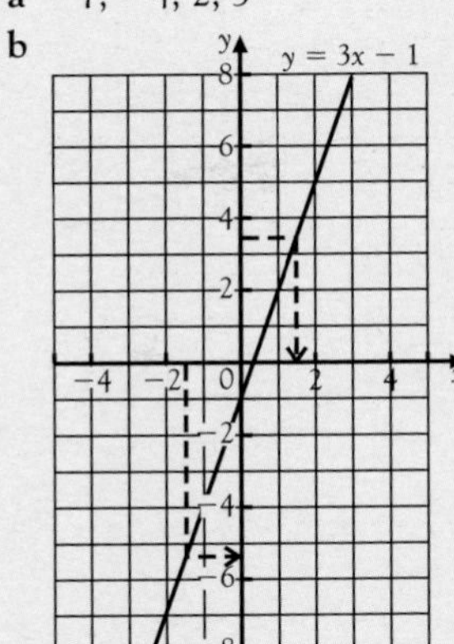

c i 1.5   ii −5.5

7 a

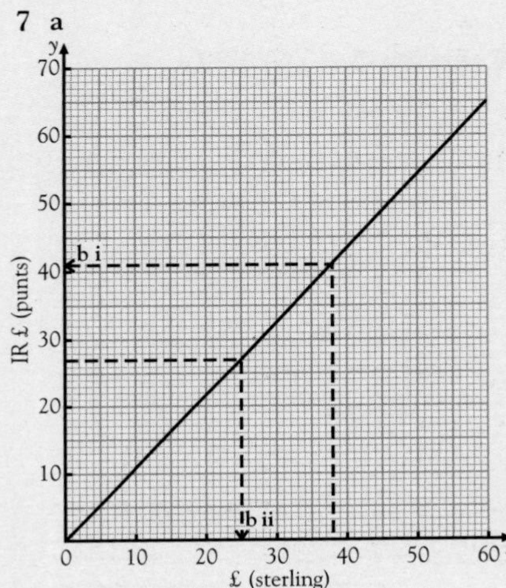

b i 41 punts   ii £25

8 a −1, 0, 1, 2

b

9 a

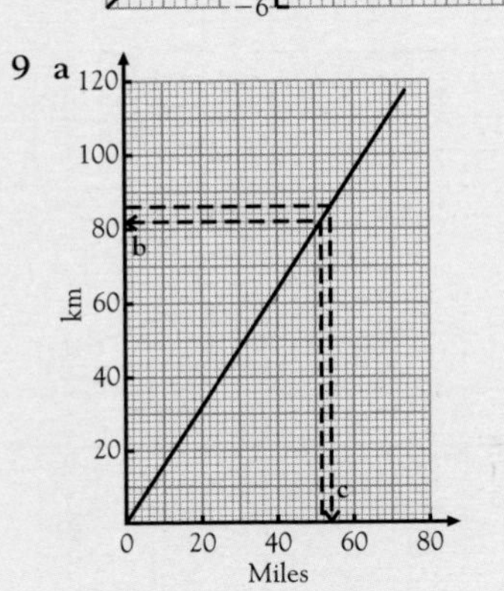

b 83 km   c 54 miles

10 a

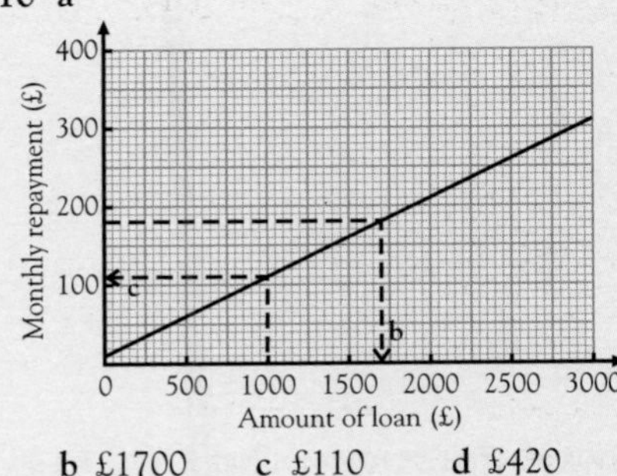

b £1700   c £110   d £420

Exercise 7.1 (p. 105)

1 a 496 cm^2, 752 cm^2

b 704 cm^2, 448 cm^2

2 a 2450 cm^2 b 1150 cm^2

3 60 cm^2

4 a 8 cm   b 2.5 cm

5 a 9 cm   b 36 cm

6 a 30 cm^2 b 16.81 cm^2 c 24.57 cm^2

7 18 cm^2

8 a 1.5 sq. ft   b £4.05

9 120 cm^2

10 40.5 m^2

11 a 12 cm   b 9.6 cm

12 A 3 cm^2, B 2.25 cm^2, C 4 cm^2, D 0.5 cm^2

13 6 cm^2 or 600 mm^2

14 15 m^2

Exercise 7.2 (p. 109)

1 28.26 m^2

2 3846.5 cm^2

3 a 6 cm   b 113.04 cm^2

c 123.84 cm^2 d 21.5%

4 1092.72 mm^2

5 b by 7 cm^2

6 a 6633.25 cm^2 b 3.341 m

7 25.625 cm^2

8 a 314 cm^2 b 2000 cm^2

9 a 122.7 cm^2 (1 d.p.)   b $\frac{1}{4}$

10 73.12 cm^2

11 4973.76 cm^2

12 a 19.2 cm^2 (1 d.p.)

b 3.4 cm^2 (1 d.p.)

c 86 cm^2 (1 d.p.)

Exercise 7.3 (p. 116)

1 a 480 cm^3 b 4 cm

2 1224 ml

3 a 1 m^3 b 25

4 160 litres

5 a 25 m^2 b 5 m^3

6 a i 6 m^3 ii 3.6 m^3 b 60%

7 Net a

8 a 44 cm^2 c 24 cm^3

9 56 250 cm^3

10 128 cm^3

11 0.477 m^3

12 a 34 cm b 44 cm^2
 d 4 cm, 3 cm, 2 cm e 24 cm^2

13 a 4200 cm^2 b Yes

14 a 36 cm^3 b 80 cm^2

15 Bronze 5500 cm^2
 Silver 8500 cm^2
 Gold 11 500 cm^2

16 9

Exam Questions: Chp 7 (p. 119)

(Use $\pi = 3.14$)

1 a 1600 m^2 b 600 m^2

2 a 399.8 m b 10 146.5 m^2

3 a 50.24 cm b 1004.8 cm^2

4 a 27 cm^3 c 54 cm^2

5 a 60 b 2400 cm^3

6 33 cm^2

7 13.5 cm

8 81 m^2

9 a 128.6 m^2 (1 d.p.)
 b 1435.6 m^2 (1 d.p.)

10 a i 35 ii 35
 b 12
 c 92 400 cm^3

11 b i 27 cm ii 58 320 cm^3
 iii 58.32 litres

12 a i 9 cubic ft b 54 cubic ft
 ii 18 cubic ft c 18 sq. ft
 iii 27 cubic ft

Exercise 8.1 (p. 124)

1 a

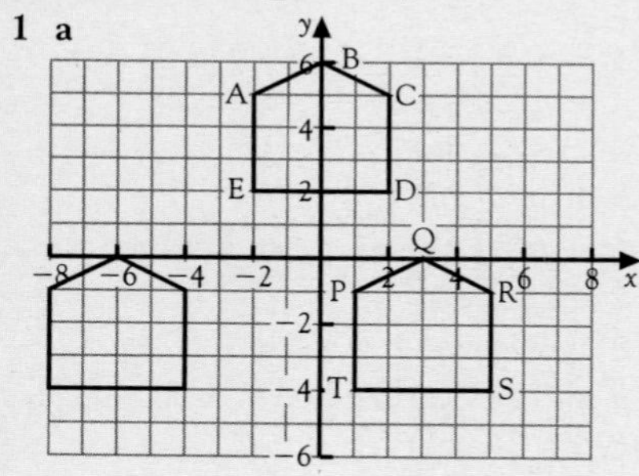

R(5, −1)

b $\begin{pmatrix} -6 \\ -6 \end{pmatrix}$

2 a $\begin{pmatrix} 14 \\ -4 \end{pmatrix}$

b

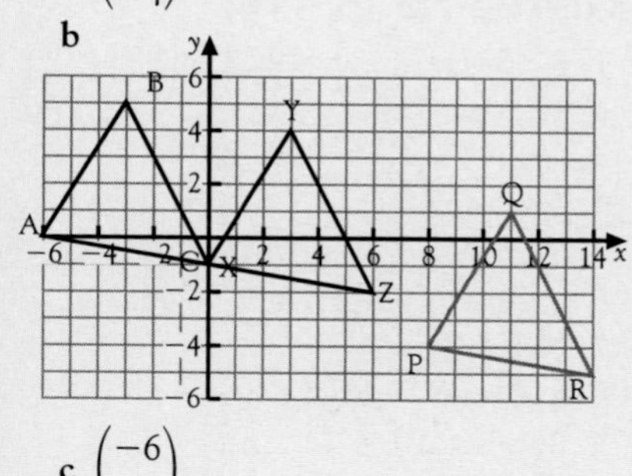

c $\begin{pmatrix} -6 \\ 1 \end{pmatrix}$

3 a Yes $\begin{pmatrix} 10 \\ 2 \end{pmatrix}$ e Yes $\begin{pmatrix} 8 \\ -7 \end{pmatrix}$

 b No f No

 c No g Yes $\begin{pmatrix} 12 \\ 4 \end{pmatrix}$

 d Yes $\begin{pmatrix} -12 \\ -4 \end{pmatrix}$ h No

4 a (4, 1) b (3, 9) c (2, 6)
 d (−4, −9) e (10, 1) f (−5, 2)

5 a $\begin{pmatrix} 1 \\ -4 \end{pmatrix}$ b $\begin{pmatrix} -3 \\ 4 \end{pmatrix}$ c $\begin{pmatrix} -1 \\ 6 \end{pmatrix}$

 d $\begin{pmatrix} 8 \\ -2 \end{pmatrix}$ e $\begin{pmatrix} -14 \\ 4 \end{pmatrix}$

Exercise 8.2 (p. 127)

1

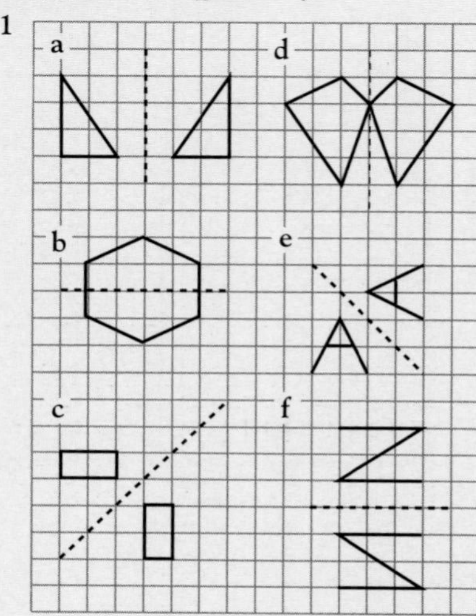

2

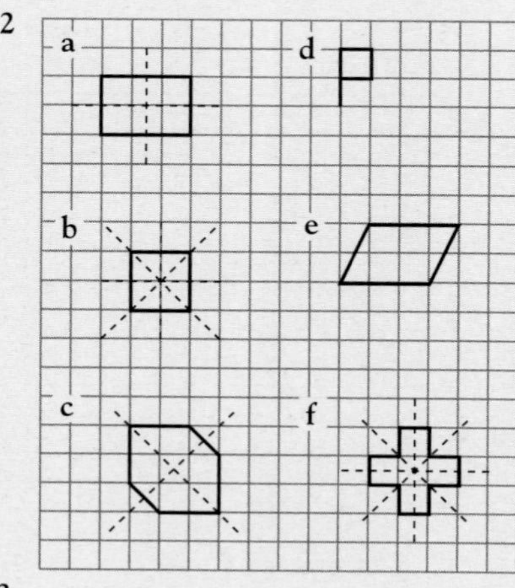

3

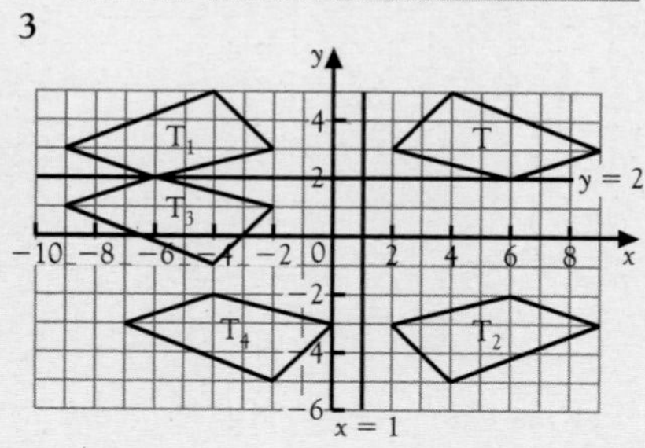

4

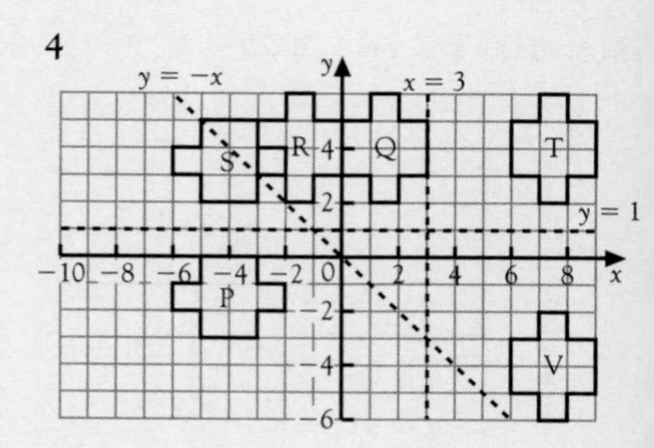

5 a $y = x$ b $x = -3$
 c $y = 0$ d $y = 1$

6 a (4, 1) b (3, 4) c (−4, −1)
 d (1, 2) e (1, 4) f (−1, 4)

Exercise 8.3 (p. 130)

1 a

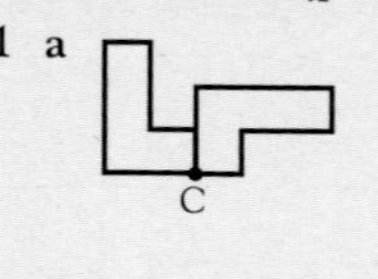

b

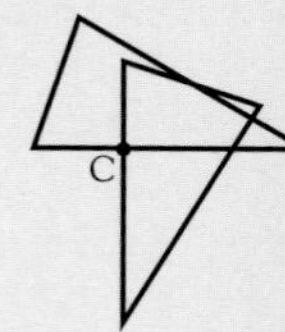

c

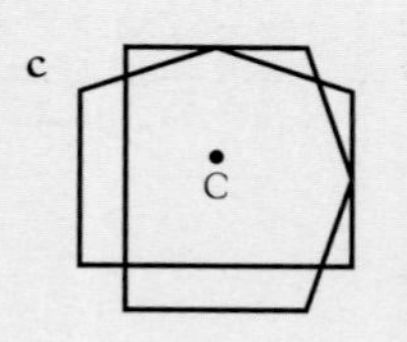

d

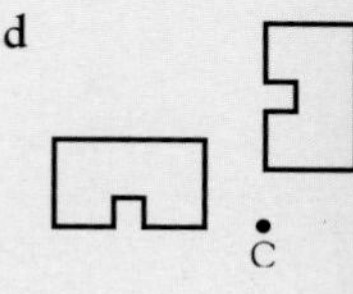

2

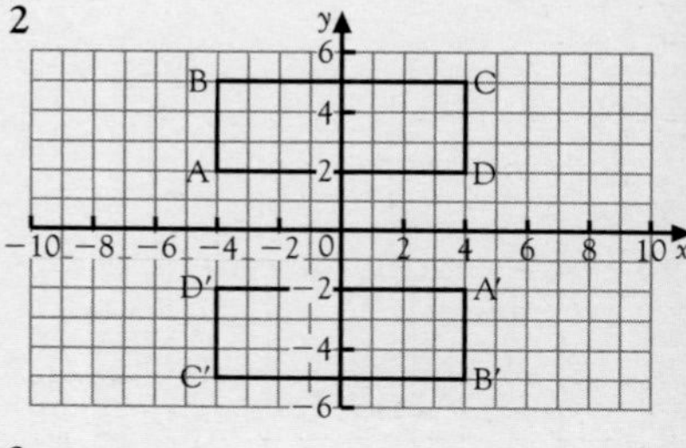

3

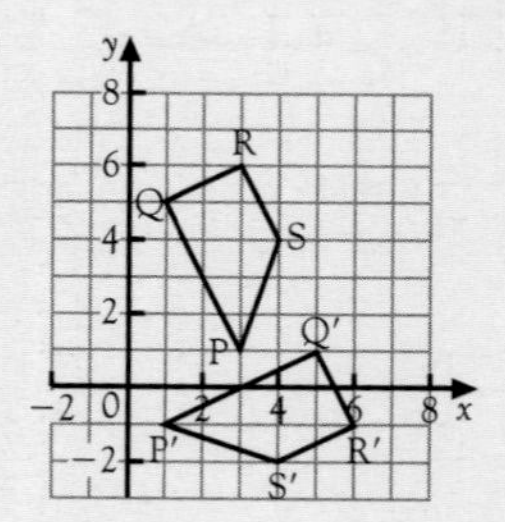

90° clockwise rotation about (1, 1)

4 D_1 (−3, 4) D_2 (3, −4)

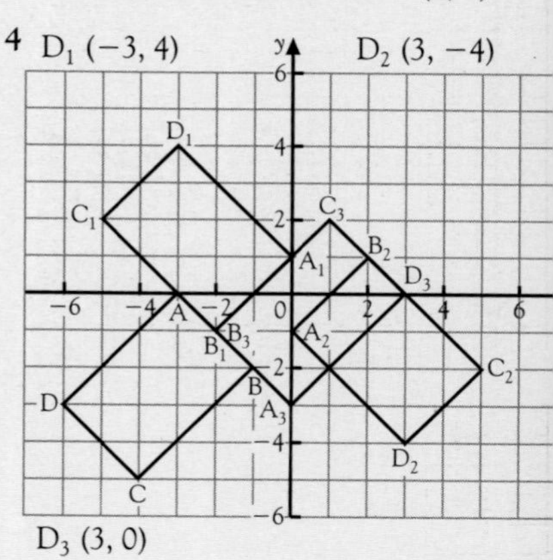

D_3 (3, 0)

Exercise 8.4 (p. 134)

1 a

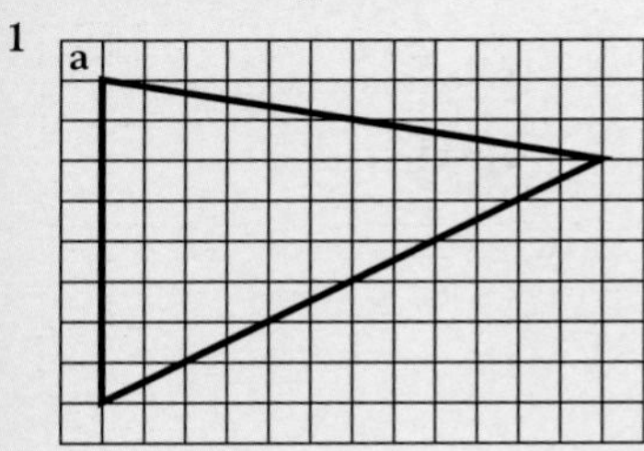

b c

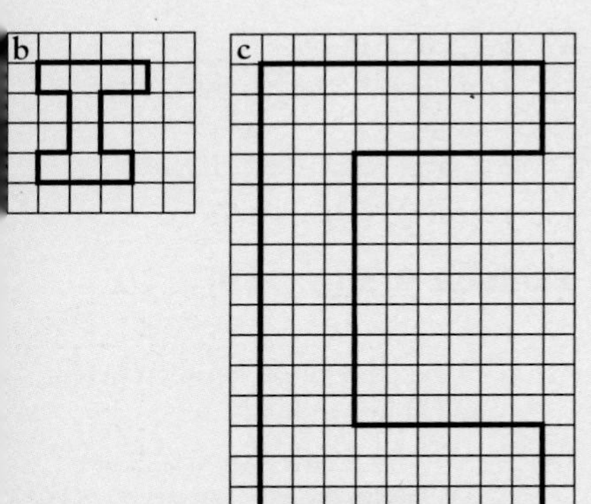

d

e

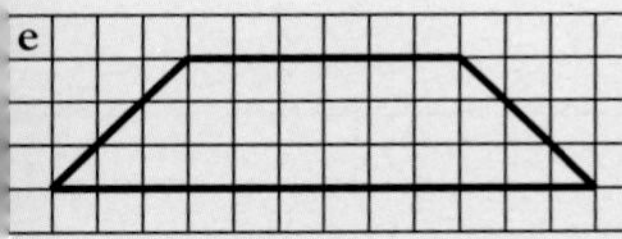

f

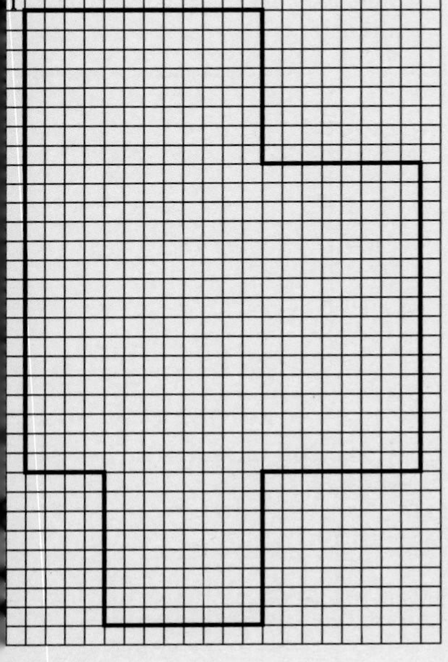

g

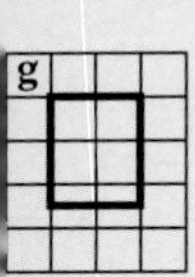
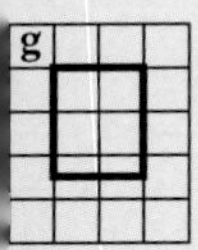

h

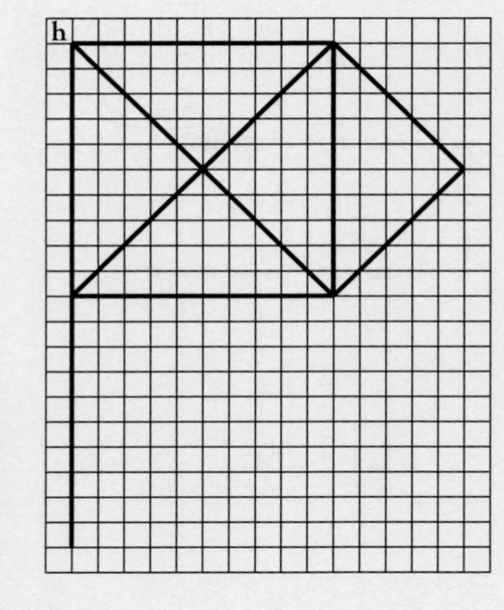

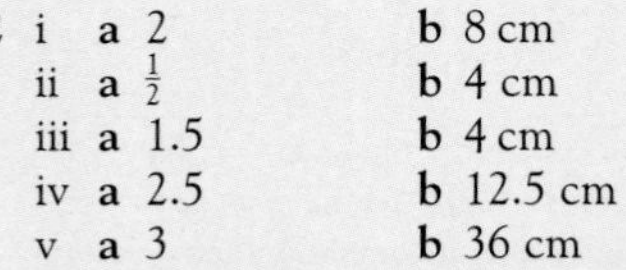

2 i a 2 b 8 cm
ii a $\frac{1}{2}$ b 4 cm
iii a 1.5 b 4 cm
iv a 2.5 b 12.5 cm
v a 3 b 36 cm

3 a

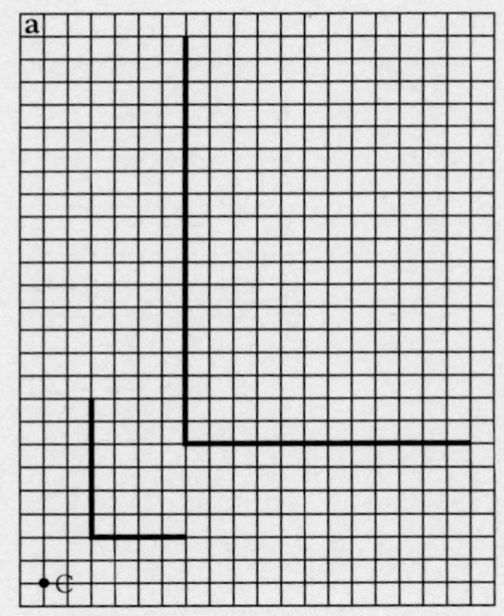

b

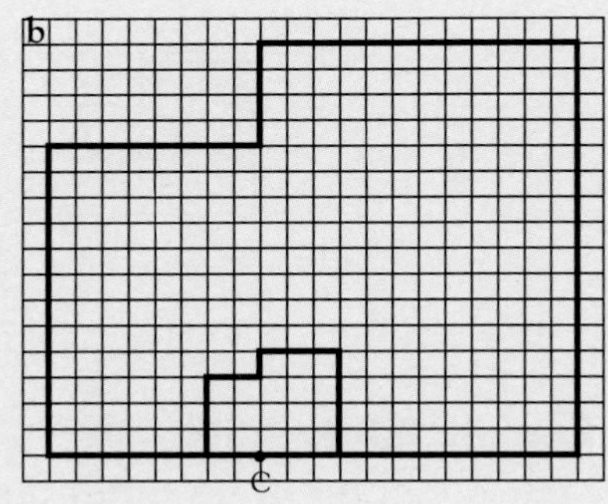

c d

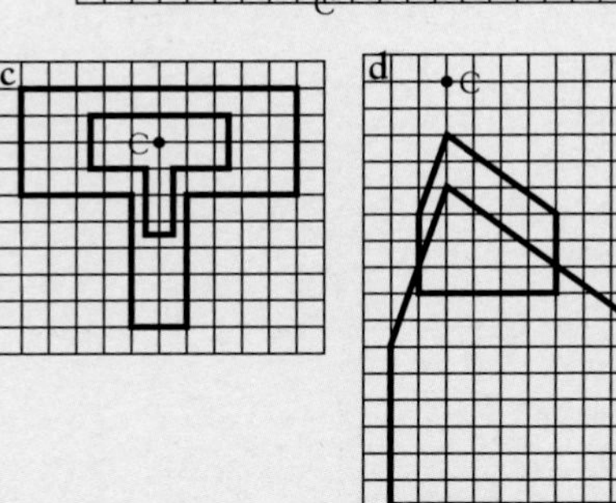

e

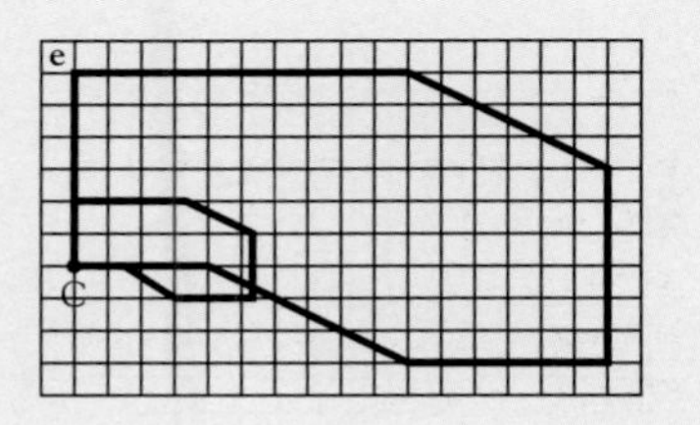

Exam Questions: Chp 8 (p. 136)

1 a $(-3, 4)$ b $(-1, -2)$

2 a

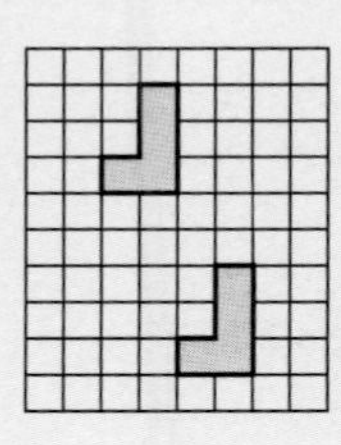

b $\begin{pmatrix} 2 \\ -5 \end{pmatrix}$

3 a i translation
ii rotation
iii enlargement
b 90° c 2

4 a i

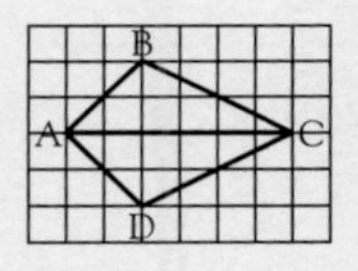

ii kite

b i

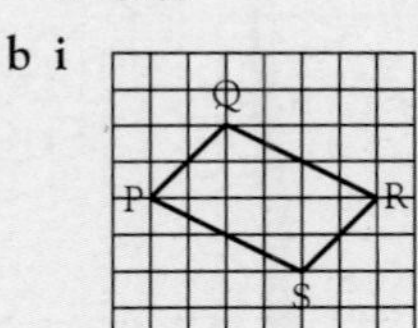

ii parallelogram

5 a Reflection in the y axis
b Plot C with vertices $(1, -1)$, $(1, -3)$, $(2, -1)$

6 a Reflection in the x axis
b Plot C with vertices $(-5, -2)$, $(-5, -4)$, $(-2, -2)$

7

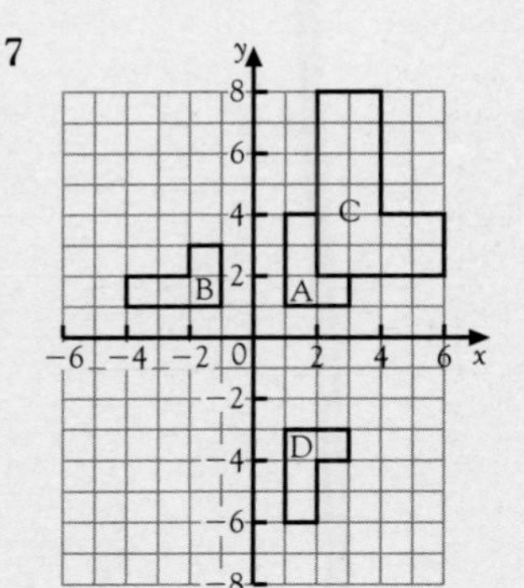

8

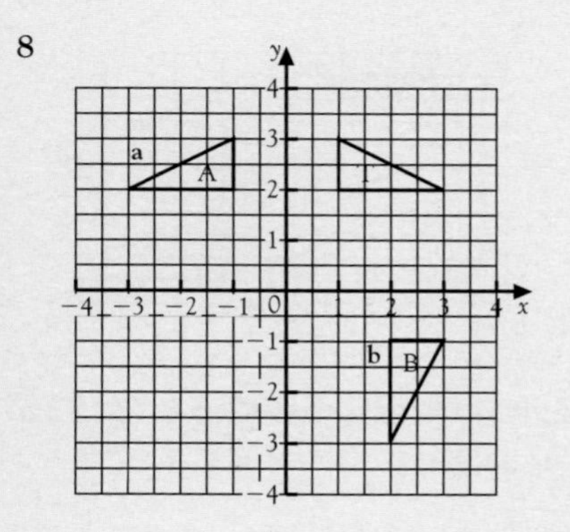

9 a C_1 (10, 11)
  b B_2 (−3, 4)
  c (8, 6)

Exercise 9.1 (p. 147)

1 Bar chart to show results of pets survey

Frequency; Dogs, Cats, Rabbits, Hamsters, Gerbils, Snakes

2 a

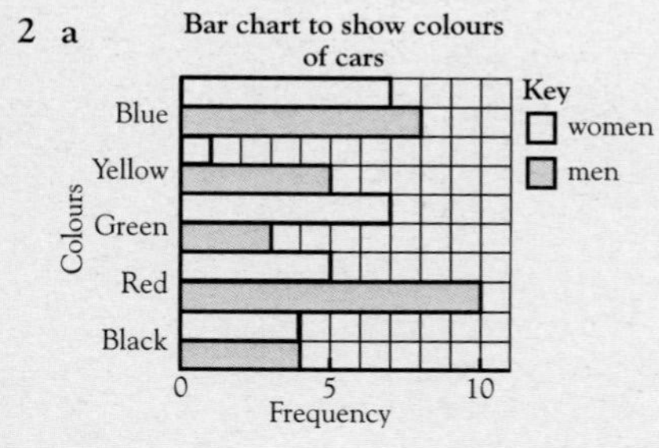

  b $33\frac{1}{3}$%
  c $\frac{1}{9}$
  d Angles: Black 60°, Red 75°, Green 105°, Yellow 15°, Blue 105°

3 a i $\frac{1}{4}$ ii 120° iii 50 g
  b i 45°
    ii Angles in pie chart: 45° water, 126° humus, 189° mineral
  c Bar chart illustrating

	Water	Humus	Mineral
Simon	30 g	40 g	50 g
Helen	15 g	42 g	63 g

4 a 12
  b 4, 12, 6, 5, 3
  c 30
  d 16
  e $\frac{4}{15}$

5 a 29
  b 3

6 112

7 a Medical treatment
  b 75
  c 0.375 ($\frac{3}{8}$)

8

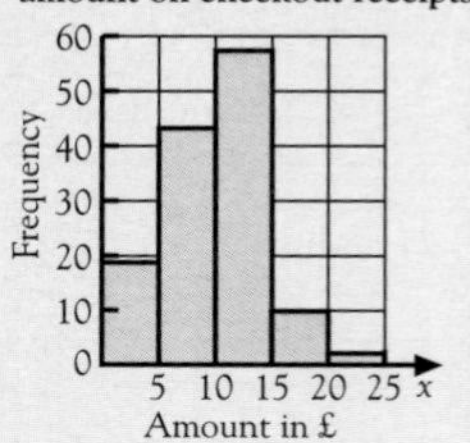

9 a Angles on pie chart: 45°, 120°, 156°, 30°, 9°
  b i $\frac{1}{12}$ ii 0 iii $\frac{17}{20}$

10 a, b

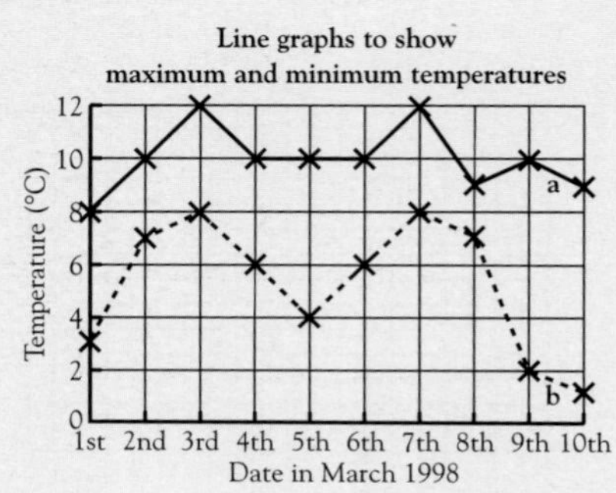

  c 8th March

11 a $\frac{1}{4}$ b 144° c 35%

12 a 4–6: 14, 7–9: 5
  b

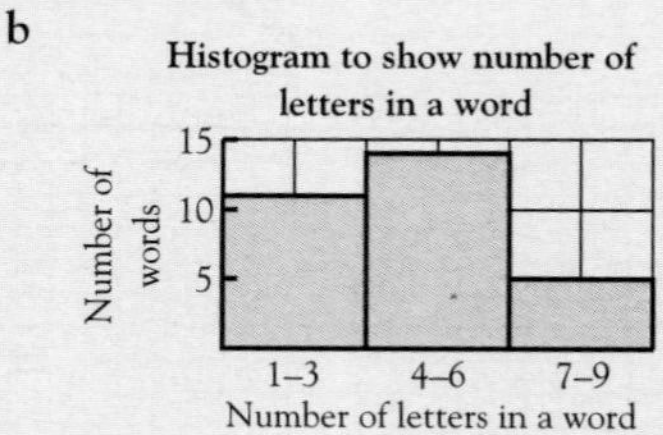

Exercise 9.2 (p. 154)

1 a 2 b 2 c 1.1 (1 d.p.)
2 a 9 b 8 c i 7 ii 7.5
3 £40.50
4 32
5 a 8 °C b 8.375 c 14
6 8.7°C (1 d.p.)
7 a 8
  b 3.1°C (1 d.p.)
8 a 1.3
  b 34.7
  c 1
9 a 29
  b 45
  c 45.8 (1 d.p.)
10 a 62
  b 2.5 (1 d.p.)
  c mode
11 a i £423 ii £200
12 a C
  b 7
  c mean
13 a i 5.6 ii 0.5
  b range 1.4 (much wider); mean 5.5 (lower)
  c Lowest mark ⩽ 5.2, highest mark ⩽ 5.8

Exercise 9.3 (p. 160)

1 a C
  b D, no correlation between height and ability in maths test
2 a

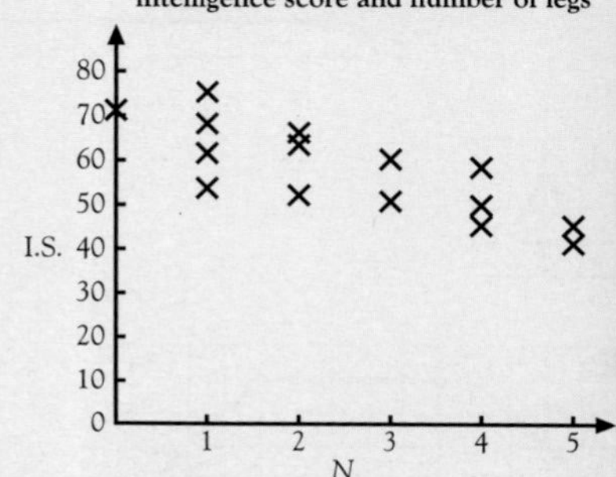

  b Some evidence of negative correlation suggesting the fewer the number of legs, the more intelligent they are.

3 a Scatter diagram to show test scores

German score; French score

  b Good positive correlation between the scores.

4 a i

  b 35 (depends on *your* line of best fit)

5

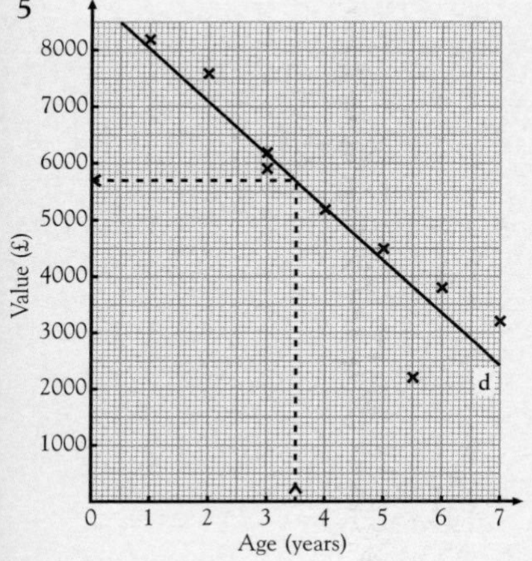

b Decreases
c It had not been looked after very well
e £5700 (depends on line of best fit)

6 a

b Strong negative correlation

Exam Questions: Chp 9 (p. 163)

1 260
2 a small 7, medium 9, large 8
b Angles 105°, 135°, 120°
3 a 69
b 15
c 82 ÷ 7 = 11.7… increase
4 a Draw scatter diagram
b 28 (approx)
5 a 10 b 9 c 8
6 a 90 b 40°
c 0.22 d $\frac{3}{25}$
7 a $\frac{5}{17}$
b 2.01 m
c 1.92 m
d 1.61 m (2 d.p.)
e It has been distorted by those who did not score. Those who scored jumped much higher than the mean value.
8 a 1000
b August
c 7000
d $3333\frac{1}{3}$
e 6
f Spring 63°, Summer 162°, Autumn 81°, Winter 54°
9 a i B ii C iii A iv A
b i 7 ii 4
iii Several answers possible, e.g. first four values 5, highest value 11

Mixed Exam Questions: End of Stage 1 (p. 169)

1 Tom's £29.38, Pete's £32, City £30
2 a i 115 000 ii 165 600 iii $\frac{3}{575}$
b i 165°, 96°, 60°, 39°, 360°
3 a $a = 128°$, $c = 38°$
b 36° (corresponding angles on parallel lines are equal)
c 108°
4 a i $\frac{9}{20}$ ii $\frac{7}{20}$ b $\frac{4}{5}$
5 Several possible answers
6 a b

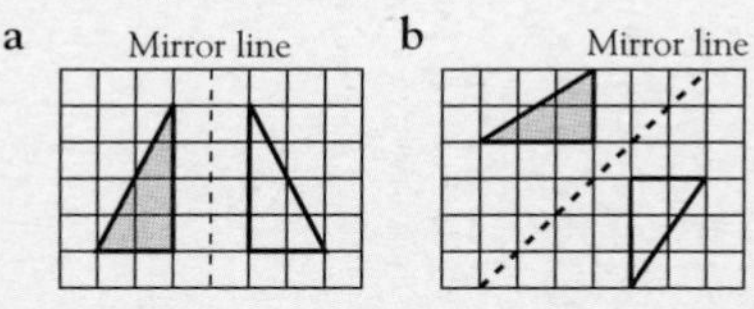

7 a 388.4 m b 8826 m² ($\pi = 3.14$)
8 a

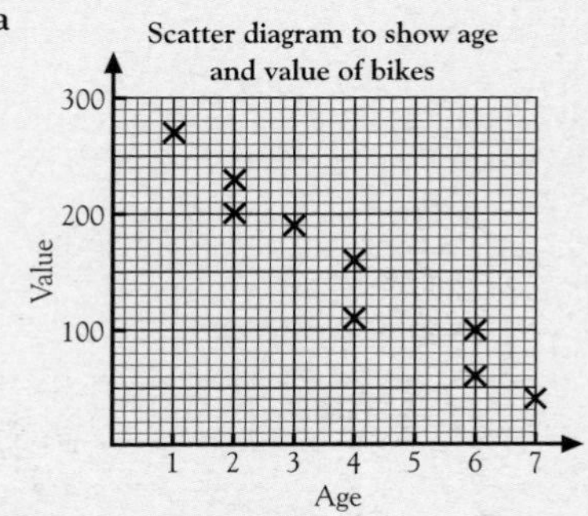

b The older the bike, the less it is worth (strong negative correlation)
c i £110 ii About $1\frac{1}{2}$ years old
9 a i £6 ii 8.4 m
b i 0.375 ii 37.5%
10 a i

Anil	Carl	Mark
w	r	b
w	b	r
r	b	w
r	w	b
b	r	w
b	w	r

ii $\frac{1}{3}$ b i $\frac{7}{20}$ ii $\frac{13}{20}$
11 a All except trapezium and kite
b All except trapezium
12 a i 10.1 cm (1 d.p.) ii 9 cm
iii 9 cm iv 9.5 cm
b i $\frac{1}{6}$ ii $\frac{1}{2}$
13 a i 11.5 mm
ii 415 mm² (nearest whole number)
b 43.3 cm (1 d.p.)
14 b 43.3 m (perpendicular distance from T to line AB)
15 a 2 b 3 c 1 d 3
16 a 42° (Z angles) b 113° (Z angles)
c XY is a line of symmetry
d 25° (angles in △ add up to 180°)
17 a D, A, B
b I watched a lot of TV and didn't do much homework.
c Generally, the more time spent watching TV, the less spent doing homework (negative correlation).
d

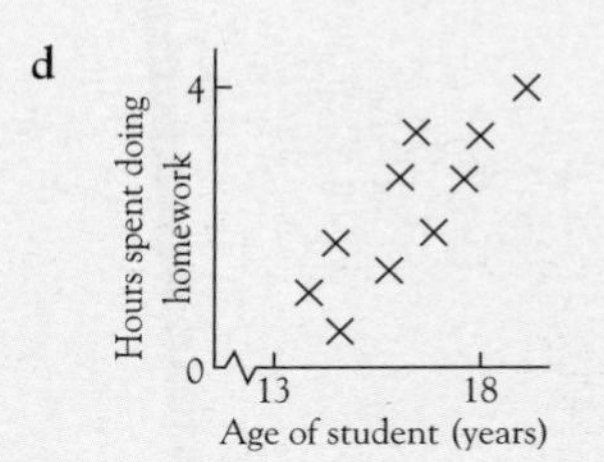

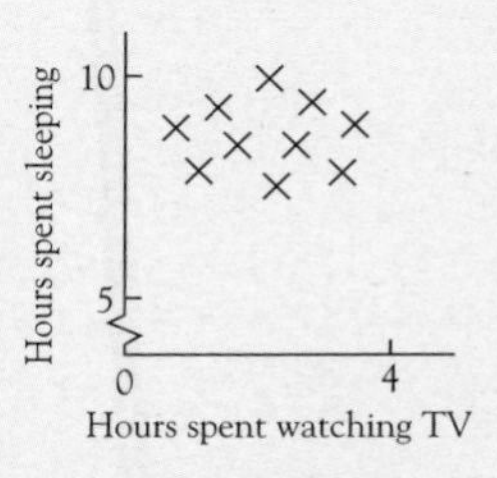

18 a 21, 29 b i 0.42 ii 0.58 c 42
19 $18n + 5g$
20 a 20 × 30 b 600
21 a 24 m³ b 68 m²
22 a Equilateral
b i 9.42 cm ii 14.13 cm
23 a $P = 2a + 4b + 2c$ b 22
24 a £450 000 b 4%
25

26 a i E ii C iii F
b rotation 180° about O

Exercise 10.1 (p. 180)

1 24, 18, 6, 60, 42
2 711, 594, 1224, 7893
3 a 24 b 60 c 36
4 a 1, 2, 3, 4, 6, 8, 12, 16, 24, 48
b 1, 2, 5, 7, 10, 14, 35, 70
c 1, 2, 3, 4, 6, 7, 12, 14, 21, 28, 42, 84
5 23, 29, 31, 37
6 a 16, 64 b 27, 64
c 35, 5, 50, 75 d 5, 3
7 a 36, 8, 64
b 36, 8, 19, 64, 1, 125
c 36, 64, 1
d 8, 64, 1, 125
e 19
8 a 49 b 64 c 16 d 3 e 5
9 a $\frac{1}{5}$ b $\frac{4}{3}$ c $\frac{3}{7}$ d 2

10 a i 1, 2, 3, 4, 6, 8, 12, 24
  ii 1, 2, 3, 4, 5, 6, 10, 12, 15, 20, 30, 60
  iii 1, 3, 5, 15, 25, 75
 b i 12 ii 15 iii 3
11 a $2^2 \times 3 \times 5$ b 2×3^3 c $2^3 \times 11$
12 41
13 $\sqrt{400}$, $(-5)^2$, $\sqrt[3]{27\,000}$, 2^5
14 b, d
15 a 10 cm b i 9 cm ii 27 cm
 c 64 cubes
16 a 1296 b 5 c 6 d 49
17 a 2, 3, 5, 7, 11, 13, 17, 19, 23, 29, 31, 37, 41, 43, 47, 53, 59, 61, 67, 71, 73, 79, 83, 89, 97
 b i 3 + 5
  ii 3 + 19 or 5 + 17 or 11 + 11
  iii 5 + 23 or 11 + 17
  iv 7 + 47 or 11 + 43 or 13 + 41 or 17 + 37
  v 3 + 97 or 11 + 89 or 17 + 83 or 29 + 71 or 41 + 59
18 120 seconds (or 2 mins)
19 a $\frac{1}{1000}$ b $\frac{1}{2}$ c 25 d $\frac{2}{3}$
20 a 1, 4, 9, 16, 25, 36, 49, 64, 81, 100
 b i 1 + 16
  ii 25 + 9 + 1
  iii 64 + 16
  iv 81 + 9 + 1 + 1 or 36 + 36 + 16 + 4
  v 100 + 4

Exercise 10.2 (p. 183)

1 a 7500 (2 s.f.) f 0.007 (1 s.f.)
 b 4 300 000 (3 sf.) g 0.483 (3 s.f.)
 c 63 (2 s.f.) h 0.093 (2 s.f.)
 d 201 000 (3 s.f.) i 0.530 (3 s.f.)
 e 2.893 (4 s.f.) j 0.00010 (2 s.f.)
2 a 296 933
 b i 200 000
  ii 330 000
  iii 300 000
3 21.4 m^2 (3 s.f.)
4 0.29 (2 s.f.)
5 8.0 cm (2 s.f.)
6 a 41 250 b 41 349

Exercise 10.3 (p. 186)

1 a 3 : 5 d 6 : 5 g 1 : 3
 b 2 : 3 e 4 : 9 h 2 : 5
 c 4 : 5 f 3 : 8 i 12 : 5
2 a 4 : 5 b 10 : 3 c 5 : 1
 d 1 : 6 e 16 : 1 f 20 : 3
3 a 2 : 15 b 6 : 7 c 6 : 1
4 a 1 : 0.8 e 1 : 2.48
 b 1 : 1.25 f 1 : 100 000
 c 1 : 3.5 g 1 : 9
 d 1 : $2\frac{1}{3}$ h 1 : $1\frac{2}{3}$
5 a 2.5 : 1 d $4\frac{1}{3}$: 1
 b $\frac{1}{3}$: 1 e 200 : 1
 c 0.4 : 1 f 4 : 1
6 6 : 7
7 7 : 8
8 a $1\frac{1}{4}$: 1 (or 1.25 : 1)
 b 1 : $\frac{4}{5}$ (or 1 : 0.8)
9 a 6 : 15 : 2
 b 3 : 1
10 5 : 2

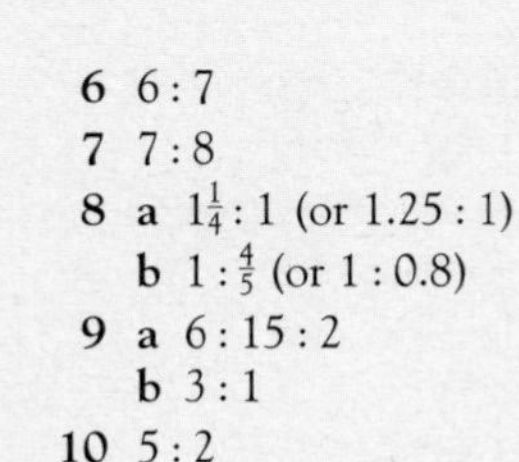

Exercise 10.4 (p. 191)

1 20 mins
2 a 30 b 42
3 28
4 6
5 28
6 12 years, 15 years
7 250 kg
8 80 ml
9 a £2.40 b 24 pence c 60 pence
10 a

No. pints	2	4	6	8	10	12
No litres	1.14	2.28	3.41	4.55	5.69	6.83

 b

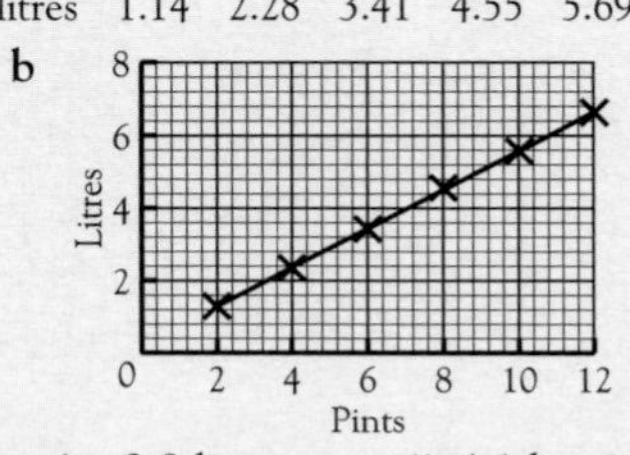

 c i 2.8 litres ii 4.1 litres
  iii 3.5 pints iv 9.1 pints
  (all to 1 d.p.)
11 25.5 kg (3 s.f.)
12 £6.80
13 a 239 000 drachma
 b £34.55
14 Macaroni 210 g
 Cheese 140 g
 Cornflour 44 g (nearest g)
 Milk 1260 ml (or 1.26 litres)
15 Motorways 81 mph
 Dual carriageways 69 mph
 Ordinary roads 56 mph
 Built up areas 38 mph (or 37 mph)
16 a 33 washes b 600 g
17 1.5 km

Exercise 10.5 (p. 194)

1 a 3 b 0.5 (or $\frac{1}{2}$)
2 a 1.25
 b X′Y′ = 8 cm XZ = 6.6 cm
3

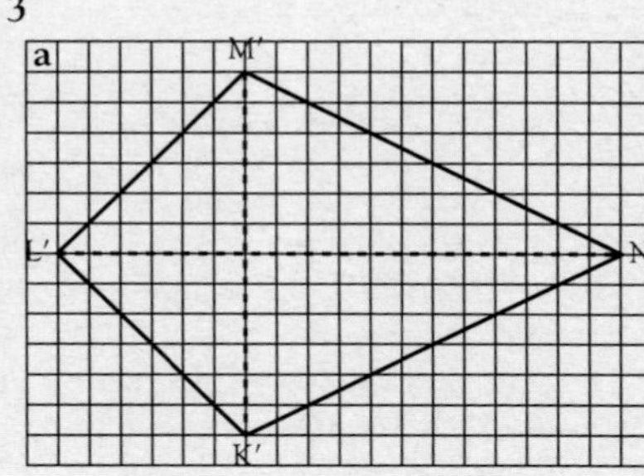

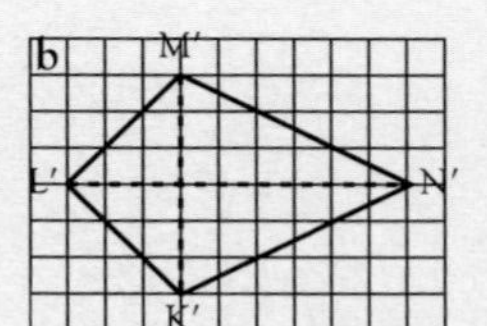

4 a i 4 m by 3 m ii Diameter 1.2 m
 b i 7 cm by 4.5 cm ii Diameter 2.4 cm
5 a 1 : 20
 b 1.8 m c 6 cm
6 a 2 larger 0.8 smaller 1.4 larger
  0.6 smaller 1 neither – same size
 b 10.5 cm by 6.02 cm
7 95 mm
 139 mm
8 a 1 : 250
 b 30 m c 72 mm
9 18 m
10

	Model	Real
Length	14 cm	70 m
Wingspan	12 cm	60 m
Height	1.2 cm	6 m
Width of fuselage	0.9 cm	4.5 m

11 a 6 cm by 9 cm
 b i 4 : 9 ii 1 : 2.25
12 a 15 cm by 12 cm
 b Ratio of lengths of picture : card = 12 : 15 = 4 : 5
  Ratio of widths of picture : card = 9 : 12 = 3 : 4
  Ratios are not the same ⇒ not a true enlargement
13 11 ft
14 a 1.2 km
 b 1.5 cm
15 a 0.004 b 1.68 cm by 1.4 cm
 c 1 : 62 500

Exercise 10.6 (p. 197)

1 8, 12
2 5, 10
3 18, 24
4 50, 30
5 30, 54
6 5, 2.5
7 9, 18, 27
8 150, 250, 200
9 100, 100, 250
10 30, 45, 90

Exercise 10.7 (p. 197)

1 a 42 b 12
2 30%
3 Jason £38.40 Kelly £25.60
4 a i Peppermints 60%
  ii Spearmints 40%
 b i 21 ii 14
5 Daffodils 75 Crocuses 45 Tulips 30
6 Angles are 24°, 72°, 96°, 168°

7 25 mins
8 a i 62.5% ii 37.5%
b i $\frac{5}{8}$ ii $\frac{3}{8}$
9 Melanie £525 Peter £455
10 a 85%
b 1.53 kg
11 Gemma £10.15 Ann £7.25
Kirsty £5.80
12 14 : 11
a $\frac{14}{25}$
b $\frac{11}{25}$
13 Tom £120 000 Dick £84 000
Harry £36 000

Exam Questions: Chp 10 (p. 199)

1 a i $\frac{7}{20}$ ii 42 b 55% c 45
2 a 6 : 5 : 4 b £7500
3 a i 6.47 g (3 s.f.) ii 6.4 g
b Bargain Pack
4 a 5 : 3 b £750
5 a 102 cm b 30
6 a i $2^3 \times 3^2$ ii $2^4 \times 5$ b 12 mins
7 a 2.5 b 3 m
8 42 mph
9 a 150 b 160 g
10 a £59.75 b £30
11 a 60 litres b 300 ml
12 a 37 170 pesetas
b £32.49 (to nearest penny)
c $300 \times 200 = 60\,000$ pesetas
13 a 287 000 (3 s.f.) b 1.61 (2 d.p.)
14 a 3 b 45 cm
15 7000 kg
16 a 6 cm b 7 cm
17 $42 = 2 \times 3 \times 7$
$63 = 3 \times 3 \times 7$
HCF = 21
18 a 23 b 4, 12, 20 c 4, 12, 20
d 4, 12 e 4, 23
19 13 cm
20 40 balloons
80 sausage rolls
30 jellies
14 packets of biscuits
21 a 25 g b 1 kg
22 16 g
23 a 20 cm² b 12 cm × 15 cm c 9

Exercise 11.1 (p. 203)

1 a $3x + 15$ g $10a - 20c$
b $x^2 - 4x$ h $18z - 24y + 30$
c $2x^2 + 14x$ i $10 - 4y$
d $30a + 24ax$ j $12a - 2b$
e $6x + 12$ k $24x - 16y + 8z$
f $3x + 5$ l $\frac{1}{4}x + \frac{1}{8}y$
2 a $5x + 10$ f $5x + 11$
b $20 + 2x$ g $14x - 6$
c $2 + 3x$ h $9x + 27$
d $16a - 4b$ i $10a + 23b + 4c$
e $x^2 + 5x + 6$

3 a $x - 8$ f $2x + 10$
b $2a - 3$ g $6x^2 + 10x$
c $2y + 8$ h $x^2 + 7x + 2$
d $7a + 2a^2$ i $14 + 2x$
e $-x$

Exercise 11.2 (p. 205)

1 64
2 70
3 44
4 20
5 a 18 b −30
c −95 d 40
6 a 10 b $18\frac{1}{3}$
c −5 d −25
7 Y = 8 years
8 $A = 10\,500$ mm² *or* 105 cm² (3 s.f.)
9 $P = 34$ m; $A = 71.25$ m²

Exercise 11.3 (p. 209)

1 $x = 2$
2 $x = -2$
3 $x = 3$
4 $x = -2$
5 $x = 0.5$
6 $x = 2$
7 $x = \frac{6}{7}$
8 $x = 6$
9 $x = 1$
10 $x = 4$
11 $x = 3.5$
12 $x = 1.5$
13 $x = 4$
14 $x = \frac{1}{3}$
15 $x = -\frac{1}{3}$
16 $x = 1$
17 $x = 2\frac{2}{3}$
18 $x = 1.5$
19 $x = 3$
20 $x = 1.5$
21 $x = 2$
22 $x = 4$
23 $x = 31$
24 $x = 7$
25 $x = -3$

Exercise 11.4 (p. 213)

1 $x < 4$ (number line −3 to 6)
2 $x \geqslant 2$ (number line 0 to 8)
3 $1 < x < 3$ (number line −1 to 5)
4 $-2 \leqslant x \leqslant 1$ (number line −4 to 3)
5 $x < -1$ or $x \geqslant 4$ (number line −3 to 7)
6 $-2.5 < x \leqslant 4.5$ (number line −4 to 6; −2.5, 4.5 marked)
7 $x = 0, 1, 2, 3$ or 4
8 $x = 4, 5$ or 6
9 $x = -1, 0, 1, 2, 3, 4$ or 5
10 $x = -5, -4, -3$ or -2
11 $x = -7, -6, -5, -4, -3, -2$ or -1
12 $x = -1, 0, 1, 2$ or 3

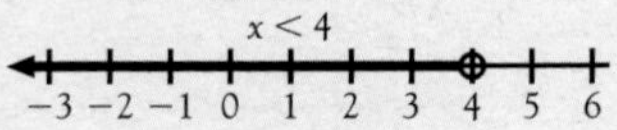

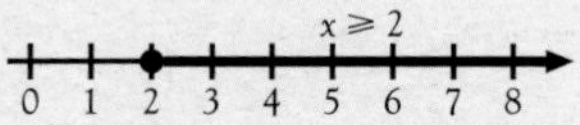

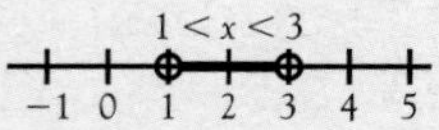

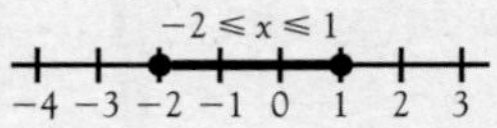

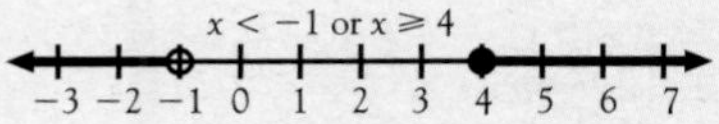

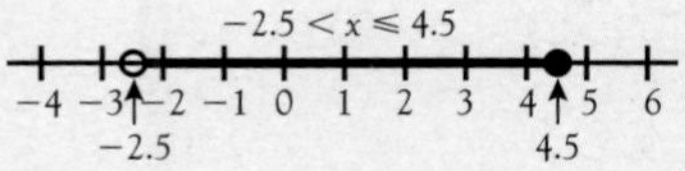

Exercise 11.5 (p. 214)

1 a $x \geqslant 6$ g $x < 3.25$
b $x < 2$ h $x \geqslant 2\frac{1}{6}$
c $x > 4$ i $x > -3$
d $x \leqslant 2$ j $x \leqslant 1\frac{3}{7}$
e $x \geqslant 2$ k $x > 5.5$
f $x < 7.5$ l $x < -2$
2 a $x \geqslant 9$ g $x > 0.5$
b $x > 2$ h $x \leqslant 5\frac{2}{3}$
c $x \leqslant 6$ i $x > -0.25$
d $x > 8$ j $x < -2.2$
e $x < 2$ k $x \geqslant 8$
f $x \geqslant 6$ l $x \geqslant 2$
3 a 7, 3 d 1, −1
b 5, 2 e 1, 0
c 3, 2 f 1, −1
4 a $71\,975 \leqslant x < 72\,025$
b $t < -18$
c $x \geqslant 15$
d $50 < c \leqslant 60$
e $64 \leqslant n \leqslant 74$
f $24.5 \leqslant L < 25.5$
5 a $3x$
b i $3x \leqslant 14$ ii $x \leqslant 4\frac{2}{3}$; 4
6 a $36 + 12x$ pence
b $36 + 12x \leqslant 200$; $x \leqslant 13\frac{2}{3}$; 13
7 a $20x$
b $20x + 45 \leqslant 300$
c $x \leqslant 12.75$; 12
8 a $7(x - 6) < 50$
b $x < 13\frac{1}{7}$
c 13
9 a i $500 + 8x$ ii $20x$
b $20x > 500 + 8x$; 42
10 a $2x - 5$
b $2x - 5 > 30$; $x > 17.5$
c 18
11 a $\dfrac{x + 8}{3} > x$
b $x < 4$ (number line −2 to 6)
c 1, 2 or 3

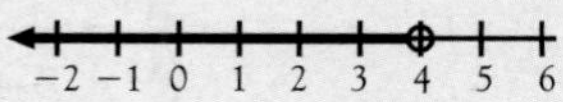

Exam Questions: Chp 11 (p. 216)

1 $y = 0.5$
2 $z = 14$
3 $q = 8$
4 $x = 2.5$
5 $x = 4.5$
6 $z = -2$
7 $y = 9$
8 6 cm^2
9 $x = -1.4$
10 $t = -0.5$
11 $P = 20$
12 $s = -3.75$
13 $A = 38\ cm^2$
14 $V = 1432.2$ ($\pi = 3.14$)
15 5, 6, 7, 8, 9 or 10
16 3
17 −1, 0, 1, 2, 3, 4 or 5
18 $x < -2.5$
19 **a** $x < 4$
b

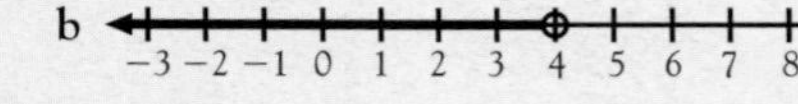

20 $x < 6.5$
21 **a** −2, −1, 0 or 1
b i

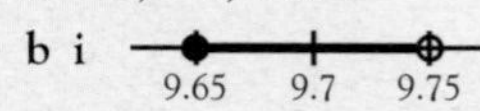

ii $9.65 \leqslant t < 9.75$
22 −3, −2, −1, 0 or 1
23 **a** −4, −3, −2, −1, 0 or 1
b $x < 0.5$
24 −2, −1, 0, 1, 2, 3, 4 or 5
25 **a** $x < 2$ **b** $x > 1$
26 −4, −3, −2, −1, 0, 1 or 2
27 $y > -0.6$
28 $n > 3$
29 **a** $x = -3$ **b** $x \geqslant -3$
30 $x < 6.5$
31 **a i** $x \leqslant 5$ **ii** $x \geqslant -1$
b $-1 \leqslant x \leqslant 5$
32 $x \leqslant 12.5$
33 **a** $n = -1, 0, 1, 2$ or 3
b 17, 18, 19, 20 or 21

Scale drawings have been reduced in the answers to Chapter 12.

Exercise 12.1 (p. 221)

1
Lamaload Reservoir	043°
Shining Tor	067°
Shutlingsloe	135°
Roach End	158°
Bosley Beacon	180°
The Cloud	201°
Jodrell Bank	270°
Alderley Edge	294°
White Nancy	339°

2
Bordeaux	027°
Le Havre	117°
Lyon	330°
Marseille	337°
Nantes	059°
Strasbourg	272°

3 325°, 190°, 080°, 210°

4 **a**

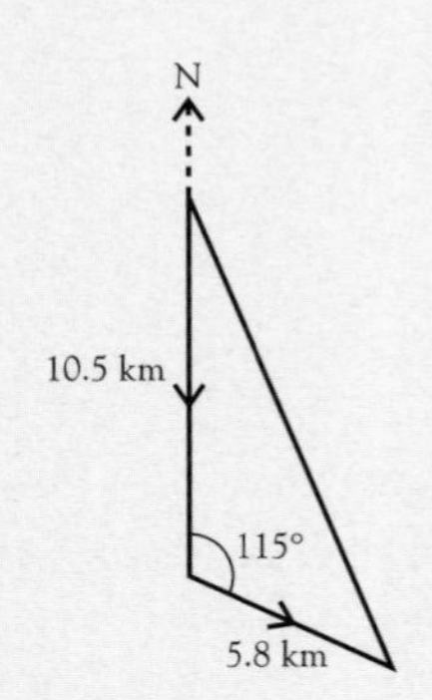

b 14 km, 158°

5 **a**

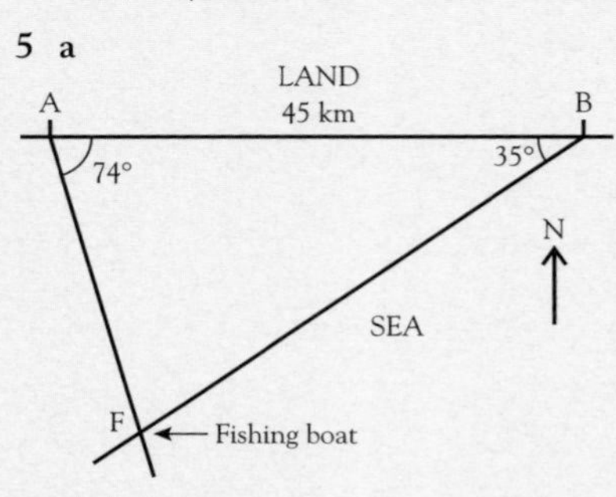

b 27 km, 46 km

6

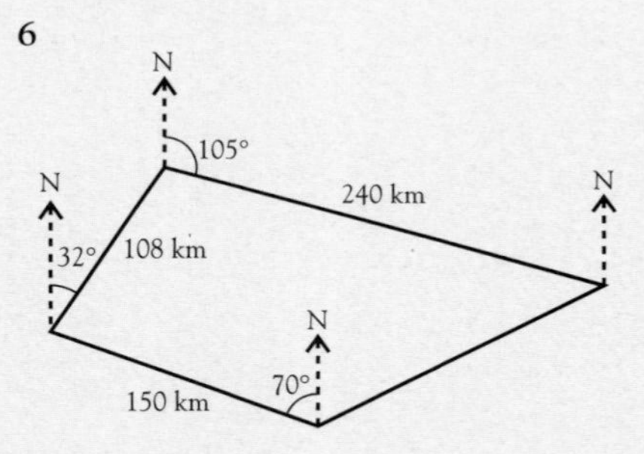

170 km, 242°

7 **a, b**

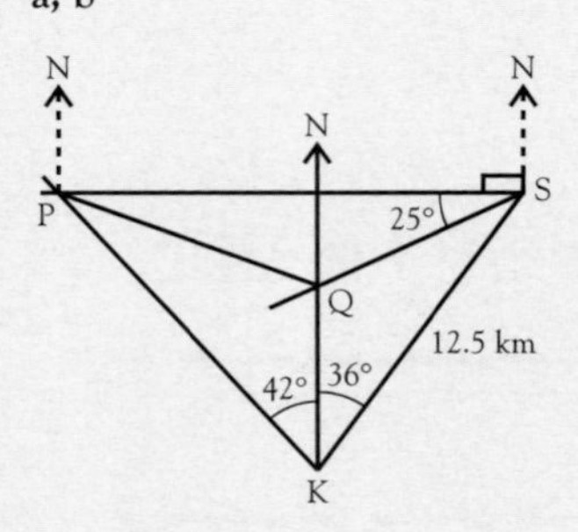

c 10 km, 110°
8 307°
9 106°
10 **a i** 242° **ii** 323° **b** 015°

Exercise 12.2 (p. 225)

1 **a, b, c**

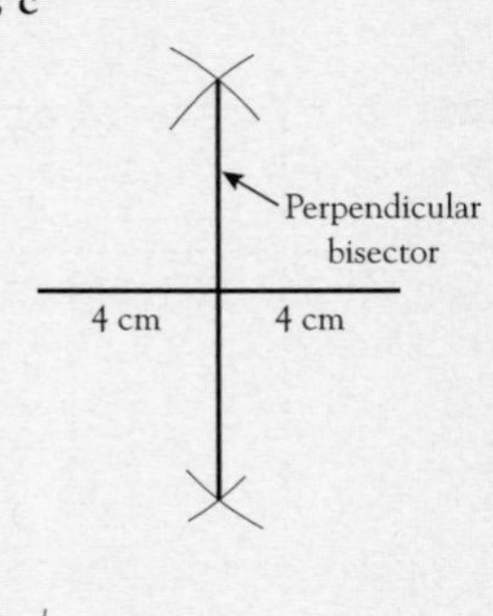

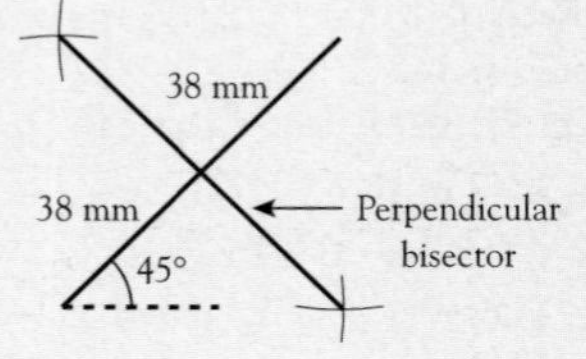

2 **a, b, c**

i

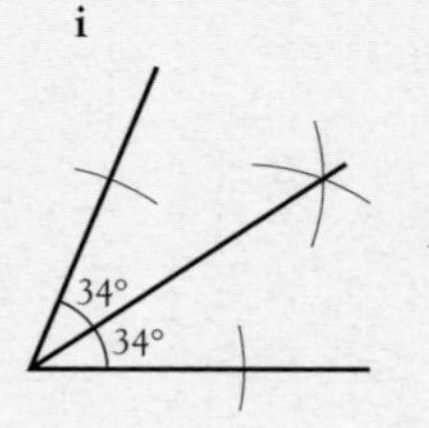

ii

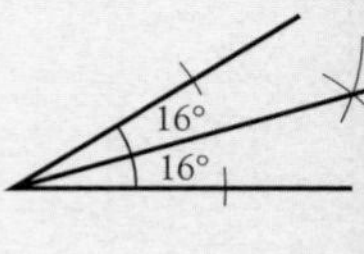

iii

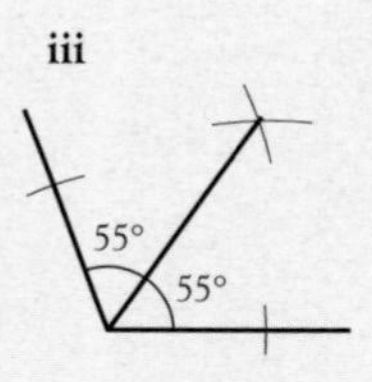

iv

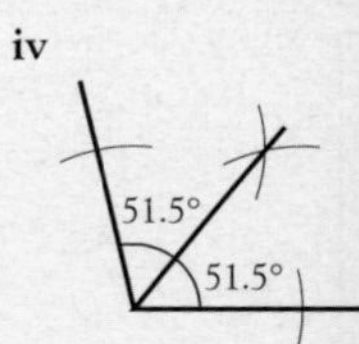

3 **a** **b**

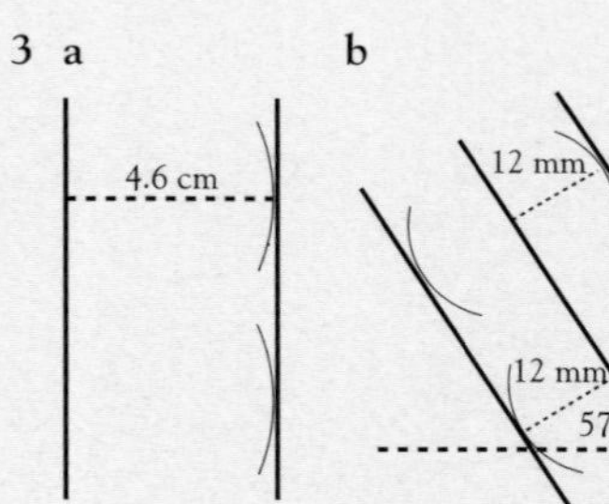

4 **a, c**

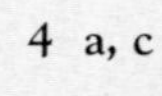

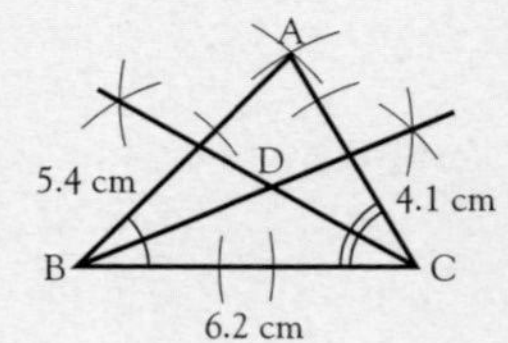

b $\hat{A} = 80°$
$\hat{B} = 41°$
$\hat{C} = 59°$
d BD = 3.9 cm
CD = 2.9 cm

5 **a, c** **b** 60°

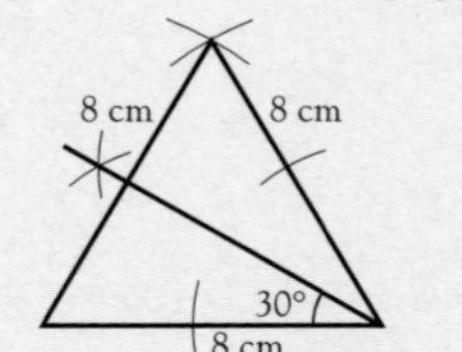

6 a, b

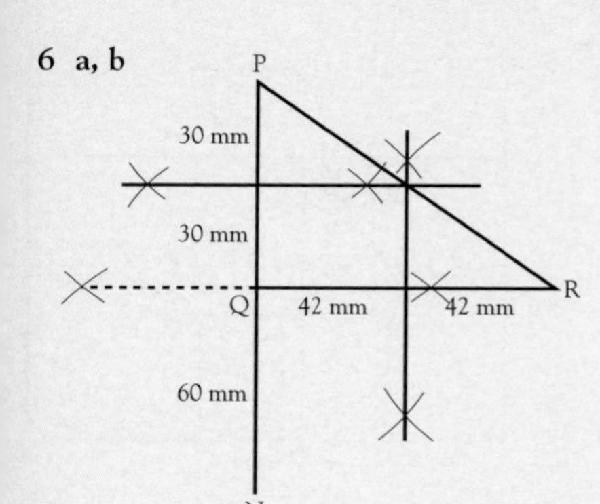

Perpendicular bisectors meet at the mid-point of PR.

7 a, b

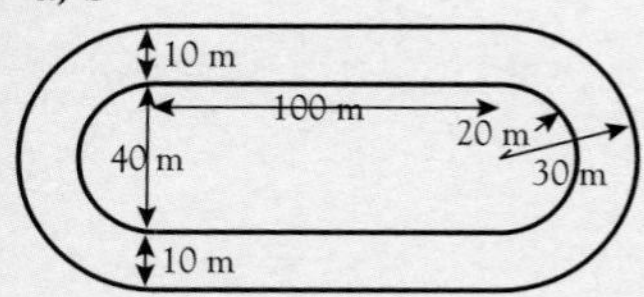

8

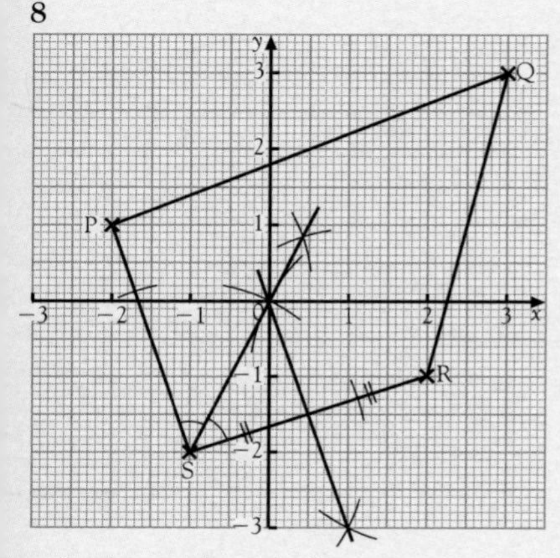

Both pass through the origin, (0, 0).

9 a, c

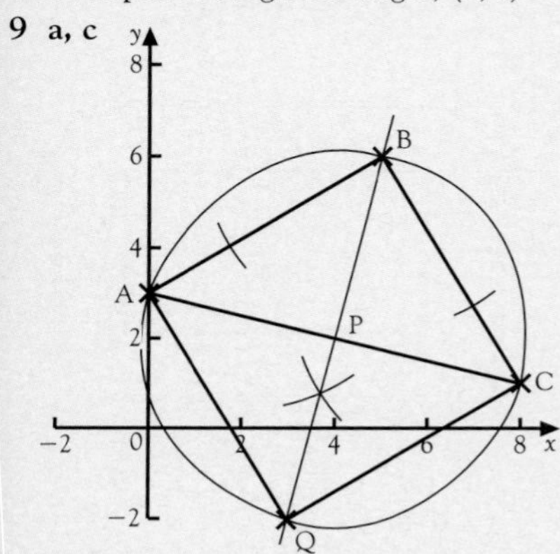

b P(4, 2) **d** Q(3, −2) **e** A square

Exercise 12.3 (p. 230)

1 Locus of the boat's position

Q

P

2

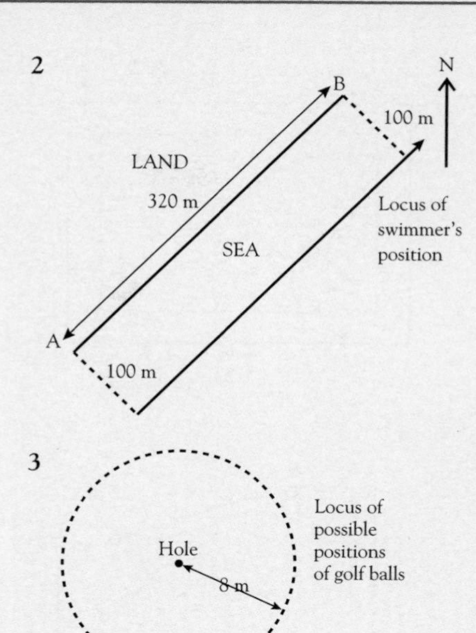

3

4 a

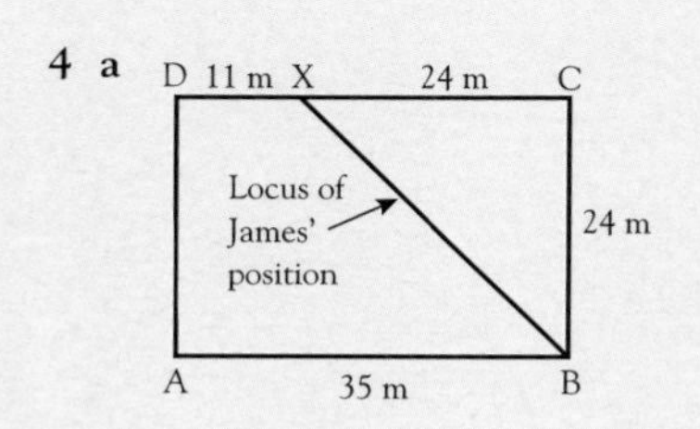

b 11 m

5 a

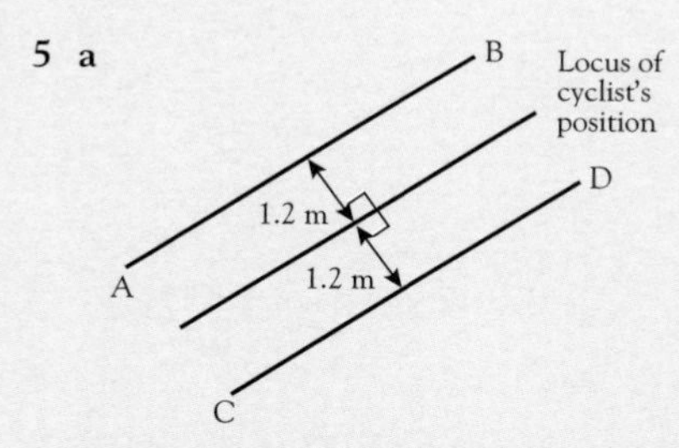

b

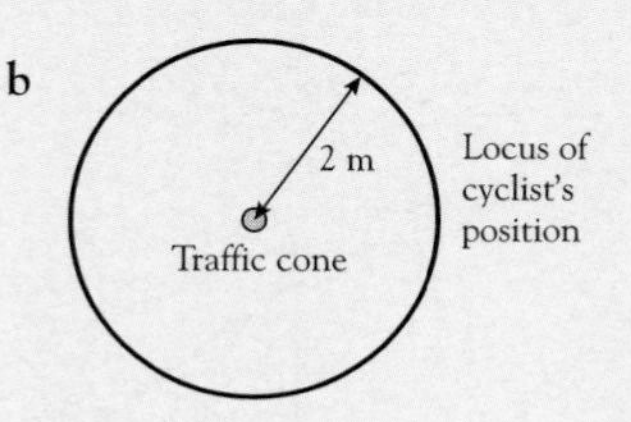

c

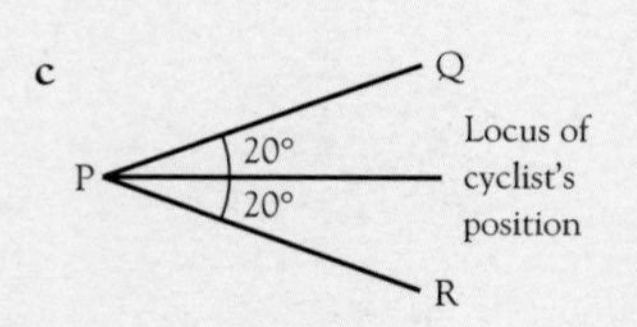

d

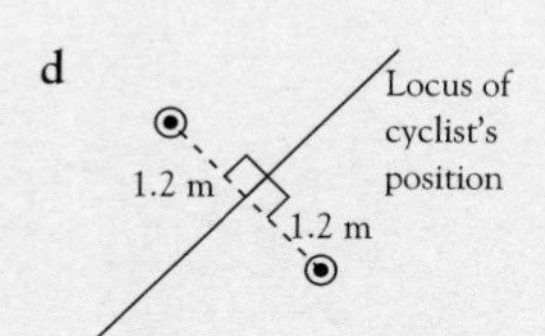

e

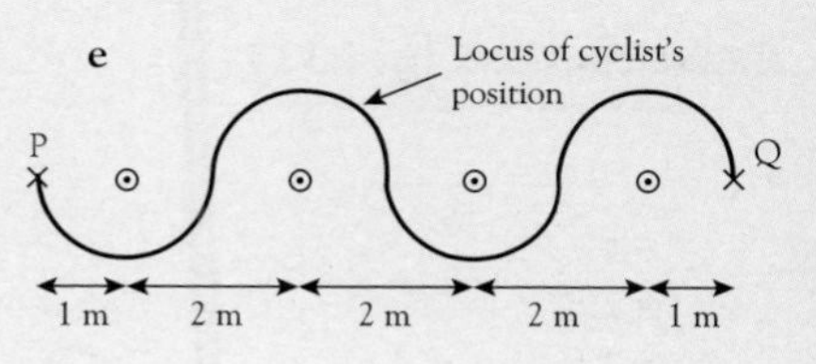

6 a, b, c

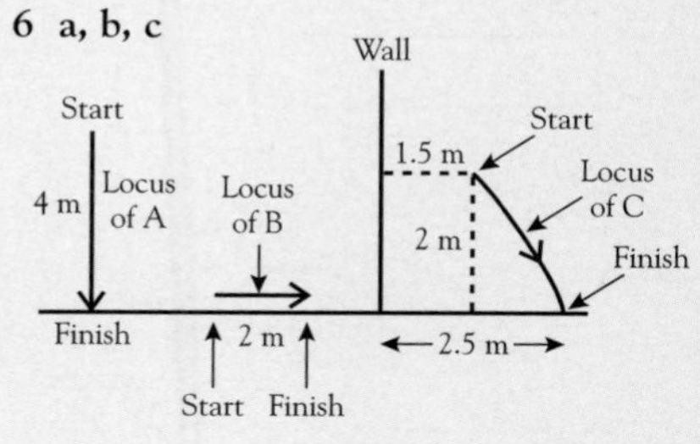

7 a

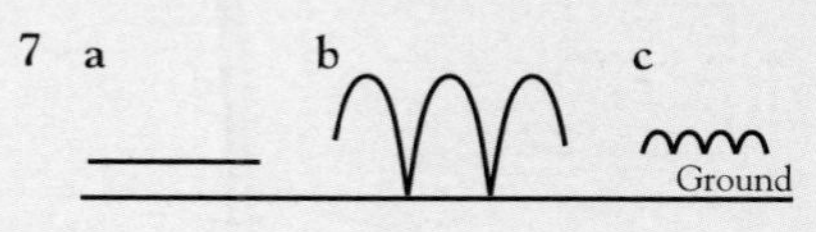

8 a, b

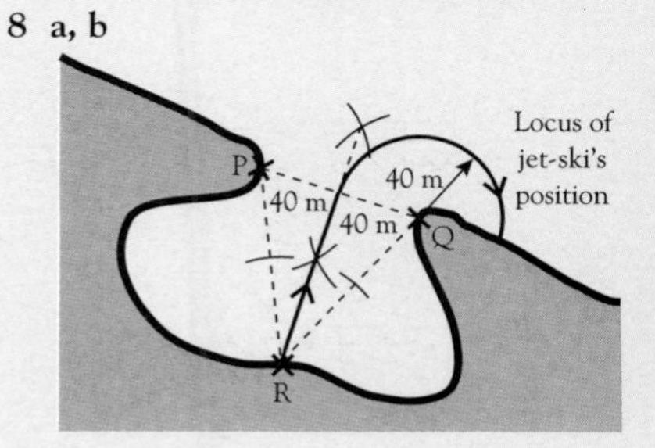

9 a 324°, 230°
d 130 m, 277°
b, c

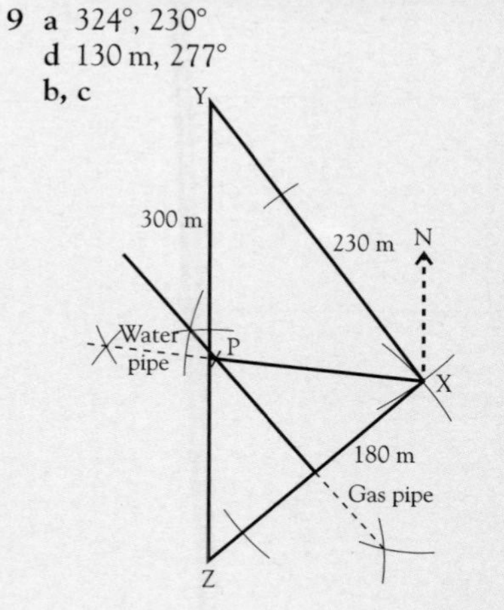

10 a 350 m, 062°
b 185 m

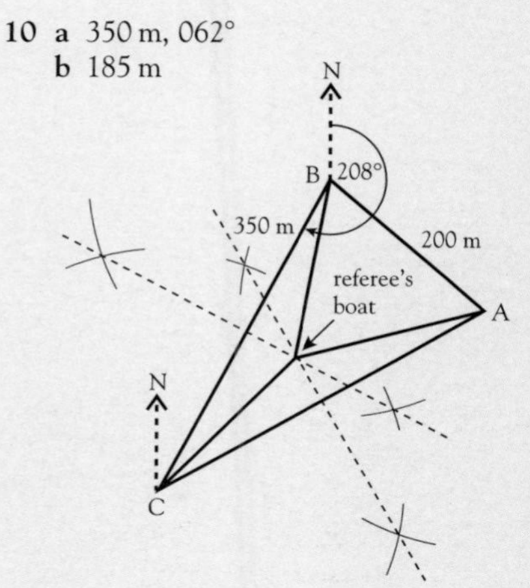

Exercise 12.4 (p. 233)

1

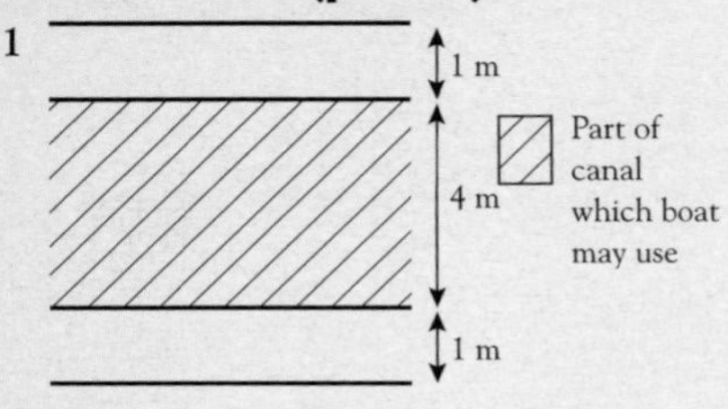

2

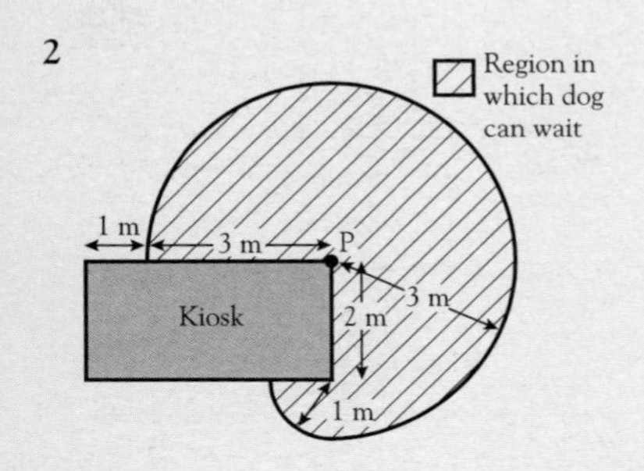

3

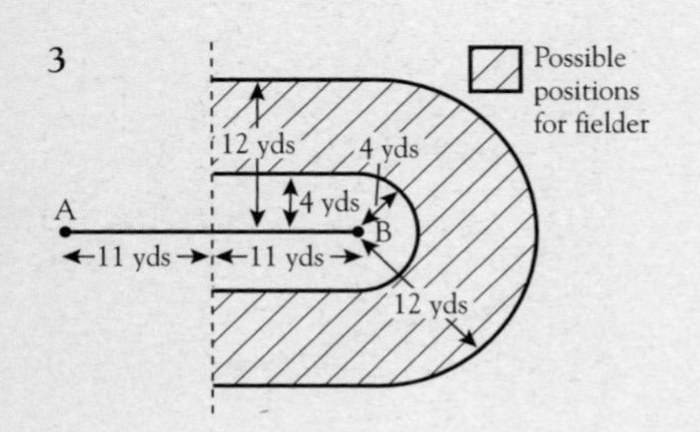

4

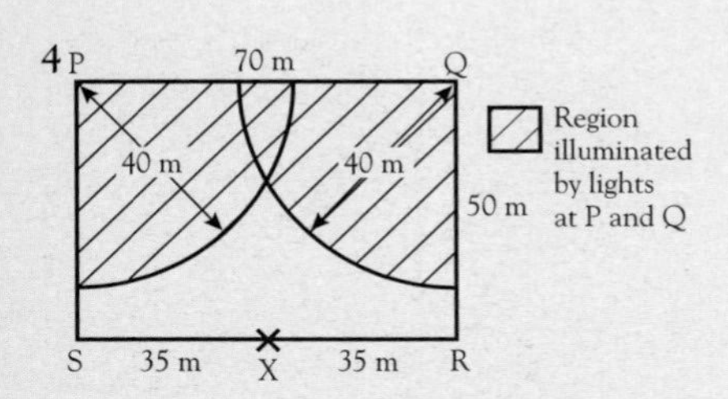

5

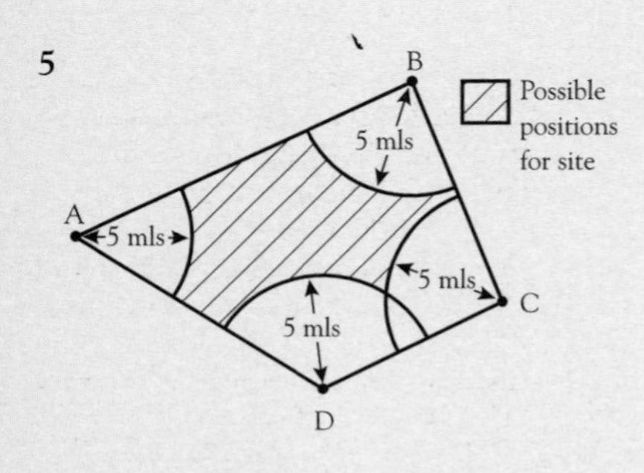

6

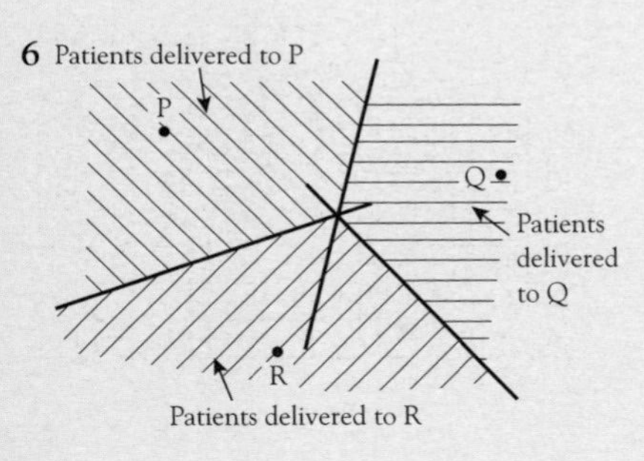

7

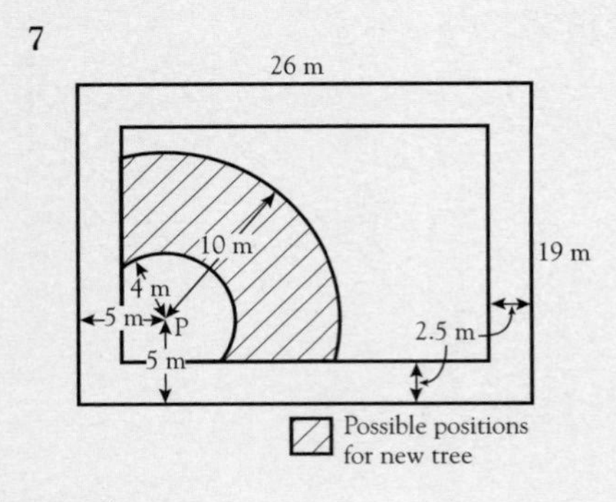

8 a, b

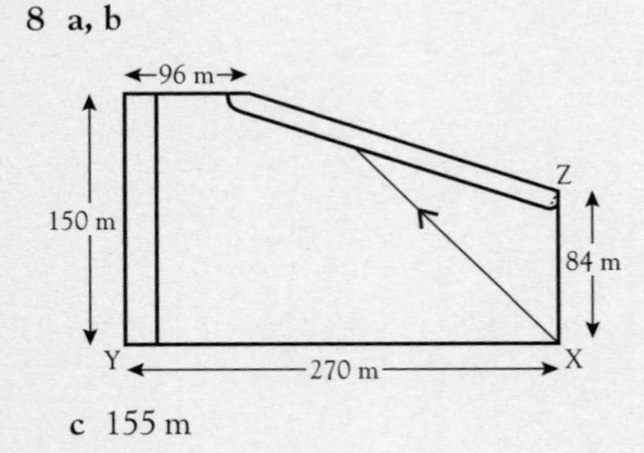

c 155 m

9 a i

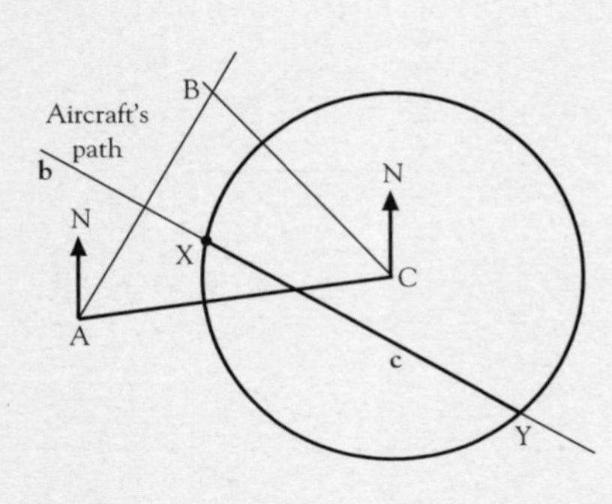

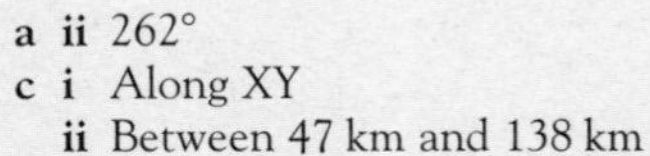
a ii 262°
c i Along XY
ii Between 47 km and 138 km

Exam Questions: Chp 12 (p. 237)

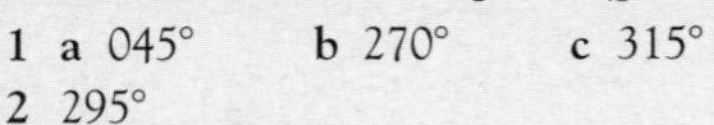
1 a 045° b 270° c 315°

2 295°

3 a 2050 metres

b

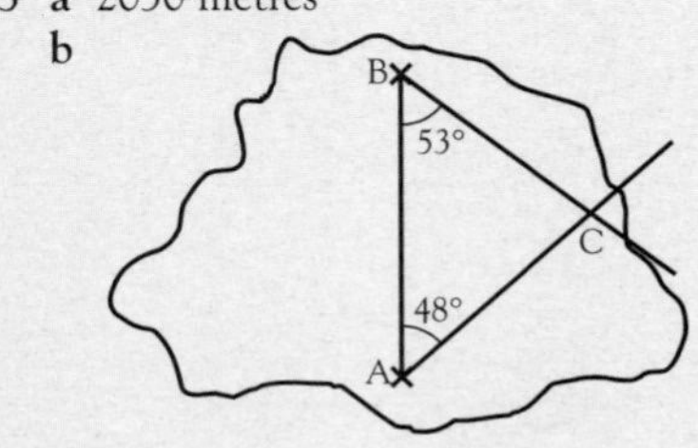

4

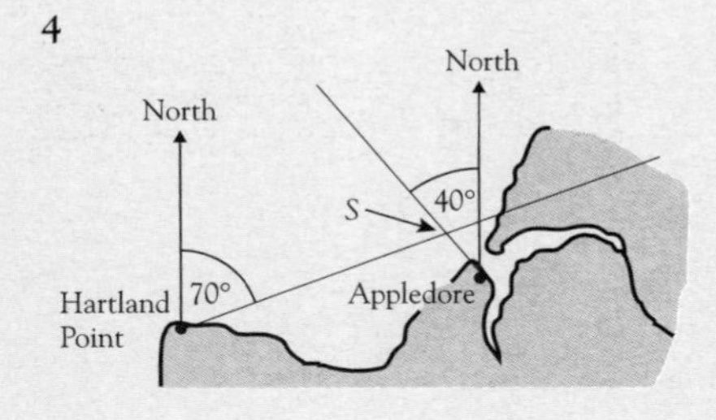

5

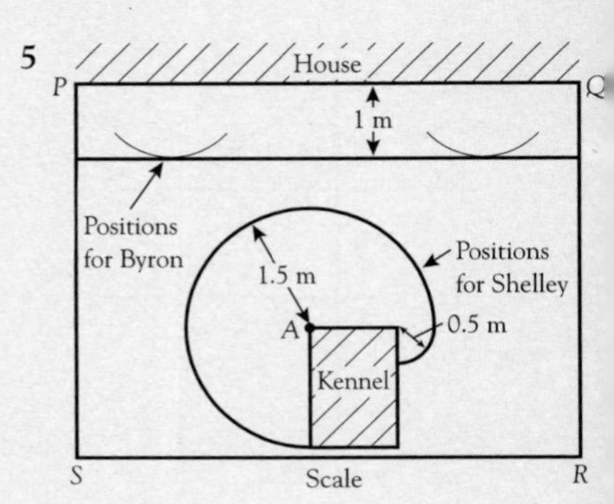

6

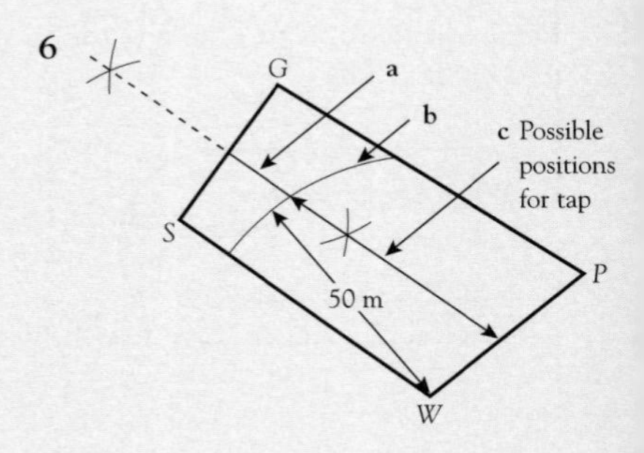

7 a, b i ii

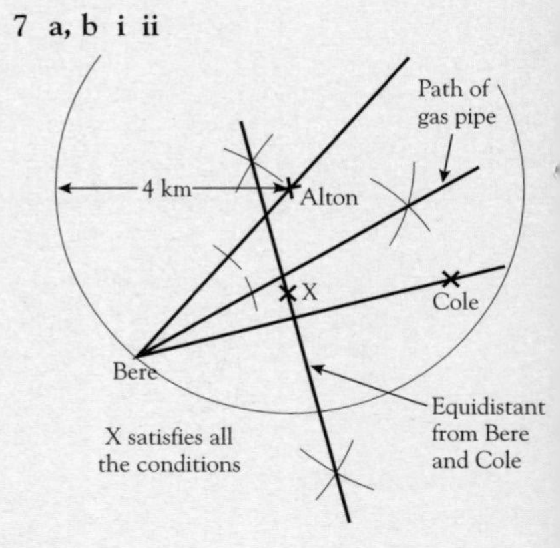

8

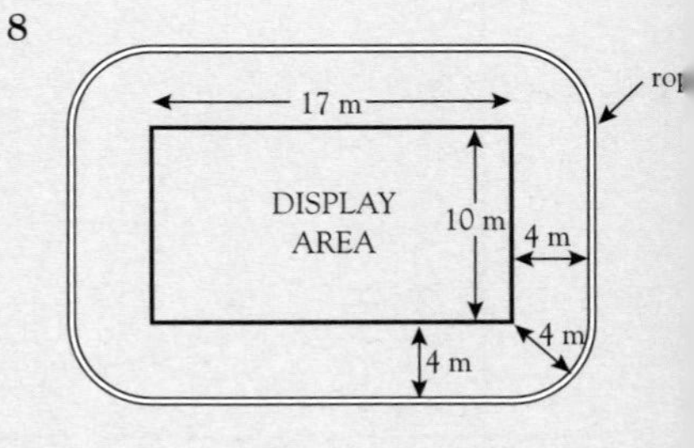

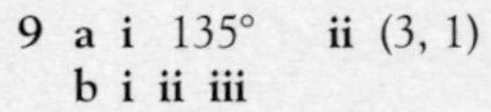
9 a i 135° ii (3, 1)

b i ii iii

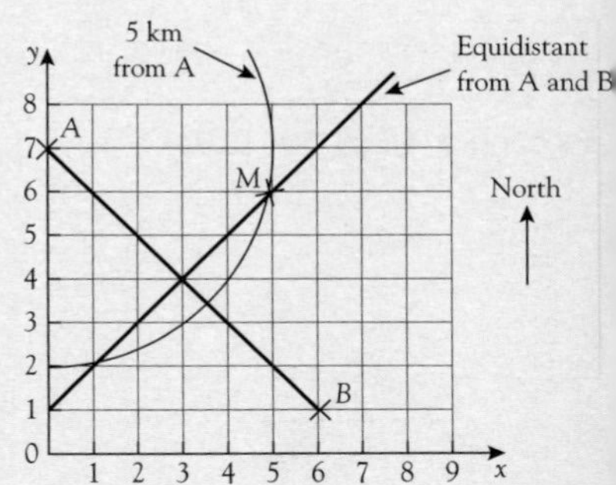

10

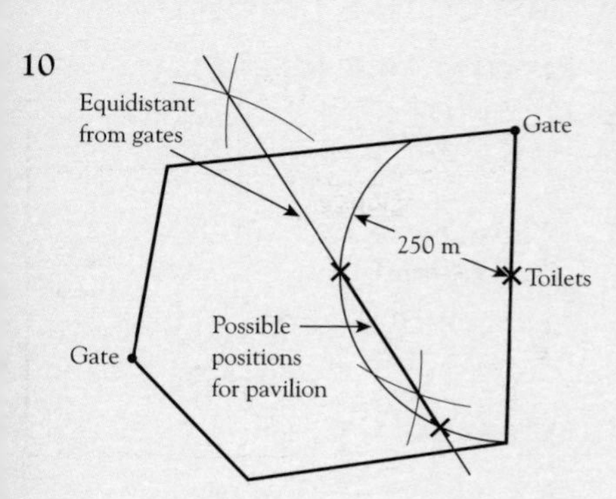

11

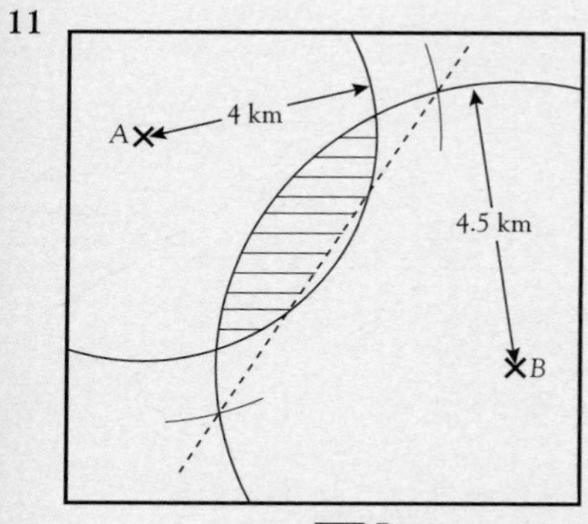

12 a, b

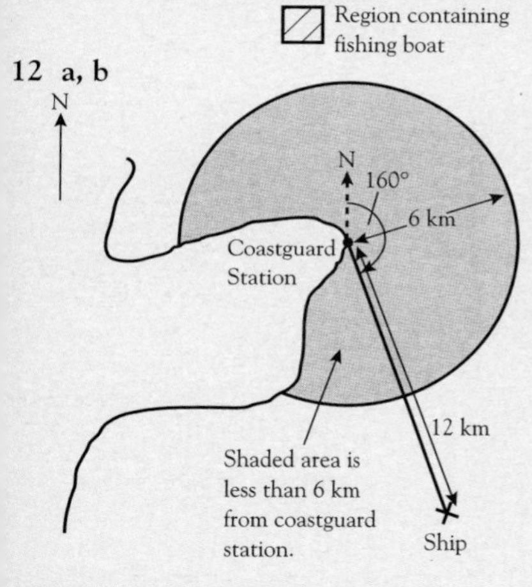

Exercise 13.1 (p. 247)

1 5 absences/class

2 a 16.2 b 20

3 5.25, 5, 4

4 7.3 hours (2 s.f.)

5 a i 8 ii 50 iii 0.84 iv 1.36
b higher, by 0.20 (2 s.f.)

6 a 75 plants b 24, 20, 10, 6
c 21 days (2 s.f.)

7 a

Amount of money, £	Mid-interval value, x	Frequency, f	$f \times x$
1–10	5.5	1	5.5
11–20	15.5	4	62
21–30	25.5	3	76.5
31–40	35.5	6	213
41–50	45.5	5	227.5
51–60	55.5	8	444
61–70	65.5	2	131
71–80	75.5	1	75.5
		$\Sigma f = 30$	$\Sigma fx = 1235$

b £41.17 (to nearest pence)

8 37.3 cm (3 s.f.)

9 a 82 vehicles
b 30.2 mph (3 s.f.)
c 43 vehicles
d 0.48 (2 s.f.)
e Yes; the probability that a vehicle will be travelling at 30 mph or over is almost one-half.

10 a 11.65 goals

b

Goals scored	Frequency f	Mid-interval value, x	$f \times x$
1–5	3	3	9
6–10	7	8	56
11–15	5	13	65
16–20	3	18	54
21–25	2	23	46
	$\Sigma f = 20$		$\Sigma fx = 230$

Mean = 11.5 goals

c Yes

Exercise 13.2 (p. 253)

1 a

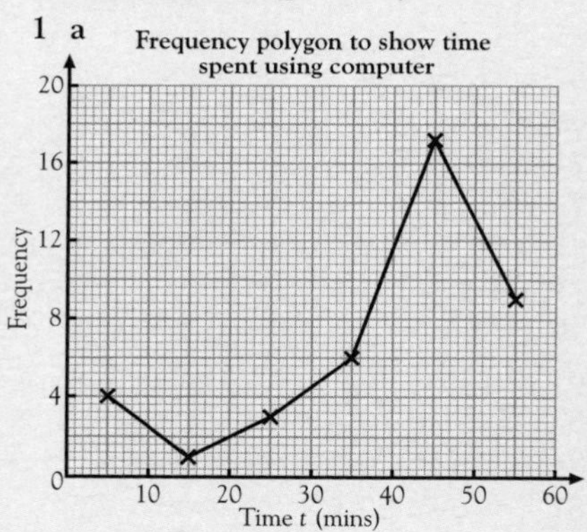

b 39.5 mins

2 a 11 videos

b

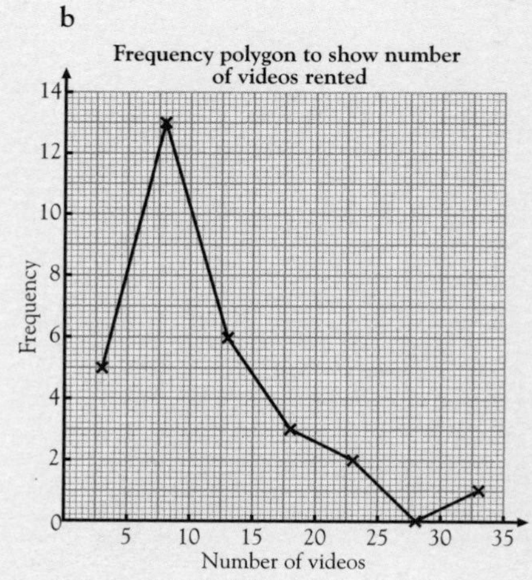

3 a Agent X: 9.42 hours
Agent Y: 4.06 hours

b

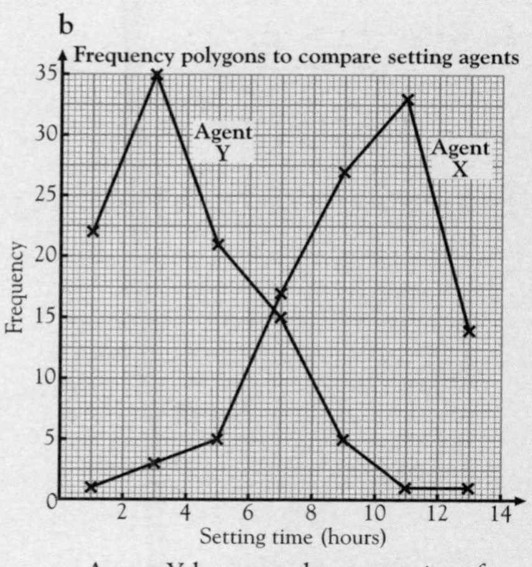

c Agent Y because the mean time for setting is much lower.

4 a

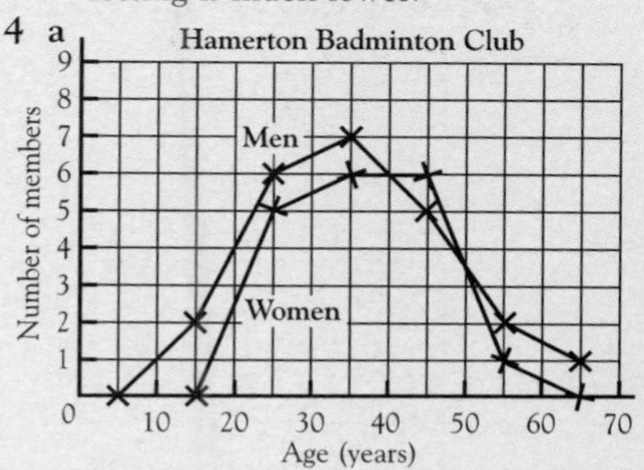

b 30–39 years c 35.9 years (3 s.f.)

5 a 9, 6, 4, 1, 0 b 27 fish
c 19.2 cm (3 s.f.)
d

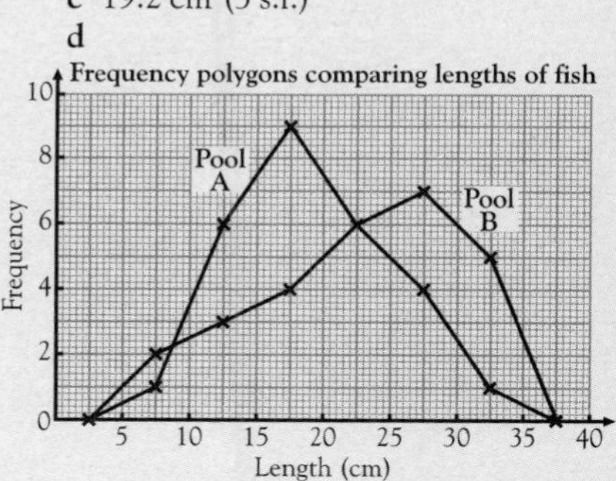

e Assuming that the farmer is rearing the fish for greatest length, he should use his usual fish food. There are a greater number of longer fish in Pool B.

f Yes; the mean for the usual brand is 3.5 cm more than for the new brand.

6 a

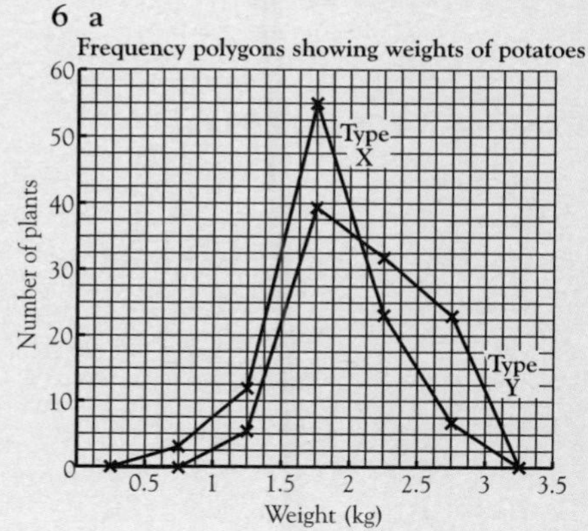

b Type Y
c i Type X
ii Range (X) = 2.5 kg
Range (Y) = 2 kg

Exam Questions: Chp 13 (p. 256)

1 £4670 (3 s.f.)
2 a 21, 7, 2
b Larger number of people for the longer viewing times and smaller number for the shorter, i.e. mean time spent viewing TV would be higher.
c 23.8 hours (3 s.f.)
3 a 13 biscuits b 6–10 biscuits
c 30 students
4 a i 0.16 ii 0.53
b

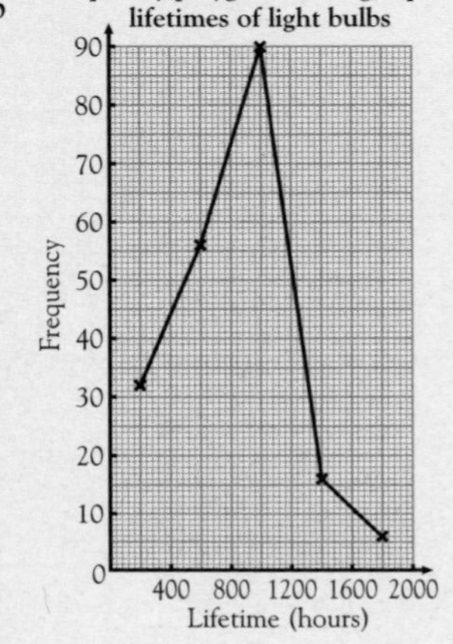

5 a $150 \leqslant p < 155$ b 156 litres (3 s.f.)
c The mean because it uses data from the whole herd (and because it is higher!).
6 73.1 kg (3 s.f.)
7 1.21 hours (3 s.f.) or 1 hour 13 minutes (to nearest min)
8 a 59.4 kg b 50 kg c girl
The total weight of the boys = 3700 kg
9 a £18.90 b 6.4 miles

Exercise 14.1 (p. 259)

1 12 4 5 7 4 10 pq
2 4 5 ac 8 x 11 x^2
3 $3a$ 6 2 9 $2x$ 12 $4abc$

Exercise 14.2 (p. 261)

1 $3(x+5)$ 2 $2(x+3)$
3 $4(x-3)$ 4 $5(a-7)$
5 $2(3x-2)$ 6 $2(7y+1)$
7 $4(2x-1)$ 8 $6(m-5)$
9 $7(a+3)$ 10 $4(2x-5)$
11 $5(3p+1)$ 12 $4(3x-2)$
13 $4(4-3a)$ 14 $3(3+4p)$
15 $3(a+b)$ 16 $2(5p-6q)$
17 $4(5x+3y)$ 18 $2(4b-3a)$
19 $2(3m-5n)$ 20 $6(3r+2s)$
21 $x(y-2)$ 22 $p(q+r)$
23 $b(a+4)$ 24 $a(5+x)$
25 $q(2p+3r)$ 26 $2s(t-2)$
27 $3x(2y-1)$ 28 $2p(2a+3b)$
29 $5x(2y+3z)$ 30 $7(2pq+rs)$
31 $x(x+2)$ 32 $3x(x-1)$
33 $a(a+4)$ 34 $y(y-18)$
35 $x(x+y)$ 36 $2x(x+2)$
37 $5p(p-2)$ 38 $2a(3a+4)$
39 $4y(3-y)$ 40 $3x(1-3x)$
41 $2x(x+3y)$ 42 $4x(2x-a)$
43 $2a(5a-2b)$ 44 $3q(5p+q)$
45 $2t(9s-t)$ 46 $y(x-8y)$
47 $q(7p+2q)$ 48 $pq(p+4)$
49 $y(3z+xy)$ 50 $ab(1-a)$
51 $ac(c+2)$ 52 $xy(x-y)$
53 $ab(ab+3)$ 54 $2pq(2p+q)$
55 $6xy(2y-x)$ 56 $3s(r^2+2st)$
57 $3y(2ay+b)$ 58 $st(5s-1)$
59 $2np(8+p)$ 60 $x^2(x+1)$
61 $x(x^2-7)$ 62 $a^2(a+8)$
63 $2p(2p^2+q)$ 64 $3x^2(2x-1)$
65 $a(3b-c+5d)$ 66 $x(x^2-2x+5)$
67 $p(p^2-p-1)$ 68 $b(b^2+b-a)$
69 70 70 130
71 720 72 116
73 65 74 2.9
75 3900 76 47 200
77 936 78 18
79 170 80 965
81 $P=2(L+W)$ 82 $I=m(v-u)$
83 $L=l(1+\alpha t)$ 84 $s=\frac{1}{2}t(u+v)$
85 $A=\pi(R^2-r^2)$ 86 $A=\frac{1}{2}h(a+b)$
87 $L=2(\pi r+100)$ $A=r(\pi r+200)$
88 $S=2\pi r(r+h)$

Exercise 14.3 (p. 264)

1 a 13 b 29
c 11 d 8
2 66
3 a 6 b −4 c 0
4 a 310 b 316
5 70
6 a 8 b 6
7 10
8 a $1\frac{1}{5}$ b 3
c 12
9 10
10 a 49 b 97
11 a 1 hour 50 mins
b 1 hour 28 mins
12 a 26 b 24.91
c −3.04
13 73.6 cm³ (3 s.f.)
14 A = 90.6 cm² (3 s.f.)
15 a 3320 b $43\frac{7}{8}$

Exercise 14.4 (p. 268)

1 a i 176 ii 260
b i 2 ii 12
2 a i 21 ii −1
b i 5 ii 17
3 a 7.5 litres b 4 coats
4 a $w=\frac{t-8}{9}$
b i 3 km ii $4\frac{1}{9}$ km
5 a i £90 ii £65
b $t=\frac{C-50}{20}$
c 3 hours
6 a $K=4020$ b $P=165$
7 a $x=\frac{y+5}{4}$ g $x=\frac{12}{y}$
b $x=4y$ h $x=\pm\sqrt{y^2+4}$
c $x=\frac{y}{3}-a$ i $x=\frac{y}{ab}$
d $x=\pm\frac{\sqrt{y}}{2}$ j $x=\pm\sqrt{2(y-10)}$
e $x=\frac{16-y}{4}$ k $x=\frac{4y}{3}$
f $x=\frac{3(8-y)}{8}$ l $x=\frac{2y^2}{5}$
8 a $P=48$ b $B=14$ c $J=3$
9 a 35 litres b 400 miles
10 a $2x(x+2h)$ b $h=\frac{S-2x^2}{4x}$
11 a i £2.60 ii 68 pence
b $n=\frac{500-p}{24}$
c 8
12 a $T=1440$ b $G=5T+A$
c $G=14\,550$
13 a i $C=15$ ii $C=100$
iii $C=-10$
b i $F=138.2$ ii $F=203$
iii $F=-112$

Exercise 14.5 (p. 273)

1 a i linear ii 27, 30; 69
b i linear ii 10, 0; −130
c i not linear
d i linear ii 3.5, 4; 10.5
e i not linear
f i linear ii 15, 19; 71
g i not linear
h i linear ii 2.4, 2.6; 5.2
i i not linear
j i linear ii 550, 500; −150
2 a 22, 27, $5n-3$
b 19, 23, $4n-1$
c 10, 7, $25-3n$
d −17, −24, $18-7n$
e 56, 65, $9n+11$
f −17, −21, $3-4n$
g 18.5, 22, $3.5n+1$
h 2, $2\frac{1}{4}$, $\frac{1}{4}n+\frac{3}{4}$
i 6.2, 6.0, $-0.2n+7.2$
(or $7.2-0.2n$)

3 8, 11, 14, 17

4 118, 116, 114; 100

5 a i 2, 4, 6, 8, 10
ii $2n$
b $2n - 1$

6 a nth term $= 4n - 3$; 77, 197
b n^{th} term $= 3n + 5$; 65, 155
c n^{th} term $= 19 - 2n$, -21, -81
d n^{th} term $= 0.4n + 0.6$; 8.6, 20.6
e n^{th} term $= 1.5n - 7$; 23, 68
f n^{th} term $= \frac{1}{4}n - 5\frac{3}{4}$; $-\frac{3}{4}$, $6\frac{3}{4}$

7 a $u_n = 4n - 3$ b 22nd

8 a $u_n = 8 - 3n$ b -142 c 15th

9 a $\frac{5}{11}, \frac{6}{13}$ b $2n + 1$ c $u_n = \dfrac{n}{2n+1}$

Exercise 14.6 (p. 276)

1 a 38 edges b $u_n = 4n + 6$

2 a $g = 16 - 2n$ b 2 bushes
c i 16 bushes ii 8 trees

3 a

Pattern	1	2	3	4	5
Matchsticks	6	11	16	21	26

b 51 matchsticks c $u_n = 5n + 1$
d Pattern 72

4 a $C = 0.8n + 1.6$
b i £8.80 ii £4.40 iii £6.16
c i 3.25 miles ii 6.75 miles

5 a $T = 8n + 17$
b i 85°C ii $10\frac{3}{8}$ mins

6 a $C = 4n + 5$
b i £45 ii £61

Exam Questions: Chp 14 (p. 278)

1 a i 16 ii $u_n = 3n - 2$
b i 5.216 ii $x = 9$

2 a 10, 13, 16
b i 22, 26, 30 ii $u_n = 4n + 2$

3 $A = 28.8$ (1 d.p.)

4 a $x(x + 3)$ b $6x - 5y$
c $C = \dfrac{F - 30}{2}$

5 a £1.28 b 2.5 cm
c $S = \sqrt{\dfrac{P - 20}{12}}$

6 a £4.95
b i $W = \dfrac{T - 20}{40}$ ii 10 kg

7 a i

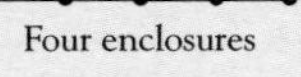

Four enclosures

ii

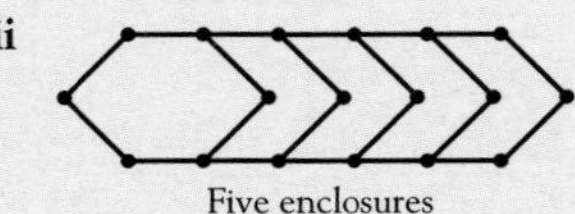

Five enclosures

b

No. of enclosures	1	2	3	4	5	6	7	8
No. of posts	6	9	12	15	18	21	24	27

c 63 posts d $u_n = 3n + 3$

8 a $f = -32.3$ (3 s.f.)
b $f \approx \dfrac{15 \times -10}{15 - 10} = -30$; reasonable

9 a $3pq(4p - 5q)$ b $n = \dfrac{C - 120}{40}$

10 $2x(2x - 3)$

11 a $h = 150$ cm
b $f = 24$ cm
c $f = -6\frac{2}{3}$ cm; impossible for a length of forearm to be negative
d $f = \dfrac{h - 90}{3}$

12 a $w = 2b + 1$ b 45 black triangles

13 a $T = -28$°C b $h = 39\,000$ feet

14 a i $8x^2 + 28x$ cm^2
ii $12x + 14$ cm
b i $P = 12$ cm ii $y = \dfrac{P + 9}{6}$
iii $3(2y - 3)$ iv $2y - 3$ cm

15 a $v = 51$ b $t = \dfrac{v - u}{g}$
c $t = 7.5$

16 a $P = -54.18$ b $P = -15\,221.25$

Exercise 15.1 (p. 283)

1 a 18 cm^2
b 36.25 cm^2
c 960 cm^2
d 16 cm^2
e 21.28 m^2
f 34.54 cm^2

2 a

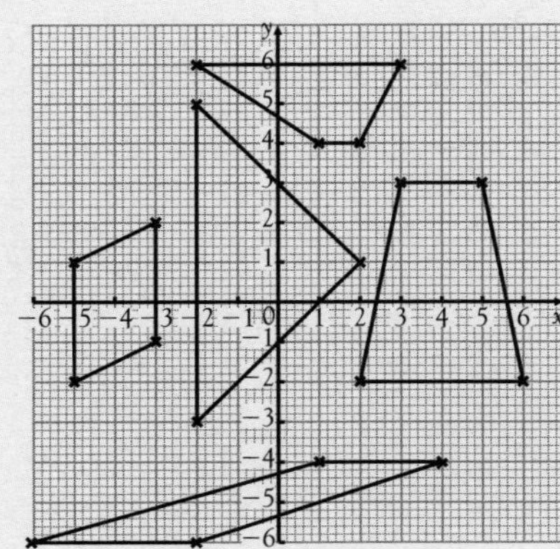

b isosceles triangle, isosceles trapezium, parallelogram, trapezium, trapezium
c 16 cm^2, 15 cm^2, 6 cm^2, 6 cm^2, 7 cm^2

3 a 4 b 54 cm^2

4 a 4 d $r = 20$ cm (2 s.f.)
b $x = 8$ cm e $y = 3$ cm
c $a = 9$ cm

5 a 388 cm^2 b 1.64 m^2 (3 s.f.)

6 52 m^2

7 a 7.1 cm (2 s.f.) b 9.0 cm (2 s.f.)

Exercise 15.2 (p. 289)

1 a 90 cm^3 b 1380 cm^3

2 a 255 cm^3 d 6.93 m^3
b 360 cm^3 e 103.43 cm^3
c 3532.5 cm^3

3 126 cm^3

4 Block B

5 a 616 mm^2
b 314 mm^2
c Silver: 0.785 m^3
Outer metal: 0.754 m^3 (all to 3 s.f.)

6 27 800 cm^3 (3 s.f.)

7 a 2.6 m^2 b 3.9 m^3
c 17.4 tonnes (3 s.f.)

8 b Yes

9 5 glasses

10 a 4 b 2 c 3

11 6 ft

12 3 m

Exercise 15.3 (p. 292)

1 a

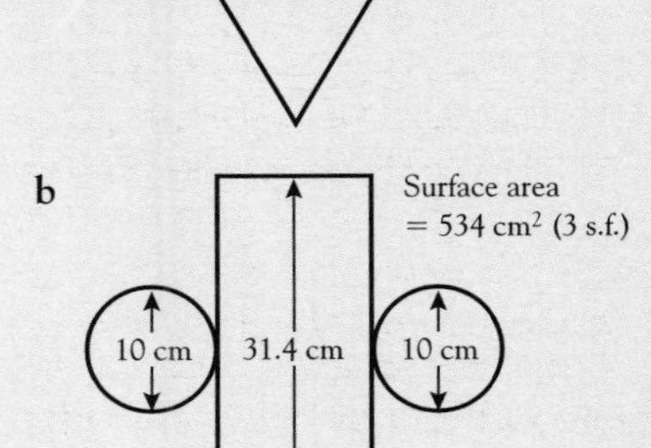

b

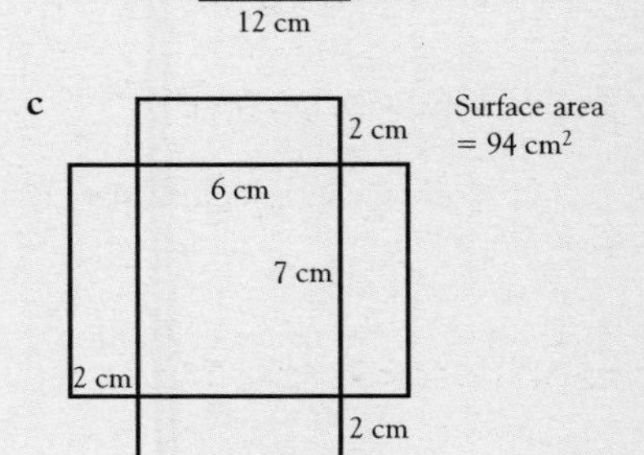

c Surface area = 94 cm^2

d

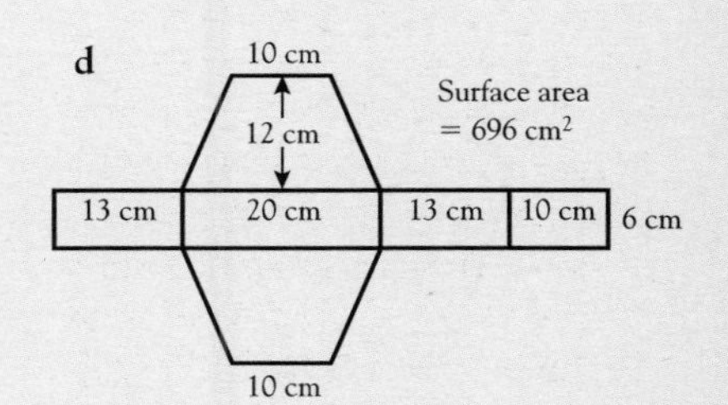

2 a $54\ \text{cm}^2$, $150\ \text{cm}^2$, $294\ \text{cm}^2$
b 9 : 25 : 49
c From top downwards : $45\ \text{cm}^2$, $116\ \text{cm}^2$, $220\ \text{cm}^2$

3 a $h = 15$ cm
b cylinder with radius of 10 cm

4 10 cm

5 $x = 25$ cm

Exam Questions: Chp 15 (p. 295)

1 $321\ \text{cm}^3$

2 a $0.2\ \text{m}^2$ b $0.7\ \text{m}^3$

3 a $60\ \text{cm}^2$ b $480\ \text{cm}^3$

4 a $39.6\ \text{cm}^2$ (3 s.f.) b $13.7\ \text{cm}^2$ (3 s.f.)

5 a $113\ \text{cm}^2$ (3 s.f.) b 17.7 cm (1 d.p.)

6 a $491\ \text{cm}^2$ (3 s.f.)
b 10 mm (to nearest mm)

7 $1530\ \text{cm}^3$ (3 s.f.)

8 a i $5024\ \text{cm}^2$ ii $753\,600\ \text{cm}^3$
b i $2100\ \text{cm}^2$ ii $147\,000\ \text{cm}^3$
c 6 loads

9 a $2640\ \text{cm}^3$ b 5.9 cm (2 s.f.)

10 a $36.9\ \text{m}^2$ b $295.2\ \text{m}^3$
c 32 cm

11 a i $1.35\ \text{m}^2$ ii $1.62\ \text{m}^3$
b 4 skips

Exercise 16.1 (p. 299)

1 i In each diagram X marks the centre of rotation.

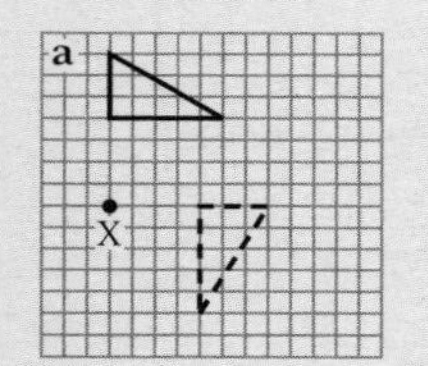

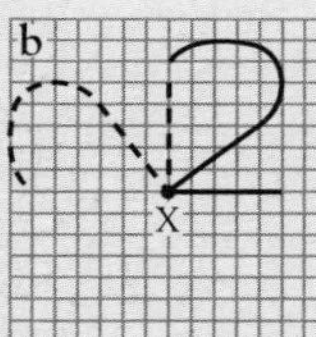

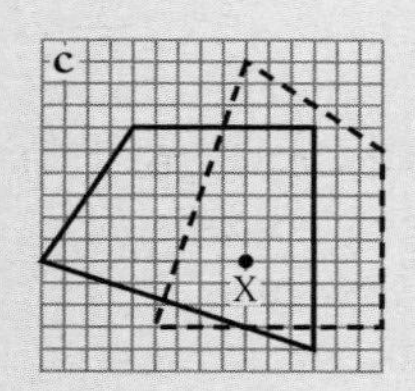

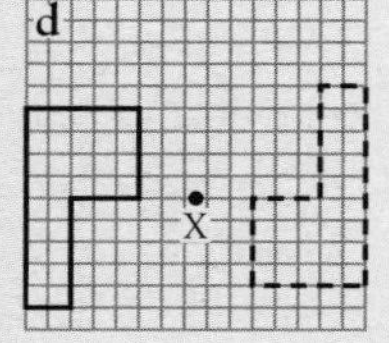

ii a clockwise, 90°
b anticlockwise, 90°
c clockwise, 90°
d clockwise or anticlockwise, 180°

2 a O
b 30°, clockwise

3 a X marks the centre of rotation.

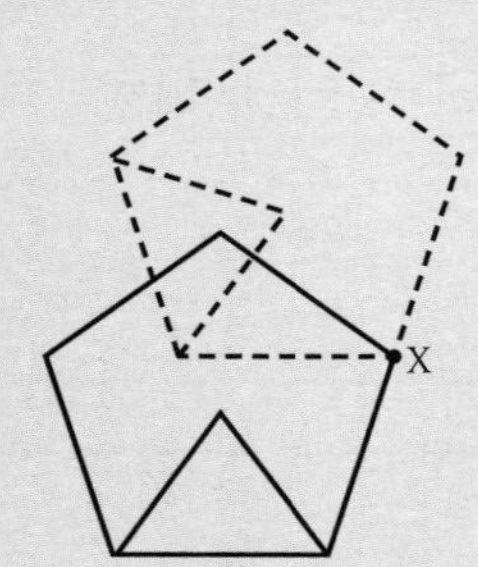

b 72° c clockwise

4 a, b

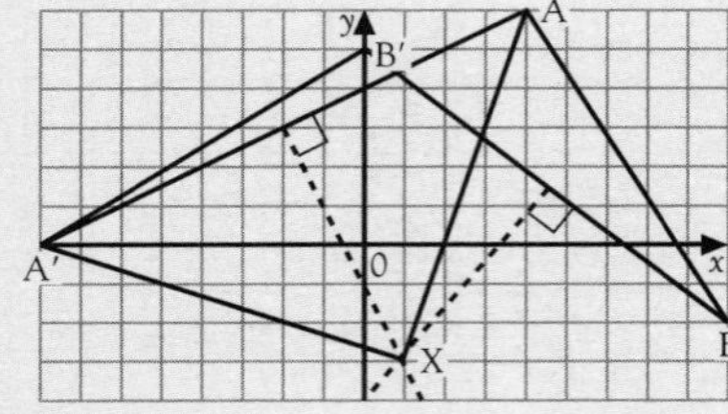

c $(1, -3)$
d 90°, anticlockwise

5 a

b 100°, clockwise

6 a

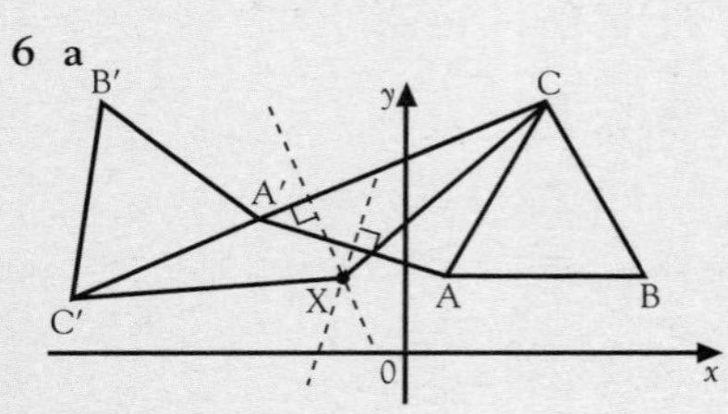

b ≈150°, anticlockwise

Exercise 16.2 (p. 304)

1 a Rotation through 180° and reflection in a vertical mirror line.
b

2 a, b, c

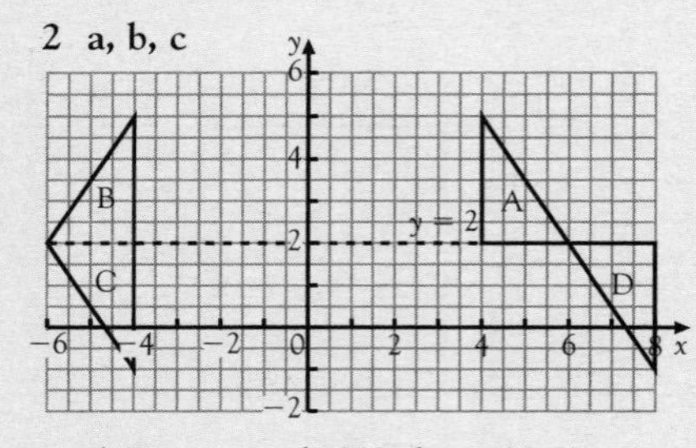

d Rotation of 180° about (6, 2).

3 a, b

c Rotation of 180° about $(2, -1)$.

4 a

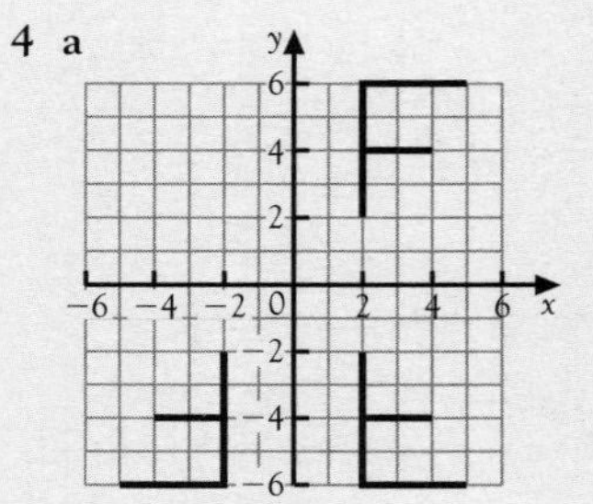

b Rotation through 180° about the origin (0, 0).

5 a, b

c Reflection in the y axis.

6 a, b

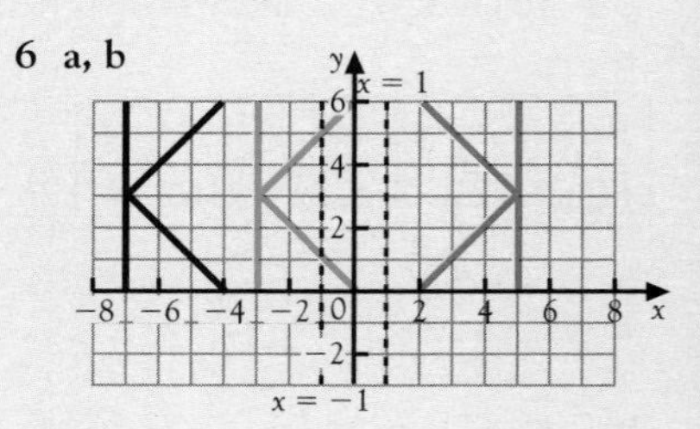

c Translation $\begin{pmatrix} 4 \\ 0 \end{pmatrix}$

7 a, b

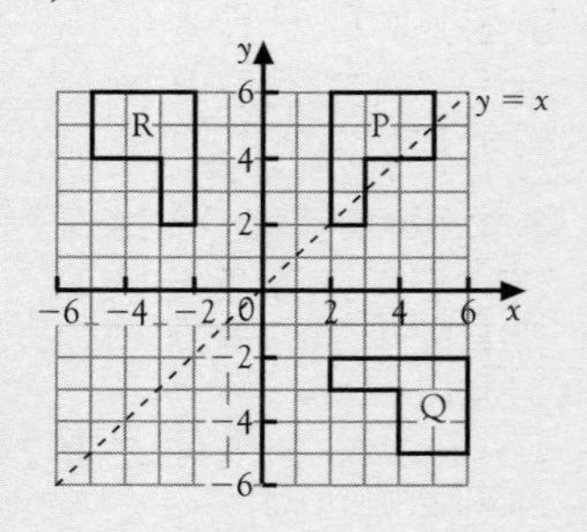

c Reflection in the y axis.

8 a, b, c

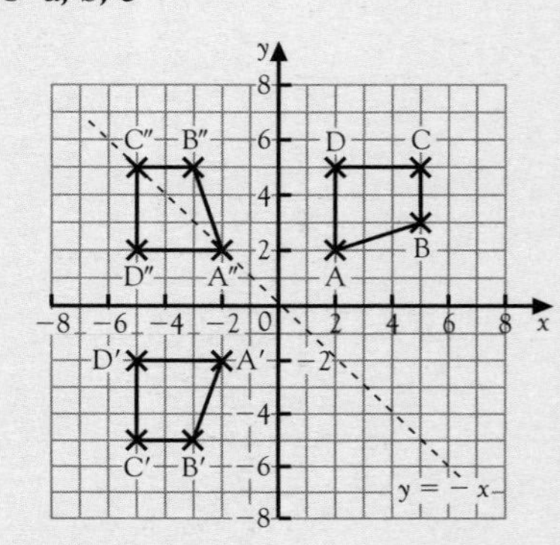

d Rotation about the origin, (0, 0), through 90° anticlockwise.

9 a, b

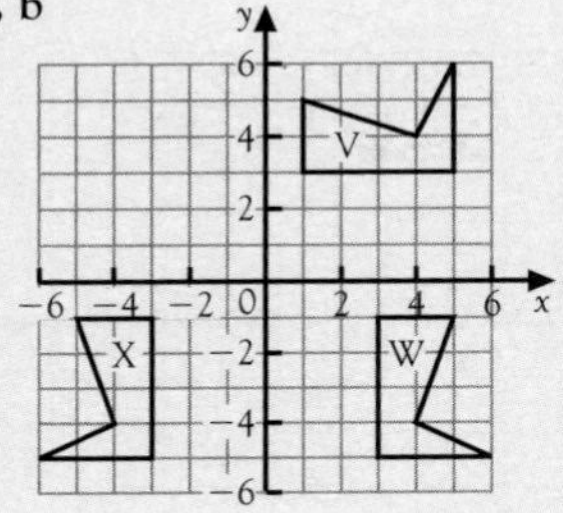

c Reflection in the line $y = -x$.

Exam Questions: Chp 16 (p. 307)

1 Reflection in the y axis.

2 Rotation about the origin, (0, 0), through 90° clockwise.

3 Rotation about the origin, (0, 0), through 90° clockwise.

4 Reflection in the line $y = -x$.

5 Rotation about the origin, (0, 0), through 90° clockwise.

6 a, b

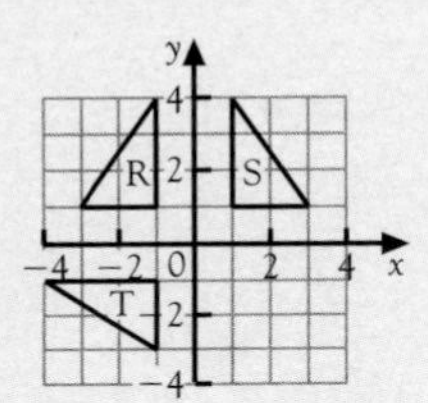

c Reflection in the line $y = -x$.

7 a, b

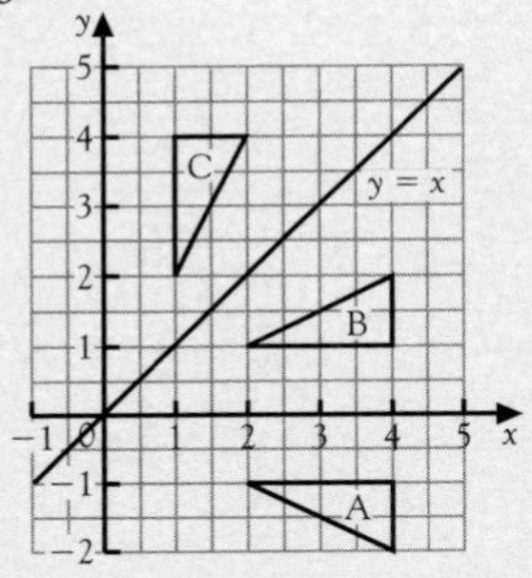

c Rotation about the origin, (0, 0), through 90° anticlockwise.

8 a Reflection in the x axis.
b Rotation about the origin, (0, 0), through 90° anticlockwise.
c Enlargement, scale factor 3, centre of enlargement (5, 4).

9 a, c

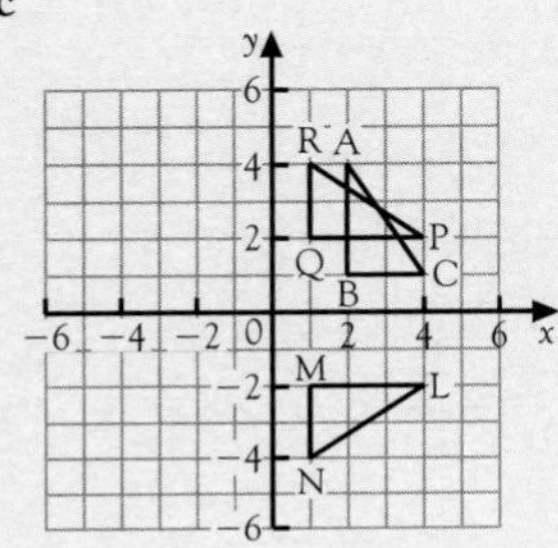

b (4, −2)
d Reflection in the line $y = x$.

10 a, b, c i

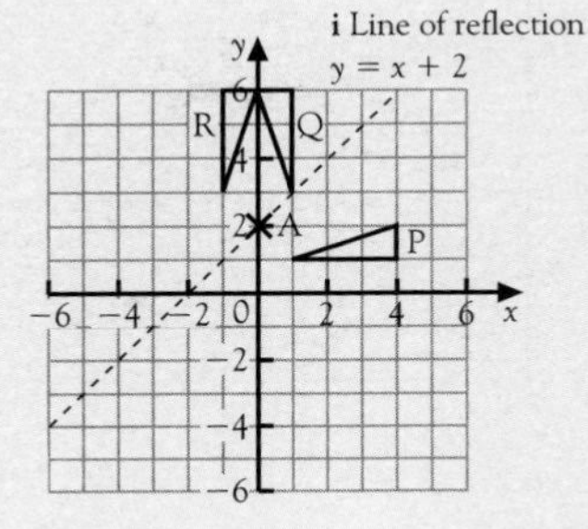

c ii $y = x + 2$

Exercise 17.1 (p. 311)

1 a $a = 4, b^2 = 9, c^2 = 25, c = 5$
b $a = 7, b^2 = 576, b = 24, c = 25$
c $a = 12, a^2 = 144, b^2 = 25, c^2 = 169$
d $a^2 = 81, a = 9, b^2 = 1600, c = 41$

2 b, d, e and f

Exercise 17.2 (p. 314)

1 a 3.61 m (3 s.f.)
b 68.4 cm (3 s.f.)
c 10.2 km (3 s.f.)

2 3.92 m (2 d.p.)

3 2.91 m (3 s.f.)

4 26.6 cm (to nearest mm)

5 a

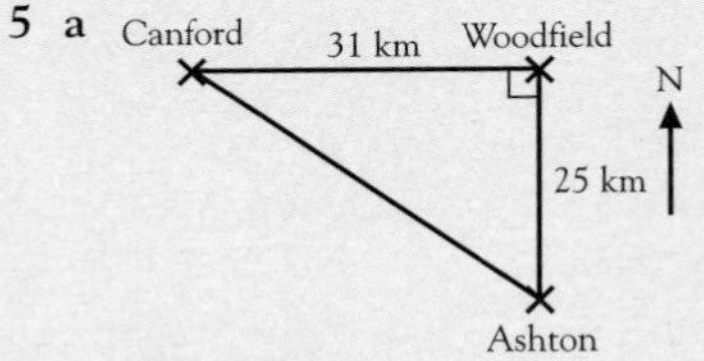

b 39.8 km (3 s.f.)

6 a 6 cm² b 5 cm c 132 cm²

7 a 3.16 cm (3 s.f.)
b 4.12 cm (3 s.f.)
c 14.6 cm (3 s.f.)

8 18.4 km (3 s.f.)

9 14.1 cm (3 s.f.)

10 8.06 ft (3 s.f.)

Exercise 17.3 (p. 318)

1 a 5.26 cm (3 s.f.)
b 96.3 mm (3 s.f.)
c 63.6 m (3 s.f.)
d 7.68 cm (3 s.f.)

2 a 9 cm
b 864 cm³

3 a 68.9 mm (3 s.f.)
b 3310 mm² (3 s.f.)

4 a 7.36 m (2 d.p.)
b 0.86 m (2 d.p.)

5 a 11.3 cm (3 s.f.)
b 190 cm³ (2 s.f.)

6 6.93 cm (3 s.f.)

7 3.61 cm (3 s.f.); 10 cm; 28.6 cm (3 s.f.)

Exercise 17.4 (p. 322)

1 18 m (2 s.f.)

2 a

a	b	c
3	4	5
5	12	13
7	24	25
9	40	41
11	60	61
13	84	85
15	112	113

b odd numbers c $c = b + 1$

3 31 km

4 a 35 cm (2 s.f.)
b 130 cm (2 s.f.)
c 320 cm² (2 s.f.)

5 a 2.2 m (1 d.p.)
b 22 m² (nearest m²)

6 a i 6.4 cm (2 s.f.)
ii 250 cm² (2 s.f.)
b i 67 mm (2 s.f.)
ii 6300 mm² (2 s.f.)

7 3.61 m (3 s.f.)

8 215 m

9 24 m (2 s.f.)

10 7.9 m (2 s.f.)

11 a 8 cm
b 24 cm²

12 a 52 mm (2 s.f.)
b 3700 m² (2 s.f.)
c 640 000 mm³ (2 s.f.)

13 a 7.7 cm (2 s.f.)
b 10 cm (2 s.f.)
c 46 cm² (2 s.f.)

14 a 9.1 cm (2 s.f.)
b 6.8 cm (2 s.f.)
c 130 cm² (2 s.f.)

Exam Questions: Chp 17 (p. 325)

1 3.0 m (2 s.f.)

2 a 3 cm b 5.2 cm (2 s.f.)
c 80 cm² (2 s.f.)

3 a 1.37 m (3 s.f.) b 6.08 m³ (3 s.f.)

4 3.20 m (3 s.f.)

5 a 416 m²
b i 11.2 m (3 s.f.) ii 112 m² (3 s.f.)
c i 4 m ii 14.5 m (3 s.f.)
iii 405 m² (3 s.f.)
d 1450 m² (3 s.f.)

Exercise 18.1 (p. 329)

1 b, e, f, h, i positive; a, d, g, j, k negative; c zero

2 a $\frac{2}{7}$ b 1
c $-\frac{1}{2}$ d $-\frac{5}{4}$

3 a −0.60 b 1.6
c 0.42 d −0.46. b is steepest

4 −2, parallel

5 a

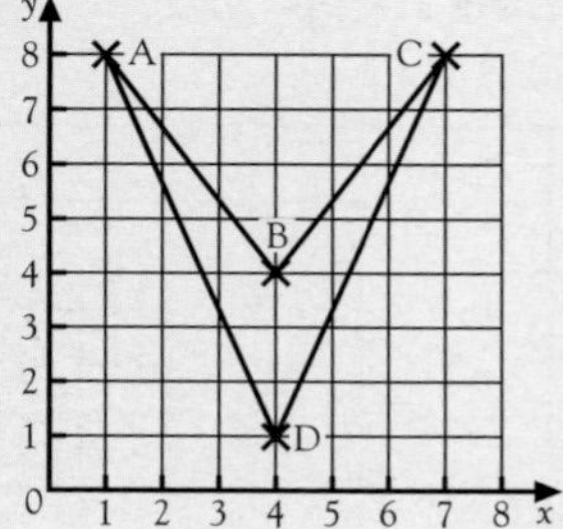

b gradient AB = $-\frac{4}{3}$
gradient BC = $\frac{4}{3}$
gradient CD = $\frac{7}{3}$
gradient DA = $-\frac{7}{3}$

6 a

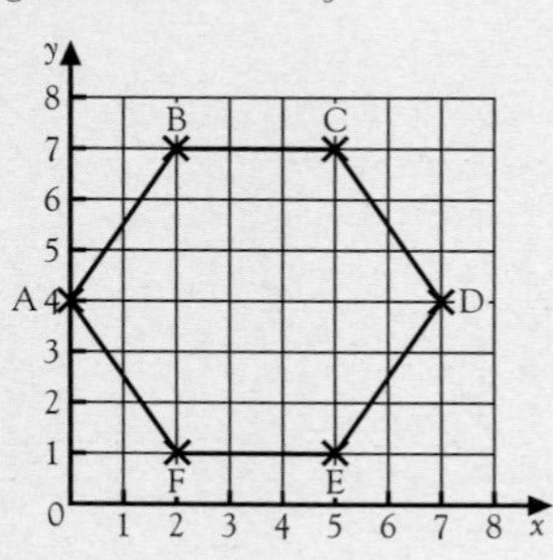

b gradient AB = $\frac{3}{2}$
BC = 0
CD = $-\frac{3}{2}$
DE = $\frac{3}{2}$
EF = 0
FA = $-\frac{3}{2}$

c AB and ED, BC and FE, AF and CD.

7 a

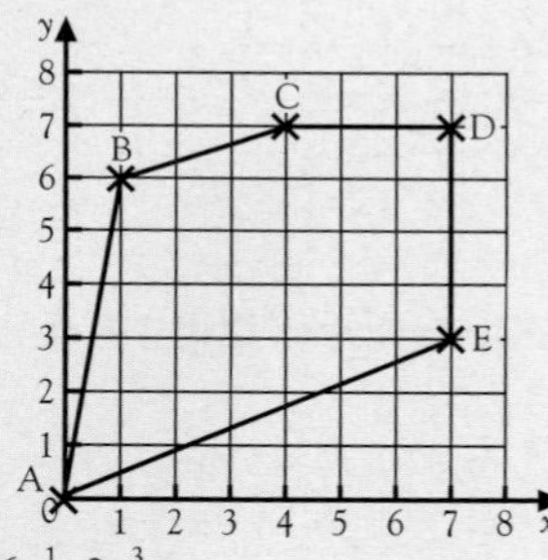

b 6, $\frac{1}{3}$, 0, $\frac{3}{7}$
c The slope of DE is infinite.

8 The slope is $\frac{2}{3}$ all along the line.

Exercise 18.2 (p. 335)

1 a 54 000 pesetas **b** 270 pesetas/£
c Exchange rate is 270 pesetas to each £1.

2 a £450 **b** 0.75 £/camper
c 75p **d** The cost would be £150

3 a 15°C **b** 4.5°C/min
c The temperature of the water increases by 4.5°C each minute.
d BC : the temperature of the water remains at 60°C for 6 mins.
CD : the water cools from 60°C to 50°C during the next 10 minutes, i.e. at a rate of 1°C/min.

4 a

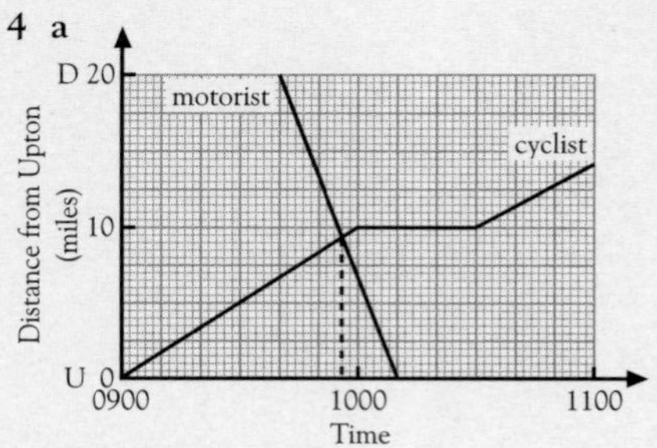

D = Dorchester
U = Upton

b 0956 a.m. (approx)

5 a 80 km/h, 0 km/h, 125 km/h
b 112.5 km/h; travelling from C to A
c 82 km/h
d 10.55 am, 120 km from A; remains stationary for $\frac{1}{2}$ hour

6 a

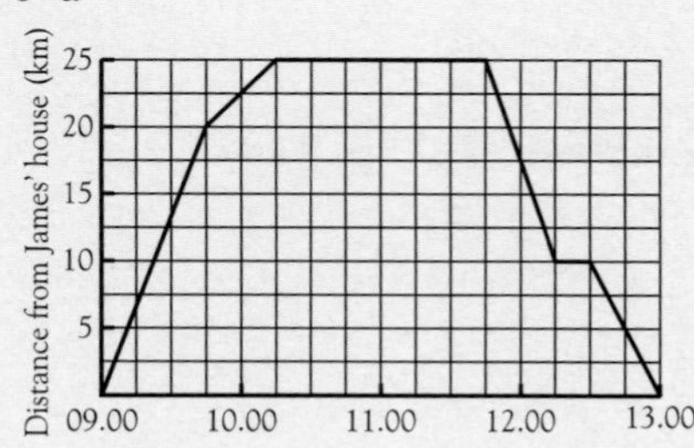

b 10.15 am, $1\frac{1}{2}$ hours
c 10 km **d** 13.00 **e** 20 km/h

7 a 2 h **b** 30 mins **c** 40 mins
d 10 h 19 mins (to nearest min)
e 20 mins (to nearest min)
f 1 h 26 mins (to nearest min)

8 a 75 km **b** 20 miles
c 4.8 km **d** 5.0 km (2 s.f.)

9 56 km/h

10 6 h 40 mins

11 40.6 mph (3 s.f.)

Exercise 18.3 (p. 343)

1 a

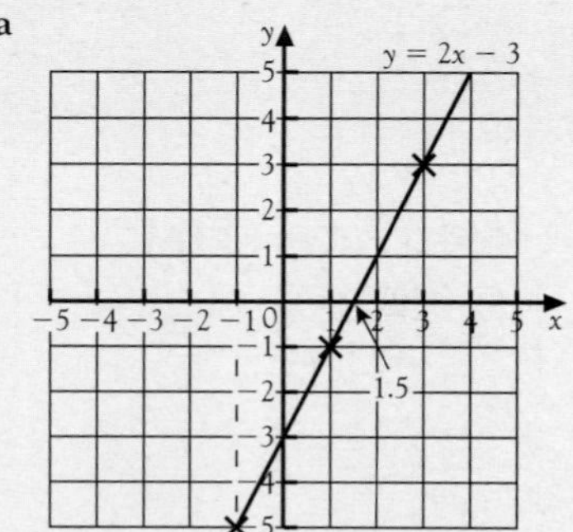

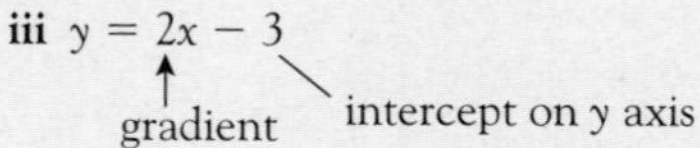

x	−1	1	3
$y = 2x - 3$	−5	−1	3

i 2
ii −3
iii $y = 2x - 3$ (↑ gradient; −3: intercept on y axis)

b

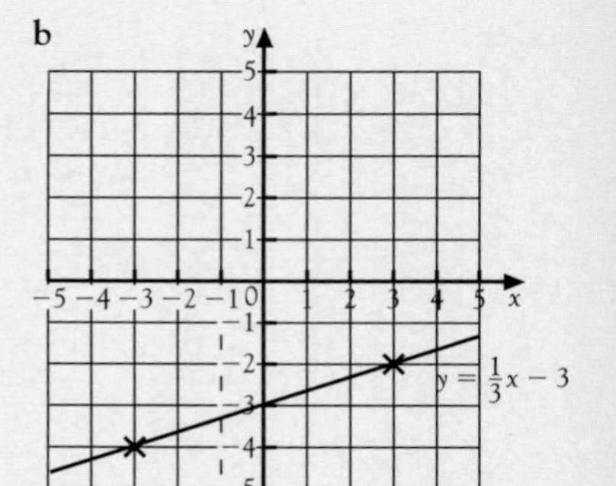

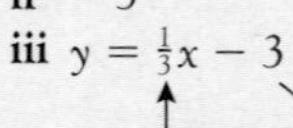

x	−3	0	3
$y = \frac{1}{3}x - 3$	−4	−3	−2

i $\frac{1}{3}$
ii −3
iii $y = \frac{1}{3}x - 3$ (↑ gradient; −3: crosses y axis at −3)

2 a and c, b and e, d and f

3 b and c have the same intercept, c and d are parallel

4 a grad = 3, intercept = 5

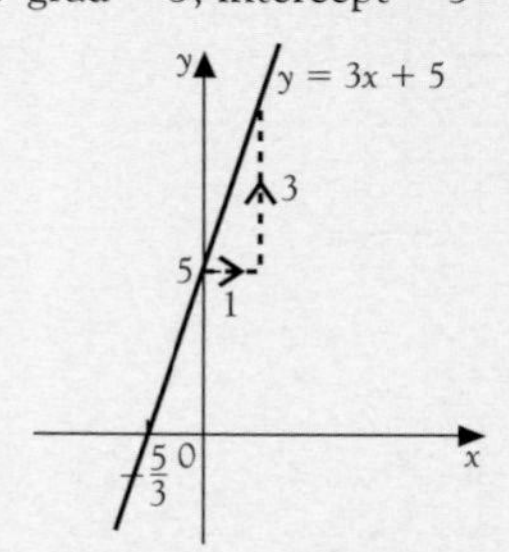

b grad = 1, intercept = −6

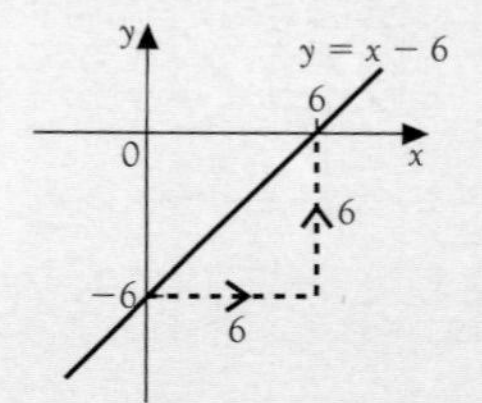

c grad = −4, intercept = 10

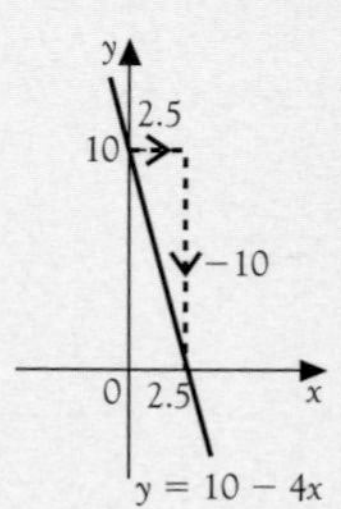

d grad = $\frac{1}{3}$, intercept = 0

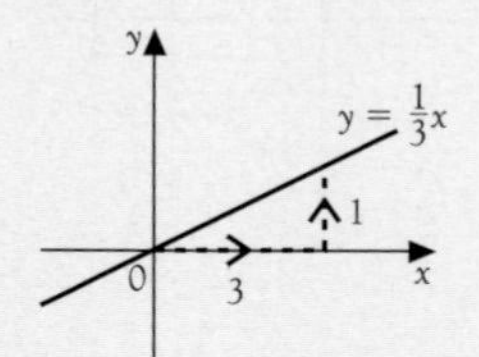

e grad = 2, intercept = −4

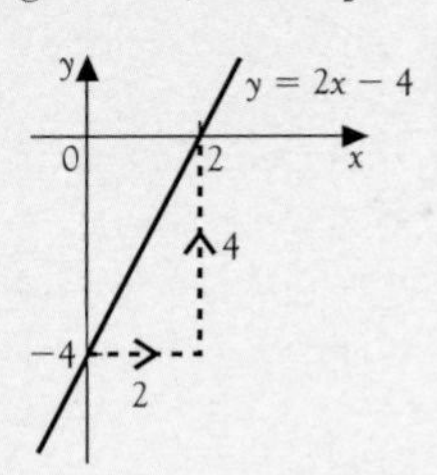

f grad = 5, intercept = 1

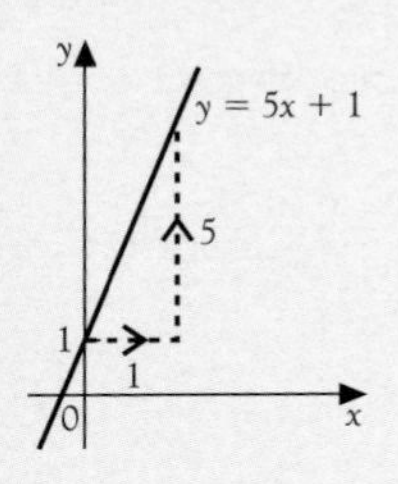

g grad = −1, intercept = 3

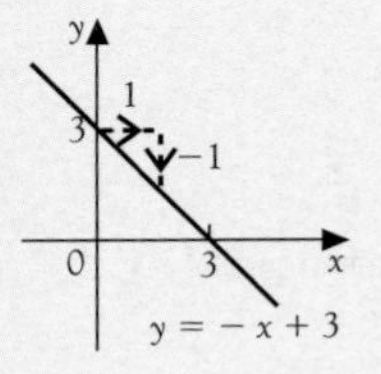

h grad = $-\frac{1}{4}$, intercept = 6

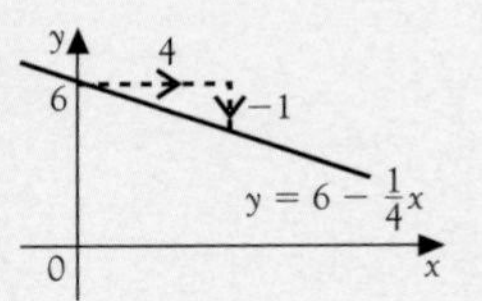

i grad = 0, intercept = 3

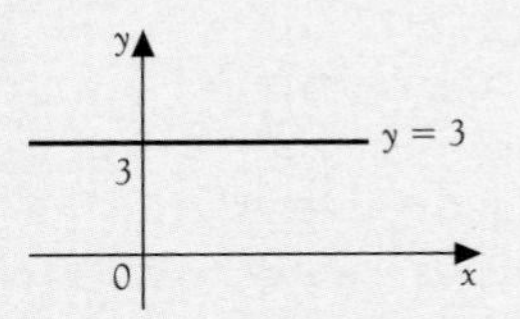

j grad = −1, intercept = 1

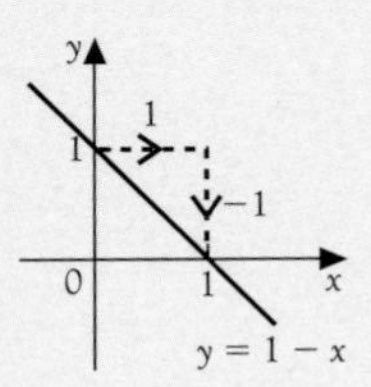

5 a $y = 4x - 3$ d $y = \frac{1}{2}x + 2$
b $y = 5 - 3x$ e $y = 2$
c $y = -\frac{2}{3}x$

6 a i 2   ii 4   iii $y = 2x + 4$
b i −3   ii 3   iii $y = 3 - 3x$
c i 1   ii 2   iii $y = x + 2$
d i −1   ii −3   iii $y = -x - 3$
e i $\frac{1}{2}$ ii 2   iii $y = \frac{1}{2}x + 2$
f i $-\frac{1}{3}$ ii 1   iii $y = -\frac{1}{3}x + 1$ *or* $y = 1 - \frac{1}{3}x$

7 a $y = -7x + 2$ $m = -7$ $c = 2$
b $y = x + 5$ $m = 1$ $c = 5$
c $y = 2x - 4$ $m = 2$ $c = -4$
d $y = -\frac{3}{2}x + 3$ $m = -\frac{3}{2}$ $c = 3$

8

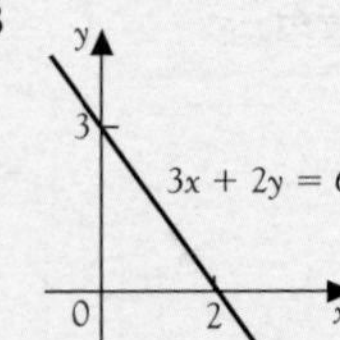

9 $m = 3, c = 1$
10 $y = -2x - 1$

Exam Questions: Chp 18 (p. 348)

1 a 0.7 d/p
b Exchange rate of 0.7 dollars per pound
2 Louise is not moving from B to C
Louise is cycling the quickest from D to E
Louise is cycling the slowest from C to D
3 a 20 litres/minute   b $V = 20t$
4 a 1.2   b 7.8 units (2 s.f.)
c Circle, centre $P(3, 1)$, radius 7.8 units
5 a 09.30   b 7 km
c 90 mins   d 4 km/h
6 a i, ii

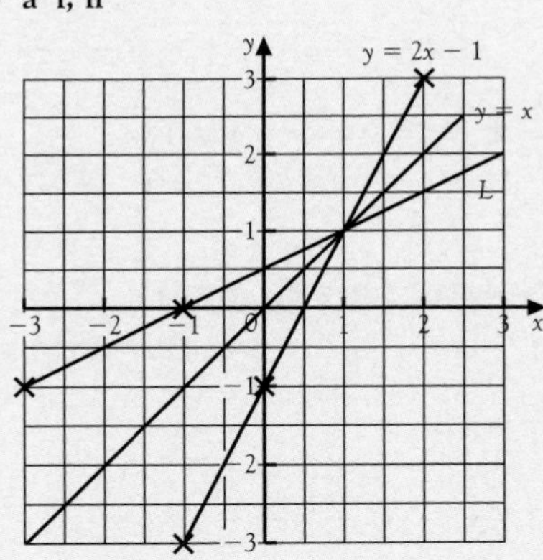

b i $\frac{1}{2}$ ii $\frac{1}{2}$ iii $y = \frac{1}{2}x + \frac{1}{2}$ *or* $2y = x + 1$

7 a The cold tap was turned on as well as the hot
b Both taps were turned off
c The plug was pulled out
8 a

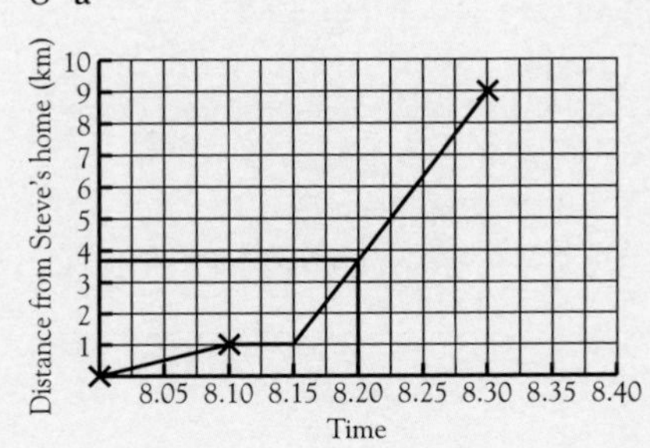

b 3.7 km
c i 36 mins   ii 07.54

Exercise 19.1 (p. 353)

1 a

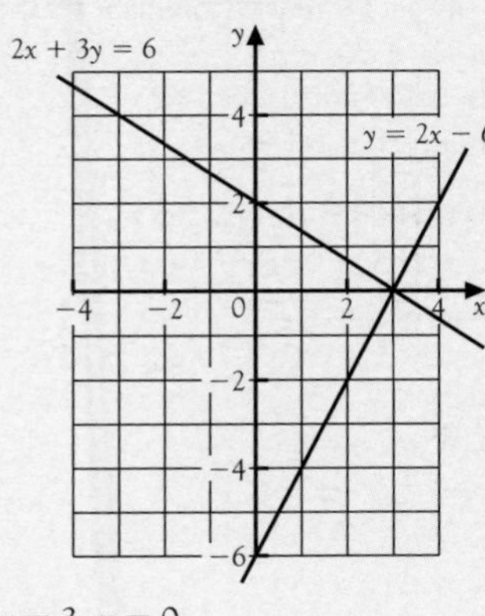

b $x = 3, y = 0$

2 a i

x	−1	2	5
y	−5	1	7

ii

x	−3	0	3
y	1	2	3

b

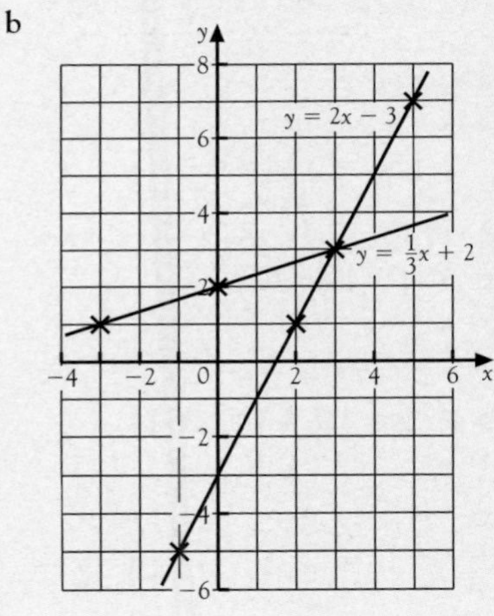

c $x = 3, y = 3$

3

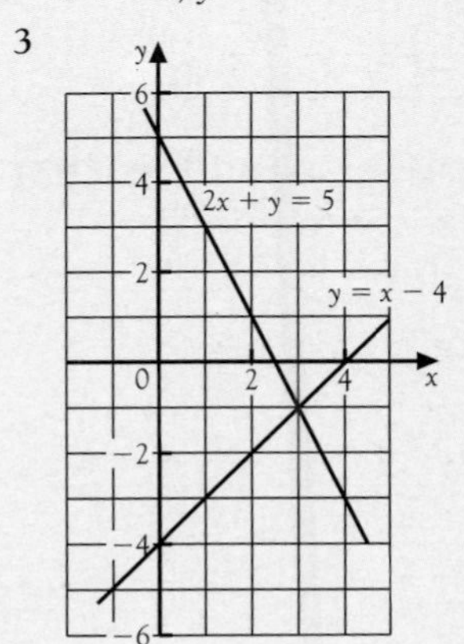

$x = 3, y = -1$

4 $x = -0.7, y = 2.3$
5 a Total cost = 2 × cost for an adult + 3 × cost for a child
b $x + 2y = 38$
c adult : £18; child : £10

6 $x + y = 150$; $x = y + 30$;
Maths: 90 marks; English: 60 marks

7 a $2x + 3y = 370$; $3x + 5y = 595$
b tea : 65p; coffee : 80p

8 man : 31; son : 7

Exercise 19.2 (p. 355)

1 a $x = 8, y = 4$
b $s = 1, t = 7$
c $x = 5, y = 6$
d $x = 2, y = 2$
e $x = 7, y = 2$
f $x = 3, y = 8$
g $p = 1, q = 3$
h $x = 2.5, y = 1.5$

2 $a = 18, b = 11$

3 $A = 1.5, M = 0.35$

4 a 30p b £2.40

5 a $x = 0, y = 2$ b $x = 3\frac{1}{3}, y = 4$

Exercise 19.3 (p. 359)

1 a $x = 7, y = 3$
b $p = 1, q = 1$
c $m = 2, n = -1$
d $a = 3, b = 0$
e $x = 6, y = 2$
f $a = 2, b = 3$
g $x = 1.5, y = 0.5$
h $b = 3.5, c = 2.5$

2 a $x = 3, y = 2$
b $a = 7, b = 3$
c $x = 8, y = 5$
d $x = 2, y = -1$
e $x = 1, y = 1.5$
f $a = 17, b = 12$
g $m = -0.5, n = 2$
h $x = 1.7, y = 0.7$

3 a £12 b £7.50

4 a $y = x + 24$ b $2y = 3x + 2$
c Kate : 46 years old
Jane : 70 years old

5 17 at £4 and 18 at £7.50

6 a $5f + 10t = 95$
b $f + t = 14$
c $9 \times$ £5 notes and $5 \times$ £10 notes

Exercise 19.4 (p. 361)

1 $x = 7, y = 11$

2 $x = 1, y = 3$

3 $x = -2, y = 1$

4 $x = 2, y = 1$

5 $x = 2, y = 5$

6 $p = 2.5, q = 5.5$

7 $a = -1, b = 2$

8 $x = 3, y = 11$

Exam Questions: Chp 19 (p. 363)

1 a $3x + 2y = 67$; $5x + y = 86$
b pen: 15p; pencil: 11p

2 £2.49

3 $x = 7, y = 3$

4 $x = 7.5, y = 2.25$

5 a $3x + 2y = 26$, $4x + y = 28$ b £6

6 a $117 = k + 4m$, $149 = k + 6m$
b $m = 16$

7 $x = 5, y = -2$

8 $x = 4.3, y = 5.7$ (1 d.p.)

9 b $a = 1.5, b = 0.5$

10 $x = 3.2, y = 1.5$ (1 d.p.)

11 a Consider the total cash received
b $x + y = 229$
c 178 articles at 50p and 51 articles at £1

12 $x = 2, y = 5$

13 $x = 4.5, y = 3.5$

14 a i

x	0	3	6
y	2	11	20

a ii, b

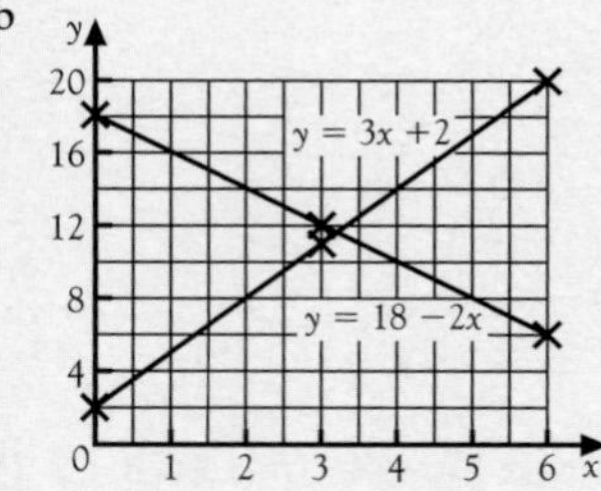

c $x = 3.2, y = 11.6$

15 a i, ii

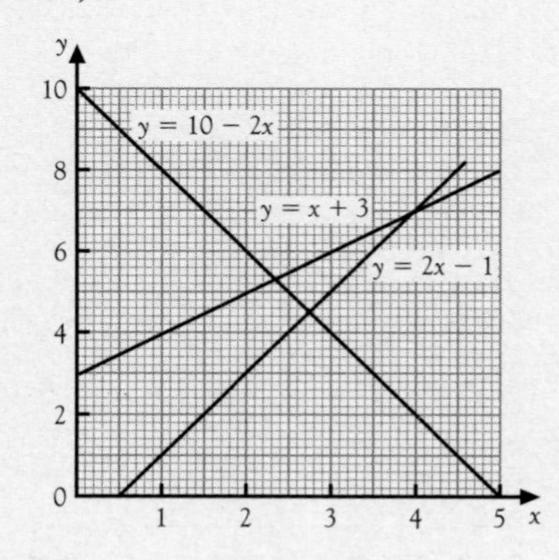

b i $x = 2.3, y = 5.3$ (1 d.p.)
ii $x = 4, y = 7$

Exercise 20.1 (p. 371)

1 $a = 20$ mm, $b = 38$ mm, $c = 8.5$ cm, $d = 8.3$ m (all to 2 s.f.)

2 1.8 m

3 1.9 m (2 s.f.)

4 530 m (nearest m)

5 70 m (nearest m)

6 66 m (nearest m)

7 7.3 cm (2 s.f.)

8 a 6.8 cm (2 s.f.) b 6.3 cm (2 s.f.)
c 22 cm^2 (2 s.f.)

Exercise 20.2 (p. 374)

1 a 34° b 56° c 48° d 28°
e 47° f 39° (all to nearest degree)

2 6° (nearest degree)

3 70° (nearest degree)

4 51° (nearest degree)

5 1.2° (2 s.f.)

6 a 25 cm b 16° (nearest degree)

7 15° (nearest degree)

8 59° (nearest degree)

9 125°, 29 km (2 s.f.)

10 56° (nearest degree)

Exercise 20.3 (p. 378)

1 63°, 63°, 54° (all to nearest degree)

2 54°, 54°, 126°, 126°
(all to nearest degree), 76 mm (2 s.f.)

3 74.6 mm, 45.8° (both to 3 s.f.)

4 a 25 m (nearest m) b 3.2 m (2 s.f.)

5 32° (nearest degree)

6 a 2.8 m (2 s.f.)
b 3.4 m (2 s.f.)

7 a 49° (nearest degree)
b 12 cm (2 s.f.)
c i 95° (nearest degree)
ii 1.8 m (2 s.f.)

8 a $h = 303$ cm (3 s.f.)
$w = 239$ cm (3 s.f.)
b 119 cm (3 s.f.)

9 a 4.6° (2 s.f.)
b 50 m^2
c 625 m^3

10 a 22°, 53° (nearest degree)
b i 37° (nearest degree)
ii 408 m (3 s.f.)

11 53 m (nearest m)

12 a 90 mm
b 106° (nearest degree)

Exam Questions: Chp 20 (p. 381)

1 10.3 m (3 s.f.)

2 a 8.2 m (2 s.f.)
b 14.8° (1 d.p.)

3 a 12.4 m (3 s.f.)
b 72° (nearest degree)

4 a 76° (nearest degree)
b 7.7 m (2 s.f.)

5 a 11.7 m (3 s.f.)
b 77° (nearest degree)

6 a 7.6 m (2 s.f.)
b 31° (nearest degree)

7 a ∠ACD is the angle standing on diameter BD and hence is a right angle
b 18.4° (1 d.p.)
c i 6.4 cm (2 s.f.)
ii 3.2 cm (2 s.f.)

8 a 8.96 m
b 38.5°
c 6.72 m (all to 3 s.f.)

9 b i 16° ii 17.8 m
c i 16.2° (3 s.f.) ii 17.9 m (3 s.f.)

10 a 42 cm^2 (2 s.f.) b 8 cm

11 a 2.41 m (3 s.f.) b 50.7° (3 s.f.)

Exercise 21.1 (p. 386)

1 $x^2 + 7x + 10$

2 $x^2 + 8x + 7$

3 $a^2 + 3a + 2$

4 $p^2 + p - 6$

5 $x^2 - x - 20$
6 $x^2 - 4x - 21$
7 $a^2 - 9a + 14$
8 $12 + 8a + a^2$
9 $12 - 7x + x^2$
10 $6p^2 - 5p - 6$
11 $6x^2 + 25x + 14$
12 $56a^2 - 31a + 3$
13 $18x^2 - 9x - 2$
14 $a^2 - 9$
15 $6x^2 - x - 2$
16 $9p^2 - 1$
17 $2x^2 - 3x - 2$
18 $21t^2 - 10t + 1$
19 $16x^2 - 25$
20 $49 - 35x + 6x^2$
21 $49x^2 - 7x - 6$
22 $15a^2 - a - 28$
23 $12y^2 - 17xy + 6x^2$
24 $30p^2 - 19pq - 4q^2$
25 $m^2 + 2m + 1$
26 $3 + 4x - 4x^2$
27 $5m^2 + 3mn - 2n^2$
28 $9t^2 + 12t + 4$
29 $12a^2 + 16ab - 3b^2$
30 $p^2 - 2pq + q^2$
31 $21x^2 - 8x - 4$
32 $25y^2 - x^2$
33 $25x^2 + 20xy + 4y^2$
34 $25x^2 - 30x + 9$
35 $35 + 4x - 4x^2$
36 $25 - a^2$
37 $4x^2 - 9y^2$
38 $6p^2 - 43pq + 7q^2$
39 $99 - 41a - 14a^2$
40 $6t^2 - 25t - 221$

Exercise 21.2 (p. 393)

1

x	−3	−2	−1	0	1	2	3
$y = 3x^2 - 10$	17	2	−7	−10	−7	2	17

a

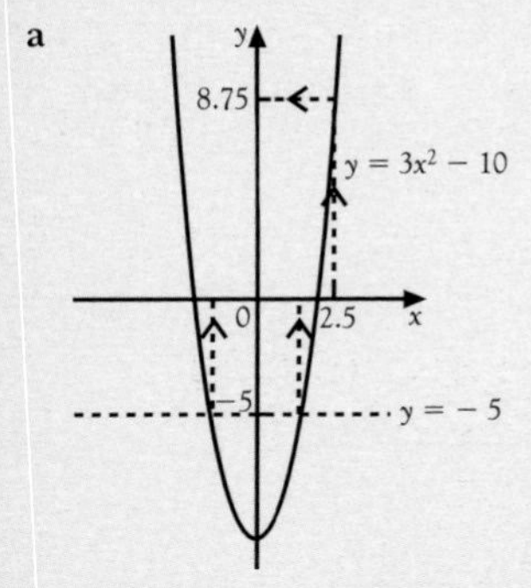

b 8.75 c ± 1.3

2 a

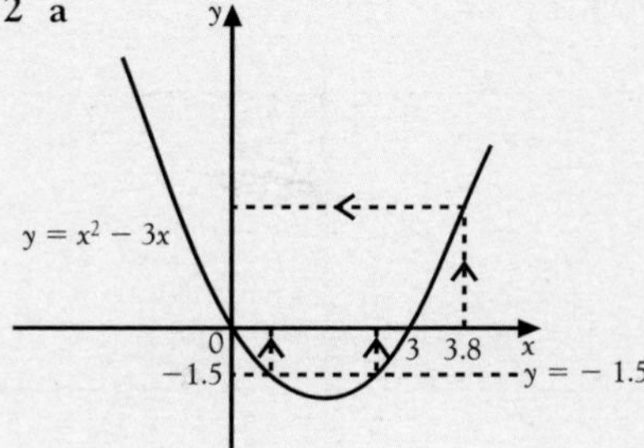

b $y \approx 3.0$ c $x \approx 0.6$ and $x \approx 2.4$
d Line of symmetry is $x = 1.5$

3 a

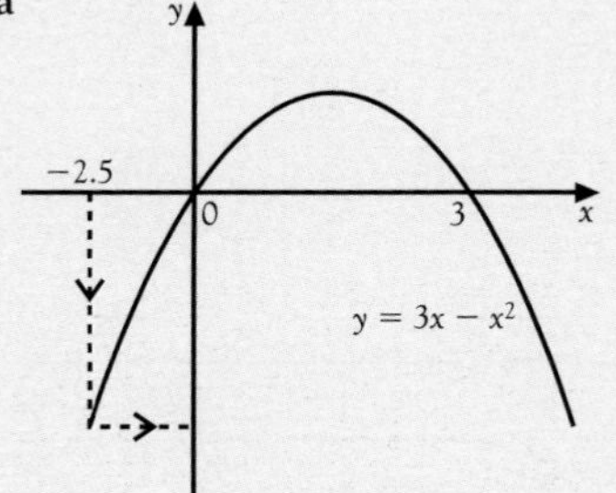

b This graph is a reflection in the x axis of graph in qu. 2
This 'way up' is expected from the sign of the x^2 term
c $y \approx -13.7$
d No values for x exist

4

x	−2	−1	0	1	2	3
$3x^2 - 2x - 1$	15	4	−1	0	7	20

a

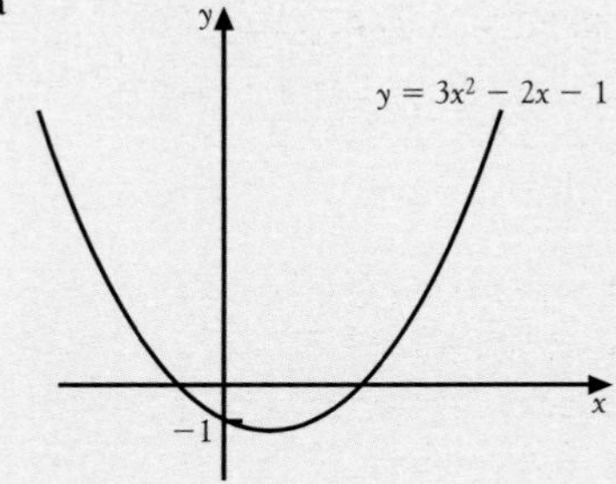

b $y \approx 11.5$
c $x \approx -1.8$ and $x \approx 2.4$
d $x = -0.3$ and $x = 1$ e $x = \frac{1}{3}$

5 a

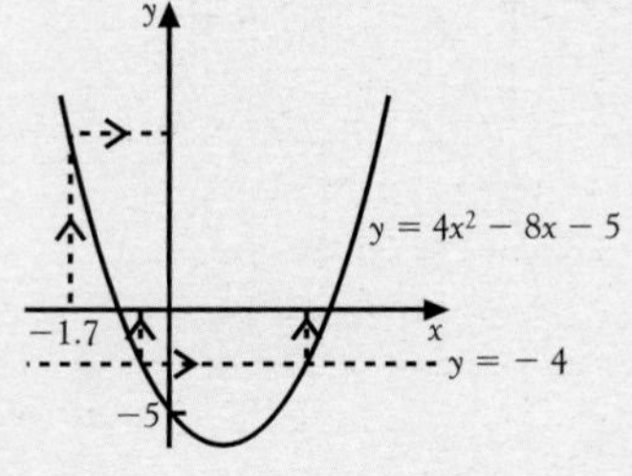

b $y \approx 20.2$ (By calculation $y = 20.16$)
c i $x = -0.5$ and $x = 2.5$
ii $x = -0.1$ and $x \approx 2.1$

6

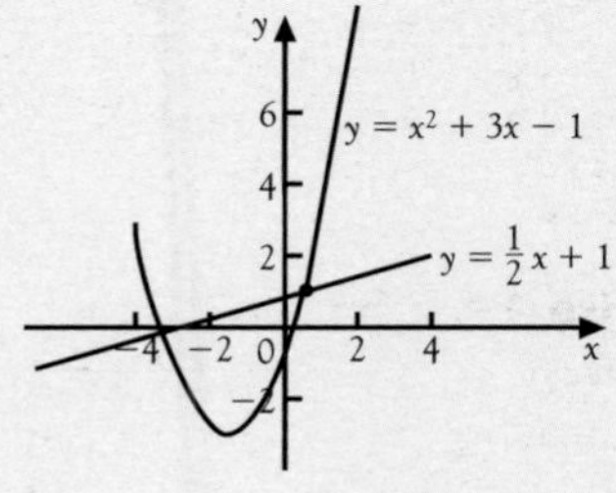

a $x = -3.3$ and $x = 0.3$
b $x = -3.1$ and $x = 0.6$,
$2x^2 + 5x - 4 = 0$

7 a $x = \pm 3$
b Crosses y axis at $(0, -9)$
c

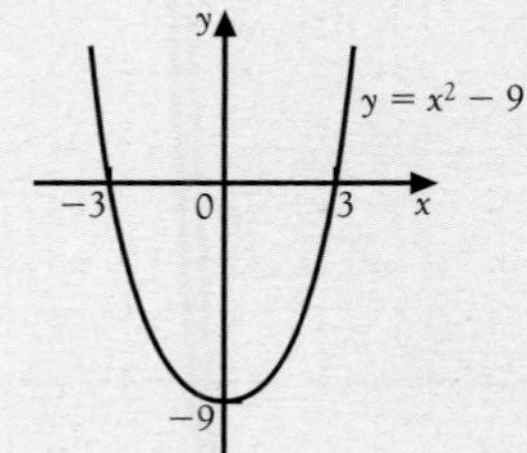

d

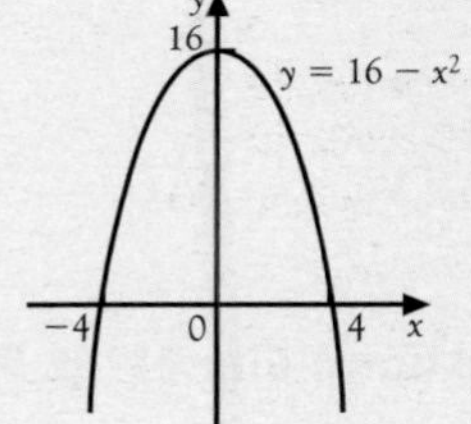

8 $A = x(10 - x)$
$\Rightarrow A = 10x - x^2$
a

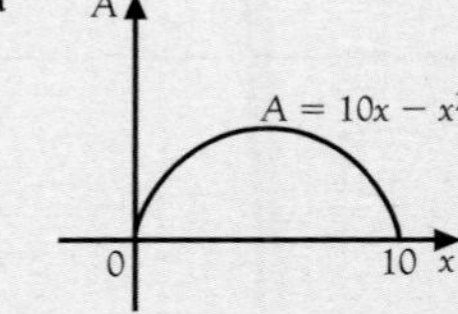

b 2.8 m by 7.2 m c $x = 5$
d $A = 25\text{ m}^2$. Square

Exercise 21.3 (p. 397)

1 $x = 1.8$ (1 d.p.)
2 $3^2 - 10 = -1 < 0$
$4^2 - 10 = 6 > 0$
Hence solution between 3 and 4
$x = 3.16$ (2 d.p.)
3 $x = 1.3$ (1 d.p.)
4 $x = 1.78$ (2 d.p.)
5 $7 \times 0.3^2 + 3 \times 0.3 = 1.53 < 2$
Hence solution between 0.3 and 0.4
$7 \times 0.4^2 + 3 \times 0.4 = 2.32 > 2$
$x = 0.36$ (2 d.p.)

6 $(x+5)^2 = x^2 + (x+3)^2$
$\Rightarrow x^2 - 4x - 16 = 0$
$x = 6.47$ cm, 9.47 cm, 11.47 cm

7 a 4 cm
c

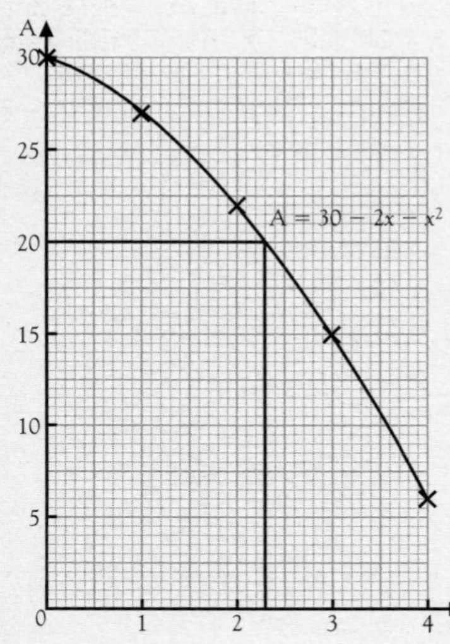

d $x \approx 2.3$ cm
e 2.32 cm (2 d.p.)

8 a $(x+4)$ years old
b i $y = x(x+4)$
ii 8 years and 6 months

Exam Questions: Chp 21 (p. 399)

1 a

x	−3	−2	−1	0	1	2	3
x^2	9	4	1	0	1	4	9
$y = x^2 - 5$	4	−1	−4	−5	−4	−1	4

b

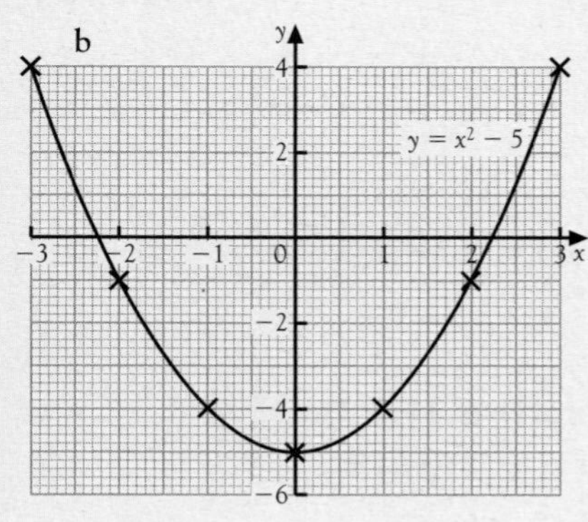

c $x = \pm 2.2$

2 $x = 12.8$ (3 s.f.)

3 a $6x(2x - 1)$
b $15x^2 - 23x - 28$

4 a

x	−3	−2	−1	0	1	2	3
y	−5	0	3	4	3	0	−5

b

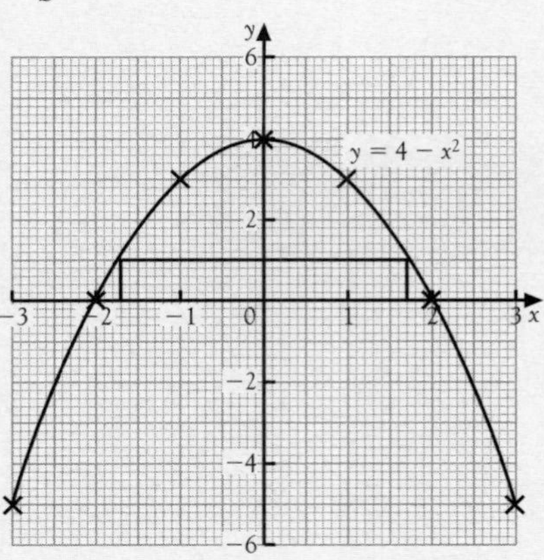

c $x = \pm 1.7$

5 a

x	−3	−2	−1	0	1	2	3
$y = x^2 - 3$	6	1	−2	−3	−2	1	6

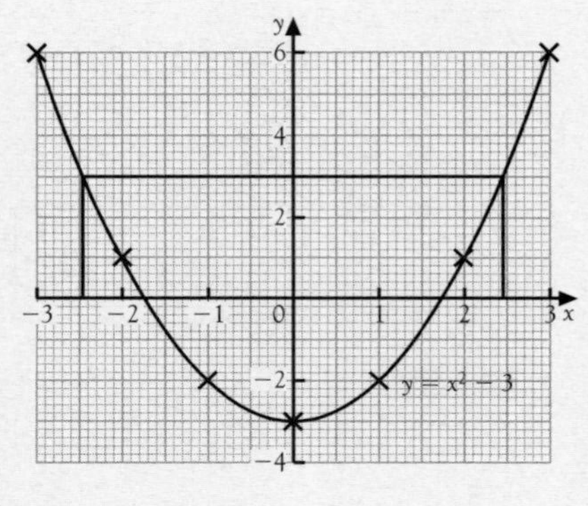

b $x = \pm 2.45$

6 a i $4x^2 + 12x$
ii $4x^2 + 12x + 9$
b 9 cm²

7 a

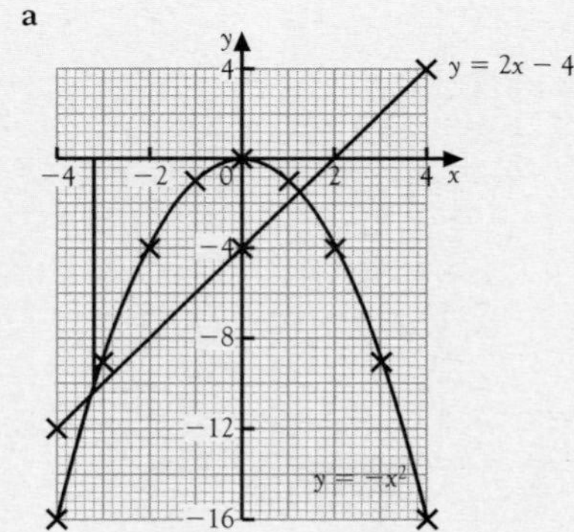

b $x = -3.2$

8 a

t	0	1	2	3	4	5	6
h	0	25	40	45	40	25	0

b i, ii

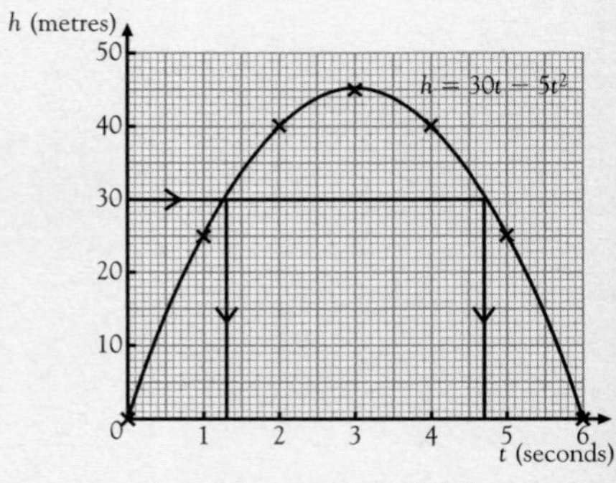

c After 1.3 seconds and 4.7 seconds

9 $x = 3.6$ (1 d.p.)

10 a

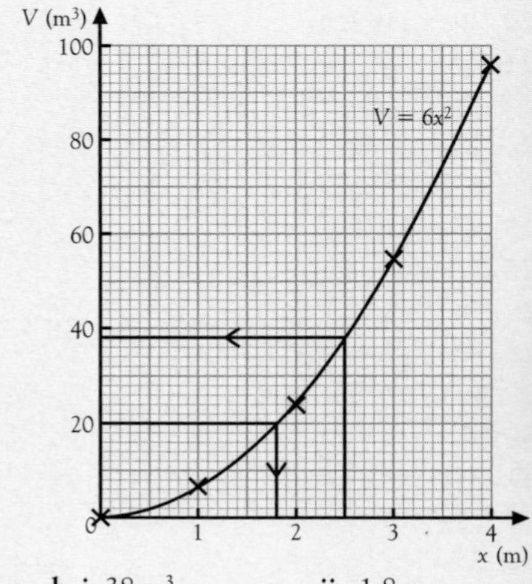

b i 38 m³ ii 1.8 m

11 $x = 3.2$ (1 d.p.)

12 a $7x - xy + 7y$ b $2p(3p - 4)$
c $2z^2 + 7z - 15$

13 a i 24 cm² ii 35 cm²
b 4.7 cm (1 d.p.)

Exercise 22.1 (p. 404)

1 a 900° b 2340°
c 3600° d 17 640°

2 a 9 b 13 c 16 d 21

3 a 12 sides, 150° b 10 sides, 144°
c 20 sides, 162° d 24 sides, 165°

4 a 120° b 140° c 160° d 168°

5 $a = 113°$, $b = 82°$, $c = 138°$, $d = 103°$

6 a 140° b 26° c 50°

7 135°

8 124°

9 112°

10

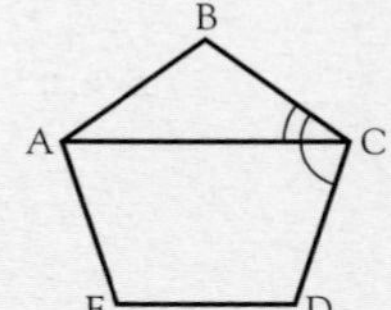

10 a i 108°
ii 36°
iii 72°
b, c AC and ED are parallel. Interior angles add up to 180°

11 135°, $22\frac{1}{2}$°, $67\frac{1}{2}$°, 90°, 45°

12 $a = 108°$
$b = 72°$
$c = 36°$

Exercise 22.2 (p. 410)

1 a 6 b 9 c 18 d 30

2 a 10°, 36 b 45°, 8
c 15°, 24 d 30°, 12

3 a i 72° ii 60° iii 45°
iv 40° v 36°
b i 108° ii 120° iii 135°
iv 140° v 144°

4 6 sides. Hexagon

5 8 sides. Octagon

6 10 sides. Decagon

7 a 69°
b 31°

8 a 135° b 117°

9 a Pentagon b 60°
c 63° (nearest degree)
d 53° (nearest degree
e 74° (nearest degree)
f 540°

10 a

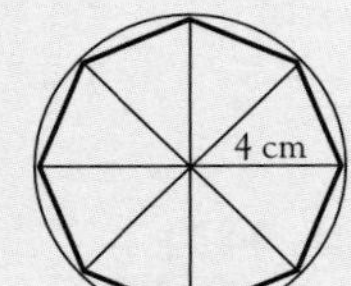

b 3.1 cm c 3.06 cm (3 s.f.)

11

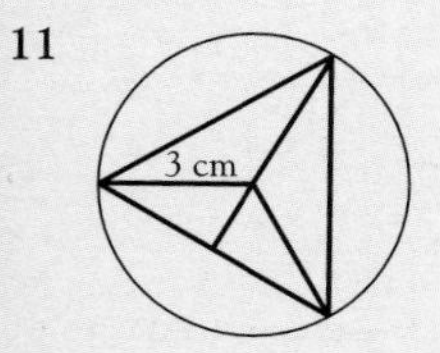

a 5.2 cm, 4.5 cm
b 5.20 cm (3 s.f.), 4.5 cm
c 11.7 cm^2 (1 d.p.)

12 a 1.3 m (2 s.f.) b 5.8 m^2 (2 s.f.)
c 1.2 m^2 (2 s.f.) d 0.3 m^3 (1 s.f.)

Exam Questions: Chp 22 (p. 413)

1 a 62° b 60° c 135°

2 a 108° b 132°, 24°, 24°

3 a 108° b △DBA or △ACD

4 a 90° b i 72° ii 108°

5 $p = 108°$, $q = 72°$

6 a Since the angles at the point where P, Q and T meet must add up to 360°, $3x = 360°$ and so $x = 120°$

b i ii 120°

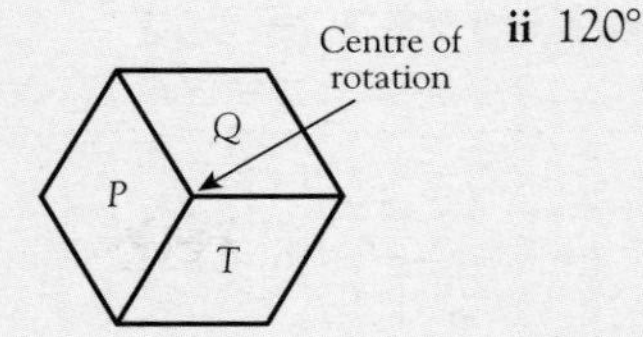

c i 2 ii 6 d Cube

7 a 360° b i 60° ii 120°
c 12 sides

8 a i 45° ii 135°

9 Not correct

Mixed Exam Questions: End of Stage 2 (p. 419)

1 £3840

2 $x = 1.5$

3 a 24.4 cm^2 (3 s.f.) b 15.2 cm (3 s.f.)

4 a 45.3 cm^2 (3 s.f.) b 113 cm^3 (3 s.f.)

5 a, i ii $x = 2.45$

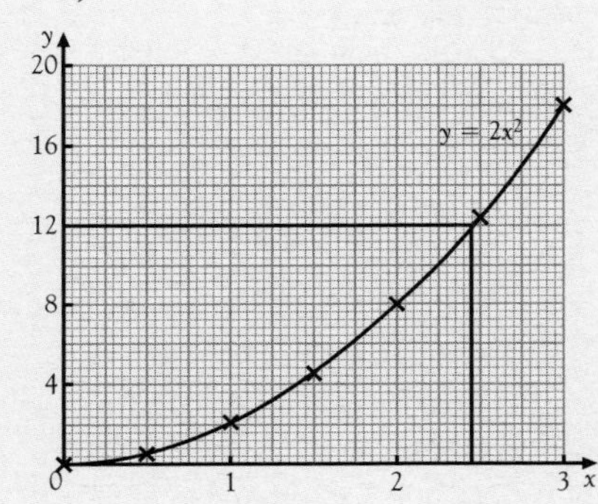

b $x = 3$, $y = 0.5$

6 a, b

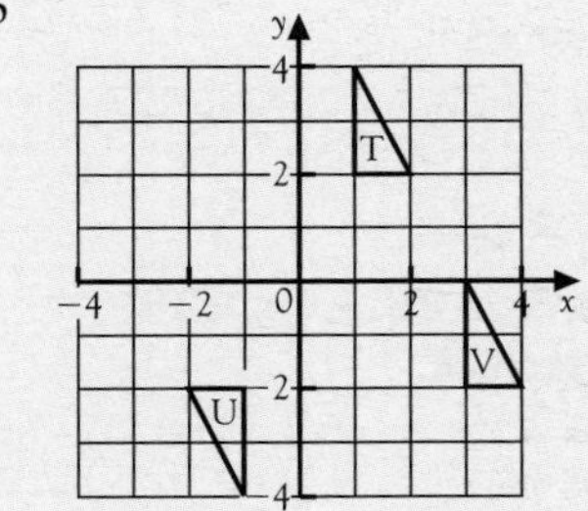

c Translation $\begin{pmatrix} 2 \\ -4 \end{pmatrix}$

7 a 28.3 km (1 d.p.)
b 298° (nearest degree)

8 a £2800
b £2880

9 $2p^2q^2(p - 2q)$

10 12 sides

11 a i $\frac{11}{67}$
ii $\frac{3}{32}$
b 43.4 s (3 s.f.)
c For: mean time for boys > mean time for girls
Against: (i) modal class is the same for both boys and girls
(ii) no boy is able to hold his breath for at least 80 seconds but 1 girl can

12 a $7x - y$
b $2x - 4y$
c $x^2 - 8x + 16$
d $16x^6$

13 a 61 pins
b To find the number of pins multiply the number of L-shapes by two, and add one

14 a 5.46 m (3 s.f.)
b 64.1° (1 d.p.)

15 a

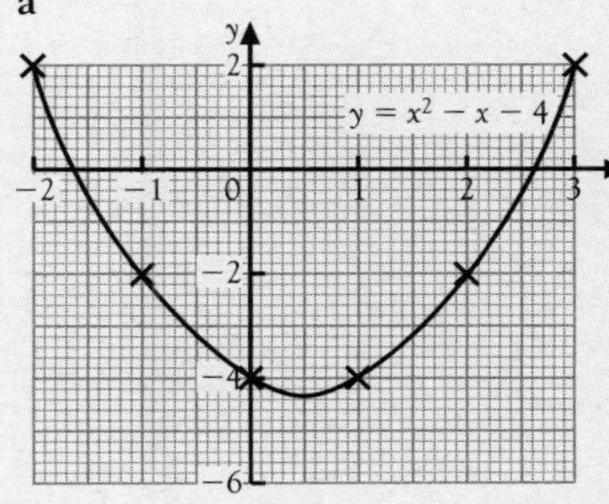

b i −4.3 ii $x = -1.6$ and $x = 2.6$

16 a £30
b £56

17 a 14.03 m^2
b 98.21 m^3
c 23.1° (1 d.p.)

18 a, b, c

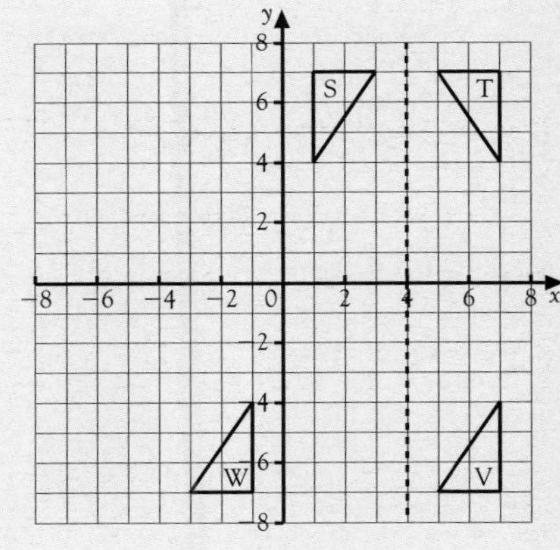

d Translation $\begin{pmatrix} 8 \\ 0 \end{pmatrix}$

19 a 462 cm^3 (3 s.f.)
b 3 cm

20 a, b i

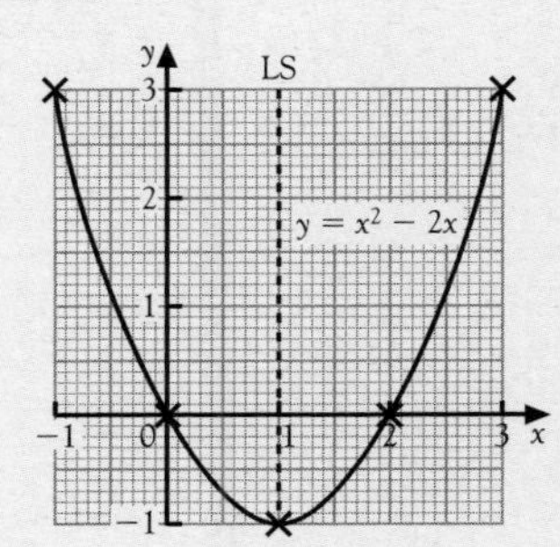

b ii $x = 1$

c $x = -3$

21 a i

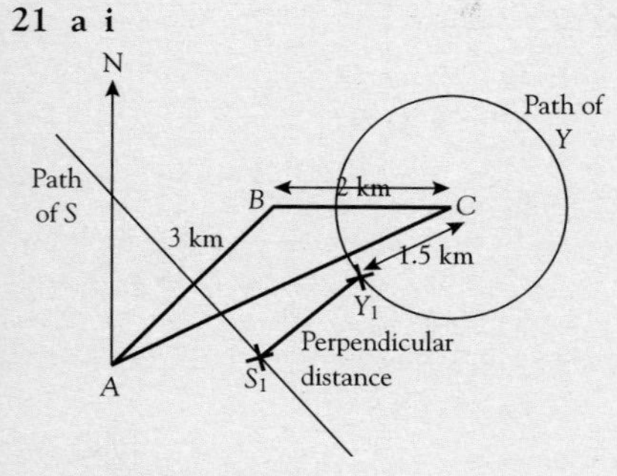

ii 4.6 km, 063°
b, c, d i (see diagram)
d ii 1.4 km

22 a $x = 7, y = 3$ b $y = 6.12$ c $x < 6.5$
23 a 612 m² b 10.1 m (3 s.f.)
24 a $2x^2 + 9x + 4$ b $2x(2x - 3)$
25 $x = 2.3$ (1 d.p.)
26 $s = 78$ km/h (2 s.f.)
27 a 72 m b 40 m c $\sqrt{200}$ s d 89.4 km/h

Exercise 23.1 (p. 426)

1 a 81 b 1024 c 216 d 1024 e 6561 f 16 807 g 1360 h 1 i 3.375 j 0.0625 k 1 889 568 l 1 000 000 000 000 m 512 n 8 o −108 p −864 q −21 609 r 128
2 a $\frac{1}{2}$ b $\frac{1}{9}$ c $\frac{1}{64}$ d $\frac{1}{100\,000}$ e $\frac{1}{625}$ f $\frac{1}{10}$
3 a 5^{-1} b 6^{-3} c 10^{-2} d 10^{-3} e t^{-2} f $2x^{-3}$ g $\frac{1}{3}y^{-4}$ h $\frac{3}{4}t^{-10}$
4 a $\frac{1}{x^3}$ b $\frac{1}{t^4}$ c $\frac{3}{y^2}$ d $\frac{4}{x^5}$ e $\frac{2}{a}$ f $\frac{1}{2x^4}$ g $\frac{1}{4m^3}$ h $\frac{3}{4t^4}$
5 a 1 b −1 c 1 d −1 e 1 f −1 g 1 h 1 i −1 j 1 k −1
6 a 5 b 3 c 3
7 a $\frac{1}{3}$ b 25 c $\frac{1}{4}$ d −14.4 e 13 f 1
8 343, 1844
9 a 5^{10} b 3^2 c 2^{-1} d 7^{-7} e 7^{-1} f 6^{-8} g 10^6 h 2^4 i 1 j y^7 k $9a^{-2}$ l 1
10 a $3^4 \times 2^9$ b $\frac{3^2}{5}$ c $30a^6b^{-4}c^4$ d $12x^{-1}y^{-1}$ e $-\frac{ab}{2}$ f x g $6ab^{-2}$ h −1 i $\frac{1}{3}$

Exercise 23.2 (p. 432)

1 a 360 000 b 47 500 c 0.54 d 58 400 000 e 0.000 14 f 723.6 g 0.000 000 098 h 2 000 000 i 0.000 082 5 j 963 k 0.047 l 0.000 006 6
2 0.000 000 000 2
3 1.99×10^{27} tonnes
4 a 3.46×10^2 b 7.26×10^3 c 3.4×10^1 d 5.67×10^{-1} e 1×10^6 f 1.7×10^{-2} g 9.657×10^5 h 1.6×10^0 i 7×10^{-4} j 2×10^{-1} k 6.4×10^{-3} l 1.4478×10^2
5 3.527×10^{-2} ounces
6 1.392×10^6 km
7 a 2.079×10^{10} b 4.75×10^5 c 5.965×10^{-6}
8 a 4.309×10^{10} b 7.7086×10^{-8} c 4.495×10^9 d -4.8957×10^{-10} e 1.428×10^{17} f 3.23×10^8 g 1.7822×10^{-6} h 1.4×10^{-4} i 8.90×10^{-3} (3 s.f.) j 4×10^{12} k 7.905×10^{-4} l 5.0085×10^{-2}
9 1 : 333 000 (3 s.f.)
10 9.45×10^{15} m (3 s.f.)
11 a 44 007 300 b 1 : 38.1
12 1.6747×10^{-24}, $1.67\,239 \times 10^{-24}$ 9.1083×10^{-28}
13 a £1.0504×10^{10}, £8.279×10^9 b £2225 million c 21.2%
14 909 years

Exercise 23.3 (p. 438)
Section A

1 a 1.15 b 1.125 c 1.08 d 1.035 e 1.5 f 1.0225
2 a 0.87 b 0.915 c 0.72 d 0.855 e 0.75 f 0.9275
3 a £120 b £77 c £85 d £28.50 e £26 f £558 g £770 h £720
4 £198
5 a £1200 b £1500
6 a £35 b £126 c £16.10 d £30.45
7 £11.28
8 1.2%
9 £3920
10 8%, £27.60
11 30p

Section B

1 a £90.45 b £841.58 c £108.24 d £436.33
2 a £367.01 b £392.70 c £481.08
3 6.9% (2 s.f.)
4 0.67% (2 s.f.)
5 4.1% (2 s.f.)
6 a £6375 b £5418.75 c £2828.62
7 a 75% b 25% c 51.9% (3 s.f.)
8 £33.73

Exercise 23.4 (p. 441)
Section A

1 a £360 b 200 m c 32 kg d £160 e £4800 f 80 cm g 30 g h 270 m i £20 j 300 kg
2 £20
3 £4400
4 £900

Section B

1 £1200
2 £16 153.85
3 £70
4 £20.85, £76.34, £2936.17
5 a £9.83 b £11.77 c £163.83 d £3574.47
6 400
7 1600
8 £7500
9 a £3000 b £750, £1050
10 a i £775 ii £1450 b £50 000
11 a £75.60 b 37%
12 £295.20, 64%
13 500 litres

Exercise 23.5 (p. 445)

1 a $\frac{1}{3}$ b $\frac{3}{4}$ c $\frac{4}{15}$ d $\frac{1}{6}$ e $\frac{1}{6}$ f $\frac{1}{10}$ g 1 h $\frac{2xy}{3}$ i $\frac{1}{4a}$ j $\frac{4}{a}$
2 a $\frac{1}{7}$ b 6 c $\frac{5}{4}$ d $\frac{8}{11}$
3 a $1\frac{1}{3}$ b $\frac{3}{4}$ c $\frac{1}{2}$ d $\frac{6y}{x}$
4 a $1\frac{3}{5}$ b $8\frac{4}{5}$ c $2\frac{1}{3}$ d $3\frac{1}{8}$
5 150
6 74
7 a i $\frac{7}{8}$ m² ii 8750 cm² b $6\frac{1}{4}$ cm²

Exam Questions: Chp 23 (p. 447)

1 a 25.586 … b i 100 ii No, power of 10 out
2 a £5.32 b 44%
3 a £61.80 b 1.225 043 ($=1.07^3$)
4 a £21 000 b £22 050 c £26 802 (nearest £)
5 637.2 (1 d.p.)
6 a $\frac{1}{12}$ b 84

7 a 54, 0.51 b 67.5 mg c 85%

8 a 2.493×10^{-23} grams
b i 2.994×10^{-23} grams
ii 3.34×10^{22} (3 s.f.)

9 a 52 000 000 b 1.2×10^{-1} cm

10 £202.54

11 a 1 b 0.003 c $32a^5b^3c^6$

12 a £64 800
b 1978
c £55 556 (nearest £)

13 a £763.75 b 14.9% (1 d.p.)

14 a i $383\,500 \leqslant d < 384\,500$
ii 3.84×10^5 km
b 500 seconds c 9.2×10^7 km

15 a 3 b 2

16 a i 4.524×10^{-6} mm²
ii 1.214×10^{-2} mm
b ii 2 m

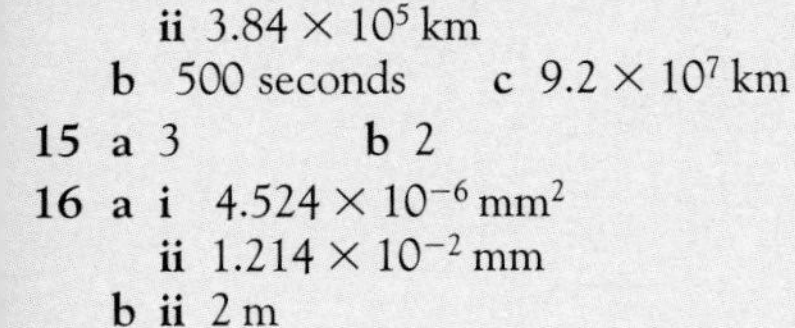

Exercise 24.1 (p. 450)

1 a No d Yes
b Yes e No
c No f Yes

2 a $\frac{1}{3}$ b $\frac{5}{12}$ c $\frac{3}{4}$ d $\frac{7}{12}$

3 a $\frac{1}{13}$ b $\frac{3}{13}$; Yes c $\frac{4}{13}$

4 a $\frac{5}{6}$ b $\frac{1}{6}$

5 a $\frac{11}{15}$ b $\frac{8}{15}$ c $\frac{14}{15}$ d $\frac{1}{15}$

6 a $\frac{2}{5}$ b $\frac{1}{10}$ c $\frac{1}{2}$

Exercise 24.2 (p. 452)

1 a $\frac{1}{12}$ b $\frac{1}{6}$

2 $\frac{1}{300}$

3 a $\frac{1}{4}$ b $\frac{1}{256}$; 9

4 a $\frac{1}{8}$ b $\frac{1}{8}$ c $\frac{1}{8}$

5 a 0.16 b 0.36 c 0.05

6 a $\frac{1}{78}$ b $\frac{1}{78}$

7 a 0.336 b 0.224

Exercise 24.3 (p. 456)

1 a

freezer $\frac{2}{3}$: computer $\frac{1}{4}$, no computer $\frac{3}{4}$
no freezer $\frac{1}{3}$: computer $\frac{1}{4}$, no computer $\frac{3}{4}$

b $\frac{1}{2}$

2

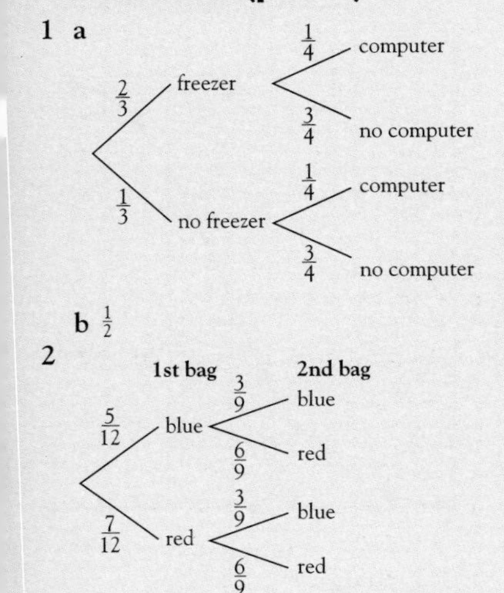

a $\frac{5}{36}$ b $\frac{17}{36}$ c $\frac{31}{36}$ d $\frac{19}{36}$

3 a

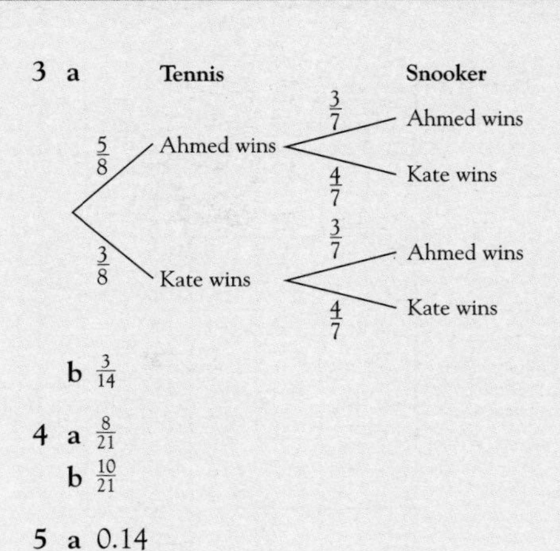

b $\frac{3}{14}$

4 a $\frac{8}{21}$
b $\frac{10}{21}$

5 a 0.14
b

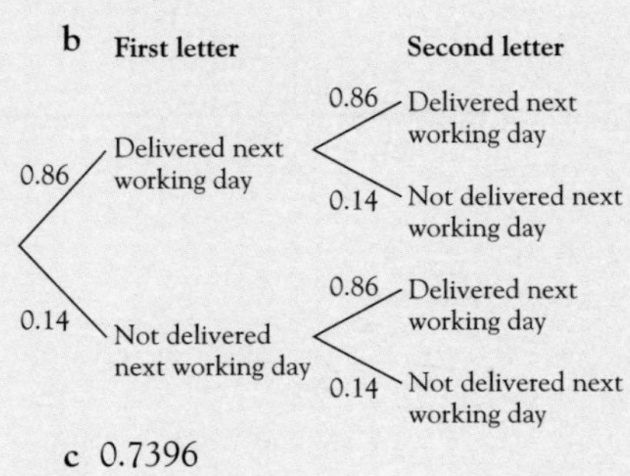

c 0.7396

6 a

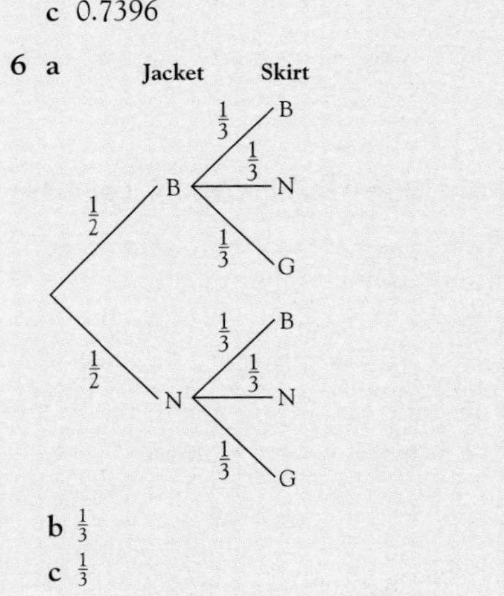

b $\frac{1}{3}$
c $\frac{1}{3}$

7

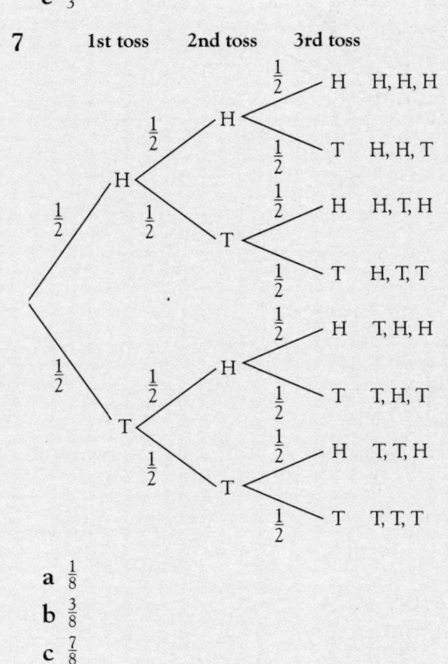

a $\frac{1}{8}$
b $\frac{3}{8}$
c $\frac{7}{8}$

8

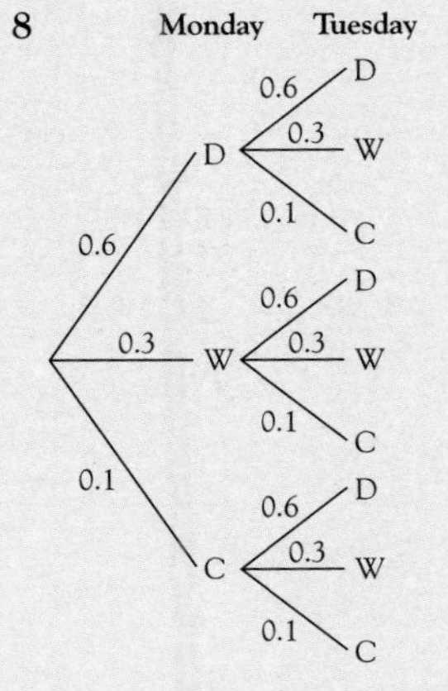

a 0.09 b 0.24 c 0.54 d 0.19

9

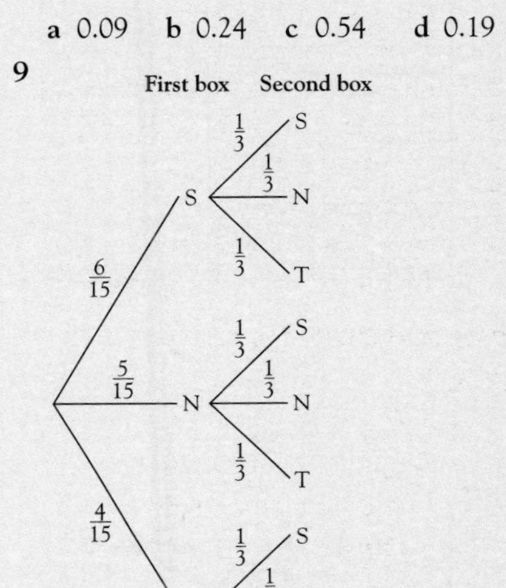

a $\frac{1}{9}$ b $\frac{2}{9}$ c $\frac{2}{3}$

10 a $\frac{1}{16}$ b $\frac{1}{6}$

11 a $\frac{1}{2}$ b $\frac{5}{12}$

12 a 0.83 b 0.0459

Exercise 24.4 (p. 459)

1 a 6 b 17 cm
c 16 d 157.5 kg

2 a 18; 14, 30.5; 16.5 b 3; 2, 5; 3
c 18 m; 14 m, 20 m; 6 m
d 96.5; 93.35, 97.85; 4.5

3 44.5 h, 11 h, 222.5 days

4 d: 116, 64 t: 135, 105

5 A: 39, 10.5 B: 36, 27

Exercise 24.5 (p. 464)

Where values are read from graphs, the answers given are approximate.

1 c.f. 1, 11, 24, 44, 48, 50
a 3 b 2, 3; 1

2 c.f. 0, 0, 0, 1, 4, 6, 10, 14, 19, 25, 29
a 8 b 6, 9; 3

3 a c.f. 0, 2, 10, 25, 48, 80, 100, 110, 115, 118, 120
b 23 c £950 d £210

4 a 3.1 mins b 1 min
c 150 d 12

5 a b Plot (0, 0), (10, 4), (20, 20), (30, 43), (40, 105), (50, 158), (60, 186), (70, 195), (80, 198), (90, 199), (100, 200)
c 39, 16 d 66 e 15

6 a Plot (5.5, 2), (10.5, 20), (15.5, 42), (20.5, 86), (25.5, 127), (30.5, 140), (35.5, 147), (40.5, 150)
b 9 hours c 28% d 11.5

7 a c.f. 2, 7, 22, 36, 45, 48, 50
b

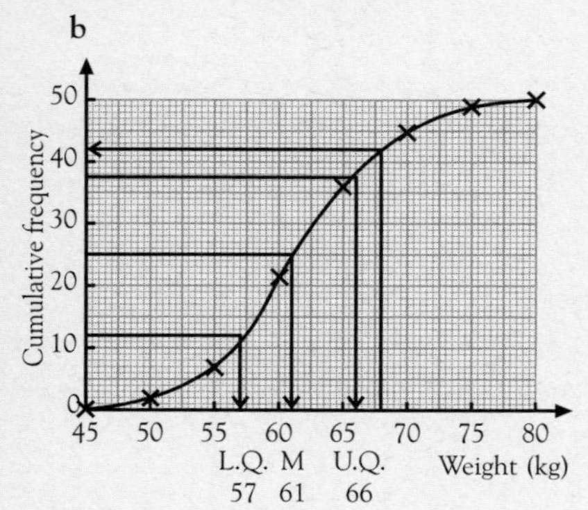

c 61 kg, 9 kg
d 16%
e Second group are much heavier

8

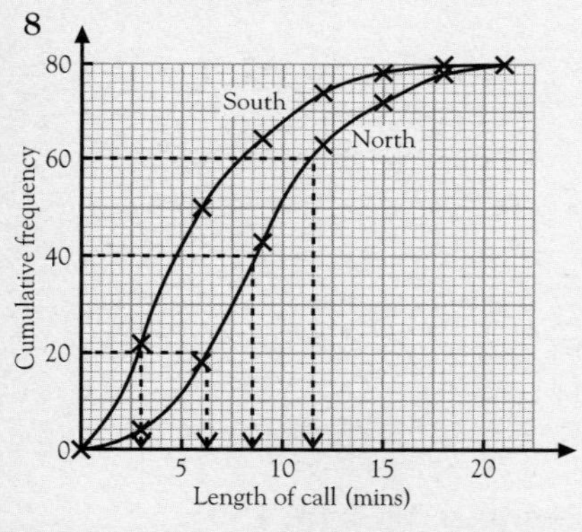

	North Street	South Street
median	8.5	5
lower quartile	6.25	2.75
upper quartile	11.5	7.75
inter-quartile range	5.25	5

North Street calls are longer but range of middle half of values is similar for both kiosks.

9 a 16, 76, 104, 114, 120
b Plot (0, 0), (10, 16), (20, 76), (30, 104), (40, 114), (50, 120)
c 17.5 minutes
d No, 22%

10 a i $\frac{7}{40}$ ii 420
b 2, 21, 45, 66, 79, 80
c Plot (40, 0), (45, 2), (50, 21), (55, 45), (60, 66), (65, 79), (70, 80)
d i 54 ii 8

11 a

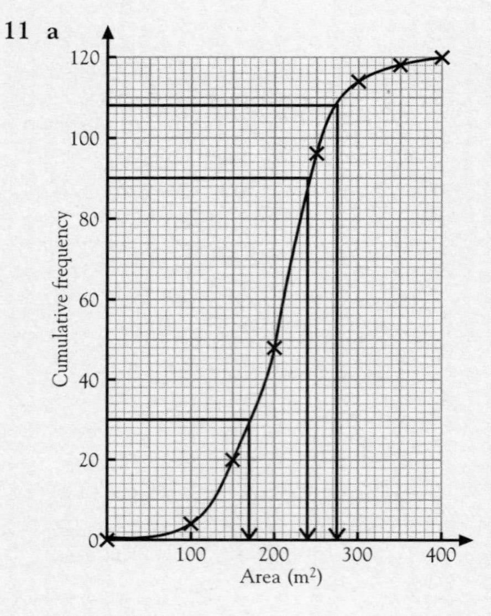

b 70 m^2
c 275 m^2

12 a c.f. 0, 14, 53, 86, 106, 116, 120
b Plot (400, 0), (500, 14), (600, 53), (700, 86), (800, 106), (900, 116), (1000, 120)
c i £620 ii £170 iii 7%

Exam Questions: Chp 24 (p. 469)

Where values are read from graphs, the answers given are approximate.

1 a $\frac{1}{81}$ b $\frac{8}{27}$ c $\frac{37}{81}$

2 a 0.17
b 0.55
c

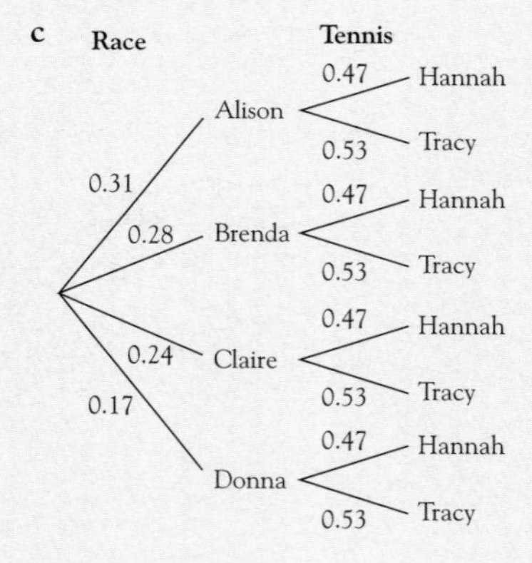

d 0.1484

3 a i $\frac{1}{6}$ ii $\frac{1}{36}$ b $\frac{13}{36}$

4 a

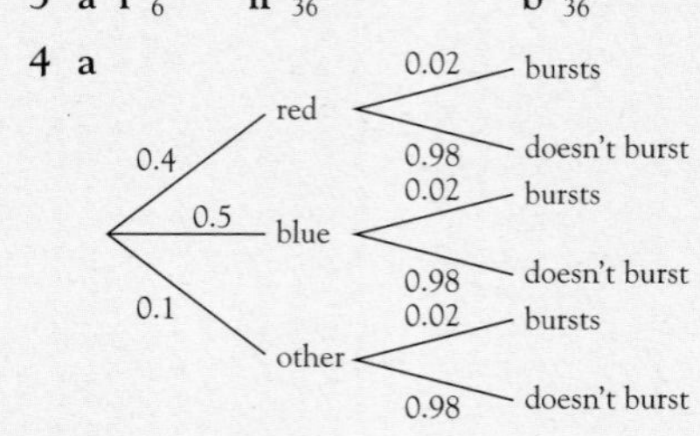

b 0.008 c 0.404

5 a

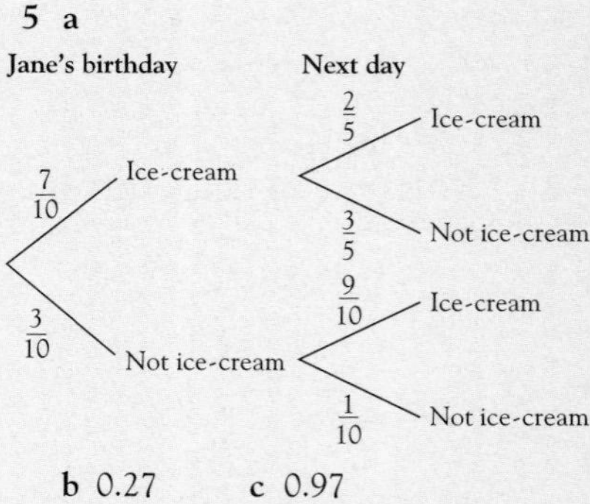

b 0.27 c 0.97

6 a 0.35 b i 0.45 ii 0.06

7 a 17 b 32 mins c 45 mins
d 25 mins e 20 mins

8 a 16 to 20
b 18
c 1, 3, 8, 15, 23, 25
d Plot (0, 0), (5, 1), (10, 3), (15, 8), (20, 15), (25, 23), (30, 25)

9 a 44, 73, 96, 107
b

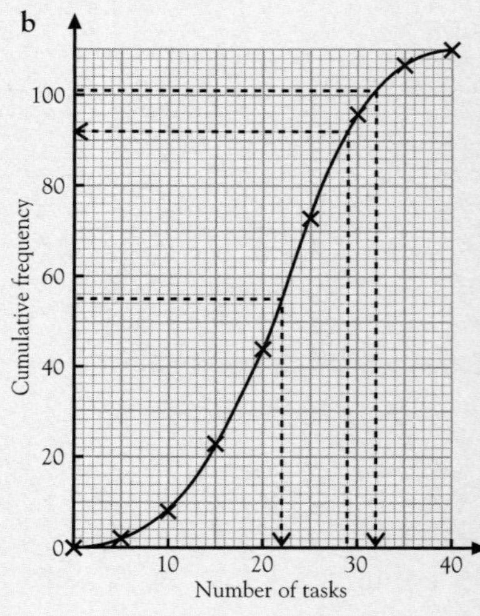

c 22 d 18 e 32

10 a c.f. 2, 9, 22, 45, 63, 68
b Plot (0, 0), (1, 2), (2, 9), (3, 22), (4, 45), (5, 63), (6, 68)

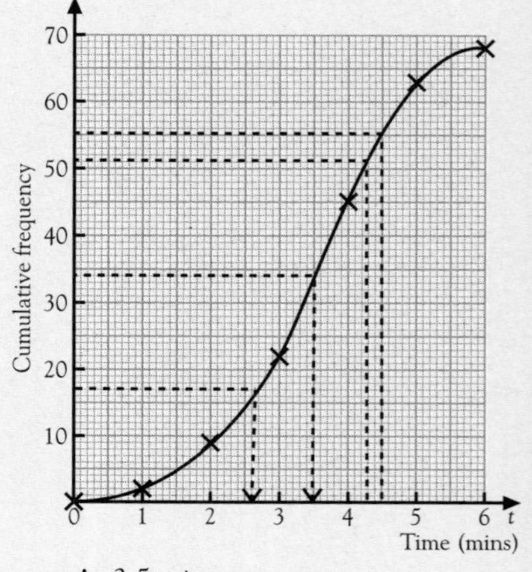

c i 3.5 mins
ii 1.7 mins
d 19%

11 a 0, 3, 10, 23, 38, 50, 58, 60
b

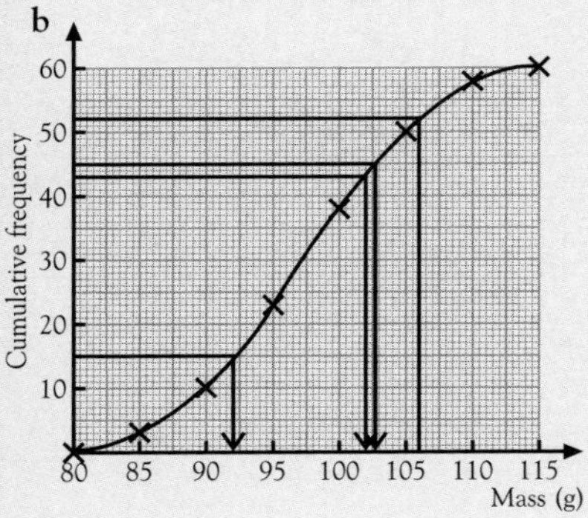

c 11 grams
d 52 apples
e 102 grams

Exercise 25.1 (p. 475)

1 a i 6, 4 **ii** 3, 4 **iii** 5, −2
iv 6, −2 **v** −3, −12 **vi** −9, 7
b i −9, −9 **ii** −8, 7 **iii** −6, −8
iv −2, 14 **v** −3, 16 **vi** −6, 6
2 a $(x + 2)(x + 3)$ **b** $(x + 5)(x + 2)$
c $(x + 7)(x + 5)$ **d** $(x - 2)(x - 6)$
e $(x - 1)(x - 3)$ **f** $(x - 3)(x - 8)$
g $(x - 3)(x - 4)$ **h** $(x - 1)(x + 6)$
i $(x - 1)(x + 5)$ **j** $(x + 2)(x - 6)$
k $(x + 2)(x - 4)$ **l** $(x + 2)(x - 11)$
m $(x - 1)(x + 12)$ **n** $(x + 1)(x - 2)$
o $(x - 4)(x - 4)$ **p** $(a + 3)(a - 6)$
q $(p + 10)(p + 10)$ **r** $(b - 6)(b - 6)$
s $(q - 8)(q - 2)$ **t** $(w + 1)(w + 21)$
u $(t - 7)(t + 8)$ **v** $(m + 7)(m - 9)$
w $(n + 6)(n - 7)$ **x** $(y + 3)(y + 3)$
y $(c + 2)(c - 5)$ **z** $(d + 8)(d - 9)$
3 a $x(x + 5)$ **b** $x(x - 3)$ **c** $y(y - 20)$
d $a(a + 2)$ **e** $5x(x - 5)$ **f** $7a(a - 3)$
4 a $(x - 3)(x + 3)$ **b** $(a - 4)(a + 4)$
c $(b - 9)(b + 9)$ **d** $(6 - w)(6 + w)$
e $(x - 11)(x + 11)$ **f** $(2 - x)(2 + x)$
5 a $2(x + 1)(x + 2)$ **b** $3(x - 2)(x - 7)$
c $3(x - 1)(x - 2)$ **d** $3(x - 2)(x + 2)$
e $7x(x + 1)$ **f** $4(x - 2)(x + 2)$
6 a $(x - 2)(x + 3)$ **b** $3a(a + 4)$
c $(n - 5)^2$ **d** $(t - 12)(t + 12)$
e $(b + 6)^2$ **f** $2(y + 4)^2$
g $25(w - 2)(w + 2)$ **h** $(x + 4)(x - 5)$
i $5(m + 1)(m + 4)$ **j** $x(x - 25)$
k $2(s - 1)^2$ **l** $9(u - 3)(u + 3)$

Exercise 25.2 (p. 478)

1 a 3, 1 **b** 5, −6
c 0, 7 **d** 3, −3
e 7, −7 **f** −5, 2
g −12, −13 **h** 9
i 0, 10 **j** 10, 5
k −7, 8 **l** −10
m −13, 12
2 a −4, −1 **b** 7, 3 **c** 2
d 4, 6 **e** −1, 8 **f** −2, −8
g 4, −4 **h** 0, −14 **i** −5
j 4, −5 **k** 7 **l** −8, 2
m 0, −6 **n** 3, −8 **o** 0, −3
p 0, 2 **q** 2, −4 **r** −3, 3
3 a 1 **b** −5, 6 **c** 0, 9
d 0, 6 **e** 4, 7 **f** 1, −17
g 3, −6 **h** 7, −7
4 a $x(x + 7) = 60$
b $x^2 + 7x - 60 = 0$
c 5
5 $x(x - 3) = 70$ Shamshir is 7, Ted is 10
6 Width is 10 cm, length is 12 cm
7 $x^2 + 2x - 63 = 0$, length is 12 inches, width is 4 inches
8 2, 832 cm^3

Exercise 25.3 (p. 483)

1 −1, 0, 3, 8, 15; 2
2 4, 13, 28, 49, 76; 6
3 5, 10, 19, 32, 49; 4
4 48, 63, 80; 2
5 −36, −73, −116; −6
6 a $n^2 - 2$ **b** $(n + 1)^2$
c $n^2 + 2n - 2$ $(= (n + 1)^2 - 3)$
7 a $3n^2 + 2n - 4$ **b** $2n^2 - 8$
c $5n^2 + n - 1$ **d** $2n^2 - 6n + 3$
e $4n^2 - 3n + 1$ **f** $n^2 - n + 1$
8

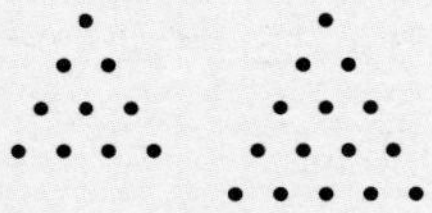

1, 3, 6, 10, 15, 21;
$\frac{1}{2}n^2 + \frac{1}{2}n$ $(= \frac{1}{2}n(n + 1))$
9 30; 3, 9, 18, 30, 45, 63;
$\frac{3}{2}n^2 + \frac{3}{2}n$ $(= \frac{3}{2}n(n + 1))$

Exam Questions: Chp 25 (p. 485)

1 b Maria 11, Aaron 14, Jason 9
2 a $(x + 2)(x - 5)$
b 0, −3
c $-1 \leqslant x < 1$
3 b −2, 6 **c** 6 cm, 8 cm, 10 cm
4 a 50 **b** 1250 **c** $2n^2$
5 a

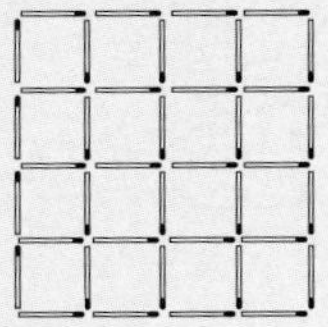

b 40, 60
c Add 24 to number for fifth shape (or nth term $= 2n^2 + 2n$)
6 $A = n(n + 1)$
7 a nth shape has $4n - 1$ triangles
b nth diagram has $n^2 + 2$ triangles
8 b i −10, 5 **ii** 25 m^2
9 a 4.5 **b ii** 7
10 $n(n + 1)$
11 a $(x - 2)(x + 6)$ **b** 2, −6
12 a i $40 = 5 \times 8 = 5 \times (5 + 3)$
ii nth number $= n \times (n + 3)$
b i $40 = 6 \times 7 - 2$
ii nth number $= (n + 1)(n + 2) - 2$
c $n \times (n + 3) = n^2 + 3n$;
$(n + 1)(n + 2) - 2 =$
$n^2 + 3n + 2 - 2 = n^2 + 3n$
13 a $4x^2 - 9$
b i $x(x + 6)$ **ii** $(x + 2)(x + 4)$
c 3, −5
14 $u_n = n^2 + 1$

Exercise 26.1 (p. 491)

1 a none **e** volume **h** length
b area **f** area **i** length
c volume **g** none **j** area
d area
2 a ii **b** iv
3 a vi **b** iv
4 a i A **ii** V **iii** L
iv A **v** None
b $[L^1] + [L^2]$ not valid formula
5 $r(r + 4h)$, $\frac{rh}{4}$, $\frac{3r^3}{h}$
6 c $[L^2]$ dimension
7 $\pi r^2 l$, $4\pi r^3$, $3(a^2 + b^2)r$
8 a iv **b** ii

Exercise 26.2 (p. 499)

1 a Yes, equiangular
b No, sides not in same ratio
2 Yes
3 No
4 Yes
5 a $A\hat{E}D = C\hat{E}B$ (opposite angles)
$A\hat{D}E = C\hat{B}E$ (3rd angle in $\triangle$)
Triangles are equiangular
b BC = 6, AE = 5
6 a $a = 12$ **b** $b = 12$
c $c = 7.5$, $d = 12$
7 a $a = 4.5$ **b** $d = 3.75$
c $b = 2.25$ **d** $e = 5.4$
8 a 12 cm **b** 10 cm
9 a $x = 5$, $y = 3$ **b** $x = 3\frac{3}{4}$, $y = 3\frac{1}{3}$
10 $x = 5\frac{1}{3}$, $y = 6\frac{2}{3}$
11 b PT = 6.25 cm, TQ = 4.8 cm
12 5.28 m
13 10 cm
14 b 3.75 m
15 12 m
16 DC = 6.25 cm, ED = 1.75 cm

Exercise 26.3 (p. 502)

1 $a = 56.3$ m, $b = 27.5$ cm, $c = 3.46$ cm, $d = 0.602$ km, $e = 9.71$ cm, $f = 3.22$ m, $g = 123$ mm, $h = 5.39$ m, $i = 185$ mm
2 8.5 m (3 s.f.)

3 a 7.83 cm (3 s.f.)
 b i 5.03 cm (3 s.f.)
 ii 7.83 cm (3 s.f.); same (if you retained all the figures on your calculator)
4 a 26.1° (1 d.p.) b 425 m (3 s.f.)
5 a 50° b 2.33 m (3 s.f.)
6 a 2.28 m (3 s.f.)
 b i 1.23 m (3 s.f.) ii 2.43 m (3 s.f.)
7 a 45°
 b A : 0 m, B : 2.2 m (1 d.p.), C : 7.5 m, D : 12.8 m (1 d.p.), E : 15 m, F : 12.8 m (1 d.p.), G : 7.5 m, H : 2.2 m (1 d.p.)

Exercise 26.4 (p. 506)

1 a 3.0 cm (2 s.f.) b 27 cm (2 s.f.)
2 a 72° b 1.31 m (3 s.f.)
 c 1.06 m (3 s.f.) d 4.08 m² (3 s.f.)
3 a 9.6 m b 3
4 a 17.6 cm (3 s.f.) b 511 cm² (3 s.f.)
 c 92 *l* (2 s.f.)
5 a $x = 54°$, $y = 68°$
 b 5.21 cm (3 s.f.)
 c 3.59 cm (3 s.f.)
 d 163 cm³ (3 s.f.)
6 a i 34.9° (1 d.p.) ii 154 cm (3 s.f.)
 b 33°
7 a 72° b 9.05 cm (3 s.f.)
 c 5.88 cm (3 s.f.) d 19 cm² (2 s.f.)
8 a 107 cm (3 s.f.) b 35 cm (2 s.f.)

Exam Questions: Chp 26 (p. 510)

1 a i 20 cm ii 106 cm iii 53°
 b 140 cm (3 s.f.)
2 a 1.86 m (3 s.f.) b 1.67 m (3 s.f.)
 c 7.12 m (3 s.f.)
3 a i △BCD ii 3 cm
 b It stays the same (3 cm)
4 No, ratios of corresponding sides are not the same
5 a 47.7 km (3 s.f.)
 b 198 km, 069° (nearest degree)
6 a $3a + 3b$ b ab c a^2b
7 b ii 15.0 m (3 s.f.)
 c 12.6 m (3 s.f.)
8 a 1.73 cm (3 s.f.)
 b i Equiangular ii 5.5 cm
9 a Equiangular b 18.75 cm
10 $\frac{1}{3}\pi h(b^2 + ab + a^2)$

Exercise 27.1 (p. 513)

1 a $-2 \leq x \leq 2$
 −2 0 2
 b $-5 < x < 5$
 −5 0 5
 c $x \leq -4, x \geq 4$
 −4 0 4
 d $x < -10, x > 10$
 −10 0 10
 e $-3 < x < 3$
 −3 0 3
 f $x \leq -1, x \geq 1$
 −1 0 1
 g $x < 0, x > 0$
 0
 h $-7 < x < 7$
 −7 0 7
 i $x < -1.5, x > 1.5$
 −1.5 0 1.5
 j $x < 0, x > 2$
 0 2
 k $-4 \leq x \leq 0$
 −4 0
 l $2 \leq x \leq 4$
 0 2 4
2 a −1, 0, 1
 b −3, −2, −1, 0, 1, 2, 3
 c −1, 0, 1
 d −2, −1, 0, 1, 2
 e 0, 1, 2, 3
 f 1, 2, 3, 4
3 −3, −2, −1, 0, 1, 2, 3
4 −5

Exercise 27.2 (p. 519)

1 a $x \geq 2$ b $y < 3$
 c $x + y \leq 2$ d $y \geq x + 1$
2 a, c, d

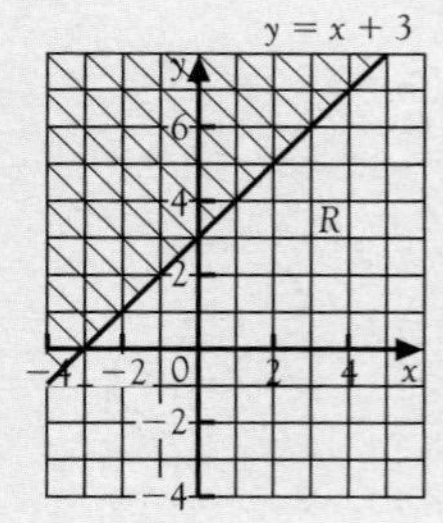

 b −1, 1, 3, 5, 7
3 a

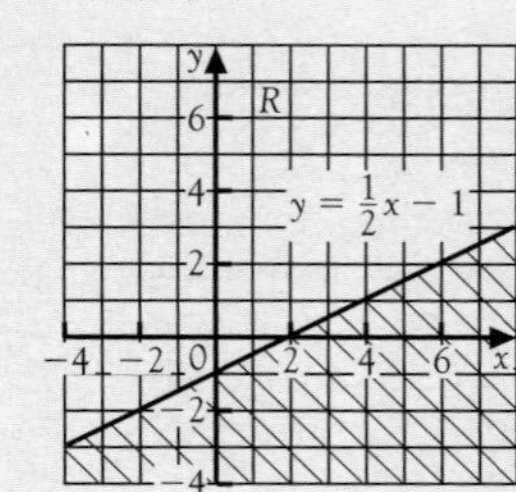

b

y = 3x + 1

R

c

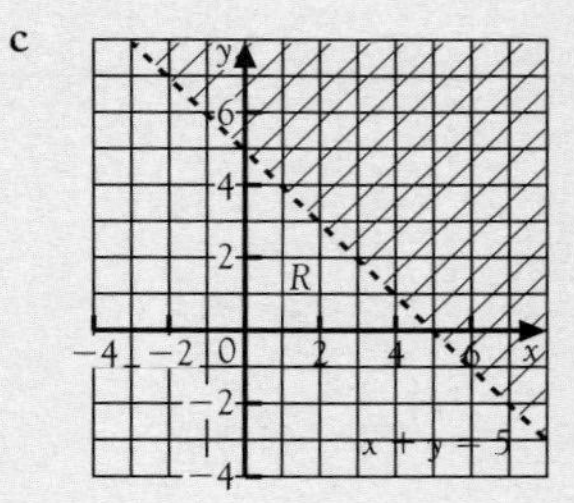

d

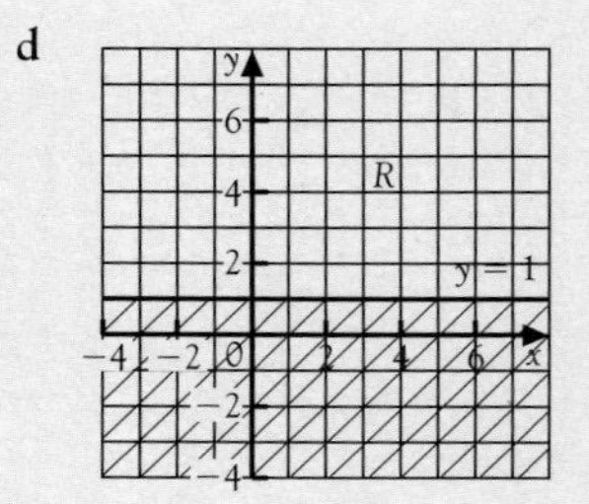

e

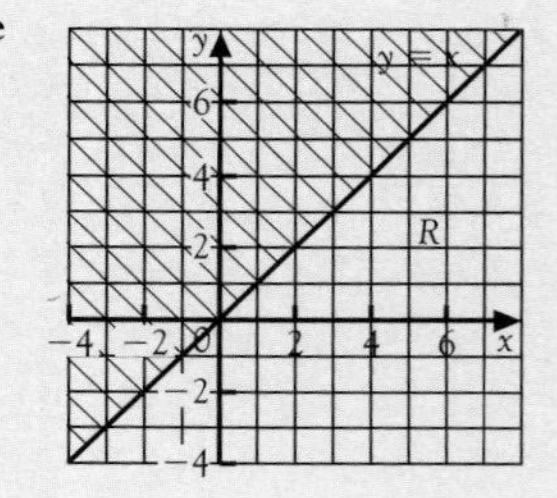

f

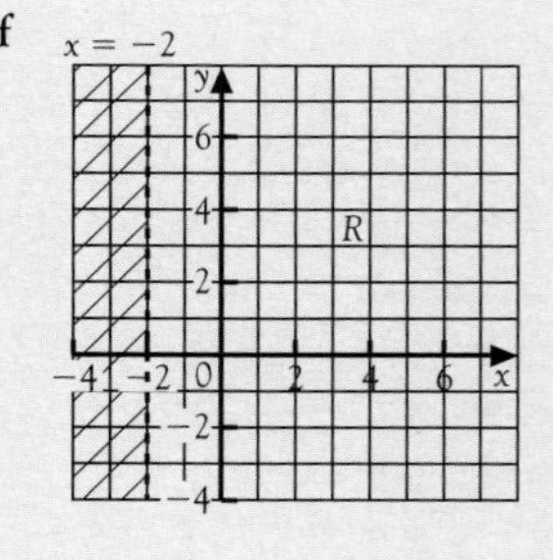

g

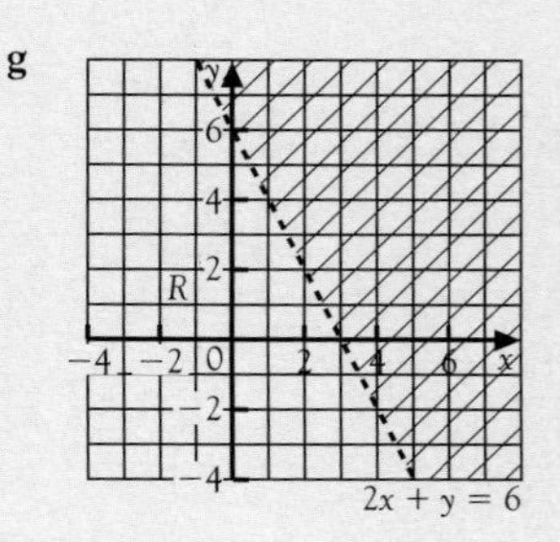

h

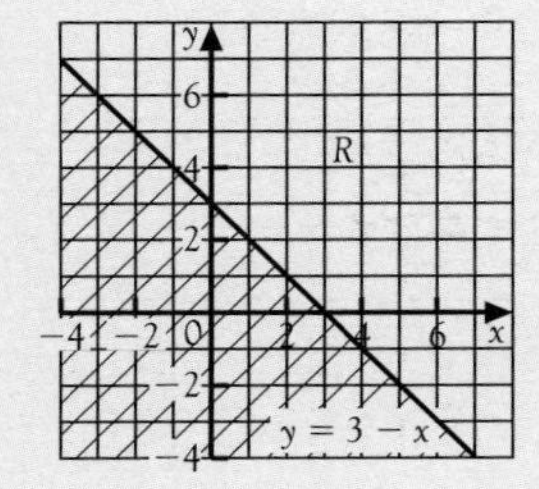

i

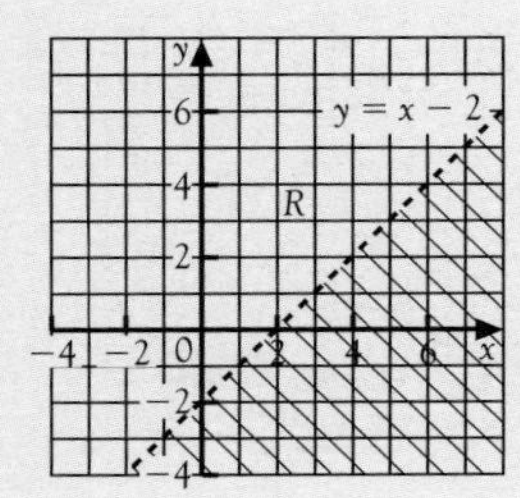

4 a

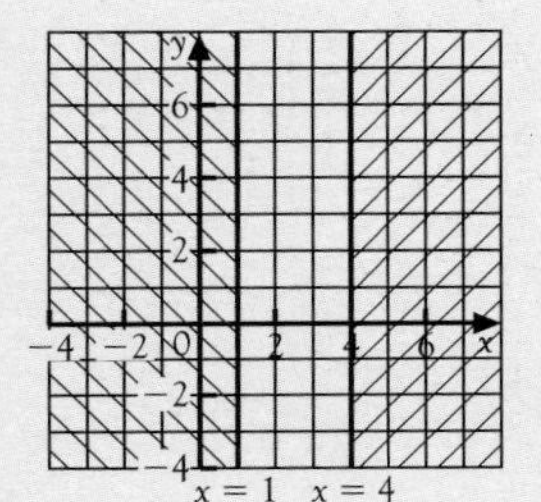

b

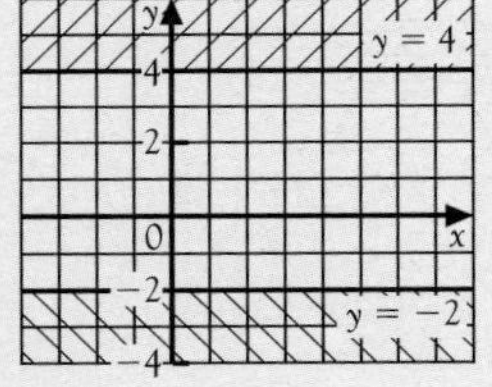

c

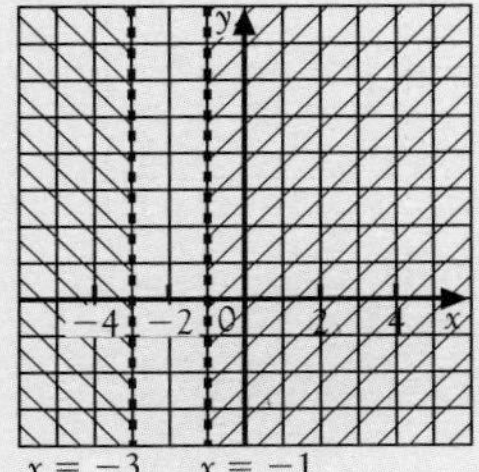

d

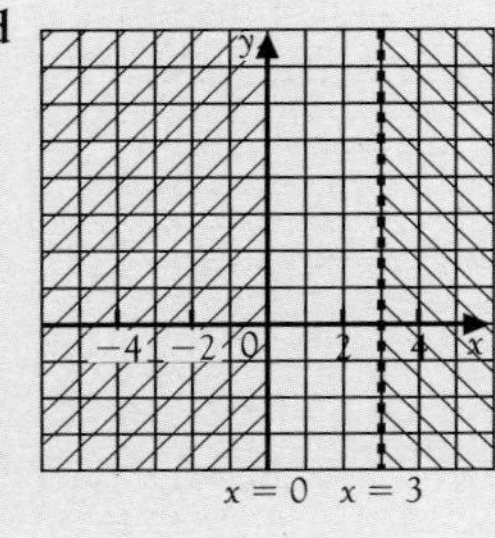

e

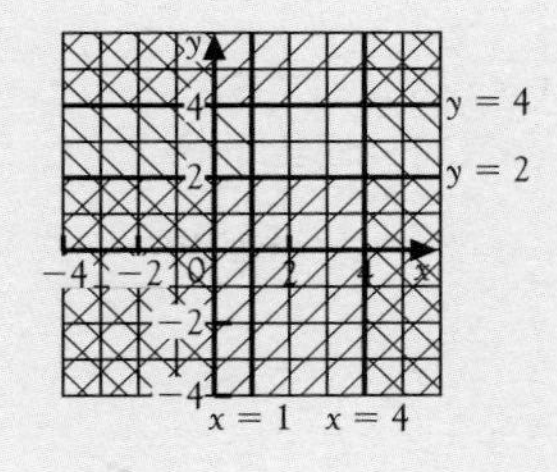

f

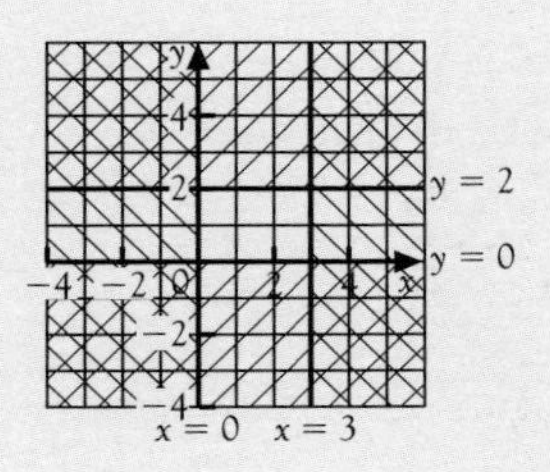

5 i

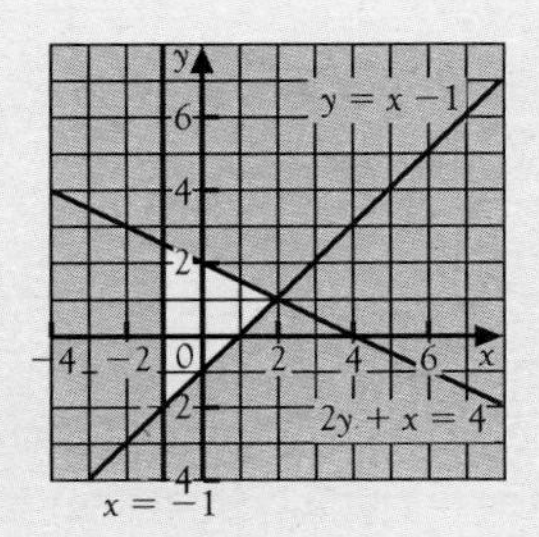

ii

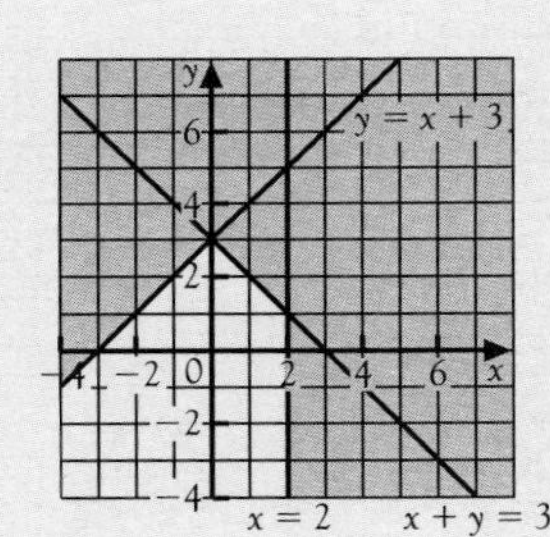

iii

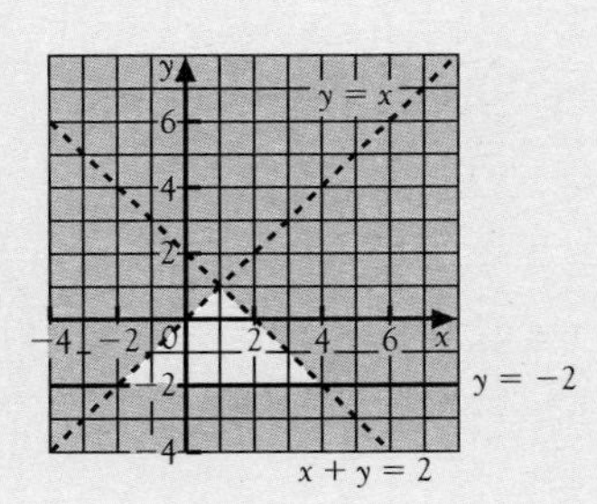

iv

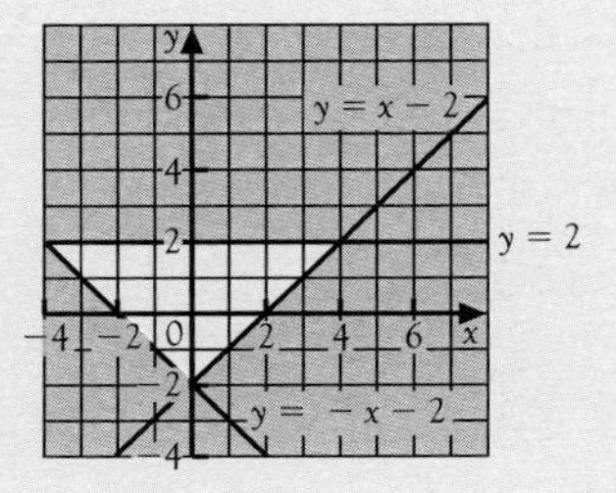

v

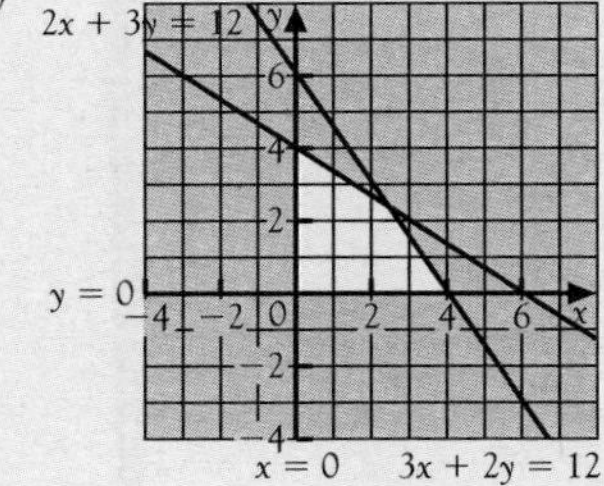

vi

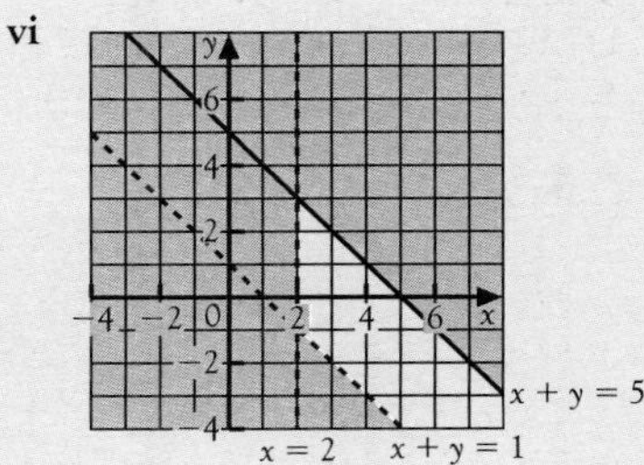

6 i

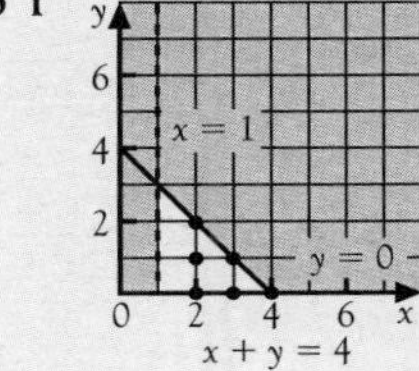

(2, 0), (2, 1), (2, 2), (3, 0), (3, 1), (4, 0)

ii

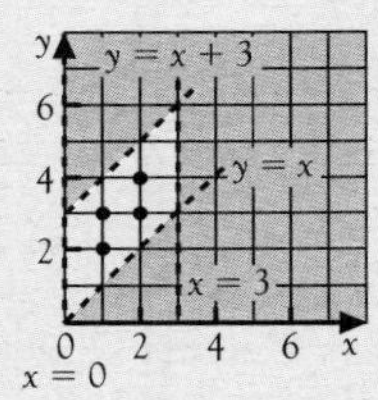

(1, 2), (1, 3), (2, 3), (2, 4)

7 a $x + y \leqslant 50$ d $y > x$
b $x \geqslant 10$ e $y \geqslant x + 20$
c $y < 30$ f $x \geqslant 2y$ (i.e. $y \leqslant \frac{1}{2}x$)

8 a $x < 250$, $x + y \leqslant 300$,
$x \geqslant 2y$, i.e. $y \leqslant \frac{1}{2}x$

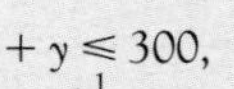

b

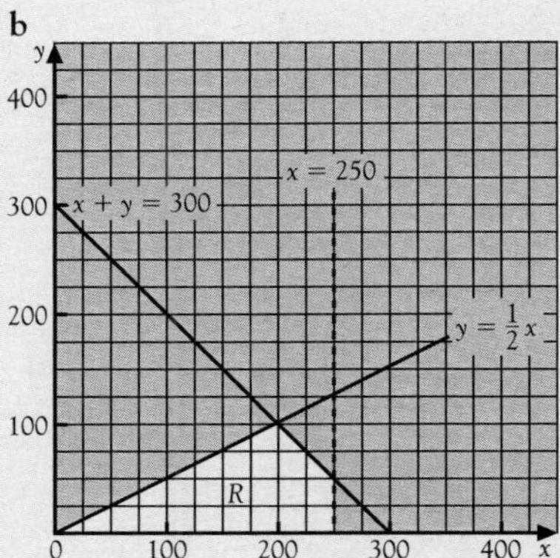

9 a Cost restriction $50x + 20y \leqslant 1400$, divide by 10
b $x + 4y \leqslant 64$

c

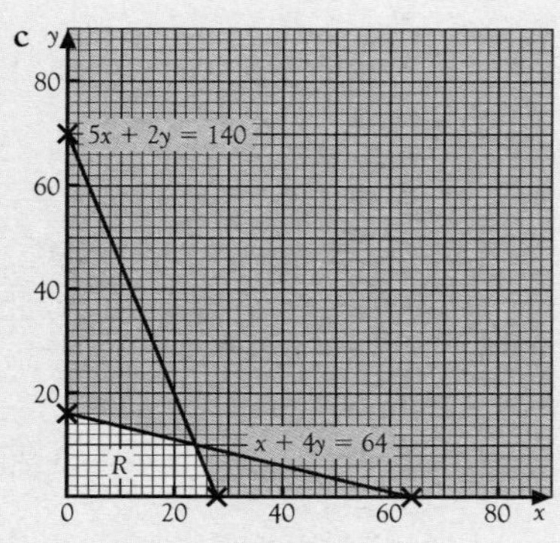

10 $x + y \leqslant 6, y \geqslant 2, x \geqslant 1$

11 a $x \geqslant 2, y \geqslant 1$

b, c, d

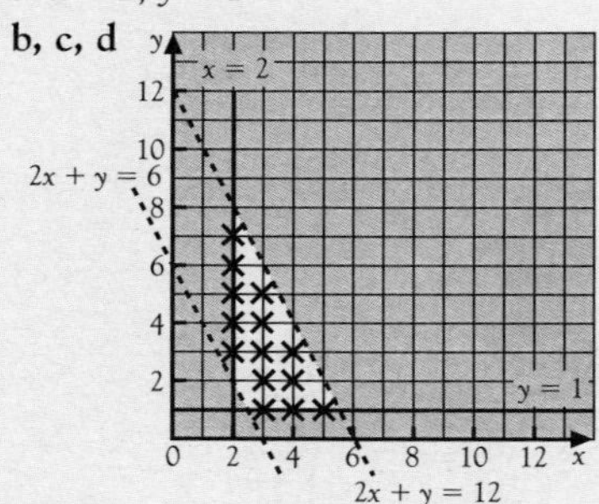

e 2 pizzas and 7 quiches (£20)

Exercise 27.3 (p. 525)

1 5.34 (2 d.p.)

2 1.56 (2 d.p.)

3 a −2.5 (1 d.p.)

b −1.3 (1 d.p.), 0.9 (1 d.p.)

4 a 1, 2, 3, 10

b

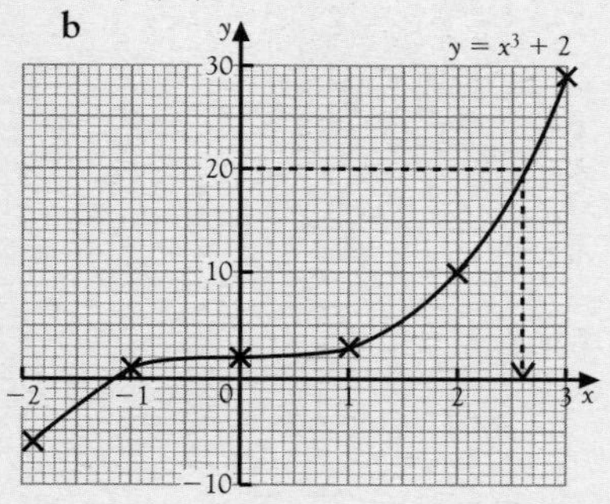

c 2.6; 2.6

5

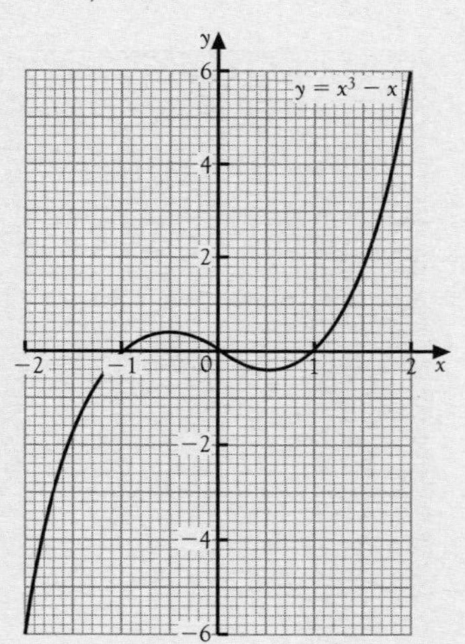

6 a −1, −1.5, −2, −3, −12, 12, 6, 2, 1, 0.5

b Sketch of answer graph

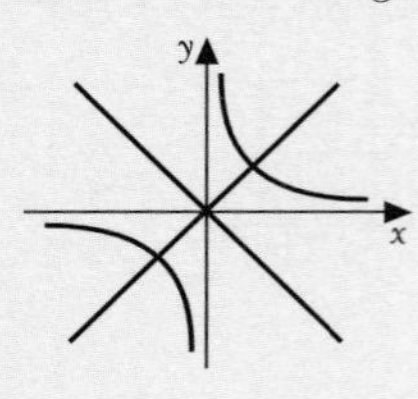

c $y = x$, $y = -x$ d 2

7 a, b Sketch of answer graph

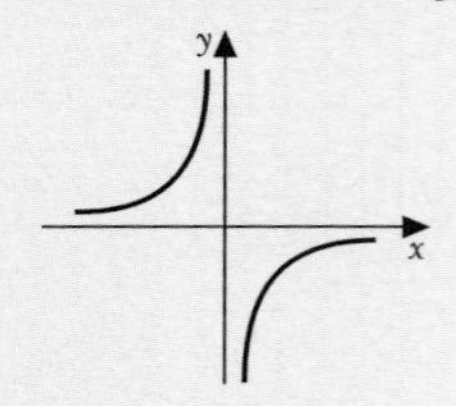

8 a D b A c C

d E e F f B

9 a 0, 81, 147, 144, 96, 63, 9, 0

b, c

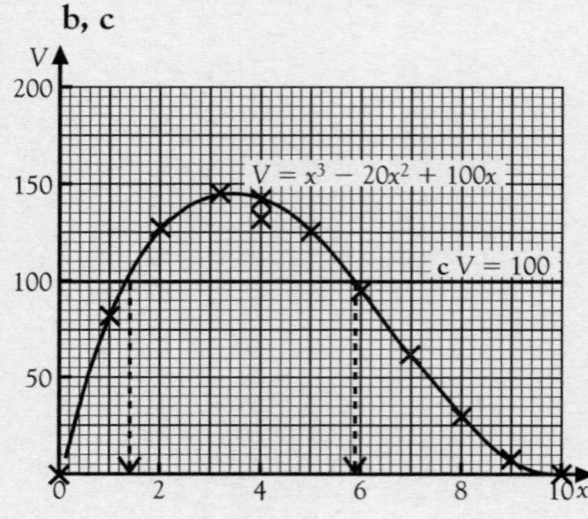

d 1.4, 5.9

10 a b

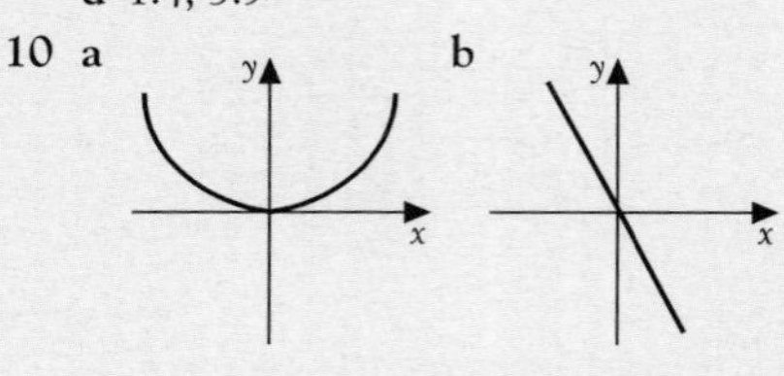

c d

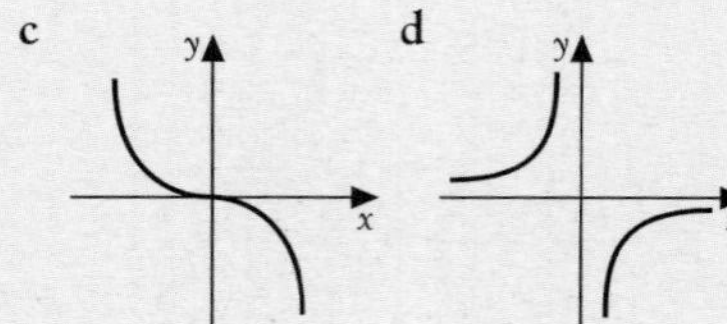

e f

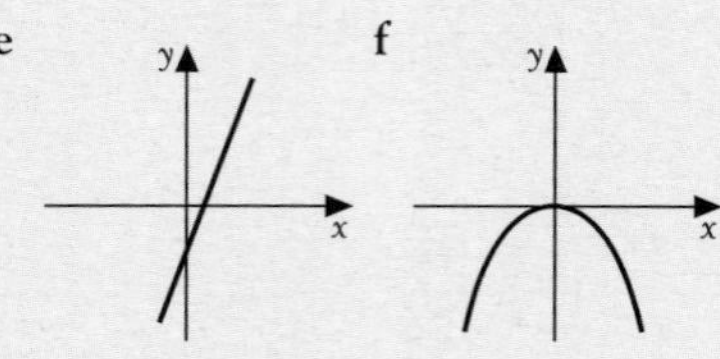

g h

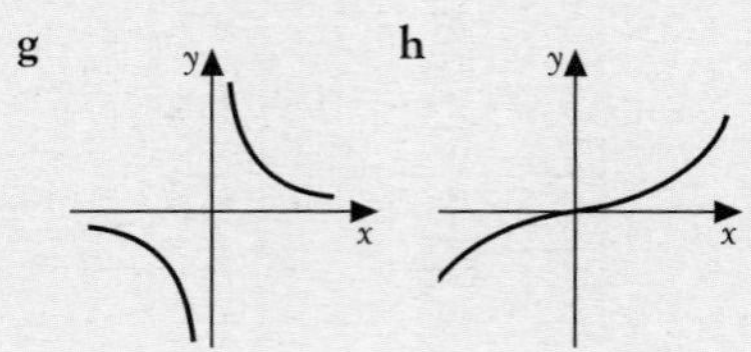

11 a 26, 12, 10, 9.75, 8, −6

b, c

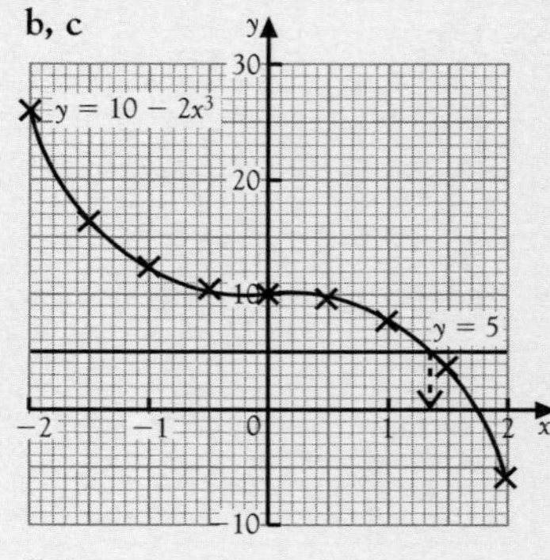

d Draw $y = 5$, $x = 1.35$

Exercise 27.4 (p. 528)

1 A–W, B–Z, C–X

2 a

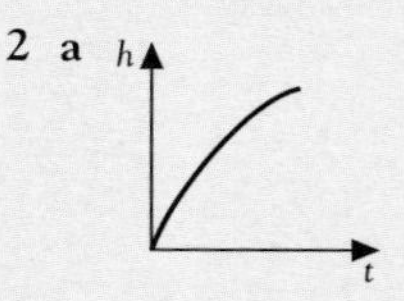

b

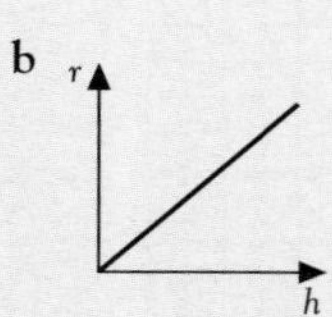

3 a R b S

c

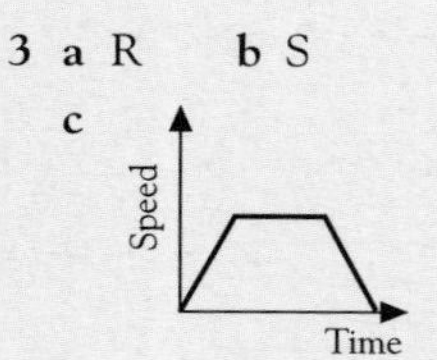

4 a 1500 people per year

b After 10 years

c 4 years

5 A

6 a BC

b CD

c AB

7 a 0, 8.75, 15, 18.75, 20, 18.75, 15, 8.75, 0

b 2 seconds

c At the start

d Increasing from zero

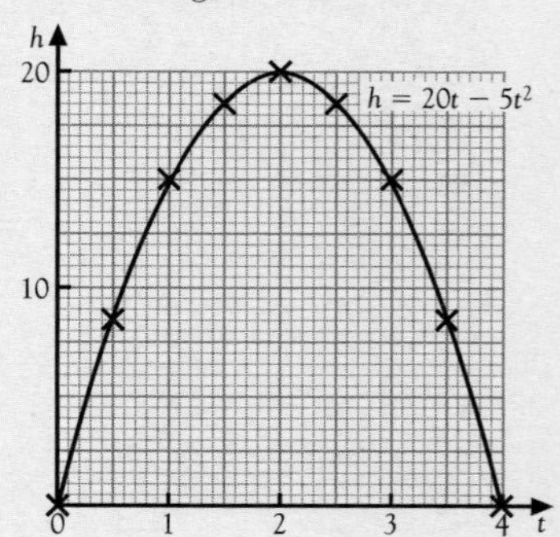

Exam Questions: Chp 27 (p. 531)

1 a $x < -\frac{1}{3}$ b

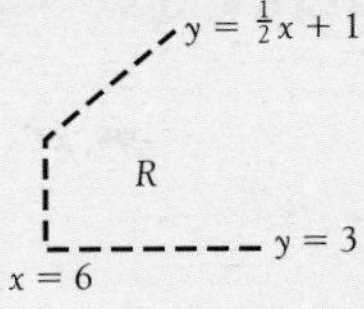

2 a $-1, 0, 1, 2$ b

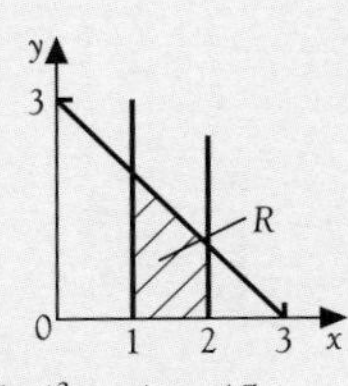

3 a $3.5^3 - 3.5 < 45$, $4^3 - 4 > 45$
b 3.7

4 a 24 b

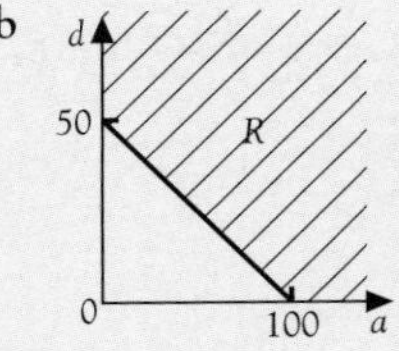

c i $a + d \leqslant 85$ ii 32
d $24 \leqslant d \leqslant 32$

5 a 6, 6, 26
b

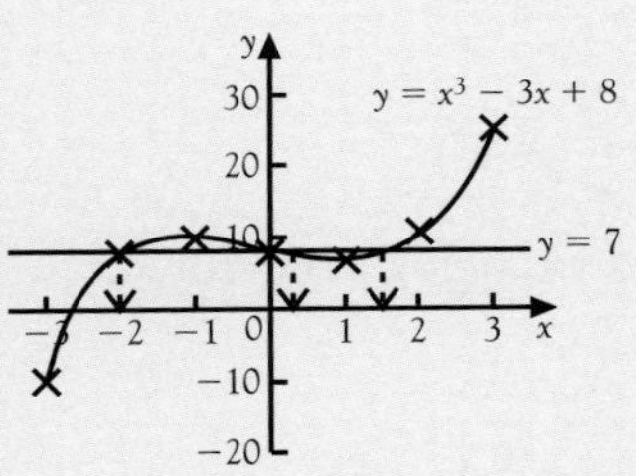

c -1.9, 0.3, 1.6 (approximate answers)

6 3.53

7 a (2, 5) b $y = 2x - 2$
c

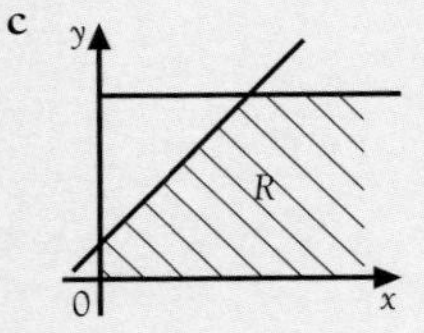

8 a $n > 3$
b $-2, -1, 0, 1, 2$

9 4.25

10 a A iii, B i, C ii, D iv
b

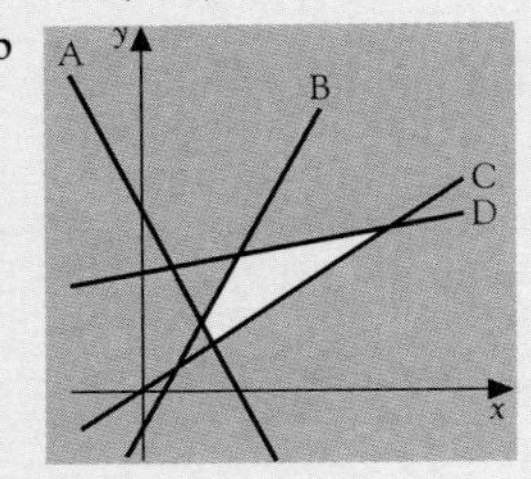

11 a A–G, B–H, C–E, D–F
b

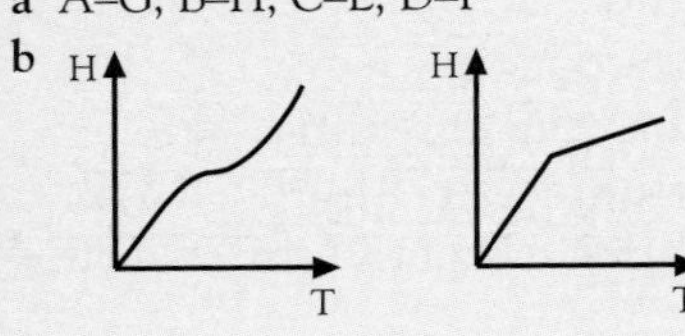

12 a Not symmetric about y axis.
b 3, 5, 7, … (positive odd number)

13 b $y \geqslant \frac{1}{5}x$
c

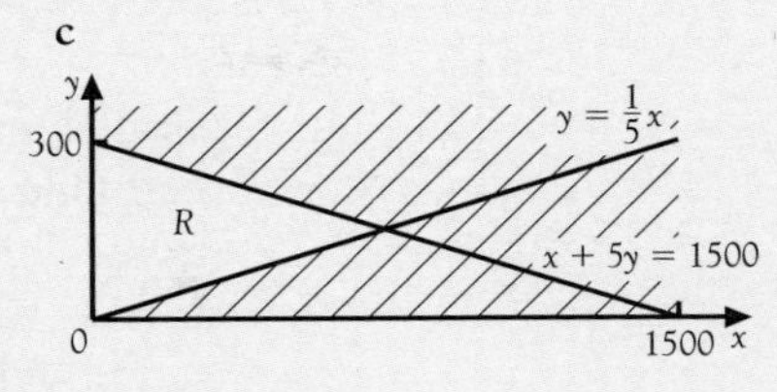

14 a 1 (B), 2 (A), 3 (E), 4 (D)
b

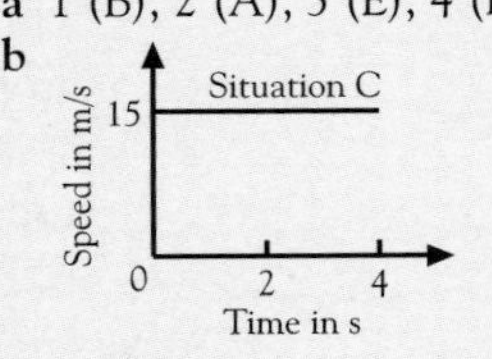

Mixed Exam Questions: End of Stage 3 (p. 537)

1 a 0.0002 seconds
b i 0.34 km (2 s.f.)
ii 1.76×10^2 (3 s.f.)

2 31.5 mpg (3 s.f.)

3 a £434.75 b £59.50

4 a 0.6 b 0.16

5 a $2n(7 - 2n)$ b 1, 2, 3

6 a Sketch of answer

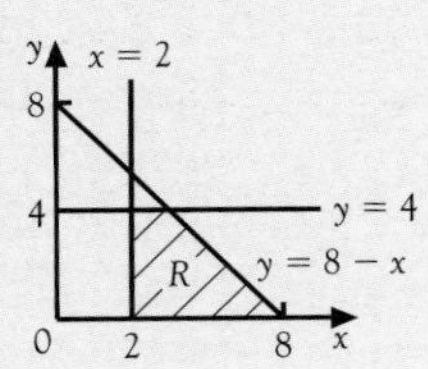

b i $x = -3$ ii $x \geqslant -3$

7 Length 4; Area 2, 3; Volume 5

8 6.52 cm

9 πab, $\pi(a + b)l$

10 a B b E

11 a 1×10^{297} b 4×10^{304}

12 a ii 226° b 170 miles
c i 28.1° (1 d.p.) ii 344° (3 s.f.)

13 a

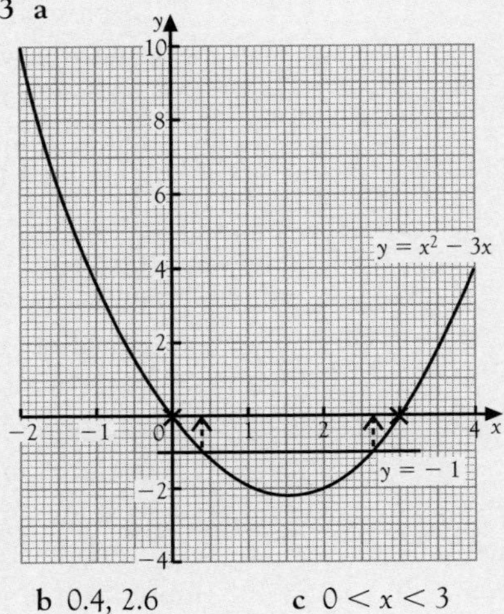

b 0.4, 2.6 c $0 < x < 3$

14 D, C, E, F, A, B

15 a i £360 ii £230.40 b £375

16 a 10, 34, 56, 77, 90
b

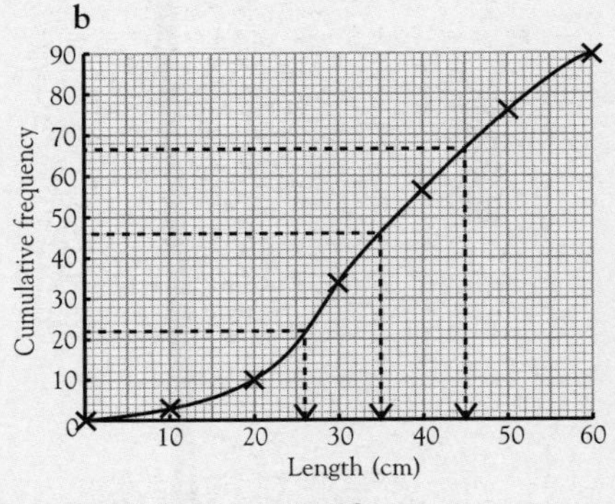

c i 35 cm ii 19 cm

17 a 20 b 12
c $n^2 - 3n - 14 = 0$ does not have any integer solutions

18 a 69.153 b 3.49

19 a 8.04×10^{23} (3 s.f.) b $T = \sqrt{\dfrac{R^3}{K}}$
c 223 days

20 a 9.6 cm b 3 cm

21 Asia

22 a £20 280 b £20 160

23 a 2, 5, 13, 22, 35, 40
b

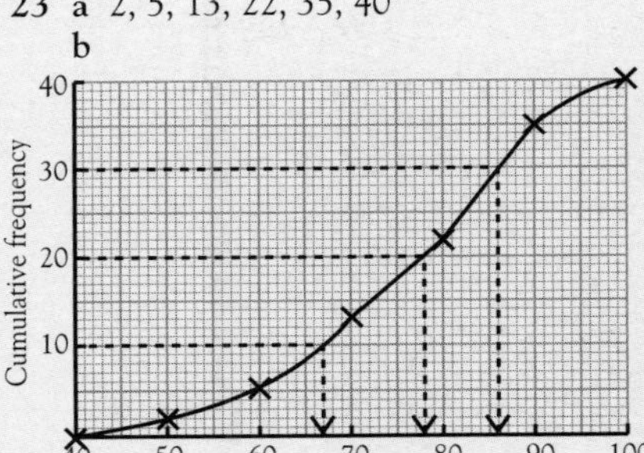

c 78 g d 19 g
e Although the medians are similar, the masses in the second sample are much more spread out, with the inter-quartile range double that of the first sample.

24 a 5×10^{101} b 5×10^{-8} m

25

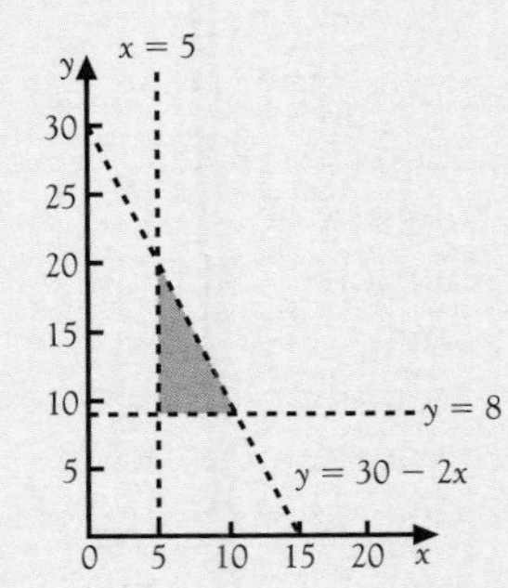

INDEX